Conversions Between U.S. Customary Units and SI Units (continued)

Customary unit		Times conversion factor		Equals SI unit	
		Accurate	Practical		
Moment of inertia (area)					
inch to fourth power	in.^{4}	416,231	416,000	millimeter to fourth power	mm^4
inch to fourth power	in.^{4}	0.416231×10^{-6}	0.416×10^{-6}	meter to fourth power	m^4
Moment of inertia (mass)					
slug foot squared	slug-ft^2	1.35582	1.36	kilogram meter squared	kg·m^2
Power					
foot-pound per second	ft-lb/s	1.35582	1.36	watt	W
foot-pound per minute	ft-lb/min	0.0225970	0.0226	watt	W
horsepower					
(550 ft-lb/s)	hp	745.701	746	watt	W
Pressure; stress					
pound per square foot	psf	47.8803	47.9	pascal (N/m^2)	Pa
pound per square inch	psi	6894.76	6890	pascal	Pa
kip per square foot	ksf	47.8803	47.9	kilopascal	kPa
kip per square inch	ksi	6.89476	6.89	megapascal	MPa
Section modulus					
inch to third power	in.^{3}	16,387.1	16,400	millimeter to third power	mm^3
inch to third power	in.^{3}	16.3871×10^{-6}	16.4×10^{-6}	meter to third power	m^3
Velocity (linear)					
foot per second	ft/s	0.3048*	0.305	meter per second	m/s
inch per second	in./s	0.0254*	0.0254	meter per second	m/s
mile per hour	mph	0.44704*	0.447	meter per second	m/s
mile per hour	mph	1.609344*	1.61	kilometer per hour	km/h
Volume					
cubic foot	ft^3	0.0283168	0.0283	cubic meter	m^3
cubic inch	in.^{3}	16.3871×10^{-6}	16.4×10^{-6}	cubic meter	m^3
cubic inch	in.^{3}	16.3871	16.4	cubic centimeter (cc)	cm^3
gallon (231 in.^{3})	gal.	3.78541	3.79	liter	L
gallon (231 in.^{3})	gal.	0.00378541	0.00379	cubic meter	m^3

*An asterisk denotes an *exact* conversion factor

Note: **To convert from SI units to USCS units, *divide* by the conversion factor**

Temperature Conversion Formulas

$$T(°C) = \frac{5}{9}[T(°F) - 32] = T(K) - 273.15$$

$$T(K) = \frac{5}{9}[T(°F) - 32] + 273.15 = T(°C) + 273.15$$

$$T(°F) = \frac{9}{5}T(°C) + 32 = \frac{9}{5}T(K) - 459.67$$

FOURTH EDITION

MECHANICS OF MATERIALS

JAMES M. GERE
Professor Emeritus, Stanford University

STEPHEN P. TIMOSHENKO (1878–1972)
Late Professor, Stanford University

PWS PUBLISHING COMPANY

I(T)P An International Thomson Publishing Company

Boston • Albany • Bonn • Cincinnati • Detroit • London • Madrid
Melbourne • Mexico City • New York • Pacific Grove • Paris
San Francisco • Singapore • Tokyo • Toronto • Washington

PWS PUBLISHING COMPANY
20 Park Plaza, Boston, MA 02116–4324

You can request permission to use material from this text through the following
phone and fax numbers: Phone: 1-800-730-2214; Fax: 1-800-730-2215

I(T)P™ International Thomson Publishing
        The trademark ITP is used under license.

Mathcad™ Engine is a registered trademark of MathSoft, Inc.

Sponsoring Editor: *Jonathan Plant*
Developmental Editor: *Mary Thomas Stone*
Technology Editor: *Leslie G. Bondaryk*
Marketing Manager: *Nathan Wilbur*
Production Coordinator: *Elise Kaiser*
Manufacturing Buyer: *Andrew Christensen*
Interior Designer/Production: *Helen Walden*
Compositor: *American Composition & Graphics, Inc.*
Interior Illustrator: *Tech Graphics/American Composition & Graphics, Inc.*
Cover Designer: *Diane Levy*
Cover Image: *David Sailors © 1995 The Stock Market, Inc. Image is of
construction on the central dome and roof structure at Washington National
Airport; Cesar Pelli, architect.*
Cover Printer: *Mid-City Lithographers*
Text Printer/Binder: *Quebecor Hawkins*

Library of Congress Cataloging-in-Publication Data
Gere, James M.
    Mechanics of materials / James M. Gere, Stephen P. Timoshenko--
4th ed.
        p.   cm.
    Includes indexes.
    ISBN 0-534-93429-3 (hc). -- ISBN 0-534-95102-3
    1. Strength of materials.   I. Timoshenko, Stephen P., 1878–
                                                        1972.
    II. Title.
    TA405.G44   1997
    624.171--dc20   CIP

Printed in the United States of America.

99 00—10 9 8 7 6 5

 This book is printed on recycled, acid-free paper.

Student Edition ISBN 0-534-93429-3
Free Copy Edition ISBN 0-534-95102-3

For more information, contact:

PWS Publishing Company
20 Park Plaza
Boston, MA 02116–4324

Nelson Canada
1120 Birchmount Road
Scarborough, Ontario
Canada M1K 5G4

International Thomson Publishing Asia
221 Henderson Road
#05-10 Henderson Building
Singapore 0315

International Thomson Publishing Europe
Berkshire House I68-I73
High Holborn
London SC1V 7AA

International Thomson Editores
Campos Eliseos 385, Piso 7
Col. Polanco
11560 Mexico D.F., Mexico

International Thomson Publishing Japan
Hirakawacho Kyowa Building, 31
2-2-1 Hirakawacho
Chiyoda-ku, Tokyo 102
Japan

Thomas Nelson Australia
102 Dodds Street
South Melbourne, 3205
Australia

International Thomson Publishing GmbH
Königswinterer Strasse 418
53227 Bonn, Germany

CONTENTS

*Asterisks denote specialized and advanced topics

iii

12 ▮ REVIEW OF CENTROIDS AND MOMENTS OF INERTIA   805

References and Historical Notes   835

Appendix A   Systems of Units and Conversion Factors   843

Appendix B   Problem Solving   857

PREFACE

By taking a course in mechanics of materials, students learn a fundamental engineering subject while at the same time developing their analytical and problem-solving abilities. During the preparation of this revised Fourth Edition, I have kept these important goals in mind. The facts and theories of mechanics are presented in a teachable and easy-to-learn manner, with ample discussions and many examples, so that students can readily master the subject matter. In addition, considerable emphasis is placed on how to analyze mechanical and structural systems, and many of the problems require that students do some original thinking.

This book covers all of the basic topics of mechanics of materials, presented at a level suitable for sophomore and junior engineering students. The principal topics are the analysis and design of structural members subjected to tension, compression, torsion, and bending, including such fundamental concepts as stress, strain, elastic behavior, inelastic behavior, and strain energy. Other topics of general interest are the transformations of stress and strain, combined loadings, stress concentrations, deflections of beams, and stability of columns. More specialized topics are thermal effects, dynamic loading, nonprismatic members, beams of two materials, shear centers, pressure vessels, and statically indeterminate beams. For completeness and occasional reference, elementary topics such as shear forces, bending moments, centroids, and moments of inertia are also presented in the book.

Much more material than can be covered in a single course is included in the book, and therefore teachers have the opportunity to select the topics that they feel are the most fundamental and relevant.

Advanced topics are identified in the contents and chapter sections by use of the asterisk (*). Teachers will also appreciate the hundreds of new problems (with over 1100 problems total) that are available for homework assignments and classroom discussions. Problems are placed at the ends of the chapters so that they are easy to find and don't break up the presentation of the subject matter. (An unusually difficult or lengthy problem is indicated by attaching one or more asterisks to the problem number.)

Both the International System of Units (SI) and the U. S. Customary System (USCS) are used in the numerical examples and problems. Discussions of both systems and a table of conversion factors are given in the appendix. For problems involving numerical solutions, odd-numbered problems are in USCS units and even-numbered problems are in SI units. The only exceptions are problems and examples involving the tabulated properties of structural-steel shapes, because the tables for these shapes are presented only in USCS units. Answers to the problems are listed near the back of the book, so that students and other problem-solvers can have the satisfaction of checking their results.

References and historical notes are also collected near the back of the book. They consist of original sources of the subject matter and biographical notes about the pioneering engineers, scientists, and mathematicians who created the subject of mechanics of materials. A separate name index makes it easy to look up any of these historical figures. The book concludes with an extensive and carefully prepared subject index, so that any topic, concept, key term, or definition can be quickly located.

This Fourth Edition of *Mechanics of Materials* has been carefully rewritten with expanded discussions, new figures, new examples and problems, and many changes in organization to make the book more useful in the classroom. All changes in organization and presentation were prompted by teachers and students who were familiar with the Third Edition.

Considerable effort has been spent in checking and proofreading the text so as to eliminate errors, but if you happen to find one, no matter how minor, please notify me at the Department of Civil Engineering, Stanford University, Stanford, California 94305-4020 U.S.A. (e-mail: jgere@cive.stanford.edu), or notify the publisher. (All correspondence will be answered.)

■ Acknowledgments

The first edition of this book, published in 1972 and written by the present author, was an outgrowth of earlier books by Professor Stephen P. Timoshenko (1878–1972), who used the title *Strength of Materials.* Timoshenko was the most outstanding pioneer in the field of applied mechanics. Through his research and his many textbooks, he revolutionized the teaching of mechanics, not only in the United States but throughout the world. (You can find a brief biography of Timoshenko in the first reference at the back of the book.)

To acknowledge everyone else who contributed to this book in some manner is clearly impossible, but I owe a major debt to my former Stanford teachers, including those giants of mechanics, Wilhelm Flügge, James Norman Goodier, Miklós Hetényi, Nicholas J. Hoff, and Donovan H. Young. I am also indebted to my Stanford colleagues—especially Tom Kane, Anne Kiremidjian, Helmut Krawinkler, Kincho Law, Peter Pinsky, Haresh Shah, Sheri Sheppard, Allison Smith, and the late Bill Weaver—who discussed mechanics and educational philosophy with me on many occasions. In addition, many helpful ideas and comments were contributed by Thalia Anagnos of San Jose State University, John Burgess of the University of Hawaii, and Aron Zaslavsky of the Technion.

The following reviewers read the entire Fourth Edition in manuscript form and provided both general and specific comments for changes and improvements. Their suggestions proved invaluable, and I am deeply appreciative of their careful and thorough reviews. My thanks to: Majid R. Chitsaz of Pennsylvania State University; Robert D. Cook of the University of Wisconsin-Madison; Janak Dave of the University of Cincinnati; Sergey Drabkin of the Polytechnic University of New York; Raghu Echempati of the University of Mississippi; Harvey Lipkin of Georgia Institute of Technology; Douglas Nims of the University of Toledo; Douglas B. Rigby of the Hong Kong University of Science and Technology; and P. D. Scarlatos of Florida Atlantic University.

In addition, the following reviewers provided comments on the Third Edition in a prerevision review. Their suggestions were crucial in shaping the Fourth Edition, and I much appreciate their thoughtful ideas. My thanks to: Hojjat Adeli of Ohio State University; Kevyan Ahdut of the University of the District of Columbia; John B. Brunski and Robert H. P. Dunn, both of Rensselaer Polytechnic Institute; Ted A. Conway of the University of Akron; Xiaomin Deng of the University of South Carolina; Arya Ebrahimpour of Pennsylvania State University; M. Elgaaly and Anisur Rahman, both of Drexel University; Ahmed Ibrahim of the State University of New York at Farmingdale; Norman F. Knight and Ramamurthy Prabhakaran, both of Old Dominion University; Gladius Lewis of the University of Memphis; Zhong Ming Liang of Purdue University; E. L. Parker of Valley Forge Military College; Edwin Powers of Catonsville Community College; Charles Rondeau of Jamestown Community College; Michael Schwartz of the University of St. Thomas; Sheri Sheppard of Stanford; R. Sierakowski of Ohio State University; L. T. D. Topoleski of the University of Maryland at Baltimore; Morteza Torkamani of the University of Pittsburgh, and Manoochehr Zoghi of the University of Dayton.

I was assisted in word processing, manuscript preparation, and proofreading by Duc Wong, who worked with great concern, care, and accuracy. In addition, the following graduate students provided expert assistance by proofreading or preparing solutions to problems: Yih-Lin Shelley Cheng, Krista Marie Donaldson, Denise M. Fennell, Jamie Hsieh, Peter I. Huang, Chao-Hua (Eric) Lin, Angela Chia-Lin Teng, and May Min-Chiao Wong.

Editing and production were handled skillfully and efficiently by the staff of PWS Publishing Company, including Jonathan Plant, Mary Thomas Stone, and Helen M. Walden. I am especially indebted to Mary Thomas Stone, who was the editor for the book and provided comments, insights, and assistance that went far beyond what I had a right to expect. The friendly and cooperative spirit shown by everyone at PWS made my work a pleasure.

Finally, I much appreciate the patience and encouragement provided by my family, especially my wife, Janice, throughout this project.

To each of these wonderful people, I am pleased to express my gratitude.

James M. Gere

■ Ancillaries

The Fourth Edition includes a bound-in 3.5" disk containing a powerful, easy-to-use computational program—Mathcad™ Engine 5.0 for Windows®—and a set of worksheets for solving problems in mechanics of materials. The worksheets are cross-referenced to appropriate text examples and problems with an icon ▣ . This icon simply indicates the type of problem or example for which the indicated worksheet(s) would be appropriate. All problems and examples in the text are meant to be solved as desired by the instructor and student and are *not* designed for specific computational tools.

Also available is a unique new CD-ROM and workbook package, *Visual Mechanics*. Developed at the University of Washington by Gregory R. Miller and Stephen C. Cooper, the package consists of a CD-ROM with two programs (called *Dr. Beam* and *Dr. Stress*), and a manual with worksheets, exercises, and examples, which focus on beam bending and analysis of stress states. The software and supporting instructional materials provide students with a hands-on "virtual lab" to visualize the behavior of beams, understand mathematical models, and explore mechanics of materials theories and design methods.

These two software-based tools are intended for optional use; the textbook can be effectively used on its own.

An Instructor's Solutions Manual with complete solutions to all of the text problems is available to adopters of the text.

If you are interested in receiving any of the above products, please call our order desk: 1-800-354-9706.

PWS Publishing Company

SYMBOLS

A	area
A_f	area of flange
A_w	area of web
a, b, c	dimensions, distances
C	centroid, constant of integration, compressive force
c	distance from neutral axis to outer surface of a beam
D	diameter
d	diameter, dimension, distance
E	modulus of elasticity
E_r	reduced modulus of elasticity
E_t	tangent modulus of elasticity
e	eccentricity, dimension, distance, unit volume change (dilatation)
F	force
f	shear flow, shape factor for plastic bending, flexibility, frequency (Hz)
f_T	torsional flexibility of a bar
G	modulus of elasticity in shear
g	acceleration of gravity
H	height, distance, force, reaction, horsepower
h	height, dimension
I	moment of inertia (or second moment) of a plane area
I_x, I_y, I_z	moments of inertia with respect to x, y, and z axes
I_{x_1}, I_{y_1}	moments of inertia with respect to x_1 and y_1 axes (rotated axes)
I_{xy}	product of inertia with respect to xy axes
$I_{x_1y_1}$	product of inertia with respect to $x_1 y_1$ axes (rotated axes)
I_p	polar moment of inertia
I_1, I_2	principal moments of inertia
J	torsion constant
K	stress-concentration factor, bulk modulus of elasticity, effective length factor for a column
k	spring constant, stiffness, symbol for $\sqrt{P/EI}$

k_T	torsional stiffness of a bar
L	length, distance
L_e	effective length of a column
ln	natural logarithm (base e)
log	common logarithm (base 10)
M	bending moment, couple, mass
M_p	plastic moment for a beam
M_y	yield moment for a beam
m	moment per unit length, mass per unit length
N	axial force
n	factor of safety, integer, revolutions per minute (rpm)
O	origin of coordinates
O'	center of curvature
P	force, concentrated load, power
P_allow	allowable load (or working load)
P_cr	critical load for a column
P_p	plastic load for a structure
P_r	reduced-modulus load for a column
P_t	tangent-modulus load for a column
P_y	yield load for a structure
p	pressure (force per unit area)
Q	force, concentrated load, first moment of a plane area
q	intensity of distributed load (force per unit distance)
R	reaction, radius
r	radius, radius of gyration ($r = \sqrt{I/A}$)
S	section modulus of the cross section of a beam, shear center
s	distance, distance along a curved line
T	tensile force, twisting couple or torque, temperature
T_p	plastic torque
T_y	yield torque
t	thickness, time, intensity of torque (torque per unit distance)
t_f	thickness of flange
t_w	thickness of web
U	strain energy
u	strain-energy density (strain energy per unit volume)
u_r	modulus of resistance
u_t	modulus of toughness
V	shear force, volume
v	deflection of a beam, velocity
v', v'', etc.	dv/dx, d^2v/dx^2, etc.
W	force, weight, work
w	load per unit of area (force per unit area)
x, y, z	rectangular axes (origin at point O)
x_c, y_c, z_c	rectangular axes (origin at centroid C)
$\bar{x}, \bar{y}, \bar{z}$	coordinates of centroid
Z	plastic modulus of the cross section of a beam

α	angle, coefficient of thermal expansion, nondimensional ratio
β	angle, nondimensional ratio, spring constant, stiffness
β_R	rotational stiffness of a spring
γ	shear strain, weight density (weight per unit volume)
$\gamma_{xy}, \gamma_{yz}, \gamma_{zx}$	shear strains in xy, yz, and zx planes
$\gamma_{x_1 y_1}$	shear strain with respect to $x_1 y_1$ axes (rotated axes)
γ_θ	shear strain for inclined axes
δ, Δ	deflection, displacement, elongation of a bar or spring
ΔT	temperature differential
δ_p	plastic displacement
δ_y	yield displacement
ϵ	normal strain
$\epsilon_x, \epsilon_y, \epsilon_z$	normal strains in x, y, and z directions
$\epsilon_{x_1}, \epsilon_{y_1}$	normal strains in x_1 and y_1 directions (rotated axes)
ϵ_θ	normal strain for inclined axes
$\epsilon_1, \epsilon_2, \epsilon_3$	principal normal strains
ϵ'	lateral strain
ϵ_T	thermal strain
ϵ_y	yield strain
θ	angle, angle of rotation of beam axis, rate of twist of a bar in torsion (angle of twist per unit length)
θ_p	angle to a principal plane or to a principal axis
θ_s	angle to a plane of maximum shear stress
κ	curvature ($\kappa = 1/\rho$)
λ	distance
ν	Poisson's ratio
ρ	radius, radius of curvature ($\rho = 1/\kappa$), radial distance in polar coordinates, mass density (mass per unit volume)
σ	normal stress
$\sigma_x, \sigma_y, \sigma_z$	normal stresses on planes perpendicular to x, y, and z axes
$\sigma_{x_1}, \sigma_{y_1}$	normal stresses on planes perpendicular to $x_1 y_1$ axes (rotated axes)
σ_θ	normal stress on an inclined plane
$\sigma_1, \sigma_2, \sigma_3$	principal normal stresses
σ_{allow}	allowable stress (or working stress)
σ_{cr}	critical stress for a column ($\sigma_{\text{cr}} = P_{\text{cr}}/A$)
σ_{pl}	proportional-limit stress
σ_r	residual stress
σ_T	thermal stress
σ_u	ultimate stress
σ_y	yield stress
τ	shear stress
$\tau_{xy}, \tau_{yz}, \tau_{zx}$	shear stresses on planes perpendicular to the x, y, and z axes and acting parallel to the y, z, and x axes
$\tau_{x_1 y_1}$	shear stress on a plane perpendicular to the x_1 axis and acting parallel to the y_1 axis (rotated axes)
τ_θ	shear stress on an inclined plane

τ_{allow}	allowable stress (or working stress) in shear
τ_u	ultimate stress in shear
τ_y	yield stress in shear
ϕ	angle, angle of twist of a bar in torsion
ψ	angle, angle of rotation
ω	angular velocity, angular frequency ($\omega = 2\pi f$)

* An asterisk attached to a section number indicates a specialized or advanced topic. One or more asterisks attached to a problem number indicate an unusually difficult or lengthy problem.

■ Greek Alphabet

A	α	Alpha	N	ν	Nu
B	β	Beta	Ξ	ξ	Xi
Γ	γ	Gamma	O	o	Omicron
Δ	δ	Delta	Π	π	Pi
E	ϵ	Epsilon	P	ρ	Rho
Z	ζ	Zeta	Σ	σ	Sigma
H	η	Eta	T	τ	Tau
Θ	θ	Theta	Υ	υ	Upsilon
I	ι	Iota	Φ	ϕ	Phi
K	κ	Kappa	X	χ	Chi
Λ	λ	Lambda	Ψ	ψ	Psi
M	μ	Mu	Ω	ω	Omega

1

TENSION, COMPRESSION, AND SHEAR

1.1 INTRODUCTION TO MECHANICS OF MATERIALS

Mechanics of materials is a branch of applied mechanics that deals with the behavior of solid bodies subjected to various types of loading. Other names for this field of study are *strength of materials* and *mechanics of deformable bodies*. The solid bodies considered in this book include bars with axial loads, shafts in torsion, beams in bending, and columns in compression.

The principal objective of mechanics of materials is to determine the stresses, strains, and displacements in structures and their components due to the loads acting on them. If we can find these quantities for all values of the loads up to the loads that cause failure, we will have a complete picture of the mechanical behavior of these structures. An understanding of mechanical behavior is essential for the safe design of all types of structures, whether airplanes and antennas, buildings and bridges, machines and motors, or ships and spacecraft. That is why mechanics of materials is a basic subject in so many engineering fields. Statics and dynamics are also essential, but those subjects deal primarily with the forces and motions associated with particles and rigid bodies. In mechanics of materials we go one step further by examining the stresses and strains inside real bodies, that is, bodies of finite dimensions that deform under loads. To determine the stresses and strains, we use the physical properties of the materials as well as numerous theoretical laws and concepts.

Theoretical analyses and experimental results have equally important roles in mechanics of materials. Often we use theories to derive formulas and equations for predicting mechanical behavior, but these

1

expressions cannot be used in practical design unless the physical properties of the materials are known. Such properties are available only after careful experiments have been carried out in the laboratory. Furthermore, many practical problems are not amenable to theoretical analysis alone, and in such cases physical testing is a necessity.

The historical development of mechanics of materials is a fascinating blend of both theory and experiment—theory has pointed the way to useful results in some instances, and experiment has done so in others. Such famous persons as Leonardo da Vinci (1452–1519) and Galileo Galilei (1564–1642) performed experiments to determine the strength of wires, bars, and beams, although they did not develop adequate theories (by today's standards) to explain their test results. By contrast, the famous mathematician Leonhard Euler (1707–1783) developed the mathematical theory of columns and calculated the critical load of a column in 1744, long before any experimental evidence existed to show the significance of his results. Without appropriate tests to back up his theories, Euler's results remained unused for over a hundred years, although today they are the basis for the design and analysis of most columns.*

When studying mechanics of materials, you will find that your efforts are divided naturally into two parts: first, understanding the logical development of the concepts, and second, applying those concepts to practical situations. The former is accomplished by studying the derivations, discussions, and examples that appear in each chapter, and the latter is accomplished by solving the problems at the ends of the chapters. Some of the problems are numerical in character, and others are symbolic (or algebraic).

An advantage of *numerical problems* is that the magnitudes of all quantities are evident at every stage of the calculations, thus providing an opportunity to judge whether the values are reasonable or not. The principal advantage of *symbolic problems* is that they lead to general-purpose formulas. A formula displays the variables that affect the final results; for instance, a quantity may actually cancel out of the solution, a fact that would not be evident from a numerical solution. Also, an algebraic solution shows the manner in which each variable affects the results, as when one variable appears in the numerator and another appears in the denominator. Furthermore, a symbolic solution provides the opportunity to check the dimensions at every stage of the work. Finally, the most important reason for solving algebraically is to obtain a general formula that can be used for many different problems. In contrast, a numerical solution applies to only one set of circumstances. Because engineers must be adept at both kinds of solutions, you will find a mixture of numeric and symbolic problems throughout this book.

Numerical problems require that you work with specific units of measurement. In keeping with current engineering practice, this book

---

*The history of mechanics of materials, beginning with Leonardo and Galileo, is given in Refs. 1-1, 1-2, and 1-3.

utilizes both the International System of Units (SI) and the U.S. Customary System (USCS). A discussion of both systems appears in Appendix A, where you will also find many useful tables, including a table of conversion factors.

All problems appear at the ends of the chapters, with the problem numbers identifying the sections to which they belong. In the case of problems requiring numerical solutions, odd-numbered problems are in USCS units and even-numbered problems are in SI units. The only exceptions are problems involving commercially available structural-steel shapes, because the properties of these shapes are tabulated in Appendix E in USCS units only.

The techniques for solving problems are discussed in detail in Appendix B. In addition to a list of sound engineering procedures, Appendix B includes sections on dimensional homogeneity and significant digits. These topics are especially important, because every equation must be dimensionally homogeneous and every numerical result must be expressed with the proper number of significant digits. In this book, final numerical results are usually presented with three significant digits when a number begins with the digits 2 through 9, and with four significant digits when a number begins with the digit 1. Intermediate values are often recorded with additional digits to avoid losing numerical accuracy due to rounding of numbers.

1.2 NORMAL STRESS AND STRAIN

The most fundamental concepts in mechanics of materials are **stress** and **strain**. These concepts can be illustrated in their most elementary form by considering a prismatic bar subjected to axial forces. A **prismatic bar** is a straight structural member having constant cross section throughout its length, and an **axial force** is a load directed along the axis of the member, resulting in either tension or compression in the bar. Examples are shown in Fig. 1-1, where the tow bar is a prismatic member in tension and the landing gear strut is a member in compression. Other examples are the members of a bridge truss, connecting rods in automobile engines, spokes of bicycle wheels, columns in buildings, and wing struts in small airplanes.

Fig. 1-1 Structural members subjected to axial loads. (The tow bar is in tension and the landing gear strut is in compression.)

Landing gear strut

Tow bar

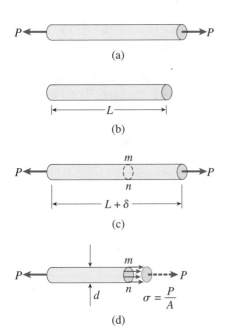

Fig. 1-2 Prismatic bar in tension:
(a) free-body diagram of a segment of the
bar, (b) segment of the bar before loading,
(c) segment of the bar after loading, and
(d) normal stresses in the bar

For discussion purposes, we will consider the tow bar of Fig. 1-1 and isolate a segment of it as a free body (Fig. 1-2a). When drawing this free-body diagram, we disregard the weight of the bar itself and assume that the only active forces are the axial forces P at the ends. Next we consider two views of the bar, the first showing the bar before the loads are applied (Fig. 1-2b) and the second showing it after the loads are applied (Fig. 1-2c). Note that the original length of the bar is denoted by the letter L, and the increase in length is denoted by the Greek letter δ (delta).

The internal stresses in the bar are exposed if we make an imaginary cut through the bar at section mn (Fig. 1-2c). Because this section is taken perpendicular to the longitudinal axis of the bar, it is called a **cross section**. We now isolate the part of the bar to the left of cross section mn as a free body (Fig. 1-2d). At the right-hand end of this free body (section mn) we show the action of the removed part of the bar (that is, the part to the right of section mn) upon the part that remains. This action consists of a continuously distributed force acting over the entire cross section. The intensity of the force (that is, the force per unit area) is called the **stress** and is denoted by the Greek letter σ (sigma). Thus, the axial force P acting at the cross section is the *resultant* of the continuously distributed stresses. (The resultant force is shown with a dashed line in Fig. 1-2d.)

Assuming that the stresses are *uniformly distributed* over cross section mn (Fig. 1-2d), we see that their resultant must be equal to the intensity σ times the cross-sectional area A of the bar. Therefore, we obtain the following expression for the magnitude of the stresses:

$$\sigma = \frac{P}{A} \tag{1-1}$$

This equation gives the intensity of uniform stress in an axially loaded, prismatic bar of arbitrary cross-sectional shape. When the bar is stretched by the forces P, the stresses are **tensile stresses**; if the forces are reversed in direction, causing the bar to be compressed, we obtain **compressive stresses**. Inasmuch as the stresses act in a direction perpendicular to the cut surface, they are called **normal stresses**. Thus, normal stresses may be either tensile or compressive. Later, in Section 1.6, we will encounter another type of stress, called *shear stress*, that acts parallel to the surface.

When a sign convention for normal stresses is required, it is customary to define tensile stresses as positive and compressive stresses as negative.

Because the normal stress σ is obtained by dividing the axial force by the cross-sectional area, it has units of force per unit of area. When USCS units are used, stress is customarily expressed in pounds per square inch (psi) or kips per square inch (ksi).* For instance, suppose

---

*One kip, or kilopound, equals 1000 lb.

that the bar of Fig. 1-2 has a diameter d of 2.0 inches and the load P has a magnitude of 6 kips. Then the stress in the bar is

$$\sigma = \frac{P}{A} = \frac{P}{\pi d^2/4} = \frac{6\text{ k}}{\pi(2.0\text{ in.})^2/4} = 1.91\text{ ksi (or 1910 psi)}$$

In this example the stress is tensile, or positive.

When SI units are used, force is expressed in newtons (N) and area in square meters (m^2). Consequently, stress has units of newtons per square meter (N/m^2), that is, pascals (Pa). However, the pascal is such a small unit of stress that it is necessary to work with large multiples, usually the megapascal (MPa). To demonstrate that a pascal is indeed small, we have only to note that it takes almost 7000 pascals to make 1 psi.* As a numerical example, the stress in the bar described in the preceding paragraph (1.91 ksi) converts to 13.2 MPa, which is 13.2×10^6 pascals. Although it is not recommended in SI, you will sometimes find stress given in newtons per square millimeter (N/mm^2), which is a unit equal to the megapascal (MPa).

The equation $\sigma = P/A$ is valid only if the stress is uniformly distributed over the cross section of the bar. This condition is realized if the axial force P acts through the centroid of the cross-sectional area, as demonstrated later in this section. When the load P does not act at the centroid, bending of the bar will result, and a more complicated analysis is necessary (see Sections 5.12 and 11.5). However, in this book (as in common practice) it is understood that axial forces are applied at the centroids of the cross sections unless specifically stated otherwise.

The uniform stress condition pictured in Fig. 1-2d exists throughout the length of the bar except near the ends. The stress distribution at the end of a bar depends upon how the load P is transmitted to the bar. If the load happens to be distributed uniformly over the end, then the stress pattern at the end will be the same as everywhere else. However, it is more likely that the load is transmitted through a pin or a bolt, producing high localized stresses called **stress concentrations**. One possibility is illustrated by the eyebar shown in Fig. 1-3. In this instance the loads P are transmitted to the bar by pins that pass through the holes (or eyes) at the ends of the bar. Thus, the forces shown in the figure are actually the resultants of bearing pressures between the pins and the eyebar, and the stress distribution around the holes is quite complex. However, as we move away from the ends and toward the middle of the bar, the stress distribution gradually approaches the uniform distribution pictured in Fig. 1-2d.

As a practical rule, the formula $\sigma = P/A$ may be used with good accuracy at any point within a prismatic bar that is at least as far away from the stress concentration as the largest lateral dimension of the bar. In other words, the stress distribution in the bar of Fig. 1-2d is uniform at

Fig. 1-3 Steel eyebar subjected to tensile loads P

---

*Conversion factors between USCS units and SI units are listed in Table A-5, Appendix A.

distances d or greater from the ends, where d is the diameter of the bar, and the stress distribution in the eyebar (Fig. 1-3) is uniform at distances b or greater from the enlarged ends, where b is the width of the bar. More detailed discussions of stress concentrations produced by axial loads are given in Section 2.10.

Of course, even when the stress is *not* uniformly distributed, the equation $\sigma = P/A$ may still be useful because it gives the *average* normal stress on the cross section.

■ Normal Strain

As already observed, a straight bar will change in length when loaded axially, becoming longer when in tension and shorter when in compression. For instance, consider again the prismatic bar of Fig. 1-2. The elongation δ of this bar (Fig. 1-2c) is the cumulative result of the stretching of all elements of the material throughout the volume of the bar. Let us assume that the material is the same everywhere in the bar. Then, if we consider half of the bar (length $L/2$), it will have an elongation equal to $\delta/2$, and if we consider one-fourth of the bar, it will have an elongation equal to $\delta/4$. Similarly, a unit length of the bar will have an elongation equal to $1/L$ times the total elongation δ. By this process we arrive at the concept of *elongation per unit length*, or **strain**, denoted by the Greek letter ϵ (epsilon) and given by the equation

$$\epsilon = \frac{\delta}{L} \tag{1-2}$$

If the bar is in tension, the strain is called a **tensile strain**, representing an elongation or stretching of the material. If the bar is in compression, the strain is a **compressive strain** and the bar shortens. Tensile strain is usually taken as positive and compressive strain as negative. The strain ϵ is called a **normal strain** because it is associated with normal stresses.

Because normal strain is the ratio of two lengths, it is a **dimensionless quantity**, that is, it has no units. Therefore, strain is expressed simply as a number, independent of any system of units. Numerical values of strain are usually very small, because bars made of structural materials undergo only small changes in length when loaded. As an example, consider a steel bar having length L equal to 2.0 m. When heavily loaded in tension, this bar might elongate by 1.4 mm, which means that the strain is

$$\epsilon = \frac{\delta}{L} = \frac{1.4 \text{ mm}}{2.0 \text{ m}} = 0.0007 = 700 \times 10^{-6}$$

In practice, the original units of δ and L are often attached to the strain itself, and then the strain is recorded in forms such as mm/m, μm/m, and in./in. For instance, the strain ϵ in the preceding illustration could be given as 700 μm/m or 700×10^{-6} in./in. Also, strain is sometimes expressed as a percent, especially when the strains are large. (In the preceding example, the strain is 0.07%.)

■ Uniaxial Stress and Strain

The definitions of normal stress and normal strain are based upon purely static and geometric considerations, which means that Eqs. (1-1) and (1-2) can be used for loads of any magnitude and for any material. The principal requirement is that the deformation of the bar be uniform throughout its volume, which in turn requires that the bar be prismatic, the loads act through the centroids of the cross sections, and the material be **homogeneous** (that is, the same throughout all parts of the bar). The resulting state of stress and strain is called **uniaxial stress and strain**. Further discussions of uniaxial stress, including stresses and strains in other than the longitudinal direction of the bar, are given later in Section 2.6. We will also analyze more complicated stress states, such as biaxial stress and plane stress, in Chapter 7.

■ Line of Action of the Axial Forces for a Uniform Stress Distribution

Throughout the preceding discussion of stress and strain in a prismatic bar, we assumed that the normal stress σ was distributed uniformly over the cross section. Now we will demonstrate that this condition is met if the line of action of the axial forces is through the centroid of the cross-sectional area.

Consider a prismatic bar of arbitrary cross-sectional shape subjected to axial forces P that produce uniformly distributed stresses σ (Fig. 1-4a). Also, let p_1 represent the point in the cross section where the line of action of the forces intersects the cross section (Fig. 1-4b). We construct a set of xy axes in the plane of the cross section and denote the coordinates of point p_1 by $\bar{x}$ and $\bar{y}$. To determine these coordinates, we observe that the moments M_x and M_y of the force P about the x and y axes, respectively, must be equal to the corresponding moments of the uniformly distributed stresses.

The moments of the force P are

$$M_x = P\bar{y} \qquad M_y = -P\bar{x} \tag{a,b}$$

in which a moment is considered positive when its vector (using the right-hand rule) acts in the positive direction of the axis.*

The moments of the distributed stresses are obtained by integrating over the cross-sectional area A. The differential force acting on an element of area dA (Fig. 1-4b) is equal to $\sigma \, dA$. The moments of this elemental force about the x and y axes are $\sigma y \, dA$ and $-\sigma x \, dA$, respectively, in which

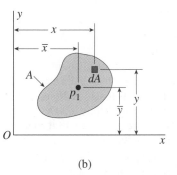

$$\sigma = \frac{P}{A}$$

(a)

(b)

Fig. 1-4 Uniform stress distribution in a prismatic bar: (a) axial forces P, and (b) cross-section of bar

---

*To visualize the right-hand rule, imagine that you grasp an axis of coordinates with your right hand so that your fingers fold around the axis and your thumb points in the positive direction of the axis. Then a moment is positive if it acts about the axis in the same direction as your fingers.

x and y denote the coordinates of the element dA. The total moments are obtained by integrating over the cross-sectional area; thus, we obtain

$$M_x = \int \sigma y \, dA \qquad M_y = -\int \sigma x \, dA \qquad \text{(c,d)}$$

These expressions give the moments produced by the stress σ.

Next, we equate the moments M_x and M_y as obtained from the force P (Eqs. a and b) to the moments obtained from the distributed stress (Eqs. c and d):

$$P\bar{y} = \int \sigma y \, dA \qquad P\bar{x} = \int \sigma x \, dA$$

Because the stress σ is uniformly distributed, we know that it is a constant over the cross-sectional area A and can be placed outside the integral signs. Also, we know that the stress is equal to P/A. Therefore, we obtain the following formulas for the coordinates of point p_1:

$$\bar{y} = \frac{\int y \, dA}{A} \qquad \bar{x} = \frac{\int x \, dA}{A} \qquad \text{(1-3a,b)}$$

These equations are the same as the equations defining the coordinates of the centroid of an area (see Eqs. 12-3a and b in Chapter 12). Therefore, we have now arrived at an important conclusion: *In order to have uniform tension or compression in a prismatic bar, the axial force must act through the centroid of the cross-sectional area.* As explained previously, we always assume these conditions are met unless it is specifically stated otherwise.

The following examples illustrate the calculation of stresses and strains in prismatic bars. In the first example we disregard the weight of the bar and in the second we include it. (In keeping with typical textbook practice, we always omit the weight of a structure when solving a problem unless instructed to include it.)

▪ **Example 1-1**

A short post constructed from a hollow circular tube of aluminum supports a compressive load of 54 kips (Fig. 1-5). The inner and outer diameters of the tube are $d_1 = 3.6$ in. and $d_2 = 5.0$ in., respectively, and its length is 40 in. The shortening of the post due to the load is measured as 0.022 in. Determine the compressive stress and strain in the post. (Disregard the weight of the post itself, and assume that the post does not buckle under the load.)

Solution   Assuming that the compressive load acts at the center of the hollow tube, we can use the equation $\sigma = P/A$ (Eq. 1-1) to calculate the normal stress. The force P equals 54 k (or 54,000 lb), and the cross-sectional area A is

$$A = \frac{\pi}{4}\left(d_2^2 - d_1^2\right) = \frac{\pi}{4}\left[(5.0 \text{ in.})^2 - (3.6 \text{ in.})^2\right] = 9.456 \text{ in.}^2$$

Therefore, the compressive stress in the post is

$$\sigma = \frac{P}{A} = \frac{54,000 \text{ lb}}{9.456 \text{ in.}^2} = 5710 \text{ psi} \qquad \blacktriangleleft$$

54 k

40 in.

Fig. 1-5 Example 1-1. Hollow aluminum post in compression

The compressive strain (from Eq. 1-2) is

$$\epsilon = \frac{\delta}{L} = \frac{0.022 \text{ in.}}{40 \text{ in.}} = 550 \times 10^{-6}$$ ⬅

Thus, the stress and strain in the post have been calculated.

Note: As explained earlier, strain is a dimensionless quantity and no units are needed. For clarity, however, units are often given. In this example, ϵ could be written as 550×10^{-6} in./in. or 550 μin./in.

■ **Example 1-2**

A circular steel rod of length L and diameter d hangs in a mine shaft and holds an ore bucket of weight W at its lower end (Fig. 1-6). (a) Obtain a formula for the maximum stress σ_{max} in the rod, taking into account the weight of the rod itself. (b) Calculate the maximum stress if $L = 40$ m, $d = 8$ mm, and $W = 1.5$ kN.

Solution   (a) The maximum axial force F_{max} in the rod occurs at the upper end and is equal to the weight W plus the weight W_0 of the rod itself. The latter is equal to the weight density γ of the steel times the volume V of the rod, or

$$W_0 = \gamma V = \gamma AL \tag{1-4}$$

in which A is the cross-sectional area of the rod. Therefore, the formula for the maximum stress (from Eq. 1-1) becomes

$$\sigma_{max} = \frac{F_{max}}{A} = \frac{W + \gamma AL}{A} = \frac{W}{A} + \gamma L \tag{1-5}$$ ⬅

(b) To calculate the maximum stress, we substitute numerical values into the preceding equation. The cross-sectional area A equals $\pi d^2/4$, where $d = 8$ mm, and the weight density γ of steel is 77.0 kN/m³ (from Table H-1 in Appendix H). Thus,

$$\sigma_{max} = \frac{1.5 \text{ kN}}{\pi(8 \text{ mm})^2/4} + (77.0 \text{ kN/m}^3)(40 \text{ m})$$

$$= 29.84 \text{ MPa} + 3.11 \text{ MPa} = 33.0 \text{ MPa}$$ ⬅

In this example, the weight of the rod contributes noticeably to the maximum stress and should not be disregarded.

Fig. 1-6 Example 1-2. Steel rod supporting a weight W

1.3   MECHANICAL PROPERTIES OF MATERIALS

The design of machines and structures so that they will function properly requires that we understand the **mechanical behavior** of the materials being used. Ordinarily, the only way to determine how materials behave when they are subjected to loads is to perform experiments in the laboratory. The usual procedure is to place small specimens of the material in testing machines, apply the loads, and then measure the resulting deformations (such as changes in length and changes in diameter). Most materials-testing laboratories are equipped with machines capable of loading specimens in a variety of ways, including both static and dynamic loading in tension and compression.

A typical **tensile-test machine** is shown in Fig. 1-7. The test specimen is installed between the two large grips of the testing machine and then loaded in tension. Measuring devices record the deformations, and the automatic control and data-processing systems (at the left in the photo) tabulate and graph the results.

A more detailed view of a **tensile-test specimen** is shown in Fig. 1-8. The ends of the circular specimen are enlarged where they fit in the grips so that failure will not occur near the grips themselves. A failure at the ends would not produce the desired information about the material, because the stress distribution near the grips is not uniform, as explained in Section 1.2. In a properly designed specimen, failure will occur in the prismatic portion of the specimen where the stress distribution is uniform

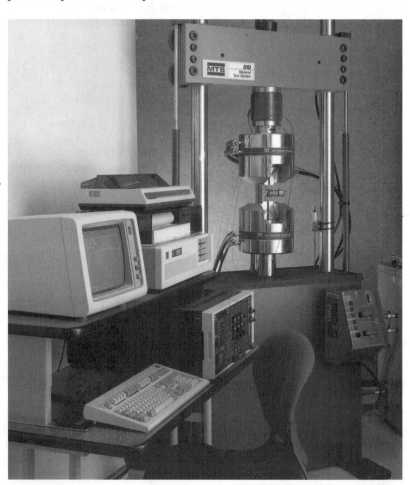

Fig. 1-7 Tensile-test machine with automatic data-processing system. (Courtesy of MTS Systems Corporation)

Gage length

Fig. 1-8 Typical tensile-test specimen with extensometer attached; the specimen has just fractured in tension. (Courtesy of MTS Systems Corporation)

and the bar is subjected only to pure tension. This situation is shown in Fig. 1-8, where the steel specimen has just fractured under load. The device at the left, which is attached by two arms to the specimen, is an **extensometer** that measures the elongation during loading.

In order that test results will be comparable, the dimensions of test specimens and the methods of applying loads have been standardized. One of the major standards organizations is the American Society for Testing and Materials (ASTM), a national technical society that publishes specifications and standards for materials and testing. Other standardizing organizations in the United States are the American Standards Association (ASA) and the National Institute of Standards and Technology (NIST), formerly the National Bureau of Standards (NBS). Similar organizations exist in other countries.

The ASTM standard tension specimen has a diameter of 0.505 in. and a **gage length** of 2.0 in. between the gage marks, which are the points where the extensometer arms are attached to the specimen (see Fig. 1-8). As the specimen is pulled, the axial load is measured and recorded, either automatically or by reading from a dial. The elongation over the gage length is measured simultaneously, either by mechanical gages of the kind shown in Fig. 1-8 or by electrical-resistance strain gages. In a **static test**, the load is applied slowly and the precise *rate* of loading is not of interest because it does not affect the behavior of the specimen. However, in a **dynamic test** the load is applied rapidly and sometimes in a cyclical manner. Since the nature of a dynamic load affects the properties of the materials, the rate of loading must also be measured.

Compression tests of metals are customarily made on small specimens in the shape of cubes or circular cylinders. Cubes are often 2.0 in. on a side, and cylinders usually have diameters of about 1 in. and lengths of 1 to 12 in. Both the load applied by the machine and the shortening of the specimen may be measured. The shortening should be measured over a gage length that is less than the total length of the specimen in order to eliminate end effects.

Concrete is tested in compression on every important construction project to ensure that the required strength has been obtained. The standard ASTM concrete test specimen is 6 in. in diameter, 12 in. in length, and 28 days old (the age of concrete is important because concrete gains strength as it cures). Similar but somewhat smaller specimens are used when performing compression tests of rock (Fig. 1-9).

■ **Stress-Strain Diagrams**

Test results generally depend upon the size of the specimen being tested. Since it is unlikely that we will be designing a structure having parts that are the same size as the test specimens, we need to express the test results in a form that can be applied to members of any size. A simple way to achieve this objective is to convert the test results to stresses and strains.

The axial stress σ in a test specimen is calculated by dividing the axial load P by the cross-sectional area A (see Eq. 1-1). When the initial area of the specimen is used in the calculation, the stress is called the **nominal stress** (other names are *conventional stress* and *engineering stress*). A more exact value of the axial stress, called the **true stress**, can be calculated by using the actual area of the bar at the cross section where failure occurs. Since the actual area in a tension test is always less than the initial area (as illustrated in Fig. 1-8), the true stress is larger than the nominal stress.

The average axial strain ϵ in the test specimen is found by dividing the measured elongation δ between the gage marks by the gage length L (see Fig. 1-8 and Eq. 1-2). If the initial gage length is used in the calculation (for instance, 2.0 in.), then the **nominal strain** is obtained. Since the distance between the gage marks increases as the tensile load is applied, we can calculate the **true strain** (or *natural strain*) at any value of the load by using the actual distance between the gage marks. In tension, true

Fig. 1-9 Rock sample being tested in compression. (Courtesy of MTS Systems Corporation)

strain is always smaller than nominal strain. However, for most engineering purposes, nominal stress and nominal strain are adequate, as explained later in this section.

After performing a tension or compression test and determining the stress and strain at various magnitudes of the load, we can plot a diagram of stress versus strain. Such a **stress-strain diagram** is a characteristic of the particular material being tested and conveys important information about the mechanical properties and type of behavior.*

The first material we will discuss is **structural steel**, also known as *mild steel* or *low-carbon steel*. Structural steel is one of the most widely used metals and is found in buildings, bridges, cranes, ships, towers,

---

*Stress-strain diagrams were originated by Jacob Bernoulli (1654–1705) and J. V. Poncelet (1788–1867); see Ref. 1-4.

vehicles, and many other types of construction. A stress-strain diagram for a typical structural steel in tension is shown in Fig. 1-10. Strains are plotted on the horizontal axis and stresses on the vertical axis. (In order to display all of the important features of this material, the strain axis in Fig. 1-10 is not drawn to scale.)

The diagram begins with a straight line from the origin O to point A, which means that the relationship between stress and strain in this initial region is not only *linear* but also *proportional*.* Beyond point A, the proportionality between stress and strain no longer exists; hence the stress at A is called the **proportional limit**. For low-carbon steels, this limit is in the range 30 to 50 ksi (210 to 350 MPa), but high-strength steels (with higher carbon content plus other alloys) can have proportional limits of more than 80 ksi (550 MPa). The slope of the straight line from O to A is called the **modulus of elasticity**. Because the slope has units of stress divided by strain, modulus of elasticity has the same units as stress. (Modulus of elasticity is discussed later in Section 1.5.)

With an increase in stress beyond the proportional limit, the strain begins to increase more rapidly for each increment in stress. Consequently, the stress-strain curve has a smaller and smaller slope, until, at point B, the curve becomes horizontal (see Fig. 1-10). Beginning at this point, considerable elongation of the test specimen occurs with no noticeable increase in the tensile force (from B to C). This phenomenon is known as **yielding** of the material, and point B is called the **yield point**. The corresponding stress is known as the **yield stress** of the steel. In the region from B to C, the material becomes **perfectly plastic**, which

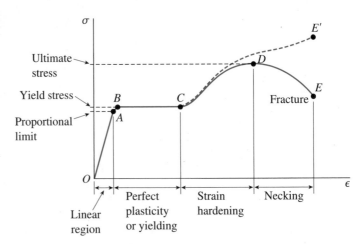

Fig. 1-10 Stress-strain diagram for a typical structural steel in tension (not to scale)

*Two variables are said to be *proportional* if their ratio remains constant. Therefore, a proportional relationship may be represented by a straight line through the origin. However, a proportional relationship is not the same as a *linear* relationship. Although a proportional relationship is linear, the converse is not necessarily true, because a relationship represented by a straight line that does *not* pass through the origin is linear but not proportional. The often-used expression "directly proportional" is synonymous with "proportional" (Ref. 1-5).

Load

Region
of
necking

Region
of
fracture

Load

Fig. 1-11 Necking of a mild-steel bar in
tension

means that it deforms without an increase in the applied load. The elongation of a mild-steel specimen in the perfectly plastic region is typically 10 to 15 times the elongation that occurs in the linear region (between the onset of loading and the proportional limit). The presence of very large strains in the plastic region (and beyond) is the reason for not plotting this diagram to scale.

After undergoing the large strains that occur during yielding in the region *BC*, the steel begins to **strain harden**. During strain hardening, the material undergoes changes in its crystalline structure, resulting in increased resistance of the material to further deformation. Elongation of the test specimen in this region requires an increase in the tensile load, and therefore the stress-strain diagram has a positive slope from *C* to *D*. The load eventually reaches its maximum value, and the corresponding stress (at point *D*) is called the **ultimate stress**. Further stretching of the bar is actually accompanied by a reduction in the load, and fracture finally occurs at a point such as *E* in Fig. 1-10.

The yield stress and ultimate stress of a material are also called the **yield strength** and **ultimate strength**, respectively. **Strength** is a general term that refers to the capacity of a structure to resist loads. For instance, the yield strength of a beam is the magnitude of the load required to cause yielding in the beam, and the ultimate strength of a truss is the maximum load it can support, that is, the failure load. However, when conducting a tension test of a particular material, we define load-carrying capacity by the stresses in the specimen rather than by the total loads acting on the specimen. As a result, the strength of a material is usually stated as a stress.

When a test specimen is stretched, **lateral contraction** occurs, as previously mentioned. The resulting decrease in cross-sectional area is too small to have a noticeable effect on the calculated values of the stresses up to about point *C* in Fig. 1-10, but beyond that point the reduction in area begins to alter the shape of the curve. In the vicinity of the ultimate stress, the reduction in area of the bar becomes clearly visible and a pronounced **necking** of the bar occurs (see Figs. 1-8 and 1-11). If the actual cross-sectional area at the narrow part of the neck is used to calculate the stress, the **true stress-strain curve** (the dashed line *CE′* in Fig. 1-10) is obtained. The total load the bar can carry does indeed diminish after the ultimate stress is reached (as shown by curve *DE*), but this reduction is due to the decrease in area of the bar and not to a loss in strength of the material itself. In reality, the material withstands an increase in true stress up to failure (point *E′*). Because most structures are expected to function at stresses below the proportional limit, the **conventional stress-strain curve** *OABCDE*, which is based upon the original cross-sectional area of the specimen and is easy to determine, provides satisfactory information for use in engineering design.

The diagram of Fig. 1-10 shows the general characteristics of the stress-strain curve for mild steel, but its proportions are not realistic

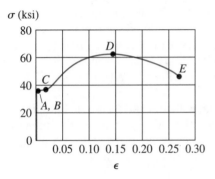

Fig. 1-12 Stress-strain diagram for a typical structural steel in tension (drawn to scale)

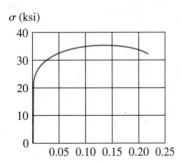

Fig. 1-13 Typical stress-strain diagram for an aluminum alloy

Fig. 1-14 Arbitrary yield stress determined by the offset method

because, as already mentioned, the strain that occurs from B to C may be more than ten times the strain occurring from O to A. Furthermore, the strains from C to E are many times greater than those from B to C. The correct relationships are portrayed in Fig. 1-12, which shows a stress-strain diagram for mild steel drawn to scale. In this figure, the strains from the zero point to point A are so small in comparison to the strains from point A to point E that they cannot be seen, and the initial part of the diagram appears to be a vertical line.

The presence of a clearly defined yield point followed by large plastic strains is an important characteristic of structural steel that is sometimes utilized in practical design (see, for instance, the discussions of elastoplastic behavior in Sections 2.12, 3.12, and 6.9). Metals such as structural steel that undergo large *permanent* strains before failure are classified as **ductile**. For instance, ductility is the property that enables a bar of steel to be bent into a circular arc or drawn into a wire without breaking. A desirable feature of ductile materials is that visible distortions occur if the loads become too large, thus providing an opportunity to take remedial action before an actual fracture occurs. Also, materials exhibiting ductile behavior are capable of absorbing large amounts of strain energy prior to fracture.

Structural steel is an alloy of iron containing about 0.2% carbon, and therefore it is classified as a low-carbon steel. With increasing carbon content, steel becomes less ductile but stronger (higher yield stress and higher ultimate stress). The physical properties of steel are also affected by heat treatment, the presence of other metals, and manufacturing processes such as rolling. Other materials that behave in a ductile manner (under certain conditions) include aluminum, copper, magnesium, lead, molybdenum, nickel, brass, bronze, monel metal, nylon, and teflon.

Although they may have considerable ductility, **aluminum alloys** typically do not have a clearly definable yield point, as shown by the stress-strain diagram of Fig. 1-13. However, they do have an initial linear region with a recognizable proportional limit. Alloys produced for structural purposes have proportional limits in the range 10 to 60 ksi (70 to 410 MPa) and ultimate stresses in the range 20 to 80 ksi (140 to 550 MPa).

When a material such as aluminum does not have an obvious yield point and yet undergoes large strains after the proportional limit is exceeded, an *arbitrary* yield stress may be determined by the **offset method**. A straight line is drawn on the stress-strain diagram parallel to the initial linear part of the curve (Fig. 1-14) but offset by some standard strain, such as 0.002 (or 0.2%). The intersection of the offset line and the stress-strain curve (point A in the figure) defines the yield stress. Because this stress is determined by an arbitrary rule and is not an inherent physical property of the material, it should be distinguished from a true yield stress by referring to it as the **offset yield stress**. For a material such as aluminum, the offset yield stress is slightly above the proportional limit. In the case of structural steel, with its abrupt transition from the linear region to the region of plastic stretching, the offset stress is essentially the same as both the yield stress and the proportional limit.

Rubber maintains a linear relationship between stress and strain up to relatively large strains (as compared to metals). The strain at the proportional limit may be as high as 0.1 or 0.2 (10% or 20%). Beyond the proportional limit, the behavior depends upon the type of rubber (Fig. 1-15). Some kinds of soft rubber will stretch enormously without failure, reaching lengths several times their original lengths. The material eventually offers increasing resistance to the load, and the stress-strain curve turns markedly upward. You can easily sense this characteristic behavior by stretching a rubber band with your hands. (Note that although rubber exhibits very large strains, it is not a ductile material because the strains are not permanent.)

The ductility of a material in tension can be characterized by its elongation and by the reduction in area at the cross section where fracture occurs. The **percent elongation** is defined as follows:

$$\text{Percent elongation} = \frac{L_f - L_0}{L_0}(100) \qquad (1\text{-}6)$$

in which L_0 is the original gage length and L_f is the distance between the gage marks at fracture. Because the elongation is not uniform over the length of the specimen but is concentrated in the region of necking, the percent elongation depends upon the gage length. Therefore, when stating the percent elongation, the gage length should always be given. For a 2 in. gage length, steel may have an elongation in the range from 3% to 40%, depending upon composition; in the case of structural steel, values of 20% or 30% are common. The elongation of aluminum alloys varies from 1% to 45%, depending upon composition and treatment.

The **percent reduction in area** measures the amount of necking that occurs and is defined as follows:

$$\text{Percent reduction in area} = \frac{A_0 - A_f}{A_0}(100) \qquad (1\text{-}7)$$

in which A_0 is the original cross-sectional area and A_f is the final area at the fracture section. For ductile steels, the reduction is about 50%.

Materials that fail in tension at relatively low values of strain are classified as **brittle**. Examples are concrete, stone, cast iron, glass, ceramics, and a variety of metallic alloys. Brittle materials fail with only little elongation after the proportional limit (the stress at point A in Fig. 1-16) is exceeded. Furthermore, the reduction in area is insignificant, and so the nominal fracture stress (point B) is the same as the true ultimate stress. High-carbon steels have very high yield stresses—over 100 ksi (700 MPa) in some cases—but they behave in a brittle manner and fracture occurs at an elongation of only a few percent.

Ordinary **glass** is a nearly ideal brittle material, because it exhibits almost no ductility whatsoever. The stress-strain curve for glass in

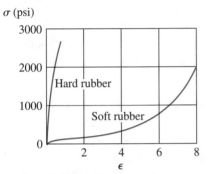

Fig. 1-15 Stress-strain curves for two kinds of rubber in tension

Fig. 1-16 Typical stress-strain diagram for a brittle material showing the proportional limit (point A) and fracture stress (point B)

tension is essentially a straight line, with failure occurring before any yielding takes place. The ultimate stress is about 10,000 psi (70 MPa) for certain kinds of plate glass, but great variations exist, depending upon the type of glass, the size of the specimen, and the presence of microscopic defects. **Glass fibers** can develop enormous strengths, and ultimate stresses over 1,000,000 psi (7 GPa) have been attained.

Many types of **plastics** are used for structural purposes because of their light weight, resistance to corrosion, and good electrical insulation properties. Their mechanical properties vary tremendously, with some plastics being brittle and others ductile. When designing with plastics it is important to realize that their properties are greatly affected by both temperature changes and the passage of time. For instance, the ultimate tensile stress of some plastics is cut in half merely by raising the temperature from 50° F to 120° F. Also, a loaded plastic may stretch gradually over time until it is no longer serviceable. For example, a bar of polyvinyl chloride subjected to a tensile load that initially produces a strain of 0.005 may have that strain doubled after one week, even though the load remains constant. (This phenomenon, known as *creep*, is discussed in the next section.)

Ultimate tensile stresses for plastics are generally in the range 2 to 50 ksi (14 to 350 MPa) and weight densities vary from 50 to 90 lb/ft³ (8 to 14 kN/m³). One type of nylon has an ultimate stress of 12 ksi (80 MPa) and weighs only 70 lb/ft³ (11 kN/m³), which is only 12% heavier than water. Because of its light weight, the strength-to-weight ratio for nylon is about the same as for structural steel (see Prob. 1.3-6).

A **filament-reinforced material** consists of a base material (or *matrix*) in which high-strength filaments, fibers, or whiskers are embedded. The resulting composite material has much greater strength than the base material. As an example, the use of glass fibers can more than double the strength of a plastic matrix. Composites are widely used in aircraft, boats, rockets, and space vehicles where high strength and light weight are needed.

Stress-strain curves for materials in **compression** differ from those in tension. Ductile metals such as steel, aluminum, and copper have proportional limits in compression very close to those in tension, and the initial regions of their compressive and tensile stress-strain diagrams are about the same. However, after yielding begins, the behavior is quite different. In a tension test, the specimen is stretched, necking may occur, and fracture ultimately takes place. When the material is compressed, it bulges outward on the sides and becomes barrel shaped, because friction between the specimen and the end plates prevents lateral expansion. With increasing load, the specimen is flattened out and offers greatly increased resistance to further shortening (which means that the stress-strain curve becomes very steep). These characteristics are illustrated in Fig. 1-17, which shows a compressive stress-strain diagram for copper. Since the actual cross-sectional area of a specimen tested in

Fig. 1-17 Stress-strain diagram for copper in compression

compression is larger than the initial area, the true stress in a compression test is smaller than the nominal stress.

Brittle materials loaded in compression typically have an initial linear region followed by a region in which the shortening increases at a slightly higher rate than does the load. The stress-strain curves for compression and tension often have similar shapes, but the ultimate stresses in compression are much higher than those in tension. Also, unlike ductile materials, which flatten out when compressed, brittle materials actually break at the maximum load.

Tables of mechanical properties of materials are given in Appendix H at the back of the book. The data in the tables are typical of the materials and are suitable for solving problems in this book. However, properties of materials and stress-strain curves vary greatly, even for the same material, because of different manufacturing processes, chemical composition, internal defects, temperature, and many other factors. Consequently, data obtained from Appendix H (or other tables of a similar nature) should not be used for specific engineering or design purposes. Instead, the manufacturers or suppliers should be consulted for information about a particular product.

1.4 ELASTICITY, PLASTICITY, AND CREEP

Stress-strain diagrams portray the behavior of engineering materials when the materials are loaded in tension or compression, as described in the preceding section. To go one step further, let us now consider what happens when the load is removed and the material is *unloaded*. Assume, for instance, that we apply a load to a tensile specimen so that the stress and strain go from the origin O to point A on the stress-strain curve of Fig. 1-18a. Suppose further that when the load is removed, the material follows exactly the same curve back to the origin O. This property of a material, by which it returns to its original dimensions during unloading, is called **elasticity**, and the material itself is said to be *elastic*. Note that the stress-strain curve from O to A need not be linear in order for the material to be elastic.

Now suppose that we load this same material to a higher level, so that point B is reached on the stress-strain curve (Fig. 1-18b). When unloading occurs from point B, the material follows line BC on the diagram. This unloading line is parallel to the initial portion of the loading curve; that is, line BC is parallel to a tangent to the stress-strain curve at the origin. When point C is reached, the load has been entirely removed, but a **residual strain**, or *permanent strain*, represented by line OC, remains in the material. As a consequence, the bar being tested is longer than it was before loading. This residual elongation of the bar is called the **permanent set**. Of the total strain OD developed during

(a)

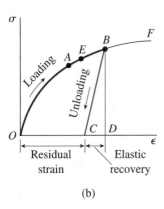

(b)

Fig. 1-18 Stress-strain diagrams illustrating: (a) elastic behavior, and (b) partially elastic behavior

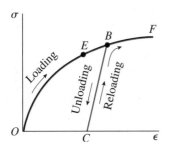

Fig. 1-19 Reloading of a material and raising of the elastic and proportional limits

loading from O to B, the strain CD has been recovered elastically and the strain OC remains as a permanent strain. Thus, during unloading the bar returns partially to its original shape, and so the material is said to be **partially elastic**.

Between points A and B on the stress-strain curve (Fig. 1-18b), there must be a point before which the material is elastic and beyond which the material is partially elastic. To find this point, we load the material to some selected value of stress and then remove the load. If there is no permanent set (that is, if the elongation of the bar returns to zero), then the material is fully elastic up to the selected value of the stress. This process of loading and unloading can be repeated for successively higher values of stress. Eventually, a stress will be reached such that not all the strain is recovered during unloading. By this procedure, it is possible to determine the stress at the upper limit of the elastic region, for instance, the stress at point E in Figs. 1-18a and b. The stress at this point is known as the **elastic limit** of the material.

Many materials, including most metals, have linear regions at the beginning of their stress-strain curves (for example, see Figs. 1-10 and 1-13). The stress at the upper limit of this linear region is the proportional limit. The elastic limit is usually the same as, or slightly above, the proportional limit. Hence, for many materials the two limits are assigned the same numerical value. In the case of mild steel, the yield stress is also very close to the proportional limit, so that for practical purposes the yield stress, the elastic limit, and the proportional limit are assumed to be equal. Of course, this situation does not hold for all materials. Rubber is an outstanding example of a material that is elastic far beyond the proportional limit.

The characteristic of a material by which it undergoes inelastic strains beyond the strain at the elastic limit is known as **plasticity**. Thus, on the stress-strain curve of Fig. 1-18a, we have an elastic region followed by a plastic region. When large deformations occur in a ductile material loaded into the plastic region, the material is said to undergo **plastic flow**.

If the material remains within the elastic range, it can be loaded, unloaded, and loaded again without significantly changing the behavior. However, when loaded into the plastic range, the internal structure of the material is altered and its properties change. For instance, we have already observed that a permanent strain exists in the specimen after unloading from the plastic region (Fig. 1-18b). Now suppose that the material is **reloaded** after such an unloading (Fig. 1-19). The new loading begins at point C on the diagram and continues upward to point B, the point at which unloading began during the first loading cycle. The material then follows the original stress-strain diagram toward point F. Thus, for the second loading, we can imagine that we have a new stress-strain curve with its origin at point C.

During the second loading, the material behaves in a linearly elastic manner from C to B, with the slope of line CB being the same as the slope of the tangent to the original loading curve at the origin O. The proportional limit is now at point B, which is at a higher stress than the original elastic limit (point E). Thus, by stretching a material such as steel or aluminum into the inelastic or plastic range, the *properties of the material are changed*—the linearly elastic region is increased, the proportional limit is raised, and the elastic limit is raised. However, the ductility is reduced because in the "new material" the amount of yielding beyond the elastic limit (from B to F) is less than in the original material (from E to F).*

■ Creep

The stress-strain diagrams described previously were obtained from tension tests involving static loading and unloading of the specimens, and the passage of time did not enter our discussions. However, when loaded for long periods of time, some materials develop additional strains and are said to **creep**. This phenomenon can manifest itself in a variety of ways. For instance, suppose that a vertical bar (Fig. 1-20a) is loaded slowly by a force P, producing an elongation equal to δ_0. Let us assume that the loading and elongation take place during a time interval of duration t_0 (Fig. 1-20b). Subsequent to time t_0, the load remains constant. However, due to creep, the bar may gradually lengthen, as shown in Fig. 1-20b, even though the load does not change. This behavior occurs with many materials, although sometimes the change is too small to be of concern.

As a second example of creep, consider a wire that is stretched between two immovable supports so that it has an initial tensile stress σ_0 (Fig. 1-21). Again, we will denote the time during which the wire is initially loaded as t_0. With the elapse of time, the stress in the wire gradually diminishes, eventually reaching a constant value, even though the supports at the ends of the wire do not move. This process, which is another manifestation of creep, is called **relaxation** of the material.

Creep is usually more important at high temperatures than at ordinary temperatures, and therefore it should always be considered in the design of engines, furnaces, and other structures that operate at elevated temperatures for long periods of time. However, materials such as steel, concrete, and wood will creep slightly even at atmospheric temperatures. For example, creep of concrete over long periods of time can create undulations in bridge decks because of sagging between the supports. One remedy is to construct the deck with an upward **camber**, which is an

Fig. 1-20 Creep in a bar under constant load

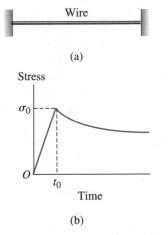

Fig. 1-21 Relaxation of stress in a wire under constant strain

---

*The study of material behavior under various environmental and loading conditions is an important branch of applied mechanics. For more detailed engineering information about materials, consult a textbook devoted solely to this subject.

initial displacement above the horizontal, so that when creep occurs, the spans lower to the level position.

1.5   LINEAR ELASTICITY, HOOKE'S LAW, AND POISSON'S RATIO

Many structural materials, including most metals, wood, plastics, and ceramics, behave both elastically and linearly when first loaded. Consequently, their stress-strain curves begin with a straight line passing through the origin. An example is the stress-strain curve for structural steel (Fig. 1-10), where the region from the origin O to the proportional limit (point A) is both linear and elastic. Other examples are the regions below *both* the proportional limits and the elastic limits on the diagrams for aluminum (Fig. 1-13), rubber (Fig. 1-15), and copper (Fig. 1-17). When a material behaves elastically and also exhibits a linear relationship between stress and strain, it is said to be **linearly elastic**. This type of behavior is extremely important in engineering for an obvious reason—by designing structures and machines to function in this region, we avoid permanent deformations due to yielding.

The linear relationship between stress and strain for a bar in simple tension or compression is expressed by the equation

$$\sigma = E\epsilon \tag{1-8}$$

in which σ is the axial stress, ϵ is the axial strain, and E is a constant of proportionality known as the **modulus of elasticity** for the material. The modulus of elasticity is the slope of the stress-strain diagram in the linearly elastic region, as mentioned previously in Section 1.3. Since strain is dimensionless, the units of E are the same as the units of stress. Typical units of E are psi or ksi in USCS units and pascals (or multiples thereof) in SI units.

The equation $\sigma = E\epsilon$ is commonly known as **Hooke's law**, named for the famous English scientist Robert Hooke (1635–1703). Hooke was the first person to investigate scientifically the elastic properties of materials, and he tested such diverse materials as metal, wood, stone, bone, and sinew. He measured the stretching of long wires supporting weights and observed that the elongations "always bear the same proportions one to the other that the weights do that made them" (Ref. 1-6). Thus, Hooke established the linear relationship between the applied loads and the resulting elongations.

Equation (1-8) is actually a very limited version of Hooke's law because it relates only to the longitudinal stresses and strains developed in simple tension or compression of a bar (*uniaxial stress*). To deal with more complicated states of stress, such as those found in most structures and machines, we must use more extensive equations of Hooke's law (see Sections 7.5 and 7.6).

The modulus of elasticity has relatively large values for materials that are very stiff, such as structural metals. Steel has a modulus of approximately 30,000 ksi (210 GPa); for aluminum, values around 10,600 ksi (73 GPa) are typical. More flexible materials have a lower modulus—values for plastics range from 100 to 2,000 ksi (0.7 to 14 GPa). Some representative values of E are listed in Table H-2, Appendix H. For most materials, the value of E in compression is nearly the same as in tension.

Modulus of elasticity is often called **Young's modulus**, after another English scientist, Thomas Young (1773–1829). In connection with an investigation of tension and compression of prismatic bars, Young introduced the idea of a "modulus of the elasticity." However, his modulus was not the same as the one in use today, because it involved properties of the bar as well as of the material (Ref. 1-7).

■ Poisson's Ratio

When a prismatic bar is loaded in tension, the axial elongation is accompanied by **lateral contraction** (that is, contraction normal to the direction of the applied load). This change in shape is pictured in Fig. 1-22, where part (a) shows the bar before loading and part (b) shows it after loading. In part (b), the dashed lines represent the shape of the bar prior to loading.

Lateral contraction is easily seen by stretching a rubber band, but in metals the changes in lateral dimensions (in the linearly elastic region) are usually too small to be visible. However, they can be detected with sensitive measuring devices.

The **lateral strain** at any point in a bar is proportional to the axial strain at that same point if the material is linearly elastic. However, in order for the lateral strains to be the same throughout the entire bar, additional conditions must be met. First, the axial force must be constant throughout the length of the bar, so that the axial strain is constant. Second, the material must be **homogeneous**, that is, it must have the same composition (and hence the same elastic properties) at every point. Of course, we already assumed that the material was homogeneous so that the stress and strain would be uniform throughout the bar (see Section 1.2). However, it is important to recognize that having a homogeneous material does not mean that the elastic properties are the same in all directions. For instance, the modulus of elasticity could be different in the axial and lateral directions. Therefore, a third condition for uniformity in the lateral strains is that the elastic properties must be the same in all directions *perpendicular* to the longitudinal axis. Materials that are either isotropic or orthotropic (see the definitions given in the next paragraph) meet this condition. When all three conditions are met, as is often the case, the lateral strains in a bar subjected to uniform tension will be the same at every point in the bar and the same in all lateral directions.

(a)

P ←   → P

(b)

Fig. 1-22 Axial elongation and lateral contraction of a bar in tension: (a) bar before loading, and (b) bar after loading. (The deformations of the bar are highly exaggerated.)

Materials having the same properties in *all* directions (axial, lateral, and in between) are said to be **isotropic**. If the properties differ in various directions, the material is **anisotropic** (or **aeolotropic**). A special case of anisotropy occurs when the properties in a particular direction are the same throughout the material and the properties in all directions perpendicular to that direction are the same (but different from the first properties); then the material is classified as **orthotropic**. Fiber-reinforced plastics and concrete reinforced with parallel steel bars are examples of composite materials that exhibit orthotropic behavior.

The ratio of the lateral strain ϵ' to the axial strain ϵ is known as **Poisson's ratio** and is denoted by the Greek letter ν (nu); thus,

$$\nu = -\frac{\text{lateral strain}}{\text{axial strain}} = -\frac{\epsilon'}{\epsilon} \tag{1-9}$$

from which

$$\epsilon' = -\nu\epsilon \tag{1-10}$$

For a bar in tension, the axial strain is positive and the lateral strain is negative (because the width of the bar decreases). For compression we have the opposite situation, with the bar becoming shorter (negative axial strain) and wider (positive lateral strain). Therefore, for ordinary materials Poisson's ratio always has a positive value. (*Note:* When using Eqs. 1-9 and 1-10, we must always keep in mind that they apply only to a bar in uniaxial stress, that is, a bar for which the only stress is the normal stress σ in the axial direction.)

Poisson's ratio is named for the famous French mathematician Siméon Denis Poisson (1781–1840), who attempted to calculate this ratio by a molecular theory of materials (Ref. 1-8). For isotropic materials, Poisson found $\nu = 1/4$. More recent calculations based upon better models of atomic structure give $\nu = 1/3$. Both of these values are close to actual measured values, which are in the range 0.25 to 0.35 for most metals and many other materials. Materials with an extremely low value of Poisson's ratio include cork, for which ν is practically zero, and concrete, for which ν is about 0.1 or 0.2. A theoretical upper limit for Poisson's ratio is 0.5, as explained in the following subsection on volume changes. Rubber comes close to this limiting value.

A table of Poisson's ratios for various materials in the linearly elastic range is given in Appendix H (see Table H-2). For most purposes, Poisson's ratio is assumed to be the same in both tension and compression.

When the strains in a material become large, Poisson's ratio changes. For instance, in the case of structural steel the ratio becomes almost 0.5 when plastic yielding occurs. Thus, Poisson's ratio remains constant only in the linearly elastic range. From a more general view-point, the ratio of lateral strain to axial strain is often called the *contraction ratio*. Of course, in the special case of linearly elastic behavior, the contraction ratio is the same as Poisson's ratio.

■ Volume Change

Because the dimensions of a bar in tension or compression change when the load is applied (Fig. 1-22), the volume of the bar changes too. The change in volume can be calculated from the axial and lateral strains, as follows. Let us take a small element of isotropic material cut from a bar in tension (Fig. 1-23). The original shape of the element (shown by the dashed lines) is a rectangular parallelepiped having sides of lengths a, b, and c in the x, y, and z directions, respectively.* The x axis is taken in the longitudinal direction of the bar, which is the direction of the normal stresses σ produced by the axial forces. The deformed shape of the element is shown by solid lines.

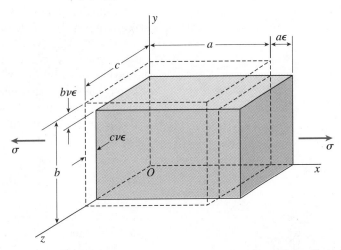

Fig. 1-23 Change in shape of an element cut from a bar in tension. (The original shape is shown by dashed lines; the deformed shape is shown by solid lines.)

The elongation of the element in the direction of loading is $a\epsilon$, where ϵ is the axial strain. Because the lateral strains are $-\nu\epsilon$ (see Eq. 1-10), the lateral dimensions decrease by $b\nu\epsilon$ and $c\nu\epsilon$ in the y and z directions, respectively. Thus, the final dimensions of the element are $a(1 + \epsilon)$, $b(1 - \nu\epsilon)$, and $c(1 - \nu\epsilon)$; therefore, the final volume is

$$V_1 = abc(1 + \epsilon)(1 - \nu\epsilon)(1 - \nu\epsilon) = V_0(1 + \epsilon)(1 - \nu\epsilon)(1 - \nu\epsilon) \quad (1\text{-}11)$$

in which V_0 is the original volume abc.

When the product on the right-hand side of Eq. (1-11) is expanded, we obtain terms involving the square and cube of the axial strain ϵ. In cases where ϵ is very small compared to unity, its square and cube are

---

*A **parallelepiped** is a prism whose bases are parallelograms; thus, a parallelepiped has six faces, each of which is a parallelogram. Opposite faces are parallel and identical parallelograms. A *rectangular parallelepiped* has all faces in the form of rectangles.

negligible in comparison to ϵ itself and may be dropped from the expression. The final volume then becomes

$$V_1 = V_0(1 + \epsilon - 2\nu\epsilon) \tag{1-12}$$

and the change in volume is

$$\Delta V = V_1 - V_0 = V_0\epsilon(1 - 2\nu) \tag{1-13}$$

The **unit volume change** e is defined as the change in volume divided by the original volume, or

$$e = \frac{\Delta V}{V_0} = \epsilon(1 - 2\nu) = \frac{\sigma}{E}(1 - 2\nu) \tag{1-14}$$

The quantity e *is* also known as the **dilatation**.

The preceding equations for volume changes may also be used for compression, in which case the axial strain ϵ is negative and the volume of the bar decreases.

From the equation for the dilatation (Eq. 1-14), we see that the maximum possible value of Poisson's ratio for ordinary materials is 0.5, because any larger value means that the dilatation e becomes negative and the volume decreases when the material is in tension, which seems physically unlikely. As already pointed out, for most materials ν is about 1/4 or 1/3 in the linearly elastic region, which means that the unit volume change is in the range $\epsilon/3$ to $\epsilon/2$.

The following example illustrates the calculation of dimensional changes in a prismatic bar.

■ Example 1-3

A steel pipe of length $L = 4.0$ ft, outside diameter $d_2 = 6.0$ in., and inside diameter $d_1 = 4.5$ in. is compressed by an axial force $P = 140$ k (Fig. 1-24). The material has modulus of elasticity $E = 30,000$ ksi and Poisson's ratio $\nu = 0.30$.

Determine the following quantities for the pipe: (a) the shortening δ, (b) the lateral strain ϵ', (c) the increase Δd_2 in the outer diameter and the increase Δd_1 in the inner diameter, (d) the increase Δt in the wall thickness, (e) the increase ΔV in the volume of material, and (f) the dilatation e.

Solution   The cross-sectional area A and longitudinal stress σ are determined as follows:

$$A = \frac{\pi}{4}(d_2^2 - d_1^2) = \frac{\pi}{4}\left[(6.0 \text{ in.})^2 - (4.5 \text{ in.})^2\right] = 12.37 \text{ in.}^2$$

$$\sigma = -\frac{P}{A} = -\frac{140 \text{ k}}{12.37 \text{ in.}^2} = -11.32 \text{ ksi (compression)}$$

Because the stress is well below the yield stress (see Table H-3, Appendix H), the material behaves linearly elastically and the axial strain may be found from Hooke's law:

$$\epsilon = \frac{\sigma}{E} = \frac{-11.32 \text{ ksi}}{30,000 \text{ ksi}} = -377.3 \times 10^{-6}$$

Fig. 1-24 Example 1-3. Steel pipe in compression

(a) Knowing the axial strain, we can now find the change in length of the pipe (see Eq. 1-2):

$$\delta = \epsilon L = (-377.3 \times 10^{-6})(4.0 \text{ ft})(12 \text{ in./ft}) = -0.018 \text{ in.} \quad \blacktriangleleft$$

The negative sign for δ indicates a shortening of the pipe.

(b) The lateral strain is obtained from Poisson's ratio (see Eq. 1-10):

$$\epsilon' = -\nu\epsilon = -(0.30)(-377.3 \times 10^{-6}) = 113.2 \times 10^{-6} \quad \blacktriangleleft$$

The positive sign for ϵ' indicates an increase in the lateral dimensions, as expected for compression.

(c) The increase in outer diameter equals the lateral strain times the diameter:

$$\Delta d_2 = \epsilon' d_2 = (113.2 \times 10^{-6})(6.0 \text{ in.}) = 0.000679 \text{ in.} \quad \blacktriangleleft$$

Similarly, the increase in inner diameter is

$$\Delta d_1 = \epsilon' d_1 = (113.2 \times 10^{-6})(4.5 \text{ in.}) = 0.000509 \text{ in.} \quad \blacktriangleleft$$

(d) The increase in wall thickness is found in the same manner as the increases in the diameters; thus,

$$\Delta t = \epsilon' t = (113.2 \times 10^{-6})(0.75 \text{ in.}) = 0.000085 \text{ in.} \quad \blacktriangleleft$$

This result can be verified by noting that the increase in wall thickness is equal to the following:

$$\Delta t = \frac{\Delta d_2 - \Delta d_1}{2} = \frac{1}{2}(0.000679 \text{ in.} - 0.000509 \text{ in.}) = 0.000085 \text{ in.}$$

as expected. Note that under compression, all three quantities increase (outer diameter, inner diameter, and thickness).

(e) The change in volume of the material is calculated from Eq. (1-13):

$$\Delta V = V_0 \epsilon (1 - 2\nu) = AL\epsilon(1 - 2\nu)$$

$$= (12.37 \text{ in.}^2)(4.0 \text{ ft})(12 \text{ in./ft})(-377.3 \times 10^{-6})(1 - 0.60) = -0.0896 \text{ in.}^3 \quad \blacktriangleleft$$

The volume change is negative, indicating a decrease in volume, as expected for compression.

(f) Finally, the dilatation (Eq. 1-14) is

$$e = \epsilon(1 - 2\nu) = (-377.3 \times 10^{-6})(1 - 0.60) = -0.00015 \quad \blacktriangleleft$$

which is a 0.015% reduction in the volume of material.

Note: The numerical results obtained in this example illustrate that the dimensional changes in structural materials under normal loading conditions are extremely small. In spite of their smallness, changes in dimensions can be important in certain kinds of analysis (such as the analysis of statically indeterminate structures) and in the experimental determination of stresses and strains.

1.6 SHEAR STRESS AND STRAIN

In the preceding sections we discussed the effects of normal stresses produced by axial loads acting on straight bars. These stresses are called "normal stresses" because they act in directions *perpendicular* to the surface of the material. Now we will consider another kind of stress, called a **shear stress**, that acts *tangential* to the surface of the material.

As an illustration of the action of shear stresses, consider the bolted connection shown in Fig. 1-25a. This connection consists of a flat bar A, a clevis C, and a bolt B that passes through holes in the bar and clevis. Under the action of the tensile loads P, the bar and clevis will press against the bolt in **bearing**, and contact stresses, called **bearing stresses**, will be developed. In addition, the bar and clevis tend to *shear* the bolt, that is, cut through it, and this tendency is resisted by shear stresses in the bolt.

To show more clearly the actions of the bearing and shear stresses, let us look at the connection in a side view (Fig. 1-25b). With this view in mind, we draw a free-body diagram of the bolt (Fig. 1-25c). The bearing stresses exerted by the clevis against the bolt appear on the left-hand side of the free-body diagram and are labeled 1 and 3. The stresses from the bar appear on the right-hand side and are labeled 2. The actual distribution of the bearing stresses is difficult to determine, so it is customary to assume that the stresses are uniformly distributed. Based upon the assumption of uniform distribution, we can calculate an **average bearing stress** σ_b by dividing the total bearing force F_b by the bearing area A_b:

$$\sigma_b = \frac{F_b}{A_b} \tag{1-15}$$

Fig. 1-25 Bolted connection in which the bolt is loaded in double shear

(a)

(b) (c) (d) (e)

The **bearing area** is defined as the projected area of the curved bearing surface. For instance, consider the bearing stresses labeled 1. The projected area A_b on which they act is a rectangle having a height equal to the thickness of the clevis and a width equal to the diameter of the bolt. Also, the bearing force F_b represented by the stresses labeled 1 is equal to $P/2$. The same area and the same force apply to the stresses labeled 3.

Now consider the bearing stresses between the flat bar and the bolt (the stresses labeled 2). For these stresses, the bearing area A_b is a rectangle with height equal to the thickness of the flat bar and width equal to the bolt diameter. The corresponding bearing force is equal to the load P.

The free-body diagram of Fig. 1-25c shows that there is a tendency to shear the bolt along cross sections *mn* and *pq*. From a free-body diagram of the portion *mnpq* of the bolt (see Fig. 1-25d), we see that shear forces V act over the cut surfaces of the bolt. In this particular example there are two planes of shear (*mn* and *pq*), and so the bolt is said to be in **double shear**. In double shear, each of the shear forces is equal to one-half of the total load transmitted by the bolt, that is, $V = P/2$. The shear forces are the resultants of the shear stresses distributed over the cross-sectional area of the bolt. For instance, the shear stresses acting on cross section *mn* are shown in Fig. 1-25e. These stresses act parallel to the cut surface. The exact distribution of the stresses is not known, but they are highest near the center and become zero at certain locations on the edges. As indicated in Fig. 1-25e, shear stresses are customarily denoted by the Greek letter τ (tau).

A bolted connection in **single shear** is shown in Fig. 1-26a, where the axial force P in the metal bar is transmitted to the flange of the steel column through a bolt. A cross-sectional view of the column (Fig. 1-26b) shows the connection in more detail. Also, a sketch of the bolt (Fig. 1-26c) shows the assumed distribution of the bearing stresses acting on the bolt. As mentioned earlier, the actual distribution of the bearing stresses is much more complex than shown in the figure. Furthermore, bearing stresses are also developed against the bolt head and against the nut. Thus, Fig. 1-26c is not a free-body diagram—only the idealized bearing stresses are shown in the figure.

By cutting through the bolt at section *mn* we obtain the bearing stress distribution shown in Fig. 1-26d. This diagram includes the shear force V (equal to the load P) acting on the cross section of the bolt. As already pointed out, this shear force is the resultant of the shear stresses that act over the cross-sectional area of the bolt.

The deformation of a bolt loaded almost to failure in single shear is shown in Fig. 1-27 (compare with Fig. 1-26c).

In the preceding discussions of bolted connections we disregarded any friction between the connecting elements. The presence of friction means that part of the load is carried by friction forces, thereby reducing the loads on the bolts. Since friction forces are unreliable and difficult to

(a)

(b)   (c)   (d)

Fig. 1-26 Bolted connection in which the bolt is loaded in single shear

estimate, it is common practice to err on the conservative side and omit them from the calculations.

The **average shear stress** on the cross section of a bolt is obtained by dividing the total shear force V by the area A of the cross section on which it acts, as follows:

$$\tau_{\text{aver}} = \frac{V}{A} \tag{1-16}$$

Fig. 1-27 Failure of a bolt in single shear

In the example of Fig. 1-26, the shear force V is equal to the load P and the area A is the cross-sectional area of the bolt. However, in the example of Fig. 1-25, the shear force V equals $P/2$.

From Eq. (1-16) we see that shear stresses, like normal stresses, represent intensity of force, or force per unit of area. Thus, the units of shear stress are the same as those for normal stress, namely, psi or ksi in USCS units and pascals in SI units.

The loading arrangements shown in Figs. 1-25 and 1-26 are examples of **direct shear** (or *simple shear*) in which the shear stresses are created by a direct action of the forces in trying to cut through the material. Direct shear arises in the design of bolts, pins, rivets, keys, welds, and glued joints. Shear stresses also arise in an indirect manner when members are subjected to tension, torsion, and bending, as discussed later in Sections 2.6, 3.3, and 5.8, respectively.

■ Equality of Shear Stresses on Perpendicular Planes

To obtain a more complete picture of the action of shear stresses, let us consider a small element of material in the form of a rectangular parallelepiped having sides of lengths a, b, and c in the x, y, and z directions, respectively (Fig. 1-28a). The front and rear faces of the element are free of stress. Now assume that a shear stress τ is distributed uniformly over the top face, which has area ac. In order for the element to be in equilibrium in the x direction, the total shear force τac on the top face must be balanced by an equal but oppositely directed shear force on the bottom face. Since the areas of the top and bottom faces are equal, it follows that the shear stresses on the two faces are equal.

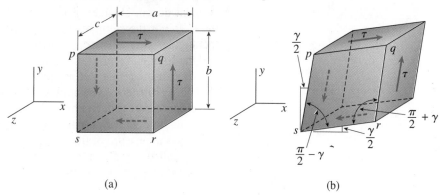

Fig. 1-28 Small element of material subjected to shear stresses and strains

(a)   (b)

The forces τac acting on the top and bottom faces (Fig. 1-28a) form a couple having a moment about the z axis of magnitude τabc, clockwise in the figure.* Equilibrium of the element requires that this moment be balanced by an equal and opposite moment resulting from shear stresses acting on the side faces of the element. Denoting the stresses on the side

---

*A **couple** consists of two parallel forces that are equal in magnitude and opposite in direction.

faces as τ_1, we see that the vertical shear forces equal $\tau_1 bc$. These forces form a counterclockwise couple of moment $\tau_1 abc$. From moment equilibrium about the z axis, we see that τabc equals $\tau_1 abc$, or $\tau_1 = \tau$. Therefore, the magnitudes of the shear stresses on all four faces of the element are equal, as shown in Fig. 1-28a.

In summary, we have arrived at the following general observations:

1. Shear stresses on opposite (and parallel) faces of an element are equal in magnitude and opposite in direction.
2. Shear stresses on adjacent (and perpendicular) faces of an element are equal in magnitude and have directions such that both stresses point toward, or both point away from, the line of intersection of the faces.

These observations were obtained for an element subjected only to shear stresses (no normal stresses), as pictured in Fig. 1-28a. This state of stress is called **pure shear** and is discussed later in greater detail (Section 3.5). However, for most purposes the preceding conclusions remain valid even when normal stresses act on the faces of the element. The reason is that the normal stresses on opposite faces of a small element usually are equal in magnitude and opposite in direction; hence they do not alter the equilibrium equations used in reaching the preceding conclusions.

■ **Shear Strain**

Shear stresses acting on an element of material (Fig. 1-28a) are accompanied by *shear strains*. As an aid in visualizing these strains, we note that the shear stresses have no tendency to elongate or shorten the element in the x, y, and z directions—in other words, the lengths of the sides of the element do not change. Instead, the shear stresses produce a change in the *shape* of the element (Fig. 1-28b). The original element, which is a rectangular parallelepiped, is deformed into an oblique parallelepiped,* and the front and rear faces become rhomboids.**

Because of this deformation, the angles between the side faces change. For instance, the angles at points q and s, which were $\pi/2$ before deformation, are reduced by a small angle γ to $\pi/2 - \gamma$ (Fig. 1-28b). At the same time, the angles at points p and r are increased to $\pi/2 + \gamma$. The angle γ is a measure of the distortion, or change in shape, of the element and is called the **shear strain**. Because shear strain is an angle, it is measured in degrees or radians.

■ **Sign Conventions for Shear Stresses and Strains**

As an aid in establishing sign conventions for shear stresses and strains, we need a scheme for identifying the various faces of a stress element (Fig. 1-28). Henceforth, we will refer to the faces oriented toward the

---

*An **oblique angle** can be either acute or obtuse, but it is *not* a right angle.

A **rhomboid is a parallelogram with oblique angles and adjacent sides *not* equal. (A *rhombus* is a parallelogram with oblique angles and all four sides equal.)

positive directions of the axes as the positive faces of the element. In other words, a positive face has its outward normal directed in the positive direction of a coordinate axis. The opposite faces are negative faces. Thus, in Fig. 1-28a, the right-hand, top, and front faces are the positive *x*, *y*, and *z* faces, respectively, and the opposite faces are the negative *x*, *y*, and *z* faces.

Using the terminology described in the preceding paragraph, we may state the sign convention for shear stresses in the following manner. *A shear stress acting on a positive face of an element is positive if it acts in the positive direction of one of the coordinate axes and negative if it acts in the negative direction of an axis. A shear stress acting on a negative face of an element is positive if it acts in the negative direction of an axis and negative if it acts in a positive direction.* Thus, all shear stresses shown in Fig. 1-28a are positive.

The sign convention for shear strains is as follows: *Shear strain in an element is positive when the angle between two positive faces (or two negative faces) is reduced. The strain is negative when the angle between two positive (or two negative) faces is increased.* Thus, the strains shown in Fig. 1-28b are positive, and we see that positive shear stresses are accompanied by positive shear strains.

■ Hooke's Law in Shear

The properties of a material in shear can be determined experimentally from direct-shear tests or from torsion tests. The latter tests are performed by twisting hollow, circular tubes, thereby producing a state of pure shear, as explained later in Section 3.5. From the results of these tests, we can plot **stress-strain diagrams in shear** (that is, diagrams of shear stress τ versus shear strain γ). These diagrams are similar in shape to tension-test diagrams (σ versus ϵ) for the same materials, although they differ in magnitudes.

From shear stress-strain diagrams, we can obtain material properties such as the proportional limit, modulus of elasticity, yield stress, and ultimate stress. These properties in shear are usually about half as large as those in tension. For instance, the yield stress for structural steel in shear is 0.5 to 0.6 times the yield stress in tension.

For many materials, the initial part of the shear stress-strain diagram is a straight line through the origin, just as it is in tension. For this linearly elastic region, the shear stress and shear strain are proportional, and therefore we have the following equation for Hooke's law in shear:

$$\tau = G\gamma \tag{1-17}$$

in which G is the **shear modulus of elasticity** (also called the *modulus of rigidity*). The shear modulus G has the same units as the tension modulus E, namely, psi or ksi in USCS units and pascals in SI units. For mild steel, typical values of G are 11,000 ksi or 75 GPa; for aluminum alloys, typical values are 4,000 ksi or 28 GPa. Additional values are listed in Table H-2, Appendix H.

The moduli of elasticity in tension and shear are related by the following equation:

$$G = \frac{E}{2(1 + \nu)} \tag{1-18}$$

in which ν is Poisson's ratio. This relationship, which is derived later in Section 3.6, shows that E, G, and ν are not independent elastic properties of the material. Because the value of Poisson's ratio for ordinary materials is between zero and one-half, we see from Eq. (1-18) that G must be from one-third to one-half of E.

The following examples illustrate some typical analyses involving the effects of shear. Example 1-4 is concerned with bearing and shear stresses in pins and bolts, Example 1-5 deals with shear stresses in a punched plate, and Example 1-6 involves finding shear stresses and shear strains in an elastomeric bearing pad subjected to a horizontal shear force.

■ Example 1-4

A steel strut S serving as a brace for a boat hoist transmits a compressive force $P = 54$ kN to the deck of a pier (Fig. 1-29a). The strut has a hollow square cross section with a wall thickness $t = 12$ mm (Fig. 1-29b), and the angle θ between the strut and the horizontal is 40°. A pin through the strut transmits the compressive force from the strut to two gussets G that are welded to the base plate B. Four anchor bolts fasten the base plate to the deck. The diameter of the pin is $d_{pin} = 18$ mm, the thickness of the gussets is $t_G = 15$ mm, the thickness of the base plate is $t_B = 8$ mm, and the diameter of the anchor bolts is $d_{bolt} = 12$ mm.

Determine the following stresses: (a) the bearing stress between the strut and the pin, (b) the shear stress in the pin, (c) the bearing stress between the pin and the gussets, (d) the bearing stress between the anchor bolts and the base plate, and (e) the shear stress in the anchor bolts. (Disregard any friction between the base plate and the deck.)

Solution   (a) *Bearing stress between strut and pin.* The average value of the bearing stress between the strut and the pin is found by dividing the force in the strut by the total bearing area of the strut against the pin. The latter is equal to twice the thickness of the strut (because bearing occurs at two locations) times the diameter of the pin (see Fig. 1-29b). Thus, the bearing stress is

$$\sigma_{b1} = \frac{P}{2td_{pin}} = \frac{54 \text{ kN}}{2(12 \text{ mm})(18 \text{ mm})} = 125 \text{ MPa} \quad \longleftarrow$$

This stress is not excessive for a strut made of structural steel, since the yield stress is probably greater than 200 MPa (see Table H-3, Appendix H).

(b) *Shear stress in pin.* As can be seen from Fig. 1-29b, the pin tends to shear on two planes, namely, the planes between the strut and the gussets.

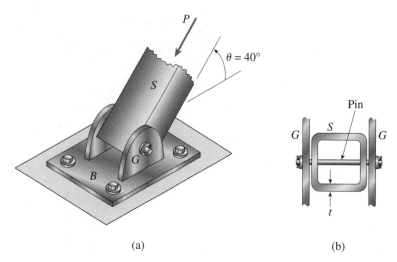

Fig. 1-29 Example 1-4. (a) Pin connection between strut S and base plate B. (b) Cross section through the strut S.

(a)   (b)

Therefore, the average shear stress in the pin (which is in double shear) is equal to the total load applied to the pin divided by twice its cross-sectional area:

$$\tau_{\text{pin}} = \frac{P}{2\pi d_{\text{pin}}^2/4} = \frac{54 \text{ kN}}{2\pi(18 \text{ mm})^2/4} = 106 \text{ MPa} \qquad \longleftarrow$$

The pin would normally be made of high-strength steel (tensile yield stress greater than 340 MPa) and could easily withstand this shear stress (the yield stress in shear is usually at least 50% of the yield stress in tension).

(c) *Bearing stress between pin and gussets.* The pin bears against the gussets at two locations, so the bearing area is twice the thickness of the gussets times the pin diameter; thus,

$$\sigma_{b2} = \frac{P}{2t_G d_{\text{pin}}} = \frac{54 \text{ kN}}{2(15 \text{ mm})(18 \text{ mm})} = 100 \text{ MPa} \qquad \longleftarrow$$

which is less than the bearing stress against the strut.

(d) *Bearing stress between anchor bolts and base plate.* The vertical component of the force P (see Fig. 1-29a) is transmitted to the pier by direct bearing between the base plate and the pier. The horizontal component, however, is transmitted through the anchor bolts. The average bearing stress between the base plate and the anchor bolts is equal to the horizontal component of the force P divided by the bearing area of four bolts. The bearing area for one bolt is equal to the thickness of the plate times the bolt diameter. Consequently, the bearing stress is

$$\sigma_{b3} = \frac{P\cos 40°}{4t_B d_{\text{bolt}}} = \frac{(54 \text{ kN})(\cos 40°)}{4(8 \text{ mm})(12 \text{ mm})} = 108 \text{ MPa} \qquad \longleftarrow$$

(e) *Shear stress in anchor bolts.* The average shear stress in the anchor bolts is equal to the horizontal component of the force P divided by the total cross-sectional area of four bolts (note that each bolt is in single shear). Therefore,

$$\tau_{\text{bolt}} = \frac{P\cos 40°}{4\pi d_{\text{bolt}}^2/4} = \frac{(54 \text{ kN})(\cos 40°)}{4\pi(12 \text{ mm})^2/4} = 91.4 \text{ MPa} \qquad \longleftarrow$$

Any friction between the base plate and the pier would reduce the load on the anchor bolts.

■ **Example 1-5**

A punch for making holes in steel plates is shown in Fig. 1-30a. Assume that a punch having a diameter of 0.75 in. is used to punch a hole in a 1/4 in. plate, as shown in the cross-sectional view (Fig. 1-30b). If a force $P = 28,000$ lb is required, what is the average shear stress in the plate and the average compressive stress in the punch?

Fig. 1-30 Example 1-5. Punching a hole in a steel plate

(a)   (b)

Solution   The average shear stress in the plate is obtained by dividing the force P by the shear area of the plate. The shear area A_s is equal to the circumference of the hole times the thickness of the plate, or

$$A_s = \pi dt = \pi(0.75 \text{ in.})(0.25 \text{ in.}) = 0.5890 \text{ in.}^2$$

in which d is the diameter of the punch and t is the thickness of the plate. Therefore, the average shear stress in the plate is

$$\tau_{\text{aver}} = \frac{P}{A_s} = \frac{28,000 \text{ lb}}{0.5890 \text{ in.}^2} = 47,500 \text{ psi} \qquad \longleftarrow$$

The average compressive stress in the punch is

$$\sigma_c = \frac{P}{A_{\text{punch}}} = \frac{P}{\pi d^2/4} = \frac{28,000 \text{ lb}}{\pi(0.75 \text{ in.})^2/4} = 63,400 \text{ psi} \qquad \longleftarrow$$

in which A_{punch} is the cross-sectional area of the punch.

Note: This analysis is highly idealized because we are disregarding impact effects that occur when a punch is rammed through a plate. (The inclusion of such effects requires advanced methods of analysis that are beyond the scope of mechanics of materials.)

■ Example 1-6

A bearing pad of the kind used to support machines and bridge girders consists of a linearly elastic material (usually an elastomer, such as rubber) capped by a steel plate (Fig. 1-31a). Assume that the thickness of the elastomer is h, the dimensions of the plate are $a \times b$, and the pad is subjected to a horizontal shear force V.

Obtain formulas for the average shear stress τ_{aver} in the elastomer and the horizontal displacement d of the plate (Fig. 1-31b).

Solution   Assume that the shear stresses in the elastomer are uniformly distributed throughout its entire volume. Then the shear stress on any horizontal plane through the elastomer equals the shear force V divided by the area of the plane (Fig. 1-31a):

$$\tau_{\text{aver}} = \frac{V}{ab} \qquad (1\text{-}19)$$

The corresponding shear strain (from Hooke's law in shear) is

$$\gamma = \frac{\tau_{\text{aver}}}{G_e} = \frac{V}{abG_e} \qquad (1\text{-}20)$$

in which G_e is the shear modulus of the elastomeric material. Finally, the horizontal displacement d is equal to $h \tan \gamma$ (from Fig. 1-31b):

$$d = h \tan \gamma = h \tan\left(\frac{V}{abG_e}\right) \qquad (1\text{-}21)$$

In most practical situations the shear strain γ is a small angle, and in such cases we may replace $\tan \gamma$ by γ and obtain

$$d = h\gamma = \frac{hV}{abG_e} \qquad (1\text{-}22)$$

Equations (1-21) and (1-22) give approximate results for the horizontal displacement of the plate because they are based upon the assumption that the shear stress and strain are constant throughout the volume of the elastomeric material. In reality the shear stress is zero at the edges of the material (because there are no shear stresses on the free vertical faces), and therefore the deformation of the material is more complex than pictured in Fig. 1-31b. However, if the length a of the plate is large compared with the thickness h of the elastomer, the preceding results are satisfactory for design purposes.

(a)

(b)

Fig. 1-31 Example 1-6. Bearing pad in shear

1.7 ALLOWABLE STRESSES AND ALLOWABLE LOADS

Engineering has been aptly described as the *application of science to the common purposes of life*. In fulfilling that mission, engineers design a seemingly endless variety of objects to serve the basic needs of society. These needs include housing, agriculture, transportation, communication, and many other aspects of modern life. Factors to be considered in design include functionality, strength, appearance, economics, and environmental protection. However, when studying mechanics of materials, our principal design interest is strength, that is, *the capacity of the object to support or transmit loads.* Objects that must sustain loads include buildings, machines, containers, trucks, aircraft, ships, and the like. For simplicity, we will refer to all such objects as **structures**; thus, *a structure is any object that must support or transmit loads.*

If structural failure is to be avoided, the loads that a structure is capable of supporting must be greater than the loads it will be subjected to when in service. The ability of a structure to resist loads is called **strength**; hence the preceding criterion can be restated as follows: *The actual strength of a structure must exceed the required strength.* The ratio of the actual strength to the required strength is called the **factor of safety** n:

$$\text{Factor of safety } n = \frac{\text{Actual strength}}{\text{Required strength}} \tag{1-23}$$

Of course, the factor of safety must be greater than 1.0 if failure is to be avoided. Depending upon the circumstances, factors of safety from slightly above 1.0 to as much as 10 are used.

The incorporation of factors of safety into design is not a simple matter, because both strength and failure have many different meanings. Strength may be measured by the load-carrying capacity of a structure, or it may be measured by the stress in the material. Failure may mean the fracture and complete collapse of a structure or it may mean that the deformations have become so large that the structure can no longer perform its intended functions. The latter kind of failure may occur at loads much smaller than those that cause actual collapse.

The determination of a factor of safety must also take into account such matters as the following: probability of accidental overloading of the structure by loads that exceed the design loads; types of loads (static or dynamic); whether the loads are applied once or are repeated; how accurately the loads are known; possibilities for fatigue failure (see Section 2.9); inaccuracies in construction; variability in the quality of workmanship; variations in properties of materials; deterioration due to corrosion or other environmental effects; accuracy of the methods of analysis; whether failure is gradual (ample warning) or sudden (no warning); consequences of failure (minor damage or major catastrophe); and other such considerations. If the factor of safety is too low, the likelihood of failure will be high and the structure will be unacceptable; if the factor is too large, the structure will be wasteful of materials and perhaps unsuitable for its function (for instance, it may be too heavy).

Because of these complexities and uncertainties, factors of safety must be determined on a probabilistic basis. They usually are established by groups of experienced engineers who write the codes and specifications used by other designers, and sometimes they are even enacted into law. The provisions of codes and specifications are intended to provide reasonable levels of safety without unreasonable costs.

Factors of safety are defined and implemented in various ways. For many structures, it is important that the material remain within the linearly elastic range in order to avoid permanent deformations when the loads are removed. Under these conditions, the factor of safety is established with respect to yielding of the structure. Yielding begins when the yield stress is reached at *any* point within the structure. Therefore, by applying a factor of safety with respect to the yield stress (or yield strength), we obtain an **allowable stress** (or *working stress*) that must not be exceeded anywhere in the structure. Thus,

$$\text{Allowable stress} = \frac{\text{Yield strength}}{\text{Factor of safety}} \qquad (1\text{-}24)$$

or, for tension and shear, respectively,

$$\sigma_{\text{allow}} = \frac{\sigma_y}{n_1} \quad \text{and} \quad \tau_{\text{allow}} = \frac{\tau_y}{n_2} \qquad (1\text{-}25\text{a,b})$$

in which σ_y and τ_y are the yield stresses and n_1 and n_2 are the factors of safety. In building design, a typical factor of safety with respect to yielding in tension is 1.67; thus, a mild steel having a yield stress of 36 ksi has an allowable stress of 21.6 ksi.

Sometimes the factor of safety is applied to the **ultimate stress** instead of the yield stress. This method is suitable for brittle materials, such as concrete and some plastics, and for materials without a clearly defined yield stress, such as wood and high-strength steels. In these cases the allowable stresses in tension and shear are

$$\sigma_{\text{allow}} = \frac{\sigma_u}{n_3} \quad \text{and} \quad \tau_{\text{allow}} = \frac{\tau_u}{n_4} \qquad (1\text{-}26\text{a,b})$$

in which σ_u and τ_u are the ultimate stresses (or ultimate strengths). Factors of safety with respect to the ultimate strength of a material are usually larger than those based upon yield strength. In the case of mild steel, a factor of safety of 1.67 with respect to yielding corresponds to a factor of approximately 2.8 with respect to the ultimate strength.

In aircraft design it is customary to speak of the **margin of safety** rather than the factor of safety. The margin of safety is defined as the factor of safety minus one:

$$\text{Margin of safety} = n - 1 \qquad (1\text{-}27)$$

Margin of safety is often expressed as a percent, in which case the value given above is multiplied by 100. Thus, a structure having an ultimate strength that is 1.75 times the required strength has a factor of safety of 1.75 and a margin of safety of 0.75 (or 75%). When the margin of safety is reduced to zero or less, the structure (presumably) will fail.

■ **Allowable Loads**

After the allowable stress has been established for a particular material and structure, the **allowable load** on that structure can be determined. The relationship between the allowable load and the allowable stress depends upon the type of structure. In this chapter we are concerned only with the most elementary kinds of structures, namely, bars in tension or compression and pins (or bolts) in direct shear and bearing. In these structures the stresses are uniformly distributed (or at least *assumed* to be uniformly distributed) over an area. For instance, in the case of a bar in tension, the stress is uniformly distributed over the cross-sectional area provided the resultant axial force acts through the centroid of the cross section. The same is true of a bar in compression provided the bar is not subject to buckling. In the case of a pin subjected to shear, we consider only the average shear stress on the cross section, which is equivalent to assuming that the shear stress is uniformly distributed. Similarly, we consider only an average value of the bearing stress acting on the projected area of the pin.

Therefore, in all four of the preceding cases the **allowable load** (also called the *permissible load* or the *safe load*) is equal to the allowable stress times the area over which it acts:

$$\text{Allowable load} = (\text{Allowable stress})(\text{Area}) \qquad (1\text{-}28)$$

For bars in direct tension and compression (no buckling), this equation becomes

$$P_{\text{allow}} = \sigma_{\text{allow}} A \qquad (1\text{-}29)$$

in which σ_{allow} is the permissible normal stress and A is the cross-sectional area of the bar. If the bar has a hole through it, the *net area* is normally used when the bar is in tension. The net area is the gross cross-sectional area minus the area removed by the hole. For compression, the gross area may be used if the hole is filled by a bolt or pin that can transmit the compressive stresses.

For pins in direct shear, Eq. (1-28) becomes

$$P_{\text{allow}} = \tau_{\text{allow}} A \qquad (1\text{-}30)$$

in which τ_{allow} is the permissible shear stress and A is the area over which the shear stresses act. If the pin is in single shear, the area is the cross-sectional area of the pin; in double shear, it is twice the cross-sectional area.

Finally, the permissible load based upon bearing is

$$P_{\text{allow}} = \sigma_b A_b \qquad (1\text{-}31)$$

in which σ_b is the allowable bearing stress and A_b is the projected area of the pin or other surface over which the bearing stresses act.

The following example illustrates how allowable loads are determined when the allowable stresses for the material are known.

■ **Example 1-7**

A steel bar serving as a vertical hanger to support heavy machinery in a factory building is attached to a support by the bolted connection shown in Fig. 1-32. The main part of the hanger has rectangular cross section with width $b_1 = 1.5$ in. and thickness $t = 0.5$ in. At the connection the hanger is enlarged to a width $b_2 = 3.0$ in. The bolt, which transfers the load from the hanger to the two gussets, has diameter $d = 1.0$ in.

Determine the allowable value of the tensile load P in the hanger based upon the following considerations: (a) The allowable stress in the main part of the hanger is 16,000 psi. (b) The allowable stress in the hanger at its cross section through the bolt is 11,000 psi. (The permissible stress at this section is lower because of the stress concentrations around the hole.) (c) The allowable bearing stress between the hanger and the bolt is 26,000 psi. (d) The allowable shear stress in the bolt is 6,500 psi. (*Note:* The factors of safety for tension, bearing, and shear were taken into account when determining the allowable stresses.)

$b_2 = 3.0$ in.

Bolt

Washer

$d = 1.0$ in.

Gusset

Hanger

$t = 0.5$ in.

$b_1 = 1.5$ in.

P

P

(a)

(b)

Fig. 1-32 Example 1-7. Vertical hanger subjected to a tensile load P: (a) front view of bolted connection, and (b) side view of connection

Solution   (a) The allowable load P_1 based upon the stress in the main part of the hanger is equal to the allowable stress in tension times the cross-sectional area of the hanger (Eq. 1-29):

$$P_1 = \sigma_{\text{allow}} A = \sigma_{\text{allow}} b_1 t = (16{,}000 \text{ psi})(1.5 \text{ in.} \times 0.5 \text{ in.}) = 12{,}000 \text{ lb}$$

A load greater than this value will overstress the main part of the hanger, that is, the actual stress will exceed the allowable stress, thereby reducing the factor of safety.

(b) At the cross section through the bolt, we must make a similar calculation but with a different allowable stress and a different area. The net cross-sectional area, that is, the area that remains after the hole is drilled through the bar, is equal to the net width times the thickness. The net width is equal to the gross width b_2 minus the diameter d of the hole. Thus, the equation for the allowable load P_2 at this section is

$$P_2 = \sigma_{allow} A = \sigma_{allow}(b_2 - d)t = (11{,}000 \text{ psi})(3.0 \text{ in.} - 1.0 \text{ in.})(0.5 \text{ in.}) = 11{,}000 \text{ lb}$$

(c) The bearing area between the hanger and the bolt is the projection of the actual contact area. This projected area is equal to the bolt diameter times the thickness of the hanger, and therefore the allowable load based upon bearing (Eq. 1-31) is

$$P_3 = \sigma_b A = \sigma_b dt = (26{,}000 \text{ psi})(1.0 \text{ in.})(0.5 \text{ in.}) = 13{,}000 \text{ lb}$$

(d) Finally, the allowable load P_4 based upon shear in the bolt is equal to the allowable shear stress times the shear area (Eq. 1-30). The shear area is twice the area of the bolt because the bolt is in double shear; thus:

$$P_4 = \tau_{allow} A = \tau_{allow}(2)(\pi d^2/4) = (6{,}500 \text{ psi})(2)(\pi)(1.0 \text{ in.})^2/4 = 10{,}200 \text{ lb}$$

Comparing the four preceding results, we see that the smallest value of the load is

$$P_{allow} = 10{,}200 \text{ lb} \qquad \longleftarrow$$

This load, which is based upon shear in the bolt, is the allowable tensile load in the hanger.

1.8 ∎ DESIGN FOR AXIAL LOADS AND DIRECT SHEAR

In the preceding section we discussed the determination of allowable loads for simple structures, and in earlier sections we saw how to find the stresses, strains, and deformations of bars. The determination of such quantities is known as **analysis**. In the context of mechanics of materials, analysis consists of determining the *response* of a structure to loads, temperature changes, and other physical actions. By the response of a structure, we mean the stresses, strains, and deformations produced by the loads. Response also refers to the load-carrying capacity of a structure; for instance, the allowable load on a structure is a form of response. A structure is said to be *known* (or *given*) when we have a complete physical description of the structure, that is, when we know all of its *properties*. The properties of a structure include the types of members and how they are arranged, the dimensions of all members, the types of supports and where they are located, the materials used, and the properties of the materials. Thus, when analyzing a structure, *the properties are given and the response is to be determined.*

The inverse process is called **design**. When designing a structure, *we must determine the properties of the structure in order that the structure will support the loads and perform its intended functions.* For instance, a common design problem is to determine the size of a member to support given loads. Designing a structure is usually a much lengthier and more difficult process than analyzing it—indeed, analyzing a structure, often more than once, is typically part of the design process.

In this section we will deal with design in its most elementary form by calculating the required sizes of simple tension and compression members as well as pins and bolts loaded in shear. In these cases the design process is quite straightforward. Knowing the loads to be transmitted and the allowable stresses in the materials, we can calculate the required areas of members from the following general relationship (compare with Eq. 1-28):

$$\text{Required area} = \frac{\text{Load to be transmitted}}{\text{Allowable stress}} \tag{1-32}$$

This equation can be applied to any structure in which the stresses are uniformly distributed over the area. (The use of this equation for finding the size of a bar in tension and the size of a bolt in shear is illustrated in Example 1-8 that follows.)

In addition to **strength** considerations, as exemplified by Eq. (1-32), the design of a structure is likely to involve **stiffness** and **stability**. Stiffness refers to the ability of the structure to resist changes in shape (for instance, to resist stretching, bending, or twisting), and stability refers to the ability of the structure to resist buckling under compressive stresses. Limitations on stiffness are sometimes necessary to prevent excessive deformations, such as large deflections of a beam that might interfere with its performance. Buckling is the principal consideration in the design of columns, which are slender compression members (Chapter 11).

Another part of the design process is **optimization**, which is the task of designing the best structure to meet a particular goal, such as minimum weight. For instance, there may be many structures that will support a given load, but in some circumstances the best structure will be the lightest one. Of course, a goal such as minimum weight usually must be balanced against more general considerations, including the aesthetic, economic, environmental, political, and technical aspects of the particular design project.

When analyzing or designing a structure, we refer to the forces that act on it as either **loads** or **reactions**. Loads are *active forces* that are applied to the structure by some external cause, such as gravity or water pressure. Reactions are *passive forces* that are induced at the supports of the structure—their magnitudes and directions are determined by the nature of the structure itself. Thus, reactions must be calculated as part of the analysis, whereas loads are known in advance.

When drawing **free-body diagrams**, it is helpful to distinguish reactions from loads or other applied forces. A common scheme is to place a slash, or slanted line, across the arrow when it represents a reactive force. This convention is illustrated in the following example and elsewhere throughout the book.

■ **Example 1-8**

The two-bar truss *ABC* shown in Fig. 1-33 has pin supports at points *A* and *C*, which are 2.0 m apart. Members *AB* and *BC* are steel bars, pin connected at joint *B*. The length of bar *BC* is 3.0 m. A sign weighing 5.4 kN is suspended from bar *BC* at points *D* and *E*, which are located 0.8 m and 0.4 m, respectively, from the ends of the bar.

Determine the required cross-sectional area of bar *AB* and the required diameter of the pin at support *C* if the allowable stresses in tension and shear are 125 MPa and 45 MPa, respectively. (*Note:* The pins at the supports are in double shear. Also, disregard the weights of members *AB* and *BC*.)

Fig. 1-33 Example 1-8. Two-bar truss *ABC* supporting a sign of weight *W*

Solution   *Reactions, forces in the bars, and shear force in the pin.* The tensile force F_{AB} in bar *AB* and the shear force V_C acting on the pin at *C* are found from equilibrium. We begin with a free-body diagram of the entire truss (Fig. 1-34a). On this diagram we show all forces acting on the truss—namely, the loads from the weight of the sign and the reactive forces exerted by the pin supports at *A* and *C*. Each reaction is shown by its horizontal and vertical components, with the resultant reaction shown by a dashed line. (Note the use of slashes across the arrows to distinguish reactions from loads.)

The horizontal component R_{AH} of the reaction at support *A* is obtained by summing moments about point *C*, as follows (counterclockwise moments are positive):

$$\Sigma M_C = 0 \qquad R_{AH}(2.0 \text{ m}) - (2.7 \text{ kN})(0.8 \text{ m}) - (2.7 \text{ kN})(2.6 \text{ m}) = 0$$

(a)

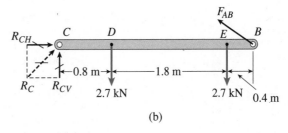

Fig. 1-34 Free-body diagrams for Example 1-8

(b)

Solving this equation, we get

$$R_{AH} = 4.590 \text{ kN}$$

Next, we sum forces in the horizontal direction and obtain

$$\Sigma F_{\text{horiz}} = 0 \qquad R_{CH} = R_{AH} = 4.590 \text{ kN}$$

To obtain the vertical component of the reaction at support C, we need a free-body diagram of member BC, as shown in Fig. 1-34b. Summing moments about joint B gives the desired reaction component:

$$\Sigma M_B = 0 \qquad -R_{CV}(3.0 \text{ m}) + (2.7 \text{ kN})(2.2 \text{ m}) + (2.7 \text{ kN})(0.4 \text{ m}) = 0$$

$$R_{CV} = 2.340 \text{ kN}$$

Now we return to the free-body diagram of the entire truss (Fig. 1-34a) and sum forces in the vertical direction to obtain the vertical component R_{AV} of the reaction at A:

$$\Sigma F_{\text{vert}} = 0 \qquad R_{AV} + R_{CV} - 2.7 \text{ kN} - 2.7 \text{ kN} = 0$$

$$R_{AV} = 3.060 \text{ kN}$$

Knowing the horizontal and vertical components of the reaction at A, we can find the reaction itself:

$$R_A = \sqrt{(R_{AH})^2 + (R_{AV})^2} = 5.516 \text{ kN}$$

Because we are disregarding the weight of bar AB, this reactive force is equal to the tensile force F_{AB} in the member:

$$F_{AB} = 5.516 \text{ kN}$$

The shear force V_C acting on the pin at C is equal to the reaction R_C (Fig. 1-34a). This force is obtained from its components R_{CH} and R_{CV}, as follows:

$$R_C = V_C = \sqrt{(R_{CH})^2 + (R_{CV})^2} = 5.152 \text{ kN}$$

Thus, we have now found the tensile force F_{AB} in bar AB and the shear force V_C acting on the pin at C.

Required areas. The required area of bar AB is calculated by dividing the tensile force by the allowable stress, inasmuch as the stress is uniformly distributed over the cross section (see Eq. 1-32):

$$A_{AB} = \frac{F_{AB}}{\sigma_{\text{allow}}} = \frac{5.516 \text{ kN}}{125 \text{ MPa}} = 44.1 \text{ mm}^2 \qquad \longleftarrow$$

Bar AB must be designed with a cross-sectional area equal to or greater than 44.1 mm^2 in order to support the weight of the sign, which is the only load we considered. For instance, if the bar is round, its calculated diameter must be at least 7.50 mm, and therefore a diameter of perhaps 8 or 10 mm would be used. (In practice, other loads besides the weight of the sign would have to be considered before making a final decision about the size of the bar. Loads that might be important include wind loads, earthquake loads, the weight of people working on the truss or on the sign, and the weight of the truss itself.)

The required area of the pin at C (double shear) is

$$A_{\text{pin}} = \frac{V_C}{2\tau_{\text{allow}}} = \frac{5.152 \text{ kN}}{2(45 \text{ MPa})} = 57.2 \text{ mm}^2$$

from which we can calculate the required diameter:

$$d_{\text{pin}} = \sqrt{4A_{\text{pin}}/\pi} = 8.54 \text{ mm} \qquad \longleftarrow$$

A pin of at least this diameter is needed to support the weight of the sign without exceeding the allowable shear stress.

Note: In this example we intentionally omitted the weight of the truss from the calculations. However, once the sizes of the members are known, their weights can be calculated and included in the free-body diagrams of Fig. 1-34. For the purpose of finding the reactions, the weights can be treated as concentrated loads acting at the midpoints of the members, although we recognize that in reality the weights are uniformly distributed along the axes of the bars.

When the weights of the bars are included, the design of member AB becomes more complicated, because it is no longer a bar in simple tension. Instead, it is a beam subjected to bending as well as tension. An analogous situation exists for member BC. Not only because of its own weight but also because of the weight of the sign, member BC is subjected to both bending and compression. The design of such members must wait until we study stresses in beams (Chapter 5).

PROBLEMS/CHAPTER 1

NORMAL STRESS AND STRAIN

1.2-1 A metal bar *ABC* having two different cross-sectional areas is loaded by an axial force *P* (see figure). Parts *AB* and *BC* are circular in cross section with diameters 1.75 in. and 1.25 in., respectively. If the normal stress in part *AB* is 5000 psi, what is the normal stress σ_{BC} in part *BC*?

Prob. 1.2-1

1.2-2 Calculate the compressive stress σ_c in the piston rod (see figure) when a force $P = 40$ N is applied to the brake pedal. (Note that the line of action of the force *P* is parallel to the piston rod. Also, the diameter of the piston rod is 5 mm, and the other dimensions shown in the figure are measured perpendicular to the line of action of the force *P*.)

Prob. 1.2-2

1.2-3 A circular aluminum tube of length $L = 20.0$ in. is loaded in compression by forces *P* (see figure). The outside and inside diameters are 2.4 in. and 2.0 in., respectively. A strain gage is placed on the outside of the bar to measure normal strains in the longitudinal direction. (a) If the measured strain is $\epsilon = 570 \times 10^{-6}$, what is the shortening δ of the bar? (b) If the compressive stress in the bar is intended to be 6 ksi, what should be the load *P*?

Prob. 1.2-3

1.2-4 The cross section of a concrete pedestal that is loaded uniformly in compression is shown in the figure. (a) Determine the average compressive stress σ_c in the concrete if the load is equal to 11.5 MN. (b) Determine the coordinates $\bar{x}$ and $\bar{y}$ of the point where the resultant load must act in order to produce uniform normal stress.

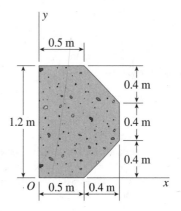

Prob. 1.2-4

1.2-5 A steel wire *ABC* supporting a lamp at its midpoint is attached to supports that are 5 ft apart (see figure). The length of the wire is 6 ft and its diameter is 20 mils (it is common practice to specify wire diameters in mils; one mil equals 0.001 in.). If the lamp weighs 13 lb, what is the tensile stress σ_t in the wire?

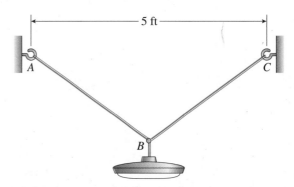

Prob. 1.2-5

1.2-6 A long retaining wall is braced by wood shores set at an angle of 30° and supported by concrete thrust blocks, as shown in part (a) of the figure. The shores are evenly spaced, 3 meters apart. For analysis purposes, the wall and shores are idealized as shown in part (b) of the figure, with the base of the wall and both ends of the shores assumed to be pinned. The pressure of the soil against the wall is assumed to be triangularly distributed, and the resultant force acting on a 3-meter length of the wall is $F = 190$ kN. If each shore has a square cross section, 150×150 mm, what is the compressive stress σ_c in the shores?

Prob. 1.2-6

1.2-7 A car weighing 16 k when fully loaded is pulled slowly up a steep inclined track by a steel cable (see figure). The cable has an effective cross-sectional area of 0.471 in.^{2}, and the angle of the incline is 31°. What is the tensile stress σ_t in the cable?

Prob. 1.2-7

1.2-8 A cable and strut assembly ABC (see figure) supports a vertical load $P = 12$ kN. The cable has an effective cross-sectional area of 160 mm^2, and the strut has an area of 340 mm^2. (a) Calculate the normal stresses σ_{AB} and σ_{BC} in the cable and strut, respectively, and indicate whether they are tension or compression. (b) If the cable elongates 1.1 mm, what is the strain? (c) If the strut shortens 0.37 mm, what is the strain?

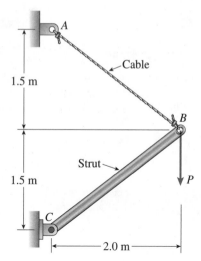

Prob. 1.2-8

1.2-9 An arrangement for pulling a pipe out of the ground is shown in the figure. A cable ABC is attached to the pipe at A and to the sturdy frame at C. A second cable BDE is then clamped to the first cable at B and attached to the frame at E. Finally, a person pulls vertically downward on the second cable at D with a force Q. The lengths of the cables are such that AB is vertical, BD is horizontal, BC is inclined by a small angle α to the vertical, and DE is inclined by a small angle β to the horizontal. Both cables have the same cross-sectional area A_c. (a) Obtain formulas for the lifting force P and the maximum tensile stresses σ_{BC} and σ_{DE} in cables ABC and BDE, respectively. (b) Calculate the lifting force P and the maximum stresses for the following data: $Q = 120$ lb, $\alpha = 6°$, $\beta = 8°$, and $A_c = 0.119$ in.^{2}

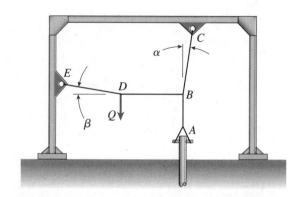

Prob. 1.2-9

1.2-10 A pump moves a piston up and down in a deep water well (see figure). The pump rod has diameter $d = 20$ mm and length $L = 110$ m. The rod is made of steel having weight density $\gamma = 77.0$ kN/m^3. The resisting force associated with the piston during the downstroke is 900 N and during the upstroke is 10,800 N. Determine the maximum tensile and compressive stresses in the pump rod due to the combined effects of the resistance forces and the weight of the rod. (Note that for simplicity we are disregarding other considerations, such as bending of the pump rod and dynamic effects.)

Prob. 1.2-10

1.2-11 A reinforced concrete slab 8 ft square and 9 in. thick is lifted by four cables attached to the corners, as shown in the figure. The cables are attached to a hook at a point 5 ft above the top of the slab. The cables have an effective cross-sectional area $A = 0.12$ in.^{2} Determine the tensile stress σ_t in the cables. (The concrete slab weighs 150 lb/ft^3.)

Prob. 1.2-11

****1.2-12** A round bar ACB of total length $2L$ (see figure) rotates about an axis through the midpoint C with constant angular speed ω (radians per second). The material of the bar has weight density γ. (a) Derive a formula for the tensile stress σ_x in the bar as a function of the distance x from the midpoint C. (b) What is the maximum tensile stress σ_{max}?

Prob. 1.2-12

MECHANICAL PROPERTIES AND STRESS-STRAIN DIAGRAMS

1.3-1 A long steel wire hangs vertically in a dry well under its own weight (see figure). (a) What is the greatest length (feet) it can have without yielding if the yield stress of the steel is 36 ksi? (b) If the same wire hangs from a ship at sea, what is the greatest length? (The weight densities of steel and sea water are available in Table H-1, Appendix H.)

Prob. 1.3-1

1.3-2 Three different materials A, B, and C are tested in tension using test specimens having diameters of 12 mm and gage lengths of 50 mm (see figure). At failure, the distances between the gage marks are found to be 54.1, 63.0, and 70.6 mm, respectively. Also, the diameters are 11.50, 9.46, and 6.01 mm, respectively, at the failure cross sections. Determine the percent elongation and percent reduction in area of each specimen, and then, using your own judgment, classify each material as brittle or ductile.

Gage length

Probs. 1.3-2 and 1.3-5

1.3-3 A symmetrical framework consisting of three pin-jointed bars is loaded by a force P (see figure). The angle between the inclined bars and the horizontal is $\alpha = 50°$. The axial strain in the middle bar is measured as 0.049. Determine the tensile stress in the outer bars if they are constructed of an aluminum alloy having the stress-strain diagram shown in Fig. 1-13.

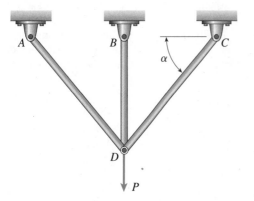

Prob. 1.3-3

1.3-4 A specimen of a methacrylate plastic is tested in tension at room temperature (see figure), producing the stress-strain data listed in the accompanying table. Plot the stress-strain curve and determine the proportional limit, modulus of elasticity (i.e., the slope of the initial part of the stress-strain curve), and yield stress at 0.2% offset. Is the material ductile or brittle?

Prob. 1.3-4

Stress-Strain Data for Problem 1.3-4

Stress (MPa)	Strain
8.0	0.0032
17.5	0.0073
25.6	0.0111
31.1	0.0129
39.8	0.0163
44.0	0.0184
48.2	0.0209
53.9	0.0260
58.1	0.0331
62.0	0.0429
62.1	fracture

1.3-5 The data shown in the accompanying table were obtained from a tensile test of high-strength steel. The test specimen had a diameter of 0.505 in. and a gage length of 2.00 in. (see figure). The total elongation between the gage marks at fracture was 0.12 in., and the minimum diameter was 0.42 in. Plot the nominal stress-strain curve for the steel and determine the proportional limit, modulus of elasticity (i.e., the slope of the initial part of the stress-strain curve), yield stress at 0.1% offset, ultimate stress, percent elongation in 2.00 in., and percent reduction in area.

Tensile-Test Data for Problem 1.3-5

Load (lb)	Elongation (in.)
1,000	0.0002
2,000	0.0006
6,000	0.0019
10,000	0.0033
12,000	0.0039
12,900	0.0043
13,400	0.0047
13,600	0.0054
13,800	0.0063
14,000	0.0090
14,400	0.0102
15,200	0.0130
16,800	0.0230
18,400	0.0336
20,000	0.0507
22,400	0.1108
22,600	fracture

1.3-6 The *strength-to-weight ratio* of a structural material is defined as its load-carrying capacity divided by its weight. For materials in tension, we use a characteristic tensile stress (as obtained from a stress-strain curve) as the measure of strength. For instance, either the yield stress or the ultimate stress could be used, depending upon the particular application. Thus, the strength-to-weight ratio $R_{s/w}$ for a material in tension is defined as

$$R_{s/w} = \frac{\sigma}{\gamma}$$

in which σ is the characteristic stress and γ is the weight density. Note that the ratio has units of length.

Using the ultimate stress σ_u as the strength parameter, calculate the strength-to-weight ratio (in units of meters) for each of the following materials: aluminum alloy 6061-T6, Douglas fir, nylon, structural steel ASTM-A572, and a titanium alloy. (Obtain the material properties from Tables H-1 and H-3 of Appendix H. When a range of values is given in a table, use the average value.)

1.3-7 The mechanical properties of structural materials are available in various reference books and reports. Go to your technical library and find an actual stress-strain diagram for a structural material of your choice, such as aluminum, copper, fiberglass, magnesium, monel, nylon, or titanium. The diagram may be in either USCS or SI units. (a) Make a photocopy of the diagram and use it to determine as many mechanical properties as you can. (b) Based upon the information from part (a) plus supplementary reading, prepare a one-page description of the material.

ELASTICITY AND PLASTICITY

1.4-1 A bar made of structural steel having the stress-strain curve shown in the figure has a length of 4 ft. The yield stress of the steel is 42 ksi and the slope of the initial linear part of the stress-strain curve (modulus of elasticity) is 30×10^3 ksi. The bar is loaded axially until it elongates 0.2 in., and then the load is removed. How does the final length of the bar compare with its original length? (*Hint:* Use the concepts illustrated in Fig. 1-18.)

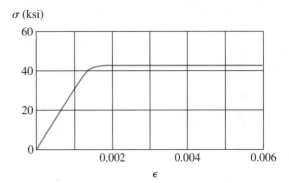

Prob. 1.4-1

1.4-2 A bar of length 1.5 m is made of a structural steel having the stress-strain curve shown in the figure. The yield stress of the steel is 250 MPa and the slope of the initial linear part of the stress-strain curve (modulus of elasticity) is 200 GPa. The bar is loaded axially until it elongates 7.5 mm, and then the load is removed. How does the final length of the bar compare with its original length? (*Hint:* Use the concepts illustrated in Fig. 1-18.)

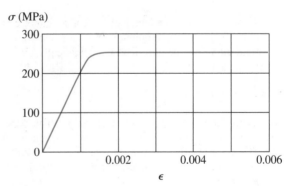

Prob. 1.4-2

1.4-3 A bar of aluminum has length $L = 4$ ft and cross-sectional area $A = 0.75$ in.^{2} The aluminum has the stress-strain curve shown in Fig. 1-13 of Section 1.3. The bar is loaded by a force $P = 24$ k and then unloaded. (a) What is the elastic recovery of strain if the initial slope of the stress-strain curve (modulus of elasticity) is 10×10^6 psi? (b) What is the residual strain? (c) What is the permanent set? (d) If the bar is reloaded, what is the proportional limit?

*1.4-4 A round bar having a length of 3 m and a diameter of 10 mm is made of an aluminum alloy for which the stress-strain curve can be described mathematically by the following equation:

$$\epsilon = \frac{\sigma}{70{,}000}\left[1 + \frac{3}{7}\left(\frac{\sigma}{270}\right)^9\right]$$

in which σ has units of megapascals (MPa). The shape of the stress-strain curve is shown in the accompanying figure. The bar is axially loaded by tensile forces $P = 20$ kN, after which the loads are removed. What is the permanent set of the bar?

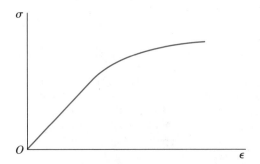

Probs. 1.4-4 and 1.4-5

*1.4-5 A round bar having a length of 5 ft and a diameter of 0.25 in. is made of an aluminum alloy for which the stress-strain curve can be described mathematically by the following equation:

$$\epsilon = \frac{\sigma}{10{,}600}\left[1 + \frac{3}{7}\left(\frac{\sigma}{60}\right)^9\right]$$

in which σ has units of kips per square inch (ksi). The shape of the stress-strain curve is shown in the accompanying figure. The bar is axially loaded by tensile forces $P = 2.8$ k, after which the loads are removed. What is the permanent set of the bar?

HOOKE'S LAW AND POISSON'S RATIO

1.5-1 A high-strength steel rod having modulus of elasticity $E = 29 \times 10^6$ psi and Poisson's ratio $\nu = 0.29$ is compressed by an axial force P (see figure). Before the load was applied, the diameter of the rod was exactly 2.000 in. In order to provide certain clearances, the diameter of the rod must not exceed 2.001 in. under load. What is the largest permissible load P?

Prob. 1.5-1

1.5-2 A prismatic bar of circular cross section is loaded by tensile forces $P = 120$ kN (see figure). The bar has length $L = 3.0$ m and diameter $d = 30$ mm. It is made of aluminum alloy (2014-T6) with modulus of elasticity $E = 73$ GPa and Poisson's ratio $\nu = 1/3$. Calculate the elongation δ, the decrease in diameter Δd, and the increase in volume ΔV of the bar.

Prob. 1.5-2

1.5-3 A high-strength steel wire, 1/8 inch in diameter, stretches 1.41 in. when a 50 ft length of it is stretched by a force of 850 lb. (a) What is the modulus of elasticity E of the steel? (b) If the diameter of the wire decreases by 0.000085 in., what is Poisson's ratio? (c) What is the unit volume change for the steel?

1.5-4 A round bar of 10 mm diameter is made of aluminum alloy 7075-T6. When the bar is stretched by an axial force P, its diameter decreases by 0.016 mm. Find the magnitude of the load P and the dilatation of the bar.

1.5-5 A bar of monel metal (length 8 in., diameter 1/4 in.) is loaded axially by a tensile force of 1500 lb (see figure). (a) Using the data in Table H-2, Appendix H, determine the increase in length and decrease in diameter of the bar. (b) Determine the increase in volume of the bar and the dilatation.

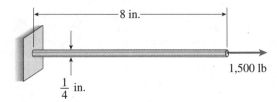

Prob. 1.5-5

1.5-6 A tensile test is performed on a brass specimen 10 mm in diameter using a gage length of 50 mm (see figure). When the tensile load P reaches a value of 20 kN, the distance between the gage marks has increased by 0.122 mm. (a) What is the modulus of elasticity E of the brass? (b) If the diameter decreases by 0.00830 mm, what is Poisson's ratio? (c) What is the dilatation of the bar?

Prob. 1.5-6

*1.5-7 A prismatic bar is loaded in tension by axial forces. Find Poisson's ratio for the material if the ratio of the unit volume change to the unit change in cross-sectional area is equal to $-2/3$.

*1.5-8 A plate of length L and width b is subjected to a uniform tensile stress σ at the ends (see figure). The material has modulus of elasticity E and Poisson's ratio ν. (a) Before the stress was applied, the slope of the diagonal line OA was b/L. What is the slope when the stress σ is acting? (b) What is the unit change in area of the face of the plate? (c) What is the unit change in cross-sectional area?

Prob. 1.5-8

*1.5-9 Derive a formula for the increase ΔV in the volume of a prismatic bar of length L hanging vertically under its own weight (W = total weight of the bar).

SHEAR STRESS AND STRAIN

1.6-1 The vertical load P acting on the wheel of a traveling crane is 13,000 lb (see figure). What is the average shear stress τ_{aver} in the 1.25 in. diameter axle?

Prob. 1.6-1

1.6-2 A block of wood is tested in direct shear using the loading frame and test specimen shown in the figure. The load P produces shear in the specimen along plane AB. The height h of plane AB is 50 mm and its width (perpendicular to the plane of the drawing) is 100 mm. If the load $P = 16$ kN, what is the average shear stress τ_{aver} in the wood?

Prob. 1.6-2

1.6-3 Two lines are inscribed at right angles on a block of material. When the block is loaded in shear, the lines are found to be at an angle of 89.75°. What is the shear strain in the material? If the modulus of elasticity is 10,600,000 psi and Poisson's ratio is 1/3, what is the shear stress in the material?

1.6-4 An angle bracket having thickness $t = 18$ mm is attached to the flange of a column with two 16 mm diameter bolts as shown in the figure. A uniformly distributed load acts on the top face of the bracket with a pressure $p = 2.0$ MPa. The top face of the bracket has length $L = 200$ mm and width $b = 70$ mm. Determine the bearing pressure σ_b between the angle bracket and the bolts and the average shear stress τ_{aver} in the bolts. (Disregard friction between the bracket and the column.)

Side view   Front view

Prob. 1.6-4

1.6-5 Three steel plates, each 3/4 in. thick, are joined by two 5/8 in. rivets as shown in the figure. (a) If the load $P = 16.0$ k, what is the maximum bearing stress σ_b on the rivets? (b) If the ultimate shear stress (average stress) in the rivets is 32 ksi, what force P_{ult} is required to cause the rivets to fail in shear? (Disregard friction between the plates.)

Prob. 1.6-5

1.6-6 A frame ACD consists of a vertical pipe CD and a brace AB constructed from two flat bars (see figure). The frame is supported by bolted connections at points A and C, which are 2 m apart. The brace is fastened to the pipe at point B (1 m above point C) by an 18 mm diameter bolt. A horizontal load P acts at point D (2 m above point C). If the load $P = 12$ kN, what is the average shear stress τ_{aver} in the bolt at B?

Prob. 1.6-6

1.6-7 A special-purpose bolt of shank diameter $d = 0.50$ in. passes through a hole in a steel plate (see figure). The hexagonal head of the bolt bears directly against the steel plate. The diameter of the circumscribed circle for the hexagon is $D = 0.80$ in. (which means that each side of the hexagon has length 0.40 in.). Also, the thickness t of the bolt head is 0.25 in. For calculation purposes, assume the tensile force P in the bolt is 1000 lb. (a) Determine the average bearing stress σ_b between the hexagonal head of the bolt and the plate. (b) Determine the average shear stress τ_{aver} in the head of the bolt.

Prob. 1.6-7

1.6-8 A steel plate of dimensions $2.5 \times 1.2 \times 0.1$ m is hoisted by a sling that has a clevis at each end (see figure). The pins through the clevises are 18 mm in diameter and located 2 m apart. Each half of the cable is at an angle of $32°$ to the vertical. For these conditions, determine the average shear stress τ_{aver} in the pins and the bearing stress σ_b between the steel plate and the pins.

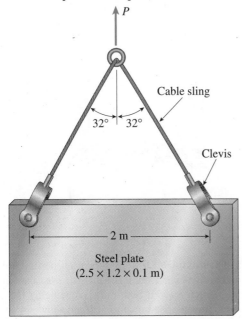

Prob. 1.6-8

1.6-9 A torque T_0 having a moment of 75 k-in. is transmitted between two flanged shafts by means of four 3/4 in. bolts (see figure). What is the average shear stress τ_{aver} in each bolt if the diameter d of the bolt circle is 6 in.?

Prob. 1.6-9

1.6-10 An elastomeric bearing pad consisting of two steel plates bonded to a chloroprene elastomer (chloroprene is an artificial rubber) is subjected to a shear force V during a static loading test (see figure). The pad has dimensions $a = 120$ mm and $b = 150$ mm, and the elastomer has thickness $t = 40$ mm. When the force V equals 5.4 kN, the top plate is found to have displaced laterally 6.0 mm with respect to the bottom plate. What is the shear modulus of elasticity G of the chloroprene?

Prob. 1.6-10

1.6-11 A flexible connection consisting of rubber pads (thickness $t = 0.5$ in.) bonded to steel plates is shown in the figure. The pads are 8.0 in. long and 6.0 in. wide. (a) Find the average shear strain γ_{aver} in the rubber if the force $P = 3400$ lb and the shear modulus for the rubber is $G = 120$ psi. (b) Find the relative horizontal displacement δ between the interior plate and the outer plates.

Prob. 1.6-11

1.6-12 A bicycle chain consists of a series of small links, each about 12 mm long between the centers of the pins (see figure). For the purpose of this problem, you must examine a bicycle chain and observe closely the construction of the links. Note particularly the cross pins, which we will assume to have a diameter of 2.5 mm. Now you must make two measurements: (1) the length of the crank arm, and (2) the radius of the chain wheel (sprocket wheel). (a) Using those dimensions, calculate the tensile force T in the chain due to a force of 800 N applied to one of the pedals. (b) Calculate the average shear stress τ_{aver} in the pins.

Prob. 1.6-12

*1.6-13 The bond between reinforcing bars and concrete is tested by means of a "pull-out test" of a bar embedded in concrete (see figure). A tensile force P is applied to the end of the bar, which has diameter d and embedment length L. (a) Assume that the shear stress (or *bond stress*) between the bar and the concrete is uniformly distributed over the length L. Then, if $P = 4000$ lb, $d = 0.5$ in., and $L = 12$ in., what average shear stress τ_{aver} is developed between the steel and concrete? (b) In reality the bond stress between the steel and concrete is smallest near the surface and largest at the interior end of the bar. Therefore, to get slightly better (but still not very accurate) results, assume that the shear stress τ is given by the equation

$$\tau = \frac{\tau_{max}}{4L^3}(4L^3 - 9Lx^2 + 6x^3)$$

in which τ_{max} is the maximum shear stress and the distance x is measured from the interior end of the bar toward the surface of the concrete. Then, if $P = 4000$ lb, $d = 0.5$ in., and $L = 12$ in., what is the maximum shear stress τ_{max}?

Prob. 1.6-13

*1.6-14 A special lever is used to twist a circular shaft by means of a square key that fits into slots (or *keyways*) in the shaft and lever, as shown in the figure. The shaft has diameter d, the key has a square cross section of dimensions $b \times b$, and the length of the key is c. The key fits half into the lever and half into the shaft (i.e., the keyways have a depth equal to $b/2$). Derive a formula for the average shear stress τ_{aver} in the key when a load P is applied at distance L from the center of the shaft. (*Hint:* Disregard the effects of any friction, assume that the bearing pressure between the key and the lever is uniformly distributed, and be sure to draw a free-body diagram of the lever.)

Prob. 1.6-14

*1.6-15 The truss *ABCDEFGH* shown in the figure is part of a wood bridge over a small stream. The truss has height h and panel length b, with both dimensions being the same. The truss members meeting at joint H are shown in detail in the second part of the figure. A single bolt of diameter $d = 1.5$ in. connects the members at this joint. We will consider the effect of only one load on the truss, namely, a load $P = 1$ k acting at the midpoint (because the load has a unit value, the stresses for any other value of the load can be obtained by multiplication). (a) What is the maximum shear force V_{max} in the bolt at joint H? (b) What is the average shear stress τ_{aver} in the bolt at the cross section of maximum shear force?

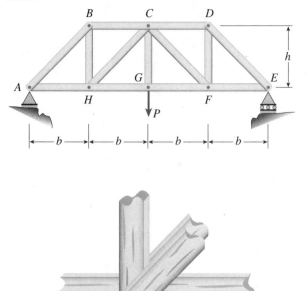

Prob. 1.6-15

*1.6-16 A shock mount constructed as shown in the figure is used to support an expensive instrument. The mount consists of an outer steel tube with inside diameter b, a central steel bar of diameter d that supports the load P, and a hollow rubber cylinder (height h) bonded to the tube and bar. (a) Obtain a formula for the shear stress τ in the rubber at a radial distance r from the center of the mount. (b) Obtain a formula for the downward displacement δ of the bar due to the load P, assuming that G is the shear modulus of elasticity of the rubber and that the steel tube and bar are rigid.

Prob. 1.6-16

ALLOWABLE LOADS

1.7-1 A tie-down on the deck of a sailboat consists of a bent bar bolted at both ends, as shown in the figure. The diameter d_b of the bolts is 1/4 in., the diameter d of the washers is 7/8 in., and the thickness t of the fiberglass deck is 3/8 in. If the allowable shear stress in the fiberglass is 300 psi and the allowable bearing pressure between the washer and the fiberglass is 550 psi, what is the allowable load P_{allow} on the tie-down?

Prob. 1.7-1

1.7-2 An aluminum bar AB is attached to a support by a pin of diameter $d = 16$ mm (see figure). The bar has thickness $t = 15$ mm and width $b = 40$ mm. If the allowable tensile stress in the bar is 125 MPa and the allowable shear stress in the pin is 75 MPa, find the allowable value of the tensile load P.

Prob. 1.7-2

1.7-3 A bolted connection between three wood members is shown in the figure. When the bolt is tightened, the wood is compressed laterally and the bolt is in tension. The allowable tensile stress in the 1/4 inch diameter bolt is 16,000 psi, and the allowable bearing stress between the 1.0 inch diameter washers and the wood is 400 psi. What is the maximum permissible tensile force T in the bolt?

Prob. 1.7-3

1.7-4 A ship's spar is attached at the base of a mast by a pin connection (see figure). The spar is a steel tube of outside diameter $d_2 = 80$ mm and inside diameter $d_1 = 70$ mm. The steel pin has diameter $d = 25$ mm, and the two plates connecting the spar to the pin have thickness $t = 12$ mm. The allowable stresses are as follows: compressive stress in the spar, 75 MPa; shear stress in the pin, 50 MPa; and bearing stress between the pin and the connecting plates, 120 MPa. Determine the allowable compressive force P_{allow} in the spar.

Prob. 1.7-4

1.7-5 A long steel wire hanging from a balloon carries a weight W at its lower end (see figure). The 1/16th inch diameter wire is 250 ft long. What is the maximum weight W_{max} that can safely be carried if the tensile yield stress for the wire is $\sigma_y = 50$ ksi and a margin of safety against yielding of 1.5 is desired? (Include the weight of the wire in the calculations.)

Prob. 1.7-5

1.7-6 What is the maximum possible value of the clamping force in the jaws of the pliers shown in the figure if $a = 95$ mm, $b = 40$ mm, and the ultimate shear stress in the 5 mm diameter pin is 310 MPa? What is the maximum permissible value of the applied load P if a factor of safety of 3 with respect to failure of the pin is to be maintained?

Prob. 1.7-6

1.7-7 Two flat bars loaded in tension by a force P are spliced using two rectangular splice plates and two 5/8 inch diameter rivets (see figure). The bars have width $b = 1.0$ in. (except at the splice, where the bars are wider) and thickness $t = 0.4$ in. The bars are made of steel having an ultimate stress in tension equal to 60 ksi. The ultimate stresses in shear and bearing for the rivet steel are 25 ksi and 80 ksi, respectively. Determine the allowable load P_{allow} if a safety factor of 2.5 is desired with respect to the ultimate load that can be carried. (Consider tension in the bars, shear in the rivets, and bearing between the rivets and the bars. Disregard friction between the plates.)

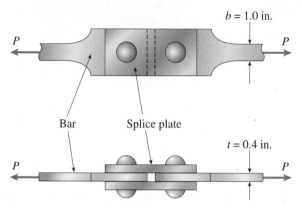

Prob. 1.7-7

1.7-8 A tubular post of outside diameter $d_2 = 90$ mm and inside diameter $d_1 = 78$ mm is guyed by two cables fitted with turnbuckles (see figure). The cables are tightened by rotating the turnbuckles, thus producing tension in the cables and compression in the post. Both cables are inclined at a 60° angle to the ground. The allowable compressive stress in the post (to prevent buckling) is $\sigma_c = 10.0$ MPa. The cables are rated for an ultimate load of 29 kN, and a factor of safety of 3 with respect to the ultimate load is required for the cables. What is the maximum permissible force T_{max} that may be developed in the cables by tightening the turnbuckles?

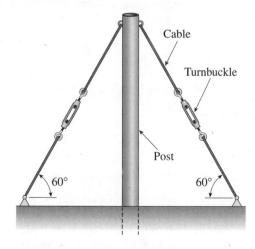

Prob. 1.7-8

1.7-9 The piston in an engine is attached to a connecting rod AB, which in turn is connected to a crank arm BC (see figure). The piston slides without friction in a cylinder and is subjected to a force P (assumed to be constant) while moving to the right in the figure. The connecting rod, which has cross-sectional area A and length L, is attached at both ends by pins. The crank arm rotates about the axle at C with the pin at B moving in a circle of radius R. The axle at C, which is supported by bearings, exerts a resisting moment M against the crank arm. (a) Obtain a formula for the maximum permissible force P_{allow} based upon an allowable compressive stress σ_c in the connecting rod. (b) Calculate the force P_{allow} for the following data: $\sigma_c = 22$ ksi, $A = 0.10$ in.^{2}, and $R = 0.28L$.

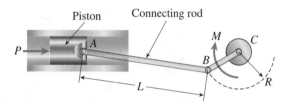

Prob. 1.7-9

1.7-10 A metal bar AB of weight W is suspended by a system of steel wires arranged as shown in the figure. The diameter of the wires is 2 mm, and the yield stress of the steel is 450 MPa. Determine the maximum permissible weight W_{max} for a factor of safety of 1.9 with respect to yielding.

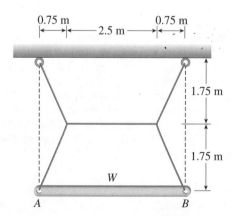

Prob. 1.7-10

1.7-11 The pin-connected truss shown in the figure supports three vertical loads of 3 k each. In addition, a horizontal load P acts to the right at joint C. Member AB has a cross-sectional area of 3.13 in.^{2}, and the allowable stress in compression is 12,000 psi. Considering only the stress in member AB, what is the allowable load P?

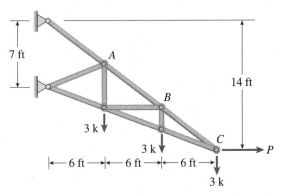

7 ft

A

14 ft

B

3 k

3 k

C

P

$\leftarrow$ 6 ft $\rightarrow$$\leftarrow$ 6 ft $\rightarrow$$\leftarrow$ 6 ft $\rightarrow$

3 k

Prob. 1.7-11

1.7-12 A solid bar of circular cross section (diameter d) has a hole of diameter $d/4$ drilled laterally through the center of the bar (see figure). The allowable average tensile stress on the net cross section of the bar is σ_{allow}. (a) Obtain a formula for the allowable load P_{allow} that the bar can carry in tension. (b) Calculate the value of P_{allow} if the bar is made of brass with diameter $d = 30$ mm and $\sigma_{\text{allow}} = 75$ MPa. (*Hint:* Use the formulas of Case 15, Appendix D.)

d

P $\rightarrow||\leftarrow \dfrac{d}{4}$

P $\rightarrow||\leftarrow \dfrac{d}{4}$

$\rightarrow| d |\leftarrow$

Prob. 1.7-12

1.7-13 A solid steel bar of diameter $d_1 = 2.25$ in. has a hole of diameter $d_2 = 1.125$ in. drilled through it (see figure). A steel pin passes through the hole and is attached to supports. Determine the maximum permissible tensile load P_{allow} in the bar if the yield stress for shear in the pin is $\tau_y = 17,000$ psi, the yield stress for tension in the bar is $\sigma_y = 36,000$ psi, and a factor of safety of 2.0 with respect to yielding is required. (*Hint:* Use the formulas of Case 15, Appendix D.)

d_1

d_2

$\leftarrow d_1 \rightarrow$

P

Prob. 1.7-13

* **1.7-14** A flexible string of length $L_s = 1.4$ m is fastened to supports at points A and B (see figure). Points A and B are at different elevations, with B being lower than A. The horizontal distance L between the supports equals 1.2 m. A load P hangs from a small roller that slides without friction along the string until it comes to rest in the equilibrium position at C. If the string has a breaking strength $S = 160$ N, and if a factor of safety $n = 2.5$ is required, what is the allowable load P?

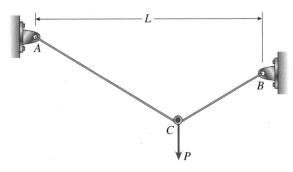

$\leftarrow$ L $\rightarrow$

A

B

C

P

Prob. 1.7-14

DESIGN FOR AXIAL LOADS AND DIRECT SHEAR

1.8-1 A compression member in a building truss is required to transmit an axial force $P = 34$ k (see figure). The thickness of the wall of the tubular member is to be 0.375 in. What is the minimum required outside diameter d_{min} if the allowable compressive stress is 9000 psi?

P

d

P

Prob. 1.8-1

1.8-2 A steel pad supporting heavy machinery and equipment rests on four short, hollow, cast iron piers (see figure). Each pier is to be designed for a compressive load of 280 kN. The ultimate strength of the cast iron in compression is 350 MPa. The designer decides to design the piers with a wall thickness t of 10 mm and a factor of safety of 4.0 with respect to the ultimate strength. Compute the minimum required outside diameter d_{min} of the piers.

Prob. 1.8-2

1.8-3 An aluminum tube serving as a brace in the fuselage of a small airplane (see figure) must resist a compressive force $P = 580$ lb. To prevent localized buckling of the wall of the tube, the ratio of the outside diameter d to the thickness t of the wall must not exceed 10. Furthermore, to keep the weight of the strut as low as possible, the cross-sectional area should be as small as possible, which means that the compressive stress should be equal to the allowable stress of 2000 psi. What is the largest permissible outside diameter d_{max}?

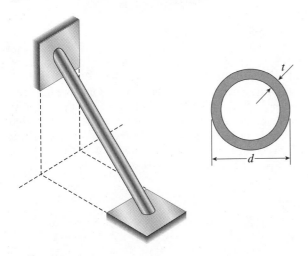

Prob. 1.8-3

1.8-4 A lifeboat hangs from two ship's davits, as shown in the figure. A pin through each davit supports two pulleys, one on each side of the davit. Cables attached to the lifeboat pass over the pulleys and wind around winches that raise and lower the lifeboat. The lower parts of the cables are vertical and the upper parts make an angle $\alpha = 18°$ with the horizontal. The tensile force in each cable is $T = 8.0$ kN when the lifeboats are fully loaded. What is the minimum

diameter d_{min} of the pins through the davits based upon an allowable shear stress of 28 MPa?

Prob. 1.8-4

1.8-5 A horizontal beam AB supported by an inclined strut CD carries a load $P = 2600$ lb at the position shown in the figure. The strut, which consists of two bars, is connected to the beam by a bolt passing through the three bars meeting at joint C. If the allowable shear stress in the bolt is 13,500 psi, what is the minimum required diameter d_{min} of the bolt?

Prob. 1.8-5

1.8-6 Two bars of rectangular cross section, each subjected to a tensile load $P = 31$ kN, are connected by a single bolt AB (see figure). The allowable shear stress in the bolt is 90 MPa and the allowable bearing stress between the bolt and the bars is 150 MPa. If the thickness t of the bars is 15 mm, determine the minimum required diameter d_{min} of the bolt.

Prob. 1.8-6

1.8-7 A pressurized cylinder has a sealed cover plate fastened with steel bolts (see figure). The pressure p of the gas in the cylinder is 280 psi, the inside diameter D of the cylinder is 10 in., and the diameter d_b of the bolts is 0.5 in. If the allowable tensile stress in the bolts is 10,000 psi, find the number n of bolts needed to fasten the cover.

Prob. 1.8-7

1.8-8 A cage for transporting workers and supplies on a construction site is hoisted by a crane (see figure). The floor of the cage is rectangular with dimensions 2.4 m by 3.6 m. Each of the four lifting cables is attached to a corner of the cage and is 4.2 m long. The mass of the cage and its contents is limited by regulations to 4800 kg. Determine the required cross-sectional area A_c of each cable if the breaking

strength of a cable is 580 MPa and a factor of safety of 3.5 with respect to failure is desired.

Prob. 1.8-8

1.8-9 A bar of rectangular cross section, 2 in. wide and 1/4 in. thick, is subjected to an axial load P (see figure). A hole of diameter d is drilled through the bar to provide for a pin support. The allowable tensile stress on the net cross section of the bar is 20 ksi, and the allowable shear stress in the pin is 11.5 ksi. (a) Determine the pin diameter d_m for which the load P will be a maximum. (b) Determine the corresponding value P_{max} of the load.

Prob. 1.8-9

1.8-10 A steel column of hollow circular cross section is supported on a circular steel base plate and a concrete pedestal (see figure). The column has outside diameter $d = 250$ mm and supports a load $P = 750$ kN. (a) If the allowable stress in the column is 55 MPa, what is the minimum required thickness t? Based upon your result, select a thickness for the column. (Select a thickness that is an even integer, such as 10, 12, 14, . . ., in units of millimeters.) (b) If the allowable bearing stress on the concrete pedestal is 11.5 MPa, what is the minimum required diameter D of the base plate if it is designed for the allowable load P_{allow} that the column with the selected thickness can support? Based upon your result, select a diameter for the base plate. (Select a thickness that is an even integer, in units of millimeters.)

Prob. 1.8-10

***1.8-11** A prismatic arm of length L rotates on a smooth horizontal surface about a vertical axis through point C with constant angular speed ω (see figure). A mass M_1 is attached to the end of the arm. Denoting by ρ the mass density of the material of the arm and by σ_t the allowable stress in tension, derive a formula for the required cross-sectional area A of the arm.

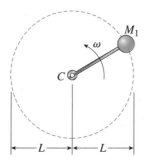

Prob. 1.8-11

***1.8-12** A flat bar of width $b = 50$ mm and thickness $t = 6$ mm is loaded in tension by a force P (see figure). The bar is attached to a support by a pin of diameter d that passes through a hole of the same size in the bar. The distance between the hole and the end of the bar is $h = 25$ mm. The allowable tensile stress on the net cross section of the bar (along planes ab and cd) is $\sigma_1 = 112$ MPa, the allowable shear stress in the pin is $\tau_2 = 64$ MPa, and the allowable shear stress in the bar along planes eb and fc is $\tau_3 = 43$ MPa. (Failure along planes eb and fc is called "tear out.") (a) Determine the pin diameter d_m for which the load P will be a maximum. (b) Determine the corresponding value P_{max} of the load.

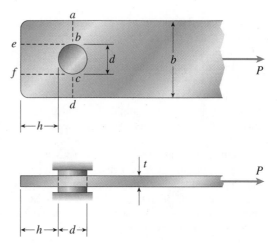

Prob. 1.8-12

***1.8-13** Two identical bars AC and BC support a vertical load P (see figure). The distance L between the supports is fixed, but the angle θ can be varied by changing the lengths of the bars. We observe that when the angle θ becomes smaller, the bars become shorter but their cross-sectional areas become larger (because the axial forces are larger). On the other hand, if the angle θ becomes larger, the bars become longer and the areas become smaller. Thus, we see that the weight of the structure (which is proportional to the volume) is a function of the angle θ. Determine the angle θ so that the structure has minimum weight without exceeding the allowable tensile stress in the bars. (*Note:* The weights of the bars are very small compared to the force P and may be disregarded.)

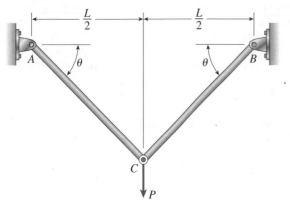

Prob. 1.8-13

**1.8-14 Two bars AC and BC of the same material support a vertical load P (see figure). The length L of the horizontal bar is fixed, but the angle θ can be varied by moving support A vertically and changing the length of bar AC to correspond with the new position of support A. The allowable stresses in the bars are the same in tension and compression. We observe that when the angle θ is reduced, bar AC becomes shorter but the cross-sectional areas of both bars increase (because the axial forces are larger). The opposite effects occur if the angle θ is increased. Thus, we see that the weight of the structure (which is proportional to the volume) depends upon the angle θ. Determine the angle θ so that the structure has minimum weight without exceeding the allowable stresses in the bars. (*Note:* The weights of the bars are small compared to the force P and may be disregarded.)

Prob. 1.8-14

***1.8-15 A uniform beam AB of length L and weight $W = 2390$ lb is to be supported at an angle of $45°$ by a slender steel strut CD of length $L/2$ (see figure). The strut is placed in such a position that it is subjected to the smallest possible compressive force. Determine the required cross-sectional area A_s of the strut if the allowable stress in compression is $\sigma_{\text{allow}} = 700$ psi. (Disregard the weight of the strut, and assume pinned conditions at points A, C, and D.)

(*Hint:* Let β be the angle between the strut and the horizontal, that is, angle ACD. Then express the compressive force F_{CD} in the strut in terms of the angle β. Next, find the value of β that makes F_{CD} a minimum. Finally, knowing the angle β, determine F_{CD} and the required area A_s.)

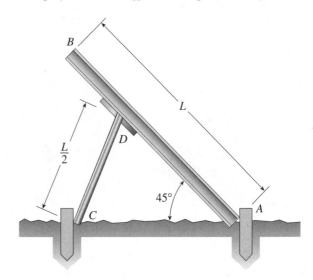

Prob. 1.8-15

2

AXIALLY LOADED MEMBERS

2.1 INTRODUCTION

Structural components subjected only to tension or compression are known as **axially loaded members.** Solid bars with straight longitudinal axes are the most common type, although cables and coil springs also carry axial loads. Examples of axially loaded bars are truss members, connecting rods in engines, spokes in bicycle wheels, columns in buildings, and struts in aircraft engine mounts. The stress-strain behavior of such members was discussed in Chapter 1, where we also obtained equations for the stresses acting on cross sections ($\sigma = P/A$) and the strains in longitudinal directions ($\epsilon = \delta/L$).

In this chapter we consider several other aspects of axially loaded members, beginning with the determination of changes in lengths caused by loads (Sections 2.2 and 2.3). The calculation of changes in length is an essential ingredient in the analysis of statically indeterminate structures, a topic we introduce in Section 2.4. Changes in length also must be calculated whenever it is necessary to control the displacements of a structure, whether for aesthetic or functional reasons. In Section 2.5, we discuss the effects of temperature on the length of a bar, and we introduce the concepts of thermal stress and thermal strain. A generalized view of the stresses in axially loaded bars is presented in Section 2.6, where we discuss the stresses on *inclined sections* (as distinct from *cross sections*) of bars. Although only normal stresses act on cross sections of axially loaded bars, both normal and shear stresses act on inclined sections.

We then introduce several additional topics of importance in mechanics of materials, namely, strain energy (Section 2.7), impact loading (Section 2.8), fatigue (Section 2.9), stress concentrations

(Section 2.10), and nonlinear behavior (Sections 2.11 and 2.12). Although these subjects are discussed in the context of members with axial loads, the discussions provide the foundation for applying the same concepts to other structural elements, such as bars in torsion and beams in bending.

2.2

CHANGES IN LENGTH OF AXIALLY LOADED MEMBERS

Fig. 2-1 Spring subjected to an axial load P

When determining the changes in length of axially loaded members, it is convenient to begin with a **coil spring** (Fig. 2-1). Springs of this type are used in large numbers in many kinds of machines and devices—for instance, there are dozens of them in every automobile. When a load is applied along the axis of a spring, as shown in Fig. 2-1, the spring gets longer or shorter depending upon the direction of the load. If the load acts away from the spring, the spring elongates and we say that the spring is loaded in *tension*. If the load acts toward the spring, the spring shortens and we say it is in *compression*. However, it should not be inferred from this terminology that the individual coils of a spring are subjected to direct tensile or compressive stresses; rather, the coils act primarily in direct shear and torsion (or twisting). Nevertheless, the overall stretching or shortening of a spring is analogous to the behavior of a bar in tension or compression, and so the same terminology is used.

The elongation of a spring is pictured in Fig. 2-2, where the upper part of the figure shows the spring in its **natural length** L (also called its *unstressed length*, *relaxed length*, or *free length*), and the lower part of the figure shows the effects of applying a tensile load. Under the action of the force P, the spring lengthens by an amount δ and its final length becomes $L + \delta$. If the material of the spring is **linearly elastic**, the load and elongation will be proportional:

$$P = k\delta \qquad \delta = fP \qquad (2\text{-}1a,b)$$

Fig. 2-2 Elongation of an axially loaded spring

in which k and f are constants of proportionality. The constant k is called the **stiffness** of the spring and is defined as the force required to produce a unit elongation, that is, $k = P/\delta$. Similarly, the constant f is known as the **flexibility** and is defined as the elongation produced by a load of unit value, that is, $f = \delta/P$. Although we used a spring in tension for this discussion, it should be obvious that Eqs. (2-1a) and (2-1b) also apply to springs in compression.

From the preceding discussion it is apparent that the stiffness and flexibility of a spring are the reciprocal of each other:

$$k = \frac{1}{f} \qquad f = \frac{1}{k} \qquad (2\text{-}2a,b)$$

The flexibility of a spring can easily be determined by measuring the elongation produced by a known load, and the stiffness can then be calculated from Eq. (2-2a). Other terms for the stiffness and flexibility of a spring are the **spring constant** and **compliance**, respectively.

The spring properties given by Eqs. (2-1) and (2-2) can be used in the analysis and design of various mechanical devices involving springs, as illustrated later in Example 2-1.

■ **Prismatic Bars**

Axially loaded bars elongate under tensile loads and shorten under compressive loads, just as springs do. To analyze this behavior, let us consider the prismatic bar shown in Fig. 2-3. A **prismatic bar** is a structural member having a straight longitudinal axis and constant cross section throughout its length. Although we often use circular bars in our illustrations, we should bear in mind that structural members may have a variety of cross-sectional shapes, such as those shown in Fig. 2-4.

The **elongation** δ of a prismatic bar subjected to a tensile load P is shown in Fig. 2-5. If the load acts through the centroid of the end cross section, the uniform normal stress at cross sections away from the ends is given by the formula $\sigma = P/A$, where A is the cross-sectional area. Furthermore, if the bar is made of a homogeneous material, the axial strain is $\epsilon = \delta/L$, where δ is the elongation and L is the length of the bar. Let us also assume that the material is linearly elastic, which means that it follows Hooke's law. Then the longitudinal stress and strain are related by the equation $\sigma = E\epsilon$, where E is the modulus of elasticity. Combining these basic relationships, we get the following equation for the elongation of the bar:

$$\delta = \frac{PL}{EA} \tag{2-3}$$

Fig. 2-3 Prismatic bar of circular section

Solid cross sections

Hollow or tubular cross sections

Thin-walled open cross sections

Fig. 2-4 Typical cross sections of structural members

This equation shows that the elongation is directly proportional to the load P and the length L and inversely proportional to the modulus of elasticity E and the cross-sectional area A. The product EA is known as the **axial rigidity** of the bar.

Although Eq. (2-3) was derived for a member in tension, it applies equally well to a member in compression, in which case δ represents the shortening of the bar. Usually we know by inspection whether a member gets longer or shorter; however, there are occasions when a **sign convention** is needed (for instance, when analyzing a statically indeterminate bar). When that happens, elongation is usually taken as positive and shortening as negative.

The change in length of a bar is normally very small in comparison to its length, especially when the material is a structural metal, such as steel or aluminum. As an example, consider an aluminum strut 75.0 in.

Fig. 2-5 Elongation of a prismatic bar in tension

long and subjected to a moderate compressive stress of 7000 psi. If the modulus of elasticity is 10,500 ksi, the shortening of the strut (from Eq. 2-3 with P/A replaced by σ) is $\delta = 0.050$ in. Consequently, the ratio of the change in length to the original length is 0.05/75, or 1/1500, and the final length is 0.999 times the original length. Under conditions similar to these, we can use the original length of a bar (instead of the final length) in calculations.

The stiffness and flexibility of a prismatic bar are defined in the same way as for a spring. The stiffness is the force required to produce a unit elongation, or P/δ, and the flexibility is the elongation due to a unit load, or δ/P. Thus, from Eq. (2-3) we see that the **stiffness** and **flexibility** of a prismatic bar are, respectively,

$$k = \frac{EA}{L} \qquad f = \frac{L}{EA} \qquad\qquad (2\text{-}4a,b)$$

Stiffnesses and flexibilities of structural members, including those given by Eqs. (2-4a) and (2-4b), have a special role in the analysis of large structures by computer-oriented methods.

▪ Cables

Cables are used to transmit large tensile forces, for example, when lifting and pulling heavy objects, guying towers, raising elevators, and supporting suspension bridges. Unlike springs and prismatic bars, cables cannot resist compression. Furthermore, they have little resistance to bending and therefore are likely to be curved rather than straight. Nevertheless, a cable is considered to be an axially loaded member because it is subjected only to tensile forces. Because the tensile forces in a cable are directed along the axis, the forces may vary in both direction and magnitude depending upon the configuration of the cable and the location of the cross section.

Cables are constructed from a large number of wires wound in some particular manner. While many arrangements are available depending upon how the cable will be used, a common type of cable, shown in Fig. 2-6, is formed by six *strands* wound helically around a central strand. Each strand is in turn constructed of many wires, also wound helically. For this reason, cables are often referred to as **wire rope**.

The cross-sectional area of a cable is equal to the total cross-sectional area of the individual wires, called the **effective area** or **metallic area**. This area is less than the area of a circle having the same diameter as the cable because there are spaces between the individual wires. For example, the actual cross-sectional area (effective area) of a particular 1.0 inch diameter cable is only 0.471 in.², whereas the area of a 1.0 in. diameter circle is 0.785 in.²

Fig. 2-6 Typical arrangement of strands and wires in a steel cable

Under the same tensile load, the elongation of a cable is greater than the elongation of a solid bar of the same material and same metallic cross-sectional area, because the wires "tighten up" in the same manner as the fibers in a rope. Thus, the modulus of elasticity (called the **effective modulus**) of a cable is less than the modulus of the material of which it is made. The effective modulus of steel cables is about 20,000 ksi (140 GPa), whereas the steel itself has a modulus of about 30,000 ksi (210 GPa). When determining the elongation of a cable from Eq. (2-3), the effective modulus should be used for E and the effective area should be used for A.

In practice, the cross-sectional dimensions and other properties of cables are obtained from their manufacturers. However, for use in solving problems in this book (and definitely *not* for use in engineering applications), we list in Table 2-1 the properties of a particular type of cable. Note that the last column contains the breaking strength, or ultimate load. The allowable load is obtained by applying a safety factor that may range from 3 to 10, depending upon how the cable is to be used. The individual wires in a cable are usually made of high-strength steel, and so the calculated tensile stress at the breaking load can be as high as 200,000 psi (1400 MPa).

Table 2-1   Properties of Steel Cables

Nominal diameter (in.)	Approximate weight (lb/ft)	Effective area (in.²)	Breaking strength (lb)
0.50	0.42	0.119	23,100
0.75	0.95	0.268	51,900
1.00	1.67	0.471	91,300
1.25	2.64	0.745	144,400
1.50	3.83	1.08	209,000
1.75	5.24	1.47	285,000
2.00	6.84	1.92	372,000

The following examples illustrate techniques for analyzing simple devices containing springs and bars. The solutions require the use of free-body diagrams, equations of equilibrium, and equations for changes in length. The problems at the end of the chapter provide many additional examples.

■ Example 2-1

A rigid L-shaped frame *ABC* consisting of a horizontal arm *AB* (length *b* = 11.0 in.) and a vertical arm *BC* (length *c* = 9.5 in.) is pivoted at point *B*, as shown in Fig. 2-7a. The pivot is attached to the outer frame *BCD*, which stands on a laboratory bench. The position of the pointer at *C* is controlled by a spring (stiffness *k* = 4.2 lb/in.) that is attached to a threaded rod. The position of the threaded rod is adjusted by turning the knurled nut. The pitch of the threads (that is, the distance from one thread to the next) is *p* = 1/16 in., which means that one full revolution of the nut will move the rod by that same amount. Initially, the nut is turned until the pointer at the end of arm *BC* is directly over the reference mark on the outer frame.

If a weight *W* = 2 lb is placed on the hanger at *A*, how many revolutions of the nut are required to bring the pointer back to the mark? (Deformations of the metal parts of the device may be disregarded because they are negligible compared to the change in length of the spring.)

(a)

(b)

Fig. 2-7 Example 2-1. L-shaped frame *ABC* pivoted at point *B*

Solution   Inspection of the device shows that the weight *W* acting downward will cause the pointer at *C* to move to the right. When the pointer moves to the right, the spring stretches by an additional amount—an amount that we can determine from the force *F* in the spring. The force *F* is found from a free-body diagram of the frame (Fig. 2-7b). Note that the reactions at the pivot are indicated with slashes across the arrows (see the discussion of reactions at the end of Section 1.8).

Taking moments about point *B* gives

$$F = \frac{Wb}{c} \tag{a}$$

The corresponding elongation δ of the spring (from Eq. 2-1a) is

$$\delta = \frac{F}{k} = \frac{Wb}{ck} \tag{b}$$

To bring the pointer back to the mark, we must turn the nut through enough revolutions to move the threaded rod to the left an amount equal to the elongation of the spring. Since each complete turn of the nut moves the rod a distance equal to the pitch p, the total movement of the rod is equal to np, where n is the number of turns. Therefore,

$$np = \delta = \frac{Wb}{ck} \tag{c}$$

from which we get the following formula for the number of revolutions of the nut:

$$n = \frac{Wb}{ckp} \tag{d} \quad \longleftarrow$$

To obtain a numerical result, we substitute the given data into Eq. (d), as follows:

$$n = \frac{Wb}{ckp} = \frac{(2 \text{ lb})(11.0 \text{ in.})}{(9.5 \text{ in.})(4.2 \text{ lb/in.})(1/16 \text{ in.})} = 8.8 \text{ revolutions} \quad \longleftarrow$$

This result shows that if we rotate the nut through 8.8 revolutions, the threaded rod will move to the left an amount equal to the elongation of the spring caused by the 2 lb load, thus returning the pointer to the reference mark.

■ Example 2-2

The contraption shown in Fig. 2-8a consists of a horizontal beam ABC supported by two vertical bars BD and CE. Bar CE is pinned at both ends but bar BD is fixed to the foundation at its lower end. The distance from A to B is 450 mm and from B to C is 225 mm. Bars BD and CE have lengths of 480 mm and 600 mm, respectively, and their cross-sectional areas are 1020 mm^2 and 520 mm^2, respectively. The bars are made of steel having a modulus of elasticity $E = 205$ GPa. Assuming that beam ABC is rigid, find the maximum allowable load P_{max} if the displacement of point A is limited to 1.0 mm.

Solution   To find the displacement of point A, we need to know the displacements of points B and C. Therefore, we must find the changes in lengths of bars BD and CE, using the general equation $\delta = PL/EA$ (Eq. 2-3). We begin by finding the forces in the bars from a free-body diagram of the beam (Fig. 2-8b). Because bar CE is pinned at both ends, it is a "two-force" member and transmits only a vertical force F_{CE} to the beam. However, bar BD can transmit both a vertical and a horizontal force. From equilibrium of beam ABC in the horizontal direction, we see that the horizontal force vanishes.

Two additional equations of equilibrium enable us to express the forces F_{BD} and F_{CE} in terms of the load P. Thus, by taking moments about point B and then summing forces in the vertical direction, we find

$$F_{CE} = 2P \qquad F_{BD} = 3P \tag{e}$$

Note that the force F_{CE} acts downward on bar ABC and the force F_{BD} acts upward. Therefore, member CE is in tension and member BD is in compression.

(a)

(b)

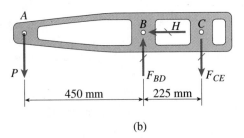

Fig. 2-8 Example 2-2. Horizontal beam
ABC supported by two vertical bars

(c)

The shortening of member *BD* is

$$\delta_{BD} = \frac{F_{BD}L_{BD}}{EA_{BD}}$$

$$= \frac{(3P)(480 \text{ mm})}{(205 \text{ GPa})(1020 \text{ mm}^2)} = 6.887P \times 10^{-6} \text{ mm} \qquad (P = \text{newtons}) \qquad \text{(f)}$$

Note that the shortening δ_{BD} is expressed in millimeters provided the load P is expressed in newtons. Similarly, the lengthening of member CE is

$$\delta_{CE} = \frac{F_{CE}L_{CE}}{EA_{CE}}$$

$$= \frac{(2P)(600 \text{ mm})}{(205 \text{ GPa})(520 \text{ mm}^2)} = 11.26P \times 10^{-6} \text{ mm} \quad (P = \text{newtons}) \quad \text{(g)}$$

Again, the displacement is expressed in millimeters provided the load P is expressed in newtons. Knowing the changes in lengths of the two bars, we can now find the displacement of point A.

A *displacement diagram* showing the relative positions of points A, B, and C is sketched in Fig. 2-8c. Line ABC represents the original alignment of the three points. After the load P is applied, member BD shortens by the amount δ_{BD} and point B moves to B'. Also, member CE elongates by the amount δ_{CE} and point C moves to C'. Because the beam ABC is assumed to be rigid, points A', B', and C' lie on a straight line.

For clarity, the displacements are highly exaggerated in the diagram. In reality, line ABC rotates through a very small angle to its new position $A'B'C'$ (see Note 2 at the end of the example).

Using similar triangles, we can now find the relationships between the displacements at points A, B, and C. From triangles $A'A''C'$ and $B'B''C'$ we get

$$\frac{A'A''}{A''C'} = \frac{B'B''}{B''C'} \quad \text{or} \quad \frac{\delta_A + \delta_{CE}}{450 + 225} = \frac{\delta_{BD} + \delta_{CE}}{225} \quad \text{(h)}$$

in which all terms are expressed in millimeters. Substituting for δ_{BD} and δ_{CE} from Eqs. (f) and (g) gives

$$\frac{\delta_A + 11.26P \times 10^{-6}}{450 + 225} = \frac{6.887P \times 10^{-6} + 11.26P \times 10^{-6}}{225}$$

Finally, we substitute for δ_A its limiting value of 1.0 mm and solve the equation for the load P. The result is

$$P = P_{\max} = 23,200 \text{ N (or 23.2 kN)} \quad \longleftarrow$$

When the load reaches this value, the downward displacement at point A is 1.0 mm.

Note 1: Since the structure behaves in a linearly elastic manner, the displacements are proportional to the magnitude of the load. For instance, if the load is one-half of $P_{\max}$, that is, if $P = 11.6$ kN, the downward displacement of point A is 0.5 mm.

Note 2: To verify our premise that line ABC rotates through a very small angle, we can calculate the angle of rotation α from the displacement diagram (Fig. 2-8c), as follows:

$$\tan \alpha = \frac{A'A''}{A''C'} = \frac{\delta_A + \delta_{CE}}{675 \text{ mm}} \quad \text{(i)}$$

The displacement δ_A of point A is 1.0 mm, and the elongation δ_{CE} of bar CE is found from Eq. (g) by substituting $P = 23,200$ N; the result is $\delta_{CE} = 0.261$ mm. Therefore, from Eq. (i) we get

$$\tan \alpha = \frac{1.0 \text{ mm} + 0.261 \text{ mm}}{675 \text{ mm}} = \frac{1.261 \text{ mm}}{675 \text{ mm}} = 0.001868$$

from which $\alpha = 0.11°$. This angle is so small that if we tried to draw the displacement diagram to scale, we would not be able to distinguish between the original line ABC and the rotated line $A'B'C'$. Thus, when working with displacement diagrams, we usually can consider the displacements to be very small quantities, thereby simplifying the geometry. In this example we were able to assume that points A, B, and C moved only vertically, whereas if the displacements were large, we would have to consider that they moved along curved paths.

2.3   CHANGES IN LENGTH OF NONUNIFORM BARS

When a prismatic bar of linearly elastic material is loaded only at the ends, we can obtain its change in length from the equation $\delta = PL/EA$, as described in the preceding section. In this section we will see how this same equation can be used in more general situations.

Suppose, for instance, that a prismatic bar is loaded by one or more axial loads acting at intermediate points along the axis (Fig. 2-9a). We can determine the change in length of this bar by adding algebraically the elongations and shortenings of the individual segments. The procedure is as follows. (1) Identify the segments of the bar (segments AB, BC, and CD) as segments 1, 2, and 3, respectively. (2) Determine the internal axial forces N_1, N_2, and N_3 in the segments from the free-body diagrams of Figs. 2-9b, c, and d. Note that the internal axial forces are denoted by the letter N to distinguish them from the external loads P. By summing forces in the vertical direction, we obtain the following expressions for the axial forces:

$$N_1 = -P_B + P_C + P_D \qquad N_2 = P_C + P_D \qquad N_3 = P_D$$

In writing these equations we used the sign convention given in the preceding section (internal axial forces are positive when in tension and negative when in compression). (3) Determine the changes in length of the segments from Eq. (2-3):

$$\delta_1 = \frac{N_1 L_1}{EA} \qquad \delta_2 = \frac{N_2 L_2}{EA} \qquad \delta_3 = \frac{N_3 L_3}{EA}$$

in which L_1, L_2, and L_3 are the lengths of the segments and EA is the axial rigidity of the bar. (4) Add δ_1, δ_2, and δ_3 to obtain δ, the change in length of the entire bar:

$$\delta = \delta_1 + \delta_2 + \delta_3$$

As already explained, the changes in lengths must be added algebraically, with elongations being positive and shortenings negative.

Fig. 2-9 (a) Bar with external loads acting at intermediate points; (b), (c), and (d) free-body diagrams showing the internal axial forces N_1, N_2, and N_3

(a)   (b)   (c)   (d)

This same method can be used when the bar consists of several prismatic segments, each having different axial forces, different dimensions, and different materials (Fig. 2-10). The change in length may be obtained from the equation

$$\delta = \sum_{i=1}^{n} \frac{N_i L_i}{E_i A_i} \qquad (2\text{-}5)$$

in which the subscript i is a numbering index for the various segments of the bar and n is the total number of segments. Note especially that N_i is not an external load but is the internal axial force in segment i.

Sometimes the axial force N and the cross-sectional area A vary continuously along the axis of a bar, as illustrated by the tapered bar of Fig. 2-11a. This bar not only has a continuously varying cross-sectional area but also a continuously varying axial force (because the load itself is distributed along the axis). A distributed axial load may be produced by centrifugal forces, friction forces, or the weight of a bar in a vertical position. Under these conditions we can no longer use Eq. (2-5) to obtain the change in length. Instead, we must determine the change in length of a differential element of the bar (Fig. 2-11a) and then integrate over the length of the bar.

Fig. 2-10 Bar consisting of prismatic segments having different axial forces, different dimensions, and different materials

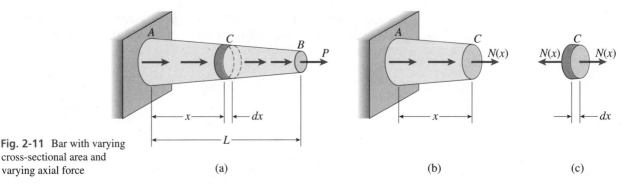

Fig. 2-11 Bar with varying cross-sectional area and varying axial force

(a) (b) (c)

We select a differential element at distance x from the left-hand end of the bar (Fig. 2-11a). The internal axial force $N(x)$ acting at this cross section (Fig. 2-11b) may be determined from equilibrium using either segment AC or segment CB as a free body. In general, this force is a function of x. Also, knowing the dimensions of the bar, we can express the cross-sectional area $A(x)$ as a function of x. The elongation $d\delta$ of the differential element (Fig. 2-11c) may be obtained from the equation $\delta = PL/EA$ by substituting $N(x)$ for P, dx for L, and $A(x)$ for A, as follows:

$$d\delta = \frac{N(x)dx}{EA(x)}$$

The elongation of the entire bar is obtained by integrating over the length:

$$\delta = \int_0^L d\delta = \int_0^L \frac{N(x)dx}{EA(x)} \tag{2-6}$$

If the expressions for $N(x)$ and $A(x)$ are not too complicated, the integral can be evaluated analytically and a formula for δ can be obtained, as illustrated later in Example 2-4. However, if formal integration is either difficult or impossible, a numerical method for evaluating the integral should be used.

Equations (2-5) and (2-6) apply only to bars made of linearly elastic materials, as shown by the presence of the modulus of elasticity E in the formulas. Also, the formula $\delta = PL/EA$ was derived using the assumption that the stress distribution is uniform over every cross section (because it is based on the formula $\sigma = P/A$). This assumption is valid for prismatic bars but not for tapered bars, and therefore Eq. (2-6) gives satisfactory results for a tapered bar only if the angle between the sides of the bar is small. As an illustration, if the angle between the sides is 20°, the stress calculated from the expression $\sigma = P/A$ (at an arbitrarily selected cross section) is 3% less than the exact stress for that same cross section (calculated by more advanced methods). For smaller angles, the error is even less. Consequently, we can say that Eq. (2-6) is satisfactory if the angle of taper is small. If the taper is large, more accurate methods of analysis are needed (Ref. 2-1).

The following examples illustrate the determination of changes in lengths of nonuniform bars.

■ Example 2-3

A vertical steel bar ABC is supported at its upper end and loaded by a force P_1 at its lower end (Fig. 2-12a). A horizontal beam BDE is pinned to the vertical bar at joint B and supported at point D. The beam carries a load P_2 at end E.

The upper part of the vertical bar (segment AB) has length $L_1 = 20.0$ in. and cross-sectional area $A_1 = 0.25$ in.^{2}; the lower part (segment BC) has length $L_2 = 30.0$ in. and area $A_2 = 0.16$ in.^{2} The modulus of elasticity E of the steel is 29.0×10^6 psi. The left- and right-hand parts of beam BDE have lengths $a = 28$ in. and $b = 25$ in., respectively.

Calculate the vertical displacement δ_C at point C if the load $P_1 = 2250$ lb and the load $P_2 = 5800$ lb. (Disregard the weights of the bar and the beam.)

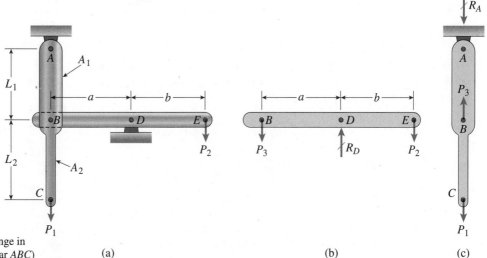

Fig. 2-12 Example 2-3. Change in length of a nonuniform bar (bar ABC)

(a)   (b)   (c)

Solution   *Axial forces in bar ABC.* From Fig. 2-12a, we see that the vertical displacement of point C is equal to the change in length of bar ABC. Therefore, we must find the axial forces in both segments of this bar. The axial force N_2 in the lower segment is equal to the load P_1. The axial force N_1 in the upper segment can be found if we know the force applied to the bar by the beam. This force can be obtained from a free-body diagram of the beam (Fig. 2-12b), in which the force acting on the beam (from the vertical bar) is denoted P_3 and the vertical reaction at support D is denoted R_D. No horizontal force acts between the bar and the beam, as can be seen from a free-body diagram of the vertical bar (Fig. 2-12c). Therefore, there is no horizontal reaction at support D of the beam.

Taking moments about point D for the free-body diagram of the beam (Fig. 2-12b) gives

$$P_3 = \frac{P_2 b}{a} = \frac{(5800 \text{ lb})(25.0 \text{ in.})}{28.0 \text{ in.}} = 5179 \text{ lb}$$

This force acts downward on the beam and upward on the vertical bar (Fig. 2-12c).

Now we can determine the downward reaction at support A of the bar (Fig. 2-12c):

$$R_A = P_3 - P_1 = 5179 \text{ lb} - 2250 \text{ lb} = 2929 \text{ lb}$$

The upper part of the vertical bar (segment AB) is subjected to an axial compressive force N_1 equal to R_A, or 2929 lb. The lower part (segment BC) carries an axial tensile force N_2 equal to P_1, or 2250 lb. (As an alternative, we can obtain these same results by using a free-body diagram of the entire structure instead of a free-body diagram of beam BDE.)

Changes in length. With tension considered positive, Eq. (2-5) yields

$$\delta = \sum_{i=1}^{n} \frac{N_i L_i}{E_i A_i} = \frac{N_1 L_1}{EA_1} + \frac{N_2 L_2}{EA_2}$$

$$= \frac{(-2929 \text{ lb})(20.0 \text{ in.})}{(29.0 \times 10^6 \text{ psi})(0.25 \text{ in.}^2)} + \frac{(2250 \text{ lb})(30.0 \text{ in.})}{(29.0 \times 10^6 \text{ psi})(0.16 \text{ in.}^2)}$$

$$= -0.00808 \text{ in.} + 0.01455 \text{ in.} = 0.0065 \text{ in.}$$

in which δ is the change in length of bar ABC. Since δ is positive, the bar elongates. The displacement of point C is equal to the change in length of the bar:

$$\delta_C = 0.0065 \text{ in.} \qquad \longleftarrow$$

This displacement is downward.

■ **Example 2-4**

A slightly tapered bar AB of solid circular cross section and length L (Fig. 2-13a) is supported at end B and subjected to a tensile load P at the free end A. The diameters of the bar at ends A and B are d_A and d_B, respectively. Determine the elongation of the bar due to the load P.

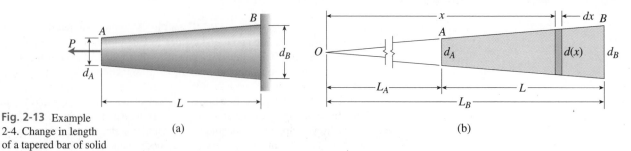

Fig. 2-13 Example 2-4. Change in length of a tapered bar of solid circular cross section

(a)   (b)

Solution The bar being analyzed in this example has a constant axial force (equal to the load P) throughout its length. However, the cross-sectional area varies continuously from one end to the other. Therefore, we must use integration (see Eq. 2-6) to determine the change in length.

Cross-sectional area. The first step in the solution is to obtain an expression for the cross-sectional area $A(x)$ at any cross section of the bar. For this purpose, we must establish an origin for the coordinate x. One possibility is to place the origin of coordinates at the free end A of the bar. However, the integrations to be performed will be slightly simplified if we locate the origin of coordinates by extending the sides of the tapered bar until they meet at point O, as shown in Fig. 2-13b.

The distances L_A and L_B from the origin O to ends A and B, respectively, are in the ratio

$$\frac{L_A}{L_B} = \frac{d_A}{d_B} \tag{a}$$

as obtained from similar triangles in Fig. 2-13b. From similar triangles we also get the ratio of the diameter $d(x)$ at distance x from the origin to the diameter d_A at the small end of the bar:

$$\frac{d(x)}{d_A} = \frac{x}{L_A} \quad \text{or} \quad d(x) = \frac{d_A x}{L_A} \tag{b}$$

Therefore, the cross-sectional area at distance x from the origin is

$$A(x) = \frac{\pi[d(x)]^2}{4} = \frac{\pi d_A^2 x^2}{4L_A^2} \tag{c}$$

Change in length. We now substitute the expression for $A(x)$ into Eq. (2-6) and obtain the elongation δ:

$$\delta = \int \frac{N(x)dx}{EA(x)} = \int_{L_A}^{L_B} \frac{Pdx(4L_A^2)}{E(\pi d_A^2 x^2)} = \frac{4PL_A^2}{\pi Ed_A^2}\int_{L_A}^{L_B} \frac{dx}{x^2} \tag{d}$$

By performing the integration (see Appendix C for integration formulas) and substituting the limits, we get

$$\delta = \frac{4PL_A^2}{\pi Ed_A^2}\left[-\frac{1}{x}\right]_{L_A}^{L_B} = \frac{4PL_A^2}{\pi Ed_A^2}\left(\frac{1}{L_A} - \frac{1}{L_B}\right)$$

This expression for δ can be simplified by noting that

$$\frac{1}{L_A} - \frac{1}{L_B} = \frac{L_B - L_A}{L_A L_B} = \frac{L}{L_A L_B}$$

Thus, the equation for δ becomes

$$\delta = \frac{4PL}{\pi Ed_A^2}\left(\frac{L_A}{L_B}\right) \tag{e}$$

Finally, we substitute $L_A/L_B = d_A/d_B$ (see Eq. a) and obtain

$$\delta = \frac{4PL}{\pi Ed_A d_B} \tag{2-7}$$ ⬅

This formula gives the elongation of a tapered bar of solid circular cross section. By substituting numerical values, we can determine the change in length for any particular bar.

Note 1: A common mistake is to assume that the elongation of a tapered bar can be determined by calculating the elongation of a prismatic bar that has the same cross-sectional area as the midsection of the tapered bar. Examination of Eq. (2-7) shows that this idea is not valid.

Note 2: The preceding formula for a tapered bar (Eq. 2-7) can be reduced to the special case of a prismatic bar by substituting $d_A = d_B = d$. The result is

$$\delta = \frac{4PL}{\pi E d^2} = \frac{PL}{EA}$$

which we know to be correct.

A general formula such as Eq. (2-7) should be checked whenever possible by verifying that it reduces to known results for simpler cases. If the reduction does not produce a correct result, the original formula is in error. If a correct result is obtained, the original formula may still be incorrect but our confidence in it increases. In other words, this type of check is a necessary but not sufficient condition for the correctness of the original formula.

2.4 ■ STATICALLY INDETERMINATE STRUCTURES

The springs, bars, and cables that we discussed in the preceding sections had one important feature in common—their reactions and internal forces could be determined solely from free-body diagrams and equations of equilibrium. Structures of this type are classified as **statically determinate**. We should note especially that the forces in a statically determinate structure can be found without knowing the properties of the materials. Consider, for instance, the bar *AB* shown in Fig. 2-14. The calculations for the internal axial forces in both parts of the bar, as well as for the reaction *R* at the base, are independent of the material of which the bar is made.

Most structures are more complex than the bar of Fig. 2-14, and their reactions and internal forces cannot be found by statics alone. This situation is illustrated in Fig. 2-15, which shows a bar *AB* fixed at *both* ends. There are now two vertical reactions (R_A and R_B) but only one useful equation of equilibrium—the equation for summing forces in the vertical direction. Since this equation contains two unknowns, it is not sufficient for finding the reactions. Structures of this kind are classified as **statically indeterminate**. To analyze such structures we must supplement the equilibrium equations with additional equations pertaining to the displacements of the structure.

To see how a statically indeterminate structure is analyzed, consider the example of Fig. 2-16a. The prismatic bar *AB* is attached to rigid supports at both ends and is axially loaded by a force *P* at an intermediate

Fig. 2-14 Statically determinate bar

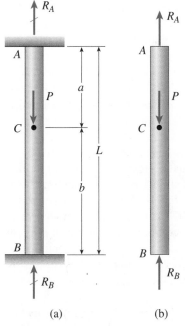

Fig. 2-15 Statically indeterminate bar

(a)   (b)

Fig. 2-16 Analysis of a statically indeterminate bar

point C. As already discussed, the reactions R_A and R_B cannot be found by statics alone, because only one **equation of equilibrium** is available:

$$\Sigma F_{\text{vert}} = 0 \qquad R_A - P + R_B = 0 \qquad \text{(a)}$$

An additional equation is needed in order to solve for the two unknown reactions.

The additional equation is based upon the observation that a bar with both ends fixed does not change in length. If we separate the bar from its supports (Fig. 2-16b), we obtain a bar that is free at both ends and loaded by the three forces, R_A, R_B, and P. These forces cause the bar to change in length by an amount δ_{AB}, which must be equal to zero:

$$\delta_{AB} = 0 \qquad \text{(b)}$$

This equation, called an **equation of compatibility**, expresses the fact that the change in length of the bar must be compatible with the conditions at the supports.

In order to solve Eqs. (a) and (b), we must now express the compatibility equation in terms of the unknown forces R_A and R_B. The relationships between the forces acting on a bar and its changes in length are known as **force-displacement relations**. These relations have various forms depending upon the properties of the material. If the material is linearly elastic, the equation $\delta = PL/EA$ can be used to obtain the force-displacement relations.

Let us assume that the bar of Fig. 2-16 has cross-sectional area A and is made of a material with modulus E. Then the changes in lengths of the upper and lower segments of the bar are, respectively,

$$\delta_{AC} = \frac{R_A a}{EA} \qquad \delta_{CB} = -\frac{R_B b}{EA} \qquad \text{(c,d)}$$

where the minus sign indicates a shortening of the bar. These force-displacement relations are now combined to give the change in length of the entire bar:

$$\delta_{AB} = \delta_{AC} + \delta_{CB} = \frac{R_A a}{EA} - \frac{R_B b}{EA}$$

Thus, the equation of compatibility (Eq. b) now becomes

$$\frac{R_A a}{EA} - \frac{R_B b}{EA} = 0 \qquad \text{(e)}$$

which contains the two reactions as unknowns.

The final step in analyzing the statically indeterminate bar is to solve simultaneously the equations of equilibrium (Eq. a) and compatibility (Eq. e). The results are

$$R_A = \frac{Pb}{L} \qquad R_B = \frac{Pa}{L} \qquad \text{(2-8a,b)}$$

With the reactions known, all other force and displacement quantities can be determined. Suppose, for instance, that we wish to find the downward displacement δ_C of point C. This displacement is equal to the elongation of segment AC:

$$\delta_C = \delta_{AC} = \frac{R_A a}{EA} = \frac{Pab}{LEA} \tag{2-9}$$

Also, we can find the stresses in the two segments of the bar directly from the internal axial forces (e.g., $\sigma_{AC} = R_A/A = Pb/AL$).

To summarize, we see from this example that the analysis of a statically indeterminate structure involves setting up and solving equations of equilibrium and equations of compatibility. The equilibrium equations relate the loads acting on the structure to the unknown forces (which may be reactions or internal forces), and the compatibility equations express conditions on the displacements of the structure. The compatibility equations are expressed in terms of the unknown forces by substituting the force-displacement relations. Finally, the equations of equilibrium and compatibility are solved simultaneously for the unknown forces.

In the engineering literature, various terms are used for the conditions expressed by the equilibrium, compatibility, and force-displacement equations. The equilibrium equations are also known as *static* or *kinetic* equations; the compatibility equations are sometimes called *geometric* equations, *kinematic* equations, or equations of *consistent deformations*; and the force-displacement relations are often referred to as *constitutive relations* (because they deal with the *constitution*, or physical properties, of the materials).

For the relatively simple structures discussed in this chapter, the preceding method of analysis is adequate. However, more formalized approaches are needed for complicated structures. Two commonly used methods, the *flexibility method* and the *stiffness method*, are described in detail in textbooks on structural analysis (Ref. 2-2). Even though these methods are normally used for large and complex structures requiring the solution of hundreds and sometimes thousands of simultaneous equations, they still are based upon the concepts described above, that is, equilibrium equations, compatibility equations, and force-displacement relations.*

The following examples illustrate the methodology for analyzing statically indeterminate structures consisting of axially loaded members. The second and third examples also illustrate the use of displacement diagrams in setting up equations of compatibility.

---

*From a historical viewpoint, it appears that Euler in 1774 was the first to analyze a statically indeterminate system; he considered the problem of a rigid table with four legs supported on an elastic foundation (Refs. 2-3 and 2-4). The next work was done by the French mathematician and engineer L. M. H. Navier, who in 1825 pointed out that statically indeterminate reactions could be found only by taking into account the elasticity of the structure (Ref. 2-5). Navier solved a statically indeterminate three-bar truss.

■ Example 2-5

A solid circular steel cylinder S is encased in a hollow circular copper tube C (Figs. 2-17a and b). The cylinder and tube are compressed between the rigid plates of a testing machine by compressive forces P. The steel cylinder has cross-sectional area A_s and modulus of elasticity E_s, the copper tube has area A_c and modulus E_c, and both parts have length L.

Determine the following quantities: (a) the compressive forces P_s in the steel cylinder and P_c in the copper tube; (b) the corresponding compressive stresses σ_s and σ_c; and (c) the shortening δ of the assembly.

(a)

(b)

(c)

(d)

Fig. 2-17 Example 2-5. Analysis of a statically indeterminate structure

Solution   (a) *Compressive forces in the steel cylinder and copper tube.* We begin by removing the upper plate of the assembly in order to expose the compressive forces P_s and P_c acting on the steel cylinder and copper tube, respectively (Fig. 2-17c). The force P_s is the resultant of the uniformly distributed stresses acting over the cross section of the steel cylinder, and the force P_c is the resultant of the stresses acting over the cross section of the copper tube.

Equation of equilibrium. A free-body diagram of the upper plate is shown in Fig. 2-17d. This plate is subjected to the force P and to the unknown compressive forces P_s and P_c ; thus, the equation of equilibrium is

$$\sum F_{\text{vert}} = 0 \qquad P_s + P_c - P = 0 \tag{f}$$

This equation, which is the only nontrivial equilibrium equation available, contains two unknowns. Therefore, we conclude that the structure is statically indeterminate.

Equation of compatibility. Because the end plates are rigid, the steel cylinder and copper tube must shorten by the same amount. Denoting the shortenings of the steel and copper parts by δ_s and δ_c, respectively, we obtain the following equation of compatibility:

$$\delta_s = \delta_c \tag{g}$$

Force-displacement relations. The changes in lengths of the cylinder and tube can be obtained from the general equation $\delta = PL/EA$. Therefore, in this example the force-displacement relations are

$$\delta_s = \frac{P_s L}{E_s A_s} \qquad \delta_c = \frac{P_c L}{E_c A_c} \tag{h,i}$$

Substituting these relations into the equation of compatibility (Eq. g) gives

$$\frac{P_s}{E_s A_s} = \frac{P_c}{E_c A_c} \tag{j}$$

This equation gives the compatibility condition in terms of the unknown forces.

Solution of equations. We now solve simultaneously the equations of equilibrium and compatibility (Eqs. f and j) and obtain the axial forces in the steel cylinder and copper tube:

$$P_s = P\left(\frac{E_s A_s}{E_s A_s + E_c A_c}\right) \qquad P_c = P\left(\frac{E_c A_c}{E_s A_s + E_c A_c}\right) \tag{2-10a,b}$$

These equations show that the compressive forces in the steel and copper parts are directly proportional to their respective axial rigidities and inversely proportional to the sum of their rigidities.

(b) *Compressive stresses in the steel cylinder and copper tube.* Knowing the axial forces, we can now obtain the compressive stresses in the two materials:

$$\sigma_s = \frac{P_s}{A_s} = \frac{PE_s}{E_s A_s + E_c A_c} \qquad \sigma_c = \frac{P_c}{A_c} = \frac{PE_c}{E_s A_s + E_c A_c} \tag{2-11a,b}$$

Note that the stresses are proportional to the moduli of elasticity of the respective materials. Therefore, the "stiffer" material has the larger stress.

(c) *Shortening of the assembly.* The shortening δ of the entire assembly can be obtained from either Eq. (h) or Eq. (i). Thus, upon substituting the forces (from Eqs. 2-10a and b), we get

$$\delta = \frac{P_s L}{E_s A_s} = \frac{P_c L}{E_c A_c} = \frac{PL}{E_s A_s + E_c A_c} \qquad (2\text{-}12) \quad \Longleftarrow$$

This result shows that the shortening of the assembly is equal to the total load divided by the sum of the stiffnesses of the two parts (recall from Eq. 2-4a that the stiffness of an axially loaded bar is $k = EA/L$).

■ Example 2-6

A horizontal rigid bar AB is pinned at end A and supported by two wires (CD and EF) at points D and F (Fig. 2-18a). A vertical load P acts at end B of the bar. The bar has length $3b$ and wires CD and EF have lengths L_1 and L_2, respectively. Also, wire CD has diameter d_1 and modulus of elasticity E_1; wire EF has diameter d_2 and modulus E_2.

(a) Obtain formulas for the allowable load P if the allowable stresses in wires CD and EF, respectively, are σ_1 and σ_2. (Disregard the weight of the bar itself.)

(b) Calculate the allowable load P for the following conditions: Wire CD is made of aluminum with modulus $E_1 = 72$ GPa, diameter $d_1 = 4.0$ mm, and length $L_1 = 0.40$ m. Wire EF is made of magnesium with modulus $E_2 = 45$ GPa, diameter $d_2 = 3.0$ mm, and length $L_2 = 0.30$ m. The allowable stresses in the aluminum and magnesium wires are $\sigma_1 = 200$ MPa and $\sigma_2 = 175$ MPa, respectively.

Solution *Equations of equilibrium.* We begin the analysis by drawing a free-body diagram of bar AB (Fig. 2-18b). In this diagram T_1 and T_2 are the unknown tensile forces in the wires and R_H and R_V are the horizontal and vertical components of the reaction at the support. We see immediately that the structure is statically indeterminate because there are four unknown forces (T_1, T_2, R_H, and R_V) but only three independent equations of equilibrium. Taking moments about point A (with counterclockwise moments being positive) yields

$$\sum M_A = 0 \qquad T_1 b + T_2(2b) - P(3b) = 0 \quad \text{or} \quad T_1 + 2T_2 = 3P \qquad (k)$$

The other two equations, obtained by summing forces in the horizontal direction and summing forces in the vertical direction, are of no benefit in finding T_1 and T_2.

Equation of compatibility. To obtain an equation pertaining to the displacements, we observe that the load P causes bar AB to rotate about the pin support at A, thereby stretching the wires. The resulting displacements are shown in the displacement diagram of Fig. 2-18c, where line AB represents the original posi-

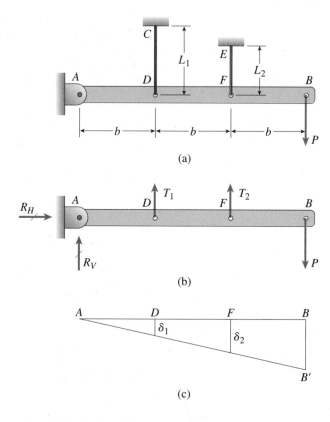

Fig. 2-18 Example 2-6. Analysis of a statically indeterminate structure

tion of the rigid bar and line AB' represents the rotated position. The displacements δ_1 and δ_2 are the elongations of the wires. Because these displacements are very small, the bar rotates through a very small angle (shown highly exaggerated in the figure) and we can make calculations on the assumption that points D, F, and B move vertically downward (instead of moving along the arcs of circles).

Because the horizontal distances AD and DF are equal, we obtain the following geometric relationship between the elongations:

$$\delta_2 = 2\delta_1 \tag{1}$$

Equation (1) is the equation of compatibility.

Force-displacement relations. Since the wires behave in a linearly elastic manner, their elongations can be expressed in terms of the unknown forces T_1 and T_2 by means of the following expressions:

$$\delta_1 = \frac{T_1 L_1}{E_1 A_1} \qquad \delta_2 = \frac{T_2 L_2}{E_2 A_2}$$

in which A_1 and A_2 are the cross-sectional areas of wires CD and EF, respectively; that is,

$$A_1 = \frac{\pi d_1^2}{4} \qquad A_2 = \frac{\pi d_2^2}{4}$$

For convenience in writing equations, let us introduce the following notation for the flexibilities of the wires (see Eq. 2-4b):

$$f_1 = \frac{L_1}{E_1 A_1} \qquad f_2 = \frac{L_2}{E_2 A_2} \qquad \text{(m,n)}$$

Then the force-displacement relations become

$$\delta_1 = f_1 T_1 \qquad \delta_2 = f_2 T_2$$

Substituting these expressions into the equation of compatibility (Eq. l) gives

$$f_2 T_2 = 2 f_1 T_1 \qquad \text{(o)}$$

We now have expressed the equation of compatibility in terms of the unknown forces.

Solution of equations. The equation of equilibrium (Eq. k) and the equation of compatibility (Eq. o) each contain the forces T_1 and T_2 as unknown quantities. Solving the two equations simultaneously yields

$$T_1 = \frac{3 f_2 P}{4 f_1 + f_2} \qquad T_2 = \frac{6 f_1 P}{4 f_1 + f_2} \qquad \text{(p,q)}$$

Knowing the forces T_1 and T_2, we can easily find the elongations of the wires from the force-displacement relations.

(a) *Allowable load P.* Now that the statically indeterminate analysis is completed and the forces in the wires are known, we can determine the permissible value of the load P. The stress σ_1 in wire CD and the stress σ_2 in wire EF are readily obtained from the forces (Eqs. p and q):

$$\sigma_1 = \frac{T_1}{A_1} = \frac{3P}{A_1}\left(\frac{f_2}{4 f_1 + f_2}\right) \qquad \sigma_2 = \frac{T_2}{A_2} = \frac{6P}{A_2}\left(\frac{f_1}{4 f_1 + f_2}\right)$$

From the first of these equations we solve for the permissible force P_1 based upon the allowable stress σ_1 in the aluminum wire:

$$P_1 = \frac{\sigma_1 A_1 (4 f_1 + f_2)}{3 f_2} \qquad \text{(r)} \quad \Longleftarrow$$

Similarly, from the second equation we get the permissible force P_2 based upon the allowable stress σ_2 in the magnesium wire:

$$P_2 = \frac{\sigma_2 A_2 (4 f_1 + f_2)}{6 f_1} \qquad \text{(s)} \quad \Longleftarrow$$

The smaller of these two values of the load is the maximum allowable load P_{allow}.

(b) *Numerical calculations for the allowable load.* Using the given data and the preceding equations, we obtain the following numerical values:

$$A_1 = \frac{\pi d_1^2}{4} = \frac{\pi (4.0 \text{ mm})^2}{4} = 12.57 \text{ mm}^2$$

$$A_2 = \frac{\pi d_2^2}{4} = \frac{\pi (3.0 \text{ mm})^2}{4} = 7.069 \text{ mm}^2$$

$$f_1 = \frac{L_1}{E_1 A_1} = \frac{0.40 \text{ m}}{(72 \text{ GPa})(12.57 \text{ mm}^2)} = 0.4420 \times 10^{-6} \text{ m/N}$$

$$f_2 = \frac{L_2}{E_2 A_2} = \frac{0.30 \text{ m}}{(45 \text{ GPa})(7.069 \text{ mm}^2)} = 0.9431 \times 10^{-6} \text{ m/N}$$

Also, the allowable stresses are

$$\sigma_1 = 200 \text{ MPa} \qquad \sigma_2 = 175 \text{ MPa}$$

Therefore, substituting into Eqs. (r) and (s) gives

$$P_1 = 2.41 \text{ kN} \qquad P_2 = 1.26 \text{ kN}$$

The first result is based upon the allowable stress σ_1 in the aluminum wire and the second is based upon the allowable stress σ_2 in the magnesium wire. The allowable load is the smaller of the two values:

$$P_{\text{allow}} = 1.26 \text{ kN} \qquad \longleftarrow$$

At this load the stress in the magnesium is 175 MPa (the allowable stress) and the stress in the aluminum is (1.26/2.41)(200 MPa) = 105 MPa. As expected, this stress is less than the allowable stress of 200 MPa.

■ **Example 2-7**

A horizontal rigid bar *ADB* of length $2b$ is pinned to a support at *A* and held by two inclined wires *CD* and *CB* (Fig. 2-19a). The wires are attached to a support at point *C*, which is located at a vertical distance b above point *A*. Both wires are made of the same material and have the same diameter.

Determine the tensile forces T_1 and T_2 in wires *CD* and *CB*, respectively, due to the vertical load *P* acting at the end of the bar.

Solution *Equation of equilibrium.* A free-body diagram of bar *AB* (Fig. 2-19b) shows that there are four unknown forces, namely, the tensile forces T_1 and T_2 in the wires and the two reaction components (R_H and R_V) at the pin support. Since there are only three independent equations of equilibrium, the structure is statically indeterminate. We can obtain an equation of equilibrium containing only T_1 and T_2 as unknowns by summing moments about point *A*:

$$\sum M_A = 0 \qquad bT_1 \sin \alpha_1 + 2bT_2 \sin \alpha_2 - 2bP = 0 \qquad \text{(t)}$$

in which α_1 and α_2 are the angles between the wires and the bar.

From the geometry of the structure (Fig. 2-19a) we see that

$$\sin \alpha_1 = \frac{CA}{CD} = \frac{b}{b\sqrt{2}} = \frac{1}{\sqrt{2}} \qquad \sin \alpha_2 = \frac{CA}{CB} = \frac{b}{b\sqrt{5}} = \frac{1}{\sqrt{5}} \qquad \text{(u)}$$

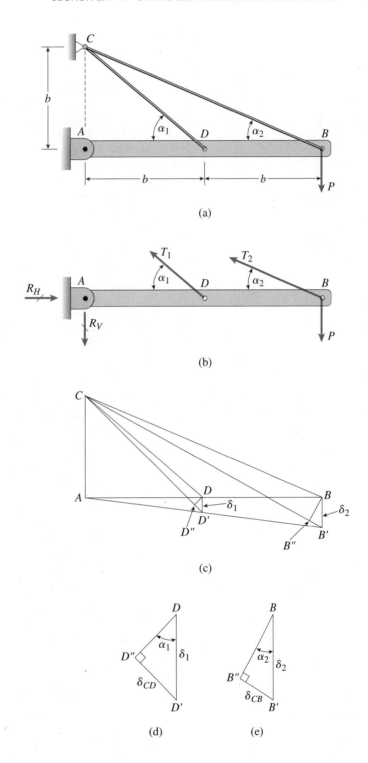

Fig. 2-19 Example 2-7. Analysis of a statically indeterminate structure

(a)

(b)

(c)

(d)   (e)

Substituting these values into Eq. (t) and rearranging, we get

$$\frac{T_1}{\sqrt{2}} + \frac{2T_2}{\sqrt{5}} = 2P \tag{v}$$

which is the final form of the equation of equilibrium.

Equation of compatibility. An equation of compatibility can be obtained by considering the displacements of points D and B. For this purpose, we draw the displacement diagram shown in Fig. 2-19c. Line AB represents the original position of bar AB and line AB' represents the rotated position. Since the angle of rotation of bar AB is very small, the displacements δ_1 and δ_2 of points D and B, respectively, can be treated as small vertical displacements. Because the distance from A to B is twice the distance from A to D, we see that

$$\delta_2 = 2\delta_1 \tag{w}$$

which is the equation of compatibility.

Force-displacement relations. Now we need to relate the forces T_1 and T_2 in the wires to the vertical displacements δ_1 and δ_2 at points D and B, respectively. Let us begin by considering wire CD, which rotates from its original position CD to its final position CD' (Fig. 2-19c). We draw a perpendicular DD'' from point D to line CD'. Because the displacements and angle changes are very small, we can assume that the distance CD'' is equal in length to the distance CD. (That is, a circular arc DD'' with point C as its center is indistinguishable from the perpendicular line DD''.)

The displacement triangle $DD'D''$ in Fig. 2-19c is redrawn for clarity in Fig. 2-19d. The distance DD' is the vertical displacement δ_1 of point D and the distance $D'D''$ is the elongation δ_{CD} of wire CD. The angle α_1 appears in the triangle as angle $D'DD''$. To prove this, we note that line DD' is perpendicular to line AB, and line DD'' is perpendicular to line CD (for small angles of rotation). Therefore, from the displacement triangle $DD'D''$ we get

$$\delta_{CD} = \delta_1 \sin \alpha_1 = \frac{\delta_1}{\sqrt{2}}$$

This equation provides the geometric relation between the elongation of wire CD and the downward displacement of point D.

The elongation δ_{CD} of wire CD is related to the force in the wire by the following force-displacement relation:

$$\delta_{CD} = \frac{T_1 L_{CD}}{EA} = \frac{T_1 b\sqrt{2}}{EA}$$

in which $L_{CD} = b\sqrt{2}$ is the length of wire CD. Combining the last two equations for δ_{CD}, we get

$$\delta_1 = \frac{2bT_1}{EA} \tag{x}$$

In a similar manner, we can relate the displacement δ_2 to the elongation δ_{CB} of wire CB (see Fig. 2-19e) and obtain

$$\delta_2 = \frac{5bT_2}{EA} \tag{y}$$

Equations (x) and (y) are the force-displacement relations that give the displacements of points D and B in terms of the unknown forces in the wires.

Substituting δ_1 and δ_2 from the force-displacement relations into the equation of compatibility (Eq. w) yields

$$5T_2 = 4T_1 \tag{z}$$

which is the equation of compatibility in terms of the unknown forces.

Solution of equations. As the final step in this example we solve simultaneously the equation of equilibrium (Eq. v) and the equation of compatibility (Eq. z). The results are

$$T_1 = \frac{10\sqrt{10}\,P}{8\sqrt{2} + 5\sqrt{5}} = 1.406P \qquad T_2 = \frac{4T_1}{5} = 1.125P \quad \longleftarrow$$

Thus, we have found the tensile forces in the wires by solving the statically indeterminate structure.

Knowing the tensile forces in the wires, we can easily determine all other force and displacement quantities. For instance, the reactions at support A can be found from equilibrium equations and the displacements δ_1 and δ_2 can be found from Eqs. (x) and (y).

Note: In this example, the final solution for the forces does not involve the axial rigidity EA of the wires. (This fact is readily apparent in a symbolic solution but might escape notice in a numerical solution.) The reason is that both wires have the same axial rigidity, and therefore EA cancels out of the solution. If each wire had a different rigidity, the rigidities would appear in the final expressions.

2.5 ■ THERMAL EFFECTS

External loads are not the only sources of stresses and strains in a structure. Changes in temperature produce expansion or contraction of the material, resulting in **thermal strains** and **thermal stresses**. A simple illustration of thermal expansion is shown in Fig. 2-20, where the block of material is unrestrained and therefore free to expand. When the block is heated, every element of the material undergoes thermal strains in all directions, and consequently the dimensions of the block increase. If we take corner A as a fixed reference point and let side AB maintain its original alignment, the block will have the shape shown by the dashed lines.

For most structural materials, thermal strain ϵ_T is proportional to the temperature change ΔT; that is,

$$\epsilon_T = \alpha(\Delta T) \tag{2-13}$$

in which α is a property of the material called the **coefficient of thermal expansion**. Since strain is a dimensionless quantity, the coefficient of thermal expansion has units equal to the reciprocal of temperature change. In SI units the dimensions of α can be expressed as either 1/K

A B

Fig. 2-20 Block of material subjected to an increase in temperature (the reciprocal of kelvins) or 1/°C (the reciprocal of degrees Celsius). The value of α is the same in both cases because a *change* in temperature is numerically the same in both kelvins and degrees Celsius. In USCS units, the dimensions of α are 1/°F (the reciprocal of degrees Fahrenheit).* Typical values of α are listed in Table H-4 of Appendix H.

When a sign convention is needed for thermal strains, we usually assume that expansion is positive and contraction is negative.

To demonstrate the relative importance of thermal strains, we will compare thermal strains with load-induced strains in the following manner. Suppose we have an axially loaded bar with longitudinal strains given by the equation $\epsilon = \sigma/E$, where σ is the stress and E is the modulus of elasticity. Then suppose we have an identical bar subjected to a temperature change ΔT, which means that the bar has thermal strains given by Eq. (2-13). Equating the two strains gives the equation $\sigma = E\alpha(\Delta T)$. From this equation we can calculate the axial stress σ that produces the same strain as does the temperature change ΔT. For instance, consider a stainless steel bar with $E = 30 \times 10^6$ psi and $\alpha = 9.6 \times 10^{-6}$/°F. A quick calculation from the preceding equation for σ shows that a change in temperature of 100°F produces the same strain as a stress of 29,000 psi. This stress is in the range of typical allowable stresses for the material. Thus, a relatively modest change in temperature produces strains of the same magnitude as the strains caused by ordinary loads, which shows that temperature effects can be important in engineering design.

Ordinary structural materials expand when heated and contract when cooled, and therefore an increase in temperature produces a positive thermal strain. Thermal strains usually are reversible, in the sense that the member returns to its original shape when its temperature returns to the original value. However, a few special metallic alloys have recently been developed that do not behave in the customary manner. Instead, over certain temperature ranges their dimensions decrease when heated and increase when cooled. Water is also an unusual material from a thermal standpoint—it expands when heated at temperatures above 4°C and also expands when cooled below 4°C. Thus, water has its maximum density at 4°C.

Now let us return to the block of material shown in Fig. 2-20. We assume that the material is homogeneous and isotropic and that the temperature increase ΔT is uniform throughout the block. We can calculate the increase in *any* dimension of the block by multiplying the original dimension by the thermal strain. For instance, if one of the dimensions is L, then that dimension will increase by the amount

$$\delta_T = \epsilon_T L = \alpha(\Delta T)L \tag{2-14}$$

---

*For a discussion of temperature units and scales, see Section A.4 of Appendix A.

Equation (2-14) is a **temperature-displacement relation**, analogous to the force-displacement relations described in the preceding section. It can be used to calculate changes in lengths of structural members subjected to uniform temperature changes, such as the elongation δ_T of the prismatic bar shown in Fig. 2-21. (The transverse dimensions of the bar also change, but these changes are not shown in the figure since they usually have no effect on the axial forces being transmitted by the bar.)

Fig. 2-21 Increase in length of a prismatic bar due to a uniform increase in temperature (Eq. 2-14)

In the preceding discussions of thermal strains, we assumed that the structure had no restraints and was able to expand or contract freely. These conditions exist when an object rests on a frictionless surface or hangs in open space. In such cases no stresses are produced by a uniform temperature change throughout the object, although nonuniform temperature changes may produce internal stresses. However, many structures have supports that prevent free expansion and contraction, in which case **thermal stresses** will develop even when the temperature change is uniform throughout the structure.

To illustrate some of these ideas about thermal effects, consider the two-bar truss *ABC* of Fig. 2-22 and assume that the temperature of bar *AB* is changed by ΔT_1 and the temperature of bar *BC* is changed by ΔT_2. Because the truss is statically determinate, both bars are free to lengthen or shorten, resulting in a displacement of joint *B*. However, there are no stresses in either bar and no reactions at the supports. This conclusion applies generally to **statically determinate structures**; that is, uniform temperature changes in the members produce thermal strains (and the corresponding changes in lengths) without producing any corresponding stresses.

A **statically indeterminate structure** may or may not develop temperature stresses, depending upon the character of the structure and the nature of the temperature changes. To illustrate some of the possibilities, consider the statically indeterminate truss shown in Fig. 2-23. Because the supports of this structure permit joint *D* to move horizontally, no stresses are developed when the *entire* truss is heated uniformly. All members increase in length in proportion to their original lengths, and hence the truss becomes slightly larger in size. However, if some (but not all) of the bars are heated, thermal stresses will develop because the statically indeterminate arrangement of the bars prevents free expansion. To visualize this condition, imagine that just one bar is heated. As this bar becomes longer, it meets resistance from the other bars, and therefore stresses develop in every member.

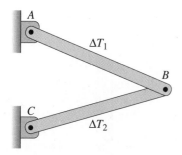

Fig. 2-22 Statically determinate truss with a uniform temperature change in each member.

Fig. 2-23 Statically indeterminate truss subjected to temperature changes

The analysis of a statically indeterminate structure with temperature changes is based upon the concepts discussed in the preceding section, namely equilibrium equations, compatibility equations, and displacement relations. The principal difference is that we now use temperature-displacement relations (Eq. 2-14) in addition to force-displacement relations (such as $\delta = PL/EA$) when performing the analysis. The following examples illustrate the procedures in detail.

■ **Example 2-8**

A prismatic bar AB of length L is held between immovable supports (Fig. 2-24a). If the temperature of the bar is raised uniformly by an amount ΔT, what thermal stress σ_T is developed in the bar? (Assume that the bar is made of linearly elastic material.)

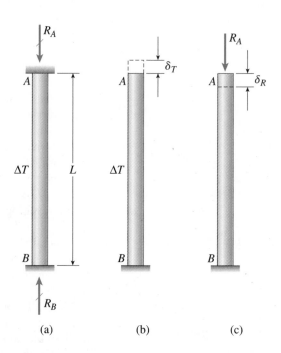

Fig. 2-24 Example 2-8. Statically indeterminate bar with uniform temperature increase ΔT

(a)   (b)   (c)

Solution   Because the temperature increases, the bar tends to elongate but is restrained by the rigid supports at A and B. Therefore, reactions R_A and R_B are developed at the supports, and the bar is subjected to uniform compressive stresses.

Equation of equilibrium. The only forces acting on the bar are the reactions shown in Fig. 2-24a. Therefore, equilibrium of forces in the vertical direction gives

$$\Sigma F_{\text{vert}} = 0 \qquad R_B - R_A = 0 \qquad\qquad (a)$$

Since this is the only nontrivial equation of equilibrium, and since it contains two unknowns, we see that the structure is statically indeterminate and an additional equation is needed.

Equation of compatibility. The equation of compatibility expresses the fact that the change in length of the bar is zero (because the supports do not move):

$$\delta_{AB} = 0 \tag{b}$$

To determine this change in length, we remove the upper support of the bar and obtain a bar that is fixed at the base and free to displace at the upper end (Figs. 2-24b and c). When only the temperature change is acting (Fig. 2-24b), the bar elongates by the amount δ_T, and when only the reaction R_A is acting, the bar shortens by the amount δ_R (Fig. 2-24c). Thus, the net change in length is $\delta_{AB} = \delta_T - \delta_R$, and the equation of compatibility becomes

$$\delta_{AB} = \delta_T - \delta_R = 0 \tag{c}$$

Displacement relations. The increase in length of the bar due to the temperature change ΔT is given by the temperature-displacement relation (Eq. 2-14):

$$\delta_T = \alpha(\Delta T)L \tag{d}$$

in which α is the coefficient of thermal expansion. The decrease in length due to the force R_A is given by the force-displacement relation:

$$\delta_R = \frac{R_A L}{EA} \tag{e}$$

in which E is the modulus of elasticity and A is the cross-sectional area. Substituting the displacement relations (d) and (e) into Eq. (c) gives the final form of the equation of compatibility:

$$\delta_T - \delta_R = \alpha(\Delta T)L - \frac{R_A L}{EA} = 0 \tag{f}$$

Solution of equations. We now solve the equations of equilibrium and compatibility (Eqs. a and f) for the reactions R_A and R_B:

$$R_A = R_B = EA\alpha(\Delta T) \tag{2-15}$$

From these results we obtain the thermal stress σ_T in the bar:

$$\sigma_T = \frac{R_A}{A} = \frac{R_B}{A} = E\alpha(\Delta T) \tag{2-16}$$

This stress is compressive when the temperature of the bar increases.

Note 1: In this example the reactions are independent of the length of the bar and the stress is independent of both the length and the cross-sectional area (see Eqs. 2-15 and 2-16). Thus, once again we see the usefulness of a symbolic solution, because these important features of the bar's behavior might not be noticed in a numerical problem.

Note 2: When determining the thermal elongation of the bar (Eq. d), we assumed that the material was homogeneous and that the increase in temperature was uniform throughout the volume of the bar. Also, when determining the decrease in length due to the reactive force (Eq. e), we assumed linearly elastic behavior of the material. These limitations should always be kept in mind when writing the temperature-displacement and force-displacement relations.

Note 3: The bar in this example has zero longitudinal displacements, not only at the fixed ends but also at every cross section. Thus, there are no axial strains in this bar, and we have the special situation of longitudinal stresses without longitudinal strains. Of course, there are transverse strains in the bar, from both the temperature change and the axial compression.

■ Example 2-9

A sleeve in the form of a circular tube of length L is placed around a bolt and fitted between washers at each end (Fig. 2-25a). The nut is then turned until it is just snug. The sleeve and bolt are made of different materials and have different cross-sectional areas. (a) If the temperature of the entire assembly is raised by an amount ΔT, what stresses σ_s and σ_b are developed in the sleeve and bolt, respectively? (b) What is the increase δ in the length L of the sleeve and bolt? (Assume that the coefficient of thermal expansion α_s of the sleeve is greater than the coefficient α_b of the bolt.)

Fig. 2-25 Example 2-9. Sleeve and bolt assembly with uniform temperature increase ΔT

Solution   Because the sleeve and bolt are of different materials, they will elongate by different amounts when heated and allowed to expand freely. However, when they are held together by the assembly, free expansion cannot occur and thermal stresses are developed in both materials. To find these stresses, we use the same concepts as in any statically indeterminate analysis—equilibrium equations, compatibility equations, and displacement relations. However, we cannot formulate these equations until we disassemble the structure.

A simple way to cut the structure is to remove the head of the bolt, thereby allowing the sleeve and bolt to expand freely under the temperature change ΔT (Fig. 2-25b). The resulting elongations of the sleeve and bolt are denoted δ_1 and δ_2, respectively, and the corresponding *temperature-displacement relations* are

$$\delta_1 = \alpha_s(\Delta T)L \qquad \delta_2 = \alpha_b(\Delta T)L \qquad \text{(g)}$$

Since α_s is greater than α_b, the elongation δ_1 is greater than δ_2, as shown in Fig. 2-25b.

The axial forces in the sleeve and bolt must be such that they shorten the sleeve and stretch the bolt until the final lengths of the sleeve and bolt are the same. These forces are shown in Fig. 2-25c, where P_s denotes the compressive force in the sleeve and P_b denotes the tensile force in the bolt. The corresponding shortening δ_3 of the sleeve and elongation δ_4 of the bolt are

$$\delta_3 = \frac{P_s L}{E_s A_s} \qquad \delta_4 = \frac{P_b L}{E_b A_b} \qquad \text{(h)}$$

in which $E_s A_s$ and $E_b A_b$ are the respective axial rigidities. Equations (h) are the *load-displacement relations.*

Now we can write an *equation of compatibility* expressing the fact that the final elongation δ is the same for both the sleeve and the bolt. The elongation of the sleeve is $\delta_1 - \delta_3$ and of the bolt is $\delta_2 + \delta_4$; therefore,

$$\delta = \delta_1 - \delta_3 = \delta_2 + \delta_4 \qquad \text{(i)}$$

Substituting the temperature-displacement and load-displacement relations (Eqs. g and h) into this equation gives

$$\delta = \alpha_s(\Delta T)L - \frac{P_s L}{E_s A_s} = \alpha_b(\Delta T)L + \frac{P_b L}{E_b A_b} \qquad \text{(j)}$$

from which we get

$$\frac{P_s L}{E_s A_s} + \frac{P_b L}{E_b A_b} = \alpha_s(\Delta T)L - \alpha_b(\Delta T)L \qquad \text{(k)}$$

which is the final form of the compatibility equation. Note that it contains the forces P_s and P_b as unknowns.

An *equation of equilibrium* is obtained from Fig. 2-25c, which is a free-body diagram of the part of the assembly remaining after the head of the bolt is removed. Summing forces in the horizontal direction gives

$$P_s = P_b \qquad \text{(l)}$$

which expresses the obvious fact that the compressive force in the sleeve is equal to the tensile force in the bolt.

We now solve simultaneously the equations of compatibility and equilibrium (Eqs. k and l) and obtain the axial forces in the sleeve and bolt:

$$P_s = P_b = \frac{(\alpha_s - \alpha_b)(\Delta T)E_s A_s E_b A_b}{E_s A_s + E_b A_b} \tag{2-17}$$

When deriving this equation, we assumed that the temperature increased and that the coefficient α_s was greater than the coefficient α_b. Under these conditions, P_s is the compressive force in the sleeve and P_b is the tensile force in the bolt.

The results will be quite different if the temperature increases but the coefficient α_s is less than the coefficient α_b. Under these conditions, a gap will open between the bolt head and the sleeve and there will be no stresses in either part of the assembly.

(a) *Stresses in the sleeve and bolt.* Expressions for the stresses σ_s and σ_b in the sleeve and bolt, respectively, are obtained by dividing the corresponding forces by the appropriate areas:

$$\sigma_s = \frac{P_s}{A_s} = \frac{(\alpha_s - \alpha_b)(\Delta T)E_s E_b A_b}{E_s A_s + E_b A_b} \tag{2-18a}$$

$$\sigma_b = \frac{P_b}{A_b} = \frac{(\alpha_s - \alpha_b)(\Delta T)E_s A_s E_b}{E_s A_s + E_b A_b} \tag{2-18b}$$

Under the assumed conditions, the stress σ_s in the sleeve is compressive and the stress σ_b in the bolt is tensile. It is interesting to note that these stresses are independent of the length of the assembly and their magnitudes are inversely proportional to their respective areas (that is, $\sigma_s/\sigma_b = A_b/A_s$).

(b) *Increase in length of the sleeve and bolt.* The elongation δ of the assembly can be found by substituting either P_s or P_b from Eq. (2-17) into Eq. (j), yielding

$$\delta = \frac{(\alpha_s E_s A_s + \alpha_b E_b A_b)(\Delta T)L}{E_s A_s + E_b A_b} \tag{2-19}$$

With the preceding formulas available, we can readily calculate the forces, stresses, and displacements of the assembly for any given set of numerical data.

Note: As a partial check on the results, we can see if Eqs. (2-17), (2-18), and (2-19) reduce to known values in simplified cases. For instance, suppose that the bolt is rigid and therefore unaffected by temperature changes. We can represent this situation by setting $\alpha_b = 0$ and letting E_b become infinitely large, thereby creating an assembly in which the sleeve is held between rigid supports. Substituting the stated values of α_b and E_b into Eqs. (2-17), (2-18a), and (2-19), we find

$$P_s = E_s A_s \alpha_s(\Delta T) \qquad \sigma_s = E_s \alpha_s(\Delta T) \qquad \delta = 0$$

These results agree with those of Example 2-8 for a bar held between rigid supports (compare with Eqs. 2-15 and 2-16, and with Eq. b).

As a second special case, suppose that the sleeve and bolt are made of the same material. Then both parts will expand freely and will lengthen the same amount when the temperature changes. No forces or stresses will be developed. To see if the derived equations predict this behavior, we will assume that both parts have properties α, E, and A. Substituting these values into Eqs. (2-17), (2-18), and (2-19), we get

$$P_s = P_b = 0 \qquad \sigma_s = \sigma_b = 0 \qquad \delta = \alpha(\Delta T)L$$

which are the expected results.

As a third special case, suppose we remove the sleeve so that only the bolt remains. The bolt is then free to expand and no forces or stresses will develop. We can represent this case by setting $A_s = 0$ in Eqs. (2-17), (2-18b), and (2-19), which then give

$$P_b = 0 \qquad \sigma_b = 0 \qquad \delta = \alpha_b(\Delta T)L$$

Again, these are the expected results.

As pointed out at the end of Section 2.3, the investigation of a special case is not sufficient to establish the correctness of a general formula. However, it certainly is a useful check. If a special case is verified, you gain confidence in your results. If a special case does *not* check, it means that the general formula has an error.

2.6 ■ STRESSES ON INCLINED SECTIONS

In our previous discussions of tension and compression in axially loaded members, the only stresses we considered were the normal stresses acting on cross sections. These stresses are pictured in Fig. 2-26, where we consider a bar AB subjected to axial loads P. When the bar is cut at an intermediate cross section by a plane mn (perpendicular to the x axis), we obtain the free-body diagram shown in Fig. 2-26b. The normal stresses acting over the cut section may be calculated from the formula $\sigma_x = P/A$ provided that the stress distribution is uniform over the entire cross-sectional area A. As explained in Chapter 1, this condition exists if the bar is prismatic, the material is homogeneous, the axial force P acts at the centroid of the cross-sectional area, and the cross section is away from any localized stress concentrations. Of course, there are no shear stresses acting on the cut section, because it is perpendicular to the longitudinal axis of the bar.

For convenience, we usually show the stresses in a two-dimensional view of the bar (Fig. 2-26c) rather than the more complex three-dimensional view (Fig. 2-26b). However, when working with two-dimensional figures we must not forget that the bar has a thickness perpendicular to the plane of the figure. This third dimension must be considered when making derivations and calculations.

The most useful way of representing the stresses in the bar of Fig. 2-26 is to isolate a small element of material, such as the element labeled C in Fig. 2-26c, and then show the stresses acting on all faces of this element. An element of this kind is called a **stress element**. The stress element at point C is a small rectangular block (it doesn't matter whether

(a)

(b)

Fig. 2-26 Prismatic bar in tension
showing the stresses acting on cross
section *mn*: (a) bar with axial forces *P*,
(b) three-dimensional view of the cut bar
showing the normal stresses, and
(c) two-dimensional view

(c)

it is a cube or a rectangular parallelepiped) with its right-hand face lying
in cross section *mn*. The dimensions of any stress element are assumed to
be infinitesimally small, but for clarity we draw the element to a large
scale, as in Fig. 2-27a. In this case, the edges of the element are parallel to
the *x*, *y*, and *z* axes, and the only stresses are the normal stresses σ_x acting
on the *x* faces (recall that the *x* faces have their normals parallel to the *x*
axis). Because it is more convenient, we usually draw a two-dimensional
view of the element (Fig. 2-27b) instead of a three-dimensional view.

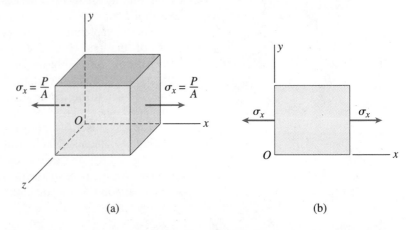

Fig. 2-27 Stress element at point *C* of
the axially loaded bar shown in
Fig. 2-26c: (a) three-dimensional view
of the element, and (b) two-dimensional
view of the element

(a)

(b)

The stress element of Fig. 2-27 provides only a limited view of the stresses in an axially loaded bar. To obtain a more complete picture, we need to investigate the stresses acting on **inclined sections**, such as the section cut by the inclined plane pq in Fig. 2-28a. Because the stresses are the same throughout the entire bar, the stresses acting over the inclined section must be uniformly distributed, as pictured in the free-body diagrams of Fig. 2-28b (three-dimensional view) and Fig. 2-28c (two-dimensional view). From the equilibrium of the free body we know that the resultant of the stresses must be a horizontal force P. (The resultant is drawn with a dashed line in Figs. 2-28b and 2-28c).

(a)

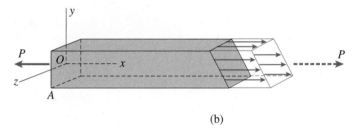

(b)

Fig. 2-28 Prismatic bar in tension showing the stresses acting on an inclined section pq: (a) bar with axial forces P, (b) three-dimensional view of the cut bar showing the stresses, and (c) two-dimensional view

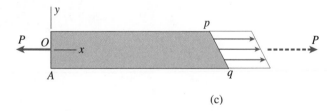

(c)

As a preliminary matter, we need a scheme for specifying the orientation of the inclined section pq. A standard method is to specify the angle θ between the x axis and the normal n to the section (Fig. 2-29a). Thus, the angle θ for the inclined section shown in the figure is approximately 30°. By contrast, cross section mn (Fig. 2-26a) has an angle θ equal to zero (because the normal to the section is the x axis). For additional examples, consider the stress element of Fig. 2-27. The angle θ for the right-hand face is 0, for the top face is 90° (a longitudinal section of the bar), for the left-hand face is 180°, and for the bottom face is 270° (or −90°).

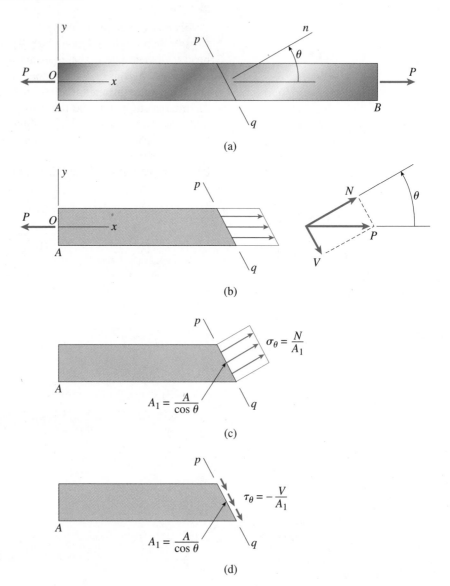

Fig. 2-29 Prismatic bar in tension showing the stresses acting on an inclined section pq

Let us now return to the task of finding the stresses acting on section pq (Fig. 2-29b). As already mentioned, the resultant of these stresses is a force P acting in the x direction. This resultant may be resolved into two components, a normal force N that is perpendicular to the inclined plane pq and a shear force V that is tangential to it. These force components are

$$N = P \cos \theta \qquad V = P \sin \theta \qquad \text{(2-20a,b)}$$

Associated with the forces N and V are normal and shear stresses that are uniformly distributed over the inclined section (Figs. 2-29c and d). The normal stress is equal to the normal force N divided by the area of the section, and the shear stress is equal to the shear force V divided by the area of the section. Thus, the stresses are

$$\sigma = \frac{N}{A_1} \qquad \tau = \frac{V}{A_1} \tag{2-21a,b}$$

in which A_1 is the area of the inclined section:

$$A_1 = \frac{A}{\cos\theta} \tag{2-22}$$

As usual, A represents the cross-sectional area of the bar. The stresses σ and τ act in the directions shown in Figs. 2-29c and d, that is, in the same directions as the normal force N and shear force V, respectively.

At this point we need to establish a standardized **notation and sign convention** for stresses acting on inclined sections. We will use a subscript θ to indicate that the stresses act on a section inclined at an angle θ (Fig. 2-30), just as we use a subscript x to indicate that the stresses act on a section perpendicular to the x axis (see Fig. 2-26). Normal stresses σ_θ are positive in tension and shear stresses τ_θ are positive when they tend to produce counterclockwise rotation of the material, as shown in Fig. 2-30.

Fig. 2-30 Sign convention for stresses acting on an inclined section. (Normal stresses are positive when in tension and shear stresses are positive when they tend to produce counterclockwise rotation.)

For a bar in tension, the normal force N produces positive normal stresses σ_θ (see Fig. 2-29c) and the shear force V produces negative shear stresses τ_θ (see Fig. 2-29d). These stresses are given by the following equations (see Eqs. 2-20, 2-21, and 2-22):

$$\sigma_\theta = \frac{N}{A_1} = \frac{P}{A}\cos^2\theta \qquad \tau_\theta = -\frac{V}{A_1} = -\frac{P}{A}\sin\theta\cos\theta$$

Introducing the notation $\sigma_x = P/A$, in which σ_x is the normal stress on a cross section, and also using the trigonometric relations

$$\cos^2\theta = \frac{1}{2}(1 + \cos 2\theta) \qquad \sin\theta\cos\theta = \frac{1}{2}(\sin 2\theta)$$

we get

$$\sigma_\theta = \sigma_x\cos^2\theta = \frac{\sigma_x}{2}(1 + \cos 2\theta) \tag{2-23a}$$

$$\tau_\theta = -\sigma_x \sin\theta \, \cos\theta = -\frac{\sigma_x}{2}(\sin 2\theta) \qquad (2\text{-}23b)$$

These equations give the normal and shear stresses acting on an inclined section oriented at an angle θ.

It is important to recognize that Eqs. (2-23a) and (2-23b) were derived only from statics, and therefore they are independent of the material. Thus, these equations are valid for any material, whether it behaves linearly or nonlinearly, elastically or inelastically.

The manner in which the stresses vary as the inclined section is cut at various angles is shown in Fig. 2-31. The horizontal axis gives the angle θ as it varies from $-90°$ to $+90°$, and the vertical axis gives the stresses σ_θ and τ_θ. Note that a positive angle θ is measured counterclockwise from the x axis (Fig. 2-30) and a negative angle is measured clockwise.

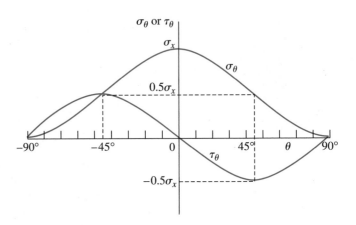

Fig. 2-31 Graph of normal stress σ_θ and shear stress τ_θ versus angle θ of the inclined section (see Fig. 2-30 and Eqs. 2-23a and b)

As shown on the graph, the normal stress σ_θ equals σ_x when $\theta = 0$. Then, as θ increases or decreases, the normal stress diminishes until at $\theta = \pm 90°$ it becomes zero, because there are no normal stresses on sections cut parallel to the longitudinal axis. The **maximum normal stress** occurs at $\theta = 0$ and is

$$\sigma_{\max} = \sigma_x \qquad (2\text{-}24)$$

Also, we note that when $\theta = \pm 45°$, the normal stress is one-half the maximum value.

The shear stress τ_θ is zero on cross sections of the bar ($\theta = 0$) as well as on longitudinal sections ($\theta = \pm 90°$). Between these extremes, the stress varies as shown on the graph, reaching the largest positive value when $\theta = -45°$ and the largest negative value when $\theta = +45°$. These **maximum shear stresses** have the same magnitude:

$$\tau_{max} = \frac{\sigma_x}{2} \tag{2-25}$$

but they tend to rotate the element in opposite directions.

The maximum stresses in a bar in tension are shown in Fig. 2-32. Two stress elements are selected—element A is oriented at $\theta = 0°$ and element B is oriented at $\theta = 45°$. Element A has the maximum normal stresses (Eq. 2-24) and element B has the maximum shear stresses (Eq. 2-25). In the case of element A (Fig. 2-32b), the only stresses are the maximum normal stresses (no shear stresses exist on any of the faces).

Fig. 2-32 Normal and shear stresses acting on stress elements oriented at $\theta = 0°$ and $\theta = 45°$ for a bar in tension

In the case of element B (Fig. 2-32c), both normal and shear stresses act on all faces (except, of course, the front and rear faces of the element). Consider, for instance, the face at 45° (the upper right-hand face). On this face the normal and shear stresses (from Eqs. 2-23a and b) are $\sigma_x/2$ and $-\sigma_x/2$, respectively. Hence, the normal stress is tension

Load

Load

Fig. 2-33 Shear failure along a 45° plane of a wood block loaded in compression

Load

Load

Fig. 2-34 Slip bands (or Lüders' bands) in a polished steel specimen loaded in tension.

(positive) and the shear stress acts clockwise (negative) against the element. The stresses on the remaining faces are obtained in a similar manner by substituting $\theta = 135°$, $-45°$, and $-135°$ into Eqs. (2-23a and b). Thus, in this special case of an element oriented at $\theta = 45°$, the normal stresses on all four faces are the same (equal to $\sigma_x/2$) and all four shear stresses have the maximum magnitude (equal to $\sigma_x/2$). Also, note that the shear stresses acting on perpendicular planes are equal in magnitude and have directions either toward, or away from, the line of intersection of the planes, as discussed in detail in Section 1.6.

If a bar is loaded in compression instead of tension, the stress σ_x will be compression and will have a negative value. Consequently, all stresses acting on stress elements will have directions opposite to those for a bar in tension. Of course, Eqs. (2-23a and b) can still be used for the calculations simply by substituting σ_x as a negative quantity.

Even though the maximum shear stress in an axially loaded bar is only one-half the maximum normal stress, the shear stress may cause failure if the material is much weaker in shear than in tension. An example of a shear failure is pictured in Fig. 2-33, which shows a block of wood that was loaded in compression and failed by shearing along a 45° plane. A similar type of behavior occurs in mild steel loaded in tension. During a tensile test of a flat bar of low-carbon steel with polished surfaces, visible *slip bands* appear on the sides of the bar at approximately 45° to the axis (Fig. 2-34). These bands indicate that the material is failing in shear along the planes on which the shear stress is maximum. Such bands were first observed by G. Piobert in 1842 and W. Lüders in 1860 (see Refs. 2-6 and 2-7), and today they are called either *Lüders' bands* or *Piobert's bands*. They begin to appear when the yield stress is reached in the bar (point *B* in Fig. 1-10).

The state of stress described in this section is called **uniaxial stress**, for the obvious reason that the bar is subjected to simple tension or compression in just one direction. The most important orientations of stress elements for uniaxial stress are $\theta = 0$ and $\theta = 45°$ (Fig. 2-32); the former has the maximum normal stress and the latter has the maximum shear stress. If sections are cut through the bar at other angles, the stresses acting on the faces of the corresponding stress elements can be determined from Eqs. (2-23a and b), as illustrated in Examples 2-10 and 2-11 that follow. Uniaxial stress is a special case of a more general stress state known as *plane stress*, which is described in detail in Chapter 7.

■ **Example 2-10**

A prismatic bar having cross-sectional area $A = 1200$ mm² is compressed by an axial load $P = 90$ kN (Fig. 2-35a). (a) Determine the stresses acting on an inclined section pq cut through the bar at an angle $\theta = 25°$. (b) Determine the complete state of stress for $\theta = 25°$ and show the stresses on a properly oriented stress element.

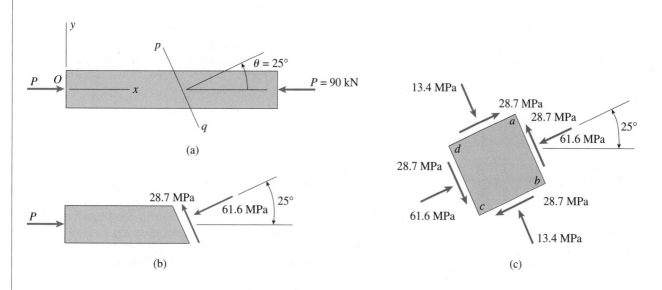

(a)

(b)

(c)

Fig. 2-35 Example 2-10. Stresses on an inclined section

Solution   (a) To find the stresses acting on a section at $\theta = 25°$, we first calculate the normal stress σ_x acting on a cross section:

$$\sigma_x = -\frac{P}{A} = -\frac{90 \text{ kN}}{1200 \text{ mm}^2} = -75 \text{ MPa}$$

where the minus sign indicates that the stress is compressive. Next, we calculate the normal and shear stresses from Eqs. (2-23a and b) with $\theta = 25°$, as follows:

$$\sigma_\theta = \sigma_x \cos^2 \theta = (-75 \text{ MPa})(\cos 25°)^2 = -61.6 \text{ MPa} \quad \Leftarrow$$

$$\tau_\theta = -\sigma_x \sin \theta \cos \theta = (75 \text{ MPa})(\sin 25°)(\cos 25°) = 28.7 \text{ MPa} \quad \Leftarrow$$

These stresses are shown acting on the inclined section in Fig. 2-35b. Note that the normal stress σ_θ is negative (compressive) and the shear stress τ_θ is positiv (counterclockwise).

(b) To determine the complete state of stress, we need to find the stres acting on all faces of a stress element oriented at 25° (Fig. 2-35c). Face ab, which $\theta = 25°$, has the same orientation as the inclined plane shown in Fig. 2-35b. Therefore, the stresses are the same as those given above.

The stresses on the opposite face *cd* are the same as those on face *ab*, which can be verified by substituting $\theta = 25° + 180° = 205°$ into Eqs. (2-23a and b).

For face *bc* we substitute $\theta = 25° - 90° = -65°$ into Eqs. (2-23a and b) and obtain

$$\sigma_\theta = -13.4 \text{ MPa} \qquad \tau_\theta = -28.7 \text{ MPa} \qquad \longleftarrow$$

These same stresses apply to the opposite face *ad*, as can be verified by substituting $\theta = 25° + 90° = 115°$ into Eqs. (2-23a and b). Note that the normal stress is compressive and the shear stress is clockwise.

The complete state of stress is shown by the stress element of Fig. 2-35c. A sketch of this kind is an excellent way to show the directions of the stresses and the orientations of the planes on which they act.

▪ Example 2-11

A compression bar having a square cross of width *b* must support a load $P = 8000$ lb (Fig. 2-36a). The two parts of the bar are connected by a glued joint (known as a *scarf joint*) along plane *pq*, which is at an angle $\alpha = 40°$ to the vertical. The bar is constructed of a structural plastic for which the allowable stresses in compression and shear are 1100 psi and 600 psi, respectively. In addition, the allowable stresses in the glued joint are 750 psi in compression and 500 psi in shear. Determine the minimum width *b* of the bar.

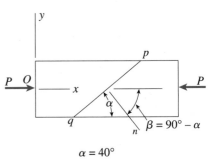

$\alpha = 40°$
$\beta = 50°$
$\theta = -\beta = -50°$

(a)   (b)

Solution   For convenience, let us rotate the bar to a horizontal position (Fig. 2-36b) that matches the figures used in deriving the equations for the stresses on an inclined section (see Figs. 2-29 and 2-30). With the bar in this position, we see that the normal n to the plane of the glued joint (plane pq) makes an angle $\beta = 90° - \alpha$, or 50°, with the axis of the bar. Since the angle θ is defined as positive when counterclockwise (Fig. 2-30), we conclude that $\theta = -50°$ for the glued joint.

The cross-sectional area of the bar is related to the load P and the stress σ_x on the cross sections by the equation

$$A = \frac{P}{\sigma_x} \tag{a}$$

Therefore, to find the required area, we first must determine the value of σ_x corresponding to each of the four allowable stresses. Then the smallest value of σ_x will determine the required area. The values of σ_x are obtained by rearranging Eqs. (2-23a and b) as follows:

$$\sigma_x = \frac{\sigma_\theta}{\cos^2\theta} \qquad \sigma_x = -\frac{\tau_\theta}{\sin\theta\cos\theta} \tag{2-26a,b}$$

We will now apply these equations to the glued joint and to the plastic.

(a) *Values of σ_x based upon the allowable stresses in the glued joint.* For compression in the glued joint we have $\sigma_\theta = -750$ psi and $\theta = -50°$. Substituting into Eq. (2-26a), we get

$$\sigma_x = \frac{-750 \text{ psi}}{(\cos -50°)^2} = -1815 \text{ psi} \tag{b}$$

For shear in the glued joint we have an allowable stress of 500 psi. However, it is not immediately evident whether τ_θ is $+500$ psi or -500 psi. One approach is to substitute both $+500$ psi and -500 psi into Eq. (2-26b) and then select the value of σ_x that is negative. The other value of σ_x will be positive (tension) and does not apply to this bar, which is in compression. Another approach is to inspect the bar itself (Fig. 2-36b) and observe from the directions of the loads that the shear stress will act clockwise against plane pq, which means that the shear stress is negative. Therefore, we substitute $\tau_\theta = -500$ psi and $\theta = -50°$ into Eq. (2-26b) and obtain

$$\sigma_x = -\frac{-500 \text{ psi}}{(\sin -50°)(\cos -50°)} = -1015 \text{ psi} \tag{c}$$

(b) *Values of σ_x based upon the allowable stresses in the plastic.* The maximum compressive stress in the plastic occurs on a cross section. Therefore, since the allowable stress in compression is 1100 psi, we know immediately that

$$\sigma_x = -1100 \text{ psi} \tag{d}$$

The maximum shear stress occurs on a plane at 45° and is numerically equal to $\sigma_x/2$ (see Eq. 2-25). Since the allowable stress in shear is 600 psi, we obtain

$$\sigma_x = -1200 \text{ psi} \tag{e}$$

This same result can be obtained from Eq. (2-26b) by substituting $\tau_\theta = 600$ psi and $\theta = 45°$.

(c) *Minimum width of the bar.* Comparing the four values of σ_x (Eqs. b, c, d, and e), we see that the smallest is $\sigma_x = -1015$ psi. Therefore, this value governs the design. Substituting into Eq. (a), and using only numerical values, we obtain the required area:

$$A = \frac{8000 \text{ lb}}{1015 \text{ psi}} = 7.88 \text{ in.}^2$$

Since the bar has a square cross section ($A = b^2$), the minimum width is

$$b_{\min} = \sqrt{A} = \sqrt{7.88 \text{ in.}^2} = 2.81 \text{ in.} \qquad \longleftarrow$$

Any width larger than $b_{\min}$ will ensure that the allowable stresses are not exceeded. For instance, we might select $b = 3.0$ in.

2.7 STRAIN ENERGY

Strain energy is a fundamental concept in applied mechanics, and strain-energy principles are widely used for determining the response of machines and structures to both static and dynamic loads. In this section we introduce the subject of strain energy in its simplest form by considering only axially loaded members subjected to static loads. More complicated structural elements are discussed in later chapters—bars in torsion in Section 3.9 and beams in bending in Section 9.8. In addition, the use of strain energy in connection with dynamic loads is described in the next section (Section 2.8) and in Section 9.10.

To illustrate the basic ideas, let us again consider a prismatic bar of length L subjected to a tensile force P (Fig. 2-37). We assume that the load is applied slowly, so that it gradually increases from zero to its maximum value P. Such a load is called a **static load** because there are no dynamic or inertial effects due to motion. The bar gradually elongates as the load is applied, eventually reaching its maximum elongation δ at the same time that the load reaches its full value P. Thereafter, the load and elongation remain unchanged.

During the loading process, the load P moves slowly through the distance δ and does a certain amount of **work**. To evaluate this work, we recall from elementary mechanics that a constant force does work equal to the product of the force and the distance through which it moves. However, in our case the force varies in magnitude from zero to its maximum value P. To find the work done by the load under these conditions, we need to know the manner in which the force varies. This information is supplied by a **load-displacement diagram**, such as the one plotted in Fig. 2-38. On this diagram the vertical axis represents the axial load and the horizontal axis represents the corresponding elongation of the bar. The shape of the curve depends upon the properties of the material.

Fig. 2-37 Prismatic bar subjected to a statically applied load

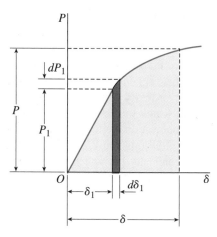

Fig. 2-38 Load-displacement diagram

Let us denote by P_1 any value of the load between zero and the maximum value P, and let us denote the corresponding elongation of the bar by δ_1. Then an increment dP_1 in the load will produce an increment $d\delta_1$ in the elongation. The work done by the load during this incremental elongation is the product of the load and the distance through which it moves, that is, the work equals $P_1 d\delta_1$. This work is represented in the figure by the area of the shaded strip below the load-displacement curve. The total work done by the load as it increases from zero to the maximum value P is the summation of all such elemental strips:

$$W = \int_0^\delta P_1 d\delta_1 \qquad (2\text{-}27)$$

In geometric terms, *the work done by the load is equal to the area below the load-displacement curve.*

When the load stretches the bar, strains are produced. The presence of these strains increases the energy level of the bar itself. Therefore, a new quantity, called **strain energy**, is defined as the energy absorbed by the bar during the loading process. From the principle of conservation of energy, we know that this strain energy is equal to the work done by the load provided no energy is added or subtracted in the form of heat. Therefore,

$$U = W = \int_0^\delta P_1 d\delta_1 \qquad (2\text{-}28)$$

in which U is the symbol for strain energy. Sometimes strain energy is referred to as **internal work** to distinguish it from the external work done by the load.

Work and energy are expressed in the same **units**. In SI, the unit of work and energy is the joule (J), which is equal to one newton meter (1 J = 1 N·m). In USCS units, work and energy are expressed in foot-pounds (ft-lb), foot-kips (ft-k), inch-pounds (in.-lb), and inch-kips (in.-k).*

If the force P (Fig. 2-37) is slowly removed from the bar, the bar will shorten. If the elastic limit of the material is not exceeded, the bar will return to its original length. If the limit is exceeded, a permanent set will remain (see Section 1.4). Thus, either all or part of the strain energy will be recovered in the form of work. This behavior is shown on the load-displacement diagram of Fig. 2-39. During loading, the work done by the load is equal to the area below the curve (area $OABCDO$). When the load is removed, the load-displacement diagram follows line BD if point B is beyond the elastic limit and a permanent elongation OD

---

*Conversion factors for work and energy are given in Appendix A, Table A-5.

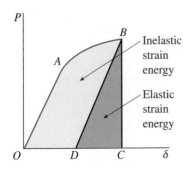

Fig. 2-39 Elastic and inelastic strain energy

remains. Thus, the strain energy recovered during unloading, called the **elastic strain energy**, is represented by the shaded triangle *BCD*. Area *OABDO* represents energy that is lost in the process of permanently deforming the bar. This energy is known as the **inelastic strain energy**.

Most structures are designed with the expectation that the material will remain within the elastic range under ordinary conditions of service. In the case of a bar in tension, the load at which the stress in the material reaches the elastic limit is represented by point *A* on the load-displacement curve (Fig. 2-39). As long as the load is below this value, all of the strain energy is recovered during unloading and no permanent elongation remains. Thus, the bar acts as an elastic spring, storing and releasing energy as the load is applied and removed.

■ Linearly Elastic Behavior

Let us now assume that the material of the bar follows Hooke's law, so that the load-displacement curve is a straight line (Fig. 2-40). Then the strain energy *U* stored in the bar (equal to the work *W* done by the load) is

$$U = W = \frac{P\delta}{2} \tag{2-29}$$

which is the area of the shaded triangle *OAB* in the figure.* Furthermore, we know that the relationship between the load *P* and the elongation δ for a linearly elastic material is given by the equation

$$\delta = \frac{PL}{EA} \tag{2-30}$$

Combining this equation with Eq. (2-29) enables us to express the strain energy of a linearly elastic bar in the following forms:

$$U = \frac{P^2L}{2EA} \qquad U = \frac{EA\delta^2}{2L} \tag{2-31a,b}$$

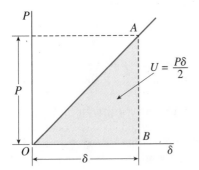

Fig. 2-40 Load-displacement diagram for a bar of linearly elastic material

The first equation expresses the strain energy as a function of the load and the second expresses it as a function of the elongation. From the first equation we see that increasing the length of a bar increases the amount of strain energy even though the load is unchanged (because more material is being strained by the load). On the other hand, increasing either the

---

*The principle that the work of the external loads is equal to the strain energy (for the case of linearly elastic behavior) was first stated by the French engineer B. P. E. Clapeyron (1799–1864) and is known as *Clapeyron's theorem* (Ref. 2-8).

modulus of elasticity or the cross-sectional area decreases the strain energy because the strains in the bar are reduced. These ideas are illustrated in Examples 2-12 and 2-13.

Strain-energy equations analogous to Eqs. (2-31a) and (2-31b) can be written for a **linearly elastic spring** by replacing the stiffness EA/L of the prismatic bar by the stiffness k of the spring. Thus,

$$U = \frac{P^2}{2k} \qquad U = \frac{k\delta^2}{2} \qquad \text{(2-32a,b)}$$

Of course, these expressions are equivalent inasmuch as $P = k\delta$ for a spring.

The total strain energy U of a bar consisting of several segments is equal to the sum of the strain energies of the individual segments. For instance, the strain energy of the bar pictured in Fig. 2-41 equals the strain energy of segment AB plus the strain energy of segment BC. This concept is expressed in general terms by the following equation:

$$U = \sum_{i=1}^{n} U_i \qquad \text{(2-33)}$$

Fig. 2-41 Bar consisting of prismatic segments having different cross-sectional areas and different axial forces

in which U_i is the strain energy of segment i of the bar and n is the number of segments. (This relation holds whether the material behaves in a linear or nonlinear manner.)

Now assume that the material of the bar is linearly elastic and that the internal axial force is constant within each segment. We can then use Eq. (2-31a) to obtain the strain energies of the segments, and Eq. (2-33) becomes

$$U = \sum_{i=1}^{n} \frac{N_i^2 L_i}{2E_i A_i} \qquad \text{(2-34)}$$

in which N_i is the axial force acting in segment i and L_i, E_i, and A_i are properties of segment i. (The use of this equation is illustrated in Examples 2-12 and 2-13 at the end of the section.)

We can obtain the strain energy of a nonprismatic bar with continuously varying axial force (Fig. 2-42) by applying Eq. (2-31a) to a differential element (shown shaded in the figure) and then integrating along the length of the bar:

$$U = \int_0^L \frac{[N(x)]^2 dx}{2EA(x)} \qquad \text{(2-35)}$$

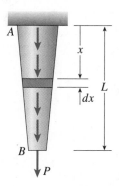

Fig. 2-42 Nonprismatic bar with varying axial force

In this equation, $N(x)$ and $A(x)$ are the axial force and cross-sectional area at distance x from the end of the bar. (Example 2-14 illustrates the use of this equation.)

The preceding expressions for strain energy (Eqs. 2-31 through 2-35) show that strain energy is *not* a linear function of the loads, not even when the material is linearly elastic. Thus, it is important to realize that *we cannot obtain the strain energy of a structure supporting more than one load by combining the strain energies obtained from the individual loads acting separately.* In the case of the nonprismatic bar shown in Fig. 2-41, the total strain energy is *not* the sum of the strain energy due to load P_1 acting alone and the strain energy due to load P_2 acting alone. Instead, we must evaluate the strain energy with all of the loads acting simultaneously, as demonstrated later in Example 2-14.

Although we considered only tension members in the preceding discussions of strain energy, all of the concepts and equations apply equally well to members in **compression**. Since the work done by an axial load is positive regardless of whether the load causes tension or compression, it follows that strain energy is always a positive quantity. This fact is also evident in the expressions for strain energy of linearly elastic bars (such as Eqs. 2-31a and 2-31b). These expressions are always positive because the load and elongation terms are squared.

Strain energy is a form of **potential energy** (or "energy of position") because it depends upon the relative locations of the particles or elements that make up the member. When a bar or a spring is compressed, its particles are crowded more closely together; when it is stretched, the distances between particles increase. In both cases the strain energy of the member increases as compared to its strain energy in the unloaded position.

■ Displacements Caused by a Single Load

The displacement of a linearly elastic structure supporting only one load can be determined from its strain energy. To illustrate the method, consider a two-bar truss (Fig. 2-43) loaded by a vertical force P. Our objective is to determine the vertical displacement δ at joint B where the load is applied.

When applied slowly to the truss, the load P does work as it moves through the vertical displacement δ. However, it does no work as it moves laterally, that is, sideways. Therefore, since the load-displacement diagram is linear (see Eq. 2-29 and Fig. 2-40), the strain energy U stored in the structure, equal to the work done by the load, is

$$U = W = \frac{P\delta}{2}$$

from which we get

$$\delta = \frac{2U}{P} \qquad (2\text{-}36)$$

Fig. 2-43 Structure supporting a single load P

This equation shows that under certain special conditions, as outlined below, the displacement of a structure can be determined directly from the strain energy.

The conditions that must be met in order to use Eq. (2-36) are as follows: (1) the structure must behave in a linearly elastic manner, and (2) only one load may act on the structure. Furthermore, the only displacement that can be determined is the displacement corresponding to the load itself (that is, the displacement must be in the direction of the load and must be at the point where the load is applied). Therefore, this method for finding displacements is extremely limited in its application and is not a good indicator of the great importance of strain-energy principles in structural mechanics. However, the method does provide an introduction to the use of strain energy. (The method is illustrated later in Example 2-15.)

■ **Strain-Energy Density**

In many situations it is convenient to use a quantity called **strain-energy density**, defined as the strain energy per unit volume of material. Expressions for strain-energy density in the case of linearly elastic materials can be obtained from the formulas for strain energy of a prismatic bar (Eqs. 2-31a and b). Since the strain energy of the bar is distributed uniformly throughout its volume, we can determine the strain-energy density by dividing the total strain energy U by the volume AL of the bar. Thus, the strain-energy density, denoted by the symbol u, can be expressed in either of these forms:

$$u = \frac{P^2}{2EA^2} \qquad u = \frac{E\delta^2}{2L^2} \qquad (2\text{-}37a,b)$$

If we replace P/A by the stress σ and δ/L by the strain ϵ, we get

$$u = \frac{\sigma^2}{2E} \qquad u = \frac{E\epsilon^2}{2} \qquad (2\text{-}38a,b)$$

These equations give the strain-energy density in a linearly elastic material in terms of the normal stress σ and the normal strain ϵ.

The expressions in Eqs. (2-38a and b) have a simple geometric interpretation. They are equal to the area $\sigma\epsilon/2$ of the triangle below the stress-strain diagram for a material that follows Hooke's law ($\sigma = E\epsilon$). In more general situations where the material does not follow Hooke's law, the strain-energy density is still equal to the area below the stress-strain curve, but the area must be evaluated in each particular case.

Strain-energy density has **units** of energy divided by volume. The SI units are joules per cubic meter (J/m^3) and the USCS units are foot-pounds per cubic foot, inch-pounds per cubic inch, and other similar units. Since all of these units reduce to units of stress (recall that 1 J = 1 N·m), we can also use units such as pascals (Pa) and pounds per square inch (psi) for strain-energy density.

The strain-energy density of the material when it is stressed to the proportional limit is called the **modulus of resilience** u_r. It is found by substituting the proportional limit σ_{pl} into Eq. (2-38a):

$$u_r = \frac{\sigma_{pl}^2}{2E} \tag{2-39}$$

For example, a mild steel having $\sigma_{pl} = 30,000$ psi and $E = 30 \times 10^6$ psi has a modulus of resilience $u_r = 15$ psi (or 103 kPa). Note that the modulus of resilience is equal to the area below the stress-strain curve up to the proportional limit. *Resilience* represents the ability of the material to absorb and release energy within the elastic range.

Another quantity, called *toughness*, refers to the ability of a material to absorb energy without fracturing. The corresponding modulus, called the **modulus of toughness** u_t, is the strain-energy density when the material is stressed to the point of failure. It is equal to the area below the entire stress-strain curve. The higher the modulus of toughness, the greater the ability of the material to absorb energy without failing. A high modulus of toughness is therefore important when the material is subject to impact loads (see Section 2.8).

The preceding expressions for strain-energy density (Eqs. 2-37 to 2-39) were derived for *uniaxial stress*, that is, for materials subjected only to tension or compression. Strain-energy formulas for more general stress states are presented in Chapter 7.

■ Example 2-12

Three round bars having the same length L but different shapes are shown in Fig. 2-44. The first bar has diameter d over its entire length, the second has diameter d over one-fifth of its length, and the third has diameter d over one-fifteenth of its length. Elsewhere, the second and third bars have diameter $2d$. All three bars are subjected to the same axial load P.

Compare the amounts of strain energy stored in the bars, assuming linearly elastic behavior. (Disregard the effects of stress concentrations and the weights of the bars.)

Solution   (a) *Strain energy U_1 of the first bar.* The strain energy of the first bar is found directly from Eq. (2-31a):

$$U_1 = \frac{P^2 L}{2EA} \tag{a} \quad \longleftarrow$$

in which $A = \pi d^2/4$.

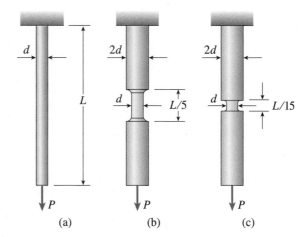

Fig. 2-44 Example 2-12. Calculation of strain energy

(b) *Strain energy U_2 of the second bar.* The strain energy is found by summing the strain energies in the three segments of the bar (see Eq. 2-34). Thus,

$$U_2 = \sum_{i=1}^{n} \frac{N_i^2 L_i}{2E_i A_i} = \frac{P^2(L/5)}{2EA} + \frac{P^2(4L/5)}{2E(4A)} = \frac{P^2 L}{5EA} = \frac{2U_1}{5} \qquad \text{(b)} \quad \Longleftarrow$$

which is only 40% of the strain energy of the first bar. Thus, increasing the cross-sectional area over part of the length has greatly reduced the amount of strain energy that can be stored in the bar.

(c) *Strain energy U_3 of the third bar.* Again using Eq. (2-34), we get

$$U_3 = \sum_{i=1}^{n} \frac{N_i^2 L_i}{2E_i A_i} = \frac{P^2(L/15)}{2EA} + \frac{P^2(14L/15)}{2E(4A)} = \frac{3P^2 L}{20EA} = \frac{3U_1}{10} \qquad \text{(c)} \quad \Longleftarrow$$

The strain energy has now decreased to 30% of the strain energy of the first bar.

Note: Comparing these results, we see that the strain energy decreases as the part of the bar with the larger area increases. The same amount of work applied to all three bars will produce the highest stress in the third bar, because the third bar has the least energy-absorbing capacity. If the region having diameter d is made even smaller, the energy-absorbing capacity will decrease further. We therefore conclude that it takes only a small amount of work to bring the tensile stress to a high value in a bar with a groove, and the narrower the groove, the more severe the condition. When the loads are dynamic and the ability to absorb energy is important, the presence of grooves is very damaging.

In the case of static loads, the maximum stresses are more important than the ability to absorb energy. In this example, all three bars have the same maximum stress P/A (provided stress concentrations are alleviated), and therefore all three bars have the same load-carrying capacity when the load is applied statically.

■ Example 2-13

The cylinder and cylinder head for a compressed air machine are clamped by bolts through the flanges of the cylinder (Fig. 2-45a). A detail of one of the bolts is shown in part (b) of the figure. The diameter d of the shank is 0.500 in. and the root diameter d_r of the threaded portion is 0.406 in. The grip g of the bolts is 1.50 in. and the threads extend a distance $t = 0.25$ in. into the grip. Under the action of repeated cycles of high and low pressure in the chamber, the bolts may eventually break.

To reduce the likelihood of the bolts failing, the designers suggest two possible modifications: (a) Machine down the shanks of the bolts so that the shank diameter is the same as the thread diameter d_r, as shown in Fig. 2-46a. (b) Replace each pair of bolts by a single long bolt, as shown in Fig. 2-46b. The long bolts are the same as the original bolts (Fig. 2-45b) except that the grip is increased to the distance $L = 13.5$ in.

Compare the energy-absorbing capacity of the three bolt configurations: (1) original bolts, (2) bolts with reduced shank diameter, and (3) long bolts. (Assume linearly elastic behavior and disregard the effects of stress concentrations.)

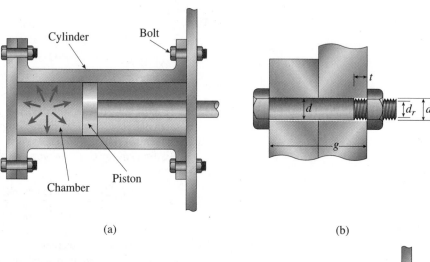

Cylinder   Bolt

Chamber   Piston

(a)   (b)

Fig. 2-45 Example 2-13. (a) Cylinder with piston and clamping bolts, and (b) detail of one bolt

(a)   (b)

Fig. 2-46 Example 2-13. Proposed modifications to the bolts: (a) Bolt with reduced shank diameter, and (b) bolt with increased length

Solution   (1) *Original bolts.* The original bolts can be idealized as bars consisting of two segments (Fig. 2-45b). The left-hand segment has length $g - t$ and diameter d, and the right-hand segment has length t and diameter d_r. The strain energy of one bolt under a tensile load P can be obtained by adding the strain energies of the two segments (Eq. 2-34):

$$U_1 = \sum_{i=1}^{n} \frac{N_i^2 L_i}{2E_i A_i} = \frac{P^2(g - t)}{2EA_s} + \frac{P^2 t}{2EA_r} \tag{d}$$

in which A_s is the cross-sectional area of the shank and A_r is the cross-sectional area at the root of the threads; thus,

$$A_s = \frac{\pi d^2}{4} \qquad A_r = \frac{\pi d_r^2}{4} \tag{e}$$

Substituting these expressions into Eq. (d), we get the following formula for the strain energy of one of the original four bolts:

$$U_1 = \frac{2P^2(g - t)}{\pi E d^2} + \frac{2P^2 t}{\pi E d_r^2} \tag{f}$$

(2) *Bolts with reduced shank diameter.* These bolts can be idealized as prismatic bars having length g and diameter d_r (Fig. 2-46a). Therefore, the strain energy of one bolt (see Eq. 2-31a) is

$$U_2 = \frac{P^2 g}{2EA_r} = \frac{2P^2 g}{\pi E d_r^2} \tag{g}$$

The ratio of the strain energies for cases (1) and (2) is

$$\frac{U_2}{U_1} = \frac{g d^2}{(g - t)d_r^2 + t d^2} \tag{h}$$

or, upon substituting numerical values,

$$\frac{U_2}{U_1} = \frac{(1.50 \text{ in.})(0.500 \text{ in.})^2}{(1.50 \text{ in.} - 0.25 \text{ in.})(0.406 \text{ in.})^2 + (0.25 \text{ in.})(0.500 \text{ in.})^2} = 1.40 \impliedby$$

Thus, using bolts with reduced shank diameters results in a 40% increase in the amount of strain energy that can be absorbed by the bolts. If implemented, this scheme should reduce the number of failures caused by the impact loads.

(3) *Long bolts.* The calculations for the long bolts (Fig. 2-46b) are the same as for the original bolts except the grip g is changed to the grip L. Therefore, the strain energy of one long bolt (compare with Eq. f) is

$$U_3 = \frac{2P^2(L - t)}{\pi E d^2} + \frac{2P^2 t}{\pi E d_r^2} \tag{i}$$

Since one long bolt replaces two of the original bolts, we must compare the strain energies by taking the ratio of U_3 to $2U_1$, as follows:

$$\frac{U_3}{2U_1} = \frac{(L - t)d_r^2 + t d^2}{2(g - t)d_r^2 + 2t d^2} \tag{j}$$

Substituting numerical values gives

$$\frac{U_3}{2U_1} = \frac{(13.5 \text{ in.} - 0.25 \text{ in.})(0.406 \text{ in.})^2 + (0.25 \text{ in.})(0.500 \text{ in.})^2}{2(1.50 \text{ in.} - 0.25 \text{ in.})(0.406 \text{ in.})^2 + 2(0.25 \text{ in.})(0.500 \text{ in.})^2} = 4.18 \quad \Longleftarrow$$

Thus, using long bolts increases the energy-absorbing capacity by 318% and achieves the greatest safety from the standpoint of strain energy.

Note: When designing bolts, designers must also consider the maximum tensile stresses, maximum bearing stresses, stress concentrations, and other matters.

■ Example 2-14

Determine the strain energy of a prismatic bar suspended from its upper end (Fig. 2-47). Consider the following loads: (a) the weight of the bar itself, and (b) the weight of the bar plus a load P at the lower end. (Assume linearly elastic behavior.)

Solution   (a) *Strain energy due to the weight of the bar itself (Fig. 2-47a).* The bar is subjected to a varying axial force, the force being zero at the lower end and maximum at the upper end. To determine the axial force, we consider an element of length dx (shown shaded in the figure) at distance x from the upper end. The internal axial force $N(x)$ acting on this element is equal to the weight of the bar below the element:

$$N(x) = \gamma A(L - x) \tag{k}$$

in which γ is the weight density of the material and A is the cross-sectional area of the bar. Substituting into Eq. (2-35) and integrating gives the total strain energy:

$$U = \int_0^L \frac{[N(x)]^2 \, dx}{2EA(x)} = \int_0^L \frac{[\gamma A(L - x)]^2 \, dx}{2EA} = \frac{\gamma^2 A L^3}{6E} \tag{2-40} \quad \Longleftarrow$$

This same result can be obtained from the strain-energy density. At any distance x from the support, the stress is

$$\sigma = \frac{N(x)}{A} = \gamma(L - x) \tag{l}$$

and, therefore, the strain-energy density (Eq. 2-38a) is

$$u = \frac{\sigma^2}{2E} = \frac{\gamma^2 (L - x)^2}{2E} \tag{m}$$

The total strain energy is now found by integrating u throughout the volume of the bar:

$$U = \int u \, dV = \int_0^L u(A \, dx) = \int_0^L \frac{\gamma^2 A(L - x)^2 \, dx}{2E} = \frac{\gamma^2 A L^3}{6E} \tag{n}$$

which agrees with Eq. (2-40).

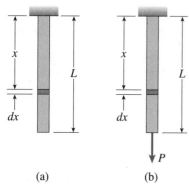

(a)   (b)

Fig. 2-47 Example 2-14. (a) Bar hanging under its own weight, and (b) bar hanging under its own weight and also supporting a load P

(b) *Strain energy due to the weight of the bar plus the load P (Fig. 2-47b).* In this case the axial force $N(x)$ acting on the element is

$$N(x) = \gamma A(L - x) + P \tag{o}$$

(compare with Eq. k). From Eq. (2-35) we now obtain

$$U = \int_0^L \frac{[\gamma A(L - x) + P]^2 \, dx}{2EA} = \frac{\gamma^2 A L^3}{6E} + \frac{\gamma P L^2}{2E} + \frac{P^2 L}{2EA} \tag{2-41}$$ ←

Note: The first term in this expression is the same as the strain energy of a bar hanging under its own weight (Eq. 2-40), and the last term is the same as the strain energy of a bar subjected only to an axial force P (Eq. 2-31a). However, the middle term contains both γ and P, showing that it depends upon both the weight of the bar and the magnitude of the applied load. Thus, this example illustrates that the strain energy of a bar subjected to two loads is *not* equal to the sum of the strain energies produced by the individual loads acting separately.

■ **Example 2-15**

Determine the vertical displacement δ_B of joint B of the truss shown in Fig. 2-48. Note that the only load acting on the truss is a vertical load P at joint B. Assume that both members of the truss have the same axial rigidity EA.

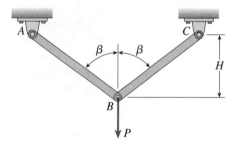

Fig. 2-48 Example 2-15. Displacement of a truss supporting a single load P

Solution   Since there is only one load acting on the truss, we can find the displacement corresponding to that load by equating the work of the load to the strain energy of the members. However, to find the strain energy we must know the forces in the members (see Eq. 2-31a).

From the equilibrium of forces acting at joint B we see that the axial force F in either bar is

$$F = \frac{P}{2 \cos \beta} \tag{p}$$

Also, from the geometry of the truss we see that the length of each bar is $L_1 = H/\cos \beta$, where H is the height of the truss and β is the angle shown in the figure. We can now obtain the strain energy of the two bars:

$$U = (2)\frac{F^2 L_1}{2EA} = \frac{P^2 H}{4EA \cos^3 \beta} \tag{q}$$

The work of the load P is

$$W = \frac{P\delta_B}{2}$$

where δ_B is the downward displacement of joint B. Equating U and W and solving for δ_B, we obtain

$$\delta_B = \frac{PH}{2EA\cos^3\beta} \qquad (2\text{-}42) \quad \Longleftarrow$$

This equation gives the vertical displacement of joint B of the truss. Note that we found the displacement using only equilibrium and strain energy—we did not need to draw a displacement diagram at joint B.

*2.8 ▌ IMPACT LOADING

Loads can be classified as static or dynamic depending upon whether they remain constant or vary with time. A **static load** is applied slowly, so that it causes no vibrational or dynamic effects in the structure. The load increases gradually from zero to its maximum value, and thereafter it remains constant. A **dynamic load** may take many forms—some loads are applied and removed suddenly (*impact loads*), others persist for long periods of time and continuously vary in intensity (*fluctuating loads*). Impact loads are produced when two objects collide or when a falling object strikes a structure. Fluctuating loads are produced by rotating machinery, traffic, wind gusts, water waves, earthquakes, and manufacturing processes.

As an example of how structures respond to dynamic loads, we will discuss the impact of an object falling onto a prismatic bar (Fig. 2-49). A collar of mass M, initially at rest, falls from a height h onto a flange at the lower end of bar AB. When the collar strikes the flange, the bar begins to elongate, creating axial stresses and strains within the bar. In a very short interval of time, probably only a few milliseconds, the flange will move downward and reach its position of maximum displacement. Thereafter, the bar shortens, then lengthens, then shortens again as the bar vibrates longitudinally and the end of the bar moves up and down. The vibrations are analogous to those that occur when a spring is stretched and then released, or when a person makes a bungee jump. The vibrations of the bar soon cease because of various damping effects, and then the bar comes to rest with the mass M supported on the flange.

The response of the bar to the falling collar is obviously very complicated, and a complete and accurate analysis requires the use of advanced mathematical techniques. However, we can make an approximate analysis by using the concept of strain energy (Section 2.7) and making several simplifying assumptions.

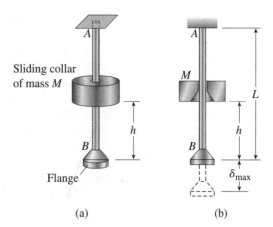

(a)   (b)

Fig. 2-49 Impact load on a prismatic bar *AB* due to a falling object of mass *M*

Let us begin by considering the energy of the system just before the collar is released (Fig. 2-49a). The potential energy of the collar with respect to the elevation of the flange is Mgh, where g is the acceleration of gravity.* This potential energy is converted into kinetic energy as the collar falls. At the instant the collar strikes the flange, its potential energy with respect to the elevation of the flange is zero and its kinetic energy is $Mv^2/2$, where $v = \sqrt{2gh}$ is its velocity.** During the ensuing impact, the kinetic energy of the collar is transformed into other forms of energy. Part of the kinetic energy is transformed into the strain energy of the stretched bar. Some of the energy is dissipated in the production of heat and in causing localized plastic deformations of the collar and flange. A small part remains as the kinetic energy of the collar, which either moves further downward (while in contact with the flange) or else bounces upward.

To make a simplified analysis of this complex situation, we will idealize the behavior by making the following assumptions. First, we assume that the collar and flange are so constructed that the collar "sticks" to the flange and moves downward with it (in other words, the collar does not rebound). This behavior is more likely to prevail when the mass of the collar is large compared to the mass of the bar. Second, we disregard all energy losses and assume that the kinetic energy of the falling mass is transformed entirely into strain energy of the bar. This assumption is conservative in the sense that it predicts larger stresses in the bar than would be predicted if we took energy losses into account. Third, we disregard any change in the potential energy of the bar itself (due to the vertical movement of elements of the bar), and we ignore the

---

*In SI units, the acceleration of gravity $g = 9.81$ m/s^2; in USCS units, $g = 32.2$ ft/s^2. For more precise values of g, or for a discussion of mass and weight, see Appendix A.

**In engineering work, velocity is usually treated as a vector quantity. However, since kinetic energy is a scalar, we will use the word "velocity" to mean the *magnitude* of the velocity, or the *speed*.

existence of strain energy in the bar due to its own weight. Both of these effects are extremely small. Fourth, we assume that the stresses in the bar remain within the linearly elastic range. Finally, we assume that the stress distribution throughout the bar is the same as when the bar is loaded statically by a force at the lower end, that is, the stresses are uniform throughout the volume of the bar. This condition is only an approximation because in reality longitudinal stress waves will travel through the bar, thereby causing variations in the stress distribution.

On the basis of the preceding assumptions, we can calculate the maximum elongation and the maximum tensile stresses produced by the impact load. (Recall that we are disregarding the weight of the bar itself and finding the stresses due solely to the falling collar.)

■ Maximum Elongation of the Bar

The maximum elongation δ_{max} (Fig. 2-49b) can be obtained from the principle of *conservation of energy* by equating the potential energy lost by the falling mass to the strain energy acquired by the bar. The potential energy lost is $W(h + \delta_{max})$, where $W = Mg$ is the weight of the collar and $h + \delta_{max}$ is the distance through which it moves. The strain energy of the bar is $EA\delta_{max}^2/2L$, where EA is the axial rigidity and L is the length of the bar (see Eq. 2-31b). Thus, we obtain the following equation:

$$W(h + \delta_{max}) = \frac{EA\delta_{max}^2}{2L} \tag{2-43}$$

This equation is quadratic in δ_{max} and can be solved for the positive root; the result is

$$\delta_{max} = \frac{WL}{EA} + \left[\left(\frac{WL}{EA}\right)^2 + 2h\left(\frac{WL}{EA}\right)\right]^{1/2} \tag{2-44}$$

Note that the maximum elongation of the bar increases if either the weight of the collar or the height of fall is increased. The elongation diminishes if the stiffness EA/L is increased.

The preceding equation can be written in simpler form by introducing the notation

$$\delta_{st} = \frac{WL}{EA} \tag{2-45}$$

in which δ_{st} is the elongation of the bar due to the weight of the collar under static loading conditions. Equation (2-44) now becomes

$$\delta_{max} = \delta_{st} + (\delta_{st}^2 + 2h\delta_{st})^{1/2} \tag{2-46}$$

From this equation we see that the elongation of the bar under the impact load is much larger than it would be if the same load were applied statically. Suppose, for instance, that the height h is 40 times the static displacement δ_{st}; the maximum elongation would then be 10 times the static elongation.

When the height h is large compared to the static elongation, we can disregard both the first and second terms on the right-hand side of Eq. (2-46) and obtain

$$\delta_{max} = \sqrt{2h\delta_{st}} = \sqrt{\frac{Mv^2L}{EA}} \qquad (2\text{-}47)$$

in which $M = W/g$ and $v = \sqrt{2gh}$ is the velocity of the falling mass when it strikes the flange. This equation can also be obtained directly from Eq. (2-43) by omitting δ_{max} on the left-hand side of the equation and solving for δ_{max}. Because of the omitted terms, values of δ_{max} calculated from Eq. (2-47) are always less than those obtained from Eq. (2-46).

■ Maximum Stress in the Bar

The maximum stress can be calculated easily from the maximum elongation because we are assuming that the stress distribution is uniform throughout the length of the bar. From the general equation $\delta = PL/EA = \sigma L/E$, we know that

$$\sigma_{max} = \frac{E\delta_{max}}{L} \qquad (2\text{-}48)$$

Substituting from Eq. (2-44), we obtain the following equation for the maximum tensile stress:

$$\sigma_{max} = \frac{W}{A} + \left[\left(\frac{W}{A}\right)^2 + \frac{2WhE}{AL}\right]^{1/2} \qquad (2\text{-}49)$$

Introducing the notation

$$\sigma_{st} = \frac{W}{A} = \frac{Mg}{A} = \frac{E\delta_{st}}{L} \qquad (2\text{-}50)$$

in which σ_{st} is the stress when the load acts statically, we can write Eq. (2-49) in the form

$$\sigma_{max} = \sigma_{st} + \left(\sigma_{st}^2 + \frac{2hE}{L}\sigma_{st}\right)^{1/2} \qquad (2\text{-}51)$$

This equation is analogous to Eq. (2-46) and again shows that an impact load produces much larger effects than when the same load is applied statically.

Again considering the case where the height h is large compared to the elongation of the bar (compare with Eq. 2-47), we obtain

$$\sigma_{max} = \sqrt{\frac{2hE\sigma_{st}}{L}} = \sqrt{\frac{Mv^2E}{AL}} \qquad (2\text{-}52)$$

From this result we see that an increase in the kinetic energy $Mv^2/2$ of the falling mass will increase the stress, whereas an increase in the volume AL of the bar will reduce the stress. This situation is quite different from static tension of the bar, where the stress is independent of the length L and the modulus of elasticity E.

The preceding equations for the maximum elongation and maximum stress apply only at the instant when the flange of the bar is at its lowest position. After the maximum elongation is reached in the bar, the bar will vibrate axially until it comes to rest at the static elongation. From then on, the elongation and stress have the values given by Eqs. (2-45) and (2-50).

▪ Impact Factor

The ratio of the dynamic response of a structure to the static response (for the same load) is known as an **impact factor**. For instance, the impact factor for the elongation of the bar of Fig. 2-49 is the ratio of the maximum elongation (Eq. 2-44, 2-46, or 2-47) to the static elongation (Eq. 2-45):

$$\text{Impact factor} = \frac{\delta_{\text{max}}}{\delta_{\text{st}}} \tag{2-53}$$

This factor represents the amount by which the static elongation is amplified due to the dynamic effects of the impact.

Equations analogous to Eq. (2-53) can be written for other impact factors, such as the impact factor for the stress in the bar (the ratio of σ_{max} to σ_{st}). When the collar falls through a considerable height, the impact factor can be very large, such as 100 or more.

▪ Suddenly Applied Load

A special case of impact occurs when a load is applied suddenly with no initial velocity. To explain this kind of loading, consider again the vertical, prismatic bar shown in Fig. 2-49 and assume that the sliding collar is lowered gently until it just touches the flange. Then the collar is suddenly released. Although in this instance no kinetic energy exists at the beginning of extension of the bar, the behavior is quite different from that of static loading of the bar. Under static loading conditions, the load is released gradually and equilibrium always exists between the applied load and the resisting force of the bar.

However, consider what happens when the collar is released suddenly from its point of contact with the flange. Initially the elongation of the bar and the stress in the bar are zero, but then the collar moves downward under the action of its own weight. During this motion the bar elongates and its resisting force gradually increases. The motion continues until at some instant the resisting force just equals W, the weight of the collar. At this particular instant the elongation of the bar is

δ_{st}. However, the collar now has a certain kinetic energy, which it acquired during the downward displacement δ_{st}. Therefore, the collar continues to move downward until its velocity is brought to zero by the resisting force in the bar. The maximum elongation for this condition is obtained from Eq. (2-46) by setting h equal to zero; thus,

$$\delta_{max} = 2\delta_{st} \qquad (2\text{-}54)$$

From this equation we see that a suddenly applied load produces an elongation twice as large as the elongation caused by the same load applied statically. Thus, the impact factor is 2.

After the maximum elongation $2\delta_{st}$ has been reached, the end of the bar will move upward and begin a series of up and down vibrations, eventually coming to rest at the static elongation produced by the weight of the collar.*

■ Energy Losses and Inelastic Effects

The preceding analyses were based upon the assumption that no energy losses occur during impact. In reality, energy losses always occur during impact, with most of the lost energy being dissipated in the form of heat and localized deformation of the materials. Because of these losses, the kinetic energy of a system immediately after an impact is less than it was before the impact. Consequently, less energy is converted into strain energy of the bar than we previously assumed. As a result, the actual displacement of the end of the bar of Fig. 2-49 is less than that predicted by our simplified analysis.

We also assumed that the stresses in the bar remain within the proportional limit. If the maximum stress exceeds this limit, the analysis becomes more complicated because the elongation of the bar is no longer proportional to the axial force. Other factors to consider are the effects of stress waves, damping, and imperfections at the contact surfaces. Therefore, we must remember that all of the formulas in this section are based upon highly idealized conditions and give only a rough approximation of the true conditions (usually overestimating the elongation).

Materials that exhibit considerable ductility beyond the proportional limit generally offer much greater resistance to impact loads than do brittle materials. Also, bars with grooves, holes, and other forms of stress concentrations (see Sections 2.9 and 2.10) are very weak against impact—a slight shock may produce fracture, even when the material itself is ductile under static loading.

■ Examples

The calculation of the maximum elongation and maximum stress caused by an impact load on a vertical bar is illustrated in Example 2-16. The

---

*Equation (2-54) was first obtained by the French mathematician and scientist J. V. Poncelet (1788–1867); see Ref. 2-9.

remaining examples (Examples 2-17 and 2-18) illustrate how the principle of conservation of energy can be used in the analysis of other kinds of impact problems. All three examples, as well as the problems at the end of the chapter, are based upon the assumptions and idealizations described earlier.

■ **Example 2-16**

A round, prismatic steel bar (E = 210 GPa) of length L = 2.0 m and diameter d = 15 mm hangs vertically from a support at its upper end (Fig. 2-50). A sliding collar of mass M = 20 kg drops from a height h = 150 mm onto the flange at the lower end of the bar without rebounding. (a) Calculate the maximum elongation of the bar due to the impact and determine the corresponding impact factor. (b) Calculate the maximum tensile stress and the corresponding impact factor.

Solution   Because the arrangement of the bar and collar in this example matches that of Fig. 2-49, we can use the equations derived previously (Eqs. 2-43 to 2-52).

(a) *Maximum elongation.* The elongation produced by the falling collar can be determined directly from Eq. (2-46). The first step is to determine the static elongation of the bar due to the weight of the collar. Since the weight of the collar is Mg, we calculate as follows:

$$\delta_{st} = \frac{WL}{EA} = \frac{MgL}{EA} = \frac{(20.0 \text{ kg})(9.81 \text{ m/s}^2)(2.0 \text{ m})}{(210 \text{ GPa})(\pi/4)(15 \text{ mm})^2} = 0.0106 \text{ mm}$$

This elongation is now substituted into Eq. (2-46) to obtain the maximum elongation:

$$\delta_{max} = \delta_{st} + (\delta_{st}^2 + 2h\delta_{st})^{1/2}$$

$$= 0.0106 \text{ mm} + [(0.0106 \text{ mm})^2 + 2(150 \text{ mm})(0.0106 \text{ mm})]^{1/2}$$

$$= 1.79 \text{ mm} \qquad \qquad \qquad \qquad \leftarrow$$

Since the height of fall is very large compared to the static elongation, we obtain nearly the same result by calculating the maximum elongation from Eq. (2-47):

$$\delta_{max} = \sqrt{2h\delta_{st}} = [2(150 \text{ mm})(0.0106 \text{ mm})]^{1/2} = 1.78 \text{ mm}$$

The impact factor is equal to the ratio of the maximum elongation to the static elongation:

$$\text{Impact factor} = \frac{\delta_{max}}{\delta_{st}} = \frac{1.79 \text{ mm}}{0.0106 \text{ mm}} = 169 \qquad \leftarrow$$

This result shows that the effects of a dynamically applied load can be very large as compared to the effects of the same load acting statically.

(b) *Maximum tensile stress.* The maximum stress produced by the falling collar is obtained from Eq. (2-48), as follows:

$$\sigma_{max} = \frac{E\delta_{max}}{L} = \frac{(210 \text{ GPa})(1.79 \text{ mm})}{2.0 \text{ m}} = 188 \text{ MPa} \qquad \leftarrow$$

d = 15 mm

M = 20 kg

L = 2.0 m

h = 150 mm

Fig. 2-50 Example 2-16. Impact load on a vertical bar

This stress may be compared with the static stress, which is

$$\sigma_{st} = \frac{W}{A} = \frac{Mg}{A} = \frac{(20 \text{ kg})(9.81 \text{ m/s}^2)}{(\pi/4)(15 \text{ mm})^2} = 1.11 \text{ MPa}$$

The ratio of σ_{max} to σ_{st} is $188/1.11 = 169$, which is the same impact factor as for the elongations. This result is expected, because the stresses are directly proportional to the corresponding elongations (see Eqs. 2-48 and 2-50).

■ **Example 2-17**

A horizontal bar AB of length L is struck at its free end by a heavy block of mass M moving horizontally with velocity v (Fig. 2-51). (a) Determine the maximum shortening δ_{max} of the bar due to the impact and determine the corresponding impact factor. (b) Determine the maximum compressive stress σ_{max} and the corresponding impact factor. (Let EA represent the axial rigidity of the bar.)

Solution   The loading on the bar in this example is quite different from the loads on the bars pictured in Figs. 2-49 and 2-50. Therefore, we must make a new analysis based upon conservation of energy.

(a) *Maximum shortening of the bar.* For this analysis we adopt the same assumptions as those described previously. Thus, we disregard all energy losses and assume that the kinetic energy of the moving block is transformed entirely into strain energy of the bar.

The kinetic energy of the block at the instant of impact is $Mv^2/2$. The strain energy of the bar when the block comes to rest at the instant of maximum shortening is $EA\delta_{max}^2/2L$, as given by Eq. (2-31b). Therefore, we can write the following equation of conservation of energy:

$$\frac{Mv^2}{2} = \frac{EA\delta_{max}^2}{2L} \tag{2-55}$$

Solving for δ_{max}, we get

$$\delta_{max} = \sqrt{\frac{Mv^2 L}{EA}} \tag{2-56} \quad \Longleftarrow$$

This equation is the same as Eq. (2-47), which we might have anticipated.

To find the impact factor, we need to know the static displacement of the end of the bar. In this case the static displacement is the shortening of the bar due to the weight of the block applied as a compressive load on the bar:

$$\delta_{st} = \frac{WL}{EA} = \frac{MgL}{EA} \tag{2-57}$$

Thus, the impact factor is

$$\text{Impact factor} = \frac{\delta_{max}}{\delta_{st}} = \sqrt{\frac{EAv^2}{Mg^2 L}} \tag{2-58} \quad \Longleftarrow$$

In realistic situations, the value determined from this equation is likely to be much larger than 1.

Fig. 2-51 Example 2-17. Impact load on a horizontal bar

(b) *Maximum compressive stress in the bar*. The maximum stress in the bar is found from the maximum shortening by means of the equation

$$\sigma_{max} = \frac{E\delta_{max}}{L} = \frac{E}{L}\sqrt{\frac{Mv^2L}{EA}} = \sqrt{\frac{Mv^2E}{AL}} \qquad (2\text{-}59) \;\longleftarrow$$

which is the same as Eq. (2-52).

The static stress σ_{st} in the bar is equal to W/A or Mg/A, which (in combination with Eq. 2-59) leads to the same impact factor as before (Eq. 2-58).

■ Example 2-18

An elevator car of weight W is supported by a cable wound on a rotating drum (Fig. 2-52). The car and cable are moving downward with constant velocity v. What maximum tensile stress is produced in the cable when the drum is suddenly locked? (Let EA represent the axial rigidity of the cable and let L represent the length of exposed cable at the instant the drum is locked.)

Solution   The analysis of the cable and car differs significantly from any of the preceding analyses because we cannot reasonably assume that the cable is unstressed before the drum is locked. In our earlier discussions and examples we assumed that the bars were unstressed because they were not subjected to a force from the striking object prior to impact. Consequently, the bars had no strain energy before impact. However, because the cable in this example supports the weight of the car, a significant amount of strain energy is present even before the drum is locked. Therefore, we must write an equation of conservation of energy appropriate to these new conditions.

As in previous analyses, we will be conservative and assume that there are no energy losses when the drum is locked. Then the total energy of the system (consisting of kinetic, potential, and strain energy) just prior to locking must be equal to the total energy at the instant the car comes to rest.

Total energy of the system at the instant before the drum is locked.

(a) *Kinetic energy.* Before locking, the kinetic energy of the moving elevator car is

$$\text{K.E.} = \frac{Wv^2}{2g} \qquad (a)$$

We will disregard the kinetic energies of the cable and rotating drum because they are relatively small compared to the kinetic energy of the car.

(b) *Potential energy.* The potential energy of the elevator car will be determined with respect to its lowest position (that is, the position of the car when the cable has its greatest elongation). Let δ_1 be the distance that the car moves downward after the drum locks. This distance is not the same as the maximum elongation of the cable, because the cable already has a static elongation δ_{st} due to the weight of the car:

$$\delta_{st} = \frac{WL}{EA} \qquad (2\text{-}60)$$

The maximum elongation δ_{max} of the cable is δ_{st} plus δ_1, and therefore the additional elongation δ_1 due to locking is $\delta_{max} - \delta_{st}$. This elongation is the same as

Drum for cable

v

L

W

Fig. 2-52 Example 2-18. Elevator car that is suddenly stopped the downward displacement δ_1 of the car after the drum locks. Consequently, the potential energy of the car prior to locking is

$$\text{P.E.} = W\delta_1 = W(\delta_{max} - \delta_{st}) \tag{b}$$

For the sake of simplicity in the analysis, we will disregard the potential energy of the cable in comparison to that of the elevator car.

(c) *Strain energy.* The strain energy of the cable before locking (from Eq. 2-31b) is

$$\text{S.E.} = \frac{EA\delta_{st}^2}{2L} \tag{c}$$

(d) *Total energy.* Combining expressions (a), (b), and (c), we obtain the total energy of the system before locking:

$$\frac{Wv^2}{2g} + W(\delta_{max} - \delta_{st}) + \frac{EA\delta_{st}^2}{2L} \tag{d}$$

in which δ_{st} is given by Eq. (2-60).

Total energy of the system at the instant the cable has its maximum elongation.

After the drum is locked, the cable stretches and reaches its maximum elongation δ_{max}. At that instant the elevator car has no kinetic energy because its velocity is zero. Its potential energy is also zero because the car is at its lowest position (downward displacement equal to δ_1). That leaves only the strain energy of the cable to be evaluated. Since the cable has its maximum elongation, its strain energy is

$$\text{S.E.} = \frac{EA\delta_{max}^2}{2L} \tag{e}$$

This quantity represents the total energy of the system after the cable has locked and the car is at its lowest point.

Conservation of energy. Using the principle of conservation of energy and equating the total energies before and after locking of the drum (expressions d and e), we obtain

$$\frac{Wv^2}{2g} + W(\delta_{max} - \delta_{st}) + \frac{EA\delta_{st}^2}{2L} = \frac{EA\delta_{max}^2}{2L} \tag{2-61}$$

This equation can be solved for the maximum elongation δ_{max} of the cable in the following way. First, we note from Eq. (2-60) that $W = EA\delta_{st}/L$. Then we substitute this expression for W into the second term of Eq. (2-61) and rearrange, obtaining

$$\frac{Wv^2}{2g} + \frac{EA\delta_{st}}{L}(\delta_{max} - \delta_{st}) = \frac{EA}{2L}(\delta_{max}^2 - \delta_{st}^2)$$

Further rearrangement of terms gives

$$\frac{Wv^2}{2g} = \frac{EA}{2L}(\delta_{max} - \delta_{st})^2$$

Next we solve this equation for the maximum elongation of the cable:

$$\delta_{max} = \delta_{st} + \sqrt{\frac{Wv^2L}{gEA}} = \delta_{st}\left(1 + \sqrt{\frac{v^2EA}{gWL}}\right) \qquad (2\text{-}62)$$

Finally, we obtain the maximum tensile stress in the cable:

$$\sigma_{max} = \frac{E\delta_{max}}{L} = \frac{W}{A}\left(1 + \sqrt{\frac{v^2EA}{gWL}}\right) = \sigma_{st}\left(1 + \sqrt{\frac{v^2EA}{gWL}}\right) \qquad (2\text{-}63) \longleftarrow$$

in which $\sigma_{st} = W/A$.

The impact factor, which is the term in parentheses in Eqs. (2-62) and (2-63), may be many times greater than unity, showing that the dynamic elongation and dynamic stress in the cable may be much greater than the corresponding static quantities.

As a special case for this example, note that if the velocity $v = 0$, the impact factor is 1 and Eqs. (2-62) and (2-63) give the static values.

*2.9 REPEATED LOADING AND FATIGUE

The behavior of a structure depends not only upon the nature of the material but also upon the character of the loads. In some situations the loads are static—they are applied gradually, act for long periods of time, and change slowly. Other loads are dynamic in character—examples are impact loads acting suddenly (Section 2.8) and repeated loads recurring for large numbers of cycles.

Some typical patterns for **repeated loads** are sketched in Fig. 2-53. The first graph (a) shows a load that is applied, removed, and applied again, always acting in the same direction. The second graph (b) shows an alternating load that reverses direction during every cycle of loading, and the third graph (c) illustrates a fluctuating load that varies about an average value. Repeated loads are commonly associated with machinery, engines, turbines, generators, shafts, propellers, airplane parts, automobile parts, and the like. Some of these structures are subjected to millions (and even billions) of loading cycles during their useful life.

A structure subjected to dynamic loads is likely to fail at a lower stress than when the same loads are applied statically, especially when the loads are repeated for a large number of cycles. In such cases failure is usually caused by **fatigue**, or **progressive fracture**. A familiar example of a fatigue failure is stressing a metal paper clip to the breaking point by repeatedly bending it back and forth. If the clip is bent only once, it does not break. But if the load is reversed by bending the clip in the opposite direction, and if the entire loading cycle is repeated several times, the clip will finally break. *Fatigue* may be defined as the deterioration of a material under repeated cycles of stress and strain, resulting in progressive cracking that eventually produces fracture.

(a)

(b)

(c)

Fig. 2-53 Types of repeated loads:
(a) load acting in one direction only,
(b) alternating or reversed load, and
(c) fluctuating load that varies about an average value

In a typical fatigue failure, a microscopic crack forms at a point of high stress (usually at a *stress concentration*, discussed in the next section) and gradually enlarges as the loads are applied repeatedly. When the crack becomes so large that the remaining material cannot resist the loads, a sudden fracture of the material occurs (Fig. 2-54). Depending upon the nature of the material, it may take anywhere from a few cycles of loading to hundreds of millions of cycles to produce a fatigue failure.

Fig. 2-54 Fatigue failure of a bar loaded repeatedly in tension; the crack spread gradually over the cross section until fracture occurred suddenly. (Courtesy of MTS Systems Corporation)

The magnitude of the load causing a fatigue failure is less than the load that can be sustained statically, as already pointed out. To determine the failure load, tests of the material must be performed. In the case of repeated loading, the material is tested at various stress levels and the number of cycles to failure is counted. For instance, a specimen of material is placed in a fatigue-testing machine and loaded repeatedly to a certain stress, say σ_1. The loading cycles are continued until failure occurs, and the number n of loading cycles to failure is noted. The test is then repeated for a different stress, say σ_2. If σ_2 is greater than σ_1, the number of cycles to failure will be smaller. If σ_2 is less than σ_1, the number will be larger. Eventually, enough data are accumulated to plot an **endurance curve**, or **S-N diagram**, in which failure stress (S) is plotted versus the number (N) of cycles to failure (Fig. 2-55). The vertical axis is usually a linear scale and the horizontal axis is usually a logarithmic scale.

The endurance curve of Fig. 2-55 shows that the smaller the stress, the larger the number of cycles to produce failure. For some materials the curve has a horizontal asymptote known as the **fatigue limit** or **endurance limit**. When it exists, this limit is the stress below which a fatigue failure will not occur regardless of how many times the load is repeated. The precise shape of an endurance curve depends upon many factors, including properties of the material, geometry of the test specimen, speed of testing, pattern of loading, and surface condition of the specimen. The results of numerous fatigue tests, made on a great variety of materials and structural components, have been reported in the engineering literature.

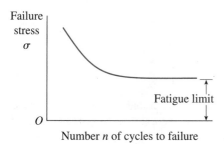

Fig. 2-55 Endurance curve, or *S-N* diagram, showing fatigue limit

Typical *S-N* diagrams for steel and aluminum are shown in Fig. 2-56. The ordinate is the failure stress, expressed as a percentage of the ultimate stress for the material, and the abscissa is the number of cycles at which failure occurred. Note that the number of cycles is plotted on a logarithmic scale. The curve for steel becomes horizontal at about 10^7 cycles, and the fatigue limit is about 50% of the ultimate tensile stress for ordinary static loading. The fatigue limit for aluminum is not as clearly defined as that for steel, but a typical value of the fatigue limit is the stress at 5×10^8 cycles, or about 25% of the ultimate stress.

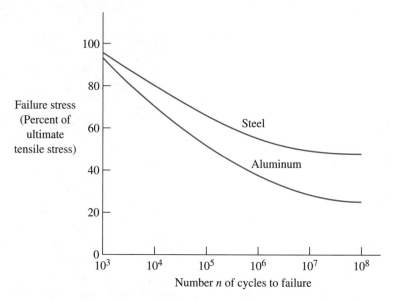

Fig. 2-56 Typical *S-N* diagrams for steel and aluminum in alternating (reversed) loading

Since fatigue failures usually begin with a microscopic crack at a point of high localized stress (that is, at a stress concentration), the condition of the surface of the material is extremely important. Highly polished specimens have higher endurance limits. Rough surfaces, especially those at stress concentrations around holes or grooves, greatly lower the endurance limit. Corrosion, which creates tiny surface irregularities, has a similar effect. For steel, ordinary corrosion may reduce the fatigue limit by more than 50%.

*2.10 STRESS CONCENTRATIONS

When determining the stresses in axially loaded bars, we customarily use the basic formula $\sigma = P/A$, in which P is the axial force in the bar and A is its cross-sectional area. This formula is based upon the assumption that the stress distribution is uniform throughout the cross section. In reality, bars often have holes, grooves, notches, keyways, shoulders, threads, or other abrupt changes in geometry that create a disruption in the otherwise uniform stress pattern. These discontinuities in geometry cause high stresses in very small regions of the bar, and these high stresses are known as **stress concentrations**. The discontinuities themselves are known as **stress raisers**.

Stress concentrations also appear at points of loading. For instance, a concentrated load is rarely distributed uniformly over a cross section. Instead, it is more likely to act over a very small area and produce high stresses in the region around its point of application. An example is a load applied through a pin connection, in which case the load is applied over the bearing area of the pin.

The stresses existing at stress concentrations can be determined either by experimental methods or by advanced methods of analysis, including the finite-element method. The results of such research for many cases of practical interest are readily available in the engineering literature (for example, Refs. 2-10 and 2-11). Some typical stress-concentration data are given in this section and later in Sections 3.11 and 5.13.

■ Saint-Venant's Principle

To illustrate the nature of stress concentrations, consider the stresses in a bar of rectangular cross section (width b, thickness t) subjected to a concentrated load P at the end (Fig. 2-57). The peak stress directly under the load may be several times the average stress P/bt, depending upon the area over which the load is applied. However, the maximum stress diminishes rapidly as we move away from the point of load application, as shown by the stress diagrams in the figure. At a distance from the end of the bar equal to the width b of the bar, the stress distribution is nearly uniform, and the maximum stress is only a few percent larger than the average stress. This observation is true for most stress concentrations, such as holes and grooves. Thus, we can make a general statement that the equation $\sigma = P/A$ gives the axial stresses only when the cross section of the bar is at least a distance b away from any concentrated load or discontinuity in shape, where b is the largest lateral dimension of the bar (such as the width or diameter).

The preceding statement about the stresses in a prismatic bar is part of a more general observation known as **Saint-Venant's principle**. With rare exceptions, this principle applies to linearly elastic bodies of all types. To understand Saint-Venant's principle, imagine that we have a body with a system of loads acting over a small part of its surface. For instance, suppose we have a prismatic bar of width b subjected to a system of several concentrated loads acting at the end (Fig. 2-58a). For simplicity, assume that the loads are symmetrical and have only a vertical resultant. Next, consider a different but statically equivalent load system acting over the same small region of the bar. ("Statically equivalent"

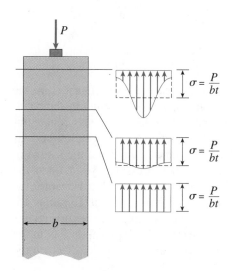

Fig. 2-57 Stress distributions near the end of a bar of rectangular cross section (width b, thickness t) subjected to a concentrated load P acting over a small area

means the two load systems have the same force resultant and same moment resultant.) For instance, the uniformly distributed load shown in Fig. 2-58b is statically equivalent to the system of concentrated loads shown in Fig. 2-58a. Saint-Venant's principle states that the stresses in the body caused by either of the two systems of loading are the same, provided we move away from the loaded region a distance at least equal to the largest dimension of the loaded region (distance *b* in our example). Thus, the stress distributions shown in Fig. 2-57 are an illustration of Saint-Venant's principle. Of course, this "principle" is not a rigorous law of mechanics but is a common-sense observation based upon theoretical and practical experience.*

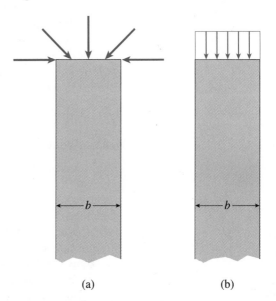

(a)   (b)

Fig. 2-58 Illustration of Saint-Venant's principle: (a) system of concentrated loads acting over a small region of a bar, and (b) statically equivalent system

Saint-Venant's principle has great practical significance in the design and analysis of bars, beams, shafts, and other structures encountered in mechanics of materials. Because the effects of stress concentrations are localized, we can use all of the standard stress formulas (such as $\sigma = P/A$) at cross sections a sufficient distance away from the source of the concentration. Close to the source, the stresses depend upon the details of the loading and the shape of the member. Furthermore, the formulas that are applicable to entire members, such as formulas for elongations, displacements, and strain energy, give satisfactory results even when stress concentrations are present. The explanation lies in the fact that stress concentrations are localized and have little effect on the overall behavior of a member.

---

*Saint-Venant's principle is named for Barré de Saint-Venant (1797–1886), a famous French mathematician and elastician (Ref. 2-12). It appears that the principle applies generally to solid bars and beams but not to all thin-walled open sections. For a discussion of the limitations of Saint-Venant's principle, see Ref. 2-13.

■ Stress-Concentration Factors

Now let us consider some particular cases of stress concentrations caused by discontinuities in the shape of a bar. We begin with a bar of rectangular cross section having a circular hole and subjected to a tensile force P (Fig. 2-59a). The bar is relatively thin with its width b much larger than its thickness t. Also, the hole has diameter d.

(a)

(b)

Fig. 2-59 Stress distribution in a flat bar with a circular hole

The normal stress acting on the cross section through the center of the hole has the distribution shown in Fig. 2-59b. The maximum stress σ_{max} occurs at the edges of the hole and may be significantly larger than the *nominal stress* $\sigma = P/ct$ at the same cross section. (Note that ct is the net area at the cross section through the hole.) The intensity of a stress concentration is usually expressed by the ratio of the maximum stress to the nominal stress, called the **stress-concentration factor K:**

$$K = \frac{\sigma_{max}}{\sigma_{nom}} \tag{2-64}$$

For a bar in tension, the nominal stress is the average stress based upon the net cross-sectional area. In other cases, a variety of stresses may be used. Thus, whenever a stress concentration factor is used, it is important to note carefully how the nominal stress is defined.

A graph of the stress-concentration factor K for a bar with a hole is given in Fig. 2-60. If the hole is tiny, the factor K equals 3, which means that the maximum stress is three times the nominal stress. As the hole becomes larger in proportion to the width of the bar, K becomes smaller and the effect of the concentration is not as severe. From Saint-Venant's principle we know that, at distances equal to the width b of the bar away from the hole in either axial direction, the stress distribution is practically uniform and equal to P divided by the gross cross-sectional area ($\sigma = P/bt$), regardless of the size of the hole.

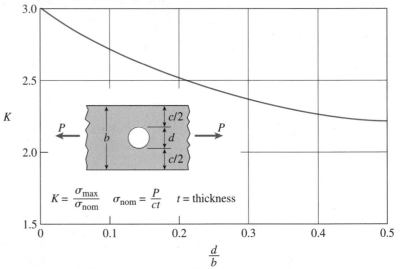

Fig. 2-60 Stress-concentration factor K for flat bars with circular holes

Stress-concentration factors for two other cases of practical interest are given in Figs. 2-61 and 2-62. These graphs are for flat bars and circular bars, respectively, that are stepped down in size, forming a *shoulder*. To reduce the stress-concentration effects, *fillets* are used to round off the re-entrant corners.* Without the fillets, the stress-concentration factors would be extremely large, as indicated at the left-hand side of each graph where K approaches infinity as the fillet radius R approaches zero. In both cases the maximum stress occurs in the smaller part of the bar in the region of the fillet.**

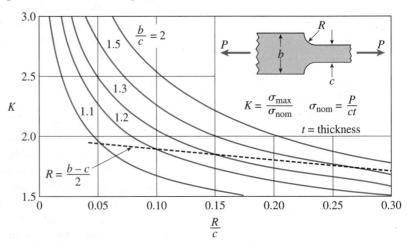

Fig. 2-61 Stress-concentration factor K for flat bars with shoulder fillets. The dashed line is for a full quarter-circular fillet.

---

*A *fillet* is a curved concave surface formed where two other surfaces meet. Its purpose is to round off what would otherwise be a sharp re-entrant corner.

**The stress-concentration factors given in the graphs are theoretical factors for bars of linearly elastic material. The graphs are plotted from the formulas given in Ref. 2-10.

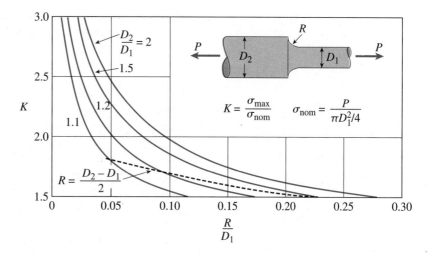

Fig. 2-62 Stress concentration factor K for round bars with shoulder fillets. The dashed line is for a full quarter-circular fillet.

■ Designing for Stress Concentrations

Because of the possibility of fatigue failures, stress concentrations are especially important when the member is subjected to repeated loading. As explained in the preceding section, cracks begin at the point of highest stress and then spread gradually through the material as the load is repeated. In practical design, the fatigue limit (Fig. 2-55) is considered to be the ultimate stress for the material when the number of cycles is extremely large. The allowable stress is obtained by applying a factor of safety with respect to this ultimate stress. Then the peak stress at the stress concentration is compared with the allowable stress.

In many situations the use of the full theoretical value of the stress-concentration factor is too severe. Fatigue tests on specimens with stress concentrations usually produce failure at higher levels of the nominal stress than those obtained by dividing the fatigue limit by K. In other words, a structural member under repeated loading is not as sensitive to the stress concentration as the value of K indicates. Consequently, a reduced stress-concentration factor is often used.

Other kinds of dynamic loads, such as impact loads, also require that stress-concentration effects be taken into account. Unless better information is available, the full stress-concentration factor should be used. Members subjected to low temperatures also are highly susceptible to failures at stress concentrations, and therefore special precautions should be taken in such cases.

The significance of stress concentrations when a member is subjected to static loading depends upon the kind of material. With ductile materials, such as structural steel, a stress concentration can often be ignored. The reason is that the material at the point of maximum stress (such as around a hole) will yield and plastic flow will occur, thus reducing the intensity of the stress concentration and making the stress distribution more nearly uniform. On the other hand, with brittle materials (such as glass) a stress concentration will remain up to the point of fracture. Therefore, we can make the general observation that with static loads and a ductile material the stress-concentration effect is not likely to be important, but with static loads and a brittle material the full stress-concentration factor should be considered.

Stress concentrations can be reduced in intensity by properly proportioning the parts. Generous fillets reduce stress concentrations at re-entrant corners. Smooth surfaces at points of high stress, such as on the inside of a hole, inhibit the formation of cracks. Proper reinforcing around holes can also be beneficial. There are many other techniques for smoothing out the stress distribution in a structural member and thereby reducing the stress-concentration factor. These techniques, which are usually studied in engineering design courses, are extremely important in the design of aircraft, ships, and machines. Many unnecessary structural failures have occurred because designers failed to recognize the effects of stress concentrations and fatigue.

*2.11 NONLINEAR BEHAVIOR

Up to this point, our discussions have dealt primarily with members and structures composed of materials that follow Hooke's law. Now we will consider the behavior of axially loaded members when the stresses exceed the proportional limit. In such cases the stresses, strains, and displacements depend upon the shape of the stress-strain curve in the region beyond the proportional limit (see Section 1.3 for some typical stress-strain diagrams).

For purposes of analysis and design, we often represent the actual stress-strain curve of a material by an **idealized stress-strain curve** that can be expressed as a mathematical function. Some examples are shown in Fig. 2-63. The first diagram (Fig. 2-63a) consists of two parts, an initial linearly elastic region followed by a nonlinear region defined by an appropriate mathematical expression. The behavior of aluminum alloys can sometimes be represented quite accurately by a curve of this type, at least in the region before the strains become excessively large (compare Fig. 2-63a with Fig. 1-13). In the second example (Fig. 2-63b), a single mathematical expression is used for the entire stress-strain curve. The best known expression of this kind is the Ramberg-Osgood stress-strain law, which is described later in more detail (see Eqs. 2-67 and 2-68).

The stress-strain diagram frequently used for structural steel is shown in Fig. 2-63c. Because steel has a linearly elastic region followed by a region of considerable yielding (see the stress-strain diagrams of Figs. 1-10 and 1-12), its behavior can be represented by two straight lines. The material is assumed to follow Hooke's law up to the yield

(a)

(b)

(c)

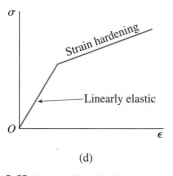

(d)

Fig. 2-63 Types of idealized material behavior: (a) elastic-nonlinear stress-strain curve, (b) general nonlinear stress-strain curve, (c) elastoplastic stress-strain curve, and (d) bilinear stress-strain curve

stress σ_y, after which it yields under constant stress, the latter behavior being known as **perfect plasticity**. The perfectly plastic region continues until the strains are 10 or 20 times larger than the yield strain. A material having a stress-strain diagram of this kind is called an **elastoplastic material** (or *elastic-plastic material*).

Eventually, as the strain becomes extremely large, the stress-strain curve for steel rises above the yield stress due to strain hardening, as explained in Section 1.3. However, by the time strain hardening begins, the displacements are so large that the structure will have lost its usefulness. Consequently, it is common practice to analyze steel structures on the basis of the elastoplastic diagram shown in Fig. 2-63c, with the same diagram being used for both tension and compression. An analysis made with these assumptions is called an **elastoplastic analysis**, or simply, **plastic analysis**, and is described in the next section.

Figure 2-63d shows a stress-strain diagram consisting of two lines having different slopes, called a **bilinear stress-strain diagram**. Note that in both parts of the diagram the relationship between stress and strain is linear, but only in the first part is the stress proportional to the strain (Hooke's law). This idealized diagram may be used to represent materials with strain hardening or it may be used as an approximation to diagrams of the general nonlinear shapes shown in Figs. 2-63a and b.

■ **Changes in Length of Bars**

The elongation or shortening of a bar can be determined if the stress-strain curve of the material is known. To illustrate the general procedure, we will consider the tapered bar *AB* shown in Fig. 2-64a. Both the cross-sectional area and the axial force vary along the length of the bar, and the material has a general nonlinear stress-strain curve (Fig. 2-64b). Because the bar is statically determinate, we can determine the axial forces at all cross sections from static equilibrium alone. Then we can find the stresses by dividing the forces by the cross-sectional areas, and we can find the strains from the stress-strain curve. Lastly, we can determine the change in length from the strains, as described in the following paragraph.

The change in length of an element dx of the bar (Fig. 2-64a) is $\epsilon\, dx$, where ϵ is the strain at distance x from the end. By integrating this expression from one end of the bar to the other, we obtain the change in length of the entire bar:

$$\delta = \int_0^L \epsilon \, dx \tag{2-65}$$

where L is the length of the bar. If the strains are expressed analytically, that is, by algebraic formulas, it may be possible to integrate Eq. (2-65) by formal mathematical means and thus obtain a formula for the change in length. If the stresses and strains are expressed numerically, that is, by a table of numerical values, we can proceed as follows. We can divide the bar into small segments of length Δx, determine the average stress and strain for each segment, and then calculate the elongation of the entire bar by summing the elongations for the individual segments. This process is equivalent to evaluating the integral in Eq. (2-65) by numerical methods instead of by formal integration.

(a)

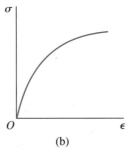

(b)

Fig. 2-64 Change in length of a tapered bar consisting of a material having a nonlinear stress-strain curve

If the strains are uniform throughout the length of the bar, as in the case of a prismatic bar with constant axial force, the integration of Eq. (2-65) is trivial and the change in length is

$$\delta = \epsilon L \tag{2-66}$$

as expected (compare with Eq. 1-2 in Section 1.2).

■ Ramberg-Osgood Stress-Strain Law

Stress-strain curves for several metals, including aluminum and magnesium, can be accurately represented by the **Ramberg-Osgood equation**:

$$\frac{\epsilon}{\epsilon_0} = \frac{\sigma}{\sigma_0} + \alpha\left(\frac{\sigma}{\sigma_0}\right)^m \tag{2-67}$$

In this equation, σ and ϵ are the stress and strain, respectively, and ϵ_0, σ_0, α, and m are constants of the material (obtained from tension tests). An alternative form of the equation is

$$\epsilon = \frac{\sigma}{E} + \frac{\sigma_0\alpha}{E}\left(\frac{\sigma}{\sigma_0}\right)^m \tag{2-68}$$

in which $E = \sigma_0/\epsilon_0$ is the modulus of elasticity in the initial part of the stress-strain curve.*

———————

*The Ramberg-Osgood stress-strain law was presented in Ref. 2-14.

A graph of Eq. (2-68) is given in Fig. 2-65 for an aluminum alloy for which the constants are as follows: $E = 10 \times 10^6$ psi, $\sigma_0 = 38,000$ psi, $\alpha = 3/7$, and $m = 10$. The equation of this particular stress-strain curve is

$$\epsilon = \frac{\sigma}{10 \times 10^6} + \frac{1}{614.0}\left(\frac{\sigma}{38,000}\right)^{10} \qquad (2\text{-}69)$$

where σ has units of psi. The calculation of the change in length of a bar, using Eq. (2-69) for the stress-strain relationship, is illustrated in Example 2-19.

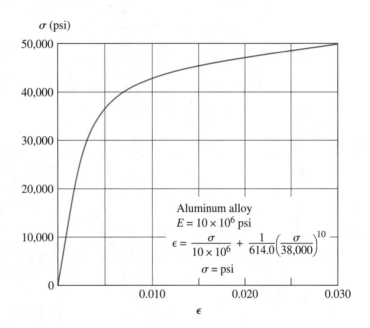

Fig. 2-65 Stress-strain curve for an aluminum alloy using the Ramberg-Osgood equation

■ Statically Indeterminate Structures

If a structure is statically indeterminate and the material behaves nonlinearly, the stresses, strains, and displacements can be found by solving the same general equations as those described in Section 2.4 for linearly elastic structures, namely, equations of equilibrium, equations of compatibility, and force-displacement relations (or equivalent stress-strain relations). The principal difference is that the force-displacement relations are now nonlinear, which means that analytical solutions cannot be obtained except in very simple situations. Instead, the equations must be solved numerically, either by using a computer program or by performing tedious manual calculations (as illustrated in Example 2-20).

Problems 2.11-9 through 2.11-14 at the end of the chapter involve statically indeterminate structures with nonlinear material characteristics.

Fig. 2-66 Example 2-19. Elongation of a bar of nonlinear material using the Ramberg-Osgood equation

■ **Example 2-19**

A prismatic bar AB of length $L = 86$ in. and cross-sectional area $A = 0.75$ in.^{2} supports two concentrated loads $P_1 = 24$ k and $P_2 = 6$ k, as shown in Fig. 2-66. The material of the bar is an aluminum alloy having a nonlinear stress-strain curve described by the following Ramberg-Osgood equation (Eq. 2-69):

$$\epsilon = \frac{\sigma}{10 \times 10^6} + \frac{1}{614.0}\left(\frac{\sigma}{38,000}\right)^{10}$$

in which σ has units of psi. (This stress-strain relation is shown graphically in Fig. 2-65.)

Determine the displacement δ_B of the lower end of the bar under each of the following conditions: (a) the load P_1 acts alone, (b) the load P_2 acts alone, and (c) the loads P_1 and P_2 act simultaneously.

Solution   (a) *Displacement due to the load P_1 acting alone.* The load P_1 produces a uniform tensile stress throughout the length of the bar equal to P_1/A, or 32,000 psi. Substituting this value into the stress-strain relation gives $\epsilon = 0.003492$. Therefore, the elongation of the bar, equal to the displacement at B, is (see Eq. 2-66):

$$\delta_B = \epsilon L = (0.003492)(86 \text{ in.}) = 0.300 \text{ in.} \qquad \longleftarrow$$

(b) *Displacement due to the load P_2 acting alone.* The stress in the upper half of the bar is P_2/A or 8,000 psi, and there is no stress in the lower half. Proceeding as in part (a), we obtain the following elongation:

$$\delta_B = \epsilon L/2 = (0.000800)(43 \text{ in.}) = 0.034 \text{ in.} \qquad \longleftarrow$$

(c) *Displacement due to both loads acting simultaneously.* The stress in the lower half of the bar is P_1/A and in the upper half is $(P_1 + P_2)/A$. The corresponding stresses are 32,000 psi and 40,000 psi, and the corresponding strains are 0.003492 and 0.006720 (from the Ramberg-Osgood equation). Therefore, the elongation of the bar is

$$\delta_B = (0.003492)(43 \text{ in.}) + (0.006720)(43 \text{ in.})$$

$$= 0.150 \text{ in.} + 0.289 \text{ in.} = 0.439 \text{ in.} \qquad \longleftarrow$$

The three calculated values of δ_B illustrate an important principle pertaining to a structure made of a material that behaves nonlinearly: *In a nonlinear structure, the displacement produced by two (or more) loads acting simultaneously is not equal to the sum of the displacements produced by the loads acting separately.*

■ **Example 2-20**

A horizontal rigid bar AB of length $3b$ is held by two identical cables and supports a load $Q = 20$ kN at end B (Fig. 2-67). The load-displacement relation for the cables is given by the following hyperbolic equation:

$$P = \frac{1.3\delta}{1 + 0.026\delta} \qquad 0 \le \delta \le 60 \text{ mm} \qquad (P = \text{kN}, \delta = \text{mm}) \quad (2\text{-}70)$$

where P is the axial force in a cable (units of kilonewtons) and δ is the elongation (units of millimeters). This load-displacement relation is shown graphically in Fig. 2-68. (The curve is a segment of a hyperbola.)

Determine the tensile forces F_1 and F_2 in cables 1 and 2, respectively, the elongations δ_1 and δ_2 of cables 1 and 2, respectively, and the downward displacement δ_B of point B.

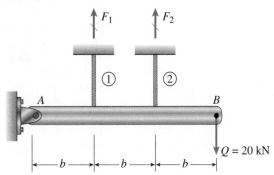

Fig. 2-67 Example 2-20. Nonlinear analysis of a statically indeterminate structure

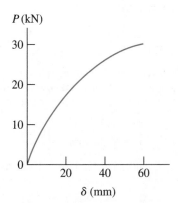

Fig. 2-68 Load-displacement curve (Eq. 2-70) for the cables of Example 2-20

Solution   Because the structure is statically indeterminate, we need to formulate an equation of equilibrium, equations of compatibility, and the load-displacement relations. These equations can then be solved for the unknown forces and displacements of the structure.

Equation of equilibrium. From the equilibrium of bar AB we obtain the following equation:

$$\sum M_A = 0 \qquad F_1(b) + F_2(2b) - Q(3b) = 0$$

or
$$F_1 + 2F_2 = 3Q = 60 \text{ kN} \tag{a}$$

The forces F_1 and F_2 are the unknowns in this equation (note that the forces are expressed in units of kilonewtons).

Equations of compatibility. When the load Q is applied to the structure, bar AB will rotate about the pin at end A. Because the bar is assumed to be rigid, the downward displacement at any point of the bar is proportional to its distance from A. Thus, the elongation δ_2 of cable 2 is twice the elongation δ_1 of cable 1:

$$\delta_2 = 2\delta_1 \tag{b}$$

and the downward displacement of point B is three times the elongation of cable 1:

$$\delta_B = 3\delta_1 \tag{c}$$

Load-displacement relations. The relationships between the unknown forces (F_1 and F_2) and the elongations of the cables (δ_1 and δ_2) are obtained from the load-displacement equation (Eq. 2-70), as follows:

$$F_1 = \frac{1.3\delta_1}{1 + 0.026\delta_1} \qquad F_2 = \frac{1.3\delta_2}{1 + 0.026\delta_2} \tag{d,e}$$

In these equations, F_1 and F_2 are expressed in units of kilonewtons and δ_1 and δ_2 are expressed in millimeters.

Solution of equations. We are now ready to solve the five equations (Eqs. a, b, c, d, and e) for the five unknown quantities F_1, F_2, δ_1, δ_2, and δ_B. Using a computer program for solving equations, we enter the five equations into the program and execute the appropriate solve command, obtaining the following results:

$$F_1 = 14.66 \text{ kN} \qquad F_2 = 22.67 \text{ kN} \qquad \longleftarrow$$

$$\delta_1 = 15.95 \text{ mm} \qquad \delta_2 = 31.91 \text{ mm} \qquad \delta_B = 47.86 \text{ mm} \qquad \longleftarrow$$

Thus, all of the unknown quantities have been found and the analysis of the statically indeterminate structure is completed. (Note that the displacements δ_1 and δ_2 are within the range of values for which Eq. 2-70 is valid.)

Alternative solution of the equations. In this particular example, we can combine some of the preceding equations and reduce the solution to a quadratic equation. The procedure is as follows. First, we substitute F_1 and F_2 from Eqs. (d) and (e) into Eq. (a), obtaining an equation with δ_1 and δ_2 as unknowns. Then we substitute for δ_2 in terms of δ_1 using Eq. (b). The result is the following equation containing δ_1 as the only unknown quantity:

$$\frac{1.3\delta_1}{1 + 0.026\delta_1} + \frac{5.2\delta_1}{1 + 0.052\delta_1} = 3Q = 60 \tag{f}$$

Next, we multiply every term in this equation by the product of the denominators. Then, upon rearranging and collecting terms, we obtain the following quadratic equation:

$$0.12168\delta_1^2 + 1.82\delta_1 - 60 = 0 \tag{g}$$

Solving by means of the quadratic formula, we get $\delta_1 = 15.95$ mm, as before.

It is now a straightforward matter to solve for the other unknowns in the solution. The elongation of cable 2 (from Eq. b) and the displacement of point B (from Eq. c) are

$$\delta_2 = 2\delta_1 = 31.90 \text{ mm} \qquad \delta_B = 3\delta_1 = 47.85 \text{ mm}$$

Also, the tensile forces F_1 and F_2 in the cables are calculated from Eqs. (d) and (e):

$$F_1 = \frac{1.3\delta_1}{1 + 0.026\delta_1} = 14.66 \text{ kN} \qquad F_2 = \frac{1.3\delta_2}{1 + 0.026\delta_2} = 22.67 \text{ kN}$$

Thus, all of the desired quantities have been calculated and the analysis is completed.

Note 1: For practical use, all of the preceding results should be rounded off to two significant digits, which is the accuracy of the original data (see, for instance, Eq. 2-70).

Note 2: Because of the nonlinear behavior of the cables, we cannot find the forces and displacements for other values of the load Q by direct proportion, as we can with a linear structure. Instead, the five equations must be solved anew, with Eq. (a) being changed in accordance with the new value of the load.

Note 3: This example illustrates how the forces and displacements of a statically indeterminate structure can be calculated from the three fundamental concepts of mechanics of materials, namely, equations of equilibrium (based upon principles of statics), equations of compatibility (based upon principles of geometry), and load-displacement relations (based upon material properties).

*2.12 ▍ ELASTOPLASTIC ANALYSIS

In the preceding section we discussed the behavior of structures when the stresses in the material exceed the proportional limit. Now we will consider a material of considerable importance in engineering design—steel, the most widely used structural metal. Mild steel (or structural steel) can be modeled as an *elastoplastic* material (see the stress-strain diagram in Fig. 2-63c). An elastoplastic material initially behaves in a linearly elastic manner with a modulus of elasticity E. After plastic yielding begins, the strains increase at a more-or-less constant stress, called the *yield stress* σ_y. The strain at the onset of yielding is known as the *yield strain* ϵ_y.

In this section we will discuss the analysis of statically indeterminate structures of elastoplastic materials. To illustrate the behavior of such structures, we will use the simple arrangement shown in Fig. 2-69a. This structure consists of three steel bars supporting a load P applied through a rigid plate. The two outer bars have length L_1, the inner bar has length L_2, and all three bars have the same cross-sectional area A. The stress-strain diagram for the steel is idealized as shown in Fig. 2-69b, and the modulus of elasticity in the linearly elastic region is $E = \sigma_y/\epsilon_y$.

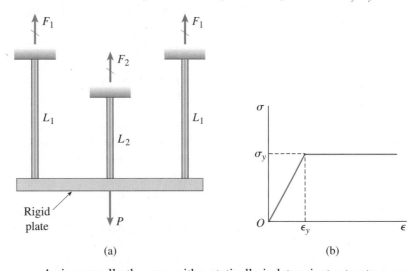

Fig. 2-69 Elastoplastic analysis of a statically indeterminate structure

(a)   (b)

As is normally the case with a statically indeterminate structure, we begin the analysis with the equations of equilibrium and compatibility. From equilibrium of the rigid plate in the vertical direction we obtain

$$2F_1 + F_2 = P \tag{a}$$

where F_1 and F_2 are the axial forces in the outer and inner bars, respectively. Because the plate moves downward as a rigid body when the load is applied, the compatibility equation is

$$\delta_1 = \delta_2 \tag{b}$$

where δ_1 and δ_2 are the elongations of the outer and inner bars, respectively. Because they depend only upon equilibrium and geometry, the two preceding equations are valid at all levels of the load P; it does not matter whether the strains fall in the linearly elastic region or in the plastic region.

When the load P is small, the stresses in the bars are less than the yield stress σ_y and the material is stressed within the linearly elastic region. Therefore, the force-displacement relationships between the bar forces and their elongations are

$$\delta_1 = \frac{F_1 L_1}{EA} \qquad \delta_2 = \frac{F_2 L_2}{EA} \tag{c}$$

Substituting in the compatibility equation (Eq. b), we get

$$F_1 L_1 = F_2 L_2 \tag{d}$$

Solving simultaneously Eqs. (a) and (d), we obtain

$$F_1 = \frac{P L_2}{L_1 + 2L_2} \qquad F_2 = \frac{P L_1}{L_1 + 2L_2} \tag{2-71a,b}$$

Thus, we have now found the forces in the bars in the linearly elastic region. The corresponding stresses are

$$\sigma_1 = \frac{F_1}{A} = \frac{P L_2}{A(L_1 + 2L_2)} \qquad \sigma_2 = \frac{F_2}{A} = \frac{P L_1}{A(L_1 + 2L_2)} \tag{2-72a,b}$$

These equations for the forces and stresses are valid provided the stresses in all three bars remain below the yield stress σ_y.

As the load P gradually increases, the stresses in the bars increase until the yield stress is reached in either the inner bar or the outer bars. Let us assume that the outer bars are longer than the inner bar, as sketched in Fig. 2-69a:

$$L_1 > L_2 \tag{e}$$

Then the inner bar is more highly stressed than the outer bars (see Eqs. 2-72a and b) and will reach the yield stress first. When that happens, the force in the inner bar is $F_2 = \sigma_y A$. The magnitude of the load P when the yield stress is first reached in any one of the bars is called the **yield load** P_y. We can determine P_y by setting F_2 equal to $\sigma_y A$ in Eq. (2-71b) and solving for the load:

$$P_y = \sigma_y A \left(1 + \frac{2L_2}{L_1} \right) \tag{2-73}$$

As long as the load P is less than P_y, the structure behaves in a linearly elastic manner and the forces in the bars can be determined from Eqs. (2-71a and b).

The downward displacement of the rigid bar at the yield load, called the **yield displacement** δ_y, is equal to the elongation of the inner bar when its stress first reaches the yield stress σ_y:

$$\delta_y = \frac{F_2 L_2}{EA} = \frac{\sigma_2 L_2}{E} = \frac{\sigma_y L_2}{E} \tag{2-74}$$

The relationship between the applied load P and the downward displacement δ of the rigid bar is portrayed in the load-displacement diagram of Fig. 2-70. The behavior of the structure up to the yield load P_y is represented by line OA.

With a further increase in the load, the forces F_1 in the outer bars increase but the force F_2 in the inner bar remains constant at the value $\sigma_y A$ because this bar is now perfectly plastic (see Fig. 2-69b). When the forces F_1 reach the value $\sigma_y A$, the outer bars also yield and therefore the structure cannot support any additional load. Instead, all three bars will elongate plastically under this constant load, called the **plastic load** P_p. The plastic load is represented by point B on the load-displacement diagram (Fig. 2-70), and the horizontal line BC represents the region of continuous plastic deformation without any increase in the load.

The plastic load P_p can be calculated from static equilibrium (Eq. a) knowing that

$$F_1 = \sigma_y A \qquad F_2 = \sigma_y A \tag{f}$$

Thus, from equilibrium we find

$$P_p = 3\sigma_y A \tag{2-75}$$

The **plastic displacement** δ_p at the instant the load just reaches the plastic load P_p is equal to the elongation of the outer bars at the instant they reach the yield stress. Therefore,

$$\delta_p = \frac{F_1 L_1}{EA} = \frac{\sigma_1 L_1}{E} = \frac{\sigma_y L_1}{E} \tag{2-76}$$

Comparing δ_p with δ_y, we see that the ratio of the plastic displacement to the yield displacement is

$$\frac{\delta_p}{\delta_y} = \frac{L_1}{L_2} \tag{2-77}$$

Also, the ratio of the plastic load to the yield load is

$$\frac{P_p}{P_y} = \frac{3L_1}{L_1 + 2L_2} \tag{2-78}$$

For example, if $L_1 = 1.5L_2$, the ratios are $\delta_p/\delta_y = 1.5$ and $P_p/P_y = 9/7 = 1.29$. In general, the ratio of the displacements is always larger than the ratio of the corresponding loads, and therefore the partially plastic region AB on the load-displacement diagram (Fig. 2-70) always has a smaller

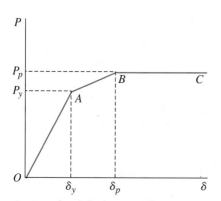

Fig. 2-70 Load-displacement diagram for the statically indeterminate structure shown in Fig. 2-69a

slope than does the elastic region *OA*. Of course, the fully plastic region *BC* has the smallest slope (zero).

To understand why the load-displacement graph is linear in the partially plastic region (line *AB*) and has a slope that is less than in the linearly elastic region, consider the following. In the partially plastic region of the structure, the outer bars still behave in a linearly elastic manner. Therefore, their elongation is a linear function of the load. Since their elongation is the same as the downward displacement of the rigid plate, the displacement of the rigid plate must also be a linear function of the load. Consequently, we have a straight line between points *A* and *B*. However, the slope of the load-displacement diagram in this region is less than in the initial linear region because the inner bar yields plastically and only the outer bars offer increasing resistance to the increasing load. In effect, the stiffness of the structure has diminished.

From the discussion associated with Eq. (2-75) we see that the calculation of the plastic load P_p requires only the use of statics, because all members have yielded and their axial forces are known. In contrast, the calculation of the yield load P_y requires a statically indeterminate analysis, which means that equilibrium, compatibility, and force-displacement equations must be solved.

After the plastic load P_p is reached, the structure continues to deform as shown by line *BC* on the load-displacement diagram (Fig. 2-70). Strain hardening occurs eventually, and then the structure is able to support additional loads. However, the presence of very large displacements usually means that the structure is no longer of use, and so the plastic load P_p is considered to be the failure load.

The preceding discussion has dealt with the behavior of a structure when the load is applied for the first time. If the load is removed before the yield load is reached, the structure will behave elastically and return to its original unstressed condition. However, if the yield load is exceeded, some members of the structure will retain a permanent set when the load is removed, thus creating a prestressed condition. Consequently, the structure will have *residual stresses* in it even though no external loads are acting. If the load is applied a second time, the structure will behave in a different manner.

■ **Example 2-21**

The structure shown in Fig. 2-71a consists of a horizontal beam *AB* (assumed to be rigid) supported by two identical bars (bars 1 and 2) made of an elastoplastic material. The bars have length *L* and cross-sectional area *A*, and the material has yield stress σ_y, yield strain ϵ_y, and modulus of elasticity $E = \sigma_y/\epsilon_y$. The beam has length 3*b* and supports a load *P* at end *B*.

(a) Determine the yield load P_y and the corresponding yield displacement δ_y at point *B*. (b) Determine the plastic load P_p and the corresponding plastic displacement δ_p at point *B*. (c) Construct a load-displacement diagram relating the load *P* to the displacement δ_B of point *B*.

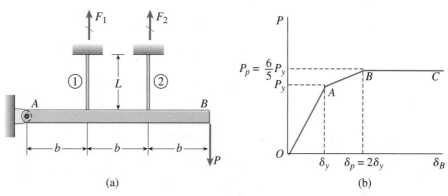

Fig. 2-71 Example 2-21. Elastoplastic analysis of a statically indeterminate structure

(a) (b)

Solution   *Equation of equilibrium.* Because the structure is statically indeterminate, we begin with the equilibrium and compatibility equations. Considering the equilibrium of beam *AB*, we take moments about point *A* and obtain

$$\sum M_A = 0 \qquad F_1(b) + F_2(2b) - P(3b) = 0$$

in which F_1 and F_2 are the axial forces in bars 1 and 2, respectively. This equation simplifies to

$$F_1 + 2F_2 = 3P \tag{g}$$

Equation of compatibility. The compatibility equation is based upon the geometry of the structure. Under the action of the load *P* the rigid beam rotates about point *A*, and therefore the downward displacement at every point along the beam is proportional to its distance from point *A*. Thus, the compatibility equation is

$$\delta_2 = 2\delta_1 \tag{h}$$

where δ_2 is the elongation of bar 2 and δ_1 is the elongation of bar 1.

(a) *Yield load and yield displacement.* When the load *P* is small and the stresses in the material are in the linearly elastic region, the force-displacement relations for the two bars are

$$\delta_1 = \frac{F_1 L}{EA} \qquad \delta_2 = \frac{F_2 L}{EA} \tag{i}$$

Combining these equations with the compatibility condition (Eq. h) gives

$$\frac{F_2 L}{EA} = 2\frac{F_1 L}{EA} \qquad \text{or} \qquad F_2 = 2F_1 \tag{j}$$

Now substituting into the equilibrium equation (Eq. g), we find

$$F_1 = \frac{3P}{5} \qquad F_2 = \frac{6P}{5} \tag{k}$$

Bar 2, which has the larger force, will be the first to reach the yield stress. At that instant the force in bar 2 will be $F_2 = \sigma_y A$. Substituting that value into the second of Eqs. (k) gives the yield load P_y:

$$P_y = \frac{5\sigma_y A}{6} \qquad\qquad (1) \quad \Longleftarrow$$

The corresponding elongation of bar 2 (from the second of Eqs. i) is $\delta_2 = \sigma_y L/E$, and therefore the yield displacement at point B is

$$\delta_y = \frac{3\delta_2}{2} = \frac{3\sigma_y L}{2E} \qquad\qquad (m) \quad \Longleftarrow$$

Both P_y and δ_y are indicated on the load-displacement diagram (Fig. 2-71b).

(b) *Plastic load and plastic displacement.* When the plastic load P_p is reached, both bars will be stretched to the yield stress and both forces F_1 and F_2 will be equal to $\sigma_y A$. It follows from equilibrium (Eq. g) that the plastic load is

$$P_p = \sigma_y A \qquad\qquad (n) \quad \Longleftarrow$$

At this load, the left-hand bar (bar 1) has just reached the yield stress; therefore, its elongation (from the first of Eqs. i) is $\delta_1 = \sigma_y L/E$, and the plastic displacement of point B is

$$\delta_p = 3\delta_1 = \frac{3\sigma_y L}{E} \qquad\qquad (o) \quad \Longleftarrow$$

The ratio of the plastic load to the yield load (see Eqs. n and l) is 6/5, and the ratio of the plastic displacement to the yield displacement (see Eqs. o and m) is 2. These values are also shown on the load-displacement diagram.

(c) *Load-displacement diagram.* The complete load-displacement behavior of the structure is pictured in Fig. 2-71b. The behavior is linearly elastic in the region from O to A, partially plastic from A to B, and fully plastic from B to C.

PROBLEMS/CHAPTER 2

CHANGES IN LENGTH OF AXIALLY LOADED MEMBERS

2.2-1 A steel wire and a copper wire have equal lengths and support equal loads P (see figure). The moduli of elasticity for the steel and copper are $E_s = 29 \times 10^6$ psi and $E_c = 18 \times 10^6$ psi, respectively. (a) If the wires have the same diameters, what is the ratio of the elongation of the copper wire to the elongation of the steel wire? (b) If the wires stretch the same amount, what is the ratio of the diameter of the copper wire to the diameter of the steel wire?

2.2-2 A heavy object is being raised from the sea floor by a tug equipped with a crane and winch (see figure). The steel cable has an effective modulus of elasticity equal to 140 GPa, and the load produces a stress in the cable equal to 290 MPa. If the loaded part of the cable is 90 m long, how much does the cable stretch when the load is picked up?

Copper
wire

Steel
wire

P

Prob. 2.2-1

P

Prob. 2.2-2

Prob. 2.2-4

2.2-3 A 1.0 in. diameter steel cable (see Table 2-1) is used in a construction yard to lift a bridge section weighing 12 tons, as shown in the figure. The cable has an effective modulus of elasticity $E = 20 \times 10^6$ psi. (a) If the cable is 45 ft long, how much will it stretch when the load is picked up? (b) If the cable is rated for a maximum load of 15 tons, what is the factor of safety with respect to failure of the cable?

Prob. 2.2-3

2.2-4 A safety valve on the top of a tank containing steam under pressure p has a discharge hole of diameter d (see figure). The valve is designed to release the steam when the pressure reaches the value p_{max}. If the natural length of the spring is L and its stiffness is k, what should be the dimension h of the valve? (Express your result as a formula for h).

2.2-5 The device shown in the figure consists of a pointer ABC supported by a spring of stiffness k. The spring is positioned at distance b from the pinned end of the pointer. The device is adjusted so that when there is no load P, the pointer reads zero on the angular scale. At what distance x should the load be placed so that the pointer will read 2° on the scale? Use the following data: $k = 8.0$ lb/in., $b = 8$ in., and $P = 3$ lb.

Prob. 2.2-5

2.2-6 A uniform bar AB of weight $W = 25$ N is supported by two springs, as shown in the figure. The spring on the left has stiffness $k_1 = 300$ N/m and natural length $L_1 = 250$ mm. The corresponding quantities for the spring on the right are $k_2 = 400$ N/m and $L_2 = 200$ mm. The distance between the springs is $L = 350$ mm, and the spring on the right is suspended from a support that is distance $h = 80$ mm below the point of support for the spring on the left. At what distance x from the left-hand spring should a load $P = 18$ N be placed in order to bring the bar to a horizontal position?

Prob. 2.2-6

2.2-7 A high-strength steel rod with diameter $d = 0.5$ in. and modulus of elasticity $E = 30 \times 10^6$ psi must transmit a tensile load $P = 6100$ lb (see figure). (a) If the length L of the rod is 14.50 ft, what is its final length? (b) What is the ratio of the length of the rod to the increase in length? (c) If the increase in length is limited to 0.15 in., what is the maximum load P_{max} that can be permitted?

Probs. 2.2-7 and 2.2-8

2.2-8 An aluminum wire having diameter $d = 2$ mm and modulus of elasticity $E = 75$ GPa is subjected to a tensile load P (see figure). (a) If the wire stretches 2.25 mm when the tensile force is 175 N, what is the length L of the wire? (b) If the maximum permissible elongation is 3.0 mm, the allowable stress in tension is 60 MPa, and the length of the wire is 3.8 m, what is the allowable load P_{max}?

2.2-9 A long, rectangular copper bar under a tensile load P hangs from a pin supported by two steel posts (see figure). The copper bar has a length of 26 ft, a cross-sectional area

of 6.5 in.^{2}, and a modulus of elasticity $E_c = 15 \times 10^6$ psi. Each steel post has a height of 35 in., a cross-sectional area of 8.4 in.^{2}, and a modulus of elasticity $E_s = 29 \times 10^6$ psi. (a) Determine the downward displacement δ of the lower end of the copper bar due to a load $P = 85$ k. (b) What is the maximum permissible load P_{max} if the displacement δ is limited to 0.30 in.?

Prob. 2.2-9

2.2-10 The three-bar truss ABC shown in the figure has a span $L = 3$ m and is constructed of steel pipes having cross-sectional area $A = 3540$ mm^2 and modulus of elasticity $E = 200$ GPa. A load P acts horizontally to the right at joint C. (a) If $P = 620$ kN, what is the horizontal displacement of joint B? (b) What is the maximum permissible load P_{max} if the displacement of joint B is limited to 1.0 mm?

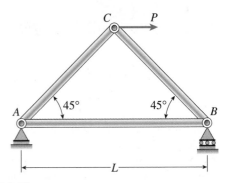

Prob. 2.2-10

2.2-11 A hollow circular column of steel (E = 30,000 ksi) is subjected to a compressive load P, as shown in the figure. The column has length L = 8.2 ft and outside diameter d = 8.0 in. The load P = 110 k. If the allowable compressive stress is 8.1 ksi and the allowable shortening of the column is 0.025 in., what is the required wall thickness t?

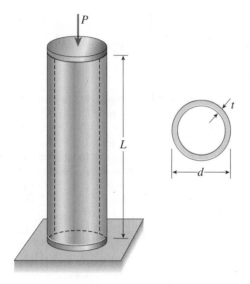

Probs. 2.2-11 and 2.2-12

2.2-12 A hollow circular column of steel (E = 210 GPa) is subjected to a compressive load P, as shown in the figure. The column has length L = 2.5 m and outside diameter d = 200 mm. The load P = 490 kN. If the allowable compressive stress is 56 MPa and the allowable shortening of the column is 0.6 mm, what is the required wall thickness t?

2.2-13 The horizontal rigid beam $ABCD$ is supported by vertical bars BE and CF and is loaded by vertical forces P_1 = 100 k and P_2 = 90 k acting at points A and D, respectively (see figure). Bars BE and CF are made of steel (E = 29.5 × 10⁶ psi) and have cross-sectional areas A_{BE} = 22.1 in.² and A_{CF} = 18.3 in.² The distances between points on the bars are shown in the figure. Determine the vertical displacements δ_A and δ_D of points A and D, respectively.

Prob. 2.2-13

2.2-14 A semicircular plate $ACBD$ of weight W = 500 N and radius R = 1.0 m is supported at points A, B, and D by three identical wires (see figure). Point C is at the center of the circular arc ADB and radius CD is perpendicular to diameter AB. Each wire has modulus of elasticity E = 210 GPa, diameter d = 2 mm, and length L = 1.2 m. Assume that the plate is rigid. A downward load P = $3W$ is placed on radius CD at point p, distance x from point C. What should be the distance x so that the plate will have a slope of 0.1° (along line CD) under the combined action of the force P and the weight W?

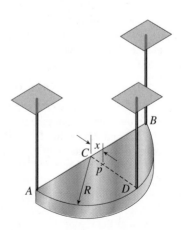

Prob. 2.2-14

NONUNIFORM BARS

2.3-1 Refer to the bar and beam assembly shown in Fig. 2-12a of Example 2-3, and note the loads P_1 and P_2 acting at points C and E, respectively. Assuming that both parts of the vertical bar ABC are made of the same material, obtain a formula for the ratio P_2/P_1 so that the vertical displacement of point C will be zero. (Express your result in terms of the cross-sectional areas A_1 and A_2 and the dimensions L_1, L_2, a, and b shown in the figure.)

2.3-2 A steel bar AD (see figure) has a cross-sectional area of 260 mm² and is loaded by forces $P_1 = 12$ kN, $P_2 = 8$ kN, and $P_3 = 6$ kN. The lengths of the segments of the bar are $a = 1.5$ m, $b = 0.6$ m, and $c = 0.9$ m. (a) Assuming that the modulus of elasticity $E = 210$ GPa, calculate the change in length δ of the bar. Does the bar elongate or shorten? (b) By what amount P should the load P_3 be increased so that end D of the bar does not move when the loads are applied?

Probs. 2.3-2 and 2.3-3

2.3-3 A steel pipe having modulus of elasticity $E = 30 \times 10^6$ psi is loaded by forces P_1, P_2, and P_3 (see figure). The pipe is 12 ft long and the distances between points of loading are $a = b = 3$ ft and $c = 6$ ft. The loads are $P_1 = 4$ k, $P_2 = 3$ k, and $P_3 = 3$ k, and the cross-sectional area of the pipe is 2.8 in.² (a) Calculate the displacements δ_B, δ_C, and δ_D of the points of loading. (b) Determine the distance x from the left-hand support to the cross section at which the displacement is zero.

2.3-4 A two-story building has steel columns AB in the first floor and BC in the second floor (see figure). The roof load P_1 equals 400 kN and the second-floor load P_2 equals 720 kN. Each column has length $a = 3.75$ m. The cross-sectional areas of the first- and second-floor columns are 11,000 mm² and 3,900 mm², respectively. (a) Assuming that $E = 206$ GPa, determine the downward displacement δ_C of point C due to the loads P_1 and P_2. (b) How much additional load P_0 can be placed at the top of the column (point C) if the downward displacement of point C is not to exceed 4.0 mm?

Prob. 2.3-4

2.3-5 A prismatic bar AD is subjected to loads P_1, P_2, and P_3 acting at points B, C, and D, respectively, as shown in the figure. Each segment of the bar is 20 in. long. The bar has cross-sectional area $A = 1.40$ in.² and is made of copper with $E = 17,000$ ksi. (a) Determine the displacement δ_D at the free end of the bar. (b) What should be the load P_3 if it is desired to reduce the displacement at end D to half of its original value?

Prob. 2.3-5

2.3-6 Bar ABC is composed of two materials and has a total length of 1 m and a diameter of 50 mm (see figure). Part AB is steel ($E_s = 210$ GPa) and part BC is aluminum ($E_a = 70$ GPa). The bar is subjected to a tensile force of 110 kN. (a) Determine the lengths L_1 and L_2 for the steel and aluminum parts, respectively, in order that both parts have the same elongation. (b) What is the total elongation δ of the bar?

Prob. 2.3-6

2.3-7 A steel bar 8 ft long has a circular cross section of diameter $d_1 = 0.75$ in. over one-half of its length and diameter $d_2 = 0.50$ in. over the other half (see figure). (a) How much will the bar elongate under a tensile load $P = 5000$

lb? (b) If the same volume of material is made into a bar of constant diameter d and length 8 ft, what will be the elongation under the same load P? (Assume $E = 30 \times 10^6$ psi.)

Prob. 2.3-7

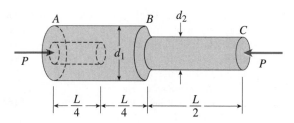

Prob. 2.3-8

2.3-8 A bar ABC of length L consists of two parts of equal lengths but different diameters (see figure). Segment AB has diameter $d_1 = 100$ mm and segment BC has diameter $d_2 = 60$ mm. Both segments have length $L/2 = 0.6$ m. A longitudinal hole of diameter d is drilled through segment AB for one-half of its length (distance $L/4 = 0.3$ m). The bar is made of plastic having modulus of elasticity $E = 4.0$ GPa. Compressive loads $P = 110$ kN act at the ends of the bar. If the shortening of the bar is limited to 8.0 mm, what is the maximum allowable hole diameter d?

2.3-9 A steel pile, driven into the earth, supports a load P entirely by friction along its sides (see figure). The friction force f per unit length of pile is assumed to be uniformly distributed over the surface of the pile. The pile has length L, cross-sectional area A, and modulus of elasticity E. (a) Derive a formula for the shortening δ of the pile in terms of P, L, E, and A. (b) Draw a diagram showing how the compressive stress σ_c varies throughout the length of the pile.

Prob. 2.3-9

2.3-10 A long steel cable AB hanging in a mine shaft supports a load $P = 22.0$ kN at its lower end (see figure). The length of the cable is $L = 450$ m and its weight per unit length is $w = 6.1$ N/m. The strain at the midheight of the cable is measured with a strain gage and found to be 0.00223. (a) What is the strain ϵ_A at the upper end of the cable? (b) What is the strain ϵ_B at the lower end of the cable? (c) What is the elongation δ of the cable?

Prob. 2.3-10

2.3-11 A prismatic bar AB of length L, cross-sectional area A, modulus of elasticity E, and weight W hangs vertically under its own weight (see figure). (a) Derive a formula for the downward displacement δ_C of point C, located at distance c from the lower end of the bar. (b) What is the elongation δ_B of the entire bar? (c) What is the ratio r of the elongation of the upper half of the bar to the elongation of the lower half of the bar?

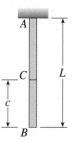

Prob. 2.3-11

2.3-12 A concrete pier AB of square cross section and height L tapers uniformly from top to bottom (see figure). The cross sections at the top and bottom have widths b and $2b$, respectively. (a) Obtain a formula for the shortening δ of the pier under a compressive load P (disregard the weight of the pier itself). (b) Calculate the shortening if $L = 4$ m, $b = 0.8$ m, $P = 1800$ kN, and the modulus of elasticity E of the concrete is 24 GPa.

* **2.3-13** A timber pile supporting a pier develops a friction force f per unit length of pile over the part of its length that is embedded in the earth (see figure). The intensity of the friction force varies linearly with the distance y from the base of the pile, that is, $f = cy$, where c is a constant. The embedment length is L_1 and the upper part of the pile has length L_2. Also, the load on the pile is P, its cross-sectional area is A, and its modulus of elasticity is E. (a) Derive a formula for the shortening δ of the pile in terms of $P, L_1, L_2, E,$ and A. (b) Draw a diagram showing how the compressive stress σ_c varies throughout the length of the pile.

* **2.3-14** A long, slender bar in the shape of a right circular cone with length L and base diameter d hangs vertically under the action of its own weight, as shown in part (a) of the figure. The bar has weight W and modulus of elasticity E. (a) Derive a formula for the elongation δ_c of the conical bar. (b) If the same amount of material is used in a prismatic bar of circular cross section and length L [see part (b) of the figure], what is the elongation?

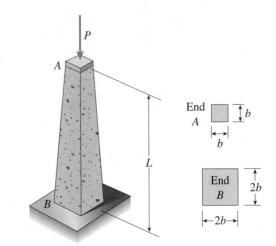

End A $\updownarrow b$

$\leftarrow b \rightarrow$

End B $\updownarrow 2b$

$\leftarrow 2b \rightarrow$

Prob. 2.3-12

Prob. 2.3-13

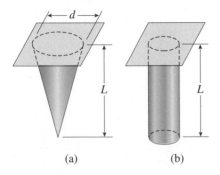

(a)   (b)

Prob. 2.3-14

* **2.3-15** A rigid circular disc of radius R rotates with constant angular speed ω about an axis through its center C (see figure). Six identical prismatic blades, each of length L, project radially outward from the disc. The material of the blades has modulus of elasticity E and mass density ρ (mass per unit of volume). Determine the elongation δ of a blade due to the centrifugal effects.

Prob. 2.3-15

**2.3-16 A tapered aluminum post *AB* of hollow circular cross section and length *L* is compressed by a load *P* (see figure). The outside diameters at the top and bottom of the post are d_A and d_B, respectively, and the wall thickness is *t*. Derive the following formula for the shortening of the post:

$$\delta = \frac{PL}{\pi E t (d_B - d_A)} \ln\left(\frac{d_B - t}{d_A - t}\right)$$

in which *E* is the modulus of elasticity.

Prob. 2.3-16

**2.3-17 The main cables of a suspension bridge [see part (a) of the figure] follow a curve that is nearly parabolic because the primary load on the cables is the weight of the bridge deck, which is uniform in intensity along the horizontal. Therefore, let us represent the central region *AOB* of one of the main cables [see part (b) of the figure] as a parabolic cable supported at points *A* and *B* and carrying a uniform load of intensity *q* along the horizontal. The span of the cable is *L*, the sag is *h*, the axial rigidity is *EA*, and the origin of coordinates is at mid-span. (a) Derive a formula for the elongation δ of cable *AOB* shown in part (b) of the figure. (b) Calculate the elongation δ of the central span of one of the main cables of the Golden Gate Bridge, for which the dimensions and properties are *L* = 4200 ft, *h* = 470 ft, *q* = 12,700 lb/ft, and *E* = 28,800,000 psi. The cable consists of 27,572 parallel wires of diameter 0.196 in. (*Hint:* Determine the tensile force *T* at any point in the cable from a free-body diagram of part of the cable; then determine the elongation of an element of the cable of length *ds*; finally, integrate along the curve of the cable to obtain an equation for the elongation δ.)

STATICALLY INDETERMINATE STRUCTURES

2.4-1 A cylindrical assembly consisting of a brass core and an aluminum collar is compressed by a load *P* (see figure). The length of the aluminum collar and brass core is 14.0 in., the diameter of the core is 1.0 in., and the outside diameter

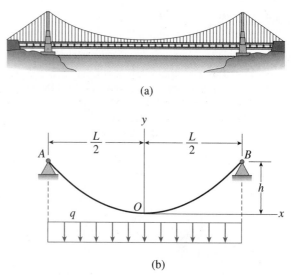

(a)

(b)

Prob. 2.3-17

of the collar is 1.6 in. Also, the moduli of elasticity of the aluminum and brass are 10.5×10^6 psi and 15.0×10^6 psi, respectively. (a) If the length of the assembly decreases by 0.1% when the load *P* is applied, what is the magnitude of the load? (b) What is the maximum permissible load P_{max} if the allowable stresses in the aluminum and brass are 12,000 psi and 18,000 psi, respectively?

Prob. 2.4-1

2.4-2 The assembly shown in the figure consists of a brass core (diameter $d_1 = 6.0$ mm) surrounded by a steel shell (inner diameter $d_2 = 7.0$ mm, outer diameter $d_3 = 9$ mm). A load P compresses the core and shell, which have length $L = 85$ mm. The moduli of elasticity of the brass and steel are 100 GPa and 200 GPa, respectively. (a) What load P will compress the assembly by 0.1 mm? (b) If the allowable stress in the steel is 180 MPa and the allowable stress in the brass is 140 MPa, what is the allowable compressive load P_{allow}?

Prob. 2.4-2

2.4-3 A reinforced concrete column of circular cross section and length $L = 12$ ft carries a compressive load $P = 200$ k (see figure). The column is constructed with eight steel reinforcing bars, each having an effective diameter of 1.0 in. The bars are confined by circular ties to prevent them from buckling outward. The effective area of the concrete has diameter $d_c = 13$ in., as shown in the figure (the outer part of the concrete may break off and therefore is not considered effective in supporting the load). Assume that the modulus of elasticity for steel is 29×10^6 psi and for concrete is 3.6×10^6 psi. (a) Calculate the stresses σ_s and σ_c in the steel and concrete, respectively. (b) Calculate the shortening δ of the column. (*Hint:* When calculating the area of the concrete, be sure to omit the area occupied by the eight reinforcing bars.)

Prob. 2.4-3

2.4-4 Three prismatic bars, two of material A and one of material B, transmit a tensile load P (see figure). The two outer bars (material A) are identical. The cross-sectional area of the middle bar (material B) is 50% larger than the cross-sectional area of one of the outer bars. Also, the modulus of elasticity of material A is twice that of material B. (a) What fraction of the load P is transmitted by the middle bar? (b) What is the ratio of the stress in the middle bar to the stress in the outer bars? (c) What is the ratio of the strain in the middle bar to the strain in the outer bars?

Prob. 2.4-4

2.4-5 Three steel cables jointly support a load of 12 k (see figure). The diameter of the middle cable is 3/4 in. and the diameter of each outer cable is 1/2 in. The tensions in the cables are adjusted so that each cable carries one-third of the load, i.e., 4 k. Later, the load is increased by 9 k to a total load of 21 k. (a) What percent of the total load is now carried by the middle cable? (b) What are the stresses σ_m

and σ_o in the middle and outer cables, respectively? (*Note:* See Table 2-1 in Section 2.2 for properties of cables.)

Prob. 2.4-5

2.4-6 A bar *ACB* having two different cross-sectional areas A_1 and A_2 is held between rigid supports at *A* and *B* (see figure). A load *P* acts at point *C*, which is distance b_1 from end *A* and distance b_2 from end *B*. (a) Obtain formulas for the reactions R_A and R_B at supports *A* and *B*, respectively, due to the load *P*. (b) Obtain a formula for the downward displacement δ_C of point *C*.

Prob. 2.4-6

2.4-7 The aluminum and steel pipes shown in the figure are fastened to rigid supports at ends *A* and *B* and to a rigid plate *C* at their junction. The aluminum pipe is twice as long as the steel pipe. Two equal and symmetrically placed loads *P* act on the plate at *C*. (a) Obtain formulas for the axial stresses σ_a and σ_s in the aluminum and steel pipes, respectively. (b) Calculate the stresses for the following data: $P = 12$ k, cross-sectional area of aluminum pipe $A_a = 8.92$ in.^{2}, cross-sectional area of steel pipe $A_s = 1.03$ in.^{2}, modulus of elasticity of aluminum $E_a = 10 \times 10^6$ psi, and modulus of elasticity of steel $E_s = 29 \times 10^6$ psi.

Prob. 2.4-7

2.4-8 The axially loaded bar *ABCD* shown in the figure is held between rigid supports. The bar has cross-sectional area A_0 from *A* to *C* and $2A_0$ from *C* to *D*. (a) Obtain formulas for the reactions R_A and R_D at the ends of the bar. (b) Determine the displacements δ_B and δ_C at points *B* and *C*, respectively. (c) Draw a diagram in which the abscissa is the distance *x* from the left-hand support and the ordinate is the horizontal displacement of the corresponding point in the bar.

Prob. 2.4-8

2.4-9 A bar AB with fixed ends and two different cross-sectional areas is loaded by two equal and opposite forces P, as shown in the figure. The cross-sectional area near the ends of the bar is A_1 and in the middle of the bar is A_2. (a) Determine the reactions R_A and R_B at the ends of the bar. (b) Determine the axial compressive force F in the middle part of the bar. (c) Under what condition is the compressive stress in the middle of the bar numerically equal to the tensile stress near the ends of the bar?

Prob. 2.4-9

2.4-10 A composite bar of square cross section with dimensions $2b \times 2b$ is constructed of two different materials having moduli of elasticity E_1 and E_2 (see figure). The two parts of the bar have the same cross-sectional dimensions. The bar is compressed by forces P acting through rigid end plates. The line of action of the loads has an eccentricity e of such magnitude that each part of the bar is stressed uniformly in compression. (a) Determine the axial forces P_1 and P_2 in the two parts of the bar. (b) Determine the eccentricity e of the loads. (c) Determine the ratio of the stresses σ_1 and σ_2 in the two parts of the bar.

Prob. 2.4-10

2.4-11 Three vertical cables of the same diameter and same material support a rigid horizontal beam at points A, B, and C (see figure). Cables B and C have length h and cable A has length $2h$. Determine the distance x between wires A and B so that the beam will remain horizontal when a load P is placed at its midpoint.

Prob. 2.4-11

2.4-12 A rigid block AB of weight W hangs on three equally spaced vertical wires, two of steel and one of aluminum (see figure). The wires also support a load P acting at the midpoint of the block. The diameter of the steel wires is 2 mm, and the diameter of the aluminum wire is 4 mm. What load P can be supported if the allowable stress in the steel wires is 220 MPa and in the aluminum wire is 80 MPa? (Assume $W = 800$ N, $E_s = 210$ GPa, and $E_a = 70$ GPa.)

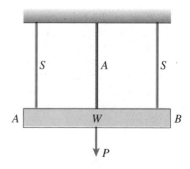

Prob. 2.4-12

2.4-13 A horizontal rigid bar AB weighing $W = 7200$ lb is supported by three slender circular rods (see figure). The two outer rods are made of aluminum ($E_1 = 10 \times 10^6$ psi) with diameters $d_1 = 0.4$ in. and lengths $L_1 = 40$ in. The inner rod is magnesium ($E_2 = 6.5 \times 10^6$ psi) with diameter d_2 and length L_2. The allowable stresses in the aluminum and magnesium are 24,000 psi and 13,000 psi, respectively. If it is desired to have all three rods loaded to their maximum allowable values, what should be the diameter d_2 and length L_2 of the middle rod?

Prob. 2.4-13

2.4-14 A steel bar ABC ($E = 200$ GPa) has cross-sectional area A_1 from A to B and cross-sectional area A_2 from B to C (see figure). The bar is supported at end A and is subjected to a load P equal to 40 kN at end C. A circular steel collar BD having cross-sectional area A_3 supports the bar at B. Determine the displacement δ_C at the lower end of the bar due to the load P, assuming that the collar fits snugly at B when there is no load. (Assume $L_1 = 2L_3 = 250$ mm, $L_2 = 225$ mm, $A_1 = 2A_3 = 960$ mm², and $A_2 = 300$ mm².)

Prob. 2.4-14

2.4-15 A rigid bar AB of length L is hinged to a wall at A and supported by two vertical wires attached at points C and D (see figure). The wires have the same cross-sectional area

A and are made of the same material (modulus E), but the wire at D is twice as long as the wire at C. (a) Find the tensile forces T_C and T_D in the wires due to the vertical load P acting at end B of the bar. (b) Find the downward displacement δ_B at end B of the bar.

Prob. 2.4-15

2.4-16 A rigid bar $ABCD$ is pinned at point B and supported by two springs at A and D (see figure). The springs at A and D have stiffnesses $k_1 = 15$ kN/m and $k_2 = 35$ kN/m, respectively. A load P acts at point C. If the angle of rotation of the bar due to the action of the load P is limited to 2°, what is the maximum permissible load P_{max}?

Prob. 2.4-16

2.4-17 Three identical and equally spaced springs support a horizontal rigid bar ABC (see figure). A load P acts at distance x from point A. (a) Plot a graph showing how the forces F_A, F_B, and F_C in the three springs vary with the distance x. (Let x vary from 0 to $2L$.) (b) For what values of

x are all three springs in compression? (c) For what values of x is the spring at A in tension?

Prob. 2.4-17

2.4-18 A rigid bar AB is supported by two cables CE and BD as shown in the figure. The cables are identical except for their lengths—cable BD has length h and cable CE has length $1.5\,h$ (which means that the length of the bar is $L = h\sqrt{5}$). Determine the tensile forces T_{BD} and T_{CE} in the cables due to the load P acting at point F.

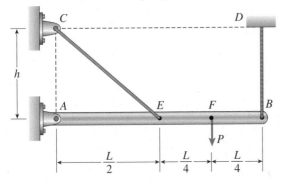

Prob. 2.4-18

2.4-19 A rigid bar BD is pinned at end B and supported by two cables AC and AD (see figure). The length of the bar is $3b$ and point A is height $1.5b$ above point B. A load P acts downward at end D of the bar. (a) Determine the tensile forces T_{AC} and T_{AD} in the cables. (b) Determine the downward displacement δ_D of point D.

Prob. 2.4-19

2.4-20 A rectangular frame of width $7b$ and height $3b$ (see figure) is pivoted at C, supported at A and D by identical vertical wires, and loaded by an inclined force P acting at B. The diameter of the wires is 2 mm. If the tensile stresses in the wires due to the load P are not to exceed 220 MPa, what is the maximum permissible load P_{max}? (Disregard any deformations of the frame.)

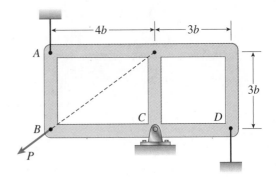

Prob. 2.4-20

2.4-21 A rigid bar EDB is pinned at support E and held by two cables AB and CD (see figure). Both cables are the same except for their lengths. A load P acts at end B of the bar. Determine the tensile forces in the cables and the angle of rotation of the bar.

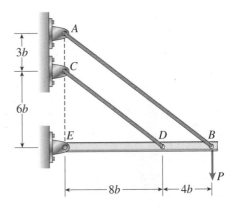

Prob. 2.4-21

*** 2.4-22** A trimetallic bar is uniformly compressed by axial forces P applied through rigid plates (see figure). The bar consists of a circular steel core surrounded by brass and copper tubes. The steel core has diameter 10 mm, the brass tube has outer diameter 15 mm, and the copper tube has outer diameter 20 mm. The corresponding moduli of elasticity are $E_s = 210$ GPa, $E_b = 100$ GPa, and $E_c = 120$ GPa.

Calculate the compressive stresses σ_s, σ_b, and σ_c in the steel, brass, and copper, respectively, when the force P equals 12 kN.

(a)   (b)

Prob. 2.4-22

THERMAL EFFECTS

2.5-1 The rails of a railroad track are welded together at their ends (to form a continuous rail and thus eliminate the clacking sound of the wheels) when the temperature is 50°F. What compressive stress σ is produced in the rails when they are heated by the sun to 120°F if the coefficient of thermal expansion $\alpha = 6.5 \times 10^{-6}/°F$ and the modulus of elasticity $E = 29 \times 10^6$ psi?

2.5-2 A 150 mm diameter steam pipe is laid in a trench at a temperature of 13°C. When steam passes through the pipe, its temperature rises to 120°C. (a) What is the increase Δd in the diameter of the pipe if the pipe is free to expand in all directions? (b) What is the axial stress σ in the pipe if the trench restrains the pipe so that it lengthens only one-third as much as it would if it could expand freely? (*Note:* The pipe is made of steel with modulus of elasticity $E = 200$ GPa and coefficient of thermal expansion $\alpha = 12 \times 10^{-6}/°C$.)

2.5-3 An aluminum pipe has a length of 190 ft at a temperature of 65°F. An adjacent steel pipe at the same temperature is 0.25 in. longer than the aluminum pipe. At what temperatures (degrees Fahrenheit) will the difference in lengths of the two pipes be 0.50 in.? (Assume that the coefficients of thermal expansion of aluminum and steel are $\alpha_a = 13 \times 10^{-6}/°F$ and $\alpha_s = 6.5 \times 10^{-6}/°F$, respectively.)

2.5-4 The apparatus shown in the figure is made of a tungsten bar AC and a magnesium bar BD that are attached to a pointer CDP by pins at C and D. Let the coefficients of thermal expansion for magnesium and tungsten be denoted by α_m and α_t, respectively. Derive a formula for the vertical displacement δ (positive upward) of point P due to a uniform temperature increase ΔT. Could such a device be used as a thermometer?

Prob. 2.5-4

2.5-5 A solid steel cylinder S is placed inside a copper tube C having the same length (see figure). The coefficient of thermal expansion α_c of the copper is larger than the coefficient α_s of the steel. After being assembled, the cylinder and tube are compressed between rigid plates by forces P. Obtain a formula for the increase in temperature ΔT that will cause all of the load to be carried by the copper tube.

Prob. 2.5-5

2.5-6 A bar AB of length L is held between rigid supports and heated nonuniformly in such a manner that the temperature increase ΔT at distance x from end A is given by the expression $\Delta T = \Delta T_1 x^2/L^2$, where ΔT_1 is the increase in temperature at end B of the bar (see figure). Determine the compressive stress σ_c in the bar. (Assume that the material has modulus of elasticity E and coefficient of thermal expansion α.)

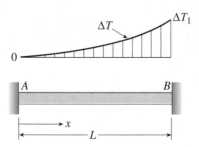

Prob. 2.5-6

2.5-7 A steel rod of diameter 0.375 in. is held snugly (but without any initial stresses) between rigid walls by the arrangement shown in the figure. Calculate the temperature drop ΔT (degrees Fahrenheit) at which the average shear stress in the 0.25 in. diameter bolt becomes 7500 psi. (For steel, use $\alpha_s = 6.5 \times 10^{-6}/°F$ and $E_s = 30 \times 10^6$ psi.)

0.25 in. diameter bolt

0.375 in.

Prob. 2.5-7

2.5-8 A circular steel rod AB (diameter $d_1 = 15$ mm, length $L_1 = 1100$ mm) has a bronze sleeve (outer diameter $d_2 = 21$ mm, length $L_2 = 400$ mm) shrunk onto it so that the two parts are securely bonded together (see figure). Calculate the total elongation δ of the steel bar due to a temperature rise $\Delta T = 350°C$. (Material properties are as follows: for steel, $E_s = 210$ GPa and $\alpha_s = 12 \times 10^{-6}/°C$; for bronze, $E_b = 110$ GPa and $\alpha_b = 20 \times 10^{-6}/°C$.)

Prob. 2.5-8

2.5-9 A plastic bar ACB having two different solid circular cross sections is held between rigid supports as shown in the figure. The diameter in the left- and right-hand parts are 2.0

in. and 3.0 in., respectively. The corresponding lengths are 9 in. and 12 in. Also, the modulus of elasticity E is 860 ksi, and the coefficient of thermal expansion α is $60 \times 10^{-6}/°F$. The bar is subjected to a uniform temperature increase of 50°F. Calculate the following quantities: (a) the compressive force P in the bar; (b) the maximum compressive stress σ_c; and (c) the displacement δ_C of point C.

Prob. 2.5-9

2.5-10 A nonprismatic bar ACB of length L is held between rigid supports (see figure). The left-hand half of the bar has cross-sectional area A_1 and the right-hand half has cross-sectional area A_2. The modulus of elasticity is E and the coefficient of thermal expansion is α. Assuming that the bar is subjected to a uniform temperature increase ΔT and that $A_2 > A_1$, derive expressions for: (a) the compressive force P in the bar, and (b) the displacement δ_C of point C (positive means the displacement is toward the left).

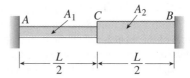

Prob. 2.5-10

2.5-11 A brass sleeve is fitted over a steel bolt (see figure), and the nut is tightened until it is just snug. The bolt has a diameter of 3/8 in., and the sleeve has inside and outside diameters of 13/32 in. and 5/8 in., respectively. Calculate the temperature rise ΔT that is required to produce a compressive stress of 3500 psi in the sleeve. (Use material properties as follows: for brass, $\alpha_b = 11 \times 10^{-6}/°F$ and $E_b = 15 \times 10^6$ psi; for steel, $\alpha_s = 6.5 \times 10^{-6}/°F$ and $E_s = 30 \times 10^6$ psi.)

Brass sleeve

Steel bolt

Prob. 2.5-11

2.5-12 A solid circular bar of aluminum fits inside a tube of copper having the same length (see figure). The outside diameter of the copper tube is 50 mm and the inside diameter is 45 mm. The diameter of the aluminum bar is 42 mm. At each end of the assembly, a metal pin of 8 mm diameter goes through both bars at right angles to the axis. Find the average shear stress in the pins if the temperature is raised 20°C. (For aluminum, $E_a = 70$ GPa and $\alpha_a = 23 \times 10^{-6}/°C$; for copper, $E_c = 120$ GPa and $\alpha_c = 17 \times 10^{-6}/°C$.)

Prob. 2.5-12

2.5-13 A rigid bar $ABCD$ is pinned at end A and supported by two cables at points B and C (see figure). The cable at B has diameter $d_B = 0.5$ in. and the cable at C has diameter $d_C = 0.75$ in. A load P acts at end D of the bar. What is the allowable load P if the temperature rises by 100°F and each cable is required to have a factor of safety of at least 5 against breaking? (*Note:* The cables have effective modulus of elasticity $E = 20 \times 10^6$ psi and coefficient of thermal expansion $\alpha = 6.5 \times 10^{-6}/°F$. Other properties of the cables can be found in Table 2-1, Section 2.2.)

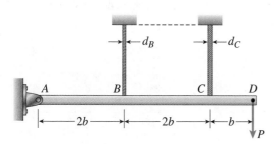

Prob. 2.5-13

2.5-14 A rigid triangular plate is pivoted at C and held by two identical horizontal wires at points A and B (see figure). Each wire has axial rigidity $EA = 540$ kN and coefficient of thermal expansion $\alpha = 23 \times 10^{-6}/°C$. (a) If a vertical load $P = 2.2$ kN acts at point D, what are the tensile forces T_A and T_B in the wires at A and B, respectively? (b) If, while the load P is acting, both wires have their temperatures raised by 100°C, what are the forces T_A and T_B? (c) What increase ΔT in temperature will cause the wire at B to become slack?

Prob. 2.5-14

STRESSES ON INCLINED SECTIONS

2.6-1 A steel bar of square cross section (1.5 in. × 1.5 in.) carries a tensile load P (see figure). The allowable stresses in tension and shear are 18,000 psi and 11,000 psi, respectively. Determine the maximum permissible load P_{max}.

Prob. 2.6-1

2.6-2 A circular steel rod of diameter d is subjected to a tensile force $P = 80$ kN (see figure). The allowable stresses in tension and shear are 120 MPa and 55 MPa, respectively. What is the minimum permissible diameter d_{min} of the rod?

Prob. 2.6-2

2.6-3 A standard brick having dimensions 8 in. × 4 in. × 2.5 in. is compressed lengthwise in a testing machine (see figure). If the ultimate shear stress for brick is 1100 psi and the ultimate compressive stress is 3800 psi, what force P_{max} is required to break the brick?

Prob. 2.6-3

2.6-4 A brass wire of diameter $d = 2$ mm is stretched tightly between rigid supports so that the tensile force is $T = 140$ N (see figure). What is the maximum permissible temperature drop ΔT if the allowable shear stress in the wire is 80 MPa? (The coefficient of thermal expansion for the wire is $19.5 \times 10^{-6}/°C$ and the modulus of elasticity is 105 GPa.)

Probs. 2.6-4 and 2.6-5

2.6-5 A steel wire of diameter $d = 1/16$ in. is stretched between rigid supports with an initial tension T of 45 lb (see figure). (a) If the temperature is then lowered by 50°F, what is the maximum shear stress τ_{max} in the wire? (b) If the allowable shear stress is 18,000 psi, what is the maximum permissible temperature drop? (Assume that the coefficient of thermal expansion is $6.5 \times 10^{-6}/°F$ and the modulus of elasticity is 30×10^6 psi.)

2.6-6 A 20 mm diameter steel bar is subjected to a tensile load $P = 25$ kN (see figure). (a) What is the maximum normal stress σ_{max} in the bar? (b) What is the maximum shear stress τ_{max}? (c) Draw a stress element oriented at 45° to the axis of the bar and show all stresses acting on the faces of this element.

Prob. 2.6-6

2.6-7 During a tension test of a mild-steel specimen (see figure), the extensometer shows an elongation of 0.00150 in. over a gage length of 2 in. Assume that the steel is stressed below the proportional limit and that the modulus of elasticity $E = 29 \times 10^6$ psi. (a) What is the maximum normal stress σ_{max} in the specimen? (b) What is the maximum shear stress τ_{max}? (c) Draw a stress element oriented at an angle of 45° to the axis of the bar and show all stresses acting on the faces of this element.

Prob. 2.6-7

2.6-8 A compression member in a truss is fabricated from a wide-flange steel section (see figure). The cross-sectional area $A = 6640$ mm² and the axial load $P = 410$ kN. Determine the normal and shear stresses acting on all faces of stress elements located in the web of the beam and oriented at: (a) an angle $\theta = 0°$, (b) an angle $\theta = 22.5°$, and (c) an angle $\theta = 45°$. In each case, show the stresses on a sketch of a properly oriented element.

Prob. 2.6-8

2.6-9 A plastic bar of diameter $d = 1.0$ in. is compressed in a testing device by a force $P = 25$ lb applied as shown in the figure. Determine the normal and shear stresses acting on all faces of stress elements oriented at: (a) an angle $\theta = 0°$, (b) an angle $\theta = 22.5°$, and (c) an angle $\theta = 45°$. In each case, show the stresses on a sketch of a properly oriented element.

Prob. 2.6-9

2.6-10 Two boards are joined by gluing along a scarf joint as shown in the figure. For practical reasons, the angle α between the plane of the joint and the faces of the boards must be 45° or less. Under a tensile load P, the normal stress in the boards is 4.9 MPa. (a) What are the normal and shear stresses acting on the glued joint if $\alpha = 20°$? (b) If the allowable shear stress on the joint is 2.25 MPa, what is the largest permissible value of the angle α? (c) For what angle α will the shear stress on the glued joint be numerically equal to twice the normal stress on the joint?

Prob. 2.6-10

2.6-11 A copper bar is held snugly (but without any initial stress) between rigid supports (see figure). The allowable stresses on the inclined plane pq, for which $\theta = 53.13°$, are specified as 7500 psi in compression and 4500 psi in shear. (a) What is the maximum permissible temperature rise ΔT if the allowable stresses on plane pq are not to be exceeded? (Assume $\alpha = 9.4 \times 10^{-6}/°F$ and $E = 18 \times 10^6$ psi.) (b) If the temperature increases by the maximum permissible amount, what are the stresses on plane pq?

Probs. 2.6-11 and 2.6-12

2.6-12 A plastic bar fits snugly between rigid supports at room temperature (20°C) but with no initial stress (see figure). When the temperature of the bar is raised to 70°C, the compressive stress on an inclined plane pq becomes 12.0 MPa. (a) What is the shear stress on plane pq? (Assume $\alpha = 100 \times 10^{-6}/°C$ and $E = 3.0$ GPa.) (b) Draw a stress element oriented to plane pq and show the stresses acting on all faces of this element.

2.6-13 A circular brass bar of diameter d is composed of two segments brazed together on a plane pq making an angle $\alpha = 36°$ with the axis of the bar (see figure). The allowable stresses in the brass are 13,000 psi in tension and 7000 psi in shear. On the brazed joint, the allowable stresses are 6000 psi in tension and 3000 psi in shear. If the bar must resist a tensile force $P = 7000$ lb, what is the minimum required diameter d_{min} of the bar?

Prob. 2.6-13

2.6-14 A prismatic bar is subjected to an axial force that produces a compressive stress of 56 MPa on a plane at an angle $\theta = 30°$ (see figure). Determine the stresses acting on all faces of a stress element oriented at $\theta = 60°$ and show the stresses on a sketch of the element.

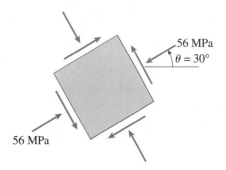

Prob. 2.6-14

2.6-15 A prismatic bar is subjected to an axial force that produces a tensile stress $\sigma_\theta = 10.4$ ksi and a shear stress $\tau_\theta = -5.2$ ksi on a certain inclined plane (see figure). Determine the stresses acting on all faces of a stress element oriented at $\theta = 50°$ and show the stresses on a sketch of the element.

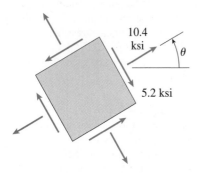

Prob. 2.6-15

2.6-16 Acting on the sides of a stress element cut from a bar in uniaxial stress are normal stresses of 60 MPa and 20 MPa, as shown in the figure. (a) Determine the angle θ and the shear stress τ_θ. (b) What are the maximum normal stress σ_{max} and the maximum shear stress τ_{max} in the material?

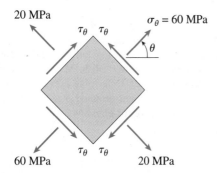

Prob. 2.6-16

* **2.6-17** The normal stress on plane pq of a prismatic bar in tension (see figure) is found to be 8220 psi. On plane rs, which makes an angle $\beta = 30°$ with plane pq, the stress is found to be 3290 psi. Determine the maximum normal stress σ_{max} and maximum shear stress τ_{max} in the bar.

Prob. 2.6-17

* **2.6-18** A tension member is to be made of two pieces of plastic glued along plane pq (see figure). For practical reasons, the angle θ must be between 25° and 45°. The allowable stresses on the glued joint in tension and shear are 13.5 MPa and 9 MPa, respectively. (a) Determine the angle θ so that the bar will carry the largest load P. (Assume that the strength of the glued joint controls the design.) (b) Determine the maximum allowable load P_{max} if the cross-sectional area of the bar is 600 mm².

Prob. 2.6-18

STRAIN ENERGY

The problems for Section 2.7 are to be solved assuming that the material behaves linearly elastically.

2.7-1 A three-story steel column in a building supports roof and floor loads as shown in the figure. The story height H is 10.5 ft, the cross-sectional area of the column is 15.4 in.², and the modulus of elasticity of the steel is 29×10^6 psi. Calculate the strain energy U of the column assuming $P_1 = 41$ k and $P_2 = P_3 = 62$ k.

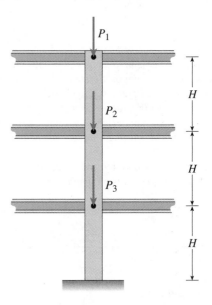

Prob. 2.7-1

2.7-2 A bar of circular cross section having two different diameters d and $2d$ is shown in the figure. Each segment of the bar has length $L/2$ and the modulus of elasticity of the material is E. (a) Obtain a formula for the strain energy U of the bar due to the load P. (b) Calculate the strain energy if the load $P = 27$ kN, the length $L = 600$ mm, the material is brass with $E = 105$ GPa, and the diameter $d = 40$ mm.

Prob. 2.7-2

2.7-3 A bar AD of length L, cross-sectional area A, and modulus of elasticity E is subjected to loads $5P$, $3P$, and P acting at points B, C, and D, respectively (see figure). Segments AB, BC, and CD have lengths $L/6$, $L/2$, and $L/3$, respectively. (a) Obtain a formula for the strain energy U of the bar. (b) Calculate the strain energy if $P = 5$ k, $L = 48$ in., $A = 2.76$ in.², and the material is aluminum with $E = 10.4 \times 10^6$ psi.

Prob. 2.7-3

2.7-4 The bar ABC shown in the figure is loaded by a force P acting at end C and by a force Q acting at the midpoint B. The bar has constant axial rigidity EA. (a) Determine the strain energy U_1 of the bar when the force P acts alone ($Q = 0$). (b) Determine the strain energy U_2 when the force Q acts alone ($P = 0$). (c) Determine the strain energy U when the forces P and Q act simultaneously upon the bar.

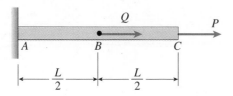

Prob. 2.7-4

2.7-5 The truss ABC shown in the figure supports a horizontal load $P_1 = 2000$ lb and a vertical load $P_2 = 4000$ lb. Both bars have cross-sectional area $A = 2.00$ in.² and are made of steel with $E = 29 \times 10^6$ psi. (a) Determine the strain energy U_1 of the truss when the load P_1 acts alone ($P_2 = 0$). (b) Determine the strain energy U_2 when the load P_2 acts alone ($P_1 = 0$). (c) Determine the strain energy U when both loads act simultaneously.

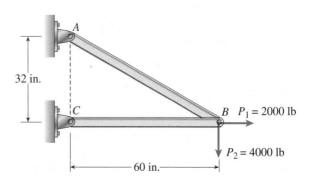

Prob. 2.7-5

2.7-6 A bar in the shape of a circular cone having length L and diameter d at the support hangs vertically under its own weight (see figure). Derive a formula for the strain energy U of the bar. (Let γ = weight density and E = modulus of elasticity of the material.)

Prob. 2.7-6

2.7-7 The truss ABC shown in the figure is subjected to a horizontal load P at joint B. Each bar has cross-sectional area A and modulus of elasticity E. (a) Determine the strain energy U of the truss if $\beta = 60°$. (b) Determine the horizontal displacement δ_B of joint B by equating the strain energy of the truss to the work done by the load.

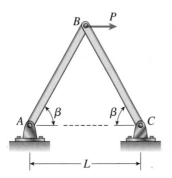

Prob. 2.7-7

2.7-8 The truss $ABCD$ shown in the figure supports a vertical load P at joint D. All bars of the truss have axial rigidity EA. (a) Determine the strain energy U of the truss. (b) Determine the vertical displacement δ_D of joint D by equating the strain energy of the truss to the work done by the load.

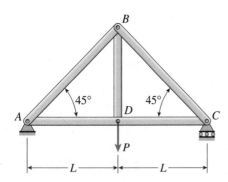

Prob. 2.7-8

2.7-9 The statically indeterminate structure shown in the figure consists of a horizontal bar AB, assumed to be rigid, supported by five equally spaced springs. The two outer springs have stiffness $2k$ and the three inner springs have stiffness k. When unstressed, the lower ends of all five springs lie along a horizontal line. Bar AB, which has weight W, causes the springs to elongate by an amount δ. (a) Determine the total strain energy U of the springs in terms of the downward displacement δ of the bar. (b) Evaluate the

displacement δ by equating the strain energy of the springs to the work done by the weight W. (c) Determine the forces F_1 and F_2 in the inner and outer springs, respectively.

Prob. 2.7-9

2.7-10 A uniformly tapered bar AB of circular cross section is subjected to a load P at the free end, as shown in the figure. The diameters at the ends are d_1 and d_2, the length is L, and the modulus of elasticity is E. (a) Determine the strain energy U of the bar. (b) Determine the elongation δ of the bar by equating the strain energy to the work done by the load.

Prob. 2.7-10

2.7-11 A tapered bar AB of rectangular cross section and length L is acted upon by a force P (see figure). The width of the bar varies uniformly from b_2 at end A to b_1 at end B. The thickness t is constant. (a) Determine the strain energy U of the bar. (b) Determine the elongation δ of the bar by equating the strain energy to the work done by the force P.

Prob. 2.7-11

2.7-12 A bar *ACB* rotates with constant angular speed ω about an axis through *C* (see figure). Determine the strain energy *U* of the bar due to the centrifugal effects. (Let *L* = length of each arm of the bar, *A* = cross-sectional area, *E* = modulus of elasticity, and ρ = mass density.)

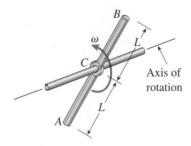

Prob. 2.7-12

2.7-13 A compressive load *P* is transmitted through a rigid plate to three magnesium-alloy bars that are identical except that initially the middle bar is slightly shorter than the other two bars (see figure). The dimensions and properties of the assembly are as follows: length *L* = 2.5 ft, cross-sectional area of each bar *A* = 4.91 in.², modulus of elasticity *E* = 6.5 × 10⁶ psi, and the gap *s* = 0.05 in. (a) Calculate the load P_1 required to close the gap. (b) Determine the downward displacement δ of the rigid plate when *P* = 150 k. (c) Calculate the total strain energy *U* of the three bars when *P* = 150 k. (d) Explain why the strain energy *U* is *not* equal to $P\delta/2$. (*Hint:* Draw a load-displacement diagram.)

Prob. 2.7-13

2.7-14 A block *B* is pushed against three springs by a force *P* (see figure). The middle spring has stiffness k_1 and the outer springs each have stiffness k_2. Initially, the springs are unstressed and the middle spring is slightly longer than the outer springs (the difference in length is denoted *s*).

(a) Draw a force-displacement diagram with the force *P* as ordinate and the displacement *x* of the block as abscissa. (b) From the diagram, determine the strain energy U_1 of the springs when *x* = 2*s*. (c) Explain why the strain energy U_1 is not equal to $P\delta/2$, where δ = 2*s*.

Prob. 2.7-14

2.7-15 The truss *ABC* shown in the figure supports a vertical load *P* at joint *B*. Bars *AB* and *BC* are identical, with modulus of elasticity *E* and cross-sectional area *A*. The angle β can be varied by changing the lengths of the bars, but joint *B* must remain at distance *L* from the vertical wall. (a) Evaluate the strain energy *U* of the truss. (b) Determine the angle β in order that the strain energy will be a minimum. (c) Determine the corresponding minimum strain energy. (d) Determine the corresponding vertical displacement δ_B of joint *B*.

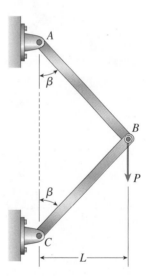

Prob. 2.7-15

* 2.7-16 A vertical load *P* is supported by a statically inde-terminate truss *ABCD* (see figure). All bars are made of the same material (modulus of elasticity *E*) and have the same cross-sectional area *A*. (a) Determine the strain energy *U* of

the truss in terms of the vertical displacement δ_D of joint D.
(b) Determine the displacement δ_D. (c) Determine the forces
F_{AD}, F_{BD}, and F_{CD} in the bars of the truss.

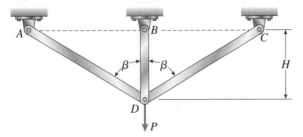

Prob. 2.7-16

**2.7-17 A bungee cord that behaves linearly elastically has
an unstressed length $L_0 = 30$ in. and a stiffness $k =$
0.8 lb/in. The cord is attached to two pegs, distance $b =$
15 in. apart, and pulled at its midpoint by a force $P = 18$ lb
(see figure). (a) How much strain energy U is stored in the
cord? (b) What is the displacement δ_C of the point where the
load is applied? (c) Compare the strain energy U with the
quantity $P\delta_C/2$. (*Note:* The elongation of the cord is *not*
small compared to its original length.)

Prob. 2.7-17

IMPACT LOADING

*The problems for Section 2.8 are to be solved on the basis
of the assumptions and idealizations described in the text
(in particular, no energy is lost during the impact and the
material behaves linearly elastically).*

2.8-1 A sliding collar of weight $W = 150$ lb falls from a
height $h = 2$ in. onto a flange at the bottom of a slender
vertical rod (see figure). The rod has length $L = 4$ ft, cross-
sectional area $A = 0.75$ in.^{2}, and modulus of elasticity $E =$
30×10^6 psi. Calculate the following quantities: (a) the
maximum downward displacement of the flange, (b) the
maximum tensile stress in the rod, and (c) the impact factor.

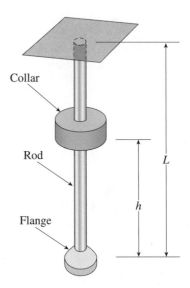

Probs. 2.8-1 and 2.8-2

2.8-2 Solve the preceding problem if the collar has a mass
$M = 80$ kg, the height $h = 0.5$ m, the length $L = 3$ m, the
cross-sectional area $A = 350$ mm^2, and the modulus of elasticity $E = 170$ GPa.

2.8-3 A small rubber ball (weight $W = 1$ oz) is attached by
a rubber cord to a wood paddle (see figure). The natural
length of the cord is $L_0 = 12$ in., its cross-sectional area is
$A = 0.0025$ in.^{2}, and its modulus of elasticity is $E = 300$
psi. After being struck by the paddle, the ball stretches the
cord to a total length $L_1 = 40$ in. What was the velocity ν of
the ball when it left the paddle? (Assume linearly elastic
behavior of the rubber cord, and disregard the potential
energy due to any change in elevation of the ball.)

Prob. 2.8-3

2.8-4 A block weighing $W = 5$ N drops inside a cylinder from a height $h = 200$ mm onto a spring having stiffness $k = 90$ N/m (see figure). (a) Determine the maximum shortening of the spring due to the impact, and (b) determine the impact factor.

Probs. 2.8-4 and 2.8-5

2.8-5 Solve the preceding problem if the block weighs $W = 1$ lb, $h = 12$ in., and $k = 0.5$ lb/in.

2.8-6 A bumper for a mine car is constructed with a spring of stiffness $k = 176$ kN /m (see figure). If a car weighing 14 kN is traveling at velocity $v = 8$ km/h when it strikes the spring, what is the maximum shortening of the spring?

Prob. 2.8-6

2.8-7 A bumping post at the end of a track in a railway yard has a spring constant $k = 35,000$ lb/in. (see figure). The maximum possible displacement d of the end of the striking plate is 1.5 ft. What is the maximum velocity v that a railway car of weight $W = 105,000$ lb can have without damaging the bumping post when it strikes it?

Prob. 2.8-7

2.8-8 A cable with a restrainer at the bottom hangs vertically from its upper end (see figure). The cable has length $L = 12$ m, effective cross-sectional area $A = 40$ mm^2, and effective modulus of elasticity $E = 130$ GPa. A slider of mass $M = 40$ kg drops from a height h onto the restrainer. If the allowable stress in the cable under an impact load is 500 MPa, what is the maximum permissible height h?

Probs. 2.8-8 and 2.8-9

2.8-9 Solve the preceding problem if the slider has weight $W = 100$ lb, $L = 50$ ft, $A = 0.065$ in.^{2}, $E = 20 \times 10^6$ psi, and the allowable stress is 70 ksi.

2.8-10 A weight W rests on top of a wall and is attached to one end of a very flexible cord having cross-sectional area A and modulus of elasticity E (see figure). The other end of the cord is attached securely to the wall. The weight is then pushed off the wall and falls freely the full length of the cord. (a) Derive a formula for the impact factor. (b) Evaluate the impact factor if the weight, when hanging statically, elongates the cord by 2.5% of its original length.

Prob. 2.8-10

2.8-11 A weight $W = 4500$ lb falls from a height $h = 1$ ft onto a vertical wood pole having length $L = 18$ ft, diameter $d = 12$ in., and modulus of elasticity $E = 1.5 \times 10^6$ psi (see figure). Determine the maximum shortening of the pole, the maximum compressive stress in the pole, and the impact factor.

Prob. 2.8-11

2.8-12 The end plate of a pressurized cylinder is clamped to the flange by six bolts as shown in part (a) of the figure. The grip of the bolts is the distance b_1. Under a time varying force $F(t)$ that does a certain amount of work W in straining the bolts, the bolts develop a maximum tensile stress σ_1. Now suppose that the connection is redesigned so that the grip of the bolts is increased to b_2, as shown in part (b) of the figure. The force $F(t)$ does the same amount of work W in straining the bolts and produces a maximum tensile stress equal to σ_2. If $b_2 = 5b_1$, what is the ratio σ_2/σ_1 of the stresses? (*Note:* Bolts subjected to dynamic loads often are designed with long grips to take advantage of the more favorable stress conditions.)

Prob. 2.8-12

2.8-13 A bungee jumper weighing 125 lb leaps from a bridge, braking her fall with a long elastic shock cord having axial rigidity $EA = 480$ lb (see figure). If the jumpoff point is 200 ft above the water, and if it is desired to maintain a clearance of 30 ft between the jumper and the water, what length L of cord should be used?

Prob. 2.8-13

2.8-14 A prismatic bar AB of length L is supported at its midpoint C by a spring of stiffness k (see figure). A heavy object of mass M drops onto end B of the bar from a height h. Using the assumption that bar AB is rigid, derive the following formula for the maximum displacement δ_B of point B due to the impact of the falling object:

$$\delta_B = \frac{4Mg}{k}\left[1 + \left(1 + \frac{kh}{2Mg}\right)^{1/2}\right]$$

Prob. 2.8-14

2.8-15 A rigid bar AB weighing $W = 2$ lb and having length $L = 12$ in. is hinged at end A and supported at B by a nylon cord BC (see figure). The cord has cross-sectional

area $A = 0.04$ in.^{2}, length $b = 9$ in., and modulus of elasticity $E = 300$ ksi. If the bar is raised to its maximum height and then released, what is the maximum stress in the cord?

Prob. 2.8-15

*** 2.8-16** The prismatic bar shown in the figure has length $L = 2.0$ m, diameter $d = 15$ mm, and modulus of elasticity $E = 210$ GPa. A spring of stiffness $k = 1.0$ MN/m is installed at the end of the bar. A sliding collar of mass $M = 20.0$ kg drops from a height $h = 150$ mm onto the spring. Determine the maximum elongation δ_{max} of the bar, the maximum tensile stress σ_{max} in the bar, and the impact factor. Compare your results with those obtained in Example 2-16 of Section 2.8 for this same bar without the spring (see Fig. 2-50).

Prob. 2.8-16

STRESS CONCENTRATIONS

The problems for Section 2.10 are to be solved by considering the stress-concentration factors and assuming linearly elastic behavior.

2.10-1 The flat bars shown in parts (a) and (b) of the figure are subjected to tensile forces $P = 4.0$ k. Each bar has thickness $t = 0.375$ in. (a) For the bar with a circular hole, determine the maximum stresses for hole diameters $d = 1$ in. and $d = 2$ in. if the width $b = 6.0$ in. (b) For the stepped bar with shoulder fillets, determine the maximum stresses for fillet radii $R = 0.25$ in. and $R = 0.5$ in. if the bar widths are $b = 4.0$ in. and $c = 2.5$ in.

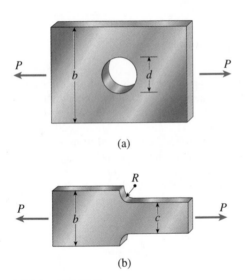

(a)

(b)

Probs. 2.10-1 and 2.10-2

2.10-2 The flat bars shown in parts (a) and (b) of the figure are subjected to tensile forces $P = 8.0$ kN. Each bar has thickness $t = 6.0$ mm. (a) For the bar with a circular hole, determine the maximum stresses for hole diameters $d = 12$ mm and $d = 20$ mm if the width $b = 60.0$ mm. (b) For the stepped bar with shoulder fillets, determine the maximum stresses for fillet radii $R = 6.0$ mm and $R = 10$ mm if the bar widths are $b = 60.0$ mm and $c = 40$ mm.

2.10-3 A flat bar of width b and thickness t has a hole of diameter d drilled through it (see figure). The hole may have

any diameter that will fit within the bar. What is the maximum permissible tensile load P_{max} if the allowable tensile stress in the material is σ_t?

Prob. 2.10-3

2.10-4 A round brass bar of diameter $d_1 = 20$ mm has upset ends of diameter $d_2 = 26$ mm (see figure). The lengths of the segments of the bar are $L_1 = 0.3$ m and $L_2 = 0.1$ m. Quarter-circular fillets are used at the shoulders of the bar, and the modulus of elasticity of the brass is $E = 100$ GPa. If the bar lengthens by 0.12 mm under a tensile load P, what is the maximum stress σ_{max} in the bar?

Probs. 2.10-4 and 2.10-5

2.10-5 Solve the preceding problem for a bar of monel metal having the following properties: $d_1 = 1.0$ in., $d_2 = 1.4$ in., $L_1 = 20.0$ in., $L_2 = 5.0$ in., and $E = 25 \times 10^6$ psi. Also, the bar lengthens by 0.0040 in. when the tensile load is applied.

2.10-6 A prismatic bar of diameter $d_0 = 20$ mm is being compared with a stepped bar of the same diameter ($d_1 = 20$ mm) that is enlarged in the middle region to a diameter $d_2 = 24$ mm (see figure). The radius of the fillets in the stepped bar is 2.0 mm. (a) Does enlarging the bar in the middle region make it stronger than the prismatic bar? Demonstrate your answer by determining the maximum permissible load P_1 for the prismatic bar and the maximum permissible load P_2 for the enlarged bar, assuming that the allowable stress for the material is 110 MPa. (b) What should be the diameter d_0 of the prismatic bar if it is to have the same maximum permissible load as does the stepped bar?

Prob. 2.10-6

2.10-7 A stepped flat bar with a hole (see figure) has widths $b_1 = 1.6$ in., $b_2 = 2.0$ in., and $b_3 = 1.4$ in. Also, the thickness $t = 0.25$ in., the hole diameter $d = 0.7$ in., and the fillets are quarter circles having the largest radii that will fit. (a) If the allowable stress is 6000 psi, what is the maximum permissible load P_{max}? (b) Assume that it is desired to increase the permissible load without changing the thickness t or the width b_2. If only one dimension can be changed (b_1, b_3, or d), which dimension would you change? Would you make the dimension smaller or larger?

Probs. 2.10-7 and 2.10-8

2.10-8 Solve the preceding problem if the dimensions are $b_1 = 75$ mm, $b_2 = 100$ mm, $b_3 = 65$ mm, $t = 15$ mm, and $d = 30$ mm. Also, the allowable stress is 40 MPa.

2.10-9 A flat bar with a hole (see figure) has widths $b = 2.4$ in. and $c = 1.6$ in. The fillets have radii equal to 0.2 in. What is the diameter d_{max} of the largest hole that can be drilled through the bar without reducing the load-carrying capacity?

Prob. 2.10-9

NONLINEAR BEHAVIOR (CHANGES IN LENGTH OF BARS)

2.11-1 A long bar AB of length L and weight density γ hangs vertically under its own weight (see figure). The stress-strain relation for the material is given by the *Ramberg-Osgood equation* (Eq. 2-68):

$$\epsilon = \frac{\sigma}{E} + \frac{\sigma_0 \alpha}{E}\left(\frac{\sigma}{\sigma_0}\right)^m$$

Derive the following formula

$$\delta = \frac{\gamma L^2}{2E} + \frac{\sigma_0 \alpha L}{(m+1)E}\left(\frac{\gamma L}{\sigma_0}\right)^m$$

for the elongation of the bar.

Prob. 2.11-1

2.11-2 A prismatic bar of length $L = 1.8$ m and cross-sectional area $A = 480$ mm^2 is loaded by forces $P_1 = 30$ kN and $P_2 = 60$ kN (see figure). The bar is constructed of magnesium alloy having a stress-strain curve described by the following *Ramberg-Osgood equation:*

$$\epsilon = \frac{\sigma}{45,000} + \frac{1}{618}\left(\frac{\sigma}{170}\right)^{10} \qquad (\sigma = \text{MPa})$$

in which σ has units of megapascals. (a) Calculate the displacement δ_C of the end of the bar when the load P_1 acts alone. (b) Calculate the displacement when the load P_2 acts alone. (c) Calculate the displacement when both loads act simultaneously.

Prob. 2.11-2

2.11-3 A hanging wire of length $L = 8$ ft and diameter $d = 0.0625$ in. supports a load P at its lower end (see figure). The wire is made of a copper alloy having the following *hyperbolic stress-strain relationship:*

$$\sigma = \frac{18,000\epsilon}{1 + 300\epsilon} \qquad 0 \le \epsilon \le 0.03 \qquad (\sigma = \text{ksi})$$

(a) Draw a stress-strain diagram for the material. (b) If the elongation of the wire is limited to 0.75 in. and the maximum stress is limited to 40 ksi, what is the allowable load P?

Prob. 2.11-3

2.11-4 A symmetric two-bar truss ABC supporting a load $P = 60$ kN (see figure) is constructed of a material having the stress-strain diagram shown. Each bar has cross-sectional $A = 265$ mm², and the distance between supports is $L = 3.0$ m. Determine the vertical displacement δ_B of joint B of the truss.

2.11-5 Two identical bars AB and BC support a vertical load P (see figure). The bars are made of an aluminum alloy having properties that may be represented approximately by the *bilinear stress-strain diagram* shown in the figure. The cross-sectional area of each bar is 2.0 in.² and the height h is 75 in. Calculate the vertical displacement δ_B of joint B for each of the following values of the load: $P = 8$ k, 16 k, 24 k, 32 k, and 40 k. From these results, plot a load-displacement diagram for the structure.

Prob. 2.11-4

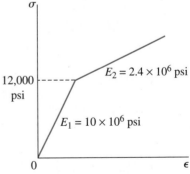

Prob. 2.11-5

2.11-6 A rigid bar AB, pinned at end A, is supported by a wire CD and loaded by a force P at end B (see figure). The wire is made of high-strength steel having modulus of elasticity $E = 210$ GPa and yield stress $\sigma_y = 820$ MPa. The length of the wire is $L = 1$ m and its diameter is $d = 3$ mm. The stress-strain diagram for the steel is defined by the *modified power law*, as follows:

$$\sigma = E\epsilon \qquad 0 \le \sigma \le \sigma_y$$

$$\sigma = \sigma_y \left(\frac{E\epsilon}{\sigma_y} \right)^n \qquad \sigma \ge \sigma_y$$

(a) Assuming $n = 0.2$, calculate the displacement δ_B at the end of the bar due to the load P. Take values of P from 2 to 6 kN in increments of 0.4 kN. (b) Plot a load-displacement diagram showing P versus δ_B.

2.11-7 A mass $M = 1.6$ slugs is attached to the end of a slender bar that rotates at high angular speed about an axis through point C (see figure). The bar has an effective length $L = 40$ in. and cross-sectional area $A = 0.10$ in.^{2} The material is a magnesium alloy having a stress-strain relation given by the *Ramberg-Osgood equation* (Eq. 2-68):

$$\epsilon = \frac{\sigma}{E} + \frac{\sigma_0 \alpha}{E} \left(\frac{\sigma}{\sigma_0} \right)^m$$

in which $E = 6.5 \times 10^6$ psi, $\sigma_0 = 25{,}000$ psi, $\alpha = 3/7$, and $m = 10$. What is the maximum permissible angular speed n (revolutions per minute) if the elongation of the bar is limited to 0.2 in.? (Disregard the mass of the bar itself.)

2.11-8 A long rod hanging vertically in a well supports a load P at its lower end (see figure). The material has the *bilinear stress-strain curve* shown in the figure, in which E_1 and E_2 represent the slopes of the two parts of the diagram. Find the elongation δ of the bar due to its own weight and the force P if the weight density $\gamma = 28$ kN/m^3, the cross-sectional area $A = 960$ mm^2, the length $L = 360$ m, and the load $P = 92$ kN.

NONLINEAR BEHAVIOR (STATICALLY INDETERMINATE STRUCTURES)

*__2.11-9__ The statically indeterminate system shown in the figure consists of a horizontal rigid bar $ABCD$ supported by two identical springs at points A and C. The springs have a nonlinear force-displacement relation given by the following *hyperbolic equation*:

$$F = \frac{60\delta}{1 + 1.2\delta} \qquad 0 \le \delta \le 2.5 \text{ in.} \qquad (F = \text{lb}, \ \delta = \text{in.})$$

Prob. 2.11-6

Prob. 2.11-7

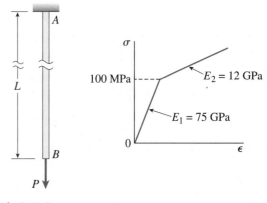

Prob. 2.11-8

where F is the axial force in a spring (units of pounds) and δ is the change in length of a spring (either elongation or shortening, in units of inches). Also, the load P equals 35 lb and the segments of the bar have lengths $a = 15$ in., $b = 10$ in., and $c = 10$ in. (a) Draw the force-displacement diagram for one of the springs. (b) Calculate the forces F_A and F_C in the springs at A and C, respectively, the changes in length δ_A and δ_C of the springs, and the angle of rotation θ of the bar.

Probs. 2.11-9 and 2.11-10

∗ 2.11-10 Solve the preceding problem if the load $P = 150$ N, the segment lengths are $a = 400$ mm, $b = 300$ mm, and $c = 370$ mm, and the force-displacement relation is

$$F = \frac{10.8\delta}{1 + 0.048\delta} \qquad 0 \le \delta \le 60 \text{ mm} \qquad (F = \text{N}, \delta = \text{mm})$$

where F is the axial force in a spring (units of newtons) and δ is the change in length of a spring (units of millimeters).

∗2.11-11 The statically indeterminate system shown in the figure consists of a rigid bar of weight $W = 750$ lb supported by three wires. The outer wires have diameter $d_1 = 100$ mils and length $L_1 = 26$ in.; the inner wire has diameter $d_2 = 100$ mils and length $L_2 = 20$ in. The wires are made of a magnesium alloy having a stress-strain relation that is represented approximately by the following *third-degree equation:*

$$\sigma = (2\epsilon \times 10^6)(3 - 150\epsilon + 2500\epsilon^2) \qquad 0 \le \epsilon \le 0.02$$

in which σ is the stress (units of pounds per square inch) and ϵ is the strain. For strains greater than 0.02, the stress is 40,000 psi. (a) Draw the stress-strain diagram for the material. (b) Determine the downward displacement δ of the bar and the forces, stresses, and strains in the wires.

Probs. 2.11-11 and 2.11-12

∗ 2.11-12 Solve the preceding problem if the weight $W = 3$ kN, the wire diameters are $d_1 = d_2 = 2.5$ mm, the lengths are $L_1 = 650$ mm and $L_2 = 500$ mm, and the stress-strain equation is

$$\sigma = 14{,}000\epsilon \, (3 - 150\epsilon + 2500\epsilon^2) \qquad 0 \le \epsilon \le 0.02$$

in which σ is the stress (units of megapascals) and ϵ is the strain. For strains greater than 0.02, the stress is 280 MPa.

∗ 2.11-13 The statically indeterminate system shown in the figure consists of a rigid bar AB supported by two wires. Wire 1 has diameter $d_1 = 150$ mils and wire 2 has diameter $d_2 = 200$ mils. Both wires have length $L = 35$ in. The wires are made of an aluminum alloy having a stress-strain relation given by the following *Ramberg-Osgood equation:*

$$\epsilon = \frac{\sigma}{10 \times 10^6} + \frac{1}{307.0}\left(\frac{\sigma}{76{,}000}\right)^{20} \qquad (\sigma = \text{psi})$$

in which σ has units of pounds per square inch. The load P is equal to 3000 lb. (a) Draw the stress-strain diagram for the material. (b) Determine the forces F_1 and F_2 in wires 1 and 2, respectively, the elongations δ_1 and δ_2 of the two wires, and the stresses and strains in the wires.

Probs. 2.11-13 and 2.11-14

*2.11-14 Solve the preceding problem if the load $P =$ 13 kN, the wire diameters are $d_1 = 4.0$ mm and $d_2 = 5.0$ mm, the length of the wires is $L = 0.8$ m, and the aluminum alloy has the following stress-strain equation:

$$\epsilon = \frac{\sigma}{70,000} + \frac{1}{314.1}\left(\frac{\sigma}{520}\right)^{20} \quad (\sigma = \text{MPa})$$

in which σ is the stress (units of megapascals) and ϵ is the strain.

ELASTOPLASTIC ANALYSIS

The problems for Section 2.12 are to be solved assuming that the material is elastoplastic with yield stress σ_y, yield strain ϵ_y, and modulus of elasticity E in the linearly elastic region.

2.12-1 A horizontal rigid bar AB supporting a load P is hung from five symmetrically placed wires, each of cross-sectional area A (see figure). The wires are fastened to a curved surface of radius R. (a) Determine the plastic load P_p if the material of the wires is elastoplastic with yield stress σ_y. (b) How is P_p changed if bar AB is flexible instead of rigid? (c) How is P_p changed if the radius R is increased?

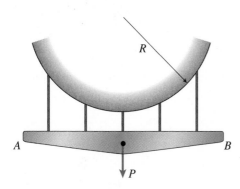

Prob. 2.12-1

2.12-2 A stepped bar ACB is held between rigid supports and loaded by an axial force P at midlength (see figure). The diameters for the two parts of the bar are $d_1 = 20$ mm and $d_2 = 25$ mm, and the material is elastoplastic with yield stress $\sigma_y = 275$ MPa. (a) Determine the yield load P_y. (b) Determine the plastic load P_p.

Prob. 2.12-2

2.12-3 A bar $ABCD$ of total weight $W = 900$ lb is suspended in a horizontal position by three steel wires arranged as shown in the figure. Each wire has diameter $d = 0.242$ in. and the steel is elastoplastic with yield stress $\sigma_y = 40,000$ psi. What load P will cause the system to collapse due to yielding of the wires?

Prob. 2.12-3

2.12-4 A circular steel rod AB of diameter $d = 12$ mm is stretched tightly between two supports so that initially the tensile stress in the rod is 50 MPa (see figure). An axial force P is then applied to the rod at an intermediate location C. (a) Determine the plastic load P_p if the material is elastoplastic with yield stress $\sigma_y = 250$ MPa. (b) How is P_p changed if the initial tensile stress is increased to 100 MPa?

Prob. 2.12-4

2.12-5 A load P is carried by a horizontal beam that is supported by four rods arranged in the symmetrical pattern shown in the figure. Each rod has cross-sectional area A and is made of an elastoplastic material having yield stress σ_y. Obtain a formula for the plastic load P_p.

2.12-6 Five bars, each having a diameter of 10 mm, support a load P as shown in the figure. Determine the plastic load P_p if the material is elastoplastic with yield stress $\sigma_y = 290$ MPa.

2.12-7 A rigid bar ACB is supported on a fulcrum at C and loaded by a force P at end B (see figure). Three identical wires made of an elastoplastic material (yield stress σ_y and modulus of elasticity E) support the bar. Each wire has cross-sectional area A and length L. (a) Determine the yield load P_y and the corresponding yield displacement δ_y at point B. (b) Determine the plastic load P_p and the corresponding displacement δ_p of point B when the load just reaches the value P_p. (c) Draw a load-displacement diagram showing the load P versus the displacement δ_B of point B.

2.12-8 The truss $ABCD$ (see figure) is constructed of three steel rods (elastoplastic material) each having cross-sectional area A and yield stress σ_y. Determine the allowable value of the load P using a factor of safety n with respect to the plastic load.

Prob. 2.12-5

Prob. 2.12-6

Prob. 2.12-8

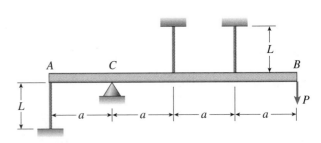

Prob. 2.12-7

*2.12-9 The structure shown in the figure consists of a horizontal rigid bar AB supported by two steel wires, one of length L and the other of length $2L$. Both wires have cross-sectional area A and are made of elastoplastic material with yield stress σ_y and modulus of elasticity E. A vertical load P acts at end B of the bar. (a) Determine the yield load P_y and the corresponding displacement δ_y of point B. (b) Determine the plastic load P_p and the corresponding displacement δ_p of point B. (c) Draw a load-displacement diagram with the load P as ordinate and the displacement δ_B of point B as abscissa.

Prob. 2.12-9

* 2.12-10 Two cables, each having a length L of approximately 40 m, support a loaded container of total weight W (see figure). The cables, which have effective cross-sectional area $A = 48.0$ mm² and effective modulus of elasticity $E = 160$ GPa, are identical except that one cable is longer than the other when they are hanging separately and unloaded. The difference in lengths is $d = 100$ mm. The cables are made of steel having an elastoplastic stress-strain diagram with $\sigma_y = 500$ MPa. Assume that the weight W is slowly increased by the addition of more material to the container. (a) Determine the weight W_y that first produces yielding of the shorter cable. Also, determine the corresponding elongation δ_y of the shorter cable. (b) Determine the weight W_p that produces yielding of both cables. Also, determine the elongation δ_p of the shorter cable when the weight W just reaches the value W_p. (c) Construct a

load-displacement diagram showing the weight W as ordinate and the elongation δ of the shorter cable as abscissa. (*Hint:* The load-displacement diagram is not a single straight line in the region $0 \le W \le W_y$.)

Prob. 2.12-10

* 2.12-11 A hollow circular tube T of length $L = 15$ in. is uniformly compressed by a force P acting through a rigid plate (see figure). The outside and inside diameters of the tube are 3.0 and 2.75 in., respectively. A concentric solid circular bar B of 1.5 in. diameter is mounted inside the tube. When no load is present, there is a clearance $c = 0.010$ in. between the bar B and the rigid plate. Both bar and tube are made of steel having an elastoplastic stress-strain diagram with $E = 29 \times 10^3$ ksi and $\sigma_y = 36$ ksi. (a) Determine the yield load P_y and the corresponding shortening δ_y of the tube. (b) Determine the plastic load P_p and the corresponding shortening δ_p of the tube. (c) Construct a load-displacement diagram showing the load P as ordinate and the shortening δ of the tube as abscissa. (*Hint:* The load-displacement diagram is not a single straight line in the region $0 \le P \le P_y$.)

Prob. 2.12-11

** 2.12-12 Three steel bars, arranged symmetrically, support a load P (see figure). The cross-sectional area of the vertical bar is twice that of the inclined bars. Assuming elastoplastic behavior, determine the angle β so that the ratio of the plastic load P_p to the yield load P_y is a maximum.

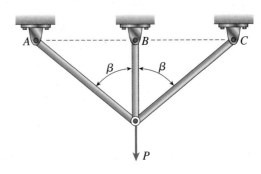

Prob. 2.12-12

3

TORSION

INTRODUCTION

In Chapters 1 and 2 we discussed the behavior of the simplest type of structural member—namely, a straight bar subjected to axial loads. Now we consider a slightly more complex type of behavior known as **torsion**. Torsion refers to the twisting of a straight bar when it is loaded by moments (or torques) that tend to produce rotation about the longitudinal axis of the bar. For instance, when you turn a screwdriver (Fig. 3-1a), your hand applies a torque T to the handle (Fig. 3-1b) and twists the shank of the screwdriver. Other examples of bars in torsion are drive shafts, axles, propeller shafts, steering rods, and drill bits.

An idealized case of torsional loading is pictured in Fig. 3-2a, which shows a straight bar supported at one end and loaded by two pairs of equal and opposite forces. The first pair consists of the forces P_1 acting near the midpoint of the bar and the second pair consists of the forces P_2 acting at the end. Each pair of forces forms a couple that tends to twist the bar about its longitudinal axis. As we know from statics, the moment of a couple is equal to the product of one of the forces and the perpendicular distance between the lines of action of the forces; thus, the first couple has a moment $T_1 = P_1 d_1$ and the second has a moment $T_2 = P_2 d_2$. Typical USCS units for moment are the pound-foot (lb-ft) and the pound-inch (lb-in.); the SI unit is the newton meter (N·m).

For convenience we often represent the moment of a couple by a vector in the form of a double-headed arrow (Fig. 3-2b). The arrow is perpendicular to the plane containing the couple, and therefore in this case both arrows are parallel to the axis of the bar. The direction (or

(a)

(b)

Fig. 3-1 Torsion of a screwdriver due to a torque T applied to the handle

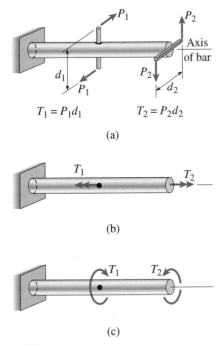

$$T_1 = P_1 d_1 \qquad T_2 = P_2 d_2$$

(a)

(b)

(c)

Fig. 3-2 Bar subjected to torsion by twisting moments T_1 and T_2

sense) of the moment is indicated by the *right-hand rule* for moment vectors—namely, using your right hand, let your fingers curl in the direction of the moment, and then your thumb will point in the direction of the vector. An alternative representation of a moment is a curved arrow acting in the direction of rotation (Fig. 3-2c). Both representations are in common use, and both are used in this book. The choice depends upon convenience and personal preference.

Moments that produce twisting of a bar, such as the moments T_1 and T_2 in Fig. 3-2, are called **torques** or **twisting moments**. Cylindrical members that transmit power through rotation are called **shafts**; for instance, the drive shaft of an automobile or the propeller shaft of a ship. Most shafts have circular cross sections, either solid or tubular.

In this chapter we begin by developing formulas for the deformations and stresses in circular bars subjected to torsion. We then analyze the state of stress known as *pure shear* and obtain the relationship between the moduli of elasticity E and G in tension and in shear, respectively. Next, we analyze rotating shafts and determine the power they transmit. Finally, we cover several additional topics related to torsion, namely, statically indeterminate members, strain energy, thin-walled tubes of noncircular cross section, stress concentrations, and nonlinear behavior.

3.2 TORSIONAL DEFORMATIONS OF A CIRCULAR BAR

We begin our discussion of torsion by considering a prismatic bar of circular cross section twisted by torques T acting at the ends (Fig. 3-3). Since every cross section of the bar is identical, and since every cross section is subjected to the same internal torque T, we say that the bar is in **pure torsion**. From considerations of symmetry, it can be proved that cross sections of the bar do not change in shape as they rotate about the longitudinal axis. In other words, all cross sections remain plane and circular and all radii remain straight. Furthermore, if the angle of rotation between one end of the bar and the other is small, neither the length of the bar nor its radius will change.

To aid in visualizing the deformation of the bar, imagine that the left-hand end of the bar shown in Fig. 3-3 is fixed in position. Then, under the action of the torque T, the right-hand end will rotate (with respect to the left-hand end) through a small angle ϕ, known as the **angle of twist**. Because of this rotation, a straight longitudinal line pq on the surface of the bar will become a helical curve pq', where q' is the position of point q after the end cross section has rotated through the angle ϕ. The angle of twist changes along the axis of the bar, and at intermediate cross sections it will have a value $\phi(x)$ that is between zero at the left-hand end and ϕ at the right-hand end. If every cross section of the bar has the same radius and is subjected to the same torque (pure torsion), the angle $\phi(x)$ will vary linearly between the ends.

Fig. 3-3 Deformations of a circular bar in pure torsion

Now consider an element of the bar between two cross sections distance dx apart (Fig. 3-4a). This element is shown isolated in Fig. 3-4b. On its outer surface we identify a small element *abcd*, with sides *ab* and *cd* that initially are parallel to the longitudinal axis. During twisting of the bar, the right-hand cross section rotates with respect to the left-hand cross section through a small angle of twist $d\phi$, so that points *b* and *c* move to *b'* and *c'*, respectively. The lengths of the sides of the element, which is now element *ab'c'd*, do not change during this small rotation. However, the angles at the corners are no longer equal to 90°. The element is therefore in a state of **pure shear** (see Section 1.6), and the magnitude of the shear strain γ_{max} (Fig. 3-4b) is equal to the decrease in the angle at point *a*, that is, the decrease in angle *bad*. From the figure, we see that the decrease in this angle is

$$\gamma_{max} = \frac{bb'}{ab}$$

where γ_{max} is measured in radians, *bb'* is the distance through which point *b* moves, and *ab* is the length of the element (equal to dx). With *r* denoting the radius of the bar, we can express the distance *bb'* as $r d\phi$, where $d\phi$ also is measured in radians. Thus, the preceding equation becomes

$$\gamma_{max} = \frac{r\,d\phi}{dx} \tag{a}$$

This equation relates the shear strain at the outer surface of the bar to the angle of twist.

(a)

Fig. 3-4 Deformation of an element of length dx cut from a bar in torsion

(b)   (c)

The quantity $d\phi/dx$ is the rate of change of the angle of twist ϕ with respect to the distance x measured along the axis of the bar. We will denote $d\phi/dx$ by the symbol θ and refer to it as the **angle of twist per unit length**, or the **rate of twist**:

$$\theta = \frac{d\phi}{dx} \tag{3-1}$$

With this notation, we can now write the equation for the shear strain at the outer surface (Eq. a) as follows:

$$\gamma_{max} = \frac{r \, d\phi}{dx} = r\theta \tag{3-2}$$

In general, both ϕ and θ vary with the distance x along the axis of the bar. However, in the special case of pure torsion, the rate of twist is constant and equal to the total angle of twist ϕ divided by the length L of the bar, that is, $\theta = \phi/L$. Therefore, for pure torsion only, we obtain

$$\gamma_{max} = r\theta = \frac{r\phi}{L} \tag{3-3}$$

This equation can be obtained directly from Fig. 3-3 by noting that the distance qq' equals $r\phi$ and that the angle γ_{max} is the angle between lines pq and pq', that is, the angle qpq'. Thus, $r\phi = \gamma_{max}L$.

The shear strains within the interior of the bar can be found by the same method used to find the shear strain γ_{max} at the surface. Because radii in the cross sections of a bar remain straight and undistorted during twisting, we see that the preceding discussion for an element $abcd$ at the outer surface (Fig. 3-4b) will also hold for a similar element situated on the surface of an interior cylinder of radius ρ (Fig. 3-4c). Thus, interior elements are also in pure shear with the corresponding shear strains given by the equation (compare with Eqs. 3-2 and 3-3):

$$\gamma = \rho\theta = \frac{\rho}{r}\gamma_{max} \tag{3-4}$$

This equation shows that the shear strains in a circular bar vary linearly with the radial distance ρ from the center, with the strain being zero at the center and reaching a maximum value at the outer surface.

A review of the preceding discussions will show that the equations for the shear strains (Eqs. 3-1 to 3-4) apply to circular tubes (Fig. 3-5) as well as to solid circular bars. Figure 3-5 shows the linear variation in shear strain between the maximum strain at the outer surface and the minimum strain at the interior surface. The minimum strain is related to the maximum strain by the equation

$$\gamma_{min} = \frac{r_1}{r_2}\gamma_{max} \tag{3-5}$$

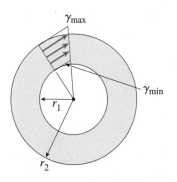

Fig. 3-5 Variation of shear strains in a circular tube

in which r_1 and r_2 are the inner and outer radii, respectively, of the tube.

Finally, we observe that the preceding equations for the strains in a circular bar are based only upon geometric concepts. Therefore, the equations are valid for any material, whether it behaves elastically or inelastically, linearly or nonlinearly. However, the equations are limited to bars having small angles of twist and small strains.

3.3 CIRCULAR BARS OF LINEARLY ELASTIC MATERIALS

Now that we have investigated the shear strains in a circular bar in torsion (Figs. 3-4 and 3-5), we are ready to determine the directions and magnitudes of the corresponding shear stresses. The directions of the stresses can be determined by inspection, as illustrated in Fig. 3-6a. We observe that the torque T tends to rotate the right-hand end of the bar counterclockwise when viewed from the right, and therefore the shear stresses τ act in the directions shown in the figure.

For clarity, the stress element on the side of the bar (Fig. 3-6a) is enlarged in Fig. 3-6b, where both the shear strain and the shear stresses are shown. As explained previously in Section 2.6, we customarily draw stress elements in two dimensions, as in Fig. 3-6b, but we must always remember that stress elements are actually three-dimensional objects with a thickness perpendicular to the plane of the figure.

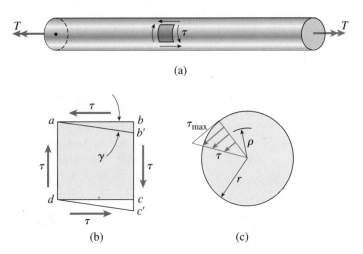

(a)

(b)　　　　　　　　(c)

Fig. 3-6 Shear stresses in a circular bar in torsion

The magnitudes of the shear stresses can be determined from the stress-strain relation for the material of the bar. If the material is linearly elastic, we can use Hooke's law in shear (see Eq. 1-17):

$$\tau = G\gamma \tag{3-6}$$

in which G is the shear modulus of elasticity and γ is the shear strain in radians. Combining this equation with the equations for the shear strains (Eqs. 3-2 and 3-4), we get

$$\tau_{max} = Gr\theta \qquad \tau = G\rho\theta = \frac{\rho}{r}\tau_{max} \tag{3-7a,b}$$

in which τ_{max} is the shear stress at the outer surface of the bar (radius r), τ is the shear stress at an interior point (radius ρ), and θ is the rate of twist (note that θ has units of radians per unit of length). These equations show that the shear stresses vary linearly with the distance from the center of the bar, as illustrated by the triangular stress diagram in Fig. 3-6c. (This linear variation of stress is a consequence of Hooke's law. If the stress-strain relation is nonlinear, other methods of analysis are needed, as discussed later in Section 3.12.)

The shear stresses acting on a cross-sectional plane are accompanied by shear stresses of the same magnitude acting on longitudinal planes (Fig. 3-7). This conclusion follows from the fact that equal shear stresses always exist on mutually perpendicular planes, as explained in Section 1.6. If the material of the bar is weaker in shear on longitudinal planes than on cross-sectional planes, as is typical of wood when the grain runs parallel to the axis of the bar, the first cracks due to torsion will appear on the surface in the longitudinal direction.

The state of pure shear at the surface of a bar (Fig. 3-6b) is equivalent to equal tensile and compressive stresses acting on an element oriented at an angle of 45°, as explained later in Section 3.5. Therefore, a rectangular element with sides at 45° to the axis of the shaft will be subjected to tensile and compressive stresses, as shown in Fig. 3-8. If a torsion bar is made of a material that is weaker in tension than in shear, failure will occur in tension along a helix inclined at 45° to the axis, as you can demonstrate by twisting a piece of classroom chalk.

Fig. 3-7 Longitudinal and transverse shear stresses in a circular bar

Fig. 3-8 Tensile and compressive stresses acting on a stress element oriented at 45° to the longitudinal axis

■ The Torsion Formula

The next step in our analysis is to determine the relationship between the shear stresses and the torque T. Once this is accomplished, we will be able to calculate the stresses and strains in a bar due to any set of applied torques.

The distribution of the shear stresses acting on a cross section is pictured in Figs. 3-6c and 3-7. Because these stresses act continuously around the cross section, they have a resultant in the form of a moment—a moment equal to the torque T acting on the bar. To determine this resultant, we consider an element of area dA located at radial distance ρ from the axis of the bar (Fig. 3-9). The shear force acting on this element is equal to $\tau\, dA$, where τ is the shear stress at radius ρ. The moment of this force about the axis of the bar is equal to the force times its distance from the center, or $\tau\rho\, dA$. Substituting for the shear stress τ from Eq. (3-7b), we can express this elemental moment as

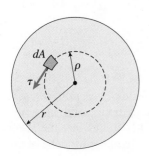

Fig. 3-9 Determination of the resultant of the shear stresses acting on a cross section

$$dM = \tau\rho\, dA = \frac{\tau_{max}}{r}\, \rho^2 dA$$

The resultant moment (equal to the torque T) is the summation over the entire cross-sectional area of all such elemental moments:

$$T = \int_A dM = \frac{\tau_{max}}{r} \int_A \rho^2\, dA = \frac{\tau_{max}}{r} I_p \tag{3-8}$$

in which

$$I_p = \int_A \rho^2 \, dA \tag{3-9}$$

is the **polar moment of inertia** of the circular cross section. For a circle of radius r and diameter d, the polar moment of inertia is

$$I_p = \frac{\pi r^4}{2} = \frac{\pi d^4}{32} \tag{3-10}$$

as given in Appendix D, Case 9. Note that moments of inertia have units of length to the fourth power.*

An expression for the maximum shear stress can be obtained by rearranging Eq. (3-8), as follows:

$$\tau_{max} = \frac{Tr}{I_p} \tag{3-11}$$

This equation, known as the **torsion formula**, shows that the shear stress is proportional to the applied torque T and inversely proportional to the polar moment of inertia I_p.

Typical units used with the torsion formula are as follows. In SI, the torque T is usually expressed in newton meters (N·m), the radius r in meters (m), the polar moment of inertia I_p in meters to the fourth power (m^4), and the shear stress τ in pascals (Pa). If USCS units are used, T is often expressed in pound-feet (lb-ft) or pound-inches (lb-in.), r in inches (in.), I_p in inches to the fourth power (in.^{4}), and τ in pounds per square inch (psi).

Substituting $r = d/2$ and $I_p = \pi d^4/32$ into the torsion formula, we get the following equation for the maximum stress:

$$\tau_{max} = \frac{16T}{\pi d^3} \tag{3-12}$$

This equation applies only to bars of *solid circular cross section*, whereas the torsion formula itself (Eq. 3-11) applies to both solid bars and circular tubes, as explained later. Equation (3-12) shows that the shear stress is inversely proportional to the cube of the diameter (or cube of the

---

*Polar moments of inertia are discussed in Section 12.6 of Chapter 12.

radius). Thus, if the diameter is doubled, the stress is reduced by a factor of eight.

Knowing both the torque and the diameter of the bar, we can use either the standard torsion formula (Eq. 3-11) or the alternative version (Eq. 3-12) to calculate the maximum shear stress. If an allowable shear stress is given, we can use Eq. (3-12) to find either the allowable torque (if the diameter is known) or the required diameter (if the torque is known). These ideas are illustrated in Examples 3-1 and 3-2 at the end of the section.

The shear stress at distance ρ from the center of the bar is

$$\tau = \frac{\rho}{r}\,\tau_{\text{max}} = \frac{T\rho}{I_p} \tag{3-13}$$

which is obtained by combining Eq. (3-7b) with the torsion formula (Eq. 3-11). Once again we see that the shear stresses vary linearly with the radial distance from the center of the bar.

■ Angle of Twist

The angle of twist of a bar of linearly elastic material can now be related to the applied torque T. Combining Eq. (3-7a) with the torsion formula, we get

$$\theta = \frac{T}{GI_p} \tag{3-14}$$

which shows that the rate of twist θ is directly proportional to the torque T and inversely proportional to the product GI_p, known as the **torsional rigidity** of the bar. For a bar in pure torsion, the total angle of twist ϕ, equal to the rate of twist times the length of the bar (that is, $\phi = \theta L$), is

$$\phi = \frac{TL}{GI_p} \tag{3-15}$$

in which ϕ is measured in radians. The use of these equations in both analysis and design is illustrated later in Examples 3-1 and 3-2.

The quantity GI_p /L, called the **torsional stiffness** of the bar, is the torque required to produce a unit angle of rotation. The **torsional flexibility** is the reciprocal of the stiffness, or L/GI_p, and is defined as the angle of rotation produced by a unit torque. Thus, we have the following expressions:

$$k_T = \frac{GI_p}{L} \qquad f_T = \frac{L}{GI_p} \qquad \text{(3-16a,b)}$$

These quantities are analogous to the axial stiffness $k = EA/L$ and axial flexibility $f = L/EA$ of a bar in tension or compression (compare with Eqs. 2-4a and 2-4b). Stiffnesses and flexibilities are important in structural analysis, especially when the structure is large and complex so that computer-oriented methods of analysis are being used.

The equation for the angle of twist (Eq. 3-15) may be used to determine the shear modulus of elasticity G for a material. By conducting a torsion test on a circular bar, we can measure the angle of twist ϕ produced by a known torque T. Then the value of G can be calculated from Eq. (3-15).

■ Circular Tubes

Circular tubes are more efficient than solid bars in resisting torsional loads. As we know, the shear stresses in a solid circular bar are maximum at the outer boundary of the cross section and zero at the center. Therefore, most of the material in a solid shaft is stressed significantly below the maximum shear stress. Furthermore, the stresses near the center of the cross section have a smaller moment arm ρ when determining the torque. By contrast, in a typical hollow tube most of the material is near the outer boundary of the cross section where both the shear stresses and the moment arms are highest (Fig. 3-10). Thus, if weight reduction and savings of material are important, it is advisable to use a circular tube. For instance, large drive shafts, propeller shafts, and generator shafts usually have hollow circular cross sections.

The analysis of the torsion of a circular tube is almost identical to that for a solid bar. The same basic expressions for the shear stresses may still be used (for instance, Eqs. 3-7a and 3-7b). Of course, the radial distance ρ is limited to the range r_1 to r_2, where r_1 is the inner radius and r_2 is the outer radius of the bar (Fig. 3-10).

The relationship between the torque T and the maximum stress is still given by Eq. (3-8), but the limits on the integral for the polar moment of inertia (Eq. 3-9) are $\rho = r_1$ and $\rho = r_2$. Therefore, the polar moment of inertia of the cross-sectional area of a tube is

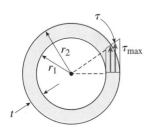

Fig. 3-10 Circular tube in torsion

$$I_p = \frac{\pi}{2}(r_2^4 - r_1^4) = \frac{\pi}{32}(d_2^4 - d_1^4) \qquad \text{(3-17)}$$

If the wall thickness t (Fig. 3-10) is small compared to the radius, the following approximate formula (see Appendix D, Case 22) may be used for the polar moment of inertia:

$$I_p \approx 2\pi r^3 t = \frac{\pi d^3 t}{4} \qquad \text{(3-18)}$$

In this equation, the quantities r and d are the *average* radius and diameter, respectively. As a measure of the accuracy of Eq. (3-18), we note that if the thickness t is less than one-fourth of the outer radius $(t/r_2 < 1/4)$, the error in the approximate formula is less than 2 percent.

The torsion formula (Eq. 3-11) may be used for a circular tube of linearly elastic material provided I_p is evaluated according to Eq. (3-17) or, if appropriate, Eq. (3-18). The same comment applies to the general equation for shear stress (Eq. 3-13), the equations for rate of twist and angle of twist (Eqs. 3-14 and 3-15), and the equations for stiffness and flexibility (Eqs. 3-16a and 3-16b).

The shear stress distribution in a tube is pictured in Fig. 3-10. From the figure, we see that the average stress in a thin tube is nearly as great as the maximum stress. This means that a hollow bar is more efficient in the use of material than is a solid bar, as demonstrated later in Examples 3-2 and 3-3.

When designing a circular tube to transmit a torque, we must be sure that the thickness t is large enough to prevent wrinkling or buckling of the wall of the tube. For instance, a maximum value of the radius to thickness ratio, such as $(r_2/t)_{max} = 10$ or 20, may be specified. Other design considerations include environmental and durability factors, which also may impose requirements for minimum wall thickness. These topics are discussed in courses and textbooks on mechanical design.

■ Limitations

The equations derived in this section are limited to bars of circular cross section (either solid or hollow) that behave in a linearly elastic manner. In other words, the loads must be such that the stresses do not exceed the proportional limit of the material. Furthermore, the equations for stresses are valid only in parts of the bars away from stress concentrations (such as holes and other abrupt changes in shape) and away from cross sections where loads are applied. (Stress concentrations in torsion are discussed later in Section 3.11.)

Finally, it is important to emphasize that the equations for the torsion of circular bars and tubes cannot be used for bars of other shapes. Noncircular bars, such as rectangular bars and bars having I-shaped cross sections, behave quite differently than do circular bars. For instance, their cross sections do *not* remain plane and their maximum stresses are not located at the farthest distances from the midpoint of the cross section. Thus, these bars require more advanced methods of analysis, such as those presented in books on theory of elasticity and advanced mechanics of materials.*

---

*The torsion theory for circular bars originated with the work of the famous French scientist C. A. de Coulomb (1736–1806); further developments were due to Thomas Young and A. Duleau (Ref. 3-1). The general theory of torsion (for bars of any shape) is due to the most famous elastician of all time, Barré de Saint-Venant (1797–1886); see Ref. 3-2.

■ Example 3-1

A solid steel bar of circular cross section (Fig. 3-11) has diameter $d = 1.5$ in., length $L = 54$ in., and shear modulus of elasticity $G = 11.5 \times 10^6$ psi. The bar is subjected to torques T acting at the ends. (a) If the torques have magnitude $T = 250$ lb-ft, what is the maximum shear stress in the bar? What is the angle of twist between the ends? (b) If the allowable shear stress is 6000 psi and the allowable angle of twist is 2.5°, what is the maximum permissible torque?

Fig. 3-11 Example 3-1

Solution   (a) Because the bar has a solid circular cross section, we can find the maximum shear stress from Eq. (3-12), as follows:

$$\tau_{max} = \frac{16T}{\pi d^3} = \frac{16(250 \text{ lb-ft})(12 \text{ in./ft})}{\pi(1.5 \text{ in.})^3} = 4530 \text{ psi} \quad \Longleftarrow$$

In a similar manner, the angle of twist is obtained from Eq. (3-15) with the polar moment of inertia given by Eq. (3-10):

$$I_p = \frac{\pi d^4}{32} = \frac{\pi(1.5 \text{ in.})^4}{32} = 0.4970 \text{ in.}^4$$

$$\phi = \frac{TL}{GI_p} = \frac{(250 \text{ lb-ft})(12 \text{ in./ft})(54 \text{ in.})}{(11.5 \times 10^6 \text{ psi})(0.4970 \text{ in.}^4)} = 0.02834 \text{ rad} = 1.62° \quad \Longleftarrow$$

Thus, the analysis of the bar under the action of the given torque is completed.

(b) The maximum permissible torque is determined either by the allowable shear stress or by the allowable angle of twist. Beginning with the shear stress, we rearrange Eq. (3-12) and calculate as follows:

$$T_1 = \frac{\pi d^3 \tau_{allow}}{16} = \frac{\pi}{16}(1.5 \text{ in.})^3(6000 \text{ psi}) = 3980 \text{ lb-in.} = 331 \text{ lb-ft}$$

Any torque larger than this value will result in a shear stress that exceeds the allowable stress of 6000 psi.

Using a rearranged Eq. (3-15), we now calculate the torque based upon the angle of twist:

$$T_2 = \frac{GI_p \phi_{allow}}{L} = \frac{(11.5 \times 10^6 \text{ psi})(0.4970 \text{ in.}^4)(2.5°)(\pi \text{ rad}/180°)}{54 \text{ in.}}$$

$$= 4618 \text{ lb-in.} = 385 \text{ lb-ft}$$

Any torque larger than T_2 will result in the allowable angle of twist being exceeded.

The maximum permissible torque is the smaller of T_1 and T_2:

$$T_{max} = 331 \text{ lb-ft}$$ ⬅

In this example, the allowable shear stress provides the limiting condition.

■ Example 3-2

$$t = \frac{d_2}{10}$$

(a)   (b)

Fig. 3-12 Example 3-2

A steel shaft is to be manufactured either as a solid circular bar or as a circular tube (Fig. 3-12). The shaft is required to transmit a torque of 1200 N·m without exceeding an allowable shear stress of 40 MPa nor an allowable rate of twist of 0.75°/m. (The shear modulus of elasticity of the steel is 78 GPa.)

(a) Determine the required diameter d_0 of the solid shaft. (b) Determine the required outer diameter d_2 of the hollow shaft if the thickness t of the shaft is specified as one-tenth of the outer diameter. (c) Determine the ratio of diameters (that is, the ratio d_2/d_0) and the ratio of weights of the hollow and solid shafts.

Solution   (a) *Solid shaft.* The required diameter d_0 is determined either from the allowable shear stress or from the allowable rate of twist. In the case of the allowable shear stress we rearrange Eq. (3-12) and obtain

$$d_0^3 = \frac{16T}{\pi \tau_{allow}} = \frac{16(1200 \text{ N·m})}{\pi(40 \text{ MPa})} = 152.8 \times 10^{-6} \text{ m}^3$$

from which we get

$$d_0 = 0.0535 \text{ m} = 53.5 \text{ mm}$$

In the case of the allowable rate of twist, we start by finding the required polar moment of inertia (see Eq. 3-14):

$$I_p = \frac{T}{G\theta_{allow}} = \frac{1200 \text{ N·m}}{(78 \text{ GPa})(0.75°/\text{m})(\pi \text{ rad}/180°)} = 1175 \times 10^{-9} \text{ m}^4$$

Since the polar moment of inertia is equal to $\pi d^4/32$, the required diameter is

$$d_0^4 = \frac{32 I_p}{\pi} = \frac{32(1175 \times 10^{-9} \text{ m}^4)}{\pi} = 11.97 \times 10^{-6} \text{ m}^4$$

or $$d_0 = 0.0588 \text{ m} = 58.8 \text{ mm}$$

Comparing the two values of d_0, we see that the rate of twist governs the design and the required diameter of the solid shaft is

$$d_0 = 58.8 \text{ mm}$$ ⬅

In a practical design, we would select a diameter slightly larger than the calculated value of d_0; for instance, 60 mm.

(b) *Hollow shaft.* Again, the required diameter is based upon either the allowable shear stress or the allowable rate of twist. We begin by noting that the outer diameter of the bar is d_2 and the inner diameter is

$$d_1 = d_2 - 2t = d_2 - 2(0.1d_2) = 0.8d_2$$

Thus, the polar moment of inertia (Eq. 3-17) is

$$I_p = \frac{\pi}{32}(d_2^4 - d_1^4) = \frac{\pi}{32}\left[d_2^4 - (0.8d_2)^4\right] = \frac{\pi}{32}(0.5904d_2^4) = 0.05796d_2^4$$

In the case of the allowable shear stress, we use the torsion formula (Eq. 3-11) as follows:

$$\tau_{allow} = \frac{Tr}{I_p} = \frac{T(d_2/2)}{0.05796d_2^4} = \frac{T}{0.1159d_2^3}$$

Rearranging, we get

$$d_2^3 = \frac{T}{0.1159\tau_{allow}} = \frac{1200 \text{ N·m}}{0.1159(40 \text{ MPa})} = 258.8 \times 10^{-6} \text{ m}^3$$

Solving for d_2 gives

$$d_2 = 0.0637 \text{ m} = 63.7 \text{ mm}$$

which is the required outer diameter based upon the shear stress.

In the case of the allowable rate of twist, we use Eq. (3-14) with θ replaced by θ_{allow} and I_p replaced by the previously obtained expression; thus,

$$\theta_{allow} = \frac{T}{G(0.05796d_2^4)}$$

from which

$$d_2^4 = \frac{T}{0.05796G\theta_{allow}}$$

$$= \frac{1200 \text{ N·m}}{0.05796(78 \text{ GPa})(0.75°/m)(\pi \text{ rad}/180°)} = 20.28 \times 10^{-6} \text{ m}^4$$

Solving for d_2 gives

$$d_2 = 0.0671 \text{ m} = 67.1 \text{ mm}$$

which is the required diameter based upon the rate of twist.

Comparing the two values of d_2, we see that the rate of twist governs the design and the required outer diameter of the hollow shaft is

$$d_2 = 67.1 \text{ mm} \qquad \leftarrow$$

The inner diameter d_1 is equal to $0.8d_2$, or 53.7 mm. (As practical values, we might select $d_2 = 70$ mm and $d_1 = 0.8d_2 = 56$ mm.)

(c) *Ratios of diameters and weights.* The ratio of the outer diameter of the hollow shaft to the diameter of the solid shaft (using the calculated values) is

$$\frac{d_2}{d_0} = \frac{67.1 \text{ mm}}{58.8 \text{ mm}} = 1.14 \qquad \leftarrow$$

Since the weights of the shafts are proportional to their cross-sectional areas, we can express the ratio of the weight of the hollow shaft to the weight of the solid shaft as follows:

$$\frac{W_{hollow}}{W_{solid}} = \frac{A_{hollow}}{A_{solid}} = \frac{\pi(d_2^2 - d_1^2)/4}{\pi d_0^2/4} = \frac{d_2^2 - d_1^2}{d_0^2}$$

$$= \frac{(67.1 \text{ mm})^2 - (53.7 \text{ mm})^2}{(58.8 \text{ mm})^2} = 0.47 \quad \longleftarrow$$

These results show that the hollow shaft uses only 47% as much material as does the solid shaft, while its outer diameter is only 14% larger.

Note: This example illustrates how to determine the required sizes of both solid and hollow bars when allowable stresses and allowable rates of twist are given. It also illustrates the fact that hollow bars are more efficient in the use of materials than are solid bars.

■ Example 3-3

A hollow shaft and a solid shaft constructed of the same material have the same length and the same outer radius R (Fig. 3-13). The inner radius of the hollow shaft is $0.6R$. (a) Assuming that both shafts are subjected to the same torque, compare their shear stresses, angles of twist, and weights. (b) Determine the strength-to-weight ratios for both shafts.

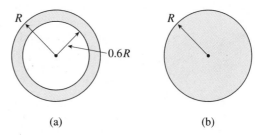

(a)   (b)

Fig. 3-13 Example 3-3

Solution   (a) *Comparisons of stresses, angles of twist, and weights.* The maximum shear stresses, given by the torsion formula (Eq. 3-11), are proportional to $1/I_p$ inasmuch as the torques and radii are the same. For the hollow shaft, we get

$$I_p = \frac{\pi R^4}{2} - \frac{\pi(0.6R)^4}{2} = 0.4352\pi R^4$$

and for the solid shaft,

$$I_p = \frac{\pi R^4}{2} = 0.5\pi R^4$$

Therefore, the ratio β_1 of the maximum shear stress in the hollow shaft to that in the solid shaft is

$$\beta_1 = \frac{T_H}{T_S} = \frac{0.5\pi R^4}{0.4352\pi R^4} = 1.15 \qquad \longleftarrow$$

where the subscripts H and S refer to the hollow shaft and the solid shaft, respectively.

The angles of twist (Eq. 3-15) are also proportional to $1/I_p$, because the torques T, lengths L, and moduli of elasticity G are the same for both shafts. Therefore, their ratio is the same as for the shear stresses:

$$\beta_2 = \frac{\phi_H}{\phi_S} = \frac{0.5\pi R^4}{0.4352\pi R^4} = 1.15 \qquad \longleftarrow$$

The weights of the shafts are proportional to their cross-sectional areas; consequently, the weight of the solid shaft is proportional to πR^2 and the weight of the hollow shaft is proportional to

$$\pi R^2 - \pi(0.6R)^2 = 0.64\pi R^2$$

Therefore, the ratio of the weight of the hollow shaft to the weight of the solid shaft is

$$\beta_3 = \frac{W_H}{W_S} = \frac{0.64\pi R^2}{\pi R^2} = 0.64 \qquad \longleftarrow$$

From the preceding ratios we again see the inherent advantage of hollow shafts. In this example, the hollow shaft has 15% greater stress and 15% greater angle of rotation than the solid shaft but 36% less weight.

(b) *Strength-to-weight ratios.* The relative efficiency of a structure is sometimes measured by its *strength-to-weight ratio*, which is defined for a bar in torsion as the allowable torque divided by the weight. The allowable torque for the hollow shaft is

$$T_H = \frac{\tau_{max}I_p}{R} = \frac{\tau_{max}\,(0.4352\pi R^4)}{R} = 0.4352\pi R^3 \tau_{max}$$

and for the solid shaft is

$$T_S = \frac{\tau_{max}I_p}{R} = \frac{\tau_{max}\,(0.5\pi R^4)}{R} = 0.5\pi R^3 \tau_{max}$$

The weights of the shafts are equal to the cross-sectional areas times the length L times the weight density γ of the material:

$$W_H = 0.64\pi R^2 L\gamma \qquad W_S = \pi R^2 L\gamma$$

Thus, the strength-to-weight ratios S_H and S_S for the hollow and solid bars, respectively, are

$$S_H = \frac{T_H}{W_H} = 0.68\frac{\tau_{max}R}{\gamma L} \qquad S_S = \frac{T_S}{W_S} = 0.5\frac{\tau_{max}\,R}{\gamma L} \qquad \longleftarrow$$

In this example, the strength-to-weight ratio of the hollow shaft is 36% greater than the strength-to-weight ratio for the solid shaft, demonstrating once again the relative efficiency of hollow shafts. For a thinner shaft, the percentage will increase; for a thicker shaft, it will decrease.

3.4 ❘ NONUNIFORM TORSION

As explained in Section 3.2, *pure torsion* refers to torsion of a prismatic bar subjected to torques acting only at the ends. **Nonuniform torsion** differs from pure torsion in that the bar need not be prismatic and the applied torques may act anywhere along the axis of the bar. Bars in nonuniform torsion can be analyzed by applying the formulas of pure torsion to finite segments of the bar and then adding the results, or by applying the formulas to differential elements of the bar and then integrating. To illustrate these procedures, we will consider three cases of nonuniform torsion. Other cases can be handled by techniques similar to those described here.

Case 1. Bar consisting of prismatic segments with constant torque throughout each segment (Fig. 3-14). The bar shown in part (a) of the figure has two different diameters and is loaded by torques acting at points *A, B, C,* and *D.* Consequently, we divide the bar into segments in such a way that each segment is prismatic and subjected to a constant torque. In this example, there are three such segments, *AB, BC,* and *CD.* Each segment is in pure torsion, and therefore all of the formulas derived in the preceding section may be applied to each part separately.

The first step in the analysis is to determine the magnitude and direction of the internal torque in each segment. Usually the torques can be determined by inspection, but if necessary they can be found by cutting sections through the bar, drawing free-body diagrams, and solving equations of equilibrium. This process is illustrated in parts (b), (c), and (d) of the figure. The first cut is made anywhere in segment *CD,* thereby exposing the internal torque T_{CD}. From the free-body diagram (Fig. 3-14b), we see that T_{CD} is equal to $-T_1 - T_2 + T_3$. From the next diagram we see that T_{BC} equals $-T_1 - T_2$, and from the last we find that T_{AB} equals $-T_1$. Thus,

$$T_{CD} = -T_1 - T_2 + T_3 \qquad T_{BC} = -T_1 - T_2 \qquad T_{AB} = -T_1 \text{ (a,b,c)}$$

Each of these torques is constant throughout the length of its segment.

When finding the shear stresses in each segment, we need only the magnitudes of these internal torques, since the directions of the stresses are not of interest. However, when finding the angle of twist for the entire bar, we need to know the direction of twist in each segment in order to combine the angles of twist correctly. Therefore, we need to establish a sign convention for the internal torques. A convenient rule in many cases is the following: *An internal torque is positive when its*

(a)

(b)

(c)

(d)

Fig. 3-14 Bar in nonuniform torsion (Case 1)

Fig. 3-15 Bar in nonuniform torsion (Case 2)

vector points away from the cut section and negative when its vector points toward the section. Thus, all of the internal torques shown in Figs. 3-14b, c, and d are pictured in their positive directions. If the calculated torque (from Eq. a, b, or c) turns out to have a positive sign, it means that the torque acts in the assumed direction; if the torque has a negative sign, it acts in the opposite direction.

The maximum shear stress in each segment of the bar is readily obtained from the torsion formula (Eq. 3-11) using the appropriate cross-sectional dimensions and internal torque. For instance, the maximum stress in segment BC (Fig. 3-14) is found using the diameter of that segment and the torque T_{BC} calculated from Eq. (b). The maximum stress in the entire bar is the largest stress from among the stresses calculated for each of the three segments.

The angle of twist for each segment is found from Eq. (3-15), again using the appropriate dimensions and torque. The total angle of twist of one end of the bar with respect to the other is then obtained by summation, as follows:

$$\phi = \phi_1 + \phi_2 + \ldots + \phi_n \qquad (3\text{-}19)$$

where ϕ_1 is the angle of twist for segment 1, ϕ_2 is the angle for segment 2, and so on, and n is the total number of segments. Since each angle of twist is found from Eq. (3-15), we can write the general formula

$$\phi = \sum_{i=1}^{n} \phi_i = \sum_{i=1}^{n} \frac{T_i L_i}{G_i I_{pi}} \qquad (3\text{-}20)$$

in which the subscript i is a numbering index for the various segments. For segment i of the bar, T_i is the internal torque (found from equilibrium, as illustrated in Figs. 3-14b, c, and d), L_i is the length, G_i is the shear modulus, and I_{pi} is the polar moment of inertia. Some of the torques (and the corresponding angles of twist) may be positive and some may be negative. By summing *algebraically* the angles of twist for all segments, we obtain the total angle of twist ϕ between the ends of the bar. The process is illustrated later in Example 3-4.

Case 2. Bar with continuously varying cross sections and constant torque (Fig. 3-15). When the torque is constant, the maximum shear stress in a solid bar always occurs at the cross section having the smallest diameter, as shown by Eq. (3-12). Furthermore, this observation usually holds for tubular bars. Therefore, by applying the torsion formula at the smallest cross section, we can quickly find the maximum stress in the bar.

To find the angle of twist, we consider an element of length dx at distance x from one end of the bar (Fig. 3-15). The differential angle of rotation $d\phi$ for this element is

$$d\phi = \frac{Tdx}{GI_p(x)} \tag{d}$$

in which $I_p(x)$ is the polar moment of inertia of the cross section at distance x from the end. The angle of twist for the entire bar is the sum of the differential angles of rotation:

$$\phi = \int_0^L d\phi = \int_0^L \frac{Tdx}{GI_p(x)} \tag{3-21}$$

If the expression for the polar moment of inertia $I_p(x)$ is not too complex, this integral can be evaluated analytically (see Example 3-5); in other cases it must be evaluated numerically.

Case 3. Bar with continuously varying cross sections and continuously varying torque (Fig. 3-16). The bar shown in part (a) of the figure is subjected to a *distributed torque* of intensity t per unit distance along the axis of the bar. As a result, the internal torque $T(x)$ varies continuously along the axis (Fig. 3-16b). The internal torque can be evaluated with the aid of a free-body diagram and an equation of equilibrium. As in Case 2, the polar moment of inertia $I_p(x)$ can be evaluated from the cross-sectional dimensions of the bar.

Knowing both the torque and polar moment of inertia as functions of x, we can use the torsion formula to determine how the shear stress varies along the axis of the bar. The cross section of maximum shear stress can then be identified, and the maximum shear stress can be determined.

The angle of twist for the bar of Fig. 3-16a can be found in the same manner as described for Case 2. The only difference is that the torque, like the polar moment of inertia, also varies along the axis. Consequently, the equation for the angle of twist becomes

$$\phi = \int_0^L d\phi = \int_0^L \frac{T(x)\,dx}{GI_p(x)} \tag{3-22}$$

This integral can be evaluated analytically in some cases, but usually it must be evaluated numerically.

(a)

(b)

Fig. 3-16 Bar in nonuniform torsion (Case 3)

■ Limitations

The analyses described in this section are valid for bars made of linearly elastic materials with circular cross sections (either solid or hollow). Also, the stresses determined from the torsion formula are valid in regions of the bar *away* from stress concentrations, which are high localized stresses that occur wherever the diameter changes abruptly and wherever concentrated torques are applied (see Section 3.11). However, stress concentrations have relatively little effect on the angle of twist, and therefore the equations for ϕ are generally valid.

Finally, we must keep in mind that the torsion formula and the formulas for angles of twist were derived for prismatic bars. We can safely apply them to bars with varying cross sections only when the changes in diameter are small and gradual. As a rule of thumb, the formulas given here are satisfactory as long as the angle of taper (the angle between the sides of the bar) is less than 10°.

■ Example 3-4

Fig. 3-17 Example 3-4

A solid steel shaft *ABCDE* (Fig. 3-17) having diameter $d = 30$ mm turns freely in bearings at its ends *A* and *E*. The shaft is driven by a gear at *C*, which applies a torque $T_2 = 450$ N·m in the direction shown in the figure. Gears at *B* and *D* are driven by the shaft and have resisting torques $T_1 = 275$ N·m and $T_3 = 175$ N·m, respectively, acting in the opposite direction to the torque T_2. Segments *BC* and *CD* have lengths $L_1 = 500$ mm and $L_2 = 400$ mm, respectively, and the shear modulus $G = 80$ GPa.

Determine the maximum shear stress in each part of the shaft and the angle of twist between gears *B* and *D*.

Solution   Each segment of the bar is prismatic and subjected to a constant torque (Case 1). Therefore, the first step in the analysis is to determine the torques acting in the segments, after which we can find the shear stresses and angles of twist.

Torques acting in the segments. The torques in the end segments (*AB* and *DE*) are zero since we are disregarding any friction in the bearings at the supports. Therefore, the end segments have no stresses and no angles of twist.

The torque T_{CD} in segment *CD* is found by cutting a section through the segment and constructing a free-body diagram, as in Fig. 3-18a. The torque is assumed to be positive, and therefore its vector points away from the cut section. From equilibrium of the free body, we obtain

$$T_{CD} = T_2 - T_1 = 450 \text{ N·m} - 275 \text{ N·m} = 175 \text{ N·m}$$

The positive sign in the result means that T_{CD} acts in the assumed positive direction.

The torque in segment *BC* is found in a similar manner, using the free-body diagram of Fig. 3-18b:

$$T_{BC} = -T_1 = -275 \text{ N·m}$$

Note that this torque has a negative sign, which means its direction is opposite to the direction shown in the figure.

Shear stresses. The maximum shear stresses in segments *BC* and *CD* are found from the modified form of the torsion formula (Eq. 3-12); thus,

$$\tau_{BC} = \frac{16 T_{BC}}{\pi d^3} = \frac{16(275 \text{ N·m})}{\pi (30 \text{ mm})^3} = 51.9 \text{ MPa} \qquad \longleftarrow$$

Fig. 3-18 Free-body diagrams for Example 3-4

$$\tau_{CD} = \frac{16T_{CD}}{\pi d^3} = \frac{16(175 \text{ N} \cdot \text{m})}{\pi (30 \text{ mm})^3} = 33.0 \text{ MPa} \quad \longleftarrow$$

Since the directions of the shear stresses are not of interest in this example, only absolute values of the torques are used in the preceding calculations.

Angles of twist. The angle of twist ϕ_{BD} between gears B and D is the algebraic sum of the angles of twist for the intervening segments of the bar, as given by Eq. (3-19); thus,

$$\phi_{BD} = \phi_{BC} + \phi_{CD}$$

When calculating the individual angles of twist, we need the moment of inertia of the cross section:

$$I_p = \frac{\pi d^4}{32} = \frac{\pi (30 \text{ mm})^4}{32} = 79,520 \text{ mm}^4$$

Now we can determine the angles of twist, as follows:

$$\phi_{BC} = \frac{T_{BC}L_1}{GI_p} = \frac{(-275 \text{ N} \cdot \text{m})(500 \text{ mm})}{(80 \text{ GPa})(79,520 \text{ mm}^4)} = -0.0216 \text{ rad}$$

$$\phi_{CD} = \frac{T_{CD}L_2}{GI_p} = \frac{(175 \text{ N} \cdot \text{m})(400 \text{ mm})}{(80 \text{ GPa})(79,520 \text{ mm}^4)} = 0.0110 \text{ rad}$$

Note that in this example the angles of twist have opposite directions. Adding algebraically, we obtain the total angle of twist:

$$\phi_{BD} = \phi_{BC} + \phi_{CD} = -0.0216 + 0.0110 = -0.0106 \text{ rad} = -0.61° \quad \longleftarrow$$

The minus sign means that gear D rotates clockwise (when viewed from the right-hand end of the shaft) with respect to gear B. However, for most purposes only the absolute value of the angle of twist is needed, and therefore it is sufficient to say that the angle of twist between gears B and D is 0.61°. The angle of twist between the two ends of a shaft is sometimes called the *wind-up*.

Notes: The procedures illustrated in this example can be used for shafts having segments of different diameters or of different materials, as long as the dimensions and properties remain constant within each segment.

Only the effects of torsion are considered in this example and in the problems at the end of the chapter. Bending effects are considered later, beginning with Chapter 4.

■ **Example 3-5**

A tapered bar AB of solid circular cross section is twisted by torques T applied at the ends (Fig. 3-19). The diameter of the bar varies linearly from d_A at the left-hand end to d_B at the right-hand end. (a) Determine the maximum shear stress in the bar. (b) Derive a formula for the angle of twist of the bar. (Assume that $d_B \geq d_A$.)

Solution (a) *Shear stresses.* Since the maximum shear stress at any cross section is given by the modified torsion formula (Eq. 3-12), we know immediately that the maximum shear stress in the bar occurs at the cross section having the smallest diameter, that is, at end A (see Fig. 3-19):

Fig. 3-19 Example 3-5. Tapered bar in torsion

$$\tau_{max} = \frac{16T}{\pi d_A^3} \quad \longleftarrow$$

(b) *Angle of twist.* Because the torque is constant and the polar moment of inertia varies continuously with the distance x from end A (Case 2), we will use Eq. (3-21) to determine the angle of twist. We begin by setting up an expression for the diameter d at distance x from end A:

$$d = d_A + \frac{d_B - d_A}{L} x \quad (3\text{-}23)$$

in which L is the length of the bar. We can now write an expression for the polar moment of inertia:

$$I_p(x) = \frac{\pi d^4}{32} = \frac{\pi}{32} \left(d_A + \frac{d_B - d_A}{L} x \right)^4 \quad (3\text{-}24)$$

Substituting this expression into Eq. (3-21), we get a formula for the angle of twist:

$$\phi = \int_0^L \frac{T dx}{G I_p(x)} = \frac{32T}{\pi G} \int_0^L \frac{dx}{\left(d_A + \frac{d_B - d_A}{L} x \right)^4} \quad (3\text{-}25)$$

To evaluate the integral in this equation, we note that it is of the form

$$\int \frac{dx}{(a + bx)^4}$$

in which

$$a = d_A \qquad b = \frac{d_B - d_A}{L} \quad (e,f)$$

With the aid of a table of integrals (see Appendix C), we find

$$\int \frac{dx}{(a + bx)^4} = -\frac{1}{3b(a + bx)^3}$$

This integral is evaluated in our case by substituting for x the limits 0 and L and substituting for a and b the expressions in Eqs. (e) and (f). Thus, the integral in Eq. (3-25) equals

$$\frac{L}{3(d_B - d_A)} \left(\frac{1}{d_A^3} - \frac{1}{d_B^3} \right) \quad (g)$$

Replacing the integral in Eq. (3-25) with this expression, we obtain

$$\phi = \frac{32TL}{3\pi G(d_B - d_A)} \left(\frac{1}{d_A^3} - \frac{1}{d_B^3} \right) \quad (3\text{-}26) \quad \longleftarrow$$

which is the desired equation for the angle of twist of the tapered bar.

A convenient form in which to write the preceding equation is

$$\phi = \frac{TL}{GI_{pA}} \left(\frac{\beta^2 + \beta + 1}{3\beta^3} \right) \tag{3-27}$$

in which

$$\beta = \frac{d_B}{d_A} \qquad I_{pA} = \frac{\pi d_A^4}{32} \tag{3-28}$$

The quantity β is the ratio of end diameters and I_{pA} is the polar moment of inertia at end A.

In the special case of a prismatic bar, we have $\beta = 1$ and Eq. (3-27) gives $\phi = TL/GI_{pA}$, as expected. For values of β greater than 1, the angle of rotation decreases because the larger diameter at end B produces an increase in the torsional stiffness (as compared to a prismatic bar).

3.5 STRESSES AND STRAINS IN PURE SHEAR

When a circular bar, either solid or hollow, is subjected to torsion, shear stresses act over the cross sections and on longitudinal planes, as explained in Section 3.3. Let us now examine in more detail the stresses and strains produced during twisting of a bar. We begin with a stress element *abcd* cut between two cross sections and between two longitudinal planes (Figs. 3-20a and b). This element is in a state of **pure shear**, because the only stresses acting on it are the shear stresses on the four side faces (see the discussion of pure shear in Section 1.6.)

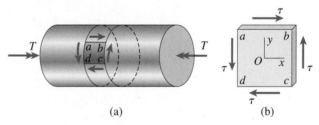

Fig. 3-20 Stresses acting on an element cut from a bar in torsion (pure shear)

(a)   (b)

The directions of the shear stresses depend upon the directions of the applied torques T. In this discussion, we assume that the torques rotate the right-hand end of the bar clockwise when viewed from the right (Fig. 3-20a); hence the shear stresses acting on the element have the directions shown in the figure. This same state of stress exists for a similar element cut from the interior of the bar, except that the magnitudes of the shear stresses are smaller because the radial distance to the element is smaller. As we will see in the next paragraph, the directions of the torques were intentionally chosen so that the resulting shear stresses are positive.

■ Stresses on Inclined Sections

A two-dimensional view of the stress element in pure shear is shown in Fig. 3-21a. As explained previously in Section 2.6, we usually draw only a two-dimensional view of a stress element, but we must always be aware that the element has a third dimension (thickness) perpendicular to the plane of the figure. In the two-dimensional view, a set of xy axes is oriented parallel to the sides of the element, with the x axis parallel to the longitudinal axis of the bar. Therefore, all stresses shown in Fig. 3-21a are positive stresses according to the sign convention for shear stresses, which is described in Section 1.6 and repeated here: A shear stress acting on a positive face of an element is positive if it acts in the positive direction of one of the coordinate axes and negative if it acts in the negative direction of an axis. Conversely, a shear stress acting on a negative face of an element is positive if it acts in the negative direction of one of the coordinate axes and negative if it acts in the positive direction of an axis.

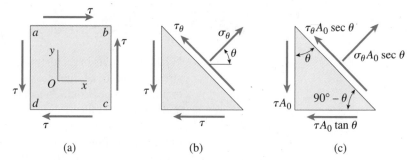

Fig. 3-21 Analysis of stresses on inclined planes: (a) element in pure shear, (b) stresses acting on a triangular stress element, and (c) forces acting on the triangular stress element (free-body diagram)

We now cut from the plane-stress element (Fig. 3-21a) a wedge-shaped (or "triangular") stress element having one face oriented at an angle θ to the x axis (Fig. 3-21b). Normal stresses σ_θ and shear stresses τ_θ act on this inclined face and are shown in their positive directions in the figure. (Recall from the discussion of stress elements in Section 2.6 that normal stresses σ_θ are positive in tension and shear stresses τ_θ are positive when they tend to produce counterclockwise rotation of the material.) The horizontal and vertical side faces of the element of Fig. 3-21b have the shear stresses τ acting on them, and the front and rear faces are free of any stresses.

The stresses acting on the inclined face may be determined from the equilibrium of the triangular element. The forces acting on its three side faces can be obtained by multiplying the stresses by the areas over which they act. For instance, the force on the left-hand face is equal to τA_0, where A_0 is the area of the vertical side face. This force acts in the negative y direction and is shown in the free-body diagram of Fig. 3-21c. Because the thickness of the element in the z direction is constant, we see that the area of the bottom face is $A_0 \tan \theta$ and the area of the inclined face is $A_0 \sec \theta$. Multiplying the stresses by the areas over which they act gives the forces acting on all faces of the element (Fig. 3-21c).

We are now ready to write two equations of equilibrium for the element, one in the direction of σ_θ and the other in the direction of τ_θ. When writing these equations, the forces acting on the left-hand and bottom faces must be resolved into components in the directions of σ_θ and τ_θ. Thus, the first equation, obtained by summing forces in the direction of σ_θ, is

$$\sigma_\theta A_0 \sec \theta = \tau A_0 \sin \theta + \tau A_0 \tan \theta \cos \theta$$

or
$$\sigma_\theta = 2\tau \sin \theta \cos \theta \qquad (3\text{-}29a)$$

The second equation is obtained by summing forces in the direction of τ_θ:

$$\tau_\theta A_0 \sec \theta = \tau A_0 \cos \theta - \tau A_0 \tan \theta \sin \theta$$

or
$$\tau_\theta = \tau(\cos^2\theta - \sin^2\theta) \qquad (3\text{-}29b)$$

These equations can be expressed in simpler forms by introducing the following trigonometric identities (see Appendix C):

$$\sin 2\theta = 2 \sin \theta \cos \theta \qquad \cos 2\theta = \cos^2 \theta - \sin^2 \theta$$

Then the equations for σ_θ and τ_θ become

$$\sigma_\theta = \tau \sin 2\theta \qquad \tau_\theta = \tau \cos 2\theta \qquad (3\text{-}30a,b)$$

Equations (3-30a and b) give the normal and shear stresses acting on any inclined plane in terms of the shear stresses τ acting on the x and y planes (Fig. 3-21a) and the angle θ defining the orientation of the inclined plane (Fig. 3-21b).

The manner in which the stresses σ_θ and τ_θ vary with the orientation of the inclined plane is shown by the graph in Fig. 3-22, which is a plot of Eqs. (3-30a and b). We see that for $\theta = 0$, which is the right-hand face of the stress element in Fig. 3-21a, the graph gives $\sigma_\theta = 0$ and $\tau_\theta = \tau$, as expected. For the top face of the element ($\theta = 90°$), we obtain $\sigma_\theta = 0$ and $\tau_\theta = -\tau$. The minus sign for the shear stress means that it acts in the negative τ_θ direction, that is, it acts to the right on face ab (Fig. 3-21a). Note that the numerically largest shear stresses occur on these planes.

From the graph we see that the normal stress σ_θ reaches a maximum value at $\theta = 45°$. At that angle, the stress is positive (tension) and equal numerically to the shear stress τ. Similarly, σ_θ has its minimum value (which is compressive) at $\theta = -45°$. At both of these 45° angles, the shear stress τ_θ is equal to zero. Thus, a stress element oriented at an angle of 45° (Fig. 3-23b) is acted upon by equal tensile and compressive stresses in perpendicular directions without any shear stresses. Note that the normal stresses shown in Fig. 3-23b correspond to an element in pure shear subjected to shear stresses acting in the directions shown in Fig. 3-23a. If the shear stresses acting on the element of Fig. 3-23a are

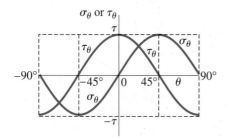

Fig. 3-22 Graph of normal stresses σ_θ and shear stresses τ_θ versus angle θ of the inclined plane

reversed in direction, the normal stresses acting on the 45° planes also will change directions.

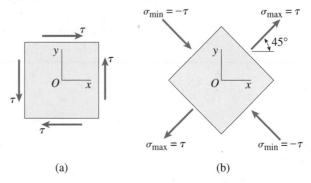

Fig. 3-23 Stress elements oriented at $\theta = 0$ and $\theta = 45°$ for pure shear

(a)   (b)

If a stress element is oriented at an angle other than 45°, both normal and shear stresses will act on the inclined faces (see Eqs. 3-30). Stress elements subjected to these more general conditions are discussed in detail in Chapter 7.

The equations derived in this section are valid for a stress element in pure shear regardless of whether the element is cut from a bar in torsion or from some other structural element. Also, since Eqs. (3-30) were derived from equilibrium only, they are valid for any material, whether or not it behaves in a linearly elastic manner.

The existence of maximum tensile stresses on planes at 45° to the x axis (Fig. 3-23b) explains why bars in torsion that are made of materials that are brittle and weak in tension fail by cracking along a 45° helical surface (Fig. 3-24). As mentioned in Section 3.3, this type of failure is readily demonstrated by twisting a piece of classroom chalk.

Fig. 3-24 Torsion failure of a brittle material by tension cracking along a 45° helical surface

■ Strains in Pure Shear

Let us now consider the strains that exist in an element in pure shear. For instance, consider the element in pure shear shown in Fig. 3-23a. The corresponding shear strains are shown in Fig. 3-25a, where the deformations are highly exaggerated. The shear strain γ is the change in angle between two lines that were originally perpendicular to each other, as discussed previously in Section 1.6. Thus, the decrease in the angle at the lower left-hand corner of the element is the shear strain γ (measured in radians). This same change in angle occurs at the upper right-hand corner, where the angle decreases, and at the other two corners, where the angles increase. However, the lengths of the sides of the element, including the thickness perpendicular to the plane of the paper, do not change when these shear deformations occur. Therefore, the element changes its shape from a rectangular parallelepiped (Fig. 3-23a) to an oblique parallelepiped (Fig. 3-25a). This change in shape is called a **shear distortion**.

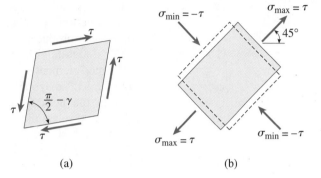

Fig. 3-25 Strains in pure shear: (a) shear distortion of an element oriented at $\theta = 0$, and (b) distortion of an element oriented at $\theta = 45°$

If the material is linearly elastic, the shear strain for the element oriented at $\theta = 0$ (Fig. 3-25a) is related to the shear stress by Hooke's law in shear:

$$\gamma = \frac{\tau}{G} \tag{3-31}$$

where, as usual, the symbol G represents the shear modulus of elasticity.

Next, consider the strains that occur in an element oriented at $\theta = 45°$ (Fig. 3-25b). The tensile stresses acting at 45° tend to elongate the element in that direction. Because of the Poisson effect, they also tend to shorten it in the perpendicular direction (the direction where $\theta = 135°$ or $-45°$). Similarly, the compressive stresses acting at 135° tend to shorten the element in that direction and elongate it in the 45° direction. These dimensional changes are shown in Fig. 3-25b, where the dashed lines show the original element. Since there are no shear distortions, the element remains a rectangular parallelepiped even though its dimensions have changed.

If the material is linearly elastic and follows Hooke's law, we can obtain an equation relating strain to stress for the element at $\theta = 45°$ (Fig. 3-25b). The tensile stress σ_{max} acting at $\theta = 45°$ produces a positive normal strain in that direction equal to σ_{max}/E. Since $\sigma_{max} = \tau$, we can also express this strain as τ/E. The stress σ_{max} also produces a negative strain in the perpendicular direction equal to $-\nu\tau/E$, where ν is Poisson's ratio. Similarly, the stress $\sigma_{min} = -\tau$ (at $\theta = 135°$) produces a negative strain equal to $-\tau/E$ in that direction and a positive strain in the perpendicular direction (the 45° direction) equal to $\nu\tau/E$. Therefore, the normal strain in the 45° direction is

$$\epsilon_{max} = \frac{\tau}{E} + \frac{\nu\tau}{E} = \frac{\tau}{E}(1 + \nu) \tag{3-32}$$

which is positive, representing elongation. The strain in the perpendicular direction is a negative strain of the same amount. In other words, pure shear produces elongation in the 45° direction and shortening in the 135° direction. These strains are consistent with the shape of the deformed element of Fig. 3-25a, because the 45° diagonal has lengthened and the 135° diagonal has shortened.

In the next section we will use the geometry of the deformed element to relate the shear strain γ (Fig. 3-25a) to the normal strain ϵ_{max} in the 45° direction (Fig. 3-25b). In so doing, we will derive the following relationship:

$$\epsilon_{max} = \frac{\gamma}{2} \tag{3-33}$$

This equation, in conjunction with Eq. (3-31), can be used to calculate the maximum shear strains and maximum normal strains in pure torsion when the shear stress τ is known.

■ **Example 3-6**

$T = 4.0$ kN·m

←60→
mm

←80→
mm

Fig. 3-26 Example 3-6. Circular tube in torsion

A circular tube with an outside diameter of 80 mm and an inside diameter of 60 mm is subjected to a torque $T = 4.0$ kN·m (Fig. 3-26). The tube is made of aluminum alloy 7075-T6. (a) Determine the maximum tensile, compressive, and shear stresses in the tube and show these stresses on sketches of properly oriented stress elements. (b) Determine the corresponding maximum strains in the tube.

Solution   (a) *Maximum stresses*. The maximum values of all three stresses (tensile, compressive, and shear) are equal numerically, although they act on different planes. Their magnitudes are found from the torsion formula:

$$\tau_{max} = \frac{Tr}{I_p} = \frac{(4000 \text{ N·m})(0.040 \text{ m})}{\dfrac{\pi}{32}\left[(0.080 \text{ m})^4 - (0.060 \text{ m})^4\right]} = 58.2 \text{ MPa}$$

The maximum shear stresses act on cross-sectional and longitudinal planes, as shown by the stress element in Fig. 3-27a, where the x axis is parallel to the longitudinal axis of the tube.

The maximum tensile and compressive stresses are

$$\sigma_t = 58.2 \text{ MPa} \qquad \sigma_c = -58.2 \text{ MPa}$$

These stresses act on planes at 45° to the axis (Fig. 3-27b).

(b) *Maximum strains.* The maximum shear strain in the tube is obtained from Eq. (3-31). The shear modulus of elasticity is obtained from Table H-2, Appendix H, as $G = 27$ GPa. Therefore, the maximum shear strain is

$$\gamma_{max} = \frac{\tau_{max}}{G} = \frac{58.2 \text{ MPa}}{27 \text{ GPa}} = 0.0022$$

This strain has the direction shown in Fig. 3-27c.

The magnitude of the maximum normal strains (from Eq. 3-33) is

$$\epsilon_{max} = \frac{\gamma_{max}}{2} = 0.0011$$

Thus, the maximum tensile and compressive strains are

$$\epsilon_t = 0.0011 \qquad \epsilon_c = -0.0011$$

These strains are pictured in Fig. 3-27d for an element with sides of unit dimensions.

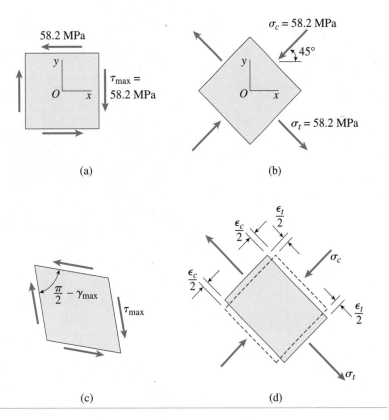

Fig. 3-27 Stress and strain elements for the tube of Example 3-6

3.6 RELATIONSHIP BETWEEN MODULI OF ELASTICITY E AND G

An important relationship between the moduli of elasticity E and G can be obtained from the equations derived in the preceding section. For this purpose, consider the stress element $abcd$ shown in Fig. 3-28a. The front face of the element is assumed to be square, with the length of each side denoted as h. When this element is subjected to pure shear by stresses τ, the front face distorts into a rhombus (Fig. 3-28b) with sides of length h and with shear strain $\gamma = \tau/G$. Because of the distortion, diagonal bd is lengthened and diagonal ac is shortened. The length of diagonal bd is equal to its initial length $\sqrt{2}\,h$ times the factor $1 + \epsilon_{max}$, where ϵ_{max} is the normal strain in the 45° direction; thus,

$$L_{bd} = \sqrt{2}\,h(1 + \epsilon_{max}) \tag{a}$$

This length can be related to the shear strain γ by considering the geometry of the deformed element (Fig. 3-28b).

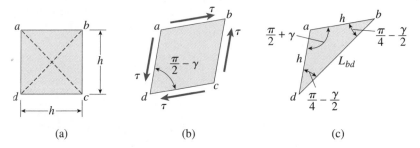

Fig. 3-28 Geometry of deformed element in pure shear

(a)   (b)   (c)

To obtain the required geometric relationships, consider triangle abd (Fig. 3-28c) which represents one-half of the rhombus pictured in Fig. 3-28b. Side bd of this triangle has length L_{bd} (Eq. a), and the other sides have length h. Angle adb of the triangle is equal to one-half of angle adc of the rhombus, or $\pi/4 - \gamma/2$. The angle abd in the triangle is the same. Therefore, angle dab of the triangle equals $\pi/2 + \gamma$. Now using the law of cosines (see Appendix C) for triangle abd, we get

$$L_{bd}^2 = h^2 + h^2 - 2h^2 \cos\left(\frac{\pi}{2} + \gamma\right)$$

Substituting for L_{bd} from Eq. (a) and simplifying, we get

$$(1 + \epsilon_{max})^2 = 1 - \cos\left(\frac{\pi}{2} + \gamma\right)$$

By expanding the term on the left-hand side, and also observing that $\cos(\pi/2 + \gamma) = -\sin \gamma$, we obtain

$$1 + 2\epsilon_{max} + \epsilon_{max}^2 = 1 + \sin \gamma$$

Because ϵ_{max} and γ are very small strains, we can disregard ϵ_{max}^2 in comparison with $2\epsilon_{max}$ and we can replace $\sin \gamma$ by γ. The resulting expression is

$$\epsilon_{max} = \frac{\gamma}{2} \tag{3-34}$$

which establishes the relationship already presented in Section 3.5 as Eq. (3-33).

The shear strain γ appearing in Eq. (3-34) is equal to τ/G by Hooke's law (Eq. 3-31) and the normal strain ϵ_{max} is equal to $\tau(1 + \nu)/E$ by Eq. (3-32). Making both of these substitutions in Eq. (3-34) yields

$$G = \frac{E}{2(1 + \nu)} \tag{3-35}$$

We see that E, G, and ν are not independent properties of a linearly elastic material. Instead, if any two of them are known, the third can be calculated from Eq. (3-35). Typical values of E, G, and ν are listed in Table H-2, Appendix H.*

3.7 | TRANSMISSION OF POWER BY CIRCULAR SHAFTS

The most important use of circular shafts is to transmit mechanical power from one device or machine to another, as in the drive shaft of an automobile, the propeller shaft of a ship, or the axle of a bicycle. The power is transmitted through the rotary motion of the shaft, and the amount of power transmitted depends upon the magnitude of the torque and the speed of rotation. A common design problem is to determine the required size of a shaft so that it will transmit a specified amount of power at a specified rotational speed without exceeding the allowable stresses for the material.

Let us suppose that a motor-driven shaft (Fig. 3-29) is rotating at an angular speed ω, measured in radians per second (rad/s). The shaft transmits a torque T to a machine (not shown in the figure) that is performing useful work. (The torque shown in the figure is the torque exerted *on the shaft* by the machine.) In general, the **work** W done by a torque of constant magnitude is equal to the product of the torque and the angle through which it rotates; that is,

*Equation (3-35) was derived originally by Poisson using the value 1/4 for ν; see Ref. 3-3.

Motor

Fig. 3-29 Shaft transmitting a constant torque T at an angular speed ω

$$W = T\psi \tag{3-36}$$

where ψ is the angle of rotation in radians.

Power is the *rate* at which work is done, or

$$P = \frac{dW}{dt} = T\frac{d\psi}{dt} \tag{3-37}$$

in which P is the symbol for power and t represents time. The rate of change $d\psi/dt$ of the angular displacement ψ is the angular speed ω, and therefore the preceding equation becomes

$$P = T\omega \qquad (\omega = \text{rad/s}) \tag{3-38}$$

This formula, which is familiar from elementary physics, gives the power transmitted by a rotating shaft transmitting a constant torque T.

The units to be used in Eq. (3-38) are as follows. If the torque T is expressed in newton meters, then the power is expressed in watts (W). One watt is equal to one newton meter per second (or one joule per second). If T is expressed in pound-feet, then the power is expressed in foot-pounds per second.*

Angular speed is often expressed as the frequency f of rotation, which is the number of revolutions per unit of time. The unit of frequency is the hertz (Hz), equal to one revolution per second (s^{-1}). Inasmuch as one revolution equals 2π radians, we obtain

$$\omega = 2\pi f \qquad (\omega = \text{rad/s}, f = \text{Hz} = s^{-1}) \tag{3-39}$$

The expression for power (Eq. 3-38) then becomes

$$P = 2\pi fT \qquad (f = \text{Hz} = s^{-1}) \tag{3-40}$$

Another commonly used unit is the number of revolutions per minute (rpm), denoted by the letter n. Therefore, we also have the following relationships:

$$n = 60f \tag{3-41}$$

---

*See Table A-1, Appendix A, for units of work and power.

and
$$P = \frac{2\pi n T}{60} \qquad (n = \text{rpm}) \qquad (3\text{-}42)$$

In Eqs. (3-40) and (3-42), the quantities P and T have the same units as in Eq. (3-38); that is, P has units of watts if T has units of newton meters, and P has units of foot-pounds per second if T has units of pound-feet.

In U.S. engineering practice, power is often expressed in horse-power (hp), a unit equal to 550 ft-lb/s. Therefore, the horsepower H being transmitted by a rotating shaft is

$$H = \frac{2\pi n T}{60(550)} = \frac{2\pi n T}{33{,}000} \qquad (n = \text{rpm}, \; T = \text{lb-ft}, \; H = \text{hp}) \qquad (3\text{-}43)$$

One horsepower is equal approximately to 746 watts.

The preceding equations relate the torque acting in a shaft to the power transmitted by the shaft. Once the torque is known, we can determine the shear stresses, shear strains, angles of twist, and other desired quantities by the methods described in Sections 3.2 through 3.5. The following examples illustrate some of the procedures for analyzing rotating shafts.

■ **Example 3-7**

A motor driving a solid circular steel shaft transmits 40 hp to a gear at B (Fig. 3-30). The allowable shear stress in the steel is 6000 psi. (a) What is the required diameter d of the shaft if it is operated at 500 rpm? (b) What is the required diameter d if it is operated at 4000 rpm?

Motor

Fig. 3-30 Example 3-7

B

Solution   (a) *Motor operating at 500 rpm.* Knowing the horsepower and the speed of rotation, we can find the torque T acting on the shaft by using Eq. (3-43). Solving that equation for T, we get

$$T = \frac{33{,}000 H}{2\pi n} = \frac{33{,}000(40 \text{ hp})}{2\pi(500 \text{ rpm})} = 420.2 \text{ lb-ft} = 5042 \text{ lb-in.}$$

This torque is transmitted by the shaft from the motor to the gear.

The maximum shear stress in the shaft can be obtained from the modified torsion formula (Eq. 3-12):

$$\tau_{\max} = \frac{16T}{\pi d^3}$$

Solving that equation for the diameter d, and also substituting τ_{allow} for τ_{max}, we get

$$d^3 = \frac{16T}{\pi\tau_{allow}} = \frac{16(5042 \text{ lb-in.})}{\pi(6000 \text{ psi})} = 4.280 \text{ in.}^3$$

from which

$$d = 1.62 \text{ in.} \qquad \longleftarrow$$

The diameter of the shaft must be at least this large if the allowable shear stress is not to be exceeded.

(b) *Motor operating at 4000 rpm.* Following the same procedure as in part (a), we obtain

$$T = \frac{33,000H}{2\pi n} = \frac{33,000(40 \text{ hp})}{2\pi(4000 \text{ rpm})} = 52.52 \text{ lb-ft} = 630.3 \text{ lb-in.}$$

$$d^3 = \frac{16T}{\pi\tau_{allow}} = \frac{16(630.3 \text{ lb-in.})}{\pi(6000 \text{ psi})} = 0.5350 \text{ in.}^3$$

$$d = 0.81 \text{ in.} \qquad \longleftarrow$$

which is one-half the diameter found in part (a).

This example shows that the higher the speed of rotation, the smaller the required size of the shaft (for the same power and the same allowable stress).

■ **Example 3-8**

A solid steel shaft ABC of 50 mm diameter (Fig. 3-31a) is driven at A by a motor that transmits 50 kW to the shaft at 10 Hz. The gears at B and C drive machinery requiring power equal to 35 kW and 15 kW, respectively. Compute the maximum shear stress τ_{max} in the shaft and the angle of twist ϕ_{AC} between the motor at A and the gear at C. (Use $G = 80$ GPa.)

(a)   (b)

Fig. 3-31 Example 3-8

Solution   *Torques acting on the shaft.* We begin the analysis by determining the torques applied to the shaft by the motor and the two gears. Since the motor supplies 50 kW at 10 Hz, it creates a torque T_A at end A of the shaft (Fig. 3-31b) that we can calculate from Eq. (3-40):

$$T_A = \frac{P}{2\pi f} = \frac{50 \text{ kW}}{2\pi(10 \text{ Hz})} = 796 \text{ N·m}$$

In a similar manner, we can calculate the torques T_B and T_C applied by the gears to the shaft:

$$T_B = \frac{P}{2\pi f} = \frac{35 \text{ kW}}{2\pi(10 \text{ Hz})} = 557 \text{ N·m}$$

$$T_C = \frac{P}{2\pi f} = \frac{15 \text{ kW}}{2\pi(10 \text{ Hz})} = 239 \text{ N·m}$$

These torques are shown in the free-body diagram of the shaft (Fig. 3-31b). Note that the torques applied by the gears are opposite in direction to the torque applied by the motor. (If we think of T_A as the "load" applied to the shaft by the motor, then the torques T_B and T_C are the "reactions" of the gears.)

The internal torques in the two segments of the shaft are now found (by inspection) from the free-body diagram of Fig. 3-31b:

$$T_{AB} = 796 \text{ N·m} \qquad T_{BC} = 239 \text{ N·m}$$

Both internal torques act in the same direction, and therefore the angles of twist in segments AB and BC are additive when finding the total angle of twist. (To be specific, both torques are positive according to the sign convention adopted in Section 3.4.)

Shear stresses and angles of twist. The shear stress and angle of twist in segment AB of the shaft are found in the usual manner from Eqs. (3-12) and (3-15):

$$\tau_{AB} = \frac{16T_{AB}}{\pi d^3} = \frac{16(796 \text{ N·m})}{\pi(50 \text{ mm})^3} = 32.4 \text{ MPa}$$

$$\phi_{AB} = \frac{T_{AB}L_{AB}}{GI_p} = \frac{(796 \text{ N·m})(1.0 \text{ m})}{(80 \text{ GPa})\left(\dfrac{\pi}{32}\right)(50 \text{ mm})^4} = 0.0162 \text{ rad}$$

The corresponding quantities for segment BC are

$$\tau_{BC} = \frac{16T_{BC}}{\pi d^3} = \frac{16(239 \text{ N·m})}{\pi(50 \text{ mm})^3} = 9.7 \text{ MPa}$$

$$\phi_{BC} = \frac{T_{BC}L_{BC}}{GI_p} = \frac{(239 \text{ N·m})(1.2 \text{ m})}{(80 \text{ GPa})\left(\dfrac{\pi}{32}\right)(50 \text{ mm})^4} = 0.0058 \text{ rad}$$

Thus, the maximum shear stress in the shaft occurs in segment AB and is equal to

$$\tau_{\text{max}} = 32.4 \text{ MPa} \qquad \longleftarrow$$

Also, the total angle of twist between the motor at A and the gear at C is

$$\phi_{AC} = \phi_{AB} + \phi_{BC} = 0.0162 \text{ rad} + 0.0058 \text{ rad} = 0.0220 \text{ rad} = 1.26° \qquad \longleftarrow$$

As explained previously, both parts of the shaft twist in the same direction, and therefore the angles of twist are added.

3.8

STATICALLY INDETERMINATE TORSIONAL MEMBERS

The bars and shafts described in the preceding sections of this chapter are *statically determinate* because all internal torques and all reactions can be obtained from free-body diagrams and equations of equilibrium. However, if additional restraints, such as fixed supports, are added to the bars, the equations of equilibrium will no longer be adequate for determining the torques. The bars are then classified as **statically indeterminate**. Torsional members of this kind can be analyzed by supplementing the equilibrium equations with compatibility equations pertaining to the rotational displacements. Thus, the general method for analyzing statically indeterminate torsional members is the same as described in Section 2.4 for statically indeterminate bars with axial loads.

The first step in the analysis is to write **equations of equilibrium**, using free-body diagrams obtained from the given physical situation. The unknown quantities in the equilibrium equations are torques, either internal torques or reaction torques. The second step in the analysis is to formulate **equations of compatibility**, based upon physical conditions pertaining to the angles of twist. As a consequence, the compatibility equations contain angles of twist as unknowns. The third step is to relate the angles of twist to the torques by **torque-displacement relations**, such as $\phi = TL/GI_p$. After introducing these relations into the compatibility equations, they too become equations containing torques as unknowns. Therefore, the last step is to solve simultaneously the equations of equilibrium and compatibility for the unknown torques.

To illustrate the method of solution, we will analyze the composite bar *AB* shown in Figs. 3-32a and b. This bar is attached to a fixed support at end *A* and loaded by a torque *T* at end *B*. It consists of a solid bar inside a tube, with both the solid bar and the tube joined to a rigid end plate at *B*. For convenience, we will identify the solid bar and tube (and their properties) by the numerals 1 and 2, respectively. For instance, the diameter of the solid bar is denoted d_1 and the outer diameter of the tube is denoted d_2. A small gap exists between the bar and the tube, and therefore the inner diameter of the tube is slightly larger than d_1.

When the torque *T* is applied at the end of the bar, the end plate rotates through a small angle ϕ and torques T_1 and T_2 are developed in the solid bar and tube, respectively (see Figs. 3-32c and d, which show the solid bar and tube separated from each other). From equilibrium we know that the sum of these torques equals the applied load, and so the *equation of equilibrium* is

$$T_1 + T_2 = T \tag{a}$$

Because this equation contains two unknowns (T_1 and T_2), we recognize that the composite bar *AB* (Fig. 3-32a) is statically indeterminate.

To obtain a second equation, we must consider the rotational

(a)

(b)

(c)

(d)

Fig. 3-32 Statically indeterminate bar in torsion

displacements of the solid bar and tube. Let us denote the angle of twist of the solid bar (Fig. 3-32c) by ϕ_1 and the angle of twist of the tube by ϕ_2 (Fig. 3-32d). These angles of twist must be equal because the bar and tube are securely joined to the end plate and rotate with it; consequently, the *equation of compatibility* is

$$\phi_1 = \phi_2 \tag{b}$$

The angles ϕ_1 and ϕ_2 are related to the torques T_1 and T_2 by *torque-displacement relations*, which in the case of linearly elastic materials are obtained from the equation $\phi = TL/GI_p$. Thus,

$$\phi_1 = \frac{T_1 L}{G_1 I_{p1}} \qquad \phi_2 = \frac{T_2 L}{G_2 I_{p2}} \tag{c,d}$$

in which G_1 and G_2 are the shear moduli of elasticity of the materials and I_{p1} and I_{p2} are the polar moments of inertia of the cross sections.

When the preceding expressions for ϕ_1 and ϕ_2 are substituted into Eq. (b), the equation of compatibility becomes

$$\frac{T_1 L}{G_1 I_{p1}} = \frac{T_2 L}{G_2 I_{p2}} \tag{e}$$

We now have two equations (Eqs. a and e) with two unknowns, so we can solve them for the torques T_1 and T_2. The results are:

$$T_1 = T\left(\frac{G_1 I_{p1}}{G_1 I_{p1} + G_2 I_{p2}}\right) \qquad T_2 = T\left(\frac{G_2 I_{p2}}{G_1 I_{p1} + G_2 I_{p2}}\right) \tag{3-44a,b}$$

With these torques known, the essential part of the statically indeterminate analysis is completed. All other quantities, such as stresses and angles of twist, can now be found from the torques.

The preceding discussion illustrates the general methodology for analyzing a statically indeterminate system in torsion. In the following example, this same approach is used to analyze a bar that is fixed at both ends.

Note: In all problems and examples of this section we assume that the bars are made of linearly elastic materials. However, the general methodology is also applicable to bars of nonlinear materials—the only change is in the torque-displacement relations.

■ **Example 3-9**

The bar ACB shown in Figs. 3-33a and b is fixed at both ends and loaded by a torque T_0 at point C. Segments AC and CB of the bar have diameters d_A and d_B, lengths L_A and L_B, and polar moments of inertia I_{pA} and I_{pB}, respectively. The material of the bar is the same throughout both segments. Obtain formulas for:

(a)

(b)

(c)

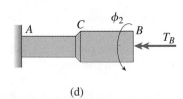

(d)

Fig. 3-33 Example 3-9. Statically indeterminate bar

(a) the reactive torques T_A and T_B at the ends, (b) the maximum shear stresses τ_{AC} and τ_{CB} in each segment of the bar, and (c) the angle of rotation ϕ_C at the cross section where the load T_0 is applied.

Solution   *Equation of equilibrium.* The load T_0 produces reactions T_A and T_B at the fixed ends of the bar, as shown in Figs. 3-33a and b. These figures serve as free-body diagrams for the bar because all torques (both the applied load and the reactions) are shown. Thus, from the equilibrium of the bar we obtain

$$T_A + T_B = T_0 \tag{f}$$

Because there are two unknowns in this equation (and no other useful equations of equilibrium), the bar is statically indeterminate.

Equation of compatibility. We now separate the bar from its support at end B and obtain a bar that is fixed at end A and free at end B (Figs. 3-33c and d). When the load T_0 acts alone on this bar, it produces an angle of twist at end B that we denote as ϕ_1 (Fig. 3-33c). Similarly, when the reactive torque T_B acts alone, it produces an angle ϕ_2 (Fig. 3-33d). The angle of twist at end B in the original bar, equal to the sum of ϕ_1 and ϕ_2, is zero. Therefore, the equation of compatibility is

$$\phi_1 + \phi_2 = 0 \tag{g}$$

Note that ϕ_1 and ϕ_2 are assumed to be positive in the direction shown in the figure.

Torque-displacement equations. The angles of twist ϕ_1 and ϕ_2 can be expressed in terms of the torques T_0 and T_B by referring to Figs. 3-33c and d and using the equation $\phi = TL/GI_p$. The equations are as follows:

$$\phi_1 = \frac{T_0 L_A}{GI_{pA}} \qquad \phi_2 = -\frac{T_B L_A}{GI_{pA}} - \frac{T_B L_B}{GI_{pB}} \tag{h,i}$$

The minus signs appear in Eq. (i) because T_B produces a rotation that is opposite in direction to the positive direction assumed in Fig. 3-33d.

We now substitute the angles of twist (Eqs. h and i) into the compatibility equation (Eq. g) and obtain

$$\frac{T_0 L_A}{GI_{pA}} - \frac{T_B L_A}{GI_{pA}} - \frac{T_B L_B}{GI_{pB}} = 0$$

or

$$\frac{T_B L_A}{I_{pA}} + \frac{T_B L_B}{I_{pB}} = \frac{T_0 L_A}{I_{pA}} \tag{j}$$

Solution of equations. The preceding equation can be solved for the torque T_B, which then can be substituted into the equation of equilibrium (Eq. f) to obtain the torque T_A. The results are

$$T_A = T_0\left(\frac{L_B I_{pA}}{L_B I_{pA} + L_A I_{pB}}\right) \qquad T_B = T_0\left(\frac{L_A I_{pB}}{L_B I_{pA} + L_A I_{pB}}\right) \tag{3-45a,b}$$

Thus, the reactive torques at the ends of the bar have been found, and the statically indeterminate part of the analysis is completed.

As a special case, note that if the bar is prismatic ($I_{pA} = I_{pB} = I_p$) the preceding results simplify to:

$$T_A = \frac{T_0 L_B}{L} \qquad T_B = \frac{T_0 L_A}{L} \qquad \text{(3-46a,b)}$$

where L is the total length of the bar. These equations are analogous to those for the reactions of an axially loaded bar with fixed ends (see Eqs. 2-8a and 2-8b).

Maximum shear stresses. The maximum shear stresses in each part of the bar are obtained directly from the torsion formula:

$$\tau_{AC} = \frac{T_A d_A}{2 I_{pA}} \qquad \tau_{CB} = \frac{T_B d_B}{2 I_{pB}}$$

Substituting from Eqs. (3-45a) and (3-45b) gives

$$\tau_{AC} = \frac{T_0 L_B d_A}{2(L_B I_{pA} + L_A I_{pB})} \qquad \tau_{CB} = \frac{T_0 L_A d_B}{2(L_B I_{pA} + L_A I_{pB})} \qquad \text{(3-47a,b)} \quad \Longleftarrow$$

By comparing the product $L_B d_A$ with the product $L_A d_B$, we can immediately determine which segment of the bar has the larger stress.

Angle of rotation. The angle of rotation ϕ_C at section C is equal to the angle of twist of either segment of the bar, since both segments rotate through the same angle at section C. Therefore, we obtain

$$\phi_C = \frac{T_A L_A}{G I_{pA}} = \frac{T_B L_B}{G I_{pB}} = \frac{T_0 L_A L_B}{G(L_B I_{pA} + L_A I_{pB})} \qquad \text{(3-48)} \quad \Longleftarrow$$

In the special case of a prismatic bar ($I_{pA} = I_{pB} = I_p$), the angle of rotation at the section where the load is applied is

$$\phi_C = \frac{T_0 L_A L_B}{G L I_p} \qquad \text{(3-49)}$$

This example illustrates not only the analysis of a statically indeterminate bar but also the determination of stresses and angles of rotation. Note also that the results obtained in this example are valid for a bar having either solid or tubular segments.

3.9 STRAIN ENERGY IN TORSION AND PURE SHEAR

When a load is applied to a structure, work is performed by the load and strain energy is developed in the structure, as described in detail in Section 2.7 for a bar subjected to axial loads. In the following discussion we will use the same basic concepts to determine the strain energy of a bar in torsion.

Consider a prismatic bar AB in pure torsion under the action of a torque T (Fig. 3-34). When the load is applied statically, the bar twists and the free end rotates through an angle ϕ. Now assume that the material of the bar is linearly elastic and follows Hooke's law. Then the relationship between the applied torque and the angle of twist will also be

Fig. 3-34 Prismatic bar in pure torsion

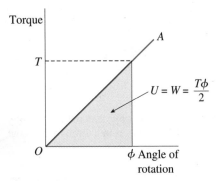

Fig. 3-35 Torque-rotation diagram for a bar in pure torsion (linearly elastic material)

linear, as shown by the torque-rotation diagram of Fig. 3-35 and as given by the equation $\phi = TL/GI_p$. The work W done by the torque as it rotates through the angle ϕ is equal to the area below the torque-rotation line OA, that is, it is equal to the area of the shaded triangle. Furthermore, from the principle of conservation of energy we know that the strain energy of the bar is equal to the work done by the load, provided no energy is gained or lost in the form of heat. Therefore, we obtain the following equation for the strain energy U of the bar:

$$U = W = \frac{T\phi}{2} \tag{3-50}$$

This equation is analogous to the equation $U = W = P\delta/2$ for a bar subjected to an axial load (see Eq. 2-29).

Using the equation $\phi = TL/GI_p$, we can now express the strain energy in the following forms:

$$U = \frac{T^2L}{2GI_p} \qquad U = \frac{GI_p\phi^2}{2L} \tag{3-51a,b}$$

The first expression is in terms of the load and the second is in terms of the angle of twist. Again, note the analogy with the corresponding equations for a bar with an axial load (see Eqs. 2-31a and b).

The SI unit for both work and energy is the joule (J), which is equal to one newton meter (1 J = 1 N·m). The basic USCS unit is the foot-pound (ft-lb), but other similar units, such as inch-pound (in.-lb) and inch-kip (in.-k), are commonly used.

■ Nonuniform Torsion

If a bar is subjected to nonuniform torsion (described in Section 3.4), we need additional formulas for the strain energy. In those cases where the bar consists of prismatic segments with constant torque in each segment (see Fig. 3-14a of Section 3.4), we can determine the strain energy of each segment and then add to obtain the total energy of the bar:

$$U = \sum_{i=1}^{n} U_i \tag{3-52}$$

in which U_i is the strain energy of segment i and n is the number of segments. For instance, if we use Eq. (3-51a) to obtain the individual strain energies, the preceding equation becomes

$$U = \sum_{i=1}^{n} \frac{T_i^2 L_i}{2G_i I_{pi}} \tag{3-53}$$

in which T_i is the internal torque in segment i and L_i, G_i, and I_{pi} are the torsional properties of the segment.

If either the cross section of the bar or the internal torque varies along the axis, as illustrated in Figs. 3-15 and 3-16 of Section 3.4, we can obtain the total strain energy by first determining the strain energy of an element and then integrating along the axis. For an element of length dx, the strain energy is (see Eq. 3-51a):

$$dU = \frac{[T(x)]^2 dx}{2GI_p(x)}$$

in which $T(x)$ is the internal torque acting on the element and $I_p(x)$ is the polar moment of inertia of the cross section at the element. Therefore, the total strain energy of the bar is

$$U = \int_0^L \frac{[T(x)]^2 dx}{2GI_p(x)} \tag{3-54}$$

Once again, the similarities of the expressions for strain energy in torsion and axial load should be noted (compare Eqs. 3-53 and 3-54 with Eqs. 2-34 and 2-35 of Section 2.7).

The use of the preceding equations for nonuniform torsion is illustrated in the examples that follow. In Example 3-10 the strain energy is found for a bar in pure torsion with prismatic segments, and in Examples 3-11 and 3-12 the strain energy is found for bars with varying torques and varying cross sectional dimensions. In addition, Example 3-12 shows how, under very limited conditions, the angle of twist of a bar can be determined from its strain energy. (For a more detailed discussion of this method, including its limitations, see the subsection "Displacements Caused by a Single Load" in Section 2.7.)

When evaluating strain energy we must keep in mind that the equations derived in this section apply only to bars of linearly elastic materials with small angles of twist. Also, we must remember the important observation stated previously in Section 2.7, namely, *the strain energy of a structure supporting more than one load cannot be obtained by adding the strain energies obtained for the individual loads acting separately.* This observation is demonstrated in Example 3-10.

■ Strain-Energy Density in Pure Shear

Because the individual elements of a bar in torsion are stressed in pure shear, it is useful to obtain expressions for the strain energy associated with the shear stresses. We begin the analysis by considering a small element of material subjected to shear stresses τ on its side faces (Fig. 3-36a). For convenience, we will assume that the front face of the element is square, with each side having length h. Although the figure shows only a two-dimensional view of the element, we recognize that the element is actually three-dimensional with thickness t perpendicular to the plane of the figure. Under the action of the shear stresses, the element is distorted so that the front face becomes a rhombus, as shown in Fig. 3-36b. The change in angle at each corner of the element is the shear strain γ.

The shear forces V acting on the side faces of the element (Fig. 3-36c) are found by multiplying the stresses by the areas ht over which they act:

$$V = \tau ht \qquad\qquad\text{(a)}$$

These forces produce work as the element deforms from its initial shape (Fig. 3-36a) to its distorted shape (Fig. 3-36b). To calculate this work we need to determine the relative distances through which the shear forces move. This task is made easier if the element in Fig. 3-36c is rotated as a rigid body until two of its faces are horizontal, as in Fig. 3-36d. During the rigid-body rotation, the net work done by the forces V is zero because the forces occur in pairs that form two equal and opposite couples.

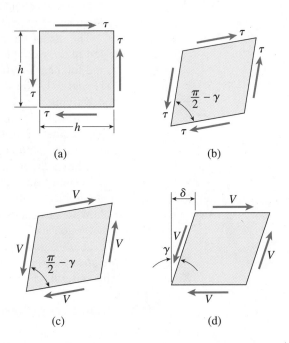

(a)

(b)

(c)

(d)

Fig. 3-36 Element in pure shear

As can be seen in Fig. 3-36d, the top face of the element is displaced horizontally through a distance δ (relative to the bottom face) as the shear force is gradually increased from zero to its final value V. The displacement δ is equal to the product of the shear strain γ (which is a small angle) and the vertical dimension of the element:

$$\delta = \gamma h \qquad\qquad\text{(b)}$$

If we assume that the material is linearly elastic and follows Hooke's law, then the work done by the forces V is equal to $V\delta/2$, which is also the strain energy stored in the element:

$$U = W = \frac{V\delta}{2} \tag{c}$$

Note that the forces acting on the side faces of the element (Fig. 3-36d) do not move along their lines of action—hence they do no work. Substituting from Eqs. (a) and (b) into Eq. (c), we get the total strain energy of the element:

$$U = \frac{\tau \gamma h^2 t}{2}$$

Because the volume of the element is $h^2 t$, the **strain-energy density** u (that is, the strain energy per unit volume) is

$$u = \frac{\tau \gamma}{2} \tag{d}$$

Finally, we substitute Hooke's law in shear ($\tau = G\gamma$) and obtain the following equations for the strain-energy density in pure shear:

$$u = \frac{\tau^2}{2G} \qquad u = \frac{G\gamma^2}{2} \tag{3-55a,b}$$

These equations are similar in form to those for uniaxial stress (Eqs. 2-38a and b of Section 2.7).

The SI unit for strain-energy density is joule per cubic meter (J/m^3), and the USCS unit is inch-pound per cubic inch (or other similar units). Since these units are the same as those for stress, we may also express strain-energy density in pascals (Pa) or pounds per square inch (psi).

In the next section (Section 3.10) we will use the equation for strain-energy density in terms of the shear stress (Eq. 3-55a) to determine the angle of twist of a thin-walled tube of arbitrary cross-sectional shape.

■ Example 3-10

A solid circular bar AB of length L is fixed at one end and free at the other (Fig. 3-37). Three different loading conditions are to be considered: (a) torque T_a acting at the free end; (b) torque T_b acting at the midpoint of the bar; and (c) torques T_a and T_b acting simultaneously. For each case of loading, obtain a formula for the strain energy stored in the bar. Then evaluate the strain energy for the following data: $T_a = 100$ N·m, $T_b = 150$ N·m, $L = 1.6$ m, $G = 80$ GPa, and $I_p = 79.52 \times 10^3$ mm⁴.

(a)

(b)

(c)

Fig. 3-37 Example 3-10

Solution   (a) *Torque T_a acting at the free end (Fig. 3-37a).* In this case the strain energy is obtained directly from Eq. (3-51a):

$$U_a = \frac{T_a^2 L}{2GI_p} \qquad (e) \Leftarrow$$

(b) *Torque T_b acting at the midpoint (Fig. 3-37b).* When the torque acts at the midpoint, we apply Eq. (3-51a) to segment AC of the bar:

$$U_b = \frac{T_b^2 (L/2)}{2GI_p} = \frac{T_b^2 L}{4GI_p} \qquad (f) \Leftarrow$$

(c) *Torques T_a and T_b acting simultaneously (Fig. 3-37c).* When both loads act on the bar, the torque in segment CB is T_a and the torque in segment AC is $T_a + T_b$. Thus, the strain energy (from Eq. 3-53) is

$$U_c = \sum_{i=1}^{n} \frac{T_i^2 L_i}{2G_i I_{pi}} = \frac{T_a^2 (L/2)}{2GI_p} + \frac{(T_a + T_b)^2 (L/2)}{2GI_p}$$

$$= \frac{T_a^2 L}{2GI_p} + \frac{T_a T_b L}{2GI_p} + \frac{T_b^2 L}{4GI_p} \qquad (g) \Leftarrow$$

A comparison of Eqs. (e), (f), and (g) shows that the strain energy produced by the two loads acting simultaneously is *not* equal to the sum of the strain energies produced by the loads acting separately. As pointed out in Section 2.7, the reason is that strain energy is a quadratic function of the loads, not a linear function.

(d) *Numerical results.* Substituting the given data into Eq. (e), we obtain

$$U_a = \frac{T_a^2 L}{2GI_p} = \frac{(100 \text{ N·m})^2 (1.6 \text{ m})}{2(80 \text{ GPa})(79.52 \times 10^3 \text{ mm}^4)} = 1.26 \text{ J} \qquad \Leftarrow$$

Recall that one joule is equal to one newton meter (1 J = 1 N·m). Proceeding in the same manner for Eqs. (f) and (g), we find

$$U_b = 2.83 \text{ J} \qquad \Leftarrow$$

$$U_c = 1.26 \text{ J} + 1.89 \text{ J} + 2.83 \text{ J} = 5.98 \text{ J} \qquad \Leftarrow$$

Note that the middle term, involving the product of the two loads, contributes significantly to the strain energy and cannot be disregarded.

■ Example 3-11

A prismatic bar AB, fixed at one end and free at the other, is loaded by a distributed torque of constant intensity t per unit distance along the axis of the bar (Fig. 3-38). (a) Derive a formula for the strain energy of the bar. (b) Evaluate the strain energy of a hollow shaft used for drilling into the earth if the data are as follows: $t = 480$ lb-in./in., $L = 12$ ft, $G = 11.5 \times 10^6$ psi, and $I_p = 17.18$ in.[4]

Solution   (a) *Strain energy of the bar.* The first step in the solution is to determine the internal torque $T(x)$ acting at distance x from the free end of the bar (Fig. 3-38). This torque is equal to the total torque acting on the segment of the bar from $x = 0$ to $x = x$; thus,

Fig. 3-38 Example 3-11

$$T(x) = tx \tag{h}$$

Substituting into Eq. (3-54), we obtain

$$U = \int_0^L \frac{[T(x)]^2 dx}{2GI_p} = \frac{1}{2GI_p} \int_0^L (tx)^2 dx = \frac{t^2 L^3}{6GI_p} \tag{3-56} \quad \longleftarrow$$

which gives the strain energy stored in the bar.

(b) *Numerical results.* To evaluate the strain energy of the shaft, we substitute the given data into Eq. (3-56):

$$U = \frac{t^2 L^3}{6GI_p} = \frac{(480 \text{ lb-in./in.})^2 (144 \text{ in.})^3}{6(11.5 \times 10^6 \text{ psi})(17.18 \text{ in.}^4)} = 580 \text{ in.-lb} \quad \longleftarrow$$

This example illustrates the use of integration to evaluate the strain energy of a bar subjected to a distributed load.

■ **Example 3-12**

A tapered bar *AB* of solid circular cross section is supported at one end and loaded by a torque *T* at the other end (Fig. 3-39). The diameter of the bar varies linearly from d_A at the left-hand end to d_B at the right-hand end. Determine the angle of rotation ϕ_A at end *A* of the bar by equating the strain energy to the work done by the load.

Solution   From the principle of conservation of energy we know that the work done by the applied torque equals the strain energy of the bar; thus, $W = U$. The work is given by the equation

Fig. 3-39 Example 3-12. Tapered bar in torsion

$$W = \frac{T\phi_A}{2} \tag{i}$$

and the strain energy *U* is given by Eq. (3-54).

To use Eq. (3-54), we need expressions for the torque $T(x)$ and the polar moment of inertia $I_p(x)$. The torque is constant along the axis of the bar and equal to the load *T*, and the polar moment of inertia is

$$I_p(x) = \frac{\pi}{32} \big[d(x) \big]^4$$

in which $d(x)$ is the diameter of the bar at distance *x* from end *A*. From the geometry of the figure, we see that

$$d(x) = d_A + \frac{d_B - d_A}{L} x \tag{j}$$

and therefore

$$I_p(x) = \frac{\pi}{32} \left(d_A + \frac{d_B - d_A}{L} x \right)^4 \tag{k}$$

Now we can substitute into Eq. (3-54), as follows:

$$U = \int_0^L \frac{[T(x)]^2 dx}{2GI_p(x)} = \frac{16T^2}{\pi G} \int_0^L \frac{dx}{\left(d_A + \dfrac{d_B - d_A}{L}x\right)^4}$$

The integral in this expression can be integrated with the aid of a table of integrals (see Appendix C). However, we already evaluated this integral in Example 3-5 of Section 3.4 (see Eq. g of that example) and found that

$$\int_0^L \frac{dx}{\left(d_A + \dfrac{d_B - d_A}{L}x\right)^4} = \frac{L}{3(d_B - d_A)}\left(\frac{1}{d_A^3} - \frac{1}{d_B^3}\right)$$

Therefore, the strain energy of the tapered bar is

$$U = \frac{16T^2L}{3\pi G(d_B - d_A)}\left(\frac{1}{d_A^3} - \frac{1}{d_B^3}\right) \tag{3-57}$$

Equating the strain energy to the work of the torque (Eq. i) and solving for ϕ_A, we get

$$\phi_A = \frac{32TL}{3\pi G(d_B - d_A)}\left(\frac{1}{d_A^3} - \frac{1}{d_B^3}\right) \tag{3-58} \quad \longleftarrow$$

This equation, which is the same as Eq. (3-26) in Example 3-5 of Section 3.4, gives the angle of rotation at end A of the tapered bar.

Note especially that the method used in this example for finding the angle of rotation is suitable only when the bar is subjected to a single load, and then only when the desired angle corresponds to that load. Otherwise, we must find angular displacements by the usual methods described in Sections 3.3, 3.4, and 3.8.

3.10 THIN-WALLED TUBES

The torsion theory described in the preceding sections is applicable to solid or hollow bars of circular cross section. Circular shapes are the most efficient shapes for resisting torsion and consequently are the most commonly used, especially in machinery. However, in lightweight structures, such as aircraft and spacecraft, thin-walled tubular members with noncircular cross sections are often required to resist torsion. In this section, we will analyze structural members of this kind.

To obtain formulas that are applicable to a variety of shapes, let us consider a thin-walled tube of arbitrary cross section (Fig. 3-40a). The tube is cylindrical in shape—that is, all cross sections are identical and the longitudinal axis is a straight line. The thickness t of the wall is not necessarily constant but may vary around the cross section. However, the thickness must be small in comparison with the total width of the tube. The tube is subjected to pure torsion by torques T acting at the ends.

■ Shear Stresses

The shear stresses τ acting on a cross section of the tube are pictured in Fig. 3-40b, which shows an element of the tube cut out between two cross sections that are distance dx apart. The stresses act parallel to the boundaries of the cross section and "flow" around the cross section. Also, the intensity of the stresses varies so slightly *across* the thickness of the tube (because the tube is assumed to be thin) that we may assume τ to be constant in that direction. However, if the thickness t is not constant, the stresses will vary in intensity as we go *around* the cross section, and the manner in which they vary must be determined from equilibrium.

To determine the magnitude of the shear stresses, we will consider a rectangular element *abcd* obtained by making two longitudinal cuts *ab* and *cd* (Figs. 3-40a and b). This element is isolated as a free body in Fig. 3-40c. Acting on the cross-sectional face *bc* are the shear stresses τ shown in Fig. 3-40b. We assume that these stresses vary in intensity as we move along the cross section from *b* to *c*; therefore, the shear stress at *b* is denoted τ_b and the stress at *c* is denoted τ_c. As we know from equilibrium, identical shear stresses act in the opposite direction on the opposite

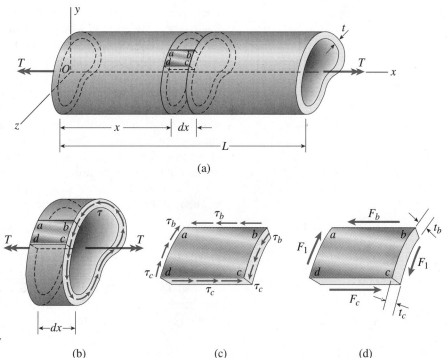

Fig. 3-40 Thin-walled tube of arbitrary cross-sectional shape

cross-sectional face *ad*, and shear stresses of the same magnitude also act on the longitudinal faces *ab* and *cd*. Thus, the constant shear stresses acting on faces *ab* and *cd* are equal to τ_b and τ_c, respectively.

The stresses acting on the longitudinal faces *ab* and *cd* produce forces F_b and F_c (Fig. 3-40d). These forces are obtained by multiplying the stresses by the areas on which they act:

$$F_b = \tau_b t_b dx \qquad F_c = \tau_c t_c dx$$

in which t_b and t_c represent the thicknesses of the tube at points *b* and *c*, respectively (see Fig. 3-40d). In addition, forces F_1 are produced by the stresses acting on faces *bc* and *ad*. From the equilibrium of the element in the longitudinal direction (the *x* direction), we see that $F_b = F_c$, or

$$\tau_b t_b = \tau_c t_c$$

Because the locations of the longitudinal cuts *ab* and *cd* were selected arbitrarily, it follows from the preceding equation that the product of the shear stress τ and the thickness *t* of the tube is the same at every point in the cross section. This product is known as the **shear flow** and is denoted by the letter *f*:

$$f = \tau t = \text{constant} \qquad (3\text{-}59)$$

This relationship shows that the largest shear stress occurs where the thickness of the tube is smallest, and vice versa. In regions where the thickness is constant, the shear stress is constant.

The next step in the analysis is to relate the shear flow *f* (and hence the shear stress τ) to the torque *T* acting on the tube. For that purpose, let us examine the cross section of the tube, as pictured in Fig. 3-41. The **median line** (also called the *centerline* or the *midline*) of the wall of the tube is shown as a dashed line in the figure. We consider an element of area of length *ds* (measured along the median line) and thickness *t*. The distance *s* defining the location of the element is measured along the median line from some arbitrarily chosen reference point.

The total shear force acting on the element of area is *f ds*, and the moment of this force about any point *O* within the tube is

$$dT = rf\,ds$$

in which *r* is the perpendicular distance from point *O* to the line of action of the force *f ds*. (Note that the line of action of the force *f ds* is tangent to the median line of the cross section at the element *ds*.) The total torque *T* produced by the shear stresses is obtained by integrating along the median line of the cross section:

$$T = f \int_0^{L_m} r\,ds \qquad (a)$$

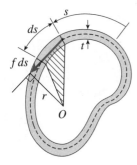

Fig. 3-41 Cross section of thin-walled tube

in which L_m denotes the length of the median line. The integral in Eq. (a) can be difficult to integrate by formal mathematical means, but fortunately it can be evaluated easily by giving it a simple geometric interpretation. The quantity $r\,ds$ represents twice the area of the shaded triangle shown in Fig. 3-41. (Note that the triangle has base length ds and height equal to r.) Therefore, the integral represents twice the area A_m enclosed by the median line of the cross section:

$$\int_0^{L_m} r\,ds = 2A_m \tag{b}$$

It follows from Eq. (a) that $T = 2fA_m$, and therefore the shear flow is

$$f = \frac{T}{2A_m} \tag{3-60}$$

Now we can eliminate the shear flow f between Eqs. (3-59) and (3-60) and obtain a **torsion formula** for thin-walled tubes:

$$\tau = \frac{T}{2tA_m} \tag{3-61}$$

Since t and A_m are properties of the cross section, the shear stresses τ can be calculated from Eq. (3-61) for any thin-walled tube subjected to a known torque T. (*Reminder:* The area A_m is the area *enclosed* by the median line—it is *not* the cross-sectional area of the tube.)

To illustrate the use of the torsion formula, consider a thin-walled **circular tube** (Fig. 3-42) of thickness t and radius r to the median line. The area enclosed by the median line is

$$A_m = \pi r^2 \tag{3-62}$$

and therefore the shear stress (constant around the cross section) is

$$\tau = \frac{T}{2\pi r^2 t} \tag{3-63}$$

This formula agrees with the stress obtained from the standard torsion formula (Eq. 3-11) when the standard formula is applied to a circular tube with thin walls using the approximate expression $I_p \approx 2\pi r^3 t$ for the polar moment of inertia (Eq. 3-18).

As a second illustration, consider a thin-walled **rectangular tube** (Fig. 3-43) having thickness t_1 on the sides and thickness t_2 on the top and bottom. Also, the height and width (measured to the median line of the cross section) are h and b, respectively. The area within the median line is

$$A_m = bh \tag{3-64}$$

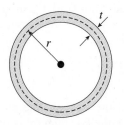

Fig. 3-42 Thin-walled circular tube

Fig. 3-43 Thin-walled rectangular tube

and thus the shear stresses in the vertical and horizontal sides, respectively, are

$$\tau_{\text{vert}} = \frac{T}{2t_1 bh} \qquad \tau_{\text{horiz}} = \frac{T}{2t_2 bh} \qquad (3\text{-}65\text{a,b})$$

If t_2 is larger than t_1, the maximum shear stress will occur in the vertical sides of the cross section.

■ **Strain Energy and Torsion Constant**

The strain energy of a thin-walled tube can be determined by first finding the strain energy of an element and then integrating throughout the volume of the bar. Consider an element of the tube having area $t\,ds$ in the cross section (see the element in Fig. 3-41) and length dx (see the element in Fig. 3-40). The volume of such an element, which is similar in shape to the element *abcd* shown in Fig. 3-40a, is $t\,ds\,dx$. Because elements of the tube are in pure shear, the strain-energy density of the element is $\tau^2/2G$, as given by Eq. (3-55a). The total strain energy of the element is equal to the strain-energy density times the volume:

$$dU = \frac{\tau^2}{2G} t\,ds\,dx = \frac{\tau^2 t^2}{2G} \frac{ds}{t}\,dx = \frac{f^2}{2G} \frac{ds}{t}\,dx$$

in which we have replaced τt by the shear flow f (a constant). The total strain energy of the tube is obtained by integrating dU throughout the volume of the tube, that is, ds is integrated from 0 to L_m around the median line and dx is integrated along the axis of the tube from 0 to L, where L is the length. Thus,

$$U = \int dU = \frac{f^2}{2G} \int_0^{L_m} \frac{ds}{t} \int_0^L dx$$

Note that the thickness t may vary around the median line and must remain with ds under the integral sign. Since the last integral is equal to the length L of the tube, the equation for the strain energy becomes

$$U = \frac{f^2 L}{2G} \int_0^{L_m} \frac{ds}{t}$$

Substituting for the shear flow from Eq. (3-60), we obtain

$$U = \frac{T^2 L}{8GA_m^2} \int_0^{L_m} \frac{ds}{t} \qquad (3\text{-}66)$$

as the equation for the strain energy of the tube in terms of the torque T.

The preceding expression for strain energy can be written in simpler form by introducing a new property of the cross section, called the **torsion constant**. For a thin-walled tube, the torsion constant (denoted by the letter J) is defined as follows:

$$J = \frac{4A_m^2}{\displaystyle\int_0^{L_m} \frac{ds}{t}} \tag{3-67}$$

With this notation, the equation for strain energy (Eq. 3-66) becomes

$$U = \frac{T^2 L}{2GJ} \tag{3-68}$$

which has the same form as the equation for strain energy in a circular bar (see Eq. 3-51a). The only difference is that the torsion constant J has replaced the polar moment of inertia I_p. Note that the torsion constant has units of length to the fourth power.

In the special case of a cross section having constant thickness t, the expression for J (Eq. 3-67) simplifies to

$$J = \frac{4tA_m^2}{L_m} \tag{3-69}$$

For each shape of cross section, we can evaluate J from either Eq. (3-67) or Eq. (3-69).

As an illustration, consider again the thin-walled **circular tube** of Fig. 3-42. Since the thickness is constant we use Eq. (3-69) and substitute $L_m = 2\pi r$ and $A_m = \pi r^2$; the result is

$$J = 2\pi r^3 t \tag{3-70}$$

which is the approximate expression for the polar moment of inertia (Eq. 3-18). Thus, in the case of a thin-walled circular tube, the polar moment of inertia is the same as the torsion constant.

As a second illustration, we will again use the **rectangular tube** of Fig. 3-43. For this cross section we have $A_m = bh$, and therefore the integral in Eq. (3-67) is

$$\int_0^{L_m} \frac{ds}{t} = 2\int_0^h \frac{ds}{t_1} + 2\int_0^b \frac{ds}{t_2} = 2\left(\frac{h}{t_1} + \frac{b}{t_2}\right)$$

Thus, the torsion constant is

$$J = \frac{2b^2 h^2 t_1 t_2}{bt_1 + ht_2} \tag{3-71}$$

Torsion constants for other cross sections can be found in a similar manner.

■ **Angle of Twist**

The angle of twist ϕ for a thin-walled tube of arbitrary cross-sectional shape (Fig. 3-44) may be determined by equating the work W done by the applied torque T to the strain energy U of the tube. Thus,

$$W = U \quad \text{or} \quad \frac{T\phi}{2} = \frac{T^2L}{2GJ}$$

from which we get the equation for the angle of twist:

$$\phi = \frac{TL}{GJ} \tag{3-72}$$

Again we observe that the equation has the same form as the corresponding equation for a circular bar (see Eq. 3-15) but with the polar moment of inertia replaced by the torsion constant. The quantity GJ is called the **torsional rigidity** of the tube.

Fig. 3-44 Angle of twist ϕ for a thin-walled tube

■ **Limitations**

The formulas developed in this section apply to prismatic members having closed tubular shapes with thin walls. If the cross section is thin-walled but open, as in the case of I-beams and channel sections, the theory given here does not apply. To emphasize this point, imagine that we take a thin-walled tube and slit it lengthwise—then the cross section becomes an open section, the shear stresses and angles of twist increase, the torsional resistance decreases, and the formulas given in this section cannot be used.

Some of the formulas given in this section are restricted to linearly elastic materials—for instance, any equation containing the shear modulus of elasticity G is in this category. However, the equations for shear flow and shear stress (Eqs. 3-60 and 3-61) are based only upon equilibrium and are valid regardless of the material properties. The entire theory is approximate and becomes less accurate as the wall thickness t increases.*

---

*The torsion theory for thin-walled tubes described in this section was developed by R. Bredt, a German engineer who presented it in 1896 (Ref. 3-4).

An important consideration in the design of any thin-walled member is the possibility that the walls will buckle. The thinner the walls and the longer the tube, the more likely it is that buckling will occur. In the case of noncircular tubes, stiffeners and diaphragms are often used to maintain the shape of the tube and prevent localized buckling. In all of our discussions and problems, we assume that buckling is prevented.

■ Example 3-13

Fig. 3-45 Example 3-13

Compare the maximum shear stress in a circular tube (Fig. 3-45) as calculated by the approximate theory for a thin-walled tube with the stress calculated by the exact torsion theory. (Note that the tube has constant thickness t and radius r to the median line of the cross section.)

Solution The shear stress obtained from the approximate theory for a thin-walled tube (Eq. 3-63) is

$$\tau_1 = \frac{T}{2\pi r^2 t} = \frac{T}{2\pi t^3 \beta^2} \tag{c}$$

in which the relation

$$\beta = \frac{r}{t} \tag{d}$$

is introduced.

The maximum stress obtained from the more accurate torsion formula (Eq. 3-11) is

$$\tau_2 = \frac{T(r + t/2)}{I_p} \tag{e}$$

where

$$I_p = \frac{\pi}{2}\left[\left(r + \frac{t}{2}\right)^4 - \left(r - \frac{t}{2}\right)^4\right]$$

After expansion, this expression simplifies to

$$I_p = \frac{\pi r t}{2}(4r^2 + t^2) \tag{f}$$

and the expression for the shear stress (Eq. e) becomes

$$\tau_2 = \frac{T(2r + t)}{\pi r t(4r^2 + t^2)} = \frac{T(2\beta + 1)}{\pi t^3 \beta(4\beta^2 + 1)} \tag{g}$$

Thus, the ratio τ_1/τ_2 of the shear stresses is

$$\frac{\tau_1}{\tau_2} = \frac{4\beta^2 + 1}{2\beta(2\beta + 1)} \tag{3-73} \quad \Longleftarrow$$

which depends only on the ratio β.

For values of β equal to 5, 10, and 20, we obtain from Eq. (3-73) the values $\tau_1/\tau_2 = 0.92$, 0.95, and 0.98, respectively. Thus, we see that the approximate formula for the shear stresses gives results that are slightly less than those obtained from the exact formula. The accuracy of the approximate formula increases as the wall of the tube becomes thinner. In the limit, as the thickness approaches zero and β approaches infinity, the ratio τ_1/τ_2 becomes 1.

■ **Example 3-14**

A circular tube and a square tube (Fig. 3-46) are constructed of the same material and subjected to the same torque. Both tubes have the same length, same wall thickness, and same cross-sectional area. What are the ratios of their shear stresses and angles of twist? (Disregard the effects of stress concentrations at the corners of the square tube.)

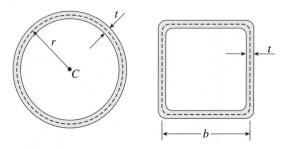

Fig. 3-46 Example 3-14

Solution   *Circular tube*. For the circular tube, the area A_{m1} enclosed by the median line of the cross section is

$$A_{m1} = \pi r^2$$

where r is the radius to the median line. Also, the torsion constant (Eq. 3-70) and cross-sectional area are

$$J_1 = 2\pi r^3 t \qquad A_1 = 2\pi r t$$

Square tube. For the square tube, the cross-sectional area is

$$A_2 = 4bt$$

where b is the length of one side, measured along the median line. Inasmuch as the areas of the tubes are the same, we obtain $b = \pi r/2$. Also, the torsion constant (Eq. 3-71) and area enclosed by the median line of the cross section are

$$J_2 = b^3 t = \frac{\pi^3 r^3 t}{8} \qquad A_{m2} = b^2 = \frac{\pi^2 r^2}{4}$$

Ratios. The ratio τ_1/τ_2 of the shear stress in the circular tube to the shear stress in the square tube is (see Eq. 3-61)

$$\frac{\tau_1}{\tau_2} = \frac{A_{m2}}{A_{m1}} = \frac{\pi^2 r^2/4}{\pi r^2} = \frac{\pi}{4} = 0.785 \qquad\qquad \text{(h)} \quad \Longleftarrow$$

The ratio of the angles of twist (see Eq. 3-72) is

$$\frac{\phi_1}{\phi_2} = \frac{J_2}{J_1} = \frac{\pi^3 r^3 t/8}{2\pi r^3 t} = \frac{\pi^2}{16} = 0.617 \qquad\qquad \text{(i)} \quad \Longleftarrow$$

These results show that the circular tube not only has a lower shear stress than does the square tube but also a greater stiffness against rotation.

*3.11 STRESS CONCENTRATIONS IN TORSION

In the previous sections of this chapter we discussed the stresses in torsional members assuming that the stress distribution varied in a smooth and continuous manner. This assumption is valid provided that there are no abrupt changes in the shape of the bar (no holes, grooves, steps, and the like) and provided that the region under consideration is away from any points of loading. If such disruptive conditions do exist, then high localized stresses will be created in the regions surrounding the discontinuities. In practical engineering work these **stress concentrations** are handled by means of **stress-concentration factors**, as explained previously in Section 2.10.

The effects of a stress concentration are confined to a small region around the discontinuity, in accord with Saint-Venant's principle (see Section 2.10). For instance, consider a stepped shaft consisting of two segments having different diameters (Fig. 3-47). The larger segment has diameter D_2 and the smaller segment has diameter D_1. The junction between the two segments forms a "step" or "shoulder" that is machined with a fillet of radius R. Without the fillet, the theoretical stress concentration factor would be infinitely large because of the abrupt 90° reentrant corner. Of course, infinite stresses cannot occur. Instead, the material at the reentrant corner would deform and partially relieve the high stress concentration. However, such a situation is very dangerous under dynamic loads, and in good design a fillet is always used. The larger the radius of the fillet, the lower the stresses.

At a distance from the shoulder approximately equal to the diameter D_2 (for instance, at cross section A-A in Fig. 3-47a) the torsional shear stresses are practically unaffected by the discontinuity. Therefore, the maximum stress τ_2 at a sufficient distance to the left of the shoulder can be found from the torsion formula using D_2 as the diameter (Fig. 3-47b). The same general comments apply at section C-C, which is distance D_1 (or greater) from the toe of the fillet. Because the diameter D_1 is less than the diameter D_2, the maximum stress τ_1 at section C-C (Fig. 3-47d) is larger than the stress τ_2.

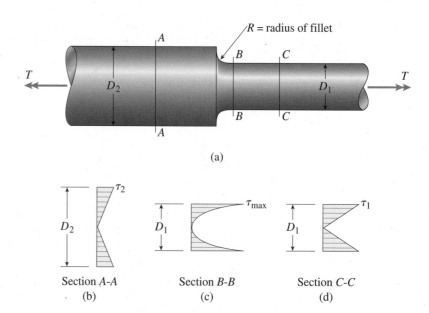

Fig. 3-47 Stepped shaft in torsion

The stress-concentration effect is greatest at section B-B, which cuts through the toe of the fillet. At this section the maximum stress is

$$\tau_{max} = K\tau_{nom} = K\frac{Tr}{I_p} = K\left(\frac{16T}{\pi D_1^3}\right) \tag{3-74}$$

In this equation, K is the stress-concentration factor and τ_{nom} (equal to τ_1) is the nominal shear stress, that is, the shear stress in the smaller part of the shaft. Values of the factor K are plotted in Fig. 3-48 as a function of the ratio R/D_1. Curves are plotted for various values of the ratio D_2/D_1. Note that when the fillet radius R becomes very small and the transition from one diameter to the other is abrupt, the value of K becomes quite large. Conversely, when R is large, the value of K approaches 1.0 and the effect of the stress concentration disappears. The dashed curve in Fig. 3-48 is for the special case of a full quarter-circular fillet, which means that $D_2 = D_1 + 2R$. (*Note:* Problems 3.11-1 through 3.11-5 provide practice in obtaining values of K from Fig. 3-48.)

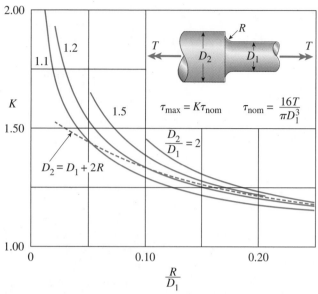

Fig. 3-48 Stress-concentration factor K for a stepped shaft in torsion. (The dashed line is for a full quarter-circular fillet.)

Many other cases of stress concentrations for circular shafts, such as a shaft with a keyway and a shaft with a hole, are available in the engineering literature (see, for example, Refs. 2-10 and 2-11).

As explained in Section 2.10, stress concentrations are important for brittle materials under static loads and for most materials under dynamic loads. As a case in point, fatigue failures are of major concern in the design of rotating shafts and axles (see Section 2.9 for a brief discussion of fatigue). The theoretical stress-concentration factors K given in this section are based upon linearly elastic behavior of the material. However, fatigue experiments show that these factors are conservative, and failures in ductile materials usually occur at larger loads than those predicted by the theoretical factors.

*3.12 | NONLINEAR TORSION OF CIRCULAR BARS

The equations derived in the preceding sections for the shear stresses in circular bars are valid only if the material follows Hooke's law. In this section we will consider the behavior of bars when the shear stresses exceed the proportional limit and Hooke's law no longer applies.

Our analysis is based upon the equations for the torsional deformations of circular bars, that is, the equations for shear strains γ and rates of twist θ (Eqs. 3-1 through 3-5). These equations are based upon the geometry of the deformed bar, and therefore they are valid for any material, whether or not it follows Hooke's law. In particular, the shear strain at distance ρ from the axis of the bar (see Fig. 3-4c) is given by Eq. (3-4), which is repeated here:

$$\gamma = \rho\theta \tag{3-75}$$

Also, the maximum shear strain (from Eq. 3-3) is

$$\gamma_{max} = r\theta \tag{3-76}$$

in which r is the radius of a solid circular bar or the outer radius of a circular tube.

Now let us assume that the shear stress-strain diagram for the material is known (Fig. 3-49a). Then, for any given value of the rate of twist θ, we can find the shear stress τ at any point in the bar. The procedure consists of obtaining the shear strain from Eq. (3-75) or Eq. (3-76) and then obtaining the corresponding shear stress from the stress-strain curve. By considering various values of the radius ρ, we can obtain the distribution of shear stresses over the entire cross section (Fig. 3-49b). Because the shear strain increases linearly with the radius ρ, the diagram of shear stresses has the same shape as the stress-strain diagram itself.

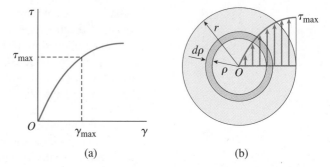

Fig. 3-49 Nonlinear torsion of a circular bar: (a) shear stress-strain diagram, and (b) distribution of shear stresses acting on the cross section

The torque T corresponding to the given value of the rate of twist θ is the moment resultant of the shear stresses acting on the cross section. To obtain this torque, we consider a ring-shaped element at distance ρ from the center and having thickness $d\rho$ (Fig. 3-49b). The elemental torque dT produced by the shear stresses acting on the ring-shaped element is

$$dT = \rho\tau(2\pi\rho\, d\rho) = 2\pi\rho^2\tau\, d\rho$$

and therefore the total torque is

$$T = \int_0^r 2\pi\rho^2\tau\, d\rho \tag{3-77}$$

From Eq. (3-75) we obtain $\rho = \gamma/\theta$ and $d\rho = d\gamma/\theta$. Substituting these relations into Eq. (3-77), and also changing the upper limit of integration to γ_{max} to correspond with the new independent variable γ, we obtain

$$T = \frac{2\pi}{\theta^3} \int_0^{\gamma_{max}} \tau\gamma^2 \, d\gamma \qquad (3\text{-}78)$$

This equation relates the torque T and the rate of twist θ.

To facilitate the evaluation of Eq. (3-78), we note that the integral on the right-hand side has a simple geometric interpretation. It is the moment of inertia with respect to the vertical axis (that is, the τ axis) of the area below the shear stress-strain curve (Fig. 3-49a). The area is located between the origin O and the maximum strain γ_{max}. Thus, we can write Eq. (3-78) in the form

$$T = \frac{2\pi}{\theta^3} I_\tau \qquad (3\text{-}79)$$

in which I_τ is the moment of inertia of the specified area, evaluated with respect to the τ axis.

The procedure for carrying out the calculations is as follows. Given a solid circular bar with a known radius and known shear stress-strain curve, we assume a value for the rate of twist θ. Then we calculate γ_{max} from Eq. (3-76) and determine (either analytically or numerically) the corresponding moment of inertia I_τ from the stress-strain curve (Fig. 3-49a). Next, we calculate the torque T from Eq. (3-79). By repeating this procedure for various values of θ, we obtain a curve giving the relationship between T and θ (Fig. 3-50). Having such a curve, we can obtain θ for any given value of the torque T. Then, with θ known, the strains, stresses, and angle of twist can also be determined for the given value of T.

Many variations of this general procedure are possible, depending upon what information is known and what quantities are to be determined. Some of the possibilities are illustrated in the problems at the end of the chapter.

Although Eqs. (3-77), (3-78), and (3-79) were derived for a solid circular bar, they can be adapted for circular tubes by changing the lower limits of the integrals from zero to the appropriate values for a tube.

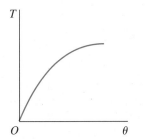

Fig. 3-50 Relationship between torque T and rate of twist θ

■ **Elastoplastic Materials**

If the material of the bar has a pronounced yield stress τ_y, the shear stress-strain diagram can be idealized as shown in Fig. 3-51a. The diagram consists of two straight lines, the first representing linearly elastic behavior and the second representing perfectly plastic behavior (discussed previously in Sections 2.11 and 2.12). The behavior of the bar depends upon the magnitude of the maximum shear strain in the bar. If the maximum strain is less than the yield strain γ_y, the material follows Hooke's law and the formulas derived in earlier sections can be used. If the maximum strain is larger than the yield strain, some of the material is plastic and an elastoplastic analysis is required. This type of analysis is described in the following paragraphs.

When the strain at the outer surface exceeds the yield strain, the stress distribution on the cross section consists of two parts, as shown in Fig. 3-51b. Yielding begins at the outer surface of the bar and moves progressively inward as the load increases. A central "core" of material remains in the linearly elastic range. If the strains become very large, the region of yielding will approach the center of the bar and the stress distribution will approach the uniform distribution shown in Fig. 3-51c. The corresponding torque T_p, called the **plastic torque**, can be found from Eq. (3-77):

$$T_p = \int_0^r 2\pi\rho^2\tau_y\,d\rho = \frac{2\pi r^3\tau_y}{3} \tag{3-80}$$

When this value of torque is reached, further twisting of the shaft will occur without an increase in the torque. Eventually, the effects of strain hardening will become noticeable and stresses greater than τ_y will develop.

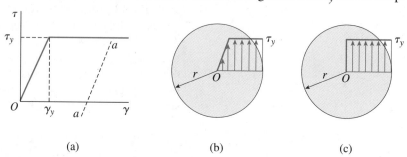

Fig. 3-51 Torsion of a solid circular bar of elastoplastic material

(a) (b) (c)

The **yield torque** T_y at which yielding first begins in the bar is found from the torsion formula (Eq. 3-12) by substituting τ_y for τ_{max}:

$$T_y = \frac{\pi d^3\tau_y}{16} = \frac{\pi r^3\tau_y}{2} \tag{3-81}$$

Comparing Eqs. (3-80) and (3-81), we see that the ratio of the plastic torque to the yield torque is

$$\frac{T_p}{T_y} = \frac{4}{3} \tag{3-82}$$

which shows that after yielding begins in the bar, it is only necessary to increase the torque by one-third to bring the bar to its ultimate load-carrying capacity.

The relationship between the applied torque T and the rate of twist θ can be obtained from Eq. (3-79) by using the shear stress-strain diagram (Fig. 3-52) and determining the moment of inertia of the shaded area about the τ axis. (Note that in Fig. 3-52 the strain γ_{max} is the maximum

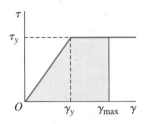

Fig. 3-52 Shear stress-strain diagram for a bar in elastoplastic torsion

shear strain in the bar, that is, the strain at the outer surface.) The result of the lengthy derivation, which is given as an exercise in Problem 3.12-6, is

$$T = \frac{\pi \tau_y}{6\theta^3}(4\gamma_{max}^3 - \gamma_y^3) \qquad (\gamma_{max} \geq \gamma_y \quad \text{and} \quad T_y \leq T \leq T_p) \qquad (3\text{-}83)$$

This equation can be put in more useful form by making the following substitutions. First, we substitute $\tau_y = 2T_y/\pi r^3$ (from Eq. 3-81); second, we substitute $\gamma_{max} = r\theta$ (Eq. 3-76); and finally, we substitute $\gamma_y = r\theta_y$, where θ_y is the rate of twist when yielding first begins in the bar, that is,

$$\theta_y = \frac{T_y}{GI_p} = \frac{2T_y}{\pi G r^4} \qquad (3\text{-}84)$$

With these three substitutions, Eq. (3-83) becomes the desired **torque-rotation** relationship, as follows:

$$\frac{T}{T_y} = \frac{1}{3}\left[4 - \left(\frac{\theta_y}{\theta}\right)^3\right] \qquad (\theta \geq \theta_y \quad \text{and} \quad T_y \leq T \leq T_p) \qquad (3\text{-}85)$$

A graph of this equation is given in Fig. 3-53. Note that as θ becomes larger and larger, the torque T approaches the plastic torque T_p.

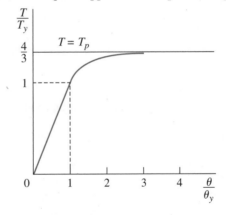

Fig. 3-53 Torque-rotation relationship for a solid circular bar of elastoplastic material (Eq. 3-85)

■ **Residual Stresses**

If a bar in torsion is loaded beyond the elastic limit and then the load is removed, **residual stresses** will remain in the bar. To illustrate how such stresses arise, consider a solid circular bar of elastoplastic material that is loaded to the plastic torque T_p. The corresponding stress distribution is shown in Fig. 3-51c. Now suppose that the torque is removed. During the unloading process, the material follows a straight line on the stress-strain diagram (line *a-a* in Fig. 3-51a), parallel to the initial straight line representing Hooke's law. Thus, the stresses produced during unloading can be obtained from the equations for linearly elastic behavior (Eqs. 3-11 through 3-13).

(a)

(b)

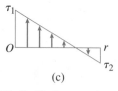

(c)

Fig. 3-54 Residual stresses in torsion (elastoplastic material)

The superposition of the shear stresses produced during loading and unloading is shown in Fig. 3-54. The stresses reached during loading are shown in the part (a) of the figure, which is the same as the stress diagram of Fig. 3-51c. The corresponding torque is $T_p = 2\pi r^3 \tau_y/3$ (Eq. 3-80). This same torque is represented by the unloading stress diagram in Fig. 3-54b, except that now the torque acts in the opposite direction and the behavior is linearly elastic. The maximum stress is $\tau_{max} = T_p r/I_p$, which becomes $\tau_{max} = 4\tau_y/3$ when we substitute for T_p and I_p. The residual stresses in the bar, obtained by superimposing the stresses due to loading and unloading, are shown in Fig. 3-54c. At the center of the bar, the residual stress is

$$\tau_1 = \tau_y \tag{3-86}$$

and at the outer edge of the cross section the residual stress is

$$\tau_2 = \tau_y - \tau_{max} = \tau_y - \frac{4\tau_y}{3} = -\frac{\tau_y}{3} \tag{3-87}$$

which is opposite in direction to the stress τ_1 (Fig. 3-54c). This same procedure for determining residual stresses can be used with stress-strain diagrams of any shape.

PROBLEMS/CHAPTER 3

TORSIONAL DEFORMATIONS

3.2-1 A circular aluminum tube subjected to pure torsion by torques T (see figure) has an outer radius r_2 equal to twice the inner radius r_1. (a) If the maximum shear strain in the tube is measured as 400×10^{-6} radians, what is the shear strain γ_1 at the inner surface? (b) If the maximum allowable rate of twist is 0.15 degrees per foot and the maximum shear strain is to be kept at 400×10^{-6} radians by adjusting the torque T, what is the minimum required outer radius $(r_2)_{min}$?

3.2-2 A circular steel tube of length $L = 0.9$ m is loaded in torsion by torques T (see figure). (a) If the inner radius of the tube is $r_1 = 40$ mm, what is the shear strain γ_1 at the inner surface if the measured angle of twist between the ends is 0.5°? (b) If the maximum allowable shear strain is 0.0005 radians and the angle of twist is to be kept at 0.5° by adjusting the torque T, what is the maximum permissible outer radius $(r_2)_{max}$?

3.2-3 Solve the preceding problem if the length $L = 50$ in., the inner radius $r_1 = 1.5$ in., the angle of twist is 0.6°, and the allowable shear strain is 0.0004 radians.

CIRCULAR BARS AND TUBES

3.3-1 While removing a wheel to change a tire, a driver applies forces $P = 25$ lb at the ends of the arms of a lug wrench (see figure). The wrench is made of steel with shear modulus of elasticity $G = 11.4 \times 10^6$ psi. Each arm of the wrench is 9.0 in. long and has a solid circular cross section of diameter $d = 0.5$ in. (a) Determine the maximum shear stress in the arm that is turning the lug nut (arm A). (b) Determine the angle of twist (in degrees) of this same arm.

Probs. 3.2-1, 3.2-2, and 3.2-3

shear is 60 MPa, what is the maximum permissible torque T_{max} that may be exerted with the wrench? Through what angle ϕ (in degrees) will the shaft twist under the action of the maximum torque? (Assume $G = 78$ GPa and disregard any bending of the shaft.)

Prob. 3.3-4

3.3-5 An aluminum bar of solid circular cross section (diameter 1.0 in. and length 3.0 ft) is twisted in a testing machine until one end rotates through an angle of 8° with respect to the other end. For this angle of twist, the torque is measured as $T = 127$ lb-ft. Calculate the maximum shear stress τ_{max} in the shaft, the shear modulus of elasticity G, and the maximum shear strain γ_{max}.

Prob. 3.3-1

3.3-2 An aluminum bar of solid circular cross section is twisted by torques T acting at the ends (see figure). The dimensions and shear modulus of elasticity are as follows: $L = 1.2$ m, $d = 30$ mm, and $G = 28$ GPa. (a) Determine the torsional stiffness of the bar. (b) If the angle of twist of the bar is 4°, what is the maximum shear stress? What is the maximum shear strain?

3.3-6 A circular tube of aluminum is subjected to torsion by torques T applied at the ends (see figure). The bar is 0.5 m long, and the inside and outside diameters are 30 mm and 40 mm, respectively. It is determined by measurement that the angle of twist is 3.57° when the torque is 600 N·m. Calculate the maximum shear stress τ_{max} in the tube, the shear modulus of elasticity G, and the maximum shear strain γ_{max}.

Probs. 3.3-2 and 3.3-3

3.3-3 A solid circular steel bar of diameter $d = 0.678$ in. and shear modulus of elasticity $G = 12 \times 10^6$ psi is subjected to pure torsion (see figure). The bar develops a maximum shear stress of 7100 psi when the angle of twist is 2.5°. (a) What is the length of the bar? (b) What is the torsional stiffness of the bar?

3.3-4 The steel shaft of a socket wrench is 8 mm in diameter and 200 mm long (see figure). If the allowable stress in

Prob. 3.3-6

3.3-7 Three identical circular disks A, B, and C are welded to the ends of three identical circular bars (see figure). The bars lie in a common plane and the disks lie in planes perpendicular to the axes of the bars. The bars are welded at their intersection to form a rigid connection. Each bar has diameter $d_1 = 0.5$ in. and each disk has diameter $d_2 = 3.0$ in. Forces P_1, P_2, and P_3 create torques acting on disks A, B, and C, respectively. If $P_1 = 28$ lb, what is the maximum shear stress τ_{max} in any of the three bars?

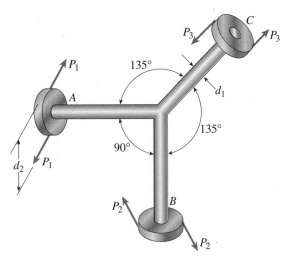

Prob. 3.3-7

3.3-8 A hollow steel shaft used in a construction auger has outer diameter $d_2 = 100$ mm and inner diameter $d_1 = 60$ mm (see figure). The steel has shear modulus of elasticity $G = 80$ GPa. For an applied torque of 6.0 kN·m, determine the following quantities: (a) shear stress τ_2 at the outer surface, (b) shear stress τ_1 at the inner surface, and (c) rate of twist θ (angle of twist per unit length). Also, draw a diagram showing how the shear stresses vary in magnitude along a radial line in the cross section.

$d_1 = 60$ mm

$d_2 = 100$ mm

Prob. 3.3-8

3.3-9 A propeller shaft for a small yacht is made of a solid steel bar 4.0 in. in diameter. The allowable stress in shear is 6500 psi, and the allowable rate of twist is 0.2°/ft. Assuming that the shear modulus of elasticity is $G = 11.5 \times 10^6$ psi, determine the maximum torque T_{max} that can be applied to the shaft.

3.3-10 The round steel axle of a large winch is subjected to a torque of 2.5 kN·m. What is the minimum required diameter d_{min} if the allowable shear stress is 60 MPa and the allowable rate of twist is 0.75°/m? (Assume that the shear modulus of elasticity is 75 GPa.)

3.3-11 A horizontal bar ABC of solid circular cross section (see figure) is held at end A, supported in a bearing at C, and loaded by a vertical force P at the end of the horizontal arm BD. The distance from A to B is $L = 15.0$ in. and the length of arm BD is $b = 4.0$ in. The bar is made of steel having shear modulus $G = 11.3 \times 10^6$ psi. Measurements show that when a certain load P is acting, the maximum shear strain is 637.5×10^{-6} rad and point D moves downward 0.102 in. What is the diameter d of the bar and what is the magnitude of the load P? (Disregard bending of the bar and consider only the effects of torsion.)

Prob. 3.3-11

3.3-12 A solid brass bar of diameter $d = 30$ mm is subjected to a torque T_1, as shown in part (a) of the figure. The allowable shear stress in the brass is 80 MPa. (a) What is the maximum permissible torque T_1? (b) If a hole of diameter 15 mm is drilled longitudinally through the bar, as shown in part (b) of the figure, what is the maximum permissible torque T_2? (c) What is the percent decrease in torque and the percent decrease in weight due to the hole?

(a)

(b)

Prob. 3.3-12

3.3-13 A hollow aluminum tube used in a roof structure has an outside diameter $d_2 = 4.0$ in. and an inside diameter $d_1 = 3.5$ in. (see figure). The tube is 8.0 ft long, and the aluminum has shear modulus $G = 4.0 \times 10^6$ psi. (a) If the tube is twisted in pure torsion by torques acting at the ends, what is the angle of twist ϕ when the maximum shear stress is 8000 psi? (b) What diameter d is required for a solid shaft (see figure) to resist the same torque with the same maximum stress? (c) What is the ratio of the weight of the hollow tube to the weight of the solid shaft?

Probs. 3.3-13, 3.3-14, and 3.3-15

3.3-14 A hollow circular shaft and a solid circular shaft (see figure) of the same material are to be designed to transmit the same torque with the same maximum shear stress. The inner diameter of the hollow shaft is 0.85 times the outer diameter; that is, $d_1 = 0.85d_2$. (a) Determine the

ratio of the outer diameter d_2 of the hollow shaft to the diameter d of the solid shaft. (b) Determine the ratio β_1 of the weight of the hollow shaft to the weight of the solid shaft. (c) Determine the ratio β_2 of the torsional stiffness of the hollow shaft to the torsional stiffness of the solid shaft.

3.3-15 A solid circular bar (diameter $d = 1.5$ in.) used as a generator shaft is to be replaced by a hollow shaft of outer diameter $d_2 = 1.75$ in. (see figure). The hollow shaft must transmit the same torque with the same maximum shear stress. (a) What is the maximum permissible inner diameter d_1 of the hollow shaft? (b) For manufacturing purposes, the inner diameter must be an integer multiple of 1/8 inch. Select the largest acceptable inner diameter d_0. Then determine the ratio β_1 of the weight of the hollow shaft to the weight of the solid shaft and the ratio β_2 of the torsional stiffness of the hollow shaft to the torsional stiffness of the solid shaft.

3.3-16 A solid circular bar of radius r_2 is subjected to a torque T (see figure). The radius r_1 divides the cross section into an inner area (radius from 0 to r_1) and an outer area (radius from r_1 to r_2). The inner area transmits a torque T_1 and the outer area transmits a torque T_2, where $T_1 + T_2 = T$. (a) Derive a formula for the ratio T_2/T of the torque transmitted by the outer area to the total torque. (b) Sketch a graph of the ratio T_2/T versus the ratio r_1/r_2 of the radii. (c) If $r_1 = r_2/2$, what percent of the total torque is transmitted by the outer area? (d) If r_1 divides the cross section into two equal areas, what percent of the total torque is transmitted by the outer area?

Prob. 3.3-16

*3.3-17 A circular tube of inner radius r_1 and outer radius r_2 is subjected to a torque produced by forces $P = 1200$ lb (see figure). The forces have their lines of action at a distance $b = 4.0$ in. from the outside of the tube. If the allowable shear stress in the tube is 7500 psi and the inner radius r_1 = 1.00 in., what is the minimum permissible outer radius r_2?

3.4-2 A circular tube of outer diameter d_3 = 70 mm and inner diameter d_2 = 60 mm is welded at the right-hand end to a fixed plate and at the left-hand end to a rigid end plate (see figure). A solid circular bar of diameter d_1 = 40 mm is inside of, and concentric with, the tube. The bar passes through a hole in the fixed plate and is welded to the end plate. The bar is 1.0 m long and the tube is half as long as the bar. A torque T = 1000 N·m acts at end A of the bar. Also, both the bar and tube are made of an aluminum alloy with shear modulus of elasticity G = 27 GPa. (a) Determine the maximum shear stresses in both the bar and tube. (b) Determine the angle of twist (in degrees) at end A of the bar.

Prob. 3.3-17

Prob. 3.4-2

WORK SHEET 2   Mathcad   **NONUNIFORM TORSION**

3.4-1 A stepped shaft ABC consisting of two solid circular segments is subjected to torques T_1 and T_2 acting in opposite directions, as shown in the figure. The larger part of the shaft has diameter d_1 = 2.25 in. and length L_1 = 30 in.; the smaller segment has diameter d_2 = 1.75 in. and length L_2 = 20 in. The material is steel with shear modulus G = 11 × 10^6 psi. If the torques are T_1 = 20,000 lb-in. and T_2 = 8,000 lb-in., calculate: (a) the maximum shear stress τ_{max} in the shaft, and (b) the angle of rotation ϕ_C (in degrees) at end C.

3.4-3 A stepped shaft $ABCD$ consisting of three solid circular segments is subjected to three torques as shown in the figure. The length of each segment is 24.0 in. and the diameters are 3.0 in., 2.5 in., and 2.0 in. The material is steel with shear modulus of elasticity G = 11.6 × 10^3 ksi. (a) Calculate the maximum shear stress τ_{max} in the shaft. (b) Calculate the angle of twist ϕ_D (in degrees) at end D.

Prob. 3.4-1

Prob. 3.4-3

3.4-4 A solid circular bar ABC consists of two segments, as shown in the figure. One segment has diameter d_1 = 50 mm and length L_1 = 1.25 m; the other segment has diameter d_2 = 40 mm and length L_2 = 1.0 m. What is the allowable torque T_{allow} if the shear stress is not to exceed 30 MPa and the angle of twist between the ends of the bar is not to exceed 1.5°? (Assume G = 80 GPa.)

Prob. 3.4-4

3.4-5 A hollow tube AE constructed of monel metal is subjected to five torques acting in the directions shown in the figure. The magnitudes of the torques are T_1 = 1000 lb-in., T_2 = T_4 = 500 lb-in., and T_3 = T_5 = 800 lb-in. The tube has an outside diameter d_2 = 1.0 in. The allowable shear stress is 12,000 psi and the allowable rate of twist is 2°/ft. Determine the maximum permissible inside diameter d_1 of the tube.

Prob. 3.4-5

3.4-6 A shaft ABC (see figure) consists of a solid bar of diameter d from A to B and a hollow bar of outside diameter $1.25d$ and inside diameter d from B to C. Both bars have length L = 0.3 m. Torques are applied to the shaft through gears at A, B, and C. These torques are T_1 = 120 N·m, T_2 = 270 N·m, and T_3 = 150 N·m, respectively, acting in the directions shown. The shaft is made of steel having shear

modulus of elasticity equal to 80 GPa. (a) What is the required diameter d if the allowable shear stress is 40 MPa? (b) What is the required diameter d if the angle of twist between any two gears is limited to 1.0°?

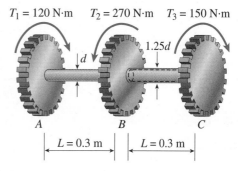

Prob. 3.4-6

3.4-7 Four gears are attached to a circular shaft and transmit the torques shown in the figure. (a) What is the required diameter d of the shaft if the allowable shear stress is 10,000 psi? (b) What is the required diameter d if the shaft is hollow with an inside diameter of 1.0 in.?

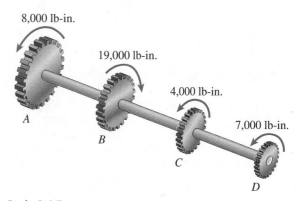

Prob. 3.4-7

3.4-8 A shaft of solid circular cross section with two different diameters is shown in the first part of the figure. A prismatic tubular shaft made of the same material, having the same length, and having the same torsional stiffness is shown in the second part of the figure. Determine the outside diameter d of the hollow shaft if its wall thickness t equals $d/10$.

Prob. 3.4-11

Prob. 3.4-8

3.4-9 Refer to the tapered bar shown in Fig. 3-19 and analyzed in Example 3-5. For what ratio d_B/d_A will the angle of twist be one-half the angle of twist for a prismatic bar of diameter d_A?

3.4-10 The tapered bar shown in Fig. 3-19 and analyzed in Example 3-5 is subjected to a torque $T = 2100$ N·m. The bar has length $L = 2.5$ m and is made of an aluminum alloy having shear modulus $G = 27$ GPa. The allowable shear stress in the bar is 52 MPa and the allowable angle of twist is 2.5°. If the diameter at end B is to be twice the diameter at end A, what is the required diameter d_A at end A?

3.4-11 A uniformly tapered tube AB of hollow circular cross section is shown in the figure. The tube has constant wall thickness t and length L. The average diameters at the ends are d_A and $d_B = 2d_A$. The polar moment of inertia may be represented by the approximate formula $I_p \approx \pi d^3 t/4$ (Eq. 3-18). Derive a formula for the angle of twist of the tube when it is subjected to torques T acting at the ends.

3.4-12 A prismatic bar AB of solid circular cross section (diameter d and length L) is fixed at the left-hand end and free at the right-hand end (see figure). The bar is loaded by a distributed torque of constant intensity t per unit length. (a) What is the maximum shear stress in the bar? (b) What is the angle of twist at end B?

Prob. 3.4-12

3.4-13 A prismatic bar AB of solid circular cross section (see figure) is loaded by a distributed torque having intensity $t(x)$ per unit length that varies linearly from a maximum value t_A at end A to zero at end B. The bar is fixed at end A and free at end B. Also, the diameter is d and the length is L. (a) What is the maximum shear stress in the bar? (b) What is the angle of twist at end B?

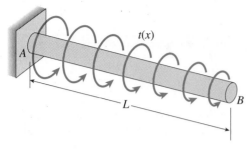

Prob. 3.4-13

*3.4-14 A horizontal circular bar of diameter d and length L is supported at end A and free at end B (see figure). Horizontal vanes are attached to the side of the bar. These vanes have lengths that vary linearly from zero at end B to a maximum length H_A at end A. The total weight of the vanes is W. (a) What is the maximum shear stress in the bar due to torsion? (b) What is the angle of twist at end B? (*Note:* The vanes also produce bending of the bar, but this bending has no effect on the angle of twist.)

Prob. 3.4-14

PURE SHEAR

3.5-1 A solid circular bar of diameter $d = 2.0$ in. (see figure) is twisted in a testing machine until the applied torque reaches the value $T = 11,000$ lb-in. At this value of torque, a strain gage oriented at 45° to the axis of the bar

gives a reading $\epsilon = 305 \times 10^{-6}$. Determine the shear modulus G of the material.

Prob. 3.5-1

3.5-2 A hollow shaft of outside diameter $d_2 = 80$ mm and inside diameter $d_1 = 50$ mm (see figure) is made of aluminum having shear modulus $G = 27$ GPa. When the shaft is subjected to a torque $T = 4.0$ kN·m, what are the values of the maximum shear strain γ_{max} and the maximum normal strain ϵ_{max}?

Probs. 3.5-2, 3.5-3, and 3.5-4

3.5-3 A steel tube ($G = 11.5 \times 10^6$ psi) has an outer diameter $d_2 = 2.0$ in. and an inner diameter $d_1 = 1.5$ in. (see figure). When twisted by a torque T, the tube develops a maximum normal strain of 170×10^{-6}. What is the magnitude of the applied torque T?

3.5-4 An aluminum tube ($G = 28$ GPa) has an outer diameter $d_2 = 100$ mm and an inner diameter $d_1 = 50$ mm (see figure). When twisted by a torque T, the tube develops a maximum normal strain of 650×10^{-6}. What is the magnitude of the applied torque T?

3.5-5 A solid circular bar of steel ($G = 11.6 \times 10^3$ ksi) transmits a torque $T = 4.0$ k-in. The allowable stresses in tension, compression, and shear are 12 ksi, 9 ksi, and 8 ksi, respectively. Also, the maximum permissible tensile strain is 300×10^{-6}. Determine the minimum required diameter d of the bar.

3.5-6 A hollow circular steel bar ($G = 80$ GPa) is twisted by a torque T that produces a maximum shear strain $\gamma_{max} = 639 \times 10^{-6}$ rad. The bar has outside and inside radii of 75 and 60 mm, respectively. What is the maximum tensile strain in the bar? What is the maximum tensile stress in the bar? What is the magnitude of the applied torque T?

3.5-7 The normal strain in the 45° direction on the surface of a circular tube (see figure) is 880×10^{-6} when the torque $T = 750$ lb-in. The tube is made of copper alloy with $G = 6.2 \times 10^6$ psi. If the outside diameter d_2 of the tube is 0.8 in., what is the inside diameter d_1?

Prob. 3.5-7

TRANSMISSION OF POWER

3.7-1 A generator shaft in a small hydroelectric plant turns at 120 rpm and delivers 50 hp (see figure). (a) If the diameter of the shaft is $d = 3.0$ in., what is the maximum shear stress τ_{max} in the shaft? (b) If the shear stress is limited to 4000 psi, what should be the diameter d of the shaft?

Prob. 3.7-1

3.7-2 A motor drives a shaft at 12 Hz and delivers 20 kW of power (see figure). (a) If the shaft has a diameter d of 30 mm, what is the maximum shear stress τ_{max} in the shaft? (b) If the maximum allowable shear stress is 40 MPa, what diameter d is required for the shaft?

Prob. 3.7-2

3.7-3 The propeller shaft of a ship has outside diameter 14 in. and inside diameter 10 in., as shown in the figure. The shaft is rated for a maximum shear stress of 9000 psi. (a) If the shaft is turning at 500 rpm, what is the maximum horse-power that can be transmitted without exceeding the allow-able stress? (b) If the rotational speed of the shaft is doubled but the power requirements remain unchanged, what happens to the shear stress in the shaft?

Prob. 3.7-3

3.7-4 The drive shaft for a truck (outer diameter 60 mm and inner diameter 40 mm) is running at 2500 rpm (see figure). (a) If the shaft transmits 150 kW, what is the maximum shear stress in the shaft? (b) If the allowable shear stress is 35 MPa, what is the maximum power that can be transmitted?

Prob. 3.7-4

3.7-5 A hollow circular shaft for use in a pumping station is being designed with an inside diameter equal to 0.8 times the outside diameter. The shaft must transmit 400 hp at 800 rpm without exceeding the allowable shear stress of 6000 psi. Determine the minimum required outside diameter d.

3.7-6 A tubular shaft being designed for use on a construction site must transmit 120 kW at 10 Hz. The inside diameter of the shaft is to be three-fourths of the outside diameter. If the allowable shear stress in the shaft is 45 MPa, what is the minimum required outside diameter d?

3.7-7 A propeller shaft of solid circular cross section and diameter d is spliced by a collar of the same material (see figure). The collar is securely bonded to both parts of the shaft. What should be the outer diameter d_1 of the collar in order that the splice can transmit the same power as the solid shaft?

Prob. 3.7-7

3.7-8 What is the maximum power that can be delivered by a hollow propeller shaft (outside diameter 50 mm, inside diameter 40 mm, and shear modulus of elasticity 80 GPa) turning at 600 rpm if the allowable shear stress is 100 MPa and the allowable rate of twist is 3°/m?

3.7-9 A motor delivers 275 hp at 1000 rpm to the end of a shaft (see figure). The gears at B and C take out 125 and 150 hp, respectively. Determine the required diameter d of the shaft if the allowable shear stress is 7500 psi and the angle of twist between the motor and gear C is limited to 1.5°. (Assume $G = 11.5 \times 10^6$ psi, $L_1 = 6$ ft, and $L_2 = 4$ ft.)

Probs. 3.7-9 and 3.7-10

3.7-10 The shaft ABC shown in the figure is driven by a motor that delivers 300 kW at a rotational speed of 32 Hz.

The gears at B and C take out 120 and 180 kW, respectively. The lengths of the two parts of the shaft are $L_1 = 1.5$ m and $L_2 = 0.9$ m. Determine the required diameter d of the shaft if the allowable shear stress is 50 MPa, the allowable angle of twist between points A and C is 4.0°, and $G = 75$ GPa.

STATICALLY INDETERMINATE TORSION MEMBERS

3.8-1 A solid circular bar with fixed ends is acted upon by torques $3T_0$ and T_0 at the locations shown in the figure. Obtain a formula for the maximum angle of twist of the bar. (*Hint:* Use Eqs. 3-46a and b of Example 3-9 to obtain the reactive torques.)

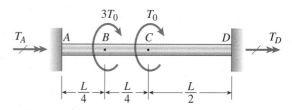

Prob. 3.8-1

3.8-2 A solid circular bar with fixed ends is acted upon by two oppositely directed torques T_0, as shown in the figure. Obtain formulas for the reactive torques T_A and T_D, the maximum angle of twist ϕ_{max}, and the angle of twist ϕ_0 at the midsection of the bar. (*Hint:* Use Eqs. 3-46a and b of Example 3-9 to obtain the reactive torques.)

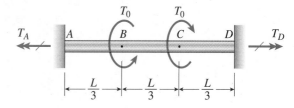

Prob. 3.8-2

3.8-3 A hollow steel shaft ACB of outside diameter 2 in. and inside diameter 1.5 in. is held against rotation at ends A and B (see figure). Horizontal forces P are applied at the ends of the vertical arm. Determine the allowable value of the forces P if the maximum permissible shear stress in the shaft is 4800 psi. (*Hint:* Use Eqs. 3-46a and b of Example 3-9 to obtain the reactive torques.)

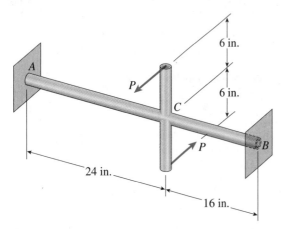

Prob. 3.8-3

3.8-4 A solid circular shaft AB of diameter d is fixed at both ends (see figure). A circular disk is attached to the shaft at the location shown. What is the largest permissible angle of rotation ϕ of the disk if the allowable shear stress is τ_{allow}? (Assume that $a > b$. Also, use Eqs. 3-46a and b of Example 3-9 to obtain the reactive torques.)

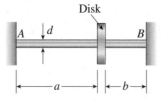

Prob. 3.8-4

3.8-5 A solid steel shaft ACB, held rigidly at both ends, has two different diameters (see figure). Assuming that the maximum permissible shear stress is 9000 psi, determine the allowable torque T_{allow} that may be applied at point C. (*Hint:* Use Eqs. 3-45a and b of Example 3-9 to obtain the reactive torques.)

Prob. 3.8-5

3.8-6 A stepped shaft of solid circular cross section (see figure) is held against rotation at the ends. If the allowable stress in shear is 55 MPa, what is the allowable torque T_{allow} that may be applied to the shaft at C? (*Hint:* Use Eqs. 3-45a and b of Example 3-9 to obtain the reactive torques.)

Prob. 3.8-6

3.8-7 A circular bar AB with fixed ends has a hole extending for half its length (see figure). The diameter of the hole is three-fourths of the outer diameter of the bar. At what distance x from the left-hand end should a torque T be applied so that the reactive torques at the supports will be equal?

Prob. 3.8-7

3.8-8 A hollow tube AB of length L and torsional rigidity GI_p is rigidly supported at end A (see figure). At end B, a horizontal bar of length $2c$ is welded to the tube. Adjacent to the ends of the horizontal bar are two identical springs, each having a spring stiffness k (force per unit displacement). The length of the gap between the ends of the springs and the ends of the bar is b. The springs are stretched, attached to the ends of the bar, and then released, thus causing the bar and tube to rotate through a small angle β. Obtain a formula for the torque T in the tube. (Disregard bending of the bar.)

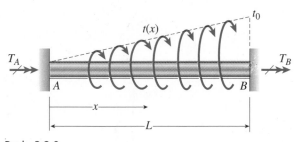

Prob. 3.8-8

3.8-9 A circular bar AB of length L is fixed at both ends and loaded by a distributed torque $t(x)$ that varies linearly in intensity from zero at end A to t_0 at end B (see figure). Obtain formulas for the fixed-end torques T_A and T_B.

Prob. 3.8-9

3.8-10 A solid steel bar of diameter 50 mm is enclosed by a steel tube of outer diameter 75 mm and inner diameter 60 mm (see figure). Both bar and tube are held rigidly at end A and joined securely to a rigid end plate at B. The composite bar, which is 750 mm long, is twisted by a torque $T = 2000$ N·m acting at end B. (a) Determine the maximum shear stresses τ_t and τ_b in the tube and bar, respectively. (b) Determine the angle of rotation ϕ of the rigid plate, assuming that the shear modulus of the steel is $G = 80$ GPa. (c) Determine the torsional stiffness k_T of the composite bar. (*Hint:* Use Eqs. 3-44a and b to find the torques.)

Prob. 3.8-10

3.8-11 A solid steel bar of diameter 2.0 in. is enclosed by a steel tube of outer diameter 3.0 in. and inner diameter 2.5 in. (see figure). Both bar and tube are held rigidly at end A and joined securely to a rigid end plate at B. The composite bar, which is 26 in. long, is twisted by a torque $T = 20$ k-in. acting at end B. (a) Determine the maximum shear stresses τ_t and τ_b in the tube and bar, respectively. (b) Determine the angle of rotation ϕ of the rigid plate, assuming the shear modulus of the steel is $G = 11.5 \times 10^6$ psi. (c) Determine the torsional stiffness k_T of the composite bar. (*Hint:* Use Eqs. 3-44a and b to find the torques.)

Prob. 3.8-11

3.8-12 A composite shaft is made by shrink-fitting a steel tube over a brass core so that they act as a unit in torsion (see figure). The shear modulus of elasticity of the steel is $G_s = 75$ GPa and of the brass is $G_b = 39$ GPa. The outer diameters are $d_1 = 25$ mm for the core and $d_2 = 40$ mm for the tube. Assume that a torque $T = 900$ N·m acts on the shaft. (a) Determine the shear stress τ_{s2} in the steel at its outer radius of 20 mm, the shear stress τ_{s1} in the steel at its inner radius of 12.5 mm, and the shear stress τ_{b1} in the brass at its outer radius of 12.5 mm. Draw a graph showing how the shear stresses vary along a radial line from the center of the shaft to the outer surface. (b) Determine the shear strains γ_{s2}, γ_{s1}, and γ_{b1} at the locations described in part (a). Draw a graph showing how the shear strains vary along a radial line from the center of the shaft to the outer surface. (*Hint:* Use Eqs. 3-44a and b to find the torques.)

Steel tube

Brass core

$d_1 = 25$ mm
$d_2 = 40$ mm

Prob. 3.8-12

3.8-13 A steel sleeve is fused onto a brass core to make a composite bar 30 in. long, as shown in the figure. The outer diameters of the sleeve and core are 2.4 in. and 2.0 in., respectively. The shear moduli of elasticity for the steel and brass are $G_s = 11,500$ ksi and $G_b = 5,600$ ksi, respectively. The bar is subjected to a torque $T = 20.0$ k-in. (a) Determine the maximum shear stresses τ_s and τ_b in the steel and brass parts, respectively. (b) Determine the angle of twist ϕ of the composite bar. (c) Determine the torsional stiffness k_T of the composite bar. (*Hint:* Use Eqs. 3-44a and b to find the torques.)

$T = 20.0$ k-in.
Steel sleeve
Brass core
$T = 20.0$ k-in.

30 in.

Prob. 3.8-13

3.8-14 A composite shaft is formed of two materials, an outer sleeve of steel ($G_s = 80$ GPa) and an inner rod of brass ($G_b = 36$ GPa), as shown in the figure. The outer diameters of the two parts are 75 mm and 60 mm. Assuming that the allowable shear stresses are $\tau_s = 65$ MPa and $\tau_b = 25$ MPa in the steel and brass, respectively, determine the maximum permissible torque T_{allow} that may be applied to the shaft. (*Hint:* Use Eqs. 3-44a and b to find the torques.)

Steel sleeve

Brass rod

60 mm
75 mm

Prob. 3.8-14

3.8-15 A steel tube S is shrink-fitted over an aluminum tube A to form a composite shaft with diameters 3 in., 5 in., and 6 in. (see figure). The allowable shear stresses in the steel and aluminum are $\tau_s = 9.0$ ksi and $\tau_a = 4.0$ ksi, respectively. Determine the allowable torque T_{allow} that may be applied to the shaft, assuming $G_s = 11,600$ ksi and $G_a = 4,000$ ksi. (*Hint:* Use Eqs. 3-44a and b to find the torques.)

S
A

3 in.
5 in.
6 in.

Prob. 3.8-15

****3.8-16** A shaft AB of diameter d_1 has two circular flanges of diameter d_2, as shown in part (a) of the figure. Torques T_0 act at the ends of the shaft. While these torques are acting, a sleeve of the same material as the shaft and with outer diameter d_3 and inner diameter d_2 is fitted over the flanges and

bonded securely to them, as shown in part (b) of the figure. Then the torques T_0 are removed. Assuming that the flanges are rigid, obtain a formula for the maximum shear stress τ_{max} in the sleeve.

(a)

(b)

Prob. 3.8-16

STRAIN ENERGY IN TORSION

3.9-1 A solid circular bar of copper ($G = 6.0 \times 10^6$ psi) with length $L = 2.0$ ft and diameter $d = 1.5$ in. is subjected to pure torsion by torques T acting at the ends. (a) Calculate the amount of strain energy U stored in the bar when the maximum shear stress is 3000 psi. (b) From the strain energy, calculate the angle of twist ϕ.

3.9-2 A solid circular bar of steel ($G = 80$ GPa) with length $L = 1.8$ m and diameter $d = 75$ mm is subjected to pure torsion by torques T acting at the ends. (a) Calculate the amount of strain energy U stored in the bar when the maximum shear stress is 55 MPa. (b) From the strain energy, calculate the angle of twist ϕ.

3.9-3 The stepped shaft shown in the figure has length $L = 100$ in., diameter $d_2 = 2.0$ in., and diameter $d_1 = 1.5$ in. The material is steel with $G = 11.8 \times 10^6$ psi. Determine the strain energy U of the shaft if the angle of twist is $3.0°$.

Probs. 3.9-3 and 3.9-4

3.9-4 A stepped shaft has length $L = 1.2$ m, diameter $d_2 = 30$ mm, and diameter $d_1 = 25$ mm (see figure). The material is brass with $G = 40$ GPa. Determine the strain energy U of the shaft if the angle of twist is 0.060 radians.

3.9-5 A circular bar of length L is fixed at one end and free at the other (see figure). The bar is loaded by a torque T at the free end and by a continuously distributed torque of constant intensity t per unit distance along the axis of the bar. (a) What is the strain energy U_1 of the bar when the load T acts alone? (b) What is the strain energy U_2 when the load t acts alone? (c) What is the strain energy U_3 when both loads act simultaneously?

Prob. 3.9-5

3.9-6 Derive a formula for the strain energy U of the circular bar shown in the figure. The intensity t of distributed torque varies linearly from zero at the free end to a maximum value t_0 at the support.

Prob. 3.9-6

3.9-7 Obtain a formula for the strain energy U of the statically indeterminate circular bar shown in the figure. (*Hint:* Use Eqs. 3-46a and b of Example 3-9, Section 3.8, to obtain the reactive torques.)

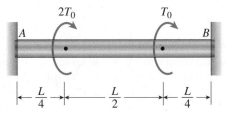

Prob. 3.9-7

3.9-8 A statically indeterminate stepped shaft ACB is fixed at ends A and B and loaded by a torque T at C (see figure). The two segments of the bar are made of the same material, have lengths L_A and L_B, and have polar moments of inertia I_{pA} and I_{pB}. Determine the angle of rotation ϕ of the cross section at C by using strain energy. (*Hint:* Use Eq. 3-51b to determine the strain energy U in terms of ϕ. Then equate the strain energy to the work done by the load T. Compare your result with Eq. 3-48 of Example 3-9, Section 3.8.)

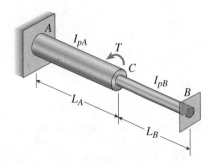

Prob. 3.9-8

3.9-9 A thin-walled hollow tube AB of conical shape has constant thickness t and average diameters d_A and d_B at the ends (see figure). (a) Determine the strain energy U of the tube when it is subjected to pure torsion by torques T. (b) Determine the angle of twist ϕ. (*Note:* Use the approximate formula $I_p \approx \pi d^3 t /4$ for a thin circular ring; see Case 22 of Appendix D.)

Prob. 3.9-9

*3.9-10 A hollow circular tube A fits over the end of a solid circular bar B, as shown in the figure. The far ends of both bars are fixed. Initially, a hole through bar B makes an angle β with a line through two holes in bar A. Then bar B is twisted until the holes are aligned, and a pin is placed through the holes. When bar B is released and the system returns to equilibrium, what is the total strain energy U of the two bars? (Let I_{pA} and I_{pB} represent the polar moments of inertia of bars A and B, respectively. The length L and shear modulus of elasticity G are the same for both bars.)

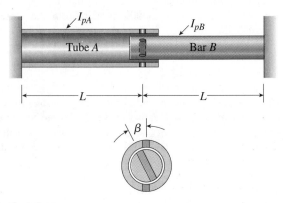

Prob. 3.9-10

*3.9-11 A heavy flywheel attached to the end of a shaft of diameter d is rotating at n revolutions per minute (see figure). If the bearing at A suddenly freezes, what will be the maximum angle of twist of the shaft? What is the corresponding maximum shear stress in the shaft? (Let L = length of the shaft, G = shear modulus of elasticity, and I_m = mass moment of inertia of the flywheel about the axis of the shaft. Also, disregard friction in the bearings at B and C and disregard the mass of the shaft.) *Hint:* Equate the kinetic energy of the rotating flywheel to the strain energy of the shaft.

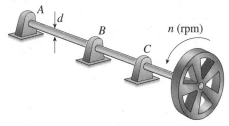

Prob. 3.9-11

THIN-WALLED TUBES

3.10-1 A hollow circular tube having 10 in. inside diameter and 1 in. wall thickness (see figure) is subjected to a torque T = 1200 k-in. Determine the maximum shear stress in the tube using: (a) the approximate theory of thin-walled tubes, and (b) the exact torsion theory. Does the approximate theory give conservative or nonconservative results?

Prob. 3.10-1

3.10-2 A solid circular bar having diameter d is to be replaced by a rectangular tube having cross-sectional dimensions $d \times 2d$ to the median line of the cross section (see figure). Determine the required thickness t_{min} of the tube so that the maximum shear stress in the tube will not exceed the maximum shear stress in the solid bar.

Prob. 3.10-2

3.10-3 A thin-walled circular tube and a solid circular bar of the same material, the same cross-sectional area, and the same length are subjected to torsion. What is the ratio of the strain energy U_1 in the tube to the strain energy U_2 in the solid bar if the maximum shear stresses are the same in both cases? (For the tube, use the approximate theory for thin-walled bars.)

3.10-4 A thin-walled aluminum tube of rectangular cross section is shown in the figure. (a) Determine the shear stress in the tube due to a torque T = 60 N·m. (b) Determine the angle of twist if the length of the tube is 0.25 m and G = 26 GPa.

Prob. 3.10-4

3.10-5 A thin-walled steel tube having an elliptical cross section with constant thickness t (see figure) is subjected to a torque T = 25 k-in. Determine the shear stress τ and the rate of twist θ if G = 12 × 10^6 psi, t = 0.2 in., a = 3 in., and b = 2 in. (*Note:* See Appendix D, Case 16, for the properties of an ellipse.)

Prob. 3.10-5

3.10-6 Calculate the shear stress τ and the angle of twist ϕ for a steel tube ($G = 76$ GPa) having the cross section shown in the figure. The tube has length $L = 1.0$ m and is subjected to a torque $T = 10$ kN·m.

Prob. 3.10-6

3.10-7 A torque T is applied to a thin-walled tube having a cross section in the shape of a regular hexagon with constant wall thickness t and side length b (see figure). Obtain formulas for the shear stress τ and the rate of twist θ.

Prob. 3.10-7

3.10-8 A tubular aluminum bar ($G = 28$ GPa) of square cross section (see figure) with outside dimensions 50 mm × 50 mm must resist a torque $T = 300$ N·m. Calculate the minimum required wall thickness t if the allowable shear stress is 20 MPa and the allowable rate of twist θ is 0.025 rad/m.

Prob. 3.10-8

3.10-9 Compare the angle of twist ϕ_1 for a thin-walled circular tube (see figure) calculated from the approximate theory for thin-walled bars with the angle of twist ϕ_2 calculated from the exact torsion theory for circular bars. Express the ratio ϕ_1/ϕ_2 in terms of the nondimensional ratio $\beta = r/t$. Calculate the ratio of angles of twist for $\beta = 5$, 10, and 20. What conclusion about the accuracy of the approximate theory do you draw from the results?

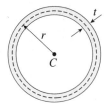

Prob. 3.10-9

3.10-10 The cross section of a thin tube having variable wall thickness t is shown in the figure. The median line of the cross section is a circle of radius r, and the thickness is given by the equation

$$t = t_0\left(1 + \sin\frac{\theta}{2}\right)$$

where t_0 is the thickness at the section where $\theta = 0$. If the tube is subjected to a torque T, what is the maximum shear stress τ_{max}? minimum shear stress τ_{min}? angle of twist ϕ?

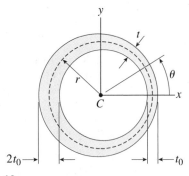

Prob. 3.10-10

3.10-11 A thin tubular shaft of circular cross section (see figure) with inside diameter 4 in. is subjected to a torque of 45,000 lb-in. If the allowable shear stress is 6000 psi, determine the required wall thickness t by using: (a) the approximate theory for a thin-walled tube, and (b) the exact torsion theory for a circular bar.

Prob. 3.10-11

3.10-12 A thin-walled rectangular tube has uniform thickness t and dimensions $a \times b$ to the median line of the cross section (see figure). How does the shear stress in the tube vary with the ratio $\beta = a/b$ if the total length L_m of the median line of the cross section and the torque T remain constant? From your results, show that the shear stress is smallest when the tube is square ($\beta = 1$).

Probs. 3.10-12 and 3.10-13

3.10-13 Repeat the preceding problem for the rate of twist θ, and then show that the rate of twist is smallest when the tube is square ($\beta = 1$).

*3.10-14 A long, thin-walled tapered tube AB of circular cross section (see figure) is subjected to a torque T. The tube has length L and constant wall thickness t. The diameters to the median lines of the cross sections at the ends A and B are d_A and d_B, respectively. Derive a formula for the angle of twist ϕ of the tube. (*Hint:* If the angle of taper is small, we may obtain approximate results by applying the formulas for a thin-walled prismatic tube to a differential element of the tapered tube and then integrating along the axis.)

Prob. 3.10-14

STRESS CONCENTRATIONS IN TORSION

The problems for Section 3.11 are to be solved by considering the stress-concentration factors.

3.11-1 A stepped shaft consisting of solid circular segments having diameters $D_1 = 2.0$ in. and $D_2 = 2.4$ in. (see figure) is subjected to a torque T. The radius of the fillet is $R = 0.1$ in. If the allowable shear stress is 12,000 psi, what is the maximum permissible torque T_{max}?

Probs. 3.11-1 through 3.11-5

3.11-2 A stepped shaft with diameters $D_1 = 40$ mm and $D_2 = 60$ mm carries a torque $T = 1100$ N·m (see figure). If the allowable shear stress is 120 MPa, what is the smallest radius R_{min} that may be used for the fillet?

3.11-3 A full quarter-circular fillet is used at the shoulder of a stepped shaft having diameter $D_2 = 1.0$ in. (see figure). A torque $T = 500$ lb-in. acts on the shaft. Determine the maximum shear stress τ_{max} in the shaft for values as follows: $D_1 = 0.7$, 0.8, and 0.9 in. Plot a graph showing τ_{max} versus D_1.

3.11-4 The stepped shaft shown in the figure is required to transmit 600 kW of power at 400 rpm. The shaft has a full quarter-circular fillet, and the smaller diameter $D_1 = 100$ mm. If the maximum allowable shear stress is 100 MPa, what is the required diameter D_2? Is this diameter an upper or a lower limit on the value of D_2?

3.11-5 A stepped shaft (see figure) has diameter $D_2 = 1.5$ in. and a full quarter-circular fillet. The allowable shear stress is 15,000 psi and the load $T = 4800$ lb-in. What is the smallest permissible diameter D_1?

NONLINEAR TORSION

3.12-1 A bar of solid circular cross section is constructed of a material that follows Hooke's law ($\tau = G\gamma$). Derive the equation for the rate of twist ($\theta = T/GI_p$) by using Eq. (3-78) for nonlinear torsion.

3.12-2 A solid circular bar (radius $= r$) subjected to a torque T is composed of a material for which the stress-strain curve in shear is given by the equation $\tau^n = B\gamma$, where n and B are constants (see figure). (a) Derive a torsion formula for the shear stress τ_{max} at the outer edge of the cross section. (b) Verify that the formula reduces to the torsion formula for linearly elastic behavior ($\tau_{max} = Tr/I_p$) when $n = 1$.

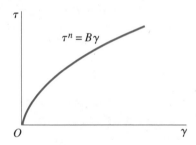

Prob. 3.12-2

3.12-3 Obtain a formula for the ratio of the plastic torque T_p to the yield torque T_y for a tubular bar of circular cross section (see figure) if the bar is made of an elastoplastic material (see Fig. 3-51a for the stress-strain diagram). Plot a graph of the ratio T_p/T_y versus the ratio r_1/r_2.

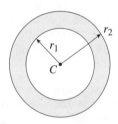

Prob. 3.12-3

3.12-4 A solid circular bar of diameter $d = 60$ mm is made of an elastoplastic material for which $G = 76$ GPa and $\tau_y = 125$ MPa. Calculate the rate of twist θ when a torque $T = 6.0$ kN·m acts on the bar.

3.12-5 A solid circular bar of diameter $d = 1.5$ in. is made of an elastoplastic material for which $G = 11.0 \times 10^6$ psi and $\tau_y = 18,000$ psi. Calculate the rate of twist θ when a torque $T = 1100$ lb-ft acts on the bar.

*__3.12-6__ Using the shear stress-strain diagram for an elastoplastic material as pictured in Fig. 3-52, derive Eq. (3-83) for the torque T.

*__3.12-7__ A hollow circular shaft has inside diameter $d_1 = 3$ in. and outside diameter $d_2 = 6$ in. (see figure). The shaft is made of steel having an elastoplastic shear stress-strain diagram (see Fig. 3-52) with $\tau_y = 24,000$ psi and $\gamma_y = 0.002$. (a) What torque T acting on the shaft will produce a maximum shear strain $\gamma_{max} = 0.003$ in the shaft? (b) What is the plastic torque T_p? (c) What is the maximum shear strain γ_p when the torque just reaches the plastic torque T_p?

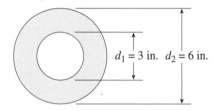

Prob. 3.12-7

__3.12-8 A solid circular bar of radius $r = 40$ mm is constructed of a material having the bilinear stress-strain diagram shown in the figure. Calculate the maximum shear stress τ_{max} in the bar when the torque $T = 6$ kN·m.

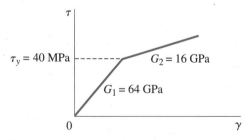

Prob. 3.12-8

4
SHEAR FORCES AND BENDING MOMENTS

4.1

INTRODUCTION

Structural members are usually classified according to the types of loads that they support. For instance, an *axially loaded bar* supports forces having their vectors directed along the axis of the bar, and a *bar in torsion* supports torques (or couples) having their moment vectors directed along the axis. In this chapter, we begin our study of **beams** (Fig. 4-1), which are structural members subjected to lateral loads, that is, forces or moments having their vectors perpendicular to the axis of the bar.

The beams shown in Fig. 4-1 are classified as *planar structures* because they lie in a single plane. If all loads act in that same plane, and if all deflections (shown by the dashed lines) occur in that plane, then we refer to that plane as the **plane of bending**.

In this chapter we discuss shear forces and bending moments in beams, and we will show how these quantities are related to each other and to the loads. Finding the shear forces and bending moments is an essential step in the design of any beam. We usually need to know not only the maximum values of these quantities, but also the manner in which they vary along the axis. Once the shear forces and bending moments are known, we can find the stresses, strains, and deflections, as discussed later in Chapters 5, 6, and 9.

Fig. 4-1 Examples of beams loaded by lateral forces

4.2

TYPES OF BEAMS, LOADS, AND REACTIONS

Beams are usually described by the manner in which they are supported. For instance, a beam with a pin support at one end and a roller support at

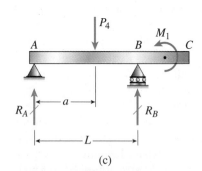

Fig. 4-2 Types of beams: (a) simple beam, (b) cantilever beam, and (c) beam with an overhang

the other (Fig. 4-2a) is called a **simply supported beam** or a **simple beam**. The essential feature of a **pin support** is that it prevents translation at the end of a beam but does not prevent rotation. Thus, end A of the beam of Fig. 4-2a cannot move horizontally or vertically but the axis of the beam can rotate in the plane of the figure. Consequently, a pin support is capable of developing a force reaction with both horizontal and vertical components (H_A and R_A), but it cannot develop a moment reaction.

At end B of the beam (Fig. 4-2a) the **roller support** prevents translation in the vertical direction but not in the horizontal direction; hence this support can resist a vertical force (R_B) but not a horizontal force. Of course, the axis of the beam is free to rotate at B just as it is at A. The vertical reactions at roller supports and pin supports may act *either* upward or downward, and the horizontal reaction at a pin support may act either to the left or to the right. In the figures, reactions are indicated by slashes across the arrows in order to distinguish them from loads, as explained previously in Section 1.8.

The beam shown in Fig. 4-2b, which is fixed at one end and free at the other, is called a **cantilever beam**. At the **fixed support** (or *clamped support*) the beam can neither translate nor rotate, whereas at the free end it may do both. Consequently, both force and moment reactions may exist at the fixed support.

The third example in the figure is a **beam with an overhang** (Fig. 4-2c). This beam is simply supported at points A and B (that is, it has a pin support at A and a roller support at B) but it also projects beyond the support at B. The overhanging segment BC is similar to a cantilever beam except that the beam axis may rotate at point B.

When drawing sketches of beams, we identify the supports by *conventional symbols*, such as those shown in Fig. 4-2. These symbols indicate the manner in which the beam is restrained, and therefore they also indicate the nature of the reactive forces and moments. However, *the symbols do not represent the actual physical construction*. For instance, consider the examples shown in Fig. 4-3. Part (a) of the figure shows a wide-flange beam supported on a concrete wall and held down by anchor bolts that pass through slotted holes in the lower flange of the beam. This connection restrains the beam against vertical movement (either upward or downward) but does not prevent horizontal movement. Also, any restraint against rotation of the longitudinal axis of the beam is small and ordinarily may be disregarded. Consequently, this type of support is usually represented by a roller, as shown in part (b) of the figure.

The second example (Fig. 4-3c) is a beam-to-column connection in which the beam is attached to the column flange by bolted angles. This type of support is usually assumed to restrain the beam against horizontal and vertical movement but not against rotation (restraint against rotation is slight because both the angles and the column can bend). Thus, this connection is likely to be represented as a pin support for the beam (Fig. 4-3d).

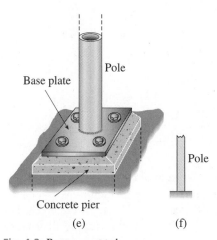

Fig. 4-3 Beam supported on a wall: (a) actual construction, and (b) representation as a roller support. Beam-to-column connection: (c) actual construction, and (d) representation as a pin support. Pole anchored to a pier: (e) actual construction, and (f) representation as a fixed support.

The last example (Fig. 4-3e) is a metal pole welded to a base plate that is anchored to a concrete pier set in the ground. Since the base of the pole is fully restrained against both translation and rotation, it is represented as a fixed support (Fig. 4-3f).

The task of representing a real structure by an *idealized model*, as illustrated by the beams shown in Fig. 4-2, is an important aspect of engineering work. The model should be simple enough to facilitate mathematical analysis and yet complex enough to represent the actual behavior of the structure with reasonable accuracy. Of course, every model is an approximation to nature. For instance, the actual supports of a beam are never perfectly rigid, and so there will always be a small amount of translation at a pin support and a small amount of rotation at a fixed support. Also, supports are never entirely free of friction, and so there will always be a small amount of restraint against translation at a roller support. In most circumstances, especially for statically determinate beams, these deviations from the idealized conditions have little effect on the action of the beam and can safely be disregarded.

■ Types of Loads

Several types of loads that act on beams are illustrated in Fig. 4-2. When a load is applied over a very small area it may be idealized as a **concentrated load**, which is a single force. Examples are the loads P_1, P_2, P_3, and P_4 in the figure. When a load is spread along the axis of a beam, it is represented as a **distributed load**, such as the load q in part (a) of the figure. Distributed loads are measured by their **intensity**, which is expressed in units of force per unit distance (for example, newtons per meter or pounds per foot). A **uniformly distributed load**, or **uniform load**, has constant intensity q per unit distance (Fig. 4-2a). A varying load has an intensity that changes with distance along the axis; for instance, the **linearly varying load** of Fig. 4-2b has an intensity that varies linearly from q_1 to q_2. Another kind of load is a **couple**, illustrated by the couple of moment M_1 acting on the overhanging beam (Fig. 4-2c).

As mentioned in Section 4.1, we assume in this discussion that the loads act in the plane of the figure, which means that all forces must have their vectors in the plane of the figure and all couples must have their moment vectors perpendicular to the plane of the figure. Furthermore, the beam itself must be symmetric about that plane, which means that every cross section of the beam must have a vertical axis of symmetry. Under these conditions, the beam will deflect only in the *plane of bending* (the plane of the figure).

■ Reactions

Finding the reactions is usually the first step in the analysis of a beam. Once the reactions are known, the shear forces and bending moments can be found, as described later in this chapter. If a beam is supported in a statically determinate manner, all reactions can be found from free-body diagrams and equations of equilibrium.

As an example, let us determine the reactions of the simple beam AB of Fig. 4-2a. This beam is loaded by an inclined force P_1, a vertical force P_2, and a uniformly distributed load of intensity q. We begin by noting that the beam has three unknown reactions: a horizontal force H_A at the pin support, a vertical force R_A at the pin support, and a vertical force R_B at the roller support. For a planar structure, such as this beam, we know from statics that we can write three independent equations of equilibrium. Thus, since there are three unknown reactions and three equations, the beam is statically determinate.

The equation of horizontal equilibrium is

$$\Sigma F_{\text{horiz}} = 0 \qquad H_A - P_1 \cos \alpha = 0$$

from which we get

$$H_A = P_1 \cos \alpha$$

This result is so obvious from an inspection of the beam that ordinarily we would not bother to write the equation of equilibrium.

To find the vertical reactions R_A and R_B we write equations of moment equilibrium about points B and A, respectively, with counterclockwise moments being positive:

$$\Sigma M_B = 0 \qquad -R_A L + (P_1 \sin \alpha)(L - a) + P_2(L - b) + qc^2/2 = 0$$

$$\Sigma M_A = 0 \qquad R_B L - (P_1 \sin \alpha)(a) - P_2 b - qc(L - c/2) = 0$$

Solving for R_A and R_B, we get

$$R_A = \frac{(P_1 \sin \alpha)(L - a)}{L} + \frac{P_2(L - b)}{L} + \frac{qc^2}{2L}$$

$$R_B = \frac{(P_1 \sin \alpha)(a)}{L} + \frac{P_2 b}{L} + \frac{qc(L - c/2)}{L}$$

As a check on these results we can write an equation of equilibrium in the vertical direction and verify that it reduces to an identity.

As a second example, consider the cantilever beam of Fig. 4-2b. The loads consist of an inclined force P_3 and a linearly varying distributed load. The latter is represented by a trapezoidal diagram of load intensity that varies from q_1 to q_2. The reactions at the fixed support are a horizontal force H_A, a vertical force R_A, and a couple M_A. Equilibrium of forces in the horizontal direction gives

$$H_A = \frac{5P_3}{13}$$

and equilibrium in the vertical direction gives

$$R_A = \frac{12P_3}{13} + \left(\frac{q_1 + q_2}{2}\right)b$$

In finding this reaction we used the fact that the resultant of the distributed load is equal to the area of the trapezoidal loading diagram.

The moment reaction M_A at the fixed support is found from an equation of equilibrium of moments. In this example we will sum moments about point A in order to eliminate both H_A and R_A from the moment equation. Also, for the purpose of finding the moment of the distributed load, we will divide the trapezoid into two triangles, as shown by the dashed line in Fig. 4-2b. Each load triangle can be replaced by its resultant, which is a force having its magnitude equal to the area of the triangle and having its line of action through the centroid of the triangle. Thus, the moment about point A of the lower triangular part of the load is

$$\left(\frac{q_1 b}{2}\right)\left(L - \frac{2b}{3}\right)$$

in which $q_1 b/2$ is the resultant force (equal to the area of the triangular load diagram) and $L - 2b/3$ is the moment arm (about point A) of the resultant.

The moment of the upper triangular portion of the load is obtained by a similar procedure, and the final equation of moment equilibrium (counterclockwise is positive) is

$$\Sigma M_A = 0 \qquad M_A - \left(\frac{12P_3}{13}\right)a - \frac{q_1 b}{2}\left(L - \frac{2b}{3}\right) - \frac{q_2 b}{2}\left(L - \frac{b}{3}\right) = 0$$

from which

$$M_A = \frac{12P_3 a}{13} + \frac{q_1 b}{2}\left(L - \frac{2b}{3}\right) + \frac{q_2 b}{2}\left(L - \frac{b}{3}\right)$$

Since this equation gives a positive result, the reactive moment M_A acts in the assumed direction, that is, counterclockwise. (The expressions for R_A and M_A can be checked by taking moments about end B of the beam and verifying that the resulting equation of equilibrium reduces to an identity.)

The beam with an overhang (Fig. 4-2c) supports a vertical force P_4 and a couple of moment M_1. Since there are no horizontal forces acting on the beam, the horizontal reaction at the pin support is nonexistent and we do not need to show it on the free-body diagram. In arriving at this conclusion, we made use of the equation of equilibrium for forces in the horizontal direction. Consequently, only two independent equations of equilibrium remain—either two moment equations or one moment equation plus the equation for vertical equilibrium. Let us arbitrarily decide to write two moment equations, the first for moments about point B and the second for moments about point A, as follows (counterclockwise moments are positive):

$$\sum M_B = 0 \qquad -R_A L + P_4(L - a) + M_1 = 0$$

$$\sum M_A = 0 \qquad -P_4 a + R_B L + M_1 = 0$$

Therefore, the reactions are

$$R_A = \frac{P_4(L - a)}{L} + \frac{M_1}{L} \qquad R_B = \frac{P_4\, a}{L} - \frac{M_1}{L}$$

Again, summation of forces in the vertical direction provides a check on these results.

The preceding discussion illustrates how the reactions of statically determinate beams are calculated from equilibrium equations. We have intentionally used symbolic examples rather than numerical examples in order to show how the individual steps are carried out. (In later examples, we often will omit the calculation of reactions because the steps are essentially the same as those described here.)

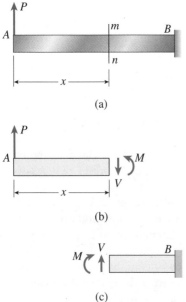

Fig. 4-4 Shear force V and bending moment M in a beam

4.3 SHEAR FORCES AND BENDING MOMENTS

When a beam is loaded by forces or couples, stresses and strains are created throughout the interior of the beam. To determine these stresses and strains, we first must find the internal forces and internal couples that act on cross sections of the beam. As an illustration of how these internal quantities are found, consider a cantilever beam AB loaded by a force P at its free end (Fig. 4-4a). We cut through the beam at a cross section mn located at distance x from the free end and isolate the left-hand part of the beam as a free body (Fig. 4-4b). The free body is held in equilibrium by the force P and by the stresses that act over the cut cross section. These stresses represent the action of the right-hand part of the beam on the left-hand part. At this stage of our discussion we do not know the distribution of the stresses acting over the cross section; all we know is that the resultant of these stresses must be such as to maintain equilibrium of the free body.

From statics, we know that the resultant of the stresses acting on the cross section can be reduced to a **shear force** V and a **bending moment** M (Fig. 4-4b). Because the load P is transverse to the axis of the beam, no axial force exists at the cross section. Both the shear force and the bending moment act in the plane of the beam, that is, the vector for the shear force lies in the plane of the figure and the vector for the moment is perpendicular to the plane of the figure.

Shear forces and bending moments, like axial forces in bars and internal torques in shafts, are the resultants of stresses distributed over the cross section. Therefore, these quantities are known collectively as **stress resultants**.

The stress resultants in statically determinate beams can be calculated from equations of equilibrium. In the case of the cantilever beam of Fig. 4-4a, we use the free-body diagram of Fig. 4-4b. Summing forces in the vertical direction and also taking moments about the cut section, we get

$$\sum F_{\text{vert}} = 0 \qquad P - V = 0 \quad \text{or} \quad V = P$$

$$\sum M = 0 \qquad M - Px = 0 \quad \text{or} \quad M = Px$$

where x is the distance from the free end of the beam to the cross section where V and M are being determined. Thus, through the use of a free-body diagram and two equations of equilibrium, we can calculate the shear force and bending moment without difficulty.

■ Sign Conventions

Let us now consider the sign conventions for shear forces and bending moments. It is customary to assume that shear forces and bending moments are positive when they act in the directions shown in Fig. 4-4b. Note that the shear force tends to rotate the material clockwise and the bending moment tends to compress the upper part of the beam and elongate the lower part. Also, in this instance, the shear force acts downward and the bending moment acts counterclockwise.

The action of these *same* stress resultants against the right-hand part of the beam is shown in Fig. 4-4c. The directions of both quantities are now reversed—the shear force acts upward and the bending moment acts clockwise. However, the shear force still tends to rotate the material clockwise and the bending moment still tends to compress the upper part of the beam and elongate the lower part.

Therefore, we must recognize that the algebraic sign of a stress resultant is determined by how it deforms the material on which it acts, rather than by its direction in space. In the case of a beam, a positive shear force acts clockwise against the material (Figs. 4-4b and c) and a negative shear force acts counterclockwise against the material. Also, a positive bending moment compresses the upper part of the beam (Figs. 4-4b and c) and a negative bending moment compresses the lower part.

To make these conventions clear, both positive and negative shear forces and bending moments are shown in Fig. 4-5. The forces and moments are shown acting on an element of a beam cut out between two cross sections that are a small distance apart.

The *deformations* of an element caused by both positive and negative shear forces and bending moments are sketched in Fig. 4-6. We see that a positive shear force tends to deform the element by causing the

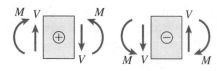

Fig. 4-5 Sign conventions for shear force V and bending moment M

(a)

(b)

Fig. 4-6 Deformations (highly exaggerated) of a beam element caused by: (a) shear forces, and (b) bending moments

right-hand face to move downward with respect to the left-hand face, and, as already mentioned, a positive bending moment compresses the upper part of a beam and elongates the lower part.

Sign conventions for stress resultants are called **deformation sign conventions** because they are based upon how the material is deformed. For instance, we previously used a deformation sign convention in dealing with axial forces in a bar. We stated that an axial force producing elongation (or tension) in a bar is positive and an axial force producing shortening (or compression) is negative. Thus, the sign of an axial force depends upon how it deforms the material, not upon its direction in space.

By contrast, when writing equations of equilibrium we use **static sign conventions**, in which forces are positive or negative according to their directions along the coordinate axes. For instance, if we are summing forces in the y direction, forces acting in the positive direction of the y axis are taken as positive and forces acting in the negative direction are taken as negative. As an example, consider Fig. 4-4b, which is a free-body diagram of part of the cantilever beam. Suppose that we are summing forces in the vertical direction and that the y axis is positive upward. Then the load P is given a positive sign in the equation of equilibrium because it acts upward. However, the shear force V (which is a *positive* shear force) is given a negative sign because it acts downward (that is, in the negative direction of the y axis). This example shows the distinction between the deformation sign convention used for the shear force and the static sign convention used in the equation of equilibrium.

The following examples illustrate the techniques for handling sign conventions and determining shear forces and bending moments in beams. The general procedure consists of constructing free-body diagrams and solving equations of equilibrium.

WORK SHEET 3 Mathcad **Example 4-1**

A simple beam AB supports two loads, a force P and a couple M_0, acting as shown in Fig. 4-7a. Find the shear force V and bending moment M in the beam at cross sections located as follows: (a) a small distance to the left of the midpoint of the beam, and (b) a small distance to the right of the midpoint of the beam.

Solution *Reactions.* The first step in the analysis of this beam is to find the reactions R_A and R_B at the supports. Taking moments about ends B and A gives two equations of equilibrium, from which we find, respectively,

$$R_A = \frac{3P}{4} - \frac{M_0}{L} \qquad R_B = \frac{P}{4} + \frac{M_0}{L} \qquad \text{(a)}$$

(a) *Shear force and bending moment to the left of the midpoint.* We cut the beam at a cross section just to the left of the midpoint and draw a free-body diagram of either half of the beam. In this example, we choose the left-hand half of the beam as the free body (Fig. 4-7b). This free body is held in equilibrium by the load P, the reaction R_A, and the two unknown stress resultants—the shear

(a)

(b)

(c)

Fig. 4-7 Example 4-1. Shear forces and bending moments in a simple beam

force V and the bending moment M, both of which are shown in their positive directions (see Fig. 4-5). The couple M_0 does not act on the free body because the beam is cut to the left of its point of application. Summing forces in the vertical direction (upward is positive) gives

$$\sum F_{\text{vert}} = 0 \qquad R_A - P - V = 0$$

from which we get the shear force:

$$V = R_A - P = -\frac{P}{4} - \frac{M_0}{L} \qquad \text{(b)} \quad \Longleftarrow$$

This result shows that when P and M_0 act in the directions shown in Fig. 4-7a, the shear force (at the selected location) is negative and acts in the opposite direction to the positive direction assumed in Fig. 4-7b.

Taking moments about an axis through the cross section where the beam is cut (see Fig. 4-7b) gives

$$\sum M = 0 \qquad -R_A\left(\frac{L}{2}\right) + P\left(\frac{L}{4}\right) + M = 0$$

in which counterclockwise moments are taken as positive. Solving for the bending moment M, we get

$$M = R_A\left(\frac{L}{2}\right) - P\left(\frac{L}{4}\right) = \frac{PL}{8} - \frac{M_0}{2} \qquad \text{(c)} \quad \Longleftarrow$$

The bending moment M may be either positive or negative, depending upon the magnitudes of the loads P and M_0. If it is positive, it acts in the direction shown in the figure; if it is negative, it acts in the opposite direction.

(b) *Shear force and bending moment to the right of the midpoint.* In this case we cut the beam at a cross section just to the right of the midpoint and again draw a free-body diagram of the part of the beam to the left of the cut section (Fig. 4-7c). The difference between this diagram and the former one is that the couple M_0 now acts on the free body. From two equations of equilibrium, the first for forces in the vertical direction and the second for moments about an axis through the cut section, we obtain

$$V = -\frac{P}{4} - \frac{M_0}{L} \qquad M = \frac{PL}{8} + \frac{M_0}{2} \qquad \text{(d,e)} \quad \Longleftarrow$$

These results show that when the cut section is shifted from the left to the right of the couple M_0, the shear force does not change (because the vertical forces acting on the free body do not change) but the bending moment increases algebraically by an amount equal to M_0 (compare Eqs. c and e).

WORK SHEET 4 Mathcad   **Example 4-2**

A cantilever beam that is free at end A and fixed at end B is subjected to a distributed load of linearly varying intensity q (Fig. 4-8a). The maximum intensity of the load occurs at the fixed support and is equal to q_0. Find the shear force V and bending moment M at distance x from the free end of the beam.

(a)

(b)

Fig. 4-8 Example 4-2. Shear force and bending moment in a cantilever beam

Solution   *Shear force.* We cut through the beam at distance x from the left-hand end and isolate part of the beam as a free body (Fig. 4-8b). Acting on the free body are the distributed load q, the shear force V, and the bending moment M. Both unknown quantities (V and M) are assumed to be positive.

The intensity of the distributed load at distance x from the end is

$$q = \frac{q_0 x}{L} \tag{4-1}$$

Therefore, the total downward load on the free body, equal to the area of the triangular loading diagram (Fig. 4-8b), is

$$\frac{1}{2}\left(\frac{q_0 x}{L}\right)(x) = \frac{q_0 x^2}{2L}$$

From an equation of equilibrium in the vertical direction we find

$$V = -\frac{q_0 x^2}{2L} \tag{4-2a}$$

At the free end A ($x = 0$) the shear force is zero, and at the fixed end B ($x = L$) the shear force has its maximum value:

$$V_{\text{max}} = -\frac{q_0 L}{2} \tag{4-2b}$$

which is numerically equal to the total downward load on the beam. The minus signs in Eqs. (4-2a) and (4-2b) show that the shear forces act in the opposite direction to that pictured in Fig. 4-8b.

Bending moment. To find the bending moment M in the beam (Fig. 4-8b), we write an equation of moment equilibrium about an axis through the cut section. Recalling that the moment of a triangular load is equal to the area of the loading diagram times the distance from its centroid to the axis of moments, we obtain the following equation of equilibrium (counterclockwise moments are positive):

$$\Sigma M = 0 \qquad M + \frac{1}{2}\left(\frac{q_0 x}{L}\right)(x)\left(\frac{x}{3}\right) = 0$$

from which we get

$$M = -\frac{q_0 x^3}{6L} \tag{4-3a}$$

At the free end of the beam ($x = 0$), the bending moment is zero, and at the fixed end ($x = L$) the moment has its numerically largest value:

$$M_{\text{max}} = -\frac{q_0 L^2}{6} \tag{4-3b}$$

The minus signs in Eqs. (4-3a) and (4-3b) show that the bending moments act in the opposite direction to that shown in Fig. 4-8b.

WORK
SHEET
3   Mathcad   **Example 4-3**

A simple beam with an overhang is supported at points A and B (Fig. 4-9a). A uniform load of intensity $q = 200$ lb/ft acts throughout the length of the beam and a concentrated load $P = 14$ k acts at a point 9 ft from the left-hand support. The span length is 24 ft and the length of the overhang is 6 ft. Calculate the shear force V and bending moment M at cross section D located 15 ft from the left-hand support.

$M_A - 14(9) + R_B(24) = 0$

Fig. 4-9 Example 4-3. Shear force and bending moment in a beam with an overhang

Solution   *Reactions.* We begin by calculating the reactions R_A and R_B from equations of equilibrium for the entire beam considered as a free body. Thus, taking moments about the supports at B and A, respectively, we find

$$R_A = 11 \text{ k} \qquad R_B = 9 \text{ k}$$

Shear force and bending moment at section D. Now we make a cut at section D and construct a free-body diagram of the left-hand part of the beam (Fig. 4-9b). When drawing this diagram, we assume that the unknown stress resultants V and M are positive. The equations of equilibrium for the free body are as follows:

$$\sum F_{vert} = 0 \qquad 11 \text{ k} - 14 \text{ k} - (0.200 \text{ k/ft})(15 \text{ ft}) - V = 0$$

$$\sum M_D = 0 \qquad -(11 \text{ k})(15 \text{ ft}) + (14 \text{ k})(6 \text{ ft}) + (0.200 \text{ k/ft})(15 \text{ ft})(7.5 \text{ ft}) + M = 0$$

in which upward forces are taken as positive in the first equation and counterclockwise moments are taken as positive in the second equation. Solving these equations, we get

$$V = -6 \text{ k} \qquad M = 58.5 \text{ k-ft}$$ ◄—

The minus sign for V means that the shear force is negative, that is, its direction is opposite to the direction shown in Fig. 4-9b. The positive sign for M means that the bending moment acts in the direction shown in the figure.

   Alternative free-body diagram. Another method of solution is to obtain V and M from a free-body diagram of the right-hand part of the beam (Fig. 4-9c). When drawing this free-body diagram, we again assume that the unknown shear force and bending moment are positive. The two equations of equilibrium are

$$\sum F_{vert} = 0 \qquad V + 9 \text{ k} - (0.200 \text{ k/ft})(15 \text{ ft}) = 0$$

$$\sum M_D = 0 \qquad -M + (9 \text{ k})(9 \text{ ft}) - (0.200 \text{ k/ft})(15 \text{ ft})(7.5 \text{ ft}) = 0$$

from which

$$V = -6 \text{ k} \qquad M = 58.5 \text{ k-ft}$$

as before. As often happens, the choice between free-body diagrams is a matter of convenience and personal preference.

4.4 ▌ RELATIONSHIPS BETWEEN LOADS, SHEAR FORCES, AND BENDING MOMENTS

We will now obtain some important relationships between loads, shear forces, and bending moments in beams. These relationships are quite useful when investigating the shear forces and bending moments throughout the entire length of a beam, and they are especially helpful when constructing shear-force and bending-moment diagrams (Section 4.5).

   As a means of obtaining the relationships, let us consider an element of a beam cut out between two cross sections that are distance dx apart (Fig. 4-10). The load acting on the top surface of the element may be a distributed load, a concentrated load, or a couple, as shown in Figs. 4-10a, b, and c, respectively. The **sign conventions** for these loads are as follows: *Distributed loads and concentrated loads are positive when they act downward on the beam and negative when they act upward. A couple acting as a load on a beam is positive when it is counterclockwise and negative when it is clockwise.* If other sign conventions are used, the only changes will be in the signs of the terms appearing in the equations derived below.

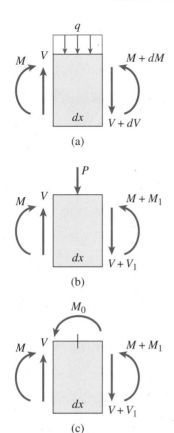

(a)

(b)

(c)

Fig. 4-10 Element of a beam used in deriving the relationships between loads, shear forces, and bending moments. (All loads and stress resultants are shown in their positive directions.)

The shear forces and bending moments acting on the sides of the element are shown in their positive directions in Fig. 4-10. In general, the shear forces and bending moments vary along the axis of the beam. Therefore, their values on the right-hand face of the element may be different from their values on the left-hand face. In the case of a distributed load (Fig. 4-10a) the increments in V and M are infinitesimal, and so we denote them by dV and dM, respectively. The corresponding stress resultants on the right-hand face are $V + dV$ and $M + dM$. In the case of a concentrated load (Fig. 4-10b) or a couple (Fig. 4-10c) the increments may be finite, and so they are denoted V_1 and M_1. The corresponding stress resultants on the right-hand face are $V + V_1$ and $M + M_1$.

For each type of loading we can write two equations of equilibrium for the element—one equation for equilibrium of forces in the vertical direction and one for equilibrium of moments. The first of these equations gives the relationship between the load and the shear force, and the second gives the relationship between the shear force and the bending moment.

■ **Distributed Loads (Fig. 4-10a)**

The first type of loading is a distributed load of intensity q, as shown in Fig. 4-10a. Equilibrium of forces in the vertical direction (upward forces are positive) gives

$$\Sigma F_{\text{vert}} = 0 \qquad V - q\,dx - (V + dV) = 0$$

or

$$\frac{dV}{dx} = -q \qquad\qquad (4\text{-}4)$$

From this equation we see that the rate of change of the shear force at any point on the axis of the beam is equal to the negative of the intensity of the distributed load at that same point. (*Note:* If the sign convention for a distributed load is reversed, so that q is positive upward instead of downward, then the minus sign is omitted in the preceding equation.)

Some useful relations are immediately obvious from Eq. (4-4). For instance, if there is no distributed load on a segment of the beam (that is, if $q = 0$), then $dV/dx = 0$ and the shear force is constant in that part of the beam. Also, if the distributed load is uniform along part of the beam ($q = $ constant), then dV/dx is also constant and the shear force changes linearly in that part of the beam.

As a demonstration of Eq. (4-4), consider the cantilever beam with a linearly varying load that we discussed in Example 4-2 of the preceding section (see Fig. 4-8). The load on the beam (from Eq. 4-1) is

$$q = \frac{q_0 x}{L}$$

which is positive because it acts downward. Also, the shear force (Eq. 4-2a) is

$$V = -\frac{q_0 x^2}{2L}$$

Taking the derivative dV/dx gives

$$\frac{dV}{dx} = \frac{d}{dx}\left(-\frac{q_0 x^2}{2L}\right) = -\frac{q_0 x}{L} = -q$$

which agrees with Eq. (4-4).

A useful relationship pertaining to the shear forces at two different cross sections of a beam can be obtained by integrating Eq. (4-4) along the axis of the beam. To obtain this relationship, we multiply both sides of Eq. (4-4) by dx and then integrate between any two points A and B on the axis of the beam; thus,

$$\int_A^B dV = -\int_A^B q\, dx \qquad \text{(a)}$$

where we are assuming that x increases as we move from point A to point B. The left-hand side of this equation equals the difference $(V_B - V_A)$ of the shear forces at B and A. The integral on the right-hand side represents the area of the loading diagram between A and B, which in turn is equal to the magnitude of the resultant of the distributed load acting between points A and B. Thus, from Eq. (a) we get

$$V_B - V_A = -\int_A^B q\, dx$$

$$= -(\text{area of the loading diagram between } A \text{ and } B) \qquad \text{(4-5)}$$

In other words, the change in shear force between two points along the axis of the beam is equal to the negative of the total downward load between those points. The area of the loading diagram may be positive (if q acts downward) or negative (if q acts upward).

Because Eq. (4-4) was derived for an element of the beam subjected *only* to a distributed load (or to no load), we cannot use Eq. (4-4) at a point where a concentrated load is applied (because the intensity of load q is not defined for a concentrated load). For the same reason, we cannot use Eq. (4-5) if a concentrated load P acts on the beam between points A and B.

Let us now consider the moment equilibrium of the beam element shown in Fig. 4-10a. Summing moments about an axis at the left-hand side of the element (the axis is perpendicular to the plane of the figure), and taking counterclockwise moments as positive, we obtain

$$\sum M = 0 \qquad -M - q\,dx\left(\frac{dx}{2}\right) - (V + dV)dx + M + dM = 0$$

Discarding products of differentials (because they are negligible compared to the other terms), we obtain the following relationship:

$$\frac{dM}{dx} = V \tag{4-6}$$

This equation shows that the rate of change of the bending moment at any point on the axis of a beam is equal to the shear force at that same point. For instance, if the shear force is zero in a region of the beam, then the bending moment is constant in that same region.

Equation (4-6) applies only in regions where distributed loads (or no loads) act on the beam. At a point where a concentrated load acts, a sudden change (or discontinuity) in the shear force occurs and the derivative dM/dx is undefined at that point.

Again using the cantilever beam of Fig. 4-8 as an example, we recall that the bending moment (Eq. 4-3a) is

$$M = -\frac{q_0 x^3}{6L}$$

Thus, the derivative dM/dx is

$$\frac{dM}{dx} = \frac{d}{dx}\left(-\frac{q_0 x^3}{6L}\right) = -\frac{q_0 x^2}{2L}$$

which is equal to the shear force in the beam (see Eq. 4-2a).

Integrating Eq. (4-6) between two points A and B on the beam axis gives

$$\int_A^B dM = \int_A^B V\,dx \tag{b}$$

The integral on the left-hand side of this equation is equal to the difference $(M_B - M_A)$ of the bending moments at points B and A. To interpret the integral on the right-hand side, we need to consider V as a function of x and visualize a shear-force diagram showing the variation of V with x. Then we see that the integral on the right-hand side represents the area below the shear-force diagram between A and B. Therefore, we can express Eq. (b) in the following manner:

$$M_B - M_A = \int_A^B V\,dx$$

$$= \text{(area of the shear-force diagram between } A \text{ and } B) \tag{4-7}$$

This equation is valid even when concentrated loads act on the beam between points A and B. However, it is not valid if a couple acts between A and B. A couple produces a sudden change in the bending moment, and the left-hand side of Eq. (b) cannot be integrated across such a discontinuity.

■ Concentrated Loads (Fig. 4-10b)

Now let us consider a concentrated load P acting on the beam element (Fig. 4-10b). From equilibrium of forces in the vertical direction, we get

$$V - P - (V + V_1) = 0 \quad \text{or} \quad V_1 = -P \tag{4-8}$$

This result means that an abrupt change in the shear force occurs at any point where a concentrated load acts. As we pass from left to right through the point of load application, the shear force decreases by an amount equal to the magnitude of the downward load P.

From equilibrium of moments about the left-hand face of the element (Fig. 4-10b), we get

$$-M - P\left(\frac{dx}{2}\right) - (V + V_1)dx + M + M_1 = 0$$

or

$$M_1 = P\left(\frac{dx}{2}\right) + V\,dx + V_1\,dx$$

Since the length dx of the element is infinitesimally small, we see from this equation that the increment M_1 in the bending moment is also infinitesimally small. *Thus, the bending moment does not change as we pass through the point of application of a concentrated load.*

Even though the bending moment M does not change at a concentrated load, its rate of change (dM/dx) undergoes an abrupt change. At the left-hand side of the element (Fig. 4-10b), the rate of change of the bending moment (see Eq. 4-6) is $dM/dx = V$. At the right-hand side, the rate of change is $dM/dx = V + V_1 = V - P$. Therefore, at the point of application of a concentrated load P, the rate of change dM/dx of the bending moment decreases abruptly by an amount equal to P.

■ Loads in the Form of Couples (Fig. 4-10c)

The last case to be considered is a load in the form of a couple M_0 (Fig. 4-10c). From equilibrium of the element in the vertical direction, we obtain $V_1 = 0$, which shows that *the shear force does not change at the point of application of a couple.*

Equilibrium of moments about the left-hand side of the element gives

$$-M + M_0 - (V + V_1)dx + M + M_1 = 0$$

Disregarding terms that contain differentials (because they are negligible compared to the finite terms), we obtain

$$M_1 = -M_0 \qquad (4\text{-}9)$$

This equation shows that the bending moment decreases by M_0 as we move from left to right through the point of load application. *Thus, the bending moment changes abruptly at the point of application of a couple.*

Equations (4-4) through (4-9) are useful when making a complete investigation of the shear forces and bending moments in a beam, as discussed in the next section.

4.5 ■ SHEAR-FORCE AND BENDING-MOMENT DIAGRAMS

When designing a beam, we usually need to know how the shear forces and bending moments vary throughout the length of the beam. Of special importance are the maximum and minimum values of these quantities. Information of this kind is provided by graphs in which the shear force and bending moment are plotted as ordinates and the distance x along the axis of the beam is plotted as the abscissa. Such graphs are called **shear-force and bending-moment diagrams**. To provide a clear understanding of these diagrams, we will explain in detail how they are constructed and interpreted for three basic loading conditions—a single concentrated load, a uniform load, and several concentrated loads. (The case of a couple acting as a load on a beam is illustrated in Example 4-7.)

■ Concentrated Loads

Let us begin with a simple beam AB supporting a concentrated load P (Fig. 4-11a). The load P acts at distance a from the left-hand support and distance b from the right-hand support. Considering the entire beam as a free body, we can readily determine the reactions of the beam from equilibrium; the results are

$$R_A = \frac{Pb}{L} \qquad R_B = \frac{Pa}{L} \qquad (4\text{-}10a,b)$$

We now cut through the beam at a section to the left of the load P and at distance x from the support at A. Then we draw a free-body diagram of the left-hand part of the beam (Fig. 4-11b). From the equations of equilibrium for this free body, we obtain the shear force V and bending moment M at distance x from the support:

$$V = R_A = \frac{Pb}{L} \qquad M = R_A x = \frac{Pbx}{L} \qquad (0 < x < a) \qquad (4\text{-}11a,b)$$

These expressions are valid only for the part of the beam to the left of the load P.

(a)

(b)

(c)

(d)

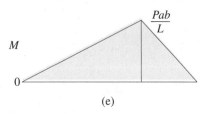

(e)

Fig. 4-11 Shear-force and bending-moment diagrams for a simple beam with a concentrated load

Next, we cut through the beam to the right of the load P (that is, in the region $a < x < L$) and again draw a free-body diagram of the left-hand part of the beam (Fig. 4-11c). From the equations of equilibrium for this free body, we obtain the following expressions for the shear force and bending moment:

$$V = R_A - P = \frac{Pb}{L} - P = -\frac{Pa}{L} \qquad (a < x < L) \qquad (4\text{-}12a)$$

$$M = R_A x - P(x - a) = \frac{Pbx}{L} - P(x - a)$$

$$= \frac{Pa}{L}(L - x) \qquad (a < x < L) \qquad (4\text{-}12b)$$

Note that these equations are valid only for the right-hand part of the beam.

The equations for the shear forces and bending moments (Eqs. 4-11 and 4-12) are plotted below the sketches of the beam. Figure 4-11d is the *shear-force diagram* and Fig. 4-11e is the *bending-moment diagram*.

From the first diagram we see that the shear force at end A of the beam ($x = 0$) is equal to the reaction R_A. Then it remains constant to the point of application of the load P ($x = a$). At that point, the shear force decreases abruptly by an amount equal to the load P. In the right-hand part of the beam, the shear force is again constant but equal numerically to the reaction at B.

As shown in the second diagram, the bending moment in the left-hand part of the beam increases linearly from zero at the support to Pab/L at the concentrated load ($x = a$). In the right-hand part, the bending moment is again a linear function of x, varying from Pab/L at $x = a$ to zero at the support ($x = L$). Thus, the maximum bending moment is

$$M_{\max} = \frac{Pab}{L} \qquad (4\text{-}13)$$

and occurs under the concentrated load.

When deriving the expressions for the shear force and bending moment to the right of the load P (Eqs. 4-12a and b), we considered the equilibrium of the left-hand part of the beam (Fig. 4-11c). This free body is acted upon by the forces R_A and P in addition to V and M. It is slightly simpler in this particular example to consider the right-hand portion of the beam as a free body, because then only one force (R_B) appears in the equilibrium equations (in addition to V and M). Of course, the final results are unchanged.

Certain characteristics of the shear-force and bending moment diagrams (Figs. 4-11d and e) may now be seen. We note first that the slope dV/dx of the shear-force diagram is zero in the regions $0 < x < a$ and $a < x < L$, which is in accord with the equation $dV/dx = -q$ (Eq. 4-4). Also, in these same regions the slope dM/dx of the bending moment diagram is equal to V (Eq. 4-6). At the point of application of the load P there is an abrupt change in the shear-force diagram (equal in magnitude to the load P) and a corresponding change in the slope of the bending-moment diagram. To the left of the load P, the slope of the moment diagram is positive and equal to Pb/L; to the right, it is negative and equal to $-Pa/L$.

Now consider the *area* of the shear-force diagram. As we move from $x = 0$ to $x = a$, the area of the shear-force diagram is $(Pb/L)a$, or Pab/L. This quantity represents the increase in bending moment between these same two points (see Eq. 4-7). From $x = a$ to $x = L$, the area of the shear-force diagram is $-Pab/L$, which means that in this region the bending moment decreases by that amount. Consequently, the bending moment is zero at end B of the beam, as expected.

If the bending moments at both ends of a beam are zero, as is usually the case with a simple beam, then the area of the shear-force diagram between the ends of the beam must be zero provided no couples act on the beam (see the discussion in Section 4.4 following Eq. 4-7).

As mentioned previously, the maximum and minimum values of the shear forces and bending moments are needed when designing beams. For a simple beam with a single concentrated load, the maximum shear force occurs at the end of the beam nearest to the concentrated load and the maximum bending moment occurs under the load itself.

■ Uniform Loads

A simple beam with a uniformly distributed load of constant intensity q is shown in Fig. 4-12a. Because the beam and its loading are symmetric, we see immediately that each of the reactions (R_A and R_B) is equal to $qL/2$. Therefore, the shear force and bending moment at distance x from the left-hand end are

$$V = R_A - qx = \frac{qL}{2} - qx \qquad (4\text{-}14a)$$

$$M = R_A x - qx\left(\frac{x}{2}\right) = \frac{qLx}{2} - \frac{qx^2}{2} \qquad (4\text{-}14b)$$

These equations, which are valid throughout the length of the beam, are plotted as shear-force and bending moment diagrams in Figs. 4-12b and c, respectively.

The shear-force diagram consists of an inclined straight line having ordinates at $x = 0$ and $x = L$ equal numerically to the reactions. The slope of the line is $-q$, as expected from Eq. (4-4). The bending-moment diagram is a parabolic curve that is symmetric about the midpoint of the beam. At each cross section the slope of the bending-moment diagram is equal to the shear force (see Eq. 4-6):

$$\frac{dM}{dx} = \frac{d}{dx}\left(\frac{qLx}{2} - \frac{qx^2}{2}\right) = \frac{qL}{2} - qx = V$$

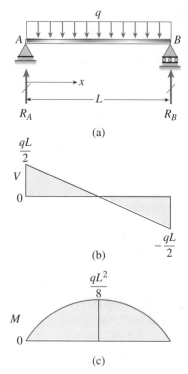

(a)

(b)

(c)

Fig. 4-12 Shear-force and bending-moment diagrams for a simple beam with a uniform load

The maximum value of the bending moment occurs at the midpoint of the beam; therefore, we substitute $x = L/2$ into the expression for M and obtain

$$M_{max} = \frac{qL^2}{8} \qquad (4\text{-}15)$$

as shown on the diagram.

The diagram of load intensity (Fig. 4-12a) has area qL, and according to Eq. (4-5) the shear force V must decrease by this amount as we move along the beam from A to B. We can see that this is indeed the case, because the shear force decreases from $qL/2$ to $-qL/2$.

The area of the shear-force diagram between $x = 0$ and $x = L/2$ is $qL^2/8$, and we see that this area represents the increase in the bending moment between those same two points (Eq. 4-7). In a similar manner, the bending moment decreases by $qL^2/8$ in the region from $x = L/2$ to $x = L$.

■ **Several Concentrated Loads**

If several concentrated loads act on a simple beam (Fig. 4-13a), expressions for the shear forces and bending moments may be determined for each segment of the beam between the points of load application. Again using free-body diagrams of the left-hand part of the beam and measuring the distance x from end A, we obtain the following equations for the first segment of the beam:

$$V = R_A \qquad M = R_A x \qquad (0 < x < a_1) \qquad (4\text{-}16a,b)$$

For the second segment, we get

$$V = R_A - P_1 \qquad M = R_A x - P_1(x - a_1) \qquad (a_1 < x < a_2) \quad (4\text{-}17a,b)$$

For the third segment of the beam, it is advantageous to consider the right-hand part of the beam rather than the left, because fewer loads act on the corresponding free body. Hence, we obtain

$$V = -R_B + P_3 \qquad (4\text{-}18a)$$

$$M = R_B(L - x) - P_3(L - b_3 - x) \qquad (a_2 < x < a_3) \qquad (4\text{-}18b)$$

Finally, for the fourth segment of the beam, we obtain

$$V = -R_B \qquad M = R_B(L - x) \qquad (a_3 < x < L) \qquad (4\text{-}19a,b)$$

Equations (4-16) through (4-19) can be used to construct the shear-force and bending-moment diagrams (Figs. 4-13b and c).

From the shear-force diagram we note that the shear force is constant in each segment of the beam and changes abruptly at every load point, with the amount of each change being equal to the load. Also, the

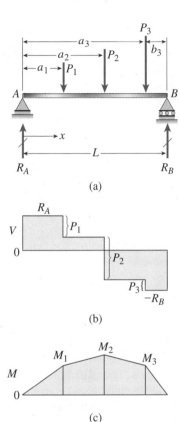

(a)

(b)

(c)

Fig. 4-13 Shear-force and bending-moment diagrams for a simple beam with several concentrated loads

bending moment in each segment is a linear function of x, and therefore the corresponding part of the bending-moment diagram is an inclined straight line. To assist in drawing these lines, we obtain the bending moments under the concentrated loads by substituting $x = a_1$, $x = a_2$, and $x = a_3$ into Eqs. (4-16b), (4-17b), and (4-18b), respectively. In this manner we obtain the following bending moments:

$$M_1 = R_A a_1 \qquad M_2 = R_A a_2 - P_1(a_2 - a_1) \qquad M_3 = R_B b_3 \quad \text{(4-20a,b,c)}$$

Knowing these values, we can readily construct the bending-moment diagram by connecting the points with straight lines.

At each discontinuity in the shear force, there is a corresponding change in the slope dM/dx of the bending-moment diagram. Also, the change in bending moment between two load points equals the area of the shear-force diagram between those same two points (see Eq. 4-7). For example, the change in bending moment between loads P_1 and P_2 is $M_2 - M_1$. Substituting from Eqs. (4-20a and b), we get

$$M_2 - M_1 = (R_A - P_1)(a_2 - a_1)$$

which is the area of the rectangular shear-force diagram between $x = a_1$ and $x = a_2$.

The maximum bending moment in a beam having only concentrated loads *must* occur under one of the loads or at a reaction. To show this, recall that the slope of the bending-moment diagram is equal to the shear force. Therefore, whenever the bending moment has a maximum or minimum value, the derivative dM/dx (and hence the shear force) must change sign. However, in a beam with only concentrated loads, the shear force can change sign only under a load.

If, as we proceed along the x axis, the shear force changes from positive to negative (as in Fig. 4-13b), then the slope in the bending-moment diagram also changes from positive to negative. Therefore, we must have a maximum bending moment at this cross section. Conversely, a change in shear force from a negative to a positive value indicates a minimum bending moment. Theoretically, the shear-force diagram can intersect the horizontal axis at several points, although this is quite unlikely. Corresponding to each such intersection point, there is a local maximum or minimum in the bending-moment diagram. The values of all local maximums and minimums must be determined in order to find the maximum positive and negative bending moments in a beam.

Note that we are using the terms "maximum" and "minimum" with their common meanings of "largest" and "smallest." Consequently, we can refer to "the maximum bending moment in a beam" regardless of whether the bending-moment diagram is described by a smooth, continuous function (as in Fig. 4-12c) or by a series of lines (as in Fig. 4-13c).

■ **General Comments**

The maximum positive and negative bending moments in a beam may occur at the following places: (1) a cross section where a concentrated load is applied and the shear force changes sign (see Figs. 4-11 and 4-13), (2) a cross section where the shear force equals zero (see Fig. 4-12), (3) a point of support where a vertical reaction is present, and (4) a cross section where a couple is applied. The discussions and examples in this section illustrate all of these possibilities.

When several loads act on a beam, the shear-force and bending-moment diagrams can be obtained by superposition (or summation) of the diagrams obtained for each of the loads acting separately. For instance, the shear-force diagram of Fig. 4-13b is actually the sum of three separate diagrams, each of the type shown in Fig. 4-11d for a single concentrated load. We can make an analogous comment for the bending-moment diagram of Fig. 4-13c. Superposition of shear-force and bending-moment diagrams is permissible because shear forces and bending moments in statically determinate beams are linear functions of the applied loads.

Computer programs are readily available for constructing shear-force and bending-moment diagrams. After you have developed an understanding of the diagrams by constructing them manually, you should feel secure in using a computer program. The experience gained from drawing the diagrams by hand will enable you to recognize errors—such as those caused by inputting incorrect data—in the computer-generated results.

WORK SHEET 3 Mathcad **Example 4-4**

Construct the shear-force and bending-moment diagrams for a simple beam with a uniform load of intensity q acting over part of the span (Fig. 4-14a).

Solution   *Reactions.* We begin the analysis by determining the reactions of the beam from the free-body diagram of the entire beam (Fig. 4-14a). The results are

$$R_A = \frac{qb(b + 2c)}{2L} \qquad R_B = \frac{qb(b + 2a)}{2L} \qquad (4\text{-}21a,b)$$

Shear forces and bending moments. To obtain the shear forces and bending moments for the entire beam, we must consider the three segments of the beam individually. For each segment we cut through the beam to expose the shear force V and bending moment M. Then we draw a free-body diagram containing V and M as unknown quantities. Lastly, we sum forces in the vertical direction to obtain the shear force and take moments about the cut section to obtain the bending moment. The results for all three segments are as follows:

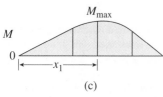

Fig. 4-14 Example 4-4. Simple beam with a uniform load over part of the span

$$V = R_A \qquad M = R_A x \qquad (0 < x < a) \qquad \text{(4-22a,b)}$$

$$V = R_A - q(x - a) \qquad M = R_A x - \frac{q(x - a)^2}{2} \qquad (a < x < a + b) \quad \text{(4-23a,b)}$$

$$V = -R_B \qquad M = R_B(L - x) \qquad (a + b < x < L) \qquad \text{(4-24a,b)}$$

These equations give the shear force and bending moment at every cross section of the beam. As a partial check on these results, we can apply Eq. (4-4) to the shear forces and Eq. (4-6) to the bending moments and verify that the equations are satisfied.

We now construct the shear-force and bending-moment diagrams (Figs. 4-14b and c) from Eqs. (4-22) through (4-24). The shear-force diagram consists of horizontal straight lines in the unloaded regions of the beam and an inclined straight line with negative slope in the loaded region, as expected from the equation $dV/dx = -q$. The bending-moment diagram consists of two inclined straight lines in the unloaded portions of the beam and a parabolic curve in the loaded portion. The inclined lines have slopes equal to R_A and $-R_B$, respectively, as expected from the equation $dM/dx = V$. Also, each of these inclined lines is tangent to the parabolic curve at the point where it meets the curve. This conclusion follows from the fact that there are no abrupt changes in the magnitude of the shear force at these points. Hence, from the equation $dM/dx = V$, we see that the slope of the bending-moment diagram does not change abruptly at these points.

Maximum bending moment. The maximum moment occurs where the shear force equals zero. This point can be found by setting the shear force V (from Eq. 4-23a) equal to zero and solving for the value of x, which we will denote by x_1. The result is

$$x_1 = a + \frac{b}{2L}(b + 2c) \qquad \text{(4-25)}$$

Now we substitute x_1 into the expression for the bending moment (Eq. 4-23b) and solve for the maximum moment. The result is

$$M_{\max} = \frac{qb}{8L^2}(b + 2c)(4aL + 2bc + b^2) \qquad \text{(4-26)}$$

The maximum bending moment occurs within the region of the uniform load.

Special cases. If the uniform load is symmetrically placed on the beam ($a = c$), then we obtain the following simplified results from Eqs. (4-25) and (4-26):

$$x_1 = \frac{L}{2} \qquad M_{\max} = \frac{qb(2L - b)}{8} \qquad \text{(4-27a,b)}$$

If the uniform load extends over the entire span, then $b = L$ and $M_{\max} = qL^2/8$, which agrees with Fig. 4-12 and Eq. (4-15).

(a)

(b)

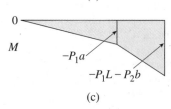

(c)

Fig. 4-15 Example 4-5. Cantilever beam with two concentrated loads

 Example 4-5

Construct the shear-force and bending-moment diagrams for a cantilever beam with two concentrated loads (Fig. 4-15a).

Solution   *Reactions.* From the free-body diagram of the entire beam we find the vertical reaction R_B (positive when upward) and the moment reaction M_B (positive when clockwise):

$$R_B = P_1 + P_2 \qquad M_B = P_1 L + P_2 b \tag{a,b}$$

Shear forces and bending moments. We obtain the shear forces and bending moments by cutting through the beam in each of the two segments, drawing the corresponding free-body diagrams, and solving the equations of equilibrium. Again measuring the distance x from the left-hand end of the beam, we get

$$V = -P_1 \qquad M = -P_1 x \qquad (0 < x < a) \tag{c,d}$$

$$V = -P_1 - P_2 \qquad M = -P_1 x - P_2(x - a) \qquad (a < x < L) \tag{e,f}$$

The corresponding shear-force and bending-moment diagrams are shown in Figs. 4-15b and c. The shear force is constant between the loads and reaches its maximum numerical value at the support, where it is equal numerically to the vertical reaction R_B (Eq. a).

The bending-moment diagram consists of two inclined straight lines, each having a slope equal to the shear force in the corresponding segment of the beam. The maximum bending moment occurs at the support and is equal numerically to the moment reaction M_B (Eq. b). It is also equal to the area of the entire shear-force diagram, as expected from Eq. (4-7).

 Example 4-6

A cantilever beam AB supporting a uniform load of constant intensity q is shown in Fig. 4-16a. Construct the shear-force and bending-moment diagrams for this beam.

Solution   *Reactions.* The reactions R_B and M_B at the fixed support are obtained from equations of equilibrium for the entire beam; thus,

$$R_B = qL \qquad M_B = \frac{qL^2}{2} \tag{4-28a,b}$$

Shear forces and bending moments. These quantities are found by cutting through the beam at distance x from the free end, drawing a free-body diagram of the left-hand part of the beam, and solving the equations of equilibrium. By this means we obtain

$$V = -qx \qquad M = -\frac{qx^2}{2} \tag{4-29a,b}$$

The shear-force and bending-moment diagrams are obtained by plotting these equations (see Figs. 4-16b and c). Note that the slope of the shear-force diagram

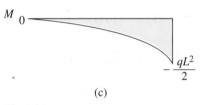

(a)

(b)

(c)

Fig. 4-16 Example 4-6. Cantilever beam with a uniform load

is equal to $-q$ (see Eq. 4-4) and the slope of the bending-moment diagram is equal to V (see Eq. 4-6).

The maximum values of the shear force and bending moment occur at the fixed support where $x = L$:

$$V_{max} = -ql \qquad M_{max} = -\frac{qL^2}{2} \qquad \text{(4-30a,b)}$$

These values are consistent with the values of the reactions R_B and M_B (Eqs. 4-28a and b).

Alternative solution. Instead of using free-body diagrams and equations of equilibrium, we can determine the shear forces and bending moments by integrating the differential relationships between load, shear force, and bending moment. The shear force V at distance x from the free end A is obtained from the load by integrating Eq. (4-5), as follows:

$$V - V_A = V - 0 = V = -\int_0^x q \, dx = -qx \qquad \text{(g)}$$

which agrees with the previous result (Eq. 4-29a).

The bending moment M at distance x from the end is obtained from the shear force by integrating Eq. (4-7):

$$M - M_A = M - 0 = M = \int_0^x V dx = \int_0^x -qx \, dx = -\frac{qx^2}{2} \qquad \text{(h)}$$

which agrees with Eq. 4-29b.

Integrating the differential relationships is quite simple in this example because the loading pattern is continuous and there are no concentrated loads or couples in the regions of integration. If concentrated loads or couples were present, discontinuities in the V and M diagrams would exist, and we cannot integrate Eq. (4-5) through a concentrated load nor can we integrate Eq. (4-7) through a couple (see Section 4.4).

WORK SHEET 4 Mathcad **Example 4-7**

A beam *ABC* with an overhang at the left-hand end is shown in Fig. 4-17a. The beam is subjected to a uniform load of intensity $q = 1.0$ k/ft on the overhang *AB* and a counterclockwise couple $M_0 = 12.0$ k-ft acting midway between the supports at *B* and *C*. Construct shear-force and bending-moment diagrams for this beam.

Solution *Reactions.* We can readily calculate the reactions R_B and R_C from a free-body diagram of the entire beam (Fig. 4-17a). In so doing, we find that R_B is upward and R_C is downward, as shown in the figure. Their numerical values are

$$R_B = 5.25 \text{ k} \qquad R_C = 1.25 \text{ k}$$

Shear forces. The shear force equals zero at the free end of the beam and equals $-qb$ (or -4.0 k) just to the left of support *B*. Since the load is uniformly distributed (that is, q is constant), the slope of the shear diagram is constant and

(a)

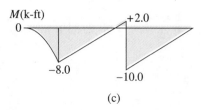

(b)

(c)

Fig. 4-17 Example 4-7. Beam with an overhang

equal to $-q$ (from Eq. 4-4). Therefore, the shear diagram is an inclined straight line with negative slope in the region from A to B (Fig. 4-17b).

Because there are no concentrated or distributed loads between the supports, the shear-force diagram is horizontal in this region. The shear force is equal to the reaction R_C, or 1.25 k, as shown in the figure. (Note that the shear force does not change at the point of application of the couple M_0.)

The numerically largest shear force occurs just to the left of support B and equals -4.0 k.

Bending moments. The bending moment is zero at the free end and decreases as we move to the right until support B is reached. The slope of the moment diagram, equal to the value of the shear force (from Eq. 4-6), is zero at the free end and -4.0 k just to the left of support B. The diagram is parabolic (second-degree) in this region, with the vertex at the end of the beam. The moment at point B is

$$M_B = -\frac{qb^2}{2} = -\frac{1}{2}(1.0 \text{ k/ft})(4.0 \text{ ft})^2 = -8.0 \text{ k-ft}$$

which is also equal to the area of the shear-force diagram between A and B (see Eq. 4-7).

The slope of the bending-moment diagram from B to C is equal to the shear force, or 1.25 k. Therefore, the bending moment just to the left of the couple M_0 is

$$-8.0 \text{ k-ft} + (1.25 \text{ k})(8.0 \text{ ft}) = 2.0 \text{ k-ft}$$

as shown on the diagram. Of course, we can get this same result by cutting through the beam just to the left of the couple, drawing a free-body diagram, and solving the equation of moment equilibrium.

The bending moment changes abruptly at the point of application of the couple M_0, as explained earlier in connection with Eq. (4-9). Because the couple acts counterclockwise, the moment decreases by an amount equal to M_0. Thus, the moment just to the right of the couple M_0 is

$$2.0 \text{ k-ft} - 12.0 \text{ k-ft} = -10.0 \text{ k-ft}$$

From that point to support C the diagram is again a straight line with slope equal to 1.25 k. Therefore, the bending moment at the support is

$$-10.0 \text{ k-ft} + (1.25 \text{ k})(8.0 \text{ ft}) = 0$$

as expected.

Maximum and minimum values of the bending moment occur where the shear force changes sign and where the couple is applied. Comparing the various high and low points on the moment diagram, we see that the numerically largest bending moment equals -10.0 k-ft and occurs just to the right of the couple M_0.

PROBLEMS/CHAPTER 4

SHEAR FORCES AND BENDING MOMENTS

4.3-1 Determine the shear force V and bending moment M at the midpoint of the simple beam AB shown in the figure.

Prob. 4.3-1

4.3-2 Calculate the shear force V and bending moment M at a cross section located 0.5 m from the fixed support A of the cantilever beam AB shown in the figure.

Prob. 4.3-2

4.3-3 Determine the shear force V and bending moment M at the midpoint of the beam with overhangs (see figure).

Prob. 4.3-3

4.3-4 Find the shear force V and bending moment M at the midpoint of the simple beam AB shown in the figure.

Prob. 4.3-4

4.3-5 Find the shear force V and bending moment M at a section located 16 ft from the left-hand end A of the beam with an overhang shown in the figure.

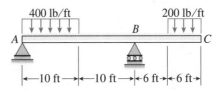

Prob. 4.3-5

4.3-6 The beam ABC shown in the figure is simply supported at A and B and has an overhang from B to C. The loads consist of a couple $M_0 = 4$ kN·m acting at end A and a concentrated load $P = 8$ kN acting at the end of the overhang. Determine the shear force V and bending moment M at a cross section located 3.0 m from the left-hand support.

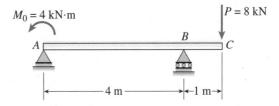

Prob. 4.3-6

4.3-7 A logger weighing 225 lb stands at the midpoint of a floating log that is 16 ft long (see figure). What is the maximum bending moment in the log due to the weight of the logger?

Prob. 4.3-7

4.3-8 A simply supported beam *AB* supports a trapezoidally distributed load (see figure). The intensity of the load varies linearly from 50 kN/m at support *A* to 30 kN/m at support *B*. Calculate the shear force *V* and bending moment *M* at the midpoint of the beam.

50 kN/m

30 kN/m

A   *B*

3 m

Prob. 4.3-8

4.3-9 A curved bar *ABC* is subjected to loads in the form of two equal and opposite forces *P*, as shown in the figure. The axis of the bar forms a semicircle of radius *r*. Determine the axial force *N*, shear force *V*, and bending moment *M* acting at a cross section defined by the angle θ.

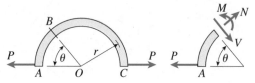

Prob. 4.3-9

4.3-10 The frame *ABC* shown in the figure is attached to a pin support at *C* and a roller support at *A*. A uniform load of intensity *q* acts on part *AB* and a triangular load of maximum intensity 2*q* acts on part *BC*. (a) Obtain an expression for the bending moment *M* in part *AB* at distance *x* from support *A*. (b) From this expression, determine the maximum positive bending moment M_{max} in part *AB*.

Prob. 4.3-10

4.3-11 A beam *ABCD* with a vertical arm *CE* is supported as a simple beam at *A* and *D* (see figure). A cable passes over a small pulley that is attached to the arm at *E*. One end of the cable is attached to the beam at point *B*. What is the force *P* in the cable if the bending moment in the beam just to the left of point *C* is equal numerically to 48 k-in.?

Prob. 4.3-11

4.3-12 The beam *ABCD* is loaded by a force *W* = 27 kN by the arrangement shown in the figure. The cable passes over a small frictionless pulley at *B* and is attached at *E* to the vertical arm. Calculate the axial force *N*, shear force *V*, and bending moment *M* at section *C*, which is just to the left of the vertical arm.

Prob. 4.3-12

4.3-13 The beam *ABCD* is held in equilibrium by uniformly distributed loads of intensities q_1 and q_2 (see figure). Find the shear force *V* and bending moment *M* at

(a) point B, and (b) the midpoint of the beam. (Assume $a = 3$ ft, $b = 8$ ft, and $q_1 = 3500$ lb/ft.)

Prob. 4.3-13

4.3-14 The beam $ABCD$ shown in the figure has overhanging ends and carries a distributed load of linearly varying intensity. For what ratio a/L will the shear force V always be zero at the midpoint of the beam?

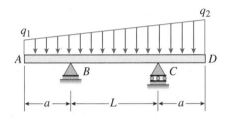

Prob. 4.3-14

*4.3-15 A beam ABC with an overhang (see figure) supports a concentrated load $P = 16$ k at the left-hand end and a uniform load of intensity q throughout its entire length. It is known that the shear force and bending moment at section X-X are equal to -34 k and -90 k-ft, respectively. Find the shear force V and bending moment M at section Y-Y.

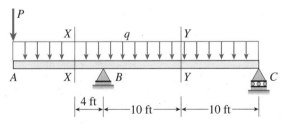

Prob. 4.3-15

*4.3-16 The centrifuge shown in the figure rotates in a horizontal plane (the xy plane) on a smooth surface about the z axis (which is vertical) with an angular acceleration α. Each of the two arms has weight w per unit length and supports a weight $W = 2.5wL$ at its end. Derive formulas for the maximum shear force and maximum bending moment in the arms, assuming $b = L/9$ and $c = L/10$.

Prob. 4.3-16

 SHEAR-FORCE AND BENDING-MOMENT DIAGRAMS

WORK SHEETS 3 & 4 Mathcad

When solving Problems 4.5-1 through 4.5-38, draw the shear-force and bending-moment diagrams approximately to scale and label all critical ordinates, including the maximum and minimum values.

4.5-1 Construct the shear-force and bending-moment diagrams for a simple beam AB supporting two equal concentrated loads (see figure).

Prob. 4.5-1

4.5-2 Construct the shear-force and bending-moment diagrams for a cantilever beam AB carrying a uniform load of intensity q over one-half of its length (see figure).

Prob. 4.5-2

4.5-3 A simple beam AB is subjected to a couple of moment M_0 acting at distance a from the left-hand support (see figure). Draw the shear-force and bending-moment diagrams for this beam.

Prob. 4.5-3

4.5-4 The simple beam AB shown in the figure is subjected to a concentrated load P and a couple $M_1 = PL/4$ acting at the positions indicated. Draw the shear-force and bending moment diagrams for this beam.

Prob. 4.5-4

4.5-5 A simply supported beam ABC supports a vertical load P by means of a bracket BDE (see figure). Draw the shear-force and bending-moment diagrams for the beam.

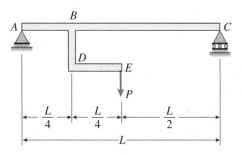

Prob. 4.5-5

4.5-6 A simple beam AB supports a uniform load of intensity $q = 5.0$ kN/m over a portion of the span (see figure). Assuming that $L = 10$ m, $a = 4$ m, and $b = 2$ m, draw the shear-force and bending-moment diagrams for this beam.

Prob. 4.5-6

4.5-7 A simple beam AB subjected to couples M_1 and $2M_1$ acting at the third points is shown in the figure. Construct the shear-force and bending-moment diagrams for this beam.

Prob. 4.5-7

4.5-8 A simple beam AB with an overhang BC is loaded by two forces P and a couple Pa through the arrangement shown in the figure. Draw the shear-force and bending-moment diagrams for beam ABC.

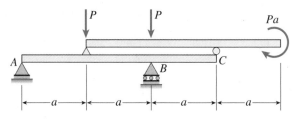

Prob. 4.5-8

4.5-9 The cantilever beam AB supports a concentrated load and a couple, as shown in the figure. Construct shear-force and bending-moment diagrams for this beam.

Prob. 4.5-9

4.5-10 Construct the shear-force and bending-moment diagrams for beam *ABC* (see figure). The beam is loaded by a weight *W* = 10.0 kN applied through a cable. The cable passes over a small frictionless pulley at *C* and is attached to the vertical arm at *D*.

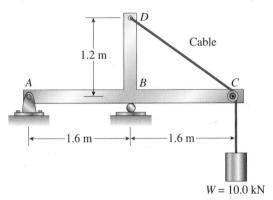

Prob. 4.5-10

4.5-11 Construct the shear-force and bending-moment diagrams for the simple beam of Problem 4.3-1.

4.5-12 Construct the shear-force and bending-moment diagrams for the cantilever beam of Problem 4.3-2.

4.5-13 Construct the shear-force and bending-moment diagrams for the overhanging beam of Problem 4.3-3.

4.5-14 Construct the shear-force and bending-moment diagrams for the simple beam shown in Problem 4.3-4.

4.5-15 Construct the shear-force and bending-moment diagrams for the beam with an overhang shown in Problem 4.3-5.

4.5-16 Construct the shear-force and bending-moment diagrams for the beam with an overhang shown in Problem 4.3-6.

4.5-17 Construct the shear-force and bending-moment diagrams for a cantilever beam *AB* with a linearly varying load of maximum intensity q_0 (see figure).

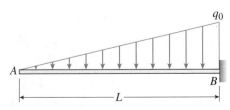

Prob. 4.5-17

4.5-18 A beam with two equal overhangs supports a uniform load of intensity *q* (see figure). The distance between supports is *a* and the total length is *L*. Find the distance *a* so that the maximum bending moment in the beam has the smallest possible numerical value. Draw the shear-force and bending-moment diagrams for this condition.

Prob. 4.5-18

4.5-19 Construct the shear-force and bending-moment diagrams for the beam with an overhang shown in the figure.

Prob. 4.5-19

4.5-20 through 4.5-30 Construct the shear-force and bending-moment diagrams for the beams shown in the figures.

Prob. 4.5-20

Prob. 4.5-21

Prob. 4.5-22

Prob. 4.5-23

Prob. 4.5-24

Prob. 4.5-25

Prob. 4.5-26

Prob. 4.5-27

Prob. 4.5-28

Prob. 4.5-29

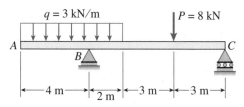

Prob. 4.5-30

4.5-31 Beam *ABCD* has two equal overhangs and supports a symmetric triangular load (see figure). Determine the distance *b* between the supports so that the bending moment at the midpoint of the beam is zero. Draw the shear-force and bending-moment diagrams for this condition.

Prob. 4.5-31

Prob. 4.5-34

4.5-32 The beam *ABC* shown in the figure consists of a cantilever segment *AB* attached to a simple span *BC* by a pin at *B*. The pin can transmit a shear force but not a bending moment. Draw the shear-force and bending-moment diagrams for the entire beam.

Prob. 4.5-32

4.5-33 The beam *ABCD* supports a distributed load of linearly varying intensity (see figure). Draw the shear-force and bending-moment diagrams for this beam.

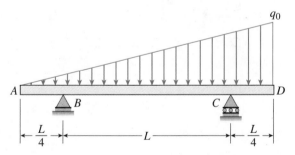

Prob. 4.5-33

4.5-34 The beam *ABCDE* shown in the figure has simple supports at *A*, *C*, and *E* and a hinge (or pin) at *D*. A load of 4 kN acts at the end of the bracket that is attached to the beam at *B*, and a load of 2 kN acts at the midpoint of segment *DE*. Draw the shear-force and bending-moment diagrams for the beam. (Note that the pin at *D* can transmit a shear force but not a bending moment.)

4.5-35 The shear-force diagram for a simple beam is shown in the figure. Determine the loading on the beam and draw the bending-moment diagram, assuming that no couples act as loads on the beam. (Note that the shear force has units of pounds.)

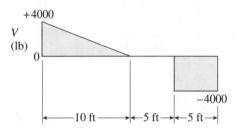

Prob. 4.5-35

4.5-36 The shear-force diagram for a beam is shown in the figure. Assuming that no couples act as loads on the beam, draw the bending-moment diagram. (Note that the shear force has units of kilonewtons.)

Prob. 4.5-36

4.5-37 Construct the shear-force and bending-moment diagrams for a simple beam with a triangular load (maximum intensity q_0) that extends for three-fourths of the span length L (see figure).

Prob. 4.5-37

4.5-38 A beam AB is subjected to concentrated loads P and $2P$ as shown in the figure. The beam rests on a foundation that exerts a continuously distributed reaction against the beam. Assuming that the distributed reaction varies linearly in intensity from A to B, determine the intensities q_A and q_B at ends A and B, respectively. Also, construct the shear-force and bending-moment diagrams for the beam.

Prob. 4.5-38

4.5-39 Two equal loads P are separated by a fixed distance d (see figure). This load combination may be positioned at any distance x from the left-hand support of the simple beam AB. (a) For what distances x will the shear force in the beam be a maximum? What is the maximum shear force V_{max}? (b) Derive a formula for the distance x that will produce the maximum bending moment M_{max} in the beam. Also, obtain an expression for M_{max}.

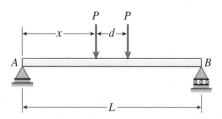

Prob. 4.5-39

4.5-40 A simple beam AB carries two connected wheel loads P and $2P$ that are distance d apart (see figure). The loads may be placed at any distance x from the left-hand support of the beam. Determine the distance x for: (a) maximum shear force in the beam, and (b) maximum bending moment in the beam. Assume $P = 7.5$ kN, $d = 1.6$ m, and $L = 8$ m. Also, determine the maximum shear force V_{max} and maximum bending moment M_{max}.

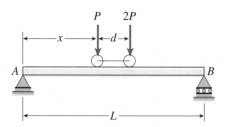

Prob. 4.5-40

4.5-41 Three wheel loads W_1, W_2, and W_3 move across a simply supported beam as shown in the figure. Determine the position of the wheels, as defined by the distance x from end A, so as to produce the maximum bending moment in the beam, assuming that $W_1 = 4$ k and $W_2 = W_3 = 16$ k. Also, determine the maximum bending moment M_{max}.

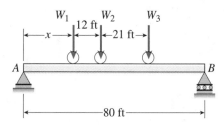

Prob. 4.5-41

$*$ **4.5-42** A small retaining wall 2.4 m high is constructed of vertical wood planks ABC, as shown in the figure. The planks are 125 mm wide and are simply supported at B and C by a horizontal steel beam and a concrete foundation wall, respectively. The steel beam is 0.6 m from the top of the wall. The soil pressure against the wall is assumed to vary linearly from zero at the top (point A) to $p = 18$ kPa at the bottom (point C). Calculate the maximum bending moment M_{max} in the planks.

Wood planks

0.6 m

Steel beam

Soil pressure

2.4 m

1.8 m

$p = 18$ kPa

Side view

Top view

Wood planks

125 mm

Prob. 4.5-42

★★4.5-43 The loads on a simple beam consist of n equally spaced forces (see figure). The total applied load is P; therefore, each force is equal to P/n. The length of the beam is L; hence the spacing between loads is $L/(n+1)$. (a) Derive general formulas for the maximum bending moment in the beam. (b) From these formulas, determine the maximum bending moment for several successive values of n ($n = 1$, 2, 3, 4, . . .). (c) Compare these results with the maximum bending moment in the beam due to a uniformly distributed load of intensity q such that $qL = P$.

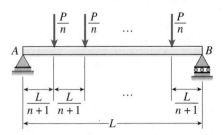

Prob. 4.5-43

★★4.5-44 A beam ABC of length L supports a triangularly distributed load of maximum intensity q_0 as shown in the figure. The beam is simply supported at points B and C, with C at the right-hand end and B at any desired distance a from the left-hand end. (a) Determine the ratio a/L so that the numerically largest bending moment in the beam is a minimum. (b) What is the maximum bending moment in the beam under this condition?

Prob. 4.5-44

5

STRESSES IN BEAMS (BASIC TOPICS)

5.1

INTRODUCTION

In the preceding chapter we saw how the loads acting on a beam create internal actions (or *stress resultants*) in the form of shear forces and bending moments. In this chapter we go one step further and investigate the *stresses* and *strains* associated with those shear forces and bending moments. Knowing the stresses and strains, we will be able to analyze and design beams subjected to a variety of loading conditions.

The loads acting on a beam cause the beam to bend (or *flex*), thereby deforming its axis into a curve. As an example, consider a cantilever beam *AB* subjected to a load *P* at the free end (Fig. 5-1a). The initially straight axis is bent into a curve (Fig. 5-1b), called the **deflection curve** of the beam.

For reference purposes, we construct a system of **coordinate axes** (Fig. 5-1b) with the origin located at a suitable point on the longitudinal axis of the beam. In this illustration, we place the origin at the fixed support. The positive *x* axis is directed to the right, and the positive *y* axis is directed upward. The *z* axis, not shown in the figure, is directed outward (that is, toward the viewer), so that the three axes form a right-handed coordinate system.

The beams considered in this chapter (like those discussed in Chapter 4) are assumed to be symmetric about the *xy* plane, which means that the *y* axis is an axis of symmetry of the cross section. In addition, all loads must act in the *xy* plane. As a consequence, the bending deflections occur in this same plane, known as the **plane of bending**. Thus, the deflection curve shown in Fig. 5-1b is a plane curve lying in the plane of bending.

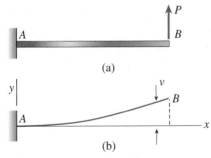

Fig. 5-1 Bending of a cantilever beam: (a) beam with load, and (b) deflection curve

The **deflection** of the beam at any point along its axis is the *displacement* of that point from its original position, measured in the y direction. We denote the deflection by the letter v to distinguish it from the coordinate y itself (see Fig. 5-1b).*

5.2 PURE BENDING AND NONUNIFORM BENDING

When analyzing beams, it is often necessary to distinguish between pure bending and nonuniform bending. **Pure bending** refers to flexure of a beam under a constant bending moment. Therefore, pure bending occurs only in regions of a beam where the shear force is zero (because $V = dM/dx$; see Eq. 4-6). In contrast, **nonuniform bending** refers to flexure in the presence of shear forces, which means that the bending moment changes as we move along the axis of the beam.

As an example of pure bending, consider a simple beam AB loaded by two couples M_1 having the same magnitude but acting in opposite directions (Fig. 5-2a). These loads produce a constant bending moment $M = M_1$ throughout the length of the beam, as shown by the bending moment diagram in part (b) of the figure. Note that the shear force V is zero at all cross sections of the beam.

Another illustration of pure bending is given in Fig. 5-3a, where the cantilever beam AB is subjected to a clockwise couple M_2 at the free end. There are no shear forces in this beam, and the bending moment M is constant throughout its length. The bending moment is negative ($M = -M_2$), as shown by the bending moment diagram in part (b) of the figure.

The symmetrically loaded simple beam of Fig. 5-4a is an example of a beam that is partly in pure bending and partly in nonuniform bending, as seen from the shear-force and bending-moment diagrams (Figs. 5-4b and c). The central region of the beam is in pure bending because the shear force is zero and the bending moment is constant. The parts of the beam near the ends are in nonuniform bending because shear forces are present and the bending moments vary.

In the following two sections we will investigate the strains and stresses in beams subjected only to pure bending. Fortunately, we can often use the results obtained for pure bending even when shear forces are present, as explained later in Section 5.8.

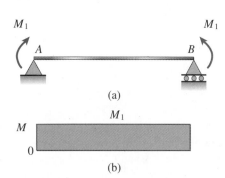

Fig. 5-2 Simple beam in pure bending ($M = M_1$)

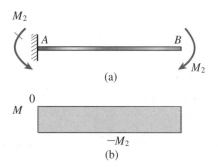

Fig. 5-3 Cantilever beam in pure bending ($M = -M_2$)

5.3 CURVATURE OF A BEAM

When loads are applied to a beam, its longitudinal axis is deformed into a curve, as illustrated previously in Fig. 5-1. The resulting strains and

---

*In applied mechanics, the traditional symbols for displacements in the x, y, and z directions are u, v, and w, respectively.

Fig. 5-4 Simple beam with central region in pure bending and end regions in nonuniform bending

Fig. 5-5 Curvature of a bent beam: (a) beam with load, and (b) deflection curve

stresses in the beam are directly related to the **curvature** of the deflection curve.

To illustrate the concept of curvature, consider again a cantilever beam subjected to a load P acting at the free end (Fig. 5-5a). The deflection curve of this beam is shown in Fig. 5-5b. For purposes of analysis, we identify two points m_1 and m_2 on the deflection curve. Point m_1 is selected at an arbitrary distance x from the y axis and point m_2 is located a small distance ds further along the curve. At each of these points we draw a line normal to the *tangent* to the deflection curve, that is, normal to the curve itself. These normals intersect at point O', which is the **center of curvature** of the deflection curve. Because most beams have very small deflections and nearly flat deflection curves, point O' is usually located much farther from the beam than is indicated in the figure.

The distance $m_1 O'$ from the curve to the center of curvature is called the **radius of curvature** ρ (Greek letter rho), and the **curvature** κ (Greek letter kappa) is defined as the reciprocal of the radius of curvature. Thus,

$$\kappa = \frac{1}{\rho} \tag{5-1}$$

Curvature is a measure of how sharply a beam is bent. If the load on a beam is small, the beam will be nearly straight, the radius of curvature will be very large, and the curvature will be small. If the load is increased, the amount of bending will increase—the radius of curvature will become smaller, and the curvature will become larger.

From the geometry of triangle $O'm_1m_2$ (Fig. 5-5b) we obtain

$$\rho \, d\theta = ds \tag{a}$$

in which $d\theta$ (measured in radians) is the infinitesimal angle between the normals and ds is the infinitesimal distance along the curve between points m_1 and m_2. Combining Eq. (a) with Eq. (5-1), we get

$$\kappa = \frac{1}{\rho} = \frac{d\theta}{ds} \tag{5-2}$$

This equation is derived in textbooks on basic calculus and holds for any curve, regardless of the amount of curvature. If the curvature is constant throughout the length of a curve, the radius of curvature will also be constant and the curve will be an arc of a circle.

The deflections of a beam are usually very small compared to its length (consider, for instance, the deflections of the structural frame of an automobile or a beam in a building). Small deflections mean that the deflection curve is nearly flat. Consequently, the distance ds along the curve may be set equal to its horizontal projection dx (see Fig. 5-5b). Under these special conditions the equation for the curvature becomes

$$\kappa = \frac{1}{\rho} = \frac{d\theta}{dx} \tag{5-3}$$

Both the curvature and the radius of curvature are functions of the distance x measured along the x axis. It follows that the position O' of the center of curvature also depends upon the distance x.

In Section 5.5 we will see that the curvature at a particular point on the axis of a beam depends upon the bending moment at that point and upon the properties of the beam itself (shape of cross section and type of material). Therefore, if the beam is prismatic and the material is homogeneous, the curvature will vary only with the bending moment. Consequently, a beam in pure bending will have constant curvature and a beam in nonuniform bending will have varying curvature.

The **sign convention for curvature** depends upon the orientation of the coordinate axes. If the x axis is positive to the right and the y axis is positive upward, as shown in Fig. 5-6, then the curvature is positive when the beam is bent concave upward (or convex downward) and the center of curvature is above the beam. Conversely, the curvature is negative when the beam is bent concave downward (or convex upward) and the center of curvature is below the beam.

In the next section we will see how the longitudinal strains in a bent beam are determined from its curvature, and in Chapter 9 we will see how curvature is related to the deflections of beams.

(a)

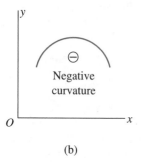

(b)

Fig. 5-6 Sign convention for curvature

5.4   LONGITUDINAL STRAINS IN BEAMS

The longitudinal strains in a beam can be found by analyzing the curvature of the beam and the associated deformations. For this purpose, let us consider a portion ab of a beam in pure bending subjected to positive bending moments M (Fig. 5-7a). We assume that the beam initially has a straight longitudinal axis (the x axis in the figure) and that its cross section is symmetric about the y axis, as shown in Fig. 5-7b.

Under the action of the bending moments, the beam deflects in the xy plane (the plane of bending) and its longitudinal axis is bent into a circular curve (curve ss in Fig. 5-7c). The beam is bent concave upward, which is positive curvature (Fig. 5-6a).

Cross sections of the beam, such as sections mn and pq in Fig. 5-7a, remain plane and normal to the longitudinal axis (Fig. 5-7c). The fact that cross sections of a beam in pure bending remain plane is so fundamental to beam theory that it is often called an assumption. However, we

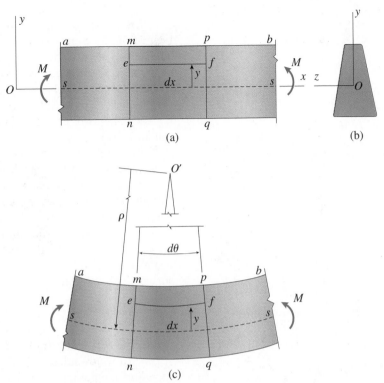

Fig. 5-7 Deformations of a beam in pure bending

could also call it a theorem, because it can be proved rigorously using only rational arguments based upon symmetry (Ref. 5-1). The basic point is that the symmetry of the beam and its loading (Figs. 5-7a and b) means that all elements of the beam (such as element *mpqn*) must deform in an identical manner, which is possible only if cross sections remain plane during bending (Fig. 5-7c). This conclusion is valid for beams of any material, whether the material is elastic or inelastic, linear or nonlinear. Of course, the material properties, like the dimensions, must be symmetric about the plane of bending. (*Note:* Even though a plane cross section in pure bending remains plane, there still may be deformations in the plane itself. Such deformations are due to the effects of Poisson's ratio, as explained at the end of this discussion.)

Because of the bending deformations shown in Fig. 5-7c, cross sections *mn* and *pq* rotate with respect to each other about axes perpendicular to the *xy* plane. Longitudinal lines on the convex (lower) part of the beam are elongated, whereas those on the concave (upper) side are shortened. Thus, the lower part of the beam is in tension and the upper part is in compression. Somewhere between the top and bottom of the beam is a surface in which longitudinal lines do not change in length. This surface, indicated by the dashed line *ss* in Figs. 5-7a and c, is called the **neutral surface** of the beam. Its intersection with any cross-sectional plane is called the **neutral axis** of the cross section; for instance, the *z* axis is the neutral axis for the cross section of Fig. 5-7b.

The planes containing cross sections *mn* and *pq* in the deformed beam (Fig. 5-7c) intersect in a line through the center of curvature *O′*. The angle between these planes is denoted *dθ*, and the distance from *O′* to the neutral surface *ss* is the radius of curvature *ρ*. The initial distance *dx* between the two planes (Fig. 5-7a) is unchanged at the neutral surface (Fig. 5-7c), hence *ρdθ* = *dx*. However, all other longitudinal lines between the two planes either lengthen or shorten, thereby creating **normal strains** ϵ_x.

To evaluate these normal strains, consider a typical longitudinal line *ef* located within the beam between planes *mn* and *pq* (Fig. 5-7a). We identify line *ef* by its distance *y* from the neutral surface in the initially straight beam. Thus, we are now assuming that the *x* axis lies along the neutral surface of the *undeformed* beam. Of course, when the beam deflects, the neutral surface moves with the beam, but the *x* axis remains fixed in position. Nevertheless, the longitudinal line *ef* in the deflected beam (Fig. 5-7c) is still located at the same distance *y* from the neutral surface. Thus, the length L_1 of line *ef* after bending takes place is

$$L_1 = (\rho - y)\, d\theta = dx - \frac{y}{\rho} dx$$

in which we have substituted $d\theta = dx/\rho$. Since the original length of line *ef* is *dx*, it follows that its elongation is $L_1 - dx$, or $-y\, dx/\rho$. The corresponding longitudinal strain is equal to the elongation divided by the initial length *dx*; therefore,

$$\epsilon_x = -\frac{y}{\rho} = -\kappa y \qquad (5\text{-}4)$$

where *κ* is the curvature.

The preceding equation shows that the longitudinal strains in the beam are proportional to the curvature and vary linearly with the distance *y* from the neutral surface. When the point under consideration is above the neutral surface, the distance *y* is positive. If the curvature is also positive (as in Fig. 5-7c), then ϵ_x will be a negative strain, representing a shortening. By contrast, if the point under consideration is below the neutral surface, the distance *y* will be negative and, if the curvature is positive, the strain ϵ_x will also be positive, representing an elongation. Note that the sign convention for ϵ_x is the same as that used for normal strains in earlier chapters, namely, elongation is positive and shortening is negative.

Equation (5-4) for the normal strains in a beam was derived solely from the geometry of the deformed beam—the properties of the material did not enter into the discussion. Therefore, *the strains in a beam in pure bending vary linearly with distance from the neutral surface regardless of the shape of the stress-strain curve of the material.*

The next step in our analysis, namely, finding the stresses from the strains, requires the use of the stress-strain curve. This step is described in the next section for linearly elastic materials and in Sections 6.9 and 6.10 for nonlinear materials.

The longitudinal strains in a beam are accompanied by transverse strains (that is, normal strains in the y and z directions) because of the effects of Poisson's ratio. However, there are no accompanying transverse stresses because beams are free to deform laterally. This stress condition is analogous to that of a prismatic bar in tension or compression, and therefore longitudinal elements in a beam in pure bending are in a state of *uniaxial stress*.

■ **Example 5-1**

A simply supported steel beam AB (Fig. 5-8a) of length $L = 16$ ft is bent by couples M_0 into a circular arc with a downward deflection δ at the midpoint (Fig. 5-8b). Measurements show that the longitudinal normal strain (elongation) on the bottom surface of the beam is 0.00125, which is the yield strain of the steel. Also, the distance from the bottom surface of the beam to the neutral surface is 6.0 in. Determine the radius of curvature ρ, the curvature κ, and the deflection δ of the beam.

Solution   *Curvature.* Since we know the longitudinal strain at the bottom surface of the beam ($\epsilon_x = 0.00125$), and since we also know the distance from the bottom surface to the neutral axis ($y = -6.0$ in.), we can use Eq. (5-4) to calculate both the radius of curvature and the curvature. Rearranging Eq. (5-4) and substituting numerical values, we get

$$\rho = -\frac{y}{\epsilon_x} = -\frac{-6 \text{ in.}}{0.00125} = 4800 \text{ in.} = 400 \text{ ft} \qquad \kappa = \frac{1}{\rho} = 0.0025 \text{ ft}^{-1} \impliedby$$

These results show that the radius of curvature is extremely large, and the curvature is very small, even when the strain in the material reaches a large value (the yield strain).

Deflection. As pointed out in Section 5.3, a constant bending moment (pure bending) produces constant curvature throughout the length of a beam. Therefore, the deflection curve is a circular arc. From Fig. 5-8b we see that the distance from the center of curvature O' to the midpoint C' of the deflected beam is the radius of curvature ρ, and the distance from O' to point C on the x axis is $\rho \cos \theta$, where θ is angle $BO'C$. This leads to the following expression for the deflection at the midpoint of the beam:

$$\delta = \rho(1 - \cos \theta) \tag{5-5}$$

For a nearly flat curve, we can assume that the distance between supports is the same as the length of the beam itself. Therefore, from triangle $BO'C$ we get

$$\sin \theta = \frac{L/2}{\rho} \tag{5-6}$$

Substituting numerical values, we obtain

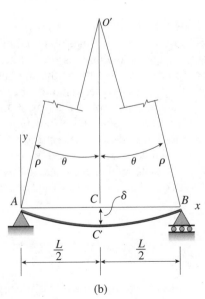

Fig. 5-8 Example 5-1. Beam in pure bending

$$\sin \theta = \frac{(16 \text{ ft})(12 \text{ in./ft})}{2(4800 \text{ in.})} = 0.0200$$

and
$$\theta = 0.0200 \text{ rad} = 1.146°$$

Note that for practical purposes we may consider $\sin \theta$ and θ (radians) to be equal numerically because θ is a very small angle.

Now we substitute into Eq. (5-5) for the deflection and obtain

$$\delta = \rho(1 - \cos \theta) = (4800 \text{ in.})(1 - 0.999800) = 0.960 \text{ in.} \quad \longleftarrow$$

This deflection is very small compared to the length of the beam, as shown by the ratio of the span length to the deflection:

$$\frac{L}{\delta} = \frac{(16 \text{ ft})(12 \text{ in./ft})}{0.960 \text{ in.}} = 200$$

Thus, we have confirmed that the deflection curve is nearly flat in spite of the large strains. Of course, in Fig. 5-8b the deflection of the beam is highly exaggerated for clarity.

Note: This method for finding the deflection of a beam has little practical value because it is limited to pure bending, which produces a circular deflected shape. More useful methods for finding beam deflections are presented in Chapter 9.

5.5 NORMAL STRESSES IN BEAMS (LINEARLY ELASTIC MATERIALS)

In the preceding section we investigated the longitudinal strains ϵ_x in a beam in pure bending (see Eq. 5-4 and Fig. 5-7). Since longitudinal elements of a beam are subjected only to tension or compression, we can now use the stress-strain curve for the material to determine the stresses from the strains. The stresses act over the entire cross section of the beam and vary in intensity depending upon the shape of the stress-strain diagram and the dimensions of the cross section. Since the x direction is longitudinal (Fig. 5-7a), we use the symbol σ_x to denote these stresses.

The most common stress-strain relationship encountered in engineering is the equation for a linearly elastic material. For such materials we substitute Hooke's law for uniaxial stress ($\sigma = E\epsilon$) into Eq. (5-4) and obtain

$$\sigma_x = E\epsilon_x = -\frac{Ey}{\rho} = -E\kappa y \tag{5-7}$$

This equation shows that the normal stresses acting on the cross section vary linearly with the distance y from the neutral surface. This stress distribution is pictured in Fig. 5-9a for the case in which the bending

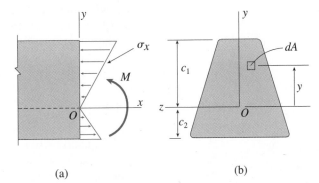

Fig. 5-9 Normal stresses in a beam of linearly elastic material: (a) side view of beam showing distribution of normal stresses, and (b) cross section of beam showing the z axis as the neutral axis of the cross section

moment M is positive and the beam bends with positive curvature. When the curvature is positive, the stresses σ_x are negative (compression) above the neutral surface and positive (tension) below it. In the figure, compressive stresses are indicated by arrows pointing toward the cross section and tensile stresses are indicated by arrows pointing away from the cross section.

In order for Eq. (5-7) to be of practical value, we must locate the origin of coordinates so that we can determine the distance y. In other words, we must locate the neutral axis of the cross section. We also need to obtain a relationship between the curvature and the bending moment—so that we can substitute into Eq. (5-7) and obtain an equation relating the stresses to the bending moment. These two objectives can be accomplished by determining the resultant of the stresses σ_x acting on the cross section.

In general, the resultant of the normal stresses consists of two stress resultants: (1) a force acting in the x direction, and (2) a bending couple acting about the z axis. However, the axial force is zero when a beam is in pure bending. Therefore, we can write the following equations of statics: (1) the resultant force in the x direction is equal to zero, and (2) the resultant moment is equal to the bending moment M. The first equation gives the location of the neutral axis and the second gives the moment-curvature relationship.

■ Location of Neutral Axis

To obtain the first equation of statics, we consider an element of area dA in the cross section (Fig. 5-9b). The element is located at distance y from the neutral axis, and therefore the stress σ_x acting on the element is given by Eq. (5-7). The *force* acting on the element is equal to $\sigma_x\, dA$ and is compressive when y is positive. Because there is no resultant force acting on the cross section, the integral of $\sigma_x\, dA$ over the area A of the entire cross section must vanish; thus, the first equation of statics is

$$\int_A \sigma_x\, dA = -\int_A E\kappa y\, dA = 0 \tag{a}$$

Because the curvature κ and modulus of elasticity E are nonzero constants at any given cross section of a bent beam, they are not involved in the integration over the cross-sectional area. Therefore, we can drop them from the equation and obtain

$$\int_A y\, dA = 0 \tag{5-8}$$

This equation states that the first moment of the area of the cross section, evaluated with respect to the z axis, is zero. In other words, the z axis must pass through the centroid of the cross section.* Since the z axis is also the neutral axis, we arrive at the following conclusion: *The neutral axis passes through the centroid of the cross-sectional area when the material follows Hooke's law and there is no axial force acting on the cross section.* This observation makes it relatively simple to determine the position of the neutral axis.

As explained in Section 5.1, our discussion is limited to beams for which the y axis is an axis of symmetry. Consequently, the y axis also passes through the centroid. Therefore, we have the following additional conclusion: *The origin O of coordinates (Fig. 5-9b) is located at the centroid of the cross-sectional area.*

Because the y axis is an axis of symmetry of the cross section, it follows that the y axis is a *principal axis* (see Chapter 12, Section 12.9, for a discussion of principal axes). Since the z axis is perpendicular to the y axis, it too is a principal axis. Thus, when a beam of linearly elastic material is subjected to pure bending, *the y and z axes are principal centroidal axes.*

■ Moment-Curvature Relationship

The second equation of statics expresses the fact that the moment resultant of the normal stresses σ_x acting over the cross section is equal to the bending moment M (Fig. 5-9a). The element of force $\sigma_x\, dA$ acting on the element of area dA (Fig. 5-9b) is in the positive direction of the x axis when σ_x is positive and in the negative direction when σ_x is negative. Since the element dA is located above the neutral axis, a positive stress σ_x acting on that element produces an element of moment equal to $\sigma_x y\, dA$. This element of moment acts opposite in direction to the positive bending moment M shown in Fig. 5-9a. Therefore, the increment dM in the bending moment is

$$dM = -\sigma_x y\, dA$$

The integral of all such elemental moments over the entire cross-sectional area A must equal the bending moment:

---

*Centroids and first moments of areas are discussed in Chapter 12.

$$M = -\int_A \sigma_x y \, dA \qquad (b)$$

or, upon substituting for σ_x from Eq. (5-7),

$$M = \int_A \kappa E y^2 \, dA = \kappa E \int_A y^2 \, dA \qquad (5-9)$$

This equation relates the curvature of the beam to the bending moment M.

Since the integral in the preceding equation is a property of the cross-sectional area, it is convenient to rewrite the equation as follows:

$$M = \kappa E I \qquad (5-10)$$

in which

$$I = \int_A y^2 \, dA \qquad (5-11)$$

This integral is the **moment of inertia** of the cross-sectional area with respect to the z axis (that is, with respect to the neutral axis). Moments of inertia are always positive and have dimensions of length to the fourth power; for instance, typical USCS units are in.^{4} and typical SI units are mm^4 when performing beam calculations.*

Equation (5-10) can now be rearranged to express the curvature in terms of the bending moment in the beam:

$$\kappa = \frac{1}{\rho} = \frac{M}{EI} \qquad (5-12)$$

Known as the **moment-curvature equation**, Eq. (5-12) shows that the curvature is directly proportional to the bending moment M and inversely proportional to the quantity EI, which is called the **flexural rigidity** of the beam. Flexural rigidity is a measure of the resistance of a beam to bending, that is, the larger the flexural rigidity, the smaller the curvature for a given bending moment.

Comparing the sign convention for bending moments (Fig. 4-5) with that for curvature (Fig. 5-6), we see that *a positive bending moment produces positive curvature and a negative bending moment produces negative curvature* (see Fig. 5-10).

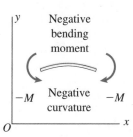

Fig. 5-10 Relationships between signs of bending moments and signs of curvatures

■ Flexure Formula

Now that we have located the neutral axis and derived the moment-curvature relationship, we can determine the stresses in terms of the

*Moments of inertia of areas are discussed in Chapter 12.

bending moment. Substituting the expression for curvature (Eq. 5-12) into the expression for the stress σ_x (Eq. 5-7), we get

$$\sigma_x = -\frac{My}{I} \tag{5-13}$$

This equation, called the **flexure formula**, shows that the stresses are directly proportional to the bending moment M and inversely proportional to the moment of inertia I of the cross section. Also, the stresses vary linearly with the distance y from the neutral axis, as previously observed. Stresses calculated from the flexure formula are called **bending stresses** or **flexural stresses.**

If the bending moment in the beam is positive, the bending stresses will be positive (tension) over the part of the cross section where y is negative, that is, over the lower part of the beam. The stresses in the upper part of the beam will be negative (compression). If the bending moment is negative, the stresses will be reversed. These relationships are shown in Fig. 5-11.

■ **Maximum Stresses at a Cross Section**

The maximum tensile and compressive bending stresses acting at any given cross section occur at points located farthest from the neutral axis. Let us denote by c_1 and c_2 the distances from the neutral axis to the extreme elements in the positive and negative y directions, respectively (see Fig. 5-9b and Fig. 5-11). Then the corresponding **maximum normal stresses** σ_1 and σ_2 (from the flexure formula) are

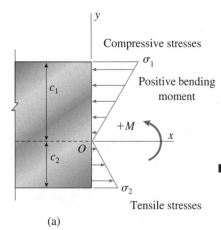

(a)

$$\sigma_1 = -\frac{Mc_1}{I} = -\frac{M}{S_1} \qquad \sigma_2 = \frac{Mc_2}{I} = \frac{M}{S_2} \tag{5-14a,b}$$

in which

$$S_1 = \frac{I}{c_1} \qquad S_2 = \frac{I}{c_2} \tag{5-15a,b}$$

The quantities S_1 and S_2 are known as the **section moduli** of the cross-sectional area. From Eqs. (5-15a and b) we see that each section modulus has dimensions of length to the third power (for example, in.[3] or mm[3]). Note that the distances c_1 and c_2 to the top and bottom of the beam are always taken as positive quantities.

The advantage of expressing the maximum stresses in terms of section moduli arises from the fact that each section modulus combines

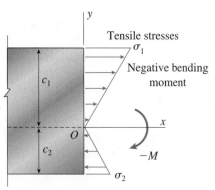

(b)

Fig. 5-11 Relationships between signs of bending moments and directions of normal stresses: (a) positive bending moment, and (b) negative bending moment the beam's relevant cross-sectional properties into a single quantity. Then this quantity can be listed in tables and handbooks as a property of the beam, which is a convenience to designers. (Design of beams using section moduli is explained in the next section.)

If the cross section of a beam is symmetric with respect to the z axis as well as the y axis (**doubly symmetric cross section**), then $c_1 = c_2 = c$ and the maximum tensile and compressive stresses are equal numerically:

$$\sigma_1 = -\sigma_2 = -\frac{Mc}{I} = -\frac{M}{S} \tag{5-16}$$

in which

$$S = \frac{I}{c} \tag{5-17}$$

is the only section modulus for the cross section. For a beam of **rectangular cross section** with width b and height h (Fig. 5-12a), the moment of inertia and section modulus are

$$I = \frac{bh^3}{12} \qquad S = \frac{bh^2}{6} \tag{5-18a,b}$$

For a **circular cross section** of diameter d (Fig. 5-12b), these properties are

$$I = \frac{\pi d^4}{64} \qquad S = \frac{\pi d^3}{32} \tag{5-19a,b}$$

Properties of other doubly symmetric shapes, such as a hollow tubular cross section, can be readily obtained from the preceding formulas.

Moments of inertia of many plane figures are listed in Appendix D for convenient reference. For other cross-sectional shapes, we must determine the location of the neutral axis, the moment of inertia, and the section moduli by direct calculation, using the techniques described in Chapter 12. (This general procedure is illustrated in Example 5-4 that follows.) The dimensions and properties of standard sizes of steel and wood beams are listed in Appendices E and F and in many engineering handbooks, as explained in more detail in the next section.

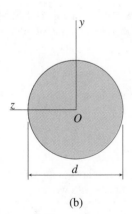

Fig. 5-12 Doubly symmetric cross-sectional shapes

■ Limitations

The analysis presented in this section is for pure bending of prismatic beams composed of homogeneous, linearly elastic materials. If a beam is subjected to nonuniform bending, the shear forces will produce warping (or out-of-plane distortion) of the cross sections. Thus, a cross section that was plane before bending is no longer plane after bending. Warping due to shear deformations greatly complicates the behavior of the beam. However, detailed investigations show that the normal stresses calculated from the flexure formula are not significantly altered by the presence of shear stresses and the associated warping (Ref. 2-1, pp. 42 and 48). Thus, we may justifiably use the theory of pure bending for calculating normal stresses in beams subjected to nonuniform bending.*

The flexure formula gives results that are accurate only in regions of the beam where the stress distribution is not disrupted by changes in the shape of the beam or by discontinuities in loading. For instance, the flexure formula is not applicable near the supports of a beam or close to a concentrated load. Such irregularities produce localized stresses, or *stress concentrations*, that are much greater than the stresses obtained from the flexure formula (see Section 5.13).

■ **Example 5-2**

Fig. 5-13 Example 5-2. Wire bent around a drum

A high-strength steel wire of diameter d is bent around a cylindrical drum of radius R_0 (Fig. 5-13). Determine the bending moment M and maximum bending stress σ_{max} in the wire, assuming $d = 4$ mm and $R_0 = 0.5$ m. (The steel wire has modulus of elasticity $E = 200$ GPa and proportional limit $\sigma_{pl} = 1200$ MPa.)

Solution   The first step in this example is to determine the radius of curvature ρ of the bent wire. Then, knowing ρ, we can find the bending moment and maximum stresses.

Radius of curvature. The radius of curvature of the bent wire is the distance from the center of the drum to the neutral axis of the cross section of the wire:

$$\rho = R_0 + \frac{d}{2} \tag{5-20}$$

Bending moment. The bending moment in the wire may be found from the moment-curvature relationship (Eq. 5-12):

$$M = \frac{EI}{\rho} = \frac{2EI}{2R_0 + d} \tag{5-21}$$

in which I is the moment of inertia of the cross-sectional area of the wire. Substituting for I in terms of the diameter d of the wire (Eq. 5-19a), we get

$$M = \frac{\pi E d^4}{32(2R_0 + d)} \tag{5-22}$$

---

*Beam theory began with Galileo Galilei (1564–1642), who investigated the behavior of various types of beams. His work in mechanics of materials is described in his famous book *Two New Sciences*, first published in 1638 (Ref. 5-2). Although Galileo made many important discoveries regarding beams, he did not obtain the stress distribution that we use today. Further progress in beam theory was made by Mariotte, Jacob Bernoulli, Euler, Parent, Saint-Venant, and others (Ref. 5-3).

This result was obtained without regard to the *sign* of the bending moment, since the direction of bending is obvious from the figure.

Maximum bending stresses. The maximum tensile and compressive stresses, which are equal numerically, are obtained from the flexure formula as given by Eq. (5-16). However, since we are interested only in the magnitude of the maximum stress, we can disregard the signs in Eq. (5-16) and write the following expression:

$$\sigma_{max} = \frac{M}{S}$$

in which S is the section modulus for a circular cross section. Substituting for M from Eq. (5-22) and for S from Eq. (5-19b), we get

$$\sigma_{max} = \frac{Ed}{2R_0 + d} \tag{5-23}$$

This same result can be obtained directly from Eq. (5-7) by replacing y with $d/2$ and substituting for ρ from Eq. (5-20).

We see by inspection of Fig. 5-13 that the stress is compressive on the lower (or inner) part of the wire and tensile on the upper (or outer) part.

Numerical values. We now substitute the given numerical data into Eqs. (5-22) and (5-23) and obtain the following results:

$$M = \frac{\pi E d^4}{32(2R_0 + d)} = \frac{\pi(200 \text{ GPa})(4 \text{ mm})^4}{32[2(0.5 \text{ m}) + 4 \text{ mm}]} = 5.01 \text{ N·m} \quad \longleftarrow$$

$$\sigma_{max} = \frac{Ed}{2R_0 + d} = \frac{(200 \text{ GPa})(4 \text{ mm})}{2(0.5 \text{ m}) + 4 \text{ mm}} = 797 \text{ MPa} \quad \longleftarrow$$

Note that σ_{max} is less than the proportional limit of the steel wire, and therefore the calculations are valid.

Note: Because the radius of the drum is large compared to the diameter of the wire, we can safely disregard d in comparison with $2R_0$ in the denominators of the expressions for M and σ_{max}. Then Eqs. (5-22) and (5-23) yield the following results:

$$M = 5.03 \text{ N·m} \qquad \sigma_{max} = 800 \text{ MPa}$$

These results are on the conservative side and differ by less than 1% from the more precise values.

■ **Example 5-3**

A simple beam AB of span length $L = 22$ ft (Fig. 5-14a) supports a uniform load of intensity $q = 1.5$ k/ft and a concentrated load $P = 12$ k. The uniform load includes an allowance for the weight of the beam. The concentrated load acts at a point 9.0 ft from the left-hand end of the beam. The beam is constructed of glued laminated wood and has a cross section of width $b = 8.75$ in. and height $h = 27$ in. (Fig. 5-14b). Determine the maximum tensile and compressive stresses in the beam due to bending.

(a)

(b)

(c)

(d)

Fig. 5-14 Example 5-3. Stresses in a simple beam

Solution   *Maximum bending moment.* We begin the analysis by constructing the shear-force and bending-moment diagrams (Figs. 5-14c and d), using the techniques described in Chapter 4. Because the beam is prismatic, the maximum bending stresses occur at the cross section of largest bending moment, which is located where the shear force changes sign. Thus, the maximum moment is

$$M_{\text{max}} = 152 \text{ k-ft}$$

Section modulus. Next, we calculate the section modulus of the cross-sectional area using Eq. (5-18b):

$$S = \frac{bh^2}{6} = \frac{1}{6}(8.75 \text{ in.})(27 \text{ in.})^2 = 1063 \text{ in.}^3$$

Maximum stresses. Finally, we determine the maximum tensile and compressive stresses σ_t and σ_c from Eq. (5-16):

$$\sigma_t = \sigma_2 = \frac{M}{S} = \frac{(152 \text{ k-ft})(12 \text{ in./ft})}{1063 \text{ in.}^3} = 1720 \text{ psi} \quad \Longleftarrow$$

$$\sigma_c = \sigma_1 = -\frac{M}{S} = -1720 \text{ psi} \quad \Longleftarrow$$

Because the bending moment is positive, the maximum tensile stress (σ_t) occurs at the bottom of the beam and the maximum compressive stress (σ_c) occurs at the top.

■ **Example 5-4**

The beam *ABC* shown in Fig. 5-15a has simple supports at *A* and *B* and an overhang from *B* to *C*. A uniform load of intensity $q = 3.0$ kN/m acts throughout the length of the beam. The beam is constructed of three steel plates (thickness $t = 12$ mm) welded to form a channel section that has width $b = 300$ mm and height $h = 80$ mm (Fig. 5-15b). For the purpose of determining properties of the cross section, we disregard fillets and welds and assume that the cross section consists of rectangles, as shown in the figure.

(a)

(b)

(c)

(d)

Fig. 5-15 Example 5-4. Stresses in a beam with an overhang

Fig. 5-16 Cross section of beam discussed in Example 5-4

Determine the maximum tensile and compressive stresses in the beam due to the uniform load.

Solution   *Bending moments.* The maximum tensile and compressive stresses always occur at cross sections where the bending moments have their largest values. To locate these cross sections, we construct the shear-force and bending-moment diagrams (Figs. 5-15c and d), using the methods described in Chapter 4. Note that both the maximum positive and maximum negative moments occur at cross sections where the shear force changes sign. These moments are equal to 1.898 kN·m and −3.375 kN·m, respectively.

Neutral axis of the cross section. We determine the position of the neutral axis (the z axis) by locating the centroid of the cross-sectional area, which is shown in Fig. 5-16. Using the techniques described in Chapter 12, Section 12.3, we divide the area into three rectangles (A_1, A_2, and A_3). Then we establish a reference axis Z-Z across the top of the cross section, and we let y_1 and y_2 be the distances from the Z-Z axis to the centroids of areas A_1 and A_2, respectively. Thus, the calculations for locating the centroid of the channel section are as follows:

Area 1: $y_1 = t/2 = 6$ mm $A_1 = (b - 2t)(t) = (276$ mm$)(12$ mm$)$
$= 3312$ mm^2

Area 2: $y_2 = h/2 = 40$ mm $A_2 = ht = (80$ mm$)(12$ mm$)$
$= 960$ mm^2

Area 3: $y_3 = y_2$ $A_3 = A_2$

$$c_1 = \frac{\sum y_i A_i}{\sum A_i} = \frac{y_1 A_1 + 2y_2 A_2}{A_1 + 2A_2}$$

$$= \frac{(6 \text{ mm})(3312 \text{ mm}^2) + 2(40 \text{ mm})(960 \text{ mm}^2)}{3312 \text{ mm}^2 + 2(960 \text{ mm}^2)} = 18.48 \text{ mm}$$

$$c_2 = h - c_1 = 80 \text{ mm} - 18.48 \text{ mm} = 61.52 \text{ mm}$$

Thus, the position of the neutral axis (the z axis) is determined.

Moment of inertia. In order to calculate the stresses from the flexure formula, we must determine the moment of inertia of the cross-sectional area about the neutral axis. These calculations require the use of the parallel-axis theorem (see Chapter 12, Section 12.5).

Beginning with area A_1, we obtain its moment of inertia I_{z_1} about the z axis from the equation

$$I_{z_1} = I_{z_c} + A_1 d_1^2 \qquad (c)$$

In this equation, I_{z_c} is the moment of inertia of area A_1 about its own centroidal axis:

$$I_{z_c} = \frac{1}{12}(b - 2t)(t)^3 = \frac{1}{12}(276 \text{ mm})(12 \text{ mm})^3 = 39,744 \text{ mm}^4$$

and d_1 is the distance from the centroidal axis of area A_1 to the z axis:

$$d_1 = c_1 - t/2 = 18.48 \text{ mm} - 6 \text{ mm} = 12.48 \text{ mm}$$

Therefore, the moment of inertia of area A_1 about the z axis (from Eq. c) is

$$I_{z_1} = 39{,}744 \text{ mm}^4 + (3312 \text{ mm}^2)(12.48 \text{ mm})^2 = 555{,}600 \text{ mm}^4$$

Proceeding in the same manner for areas A_2 and A_3, we get

$$I_{z_2} = I_{z_3} = 956{,}600 \text{ mm}^4$$

Thus, the centroidal moment of inertia I_z of the entire cross-sectional area is

$$I_z = I_{z_1} + I_{z_2} + I_{z_3} = 2.469 \times 10^6 \text{ mm}^4$$

Section moduli. The section moduli for the top and bottom of the beam, respectively, are

$$S_1 = \frac{I_z}{c_1} = 133{,}600 \text{ mm}^3 \qquad S_2 = \frac{I_z}{c_2} = 40{,}100 \text{ mm}^3$$

(see Eqs. 5-15a and b). With the cross-sectional properties determined, we can now proceed to calculate the maximum stresses from Eqs. (5-14a and b).

Maximum stresses. At the cross section of maximum positive bending moment, the largest tensile stress occurs at the bottom of the beam (σ_2) and the largest compressive stress occurs at the top (σ_1). Thus, from Eqs. (5-14b) and (5-14a), respectively, we get

$$\sigma_t = \sigma_2 = \frac{M}{S_2} = \frac{1.898 \text{ kN·m}}{40{,}100 \text{ mm}^3} = 47.3 \text{ MPa}$$

$$\sigma_c = \sigma_1 = -\frac{M}{S_1} = -\frac{1.898 \text{ kN·m}}{133{,}600 \text{ mm}^3} = -14.2 \text{ MPa}$$

Similarly, the largest stresses at the section of maximum negative moment are

$$\sigma_t = \sigma_1 = -\frac{M}{S_1} = -\frac{-3.375 \text{ kN·m}}{133{,}600 \text{ mm}^3} = 25.3 \text{ MPa}$$

$$\sigma_c = \sigma_2 = \frac{M}{S_2} = \frac{-3.375 \text{ kN·m}}{40{,}100 \text{ mm}^3} = -84.2 \text{ MPa}$$

A comparison of these four stresses shows that the largest tensile stress in the beam is 47.3 MPa and occurs at the bottom of the beam at the cross section of maximum positive bending moment; thus,

$$(\sigma_t)_{max} = 47.3 \text{ MPa} \qquad \longleftarrow$$

The largest compressive stress is -84.2 MPa and occurs at the bottom of the beam at the section of maximum negative moment:

$$(\sigma_c)_{max} = -84.2 \text{ MPa} \qquad \longleftarrow$$

Thus, we have determined the maximum stresses due to the uniform load acting on the beam.

5.6 ■ DESIGN OF BEAMS FOR BENDING STRESSES

The process of designing a beam requires that many factors be considered, including the type of structure (airplane, automobile, bridge, building, or whatever), the materials to be used, the loads to be supported, the environmental conditions to be encountered, and the costs to be paid. However, from the standpoint of strength, the task eventually reduces to selecting a shape and size of beam such that the actual stresses in the beam do not exceed the allowable stresses for the material. In the following discussion, we will consider only the bending stresses (that is, the stresses obtained from the flexure formula, Eq. 5-13). Later, we will consider the effects of shear stresses (Sections 5.8, 5.9, and 5.10) and stress concentrations (Section 5.13).

When designing a beam to resist bending stresses, we usually begin by calculating the required section modulus. For instance, if the beam has a doubly symmetric cross section and the allowable stresses are the same for both tension and compression, we can calculate the required modulus by dividing the maximum bending moment by the allowable bending stress for the material (see Eq. 5-16):

$$S = \frac{M_{\max}}{\sigma_{\text{allow}}} \tag{5-24}$$

The allowable stress is based upon the properties of the material and the desired factor of safety. To ensure that this stress is not exceeded, we must choose a beam that provides a section modulus at least as large as that obtained from Eq. (5-24).

If the cross section is not doubly symmetric, or if the allowable stresses are different for tension and compression, we probably need to determine two required section moduli—one based upon tension and the other based upon compression. Then we must provide a beam that satisfies both criteria.

To minimize weight and save material, we usually select a beam that has the least cross-sectional area while still providing the required section moduli (and also meeting any other design requirements that may be imposed).

Beams are constructed in a great variety of shapes and sizes to suit a myriad of purposes. For instance, large steel beams are fabricated by welding (Fig. 5-17), aluminum beams are extruded as round or rectangular tubes, wood beams are cut and glued to fit special requirements, and reinforced concrete beams are cast in any desired shape by proper construction of the forms. In addition, beams of steel, aluminum, plastic, and wood can be ordered in standard shapes and sizes from catalogs supplied by dealers

Fig. 5-17 Welding three large steel plates into a single solid section (Courtesy of The Lincoln Electric Company)

and manufacturers. Readily available shapes include wide-flange beams, I-beams, angles, channels, rectangular beams, and tubes.

■ Beams of Standardized Shapes and Sizes

The dimensions and properties of many kinds of beams are listed in engineering handbooks. For instance, the shapes and sizes of structural-steel beams are standardized by the American Institute of Steel Construction (AISC), which publishes a manual giving their properties (Ref. 5-4). The tables in the AISC manual list cross-sectional dimensions and important properties such as weight per foot of length, cross-sectional area, moment of inertia, and section modulus. Properties of aluminum and wood beams are tabulated in a similar manner and are available in publications of the Aluminum Association (Ref. 5-5) and the American Forest and Paper Association (Ref. 5-6). Abridged tables of steel beams and wood beams are given later in this book for use in solving problems (see Appendices E and F).

Structural steel sections are given a designation such as W 30 × 211, which means that the section is of W shape (also called a wide-flange shape) with a nominal depth of 30 in. and a weight of 211 lb per ft of length (see Table E-1, Appendix E). Analogous designations are used for S shapes (also called I-beams) and C shapes (also called channels), as shown in Tables E-2 and E-3. Angle sections, or L shapes, are designated by the lengths of the two legs and the thickness (see Tables E-4 and E-5). For example, L 8 × 6 × 1 denotes an angle with unequal legs, one of length 8 in. and the other of length 6 in., with a thickness of 1 in. All standardized steel sections are manufactured by *rolling*, a process in which a billet of hot steel is passed back and forth between rolls until it is formed into the desired shape.

Aluminum structural sections are usually made by the process of *extrusion*, in which a hot billet is pushed, or extruded, through a shaped die. Since dies are relatively easy to make and the material is workable, aluminum beams can be extruded in almost any desired shape. Standard shapes of wide-flange beams, I-beams, channels, angles, tubes, and other sections are listed in the *Aluminum Construction Manual* (Ref. 5-5). In addition, custom-made shapes can be ordered.

Most wood beams have rectangular cross sections and are designated by nominal dimensions, such as 4 × 8 (inches). These dimensions represent the rough-cut size of the lumber. The net dimensions (or actual dimensions) of a wood beam are smaller than the nominal dimensions if the sides of the rough lumber have been planed, or *surfaced*, to make them smooth. Thus, a 4 × 8 wood beam has actual dimensions 3.5 × 7.25 in. after it has been surfaced. Of course, the net dimensions of surfaced lumber should be used in all engineering computations. Therefore, net dimensions and the corresponding properties are given in Appendix F.

■ Relative Efficiency of Various Beam Shapes

One of the objectives in designing a beam is to use the material as efficiently as possible within the constraints imposed by function, appearance, manufacturing costs, and the like. From the standpoint of strength alone, efficiency in bending depends primarily upon the shape of the cross section. In particular, the most efficient beam is one in which the material is located as far as practical from the neutral axis. The farther a given amount of material is from the neutral axis, the larger the section modulus becomes—and the larger the section modulus, the larger the bending moment that can be resisted (for a given allowable stress).

As an illustration, consider a cross section in the form of a rectangle of width b and height h (Fig. 5-18a). The section modulus (from Eq. 5-18b) is

$$S = \frac{bh^2}{6} = \frac{Ah}{6} = 0.167Ah \qquad (5\text{-}25)$$

where A denotes the cross-sectional area. This equation shows that a rectangular cross section of given area becomes more efficient as the height h is increased (and the width b is decreased to keep the area constant). Of course, there is a practical limit to the increase in height, because the beam becomes laterally unstable when the ratio of height to width becomes too large. Thus, a beam of very narrow rectangular section will fail due to lateral (sideways) buckling rather than to insufficient strength of the material.

Next, let us compare a solid circular cross section of diameter d (Fig. 5-18b) with a square cross section of the same area. The side h of a square having the same area as the circle is $h = (d/2)\sqrt{\pi}$. The corresponding section moduli (from Eqs. 5-18b and 5-19b) are

$$S_{\text{square}} = \frac{h^3}{6} = \frac{\pi\sqrt{\pi}d^3}{48} = 0.1160d^3$$

$$S_{\text{circle}} = \frac{\pi d^3}{32} = 0.0982d^3 \qquad (5\text{-}26\text{a,b})$$

from which we get

$$\frac{S_{\text{square}}}{S_{\text{circle}}} = 1.18 \qquad (5\text{-}27)$$

This result shows that a beam of square cross section is more efficient in resisting bending than is a circular beam of the same area. The reason, of course, is that a circle has a relatively larger amount of material located near the neutral axis. This material is less highly stressed, and therefore it does not contribute as much to the strength of the beam.

The ideal cross-sectional shape for a beam of given cross-sectional area A and height h would be obtained by placing one-half of the area at a

Fig. 5-18 Cross-sectional shapes of beams

distance $h/2$ above the neutral axis and the other half at distance $h/2$ below the neutral axis, as shown in Fig. 5-18c. For this ideal shape, we obtain

$$I = 2\left(\frac{A}{2}\right)\left(\frac{h}{2}\right)^2 = \frac{Ah^2}{4} \qquad S = \frac{I}{h/2} = 0.5Ah \qquad (5\text{-}28\text{a,b})$$

These theoretical limits are approached in practice by wide-flange sections and I-sections, which have most of their material in the flanges (Fig. 5-18d). Because of the necessity of putting part of the material in the web of the beam, the ideal conditions can never be realized. Thus, for standard wide-flange beams, the section modulus is approximately

$$S \approx 0.35Ah \qquad (5\text{-}29)$$

which is much larger than the section modulus for a rectangular cross section of the same area and height (see Eq. 5-25). The obvious reason is that a wide-flange beam has most of the material in the flanges, at the greatest available distance from the neutral axis.

Another desirable feature of a wide-flange beam is its greater width, and hence greater stability with respect to sideways buckling, when compared to a rectangular beam of the same height and section modulus. On the other hand, there are practical limits to how thin we can make the web of a wide-flange beam. If the web is too thin, it will be susceptible to localized buckling or it may be overstressed in shear, a topic that is discussed in Section 5.10.

The following examples illustrate the process of selecting a beam on the basis of the allowable stresses. In these examples, only the effects of bending stresses (obtained from the flexure formula) are considered.

Note: When solving examples and problems that require the selection of a beam from the tables in the appendix, we use the following rule: If several choices are available in a table, select the *lightest* beam that will provide the required section modulus.

■ **Example 5-5**

A simply supported wood beam having a span length $L = 12$ ft carries a uniform load $q = 420$ lb/ft (Fig. 5-19). The allowable bending stress is 1800 psi, the wood weighs 35 lb/ft^3, and the beam is supported laterally against sideways buckling and tipping. Select a suitable size for the beam from the table in Appendix F.

Solution   Since we do not know in advance how much the beam weighs, we will proceed by trial-and-error as follows: (1) Calculate the required section modulus based upon the given uniform load. (2) Select a trial size for the beam. (3) Add the weight of the beam to the uniform load and calculate a new required section modulus. (4) Check to see that the selected beam is still satisfactory. If it is not, select a larger beam and repeat the process.

Fig. 5-19 Example 5-5. Design of a simply supported wood beam

(1) The maximum bending moment in the beam occurs at the midpoint (see Eq. 4-15):

$$M_{max} = \frac{qL^2}{8} = \frac{(420 \text{ lb/ft})(12 \text{ ft})^2(12 \text{ in./ft})}{8} = 90{,}720 \text{ lb-in.}$$

The required section modulus (Eq. 5-24) is

$$S = \frac{M_{max}}{\sigma_{allow}} = \frac{90{,}720 \text{ lb-in.}}{1800 \text{ psi}} = 50.40 \text{ in.}^3$$

(2) From the table in Appendix F we see that the lightest beam that supplies a section modulus of at least 50.40 in.^{3} about axis 1-1 is a 3 × 12 in. beam (nominal dimensions). This beam has a section modulus equal to 52.73 in.^{3} and weighs 6.8 lb/ft.

(3) The uniform load on the beam now becomes 426.8 lb/ft, and the corresponding required section modulus is

$$S = (50.40 \text{ in.}^3)\left(\frac{426.8 \text{ lb/ft}}{420 \text{ lb/ft}}\right) = 51.22 \text{ in.}^3$$

(4) The previously selected beam has a section modulus of 52.73 in.^{3}, which is larger than the required modulus of 51.22 in.^{3} Therefore, a 3 × 12 in. beam is satisfactory.   ←

Note: If the weight density of the wood is other than 35 lb/ft^3, we can obtain the weight of the beam per linear foot by multiplying the value in the last column in Appendix F by the ratio of the actual weight density to 35 lb/ft^3.

■ **Example 5-6**

A vertical post 2.5-meters high must support a lateral load $P = 12$ kN at its upper end (Fig. 5-20). Two plans are proposed—a solid wood post and a hollow aluminum tube. (a) What is the minimum required diameter d_1 of the wood post if the allowable bending stress in the wood is 15 MPa? (b) What is the minimum required outer diameter d_2 of the aluminum tube if its wall thickness is to be one-eighth of the outer diameter and the allowable bending stress in the aluminum is 50 MPa?

Solution   *Maximum bending moment.* The maximum moment occurs at the base of the post and is equal to the load P times the height h; thus,

$$M_{max} = Ph = (12 \text{ kN})(2.5 \text{ m}) = 30.0 \text{ kN·m}$$

(a) *Wood post.* The required section modulus S_1 for the wood post (see Eqs. 5-19b and 5-24) is

$$S_1 = \frac{\pi d_1^3}{32} = \frac{M_{max}}{\sigma_{allow}} = \frac{30.0 \text{ kN·m}}{15 \text{ MPa}} = 0.0020 \text{ m}^3 = 2 \times 10^6 \text{ mm}^3$$

Solving for the diameter, we get

$$d_1 = 273 \text{ mm}   ←$$

Fig. 5-20 Example 5-6. (a) Solid wood post, and (b) aluminum tube

The diameter selected for the wood post must be equal to or larger than 273 mm if the allowable stress is not to be exceeded.

(b) *Aluminum tube.* To determine the section modulus S_2 for the tube, we first must find the moment of inertia I_2 of the cross section (see Eq. 5-17). The wall thickness of the tube is $d_2/8$, and therefore the inner diameter is $d_2 - d_2/4$, or $0.75d_2$. Thus, the moment of inertia (see Eq. 5-19a) is

$$I = \frac{\pi}{64}\left[d_2^4 - (0.75d_2)^4\right] = 0.03356d_2^4$$

The section modulus is now obtained as follows:

$$S_2 = \frac{I_2}{c} = \frac{0.03356d_2^4}{d_2/2} = 0.06712d_2^3$$

The required section modulus is obtained from Eq. (5-24):

$$S_2 = \frac{M_{max}}{\sigma_{allow}} = \frac{30.0\text{ kN·m}}{50\text{ MPa}} = 0.0006\text{ m}^3 = 600 \times 10^3\text{ mm}^3$$

By equating the two preceding expressions for the section modulus, we can solve for the required outer diameter:

$$d_2 = \left(\frac{600 \times 10^3\text{ mm}^3}{0.06712}\right)^{1/3} = 208\text{ mm} \qquad \longleftarrow$$

The corresponding inner diameter is 0.75(208 mm), or 156 mm.

■ Example 5-7

A simple beam AB of span length 21 ft must support a uniform load $q = 2000$ lb/ft distributed along the beam in the manner shown in Fig. 5-21. Considering both the uniform load and the weight of the beam, and also using an allowable bending stress of 18,000 psi, select a structural steel beam of wide-flange shape to support the loads.

Solution   In this example we will proceed as follows: (1) Find the maximum bending moment in the beam due to the uniform load. (2) Knowing the maximum moment, find the required section modulus. (3) Select a trial wide-flange beam from Table E-1 in Appendix E and obtain the weight of the beam. (4) With the weight known, calculate a new value of the bending moment and a new value of the section modulus. (5) Determine whether the selected beam is still satisfactory. If it is not, select a new beam size and repeat the process until a satisfactory size of beam has been found.

Maximum bending moment. To assist in locating the cross section of maximum bending moment, we construct the shear-force diagram (Fig. 5-21) using the methods described in Chapter 4. As part of that process, we determine the reactions at the supports:

$$R_A = 18,860\text{ lb} \qquad R_B = 17,140\text{ lb}$$

The distance x_1 from the left-hand support to the cross section of zero shear force is obtained from the equation

Fig. 5-21 Example 5-7. Design of a simple beam with partial uniform loads

$$V = R_A - qx_1 = 0$$

which is valid in the range $0 \leq x \leq 12$ ft. Solving for x_1, we get

$$x_1 = \frac{R_A}{q} = \frac{18{,}860 \text{ lb}}{2{,}000 \text{ lb/ft}} = 9.430 \text{ ft}$$

which is less than 12 ft, and therefore the calculation is valid.

The maximum bending moment occurs at the cross section where the shear force is zero; therefore,

$$M_{\text{max}} = R_A x_1 - \frac{qx_1^2}{2} = 88{,}920 \text{ lb-ft}$$

Required section modulus. The required section modulus (based only upon the load q) is obtained from Eq. (5-24):

$$S = \frac{M_{\text{max}}}{\sigma_{\text{allow}}} = \frac{(88{,}920 \text{ lb-ft})(12 \text{ in./ft})}{18{,}000 \text{ psi}} = 59.3 \text{ in.}^3$$

Trial beam. We now turn to Table E-1 and select the lightest wide-flange beam having a section modulus greater than 59.3 in.^{3} The lightest beam that provides this section modulus is W 12×50 with $S = 64.7$ in.^{3} This beam weighs 50 lb/ft. (Recall that the tables in Appendix E are abridged, and therefore a lighter beam may actually be available.)

We now recalculate the reactions, maximum bending moment, and required section modulus with the beam loaded by both the uniform load q and its own weight. Under these combined loads the reactions are

$$R_A = 19{,}380 \text{ lb} \qquad R_B = 17{,}670 \text{ lb}$$

and the distance to the cross section of zero shear becomes

$$x_1 = 9.455 \text{ ft}$$

The maximum bending moment increases to 91,600 lb-ft, and the new required section modulus is

$$S = \frac{M_{\text{max}}}{\sigma_{\text{allow}}} = \frac{(91{,}600 \text{ lb-ft})(12 \text{ in./ft})}{18{,}000 \text{ psi}} = 61.1 \text{ in.}^3$$

Thus, we see that the W 12×50 beam with section modulus $S = 64.7$ in.^{3} is still satisfactory.   ⬅

If the new required section modulus exceeded that of the W 12×50 beam, a new beam with a larger section modulus would be selected and the process repeated.

■ Example 5-8

A temporary wood dam is constructed of horizontal planks A supported by vertical wood posts B that are sunk into the ground so that they act as cantilever beams (Fig. 5-22). The posts are of square cross section (dimensions $b \times b$) and spaced at distance $s = 0.8$ m, center to center. Assume that the water level

(a) Top view   (b) Side view

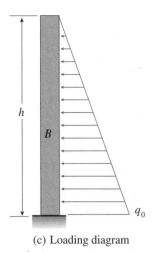

(c) Loading diagram

Fig. 5-22 Example 5-8. Wood dam with horizontal planks A supported by vertical posts B

behind the dam is at its full height $h = 2.0$ m. Determine the minimum required dimension b of the posts if the allowable bending stress in the wood is $\sigma_{\text{allow}} = 8.0$ MPa.

Solution   Each post is subjected to a triangularly distributed load produced by the water pressure acting against the planks. Consequently, the loading diagram for each post is triangular (Fig. 5-22c). The maximum intensity q_0 of the load on the posts is equal to the water pressure at depth h times the spacing s of the posts:

$$q_0 = \gamma h s \qquad \text{(a)}$$

in which γ is the specific weight of water. Note that q_0 has units of force per unit distance, γ has units of force per unit volume, and both h and s have units of length.

Since each post is a cantilever beam, the maximum bending moment occurs at the base and is given by the following expression:

$$M_{\text{max}} = \frac{q_0 h}{2}\left(\frac{h}{3}\right) = \frac{\gamma h^3 s}{6} \qquad \text{(b)}$$

Therefore, the required section modulus (Eq. 5-24) is

$$S = \frac{M_{\text{max}}}{\sigma_{\text{allow}}} = \frac{\gamma h^3 s}{6\sigma_{\text{allow}}} \qquad \text{(c)}$$

For a beam of square cross section, the section modulus is $S = b^3/6$ (see Eq. 5-18b). Substituting this expression for S into Eq. (c), we get a formula for the minimum dimension b of the posts:

$$b^3 = \frac{\gamma h^3 s}{\sigma_{\text{allow}}} \qquad \text{(d)} \quad \Longleftarrow$$

Numerical values. We now substitute the given numerical values into Eq. (d) and obtain

$$b^3 = \frac{(9.81 \text{ kN/m}^3)(2.0 \text{ m})^3(0.8 \text{ m})}{8.0 \text{ MPa}} = 0.007848 \text{ m}^3 = 7.848 \times 10^6 \text{ mm}^3$$

from which

$$b = 199 \text{ mm} \qquad \Longleftarrow$$

Thus, the minimum required dimension b of the posts is 199 mm. Any larger dimension, such as 200 mm, will ensure that the actual bending stress is less than the allowable stress.

5.7   NONPRISMATIC BEAMS

The beam theories described in this chapter were derived for prismatic beams, that is, straight beams having the same cross sections throughout their lengths. However, nonprismatic beams are commonly used to reduce weight and improve appearance. Such beams are found in automobiles, airplanes, machinery, bridges, buildings, tools, and many other applications (Fig. 5-23). Fortunately, the flexure formula (Eq. 5-13) gives reasonably accurate values for the bending stresses in nonprismatic beams whenever the changes in cross-sectional dimensions are gradual, as in the examples shown in Fig. 5-23.

Fig. 5-23 Examples of nonprismatic beams: (a) streetlamp, (b) bridge with tapered girders and piers, (c) wheel strut of a small airplane, and (d) wrench handle

Now consider how the bending stresses vary as we move along the axis of a beam. In a prismatic beam the section modulus S is constant, and therefore the stresses vary in direct proportion to the bending moment M (because $\sigma = M/S$). However, in a nonprismatic beam the section modulus also varies along the axis. Consequently, we cannot assume that the maximum stresses occur at the cross section with the largest bending moment—sometimes they occur elsewhere, as illustrated in the following example.

■ **Example 5-9**

A tapered cantilever beam AB of solid circular cross section supports a load P at the free end (Fig. 5-24). The diameter d_B at the large end is twice the diameter d_A at the small end:

$$\frac{d_B}{d_A} = 2$$

Determine the bending stress σ_B at the fixed support and the maximum bending stress σ_{max}.

Fig. 5-24 Example 5-9. Tapered cantilever beam of circular cross section

Solution   If the angle of taper of the beam is small, the bending stresses obtained from the flexure formula will differ only slightly from the exact values. As a guideline concerning accuracy, we note that if the angle between line *AB* (Fig. 5-24) and the longitudinal axis of the beam is about 20°, the error in calculating the normal stresses from the flexure formula is about 10%. Of course, as the angle of taper decreases, the error becomes smaller.

Section modulus. The section modulus at any cross section of the beam can be expressed as a function of the distance x measured along the axis of the beam. Since the section modulus depends upon the diameter (Eq. 5-19b), we first must express the diameter in terms of x, as follows:

$$d_x = d_A + (d_B - d_A)\frac{x}{L} \tag{5-30}$$

in which d_x is the diameter at distance x from the free end. Therefore, the section modulus at distance x from the end is

$$S_x = \frac{\pi d_x^3}{32} = \frac{\pi}{32}\left[d_A + (d_B - d_A)\frac{x}{L}\right]^3 \tag{5-31}$$

Bending stresses. Since the bending moment equals Px, the maximum normal stress at any cross section is given by the equation

$$\sigma_1 = \frac{M_x}{S_x} = \frac{32Px}{\pi[d_A + (d_B - d_A)(x/L)]^3} \tag{5-32}$$

We can see by inspection of the beam that the stress σ_1 is tensile at the top of the beam and compressive at the bottom.

Note that Eqs. (5-30), (5-31), and (5-32) are valid for any values of d_A and d_B, provided the angle of taper is small. In the following discussion, we consider the case where $d_B = 2d_A$.

Maximum stress at the fixed support when $d_B = 2d_A$. The maximum stress at the section of largest bending moment (end B of the beam) can be found from Eq. (5-32) by substituting $x = L$ and $d_B = 2d_A$; the result is

$$\sigma_B = \frac{4PL}{\pi d_A^3} \tag{a} \quad \Longleftarrow$$

Maximum stress in the beam when $d_B = 2d_A$. The maximum stress at a cross section at distance x from the end (Eq. 5-32) for the case where $d_B = 2d_A$ is

$$\sigma_1 = \frac{32Px}{\pi d_A^3(1 + x/L)^3} \tag{b}$$

To determine the location of the cross section having the largest bending stress in the beam, we need to find the value of x that makes σ_1 a maximum. Taking the derivative $d\sigma_1/dx$ and equating it to zero, we can solve for the value of x that makes σ_1 a maximum; the result is

$$x = \frac{L}{2}$$

The corresponding maximum stress, obtained by substituting $x = L/2$ into Eq. (b), is

$$\sigma_{max} = \frac{128PL}{27\pi d_A^3} = 4.741\frac{PL}{\pi d_A^3} \qquad\qquad \text{(c)} \quad \Leftarrow$$

In this particular example, the maximum stress occurs at the midpoint of the beam and is 19% greater than the stress at the built-in end.

Note: If the taper of the beam is reduced, the cross section of maximum normal stress moves from the midpoint toward the fixed support. For very small angles of taper, the maximum stress occurs at end B.

■ Example 5-10

Fig. 5-25 Example 5-10. Fully stressed beam having constant maximum normal stress (theoretical shape with shear stresses disregarded)

A cantilever beam of length L is being designed to support a concentrated load P at the free end (Fig. 5-25). The cross section of the beam is rectangular with constant width b. To assist them in designing this beam, the designers would like to know how the height of an idealized beam should vary in order that the maximum normal stress at every cross section will be equal to the allowable stress σ_{allow}. (A beam meeting this condition is called a **fully stressed beam** or a *beam of constant strength.*) Considering only the bending stresses obtained from the flexure formula, determine the height of such a beam.

Solution   The bending moment and section modulus at distance x from the free end of the beam are

$$M = Px \qquad S = \frac{bh_x^2}{6}$$

where h_x is the height of the beam at distance x. Substituting in the flexure formula, we obtain

$$\sigma_{allow} = \frac{M}{S} = \frac{Px}{bh_x^2/6} = \frac{6Px}{bh_x^2} \qquad\qquad \text{(d)}$$

Solving for the height of the beam, we get

$$h_x = \sqrt{\frac{6Px}{b\sigma_{allow}}} \qquad\qquad \text{(e)} \quad \Leftarrow$$

At the fixed end of the beam ($x = L$), the height h_B is

$$h_B = \sqrt{\frac{6PL}{b\sigma_{allow}}} \qquad\qquad \text{(f)}$$

and therefore we can express the height h_x in the following form:

$$h_x = h_B \sqrt{\frac{x}{L}} \qquad\qquad \text{(g)} \quad \Longleftarrow$$

This last equation shows that the height of the fully stressed beam varies with the square root of x. Consequently, the idealized beam has the parabolic shape shown in Fig. 5-25.

Note: At the loaded end of the beam ($x = 0$) the theoretical height is zero, because there is no bending moment at that point. Of course, a beam of this shape is not practical because it is incapable of supporting the shear forces near the end of the beam. Nevertheless, knowing the properties of a fully stressed beam can be an important aid to the engineer when designing structures for constant maximum stress (and hence for minimum weight). A familiar example of a structure with varying cross section that is designed to maintain constant maximum stress (within practical limits) is a leaf spring, such as found in automobiles and trucks.

5.8   SHEAR STRESSES IN BEAMS OF RECTANGULAR CROSS SECTION

When a beam is in pure bending, the only stress resultants are the bending moments and the only stresses are the normal stresses acting on the cross sections. However, most beams are subjected to loads that produce both bending moments and shear forces (nonuniform bending). In these cases, both normal and shear stresses are developed in the beam. The normal stresses are calculated from the flexure formula (see Section 5.5), provided the beam is constructed of a linearly elastic material. The shear stresses are discussed in this and the following two sections.

Consider a beam of rectangular cross section (width b and height h) subjected to a positive shear force V (Fig. 5-26a). It is reasonable to assume that the shear stresses τ acting on the cross section are parallel to the shear force, that is, parallel to the vertical sides of the cross section. It is also reasonable to assume that the shear stresses are uniformly distributed across the width of the beam, although they may vary over the height. Using these two assumptions, we can determine the intensity of the shear stress at any point on the cross section.

For purposes of analysis, we isolate a small element mn of the beam (Fig. 5-26a) by cutting between two adjacent cross sections and between two horizontal planes. According to our assumptions, the shear stresses τ acting on the front face of this element are vertical and uniformly distributed from one side of the beam to the other. Also, from the discussion of shear stresses in Section 1.6, we know that shear stresses acting on one side of an element are accompanied by shear stresses of equal magnitude acting on perpendicular faces of the element (see Figs. 5-26b and c). Thus, there are horizontal shear stresses acting between horizontal layers of the beam as well as vertical shear stresses acting on the cross sections. At any point in the beam, these complementary shear stresses are equal in magnitude.

(a)

(b)   (c)

Fig. 5-26 Shear stresses in a beam of rectangular cross section

The equality of the horizontal and vertical shear stresses acting on an element leads to an important conclusion regarding the shear stresses at the top and bottom of the beam. If we imagine that the element *mn* (Fig. 5-26a) is located at either the top or the bottom, we see that the horizontal shear stresses must vanish, because there are no stresses on the outer surfaces of the beam. It follows that the vertical shear stresses must also vanish at those locations; in other words, $\tau = 0$ where $y = \pm h/2$.

The existence of horizontal shear stresses in a beam can be demonstrated by a simple experiment. Place two identical rectangular beams on simple supports and load them by a force P, as shown in Fig. 5-27a. If friction between the beams is small, the beams will bend independently (Fig. 5-27b). Each beam will be in compression above its own neutral axis and in tension below its neutral axis, and therefore the bottom surface of the upper beam will slide with respect to the top surface of the lower beam.

Now suppose that the two beams are glued along the contact surface, so that they become a single solid beam. When this beam is loaded, horizontal shear stresses must develop along the glued surface in order to prevent the sliding shown in Fig. 5-27b. Because of the presence of these shear stresses, the single solid beam is much stiffer and stronger than the two separate beams.

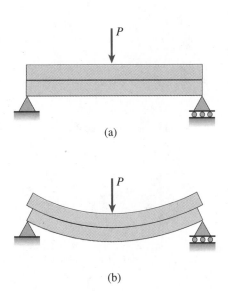

Fig. 5-27 Bending of two separate beams

■ Derivation of Shear Formula

We are now ready to derive a formula for the shear stresses τ in a rectangular beam. However, instead of evaluating the vertical shear stresses acting on a cross section, it is easier to evaluate the horizontal shear stresses acting between layers of the beam. Of course, the vertical shear stresses have the same magnitudes as the horizontal shear stresses.

With this procedure in mind, let us consider a beam in nonuniform bending (Fig. 5-28a). We take two adjacent cross sections mn and m_1n_1, distance dx apart, and isolate an element mm_1n_1n. The bending moment and shear force acting on the left-hand face of the element are denoted M and V, respectively. Since both the bending moment and shear force may change as we move along the axis of the beam, the corresponding quantities on the right-hand face are denoted $M + dM$ and $V + dV$.

Because of the presence of the bending moments and shear forces (Fig. 5-28a), the element is subjected to normal and shear stresses on both cross-sectional faces. However, only the normal stresses are needed in the following derivation, and therefore only the normal stresses are shown in Fig. 5-28b. On cross sections mn and m_1n_1 the normal stresses are, respectively,

$$\sigma_1 = -\frac{My}{I} \quad \text{and} \quad \sigma_2 = -\frac{(M + dM)y}{I} \qquad \text{(a,b)}$$

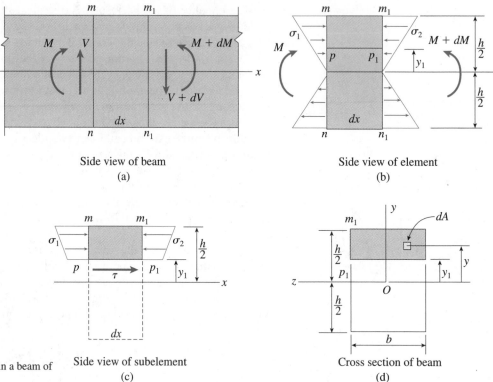

Fig. 5-28 Shear stresses in a beam of rectangular cross section

Side view of beam
(a)

Side view of element
(b)

Side view of subelement
(c)

Cross section of beam
(d)

as given by the flexure formula (Eq. 5-13). In these expressions, y is the distance from the neutral axis and I is the moment of inertia of the cross-sectional area about the neutral axis.

Next, we isolate a subelement mm_1p_1p by passing a horizontal plane pp_1 through element mm_1n_1n (Fig. 5-28b). The plane pp_1 is at distance y_1 from the neutral surface of the beam. The subelement is shown separately in Fig. 5-28c. We note that its top face is part of the upper surface of the beam and thus is free from stress. Its bottom face (which is parallel to the neutral surface and distance y_1 from it) is acted upon by the horizontal shear stresses τ existing at this level in the beam. Its cross-sectional faces mp and m_1p_1 are acted upon by the bending stresses σ_1 and σ_2, respectively, produced by the bending moments. Vertical shear stresses also act on the cross-sectional faces; however, these stresses do not affect the equilibrium of the subelement in the horizontal direction (the x direction), so they are not shown in Fig. 5-28c.

If the bending moments at cross sections mn and m_1n_1 (Fig. 5-28b) are equal (that is, if the beam is in pure bending), the normal stresses σ_1 and σ_2 acting over the sides mp and m_1p_1 of the subelement (Fig. 5-28c) also will be equal. Under these conditions, the subelement will be in equilibrium under the action of the normal stresses alone, and therefore the shear stresses τ acting on the bottom face pp_1 will vanish. This conclusion is obvious inasmuch as a beam in pure bending has no shear force and hence no shear stresses.

If the bending moments vary along the x axis (nonuniform bending), we can determine the shear stress τ acting on the bottom face of the subelement (Fig. 5-28c) by considering the equilibrium of the element in the x direction. We begin by identifying an element of area dA in the cross section at distance y from the neutral axis (Fig. 5-28d). The force acting on this element is σdA, in which σ is the normal stress obtained from the flexure formula. If the element of area is located on the left-hand face mp of the subelement (where the bending moment is M), the normal stress is given by Eq. (a), and therefore the element of force is

$$\sigma_1 dA = \frac{My}{I}dA$$

Note that we are using only absolute values in this equation because the directions of the stresses are obvious from the figure. Summing these elements of force over the area of face mp of the subelement gives the total horizontal force F_1 acting on that face:

$$F_1 = \int \sigma_1 dA = \int \frac{My}{I}dA \qquad \text{(c)}$$

Note that this integration is performed over the area of the shaded part of the cross section shown in Fig. 5-28d, that is, over the area of the cross section from $y = y_1$ to $y = h/2$. The force F_1 is shown in Fig. 5-29 on a partial free-body diagram of the subelement (vertical forces have been omitted).

In a similar manner, we find that the total force F_2 acting on the right-hand face $m_1 p_1$ of the subelement (Fig. 5-29) is

$$F_2 = \int \sigma_2 dA = \int \frac{(M + dM)y}{I}dA \qquad \text{(d)}$$

Knowing the forces F_1 and F_2, we can now determine the horizontal force F_3 acting on the bottom face of the subelement.

Since the subelement is in equilibrium, we can sum forces in the x direction and obtain

$$F_3 = F_2 - F_1 \qquad \text{(e)}$$

or $\qquad F_3 = \int \frac{(M + dM)y}{I}dA - \int \frac{My}{I}dA = \int \frac{(dM)y}{I}dA$

The quantities dM and I in the last term can be moved outside the integral sign because they are constants at any given cross section and are not involved in the integration. Thus, the expression for the force F_3 becomes

$$F_3 = \frac{dM}{I}\int y\,dA \qquad \text{(5-33)}$$

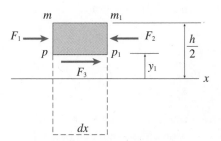

Fig. 5-29 Partial free-body diagram of subelement showing all horizontal forces

If the shear stresses τ are uniformly distributed across the width b of the beam, the force F_3 is

$$F_3 = \tau b \, dx \tag{5-34}$$

in which $b \, dx$ is the area of the bottom face of the subelement.

Combining Eqs. (5-33) and (5-34) and solving for the shear stress τ, we get

$$\tau = \frac{dM}{dx}\left(\frac{1}{Ib}\right)\int y \, dA \tag{5-35}$$

The quantity dM/dx is equal to the shear force V (see Eq. 4-6), and therefore the preceding expression becomes

$$\tau = \frac{V}{Ib}\int y \, dA \tag{5-36}$$

The integral in this equation is evaluated over the shaded part of the cross section (Fig. 5-28d), as already explained. Thus, the integral is the first moment of the shaded area with respect to the neutral axis (the z axis). In other words, *the integral is the first moment of the cross-sectional area above the level at which the shear stress τ is being evaluated.* This first moment is usually denoted by the symbol Q:

$$Q = \int y \, dA \tag{5-37}$$

With this notation, the equation for the shear stress becomes

$$\tau = \frac{VQ}{Ib} \tag{5-38}$$

This equation, known as the **shear formula**, can be used to determine the shear stress τ at any point in the cross section of a rectangular beam. Note that for a specific cross section, the shear force V, moment of inertia I, and width b are constants. However, the first moment Q (and hence the shear stress τ) varies with the distance y_1 from the neutral axis.

If the level at which the shear stress is to be determined is above the neutral axis, as shown in Fig. 5-28d, it is natural to obtain Q by calculating the first moment of the cross-sectional area *above* that level (the shaded area in the figure). However, as an alternative, we could calculate the first moment of the remaining cross-sectional area, that is, the area *below* the shaded area. Its first moment is equal to the negative of Q. The explanation lies in the fact that the first moment of the entire cross-sectional area with respect to the neutral axis is equal to zero (because the neutral axis passes through the centroid). Therefore, the value of Q for the area below the level y_1 is the negative of Q for the area above that level. As a matter of convenience, we usually use the area above the level y_1 when the point where we are finding the shear stress is in the upper part of the beam, and we use the area below the level y_1 when the point is in the lower part of the beam.

Furthermore, we usually don't bother with sign conventions for V and Q. Instead, we treat all terms in the shear formula as positive quantities and determine the direction of the shear stresses by inspection, since the stresses act in the same direction as the shear force V itself. This procedure for determining shear stresses is illustrated later in Example 5-11.

■ Distribution of Shear Stresses in a Rectangular Beam

We are now ready to determine the distribution of the shear stresses in a beam of rectangular cross section (Fig. 5-30a). The first moment Q of the shaded part of the cross-sectional area is obtained by multiplying the area by the distance from its own centroid to the neutral axis:

$$Q = b\left(\frac{h}{2} - y_1\right)\left(y_1 + \frac{h/2 - y_1}{2}\right) = \frac{b}{2}\left(\frac{h^2}{4} - y_1^2\right) \qquad (f)$$

Of course, this same result can be obtained by integration using Eq. (5-37):

$$Q = \int y\, dA = \int_{y_1}^{h/2} yb\, dy = \frac{b}{2}\left(\frac{h^2}{4} - y_1^2\right)$$

Substituting the expression for Q into the shear formula (Eq. 5-38), we get

$$\tau = \frac{V}{2I}\left(\frac{h^2}{4} - y_1^2\right) \qquad (5\text{-}39)$$

This equation shows that the shear stresses in a rectangular beam vary quadratically with the distance y_1 from the neutral axis. Thus, when plotted along the height of the beam, τ varies as shown in Fig. 5-30b. Note that the shear stress is zero when $y_1 = \pm h/2$.

The maximum value of the shear stress occurs at the neutral axis ($y_1 = 0$) where the first moment Q has its maximum value. Substituting $y_1 = 0$ into Eq. (5-39), we get

$$\tau_{max} = \frac{Vh^2}{8I} = \frac{3V}{2A} \qquad (5\text{-}40)$$

in which $A = bh$ is the cross-sectional area. Thus, the maximum shear stress in a beam of rectangular cross section is 50% larger than the average shear stress (which is equal to V/A).

Note again that the preceding equations for the shear stresses can be used to calculate either the shear stresses acting on the cross sections or

(a)

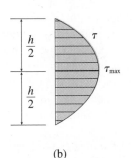

(b)

Fig. 5-30 Distribution of shear stresses in a beam of rectangular cross section: (a) cross section of beam, and (b) diagram showing the parabolic distribution of shear stresses over the height of the beam the horizontal shear stresses acting between horizontal layers of the beam.*

■ **Limitations**

The formulas for shear stresses presented in this section are subject to the same restrictions as the flexure formula from which they are derived. Thus, they are valid only for beams of linearly elastic materials with small deflections.

In the case of rectangular beams, the accuracy of the shear formula depends upon the height-to-width ratio of the cross section. The formula may be considered as exact for very narrow beams (height h much larger than the width b). However, it becomes less accurate as b increases relative to h. For instance, when the beam is square ($b = h$), the true maximum shear stress is about 13% larger than the value given by Eq. (5-40). (For a more complete discussion of the limitations of the shear formula, see Ref. 5-9.)

A common error is to apply the shear formula (Eq. 5-38) to cross-sectional shapes for which it is not applicable. For instance, it is not applicable to sections of triangular or semicircular shape. To avoid misusing the formula, we must keep in mind the following assumptions that underlie the derivation: (1) The edges of the cross section must be parallel to the y axis (so that the shear stresses act parallel to the y axis), and (2) the shear stresses must be uniform across the width of the cross section. These assumptions are fulfilled only in certain cases, such as those discussed in this and the next two sections.

Finally, the shear formula applies only to prismatic beams. If a beam is nonprismatic (for instance, if the beam is tapered), the shear stresses are quite different from those predicted by the formulas given here (see Refs. 5-9 and 5-10).

■ **Effects of Shear Strains**

Because the shear stress τ varies parabolically over the height of a rectangular beam, it follows that the shear strain $\gamma = \tau/G$ also varies parabolically. As a result of these shear strains, cross sections of the beam that were originally plane surfaces become warped. This warping is shown in Fig. 5-31, where cross sections mn and pq, originally plane, have become curved surfaces m_1n_1 and p_1q_1, with the maximum shear strain occurring at the neutral surface. At points m_1, p_1, n_1, and q_1 the shear strain is zero, and therefore the curves m_1n_1 and p_1q_1 are perpendicular to the upper and lower surfaces of the beam.

If the shear force V is constant along the axis of the beam, warping is the same at every cross section. Therefore, stretching and shortening of longitudinal elements due to the bending moments is unaffected by the

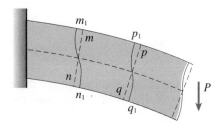

Fig. 5-31 Warping of the cross sections of a beam due to shear strains

---

*The shear-stress analysis presented in this section was developed by the Russian engineer D. J. Jourawski; see Refs. 5-7 and 5-8.

shear strains, and the distribution of the normal stresses is the same as in pure bending. Moreover, detailed investigations using advanced methods of analysis show that warping of cross sections due to shear strains does not substantially affect the longitudinal strains even when the shear force varies continuously along the length. Thus, under most conditions it is justifiable to use the flexure formula (Eq. 5-13) for nonuniform bending, even though the formula was derived for pure bending.

■ **Example 5-11**

A metal beam with span $L = 3$ ft is simply supported at points A and B (Fig. 5-32a). The uniform load on the beam (including its own weight) is $q = 160$ lb/in. The cross section of the beam is rectangular (Fig. 5-32b) with width $b = 1$ in. and height $h = 4$ in. The beam is adequately supported against sideways buckling.

Determine the normal stress σ_C and shear stress τ_C at point C, which is located 1 in. below the top of the beam and 8 in. from the right-hand support. Show these stresses on a sketch of a stress element at point C.

Solution   *Shear force and bending moment.* The shear force V_C and bending moment M_C at the cross section through point C are found by the methods described in Chapter 4. The results are

$$M_C = 17{,}920 \text{ lb-in.} \qquad V_C = -1{,}600 \text{ lb}$$

The signs of these quantities are based upon the standard sign conventions for bending moments and shear forces (see Fig. 4-5).

Moment of inertia. The moment of inertia of the cross-sectional area about the neutral axis (the z axis in Fig. 5-32b) is

$$I = \frac{bh^3}{12} = \frac{1}{12}(1.0 \text{ in.})(4.0 \text{ in.})^3 = 5.333 \text{ in.}^4$$

Normal stress at point C. The normal stress at point C is found from the flexure formula (Eq. 5-13) with the distance y from the neutral axis equal to 1.0 in.; thus,

$$\sigma_C = -\frac{My}{I} = -\frac{(17{,}920 \text{ lb-in.})(1.0 \text{ in.})}{5.333 \text{ in.}^4} = -3360 \text{ psi} \qquad \Longleftarrow$$

The minus sign indicates that the stress is compressive, as expected.

Shear stress at point C. To obtain the shear stress at point C, we need to evaluate the first moment Q_C of the cross-sectional area above point C (Fig. 5-32b). This first moment is equal to the product of the area (denoted A_C) and its centroidal distance (denoted y_C) from the z axis; thus,

$$A_C = (1.0 \text{ in.})(1.0 \text{ in.}) = 1.0 \text{ in.}^2 \qquad y_C = 1.5 \text{ in.} \qquad Q_C = A_C y_C = 1.5 \text{ in.}^3$$

Now we substitute numerical values into the shear formula (Eq. 5-38) and obtain the magnitude of the shear stress:

$$\tau_C = \frac{V_C Q_C}{Ib} = \frac{(1{,}600 \text{ lb})(1.5 \text{ in.}^3)}{(5.333 \text{ in.}^4)(1.0 \text{ in.})} = 450 \text{ psi} \qquad \Longleftarrow$$

Fig. 5-32 Example 5-11. (a) Simple beam with uniform load, (b) cross section of beam, and (c) stress element showing the normal and shear stresses at point C

The direction of this stress can be established by inspection, because it acts in the same direction as the shear force. In this example, the shear force acts upward on the left-hand part of the beam when the beam is cut by a section through point C. The best way to show the directions of both the normal and shear stresses is to draw a stress element, as follows.

Stress element at point C. The stress element, shown in Fig. 5-32c, is cut from the side of the beam at point C (Fig. 5-32a). Compressive stresses $\sigma_C = 3360$ psi act on the cross-sectional faces of the element and shear stresses $\tau_C = 450$ psi act on the top and bottom faces as well as the cross-sectional faces.

■ **Example 5-12**

A wood beam AB supporting two concentrated loads P (Fig. 5-33a) has a rectangular cross section of width $b = 100$ mm and height $h = 150$ mm (Fig. 5-33b). The distances from the ends of the beam to the loads are $a = 0.5$ m. Determine the maximum permissible value P_{max} of the loads if the allowable stress in bending is $\sigma_{allow} = 11$ MPa (for both tension and compression) and the allowable stress in horizontal shear is $\tau_{allow} = 1.2$ MPa. (Disregard the weight of the beam itself.)

Note: Wood beams are much weaker in *horizontal shear* (shear parallel to the longitudinal fibers in the wood) than in *cross-grain shear* (shear on the cross sections). Consequently, the allowable stress in cross-grain shear is not needed in this example.

Solution The maximum shear force occurs at the supports and the maximum bending moment occurs throughout the region between the loads. Their values are

$$V_{max} = P \qquad M_{max} = Pa$$

Also, the section modulus S and cross-sectional area A are

$$S = \frac{bh^2}{6} \qquad A = bh$$

The maximum normal and shear stresses in the beam are obtained from the flexure and shear formulas (Eqs. 5-16 and 5-40):

$$\sigma_{max} = \frac{M_{max}}{S} = \frac{6Pa}{bh^2} \qquad \tau_{max} = \frac{3V_{max}}{2A} = \frac{3P}{2bh}$$

Therefore, the maximum permissible values of the load P in bending and shear, respectively, are

$$P_{bending} = \frac{\sigma_{allow} bh^2}{6a} \qquad P_{shear} = \frac{2\tau_{allow} bh}{3}$$

Substituting numerical values into these formulas, we get

$$P_{bending} = \frac{(11 \text{ MPa})(100 \text{ mm})(150 \text{ mm})^2}{6(0.5 \text{ m})} = 8.25 \text{ kN}$$

(a)

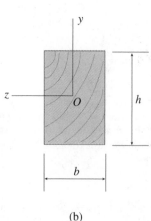

(b)

Fig. 5-33 Example 5-12. Wood beam with concentrated loads

$$P_{shear} = \frac{2(1.2 \text{ MPa})(100 \text{ mm})(150 \text{ mm})}{3} = 12.0 \text{ kN}$$

Thus, the bending stress governs the design, and the maximum permissible load is

$$P_{max} = 8.25 \text{ kN} \qquad \longleftarrow$$

A more complete analysis of this beam would require that the weight of the beam be taken into account, thus reducing the permissible load.

Note: In this example, the maximum normal stresses and maximum shear stresses do not occur at the same locations in the beam—the normal stress is maximum in the middle region of the beam at the top and bottom of the cross section, and the shear stress is maximum near the supports at the neutral axis of the cross section. For most beams, the bending stresses (not the shear stresses) control the allowable load.

Although wood is not a homogeneous material and often departs from linearly elastic behavior, we can still obtain approximate results from the flexure and shear formulas. These approximate results are usually adequate for designing wood beams.

5.9   SHEAR STRESSES IN BEAMS OF CIRCULAR CROSS SECTION

When a beam has a **circular cross section** (Fig. 5-34), we can no longer assume that the shear stresses act parallel to the y axis. For instance, we can easily prove that at point m (on the boundary of the cross section) the shear stress τ must act *tangent* to the boundary. This observation follows from the fact that the outer surface of the beam is free of stress, and therefore the shear stress acting on the cross section can have no component in the radial direction.

Although there is no simple way to find the shear stresses acting throughout the entire cross section, we can readily determine the shear stresses at the neutral axis (where the stresses are the largest) by making some reasonable assumptions about the stress distribution. We assume that the stresses act parallel to the y axis and have constant intensity across the width of the beam (from point p to point q in Fig. 5-34). Since these assumptions are the same as those used in deriving the shear formula $\tau = VQ/Ib$ (Eq. 5-38), we can use the shear formula to calculate the stresses at the neutral axis.

For use in the shear formula, we need the following properties of a circular cross section having radius r:

$$I = \frac{\pi r^4}{4} \qquad Q = A\bar{y} = \left(\frac{\pi r^2}{2}\right)\left(\frac{4r}{3\pi}\right) = \frac{2r^3}{3} \qquad b = 2r \quad (5\text{-}41\text{a,b})$$

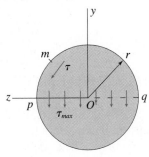

Fig. 5-34 Shear stresses acting on the cross section of a circular beam

The expression for the moment of inertia I is taken from Case 9 of Appendix D, and the expression for the first moment Q is based upon the formulas for a semicircle (Case 11, Appendix D). Substituting these expressions into the shear formula, we obtain

$$\tau_{max} = \frac{VQ}{Ib} = \frac{V(2r^3/3)}{(\pi r^4/4)(2r)} = \frac{4V}{3\pi r^2} = \frac{4V}{3A} \tag{5-42}$$

in which $A = \pi r^2$ is the area of the cross section. This equation shows that the maximum shear stress in a circular beam is equal to 4/3 times the average shear stress V/A.

If a beam has a **hollow circular cross section** (Fig. 5-35), we may again assume with reasonable accuracy that the shear stresses at the neutral axis are parallel to the y axis and uniformly distributed across the section. Consequently, we may again use the shear formula to find the maximum stresses. The required properties of the hollow circular section are

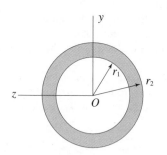

Fig. 5-35 Hollow circular cross section

$$I = \frac{\pi}{4}\left(r_2^4 - r_1^4\right) \qquad Q = \frac{2}{3}\left(r_2^3 - r_1^3\right) \qquad b = 2(r_2 - r_1) \tag{5-43a,b,c}$$

in which r_1 and r_2 are the inner and outer radii of the cross section. Therefore, the maximum stress is

$$\tau_{max} = \frac{VQ}{Ib} = \frac{4V}{3A}\left(\frac{r_2^2 + r_2 r_1 + r_1^2}{r_2^2 + r_1^2}\right) \tag{5-44}$$

in which

$$A = \pi(r_2^2 - r_1^2)$$

is the area of the cross section. Note that if $r_1 = 0$, Eq. (5-44) reduces to Eq. (5-42) for a solid circular beam.

Although the preceding theory for shear stresses in beams of circular cross section is approximate, it gives results differing by only a few percent from those obtained using the exact theory of elasticity (see Refs. 5-9 and 5-11). Consequently, Eqs. (5-42) and (5-44) can be used to determine the maximum shear stresses in circular beams under ordinary circumstances.

5.10 SHEAR STRESSES IN THE WEBS OF BEAMS WITH FLANGES

When a beam of wide-flange shape (Fig. 5-36a) is subjected to shear forces as well as bending moments (nonuniform bending), both normal and shear stresses are developed on the cross sections. The distribution of the shear stresses in a wide-flange beam is more complicated than in a rectangular beam. For instance, the shear stresses in the flanges of the

(a)

(b)

Fig. 5-36 (a) Beam of wide-flange shape, and (b) directions of the shear stresses acting on a cross section

beam act in both vertical and horizontal directions (the y and z directions), as shown by the small arrows in Fig. 5-36b. The horizontal shear stresses, which are much larger than the vertical shear stresses, are discussed later in Section 6.7.

The shear stresses in the web of a wide-flange beam act only in the vertical direction, with the largest stress occurring at the neutral axis. These stresses can be found by the same techniques we used for finding shear stresses in rectangular beams.

■ Shear Stresses in the Web

Let us begin the analysis by determining the shear stresses at level ef in the web of a wide-flange beam (Fig. 5-37a). We will make the same assumptions as those we made for a rectangular beam; that is, we assume that the shear stresses act parallel to the y axis and are uniformly distributed across the thickness of the web. Then the shear formula $\tau = VQ/Ib$ will still apply. However, the width b is now the thickness t of the web, and the area used in calculating the first moment Q is the area between line ef and the top edge of the cross section (indicated by the shaded area of Fig. 5-37a).

When finding the first moment Q of the shaded area, we will disregard the effects of the small fillets at the juncture of the web and flange (points b and c in Fig. 5-37a). The error in ignoring the areas of these fillets is very small. Then we will divide the shaded area into two rectangles. The first rectangle is the upper flange itself, which has area

$$A_1 = b\left(\frac{h}{2} - \frac{h_1}{2}\right) \tag{a}$$

in which b is the width of the flange, h is the overall height of the beam, and h_1 is the distance between the insides of the flanges. The second

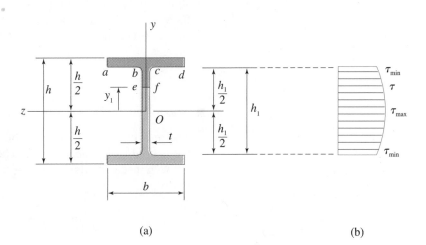

Fig. 5-37 Shear stresses in the web of a wide-flange beam. (a) Cross section of beam, and (b) distribution of vertical shear stresses in the web

(a)   (b)

rectangle is the part of the web between *ef* and the flange, that is, rectangle *efcb*, which has area

$$A_2 = t\left(\frac{h_1}{2} - y_1\right) \tag{b}$$

in which *t* is the thickness of the web and y_1 is the distance from the neutral axis to level *ef*.

The first moments of areas A_1 and A_2, evaluated about the neutral axis, are obtained by multiplying these areas by the distances from their respective centroids to the *z* axis. Adding these first moments gives the first moment *Q* of the combined area:

$$Q = A_1\left(\frac{h_1}{2} + \frac{h/2 - h_1/2}{2}\right) + A_2\left(y_1 + \frac{h_1/2 - y_1}{2}\right)$$

Upon substituting for A_1 and A_2 from Eqs. (a) and (b) and then simplifying, we get

$$Q = \frac{b}{8}(h^2 - h_1^2) + \frac{t}{8}(h_1^2 - 4y_1^2) \tag{5-45}$$

Therefore, the shear stress τ in the web of the beam at distance y_1 from the neutral axis is

$$\tau = \frac{VQ}{It} = \frac{V}{8It}\left[b(h^2 - h_1^2) + t(h_1^2 - 4y_1^2)\right] \tag{5-46}$$

in which the moment of inertia of the cross section is

$$I = \frac{bh^3}{12} - \frac{(b-t)h_1^3}{12} = \frac{1}{12}(bh^3 - bh_1^3 + th_1^3) \tag{5-47}$$

Since all quantities in Eq. (5-46) are constants except y_1, we see immediately that τ varies quadratically throughout the height of the web, as shown by the graph in Fig. 5-37b. Note that the graph is drawn only for the web and does not include the flanges. The reason is simple enough— Eq. (5-46) cannot be used to determine the vertical shear stresses in the flanges of the beam (see the discussion titled "Limitations" later in this section).

■ Maximum and Minimum Shear Stresses

The maximum shear stress in the web of a wide-flange beam occurs at the neutral axis, where $y_1 = 0$. The minimum shear stress occurs where the web meets the flanges ($y_1 = \pm h_1/2$). These stresses, found from Eq. (5-46), are

$$\tau_{\text{max}} = \frac{V}{8It}(bh^2 - bh_1^2 + th_1^2) \qquad \tau_{\text{min}} = \frac{Vb}{8It}(h^2 - h_1^2) \tag{5-48a,b}$$

Both τ_{max} and τ_{min} are labeled on the graph of Fig. 5-37b. For typical wide-flange beams, the maximum stress in the web is from 10% to 60% greater than the minimum stress.

Although it may not be apparent from the preceding discussion, the stress τ_{max} given by Eq. (5-48a) not only is the largest shear stress in the web but also is the largest shear stress anywhere in the beam.

■ Shear Force in the Web

The vertical shear force carried by the web alone may be determined by multiplying the area of the shear-stress diagram (Fig. 5-37b) by the thickness t of the web. The shear-stress diagram consists of two parts, a rectangle of area $h_1\tau_{min}$ and a parabolic segment of area

$$\frac{2}{3}(h_1)(\tau_{max} - \tau_{min})$$

By adding these two areas, multiplying by the thickness t of the web, and then combining terms, we get the total shear force in the web:

$$V_{web} = \frac{th_1}{3}(2\tau_{max} + \tau_{min}) \tag{5-49}$$

For beams of typical proportions, the shear force in the web is 90% to 98% of the total shear force V acting on the cross section; the remainder is carried by shear in the flanges.

Since the web resists most of the shear force, designers often calculate an approximate value of the maximum shear stress by dividing the total shear force by the area of the web. The result is the average shear stress in the web, assuming that the web carries *all* of the shear force:

$$\tau_{aver} = \frac{V}{th_1} \tag{5-50}$$

For typical wide-flange beams, the average stress calculated in this manner is within 10% (plus or minus) of the maximum shear stress calculated from Eq. (5-48a). Thus, Eq. (5-50) provides a simple way to estimate the maximum shear stress.

Note: The average shear stress obtained from Eq. (5-50) is not the same as the average shear stress obtained when the stresses (shown in Fig. 5-37b) are calculated from the shear formula $\tau = VQ/Ib$. The discrepancy comes from the fact that the shear formula takes into account the area of the flanges, but Eq. (5-50) does not. It is for this reason that τ_{aver} from Eq. (5-50) may be larger or smaller than τ_{max} as determined by the shear formula.

■ **Example**

As an illustration of the calculation of shear stresses, consider a wide-flange beam of W 12 × 35 shape subjected to a shear force $V = 10,000$ lb. The cross-sectional dimensions of this beam (see Fig. 5-37a and Table E-1, Appendix E) are as follows:

$$b = 6.56 \text{ in.} \qquad t = 0.30 \text{ in.} \qquad h = 12.50 \text{ in.} \qquad h_1 = 11.46 \text{ in.}$$

The moment of inertia of the cross section (from Eq. 5-47) is

$$I = \frac{1}{12}(bh^3 - bh_1^3 + th_1^3) = 283 \text{ in.}^4$$

The moment of inertia listed in Table E-1 is 285 in.^{4}, which is slightly larger than the value calculated above because it includes the areas of the fillets.

To determine the maximum and minimum shear stresses, we substitute the preceding numerical values into Eqs. (5-48a and b) and obtain

$$\tau_{max} = \frac{V}{8It}(bh^2 - bh_1^2 + th_1^2) = 2990 \text{ psi}$$

$$\tau_{min} = \frac{Vb}{8It}(h^2 - h_1^2) = 2410 \text{ psi}$$

In this case, the ratio of τ_{max} to τ_{min} is 1.24, that is, the maximum stress in the web is 24% larger than the minimum stress.

The average shear stress in the web of the beam (from Eq. 5-50) is

$$\tau_{aver} = \frac{V}{th_1} = 2910 \text{ psi}$$

which is only 3% less than the maximum stress. Thus, the average stress (when calculated by Eq. 5-50) is an easily obtained approximation to the maximum stress and is satisfactory for most purposes.

The shear force in the web (Eq. 5-49) is

$$V_{web} = \frac{th_1}{3}(2\tau_{max} + \tau_{min}) = 9610 \text{ lb}$$

which means that the web of this particular beam resists 96% of the total shear force.

■ **Limitations**

The elementary shear theory presented in this section is suitable for determining the vertical shear stresses in the web of a wide-flange beam. However, when investigating vertical shear stresses in the flanges, we can no longer assume that the shear stresses are constant across the width of the section, that is, across the width b of the flanges (Fig. 5-37a). Hence, we cannot use the shear formula to determine these stresses.

To emphasize this point, consider the junction of the web and upper flange ($y_1 = h_1/2$), where the width of the section changes abruptly from t to b. The shear stresses on the free surfaces ab and cd (Fig. 5-37a) must be zero, whereas the shear stress across the web at line bc is τ_{min}. These observations indicate that the distribution of shear stresses at the junction of the web and the flange is quite complex and cannot be investigated by elementary methods. The stress analysis is further complicated by the use of fillets at the reentrant corners (corners b and c). The fillets are necessary to prevent the stresses from becoming dangerously large, but they also alter the stress distribution across the web.

Thus, we conclude that the shear formula cannot be used to determine the vertical shear stresses in the flanges. However, the shear formula does give good results for the shear stresses acting *horizontally* in the flanges (Fig. 5-36b), as discussed later in Section 6.7.

The method described above for determining shear stresses in the webs of wide-flange beams can also be used for other sections having thin webs. The following example illustrates the procedure for a T-beam.

■ **Example 5-13**

A beam having a T-shaped cross section (Fig. 5-38a) is subjected to a vertical shear force $V = 10,000$ lb. The cross-sectional dimensions are $b = 4$ in., $t = 1.0$ in., $h = 8.0$ in., and $h_1 = 7.0$ in. Determine the shear stress τ_1 at the top of the web (level nn) and the maximum shear stress τ_{max}. (Disregard the areas of the fillets.)

Fig. 5-38 Example 5-13. Shear stresses in the web of a T-beam

(a)   (b)

Solution   *Location of neutral axis.* The neutral axis of the T-beam is located by calculating the distances c_1 and c_2 from the top and bottom of the beam to the centroid of the cross section (Fig. 5-38a). First, we divide the cross section into two rectangles, the flange and the web (see the dashed line in Fig. 5-38a). Then we calculate the first moment Q_{aa} of these two rectangles with respect to line aa at the bottom of the beam. The distance c_2 is equal to Q_{aa} divided by the area A of the entire cross section (see Chapter 12, Section 12.3, for methods for locating centroids of composite areas). The calculations are as follows:

$$A = \sum A_i = b(h - h_1) + th_1 = 11.0 \text{ in.}^2$$

$$Q_{aa} = \sum y_i A_i = \left(\frac{h + h_1}{2}\right)(b)(h - h_1) + \frac{h_1}{2}(th_1) = 54.5 \text{ in.}^3$$

$$c_2 = \frac{Q_{aa}}{A} = \frac{54.5 \text{ in.}^3}{11.0 \text{ in.}^2} = 4.955 \text{ in.}   \qquad c_1 = h - c_2 = 3.045 \text{ in.}$$

Moment of inertia. The moment of inertia I of the entire cross-sectional area (with respect to the neutral axis) can be found by determining the moment of inertia I_{aa} about line aa at the bottom of the beam and then using the parallel-axis theorem (see Section 12.5):

$$I = I_{aa} - Ac_2^2$$

The calculations are as follows:

$$I_{aa} = \frac{bh^3}{3} - \frac{(b - t)h_1^3}{3} = 339.67 \text{ in.}^4   \qquad Ac_2^2 = 270.02 \text{ in.}^4   \qquad I = 69.65 \text{ in.}^4$$

Shear stress at top of web. To find the shear stress τ_1 at the top of the web (along line nn) we need to calculate the first moment Q_1 of the area above level nn. This first moment is equal to the area of the flange times the distance from the neutral axis to the centroid of the flange:

$$Q_1 = b(h - h_1)\left(c_1 - \frac{h - h_1}{2}\right)$$

$$= (4 \text{ in.})(1 \text{ in.})(3.045 \text{ in.} - 0.5 \text{ in.}) = 10.18 \text{ in.}^3$$

Of course, we get the same result if we calculate the first moment of the area *below* level nn:

$$Q_1 = th_1\left(c_2 - \frac{h_1}{2}\right) = (1 \text{ in.})(7 \text{ in.})(4.955 \text{ in.} - 3.5 \text{ in.}) = 10.18 \text{ in.}^3$$

Substituting into the shear formula, we find

$$\tau_1 = \frac{VQ_1}{It} = \frac{(10,000 \text{ lb})(10.18 \text{ in.}^3)}{(69.65 \text{ in.}^4)(1 \text{ in.})} = 1460 \text{ psi} \qquad \longleftarrow$$

This stress exists both as a vertical shear stress acting on the cross section and as a horizontal shear stress acting on the horizontal plane between the flange and the web.

Maximum shear stress. The maximum shear stress occurs in the web at the neutral axis. Therefore, we calculate the first moment Q_{max} of the cross-sectional area below the neutral axis:

$$Q_{max} = tc_2\left(\frac{c_2}{2}\right) = (1 \text{ in.})(4.955 \text{ in.})\left(\frac{4.955 \text{ in.}}{2}\right) = 12.28 \text{ in.}^3$$

As previously indicated, we would get the same result if we calculated the first moment of the area above the neutral axis, but those calculations would be slighter longer.

Substituting into the shear formula, we obtain

$$\tau_{max} = \frac{VQ_{max}}{It} = \frac{(10,000 \text{ lb})(12.28 \text{ in.}^3)}{(69.65 \text{ in.}^4)(1 \text{ in.})} = 1760 \text{ psi} \quad \longleftarrow$$

which is the maximum shear stress in the beam.

The parabolic distribution of shear stresses in the web is shown in Fig. 5-38b.

5.11 BUILT-UP BEAMS AND SHEAR FLOW

Built-up beams are fabricated from two or more pieces of material joined together to form a single beam. Such beams can be constructed in a great variety of shapes to meet special architectural or structural needs and to provide larger cross sections than are ordinarily available. Figure 5-39 shows some typical cross sections of built-up beams. Part (a) of the figure shows a wood **box beam** constructed of two planks, which serve as flanges, and two plywood webs. The pieces are joined together with nails, screws, or glue in such a manner that the entire beam acts as a single unit. Box beams are also constructed of other materials, including steel, plastics, and composites.

The second example is a glued laminated beam (called a **glulam beam**) made of boards glued together to form a much larger beam than could be cut from a tree as a single member. Glulam beams are widely used in the construction of small buildings.

The third example is a steel **plate girder** of the type commonly used in bridges and large buildings. These girders, consisting of three steel plates joined by welding, can be fabricated in much larger sizes than are available with ordinary wide-flange or I-beams.

Built-up beams must be designed so that the beam behaves as a single member. Consequently, the design calculations involve two phases. In the first phase, the beam is designed as though it were made of one piece, taking into account both bending and shear stresses. In the second phase, the *connections* between the parts (such as nails, bolts, welds, and glue) are designed to ensure that the beam does indeed behave

(a)

(b)   (c)

Fig. 5-39 Cross sections of typical built-up beams: (a) wood box beam, (b) glulam beam, and (c) plate girder as a single entity. In particular, the connections must be strong enough to transmit the horizontal shear forces acting between the parts of the beam. To obtain these forces, we make use of the concept of *shear flow*.

■ Shear Flow

To obtain a formula for the horizontal shear forces acting between parts of a beam, let us return to the derivation of the shear formula (see Figs. 5-28 and 5-29 of Section 5.8). In that derivation, we cut an element mm_1n_1n from a beam (Fig. 5-40a) and investigated the horizontal equilibrium of a subelement mm_1p_1p (Fig. 5-40b). From the horizontal equilibrium of the subelement, we determined the force F_3 (Fig. 5-40c) acting on its lower surface:

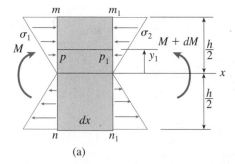

$$F_3 = \frac{dM}{I}\int y\,dA \tag{5-51}$$

This equation is repeated from Eq. (5-33) of Section 5.8.

Let us now define a new quantity called the **shear flow** f. Shear flow is the *horizontal shear force per unit distance along the longitudinal axis of the beam*. Since the force F_3 acts along the distance dx, the shear force per unit distance is equal to F_3 divided by dx; thus,

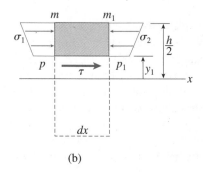

$$f = \frac{F_3}{dx} = \frac{dM}{dx}\left(\frac{1}{I}\right)\int y\,dA$$

Replacing dM/dx by the shear force V and denoting the integral by Q, we obtain the following **shear-flow formula**:

$$f = \frac{VQ}{I} \tag{5-52}$$

This equation gives the shear flow acting on the horizontal plane pp_1 shown in Fig. 5-40a. The terms V, Q, and I have the same meanings as in the shear formula.

If the shear stresses on plane pp_1 are uniformly distributed, as we assumed for rectangular beams and wide-flange beams, the shear flow f equals τb. In that case, the shear-flow formula reduces to the shear formula (Eq. 5-38). However, the derivation of Eq. (5-51) for the force F_3 does not involve any assumption about the distribution of shear stresses in the beam. Instead, the force F_3 is found solely from the horizontal equilibrium of the subelement (Fig. 5-40c). Therefore, we can now interpret the subelement and the force F_3 in more general terms than before.

The subelement may be *any* prismatic block of material between cross sections mn and m_1n_1 (Fig. 5-40a). It does not have to be obtained

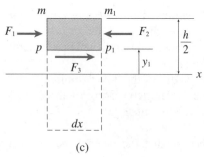

Fig. 5-40 Horizontal shear stresses and shear forces in a beam. (*Note:* These figures are repeated from Figs. 5-28 and 5-29.)

by making a single horizontal cut (such as pp_1) through the beam. Also, since the force F_3 is the total horizontal shear force acting between the subelement and the rest of the beam, it may be distributed anywhere over the sides of the subelement, not just on its lower surface. These same comments apply to the shear flow f, since it is merely the force F_3 per unit distance.

Let us now return to the shear-flow formula $f = VQ/I$ (Eq. 5-52). The terms V and I have their usual meanings and are not affected by the choice of subelement. However, the first moment Q is a property of the cross-sectional face of the subelement. To illustrate how Q is determined, we will consider three specific examples of built-up beams (Fig. 5-41).

(a)

(b)

■ Areas Used When Calculating the First Moment Q

The first example of a built-up beam is a welded steel plate girder (Fig. 5-41a). The welds must transmit the horizontal shear forces that act between the flanges and the web. At the upper flange, the horizontal shear force (per unit distance along the axis of the beam) is the shear flow along the contact surface aa. This shear flow may be calculated by taking Q as the first moment of the cross-sectional area above the contact surface aa. In other words, Q is the first moment of the flange area (shown shaded in Fig. 5-41a), calculated with respect to the neutral axis. After calculating the shear flow, we can readily determine the amount of welding needed to resist the shear force, because the strength of a weld is usually specified in terms of force per unit distance along the weld.

The second example is a wide-flange beam that is strengthened by riveting a channel section to each flange (Fig. 5-41b). The horizontal shear force acting between each channel and the main beam must be transmitted by the rivets. This force is calculated from the shear-flow formula using Q as the first moment of the entire channel (shown shaded in the figure). The resulting shear flow is the longitudinal force per unit distance acting along the contact surface bb, and the rivets must be of adequate size and longitudinal spacing to resist this force.

The last example is a wood box beam with two flanges and two webs that are connected by nails or screws (Fig. 5-41c). The total horizontal shear force between the upper flange and the webs is the shear flow acting along *both* contact surfaces cc and dd, and therefore the first moment Q is calculated for the upper flange (the shaded area). In other words, the shear flow calculated from the formula $f = VQ/I$ is the total shear flow along all contact surfaces that surround the area for which Q is computed. In this case, the shear flow f is resisted by the combined action of the nails on *both* sides of the beam, that is, at both cc and dd, as illustrated in the following example.

(c)

Fig. 5-41 Areas used when calculating the first moment Q

■ **Example 5-14**

A wood box beam (Fig. 5-42) is constructed of two boards, each 40 mm ×
180 mm in cross section, that serve as flanges and two plywood webs, each
15 mm thick. The total height of the beam is 280 mm. The plywood is fastened
to the flanges by wood screws having an allowable load in shear of $F = 1100$ N
each. If the shear force V acting on the cross section is 10.5 kN, determine the
maximum permissible longitudinal spacing s of the screws (Fig. 5-42b).

Cross section   Side view
(a)   (b)

Fig. 5-42 Example 5-14. Wood box beam

Solution   *Shear flow.* The horizontal shear force transmitted between the
upper flange and the two webs can be found from the shear-flow formula
$f = VQ/I$, in which Q is the first moment of the cross-sectional area of the flange.
To find this first moment, we multiply the area A_f of the flange by the distance d_f
from its centroid to the neutral axis:

$$A_f = 40 \text{ mm} \times 180 \text{ mm} = 7200 \text{ mm}^2 \qquad d_f = 120 \text{ mm}$$

$$Q = A_f d_f = (7200 \text{ mm}^2)(120 \text{ mm}) = 864 \times 10^3 \text{ mm}^3$$

The moment of inertia of the entire cross-sectional area about the neutral axis is
equal to the moment of inertia of the outer rectangle minus the moment of inertia
of the "hole" (the inner rectangle):

$$I = \frac{1}{12}(210 \text{ mm})(280 \text{mm})^3 - \frac{1}{12}(180 \text{ mm})(200 \text{ mm})^3 = 264.2 \times 10^6 \text{ mm}^4$$

Substituting V, Q, and I into the shear-flow formula (Eq. 5-52), we obtain

$$f = \frac{VQ}{I} = \frac{(10,500 \text{ N})(864 \times 10^3 \text{ mm}^3)}{264.2 \times 10^6 \text{ mm}^4} = 34.3 \text{ N/mm}$$

which is the horizontal shear force per millimeter of length that must be transmitted between the flange and the two webs.

Spacing of screws. Since the longitudinal spacing of the screws is s, and
since there are two lines of screws (one on each side of the flange), it follows that
the load capacity of the screws is $2F$ per distance s along the beam. Therefore,
the capacity of the screws per unit distance along the beam is $2F/s$. Equating
$2F/s$ to the shear flow f and solving for the spacing s, we get

$$s = \frac{2F}{f} = \frac{2(1100 \text{ N})}{34.3 \text{ N/mm}} = 64.1 \text{ mm}$$

This value of s is the maximum permissible spacing of the screws, based upon the allowable load per screw. Any spacing greater than 64.1 mm would overload the screws. For convenience in fabrication, and to be on the safe side, a spacing such as $s = 60$ mm would be selected.

5.12 ■ BEAMS WITH AXIAL LOADS

(a)

(b)

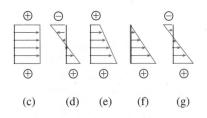

(c)   (d)   (e)   (f)   (g)

Fig. 5-43 Normal stresses in a cantilever beam subjected to both bending and axial loads: (a) beam with load P acting at the free end, (b) stress resultants acting on a cross section at distance x from the support, (c) tensile stresses due to the axial force N acting alone, (d) tensile and compressive stresses due to the bending moment M acting alone, and (e), (f), (g) possible final stress distributions due to the combined effects of N and M

Structural members are often subjected to the simultaneous action of bending loads and axial loads. This happens, for instance, in aircraft frames, columns in buildings, machinery, parts of ships, and spacecraft. If the members are not too slender, the combined stresses can be obtained by superposition of the bending stresses and the axial stresses.

To see how this is accomplished, consider the cantilever beam shown in Fig. 5-43a. The only load on the beam is an inclined force P acting through the centroid of the end cross section. This load can be resolved into two components, a lateral load Q and an axial load S. These loads produce stress resultants in the form of bending moments M, shear forces V, and axial forces N throughout the beam (Fig. 5-43b). On a typical cross section, distance x from the support, these stress resultants are

$$M = Q(L - x) \qquad V = -Q \qquad N = S$$

in which L is the length of the beam. The stresses associated with each of these stress resultants can be determined at any point in the cross section by means of the appropriate formula ($\sigma = -My/I$, $\tau = VQ/Ib$, and $\sigma = N/A$).

Since both the axial force N and bending moment M produce normal stresses, we need to combine those stresses to obtain the final stress distribution. The axial force (when acting alone) produces a uniform stress distribution $\sigma = N/A$ over the entire cross section, as shown by the stress diagram in Fig. 5-43c. In this particular example, the stress σ is tensile, as indicated by the plus signs attached to the diagram.

The bending moment produces a linearly varying stress $\sigma = -My/I$ (Fig. 5-43d) with compression on the upper part of the beam and tension on the lower part. The distance y is measured from the z axis, which passes through the centroid of the cross section.

The final distribution of normal stresses is obtained by superposing the stresses produced by the axial force and the bending moment, as follows:

$$\sigma = \frac{N}{A} - \frac{My}{I} \qquad (5\text{-}53)$$

Note that N is positive when it produces tension and M is positive according to the bending-moment sign convention (positive bending moment produces compression in the upper part of the beam and tension in the lower part). Also, the y axis is positive upward. As long as we use these sign conventions in Eq. (5-53), the normal stress σ will be positive for tension and negative for compression.

The final stress distribution depends upon the relative algebraic values of the terms in Eq. (5-53). For our particular example, the three possibilities are shown in Figs. 5-43e, f, and g. If the bending stress at the top of the beam (Fig. 5-43d) is numerically less than the axial stress (Fig. 5-43c), the entire cross section will be in tension, as shown in Fig. 5-43e. If the bending stress at the top equals the axial stress, the distribution will be triangular (Fig. 5-43f), and if the bending stress is numerically larger than the axial stress, the cross section will be partially in compression and partially in tension (Fig. 5-43g). Of course, if the axial force is a compressive force, or if the bending moment is reversed in direction, the stress distributions will change accordingly.

Whenever bending and axial loads act simultaneously, the neutral axis (that is, the line in the cross section where the normal stress is zero) no longer passes through the centroid of the cross section. As shown in Figs. 5-43e, f, and g, respectively, the neutral axis may be outside the cross section, at the edge of the section, or within the section.

The use of Eq. (5-53) to determine the stresses in a beam with axial loads is illustrated later in Example 5-15.

■ Eccentric Axial Loads

An **eccentric axial load** is an axial force that does *not* act through the centroid of the cross section. An example is shown in Fig. 5-44a, where the cantilever beam AB is subjected to a tensile load P acting at distance e from the x axis (the x axis passes through the centroids of the cross sections). The distance e, called the *eccentricity* of the load, is positive in the positive direction of the y axis.

The eccentric load P is statically equivalent to an axial force P acting along the x axis and a bending moment Pe acting about the z axis (Fig. 5-44b). Note that the moment Pe is a negative bending moment.

A cross-sectional view of the beam (Fig. 5-44c) shows the y and z axes passing through the centroid C of the cross section. The eccentric load P intersects the y axis, which is an axis of symmetry.

Since the axial force N at any cross section is equal to P, and since the bending moment M is equal to $-Pe$, the normal stress at any point in the bar (from Eq. 5-53) is

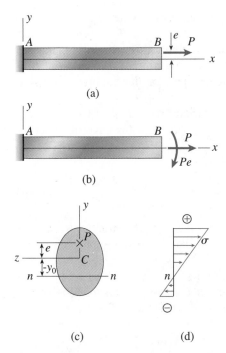

Fig. 5-44 (a) Cantilever beam with an eccentric axial load P, (b) equivalent loads P and Pe, (c) cross section of beam, and (d) distribution of normal stresses over the cross section

$$\sigma = \frac{P}{A} + \frac{Pey}{I} \qquad (5\text{-}54)$$

in which A is the area of the cross section and I is the moment of inertia about the z axis. The stress distribution obtained from Eq. (5-54), for the case where both P and e are positive, is shown in Fig. 5-44d.

The position of the neutral axis nn (Fig. 5-44c) can be obtained from Eq. (5-54) by setting the stress σ equal to zero and solving for the coordinate y, which we now denote as y_0. The result is

$$y_0 = -\frac{I}{Ae} \qquad (5\text{-}55)$$

The coordinate y_0 is measured from the z axis (which is the neutral axis under pure bending) to the line nn of zero stress (the neutral axis under combined bending and axial load). Because y_0 is positive in the direction of the y axis (upward in Fig. 5-44c), it is labeled $-y_0$ when it is shown downward in the figure.

From Eq. (5-55) we see that the neutral axis lies below the z axis when e is positive and above the z axis when e is negative. If the eccentricity is reduced, the distance y_0 increases and the neutral axis moves away from the centroid. In the limit, as e approaches zero, the load acts at the centroid, the neutral axis is at an infinite distance, and the stress distribution is uniform. If the eccentricity is increased, the distance y_0 decreases and the neutral axis moves toward the centroid. In the limit, as e becomes extremely large, the load acts at an infinite distance, the neutral axis passes through the centroid, and the stress distribution is the same as in pure bending.

Eccentric axial loads are analyzed in some of the problems at the end of this chapter (see Problems 5.12-13 to 5.12-19).

■ The Core of a Cross Section

If the axial load is applied with a small eccentricity, the neutral axis may lie outside the beam, as explained in the preceding paragraph. When that happens, the normal stresses will have the same sign throughout the cross section, and the beam will be entirely in tension or entirely in compression.

This condition is important, for instance, when a compressive load acts on a material that is very weak in tension, such as concrete, stone, or ceramics. With such materials we must be sure that the load produces only compression on the cross sections. To see how this is accomplished, consider the example of Fig. 5-45a. In this example, a compressive axial load P acts with eccentricity e on a rectangular column of width b and height h. When e is small, the neutral axis nn lies outside the column in the negative y direction (Fig. 5-45b) and the entire cross section is in compression. As the load moves in the positive y direction and the eccen-

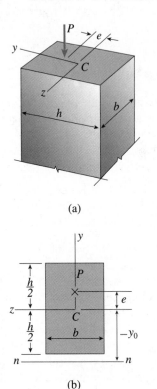

(a)

(b)

Fig. 5-45 Compressive load P acting on a rectangular column

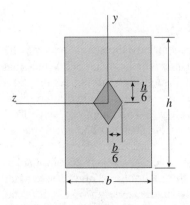

Fig. 5-46 The core of a rectangular cross section

tricity e increases, the neutral axis moves closer to the edge of the cross section. When the neutral axis reaches the edge of the cross section ($y_0 = -h/2$, or $-y_0 = h/2$), the stress distribution becomes triangular but is still entirely in compression. With a further increase in e, the neutral axis moves into the cross section and part of the column is now in tension.

The maximum value of the eccentricity so that all of the column is in compression occurs when the neutral axis is at the edge of the cross section, that is, when y_0 equals $-h/2$. The corresponding value e_1 of the eccentricity is obtained from Eq. (5-55):

$$e_1 = -\frac{I}{Ay_0} = -\frac{I}{A(-h/2)} = \frac{bh^3/12}{(bh)(h/2)} = \frac{h}{6} \qquad (5\text{-}56)$$

Thus, as long as the eccentricity is less than $h/6$, the entire cross section will be in compression. (Keep in mind that this result applies only to a beam of rectangular cross section.)

A similar analysis can be made for a compressive load acting along the negative y axis. In this case the neutral axis is on the side of the column where y is positive and the limiting value of the eccentricity is $e_1 = -h/6$.

We therefore reach the following conclusion: *If the load P is applied on the y axis with an eccentricity between h/6 and −h/6, the entire cross section will be in compression.* Since this range of values means that P is confined to the middle third of the height h, the preceding statement is known as the *middle-third rule*. The middle-third rule applies only to a rectangular cross section; for cross sections of other shapes, the limiting eccentricities will have different values.

An analysis similar to the one described above can be made for the rectangular column of Fig. 5-45 when the compressive load P acts at a point on the z axis. We then find that the entire column is in compression as long as the load acts with an eccentricity between $b/6$ and $-b/6$. Thus, we conclude that there is a small area surrounding the centroid C such that the entire member is in compression as long as the load acts within that region. This region is called the *core* or the *kern* of the cross section. With the aid of the middle-third rule and some additional analysis, it is possible to show that the core of a rectangle is a rhombus with diagonals of lengths $h/3$ and $b/3$ (Fig. 5-46).* The cores of other shapes are determined in Problems 5.12-20 to 5.12-24 at the end of the chapter.

■ **Limitations**

The preceding analysis of beams with axial loads is based upon the assumption that the bending moments can be calculated without consid-

---

*The concept of the core of a cross section was introduced by the French engineer J. A. C. Bresse in 1854; see Ref. 5-12.

ering the deflections of the beams. In other words, when determining the bending moment M for use in Eq. (5-53), we must be able to use the original dimensions of the beam—the dimensions *before* any deformations or deflections occur. The use of the original dimensions is valid provided the beams are relatively stiff in bending, so that the deflections are very small.

Thus, when analyzing a beam with axial loads, it is important to distinguish between a *stocky* beam, which is relatively short and therefore highly resistant to bending, and a *slender* beam, which is relatively long and therefore very flexible. In the case of a stocky beam, the lateral deflections are so small as to have no significant effect on the line of action of the axial forces. As a consequence, the bending moments will not depend upon the deflections and the stresses can be found from Eq. (5-53).

In the case of a slender beam, the lateral deflections (even though small in magnitude) are large enough to alter significantly the line of action of the axial forces. When that happens, an additional bending moment, equal to the product of the axial force and the lateral deflection, is created at every cross section. In other words, there is an interaction, or coupling, between the axial effects and the bending effects. This type of behavior is discussed in Chapter 11 on columns.

The distinction between a stocky beam and a slender beam is obviously not a precise one. In general, the only way to know whether interaction effects are important is to analyze the beam with and without the interaction and notice whether the results differ significantly. However, this procedure may require considerable calculating effort. Therefore, as a guideline for practical use, we usually consider a beam with a length-to-height ratio of 10 or less to be a stocky beam. Only stocky beams are considered in the examples and problems of this section.

■ **Example 5-15**

A tubular beam ACB of length $L = 60$ in. is pin-supported at its ends and loaded by an inclined force P at midlength (Fig. 5-47a). The distance from the point of application of the load P to the longitudinal axis of the tube is $d = 5.5$ in. The cross section of the tube is square (Fig. 5-47b) with outer dimension $b = 6.0$ in., area $A = 20.0$ in.^{2}, and moment of inertia $I = 86.67$ in.^{4} Determine the maximum tensile and compressive stresses in the beam due to a load $P = 1000$ lb.

Solution   *Beam and loading.* We begin by representing the beam and its load in idealized form for purposes of analysis (Fig. 5-48a). Since the support at end A resists both horizontal and vertical displacement, it is represented as a pin support. The support at B prevents vertical displacement but offers no resistance to horizontal displacement, so it is shown as a roller support.

The inclined load P is resolved into horizontal and vertical components P_h and P_v, respectively:

(a)

(b)

Fig. 5-47 Example 5-15

(a)

(b)

(c)

(d)

Fig. 5-48 Solution to Example 5-15.
(a) Idealized beam and loading, (b) axial-force diagram, (c) shear-force diagram, and
(d) bending-moment diagram

$$P_h = P \sin 60° = (1000 \text{ lb})(\sin 60°) = 866 \text{ lb}$$

$$P_v = P \cos 60° = (1000 \text{ lb})(\cos 60°) = 500 \text{ lb}$$

The horizontal component P_h is shifted to the axis of the beam by the addition of a moment M_0 (Fig. 5-48a):

$$M_0 = P_h d = (866.0 \text{ lb})(5.5 \text{ in.}) = 4760 \text{ lb-in.}$$

Note that the loads P_h, P_v, and M_0 acting at the midpoint C of the beam are statically equivalent to the original load P.

Reactions and stress resultants. The reactions of the beam (R_h, R_A, and R_B) are shown in Fig. 5-48a. Also, the diagrams of axial force N, shear force V, and bending moment M are shown in Figs. 5-48b, c, and d, respectively. All of these quantities are found from free-body diagrams and equations of equilibrium, using the techniques described in Chapter 4.

Stresses in the beam. The maximum tensile stress in the beam occurs at the bottom of the beam ($y = -3.0$ in.) just to the left of the midpoint C. We arrive at this conclusion by noting that at this point in the beam the tensile stress due to the axial force *adds* to the tensile stress due to the largest bending moment. Thus, from Eq. (5-53), we get

$$(\sigma_t)_{\text{max}} = \frac{N}{A} - \frac{My}{I} = \frac{866 \text{ lb}}{20.0 \text{ in.}^2} - \frac{(9870 \text{ lb-in.})(-3.0 \text{ in.})}{86.67 \text{ in.}^4}$$

$$= 43 \text{ psi} + 342 \text{ psi} = 385 \text{ psi} \quad \longleftarrow$$

The maximum compressive stress occurs either at the top of the beam ($y = 3.0$ in.) to the left of point C or at the top of the beam to the right of point C. These two stresses are calculated as follows:

$$(\sigma_c)_{\text{left}} = \frac{N}{A} - \frac{My}{I} = \frac{866 \text{ lb}}{20.0 \text{ in.}^2} - \frac{(9870 \text{ lb-in.})(3.0 \text{ in.})}{86.67 \text{ in.}^4}$$

$$= 43 \text{ psi} - 342 \text{ psi} = -299 \text{ psi}$$

$$(\sigma_c)_{\text{right}} = \frac{N}{A} - \frac{My}{I} = 0 - \frac{(5110 \text{ lb-in.})(3.0 \text{ in.})}{86.67 \text{ in.}^4} = -177 \text{ psi}$$

Thus, the maximum compressive stress is

$$(\sigma_c)_{max} = -299 \text{ psi} \qquad \longleftarrow$$

and occurs at the top of the beam to the left of point C.

This example shows how the normal stresses in a beam due to combined bending and axial load can be determined. The shear stresses acting on cross sections of the beam (due to the shear forces V) can be determined independently of the normal stresses, as described earlier in this chapter. Later, in Chapter 7, we will see how to determine the stresses on inclined planes when we know both the normal and shear stresses acting on cross-sectional planes.

STRESS CONCENTRATIONS IN BENDING

The flexure and shear formulas discussed in earlier sections of this chapter are valid for beams without holes, notches, or other abrupt changes in dimensions. Whenever such discontinuities exist, high localized stresses are produced. These *stress concentrations* can be extremely important when a member is made of brittle material or is subjected to dynamic loads. (See Chapter 2, Section 2.10, for a discussion of the conditions under which stress concentrations are important.)

For illustrative purposes, two cases of stress concentrations in beams are described in this section. The first case is a beam of rectangular cross section with a hole at the neutral axis (Fig. 5-49). The beam has height h and thickness b (perpendicular to the plane of the figure) and is in pure bending under the action of bending moments M. When the diameter d of the hole is small compared to the height h, the stress distribution on the cross section through the hole is approximately as shown by the diagram in Fig. 5-49a. At point B on the edge of the hole the stress is much larger than the stress that would exist at that point if the hole were not present. (The dashed line in the figure shows the stress distribution with no hole.) However, as we go toward the outer edges of the beam (toward point A), the stress distribution varies linearly with distance from the neutral axis and is only slightly affected by the presence of the hole.

When the hole is relatively large, the stress pattern is approximately as shown in Fig. 5-49b. There is a large increase in stress at point B and only a small change in stress at point A as compared to the stress distribution in the beam without a hole (again shown by the dashed line). The stress at point C is larger than the stress at A but smaller than the stress at B.

Extensive investigations have shown that the stress at the edge of the hole (point B) is approximately twice the *nominal stress* at that point. The nominal stress is calculated from the flexure formula in the standard way, that is, $\sigma = My/I$, in which y is the distance $d/2$ from the neutral axis to point B and I is the moment of inertia of the net cross section at the

(a)

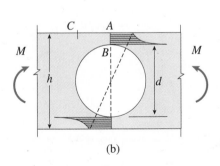

(b)

Fig. 5-49 Stress distributions in a beam in pure bending with a circular hole at the neutral axis. (The beam has a rectangular cross section with height h and thickness b.)

hole. Thus, we have the following approximate formula for the stress at point B:)

$$\sigma_B \approx 2\frac{My}{I} = \frac{12Md}{b(h^3 - d^3)} \tag{5-57}$$

At the outer edge of the beam (at point C), the stress is approximately equal to the *nominal* stress (not the actual stress) at point A (where $y = h/2$):

$$\sigma_C \approx \frac{My}{I} = \frac{6Mh}{b(h^3 - d^3)} \tag{5-58}$$

From the last two equations we see that the ratio σ_B/σ_C is approximately $2d/h$. Hence we conclude that when the ratio d/h of hole diameter to height of beam exceeds 1/2, the largest stress occurs at point B. When d/h is less than 1/2, the largest stress is at point C.

The second case we will discuss is a rectangular beam with notches (Fig. 5-50). The beam shown in the figure is subjected to pure bending and has height h and thickness b (perpendicular to the plane of the figure). Also, the net height of the beam (that is, the distance between the bases of the notches) is h_1 and the radius at the base of each notch is R. The maximum stress in this beam occurs at the base of the notches and may be much larger than the nominal stress at that same point. The nominal stress is calculated from the flexure formula with $y = h_1/2$ and $I = bh_1^3/12$; thus,

$$\sigma_{\text{nom}} = \frac{My}{I} = \frac{6M}{bh_1^2} \tag{5-59}$$

The maximum stress is equal to the stress-concentration factor K times the nominal stress:

$$\sigma_{\text{max}} = K\sigma_{\text{nom}} \tag{5-60}$$

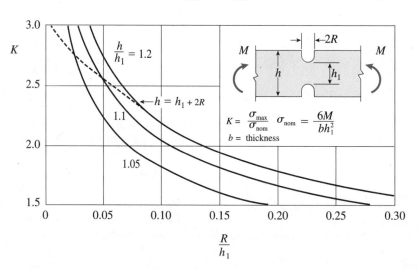

Fig. 5-50 Stress-concentration factor K for a notched beam of rectangular cross section in pure bending (h = height of beam; b = thickness of beam, perpendicular to the plane of the figure). The dashed line is for semicircular notches ($h = h_1 + 2R$).

The stress-concentration factor K is plotted in Fig. 5-50 for a few values of the ratio h/h_1. Note that when the notch becomes "sharper," that is, the ratio R/h_1 becomes smaller, the stress-concentration factor increases.*

The effects of the stress concentrations are confined to small regions around the holes and notches, as explained in the discussion of Saint-Venant's principle in Section 2.10. At a distance equal to h or greater from the hole or notch, the stress-concentration effect is negligible and the ordinary formulas for stresses may be used.

---

*Fig. 5-50 is plotted from the formulas given in Ref. 2-10, which contains formulas for many other types of stress-concentrations. For additional data on stress concentrations, including formulas and graphs, see Ref. 2-11.

PROBLEMS/CHAPTER 5

LONGITUDINAL STRAINS IN BEAMS

5.4-1 Determine the maximum normal strain ϵ_{max} produced in a steel wire of diameter $d = 1/16$ in. when it is bent around a cylindrical drum of radius $r = 24$ in. (see figure).

Prob. 5.4-1

5.4-2 A copper wire having diameter $d = 4$ mm is bent into a circle and held with the ends just touching (see figure). If the maximum permissible strain in the copper is $\epsilon_{max} = 0.0033$, what is the shortest length L of wire that can be used?

Prob. 5.4-2

5.4-3 A 4 in. outside diameter polyethylene pipe designed to carry chemical wastes is placed in a trench that makes a quarter-circular 90° bend (see figure). The curved section of the pipe is 32 ft long. Determine the maximum compressive strain ϵ_{max} in the pipe.

Prob. 5.4-3

5.4-4 A cantilever beam AB is loaded by a couple M_0 at its free end (see figure). The length of the beam is $L = 1.2$ m and the longitudinal normal strain at the top surface is 0.0008. The distance from the top surface of the beam to the neutral surface is 50 mm. Calculate the radius of curvature ρ, the curvature κ, and the vertical deflection δ at the end of the beam.

Prob. 5.4-4

5.4-5 A strip of steel of length $L = 36$ in. and thickness $t = 0.4$ in. is bent by couples M_0 (see figure). The deflection δ at the midpoint of the strip (measured from a line joining its end points) is found to be 0.29 in. Determine the longitudinal normal strain ϵ at the top surface of the strip.

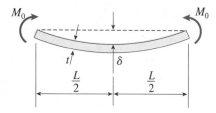

Prob. 5.4-5

5.4-6 A bar of rectangular cross section is loaded and supported as shown in the figure. The distance between supports is $L = 1.5$ m and the height of the bar is $h = 120$ mm. The deflection δ at the midpoint is measured as 2.5 mm. What is the maximum normal strain ϵ at the top and bottom of the bar?

Prob. 5.4-6

NORMAL STRESSES IN BEAMS

5.5-1 A steel wire ($E = 30 \times 10^6$ psi) of diameter $d = 1/32$ in. is bent around a pulley of radius $r = 9\text{-}3/4$ in. (see figure). (a) What is the maximum stress σ_{max} in the wire? (b) Does the stress increase or decrease if the radius of the pulley is increased?

Prob. 5.5-1

5.5-2 A thin strip of hard copper ($E = 113$ GPa) having length $L = 2$ m and thickness $t = 2$ mm is bent into a circle and held with the ends just touching (see figure). (a) Calculate the maximum bending stress σ_{max} in the strip. (b) Does the stress increase or decrease if the thickness of the strip is increased?

Prob. 5.5-2

5.5-3 A thin, high-strength steel rule ($E = 28 \times 10^6$ psi) having thickness $t = 0.06$ in. and length $L = 14$ in. is bent by couples M_0 into a circular arc subtending a central angle $\alpha = 67.8°$ (see figure). (a) What is the maximum bending stress σ_{max} in the rule? (b) Does the stress increase or decrease if the central angle is increased?

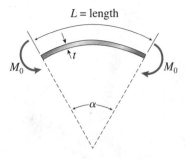

Prob. 5.5-3

5.5-4 A simply supported wood beam AB with span length $L = 3.5$ m carries a uniform load of intensity $q = 6.4$ kN/m (see figure). Calculate the maximum bending stress σ_{max} due to the load q if the beam has a rectangular cross section with width $b = 150$ mm and height $h = 280$ mm.

Prob. 5.5-4

5.5-5 A seesaw weighing 3 lb/ft of length is occupied by two children, each weighing 90 lb (see figure). The center of gravity of each child is 8 ft from the fulcrum. The board is 19 ft long, 8 in. wide, and 1.5 in. thick. What is the maximum bending stress in the board?

Prob. 5.5-5

5.5-6 A freight-car axle AB is loaded as shown in the figure, with the forces P representing the car loads (transmitted through the axle boxes) and the forces R representing the rail loads (transmitted through the wheels). The diameter of the axle is $d = 80$ mm, the wheel gauge is $L = 1.45$ m, and the distance between loads is $b = 200$ mm. Calculate the maximum bending stress σ_{max} in the axle if $P = 9.3$ kN.

Prob. 5.5-6

5.5-7 Each girder of the lift bridge (see figure) is 170 ft long and simply supported at the ends. The design load for each girder is a uniform load of intensity 1.2 k/ft. The girders are fabricated by welding steel plates so as to form an I-shaped cross section having section modulus $S = 2800$ in.^{3} What is the maximum bending stress σ_{max} in a girder due to the uniform load?

Prob. 5.5-7

5.5-8 The horizontal beam ABC of an oil-well pump has the cross section shown in the figure. If the vertical pumping force acting at end C is 38 kN, and if the distance from the line of action of that force to point B is 4.5 m, what is the maximum bending stress in the beam due to the pumping force?

Prob. 5.5-8

5.5-9 During construction of a highway bridge, the main girders are cantilevered outward from one pier toward the next (see figure). Each girder has a cantilever length of 170 ft and an I-shaped cross section with dimensions as shown in the figure. The load on each girder (during construction) is assumed to be 750 lb/ft, which includes the weight of the girder. Determine the maximum bending stress in a girder due to this load.

Prob. 5.5-9

Prob. 5.5-10

5.5-11 A railroad tie (or *sleeper*) is subjected to two concentrated loads, each of magnitude $P = 40$ k, acting as shown in the figure. The reaction q of the ballast is assumed to be uniformly distributed over the length of the tie, which has cross-sectional dimensions $b = 12$ in. and $h = 10$ in. Calculate the maximum bending stress σ_{max} in the tie due to the loads P, assuming $L = 57$ in. and $a = 19.5$ in.

5.5-10 A fiberglass pipe is hoisted by a crane using the arrangement shown in the figure. The outer diameter of the pipe is 150 mm, its thickness is 6 mm, and its weight density is 18 kN/m³. The length of the pipe is $L = 13$ m and the distance between lifting points is $s = 4$ m. Determine the maximum bending stress in the pipe due to its own weight.

Prob. 5.5-11

5.5-12 A small dam of height $h = 2.4$ m is constructed of vertical wood beams AB of thickness $t = 160$ mm, as shown in the figure. Consider the beams to be simply supported at the top and bottom. Determine the maximum bending stress σ_{max} in the beams, assuming that the weight density of water is $\gamma = 9.81$ kN/m^3.

Prob. 5.5-12

5.5-13 Obtain formulas for the maximum normal stress σ_{max} (due to pure bending by moments M) for beams having cross sections as follows (see figure): (a) the core of a circle of diameter d with the angle $\beta = 45°$, and (b) a regular hexagon with sides of length b.

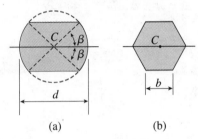

(a)   (b)

Prob. 5.5-13

5.5-14 Obtain formulas for the maximum tensile stress σ_t (due to pure bending by positive bending moments M) for beams having cross sections as follows (see figure): (a) a semicircle of diameter d, and (b) an isosceles trapezoid with bases $b_1 = b$ and $b_2 = 3b/2$, and altitude h.

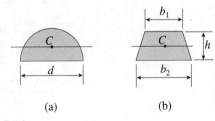

(a)   (b)

Prob. 5.5-14

5.5-15 A simple beam AB of span length $L = 24$ ft is subjected to two wheel loads acting at distance $d = 5$ ft apart (see figure). Each wheel transmits a load $P = 3$ k, and the carriage may occupy any position on the beam. Determine the maximum bending stress σ_{max} due to the wheel loads if the beam is an I-beam having section modulus $S = 16.2$ in.^{3}

Prob. 5.5-15

5.5-16 A stile crossing a pipeline at a chemical plant is supported by two fiberglass frames, as shown in part (a) of the figure. Each frame has a span $L = 5.46$ m and height $h = 0.82$ m [see part (b) of the figure]. The slope of the inclined members of the frame is 2 on 3. The cross section of the frame is I-shaped with width 100 mm, height 150 mm, and thickness 10 mm [see part (c) of the figure]. Determine the maximum bending stress at the midsection of the frame due to a uniform load $q = 2.9$ kN/m acting on the horizontal part of the frame.

(a)

(b)   (c)

Prob. 5.5-16

5.5-17 A cantilever beam AB, loaded by a uniform load and a concentrated load (see figure), is constructed of a channel section. Find the maximum tensile stress σ_t and maximum compressive stress σ_c if the cross section has the dimensions indicated and the moment of inertia about the z axis (the neutral axis) is $I = 2.81$ in.^{4} (*Note:* The uniform load represents the weight of the beam.)

Prob. 5.5-17

5.5-18 A cantilever beam AB of triangular cross section has length $L = 0.8$ m, width $b = 80$ mm, and height $h = 120$ mm (see figure). The beam is made of brass weighing 85 kN/m^3. (a) Determine the maximum tensile stress σ_t and maximum compressive stress σ_c due to the beam's own weight. (b) If the width b is doubled, what happens to the stresses? (c) If the height h is doubled, what happens to the stresses?

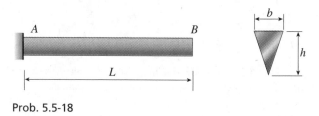

Prob. 5.5-18

5.5-19 A beam ABC with an overhang from B to C supports a uniform load of 200 lb/ft throughout its length (see figure). The beam is a channel section with dimensions as shown in the figure. The moment of inertia about the z axis (the neutral axis) equals 5.14 in.^{4} Calculate the maximum tensile stress σ_t and maximum compressive stress σ_c due to the uniform load.

Prob. 5.5-19

5.5-20 A frame ABC travels horizontally with an acceleration a_0 (see figure). Obtain a formula for the maximum stress σ_{max} in the vertical arm AB, which has length L, thickness t, and mass density ρ.

Prob. 5.5-20

5.5-21 A beam of T-section is supported and loaded as shown in the figure. The cross section has width $b = 2\ 1/2$ in., height $h = 3$ in., and thickness $t = 1/2$ in. Determine the maximum tensile and compressive stresses in the beam.

5.5-22 A cantilever beam AB with a rectangular cross section has a longitudinal hole drilled throughout its length (see figure). The beam supports a load $P = 600$ N. The cross section is 25 mm wide and 50 mm high, and the hole has a diameter of 10 mm. Find the bending stresses at the top of the beam, at the top of the hole, and at the bottom of the beam.

Prob. 5.5-22

* **5.5-23** A small dam of height $h = 6$ ft is constructed of vertical wood beams AB, as shown in the figure. The wood beams, which have thickness $t = 2.5$ in., are simply supported by horizontal steel beams at A and B. Construct a graph showing the maximum bending stress σ_{max} in the wood beams versus the depth d of the water above the lower support at B. Plot the stress σ_{max} (psi) as the ordinate of the graph and the depth d (ft) as the abscissa. (*Note:* The weight density γ of water equals 62.4 lb/ft^3.)

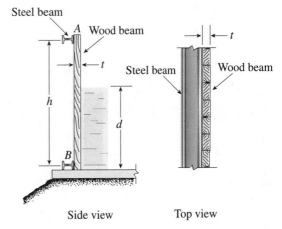

Side view   Top view

Prob. 5.5-23

Prob. 5.5-21

DESIGN OF BEAMS

5.6-1 A fiberglass bracket *ABCD* of solid circular cross section has the shape and dimensions shown in the figure. A vertical load $P = 8$ lb acts at the free end *D*. Determine the minimum permissible diameter d_{min} of the bracket if the allowable bending stress in the material is 3500 psi and $b = 1.5$ in. (Disregard the weight of the bracket itself.)

Prob. 5.6-1

5.6-2 The cross section of a narrow-gage railway bridge is shown in part (a) of the figure. The bridge is constructed with longitudinal steel girders that support the wood ties. The girders are restrained against lateral buckling by diagonal bracing, as indicated by the dashed lines. The spacing of the girders is $s_1 = 0.8$ m and the spacing of the rails is $s_2 = 0.6$ m. The load transmitted by each rail to a single tie is $P = 18$ kN. The cross section of a tie, shown in part (b) of the figure, has width $b = 120$ mm and depth *d*. Determine the minimum value of *d* based upon an allowable bending stress of 10 MPa in the tie. (Disregard the weight of the tie itself.)

(a)

(b)

Prob. 5.6-2

5.6-3 A cantilever beam of length $L = 6$ ft supports a uniform load of intensity $q = 250$ lb/ft and a concentrated load $P = 3000$ lb (see figure). Calculate the required section modulus *S* if $\sigma_{allow} = 16,000$ psi. Then select a suitable wide-flange beam (W shape) from Table E-1, Appendix E, and recalculate *S* taking into account the weight of the beam. Select a new beam size if necessary.

Prob. 5.6-3

5.6-4 A simple beam of length $L = 15$ ft carries a uniform load of intensity $q = 900$ lb/ft and a concentrated load $P = 8000$ lb (see figure). Assuming $\sigma_{allow} = 18,000$ psi, calculate the required section modulus *S*. Then select a 12-inch wide-flange beam (W shape) from Table E-1, Appendix E, and recalculate *S* taking into account the weight of the beam. Select a new 12-inch beam if necessary.

Prob. 5.6-4

5.6-5 A simple beam *AB* is loaded as shown in the figure. Calculate the required section modulus *S* if $\sigma_{allow} = 16,000$ psi, $L = 24$ ft, $P = 2000$ lb, and $q = 300$ lb/ft. Then select a suitable I-beam (S shape) from Table E-2, Appendix E, and recalculate *S* taking into account the weight of the beam. Select a new beam size if necessary.

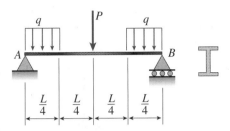

Prob. 5.6-5

5.6-6 A pontoon bridge (see figure) is constructed of two longitudinal wood beams, known as *balks*, that span between adjacent pontoons and support the transverse floor beams, which are called *chesses*. For purposes of design, assume that a uniform floor load of 10 kPa acts over the chesses. (This load includes an allowance for the weights of the chesses and balks.) Also, assume that the chesses are 2.4 m long and that the balks are simply supported with a span of 3.6 m. The allowable bending stress in the wood is 17.5 MPa. If the balks have a square cross section, what is their minimum required width b_{min}?

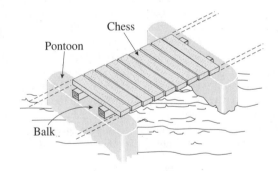

Prob. 5.6-6

5.6-7 A floor system in a small building consists of wood planks supported by 2 in. (nominal width) joists spaced at distance s, measured from center to center (see figure). The span length L of each joist is 10.5 ft, the spacing s of the joists is 16 in., and the allowable bending stress in the wood is 1200 psi. The uniform floor load is 100 lb/ft², which includes an allowance for the weight of the floor system itself. Calculate the required section modulus S for the joists, and then select a suitable joist size (surfaced lumber) from Appendix F, assuming that each joist may be represented as a simple beam carrying a uniform load.

5.6-8 The wood joists supporting a plank floor (see figure) are 40 mm × 180 mm in cross section (actual dimensions) and have a span length $L = 4.0$ m. The floor load is 2.4 kPa, which includes the weight of the joists and the floor. Calculate the maximum permissible spacing s of the joists if the allowable bending stress is 8.5 MPa. (Assume that each joist may be represented as a simple beam carrying a uniform load.)

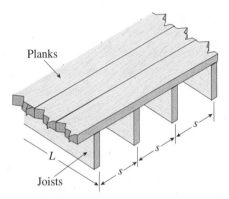

Probs. 5.6-7 and 5.6-8

5.6-9 A beam ABC with an overhang from B to C is constructed of a C 10 × 30 channel section (see figure). The beam supports its own weight (30 lb/ft) plus a uniform load of intensity q acting on the overhang. The allowable stresses in tension and compression are 20 ksi and 12 ksi, respectively. Determine the allowable uniform load q_{allow} if the distance L equals 4.0 ft.

Prob. 5.6-9

5.6-10 A so-called "trapeze bar" in a hospital room provides a means for patients to exercise while in bed (see figure). The bar is 2.1 m long and has a cross section in the shape of a regular octagon. The design load is 1100 N applied at the midpoint of the bar, and the allowable bending stress is 160 MPa. Determine the minimum width s of the bar. (Assume that the ends of the bar are simply supported and that the weight of the bar is negligible.)

Prob. 5.6-10

5.6-11 A two-axle carriage that is part of an overhead traveling crane in a testing laboratory moves slowly across a simple beam AB (see figure). The load transmitted to the beam from the front axle is 2.4 k and from the rear axle is 4.8 k. The weight of the beam itself may be disregarded. (a) Determine the minimum required section modulus S for the beam if the allowable bending stress is 18 ksi, the length of the beam is 16 ft, and the wheelbase of the carriage is 5 ft. (b) Select a suitable I-beam (S shape) from Table E-2, Appendix E.

Prob. 5.6-11

5.6-12 A cantilever beam AB of circular cross section and length $L = 500$ mm supports a load $P = 200$ N acting at the free end (see figure). The beam is made of steel with an allowable bending stress of 80 MPa. Determine the required diameter d_{min} of the beam, considering the effect of the beam's own weight.

Prob. 5.6-12

5.6-13 A compound beam $ABCD$ (see figure) is supported at points A, B, and D and has a splice (represented by the pin connection) at point C. The distance $a = 8$ ft and the beam is a W 16×57 wide-flange shape with an allowable bending stress of 16,000 psi. Find the allowable uniform load q that may be placed on top of the beam, taking into account the weight of the beam itself.

Prob. 5.6-13

5.6-14 A small balcony constructed of wood is supported by three identical cantilever beams (see figure). Each beam has length $L_1 = 2.1$ m, width b, and height $h = 4b/3$. The dimensions of the balcony floor are $L_1 \times L_2$, with $L_2 = 2.5$ m. The design load is 5.5 kPa acting over the entire floor area. (This load accounts for all loads except the weights of the cantilever beams, which have a weight density $\gamma = 5.5$ kN/m³.) The allowable bending stress in the cantilevers is 15 MPa. Assuming that the middle cantilever supports 50% of the load and each outer cantilever supports 25% of the load, determine the required dimensions b and h.

Prob. 5.6-14

5.6-15 A beam having a cross section in the form of an unsymmetric wide-flange shape (see figure) is subjected to a negative bending moment acting about the z axis. Determine the width b of the top flange in order that the stresses at the top and bottom of the beam will be in the ratio 4:3, respectively.

Prob. 5.6-15

5.6-16 A beam having a cross section in the form of a channel (see figure) is subjected to a bending moment acting about the z axis. Calculate the thickness t of the channel in order that the bending stresses at the top and bottom of the beam will be in the ratio 7:3, respectively.

Prob. 5.6-16

5.6-17 Determine the ratios of the weights of three beams that have the same length, are made of the same material, are subjected to the same maximum bending moment, and have the same maximum bending stress if their cross sections are: (1) a rectangle with height equal to twice the width, (2) a square, and (3) a circle (see figures).

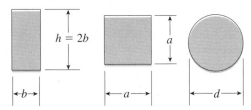

Prob. 5.6-17

5.6-18 The cross section of a rectangular beam having width b and height h is shown in part (a) of the figure. For reasons unknown to the beam designer, it is planned to add structural projections of width b/9 and height d to the top and bottom of the beam [see part (b) of the figure]. For what values of d is the bending-moment capacity of the beam increased? For what values is it decreased?

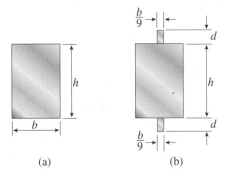

(a)   (b)

Prob. 5.6-18

5.6-19 A horizontal shelf AD of length L = 30 in., width b = 12 in., and thickness t = 0.5 in. is supported by brackets at B and C [see part (a) of the figure]. The brackets are adjustable and may be placed in any desired positions between the ends of the shelf. A uniform load of intensity q, which includes the weight of the shelf itself, acts on the shelf [see part (b) of the figure]. Determine the maximum permissible value of the load q if the allowable bending stress in the shelf is $\sigma_{\text{allow}} = 830$ psi.

(a)

(b)

Prob. 5.6-19

5.6-20 A steel beam *ABC* is simply supported at *A* and *B* and has an overhang *BC* of length $L = 150$ mm (see figure). The beam supports a uniform load of intensity $q = 3.5$ kN/m over its entire length of 450 mm. The cross section of the beam is rectangular with width *b* and height 2*b*. The allowable bending stress in the steel is $\sigma_{allow} = 60$ MPa and its weight density is $\gamma = 77.0$ kN/m³. (a) Disregarding the weight of the beam, calculate the required width *b* of the rectangular cross section. (b) Taking into account the weight of the beam, calculate the required width *b*.

Prob. 5.6-20

5.6-21 A steel plate (called a *cover plate*) having cross-sectional dimensions 4.0 in. × 0.5 in. is welded along the full length of the top flange of a W 12 × 35 wide-flange beam (see figure, which shows the beam cross section). What is the percent increase in section modulus (as compared to the wide-flange beam alone)?

4.0 × 0.5 in. cover plate

W 12 × 35

Prob. 5.6-21

5.6-22 An extruded aluminum beam has the doubly symmetric cross-sectional shape shown in part (a) of the figure. For purposes of a bending analysis, the fillets and rounded corners may be disregarded, as shown in part (b). The cross section has the following dimensions: $b = 220$ mm, $h = 200$ mm, $h_1 = 80$ mm, $t_1 = t_2 = 10$ mm, and $t_3 = 20$ mm. (a) Is the beam stronger when bent about axis 1-1 as the neutral axis or when bent about axis 2-2 as the neutral axis? (b) What is the ratio β of the larger section modulus to the smaller section modulus?

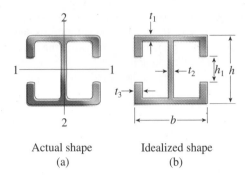

Actual shape (a) Idealized shape (b)

Prob. 5.6-22

5.6-23 Water pressure acts against an inclined panel *ABC* that serves as a barrier (see figure). The panel is pivoted at point *B*, which is height *h* above the base, and presses against the base at *A* when the water level is not too high (note that the panel will rotate about the pin at *B* if the depth *d* of the water exceeds a certain maximum depth d_{max}). The panel has thickness *t* and is inclined at an angle α to the horizontal. Derive a formula for the minimum required thickness t_{min} of the panel if the maximum bending stress in the panel is not to exceed an allowable stress σ_{allow}. (Note that the maximum stress in the panel occurs when the depth of the water reaches the maximum possible depth d_{max}. When deriving the formula, consider only the effects of bending in the panel, disregard the weight of the panel itself, and let γ be the weight density of water.)

Prob. 5.6-23

5.6-24 A beam of square cross section ($a = $ length of each side) is bent in the plane of a diagonal (see figure). By removing a small amount of material at the top and bottom corners, as shown by the shaded triangles in the figure, we can increase the section modulus and obtain a stronger beam, even though the area of the cross section is reduced. (a) Determine the ratio β defining the areas that should be removed in order to obtain the strongest cross section in bending. (b) By what percent is the section modulus increased when the areas are removed?

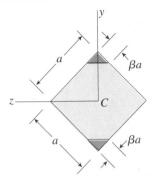

Prob. 5.6-24

5.6-25 A retaining wall 5 ft high is constructed of horizontal wood planks 3 in. thick (actual dimension) that are supported by vertical wood piles of 12 in. diameter (actual dimension), as shown in the figure. The lateral earth pressure is $p_1 = 100$ lb/ft² at the top of the wall and $p_2 = 400$ lb/ft² at the bottom. Assuming that the allowable stress in the wood is 1200 psi, calculate the maximum permissible spacing s of the piles. (*Hint:* Observe that the spacing of the piles may be governed by the load-carrying capacity of either the planks or the piles. Consider the piles to act as cantilever beams subjected to a trapezoidal distribution of load, and consider the planks to act as simple beams between the piles. To be on the safe side, assume that the pressure on the bottom plank is uniform and equal to the maximum pressure.)

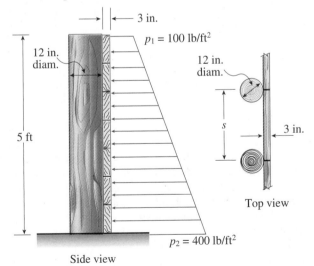

Prob. 5.6-25

NONPRISMATIC BEAMS

5.7-1 A beam *ABC* in a piece of shop equipment is tied down at *C* and rests in a cradle at *B* (see figure). The cross sections are circular with constant diameter d_B between *B* and *C* and diameter d_A at the free end *A*. Between points *A* and *B*, the diameter varies linearly with the distance *x*. A concentrated load *P* acts at end *A*. At what distance *x* from the free end does the maximum bending stress due to the load *P* occur if $d_B/d_A = 2.25$? What is the magnitude σ_{max} of the maximum bending stress? What is the ratio of the maximum stress to the largest stress σ_B at point *B*?

Prob. 5.7-1

5.7-2 A tapered cantilever beam *AB* of length *L* has square cross sections and supports a concentrated load *P* at the free end (see figure). The width and height of the beam vary linearly from h_A at the free end to h_B at the fixed end. Determine the distance *x* from the free end to the cross section of maximum bending stress if $h_B = 2.5h_A$. What is the magnitude σ_{max} of the maximum bending stress? What is the ratio of the maximum stress to the largest stress σ_B at the support?

Prob. 5.7-2

5.7-3 A tall signboard is supported by two vertical beams consisting of thin-walled, tapered circular tubes (see figure). For purposes of analysis, each beam may be represented as a cantilever *AB* of length *L* = 25 ft subjected to a lateral load *P* = 500 lb at the free end. The tubes have constant thickness *t* = 0.375 in. and average diameters d_A = 3.5 in. and d_B = 10.5 in. at ends *A* and *B*, respectively. Because the thickness is small compared to the diameters, the moment of inertia at any cross section may be obtained from the formula $I = \pi d^3 t/8$ (see *Case* 22, Appendix D), and therefore the section modulus may be obtained from the formula $S = \pi d^2 t/4$. At what distance x from the free end does the maximum bending stress occur? What is the magnitude σ_{max} of the maximum bending stress? What is the ratio of the maximum stress to the largest stress σ_B at the support?

Prob. 5.7-3

5.7-4 A tapered cantilever beam *AB* having rectangular cross sections is subjected to a concentrated load $P = 400$ N and a couple $M_0 = 160$ N·m acting at the free end (see figure). The width b of the beam is constant and equal to 25 mm, but the height varies linearly from $h_A = 50$ mm at the loaded end to $h_B = 75$ mm at the support. At what distance x from the free end does the maximum bending stress σ_{max} occur? What is the magnitude σ_{max} of the maximum bending stress? What is the ratio of the maximum stress to the largest stress σ_B at the support?

Prob. 5.7-4

* 5.7-5 The spokes in a large flywheel are modeled as beams fixed at one end and loaded by a force P and a couple M_0 at the other (see figure). The cross sections of the spokes are elliptical with major and minor axes (height and width, respectively) having the lengths shown in the figure. The cross-sectional dimensions vary linearly from end *A* to end *B*. Considering only the effects of bending due to the loads P and M_0, determine the following quantities: (a) the largest bending stress σ_A at end *A*; (b) the largest bending stress σ_B at end *B*; (c) the distance x to the cross section of maximum bending stress; and (d) the magnitude σ_{max} of the maximum bending stress.

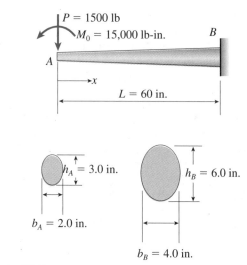

Prob. 5.7-5

** 5.7-6 Refer to the tapered cantilever beam of circular cross section shown in Fig. 5-24 of Section 5.7. (a) Considering only the effects of bending due to the load P, determine the ratio d_B/d_A of the end diameters such that the volume (and hence the weight) of the beam is a minimum. (Assume that the allowable stress in bending is σ_{allow}.) (b) Obtain a formula for the minimum volume V_{min}. (c) Determine the ratio of the minimum volume V_{min} to the volume of a prismatic beam ($d_B = d_A$).

FULLY STRESSED BEAMS

Problems 5.7-7 to 5.7-10 pertain to fully stressed beams of rectangular cross section. Consider only the bending stresses obtained from the flexure formula and disregard the weights of the beams.

5.7-7 A cantilever beam *AB* having rectangular cross sections with constant width *b* and varying height h_x is subjected to a uniform load of intensity *q* (see figure). How should the height h_x vary as a function of *x* (measured from the free end of the beam) in order to have a fully stressed beam? (Express h_x in terms of the height h_B at the fixed end of the beam.)

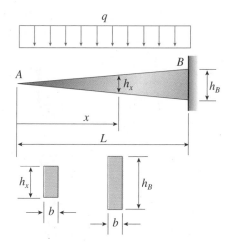

Prob. 5.7-7

5.7-8 A cantilever beam *AB* having rectangular cross sections with varying width b_x and varying height h_x is subjected to a uniform load of intensity *q* (see figure). If the width varies linearly with *x* according to the equation $b_x = b_B x/L$, how should the height h_x vary as a function of *x* in order to have a fully stressed beam? (Express h_x in terms of the height h_B at the fixed end of the beam.)

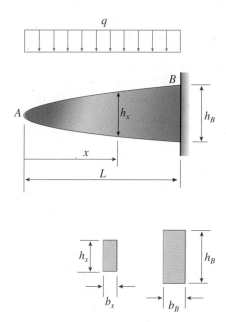

Prob. 5.7-8

5.7-9 A cantilever beam *AB* having rectangular cross sections with constant height *h* and varying width b_x is subjected to a concentrated load *P* at the free end (see figure). How should the width b_x vary as a function of *x* in order to have a fully stressed beam? (Express b_x in terms of the width b_B at the fixed end of the beam.)

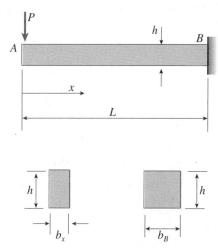

Prob. 5.7-9

5.7-10 A simple beam AB supporting a concentrated load P (see figure) has rectangular cross sections with constant width b and varying height h_x. The concentrated load may act anywhere along the span of the beam. How should the height h_x vary as a function of x (measured from the midpoint of the beam) in order to minimize the weight of the beam? (Express h_x in terms of the height h_1 at the midpoint of the beam.)

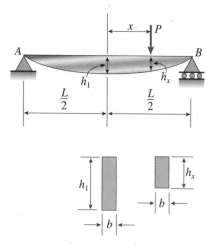

Prob. 5.7-10

SHEAR STRESSES IN RECTANGULAR BEAMS

5.8-1 The shear stresses τ in a rectangular beam are given by Eq. (5-39):

$$\tau = \frac{V}{2I}\left(\frac{h^2}{4} - y_1^2\right)$$

in which V is the shear force, I is the moment of inertia of the cross-sectional area, h is the height of the beam, and y_1 is the distance from the neutral axis to the point where the shear stress is being determined (see Fig. 5-30). By integrating over the cross-sectional area, show that the resultant of the shear stresses is equal to the shear force V.

5.8-2 Calculate the maximum shear stress τ_{max} and the maximum bending stress σ_{max} in a simply supported wood beam (see figure) carrying a uniform load of 25 kN/m (which includes the weight of the beam) if the length is 1.75 m and the cross section is rectangular with width 200 mm and height 250 mm.

Prob. 5.8-2

5.8-3 Two wood beams, each of square cross section (3.5 in. × 3.5 in., actual dimensions) are glued together to form a solid beam of dimensions 3.5 in. × 7.0 in. (see figure). The beam is simply supported with a span of 6 ft. What is the maximum load P_{max} that may act at the midpoint if the allowable shear stress in the glued joint is 200 psi? (Include the effects of the beam's own weight, assuming that the wood weighs 0.02 lb/in.³).

Prob. 5.8-3

5.8-4 A cantilever beam of length $L = 2$ m supports a load $P = 10$ kN (see figure). The beam is made of wood with cross-sectional dimensions 150 mm × 200 mm. Calculate the shear stresses due to the load P at points located 25 mm, 50 mm, 75 mm, and 100 mm from the top surface of the beam. From these results, plot a graph showing the distribution of shear stresses from top to bottom of the beam.

Prob. 5.8-4

5.8-5 A simple beam of length $L = 16$ in. and cross-sectional dimensions $b = 0.5$ in. and $h = 2$ in. (see figure) supports a uniform load of intensity $q = 200$ lb/in., which includes the weight of the beam. Calculate the shear stresses in the beam (at the cross section of maximum shear force) at points located 1/4 in., 1/2 in., 3/4 in., and 1 in. from the top surface of the beam. From these calculations, plot a graph showing the distribution of shear stresses from top to bottom of the beam.

q = 200 lb/in.

$L = 16$ in.

$h = 2$ in.

$b = 0.5$ in.

Prob. 5.8-5

5.8-6 A beam of rectangular cross section (width b and height h) supports a uniformly distributed load along its entire length L. The allowable stresses in bending and shear are σ_{allow} and τ_{allow}, respectively. (a) If the beam is simply supported, what is the span length L_0 below which the shear stress governs the allowable load and above which the bending stress governs? (b) If the beam is supported as a cantilever, what is the length L_0 below which the shear stress governs the allowable load and above which the bending stress governs?

5.8-7 A laminated wood beam on simple supports is built up by gluing together three 2 in. × 4 in. boards (actual dimensions) to form a solid beam 4 in. × 6 in. in cross section, as shown in the figure. The allowable shear stress in the glued joints is 50 psi and the allowable bending stress in the wood is 1600 psi. If the beam is 6 ft long, what is the allowable load P acting at the midpoint of the beam? (Disregard the weight of the beam.)

3 ft   P

$L = 6$ ft

2 in.
2 in.
2 in.

4 in.

Prob. 5.8-7

5.8-8 In Example 5-8 of Section 5.6, we analyzed the vertical wood posts B supporting a small dam (see Fig. 5-22). Only bending stresses were considered in that example. Therefore, let us now consider the shear stresses. If the allowable shear stress in the posts is $\tau_{allow} = 0.9$ MPa, what is the required dimension b of the posts?

5.8-9 A wood beam AB on simple supports with span length equal to 9 ft is subjected to a uniform load of intensity 120 lb/ft acting along the entire length of the beam and a concentrated load of magnitude 8800 lb acting at a point 3 ft from the right-hand support (see figure). The allowable stresses in bending and shear, respectively, are 2500 psi and 150 psi. (a) From the table in Appendix F, select the lightest beam that will support the loads (disregard the weight of the beam). (b) Taking into account the weight of the beam (weight density = 35 lb/ft³), verify that the selected beam is satisfactory, or, if it is not, select a new beam.

8800 lb   3 ft
120 lb/ft

A B

9 ft

Prob. 5.8-9

5.8-10 A laminated plastic beam of square cross section is built up by gluing together three strips, each 10 mm × 30 mm in cross section (see figure). The beam has a total weight of 3.2 N and is simply supported with span length $L = 320$ mm. Considering the weight of the beam, calculate the maximum permissible load P that may be placed at the midpoint if: (a) the allowable shear stress in the glued joints is 0.3 MPa, and (b) the allowable bending stress in the plastic is 8 MPa.

P
q

10 mm
10 mm 30 mm
10 mm

30 mm

L

Prob. 5.8-10

5.8-11 A simply supported wood beam of rectangular cross section and span length 4 ft carries a concentrated load P at midspan in addition to its own weight (see figure). The cross section has nominal dimensions 6 in. $\times$ 10 in. (see Appendix F for actual dimensions and weight of the beam). Calculate the maximum permissible value of the load P if: (a) the allowable bending stress is 1250 psi, and (b) the allowable shear stress is 140 psi.

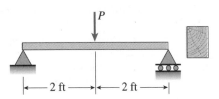

Prob. 5.8-11

***5.8-12** A wood beam ABC with simple supports at A and B and an overhang BC has height $h = 280$ mm (see figure). The length of the main span of the beam is $L = 3.6$ m and the length of the overhang is $L/3 = 1.2$m. The beam supports a concentrated load $3P = 15$ kN at the midpoint of the main span and a load $P = 5$ kN at the free end of the overhang. The wood has weight density $\gamma = 5.5$ kN/m^3. (a) Determine the required width b based upon an allowable bending stress of 8.2 MPa. (b) Determine the required width b based upon an allowable shear stress of 0.7 MPa.

Prob. 5.8-12

****5.8-13** A square wood platform, 8 ft $\times$ 8 ft in area, rests on 6-in. wide masonry walls (see figure). The deck of the platform is constructed of 2 in. nominal thickness tongue-and-groove planks (actual thickness 1.5 in.; see Appendix F) supported on two 8-ft long beams. The beams have 4 in. $\times$ 6 in. nominal dimensions (actual dimensions 3.5 in. $\times$ 5.5 in.) and are supported at their ends by the walls. The structure is designed to support a uniformly distributed load w (lb/ft^2) acting over the entire top surface of the platform. The allowable stress in bending for both planks and beams is 2400 psi and the allowable shear stress

is 100 psi. When analyzing the planks, it is assumed that their reactions are uniformly distributed over the top surfaces of the supporting beams, and when analyzing the beams it is assumed that their reactions are uniformly distributed over the area of contact with the supporting walls. The weights of the planks and beams are to be taken into account in the analysis, and the wood is assumed to have a weight density $\gamma = 30$ lb/ft^3. (a) Determine the allowable loads w_1, w_2, w_3, and w_4 (lb/ft^2) based upon the following parameters, respectively: (1) bending stress in the planks, (2) shear stress in the planks, (3) bending stress in the beams, and (4) shear stress in the beams. (b) Which of the preceding values becomes the allowable load w_{allow} on the platform? (*Hints:* Use care in constructing the loading diagrams, noting especially that the reactions are distributed loads instead of concentrated loads. Also, note that the maximum shear forces occur at the inside faces of the supports.)

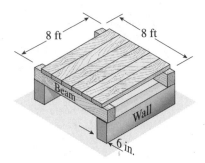

Prob. 5.8-13

SHEAR STRESSES IN CIRCULAR BEAMS

5.9-1 A wood pole of solid circular cross section ($d =$ diameter) is subjected to a horizontal force $P = 500$ lb (see figure). The length of the pole is $L = 6$ ft, and the allowable stresses in the wood are 2100 psi in bending and 150 psi in shear. Determine the minimum required diameter of the pole based upon: (a) the allowable bending stress, and (b) the allowable shear stress.

Prob. 5.9-1

5.9-2 A simple log bridge in a remote area consists of two parallel logs with planks across them (see figure). The logs are Douglas fir with average diameter 300 mm. A truck moves slowly across the bridge, which spans 2.5 m. Assume that the weight of the truck is equally distributed between the two logs. Because the wheelbase of the truck is greater than 2.5 m, only one set of wheels is on the bridge at a time. Thus, the wheel load on one log is equivalent to a concentrated load W acting at any position along the span. In addition, the weight of one log and the planks it supports is equivalent to a uniform load of 850 N/m acting on the log. Determine the maximum permissible wheel load W based upon: (a) an allowable bending stress of 7.5 MPa, and (b) an allowable shear stress of 0.8 MPa.

Prob. 5.9-2

Probs. 5.9-3 and 5.9-4

*5.9-4 Solve the preceding problem for a sign and poles having the following dimensions: $h_1 = 6.0$ m, $h_2 = 1.5$ m, $b = 3.0$ m, and $t = d/10$. The design wind pressure is 3.6 kPa, and the allowable stresses in the aluminum are 50 MPa in bending and 14 MPa in shear.

SHEAR STRESSES IN BEAMS WITH FLANGES

5.10-1 through 5.10-6 A wide-flange beam (see figure) having the cross section described below is subjected to a shear force V. Using the dimensions of the cross section, calculate the moment of inertia and then determine the following quantities: (a) The maximum shear stress τ_{max} in the web. (b) The minimum shear stress τ_{min} in the web. (c) The average shear stress τ_{aver} (obtained by dividing the shear force by the area of the web) and the ratio τ_{max}/τ_{aver}. (d) The shear force V_{web} carried in the web and the ratio V_{web}/V. (*Note:* Disregard the fillets at the junctions of the web and flanges and consider the cross section to consist of three rectangles when calculating the properties of the cross section.)

*5.9-3 A sign for a service station is supported by two aluminum poles of hollow circular cross section, as shown in the figure. The poles are being designed to resist a wind pressure of 75 lb/ft² against the full area of the sign. The dimensions of the poles and sign are $h_1 = 20$ ft, $h_2 = 5$ ft, and $b = 10$ ft. To prevent buckling of the walls of the poles, the thickness t is specified as one-tenth the outside diameter d. (a) Determine the minimum required diameter of the poles based upon an allowable bending stress of 7500 psi in the aluminum. (b) Determine the minimum required diameter based upon an allowable shear stress of 2000 psi.

Probs. 5.10-1 through 5.10-6

5.10-1 Dimensions of cross section: $b = 8$ in., $t = 0.4$ in., $h = 12$ in., $h_1 = 10.7$ in., and $V = 20,000$ lb.

5.10-2 Dimensions of cross section: $b = 160$ mm, $t = 8$ mm, $h = 300$ mm, $h_1 = 274$ mm, and $V = 50$ kN.

5.10-3 Wide-flange shape, W 8×28 (see Table E-1, Appendix E); $V = 10$ k.

5.10-4 Dimensions of cross section: $b = 220$ mm, $t = 12$ mm, $h = 600$ mm, $h_1 = 570$ mm, and $V = 200$ kN.

5.10-5 Wide-flange shape, W 24×94 (see Table E-1, Appendix E); $V = 50$ k.

5.10-6 Dimensions of cross section: $b = 120$ mm, $t = 7$ mm, $h = 350$ mm, $h_1 = 330$ mm, and $V = 60$ kN.

5.10-7 A cantilever beam AB of length $L = 6$ ft supports a uniform load of intensity q that includes the weight of the beam (see figure). The beam is a steel W 10×30 wide-flange shape (see Table E-1, Appendix E). Calculate the maximum permissible load q based upon: (a) an allowable bending stress $\sigma_{allow} = 18$ ksi, and (b) an allowable shear stress $\tau_{allow} = 9$ ksi. (*Note:* Obtain the moment of inertia and section modulus of the beam from Table E-1.)

Prob. 5.10-7

5.10-8 A bridge girder AB on a simple span of length $L = 12$ m supports a uniform load of intensity q that includes the weight of the girder (see figure). The girder is constructed of three plates welded to form the cross section shown. Determine the maximum permissible load q based upon: (a) an allowable bending stress $\sigma_{allow} = 90$ MPa, and (b) an allowable shear stress $\tau_{allow} = 50$ MPa.

Prob. 5.10-8

5.10-9 A simple beam with an overhang supports a uniform load of intensity $q = 1200$ lb/ft and a concentrated load $P = 3000$ lb (see figure). The allowable stresses in bending and shear are 18 ksi and 11 ksi, respectively. Select from Table E-2, Appendix E, the lightest I-beam (S shape) that will support the given loads. (*Note:* The uniform load q includes an allowance for the weight of the beam. *Hint:* Select a beam based upon the bending stress and then calculate the maximum shear stress. If the beam is overstressed in shear, select a heavier beam and repeat.)

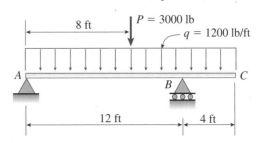

Prob. 5.10-9

5.10-10 A hollow steel box beam has the rectangular cross section shown in the figure. Determine the maximum allowable shear force V that may act on the beam if the allowable shear stress is 40 MPa.

Prob. 5.10-10

5.10-11 A steel wide-flange beam (W 16 × 77 shape) is reinforced by two cover plates, each 6 in. wide and 1/2 in. thick (see figure). Determine the maximum allowable shear force V that may act on the beam if the allowable shear stress is 6000 psi.

Prob. 5.10-11

5.10-12 The T-beam shown in the figure has cross-sectional dimensions as follows: $b = 220$ mm, $t = 15$ mm, $h = 300$ mm, and $h_1 = 275$ mm. The beam is subjected to a shear force $V = 70$ kN. Determine the maximum shear stress τ_{max} in the web of the beam.

Probs. 5.10-12 and 5.10-13

5.10-13 Calculate the maximum shear stress τ_{max} in the web of the T-beam shown in the figure if $b = 10$ in., $t = 0.6$ in., $h = 8$ in., $h_1 = 7$ in., and the shear force $V = 6000$ lb.

*** 5.10-14** An unsymmetric I-beam (see figure) has cross-sectional dimensions as follows: $b_1 = 150$ mm, $b_2 = 250$ mm, $t = 10$ mm, $h = 320$ mm, and $t_1 = 20$ mm. The beam is subjected to a shear force $V = 70$ kN. (a) Calculate the maximum and minimum shear stresses τ_{max} and τ_{min} in the web of the beam. (b) Calculate the average shear stress τ_{aver} in the web by dividing V by the area of the web, and then obtain the ratio τ_{max}/τ_{aver}. (c) Calculate the shear force V_{web} carried in the web and obtain the ratio V_{web}/V.

Prob. 5.10-14

BUILT-UP BEAMS

5.11-1 A prefabricated wood I-beam serving as a floor joist has the cross section shown in the figure. The allowable load in shear for the glued joints between the web and the flanges is 70 lb/in. in the longitudinal direction. Determine the maximum allowable shear force V_{max} for the beam.

Prob. 5.11-1

5.11-2 A welded steel girder having the cross section shown in the figure is fabricated of two 250 mm × 25 mm flange plates and a 600 mm × 15 mm web plate. The plates are joined by four fillet welds that run continuously for the length of the girder. Each weld has an allowable load in shear of 500 kN/m. Calculate the maximum allowable shear force V_{max} for the girder.

Prob. 5.11-2

5.11-3 A welded steel girder having the cross section shown in the figure is fabricated of two 16 in. × 1 in. flange plates and a 60 in. × 3/8 in. web plate. The plates are joined by four longitudinal fillet welds that run continuously throughout the length of the girder. If the girder is subjected to a shear force of 250 kips, what force F (per inch of length of weld) must be resisted by each weld?

Prob. 5.11-3

5.11-4 A box beam of wood is constructed of two 250 mm × 40 mm boards and two 250 mm×25 mm boards (see figure). The boards are nailed at a longitudinal spacing $s = 100$ mm. If each nail has an allowable shear force $F = 750$ N, what is the maximum allowable shear force V_{max}?

Prob. 5.11-4

5.11-5 A box beam constructed of four wood boards of size 6 in. × 1 in. (actual dimensions) is shown in the figure. The boards are joined by screws for which the allowable load in shear is $F = 210$ lb per screw. Calculate the maximum permissible longitudinal spacing s_{max} of the screws if the shear force V is 1150 lb.

Prob. 5.11-5

5.11-6 Two wood box beams (beams A and B) have the same outside dimensions (200 mm × 360 mm) and the same thickness ($t = 20$ mm) throughout, as shown in the figure. Both beams are formed by nailing, with each nail having an allowable shear load of 250 N. The beams are designed for a shear force $V = 3.2$ kN. (a) What is the maximum longitudinal spacing s_A for the nails in beam A? (b) What is the maximum longitudinal spacing s_B for the nails in beam B? (c) Which beam is more efficient in resisting the shear force?

Prob. 5.11-6

5.11-7 The T-beam shown in the figure is fabricated by welding together two steel plates. If the allowable load for each weld is 2.4 k/in. in the longitudinal direction, what is the maximum allowable shear force V?

Prob. 5.11-7

5.11-8 A beam of T cross section is formed by nailing together two boards having the dimensions shown in the figure. If the total shear force V acting on the cross section is 872 N and each nail may carry 400 N in shear, what is the maximum allowable nail spacing s?

Prob. 5.11-8

5.11-9 A steel beam is built up from a W 16 × 77 wide-flange beam and two 10 in. × 1/2 in. cover plates (see figure). The allowable load in shear on each bolt is 2.8 kips. What is the required bolt spacing s in the longitudinal direction if the shear force $V = 40$ kips?

Prob. 5.11-9

5.11-10 A built-up beam is fabricated from a W 12 × 50 wide-flange beam and two C 12 × 30 channels (see figure). The beam and channels are joined by bolts spaced at 7.5 in. longitudinally. What is the maximum allowable shear force V if each bolt may carry 2.4 kips in shear?

Prob. 5.11-10

BEAMS WITH AXIAL LOADS

When solving the problems for Section 5.12, assume that the bending moments are not affected by the presence of lateral deflections.

5.12-1 While drilling a hole with a brace and bit, you exert a downward force $P = 25$ lb on the handle of the brace (see figure). The diameter of the crank arm is $d = 7/16$ in. and its lateral offset is $b = 4\text{-}7/8$ in. Determine the maximum tensile and compressive stresses σ_t and σ_c, respectively, in the crank.

Prob. 5.12-1

Prob. 5.12-2

5.12-3 A rigid frame ABC is formed by welding two steel pipes at B (see figure). Each pipe has cross-sectional area $A = 16.1$ in.^{2}, moment of inertia $I = 212$ in.^{4}, and outside diameter $d = 10.75$ in. Find the maximum tensile and compressive stresses σ_t and σ_c, respectively, in the frame due to the load $P = 3200$ lb if $L = 6.0$ ft and $H = 4.5$ ft.

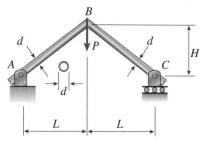

Prob. 5.12-3

5.12-2 An aluminum pole for a street light weighs 2300 N and supports an arm that weighs 330 N (see figure). The center of gravity of the arm is 1.2 m from the axis of the pole. The outside diameter of the pole (at its base) is 225 mm and its thickness is 18 mm. Determine the maximum tensile and compressive stresses σ_t and σ_c, respectively, in the pole (at its base) due to the weights.

5.12-4 A curved bar ABC having a circular axis (radius $r = 300$ mm) is loaded by forces $P = 1.6$ kN (see figure). The cross section of the bar is rectangular with height h and thickness t. If the allowable tensile stress in the bar is 80 MPa and the height $h = 30$ mm, what is the minimum required thickness t_{min}?

Prob. 5.12-4

Prob. 5.12-6

5.12-5 A palm tree inclined at an angle of approximately 60° weighs about 1000 lb (see figure). The weight of the tree may be resolved into two resultant forces, a force $P_1 = 900$ lb acting at a point 12 ft from the base and a force $P_2 = 100$ lb acting at the top of the tree, which is 30 ft long. The diameter at the base of the tree is approximately 14 in. Calculate the maximum tensile and compressive stresses σ_t and σ_c, respectively, at the base of the tree due to its weight.

5.12-7 Because of foundation settlement, a circular tower is leaning at an angle α to the vertical (see figure). The structural core of the tower is a circular cylinder of height h, outer diameter d_2, and inner diameter d_1. For simplicity in the analysis, assume that the weight of the tower is uniformly distributed along the height. Obtain a formula for the maximum permissible angle α if there is to be no tensile stress in the tower.

Prob. 5.12-5

Prob. 5.12-7

5.12-6 A vertical pole of aluminum is fixed at the base and pulled at the top by a cable having a tensile force T (see figure). The cable is attached at the outer surface of the pole and makes an angle $\alpha = 28°$ at the point of attachment. The pole has length $L = 2.0$ m and a hollow circular cross section with outer diameter $d_2 = 250$ mm and inner diameter $d_1 = 200$ mm. Determine the allowable tensile force T_{allow} in the cable if the allowable compressive stress in the aluminum pole is 80 MPa.

5.12-8 A steel bar of solid circular cross section is subjected to an axial tensile force $T = 26$ kN and a bending moment $M = 2.7$ kN·m (see figure). Based upon an allowable stress in tension of 125 MPa, determine the required diameter d of the bar. (Disregard the weight of the bar itself.)

Prob. 5.12-8

Prob. 5.12-10

5.12-9 A cylindrical brick chimney of height H weighs $w = 825$ lb/ft of height (see figure). The inner and outer diameters are $d_1 = 3$ ft and $d_2 = 4$ ft, respectively. The wind pressure against the side of the chimney is $p = 10$ lb/ft² of projected area. Determine the maximum height H if there is to be no tension in the brickwork.

Prob. 5.12-9

5.12-10 A plain concrete wall (i.e., a wall with no steel reinforcement) rests on a secure foundation and serves as a small dam (see figure). The height of the wall is $h = 2$ m and the thickness of the wall is $t = 0.3$ m. (a) Determine the maximum tensile and compressive stresses σ_t and σ_c, respectively, at the base of the wall when the water level reaches the top ($d = h$). Assume plain concrete has weight density $\gamma_c = 23$ kN/m³. (b) Determine the maximum permissible depth d_{max} of the water if there is to be no tension in the concrete.

5.12-11 A flying buttress transmits a load $P = 5400$ lb, acting at an angle of $60°$ to the horizontal, to the top of a vertical buttress AB (see figure). The vertical buttress has height $h = 16$ ft and rectangular cross section of thickness $t = 5$ ft and width $b = 3$ ft (perpendicular to the plane of the figure). The stone used in the construction weighs $\gamma = 165$ lb/ft.³ What is the required weight W of the pedestal and statue above the vertical buttress (that is, above section A) to avoid any tensile stresses in the vertical buttress?

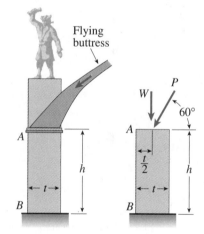

Prob. 5.12-11

* **5.12-12** A solid circular bar AB of diameter $d = 125$ mm is hinged at support B and rests against a smooth vertical surface at end A (see figure). The bar has length $L = 2$ m and is made of steel (weight density $\gamma = 77.0$ kN/m³). If the angle of inclination is $\alpha = 75°$, what is the maximum compressive stress σ_{max} in the bar? (*Hint:* Determine the

compressive stress at a cross section located at distance x, measured along the axis of the bar, from support B. Then take the derivative with respect to x and locate the cross section having the maximum stress.)

Prob. 5.12-12

ECCENTRIC AXIAL LOADS

5.12-13 Two cables, each carrying a tensile force $P = 1200$ lb, are bolted to a block of steel (see figure). The block has thickness $t = 1$ in. and width $b = 3$ in. (a) If the diameter d of the cable is 0.25 in., what are the maximum tensile and compressive stresses σ_t and σ_c, respectively, in the block? (b) If the diameter of the cable is increased (without changing the force P), what happens to the maximum tensile and compressive stresses?

Prob. 5.12-13

5.12-14 A circular post and a rectangular post are each compressed by loads that produce a resultant force P acting at the edge of the cross section (see figure). The diameter of the circular post and the depth of the rectangular post are the same. (a) For what width b of the rectangular post will the maximum tensile stresses be the same in both posts? (b) Under the conditions described in part (a), which post has the larger compressive stress?

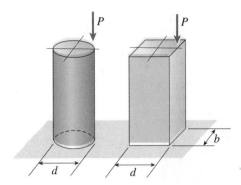

Prob. 5.12-14

5.12-15 A bar AB supports a load P acting at the centroid of the end cross section (see figure). In the middle region of the bar the cross-sectional area is reduced by removing one-half of the bar. (a) If the end cross sections are square with sides of length b, what are the maximum tensile and compressive stresses σ_t and σ_c, respectively, at cross section mn within the reduced region? (b) If the end cross sections are circular with diameter b, what are the maximum stresses σ_t and σ_c?

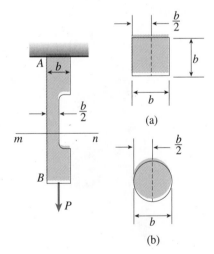

Prob. 5.12-15

5.12-16 A short column of wide-flange shape is subjected to a compressive load that produces a resultant force $P = 60$ kN acting at the midpoint of one flange (see figure). (a) Determine the maximum tensile and compressive stresses σ_t and σ_c, respectively, in the column. (b) Locate the neutral axis under this loading condition.

Prob. 5.12-16

5.12-17 A short column constructed of a W 10 × 30 wide-flange shape is subjected to a resultant compressive load $P = 12$ k having its line of action at the midpoint of one flange (see figure). (a) Determine the maximum tensile and compressive stresses σ_t and σ_c, respectively, in the column. (b) Locate the neutral axis under this loading condition.

Prob. 5.12-17

5.12-18 The T-beam shown in the figure is subjected to a compressive force P that has its line of action through the midpoint of the flange. (a) Determine the equation of the neutral axis under this loading condition. (b) If the allowable stresses in tension and compression are 100 MPa and

60 MPa, respectively, find the maximum permissible load P_{max}.

Prob. 5.12-18

5.12-19 A short length of a C 8 × 11.5 channel is subjected to an axial compressive force P that has its line of action through the midpoint of the web of the channel (see figure). (a) Determine the equation of the neutral axis under this loading condition. (b) If the allowable stresses in tension and compression are 10,000 psi and 8,000 psi, respectively, find the maximum permissible load P_{max}.

Prob. 5.12-19

CORE OF A CROSS SECTION

5.12-20 A hollow circular cross section with inner radius r_1 and outer radius r_2 is shown in the figure. (a) Determine the radius r_c of the core of the section. (b) From the result for part (a), obtain the radius r_c of the core of a solid circular cross section. (c) From the result for part (a), obtain the radius r_c of the core of a cross section in which r_1 approaches r_2 and the area becomes a thin circular ring.

Prob. 5.12-20

5.12-21 Determine the core of a W 10 × 30 structural steel shape (see figure).

Probs. 5.12-21 and 5.12-22

5.12-22 Determine the core of a W 12 × 87 structural steel shape (see figure).

5.12-23 The doubly symmetric, hollow square cross section shown in the figure has outer dimension b and inner dimension c. Determine the core of the cross section.

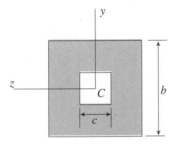

Prob. 5.12-23

5.12-24 The cross section of a square chimney ($b =$ length of each side) with a circular flue of diameter d is shown in the figure. What is the smallest diameter d (in terms of b) such that the core lies entirely within the hole?

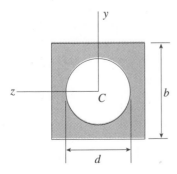

Prob. 5.12-24

STRESS CONCENTRATIONS

The problems for Section 5.13 are to be solved considering the stress-concentration factors.

5.13-1 The beams shown in the figure are subjected to bending moments $M = 2100$ lb-in. Each beam has a rectangular cross section with height $h = 1.5$ in. and width $b = 0.375$ in. (perpendicular to the plane of the figure). (a) For the beam with a hole at midheight, determine the maximum stresses for hole diameters $d = 0.25, 0.50, 0.75,$ and 1.00 in. (b) For the beam with two identical notches (inside height $h_1 = 1.25$ in.), determine the maximum stresses for notch radii $R = 0.05, 0.10, 0.15,$ and 0.20 in.

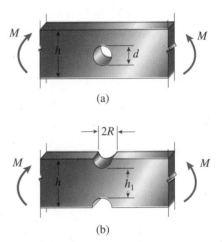

(a)

(b)

Probs. 5.13-1 through 5.13-4

5.13-2 The beams shown in the figure are subjected to bending moments $M = 250$ N·m. Each beam has a rectangular cross section with height $h = 44$ mm and width $b = 10$ mm (perpendicular to the plane of the figure). (a) For the beam with a hole at midheight, determine the maximum stresses for hole diameters $d = 10, 16, 22,$ and 28 mm. (b) For the beam with two identical notches (inside height $h_1 = 40$ mm), determine the maximum stresses for notch radii $R = 2, 4, 6,$ and 8 mm.

5.13-3 A rectangular beam with semicircular notches, as shown in part (b) of the figure, has dimensions $h = 0.88$ in. and $h_1 = 0.80$ in. The maximum allowable bending stress in the metal beam is $\sigma_{max} = 60$ ksi, and the bending moment is $M = 600$ lb-in. Determine the minimum permissible width b_{min} of the beam.

5.13-4 A rectangular beam with semicircular notches, as shown in part (b) of the figure, has dimensions $h = 120$ mm and $h_1 = 100$ mm. The maximum allowable bending stress in the plastic beam is $\sigma_{max} = 6$ MPa, and the bending moment is $M = 150$ N·m. Determine the minimum permissible width b_{min} of the beam.

5.13-5 A rectangular beam with notches and a hole (see figure) has dimensions $h = 5.5$ in., $h_1 = 5$ in., and width $b = 1.6$ in. The beam is subjected to a bending moment $M = 130$ k-in., and the maximum allowable bending stress in the material (steel) is $\sigma_{max} = 42,000$ psi. (a) What is the smallest radius R_{min} that should be used in the notches? (b) What is the diameter d_{max} of the largest hole that should be drilled at the midheight of the beam?

Prob. 5.13-5

6
STRESSES IN BEAMS (ADVANCED TOPICS)

6.1 INTRODUCTION

In this chapter we continue our study of the bending of beams by examining several specialized topics, including the analysis of composite beams (that is, beams of more than one material), beams with inclined loads, unsymmetric beams, shear stresses in thin-walled beams, shear centers, elastoplastic bending, and nonlinear bending. These subjects are based upon the fundamental topics discussed previously in Chapter 5—topics such as curvature, normal stresses in beams (including the flexure formula), and shear stresses in beams. Later, in Chapters 9 and 10, we will discuss two additional subjects of fundamental importance in beam design—deflections of beams and statically indeterminate beams.

6.2 COMPOSITE BEAMS

Beams that are fabricated of more than one material are called **composite beams**. Examples are bimetallic beams (used in thermostats), plastic coated pipes, and wood beams with steel reinforcing plates (Fig. 6-1). Many other types of composite beams have been developed in recent years, primarily to save material and reduce weight. For instance, **sandwich beams** are widely used in the aviation and aerospace industries, where light weight plus high strength and rigidity are required. Such familiar objects as skis, doors, wall panels, book shelves, and cardboard boxes are also manufactured in sandwich style. A typical sandwich beam

(a)

(b)

(c)

Fig. 6-1 Examples of composite beams:
(a) bimetallic beam, (b) plastic coated
steel pipe, and (c) wood beam reinforced
with a steel plate

(Fig. 6-2) consists of two thin *faces* of relatively high-strength material (such as aluminum) separated by a thick *core* of lightweight, low-strength material. Since the faces are at the greatest distance from the neutral axis (where the bending stresses are highest), they function somewhat like the flanges of an I-beam. The core serves as a filler and provides support for the faces, stabilizing them against wrinkling or buckling. Lightweight plastics and foams, as well as honeycombs and corrugations, are often used for cores.

■ **Strains and Stresses**

The strains in composite beams are determined from the same basic axiom that we used for finding the strains in beams of one material, namely, cross sections remain plane during bending. This axiom is valid for pure bending regardless of the nature of the material (see Section 5.4). Therefore, the longitudinal strains ϵ_x in a composite beam vary linearly from top to bottom of the beam, as expressed by Eq. (5-4), which is repeated here:

$$\epsilon_x = -\frac{y}{\rho} = -\kappa y \tag{6-1}$$

In this equation, y is the distance from the neutral axis, ρ is the radius of curvature, and κ is the curvature.

Beginning with the linear strain distribution represented by Eq. (6-1), we can determine the strains and stresses in any composite beam. To show how this is accomplished, consider the composite beam shown in Figs. 6-3a and 6-3b. This beam consists of two parts, labeled 1 and 2 in the figure, which are securely bonded so that they act as a single solid beam. As in our previous discussions of beams (Chapter 5), we assume that the xy plane is a plane of symmetry and that the xz plane is the neutral plane of the beam. However, the neutral axis (the z axis) does *not* pass through the centroid of the cross-sectional area when the beam is made of two different materials.

If the beam is bent with positive curvature, the strains ϵ_x will vary as shown in Fig. 6-3c, where ϵ_A is the compressive strain at the top of the beam (represented by the letter A), ϵ_B is the tensile strain at the bottom (represented by the letter B), and ϵ_C is the strain at the contact surface (represented by the letter C). The normal stresses acting on the cross section can be obtained from the strains by using the stress-strain relationships for the two materials. Let us assume that both materials behave in a linearly elastic manner so that Hooke's law for uniaxial stress is valid. Then the stresses in the materials are obtained by multiplying the strains by the appropriate modulus of elasticity. Denoting the moduli of elasticity for materials 1 and 2 as E_1 and E_2, respectively, and also assuming that $E_2 > E_1$, we obtain the stress diagram shown in Fig. 6-3d. The compressive stress at the top of the beam is $\sigma_A = E_1\epsilon_A$ and the tensile stress at the bottom is $\sigma_B = E_2\epsilon_B$. At the contact surface the stresses in the two materials are different because their moduli are different. In material 1 the stress is $\sigma_{1C} = E_1\epsilon_C$ and in material 2 it is $\sigma_{2C} = E_2\epsilon_C$.

Using Hooke's law and Eq. (6-1), we can express the normal stresses at distance y from the neutral axis in terms of the curvature:

$$\sigma_{x1} = -E_1\kappa y \qquad \sigma_{x2} = -E_2\kappa y \qquad (6\text{-}2a,b)$$

in which σ_{x1} is the stress in material 1 and σ_{x2} is the stress in material 2. With the aid of these equations, we can locate the neutral axis and obtain the moment-curvature relationship.

Fig. 6-2 Sandwich beams with: (a) plastic core, (b) honeycomb core, and (c) corrugated core

Fig. 6-3 (a) Composite beam of two materials, (b) cross section of beam, (c) distribution of strains ϵ_x throughout the height of the beam, and (d) distribution of stresses σ_x in the beam for the case where $E_2 > E_1$

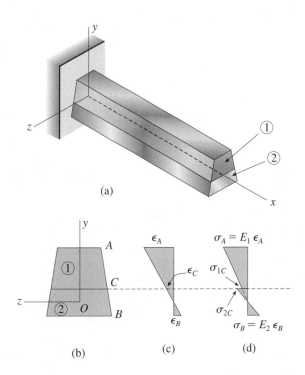

■ Neutral Axis

The position of the neutral axis is found from the condition that the resultant axial force acting on the cross section is zero (see Section 5.5); therefore,

$$\int_1 \sigma_{x1}\,dA + \int_2 \sigma_{x2}\,dA = 0$$

where it is understood that the first integral is evaluated over the cross-sectional area of material 1 and the second integral is evaluated over the cross-sectional area of material 2. Replacing σ_{x1} and σ_{x2} in the preceding equation by their expressions from Eqs. (6-2a) and (6-2b), we get

$$-\int_1 E_1 \kappa y\,dA - \int_2 E_2 \kappa y\,dA = 0$$

Since the curvature is a constant at any given cross section, it is not involved in the integrations and can be cancelled from the equation; thus,

$$E_1 \int_1 y\,dA + E_2 \int_2 y\,dA = 0 \qquad (6\text{-}3)$$

The integrals in this equation represent the first moments of the two parts of the cross-sectional area with respect to the neutral axis. (If there are more than two materials—a rare condition—additional terms are required in the equation.)

Equation (6-3) is a generalized form of the analogous equation for a beam of one material (Eq. 5-8). The details of the procedure for locating the neutral axis with the aid of Eq. (6-3) are illustrated later in Example 6-1.

If the cross section of a beam is *doubly symmetric*, as in the case of a wood beam with steel cover plates on the top and bottom (Fig. 6-4), the neutral axis is located at the midheight of the cross section and Eq. (6-3) is not needed.

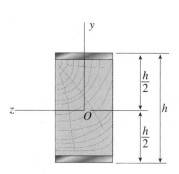

Fig. 6-4 Doubly symmetric cross section

■ Moment-Curvature Relationship

The moment-curvature relationship for a composite beam (Fig. 6-3) may be determined from the condition that the moment resultant of the bending stresses is equal to the bending moment M acting at the cross section. Following the same steps as for a beam of one material (see Eqs. 5-9 through 5-12), and also using Eqs. (6-2a) and (6-2b), we obtain

$$M = -\int_A \sigma_x y\,dA = -\int_1 \sigma_{x1} y\,dA - \int_2 \sigma_{x2} y\,dA$$

$$= \kappa E_1 \int_1 y^2\,dA + \kappa E_2 \int_2 y^2\,dA$$

This equation can be written in the simpler form

$$M = \kappa(E_1 I_1 + E_2 I_2) \qquad (6\text{-}4)$$

in which I_1 and I_2 are the moments of inertia about the neutral axis (the z axis) of the cross-sectional areas of materials 1 and 2, respectively. Note that $I = I_1 + I_2$, where I is the moment of inertia of the *entire* cross-sectional area about the neutral axis.

Equation (6-4) can now be solved for the curvature in terms of the bending moment:

$$\kappa = \frac{1}{\rho} = \frac{M}{E_1 I_1 + E_2 I_2} \tag{6-5}$$

This equation is the moment-curvature relationship for a beam of two materials (compare with Eq. 5-12 for a beam of one material). The denominator on the right-hand side is the **flexural rigidity** of the composite beam.

■ **Normal Stresses**

The normal stresses (or bending stresses) in the beam are obtained by substituting the expression for curvature (Eq. 6-5) into the expressions for σ_{x1} and σ_{x2} (Eqs. 6-2a) and (6-2b); thus,

$$\sigma_{x1} = -\frac{MyE_1}{E_1 I_1 + E_2 I_2} \qquad \sigma_{x2} = -\frac{MyE_2}{E_1 I_1 + E_2 I_2} \tag{6-6a,b}$$

These expressions, known as the *flexure formulas for a composite beam*, give the normal stresses in materials 1 and 2, respectively. If the two materials have the same modulus of elasticity ($E_1 = E_2 = E$), then both equations reduce to the flexure formula for a beam of one material (Eq. 5-13).

The analysis of composite beams, using Eqs. (6-3) through (6-6), is illustrated in Examples 6-1 and 6-2 at the end of this section.

■ **Approximate Theory for Bending of Sandwich Beams**

Sandwich beams having doubly symmetric cross sections and composed of two linearly elastic materials (Fig. 6-5) can be analyzed for bending using Eqs. (6-5) and (6-6), as described above. However, we can also develop an approximate theory for bending of sandwich beams by introducing some simplifying assumptions. If the material of the faces (material 1) has a much larger modulus of elasticity than does the material of the core (material 2), it is reasonable to disregard the normal stresses in the core and assume that the faces resist all of the longitudinal bending stresses. This assumption is equivalent to saying that the modulus of elasticity E_2 of the core is zero. Under these conditions the flexure formula for material 2 (Eq. 6-6b) gives $\sigma_{x2} = 0$ (as expected), and the flexure formula for material 1 (Eq. 6-6a) gives

$$\sigma_{x1} = -\frac{My}{I_1} \tag{6-7}$$

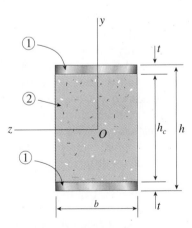

Fig. 6-5 Cross section of a sandwich beam having two axes of symmetry (doubly symmetric cross section)

which is similar to the ordinary flexure formula (Eq. 5-13). The quantity I_1 is the moment of inertia of the two faces evaluated with respect to the neutral axis; thus,

$$I_1 = \frac{b}{12}\left(h^3 - h_c^3\right) \tag{6-8}$$

in which b is the width of the beam, h is the overall height of the beam, and h_c is the height of the core. Note that $h_c = h - 2t$, where t is the thickness of the faces.

The maximum normal stresses in the sandwich beam occur at the top and bottom of the cross section where $y = h/2$ and $-h/2$, respectively. Thus, from Eq. (6-7), we obtain

$$\sigma_{\text{top}} = -\frac{Mh}{2I_1} \qquad \sigma_{\text{bottom}} = \frac{Mh}{2I_1} \tag{6-9a,b}$$

If the bending moment M is positive, the upper face is in compression and the lower face is in tension.

If the faces are thin compared to the thickness of the core (that is, if t is small compared to h_c), we can disregard the shear stresses in the faces and assume that the core carries all of the shear stresses. Under these conditions the average shear stress and average shear strain in the core are, respectively,

$$\tau_{\text{aver}} = \frac{V}{bh_c} \qquad \gamma_{\text{aver}} = \frac{V}{bh_c G_c} \tag{6-10a,b}$$

where V is the shear force acting on the cross section and G_c is the shear modulus of elasticity for the core material. (Although the maximum shear stress and maximum shear strain are larger than the average values, the average values are often used for design purposes.)

■ Limitations

Throughout the preceding discussion of composite beams, we assumed that both materials followed Hooke's law and that the two parts of the beam were adequately bonded so that they acted as a single unit. Thus, our analysis is highly idealized and represents only a first step in under-standing the behavior of composite beams and composite materials. Methods for dealing with nonhomogeneous and nonlinear materials, bond stresses between the parts, shear stresses on the cross sections, buckling of the faces, and other such matters are treated in reference books dealing specifically with composite construction.

Reinforced concrete beams are one of the most complex types of composite construction (Fig. 6-6), and their behavior differs significantly from that of the composite beams discussed in this section. Concrete is strong in compression but extremely weak in tension. Consequently, its tensile strength is usually disregarded entirely. Under those conditions, the formulas given in this section do not apply. Furthermore, reinforced concrete beams are not designed on the basis of linearly elastic behavior —instead, more realistic design methods (based upon load-carrying capacity instead of allowable stresses) are used. Since the design of rein-

Fig. 6-6 Reinforced concrete beam with main reinforcing bars and vertical stirrups

forced concrete members is a highly specialized matter, we will not discuss it here.

■ Example 6-1

A composite beam (Fig. 6-7) is constructed from a wood beam (4.0 in. × 6.0 in. actual dimensions) and a steel reinforcing plate (4.0 in. wide and 0.5 in. thick). The wood and steel are securely fastened to act as a single beam. The beam is subjected to a positive bending moment $M = 60$ k-in. Calculate the largest tensile and compressive stresses in the wood (material 1) and the maximum and minimum stresses in the steel (material 2) if $E_1 = 1500$ ksi and $E_2 = 30,000$ ksi.

Solution   *Neutral axis.* The first step in the analysis is to locate the neutral axis of the cross section. For that purpose, let us denote the distances from the neutral axis to the top and bottom of the beam as h_1 and h_2, respectively. To obtain these distances, we use Eq. (6-3). The integrals in that equation are evaluated by taking the first moments of areas 1 and 2 about the z axis, as follows:

$$\int_1 y\,dA = \bar{y}_1 A_1 = (h_1 - 3 \text{ in.})(4 \text{ in.} \times 6 \text{ in.}) = (h_1 - 3 \text{ in.})(24 \text{ in.}^2)$$

$$\int_2 y\,dA = \bar{y}_2 A_2 = -(6.25 \text{ in.} - h_1)(4 \text{ in.} \times 0.5 \text{ in.}) = (h_1 - 6.25 \text{ in.})(2 \text{ in.}^2)$$

in which A_1 and A_2 are the areas of parts 1 and 2 of the cross section, $\bar{y}_1$ and $\bar{y}_2$ represent the y coordinates of the centroids of the respective areas, and h_1 has units of inches.

Substituting the preceding expressions into Eq. (6-3) gives the equation for locating the neutral axis, as follows:

$$E_1 \int_1 y\,dA + E_2 \int_2 y\,dA = 0$$

or $(1500 \text{ ksi})(h_1 - 3 \text{ in.})(24 \text{ in.}^2) + (30,000 \text{ ksi})(h_1 - 6.25 \text{ in.})(2 \text{ in.}^2) = 0$

Solving this equation, we obtain the distance h_1 from the neutral axis to the top of the beam:

$$h_1 = 5.031 \text{ in.}$$

Also, the distance h_2 from the neutral axis to the bottom of the beam is

$$h_2 = 6.5 \text{ in.} - h_1 = 1.469 \text{ in.}$$

Thus, the position of the neutral axis is established.

Moments of inertia. The moments of inertia I_1 and I_2 of areas A_1 and A_2 with respect to the neutral axis can be found by using the parallel-axis theorem (see Section 12.5 of Chapter 12). Beginning with area 1 (Fig. 6-7), we get

$$I_1 = \frac{1}{12}(4 \text{ in.})(6 \text{ in.})^3 + (4 \text{ in.})(6 \text{ in.})(h_1 - 3 \text{ in.})^2 = 171.0 \text{ in.}^4$$

Similarly, for area 2 we get

$$I_2 = \frac{1}{12}(4 \text{ in.})(0.5 \text{ in.})^3 + (4 \text{ in.})(0.5 \text{ in.})(h_2 - 0.25 \text{ in.})^2 = 3.01 \text{ in.}^4$$

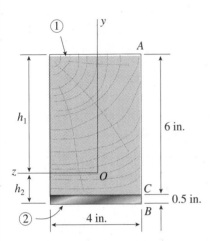

Fig. 6-7 Example 6-1. Composite beam of wood and steel

To check these calculations, we can determine the moment of inertia I of the entire cross-sectional area about the z axis as follows:

$$I = \frac{1}{3}(4 \text{ in.})h_1^3 + \frac{1}{3}(4 \text{ in.})h_2^3 = 169.8 + 4.2 = 174.0 \text{ in.}^4$$

which agrees with the sum of I_1 and I_2.

Normal stresses. The stresses in materials 1 and 2 are calculated from the flexure formulas for composite beams (Eqs. 6-6a and b). The largest compressive stress in material 1 occurs at the top of the beam (A) where $y = h_1 = 5.031$ in. Denoting this stress by σ_{1A} and using Eq. (6-6a), we get

$$\sigma_{1A} = -\frac{Mh_1E_1}{E_1I_1 + E_2I_2}$$

$$= -\frac{(60 \text{ k-in.})(5.031 \text{ in.})(1500 \text{ ksi})}{(1500 \text{ ksi})(171.0 \text{ in.}^4) + (30{,}000 \text{ ksi})(3.01 \text{ in.}^4)} = -1310 \text{ psi} \quad \Longleftarrow$$

The largest tensile stress in material 1 occurs at the contact plane between the two materials (C) where $y = -(h_2 - 0.5 \text{ in.}) = -0.969$ in. Proceeding as in the previous calculation, we get

$$\sigma_{1C} = -\frac{(60 \text{ k-in.})(-0.969 \text{ in.})(1500 \text{ ksi})}{(1500 \text{ ksi})(171.0 \text{ in.}^4) + (30{,}000 \text{ ksi})(3.01 \text{ in.}^4)} = 251 \text{ psi} \quad \Longleftarrow$$

Thus, we have found the largest tensile and compressive stresses in the wood.

The steel plate (material 2) is located below the neutral axis, and therefore it is entirely in tension. The maximum tensile stress occurs at the bottom of the beam (B) where $y = -h_2 = -1.469$ in. Hence, from Eq. (6-6b) we get

$$\sigma_{2B} = -\frac{M(-h_2)E_2}{E_1I_1 + E_2I_2}$$

$$= -\frac{(60 \text{ k-in.})(-1.469 \text{ in.})(30{,}000 \text{ ksi})}{(1500 \text{ ksi})(171.0 \text{ in.}^4) + (30{,}000 \text{ ksi})(3.01 \text{ in.}^4)} = 7620 \text{ psi} \quad \Longleftarrow$$

The minimum tensile stress in material 2 occurs at the contact plane (C) where $y = -0.969$ in. Thus,

$$\sigma_{2C} = -\frac{(60 \text{ k-in.})(-0.969 \text{ in.})(30{,}000 \text{ ksi})}{(1500 \text{ ksi})(171.0 \text{ in.}^4) + (30{,}000 \text{ ksi})(3.01 \text{ in.}^4)} = 5030 \text{ psi} \quad \Longleftarrow$$

These stresses are the maximum and minimum stresses in the steel.

Note: At the contact plane the ratio of the stress in the steel to the stress in the wood is

$$\frac{\sigma_{2C}}{\sigma_{1C}} = \frac{5030 \text{ psi}}{251 \text{ psi}} = 20$$

which is equal to the ratio E_2/E_1 of the moduli of elasticity (as expected). Although the strains in the steel and wood are the same at the contact plane, the stresses are different because of the different moduli.

(figure labels)

y

① $t = 5$ mm

②

$h_c = 150$ mm

z ─── O ①

$\dfrac{h}{2}$ $h = 160$ mm

$b = 200$ mm $t = 5$ mm

Fig. 6-8 Example 6-2. Cross section of sandwich beam having aluminum-alloy faces and a plastic core

■ **Example 6-2**

A sandwich beam having aluminum-alloy faces enclosing a plastic core (Fig. 6-8) is subjected to a bending moment $M = 3.0$ kN·m. The thickness of the faces is $t = 5$ mm and their modulus of elasticity is $E_1 = 72$ GPa. The height of the plastic core is $h_c = 150$ mm and its modulus of elasticity is $E_2 = 800$ MPa. The overall dimensions of the beam are $h = 160$ mm and $b = 200$ mm. Determine the maximum tensile and compressive stresses in the faces and the core using: (a) the general theory for composite beams, and (b) the approximate theory for sandwich beams.

Solution   *Neutral axis.* Because the cross section is doubly symmetric, the neutral axis is located at midheight (see the z axis in Fig. 6-8).

Moments of inertia. The moment of inertia I_1 of the cross-sectional area of the faces is

$$I_1 = \frac{b}{12}(h^3 - h_c^3) = \frac{200 \text{ mm}}{12}\left[\left(160 \text{ mm}\right)^3 - \left(150 \text{ mm}\right)^3\right] = 12.017 \times 10^6 \text{ mm}^4$$

and the moment of inertia I_2 of the plastic core is

$$I_2 = \frac{b}{12}\left(h_c^3\right) = \frac{200 \text{ mm}}{12}\left(150 \text{ mm}\right)^3 = 56.250 \times 10^6 \text{ mm}^4$$

(a) *Normal stresses calculated from the general theory for composite beams.* To calculate these stresses, we use Eqs. (6-6a and b). As a preliminary matter, we will evaluate the term in the denominator of those equations (that is, the flexural rigidity of the composite beam):

$$E_1 I_1 + E_2 I_2 = (72 \text{ GPa})(12.017 \times 10^6 \text{ mm}^4) + (800 \text{ MPa})(56.250 \times 10^6 \text{ mm}^4)$$

$$= 910{,}200 \text{ N·m}^2$$

The maximum tensile and compressive stresses in the aluminum faces are found from Eq. (6-6a):

$$(\sigma_1)_{\text{max}} = \pm \frac{M(h/2)(E_1)}{E_1 I_1 + E_2 I_2}$$

$$= \pm \frac{(3.0 \text{ kN·m})(80 \text{ mm})(72 \text{ GPa})}{910{,}200 \text{ N·m}^2} = \pm 19.0 \text{ MPa} \qquad \longleftarrow$$

The corresponding quantities for the plastic core (from Eq. 6-6b) are

$$(\sigma_2)_{\text{max}} = \pm \frac{M(h_c/2)(E_2)}{E_1 I_1 + E_2 I_2}$$

$$= \pm \frac{(3.0 \text{ kN·m})(75 \text{ mm})(800 \text{ MPa})}{910{,}200 \text{ N·m}^2} = \pm 0.198 \text{ MPa} \qquad \longleftarrow$$

The maximum stresses in the faces are 96 times greater than the maximum stresses in the core, which is not surprising since the modulus of elasticity of the aluminum is 90 times greater than that of the plastic.

(b) *Normal stresses calculated from the approximate theory for sandwich beams.* In the approximate theory we disregard the normal stresses in the core and assume that the faces transmit the entire bending moment. Then the maximum tensile and compressive stresses in the faces can be found from Eq. (6-7), as follows:

$$(\sigma_1)_{\text{max}} = \pm \frac{My}{I_1} = \pm \frac{M(h/2)}{I_1}$$

$$= \pm \frac{(3.0 \text{ kN·m})(80 \text{ mm})}{12.017 \times 10^6 \text{ mm}^4} = \pm 20.0 \text{ MPa} \qquad \longleftarrow$$

As expected, this theory is conservative and gives slightly higher stresses in the aluminum faces than does the general theory for composite beams.

6.3   TRANSFORMED-SECTION METHOD

The transformed-section method is an alternative procedure for analyzing the bending stresses in a composite beam. The method is based upon the theories and equations developed in the preceding section, and therefore it is subject to the same limitations (for instance, it is valid only for linearly elastic materials) and gives the same results. Although the transformed-section method does not reduce the calculating effort, many designers find that it provides a convenient way to visualize and organize the calculations.

The method consists of transforming the cross section of a composite beam into an equivalent cross section of an imaginary beam that is composed of only one material. This new cross section is called the **transformed section**. Then the imaginary beam with the transformed section is analyzed in the customary manner for a beam of one material. As a final step, the stresses in the transformed beam are converted to those in the original beam.

If the transformed beam is to be equivalent to the original beam, *its neutral axis must be located in the same place and its moment-resisting capacity must be the same.* To show how these two requirements are met, consider again a composite beam of two materials (Fig. 6-9a). The **neutral axis** of the cross section is obtained from Eq. (6-3), which is repeated here:

$$E_1 \int_1 y\,dA + E_2 \int_2 y\,dA = 0 \qquad (6-11)$$

In this equation, the integrals represent the first moments of the two parts of the cross section with respect to the neutral axis. Let us now introduce the notation

(a)

(b)

Fig. 6-9 Composite beam of two materials: (a) actual cross section, and (b) transformed section consisting only of material 1

$$n = \frac{E_2}{E_1} \tag{6-12}$$

where n is the **modular ratio**. With this notation, we can rewrite Eq. (6-11) in the form

$$\int_1 y \, dA + \int_2 yn \, dA = 0 \tag{6-13}$$

Since Eqs. (6-11) and (6-13) are equivalent, the preceding equation shows that the neutral axis is unchanged if each element of area dA in material 2 is multiplied by the factor n, provided that the y coordinate for each such element of area is not changed. Therefore, we can create a new cross section consisting of two parts: (1) area 1 with its dimensions unchanged, and (2) area 2 with its *width* (that is, its dimension parallel to the neutral axis) multiplied by n. This new cross section (the transformed section) is shown in Fig. 6-9b for the case where $E_2 > E_1$ (and therefore $n > 1$). Its neutral axis is in the same position as the neutral axis of the original beam. (Note that all dimensions perpendicular to the neutral axis remain the same.)

Since the stress in the material (for a given strain) is proportional to the modulus of elasticity ($\sigma = E\epsilon$), we see that multiplying the width of material 2 by $n = E_2/E_1$ is equivalent to transforming it to material 1. For instance, suppose that $n = 10$. Then the area of part 2 of the cross section is now 10 times wider than before. If we imagine that this part of the beam is now material 1, we see that it will carry the same force as before because its modulus is *reduced* by a factor of 10 (from E_2 to E_1) at the same time that its area is *increased* by a factor of 10. Thus, the new section (the transformed section) consists only of material 1.

The **moment-curvature relationship** for the transformed beam must be the same as for the original beam. To show that this is indeed the case, we note that the stresses in the transformed beam (since it consists only of material 1) are given by Eq. (5-7) of Section 5.5:

$$\sigma_x = -E_1\kappa y$$

Using this equation, and also following the same procedure as for a beam of one material (see Section 5.5), we can obtain the moment-curvature relation for the transformed beam:

$$M = -\int_A \sigma_x y \, dA = -\int_1 \sigma_x y \, dA - \int_2 \sigma_x y \, dA$$

$$= E_1\kappa \int_1 y^2 dA + E_1\kappa \int_2 y^2 dA = \kappa(E_1 I_1 + E_2 n I_2)$$

or
$$M = \kappa(E_1 I_1 + E_2 I_2) \tag{6-14}$$

This equation is the same as Eq. (6-4), thereby demonstrating that the moment-curvature relationship for the transformed beam is the same as for the original beam.

Since the transformed beam consists of only one material, the **bending stresses** can be found from the standard flexure formula (Eq. 5-13). Thus, the normal stresses in the beam transformed to material 1 (Fig. 6-9b) are

$$\sigma_{x1} = -\frac{My}{I_T} \tag{6-15}$$

where I_T is the moment of inertia of the transformed section with respect to the neutral axis. By substituting into this equation, we can calculate the stresses at any point in the *transformed* beam. (As explained below, the stresses in the transformed beam match those in the original beam in the part of the original beam consisting of material 1; however, in the part of the original beam consisting of material 2, the stresses are different from those in the transformed beam.)

We can easily verify Eq. (6-15) by noting that the moment of inertia of the transformed section (Fig. 6-9b) is related to the moment of inertia of the original section (Fig. 6-9a) by the following relation:

$$I_T = I_1 + nI_2 = I_1 + \frac{E_2}{E_1} I_2 \tag{6-16}$$

Substituting this expression for I_T into Eq. (6-15) gives

$$\sigma_{x1} = -\frac{MyE_1}{E_1I_1 + E_2I_2} \tag{a}$$

which is the same as Eq. (6-6a), thus demonstrating that the stresses in material 1 in the original beam are the same as the stresses in the corresponding part of the transformed beam.

As mentioned above, the stresses in material 2 in the original beam are *not* the same as the stresses in the corresponding part of the transformed beam. Instead, the stresses in the transformed beam (Eq. 6-15) must be multiplied by the modular ratio n to obtain the stresses in material 2 of the original beam:

$$\sigma_{x2} = -\frac{My}{I_T} n \tag{6-17}$$

We can verify this formula by noting that when Eq. (6-16) for I_T is substituted into Eq. (6-17), we get

$$\sigma_{x2} = -\frac{MynE_1}{E_1I_1 + E_2I_2} = -\frac{MyE_2}{E_1I_1 + E_2I_2} \tag{b}$$

which is the same as Eq. (6-6b).

In this discussion of the transformed-section method we chose to transform the original beam to a beam consisting entirely of material 1. It is also possible to transform the beam to material 2. In that case the stresses in the original beam in material 2 will be the same as the stresses in the corresponding part of the transformed beam. However, the stresses in material 1 in the original beam must be obtained by multiplying the stresses in the corresponding part of the transformed beam by the modular ratio n, which in this case is defined as $n = E_1/E_2$.

It is also possible to transform the original beam into a material having any arbitrary modulus of elasticity E, in which case all parts of the beam must be transformed to the fictitious material. Of course, the calculations are simpler if we transform to one of the original materials. Finally, with a little ingenuity it is possible to extend the transformed-section method to composite beams of more than two materials.

■ **Example 6-3**

The composite beam shown in Fig. 6-10a is formed of a wood beam (4.0 in. × 6.0 in. actual dimensions) and a steel reinforcing plate (4.0 in. wide and 0.5 in. thick). The beam is subjected to a positive bending moment $M = 60$ k-in. Using the transformed-section method, calculate the largest tensile and compressive stresses in the wood (material 1) and the maximum and minimum stresses in the steel (material 2) if $E_1 = 1500$ ksi and $E_2 = 30,000$ ksi. (*Note:* This is the same beam that was analyzed previously in Example 6-1 of Section 6.2.)

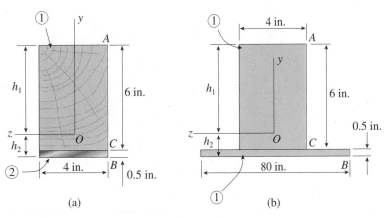

Fig. 6-10 Example 6-3. Composite beam of Example 6-1 analyzed by the transformed-section method: (a) original beam, and (b) transformed section (material 1)

Solution   *Transformed section.* We will transform the original beam into a beam of material 1, which means that the modular ratio is defined as

$$n = \frac{E_2}{E_1} = \frac{30,000 \text{ ksi}}{1,500 \text{ ksi}} = 20$$

The part of the beam made of wood (material 1) is not altered but the part made of steel (material 2) has its width multiplied by the modular ratio. Thus, the width of this part of the beam becomes

$$n(4 \text{ in.}) = 20(4 \text{ in.}) = 80 \text{ in.}$$

in the transformed section (Fig. 6-10b).

Neutral axis. Because the transformed beam consists of only one material, the neutral axis passes through the centroid of the cross-sectional area. Therefore, with the top edge of the cross-section serving as a reference line, and with the distance y_i measured positive downward, we can calculate the distance h_1 to the centroid as follows:

$$h_1 = \frac{\sum y_i A_i}{\sum A_i} = \frac{(3 \text{ in.})(4 \text{ in.})(6 \text{ in.}) + (6.25 \text{ in.})(80 \text{ in.})(0.5 \text{ in.})}{(4 \text{ in.})(6 \text{ in.}) + (80 \text{ in.})(0.5 \text{ in.})}$$

$$= \frac{322.0 \text{ in.}^3}{64.0 \text{ in.}^2} = 5.031 \text{ in.}$$

Also, the distance h_2 from the lower edge of the section to the centroid is

$$h_2 = 6.5 \text{ in.} - h_1 = 1.469 \text{ in.}$$

Thus, the location of the neutral axis is determined.

Moment of inertia of the transformed section. Using the parallel-axis theorem (see Section 12.5 of Chapter 12), we can calculate the moment of inertia I_T of the entire cross-sectional area with respect to the neutral axis as follows:

$$I_T = \frac{1}{12}(4 \text{ in.})(6 \text{ in.})^3 + (4 \text{ in.})(6 \text{ in.})(h_1 - 3 \text{ in.})^2$$

$$+ \frac{1}{12}(80 \text{ in.})(0.5 \text{ in.})^3 + (80 \text{ in.})(0.5 \text{ in.})(h_2 - 0.25 \text{ in.})^2$$

$$= 171.0 \text{ in.}^4 + 60.3 \text{ in.}^4 = 231.3 \text{ in.}^4$$

Normal stresses in the wood (material 1). The stresses in the transformed beam (Fig. 6-10b) at the top of the cross section (A) and at the contact plane between the two parts (C) are the same as in the original beam (Fig. 6-10a). These stresses can be found from the flexure formula (Eq. 6-15), as follows:

$$\sigma_{1A} = -\frac{My}{I_T} = -\frac{(60 \text{ k-in.})(5.031 \text{ in.})}{231.3 \text{ in.}^4} = -1310 \text{ psi} \qquad \longleftarrow$$

$$\sigma_{1C} = -\frac{My}{I_T} = -\frac{(60 \text{ k-in.})(-0.969 \text{ in.})}{231.3 \text{ in.}^4} = 251 \text{ psi} \qquad \longleftarrow$$

These are the largest tensile and compressive stresses in the wood (material 1) in the original beam. The stress σ_{1A} is compressive and the stress σ_{1C} is tensile.

Normal stresses in the steel (material 2). The maximum and minimum stresses in the steel plate are found by multiplying the corresponding stresses in the transformed beam by the modular ratio n (Eq. 6-17). The maximum stress occurs at the lower edge of the cross section (B) and the minimum stress occurs at the contact plane (C):

$$\sigma_{2B} = -\frac{My}{I_T}n = -\frac{(60 \text{ k-in.})(-1.469 \text{ in.})}{231.3 \text{ in.}^4}(20) = 7620 \text{ psi} \quad \longleftarrow$$

$$\sigma_{2C} = -\frac{My}{I_T}n = -\frac{(60 \text{ k-in.})(-0.969 \text{ in.})}{231.3 \text{ in.}^4}(20) = 5030 \text{ psi} \quad \longleftarrow$$

Both of these stresses are tensile.

Note that the stresses calculated by the transformed-section method agree with those found in Example 6-1 by direct application of the formulas for a composite beam.

Fig. 6-11 Beam with a lateral load acting in a plane of symmetry

Fig. 6-12 Doubly symmetric beam with an inclined load

6.4 ■ DOUBLY SYMMETRIC BEAMS WITH INCLINED LOADS

In our previous discussions of bending we dealt with beams possessing a longitudinal plane of symmetry (the xy plane in Fig. 6-11) and supporting lateral loads acting in that same plane. Under these conditions the bending stresses can be obtained from the flexure formula (Eq. 5-13) provided the material is homogeneous and linearly elastic. In this section, we will extend those ideas and consider what happens when the beam is subjected to loads that do not act in the plane of symmetry, that is, *inclined loads* (Fig. 6-12). We will limit our discussion to beams that have a doubly symmetric cross section, that is, both the xy and xz planes are planes of symmetry. Also, the inclined loads must act through the centroid of the cross section to avoid twisting the beam about the longitudinal axis.

We can determine the bending stresses in the beam shown in Fig. 6-12 by resolving the inclined load into two components, one acting in each plane of symmetry. Then the bending stresses can be obtained from the flexure formula for each load component acting separately, and the final stresses can be obtained by superposing the separate stresses.

■ Sign Convention for Bending Moments

As a preliminary matter, we will establish sign conventions for the bending moments acting on cross sections of a beam.* For this purpose, we cut through the beam and consider a typical cross section (Fig. 6-13). The bending moments M_y and M_z acting about the y and z axes, respectively, are represented as vectors using double-headed arrows. The moments are positive when their vectors point in the positive directions of the corresponding axes, and the right-hand rule for vectors gives the

---

*The directions of the normal and shear stresses in a beam are usually apparent from an inspection of the beam and its loading, and therefore we often calculate stresses by ignoring sign conventions and using only absolute values. However, when deriving general formulas we need to maintain rigorous sign conventions to avoid ambiguity in the equations.

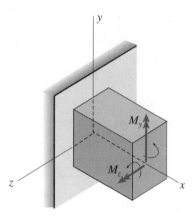

Fig. 6-13 Sign conventions for bending moments M_y and M_z

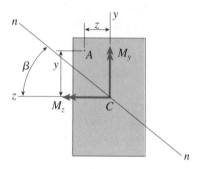

Fig. 6-14 Cross section of beam subjected to bending moments M_y and M_z

direction of rotation (indicated by the curved arrows in the figure). Thus, a positive bending moment M_y produces compression on the right-hand side of the beam (the negative z side) and tension on the left-hand side (the positive z side). Similarly, a positive moment M_z produces compression on the upper part of the beam (where y is positive) and tension on the lower part (where y is negative). Also, it is important to note that the bending moments shown in Fig. 6-13 act on the positive x face of a segment of the beam, that is, on a face having its outward normal in the positive direction of the x axis.

■ **Bending Stresses**

The normal stresses associated with the bending moments M_y and M_z acting separately are obtained from the flexure formula (Eq. 5-13). These stresses are then superposed to give the stresses produced by both moments acting simultaneously. For instance, consider the stresses at a point in the cross section having positive coordinates y and z (point A in Fig. 6-14). A positive moment M_y produces tension at this point and a positive moment M_z produces compression; thus, the normal stress at point A is

$$\sigma_x = \frac{M_y z}{I_y} - \frac{M_z y}{I_z}$$

(6-18)

in which I_y and I_z are the moments of inertia of the cross-sectional area with respect to the y and z axes, respectively. Using this equation, we can find the normal stress at any point in the cross section by substituting the appropriate algebraic values.

■ **Neutral Axis**

The equation of the neutral axis can be determined by equating the normal stress σ_x (Eq. 6-18) to zero:

$$\frac{M_y}{I_y} z - \frac{M_z}{I_z} y = 0$$

(6-19)

This equation shows that the neutral axis nn is a straight line passing through the centroid C (Fig. 6-14). The angle β between the neutral axis and the z axis is determined as follows:

$$\tan \beta = \frac{y}{z} = \frac{M_y I_z}{M_z I_y}$$

(6-20)

Depending upon the magnitudes and directions of the bending moments, the angle β may vary from $-90°$ to $+90°$. Knowing the orientation of the neutral axis is useful when determining the points in the cross section where the normal stresses are the largest. (Since the stresses vary linearly with distance from the neutral axis, the maximum stresses occur at points located farthest from the neutral axis.)

■ Relationship Between the Neutral Axis and the Inclination of the Loads

As we have just seen, the orientation of the neutral axis with respect to the z axis is determined by the bending moments and the moments of inertia (Eq. 6-20). Now we wish to determine the orientation of the neutral axis relative to the angle of inclination of the loads acting on the beam. For this purpose, we will use the cantilever beam shown in Fig. 6-15a as an example. The beam is loaded by a force P acting in the plane of the end cross section and inclined at an angle θ to the positive y axis. This particular orientation of the load is selected because it means that both bending moments (M_y and M_z) are positive when θ is between 0 and 90°.

The load P can be resolved into components $P\cos\theta$ in the positive y direction and $P\sin\theta$ in the negative z direction. Therefore, the bending moments M_y and M_z (Fig. 6-15b) acting on a cross section located at distance x from the fixed support are

$$M_y = (P\sin\theta)(L-x) \qquad M_z = (P\cos\theta)(L-x) \quad \text{(6-21a,b)}$$

in which L is the length of the beam. The ratio of these moments is

$$\frac{M_y}{M_z} = \tan\theta \qquad (6\text{-}22)$$

which shows that the resultant moment vector M is at the angle θ with the z axis (Fig. 6-15b). Consequently, the resultant moment vector is *perpendicular* to the longitudinal plane containing the force P.

The angle β between the neutral axis nn and the z axis (Fig. 6-15b) is obtained from Eq. (6-20):

$$\tan\beta = \frac{M_y I_z}{M_z I_y} = \frac{I_z}{I_y}\tan\theta \qquad (6\text{-}23)$$

which shows that the angle β is generally *not* equal to the angle θ. Thus, *except in special cases, the neutral axis is not perpendicular to the longitudinal plane containing the load.*

Exceptions to this general rule occur in three special cases: (1) When the load lies in the xy plane ($\theta = 0$ or $180°$), which means that the z axis is the neutral axis. (2) When the load lies in the xz plane ($\theta = \pm 90°$), which means that the y axis is the neutral axis. (3) When the principal moments of inertia are equal, that is, when $I_y = I_z$. In this case,

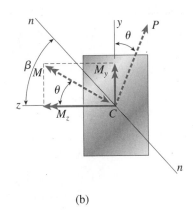

(a)

(b)

Fig. 6-15 Doubly symmetric beam with an inclined load P acting at an angle θ to the positive y axis all axes through the centroid are principal axes and all have the same moment of inertia. The plane of loading, no matter what its direction, is always a principal plane, and the neutral axis is always perpendicular to it. (This situation occurs with square, circular, and certain other cross sections, as described in Section 12.9 of Chapter 12.)

The fact that the neutral axis is not necessarily perpendicular to the plane of the load can greatly affect the stresses in a beam, especially if the ratio of the principal moments of inertia is very large. Under these conditions the stresses in the beam are very sensitive to slight changes in the direction of the load and to irregularities in the alignment of the beam itself. This characteristic of certain beams is illustrated later in Example 6-5.

■ **Example 6-4**

A wood beam *AB* of rectangular cross section serving as a roof purlin (Figs. 6-16a and b) is simply supported by the top chords of two adjacent roof trusses. The beam supports the weight of the roof sheathing and the roofing material, plus its own weight and any additional loads that affect the roof (such as wind, snow, and earthquake loads). In this example, we will consider only the effects of a uniformly distributed load of intensity $q = 3.0$ kN/m acting in the vertical direction through the centroids of the cross sections (Fig. 6-16c). The load acts along the entire length of the beam and includes the weight of the beam. The top chords of the trusses have a slope of 1 on 2 ($\alpha = 26.57°$), and the beam has width $b = 100$ mm, height $h = 150$ mm, and span $L = 1.6$ m.

Determine the maximum tensile and compressive stresses in the beam and locate the neutral axis.

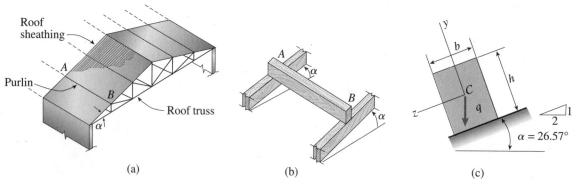

(a)   (b)   (c)

Fig. 6-16 Example 6-4. Wood beam of rectangular cross section serving as a roof purlin

Solution   *Loads and bending moments.* The uniform load q acting in the vertical direction can be resolved into components in the y and z directions (Fig. 6-17a):

$$q_y = q \cos \alpha \qquad q_z = q \sin \alpha \qquad (6\text{-}24a,b)$$

The maximum bending moments occur at the midpoint of the beam and are found from the general formula $M = qL^2/8$; hence,

$$M_y = \frac{q_z L^2}{8} = \frac{qL^2 \sin \alpha}{8} \qquad M_z = \frac{q_y L^2}{8} = \frac{qL^2 \cos \alpha}{8} \qquad (6\text{-}25a,b)$$

Both of these moments are positive because their vectors are in the positive directions of the y and z axes (Fig. 6-17b).

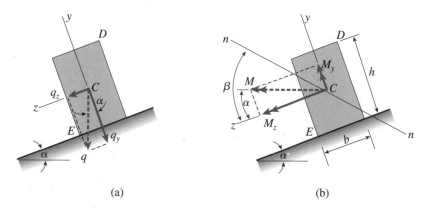

Fig. 6-17 Solution to Example 6-4.
(a) Components of the uniform load, and
(b) bending moments acting on a cross section

(a)   (b)

Moments of inertia. The moments of inertia of the cross-sectional area about the y and z axes are as follows:

$$I_y = \frac{hb^3}{12} \qquad I_z = \frac{bh^3}{12} \qquad (6\text{-}26a,b)$$

Bending stresses. The stresses at the midsection of the beam are obtained from Eq. (6-18) with the bending moments given by Eqs. (6-25) and the moments of inertia given by Eqs. (6-26):

$$\sigma_x = \frac{M_y z}{I_y} - \frac{M_z y}{I_z} = \frac{qL^2 \sin \alpha}{8hb^3/12} z - \frac{qL^2 \cos \alpha}{8bh^3/12} y$$

$$= \frac{3qL^2}{2bh}\left(\frac{\sin \alpha}{b^2} z - \frac{\cos \alpha}{h^2} y\right) \qquad (6\text{-}27)$$

The stress at any point in the cross section can be obtained from this equation by substituting the coordinates y and z of the point.

From the orientation of the cross section and the directions of the loads and bending moments (Fig. 6-17), it is apparent that the maximum compressive stress occurs at point D (where $y = h/2$ and $z = -b/2$) and the maximum tensile stress occurs at point E (where $y = -h/2$ and $z = b/2$). Substituting these coordinates into Eq. (6-27) and then simplifying, we obtain expressions for the maximum and minimum stresses in the beam:

$$\sigma_E = -\sigma_D = \frac{3qL^2}{4bh}\left(\frac{\sin \alpha}{b} + \frac{\cos \alpha}{h}\right) \qquad (6\text{-}28) \quad \Longleftarrow$$

Numerical values. The maximum tensile and compressive stresses can be calculated from the preceding equation by substituting the given data:

$q = 3.0 \text{ kN/m} \qquad L = 1.6 \text{ m} \qquad b = 100 \text{ mm} \qquad h = 150 \text{ mm} \qquad a = 26.57°$

CHAPTER 6 ■ STRESSES IN BEAMS (ADVANCED TOPICS)

The results are

$$\sigma_E = -\sigma_D = 4.01 \text{ MPa} \qquad \longleftarrow$$

Neutral axis. In addition to finding the stresses in the beam, it is often useful to locate the neutral axis. The equation of the line is obtained by setting the stress (Eq. 6-27) equal to zero:

$$\frac{\sin \alpha}{b^2}z - \frac{\cos \alpha}{h^2}y = 0 \qquad (6\text{-}29)$$

The neutral axis is shown in Fig. 6-17b as line *nn*. The angle β from the z axis to the neutral axis is obtained from Eq. (6-29) as follows:

$$\tan \beta = \frac{y}{z} = \frac{h^2}{b^2}\tan \alpha \qquad (6\text{-}30)$$

Substituting numerical values, we get

$$\tan \beta = \frac{h^2}{b^2}\tan \alpha = \frac{(150 \text{ mm})^2}{(100 \text{ mm})^2}\left(\tan 26.57°\right) = 1.125 \qquad \beta = 48.4°$$

Since the angle β is not equal to the angle α, the neutral axis is inclined to the plane of loading (which is vertical).

From the orientation of the neutral axis (Fig. 6-17b), we see that points D and E are the farthest from the neutral axis, thus confirming our assumption that the maximum stresses occur at those points. The part of the beam above and to the right of the neutral axis is in compression, and the part to the left and below the neutral axis is in tension.

■ **Example 6-5**

A 12-foot long cantilever beam (Fig. 6-18a) is constructed from an S 24 × 80 section (see Table E-2 of Appendix E for the dimensions and properties of this beam). A load $P = 10$ k acts in the vertical direction at the end of the beam.

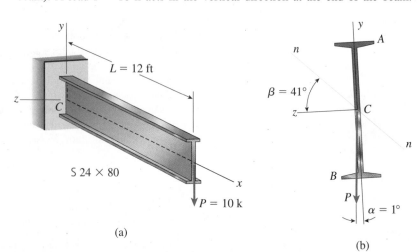

Fig. 6-18 Example 6-5. Cantilever beam with moment of inertia I_z much larger than I_y

(a)

(b)

Because the beam is very narrow compared to its height (Fig. 6-18b), its moment of inertia about the z axis is much larger than its moment of inertia about the y axis.

(a) Determine the maximum bending stresses in the beam if the y axis of the cross section is vertical and therefore aligned perfectly with the load P (Fig. 6-18a).

(b) Determine the maximum bending stresses if the beam is inclined at a small angle $\alpha = 1°$ to the load P (Fig. 6-18b). (A small inclination can be caused by imperfections in the fabrication of the beam, misalignment of the beam during construction, or movement of the supporting structure.)

Solution (a) *Maximum bending stresses when the load is aligned with the y axis.* If the beam and load are in perfect alignment, the z axis is the neutral axis and the maximum stresses in the beam (at the support) are obtained from the flexure formula:

$$\sigma_{max} = \frac{My}{I_z} = \frac{PL(h/2)}{I_z}$$

in which $M = PL$ is the bending moment at the support, h is the height of the beam, and I_z is the moment of inertia about the z axis. Substituting numerical values, we obtain

$$\sigma_{max} = \frac{(10 \text{ k})(12 \text{ ft})(12 \text{ in./ft})(12.00 \text{ in.})}{2100 \text{ in.}^4} = 8230 \text{ psi} \qquad \longleftarrow$$

This stress is tensile at the top of the beam and compressive at the bottom of the beam.

(b) *Maximum bending stresses when the load is inclined to the y axis.* We now assume that the beam has a small inclination (Fig. 6-18b), so that the angle between the y axis and the load is $\alpha = 1°$.

The components of the load P are $P\cos\alpha$ in the negative y direction and $P\sin\alpha$ in the positive z direction. Therefore, the bending moments at the support are

$$M_y = -(P\sin\alpha)L = -(10 \text{ k})(\sin 1°)(12 \text{ ft})(12 \text{ in./ft}) = -25.13 \text{ k-in.}$$

$$M_z = -(P\cos\alpha)L = -(10 \text{ k})(\cos 1°)(12 \text{ ft})(12 \text{ in./ft}) = -1440 \text{ k-in.}$$

The angle β giving the orientation of the neutral axis nn is obtained from Eq. (6-20):

$$\tan\beta = \frac{y}{z} = \frac{M_y I_z}{M_z I_y} = \frac{(-25.13 \text{ k-in.})(2100 \text{ in.}^4)}{(-1440 \text{ k-in.})(42.2 \text{ in.}^4)} = 0.8684 \qquad \beta = 41°$$

This calculation shows that the neutral axis is inclined at an angle of 41° from the z axis even though the plane of the load is inclined only 1° from the y axis. The sensitivity of the position of the neutral axis to the angle of the load is a consequence of the large I_z/I_y ratio.

From the position of the neutral axis (Fig. 6-18b), we see that the maximum stresses in the beam occur at points A and B, which are located at the farthest distances from the neutral axis. The coordinates of point A are

$$z_A = -3.50 \text{ in.} \qquad y_A = 12.0 \text{ in.}$$

Therefore, the tensile stress at point A (see Eq. 6-18) is

$$\sigma_A = \frac{M_y z_A}{I_y} - \frac{M_z y_A}{I_z}$$

$$= \frac{(-25.13 \text{ k-in.})(-3.50 \text{ in.})}{42.2 \text{ in.}^4} - \frac{(-1440 \text{ k-in.})(12.0 \text{ in.})}{2100 \text{ in.}^4}$$

$$= 2080 \text{ psi} + 8230 \text{ psi} = 10{,}310 \text{ psi} \qquad \longleftarrow$$

The stress at B has the same magnitude but is a compressive stress:

$$\sigma_B = -10{,}310 \text{ psi} \qquad \longleftarrow$$

These stresses are 25% larger than the stress $\sigma_{max} = 8230$ psi for the same beam with a perfectly aligned load. Furthermore, the inclined load produces a lateral deflection in the z direction, whereas the perfectly aligned load does not.

This example shows that beams with I_z much larger than I_y may develop large stresses if the beam or its loads deviate even a small amount from their planned alignment. Therefore, such beams should be used with caution, because they are highly susceptible to overstress and to lateral (that is, sideways) bending and buckling. The remedy is to provide adequate lateral support for the beam, thereby preventing sideways bending. For instance, wood floor joists in buildings are supported laterally by installing bridging or blocking between the joists.

6.5 BENDING OF UNSYMMETRIC BEAMS

In our previous discussions of bending, we assumed that the beams had cross sections with at least one axis of symmetry. Now we will abandon that restriction and consider beams having unsymmetric cross sections. We begin by investigating beams in pure bending, and then in later sections (Sections 6.6, 6.7, and 6.8) we will consider the effects of lateral loads. As in earlier discussions, it is assumed that the beams are made of linearly elastic materials.

Suppose that a beam having an unsymmetric cross section is subjected to a bending moment M acting at the end cross section (Fig. 6-19a). We would like to know the stresses in the beam and the position of the neutral axis. Unfortunately, at this stage of the analysis there is no direct way of determining these quantities. Therefore, we will use an indirect approach—instead of starting with a bending moment and trying to find the neutral axis, we will start with an assumed neutral axis and find the associated bending moment.

We begin by constructing two perpendicular axes (the y and z axes) at an arbitrarily selected point in the plane of the cross section (Fig. 6-19b). The axes may have any orientation, but for convenience we will orient them horizontally and vertically. Next, we *assume* that the beam is

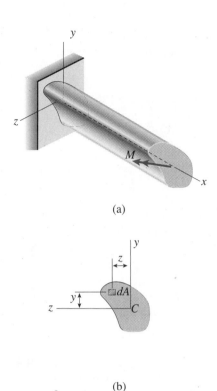

(a)

(b)

Fig. 6-19 Beam with an unsymmetric cross section

bent in such a manner that the z axis is the neutral axis of the cross section. Consequently, the beam deflects in the xy plane, which becomes the plane of bending. Under these conditions, the normal stress acting on an element of area dA located at distance y from the neutral axis (see Fig. 6-19b and Eq. 5-7 of Chapter 5) is

$$\sigma_x = -E\kappa_y y \qquad (6\text{-}31)$$

The minus sign is needed because the part of the beam above the z axis (the neutral axis) is in compression when the curvature is positive. (The sign convention for curvature when the beam is bent in the xy plane is shown in Fig. 6-20a.)

The force acting on the element of area dA is $\sigma_x dA$, and the resultant force acting on the entire cross section is the integral of this elemental force over the cross-sectional area A. Since the beam is in pure bending, the resultant force must be zero; hence,

$$\int_A \sigma_x\, dA = -\int_A E\kappa_y y\, dA = 0$$

The modulus of elasticity and the curvature are constants at any given cross section, and therefore

$$\int_A y\, dA = 0 \qquad (6\text{-}32)$$

This equation shows that the z axis (the neutral axis) passes through the centroid C of the cross section.

Now assume that the beam is bent in such a manner that the y axis is the neutral axis and the xz plane is the plane of bending. Then the normal stress acting on the element of area dA (Fig. 6-19b) is

$$\sigma_x = -E\kappa_z z \qquad (6\text{-}33)$$

The sign convention for the curvature κ_z in the xz plane is shown in Fig. 6-20b. The minus sign is needed in Eq. (6-33) because positive curvature in the xz plane produces compression on the element dA. The resultant force for this case is

$$\int_A \sigma_x\, dA = -\int_A E\kappa_z z\, dA = 0$$

from which we get

$$\int_A z\, dA = 0 \qquad (6\text{-}34)$$

and again we see that the neutral axis must pass through the centroid. Thus, we have established that *the origin of the y and z axes for an unsymmetric beam must be placed at the centroid C.*

Now let us consider the moment resultant of the stresses σ_x. Once again we assume that bending takes place with the z axis as the neutral

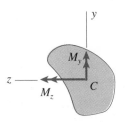

Fig. 6-20 Sign conventions for curvatures κ_y and κ_z in the xy and xz planes, respectively

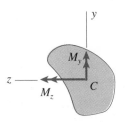

Fig. 6-21 Bending moments M_y and M_z acting about the y and z axes, respectively

axis, in which case the stresses σ_x are given by Eq. (6-31). The corresponding bending moments M_z and M_y about the z and y axes, respectively (Fig. 6-21), are

$$M_z = -\int_A \sigma_x y \, dA = \kappa_y E \int_A y^2 dA = \kappa_y E I_z \qquad (6\text{-}35\text{a})$$

$$M_y = \int_A \sigma_x z \, dA = -\kappa_y E \int_A yz \, dA = -\kappa_y E I_{yz} \qquad (6\text{-}35\text{b})$$

In these equations, I_z is the moment of inertia of the cross-sectional area with respect to the z axis and I_{yz} is the *product of inertia* with respect to the y and z axes.*

From Eqs. (6-35a) and (6-35b) we can draw the following conclusions: (1) If the z axis is selected in an arbitrary direction through the centroid, it will be the neutral axis *only* if moments M_y and M_z act about the y and z axes and *only* if these moments are in the ratio established by Eqs. (6-35a) and (6-35b). (2) If the z axis is selected as a *principal axis*, then the product of inertia I_{yz} equals zero and the only bending moment is M_z. In that case, the z axis is the neutral axis, bending takes place in the xy plane, and the moment M_z acts in that same plane. Thus, bending occurs in a manner analogous to that of a symmetric beam.

In summary, an unsymmetric beam bends in the same general manner as a symmetric beam provided the z axis is a *principal centroidal axis* and the only bending moment is the moment M_z acting about that same axis.

If we now assume that the y axis is the neutral axis, we will arrive at similar conclusions. The stresses σ_x are given by Eq. (6-33) and the bending moments are

$$M_y = \int_A \sigma_x z \, dA = -\kappa_z E \int_A z^2 dA = -\kappa_z E I_y \qquad (6\text{-}36\text{a})$$

$$M_z = -\int_A \sigma_x y \, dA = \kappa_z E \int_A yz \, dA = \kappa_z E I_{yz} \qquad (6\text{-}36\text{b})$$

in which I_y is the moment of inertia with respect to the y axis. Again we observe that if the neutral axis (the y axis in this case) is oriented arbitrarily, moments M_y and M_z must exist. However, if the y axis is a principal axis, the only moment is M_y and we have ordinary bending in the xz plane. Therefore, we can state that an unsymmetric beam bends in the same general manner as a symmetric beam when the y axis is a *principal centroidal axis* and the only bending moment is the moment M_y acting about that same axis.

One further observation—since the y and z axes are orthogonal, we know that if *either* axis is a principal axis, then the other axis is automatically a principal axis.

---

*Products of inertia are discussed in Section 12.7 of Chapter 12.

We have now arrived at the following important conclusion: *When an unsymmetric beam is in pure bending, the plane in which the bending moment acts is perpendicular to the neutral surface only if the y and z axes are principal centroidal axes of the cross section and the bending moment acts in one of the two principal planes (the xy plane or the xz plane)*. In such a case, the principal plane in which the bending moment acts becomes the plane of bending and the usual bending theory (including the flexure formula) is valid.

Having arrived at this conclusion, we now have a direct method for finding the stresses in an unsymmetric beam subjected to a bending moment acting in an arbitrary direction.

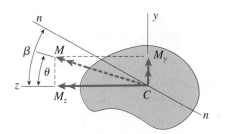

Fig. 6-22 Unsymmetric cross section with the bending moment M resolved into components M_y and M_z acting about the principal centroidal axes

■ Procedure for Analyzing an Unsymmetric Beam

We will now describe a general procedure for analyzing an unsymmetric beam subjected to any bending moment M (Fig. 6-22). We begin by locating the centroid C of the cross section and constructing a set of principal axes at that point (the y and z axes in the figure).* Next, the bending moment M is resolved into components M_y and M_z, positive in the directions shown in the figure. These components are

$$M_y = M \sin \theta \qquad M_z = M \cos \theta \qquad \text{(6-37a,b)}$$

in which θ is the angle between the moment vector M and the z axis (Fig. 6-22). Since each component acts in a principal plane, it produces pure bending in that same plane. Thus, the usual formulas for pure bending apply, and we can readily find the stresses due to the moments M_y and M_z acting separately. The bending stresses obtained from the moments acting separately are then superposed to obtain the stresses produced by the original bending moment M. (Note that this general procedure is similar to that described in the preceding section for analyzing doubly symmetric beams with inclined loads.)

The superposition of the bending stresses in order to obtain the resultant stress at any point in the cross section is given by Eq. (6-18):

$$\sigma_x = \frac{M_y z}{I_y} - \frac{M_z y}{I_z} = \frac{(M \sin \theta)z}{I_y} - \frac{(M \cos \theta)y}{I_z} \qquad \text{(6-38)}$$

in which y and z are the coordinates of the point under consideration.

Also, the equation of the neutral axis nn (Fig. 6-22) is obtained by setting σ_x equal to zero and simplifying:

$$\frac{\sin \theta}{I_y}z - \frac{\cos \theta}{I_z}y = 0 \qquad \text{(6-39)}$$

---

*Principal axes are discussed in Sections 12.8 and 12.9 of Chapter 12.

The angle β between the neutral axis and the z axis can be obtained from the preceding equation, as follows:

$$\tan \beta = \frac{y}{z} = \frac{I_z}{I_y} \tan \theta \qquad (6\text{-}40)$$

This equation shows that in general the angles β and θ are not equal, hence the neutral axis is generally not perpendicular to the plane in which the applied couple M acts. The only exceptions are the three special cases described in the preceding section (Section 6.4) in the discussion following Eq. (6-23).

In this section we have focused our attention on unsymmetric beams. Of course, symmetric beams are special cases of unsymmetric beams, and therefore the discussions of this section also apply to symmetric beams. If a beam is singly symmetric, the axis of symmetry is one of the centroidal principal axes of the cross section; the other principal axis is perpendicular to the axis of symmetry at the centroid. If a beam is doubly symmetric, the two axes of symmetry are centroidal principal axes.

In a strict sense the discussions of this section apply only to pure bending, which means that no shear forces act on the cross sections. When shear forces do exist, the possibility arises that the beam will twist about the longitudinal axis. However, twisting is avoided when the shear forces act through the *shear center*, which is described in the next section.

The following examples illustrate the analysis of beams having either one axis of symmetry or no axis of symmetry.

■ **Example 6-6**

A channel section (C 10×15.3) is subjected to a bending moment $M = 15$ k-in. oriented at an angle $\theta = 10°$ to the z axis (Fig. 6-23a). Calculate the bending stresses σ_A and σ_B at points A and B, respectively, and determine the position of the neutral axis.

Solution   *Properties of the cross section.* The centroid C is located on the axis of symmetry (the z axis) at a distance

$$c = 0.634 \text{ in.}$$

from the back of the angle (Fig. 6-24).* The y and z axes are principal centroidal axes with moments of inertia

$$I_y = 2.28 \text{ in.}^4 \qquad I_z = 67.4 \text{ in.}^4$$

Also, the coordinates of points A and B are as follows:

---

*See Table E-3, Appendix E, for dimensions and properties of channel sections.

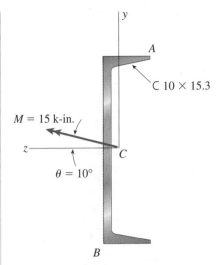

Fig. 6-23 Example 6-6. Channel section subjected to a bending moment M acting at an angle θ to the z axis

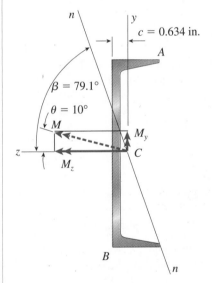

Fig. 6-24 Solution to Example 6-6

$$y_A = 5.00 \text{ in.} \qquad z_A = -2.600 \text{ in.} + 0.634 \text{ in.} = -1.966 \text{ in.}$$

$$y_B = -5.00 \text{ in.} \qquad z_B = 0.634 \text{ in.}$$

Bending moments. The bending moments about the y and z axes (Fig. 6-24) are

$$M_y = M \sin \theta = (15 \text{ k-in.})(\sin 10°) = 2.605 \text{ k-in.}$$

$$M_z = M \cos \theta = (15 \text{ k-in.})(\cos 10°) = 14.77 \text{ k-in.}$$

Bending stresses. We now calculate the stress at point A from Eq. (6-38):

$$\sigma_A = \frac{M_y z_A}{I_y} - \frac{M_z y_A}{I_z}$$

$$= \frac{(2.605 \text{ k-in.})(-1.966 \text{ in.})}{2.28 \text{ in.}^4} - \frac{(14.77 \text{ k-in.})(5.00 \text{ in.})}{67.4 \text{ in.}^4}$$

$$= -2246 \text{ psi} - 1096 \text{ psi} = -3340 \text{ psi} \qquad \longleftarrow$$

By a similar calculation we obtain the stress at point B:

$$\sigma_B = \frac{M_y z_B}{I_y} - \frac{M_z y_B}{I_z}$$

$$= \frac{(2.605 \text{ k-in.})(0.634 \text{ in.})}{2.28 \text{ in.}^4} - \frac{(14.77 \text{ k-in.})(-5.00 \text{ in.})}{67.4 \text{ in.}^4}$$

$$= 724 \text{ psi} + 1096 \text{ psi} = 1820 \text{ psi} \qquad \longleftarrow$$

These stresses are the maximum compressive and tensile stresses in the beam.

Neutral axis. The angle β that locates the neutral axis (Eq. 6-40) is found as follows:

$$\tan \beta = \frac{I_z}{I_y} \tan \theta = \frac{67.4 \text{ in.}^4}{2.28 \text{ in.}^4} \tan 10° = 5.212 \qquad \beta = 79.1°$$

The neutral axis nn is shown in Fig. 6-24, and we see that points A and B are located at the farthest distances from the neutral axis, thus confirming that σ_A and σ_B are the largest stresses in the beam.

In this example, the angle β between the z axis and the neutral axis is much larger than the angle θ (Fig. 6-24) because the ratio I_z/I_y is large. The angle β varies from 0 to 79.1° as the angle θ varies from 0 to 10°. As discussed previously in Example 6-5 of Section 6.4, beams with large I_z/I_y ratios are very sensitive to the direction of loading. Thus, beams of this kind should be provided with lateral support to prevent excessive lateral deflections.

■ **Example 6-7**

An angle section with unequal legs (L 6 × 4 × 1/2)) is shown in Fig. 6-25. The angle is subjected to a bending moment $M = 10$ k-in. having its moment vector directed along axis 1-1, which is parallel to the short leg of the angle section. Determine the maximum tensile and compressive stresses in the angle.

Fig. 6-25 Example 6-7. Angle section with unequal legs (unsymmetric section) subjected to a bending moment M acting about axis 1-1

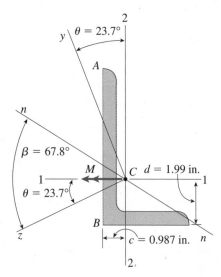

Fig. 6-26 Solution to Example 6-7

Solution   *Properties of the cross section.* The centroid C of the angle section (Fig. 6-26) is located at distance $c = 0.987$ in. from the back of the long leg of the angle and distance $d = 1.99$ in. from the back of the short leg.* The principal centroidal axes are shown as axes 3-3 and 4-4 in Table E-5 of the Appendix; however, we will label them as the y and z axes to be consistent with our earlier discussions. The angle θ between axis 2-2 and the y axis (Fig. 6-26) is the same as the angle α between axes 2-2 and 3-3 in Table E-5; thus,

$$\tan \theta = \tan \alpha = 0.440 \qquad \theta = 23.7°$$

Since in our example the moment vector M is directed along axis 1-1, it follows that the angle between the z axis and the vector M is also equal to θ.

The principal moments of inertia I_y and I_z can be found from the data in Table E-5. The table gives the radius of gyration about axis 3-3 (the y axis); therefore,

$$I_y = I_{33} = Ar_{min}^2 = (4.75 \text{ in.}^2)(0.870 \text{ in.})^2 = 3.60 \text{ in.}^4$$

The other principal moment of inertia (I_z) is obtained from the fact that the sum of the moments of inertia for all pairs of perpendicular axes is a constant (see Eq. 12-29 in Chapter 12):

$$I_y + I_z = I_{11} + I_{22}$$

from which

$$I_z = I_{11} + I_{22} - I_y = 17.4 \text{ in.}^4 + 6.27 \text{ in.}^4 - 3.60 \text{ in.}^4 = 20.1 \text{ in.}^4$$

If a table of properties were not available, it would be necessary to calculate the orientation of the principal axes and the magnitudes of the principal moments of inertia using the transformation equations given in Section 12.9 of Chapter 12.

Orientation of the neutral axis. The angle β between the z axis and the neutral axis nn is determined as follows (see Eq. 6-40):

$$\tan \beta = \frac{I_z}{I_y} \tan \theta = \frac{20.1 \text{ in.}^4}{3.60 \text{ in.}^4} \tan 23.7° = 2.45 \qquad \beta = 67.8°$$

By inspection of Fig. 6-26, we see that point B is located farthest from the neutral axis on the tensile side of the beam (below and to the left of the neutral axis). Therefore, the maximum tensile stress occurs at B. The maximum compressive stress occurs somewhere on the curved surface near point A. However, it is not worth the effort to locate this point precisely (our calculations are only valid to two or three significant digits), so we will consider the stress at A to be the maximum compressive stress for practical purposes.

Coordinates of points A and B. The coordinates of points A and B are needed in order to determine the stresses. Although the coordinates of these points with respect to the 1-1 and 2-2 axes are readily obtained from the data in Table E-5, the coordinates with respect to the y and z axes are more difficult to determine.

To make the task easier, we will derive general formulas for calculating the coordinates. In Fig. 6-27a are shown two sets of axes (axes 1-2 and axes yz) with

*See Table E-5, Appendix E, for dimensions and properties of angle sections with unequal legs.

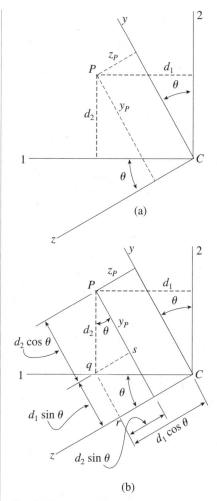

(a)

(b)

Fig. 6-27 Coordinates of point P with respect to the yz axes when the coordinates with respect to the 1-2 axes are known (Eqs. 6-41)

an angle θ between them. Point P has known coordinates d_1 and d_2 with respect to the 1-2 axes, and we wish to determine the coordinates y_P and z_P with respect to the yz axes. We can obtain the desired formulas from the geometric relationships shown in Fig. 6-27b. Let's begin with the coordinate y_P, which is equal to the distance Ps plus the distance qr. The distance Ps (from triangle Psq) equals $d_2 \cos \theta$, and the distance qr (from triangle qCr) equals $d_1 \sin \theta$. In a similar manner, we see that the coordinate z_P is equal to the distance Cr minus the distance qs. The distance Cr (from triangle qCr) equals $d_1 \cos \theta$, and the distance qs (from triangle Psq) equals $d_2 \sin \theta$. Thus, we obtain the following expressions for the coordinates y_P and z_P:

$$y_P = d_1 \sin\theta + d_2 \cos\theta \qquad z_P = d_1 \cos\theta - d_2 \sin\theta \qquad (6\text{-}41\text{a,b})$$

Knowing the coordinates d_1 and d_2 of a point with respect to the 1-2 axes, we can use the preceding equations to obtain the coordinates y_P and z_P with respect to the yz axes.

Returning now to the angle section of Fig. 6-26, we obtain the coordinates of point A with respect to the 1-2 axes by using the dimensions given in Table E-5:

$$d_1 = c = 0.987 \text{ in.} \qquad d_2 = 6.00 \text{ in.} - d = 6.00 \text{ in.} - 1.99 \text{ in.} = 4.01 \text{ in.}$$

The y and z coordinates of point A are now obtained from Eqs. (6-41a) and (6-41b):

$$y_A = (0.987 \text{ in.})(\sin 23.7°) + (4.01 \text{ in.})(\cos 23.7°) = 4.07 \text{ in.}$$

$$z_A = (0.987 \text{ in.})(\cos 23.7°) - (4.01 \text{ in.})(\sin 23.7°) = -0.71 \text{ in.}$$

In a similar manner, we find for point B the following coordinates:

$$d_1 = c = 0.987 \text{ in.} \qquad d_2 = -d = -1.99 \text{ in.}$$

$$y_B = (0.987 \text{ in.})(\sin 23.7°) + (-1.99 \text{ in.})(\cos 23.7°) = -1.43 \text{ in.}$$

$$z_B = (0.987 \text{ in.})(\cos 23.7°) - (-1.99 \text{ in.})(\sin 23.7°) = 1.70 \text{ in.}$$

An alternative method for finding the coordinates of points A and B is to make a scale drawing of the angle and the axes and then measure the desired distances.

Bending stresses at points A and B. The components of the moment M about the y and z axes (see Fig. 6-26) are

$$M_y = M \sin\theta = (10 \text{ k-in.})(\sin 23.7°) = 4.02 \text{ k-in.}$$

$$M_z = M \cos\theta = (10 \text{ k-in.})(\cos 23.7°) = 9.16 \text{ k-in.}$$

Therefore, the stress at point A (from Eq. 6-38) is

$$\sigma_A = \frac{M_y z_A}{I_y} - \frac{M_z y_A}{I_z}$$

$$= \frac{(4.02 \text{ k-in.})(-0.71 \text{ in.})}{3.60 \text{ in.}^4} - \frac{(9.16 \text{ k-in.})(4.07 \text{ in.})}{20.1 \text{ in.}^4}$$

$$= -793 \text{ psi} - 1855 \text{ psi} = -2650 \text{ psi}$$

which is the maximum compressive stress in the angle. By a similar calculation we obtain the stress at point B:

$$\sigma_B = \frac{M_y z_B}{I_y} - \frac{M_z y_B}{I_z}$$

$$= \frac{(4.02 \text{ k-in.})(1.70 \text{ in.})}{3.60 \text{ in.}^4} - \frac{(9.16 \text{ k-in.})(-1.43 \text{ in.})}{20.1 \text{ in.}^4}$$

$$= 1898 \text{ psi} + 652 \text{ psi} = 2550 \text{ psi} \quad \longleftarrow$$

which is the maximum tensile stress in the angle.

Thus, we have determined the maximum tensile and compressive stresses in an unsymmetric angle due to a bending couple M. The same general procedure can be used for other shapes of unsymmetric beams.

6.6 ▌ THE SHEAR-CENTER CONCEPT

In the preceding sections of this chapter we were concerned with determining the bending stresses in beams under a variety of special conditions. For instance, in Section 6.4 we considered symmetrical beams with inclined loads, and in Section 6.5 we considered unsymmetrical beams. However, lateral loads acting on a beam produce shear forces as well as bending moments, and therefore in this and the next two sections we will examine the effects of shear.

In Chapter 5 we saw how to determine the shear stresses in beams when the loads act in a plane of symmetry, and we derived the shear formula for calculating those stresses for certain shapes of beams. Now we will examine the shear stresses in beams when the lateral loads act in a plane that is *not* a plane of symmetry. We will find that the loads must be applied at a particular point in the cross section, called the *shear center*, if the beam is to bend without twisting.

Consider a cantilever beam of singly symmetric cross section supporting a load P at the free end (Fig. 6-28a). A beam having the cross section shown in Fig. (6-28b) is called an *unbalanced I-beam*. Beams of I-shape, whether balanced or unbalanced, are usually loaded in the plane of symmetry (the xz plane), but in this case the line of action of the force P is perpendicular to that plane. Since the origin of coordinates is taken at the centroid C of the cross section, and since the z axis is an axis of symmetry of the cross section, both the y and z axes are principal centroidal axes.

Let us assume that under the action of the load P the beam bends with the xz plane as the neutral plane, which means that the xy plane is the plane of bending. Under these conditions, two stress resultants exist at intermediate cross sections of the beam (Fig. 6-28b): a bending

(a)

(b)

Fig. 6-28 Cantilever beam with singly symmetric cross section: (a) beam with load, and (b) cross section of beam showing centroid C and shear center S

moment M_0 acting about the z axis and having its moment vector in the negative direction of the z axis, and a shear force of magnitude P acting in the negative y direction. For a given beam and loading, both M_0 and P are known quantities.

The normal stresses acting on the cross section have a resultant that is the bending moment M_0, and the shear stresses have a resultant that is the shear force (equal to P). If the material follows Hooke's law, the normal stresses vary linearly with the distance from the neutral axis (the z axis) and can be calculated from the flexure formula. Since the shear stresses acting on a cross section are determined from the normal stresses solely from equilibrium considerations (see the derivation of the shear formula in Chapter 5), it follows that the distribution of shear stresses over the cross section is also determined. The resultant of these shear stresses is a vertical force equal in magnitude to the force P and having its line of action through some point S lying on the z axis (Fig. 6-28b). This point is known as the **shear center** (also called the *center of flexure*) of the cross section.

In summary, by assuming that the z axis is the neutral axis, we can determine not only the distribution of the normal stresses but also the distribution of the shear stresses and the position of the resultant shear force. Therefore, we now recognize that a load P applied at the end of the beam (Fig. 6-28a) must act through a particular point (the shear center) if bending is to occur with the z axis as the neutral axis.

If the load is applied at some other point on the z axis (say, at point A in Fig. 6-29), it can be replaced by a statically equivalent system consisting of a force P acting at the shear center and a torque T. The force acting at the shear center produces bending about the z axis and the torque produces torsion. Therefore, we now recognize that *a lateral load acting on a beam will produce bending without twisting only if it acts through the shear center.*

The shear center (like the centroid) lies on any axis of symmetry, and therefore the shear center S and the centroid C coincide for a **doubly symmetric cross section** (Fig. 6-30a). A load P acting through the centroid produces bending about the y and z axes without torsion, and the corresponding bending stresses can be found by the method described in Section 6.4 for doubly symmetric beams.

If a beam has a **singly symmetric cross section** (Fig. 6-30b), both the centroid and the shear center lie on the axis of symmetry. A load P acting through the shear center can be resolved into components in the y and z directions. The component in the y direction will produce bending in the xy plane with the z axis as the neutral axis, and the component in the z direction will produce bending (without torsion) in the xz plane with the y axis as the neutral axis. The bending stresses produced by these components can be superposed to obtain the stresses caused by the original load.

Finally, if a beam has an **unsymmetric cross section** (Fig. 6-31), the bending analysis proceeds as follows (provided the load acts through the

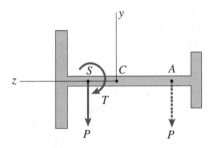

Fig. 6-29 Singly symmetric beam with load P applied at point A

(a)

(b)

Fig. 6-30 (a) Doubly symmetric beam with a load P acting through the centroid (and shear center), and (b) singly symmetric beam with a load P acting through the shear center

Fig. 6-31 Unsymmetric beam with a load P acting through the shear center S

shear center). First, locate the centroid C of the cross section and determine the orientation of the principal centroidal axes y and z. Then resolve the load into components (acting at the shear center) in the y and z directions and determine the bending moments M_y and M_z about the principal axes. Lastly, calculate the bending stresses using the method described in Section 6.5 for unsymmetric beams.

Now that we have explained the *significance* of the shear center and its use in beam analysis, it is natural to ask, "How do we *locate* the shear center?" For doubly symmetric shapes the answer, of course, is simple— it is at the centroid. For singly symmetric shapes the shear center lies on the axis of symmetry, but the precise location on that axis may not be easy to determine. Locating the shear center is even more difficult if the cross section is unsymmetric. In such cases, the task requires more advanced methods than are appropriate for this book. (Some engineering handbooks give formulas for locating shear centers; e.g., see Ref. 2-10.)

Beams of *thin-walled open cross sections*, such as wide-flange beams, channels, angles, T-beams, and Z-sections, are a special case. Not only are they in common use for structural purposes, they also are very weak in torsion. Consequently, it is especially important to locate their shear centers. Cross sections of this type are considered in the following two sections—in Section 6.7 we discuss how to find the shear stresses in such beams, and in Section 6.8 we show how to locate their shear centers.

6.7 SHEAR STRESSES IN BEAMS OF THIN-WALLED OPEN CROSS SECTIONS

The distribution of shear stresses in rectangular beams, circular beams, and in the webs of beams with flanges was described previously in Sections 5.8, 5.9, and 5.10, and we derived the shear formula (Eq. 5-38) for calculating the stresses:

$$\tau = \frac{VQ}{Ib} \tag{6-42}$$

In this formula, V represents the shear force acting on the cross section, I is the moment of inertia of the cross-sectional area (with respect to the neutral axis), b is the width of the beam at the location where the shear stress is to be determined, and Q is the first moment of the cross-sectional area outside of the location where the stress is being found.

Now we will consider the shear stresses in a special class of beams known as beams of **thin-walled open cross section**. Beams of this type are distinguished by two features: (1) the wall thickness is small compared to the height and width of the cross section, and (2) the cross section is open, as in the case of an I-beam or channel beam, rather than closed, as in the case of a hollow box beam. Examples are shown in Fig. 6-32. Beams of this type are also called **structural sections** or **profile sections**.

Fig. 6-32 Typical beams of thin-walled open cross section (wide-flange beam or I-beam, channel beam, angle section, Z-section, and T-beam)

We can determine the shear stresses in thin-walled beams of open cross section by using the same techniques we used when deriving the shear formula (Eq. 6-42). To keep the derivation as general as possible, we will consider a beam having its cross-sectional centerline *mm* of arbitrary shape (Fig. 6-33a). The *y* and *z* axes are principal centroidal axes of the cross section, and the load *P* acts parallel to the *y* axis through the shear center *S* (Fig. 6-33b). Therefore, bending will occur in the *xy* plane with the *z* axis as the neutral axis. Under these conditions, we can obtain the normal stress at any point in the beam from the flexure formula:

$$\sigma_x = -\frac{M_z y}{I_z} \tag{6-43}$$

where M_z is the bending moment about the *z* axis (positive as defined in Section 6.4 and Fig. 6-13) and *y* is the coordinate of the point under consideration.

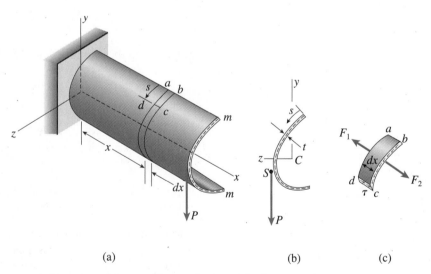

Fig. 6-33 Shear stresses in a beam of thin-walled open cross section. (The *y* and *z* axes are principal centroidal axes.)

(a)　　　　　　　(b)　　　(c)

Now consider a volume element *abcd* cut out between two cross sections distance *dx* apart (Fig. 6-33a). Note that the element begins at the edge of the cross section and has length *s* measured along the centerline *mm*. To determine the shear stresses, we isolate the element as shown in Fig. 6-33c. The resultant of the normal stresses acting on face *ad* is the force F_1 and the resultant on face *bc* is the force F_2. Since the normal stresses acting on face *ad* are larger than those acting on face *bc* (because the bending moment is larger), the force F_1 will be larger than F_2. Therefore, shear stresses τ must act along face *cd* in order for the element to be in equilibrium. These shear stresses act parallel to the top and bottom surfaces of the element and must be accompanied by complementary shear stresses acting on the cross-sectional faces *ad* and *bc*, as shown in the figure.

To evaluate these shear stresses, we sum forces in the *x* direction for element *abcd* (Fig. 6-33c); thus,

$$\tau t \, dx + F_2 - F_1 = 0 \quad \text{or} \quad \tau t \, dx = F_1 - F_2 \tag{a}$$

where *t* is the thickness of the cross section at face *cd* of the element. In other words, *t* is the thickness of the cross section at distance *s* from the free edge (Fig. 6-33b). Next, we obtain an expression for the force F_1 by using Eq. (6-43):

$$F_1 = \int_0^s \sigma_x dA = -\frac{M_{z1}}{I_z} \int_0^s y \, dA \tag{b}$$

where *dA* is an element of area on side *ad* of the volume element *abcd*, *y* is the coordinate to the element *dA*, and M_{z1} is the bending moment at the cross section. An analogous expression is obtained for the force F_2:

$$F_2 = \int_0^s \sigma_x dA = -\frac{M_{z2}}{I_z} \int_0^s y \, dA \tag{c}$$

Substituting these expressions for F_1 and F_2 into Eq. (a), we get

$$\tau = \left(\frac{M_{z2} - M_{z1}}{dx}\right) \frac{1}{I_z t} \int_0^s y \, dA \tag{d}$$

The quantity $(M_{z2} - M_{z1})/dx$ is the rate of change dM/dx of the bending moment and is equal to the shear force acting on the cross section (see Eq. 4-6):

$$\frac{dM}{dx} = \frac{M_{z2} - M_{z1}}{dx} = V_y \tag{e}$$

The shear force V_y is parallel to the *y* axis and positive in the negative direction of the *y* axis, that is, positive in the direction of the force *P* (Fig. 6-33). This convention is consistent with the sign convention previously adopted in Chapter 4 (see Fig. 4-5 for the sign convention for shear forces).

Substituting from Eq. (e) into Eq. (d), we get the following equation for the shear stress τ:

$$\tau = \frac{V_y}{I_z t} \int_0^s y\, dA \qquad (6\text{-}44)$$

This equation gives the shear stresses at any point in the cross section at distance s from the free edge. The integral on the right-hand side represents the first moment with respect to the z axis (the neutral axis) of the area of the cross section from $s = 0$ to $s = s$. Denoting this first moment by Q_z, we can write the equation for the **shear stresses** τ in the simpler form

$$\tau = \frac{V_y Q_z}{I_z t} \qquad (6\text{-}45)$$

which is analogous to the standard shear formula (Eq. 6-42). The shear stresses are directed along the centerline of the cross section and act parallel to the edges of the section. Furthermore, we tacitly assumed that these stresses have constant intensity across the thickness t of the wall, which is a valid assumption when the thickness is small. (Note that the wall thickness need not be constant but may vary as a function of the distance s.)

The **shear flow** at any point in the cross section, equal to the product of the shear stress and the thickness at that point, is

$$f = \tau t = \frac{V_y Q_z}{I_z} \qquad (6\text{-}46)$$

Because V_y and I_z are constants, the shear flow is directly proportional to Q_z. At the top and bottom edges of the cross section, Q_z is zero and hence the shear flow is also zero. The shear flow varies continuously between these end points and reaches its maximum value where Q_z is maximum, which is at the neutral axis.

Now suppose that the beam shown in Fig. 6-33 is bent by loads that act parallel to the z axis and through the shear center. Then the beam will bend in the xz plane and the y axis will be the neutral axis. In this case we can repeat the same type of analysis and arrive at the following equations for the shear stresses and shear flow (compare with Eqs. 6-45 and 6-46):

$$\tau = \frac{V_z Q_y}{I_y t} \qquad f = \tau t = \frac{V_z Q_y}{I_y} \qquad (6\text{-}47\text{a,b})$$

In these equations, V_z is the shear force parallel to the z axis and Q_y is the first moment with respect to the y axis.

In summary, we have derived expressions for the shear stresses in beams of thin-walled open cross sections with the stipulations that the shear force must act through the shear center and must be parallel to one of the principal centroidal axes. If the shear force is inclined to the y and z axes (but still acts through the shear center), it can be resolved into components parallel to the principal axes. Then two separate analyses can be made, and the results can be superimposed.

To illustrate the use of the shear-stress equation, we will consider the shear stresses in a wide-flange beam. Later, in Section 6.8, we will use the shear-stress equation to locate the shear centers of several thin-walled beams with open cross sections.

■ Shear Stresses in a Wide-Flange Beam

Let us investigate the shear stresses in the wide-flange beam of Fig. 6-34a. This beam is loaded by a force P acting in the plane of the web, that is, through the shear center, which coincides with the centroid of the cross section. The cross-sectional dimensions are shown in Fig. 6-34b, where we note that b is the flange width, h is the height between center-lines of the flanges, t_f is the flange thickness, and t_w is the web thickness.

We begin by considering the shear stresses at section bb in the right-hand part of the top flange (Fig. 6-34b). Since the distance s has its origin at the edge of the section (point a), the cross-sectional area between point a and section bb is st_f. Also, the distance from the centroid of this area to the neutral axis is $h/2$, and therefore its first moment Q_z is equal to $st_f h/2$. Thus, the shear stress τ_f in the flange at section bb (from Eq. 6-45) is

$$\tau_f = \frac{P(st_f h/2)}{I_z t_f} = \frac{shP}{2I_z} \tag{6-48}$$

The direction of this stress can be determined by examining the forces acting on element A, which is cut out of the flange between point a and section bb (see Figs. 6-34a and b). The element is drawn to a larger scale in Fig. 6-34c in order to show clearly the forces and stresses acting on it. We recognize immediately that the tensile force F_1 is larger than the force F_2, because the bending moment is larger on the rear face of the element than it is on the front face. It follows that the shear stress on the left-hand face of element A must act toward the reader if the element is to be in equilibrium. From this observation it follows that the shear stresses on the front face of element A must act toward the left.

Returning now to Fig. 6-34b, we see that we have completely determined the magnitude and direction of the shear stress at section bb, which may be located anywhere between point a and the junction of the flange and web. Thus, the shear stresses throughout the upper right-hand part of the flange are horizontal, act to the left, and have a magnitude given by Eq. (6-48). As seen from that equation, the shear stresses

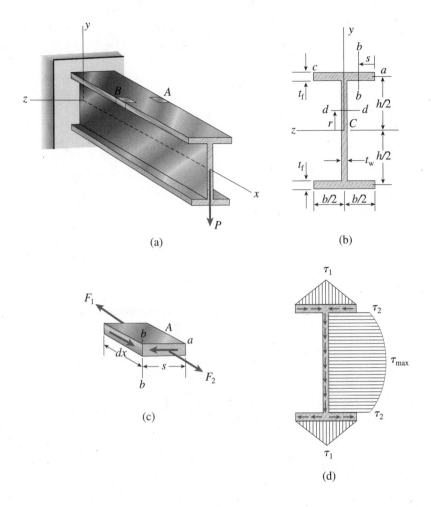

Fig. 6-34 Shear stresses in a wide-flange beam

increase linearly with the distance s. The variation of the stresses is shown graphically in Fig. 6-34d, and we see that the stresses vary from zero at point a (where $s = 0$) to a maximum value τ_1 at $s = b/2$:

$$\tau_1 = \frac{bhP}{4I_z} \tag{6-49}$$

The corresponding shear flow is

$$f_1 = \tau_1 t_f = \frac{bht_fP}{4I_z} \tag{6-50}$$

Note that we have calculated the shear stress and shear flow at the junction of the *centerlines* of the flange and web, using only centerline dimensions of the cross section in the calculations. This approximate procedure simplifies the calculations and is satisfactory for thin-walled cross sections.

By beginning at point c on the left-hand part of the top flange (Fig. 6-34b) and measuring s toward the right, we can repeat the same type of analysis. We will find that the magnitude of the shear stresses is again given by Eqs. (6-48) and (6-49). However, by cutting out an element B (Fig. 6-34a) and considering its equilibrium, we find that the shear stresses on the cross section now act toward the right, as shown in Fig. 6-34d.

The next step is to determine the shear stresses acting in the web. Considering a horizontal cut at the top of the web (at the junction of the flange and web), we find the first moment about the neutral axis to be $Q_z = bt_f h/2$, so that the corresponding shear stress is

$$\tau_2 = \frac{bht_f P}{2I_z t_w} \tag{6-51}$$

The associated shear flow is

$$f_2 = \tau_2 t_w = \frac{bht_f P}{2I_z} \tag{6-52}$$

Note that the shear flow f_2 is equal to twice the shear flow f_1, which is expected since the shear flows in the two halves of the upper flange combine to produce the shear flow at the top of the web.

The shear stresses in the web act downward and increase in magnitude until the neutral axis is reached. At section dd, located at distance r from the neutral axis (Fig. 6-34b), the shear stress τ_w in the web is calculated as follows:

$$Q_z = \frac{bt_f h}{2} + \left(\frac{h}{2} - r\right)(t_w)\left(\frac{h/2 + r}{2}\right) = \frac{bt_f h}{2} + \frac{t_w}{2}\left(\frac{h^2}{4} - r^2\right)$$

$$\tau_w = \left(\frac{bt_f h}{t_w} + \frac{h^2}{4} - r^2\right)\frac{P}{2I_z} \tag{6-53}$$

When $r = h/2$ this equation reduces to Eq. (6-51), and when $r = 0$ it gives the maximum shear stress:

$$\tau_{max} = \left(\frac{bt_f}{t_w} + \frac{h}{4}\right)\frac{Ph}{2I_z} \tag{6-54}$$

Again it should be noted that we have made all calculations on the basis of the centerline dimensions of the cross section. For this reason, the shear stresses in the web of a wide-flange beam calculated from Eq. (6-53) may be slightly different from those obtained by the more exact analysis made in Chapter 5 (see Eq. 5-46 of Section 5.10).

The shear stresses in the web vary parabolically, as shown in Fig. 6-34d, although the variation is not large. The ratio of τ_{max} to τ_2 is

$$\frac{\tau_{max}}{\tau_2} = 1 + \frac{ht_w}{4bt_f} \tag{6-55}$$

For instance, if we take typical values of $h = 2b$ and $t_f = 2t_w$, the ratio is $\tau_{max}/\tau_2 = 1.25$.

As the final step, we can investigate the shear stresses in the lower flange using the same methods we used for the top flange. We will find that the magnitudes of the stresses are the same as in the top flange, but the directions are as shown in Fig. 6-34d.

From Fig. 6-34d we see that the shear stresses on the cross section "flow" inward from the outermost edges of the top flange, then down through the web, and finally outward to the edges of the bottom flange. Because this flow is always continuous in any structural section, it serves as a convenient method for determining the directions of the stresses. For instance, if the shear force acts downward on the beam of Fig. 6-34a, we know immediately that the shear flow in the web must also be downward. Knowing the direction of the shear flow in the web, we also know the directions of the shear flows in the flanges because of the required continuity in the flow. Using this simple technique to get the directions of the shear stresses is easier than visualizing the directions of the forces acting on elements such as A (Fig. 6-34c) cut out from the beam.

The resultant of all the shear stresses acting on the cross section is clearly a vertical force, because the horizontal stresses in the flanges produce no resultant. The shear stresses in the web have a resultant R, which can be found by integrating the shear stresses over the height of the web, as follows:

$$R = \int \tau \, dA = 2 \int_0^{h/2} \tau t_w \, dr$$

Substituting from Eq. (6-53), we get

$$R = 2t_w \int_0^{h/2} \left(\frac{bt_f h}{t_w} + \frac{h^2}{4} - r^2 \right)\left(\frac{P}{2I_z} \right) dr = \left(\frac{bt_f}{t_w} + \frac{h}{6} \right)\frac{h^2 t_w P}{2I_z} \qquad (6\text{-}56)$$

The moment of inertia I_z can be calculated as follows (using centerline dimensions):

$$I_z = \frac{t_w h^3}{12} + \frac{bt_f h^2}{2} \qquad (6\text{-}57)$$

in which the first term is the moment of inertia of the web and the second term is the moment of inertia of the flanges. When this expression for I_z is substituted into Eq. (6-56), we get $R = P$, which demonstrates that the resultant of the shear stresses acting on the cross section is equal to the load. Furthermore, the line of action of the resultant is in the plane of the web, and therefore the resultant passes through the shear center.

The preceding analysis shows how the shear stresses in a thin-walled open section can be found. The results also provide a more complete picture of the shear stresses acting on the cross sections of a wide-flange or I-beam (recall that in our previous discussions in Chapter

5, we only investigated the shear stresses in the web). Additional examples are given in the next section, where the shear stresses in a channel section and an angle section are determined as part of the process of locating the shear center.

6.8 SHEAR CENTERS OF THIN-WALLED OPEN SECTIONS

In Section 6.7 we developed methods for finding the shear stresses in beams of thin-walled open cross section. Now we will use those methods to locate the shear centers of several shapes of beams. Only beams with singly symmetric or unsymmetric cross sections will be considered, because we already know that the shear center of a doubly symmetric cross section is located at the centroid.

The procedure for locating the shear center consists of two principal parts: first, evaluating the shear stresses acting on the cross section when bending occurs about one of the principal axes, and second, determining the resultant of those stresses. The shear center is located on the line of action of the resultant. By considering bending about *both* principal axes, we can determine the position of the shear center.

As in Section 6.7, we will use only centerline dimensions when deriving formulas and making calculations. This procedure is satisfactory if the beam is thin walled, that is, if the thickness of the beam is small compared to the other dimensions of the cross section.

■ Channel Section

The first beam to be analyzed is a singly symmetric channel section (Fig. 6-35a). From the discussions in Section 6.6 we know immediately that the shear center is located on the axis of symmetry (the z axis). To

Fig. 6-35 Shear center S of a channel section

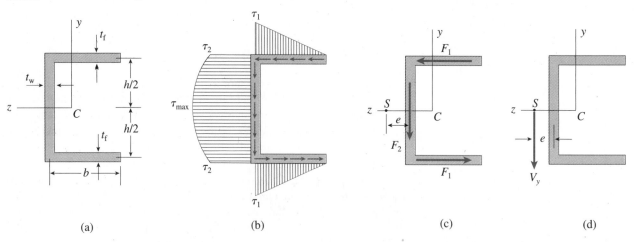

(a)   (b)   (c)   (d)

find the position of the shear center on the z axis, we assume that the beam is bent about the z axis as the neutral axis, and then we determine the line of action of the resultant shear force V_y acting parallel to the y axis. The shear center is located where the line of action of V_y intersects the z axis. (Note that the origin of axes is at the centroid C, so that both the y and z axes are principal centroidal axes.)

Based upon the discussions in Section 6.7, we conclude that the shear stresses in a channel vary linearly in the flanges and parabolically in the web (Fig. 6-35b). We can find the resultant of those stresses if we know the maximum stress τ_1 in the flange, the stress τ_2 at the top of the web, and the maximum stress $\tau_{\max}$ in the web.

To find the stress τ_1 in the flange, we use Eq. (6-45) with Q_z equal to the first moment of the flange area about the z axis:

$$Q_z = \frac{bt_f h}{2} \tag{a}$$

in which b is the flange width, t_f is the flange thickness, and h is the height of the beam. (Note again that the dimensions b and h are measured along the centerline of the section.) Thus, the stress τ_1 in the flange is

$$\tau_1 = \frac{V_y Q_z}{I_z t_f} = \frac{bhV_y}{2I_z} \tag{6-58}$$

where I_z is the moment of inertia about the z axis.

The stress τ_2 at the top of the web is obtained in a similar manner but with the thickness equal to the web thickness instead of the flange thickness:

$$\tau_2 = \frac{V_y Q_z}{I_z t_w} = \frac{bt_f h V_y}{2t_w I_z} \tag{6-59}$$

Also, at the neutral axis the first moment of area is

$$Q_z = \frac{bt_f h}{2} + \frac{ht_w}{2}\left(\frac{h}{4}\right) = \left(bt_f + \frac{ht_w}{4}\right)\frac{h}{2} \tag{b}$$

Therefore, the maximum stress is

$$\tau_{\max} = \frac{V_y Q_z}{I_z t_w} = \left(\frac{bt_f}{t_w} + \frac{h}{4}\right)\frac{hV_y}{2I_z} \tag{6-60}$$

The stresses τ_1 and τ_2 in the lower half of the beam are equal to the corresponding stresses in the upper half (Fig. 6-35b).

The horizontal shear force F_1 in either flange (Fig. 6-35c) can be found from the triangular stress diagrams. Each force is equal to the area of the stress triangle multiplied by the thickness of the flange:

$$F_1 = \left(\frac{\tau_1 b}{2}\right)(t_f) = \frac{hb^2 t_f V_y}{4I_z} \tag{6-61}$$

The vertical force F_2 in the web must be equal to the shear force V_y, since the forces in the flanges have no vertical components. As a check, we can verify that $F_2 = V_y$ by considering the parabolic stress diagram of Fig. 6-35b. The diagram is made up of two parts—a rectangle of area $\tau_2 h$ and a parabolic segment of area

$$\frac{2}{3}(\tau_{max} - \tau_2)h$$

Thus, the shear force F_2, equal to the area of the stress diagram times the web thickness t_w, is

$$F_2 = \tau_2 h t_w + \frac{2}{3}(\tau_{max} - \tau_2)h t_w$$

Substituting the expressions for τ_2 and τ_{max} (Eqs. 6-59 and 6-60) into the preceding equation, we obtain

$$F_2 = \left(\frac{t_w h^3}{12} + \frac{bh^2 t_f}{2}\right)\frac{V_y}{I_z} \tag{c}$$

Finally, we note that the expression for the moment of inertia is

$$I_z = \frac{t_w h^3}{12} + \frac{bh^2 t_f}{2} \tag{6-62}$$

in which we again base the calculations upon centerline dimensions. Substituting this expression for I_z into Eq. (c) for F_2, we get

$$F_2 = V_y \tag{d}$$

as expected.

The three forces acting on the cross section (Fig. 6-35c) have a resultant V_y that intersects the z axis at the shear center S (Fig. 6-35d). Hence, the moment of the three forces about any point in the cross section must be equal to the moment of the force V_y about that same point. This moment relationship provides an equation from which the position of the shear center may be found. As an illustration, let us select the shear center itself as the center of moments. In that case, the moment of the three forces (Fig. 6-35c) is $F_1 h - F_2 e$, where e is the distance from the centerline of the web to the shear center, and the moment of the resultant force V_y is zero (Fig. 6-35d). Equating these moments gives

$$F_1 h - F_2 e = 0 \tag{6-63}$$

Substituting for F_1 from Eq. (6-61) and for F_2 from Eq. (d), and then solving for e, we get

$$e = \frac{b^2 h^2 t_f}{4I_z} \tag{6-64}$$

When the expression for I_z (Eq. 6-62) is substituted, Eq. (6-64) becomes

$$e = \frac{3b^2 t_f}{h t_w + 6 b t_f} \tag{6-65}$$

Thus, we have determined the position of the shear center of a channel section.

As explained in Section 6.6, a channel beam will undergo bending without twisting whenever it is loaded by forces acting through the shear center. If the loads act parallel to the y axis but through some point other than the shear center (for example, if the loads act in the plane of the web), they can be replaced by a statically equivalent force system consisting of loads through the shear center and twisting couples. We then have a combination of bending and torsion of the beam. If the loads act along the z axis, we have simple bending about the y axis. If the loads act in skew directions through the shear center, they can be replaced by statically equivalent loads acting parallel to the y and z axes.

■ Angle Section

The next shape to be considered is an equal-leg angle section (Fig. 6-36a), in which each leg of the angle has length b and thickness t. The z axis is an axis of symmetry and the origin of coordinates is at the centroid C; therefore, both the y and z axes are principal centroidal axes.

To locate the shear center, we will follow the same general procedure as that described for a channel section, because we wish to determine the distribution of the shear stresses as part of the analysis. However, as we will see later, the shear center of an angle section can be determined by inspection.

We begin by assuming that the section is subjected to a shear force V_y acting parallel to the y axis. Then we use Eq. (6-45) to find the

Fig. 6-36 Shear center of an equal-leg angle section

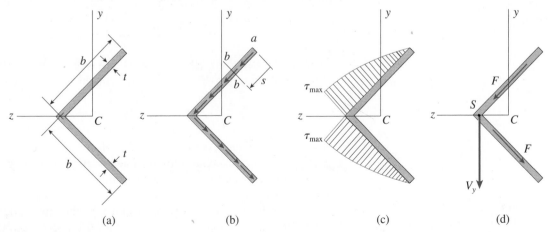

(a)   (b)   (c)   (d)

corresponding shear stresses in the legs of the angle. For this purpose we need the first moment of the cross-sectional area between point a at the outer edge of the beam (Fig. 6-36b) and section bb located at distance s from point a. The area is equal to st and its centroidal distance from the neutral axis is

$$\frac{b - s/2}{\sqrt{2}}$$

Thus, the first moment of the area is

$$Q_z = st\left(\frac{b - s/2}{\sqrt{2}}\right) \tag{6-66}$$

Substituting into Eq. (6-45), we get the following expression for the shear stress at distance s from the edge of the cross section:

$$\tau = \frac{V_y Q_z}{I_z t} = \frac{V_y s}{I_z \sqrt{2}}\left(b - \frac{s}{2}\right) \tag{6-67}$$

The moment of inertia I_z can be obtained from Case 24 of Appendix D with $\beta = 45°$:

$$I_z = 2I_{BB} = 2\left(\frac{tb^3}{6}\right) = \frac{tb^3}{3} \tag{6-68}$$

Substituting this expression for I_z into Eq. (6-67), we get

$$\tau = \frac{3V_y s}{b^3 t \sqrt{2}}\left(b - \frac{s}{2}\right) \tag{6-69}$$

This equation gives the shear stress at any point along the leg of the angle. The stress varies quadratically with s, as shown in Fig. 6-36c.

The maximum value of the stress occurs at the intersection of the legs of the angle and is obtained from Eq. (6-69) by substituting $s = b$:

$$\tau_{max} = \frac{3V_y}{2bt\sqrt{2}} \tag{6-70}$$

The shear force F in each leg (Fig. 6-36d) is equal to the area of the parabolic stress diagram (Fig. 6-36c) times the thickness t of the legs:

$$F = \frac{2}{3}(\tau_{max}b)(t) = \frac{V_y}{\sqrt{2}} \tag{6-71}$$

Since the horizontal components of the forces F cancel each other, only the vertical components remain. Each vertical component is equal to $F/\sqrt{2}$, or $V_y/2$, and therefore the resultant vertical force is equal to the shear force V_y.

Since the resultant force passes through the intersection point of the lines of action of the two forces F (Fig. 6-36d), we see that the shear center S is located at the junction of the two legs of the angle.

■ Sections Consisting of Two Intersecting Narrow Rectangles

In the preceding discussion of an angle section we evaluated the shear stresses and the forces in the legs in order to illustrate the general methodology for analyzing thin-walled open sections. However, if our sole objective had been to locate the shear center, it would not have been necessary to evaluate the stresses and forces (Eqs. 6-69 and 6-71). Since the shear stresses are parallel to the centerlines of the legs (Fig. 6-36b), we would have known immediately that their resultants are two forces F (Fig. 6-36d). The resultant of those two forces is a single force that passes through their point of intersection. Consequently, this point is the shear center. Thus, we can determine the location of the shear center of an equal-leg angle section by a simple line of reasoning (without making *any* calculations).

The same line of reasoning is valid for all cross sections consisting of two thin, intersecting rectangles (Fig. 6-37). In each case the resultants of the shear stresses are forces that intersect at the junction of the rectangles. Therefore, the shear center S is located at that point.

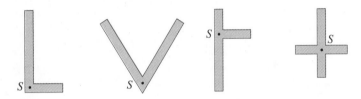

Fig. 6-37 Shear centers of sections consisting of two thin intersecting rectangles

■ Z-Section

Let us now determine the location of the shear center of a Z-section having thin walls (Fig. 6-38a). The section has no axes of symmetry but is symmetric about the centroid C (see Section 12.2 of Chapter 12 for a discussion of *symmetry about a point*). The y and z axes are principal axes through the centroid.

We begin by assuming that a shear force V_y acts parallel to the y axis and causes bending about the z axis as the neutral axis. Then the shear stresses in the flanges and web will be directed as shown in Fig. 6-38a. From symmetry considerations we conclude that the forces F_1 in the two flanges must be equal to each other (Fig. 6-38b). The resultant of the three forces acting on the cross section (F_1 in the flanges and F_2 in the web) must be equal to the shear force V_y. The forces F_1 have a resultant $2F_1$ acting through the centroid and parallel to the flanges. This force intersects the force F_2 at the centroid C, and therefore we conclude that the line of action of the shear force V_y is through the centroid.

If the beam is subjected to a shear force V_z parallel to the z axis, we arrive at a similar conclusion, namely, that the shear force acts through the centroid. Since the shear center is located at the intersection of the

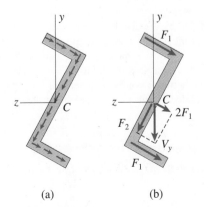

(a)   (b)

Fig. 6-38 Shear center of a thin-walled Z-section

lines of action of the two shear forces, we conclude that the shear center of the Z-section coincides with the centroid.

This conclusion applies to any Z-section that is symmetric about the centroid, that is, any Z-section having identical flanges (same width and same thickness). Note, however, that the thickness of the web does not have to be the same as the thickness of the flanges.

The locations of the shear centers for many other structural shapes are given in the problems at the end of this chapter.*

■ **Example 6-8**

A thin-walled semicircular cross section of radius r and thickness t is shown in Fig. 6-39a. Determine the distance e from the center O of the circular arcs to the shear center S.

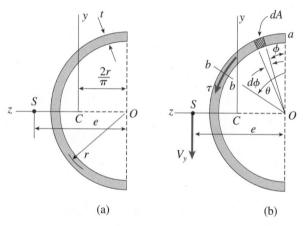

Fig. 6-39 Example 6-8. Shear center of a thin-walled semicircular section

(a)   (b)

Solution   We know immediately that the shear center is located somewhere on the axis of symmetry (the z axis). To determine the exact position, we assume that the beam is bent by a shear force V_y acting parallel to the y axis and producing bending about the z axis as the neutral axis (Fig. 6-39b).

The first step is to determine the shear stresses τ acting on the cross section (Fig. 6-39b). We consider a section bb defined by the distance s measured along the centerline of the cross section from point a. The central angle subtended between point a and section bb is denoted θ. Therefore, the distance s equals $r\theta$, where r is the radius of the centerline and θ is measured in radians. To evaluate the first moment of the cross-sectional area between point a and section bb, we identify an element of area dA (shown shaded in the figure) and integrate as follows:

$$Q_z = \int y \, dA = \int_0^{\theta} (r \cos \phi)(tr \, d\phi) = r^2 t \sin \theta \qquad \text{(e)}$$

in which ϕ is the angle to the element of area and t is the thickness of the section. Thus, the shear stress τ at section bb is

---

*The first determination of a shear center was made by Timoshenko in 1913 (Ref. 6-1). For the historical development of the shear-center concept, see Refs. 6-1 through 6-6.

$$\tau = \frac{V_y Q_z}{I_z t} = \frac{V_y r^2 \sin \theta}{I_z} \qquad \text{(f)}$$

Substituting $I_z = \pi r^3 t / 2$ (see Case 23 of Appendix D), we get

$$\tau = \frac{2 V_y \sin \theta}{\pi r t} \qquad \text{(6-72)}$$

When $\theta = 0$ or $\theta = \pi$, this expression gives $\tau = 0$, as expected. When $\theta = \pi/2$, it gives the maximum shear stress.

The resultant of the shear stresses must be the vertical shear force V_y. Therefore, the moment M_0 of the shear stresses about the center O must equal the moment of the force V_y about that same point:

$$M_0 = V_y e \qquad \text{(g)}$$

To evaluate M_0, we begin by noting that the shear stress τ acting on the element of area dA (Fig. 6-39b) is

$$\tau = \frac{2 V_y \sin \phi}{\pi r t}$$

as found from Eq. (6-72). The corresponding force is $\tau \, dA$, and the moment of this force is

$$dM_0 = r(\tau \, dA) = \frac{2 V_y \sin \phi \, dA}{\pi t}$$

Since $dA = t r d\phi$, this expression becomes

$$dM_0 = \frac{2 r V_y \sin \phi \, d\phi}{\pi}$$

Therefore, the moment produced by the shear stresses is

$$M_0 = \int dM_0 = \int_0^\pi \frac{2 r V_y \sin \phi \, d\phi}{\pi} = \frac{4 r V_y}{\pi} \qquad \text{(h)}$$

It follows from Eq. (g) that the distance e to the shear center is

$$e = \frac{M_0}{V_y} = \frac{4 r}{\pi} \approx 1.27 r \qquad \text{(6-73)} \quad \blacktriangleleft$$

This result shows that the shear center S is located outside of the semicircular section.

Note: The distance from the center O of the circular arcs to the centroid C of the cross section (Fig. 6-39a) is $2r/\pi$ (from Case 23 of Appendix D), which is one-half of the distance e. Thus, the centroid is located midway between the shear center and the center of the arcs.

The location of the shear center in a more general case of a thin-walled circular section is determined in Problem 6.8-12.

6.9 ■ ELASTOPLASTIC BENDING

In our previous discussions of bending we assumed that the beams were made of materials that followed Hooke's law (linearly elastic materials). Now we will consider the bending of beams when the stresses exceed the proportional limit of the material. When that happens, the distribution of the stresses is no longer linear but varies according to the shape of the stress-strain curve. In this section we will analyze beams of elastoplastic materials, and in the next section we will consider more general kinds of stress-strain curves.

Elastoplastic materials were discussed earlier when we analyzed axially loaded bars and bars in torsion (see Sections 2.12 and 3.12). As explained in those sections, elastoplastic materials follow Hooke's law up to the yield stress σ_y and then yield plastically under constant stress (see the stress-strain diagram of Fig. 6-40). From the figure, we see that an elastoplastic material has a region of linear elasticity between regions of perfect plasticity. Throughout this section, we will assume that the material has the same yield stress σ_y and same yield strain ϵ_y in both tension and compression.

Structural steels are the best examples of elastoplastic materials because they have sharply defined yield points and undergo large strains during yielding. The assumption of perfect plasticity after the yield stress is reached means that we are disregarding the effects of strain hardening. Since strain hardening provides an increase in strength, this assumption is on the side of safety.

Let us consider a beam of elastoplastic material subjected to a bending moment M that causes bending in the xy plane (Fig. 6-41). When the bending moment is small, the maximum stress in the beam is less than the yield stress σ_y, and therefore the beam is in the same condition as a beam in ordinary elastic bending with a linear stress distribution, as shown in Fig. 6-42b. Under these conditions, the neutral axis passes through the centroid of the cross section and the normal stresses are obtained from the flexure formula ($\sigma = -My/I$). Since the bending moment is positive, the stresses are compressive above the z axis and tensile below it. These conditions exist until the stress in the beam at the point farthest from the neutral axis reaches the yield stress σ_y, either in tension or in compression (Fig. 6-42c). The bending moment in the beam when the maximum stress just reaches the yield stress, called the **yield moment** M_y, can be obtained from the flexure formula:

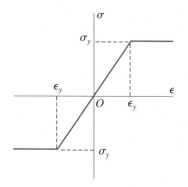

Fig. 6-40 Idealized stress-strain diagram for an elastoplastic material

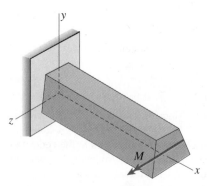

Fig. 6-41 Beam of elastoplastic material subjected to a positive bending moment M

$$M_y = \frac{\sigma_y I}{c} = \sigma_y S \qquad (6\text{-}74)$$

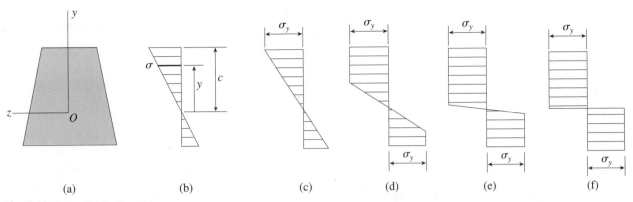

Fig. 6-42 Stress distributions in a beam of elastoplastic material

in which c is the distance to the point farthest from the neutral axis and S is the corresponding section modulus.

If we now increase the bending moment above the yield moment M_y, the strains in the beam will continue to increase and the maximum strain will exceed the yield strain ϵ_y. However, because of perfectly plastic yielding, the maximum stress will remain constant and equal to σ_y, as pictured in Fig. 6-42d. Note that the outer regions of the beam have become fully plastic while a central core (called the *elastic core*) remains linearly elastic. If the z axis is not an axis of symmetry (singly symmetric cross section), the neutral axis moves away from the centroid when the yield moment is exceeded. This shift in the location of the neutral axis is not large, and in the case of the trapezoidal cross section of Fig. 6-42, it is too small to be seen in the figure. If the cross section is doubly symmetric, the neutral axis passes through the centroid even when the yield moment is exceeded.

As the bending moment increases still further, the plastic region enlarges and moves inward toward the neutral axis until the condition shown in Fig. 6-42e is reached. At this stage the maximum strain in the beam (at the farthest distance from the neutral axis) is perhaps 10 or 15 times the yield strain ϵ_y and the elastic core has almost disappeared. Thus, for practical purposes the beam has reached its ultimate moment-resisting capacity, and we can idealize the ultimate stress distribution as consisting of two rectangular parts (Fig. 6-42f). The bending moment corresponding to this idealized stress distribution, called the **plastic moment** M_p, represents the maximum moment that can be sustained by a beam of elastoplastic material.

To find the plastic moment M_p, we begin by locating the **neutral axis** of the cross section under fully plastic conditions. For this purpose, consider the cross section shown in Fig. 6-43a and let the z axis be the neutral axis. Every point in the cross section above the neutral axis is subjected to a compressive stress σ_y (Fig. 6-43b), and every point below the neutral axis is subjected to a tensile stress σ_y. The resultant compressive force C is equal to σ_y times the cross-sectional area A_1 above the neutral axis (Fig. 6-43a), and the resultant tensile force T equals σ_y times the area A_2 below the neutral axis. Since the resultant force acting on the cross section is zero, it follows that

$$T = C \quad \text{or} \quad A_1 = A_2 \tag{a}$$

Because the total area A of the cross section is equal to $A_1 + A_2$, we see that

$$A_1 = A_2 = \frac{A}{2} \tag{6-75}$$

Therefore, under fully plastic conditions, *the neutral axis divides the cross section into two equal areas.* As a result, the location of the neutral axis for the plastic moment M_p may be different from its location for linearly elastic bending. For instance, in the case of a trapezoidal cross section that is narrower at the top than at the bottom (Fig. 6-43a), the neutral axis for fully plastic bending is slightly below the neutral axis for linearly elastic bending.

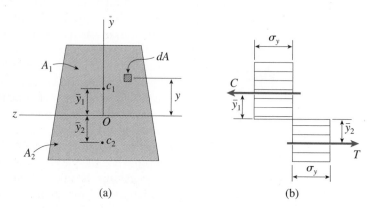

Fig. 6-43 Location of the neutral axis and determination of the plastic moment M_p under fully plastic conditions

(a)   (b)

Since the plastic moment M_p is the moment resultant of the stresses acting on the cross section, it can be found by integrating over the cross-sectional area A (Fig. 6-43a):

$$M_p = -\int_A \sigma y \, dA = -\int_{A_1} (-\sigma_y) y \, dA - \int_{A_2} \sigma_y y \, dA$$

$$= \sigma_y(\bar{y}_1 A_1) - \sigma_y(-\bar{y}_2 A_2) = \frac{\sigma_y A(\bar{y}_1 + \bar{y}_2)}{2}$$

in which y is the coordinate (positive upward) of the element of area dA and $\bar{y}_1$ and $\bar{y}_2$ are the distances from the neutral axis to the centroids c_1 and c_2 of areas A_1 and A_2, respectively.

An easier way to obtain the plastic moment is to evaluate the moments about the neutral axis of the forces C and T (Fig. 6-43b):

$$M_p = C\bar{y}_1 + T\bar{y}_2$$

Replacing T and C by $\sigma_y A/2$, we get

$$M_p = \frac{\sigma_y A(\bar{y}_1 + \bar{y}_2)}{2} \tag{6-76}$$

The procedure for any particular beam is to divide the cross section into two equal areas, locate the centroid of each half, and then use Eq. (6-76) to calculate M_p.

The expression for the plastic moment can be written in a form similar to that for the yield moment (Eq. 6-74), as follows:

$$M_p = \sigma_y Z \tag{6-77}$$

in which

$$Z = \frac{A(\bar{y}_1 + \bar{y}_2)}{2} \tag{6-78}$$

is the **plastic modulus** (or the *plastic section modulus*) for the cross section. The plastic modulus may be interpreted geometrically as the first moment (evaluated with respect to the neutral axis) of the area of the cross section above the neutral axis plus the first moment of the area below the neutral axis.

The ratio of the plastic moment to the yield moment is solely a function of the shape of the cross section and is called the **shape factor** f:

$$f = \frac{M_p}{M_y} = \frac{Z}{S} \tag{6-79}$$

This factor is a measure of the reserve strength of the beam after yielding first begins. It is highest when most of the material is located near the neutral axis (for instance, a beam having a solid circular section), and lowest when most of the material is away from the neutral axis (for instance, a beam having a wide-flange section). Values of f for cross sections of rectangular, wide-flange, and circular shapes are given in the remainder of this section. Other shapes are considered in the problems at the end of the chapter.

■ **Beams of Rectangular Cross Section**

Now let us determine the properties of a beam of rectangular cross section (Fig. 6-44) when the material is elastoplastic. The section modulus is $S = bh^2/6$, and therefore the yield moment (Eq. 6-74) is

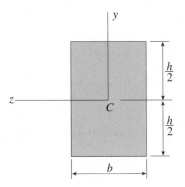

Fig. 6-44 Rectangular cross section

$$M_y = \frac{\sigma_y b h^2}{6} \tag{6-80}$$

in which b is the width and h is the height of the cross section.

Because the cross section is doubly symmetric, the neutral axis passes through the centroid even when the beam is loaded into the plastic range. Consequently, the distances to the centroids of the areas above and below the neutral axis are

$$\bar{y}_1 = \bar{y}_2 = \frac{h}{4}$$

Therefore, the plastic modulus (Eq. 6-78) is

$$Z = \frac{A(\bar{y}_1 + \bar{y}_2)}{2} = \frac{bh}{2}\left(\frac{h}{4} + \frac{h}{4}\right) = \frac{bh^2}{4} \tag{b}$$

and the plastic moment (Eq. 6-77) is

$$M_p = \frac{\sigma_y b h^2}{4} \tag{6-81}$$

Finally, the shape factor for a rectangular cross section is

$$f = \frac{M_p}{M_y} = \frac{Z}{S} = \frac{3}{2} \tag{6-82}$$

which means that the plastic moment for a rectangular beam is 50% greater than the yield moment.

Next, we consider the stresses in a rectangular beam when the bending moment M is greater than the yield moment but has not yet reached the plastic moment. The outer parts of the beam will be at the yield stress σ_y and the inner part (the elastic core) will have a linearly varying stress distribution (Figs. 6-45a and b). The fully plastic zones are shaded in Fig. 6-45a, and the distances from the neutral axis to the outer edges of the elastic core are denoted by e.

Fig. 6-45 Stress distribution in a beam of rectangular cross section with an elastic core ($M_y \leq M \leq M_p$)

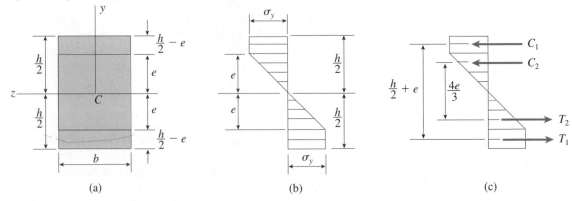

(a)   (b)   (c)

The stresses acting on the cross section have the force resultants C_1, C_2, T_1, and T_2, as shown in Fig. 6-45c. The forces C_1 and T_1 in the plastic zones are each equal to the yield stress times the cross-sectional area of the zone:

$$C_1 = T_1 = \sigma_y b \left(\frac{h}{2} - e \right) \tag{c}$$

The forces C_2 and T_2 in the elastic core are each equal to the area of the stress diagram times the width b of the beam:

$$C_2 = T_2 = \frac{\sigma_y e}{2} b \tag{d}$$

Thus, the bending moment (see Fig. 6-45c) is

$$M = C_1 \left(\frac{h}{2} + e \right) + C_2 \left(\frac{4e}{3} \right) = \sigma_y b \left(\frac{h}{2} - e \right) \left(\frac{h}{2} + e \right) + \frac{\sigma_y b e}{2} \left(\frac{4e}{3} \right)$$

$$= \frac{\sigma_y b h^2}{6} \left(\frac{3}{2} - \frac{2e^2}{h^2} \right) = M_y \left(\frac{3}{2} - \frac{2e^2}{h^2} \right) \tag{6-83}$$

This equation is valid for $M_y \le M \le M_p$. Note that when $e = h/2$, the equation gives $M = M_y$, and when $e = 0$, it gives $M = 3M_y/2$, which is the plastic moment M_p.

Equation (6-83) can be used to determine the bending moment when the dimensions of the elastic core are known. However, a more common requirement is to determine the size of the elastic core when the bending moment is known. Therefore, we solve Eq. (6-83) for e in terms of the bending moment:

$$e = h \sqrt{\frac{1}{2} \left(\frac{3}{2} - \frac{M}{M_y} \right)} \tag{6-84}$$

Again we note the limiting conditions: when $M = M_y$, the equation gives $e = h/2$, and when $M = M_p = 3M_y/2$, it gives $e = 0$, which is the fully plastic condition.

■ Beams of Wide-Flange Shape

For a doubly symmetric wide-flange beam (Fig. 6-46), the plastic modulus Z (Eq. 6-78) is calculated by taking the first moment about the neutral axis of the area of one flange and the upper half of the web and then multiplying by 2. The result is

$$Z = 2 \left[(b t_f) \left(\frac{h}{2} - \frac{t_f}{2} \right) + (t_w) \left(\frac{h}{2} - t_f \right) \left(\frac{1}{2} \right) \left(\frac{h}{2} - t_f \right) \right]$$

$$= b t_f (h - t_f) + t_w \left(\frac{h}{2} - t_f \right)^2 \tag{6-85a}$$

Fig. 6-46 Cross section of a wide-flange beam

With a little rearranging, we can express Z in this alternative form:

$$Z = \frac{1}{4}\left[bh^2 - (b - t_w)(h - 2t_f)^2\right]$$ (6-85b)

After calculating the plastic modulus from Eq. (6-85a or b), we can obtain the plastic moment M_p from Eq. (6-77).

Values of Z for commercially available shapes of wide-flange beams are listed in the AISC manual (Ref. 5-4). The shape factor f for wide-flange beams is typically in the range 1.1 to 1.2, depending upon the proportions of the cross section.

Other shapes of elastoplastic beams can be analyzed in a manner similar to that described above for rectangular and wide-flange beams (see the following examples and the problems at the end of the chapter).

■ **Example 6-9**

Determine the yield moment, plastic modulus, plastic moment, and shape factor for a beam of circular cross section with diameter d (Fig. 6-47).

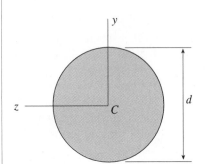

Fig. 6-47 Example 6-9. Cross section of a circular beam (elastoplastic material)

Solution As a preliminary matter, we note that since the cross section is doubly symmetric, the neutral axis passes through the center of the circle for both linearly elastic and elastoplastic behavior.

The yield moment M_y is found from the flexure formula (Eq. 6-74) as follows:

$$M_y = \frac{\sigma_y I}{c} = \frac{\sigma_y(\pi d^4/64)}{d/2} = \sigma_y\left(\frac{\pi d^3}{32}\right)$$ (6-86) ◀

The plastic modulus Z is found from Eq. (6-78) in which A is the area of the circle and $\bar{y}_1$ and $\bar{y}_2$ are the distances to the centroids c_1 and c_2 of the two halves of the circle (Fig. 6-48). Thus, from Cases 9 and 10 of Appendix D, we get

$$A = \frac{\pi d^2}{4} \qquad \bar{y}_1 = \bar{y}_2 = \frac{2d}{3\pi}$$ (e,f)

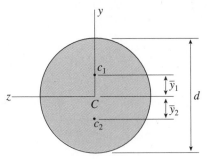

Fig. 6-48 Solution to Example 6-9

Now substituting into Eq. (6-78) for the plastic modulus, we find

$$Z = \frac{A(\bar{y}_1 + \bar{y}_2)}{2} = \frac{d^3}{6}$$ (g)

Therefore, the plastic moment M_p (Eq. 6-77) is

$$M_p = \sigma_y Z = \frac{\sigma_y d^3}{6}$$ (6-87) ◀

and the shape factor f (Eq. 6-79) is

$$f = \frac{M_p}{M_y} = \frac{16}{3\pi} \approx 1.70$$ (6-88) ◀

This result shows that the maximum bending moment for a circular beam of elastoplastic material is about 70% larger than the bending moment when the beam first begins to yield.

■ Example 6-10

A doubly symmetric hollow box beam (Fig. 6-49) of elastoplastic material ($\sigma_y = 220$ MPa) is subjected to a bending moment M of such magnitude that the flanges yield but the webs remain linearly elastic. Determine the magnitude of the moment M if the dimensions of the cross section are $b = 150$ mm, $b_1 = 130$ mm, $h = 200$ mm, and $h_1 = 160$ mm.

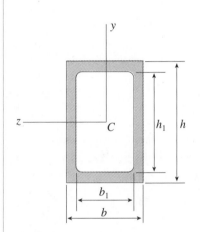

Fig. 6-49 Example 6-10. Cross section of a hollow box beam (elastoplastic material)

Solution   The cross section of the beam and the distribution of the normal stresses are shown in Figs. 6-50a and b, respectively. From the figure, we see that the stresses in the webs increase linearly with distance from the neutral axis and the stresses in the flanges equal the yield stress σ_y. Therefore, the bending moment M acting on the cross section consists of two parts: (1) a moment M_1 corresponding to the elastic core, and (2) a moment M_2 produced by the yield stresses σ_y in the flanges.

The bending moment supplied by the core is found from the flexure formula (Eq. 6-74) with the section modulus calculated for the webs alone; thus,

$$S = \frac{(b - b_1)h_1^2}{6} \tag{h}$$

and

$$M_1 = \sigma_y S = \frac{\sigma_y(b - b_1)h_1^2}{6} \tag{i}$$

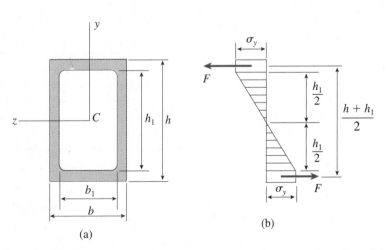

Fig. 6-50 Solution to Example 6-10

(a)

(b)

To find the moment supplied by the flanges, we note that the resultant force F in each flange (Fig. 6-50b) is equal to the yield stress multiplied by the area of the flange:

$$F = \sigma_y b\left(\frac{h - h_1}{2}\right) \tag{j}$$

The force in the top flange is compressive and the force in the bottom flange is tensile if the bending moment M is positive. Together, the two forces create the bending moment M_2:

$$M_2 = F\left(\frac{h + h_1}{2}\right) = \frac{\sigma_y b(h^2 - h_1^2)}{4} \tag{k}$$

Hence, the total moment acting on the cross section is

$$M = M_1 + M_2 = \frac{\sigma_y}{12}\left[3bh^2 - (b + 2b_1)h_1^2\right] \tag{6-89} \quad \longleftarrow$$

Substituting the given numerical values, we obtain

$$M = 138 \text{ kN·m} \quad \longleftarrow$$

Note: The yield moment M_y and the plastic moment M_p for the beam in this example have the following values (determined in Problem 6.9-14):

$$M_y = 122 \text{ kN·m} \qquad M_p = 147 \text{ kN·m}$$

The bending moment M is between these values, as expected.

6.10 ■ NONLINEAR BENDING

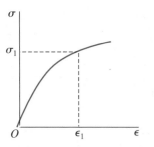

Fig. 6-51   Nonlinear stress-strain curve

In the preceding section we dealt with the special case of an elastoplastic material, an idealization that represents very closely the behavior of structural steel. Now, however, we will consider materials having more general kinds of stress-strain curves (Fig. 6-51). Initially, we will assume that the material has a stress-strain curve that is the same in both tension and compression (symmetric stress-strain curve), and later we will consider materials with nonsymmetric curves. Also, we will limit our analyses to beams with rectangular cross sections, thereby simplifying the calculations while still illustrating the general principles of nonlinear analysis.

■ Beams with Rectangular Cross Sections and Symmetric Stress-Strain Curves

When the cross section of the beam is doubly symmetric and the stress-strain curve is the same for both tension and compression, we know immediately that the neutral axis for bending passes through the centroid of the cross section, as shown for the rectangular cross section of Fig. 6-52a. In addition, we know that the strains in a beam subjected to pure bending vary linearly throughout the height of the beam regardless of the nature of the material (see Section 5.4). Therefore, the strains in the beam of Fig. 6-52a will vary as shown in Fig. 6-52b when the beam is subjected to a positive bending moment (compression in the upper half of the beam and tension in the lower half). The strains are zero at the neutral axis and equal to ϵ_1 at the top and bottom of the beam.

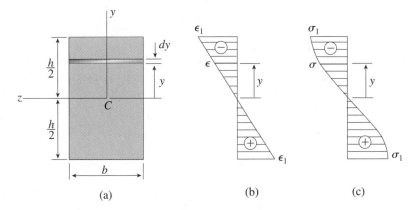

Fig. 6-52 Rectangular beam subjected to nonlinear bending: (a) cross section of beam with neutral axis through the centroid, (b) strains in the beam, and (c) stresses in the beam

The stresses in the beam are found from the strains by using the stress-strain relationship. Since the strains vary linearly with distance from the neutral axis, the stress distribution over the height of the beam (Fig. 6-52c) has the same shape as the stress-strain curve itself (Fig. 6-51). The maximum stresses σ_1 in the beam correspond to the strain ϵ_1 on the stress-strain curve.

Let us now determine the bending moment M acting on the cross section. We begin by identifying a thin element of area located at distance y from the neutral axis (Fig. 6-52a). The element has width b and height dy, so its area is $b\,dy$. The force acting on this area is equal to $\sigma b\,dy$, and the moment of that force about the z axis is $\sigma by\,dy$. Thus, the bending moment is

$$M = 2\int_0^{h/2} \sigma by\,dy \tag{a}$$

Only absolute values are used in this and later equations because the directions of the stresses and the sense of the bending moment are always obvious. For instance, we know that a positive bending moment produces compressive stresses in the upper part of the beam and tensile stresses in the lower part.

The stress σ appearing in Eq. (a) is determined from the stress-strain curve and therefore is a function of the strain ϵ. However, the integration is performed with respect to the variable y. Consequently, we need to make a change of variables in order to perform the integration. We can relate the variables ϵ and y by observing that the strain ϵ at any distance y from the neutral axis (Fig. 6-52b) is

$$\epsilon = \epsilon_1 \frac{2y}{h} \tag{b}$$

and therefore

$$y = \frac{h\epsilon}{2\epsilon_1} \qquad dy = \frac{hd\epsilon}{2\epsilon_1} \tag{c,d}$$

Substituting from these equations into Eq. (a), and also changing the upper limit of integration from $h/2$ to the corresponding value ϵ_1 of the strain, we get

$$M = 2 \int_0^{h/2} \sigma b y \, dy = 2b \int_0^{\epsilon_1} \sigma \left(\frac{h\epsilon}{2\epsilon_1} \right) \left(\frac{h \, d\epsilon}{2\epsilon_1} \right)$$

or

$$M = \frac{bh^2}{2\epsilon_1^2} \int_0^{\epsilon_1} \sigma \epsilon \, d\epsilon \tag{6-90a}$$

This equation relates the maximum strain ϵ_1 in the beam to the bending moment M, provided the stress σ is known as a function of the strain ϵ.

Equation (6-90a) can be given a simple geometric interpretation. The integral is the first moment of the area below the stress-strain curve (Fig. 6-51) from $\epsilon = 0$ to $\epsilon = \epsilon_1$, evaluated with respect to the vertical axis (the stress axis). Thus,

$$M = \frac{bh^2}{2\epsilon_1^2} Q_s \tag{6-90b}$$

where Q_s is the first moment just described.

If the stress-strain relation is simple enough to be expressed analytically, it may be possible to obtain a formula for the first moment Q_s. In that case, Eq. (6-90b) becomes a formula relating the bending moment M to the maximum strain ϵ_1. We can then substitute the stress-strain relation and obtain a formula relating the moment to the maximum stress σ_1. This last relationship is a *flexure formula* for a rectangular beam of nonlinear material. An illustration of such a formula is given later in Example 6-11.

However, if the stress-strain relation is more complex and cannot be integrated analytically, or if the stress-strain relation is given in numerical terms, then the first moment Q_s must be evaluated numerically. The procedure is to select a value of the maximum strain ϵ_1, determine the corresponding value of Q_s from the stress-strain curve, and then evaluate M from Eq. (6-90b). The result is a single value of M corresponding to the selected value of the maximum strain. By repeating these calculations for a series of values of the strain, the relationship between M and ϵ_1 is obtained as a table of numerical values. Since stress and strain are related by the stress-strain curve, we can also obtain a table of values giving the relationship between M and the maximum stress σ_1. Finally, graphs of M versus ϵ_1 and M versus σ_1 can be plotted from the tables of data. (Note that the graph of M versus σ_1 takes the place of a flexure formula.)

As a special case, consider a beam of **linearly elastic material** for which the stress-strain relation (Fig. 6-53) is

$$\sigma = E\epsilon \tag{e}$$

The first moment Q_s is

$$Q_s = \frac{\sigma_1 \epsilon_1}{2}\left(\frac{2\epsilon_1}{3}\right) = \frac{\sigma_1 \epsilon_1^2}{3} \tag{f}$$

and therefore the bending moment is

$$M = \frac{bh^2}{2\epsilon_1^2}Q_s = \frac{bh^2}{2\epsilon_1^2}\left(\frac{\sigma_1 \epsilon_1^2}{3}\right) = \frac{\sigma_1 bh^2}{6} = \sigma_1 S \tag{g}$$

where S is the section modulus. This equation is the familiar flexure formula for a rectangular beam of linearly elastic material.

■ Beams with Rectangular Cross Sections and Nonsymmetric Stress-Strain Curves

Now let us consider what happens when the tension and compression parts of the stress-strain curve are different (Fig. 6-54). We immediately recognize that the neutral axis will no longer pass through the centroid of the cross section (Fig. 6-55a). The linear strain distribution has the shape shown in Fig. 6-55b, where ϵ_1 is the compressive strain at the top of the beam and ϵ_2 is the tensile strain at the bottom. The stress distribution (Fig. 6-55c) has the same shape as the stress-strain curve.

Fig. 6-53 Stress-strain diagram for a linearly elastic material

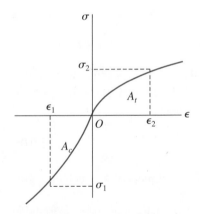

Fig. 6-54 Nonlinear stress-strain curve with different properties in tension and compression

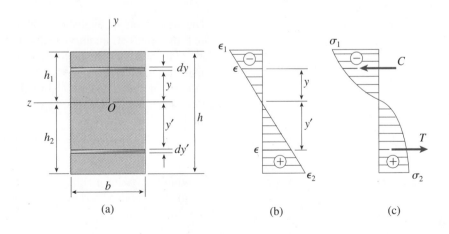

Fig. 6-55 Rectangular beam subjected to nonlinear bending: (a) cross section of beam, (b) strains in the beam, and (c) stresses in the beam

To locate the neutral axis, we use the condition that the resultant force acting on the cross section must be zero. We again consider an element of area having width b and height dy (Fig. 6-55a), and we note that the force acting on this element is $\sigma b\, dy$. If we integrate this force over the upper half of the beam, we obtain the resultant compressive force C (Fig. 6-55c):

$$C = \int_0^{h_1} \sigma b\, dy \qquad \text{(h)}$$

in which h_1 is the distance from the neutral axis to the top of the beam. (As already mentioned, only absolute values are used in this and other expressions because the directions of the stresses and forces can be determined by inspection.)

Following the same steps that we used in deriving Eq. (6-90a), we now change the independent variable from y to ϵ by means of the following equations (see Fig. 6-55b):

$$\epsilon = \epsilon_1 \frac{y}{h_1} \qquad y = \frac{h_1 \epsilon}{\epsilon_1} \qquad dy = \frac{h_1 d\epsilon}{\epsilon_1} \qquad \text{(i,j,k)}$$

Therefore, the compressive force C is

$$C = \int_0^{\epsilon_1} \sigma b \left(\frac{h_1 d\epsilon}{\epsilon_1} \right) = \frac{b h_1}{\epsilon_1} \int_0^{\epsilon_1} \sigma d\epsilon \qquad \text{(6-91a)}$$

The last integral in this expression is the area A_c between the compressive part of the stress-strain curve and the strain axis (Fig. 6-54). Thus, an alternative form of the preceding equation is

$$C = \frac{b h_1}{\epsilon_1} A_c \qquad \text{(6-91b)}$$

The area A_c can be evaluated analytically or numerically, depending upon how the stress-strain curve is defined.

The resultant tensile force T in the lower part of the beam is found in a similar manner. However, for convenience we now take a new coordinate y' in the negative y direction (Fig. 6-55a), and we identify an element of area $b\, dy'$. The force on this element is $\sigma b\, dy'$, and therefore the force T is

$$T = \int_0^{h_2} \sigma b\, dy' \qquad \text{(l)}$$

in which h_2 is the distance from the neutral axis to the bottom of the beam. To change variables, we use the following relations (see Fig. 6-55b):

$$\epsilon = \epsilon_2 \frac{y'}{h_2} \qquad y' = \frac{h_2 \epsilon}{\epsilon_2} \qquad dy' = \frac{h_2 d\epsilon}{\epsilon_2} \qquad \text{(m,n,o)}$$

Thus, the tensile force T is

$$T = \int_0^{\epsilon_2} \sigma b \left(\frac{h_2 d\epsilon}{\epsilon_2} \right) = \frac{bh_2}{\epsilon_2} \int_0^{\epsilon_2} \sigma d\epsilon \qquad \text{(6-92a)}$$

Again we note that the integral has a simple geometric interpretation—it is the area A_t between the tensile part of the stress-strain curve and the strain axis (Fig. 6-54). Thus,

$$T = \frac{bh_2}{\epsilon_2} A_t \qquad \text{(6-92b)}$$

Now that we have obtained expressions for the forces C and T, we can locate the neutral axis.

The condition that the resultant force acting on the cross section is zero gives the following equations:

$$C = T \quad \text{or} \quad \frac{bh_1}{\epsilon_1} A_c = \frac{bh_2}{\epsilon_2} A_t \qquad \text{(p)}$$

From the linear strain distribution (Fig. 6-55b), we see that

$$\frac{h_1}{\epsilon_1} = \frac{h_2}{\epsilon_2} \qquad \text{(6-93)}$$

and therefore Eq. (p) reduces to

$$A_c = A_t \qquad \text{(6-94)}$$

In other words, *the neutral axis is so located that the compression area of the stress-strain diagram is equal to the tension area (Fig. 6-54).* Thus, for any assumed value of ϵ_1, we can determine the corresponding value of ϵ_2 from Eq. (6-94) and the stress-strain curve.

Knowing ϵ_1 and ϵ_2, we can obtain the ratio of h_1 to h_2 from Eq. (6-93). In addition, we know that

$$h_1 + h_2 = h \qquad \text{(6-95)}$$

Consequently, we can find h_1 and h_2 by solving simultaneously Eqs. (6-93) and (6-95); the results are

$$h_1 = \frac{h}{1 + \epsilon_2/\epsilon_1} \qquad h_2 = \frac{h}{1 + \epsilon_1/\epsilon_2} \qquad \text{(6-96a,b)}$$

In summary, the neutral axis is located by assuming a value of the strain ϵ_1, finding the corresponding strain ϵ_2 from Eq. (6-94) and the stress-strain curve, and then finding h_1 and h_2 from Eqs. (6-96).

The next step is to find the bending moment M acting on the cross section. We will separate the moment into two parts: a moment M_c produced by the compressive stresses and a moment M_t produced by the tensile stresses. To find the moment M_c, we again consider the element of area $b\,dy$ located at distance y from the neutral axis (Fig. 6-55a). The force acting on this element is $\sigma b\,dy$, and its moment about the neutral axis is $\sigma by\,dy$; thus,

$$M_c = \int_0^{h_1} \sigma by\,dy$$

Substituting from Eqs. (j) and (k), and also changing the upper limit, we get

$$M_c = \int_0^{\epsilon_1} \sigma b\left(\frac{h_1\epsilon}{\epsilon_1}\right)\left(\frac{h_1 d\epsilon}{\epsilon_1}\right) = \frac{bh_1^2}{\epsilon_1^2} \int_0^{\epsilon_1} \sigma\epsilon\,d\epsilon \qquad (q)$$

Proceeding in a similar manner for the tensile stresses, we get

$$M_t = \int_0^{h_2} \sigma by'\,dy'$$

Substituting from Eqs. (n) and (o) yields

$$M_t = \int_0^{\epsilon_1} \sigma b\left(\frac{h_2\epsilon}{\epsilon_2}\right)\left(\frac{h_2 d\epsilon}{\epsilon_2}\right) = \frac{bh_2^2}{\epsilon_2^2} \int_0^{\epsilon_2} \sigma\epsilon\,d\epsilon \qquad (r)$$

Therefore, the total moment is

$$M = M_c + M_t = \frac{bh_1^2}{\epsilon_1^2} \int_0^{\epsilon_1} \sigma\epsilon\,d\epsilon + \frac{bh_2^2}{\epsilon_2^2} \int_0^{\epsilon_2} \sigma\epsilon\,d\epsilon \qquad (6\text{-}97a)$$

The first integral represents the first moment of the area A_c evaluated with respect to the stress axis (Fig. 6-54), and the second integral represents the first moment of the area A_t evaluated with respect to that same axis. Denoting these first moments by Q_c and Q_t, respectively, the preceding equation becomes

$$M = \frac{bh_1^2}{\epsilon_1^2} Q_c + \frac{bh_2^2}{\epsilon_2^2} Q_t \qquad (6\text{-}97b)$$

From Eq. (6-93) we see that the two fractions in the preceding equation are equal; hence,

$$M = \frac{bh_1^2}{\epsilon_1^2}(Q_c + Q_t) = \frac{bh_2^2}{\epsilon_2^2}(Q_c + Q_t) \qquad (6\text{-}98)$$

Thus, we now have an equation for determining the bending moment in the beam, assuming that we have already located the neutral axis and determined ϵ_1, ϵ_2, h_1, and h_2. All that remains is to determine the first moments Q_c and Q_t from the stress-strain diagram and substitute into Eq. (6-98).

The details of the procedure depend upon the nature of the stress-strain information. If the stress-strain curve can be expressed in a simple analytical form, we can derive equations that locate the neutral axis and relate the bending moment to the maximum stresses. This procedure is illustrated in Example 6-11. Otherwise, the solution must be carried out in numerical terms, in which case we calculate a value of M (from Eq. 6-98) for each assumed value of ϵ_1. Since we know the values of ϵ_2, σ_1, and σ_2 corresponding to each value of ϵ_1, we automatically have the relationships between the bending moment M and the maximum and minimum stresses.

■ Example 6-11

A beam of rectangular cross section with width b and height h (Fig. 6-56a) is constructed of material having a stress-strain diagram consisting of two straight lines (Fig. 6-56b). The moduli of elasticity in compression and tension are E_1 and E_2, respectively. Assuming that the beam is subjected to a positive bending moment M acting about the z axis, determine the location of the neutral axis and the maximum compressive and tensile stresses in the beam.

Solution   *Neutral axis.* The neutral axis is located at distance h_1 from the top of the beam and distance h_2 from the bottom (Fig. 6-57a). We can find these distances from the fact that the compressive and tensile parts of the stress-strain diagram (Fig. 6-57b) must have equal areas; that is,

$$A_c = A_t \quad \text{or} \quad \frac{\sigma_1\epsilon_1}{2} = \frac{\sigma_2\epsilon_2}{2} \tag{s}$$

in which ϵ_1 and ϵ_2 are the strains at the top and bottom of the beam, respectively (Fig. 6-57c), and σ_1 and σ_2 are the corresponding maximum stresses (Fig. 6-57d). The equations of the stress-strain diagram are

$$\sigma_1 = E_1\epsilon_1 \qquad \sigma_2 = E_2\epsilon_2 \tag{t,u}$$

and therefore Eq. (s) becomes

$$E_1\epsilon_1^2 = E_2\epsilon_2^2 \quad \text{or} \quad \frac{\epsilon_1}{\epsilon_2} = \frac{\sqrt{E_2}}{\sqrt{E_1}} \tag{v}$$

The distances h_1 and h_2 are now obtained from Eqs. (6-96a and b):

$$h_1 = \frac{h}{1 + \epsilon_2/\epsilon_1} = \frac{h\sqrt{E_2}}{\sqrt{E_1} + \sqrt{E_2}} \tag{6-99a} \quad \Longleftarrow$$

$$h_2 = \frac{h}{1 + \epsilon_1/\epsilon_2} = \frac{h\sqrt{E_1}}{\sqrt{E_1} + \sqrt{E_2}} \tag{6-99b} \quad \Longleftarrow$$

Thus, the position of the neutral axis is determined. Note that if E_1 is larger than E_2, the distance h_2 is larger than h_1.

(a)

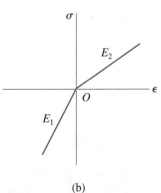

(b)

Fig. 6-56 Example 6-11. Rectangular beam with stress-strain diagram consisting of two straight lines

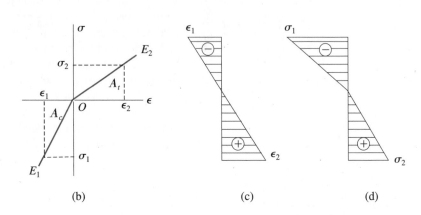

(a)   (b)   (c)   (d)

Fig. 6-57 Solution to Example 6-11:
(a) location of neutral axis,
(b) stress-strain diagram,
(c) distribution of strains, and
(d) distribution of stresses

Maximum stresses. To obtain the relation between the bending moment and the stresses, we begin by evaluating the first moments of the two parts of the stress-strain diagram (see Fig. 6-57b):

$$Q_c = A_c\left(\frac{2\epsilon_1}{3}\right) = \left(\frac{\sigma_1\epsilon_1}{2}\right)\left(\frac{2\epsilon_1}{3}\right) = \frac{\sigma_1\epsilon_1^2}{3} \tag{w}$$

$$Q_t = A_t\left(\frac{2\epsilon_2}{3}\right) = \left(\frac{\sigma_2\epsilon_2}{2}\right)\left(\frac{2\epsilon_2}{3}\right) = \frac{\sigma_2\epsilon_2^2}{3} \tag{x}$$

Substituting into Eq. (6-98) gives

$$M = \frac{bh_1^2}{\epsilon_1^2}(Q_c + Q_t) = \frac{bh_1^2}{3\epsilon_1^2}(\sigma_1\epsilon_1^2 + \sigma_2\epsilon_2^2)$$

Replacing σ_2 by $\sigma_1\epsilon_1/\epsilon_2$ (from Eq. s), and also using Eq. (v) for the ratio ϵ_1/ϵ_2, we get

$$M = \frac{\sigma_1 bh_1^2}{3}\left(1 + \frac{\sqrt{E_1}}{\sqrt{E_2}}\right) \quad\text{or}\quad \sigma_1 = \frac{3M}{bh_1^2}\left(\frac{\sqrt{E_2}}{\sqrt{E_1} + \sqrt{E_2}}\right) \tag{y,z}$$

Finally, we substitute from Eq. (6-99a) and obtain

$$\sigma_1 = \frac{3M}{bh^2}\left(\frac{\sqrt{E_1} + \sqrt{E_2}}{\sqrt{E_2}}\right) \tag{6-100a}$$ ◀

By a similar series of substitutions, we find the stress σ_2:

$$\sigma_2 = \frac{3M}{bh^2}\left(\frac{\sqrt{E_1} + \sqrt{E_2}}{\sqrt{E_1}}\right) \tag{6-100b}$$ ◀

Equations (6-100a) and (6-100b) are the flexure formulas for a rectangular beam constructed of a material having modulus E_1 in compression and E_2 in tension. The stress σ_1 is the maximum compressive stress, and the stress σ_2 is the maximum tensile stress.

Special case: If the moduli in compression and tension are the same $(E_1 = E_2)$, then Eqs. (6-99) and (6-100) give

$$h_1 = h_2 = \frac{h}{2} \qquad \sigma_1 = \sigma_2 = \frac{6M}{bh^2} = \frac{M}{S}$$

as expected.

PROBLEMS / CHAPTER 6

COMPOSITE BEAMS

When solving the problems for Section 6.2, assume that the component parts of the beams are securely bonded by adhesives or connected by fasteners. Also, be sure to use the theory for composite beams described in this section.

6.2-1 A composite beam consisting of fiberglass faces and a core of particle board has the cross-section shown in the figure. The width of the beam is 2.0 in., the thickness of the faces is 0.10 in., and the thickness of the core is 0.50 in. The beam is subjected to a bending moment of 500 lb-in. acting about the z axis. Find the maximum stresses σ_{face} and σ_{core} in the faces and the core, respectively, if their respective moduli of elasticity are 4×10^6 psi and 1.5×10^6 psi.

Prob. 6.2-1

6.2-2 A wood beam of dimensions 200×300 mm is reinforced on its sides by steel plates 12 mm thick (see figure). The moduli of elasticity for the steel and wood are $E_s = 204$ GPa and $E_w = 8.5$ GPa, respectively. Also, the corresponding allowable stresses are $\sigma_s = 140$ MPa and $\sigma_w = 10$ MPa. Calculate the maximum permissible bending moment M_{max} when the beam is bent about the z axis.

Prob. 6.2-2

6.2-3 A hollow box beam is constructed with webs of Douglas-fir plywood and flanges of pine as shown in the figure. The plywood is 1 in. thick and 12 in. wide; the flanges are 2 in. × 4 in. (actual size). The modulus of elasticity for the plywood is 1,600,000 psi and for the pine is 1,200,000 psi. If the allowable stresses are 2100 psi for the plywood and 1800 psi for the pine, find the allowable bending moment M_{max} when the beam is bent about the z axis.

Prob. 6.2-3

6.2-4 A round steel tube of outside diameter d and an aluminum core of diameter $d/2$ are bonded to form a composite beam as shown in the figure. Derive a formula for the allowable bending moment M that can be carried by the beam based upon an allowable stress σ_s in the steel. (Assume that the moduli of elasticity for the steel and aluminum are E_s and E_a, respectively.)

Prob. 6.2-4

6.2-5 A simple beam on a 12 ft span supports a uniform load of intensity 1000 lb/ft (see figure). The beam consists of a wood member (4 in. × 11.5 in. in cross section) that is reinforced by 0.25 in. thick steel plates on top and bottom. The moduli of elasticity for the steel and wood are $E_s = 30 \times 10^6$ psi and $E_w = 1.5 \times 10^6$ psi, respectively. Calculate the maximum stresses σ_s in the steel plates and the maximum stresses σ_w in the wood member due to the uniform load.

Prob. 6.2-5

6.2-6 A plastic-lined steel pipe has the cross-sectional shape shown in the figure. The steel pipe has outer diameter $d_3 = 100$ mm and inner diameter $d_2 = 94$ mm. The plastic liner has inner diameter $d_1 = 82$ mm. The modulus of elas-

ticity of the steel is 75 times the modulus of the plastic. Determine the allowable bending moment M_{allow} if the allowable stress in the steel is 35 MPa and in the plastic is 0.50 MPa.

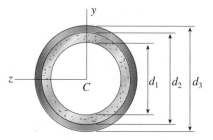

Prob. 6.2-6

6.2-7 The cross section of a sandwich beam consisting of aluminum alloy faces and a foam core is shown in the figure. The width b of the beam is 8.0 in., the thickness t of the faces is 0.25 in., and the height h_c of the core is 5.5 in. (total height $h = 6.0$ in.). The moduli of elasticity are 10.5×10^6 psi for the aluminum faces and 12,000 psi for the foam core. A bending moment $M = 40$ k-in. acts about the z axis. Determine the maximum stresses in the faces and the core using: (a) the general theory for composite beams, and (b) the approximate theory for sandwich beams.

Probs. 6.2-7 and 6.2-8

6.2-8 The cross section of a sandwich beam consisting of fiberglass faces and a lightweight plastic core is shown in the figure. The width b of the beam is 50 mm, the thickness t of the faces is 4 mm, and the height h_c of the core is 92 mm (total height $h = 100$ mm). The moduli of elasticity are 75 GPa for the fiberglass and 1200 MPa for the plastic. A bending moment $M = 275$ N·m acts about the z axis. Determine the maximum stresses in the faces and the core using: (a) the general theory for composite beams, and (b) the approximate theory for sandwich beams.

6.2-9 A bimetallic beam used in a temperature-control switch consists of strips of aluminum and copper bonded together (see figure). The width of the beam is 1 in., and each strip has a thickness of 1/16 in. Under the action of a bending moment $M = 12$ lb-in. acting about the z axis, what are the maximum stresses σ_a and σ_c in the aluminum and copper, respectively? (Assume $E_a = 10,500,000$ psi and $E_c = 16,800,000$ psi.)

Prob. 6.2-9

6.2-10 A simply supported composite beam 3 m long carries a uniformly distributed load of intensity $q = 3.2$ kN/m (see figure). The beam is constructed of a wood member, 100 mm wide by 150 mm deep, reinforced on its lower side by a steel plate 8 mm thick and 100 mm wide. Find the maximum bending stresses σ_w and σ_s in the wood and steel, respectively, due to the uniform load if the moduli of elasticity are $E_w = 10$ GPa for the wood and $E_s = 210$ GPa for the steel.

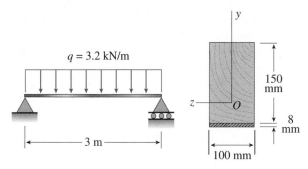

Prob. 6.2-10

TRANSFORMED-SECTION METHOD

When solving the problems for Section 6.3, assume that the component parts of the beams are securely bonded by

adhesives or connected by fasteners. Also, be sure to use the transformed-section method in the solutions.

6.3-1 A wood beam 8 in. wide and 12 in. deep is reinforced on top and bottom by 0.5 in. thick steel plates (see figure). Find the allowable bending moment M_{max} about the z axis if the allowable stress in the wood is 1,050 psi and in the steel is 18,000 psi. (Assume that the ratio of the moduli of elasticity of steel and wood is 20.)

Prob. 6.3-1

6.3-2 A simple beam of span length 3.2 m carries a uniform load of intensity 48 kN/m. The cross section of the beam is a hollow box with wood flanges and steel side plates, as shown in the figure. The wood flanges are 75 mm by 100 mm in cross section, and the steel plates are 300 mm deep. What is the required thickness t of the steel plates if the allowable stresses are 120 MPa for the steel and 6.5 MPa for the wood? (Assume that the moduli of elasticity for the steel and wood are 210 GPa and 10 GPa, respectively, and disregard the weight of the beam.)

Prob. 6.3-2

6.3-3 A simple beam that is 16 ft long supports a uniform load of intensity q. The beam is constructed of an S 8 × 18.4 section (I-beam section) reinforced with wood beams that are securely fastened to the flanges (see the cross section shown in the figure). The wood beams are 2 in. deep and 4 in. wide. The modulus of elasticity of the steel is 20 times that of the wood. If the allowable stresses in the steel and wood are 16,000 psi and 1,200 psi, respectively, what is the allowable load q_{allow}? (Disregard the weight of the beam.)

Prob. 6.3-3

6.3-4 The composite beam shown in the figure is simply supported and carries a total uniform load of 40 kN/m on a span length of 5 m. The beam is built of a wood member having cross-sectional dimensions 150 mm × 250 mm and two steel plates of cross-sectional dimensions 50 mm × 150 mm. Determine the maximum stresses σ_s and σ_w in the steel and wood, respectively, if the moduli of elasticity are E_s = 209 GPa and E_w = 11 GPa. (Disregard the weight of the beam.)

Prob. 6.3-4

6.3-5 The cross section of a beam made of thin strips of aluminum separated by a lightweight plastic is shown in the figure. The beam has width b = 3.0 in., the aluminum strips have thickness t = 0.1 in., and the plastic segments have heights d = 1.2 in. and $3d$ = 3.6 in. The total height of the

beam is h = 6.4 in. The moduli of elasticity for the aluminum and plastic are E_a = 11 × 10⁶ psi and E_p = 440 × 10³ psi, respectively. Determine the maximum stresses σ_a and σ_p in the aluminum and plastic, respectively, due to a bending moment of 10 k-in.

Probs. 6.3-5 and 6.3-6

6.3-6 Solve the preceding problem if the beam has width b = 75 mm, the aluminum strips have thickness t = 3 mm, the plastic segments have heights d = 40 mm and $3d$ = 120 mm, and the total height of the beam is h = 212 mm. Also, the moduli of elasticity are E_a = 75 GPa and E_p = 3 GPa, respectively. Determine the maximum stresses σ_a and σ_p in the aluminum and plastic, respectively, due to a bending moment of 1.0 kN·m.

6.3-7 A composite beam constructed of a wood beam reinforced by a steel plate has the cross-sectional dimensions shown in the figure. The beam is simply supported with a span length of 12 ft and supports a uniformly distributed load of intensity q = 275 lb/ft. Calculate the maximum bending stresses σ_s and σ_w in the steel and wood, respectively, due to the uniform load if E_s/E_w = 20.

Prob. 6.3-7

6.3-8 The cross section of a composite beam made of aluminum and steel is shown in the figure. The moduli of elasticity are $E_a = 75$ GPa and $E_s = 200$ GPa. Under the action of a bending moment that produces a maximum stress of 50 MPa in the aluminum, what is the maximum stress σ_s in the steel?

Prob. 6.3-8

6.3-9 A composite beam is constructed of a wood beam 6 in. wide and 8 in. deep reinforced on the lower side by a 0.5 in. by 6 in. steel plate (see figure). The modulus of elasticity for the wood is $E_w = 1.2 \times 10^6$ psi and for the steel is $E_s = 30 \times 10^6$ psi. Find the allowable bending moment M_{allow} for the beam if the allowable stress in the wood is $\sigma_w = 1,500$ psi and in the steel is $\sigma_s = 15,000$ psi.

Prob. 6.3-9

6.3-10 The cross section of a bimetallic strip is shown in the figure. Assuming that the moduli of elasticity for metals A and B are $E_A = 196$ GPa and $E_B = 105$ GPa, respectively, determine the smaller of the two section moduli for the beam. (Recall that section modulus is equal to bending moment divided by maximum bending stress.) In which material does the maximum stress occur?

Prob. 6.3-10

6.3-11 A W 12 × 50 steel wide-flange beam and a segment of a 4-inch thick concrete slab (see figure) jointly resist a positive bending moment of 100 k-ft. The beam and slab are joined by shear connectors that are welded to the steel beam. (These connectors resist the horizontal shear at the contact surface.) The moduli of elasticity of the steel and the concrete are in the ratio 12 to 1. Determine the maximum stresses σ_s and σ_c in the steel and concrete, respectively.

Prob. 6.3-11

6.3-12 A wood beam reinforced by an aluminum channel section is shown in the figure. The beam has a cross section of dimensions 150 mm by 250 mm, and the channel has a

Prob. 6.3-12

uniform thickness of 6 mm. If the allowable stresses in the wood and aluminum are 8 MPa and 40 MPa, respectively, and if their moduli of elasticity are in the ratio 1 to 6, what is the maximum allowable bending moment for the beam?

6.3-13 A 6 in. × 8 in. wood beam is reinforced by a 0.5 in. × 6 in. steel plate on the bottom and a 1 in. × 2 in. steel bar on the top (see figure). If the allowable stresses in the wood and steel are 2,000 psi and 16,000 psi, respectively, and if their moduli of elasticity are in the ratio 1 to 20, what is the maximum allowable bending moment for the beam?

Prob. 6.3 -13

BEAMS WITH INCLINED LOADS

When solving the problems for Section 6.4, be sure to draw a sketch of the cross section showing the orientation of the neutral axis and the locations of the points where the stresses are being found.

6.4-1 A beam of rectangular cross section supports an inclined load P having its line of action along a diagonal of the cross section (see figure). Show that the neutral axis lies along the other diagonal.

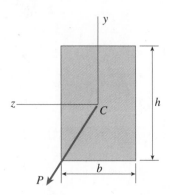

Prob. 6.4-1

6.4-2 A wood beam of rectangular cross section (see figure) is simply supported on a span of length L. The longitudinal axis of the beam is horizontal, and the cross section is tilted at an angle α. The load on the beam is a vertical uniform load of intensity q acting through the centroid C. Determine the orientation of the neutral axis and calculate the maximum tensile stress σ_{max} if $b = 75$ mm, $h = 150$ mm, $L = 3.0$ m, $\alpha = 30°$, and $q = 2.0$ kN/m.

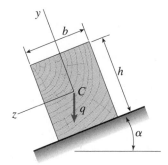

Probs. 6.4-2 and 6.4-3

6.4-3 Solve the preceding problem for the following data: $b = 6$ in., $h = 8$ in., $L = 10$ ft, $\tan \alpha = 1/3$, and $q = 200$ lb/ft.

6.4-4 A simply supported wide-flange beam of span length L carries a vertical concentrated load P acting through the centroid C at the midpoint of the span (see figure). The beam is attached to supports inclined at an angle α to the horizontal. Determine the orientation of the neutral axis and calculate the maximum stresses at the outside corners of the cross section (points A, B, D, and E) due to the load P. Data for the beam are as follows: W 10 × 30 section, $L = 8.5$ ft, $P = 5$ k, and $\alpha = 26.57°$.

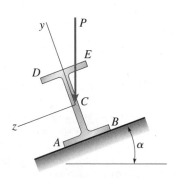

Probs. 6.4-4 and 6.4-5

6.4-5 Solve the preceding problem using the following data: W 8 × 21 section, $L = 8$ ft, $P = 3.8$ k, and $\alpha = 20°$.

6.4-6 A wood cantilever beam of rectangular cross section and length L supports an inclined load P at its free end (see figure). Determine the orientation of the neutral axis and calculate the maximum tensile stress σ_{max} due to the load P. Data for the beam are as follows: $b = 75$ mm, $h = 150$ mm, $L = 1.5$ m, $P = 800$ N, and $\alpha = 36°$.

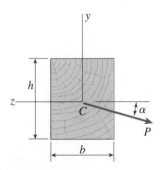

Probs. 6.4-6 and 6.4-7

6.4-7 Solve the preceding problem for a cantilever beam with data as follows: $b = 4$ in., $h = 8$ in., $L = 6$ ft, $P = 500$ lb, and $\alpha = 45°$.

6.4-8 A steel beam of I-section (see figure) is simply supported at the ends. Two equal and oppositely directed bending moments M_0 act at the ends of the beam, so that the beam is in pure bending. The moments act in plane mm, which is oriented at an angle α to the xy plane. Determine the orientation of the neutral axis and calculate the maximum tensile stress σ_{max} due to the moments M_0. Data for the beam are as follows: S 8 × 18.4 section, $M_0 = 30$ k-in., and $\alpha = 30°$.

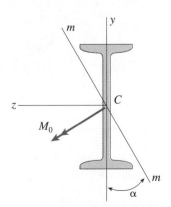

Prob. 6.4-8

6.4-9 A cantilever beam of wide-flange cross section and length L supports an inclined load P at its free end (see figure). Determine the orientation of the neutral axis and calculate the maximum tensile stress σ_{max} due to the load P. Data for the beam are as follows: W 10 × 45 section, $L = 6$ ft, $P = 2.5$ k, and $\alpha = 55°$.

Probs. 6.4-9 and 6.4-10

6.4-10 Solve the preceding problem using the following data: W 8 × 35 section, $L = 6$ ft, $P = 2.0$ k, and $\alpha = 60°$.

6.4-11 A cantilever beam of W 12 × 14 section and length $L = 9$ ft supports a slightly inclined load $P = 500$ lb at the free end (see figure). (a) Plot a graph of the stress σ_A at point A as a function of the angle of inclination α. (b) Plot a graph of the angle β, which locates the neutral axis nn, as a function of the angle α. (When plotting the graphs, let α vary from 0 to 10°.)

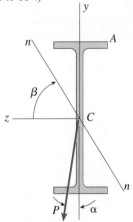

Prob. 6.4-11

BENDING OF UNSYMMETRIC BEAMS

When solving the problems for Section 6.5, be sure to draw a sketch of the cross section showing the orientation of the neutral axis and the locations of the points where the stresses are being found.

6.5-1 A beam of channel section is subjected to a bending moment M having its vector at an angle θ to the z axis (see figure). Determine the orientation of the neutral axis and calculate the maximum tensile stress σ_t and maximum compressive stress σ_c in the beam. Use the following data: C 8×11.5 section, $M = 20$ k-in., $\tan \theta = 1/3$.

Probs. 6.5-1 and 6.5-2

6.5-2 Solve the preceding problem for a C 6×13 channel section with $M = 5.0$ k-in. and $\theta = 15°$.

6.5-3 An angle section with equal legs is subjected to a bending moment M having its vector directed along the 1-1 axis, as shown in the figure. Determine the orientation of the neutral axis and calculate the maximum tensile stress σ_t and maximum compressive stress σ_c if the angle is an L $6 \times 6 \times 3/4$ section and $M = 20$ k-in.

Probs. 6.5-3 and 6.5-4

6.5-4 Solve the preceding problem for an L $4 \times 4 \times 1/2$ angle section with $M = 6.0$ k-in.

* 6.5-5 The cross section of a beam has the shape of an equilateral triangle with sides of length b (see figure). The beam is subjected to a bending moment M having its vector at an angle θ to the z axis. (a) Determine the orientation of the neutral axis. (b) Derive formulas for the stresses σ_A, σ_B, and σ_D at points A, B, and D, respectively, in terms of M, b, and θ. (c) Plot a graph of the maximum tensile stress σ_t as a function of the angle θ for θ varying from 0 to 90°. Let the ordinate of the graph be the nondimensional quantity $\sigma_{max}b^3/32M$.

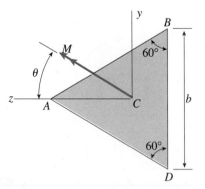

Prob. 6.5-5

* 6.5-6 A beam of semicircular cross section of radius r is subjected to a bending moment M having its vector at an angle θ to the z axis (see figure). Determine the maximum tensile stress σ_t and the maximum compressive stress σ_c in the beam for $\theta = 0, 45°$, and 90°.

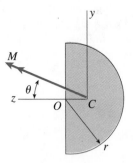

Prob. 6.5-6

* 6.5-7 An angle section with unequal legs is subjected to a bending moment M having its vector directed along axis 1-1, as shown in the figure. Determine the orientation of the neutral axis and calculate the maximum tensile stress σ_t and maximum compressive stress σ_c in the beam if the angle is an L $8 \times 6 \times 1$ section and $M = 25$ k-in.

Probs. 6.5-7 and 6.5-8

*6.5-8 Solve the preceding problem for an L $7 \times 4 \times 1/2$ angle section with $M = 15$ k-in.

SHEAR STRESSES IN THIN-WALLED BEAMS

When solving the problems for Section 6.7, assume that the cross sections are thin-walled and use centerline dimensions for all calculations and derivations.

6.7-1 A simple beam of wide-flange cross section supports a uniform load of intensity $q = 3.0$ k/ft on a span of length $L = 10$ ft (see figure). The dimensions of the cross section are $h = 10.5$ in., $b = 7$ in., and $t_f = t_w = 0.4$ in. (a) Calculate the maximum shear stress τ_{max} on cross section A-A located at distance $d = 2.0$ ft from the end of the beam. (b) Calculate the shear stress τ_B at point B on the cross section. Point B is located at a distance $a = 2.0$ in. from the edge of the lower flange.

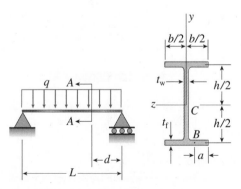

Probs. 6.7-1 and 6.7-2

6.7-2 Solve the preceding problem for the following data: $L = 3$ m, $q = 40$ kN/m, $h = 260$ mm, $b = 170$ mm, $t_f = 12$ mm, $t_w = 10$ mm, $d = 0.6$ m, and $a = 60$ mm.

6.7-3 A beam having the doubly symmetric cross section shown in the figure is subjected to a shear force $P = 4.0$ k acting through the centroid and parallel to the y axis. (a) Calculate the maximum shear stress τ_{max}. (b) What fraction of the shear force P is resisted by each of the four vertical rectangular areas in the cross section?

Prob. 6.7-3

6.7-4 A T-beam is subjected to a shear force P acting through the shear center S parallel to the y axis (see figure). The flange has width b and thickness t. Investigate the shear stresses acting on the cross section, as follows: (a) Obtain a formula for the maximum shear stress τ_{max}. (b) Plot a graph showing how the shear stress τ varies over the height of the flange. (c) Demonstrate that the resultant of the shear stresses is equal to P.

Prob. 6.7-4

6.7-5 A singly symmetric wide-flange beam is subjected to a shear force P acting through the shear center S parallel to the y axis (see figure). Obtain a formula for the maximum shear stress τ_{max} in the beam, assuming $b_1 > b_2$.

Prob. 6.7-5

6.7-6 An equal-leg angle section is subjected to a shear force P acting through the shear center S and parallel to the y axis, which is an axis of symmetry (see figure). Each leg of the angle has centerline length b and thickness t. Investigate the shear stresses acting on the cross section, as follows: (a) Derive a formula for the maximum shear stress τ_{max} in the angle. (b) Plot a graph showing how the shear stress τ varies along the centerline of the cross section. (c) Demonstrate that the resultant of the shear stresses is equal to P.

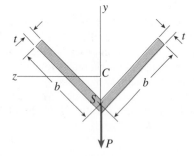

Prob. 6.7-6

6.7-7 A beam of channel section is subjected to a shear force $P = 12$ k acting along the y axis, which is an axis of symmetry (see figure). The dimensions of the cross section are $h = 18$ in., $b = 6$ in., and $t = 1$ in. Investigate the shear stresses acting on the cross section, as follows: (a) Calculate the maximum shear stress τ_{max}. (b) Plot graphs showing how the shear stress τ varies over the height b of the flanges and along the length h of the web. (c) Demonstrate that the resultant of the shear stresses is equal to P.

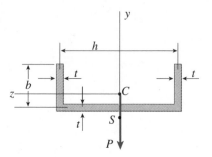

Probs. 6.7-7 and 6.7-8

6.7-8 Solve the preceding problem if $P = 20$ kN and the dimensions of the cross section are $h = 225$ mm, $b = 90$ mm, and $t = 20$ mm.

* **6.7-9** An unbalanced I-beam is turned sideways and subjected to a shear force $P = 20$ k acting in the direction of the y axis, which is an axis of symmetry (see figure). The dimensions of the cross section are $h = 16$ in., $b = 9$ in., $b_1 = 6$ in., $b_2 = 3$ in., $t_f = 1$ in., and $t_w = 0.75$ in. Investigate the shear stresses acting on the cross section, as follows: (a) Calculate the maximum shear stress τ_{max}. (b) Plot a graph showing how the shear stress τ varies over the height b of the flanges and the length h of the web. (c) Demonstrate that the resultant of the shear stresses is equal to P.

Probs. 6.7-9 and 6.7-10

* **6.7-10** Solve the preceding problem if $P = 120$ kN and the dimensions of the cross section are $h = 300$ mm, $b = 160$ mm, $b_1 = 120$ mm, $b_2 = 40$ mm, $t_f = 20$ mm, and $t_w = 15$ mm.

SHEAR CENTERS OF THIN-WALLED OPEN SECTIONS

When locating the shear centers in the problems for Section 6.8, assume that the cross sections are thin-walled and use centerline dimensions for all calculations and derivations.

6.8-1 Calculate the distance e from the centerline of the web of a C 12×20.7 channel section to the shear center S (see figure). (*Note:* For purposes of analysis, consider the flanges to be rectangles with thickness t_f equal to the average flange thickness given in Table E-3, Appendix E.)

Probs. 6.8-1 and 6.8-2

6.8-2 Calculate the distance e from the centerline of the web of a C 8×18.75 channel section to the shear center S (see figure). (*Note:* For purposes of analysis, consider the flanges to be rectangles with thickness t_f equal to the average flange thickness given in Table E-3, Appendix E.)

6.8-3 The cross section of an unbalanced wide-flange beam is shown in the figure. Derive the following formula for the distance h_1 from the centerline of one flange to the shear center S:

$$h_1 = \frac{t_2 b_2^3 h}{t_1 b_1^3 + t_2 b_2^3}$$

Also, check the formula for the special cases of a T-beam ($b_2 = t_2 = 0$) and a balanced wide-flange beam ($t_2 = t_1$ and $b_2 = b_1$).

Prob. 6.8-3

6.8-4 The cross section of an unbalanced wide-flange beam is shown in the figure. Derive the following formula for the distance e from the centerline of the web to the shear center S:

$$e = \frac{3t_f(b_2^2 - b_1^2)}{ht_w + 6t_f(b_1 + b_2)}$$

Also, check the formula for the special cases of a channel section ($b_1 = 0$ and $b_2 = b$) and a doubly symmetric beam ($b_1 = b_2 = b/2$).

Prob. 6.8-4

6.8-5 The cross section of a channel beam with double flanges and constant thickness throughout the section is shown in the figure. Derive the following formula for the distance e from the centerline of the web to the shear center S:

$$e = \frac{3b^2(h_1^2 + h_2^2)}{h_2^3 + 6b(h_1^2 + h_2^2)}$$

Prob. 6.8-5

6.8-6 The cross section of a slit circular tube of constant thickness is shown in the figure. Show that the distance e from the center of the circle to the shear center S is equal to $2r$.

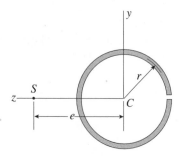

Prob. 6.8-6

6.8-7 The cross section of a slit square tube of constant thickness is shown in the figure. Derive the following formula for the distance e from the corner of the cross section to the shear center S:

$$e = \frac{b}{2\sqrt{2}}$$

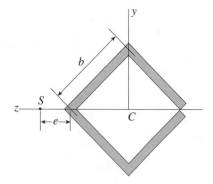

Prob. 6.8-7

6.8-8 The cross section of a slit rectangular tube of constant thickness is shown in the figure. Derive the following formula for the distance e from the centerline of the wall of the tube to the shear center S:

$$e = \frac{b(2h + 3b)}{2(h + 3b)}$$

Prob. 6.8-8

6.8-9 A U-shaped cross section of constant thickness is shown in the figure. Derive the following formula for the distance e from the center of the semicircle to the shear center S:

$$e = \frac{2(2r^2 + b^2 + \pi br)}{4b + \pi r}$$

Also, plot a graph showing how the distance e (expressed as the nondimensional ratio e/r) varies as a function of the ratio b/r. (Let b/r range from 0 to 2.)

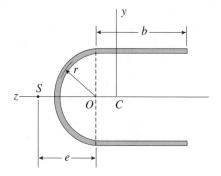

Prob. 6.8-9

* 6.8-10 Derive the following formula for the distance e from the centerline of the wall to the shear center S for the C-section of constant thickness shown in the figure:

$$e = \frac{3bh^2(b + 2a) - 8ba^3}{h^2(h + 6b + 6a) + 4a^2(2a - 3h)}$$

Also, check the formula for the special cases of a channel section ($a = 0$) and a slit rectangular tube ($a = h/2$).

Prob. 6.8-10

*6.8-11 Derive the following formula for the distance e from the centerline of the wall to the shear center S for the hat section of constant thickness shown in the figure:

$$e = \frac{3bh^2(b + 2a) - 8ba^3}{h^2(h + 6b + 6a) + 4a^2(2a + 3h)}$$

Also, check the formula for the special case of a channel section ($a = 0$).

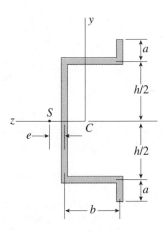

Prob. 6.8-11

*6.8-12 A cross section in the shape of a circular arc of constant thickness is shown in the figure. Derive the following formula for the distance e from the center of the arc to the shear center S:

$$e = \frac{2r(\sin \beta - \beta \cos \beta)}{\beta - \sin \beta \cos \beta}$$

in which β is in radians. Also, plot a graph showing how the distance e varies as β goes from 0 to π.

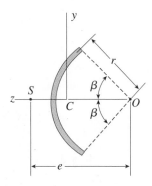

Prob. 6.8-12

ELASTOPLASTIC BENDING

The problems for Section 6.9 are to be solved using the assumption that the material is elastoplastic with yield stress σ_y.

6.9-1 Determine the shape factor f for a cross section in the shape of a double trapezoid having the dimensions shown in the figure.

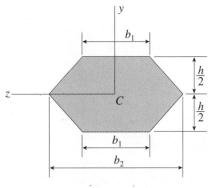

Prob. 6.9-1

6.9-2 (a) Determine the shape factor f for a hollow circular cross section having inner radius r_1 and outer radius r_2 (see figure). (b) If the section is very thin, what is the shape factor?

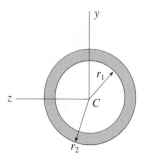

Prob. 6.9-2

6.9-3 A cantilever beam of length $L = 60$ in. supports a uniform load of intensity q (see figure). The beam is made of steel ($\sigma_y = 40$ ksi) and has a rectangular cross section of width $b = 4.0$ in. and height $h = 6.0$ in. What load intensity q will produce a fully plastic condition in the beam?

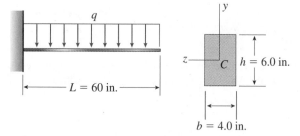

Prob. 6.9-3

6.9-4 A steel beam of rectangular cross section is 50 mm wide and 80 mm high (see figure). The yield stress of the steel is 210 MPa. (a) What percent of the cross-sectional area is occupied by the elastic core if the beam is subjected to a bending moment of 13.0 kN·m acting about the z axis? (b) What is the magnitude of the bending moment that will cause 50% of the cross section to yield?

Prob. 6.9-4

6.9-5 Calculate the shape factor f for the wide-flange beam shown in the figure if $h = 12.0$ in., $b = 6.0$ in., $t_f = 0.6$ in., and $t_w = 0.4$ in.

Probs. 6.9-5 and 6.9-6

6.9-6 Solve the preceding problem for a wide-flange beam with $h = 400$ mm, $b = 160$ mm, $t_f = 12$ mm, and $t_w = 8$ mm.

6.9-7 Determine the section modulus S, plastic modulus Z, and shape factor f for a W 12×87 wide-flange beam. (*Note:* Obtain all beam properties by performing calculations using the cross-sectional dimensions given in Table E-1, Appendix E.)

6.9-8 Solve the preceding problem for a W 10×60 wide-flange beam.

6.9-9 Determine the yield moment M_y and plastic moment M_p for a W 16×77 wide-flange beam if $\sigma_y = 36$ ksi. (*Note:* Obtain all beam properties by performing calculations using the cross-sectional dimensions given in Table E-1, Appendix E.)

6.9-10 Solve the preceding problem for a W 8 × 21 wide-flange beam.

6.9-11 A hollow box beam with height $h = 16$ in., width $b = 8$ in., and constant wall thickness $t = 0.75$ in. is shown in the figure. The beam is constructed of steel with yield stress $\sigma_y = 32$ ksi. Determine the yield moment M_y and plastic moment M_p.

Probs. 6.9-11 and 6.9-12

6.9-12 Solve the preceding problem for a box beam with dimensions $h = 0.4$ m, $b = 0.2$ m, and $t = 20$ mm. The yield stress of the steel is 230 MPa.

6.9-13 A hollow box beam with height $h = 9.0$ in., inside height $h_1 = 7.5$ in., width $b = 5.0$ in., and inside width $b_1 = 4.0$ in. is shown in the figure. Assuming that the beam is constructed of steel with yield stress $\sigma_y = 33$ ksi, calculate the yield moment M_y and plastic moment M_p.

Probs. 6.9-13, 6.9-14, 6.9-15, and 6.9-16

6.9-14 Solve the preceding problem for a box beam with dimensions $h = 200$ mm, $h_1 = 160$ mm, $b = 150$ mm, and $b_1 = 130$ mm. Assume that the beam is constructed of steel with yield stress $\sigma_y = 220$ MPa.

6.9-15 The hollow box beam shown in the figure is subjected to a bending moment M of such magnitude that the flanges yield but the webs remain linearly elastic. (a) Calculate the magnitude of the moment M if the dimensions of the cross section are $h = 14$ in., $h_1 = 12.5$ in., $b = 8$ in., and $b_1 = 7$ in. Also, the yield stress is $\sigma_y = 42$ ksi. (b) What percent of the moment M is produced by the elastic core?

6.9-16 Solve the preceding problem for a box beam with dimensions $h = 400$ mm, $h_1 = 360$ mm, $b = 200$ mm, and $b_1 = 160$ mm, and with yield stress $\sigma_y = 220$ MPa.

6.9-17 A W 12 × 50 wide-flange beam is subjected to a bending moment M of such magnitude that the flanges yield but the web remains linearly elastic. (a) Calculate the magnitude of the moment M if the yield stress is $\sigma_y = 36$ ksi. (b) What percent of the moment M is produced by the elastic core?

6.9-18 Solve the preceding problem for a wide-flange beam if the height of the beam is 210 mm, the width of the flange is 130 mm, the thickness of the flanges is 10 mm, the thickness of the web is 6.5 mm, and the yield stress is 200 MPa.

6.9-19 A singly symmetric beam of T-section (see figure) has cross-sectional dimensions $b = 6$ in., $a = 8$ in., $t_w = 1.5$ in., and $t_f = 1.5$ in. Calculate the plastic modulus Z and the shape factor f.

Probs. 6.9-19 and 6.9-20

6.9-20 Solve the preceding problem for a T-beam with dimensions $b = 140$ mm, $a = 200$ mm, $t_w = 20$ mm, and $t_f = 25$ mm.

6.9-21 A wide-flange beam of unbalanced cross section has the dimensions shown in the figure. Determine the plastic moment M_p if $\sigma_y = 36$ ksi.

Prob. 6.9-21

6.9-22 Determine the plastic moment M_p for a beam having the cross section shown in the figure if $\sigma_y = 210$ MPa.

Prob. 6.9-22

NONLINEAR BENDING

6.10-1 A rectangular beam of structural plastic is 3.0 in. wide and 6.0 in. high (see figure) and subjected to a positive bending moment M. The stress-strain diagram for the plastic is represented by two straight lines such that the modulus of elasticity in compression is $E_1 = 2.8 \times 10^6$ psi and the

modulus in tension is $E_2 = 1.5 \times 10^6$ psi. If the maximum permissible stresses in tension and compression are 900 psi and 1200 psi, respectively, what is the allowable bending moment M_{allow}?

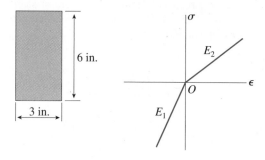

Prob. 6.10-1

6.10-2 A cantilever beam of rectangular cross section and length L supports a uniformly distributed load of intensity q (see figure). The material of the beam has a modulus of elasticity E_1 in compression and E_2 in tension. Numerical values are as follows: $L = 1.0$ m, $q = 12.0$ kN/m, $b = 60$ mm, $h = 140$ mm, $E_1 = 200$ GPa, and $E_2 = 70$ GPa. Determine the maximum tensile stress σ_t and maximum compressive stress σ_c.

Probs. 6.10-2 and 6.10-3

6.10-3 Solve the preceding problem for the following conditions: $L = 3.0$ ft, $q = 75$ lb/in., $b = 2.5$ in., $h = 6.0$ in., $E_1 = 24 \times 10^6$ psi, and $E_2 = 18 \times 10^6$ psi.

6.10-4 A beam of rectangular cross section is made of a material represented by the bilinear stress-strain diagram shown in the figure. The moduli of elasticity are E_1 and E_2, and the properties are the same in both tension and compression. The following numerical data apply to this beam: $b = 40$ mm, $h = 120$ mm, $E_1 = 70$ GPa, $E_2 = 30$ GPa, and $\sigma_y = 160$ MPa. If the maximum stress in the beam is $\sigma_{max} = 200$ MPa, what is the bending moment M?

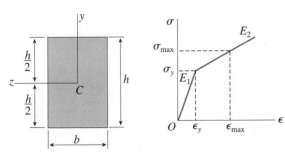

Probs. 6.10-4 and 6.10-5

6.10-5 Solve the preceding problem using the following data: $b = 2.0$ in., $h = 6.0$ in., $E_1 = 10 \times 10^6$ psi, $E_2 = 4 \times 10^6$ psi, $\sigma_y = 24{,}000$ psi, and $\sigma_{max} = 30{,}000$ psi.

6.10-6 A beam of rectangular cross section (width b, height h) is made of material having a stress-strain curve in tension (see figure) given by the equation

$$\sigma = B_1\epsilon - B_2\epsilon^2 \qquad (0 \le \epsilon \le \epsilon_{max})$$

in which B_1 and B_2 are constants and $\epsilon_{max} = B_1/2B_2$. The stress-strain curve in compression is the same as in tension. Derive a formula for the bending moment M if the strains at the top and bottom of the beam equal ϵ_1.

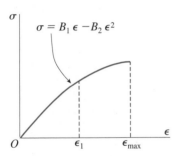

Prob. 6.10-6

6.10-7 The tension stress-strain law for the material of a beam (see figure) is assumed to be of the form $\sigma = B\epsilon^n$, in which B and n are constants ($0 \le n \le 1$). The stress-strain curve for compression is the same as for tension. The cross section of the beam is rectangular with width b and height h. (a) Derive the following formula for the maximum stress in the beam due to a bending moment M:

$$\sigma_1 = \frac{Mc}{I}\left(\frac{n+2}{3}\right)$$

in which $c = h/2$ and $I = bh^3/12$. (b) Derive the following formula for the stress σ in the beam at distance y from the neutral axis:

$$\frac{\sigma}{\sigma_1} = \left(\frac{2y}{h}\right)^n$$

Also, plot a graph showing the stress distribution in the beam for various values of n. (Plot the graph in nondimensional form, σ/σ_1 versus $2y/h$, and choose $n = 1, 0.5, 0.25$, and 0.)

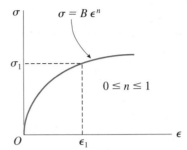

Prob. 6.10-7

6.10-8 The stress distribution on the cross section of a rectangular beam having width b and height h is given by the equation

$$\frac{\sigma}{\sigma_1} = 1 - \left(1 - \frac{2y}{h}\right)^m$$

in which σ is the stress at distance y from the neutral axis, σ_1 is the stress at the top and bottom of the beam, and m is a constant ($m \ge 1$). The stress distribution is the same in both the tension and compression parts of the beam (see figure). (a) Derive the following formula for the bending moment M acting on the cross section:

$$M = \frac{\sigma_1 I}{c}\left[\frac{3m(m+3)}{2(m+1)(m+2)}\right] \qquad (m \ge 1)$$

in which $c = h/2$ and $I = bh^3/12$. (b) Plot a graph showing the stress distribution in the beam for various values of m. (Plot the graph in nondimensional form, σ/σ_1 versus $2y/h$, and choose $m = 1, 2, 4, 10$.)

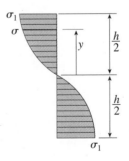

Prob. 6.10-8

6.10-9 A beam of rectangular cross section with width $b = 2.0$ in. and height $h = 4.5$ in. is constructed of high-strength steel having a stress-strain diagram defined by the numerical data listed in the accompanying table. The diagram is the same for both tension and compression. Calculate the value of the bending moment M if the maximum stress in the beam is 105 ksi.

Stress-strain data for Problem 6.10-9

Stress (ksi)	Strain
0	0
70	0.0024
75	0.0030
80	0.0038
90	0.0063
100	0.0105
110	0.0170
120	0.0274

*6.10-10 Determine the maximum stress σ_{max} in the beam of Problem 6.10-4 if the bending moment $M = 18$ kN·m.

*6.10-11 Determine the maximum stress σ_{max} in the beam of Problem 6.10-5 if the bending moment $M = 320$ k-in.

**6.10-12 A beam of rectangular cross section (width $b = 25$ mm and height $h = 75$ mm) is made of material having the idealized nonsymmetric stress-strain curve shown in the figure. The slopes of the lines are $E_1 = 6.5$ GPa and $E_2 = 2.85$ GPa, and the tensile yield strain is $\epsilon_y = 0.0022$. What is the bending moment M in the beam when the maximum tensile stress is 20 MPa?

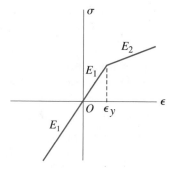

Prob. 6.10-12

7

ANALYSIS OF STRESS AND STRAIN

INTRODUCTION

Normal and shear stresses in beams, shafts, and bars can be calculated from the basic formulas discussed in the preceding chapters. For instance, the stresses in a beam are given by the flexure and shear formulas ($\sigma = My/I$ and $\tau = VQ/Ib$), and the stresses in a shaft are given by the torsion formula ($\tau = T\rho/I_p$). However, the stresses calculated from these formulas act on cross sections of the members, and sometimes larger stresses occur on inclined sections. Therefore, we will begin our analysis of stresses and strains by discussing methods for finding the normal and shear stresses acting on inclined sections cut through the members.

We have already derived expressions for the normal and shear stresses acting on inclined sections in both uniaxial stress and pure shear (see Sections 2.6 and 3.5, respectively). In the case of uniaxial stress, we found that the maximum shear stresses occur on planes inclined at 45° to the axis, whereas the maximum normal stresses occur on the cross sections. In the case of pure shear, we found that the maximum tensile and compressive stresses occur on 45° planes. In an analogous manner, the stresses on inclined sections cut through a beam may be larger than the stresses acting on a cross section. To calculate such stresses, we need to determine the stresses acting on inclined planes under a more general stress state known as *plane stress* (Section 7.2).

In our discussions of plane stress we will use *stress elements* to represent the state of stress at a point in a body. Stress elements were discussed previously in a specialized context (see Sections 2.6 and 3.5), but now we will use them in a more formalized manner. We will begin our

473

analysis by considering an element on which the stresses are known, and then we will derive the *transformation relationships* that give the stresses acting on the sides of an element oriented in a different direction.

When working with stress elements, we must always keep in mind that only one intrinsic *state of stress* exists at a point in a stressed body, regardless of the orientation of the element being used to portray that state of stress. When we have two elements with different orientations at the same point in the body, the stresses acting on the faces of the two elements are different, but they still represent the same state of stress, namely, the stress at the point under consideration. This situation is analogous to the representation of a force vector by its components—although the components are different when the coordinate axes are rotated to a new position, the force itself is the same.

Furthermore, we must always keep in mind that stresses are *not* vectors. This fact can sometimes be confusing, because we customarily represent stresses by arrows just as we represent force vectors by arrows. Although the arrows used to represent stresses have magnitude and direction, they are not vectors because they do not combine according to the parallelogram law of addition. Instead, stresses are much more complex quantities than are vectors, and in mathematics they are called *tensors*. Other tensor quantities in mechanics are strains and moments of inertia.

7.2 | PLANE STRESS

Fig. 7-1 Elements in plane stress: (a) three-dimensional view of an element oriented to the *xyz* axes, (b) two-dimensional view of the same element, and (c) two-dimensional view of an element oriented to the $x_1y_1z_1$ axes

The stress conditions that we encountered in earlier chapters when analyzing bars in tension and compression, shafts in torsion, and beams in bending are examples of a state of stress called **plane stress**. To explain plane stress, we will consider the stress element shown in Fig. 7-1a. This element is infinitesimal in size and can be sketched either

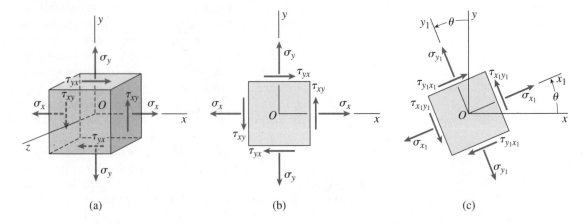

(a)   (b)   (c)

as a cube or as a rectangular parallelepiped. The *xyz* axes are parallel to the edges of the element, and the faces of the element are designated by the directions of their outward normals, as explained previously in Section 1.6. For instance, the right-hand face of the element is referred to as the positive *x* face, and the left-hand face (hidden from the viewer) is referred to as the negative *x* face. Similarly, the top face is the positive *y* face, and the front face is the positive *z* face.

When the material is in plane stress in the *xy* plane, only the *x* and *y* faces of the element are subjected to stresses, and all stresses act parallel to the *x* and *y* axes, as shown in Fig. 7-1a. This stress condition is very common because it exists at the surface of any stressed body, except at points where external loads act on the surface. When the element shown in Fig. 7-1a is located at the free surface of a body, the *z* face is in the plane of the surface (no stresses) and the *z* axis is normal to the surface.

The symbols for the stresses shown in Fig. 7-1a have the following meanings. A **normal stress** σ has a subscript that identifies the face on which the stress acts; for instance, the stress σ_x acts on the *x* face of the element and the stress σ_y acts on the *y* face of the element. Since the element is infinitesimal in size, equal normal stresses act on the opposite faces. The *sign convention* for normal stresses is the familiar one, namely, tension is positive and compression is negative.

A **shear stress** τ has two subscripts—the first subscript denotes the face on which the stress acts, and the second gives the direction on that face. Thus, the stress τ_{xy} acts on the *x* face in the direction of the *y* axis (Fig. 7-1a), and stress τ_{yx} acts on the *y* face in the direction of the *x* axis.

The *sign convention* for shear stresses is as follows. A shear stress is positive when it acts on a positive face of an element in the positive direction of an axis, and it is negative when it acts on a positive face of an element in the negative direction of an axis. Therefore, the stresses τ_{xy} and τ_{yx} shown on the positive *x* and *y* faces in Fig. 7-1a are positive shear stresses. Similarly, on a negative face of the element, a shear stress is positive when it acts in the negative direction of an axis. Hence, the stresses τ_{xy} and τ_{yx} shown on the negative *x* and *y* faces of the element are also positive.

This sign convention for shear stresses is easy to remember if we state it as follows: A shear stress is positive when the directions associated with its subscripts are plus-plus or minus-minus; the stress is negative when the directions are plus-minus or minus-plus.

The preceding sign convention for shear stresses is consistent with the equilibrium of the element, because we know that shear stresses on opposite faces of an infinitesimal element must be equal in magnitude and opposite in direction. Hence, according to our sign convention, a positive stress τ_{xy} acts upward on the positive face (Fig. 7-1a) and downward on the negative face. In a similar manner, the stresses τ_{yx} acting on the top and bottom faces of the element are positive although they have opposite directions.

We also know that shear stresses on perpendicular planes are equal in magnitude and have directions such that both stresses point toward, or both point away from, the line of intersection of the faces. Inasmuch as τ_{xy} and τ_{yx} are positive in the directions shown in the figure, they are consistent with this observation. Therefore, we note that

$$\tau_{xy} = \tau_{yx} \tag{7-1}$$

This relationship was derived previously from equilibrium of the element (see Section 1.6).

For convenience in sketching plane-stress elements, we usually draw only a two-dimensional view of the element, as shown in Fig. 7-1b. Although a figure of this kind is adequate for showing all stresses acting on the element, we must still keep in mind that the element is a solid body with a thickness perpendicular to the plane of the figure.

■ Stresses on Inclined Sections

We are now ready to consider the stresses acting on inclined sections, assuming that the stresses σ_x, σ_y, and τ_{xy} (Figs. 7-1a and b) are known. To portray the stresses acting on an inclined section, we consider a new stress element (Fig. 7-1c) that is located at the same point in the material as the original element (Fig. 7-1b). However, the new element has faces that are parallel and perpendicular to the inclined direction. Associated with this new element are axes x_1, y_1, and z_1, such that the z_1 axis coincides with the z axis and the x_1y_1 axes are rotated counterclockwise through an angle θ with respect to the xy axes. The normal and shear stresses acting on this new element are denoted σ_{x_1}, σ_{y_1}, $\tau_{x_1y_1}$, and $\tau_{y_1x_1}$, using the same subscript designations and sign conventions described above for the stresses acting on the xy element. The previous conclusions regarding the shear stresses still apply, so that

$$\tau_{x_1y_1} = \tau_{y_1x_1} \tag{7-2}$$

From this equation and the equilibrium of the element, we see that *the shear stresses acting on all four side faces of an element in plane stress are known if we determine the shear stress acting on any one of those faces.*

The stresses acting on the inclined x_1y_1 element (Fig. 7-1c) can be expressed in terms of the stresses on the xy element (Fig. 7-1b) by using equations of equilibrium. For this purpose, we choose a wedge-shaped stress element (Fig. 7-2a) having an inclined face that is the same as the x_1 face of the inclined element. The other two side faces of the wedge are parallel to the x and y axes.

(a) Stresses

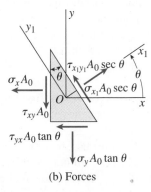

(b) Forces

Fig. 7-2 Wedge-shaped stress element in plane stress: (a) stresses acting on the element, and (b) forces acting on the element

In order to write equations of equilibrium for the wedge, we need to construct a free-body diagram showing the forces acting on the faces. Let us denote the area of the left-hand side face (that is, the negative x face) as A_0. Then the normal and shear forces acting on that face are $\sigma_x A_0$ and $\tau_{xy} A_0$, as shown in the free-body diagram of Fig. 7-2b. The area of the bottom face (or negative y face) is $A_0 \tan \theta$, and the area of the inclined face (or positive x_1 face) is $A_0 \sec \theta$. Thus, the normal and shear forces acting on these faces have the magnitudes and directions shown in the figure.

The forces acting on the left-hand and bottom faces can be resolved into orthogonal components acting in the x_1 and y_1 directions. Then we can obtain two equations of equilibrium by summing forces in those directions. The first equation, obtained by summing forces in the x_1 direction, is

$$\sigma_{x_1} A_0 \sec \theta - \sigma_x A_0 \cos \theta - \tau_{xy} A_0 \sin \theta$$
$$- \sigma_y A_0 \tan \theta \sin \theta - \tau_{yx} A_0 \tan \theta \cos \theta = 0$$

In the same manner, summation of forces in the y_1 direction gives

$$\tau_{x_1 y_1} A_0 \sec \theta + \sigma_x A_0 \sin \theta - \tau_{xy} A_0 \cos \theta$$
$$- \sigma_y A_0 \tan \theta \cos \theta + \tau_{yx} A_0 \tan \theta \sin \theta = 0$$

Using the relationship $\tau_{xy} = \tau_{yx}$, and also simplifying and rearranging, we obtain the following two equations:

$$\sigma_{x_1} = \sigma_x \cos^2 \theta + \sigma_y \sin^2 \theta + 2\tau_{xy} \sin \theta \cos \theta \qquad (7\text{-}3a)$$

$$\tau_{x_1 y_1} = -(\sigma_x - \sigma_y) \sin \theta \cos \theta + \tau_{xy} (\cos^2 \theta - \sin^2 \theta) \qquad (7\text{-}3b)$$

Equations (7-3a) and (7-3b) give the normal and shear stresses acting on the x_1 plane in terms of the angle θ and the stresses σ_x, σ_y, and τ_{xy} acting on the x and y planes.

For the special case when $\theta = 0$, we note that Eqs. (7-3a) and (7-3b) give $\sigma_{x_1} = \sigma_x$ and $\tau_{x_1 y_1} = \tau_{xy}$, as expected. Also, when $\theta = 90°$, the equations give $\sigma_{x_1} = \sigma_y$ and $\tau_{x_1 y_1} = -\tau_{xy} = -\tau_{yx}$. In the latter case, since the x_1 axis is vertical when $\theta = 90°$, the stress $\tau_{x_1 y_1}$ will be positive when it acts to the left. However, the stress τ_{yx} acts to the right, and therefore $\tau_{x_1 y_1} = -\tau_{yx}$.

Equations (7-3a) and (7-3b) can be expressed in a more convenient form by introducing the following trigonometric identities (see Appendix C):

$$\cos^2 \theta = \frac{1}{2}(1 + \cos 2\theta) \qquad \sin^2 \theta = \frac{1}{2}(1 - \cos 2\theta)$$

$$\sin \theta \cos \theta = \frac{1}{2} \sin 2\theta$$

When these substitutions are made, the equations become

$$\sigma_{x_1} = \frac{\sigma_x + \sigma_y}{2} + \frac{\sigma_x - \sigma_y}{2} \cos 2\theta + \tau_{xy} \sin 2\theta \qquad (7\text{-}4a)$$

$$\tau_{x_1 y_1} = -\frac{\sigma_x - \sigma_y}{2} \sin 2\theta + \tau_{xy} \cos 2\theta \qquad (7\text{-}4b)$$

These equations are known as the **transformation equations for plane stress** because they transform the stress components from one set of axes to another. However, as explained previously, the intrinsic state of stress at the point under consideration is the same whether represented by stresses acting on the xy element (Fig. 7-1b) or by stresses acting on the inclined $x_1 y_1$ element (Fig. 7-1c).

Since the transformation equations were derived solely from equilibrium of an element, they are applicable to stresses in any kind of material, whether linear or nonlinear, elastic or inelastic.

An important observation concerning the normal stresses can be obtained from the transformation equations. As a preliminary matter, we note that the normal stress σ_{y_1} acting on the y_1 face of the inclined element (Fig. 7-1c) can be obtained from Eq. (7-4a) by substituting $\theta + 90°$ for θ. The result is the following equation for σ_{y_1}:

$$\sigma_{y_1} = \frac{\sigma_x + \sigma_y}{2} - \frac{\sigma_x - \sigma_y}{2} \cos 2\theta - \tau_{xy} \sin 2\theta \qquad (7\text{-}5)$$

Summing the expressions for σ_{x_1} and σ_{y_1} (Eqs. 7-4a and 7-5), we obtain the following equation for plane stress:

$$\sigma_{x_1} + \sigma_{y_1} = \sigma_x + \sigma_y \qquad (7\text{-}6)$$

This equation shows that the sum of the normal stresses acting on perpendicular faces of plane-stress elements (at a given point in a stressed body) is constant and independent of the angle θ.

The manner in which the normal and shear stresses vary is shown in Fig. 7-3, which is a graph of σ_{x_1} and $\tau_{x_1 y_1}$ versus the angle θ (from Eqs. 7-4a and 7-4b). The graph is plotted for the particular case of $\sigma_y = 0.2\sigma_x$ and $\tau_{xy} = 0.8\sigma_x$. We see from the plot that the stresses vary continuously as the orientation of the element is changed. At certain angles, the normal stress reaches a maximum or minimum value; at other angles, it becomes zero. Similarly, the shear stress has maximum, minimum, and zero values at certain angles. A detailed investigation of these maximum and minimum values is made in the next section.

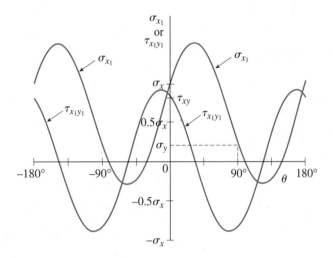

Fig. 7-3 Graph of normal stress σ_{x_1} and shear stress $\tau_{x_1y_1}$ versus the angle θ (for $\sigma_y = 0.2\sigma_x$ and $\tau_{xy} = 0.8\sigma_x$)

■ Special Cases of Plane Stress

The general case of plane stress reduces to simpler states of stress under special conditions. For instance, if all stresses acting on the xy element (Fig. 7-1b) are zero except for the normal stress σ_x, then the element is in **uniaxial stress** (Fig. 7-4). The corresponding transformation equations, obtained by setting σ_y and τ_{xy} equal to zero in Eqs. (7-4a) and (7-4b), are

$$\sigma_{x_1} = \frac{\sigma_x}{2}(1 + \cos 2\theta) \qquad \tau_{x_1y_1} = -\frac{\sigma_x}{2}(\sin 2\theta) \qquad \text{(7-7a,b)}$$

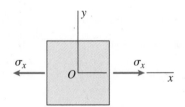

Fig. 7-4 Element in uniaxial stress

These equations agree with the equations derived previously in Section 2.6 (see Eqs. 2-23a and 2-23b), except that now we are using a more generalized notation for the stresses acting on an inclined plane.

Another special case is **pure shear** (Fig. 7-5), for which the transformation equations are obtained by substituting $\sigma_x = 0$ and $\sigma_y = 0$ into Eqs. (7-4a) and (7-4b):

$$\sigma_{x_1} = \tau_{xy} \sin 2\theta \qquad \tau_{x_1y_1} = \tau_{xy} \cos 2\theta \qquad \text{(7-8a,b)}$$

Again, these equations correspond to those derived earlier (see Eqs. 3-30a and 3-30b in Section 3.5).

Finally, we note the special case of **biaxial stress**, in which the xy element is subjected to normal stresses in both the x and y directions but without any shear stresses (Fig. 7-6). The equations for biaxial stress are

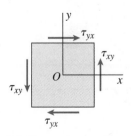

Fig. 7-5 Element in pure shear

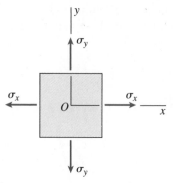

Fig. 7-6 Element in biaxial stress

obtained from Eqs. (7-4a) and (7-4b) simply by dropping the terms containing τ_{xy}:

$$\sigma_{x_1} = \frac{\sigma_x + \sigma_y}{2} + \frac{\sigma_x - \sigma_y}{2} \cos 2\theta \qquad (7\text{-}9a)$$

$$\tau_{x_1 y_1} = -\frac{\sigma_x - \sigma_y}{2} \sin 2\theta \qquad (7\text{-}9b)$$

Biaxial stress occurs in many kinds of structures, including thin-walled pressure vessels (see Sections 8.2 and 8.3).

WORK SHEETS 5 & 6 Mathcad **Example 7-1**

An element in plane stress is subjected to stresses $\sigma_x = 16,000$ psi, $\sigma_y = 6,000$ psi, and $\tau_{xy} = \tau_{yx} = 4,000$ psi, as shown in Fig. 7-7a. Determine the stresses acting on an element inclined at an angle $\theta = 45°$.

Solution   To determine the stresses acting on an inclined element, we will use the transformation equations (Eqs. 7-4a and 7-4b). From the given numerical data, we obtain the following values for substitution into those equations:

$$\frac{\sigma_x + \sigma_y}{2} = 11,000 \text{ psi} \qquad \frac{\sigma_x - \sigma_y}{2} = 5,000 \text{ psi} \qquad \tau_{xy} = 4,000 \text{ psi}$$

$$\sin 2\theta = \sin 90° = 1 \qquad \cos 2\theta = \cos 90° = 0$$

Substituting these values into Eqs. (7-4a) and (7-4b), we get

$$\sigma_{x_1} = \frac{\sigma_x + \sigma_y}{2} + \frac{\sigma_x - \sigma_y}{2} \cos 2\theta + \tau_{xy} \sin 2\theta$$

$$= 11,000 \text{ psi} + (5,000 \text{ psi})(0) + (4,000 \text{ psi})(1) = 15,000 \text{ psi} \quad \longleftarrow$$

Fig. 7-7 Example 7-1. (a) Element in plane stress, and (b) element inclined at an angle $\theta = 45°$

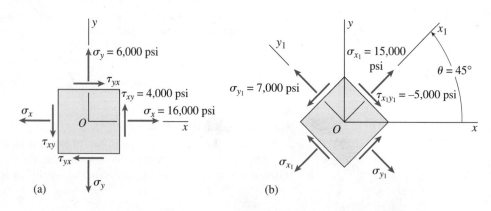

$$= -(5,000 \text{ psi})(1) + (4,000 \text{ psi})(0) = -5,000 \text{ psi} \quad \longleftarrow$$

In addition, the stress σ_{y_1} may be obtained from Eq. (7-5):

$$\sigma_{y_1} = \frac{\sigma_x + \sigma_y}{2} - \frac{\sigma_x - \sigma_y}{2} \cos 2\theta - \tau_{xy} \sin 2\theta$$

$$= 11,000 \text{ psi} - (5,000 \text{ psi})(0) - (4,000 \text{ psi})(1) = 7,000 \text{ psi} \quad \longleftarrow$$

From these results we can readily obtain the stresses acting on all sides of an element oriented at $\theta = 45°$, as shown in Fig. 7-7b. The arrows show the true directions in which the stresses act. Note especially the directions of the shear stresses, all of which have the same magnitude. Also, observe that the sum of the normal stresses remains constant and equal to 22,000 psi (see Eq. 7-6).

The stresses shown in Fig. 7-7b represent the same intrinsic state of stress as do the stresses shown in Fig. 7-7a. However, the stresses have different values because the elements on which they act have different orientations.

■ Example 7-2

A plane-stress condition exists at a point on the surface of a loaded structure, where the stresses have the magnitudes and directions shown on the stress element of Fig. 7-8a. Determine the stresses acting on an element that is oriented at a clockwise angle of 15° with respect to the original element.

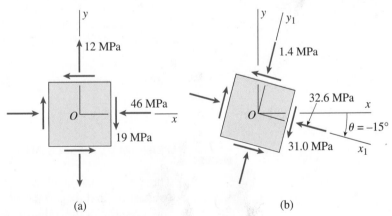

Fig. 7-8 Example 7-2. (a) Element in plane stress, and (b) element inclined at an angle $\theta = -15°$

(a)   (b)

Solution   The stresses acting on the original element (Fig. 7-8a) have the following values:

$$\sigma_x = -46 \text{ MPa} \qquad \sigma_y = 12 \text{ MPa} \qquad \tau_{xy} = -19 \text{ MPa}$$

An element oriented at a clockwise angle of 15° is shown in Fig. 7-8b, where the x_1 axis is at an angle $\theta = -15°$ with respect to the x axis. (As an alternative, the x_1 axis could be placed at a positive angle $\theta = 75°$.)

We can readily calculate the stresses on the x_1 face of the element oriented at $\theta = -15°$ by using the transformation equations (Eqs. 7-4a and 7-4b). The calculations proceed as follows:

$$\frac{\sigma_x + \sigma_y}{2} = -17 \text{ MPa} \qquad \frac{\sigma_x - \sigma_y}{2} = -29 \text{ MPa}$$

$$\sin 2\theta = \sin(-30°) = -0.5 \qquad \cos 2\theta = \cos(-30°) = 0.8660$$

Substituting into the transformation equations, we get

$$\sigma_{x_1} = \frac{\sigma_x + \sigma_y}{2} + \frac{\sigma_x - \sigma_y}{2} \cos 2\theta + \tau_{xy}\sin 2\theta$$

$$= -17 \text{ MPa} + (-29 \text{ MPa})(0.8660) + (-19 \text{ MPa})(-0.5)$$

$$= -32.6 \text{ MPa} \qquad \Longleftarrow$$

$$\tau_{x_1y_1} = -\frac{\sigma_x - \sigma_y}{2} \sin 2\theta + \tau_{xy}\cos 2\theta$$

$$= -(-29 \text{ MPa})(-0.5) + (-19 \text{ MPa})(0.8660)$$

$$= -31.0 \text{ MPa} \qquad \Longleftarrow$$

Also, the normal stress acting on the y_1 face (Eq. 7-5) is

$$\sigma_{y_1} = \frac{\sigma_x + \sigma_y}{2} - \frac{\sigma_x - \sigma_y}{2} \cos 2\theta - \tau_{xy}\sin 2\theta$$

$$= -17 \text{ MPa} - (-29 \text{ MPa})(0.8660) - (-19 \text{ MPa})(-0.5)$$

$$= -1.4 \text{ MPa} \qquad \Longleftarrow$$

This stress can be verified by substituting $\theta = 75°$ into Eq. (7-4a). As a further check on the results, we note that $\sigma_{x_1} + \sigma_{y_1} = \sigma_x + \sigma_y$.

The stresses acting on the inclined element are shown in Fig. 7-8b, where the arrows indicate the true directions of the stresses. Again we note that both stress elements shown in Fig. 7-8 represent the same state of stress.

7.3 PRINCIPAL STRESSES AND MAXIMUM SHEAR STRESSES

The transformation equations for plane stress show that the normal stresses σ_{x_1} and the shear stresses τ_{x1y1} vary continuously as the axes are rotated through the angle θ. This variation is pictured in Fig. 7-3 for a particular combination of stresses. From the figure, we see that both the normal and shear stresses reach maximum and minimum values at 90° intervals. Not surprisingly, these maximum and minimum values are usually needed for design purposes. For instance, fatigue failures of structures such as machines and aircraft are often associated with the maximum stresses, and hence their magnitudes and orientations should be determined as part of the design process.

■ **Principal Stresses**

The maximum and minimum normal stresses, called the **principal stresses**, can be found from the transformation equation for the normal stress σ_{x_1} (Eq. 7-4a). By taking the derivative of σ_{x_1} with respect to θ and setting it equal to zero, we obtain an equation from which we can find the values of θ at which σ_{x_1} is a maximum or minimum. The equation for the derivative is

$$\frac{d\sigma_{x_1}}{d\theta} = -(\sigma_x - \sigma_y)\sin 2\theta + 2\tau_{xy} \cos 2\theta = 0 \qquad (7\text{-}10)$$

from which we get

$$\tan 2\theta_p = \frac{2\tau_{xy}}{\sigma_x - \sigma_y} \qquad (7\text{-}11)$$

The subscript p indicates that the angle θ_p defines the orientation of the **principal planes**, that is, the planes on which the principal stresses act.

Two values of the angle $2\theta_p$ in the range from 0 to 360° can be obtained from Eq. (7-11). These values differ by 180°, with one value between 0 and 180° and the other between 180° and 360°. Therefore, the angle θ_p has two values that differ by 90°, one value between 0 and 90° and the other between 90° and 180°. The two values of θ_p are known as the **principal angles**. For one of these angles, the normal stress σ_{x_1} is a *maximum* principal stress; for the other, it is a *minimum* principal stress. Because the principal angles differ by 90°, we see that *the principal stresses occur on mutually perpendicular planes.*

The principal stresses can be calculated by substituting each of the two values of θ_p into the first stress-transformation equation (Eq. 7-4a) and solving for σ_{x_1}. By determining the principal stresses in this manner, we not only obtain the values of the principal stresses but we also learn which principal stress is associated with which principal angle.

We can also obtain general formulas for the principal stresses. To do so, refer to the right triangle in Fig. 7-9, which is constructed from Eq. (7-11). Note that the hypotenuse of the triangle, obtained from the Pythagorean theorem, is

$$R = \sqrt{\left(\frac{\sigma_x - \sigma_y}{2}\right)^2 + \tau_{xy}^2} \qquad (7\text{-}12)$$

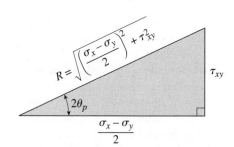

Fig. 7-9 Geometric representation of Eq. (7-11)

The quantity R is always a positive number and, like the other two sides of the triangle, has units of stress. From the triangle we obtain two additional relations:

$$\cos 2\theta_p = \frac{\sigma_x - \sigma_y}{2R} \qquad \sin 2\theta_p = \frac{\tau_{xy}}{R} \qquad (7\text{-}13a,b)$$

Now we substitute these expressions for $\cos 2\theta_p$ and $\sin 2\theta_p$ into Eq. (7-4a) and obtain the algebraically larger of the two principal stresses, denoted by σ_1:

$$\sigma_1 = \frac{\sigma_x + \sigma_y}{2} + \sqrt{\left(\frac{\sigma_x - \sigma_y}{2}\right)^2 + \tau_{xy}^2} \qquad (7\text{-}14)$$

The smaller of the principal stresses, denoted by σ_2, may be found from the condition that the sum of the normal stresses on perpendicular planes is constant (see Eq. 7-6):

$$\sigma_1 + \sigma_2 = \sigma_x + \sigma_y \qquad (7\text{-}15)$$

Substituting the expression for σ_1 into Eq. (7-15) and solving for σ_2, we get

$$\sigma_2 = \frac{\sigma_x + \sigma_y}{2} - \sqrt{\left(\frac{\sigma_x - \sigma_y}{2}\right)^2 + \tau_{xy}^2} \qquad (7\text{-}16)$$

This equation has the same form as the equation for σ_1 but differs by the presence of the minus sign before the square root.

The preceding formulas for σ_1 and σ_2 can be combined into a single formula for the **principal stresses**:

$$\sigma_{1,2} = \frac{\sigma_x + \sigma_y}{2} \pm \sqrt{\left(\frac{\sigma_x - \sigma_y}{2}\right)^2 + \tau_{xy}^2} \qquad (7\text{-}17)$$

The plus sign gives the algebraically larger principal stress and the minus sign gives the algebraically smaller principal stress.

Let us now denote the two angles defining the principal planes as θ_{p_1} and θ_{p_2}, corresponding to the principal stresses σ_1 and σ_2, respectively. Both angles can be determined from the equation for $\tan 2\theta_p$ (Eq. 7-11). However, we cannot tell from that equation which angle is θ_{p_1} and which is θ_{p_2}. A simple procedure for making this determination is to take one of the values and substitute it into the equation for σ_{x_1} (Eq. 7-4a). The resulting value of σ_{x_1} will be recognized as either σ_1 or σ_2 (assuming we have already found σ_1 and σ_2 from Eq. 7-17), thus correlating the two principal angles with the two principal stresses.

Another method for correlating the principal angles and principal stresses is to use Eqs. (7-13a) and (7-13b) to find θ_p, since the only angle that satisfies *both* of those equations is θ_{p_1}. Thus, we can rewrite those equations as follows:

$$\cos 2\theta_{p_1} = \frac{\sigma_x - \sigma_y}{2R} \qquad \sin 2\theta_{p_1} = \frac{\tau_{xy}}{R} \qquad \text{(7-18a,b)}$$

Only one angle exists between 0 and 360° that satisfies both of these equations. Thus, the value of θ_{p_1} can be determined uniquely from Eqs. (7-18a) and (7-18b). The angle θ_{p_2}, corresponding to σ_2, defines a plane that is perpendicular to the plane defined by θ_{p_1}. Therefore, θ_{p_2} can be taken as 90° larger or 90° smaller than θ_{p_1}.

An important characteristic of the principal planes can be obtained from the transformation equation for the shear stresses (Eq. 7-4b). If we set the shear stress $\tau_{x_1y_1}$ equal to zero, we get an equation that is the same as Eq. (7-10). Therefore, if we solve that equation for the angle 2θ, we get the same expression for $\tan 2\theta$ as before (Eq. 7-11). In other words, the angles to the planes of zero shear stress are the same as the angles to the principal planes. Thus, we can make the following important observation concerning the principal planes: *the shear stresses are zero on the principal planes*.

The principal planes for elements in **uniaxial stress** and **biaxial stress** are the x and y planes themselves (Fig. 7-10), because $\tan 2\theta_p = 0$ (see Eq. 7-11) and the two values of θ_p are 0 and 90°. We also know that the x and y planes are the principal planes from the fact that the shear stresses are zero on those planes.

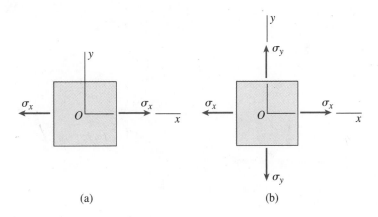

Fig. 7-10 Elements in uniaxial and biaxial stress

(a)   (b)

For an element in **pure shear** (Fig. 7-11a), the principal planes are oriented at 45° to the x axis (Fig. 7-11b), because $\tan 2\theta_p$ is infinite and the two values of θ_p are 45° and 135°. If τ_{xy} is positive, the principal stresses are $\sigma_1 = \tau_{xy}$ and $\sigma_2 = -\tau_{xy}$ (see Section 3.5 for a discussion of pure shear).

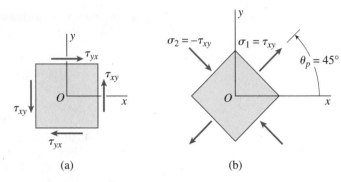

Fig. 7-11 Elements in pure shear

(a) (b)

The preceding discussion of principal stresses refers only to rotation of axes in the xy plane, that is, rotation about the z axis (Fig. 7-12a). Therefore, the two principal stresses determined from Eq. (7-17) are called the **in-plane principal stresses**. However, we must not overlook the fact that the stress element is actually three-dimensional and has three (not two) principal stresses acting on three mutually perpendicular planes. By making a more complete three-dimensional analysis, it can be shown that the three principal planes for a plane-stress element are the two principal planes already described plus the z face of the element. These principal planes are shown in Fig. 7-12b, where a stress element has been oriented at the principal angle θ_{p_1}, which corresponds to the principal stress σ_1. The principal stresses σ_1 and σ_2 are given by Eq. (7-17), and the third principal stress (σ_3) equals zero. By definition, σ_1 is algebraically larger than σ_2, but σ_3 may be algebraically larger than, between, or smaller than σ_1 and σ_2. Note again that there are no shear stresses on any of the principal planes.*

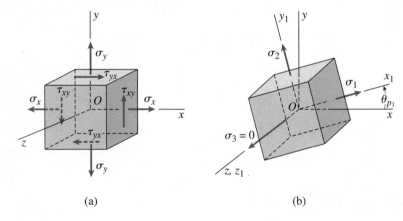

Fig. 7-12 Elements in plane stress: (a) original element, and (b) element oriented to the three principal planes and three principal stresses

(a) (b)

---

*The determination of principal stresses is an example of a type of mathematical analysis known as *eigenvalue analysis*, which is described in books on matrix algebra. The stress-transformation equations and the concept of principal stresses are due to the French mathematicians A. L. Cauchy (1789–1857) and Barré de Saint-Venant (1797–1886) and to the Scottish scientist and engineer W. J. M. Rankine (1820–1872); see Refs. 7-1, 7-2, and 7-3, respectively.

■ Maximum Shear Stresses

Having found the principal stresses and their directions for an element in plane stress, we now consider the determination of the maximum shear stresses and the planes on which they act. The shear stresses $\tau_{x_1 y_1}$ acting on inclined planes are given by the second transformation equation (Eq. 7-4b). Taking the derivative of $\tau_{x_1 y_1}$ with respect to θ and setting it equal to zero, we obtain

$$\frac{d\tau_{x_1 y_1}}{d\theta} = -(\sigma_x - \sigma_y)\cos 2\theta - 2\tau_{xy}\sin 2\theta = 0 \qquad (7\text{-}19)$$

from which

$$\tan 2\theta_s = -\frac{\sigma_x - \sigma_y}{2\tau_{xy}} \qquad (7\text{-}20)$$

The subscript s indicates that the angle θ_s defines the orientation of the planes of maximum positive and negative shear stresses. Equation (7-20) yields one value of θ_s between 0 and 90° and another between 90° and 180°. Furthermore, these two values differ by 90°, and therefore the maximum shear stresses occur on perpendicular planes. Because shear stresses on perpendicular planes are equal in absolute value, the maximum positive and negative shear stresses differ only in sign.

Comparing Eq. (7-20) for θ_s with Eq. (7-11) for θ_p shows that

$$\tan 2\theta_s = -\frac{1}{\tan 2\theta_p} = -\cot 2\theta_p \qquad (7\text{-}21)$$

From this equation we can obtain a relationship between the angles θ_s and θ_p. First, we rewrite the preceding equation in the form

$$\frac{\sin 2\theta_s}{\cos 2\theta_s} + \frac{\cos 2\theta_p}{\sin 2\theta_p} = 0$$

Multiplying by the terms in the denominator, we get

$$\sin 2\theta_s \sin 2\theta_p + \cos 2\theta_s \cos 2\theta_p = 0$$

which is equivalent to the following expression (see Appendix C):

$$\cos(2\theta_s - 2\theta_p) = 0$$

Therefore,

$$2\theta_s - 2\theta_p = \pm 90°$$

and

$$\theta_s = \theta_p \pm 45° \qquad (7\text{-}22)$$

This equation shows that *the planes of maximum shear stress occur at 45° to the principal planes.*

The plane of the maximum positive shear stress τ_{max} is defined by the angle θ_{s_1}, for which the following equations apply:

$$\cos 2\theta_{s_1} = \frac{\tau_{xy}}{R} \qquad \sin 2\theta_{s_1} = -\frac{\sigma_x - \sigma_y}{2R} \qquad \text{(7-23a,b)}$$

in which R is given by Eq. (7-12). Also, the angle θ_{s_1} is related to the angle θ_{p_1} (see Eqs. 7-18a and 7-18b) as follows:

$$\theta_{s_1} = \theta_{p_1} - 45° \qquad \text{(7-24)}$$

The corresponding maximum shear stress is obtained by substituting the expressions for $\cos 2\theta_{s_1}$ and $\sin 2\theta_{s_1}$ into the second transformation equation (Eq. 7-4b), yielding

$$\tau_{max} = \sqrt{\left(\frac{\sigma_x - \sigma_y}{2}\right)^2 + \tau_{xy}^2} \qquad \text{(7-25)}$$

The maximum negative shear stress τ_{min} has the same magnitude but opposite sign.

Another expression for the maximum shear stress can be obtained from the principal stresses σ_1 and σ_2, both of which are given by Eq. (7-17). Subtracting the expression for σ_2 from that for σ_1, and then comparing with Eq. (7-25), we see that

$$\tau_{max} = \frac{\sigma_1 - \sigma_2}{2} \qquad \text{(7-26)}$$

Thus, *the maximum shear stress is equal to one-half the difference of the principal stresses.*

The planes of maximum shear stress also contain normal stresses. The normal stress acting on the planes of maximum positive shear stress can be determined by substituting the expressions for the angle θ_{s_1} (Eqs. 7-23a and 7-23b) into the equation for σ_{x_1} (Eq. 7-4a). The resulting stress is equal to the average of the normal stresses on the x and y planes:

$$\sigma_{aver} = \frac{\sigma_x + \sigma_y}{2} \qquad \text{(7-27)}$$

This same normal stress acts on the planes of maximum negative shear stress.

In the particular cases of **uniaxial stress** and **biaxial stress** (Fig. 7-10), the planes of maximum shear stress occur at 45° to the x and y axes. In the case of **pure shear** (Fig. 7-11), the maximum shear stresses occur on the x and y planes.

The preceding analysis of shear stresses has dealt only with **in-**

plane shear stresses, that is, stresses acting in the xy plane. To obtain the maximum in-plane shear stresses (Eqs. 7-25 and 7-26), we considered elements that were obtained by rotating the xyz axes about the z axis, which is a principal axis (Fig. 7-12a). We found that the maximum shear stresses occur on planes at 45° to the principal planes. The principal planes for the element of Fig. 7-12a are shown in Fig. 7-12b, where σ_1 and σ_2 are the principal stresses. Therefore, the maximum in-plane shear stresses are found on an element obtained by rotating the $x_1 y_1 z_1$ axes (Fig. 7-12b) about the z_1 axis through an angle of 45°. These stresses are given by Eq. (7-25) or Eq. (7-26).

We can also obtain maximum shear stresses by 45° rotations about the other two principal axes (the x_1 and y_1 axes in Fig. 7-12b). As a result, we obtain three sets of **maximum positive and maximum negative shear stresses** (compare with Eq. 7-26):

$$(\tau_{\text{max}})_{x_1} = \pm \frac{\sigma_2}{2} \quad (\tau_{\text{max}})_{y_1} = \pm \frac{\sigma_1}{2} \quad (\tau_{\text{max}})_{z_1} = \pm \frac{\sigma_1 - \sigma_2}{2} \quad (7\text{-}28a,b,c)$$

in which the subscripts indicate the principal axes about which the 45° rotations take place.

The algebraic values of σ_1 and σ_2 determine which of the preceding expressions gives the numerically largest shear stress. If σ_1 and σ_2 have the same sign, then one of the first two expressions is numerically largest; if they have opposite signs, the last expression is largest.

WORK SHEETS 5 & 6   Mathcad   **Example 7-3**

Fig. 7-13 Example 7-3. (a) Element in plane stress, (b) principal stresses, and (c) maximum shear stresses

An element in plane stress is subjected to stresses $\sigma_x = 12{,}300$ psi, $\sigma_y = -4{,}200$ psi, and $\tau_{xy} = -4{,}700$ psi, as shown in Fig. 7-13a. (a) Determine the principal stresses and show them on a sketch of a properly oriented element.

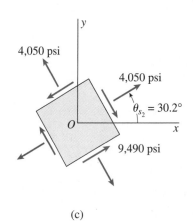

(a)   (b)   (c)

(b) Determine the maximum shear stresses and show them on a sketch of a properly oriented element. (Consider only the in-plane stresses.)

Solution   (a) *Principal stresses.* The principal angles θ_p that locate the principal planes can be obtained from Eq. (7-11):

$$\tan 2\theta_p = \frac{2\tau_{xy}}{\sigma_x - \sigma_y} = \frac{2(-4{,}700 \text{ psi})}{12{,}300 \text{ psi} - (-4{,}200 \text{ psi})} = -0.5697$$

Solving for the angles, we get the following two sets of values:

$$2\theta_p = 150.3° \quad \text{and} \quad \theta_p = 75.2°$$

$$2\theta_p = 330.3° \quad \text{and} \quad \theta_p = 165.2°$$

The principal stresses may be obtained by substituting the two values of $2\theta_p$ into the transformation equation for σ_{x_1} (Eq. 7-4a). As a preliminary calculation, we determine the following quantities:

$$\frac{\sigma_x + \sigma_y}{2} = \frac{12{,}300 \text{ psi} - 4{,}200 \text{ psi}}{2} = 4{,}050 \text{ psi}$$

$$\frac{\sigma_x - \sigma_y}{2} = \frac{12{,}300 \text{ psi} + 4{,}200 \text{ psi}}{2} = 8{,}250 \text{ psi}$$

Now we substitute the first value of $2\theta_p$ into Eq. (7-4a) and obtain

$$\sigma_{x_1} = \frac{\sigma_x + \sigma_y}{2} + \frac{\sigma_x - \sigma_y}{2} \cos 2\theta + \tau_{xy} \sin 2\theta$$

$$= 4{,}050 \text{ psi} + (8{,}250 \text{ psi})(\cos 150.3°) - (4{,}700 \text{ psi})(\sin 150.3°)$$

$$= -5{,}440 \text{ psi}$$

In a similar manner, we substitute the second value of $2\theta_p$ and obtain $\sigma_{x_1} = 13{,}540$ psi. Thus, the principal stresses and their corresponding principal angles are

$$\sigma_1 = 13{,}540 \text{ psi} \quad \text{and} \quad \theta_{p_1} = 165.2° \qquad \longleftarrow$$

$$\sigma_2 = -5{,}440 \text{ psi} \quad \text{and} \quad \theta_{p_2} = 75.2° \qquad \longleftarrow$$

Note that θ_{p_1} and θ_{p_2} differ by 90° and that $\sigma_1 + \sigma_2 = \sigma_x + \sigma_y$. The principal stresses are shown on a properly oriented element in Fig. 7-13b. Of course, no shear stresses act on the principal planes.

   Alternative solution for the principal stresses. The principal stresses may also be calculated directly from Eq. (7-17):

$$\sigma_{1,2} = \frac{\sigma_x + \sigma_y}{2} \pm \sqrt{\left(\frac{\sigma_x - \sigma_y}{2}\right)^2 + \tau_{xy}^2}$$

$$= 4{,}050 \text{ psi} \pm \sqrt{(8{,}250 \text{ psi})^2 + (-4{,}700 \text{ psi})^2}$$

$$\sigma_{1,2} = 4{,}050 \text{ psi} \pm 9{,}490 \text{ psi}$$

Therefore,

$$\sigma_1 = 13{,}540 \text{ psi} \qquad \sigma_2 = -5{,}440 \text{ psi}$$

The angle θ_{p_1} to the plane on which σ_1 acts is obtained from Eqs. (7-18a) and (7-18b):

$$\cos 2\theta_{p_1} = \frac{\sigma_x - \sigma_y}{2R} = \frac{8{,}250 \text{ psi}}{9{,}490 \text{ psi}} = 0.869$$

$$\sin 2\theta_{p_1} = \frac{\tau_{xy}}{R} = \frac{-4{,}700 \text{ psi}}{9{,}490 \text{ psi}} = -0.495$$

in which R is given by Eq. (7-12) and is equal to the square-root term in the preceding calculation for the principal stresses σ_1 and σ_2. The only angle between 0 and 360° having the specified sine and cosine is $2\theta_{p_1} = 330.3°$; hence, $\theta_{p_1} = 165.2°$. This angle is associated with the algebraically larger principal stress $\sigma_1 = 13{,}540$ psi. The other angle is 90° larger or smaller than θ_{p_1}; hence, $\theta_{p_2} = 75.2°$. This angle corresponds to the smaller principal stress $\sigma_2 = -5{,}440$ psi. Note that these results for the principal stresses and principal angles agree with those found previously.

(b) *Maximum shear stresses.* The maximum in-plane shear stresses are given by Eq. (7-25):

$$\tau_{\max} = \sqrt{\left(\frac{\sigma_x - \sigma_y}{2}\right)^2 + \tau_{xy}^2}$$

$$= \sqrt{(8{,}250 \text{ psi})^2 + (-4{,}700 \text{ psi})^2} = 9{,}490 \text{ psi} \qquad \longleftarrow$$

The angle θ_{s_1} to the plane having the maximum positive shear stress is calculated from Eq. (7-24):

$$\theta_{s_1} = \theta_{p_1} - 45° = 165.2° - 45° = 120.2° \qquad \longleftarrow$$

It follows that the maximum negative shear stress acts on the plane for which $\theta_{s_2} = 120.2° - 90° = 30.2°$.

The normal stresses acting on the planes of maximum shear stresses are calculated from Eq. (7-27):

$$\sigma_{\text{aver}} = \frac{\sigma_x + \sigma_y}{2} = 4{,}050 \text{ psi} \qquad \longleftarrow$$

Finally, the maximum shear stresses and associated normal stresses are shown on the stress element of Fig. 7-13c.

As an alternative approach to finding the maximum shear stresses, we can use Eq. (7-20) to determine the two values of the angles θ_s, and then we can use the second transformation equation (Eq. 7-4b) to obtain the corresponding shear stresses.

7.4 MOHR'S CIRCLE FOR PLANE STRESS

The transformation equations for plane stress can be represented in graphical form by a plot known as **Mohr's circle**.[*] This graphical representation is extremely useful because it enables you to visualize the relationships between the normal and shear stresses acting on various inclined planes at a point in a stressed body. It also provides a means for calculating principal stresses, maximum shear stresses, and stresses on inclined planes. Furthermore, Mohr's circle is valid not only for stresses but also for other quantities of a similar mathematical nature, including strains and moments of inertia.

■ Equations of Mohr's Circle

The equations of Mohr's circle can be derived from the transformation equations for plane stress (Eqs. 7-4a and 7-4b). The two equations are repeated here, but with a slight rearrangement of the first equation:

$$\sigma_{x_1} - \frac{\sigma_x + \sigma_y}{2} = \frac{\sigma_x - \sigma_y}{2}\cos 2\theta + \tau_{xy}\sin 2\theta \qquad (7\text{-}29\text{a})$$

$$\tau_{x_1 y_1} = -\frac{\sigma_x - \sigma_y}{2}\sin 2\theta + \tau_{xy}\cos 2\theta \qquad (7\text{-}29\text{b})$$

From analytic geometry, we might recognize that these two equations are the equations of a circle in parametric form. The angle 2θ is the parameter and the stresses σ_{x_1} and τ_{x1y1} are the coordinates. However, it is not necessary to recognize the nature of the equations at this stage—if we eliminate the parameter, the significance of the equations will become apparent.

To eliminate the parameter 2θ, we square both sides of each equation and then add the two equations. The equation that results is

$$\left(\sigma_{x_1} - \frac{\sigma_x + \sigma_y}{2}\right)^2 + \tau_{x_1 y_1}^2 = \left(\frac{\sigma_x - \sigma_y}{2}\right)^2 + \tau_{xy}^2 \qquad (7\text{-}30)$$

This equation can be written in simpler form by using the following notation from Section 7.3 (see Eqs. 7-27 and 7-12, respectively):

$$\sigma_{\text{aver}} = \frac{\sigma_x + \sigma_y}{2} \qquad R = \sqrt{\left(\frac{\sigma_x - \sigma_y}{2}\right)^2 + \tau_{xy}^2} \qquad (7\text{-}31\text{a,b})$$

Equation (7-30) now becomes

$$(\sigma_{x_1} - \sigma_{\text{aver}})^2 + \tau_{x_1 y_1}^2 = R^2 \qquad (7\text{-}32)$$

---

[*]Mohr's circle is named after the famous German civil engineer Otto Christian Mohr (1835–1918), who developed the circle in 1882 (Ref. 7-4).

which is the equation of a circle in standard algebraic form. The coordinates are σ_{x_1} and $\tau_{x_1y_1}$, the radius is R, and the center of the circle has coordinates $\sigma_{x1} = \sigma_{\text{aver}}$ and $\tau_{x1y1} = 0$.

Mohr's circle can be plotted from Eqs. (7-29) and (7-32) in two different ways. In the first form of Mohr's circle, we plot the normal stress σ_{x_1} positive to the right and the shear stress $\tau_{x_1y_1}$ positive downward, as shown in Fig. 7-14a. The advantage of plotting shear stresses positive downward is that the angle 2θ on Mohr's circle is positive when counterclockwise, which agrees with the positive direction of 2θ in the derivation of the transformation equations (see Figs. 7-1 and 7-2). In the second form of Mohr's circle, $\tau_{x_1y_1}$ is plotted positive upward but the angle 2θ is now positive clockwise (Fig. 7-14b), which is opposite to its usual positive direction.

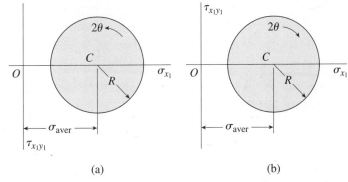

Fig. 7-14 Two forms of Mohr's circle: (a) $\tau_{x_1y_1}$ is positive downward and the angle 2θ is positive counterclockwise, and (b) $\tau_{x_1y_1}$ is positive upward and the angle 2θ is positive clockwise. (*Note:* This book uses the first form.)

Both forms of Mohr's circle are mathematically correct, and either one can be used. However, it is easier to visualize the orientation of the stress element if the positive direction of the angle 2θ is the same in Mohr's circle as it is for the element itself. Furthermore, a counterclockwise rotation agrees with the customary right-hand rule for rotation. Therefore, we will opt for the first form of Mohr's circle (Fig. 7-14a) in which *positive shear stress is plotted downward and a positive angle 2θ is plotted counterclockwise.*

■ Construction of Mohr's Circle

Mohr's circle can be constructed in a variety of ways, depending upon which stresses are known and which are unknown. Also, personal preferences often determine how one proceeds. For our immediate purpose, which is to show the basic properties of the circle, let us assume that we know the stresses σ_x, σ_y, and τ_{xy} acting on the x and y planes of an element in plane stress (Fig. 7-15a). As we will see, this information is sufficient to construct the circle. Then, with the circle drawn, we can determine the stresses σ_{x_1}, σ_{y_1}, and $\tau_{x_1y_1}$ acting on an inclined element (Fig. 7-15b). We can also obtain the principal stresses and maximum shear stresses from the circle.

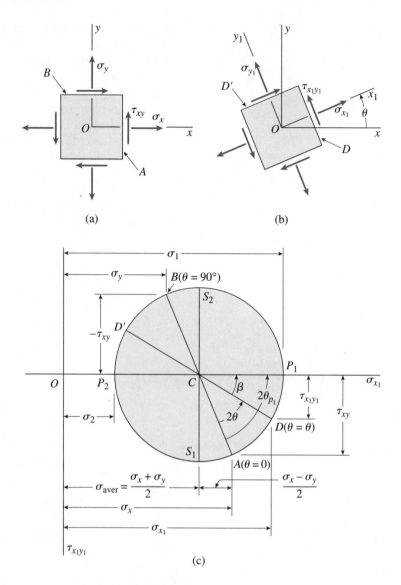

Fig. 7-15 Construction of Mohr's circle for plane stress

With σ_x, σ_y, and τ_{xy} known, the procedure for constructing Mohr's circle is as follows (see Fig. 7-15c):

(1) Draw a set of coordinate axes with σ_{x_1} as abscissa (positive to the right) and $\tau_{x_1y_1}$ as ordinate (positive downward).

(2) Locate the center C of the circle at the point having coordinates $\sigma_{x_1} = \sigma_{\text{aver}}$ and $\tau_{x_1y_1} = 0$ (see Eqs. 7-31a and 7-32).

(3) Locate point A, representing the stress conditions on the x face of the element shown in Fig. 7-15a, by plotting its coordinates $\sigma_{x_1} = \sigma_x$ and $\tau_{x_1y_1} = \tau_{xy}$. Note that point A corresponds to $\theta = 0$. Also, note

that the x face of the element (Fig. 7-15a) is labeled "A" to show its correspondence with point A in the diagram.

(4) Locate point B, representing the stress conditions on the y face of the element shown in Fig. 7-15a, by plotting its coordinates $\sigma_{x_1} = \sigma_y$ and $\tau_{x_1 y_1} = -\tau_{xy}$. Note that point B corresponds to $\theta = 90°$. In addition, the y face of the element (Fig. 7-15a) is labeled "B" to show its correspondence with point B in the diagram.

(5) Draw a line from point A to point B. This line is a diameter of the circle and passes through the center C. Points A and B, representing the stresses on planes at $90°$ to each other (Fig. 7-15a), are at opposite ends of the diameter (and therefore are $180°$ apart on the circle).

(6) Using point C as the center, draw Mohr's circle through points A and B. The circle drawn in this manner has radius R (Eq. 7-31b), as shown in the next paragraph.

Now that we have drawn the circle, we can verify by geometry that lines CA and CB are radii and have lengths equal to R. We note that the abscissas of points C and A are $(\sigma_x + \sigma_y)/2$ and σ_x, respectively. The difference in these abscissas is $(\sigma_x - \sigma_y)/2$, as dimensioned in the figure. Also, the ordinate to point A is τ_{xy}. Therefore, line CA is the hypotenuse of a right triangle having one side of length $(\sigma_x - \sigma_y)/2$ and the other side of length τ_{xy}. Taking the square root of the sum of the squares of these two sides gives the radius R:

$$R = \sqrt{\left(\frac{\sigma_x - \sigma_y}{2}\right)^2 + \tau_{xy}^2}$$

which is the same as Eq. (7-31b). By a similar procedure, we can show that the length of line CB is also equal to the radius R of the circle.

■ Stresses on an Inclined Element

Now we will consider the stresses σ_{x_1}, σ_{y_1}, and $\tau_{x_1 y_1}$ acting on the faces of a plane-stress element oriented at an angle θ from the x axis (Fig. 7-15b). If the angle θ is known, these stresses can be determined from Mohr's circle. The procedure is as follows.

On the circle (Fig. 7-15c), we measure an angle 2θ counterclockwise from radius CA, because point A corresponds to $\theta = 0$ and is the reference point from which we measure angles. The angle 2θ locates point D, which (as shown in the next paragraph) has coordinates σ_{x_1} and $\tau_{x_1 y_1}$. Therefore, point D on the circle represents the stresses on the x_1 face of the element of Fig. 7-15b. Consequently, this face of the element is labeled "D" in the figure.

Note that an angle 2θ on Mohr's circle corresponds to an angle θ on a stress element. For instance, point D on the circle is at an angle 2θ from point A, but the x_1 face of the element shown in Fig. 7-15b (the face labeled "D") is at an angle θ from the x face of the element shown in Fig.

7-15a (the face labeled "A"). Similarly, points A and B are $180°$ apart on the circle, but the corresponding faces of the element (Fig. 7-15a) are $90°$ apart.

To show that the coordinates σ_{x_1} and $\tau_{x_1 y_1}$ of point D on the circle are indeed given by the stress-transformation equations (Eqs. 7-4a and 7-4b), we again use the geometry of the circle. Let β be the angle between the radial line CD and the σ_{x_1} axis. Then, from the geometry of the figure, we obtain the following expressions for the coordinates of point D:

$$\sigma_{x_1} = \frac{\sigma_x + \sigma_y}{2} + R \cos \beta \qquad \tau_{x_1 y_1} = R \sin \beta \qquad \text{(7-33a,b)}$$

Noting that the angle between the radius CA and the horizontal axis is $2\theta + \beta$, we get

$$\cos (2\theta + \beta) = \frac{\sigma_x - \sigma_y}{2R} \qquad \sin (2\theta + \beta) = \frac{\tau_{xy}}{R}$$

Expanding the cosine and sine expressions (see Appendix C) gives

$$\cos 2\theta \cos \beta - \sin 2\theta \sin \beta = \frac{\sigma_x - \sigma_y}{2R} \qquad \text{(a)}$$

$$\sin 2\theta \cos \beta + \cos 2\theta \sin \beta = \frac{\tau_{xy}}{R} \qquad \text{(b)}$$

Multiplying the first of these equations by $\cos 2\theta$ and the second by $\sin 2\theta$ and then adding, we obtain

$$\cos \beta = \frac{1}{R} \left(\frac{\sigma_x - \sigma_y}{2} \cos 2\theta + \tau_{xy} \sin 2\theta \right) \qquad \text{(c)}$$

Also, multiplying Eq. (a) by $\sin 2\theta$ and Eq. (b) by $\cos 2\theta$ and then subtracting, we get

$$\sin \beta = \frac{1}{R} \left(-\frac{\sigma_x - \sigma_y}{2} \sin 2\theta + \tau_{xy} \cos 2\theta \right) \qquad \text{(d)}$$

When these expressions for $\cos \beta$ and $\sin \beta$ are substituted into Eqs. (7-33a) and (7-33b), we obtain the stress-transformation equations for σ_{x_1} and $\tau_{x_1 y_1}$ (Eqs. 7-4a and 7-4b). Thus, we have shown that point D on Mohr's circle, defined by the angle 2θ, represents the stress conditions on the x_1 face of the stress element defined by the angle θ (Fig. 7-15b).

Point D', which is diametrically opposite point D on the circle, is located by an angle 2θ (measured from line CA) that is $180°$ greater than the angle 2θ to point D. Therefore, point D' on the circle represents the stresses on a face of the stress element (Fig. 7-15b) at $90°$ from the face represented by point D. Thus, point D' on the circle gives the stresses σ_{y_1} and $-\tau_{x_1 y_1}$ on the y_1 face of the stress element (the face labeled "D'").

From this discussion we see how the stresses represented by points on Mohr's circle are related to the stresses acting on an element. The stresses on an inclined plane defined by the angle θ (Fig. 7-15b) are found on the circle at the point where the angle from the reference point (point A) is 2θ. Thus, as we rotate the x_1y_1 axes counterclockwise through an angle θ (Fig. 7-15b), the point on Mohr's circle corresponding to the x_1 face moves counterclockwise through an angle 2θ. Similarly, if we rotate the axes clockwise, the point on the circle moves clockwise through an angle twice as large.

■ Principal Stresses

The determination of principal stresses is probably the most important application of Mohr's circle. Note that as we move around Mohr's circle (Fig. 7-15c), we encounter point P_1 where the normal stress reaches its algebraically largest value and the shear stress is zero. Hence, point P_1 represents a **principal stress** and a **principal plane**. The abscissa σ_1 of point P_1 gives the algebraically larger principal stress and its angle $2\theta_{p_1}$ from the reference point A (where $\theta = 0$) gives the orientation of the principal plane. The other principal plane, associated with the algebraically smallest normal stress, is represented by point P_2, diametrically opposite point P_1.

From the geometry of the circle, we see that the algebraically larger principal stress is

$$\sigma_1 = \overline{OC} + \overline{CP_1} = \frac{\sigma_x + \sigma_y}{2} + R$$

which, upon substitution of the expression for R (Eq. 7-31b), agrees with the earlier equation for this stress (Eq. 7-14). In a similar manner, we can verify the expression for the algebraically smaller principal stress σ_2.

The principal angle θ_{p_1} between the x axis (Fig. 7-15a) and the plane of the algebraically larger principal stress is one-half the angle $2\theta_{p_1}$, which is the angle on Mohr's circle between radii CA and CP_1. The cosine and sine of the angle $2\theta_{p_1}$ can be obtained by inspection from the circle:

$$\cos 2\theta_{p_1} = \frac{\sigma_x - \sigma_y}{2R} \qquad \sin 2\theta_{p_1} = \frac{\tau_{xy}}{R}$$

These equations agree with Eqs. (7-18a) and (7-18b), and so once again we see that the geometry of the circle matches the equations derived earlier. On the circle, the angle $2\theta_{p_2}$ to the other principal point (point P_2) is 180° larger than $2\theta_{p_1}$; hence, $\theta_{p_2} = \theta_{p_1} + 90°$, as expected.

■ Maximum Shear Stresses

Points S_1 and S_2, representing the planes of maximum positive and maximum negative shear stresses, respectively, are located at the bottom and top of Mohr's circle (Fig. 7-15c). These points are at angles $2\theta = 90°$ from points P_1 and P_2, which agrees with the fact that the planes of maximum shear stress are oriented at 45° to the principal planes.

The maximum shear stresses are numerically equal to the radius R of the circle (compare Eq. 7-31b for R with Eq. 7-25 for τ_{max}). Also, the normal stresses on the planes of maximum shear stress are equal to the abscissa of point C, which is the average normal stress σ_{aver} (see Eq. 7-31a).

■ Alternative Sign Convention for Shear Stresses

An alternative sign convention for shear stresses is sometimes used when constructing Mohr's circle. In this convention, the direction of a shear stress acting on an element of the material is indicated by the sense of the rotation that it tends to produce (Figs. 7-16a and b). If the shear stress τ tends to rotate the stress element clockwise, it is called a clockwise shear stress, and if it tends to rotate it counterclockwise, it is called a counterclockwise stress. Then, when constructing Mohr's circle, clockwise shear stresses are plotted upward and counterclockwise shear stresses are plotted downward (Fig. 7-16c).

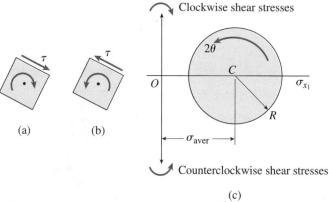

Fig. 7-16 Alternative sign convention for shear stresses: (a) shear stress τ tending to rotate the element clockwise, (b) shear stress τ tending to rotate the element counterclockwise, and (c) axes for Mohr's circle (clockwise shear stresses are plotted upward and counterclockwise shear stresses are plotted downward)

It is important to realize that the alternative sign convention produces a circle that is identical to the circle already described (Fig. 7-15c). The reason is that a positive shear stress $\tau_{x_1y_1}$ is also a counterclockwise shear stress, and both are plotted downward. Also, a negative shear stress $\tau_{x_1y_1}$ is a clockwise shear stress, and both are plotted upward.

Thus, the alternative sign convention merely provides a different (and possibly preferable) point of view. Instead of thinking of the vertical axis as having negative shear stresses plotted upward and positive shear stresses plotted downward (which is a bit awkward), we can think of the vertical axis as having clockwise shear stresses plotted upward and counterclockwise shear stresses plotted downward.

■ General Comments about the Circle

From the preceding discussions in this section, it is apparent that we can find the stresses acting on any inclined plane, as well as the principal stresses and maximum shear stresses, from Mohr's circle. However, only rotations of axes in the xy plane (that is, rotations about the z axis) are considered, and therefore *all stresses on Mohr's circle are in-plane stresses*.

For convenience, the circle of Fig. 7-15 was drawn with σ_x, σ_y, and τ_{xy} as positive stresses, but the same procedures may be followed if one or more of the stresses is negative. If one of the normal stresses is negative, part or all of the circle will be located to the left of the origin, as illustrated in Example 7-6 that follows.

Point A, representing the stresses on the plane $\theta = 0$, may be situated anywhere around the circle. However, the angle 2θ is always measured counterclockwise from the radius CA, regardless of where point A is located.

In the special cases of *uniaxial stress*, *biaxial stress*, and *pure shear*, the construction of Mohr's circle is simpler than in the general case of plane stress. These special cases are illustrated in Example 7-4 and in Problems 7.4-1 through 7.4-8.

Besides using Mohr's circle to obtain the stresses on inclined planes when the stresses on the x and y planes are known, we can also use the circle in the opposite manner. If we know the stresses σ_{x_1}, σ_{y_1}, and $\tau_{x_1 y_1}$ acting on an inclined element oriented at a known angle θ, we can easily construct the circle and determine the stresses σ_x, σ_y, and τ_{xy} for the angle $\theta = 0$. The procedure is to locate points D and D' from the known stresses and then draw the circle using line DD' as a diameter. By measuring the angle 2θ in a negative sense from radius CD, we can locate point A, corresponding to the x face of the element. Then we can locate point B by constructing a diameter from A. Finally, we can determine the coordinates of points A and B and thereby obtain the stresses acting on the element for which $\theta = 0$.

If desired, we can construct Mohr's circle to scale and measure values of stress from the drawing. However, it is usually preferable to obtain the stresses by numerical calculations, either directly from the various equations or by using trigonometry and the geometry of the circle.

Mohr's circle makes it possible to visualize the relationships between stresses acting on planes at various angles, and it also serves as a simple memory device for calculating stresses. Although many graphical techniques are no longer used in engineering work, Mohr's circle remains valuable because it provides a simple and clear picture of an otherwise complicated analysis.

Mohr's circle is also applicable to the transformations for plain strain and moments of inertia of plane areas, because these quantities follow the same transformation laws as do stresses (see Sections 7.7, 12.8, and 12.9).

Example 7-4

At a point on the surface of a pressurized cylinder, the material is subjected to biaxial stresses $\sigma_x = 90$ MPa and $\sigma_y = 20$ MPa, as shown on the stress element of Fig. 7-17a. Using Mohr's circle, determine the stresses acting on an element inclined at an angle $\theta = 30°$. (Consider only the in-plane stresses, and show the results on a sketch of a properly oriented element.)

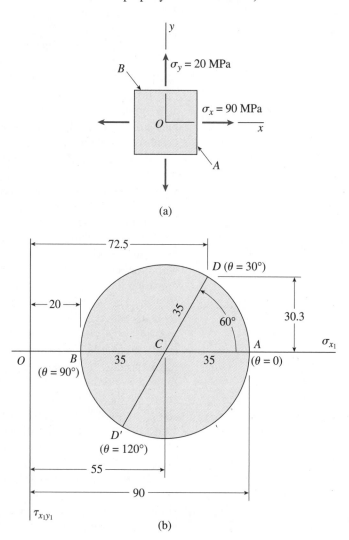

Fig. 7-17 Example 7-4. (a) Element in plane stress, and (b) the corresponding Mohr's circle. (*Note:* All stresses on the circle have units of MPa.)

Solution   *Construction of Mohr's circle.* We begin by setting up the axes for the normal and shear stresses, with σ_{x_1} positive to the right and $\tau_{x_1 y_1}$ positive downward, as shown in Fig. 7-17b. Then we place the center C of the circle on the σ_{x_1} axis at the point where the stress equals the average normal stress (Eq. 7-31a):

$$\sigma_{aver} = \frac{\sigma_x + \sigma_y}{2} = \frac{90\ \text{MPa} + 20\ \text{MPa}}{2} = 55\ \text{MPa}$$

Point A, representing the stresses on the x face of the element ($\theta = 0$), has coordinates

$$\sigma_{x_1} = 90\ \text{MPa} \qquad \tau_{x_1 y_1} = 0$$

Similarly, the coordinates of point B, representing the stresses on the y face ($\theta = 90°$), are

$$\sigma_{x_1} = 20\ \text{MPa} \qquad \tau_{x_1 y_1} = 0$$

Now we draw the circle through points A and B with center at C and radius R (see Eq. 7-31b) equal to

$$R = \sqrt{\left(\frac{\sigma_x - \sigma_y}{2}\right)^2 + \tau_{xy}^2} = \sqrt{\left(\frac{90\ \text{MPa} - 20\ \text{MPa}}{2}\right)^2 + 0} = 35\ \text{MPa}$$

Stresses on an element inclined at $\theta = 30°$. The stresses acting on a plane oriented at an angle $\theta = 30°$ are given by the coordinates of point D, which is at an angle $2\theta = 60°$ from point A (Fig. 7-17b). By inspection of the circle, we see that the coordinates of point D are

(Point D) $\sigma_{x_1} = \sigma_{aver} + R\cos 60°$

$$= 55\ \text{MPa} + (35\ \text{MPa})(\cos 60°) = 72.5\ \text{MPa} \qquad \longleftarrow$$

$$\tau_{x_1 y_1} = -R\sin 60° = -(35\ \text{MPa})(\sin 60°) = -30.3\ \text{MPa} \qquad \longleftarrow$$

In a similar manner, we can find the stresses represented by point D', which corresponds to an angle $\theta = 120°$ (or $2\theta = 240°$):

(Point D') $\sigma_{x_1} = \sigma_{aver} - R\cos 60°$

$$= 55\ \text{MPa} - (35\ \text{MPa})(\cos 60°) = 37.5\ \text{MPa} \qquad \longleftarrow$$

$$\tau_{x_1 y_1} = R\sin 60° = (35\ \text{MPa})(\sin 60°) = 30.3\ \text{MPa} \qquad \longleftarrow$$

These results are shown in Fig. 7-18 on a sketch of an element oriented at an angle $\theta = 30°$, with all stresses shown in their true directions. Note that the sum of the normal stresses on the inclined element is equal to $\sigma_x + \sigma_y$, or 110 MPa.

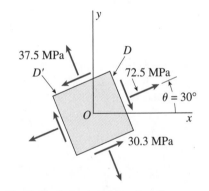

Fig. 7-18 Example 7-4 (continued). Stresses acting on an element oriented at an angle $\theta = 30°$.

WORK SHEETS 5 & 6   Mathcad

Example 7-5

An element in plane stress at the surface of a large machine is subjected to stresses $\sigma_x = 15{,}000$ psi, $\sigma_y = 5{,}000$ psi, and $\tau_{xy} = 4{,}000$ psi, as shown in Fig. 7-19a. Using Mohr's circle, determine the following quantities: (a) the stresses acting on an element inclined at an angle $\theta = 40°$, (b) the principal stresses, and (c) the maximum shear stresses. (Consider only the in-plane stresses, and show all results on sketches of properly oriented elements.)

Solution *Construction of Mohr's circle.* The first step in the solution is to set up the axes for Mohr's circle, with σ_{x_1} positive to the right and $\tau_{x_1y_1}$ positive downward (Fig. 7-19b). The center C of the circle is located on the σ_{x_1} axis at the point where σ_{x_1} equals the average normal stress (Eq. 7-31a):

$$\sigma_{aver} = \frac{\sigma_x + \sigma_y}{2} = \frac{15{,}000 \text{ psi} + 5{,}000 \text{ psi}}{2} = 10{,}000 \text{ psi}$$

Point A, representing the stresses on the x face of the element ($\theta = 0$), has coordinates

$$\sigma_{x_1} = 15{,}000 \text{ psi} \qquad \tau_{x_1y_1} = 4{,}000 \text{ psi}$$

Similarly, the coordinates of point B, representing the stresses on the y face ($\theta = 90°$), are

$$\sigma_{x_1} = 5{,}000 \text{ psi} \qquad \tau_{x_1y_1} = -4{,}000 \text{ psi}$$

The circle is now drawn through points A and B with center at C. The radius of the circle, from Eq. (7-31b), is

$$R = \sqrt{\left(\frac{\sigma_x - \sigma_y}{2}\right)^2 + \tau_{xy}^2}$$

$$= \sqrt{\left(\frac{15{,}000 \text{ psi} - 5{,}000 \text{ psi}}{2}\right)^2 + (4{,}000 \text{ psi})^2} = 6{,}403 \text{ psi}$$

Fig. 7-19 Example 7-5. (a) Element in plane stress, and (b) the corresponding Mohr's circle. (*Note:* All stresses on the circle have units of psi.)

(a) *Stresses on an element inclined at $\theta = 40°$.* The stresses acting on a plane oriented at an angle $\theta = 40°$ are given by the coordinates of point D, which is at an angle $2\theta = 80°$ from point A (Fig. 7-19b). To evaluate these coordinates, we need to know the angle between line CD and the σ_{x_1} axis (that is, angle

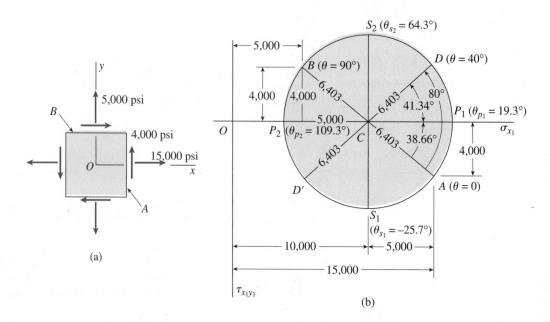

(a)

(b)

DCP_1), which in turn requires that we know the angle between line CA and the σ_{x_1} axis (angle ACP_1). These angles are found from the geometry of the circle, as follows:

$$\tan \overline{ACP_1} = \frac{4{,}000 \text{ psi}}{5{,}000 \text{ psi}} = 0.8 \qquad \overline{ACP_1} = 38.66°$$

$$\overline{DCP_1} = 80° - \overline{ACP_1} = 80° - 38.66° = 41.34°$$

Knowing these angles, we can determine the coordinates of point D directly from the figure:

(Point D) $\sigma_{x_1} = 10{,}000 \text{ psi} + (6{,}403 \text{ psi})(\cos 41.34°) = 14{,}810 \text{ psi}$ ←

$\tau_{x_1 y_1} = -(6{,}403 \text{ psi})(\sin 41.34°) = -4{,}230 \text{ psi}$ ←

In an analogous manner, we can find the stresses represented by point D', which corresponds to a plane inclined at an angle $\theta = 130°$ (or $2\theta = 260°$):

(Point D') $\sigma_{x_1} = 10{,}000 \text{ psi} - (6{,}403 \text{ psi})(\cos 41.34°) = 5{,}190 \text{ psi}$ ←

$\tau_{x_1 y_1} = (6{,}403 \text{ psi})(\sin 41.34°) = 4{,}230 \text{ psi}$ ←

These stresses are shown in Fig. 7-20a on a sketch of an element oriented at an angle $\theta = 40°$ (all stresses are shown in their true directions). Also, note that the sum of the normal stresses is equal to $\sigma_x + \sigma_y$, or 20,000 psi.

(b) *Principal stresses.* The principal stresses are represented by points P_1 and P_2 on Mohr's circle (Fig. 7-19b). The algebraically larger principal stress (point P_1) is

$$\sigma_1 = 10{,}000 \text{ psi} + 6{,}400 \text{ psi} = 16{,}400 \text{ psi}$$ ←

as seen by inspection of the circle. The angle $2\theta_{p_1}$ to point P_1 from point A is the angle ACP_1 on the circle, that is,

$$\overline{ACP_1} = 2\theta_{p_1} = 38.66° \qquad \theta_{p_1} = 19.3°$$ ←

Thus, the plane of the algebraically larger principal stress is oriented at an angle $\theta_{p_1} = 19.3°$, as shown in Fig. 7-20b.

Fig. 7-20 Example 7-5 (continued). (a) Stresses acting on an element oriented at $\theta = 40°$, (b) principal stresses, and (c) maximum shear stresses

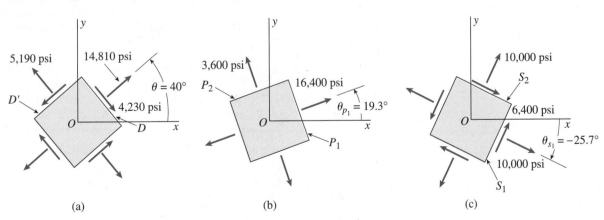

(a)   (b)   (c)

The algebraically smaller principal stress (represented by point P_2) is obtained from the circle in a similar manner:

$$\sigma_2 = 10{,}000 \text{ psi} - 6{,}400 \text{ psi} = 3{,}600 \text{ psi} \qquad \longleftarrow$$

The angle $2\theta_{p_2}$ to point P_2 on the circle is $38.66° + 180° = 218.66°$; thus, the second principal plane is defined by the angle $\theta_{p_2} = 109.3°$. The principal stresses and principal planes are shown in Fig. 7-20b, and again we note that the sum of the normal stresses is equal to 20,000 psi.

(c) *Maximum shear stresses.* The maximum shear stresses are represented by points S_1 and S_2 on Mohr's circle; therefore, the maximum in-plane shear stress (equal to the radius of the circle) is

$$\tau_{\max} = 6{,}400 \text{ psi} \qquad \longleftarrow$$

The angle ACS_1 from point A to point S_1 is $90° - 38.66° = 51.34°$, and therefore the angle $2\theta_{s_1}$ for point S_1 is

$$2\theta_{s_1} = -51.34°$$

This angle is negative because it is measured clockwise on the circle. The corresponding angle θ_{s_1} to the plane of the maximum positive shear stress is one-half that value, or $\theta_{s_1} = -25.7°$, as shown in Figs. 7-19b and 7-20c. The maximum negative shear stress (point S_2 on the circle) has the same numerical value as the maximum positive stress (6,400 psi).

The normal stresses acting on the planes of maximum shear stress are equal to σ_{aver}, which is the abscissa of the center C of the circle (10,000 psi). These stresses are also shown in Fig. 7-20c. Note that the planes of maximum shear stress are oriented at $45°$ to the principal planes.

WORK
SHEETS
5 & 6   **Example 7-6**

At a point on the surface of a generator shaft the stresses are $\sigma_x = -50$ MPa, $\sigma_y = 10$ MPa, and $\tau_{xy} = -40$ MPa, as shown in Fig. 7-21a. Using Mohr's circle, determine the following quantities: (a) the stresses acting on an element inclined at an angle $\theta = 45°$, (b) the principal stresses, and (c) the maximum shear stresses. (Consider only the in-plane stresses, and show all results on sketches of properly oriented elements.)

Solution *Construction of Mohr's circle.* The axes for the normal and shear stresses are shown in Fig. 7-21b, with σ_{x_1} positive to the right and $\tau_{x_1y_1}$ positive downward. The center C of the circle is located on the σ_{x_1} axis at the point where the stress equals the average normal stress (Eq. 7-31a):

$$\sigma_{\text{aver}} = \frac{\sigma_x + \sigma_y}{2} = \frac{-50 \text{ MPa} + 10 \text{ MPa}}{2} = -20 \text{ MPa}$$

Point A, representing the stresses on the x face of the element ($\theta = 0$), has coordinates

$$\sigma_{x_1} = -50 \text{ MPa} \qquad \tau_{x_1y_1} = -40 \text{ MPa}$$

Similarly, the coordinates of point B, representing the stresses on the y face ($\theta = 90°$), are

$$\sigma_{x_1} = 10 \text{ MPa} \qquad \tau_{x_1 y_1} = 40 \text{ MPa}$$

The circle is now drawn through points A and B with center at C and radius R (from Eq. 7-31b) equal to:

$$R = \sqrt{\left(\frac{\sigma_x - \sigma_y}{2}\right)^2 + \tau_{xy}^2}$$

$$= \sqrt{\left(\frac{-50 \text{ MPa} - 10 \text{ MPa}}{2}\right)^2 + (-40 \text{ MPa})^2} = 50 \text{ MPa}$$

(a) *Stresses on an element inclined at $\theta = 45°$.* The stresses acting on a plane oriented at an angle $\theta = 45°$ are given by the coordinates of point D, which is at an angle $2\theta = 90°$ from point A (Fig. 7-21b). To evaluate these coordinates, we need to know the angle between line CD and the negative σ_{x_1} axis (that is, angle DCP_2), which in turn requires that we know the angle between line CA and the negative σ_{x_1} axis (angle ACP_2). These angles are found from the geometry of the circle as follows:

$$\tan \overline{ACP_2} = \frac{40 \text{ MPa}}{30 \text{ MPa}} = \frac{4}{3} \qquad \overline{ACP_2} = 53.13°$$

$$\overline{DCP_2} = 90° - \overline{ACP_2} = 90° - 53.13° = 36.87°$$

Knowing these angles, we can obtain the coordinates of point D directly from the figure:

(Point D) $\qquad \sigma_{x_1} = -20 \text{ MPa} - (50 \text{ MPa})(\cos 36.87°) = -60 \text{ MPa}$ ⬅

$$\tau_{x_1 y_1} = (50 \text{ MPa})(\sin 36.87°) = 30 \text{ MPa}$$ ⬅

Fig. 7-21 Example 7-6. (a) Element in plane stress, and (b) the corresponding Mohr's circle. (*Note:* All stresses on the circle have units of MPa.)

(a)

(b)

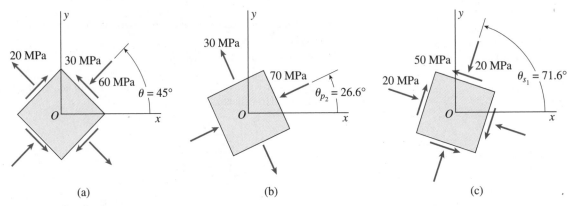

Fig. 7-22 Example 7-6 (continued).
(a) Stresses acting on an element oriented at $\theta = 45°$, (b) principal stresses, and (c) maximum shear stresses

In an analogous manner, we can find the stresses represented by point D', which corresponds to a plane inclined at an angle $\theta = 135°$ (or $2\theta = 270°$):

$$\text{(Point } D') \qquad \sigma_{x_1} = -20 \text{ MPa} + (50 \text{ MPa})(\cos 36.87°) = 20 \text{ MPa} \quad \longleftarrow$$

$$\tau_{x_1 y_1} = (-50 \text{ MPa})(\sin 36.87°) = -30 \text{ MPa} \quad \longleftarrow$$

These stresses are shown in Fig. 7-22a on a sketch of an element oriented at an angle $\theta = 45°$ (all stresses are shown in their true directions). Also, note that the sum of the normal stresses is equal to $\sigma_x + \sigma_y$, or -40 MPa.

(b) *Principal stresses.* The principal stresses are represented by points P_1 and P_2 on Mohr's circle. The algebraically larger principal stress (represented by point P_1) is

$$\sigma_1 = -20 \text{ MPa} + 50 \text{ MPa} = 30 \text{ MPa} \quad \longleftarrow$$

as seen by inspection of the circle. The angle $2\theta_{p_1}$ to point P_1 from point A is the angle ACP_1 measured counterclockwise on the circle, that is,

$$\overline{ACP_1} = 2\theta_{p_1} = 53.13° + 180° = 233.13° \qquad \theta_{p_1} = 116.6° \quad \longleftarrow$$

Thus, the plane of the algebraically larger principal stress is oriented at an angle $\theta_{p_1} = 116.6°$.

The algebraically smaller principal stress (point P_2) is obtained from the circle in a similar manner:

$$\sigma_2 = -20 \text{ MPa} - 50 \text{ MPa} = -70 \text{ MPa} \quad \longleftarrow$$

The angle $2\theta_{p_2}$ to point P_2 on the circle is $53.13°$; thus, the second principal plane is defined by the angle $\theta_{p_2} = 26.6°$.

The principal stresses and principal planes are shown in Fig. 7-22b, and again we note that the sum of the normal stresses is equal to $\sigma_x + \sigma_y$, or -40 MPa.

(c) *Maximum shear stresses.* The maximum positive and negative shear stresses are represented by points S_1 and S_2 on Mohr's circle (Fig. 7-21b). Their magnitudes, equal to the radius of the circle, are

$$\tau_{\max} = 50 \text{ MPa} \quad \longleftarrow$$

The angle ACS_1 from point A to point S_1 is $90° + 53.13° = 143.13°$, and therefore the angle $2\theta_{s_1}$ for point S_1 is

$$2\theta_{s_1} = 143.13°$$

The corresponding angle θ_{s_1} to the plane of the maximum positive shear stress is one-half that value, or $\theta_{s_1} = 71.6°$, as shown in Fig. 7-22c. The maximum negative shear stress (point S_2 on the circle) has the same numerical value as the positive stress (50 MPa).

The normal stresses acting on the planes of maximum shear stress are equal to σ_{aver}, which is the coordinate of the center C of the circle (-20 MPa). These stresses are also shown in Fig. 7-22c. Note that the planes of maximum shear stress are oriented at $45°$ to the principal planes.

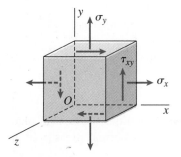

Fig. 7-23 Element of material in plane stress ($\sigma_z = 0$)

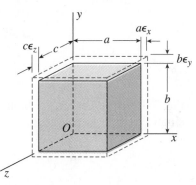

Fig. 7-24 Element of material subjected to normal strains ϵ_x, ϵ_y, and ϵ_z

7.5 | HOOKE'S LAW FOR PLANE STRESS

The stresses acting on inclined planes when the material is subjected to plane stress (Fig. 7-23) were discussed in Sections 7.2, 7.3, and 7.4. The stress-transformation equations derived in those discussions were obtained solely from equilibrium, and therefore the properties of the materials were not needed. Now, in this section, we will investigate the *strains* in the material, which means that the material properties must be considered. However, we will limit our discussion to materials that meet two important conditions: first, the material is uniform throughout the body and has the same properties in all directions (homogeneous and isotropic material), and second, the material follows Hooke's law (linearly elastic material). Under these conditions, we can readily obtain the relationships between the stresses and strains in the body.

Let us begin by considering the normal strains ϵ_x, ϵ_y, and ϵ_z in plane stress. The effects of these strains are pictured in Fig. 7-24, which shows the changes in dimensions of an infinitesimally small element having edges of lengths a, b, and c. All three strains are shown positive (elongation) in the figure. The strains can be expressed in terms of the stresses (Fig. 7-23) by superimposing the effects of the individual stresses. For instance, the strain ϵ_x in the x direction due to the stress σ_x is equal to σ_x/E, where E is the modulus of elasticity. Also, the strain ϵ_x due to the stress σ_y is equal to $-\nu\sigma_y/E$, where ν is Poisson's ratio (see Section 1.5). Of course, the shear stress τ_{xy} produces no normal strains in the x, y, or z directions. Thus, the resultant strain in the x direction is

$$\epsilon_x = \frac{1}{E}(\sigma_x - \nu\sigma_y) \tag{7-34a}$$

In a similar manner, we obtain the strains in the y and z directions:

$$\epsilon_y = \frac{1}{E}(\sigma_y - \nu\sigma_x) \qquad \epsilon_z = -\frac{\nu}{E}(\sigma_x + \sigma_y) \qquad (7\text{-}34\text{b,c})$$

These equations may be used to find the normal strains (in plane stress) when the stresses are known.

The shear stress τ_{xy} (Fig. 7-23) causes a distortion of the element such that each z face becomes a rhombus (Fig. 7-25). The shear strain γ_{xy} is the decrease in angle between the x and y faces of the element and is related to the shear stress by Hooke's law in shear, as follows:

$$\gamma_{xy} = \frac{\tau_{xy}}{G} \qquad (7\text{-}35)$$

where G is the shear modulus of elasticity. Note that the normal stresses σ_x and σ_y have no effect on the shear strain γ_{xy}. Consequently, Eqs. (7-34) and (7-35) give the strains (in plane stress) when all stresses (σ_x, σ_y, and τ_{xy}) act simultaneously.

The first two equations (Eqs. 7-34a and 7-34b) give the strains ϵ_x and ϵ_y in terms of the stresses. These equations can be solved simultaneously for the stresses in terms of the strains:

$$\sigma_x = \frac{E}{1 - \nu^2}(\epsilon_x + \nu\epsilon_y) \qquad \sigma_y = \frac{E}{1 - \nu^2}(\epsilon_y + \nu\epsilon_x) \qquad (7\text{-}36\text{a,b})$$

In addition, we have the following equation for the shear stress in terms of the shear strain:

$$\tau_{xy} = G\gamma_{xy} \qquad (7\text{-}37)$$

Equations (7-36) and (7-37) may be used to find the stresses (in plane stress) when the strains are known. Of course, the normal stress σ_z in the z direction is equal to zero.

Equations (7-34) through (7-37) are known collectively as **Hooke's law for plane stress.** They contain three material constants (E, G, and ν), but only two are independent because of the relationship

$$G = \frac{E}{2(1 + \nu)} \qquad (7\text{-}38)$$

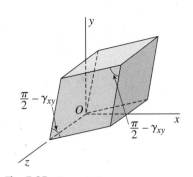

Fig. 7-25 Shear strain γ_{xy}

■ **Special Cases of Hooke's Law**

In the special case of **biaxial stress** (Fig. 7-10b), we have $\tau_{xy} = 0$, and therefore Hooke's law for plane stress simplifies to

$$\epsilon_x = \frac{1}{E}(\sigma_x - \nu\sigma_y) \qquad \epsilon_y = \frac{1}{E}(\sigma_y - \nu\sigma_x)$$

$$\epsilon_z = -\frac{\nu}{E}(\sigma_x + \sigma_y) \qquad (7\text{-}39\text{a,b,c})$$

$$\sigma_x = \frac{E}{1 - \nu^2} (\epsilon_x + \nu\epsilon_y) \qquad \sigma_y = \frac{E}{1 - \nu^2} (\epsilon_y + \nu\epsilon_x) \quad \text{(7-40a,b)}$$

These equations are the same as Eqs. (7-34) and (7-36) because the effects of normal and shear stresses are independent of each other.

For **uniaxial stress**, with $\sigma_y = 0$ (Fig. 7-10a), the equations of Hooke's law simplify even further:

$$\epsilon_x = \frac{\sigma_x}{E} \qquad \epsilon_y = \epsilon_z = -\frac{\nu\sigma_x}{E} \qquad \sigma_x = E\epsilon_x \quad \text{(7-41a,b,c)}$$

Finally, we consider **pure shear** (Fig. 7-11a), which means that $\sigma_x = \sigma_y = 0$. Then we obtain

$$\epsilon_x = \epsilon_y = \epsilon_z = 0 \qquad \gamma_{xy} = \frac{\tau_{xy}}{G} \quad \text{(7-42a,b)}$$

In all three of these special cases, the normal stress σ_z is equal to zero.

■ Unit Volume Change

The unit volume change at a point in a strained body can be found by considering the deformed element of Fig. 7-24. The original volume of this element is $V_0 = abc$, and its final volume is

$$V_1 = abc\,(1 + \epsilon_x)(1 + \epsilon_y)(1 + \epsilon_z) \quad \text{(7-43)}$$

Upon expanding the terms on the right-hand side, we obtain

$$V_1 = abc\,(1 + \epsilon_x + \epsilon_y + \epsilon_z + \epsilon_x\epsilon_y + \epsilon_x\epsilon_z + \epsilon_y\epsilon_z + \epsilon_x\epsilon_y\epsilon_z) \quad \text{(7-44)}$$

The preceding two equations are valid for strains of any size.

If we only consider structures having very small strains (as is usually the case), then we can disregard the products of small strains in comparison with the strains themselves. Thus, the final volume (Eq. 7-44) becomes

$$V_1 = abc\,(1 + \epsilon_x + \epsilon_y + \epsilon_z) \quad \text{(7-45)}$$

Therefore, the change in volume is

$$\Delta V = V_1 - V_0 = abc\,(\epsilon_x + \epsilon_y + \epsilon_z) \quad \text{(7-46)}$$

and the unit volume change (or *dilatation*) becomes

$$e = \frac{\Delta V}{V_0} = \epsilon_x + \epsilon_y + \epsilon_z \quad \text{(7-47)}$$

This equation gives the dilatation in terms of the normal strains and is valid for any material, whether or not it follows Hooke's law, provided the strains are small. Also, note that this equation is not limited to plane stress, but is valid for any stress conditions. Finally, we observe once again that shear strains produce no change in volume.

When the material follows Hooke's law, we can substitute Eqs. (7-34a, b, and c) into Eq. (7-47) and obtain the following expression for the unit volume change (in plane stress) in terms of the stresses:

$$e = \frac{\Delta V}{V_0} = \frac{1 - 2\nu}{E}(\sigma_x + \sigma_y) \tag{7-48}$$

Having this expression for e, we can find the volume change for any object subjected to plane stress by integrating throughout its volume.

■ Strain-energy Density

The strain-energy density u is the strain energy stored in a unit volume of the material (see the discussions in Sections 2.7 and 3.9). For an element in plane stress, we can obtain the strain-energy density by referring to the elements pictured in Figs. 7-24 and 7-25. Because the normal and shear strains occur independently, we can add the strain energies from these two elements to obtain the total energy.

Let us begin by finding the strain energy associated with the normal strains (Fig. 7-24). Since the stress acting on the x face of the element is σ_x (see Fig. 7-23), we find that the force acting on the x face of the element (Fig. 7-24) is equal to $\sigma_x bc$. Of course, as the loads are applied to the structure, this force increases gradually from zero to its maximum value. At the same time, the x face of the element moves through the distance $a\epsilon_x$. Therefore, the work done by this force is

$$\frac{1}{2}(\sigma_x bc)(a\epsilon_x)$$

provided Hooke's law holds for the material. Similarly, the force $\sigma_y ac$ acting on the y face does work equal to

$$\frac{1}{2}(\sigma_y ac)(b\epsilon_y)$$

The sum of these two terms gives the strain energy stored in the element:

$$\frac{abc}{2}(\sigma_x \epsilon_x + \sigma_y \epsilon_y)$$

Thus, the strain-energy density (strain energy per unit volume) due to the normal stresses and strains is

$$u_1 = \frac{1}{2}(\sigma_x \epsilon_x + \sigma_y \epsilon_y) \tag{a}$$

The strain-energy density associated with the shear strains (Fig. 7-25) was evaluated previously in Section 3.9 (see Eq. d of that section):

$$u_2 = \frac{\tau_{xy}\gamma_{xy}}{2} \tag{b}$$

By combining the strain-energy densities for the normal and shear strains, we obtain the following formula for the strain-energy density in plane stress:

$$u = \frac{1}{2} (\sigma_x \epsilon_x + \sigma_y \epsilon_y + \tau_{xy} \gamma_{xy}) \tag{7-49}$$

Substituting for the strains from Eqs. (7-34) and (7-35), we obtain the strain-energy density in terms of stresses alone:

$$u = \frac{1}{2E} (\sigma_x^2 + \sigma_y^2 - 2\nu\sigma_x\sigma_y) + \frac{\tau_{xy}^2}{2G} \tag{7-50}$$

In a similar manner, we can substitute for the stresses from Eqs. (7-36) and (7-37) and obtain the strain-energy density in terms of strains alone:

$$u = \frac{E}{2(1 - \nu^2)} (\epsilon_x^2 + \epsilon_y^2 + 2\nu\epsilon_x\epsilon_y) + \frac{G\gamma_{xy}^2}{2} \tag{7-51}$$

To obtain the strain-energy density in the special case of *biaxial stress*, we simply drop the shear terms in Eqs. (7-49), (7-50), and (7-51).

For the special case of *uniaxial stress*, we substitute the following values

$$\sigma_y = 0 \qquad \tau_{xy} = 0 \qquad \epsilon_y = -\nu\epsilon_x \qquad \gamma_{xy} = 0$$

into Eqs. (7-50) and (7-51) and obtain, respectively,

$$u = \frac{\sigma_x^2}{2E} \qquad u = \frac{E\epsilon_x^2}{2}$$

These equations agree with Eqs. (2-38a) and (2-38b) of Section 2.7.

Also, for *pure shear* we substitute

$$\sigma_x = \sigma_y = 0 \qquad \epsilon_x = \epsilon_y = 0$$

into Eqs. (7-50) and (7-51) and obtain

$$u = \frac{\tau_{xy}^2}{2G} \qquad u = \frac{G\gamma_{xy}^2}{2}$$

These equations agree with Eqs. (3-55a) and (3-55b) of Section 3.9.

7.6 TRIAXIAL STRESS

An element of material subjected to normal stresses σ_x, σ_y, and σ_z acting in three mutually perpendicular directions is said to be in a state of **triaxial stress** (Fig. 7-26a). Since there are no shear stresses on the x, y, and z faces, the stresses σ_x, σ_y, and σ_z are the *principal stresses* in the material.

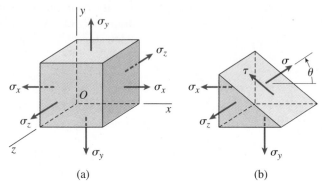

Fig. 7-26 Element in triaxial stress

(a)   (b)

If an inclined plane parallel to the z axis is cut through the element (Fig. 7-26b), the only stresses on the inclined face are the normal stress σ and shear stress τ, both of which act parallel to the xy plane. These stresses are analogous to the stresses σ_{x_1} and $\tau_{x_1y_1}$ encountered in our earlier discussions of plane stress (see, for instance, Fig. 7-2a). Because the stresses σ and τ (Fig. 7-26b) are found from equations of force equilibrium in the xy plane, they are independent of the normal stress σ_z. Therefore, we can use the transformation equations of plane stress, as well as Mohr's circle for plane stress, when determining the stresses σ and τ. The same general conclusion holds for the normal and shear stresses acting on inclined planes cut through the element parallel to the x and y axes.

■ **Maximum Shear Stresses**

From our previous discussions of plane stress, we know that the maximum shear stresses occur on planes oriented at 45° to the principal planes. Therefore, for a material in triaxial stress (Fig. 7-26a), the maximum shear stresses occur on elements oriented at angles of 45° to the x, y, and z axes. For example, consider an element obtained by a 45° rotation about the z axis. The maximum positive and negative shear stresses acting on this element are

$$(\tau_{\max})_z = \pm \frac{\sigma_x - \sigma_y}{2} \tag{7-52a}$$

Similarly, by rotating about the x and y axes through angles of 45°, we obtain the following maximum shear stresses:

$$(\tau_{\max})_x = \pm \frac{\sigma_y - \sigma_z}{2} \qquad (\tau_{\max})_y = \pm \frac{\sigma_x - \sigma_z}{2} \tag{7-52b,c}$$

The absolute maximum shear stress is the numerically largest of the stresses determined from Eqs. (7-52a, b, and c). It is equal to one-half the difference between the algebraically largest and algebraically smallest of the three principal stresses.

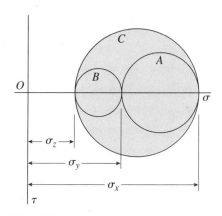

Fig. 7-27 Mohr's circles for an element in triaxial stress

The stresses acting on elements oriented at various angles to the x, y, and z axes can be visualized with the aid of Mohr's circles. For elements oriented by rotations about the z axis, the corresponding circle is labeled A in Fig. 7-27. Note that this circle is drawn for the case in which $\sigma_x > \sigma_y$ and both σ_x and σ_y are tensile stresses.

In a similar manner, we can construct circles B and C for elements oriented by rotations about the x and y axes, respectively. The radii of the circles represent the maximum shear stresses given by Eqs. (7-52a, b, and c), and the absolute maximum shear stress is equal to the radius of the largest circle. The normal stresses acting on the planes of maximum shear stresses have magnitudes given by the abscissas of the centers of the respective circles.

In the preceding discussion of triaxial stress we only considered stresses acting on planes obtained by rotating about the x, y, and z axes. Thus, every plane we considered is parallel to one of the axes. For instance, the inclined plane of Fig. 7-26b is parallel to the z axis, and its normal is parallel to the xy plane. Of course, we can also cut through the element in skew directions, so that the resulting inclined planes are skew to all three coordinate axes. The normal and shear stresses acting on such planes can be obtained by a more complicated three-dimensional analysis. However, the normal stresses acting on skew planes are intermediate in value between the algebraically maximum and minimum principal stresses, and the shear stresses on those planes are smaller (in absolute value) than the absolute maximum shear stress obtained from Eqs. (7-52a, b, and c).

■ **Hooke's Law for Triaxial Stress**

If the material follows Hooke's law, we can obtain the relationships between the normal stresses and normal strains by using the same procedure as for plane stress (see Section 7.5). The strains produced by the stresses σ_x, σ_y, and σ_z acting independently are superimposed to obtain the resultant strains. Thus, we readily arrive at the following equations for the strains in triaxial stress:

$$\epsilon_x = \frac{\sigma_x}{E} - \frac{\nu}{E}(\sigma_y + \sigma_z) \tag{7-53a}$$

$$\epsilon_y = \frac{\sigma_y}{E} - \frac{\nu}{E}(\sigma_z + \sigma_x) \tag{7-53b}$$

$$\epsilon_z = \frac{\sigma_z}{E} - \frac{\nu}{E}(\sigma_x + \sigma_y) \tag{7-53c}$$

In these equations, the standard sign conventions are used; that is, tensile stress σ and extensional strain ϵ are positive.

The preceding equations can be solved simultaneously for the stresses in terms of the strains:

$$\sigma_x = \frac{E}{(1 + \nu)(1 - 2\nu)}\left[(1 - \nu)\epsilon_x + \nu(\epsilon_y + \epsilon_z)\right] \qquad (7\text{-}54a)$$

$$\sigma_y = \frac{E}{(1 + \nu)(1 - 2\nu)}\left[(1 - \nu)\epsilon_y + \nu(\epsilon_z + \epsilon_x)\right] \qquad (7\text{-}54b)$$

$$\sigma_z = \frac{E}{(1 + \nu)(1 - 2\nu)}\left[(1 - \nu)\epsilon_z + \nu(\epsilon_x + \epsilon_y)\right] \qquad (7\text{-}54c)$$

Equations (7-53) and (7-54) represent **Hooke's law for triaxial stress**.

In the special case of **biaxial stress** (Fig. 7-10b), we can obtain the equations of Hooke's law by substituting $\sigma_z = 0$ into the above equations. The resulting equations reduce to Eqs. (7-39) and (7-40) of Section 7.5.

■ **Unit Volume Change**

The unit volume change (or *dilatation*) for an element in triaxial stress is obtained in the same manner as for plane stress (see Section 7.5). If the element is subjected to strains ϵ_x, ϵ_y, and ϵ_z, we may use Eq. (7-47) for the unit volume change:

$$e = \epsilon_x + \epsilon_y + \epsilon_z \qquad (7\text{-}55)$$

This equation is valid for any material provided the strains are small.

If Hooke's law holds for the material, we can substitute for the strains ϵ_x, ϵ_y, and ϵ_z from Eqs. (7-53a, b, and c) and obtain

$$e = \frac{1 - 2\nu}{E}(\sigma_x + \sigma_y + \sigma_z) \qquad (7\text{-}56)$$

Equations (7-55) and (7-56) give the unit volume change in triaxial stress in terms of the strains and stresses, respectively.

■ **Strain-energy Density**

The strain-energy density for an element in triaxial stress is obtained by the same method used for plane stress (see Section 7.5). When stresses σ_x and σ_y act alone (biaxial stress), the strain-energy density (from Eq. 7-49 with the shear term discarded) is

$$u = \frac{1}{2}(\sigma_x \epsilon_x + \sigma_y \epsilon_y)$$

When the element is in triaxial stress and subjected to stresses σ_x, σ_y, and σ_z, the expression for strain-energy density becomes

$$u = \frac{1}{2}(\sigma_x \epsilon_x + \sigma_y \epsilon_y + \sigma_z \epsilon_z) \qquad (7\text{-}57a)$$

Substituting for the strains from Eqs. (7-53a, b, and c), we obtain the strain-energy density in terms of the stresses:

$$u = \frac{1}{2E}(\sigma_x^2 + \sigma_y^2 + \sigma_z^2) - \frac{\nu}{E}(\sigma_x \sigma_y + \sigma_x \sigma_z + \sigma_y \sigma_z) \quad (7\text{-}57b)$$

In a similar manner, but using Eqs. (7-54a, b, and c), we can express the strain-energy density in terms of the strains:

$$u = \frac{E}{2(1 + \nu)(1 - 2\nu)}[(1 - \nu)(\epsilon_x^2 + \epsilon_y^2 + \epsilon_z^2)$$

$$+ 2\nu(\epsilon_x \epsilon_y + \epsilon_x \epsilon_z + \epsilon_y \epsilon_z)] \qquad (7\text{-}57c)$$

When calculating from these expressions, we must be sure to substitute the stresses and strains with their proper algebraic signs.

■ Spherical Stress

A special type of triaxial stress, called **spherical stress,** occurs whenever all three normal stresses are equal (Fig. 7-28):

$$\sigma_x = \sigma_y = \sigma_z = \sigma_0 \qquad (7\text{-}58)$$

Under these stress conditions, *any* plane cut through the element will be subjected to the same normal stress σ_0 and will be free of shear stress. Thus, we have equal normal stresses in every direction and no shear stresses anywhere in the material. Every plane is a principal plane, and the three Mohr's circles shown in Fig. 7-27 reduce to a single point.

The normal strains in spherical stress are also the same in all directions, provided the material is homogeneous and isotropic. If Hooke's law applies, the normal strains are

$$\epsilon_0 = \frac{\sigma_0}{E}(1 - 2\nu) \qquad (7\text{-}59)$$

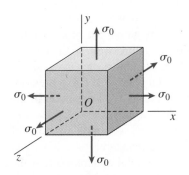

Fig. 7-28 Element in spherical stress

as obtained from Eqs. (7-53a, b, and c).

Since there are no shear strains, an element in the shape of a cube changes in size but remains a cube. In general, any body subjected to spherical stress will maintain its relative proportions but will expand or contract in volume depending upon whether σ_0 is tensile or compressive.

The expression for the unit volume change can be obtained from Eq. (7-55) by substituting for the strains from Eq. (7-59). The result is

$$e = 3\epsilon_0 = \frac{3\sigma_0(1 - 2\nu)}{E} \qquad (7\text{-}60)$$

Equation (7-60) is usually expressed in more compact form by introducing a new quantity K called the **volume modulus of elasticity**, or **bulk modulus of elasticity**, which is defined as follows:

$$K = \frac{E}{3(1 - 2\nu)} \tag{7-61}$$

With this notation, the expression for the unit volume change becomes

$$e = \frac{\sigma_0}{K} \tag{7-62}$$

and the volume modulus is

$$K = \frac{\sigma_0}{e} \tag{7-63}$$

Thus, the volume modulus can be defined as the ratio of the spherical stress to the volumetric strain, which is analogous to the definition of the modulus E in uniaxial stress. Note that the preceding formulas for e and K are based upon the assumption that the strains are small and that Hooke's law holds for the material.

From Eq. (7-61) for K, we see that if Poisson's ratio ν equals 1/3, the moduli K and E are numerically equal. If $\nu = 0$, then K has the value $E/3$, and if $\nu = 0.5$, K becomes infinite, which corresponds to a rigid material having no change in volume (that is, the material is incompressible). Thus, the theoretical maximum value of Poisson's ratio is 0.5.

The preceding formulas for spherical stress were derived for an element subjected to uniform tension in all directions, but of course the formulas also apply to an element in uniform compression. In the case of uniform compression, the stresses and strains have negative signs. Uniform compression occurs when the material is subjected to a pressure p; for example, an object submerged in sea water or rock deep within the earth. This state of stress is often called *hydrostatic stress*.

Although uniform compression is relatively common, a state of uniform tension is difficult to achieve. It can be realized by suddenly and uniformly heating the outer surface of a solid metal sphere, so that the outer layers are at a higher temperature than the interior. The tendency of the outer layers to expand produces uniform tension in all directions at the center.

7.7 PLANE STRAIN

The strains at a point in a loaded structure vary according to the orientation of the axes, in a manner similar to that for stresses. In this section, we will derive the transformation equations that relate the strains in inclined directions to the strains in the reference directions. These transformation equations are widely used in laboratory and field investigations involving measurements of strains.

Strains are customarily measured by *strain gages*; for example, gages are placed in aircraft to measure structural behavior during flight, and gages are placed in buildings to measure the effects of earthquakes. Since each gage measures the strain in one particular direction, it is usually necessary to calculate the strains in other directions by means of the transformation equations.

■ Plane Strain Versus Plane Stress

Let us begin by explaining what is meant by plane strain and how it relates to plane stress. Consider a small element of material having sides of lengths a, b, and c in the x, y, and z directions, respectively (Fig. 7-29a). If the only deformations are those in the xy plane, then three strain components may exist—the normal strain ϵ_x in the x direction (Fig. 7-29b), the normal strain ϵ_y in the y direction (Fig. 7-29c), and the shear strain γ_{xy} (Fig. 7-29d). An element of material subjected to these strains (and *only* these strains) is said to be in a state of **plane strain**. It follows that an element in plane strain has no normal strain ϵ_z in the z direction and no shear strains γ_{xz} and γ_{yz} in the xz and yz planes, respectively.

Fig. 7-29 Strain components ϵ_x, ϵ_y, and γ_{xy} in the xy plane (plane strain)

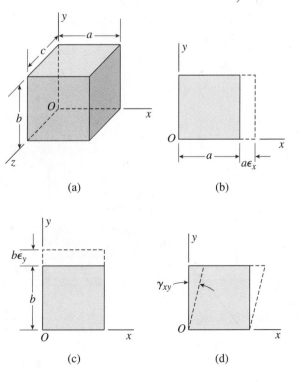

(a)

(b)

(c)

(d)

Thus, plane strain is defined by the following conditions:

$$\epsilon_z = 0 \qquad \gamma_{xz} = 0 \qquad \gamma_{yz} = 0 \qquad (7\text{-}64\text{a,b,c})$$

The remaining strains (ϵ_x, ϵ_y, and γ_{xy}) may have nonzero values.

From the preceding definition, we see that plane strain occurs when the front and rear faces of an element of material (Fig. 7-29a) are fully restrained against displacement—an idealized condition that is seldom reached in actual structures. However, this does not mean that the transformation equations of plane strain are not useful. It turns out that they are extremely useful because they also apply to plane stress, as explained below.

The definition of plane strain (Eqs. 7-64a, b, and c) is analogous to that for plane stress. In plane stress, the following stresses must be zero:

$$\sigma_z = 0 \qquad \tau_{xz} = 0 \qquad \tau_{yz} = 0 \qquad (7\text{-}65\text{a,b,c})$$

whereas the remaining stresses (σ_x, σ_y, and τ_{xy}) may have nonzero values. A comparison of the stresses and strains in plane stress and plane strain is given in Fig. 7-30.

It should not be inferred from the similarities in the definitions of plane stress and plane strain that both occur simultaneously. In general, an element in plane stress will undergo a strain in the z direction (Fig. 7-30); hence, it is *not* in plane strain. Also, an element in plane strain usually will have stresses σ_z acting on it because of the requirement that

Fig. 7-30 Comparison of plane stress and plane strain

	Plane stress	Plane strain
Stresses	$\sigma_z = 0 \qquad \tau_{xz} = 0 \qquad \tau_{yz} = 0$ σ_x, σ_y, and τ_{xy} may have nonzero values	$\tau_{xz} = 0 \qquad \tau_{yz} = 0$ σ_x, σ_y, σ_z, and τ_{xy} may have nonzero values
Strains	$\gamma_{xz} = 0 \qquad \gamma_{yz} = 0$ ϵ_x, ϵ_y, ϵ_z, and γ_{xy} may have nonzero values	$\epsilon_z = 0 \qquad \gamma_{xz} = 0 \qquad \gamma_{yz} = 0$ ϵ_x, ϵ_y, and γ_{xy} may have nonzero values

$\epsilon_z = 0$; therefore, it is *not* in plane stress. Thus, under ordinary conditions plane stress and plane strain do not occur simultaneously.

An exception occurs when an element in plane stress is subjected to equal and opposite normal stresses (that is, when $\sigma_x = -\sigma_y$) and Hooke's law holds for the material. In this special case, there is no normal strain in the z direction, as shown by Eq. (7-34c), and therefore the element is in a state of plane strain as well as plane stress. Another special case, albeit a hypothetical one, is when a material has Poisson's ratio equal to zero ($\nu = 0$); then every plane stress element is also in plane strain because $\epsilon_z = 0$ (Eq. 7-34c).*

The stress-transformation equations derived for plane stress in the xy plane (Eqs. 7-4a and 7-4b) are valid even when a normal stress σ_z is present. The explanation lies in the fact that the stress σ_z does not enter the equations of equilibrium used in deriving Eqs. (7-4a) and (7-4b). Therefore, *the transformation equations for plane stress can also be used for the stresses in plane strain.*

An analogous situation exists for plane strain. Although we will derive the strain-transformation equations for the case of plane strain in the xy plane, the equations are valid even when a strain ϵ_z exists. The reason is simple enough—the strain ϵ_z does not affect the geometric relationships used in the derivations. Therefore, *the transformation equations for plane strain can also be used for the strains in plane stress.*

Finally, we should recall that the transformation equations for plane stress were derived solely from equilibrium and therefore are valid for any material, whether linearly elastic or not. The same conclusion applies to the transformation equations for plane strain—since they are derived solely from geometry, they are independent of the material properties.

■ Transformation Equations for Plane Strain

In the derivation of the transformation equations for plane strain, we will use the coordinate axes shown in Fig. 7-31. We will assume that the normal strains ϵ_x and ϵ_y and the shear strain γ_{xy} associated with the xy axes are known (Fig. 7-29). The objectives of our analysis are to determine the normal strain ϵ_{x_1} and the shear strain $\gamma_{x_1y_1}$ associated with the x_1y_1 axes, which are rotated counterclockwise through an angle θ from the xy axes. (It is not necessary to derive a separate equation for the normal strain ϵ_{y_1} because it can be obtained from the equation for ϵ_{x_1} by substituting $\theta + 90°$ for θ.)

(1) *Normal strain ϵ_{x_1}*. To determine the normal strain ϵ_{x_1} in the x_1 direction, we consider a small element of material selected so that the x_1

Fig. 7-31 Axes x_1 and y_1 rotated through an angle θ from the xy axes

---

*In the discussions of this chapter we are omitting the effects of temperature changes and prestrains, both of which produce additional deformations that may alter some of our conclusions.

axis is along a diagonal of the z face of the element and the x and y axes are along the sides of the element (Fig. 7-32a). The figure shows a two-dimensional view of the element, with the z axis toward the viewer. Of course, the element is actually three dimensional, as in Fig. 7-29a, with a dimension in the z direction.

Consider first the strain ϵ_x in the x direction (Fig. 7-32a). This strain produces an elongation in the x direction equal to $\epsilon_x\,dx$, where dx is the length of the corresponding side of the element. As a result of this elongation, the diagonal of the element increases in length by an amount

$$\epsilon_x\,dx\cos\theta \tag{a}$$

as shown in Fig. 7-32a.

Next, consider the strain ϵ_y in the y direction (Fig. 7-32b). This strain produces an elongation in the y direction equal to $\epsilon_y\,dy$, where dy is the length of the side of the element parallel to the y axis. As a result of this elongation, the diagonal of the element increases in length by an amount

$$\epsilon_y\,dy\sin\theta \tag{b}$$

which is shown in Fig. 7-32b.

Fig. 7-32 Deformations of an element in plane strain due to: (a) normal strain ϵ_x, (b) normal strain ϵ_y, and (c) shear strain γ_{xy}

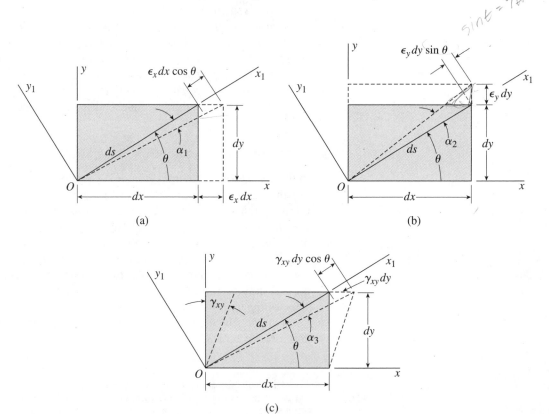

(a)

(b)

(c)

Finally, consider the shear strain γ_{xy} in the xy plane (Fig. 7-32c). This strain produces a distortion of the element such that the angle at the lower left corner of the element decreases by an amount equal to the shear strain. Consequently, the upper face of the element moves to the right (with respect to the lower face) by an amount $\gamma_{xy}\,dy$. This deformation results in an increase in the length of the diagonal equal to

$$\gamma_{xy}\,dy\cos\theta \tag{c}$$

as shown in Fig. 7-32c.

The total increase Δd in the length of the diagonal is the sum of the preceding three expressions; thus,

$$\Delta d = \epsilon_x\,dx\cos\theta + \epsilon_y\,dy\sin\theta + \gamma_{xy}\,dy\cos\theta \tag{d}$$

The normal strain ϵ_{x_1} in the x_1 direction is equal to this increase in length divided by the initial length ds of the diagonal:

$$\epsilon_{x_1} = \frac{\Delta d}{ds} = \epsilon_x\,\frac{dx}{ds}\cos\theta + \epsilon_y\,\frac{dy}{ds}\sin\theta + \gamma_{xy}\,\frac{dy}{ds}\cos\theta \tag{e}$$

Observing that $dx/ds = \cos\theta$ and $dy/ds = \sin\theta$, we obtain the following equation for the normal strain:

$$\epsilon_{x_1} = \epsilon_x\cos^2\theta + \epsilon_y\sin^2\theta + \gamma_{xy}\sin\theta\cos\theta \tag{7-66}$$

Thus, we have obtained an expression for the normal strain in the x_1 direction in terms of the strains ϵ_x, ϵ_y, and γ_{xy} associated with the xy axes. As mentioned previously, the normal strain ϵ_{y_1} in the y_1 direction is obtained from this equation by substituting $\theta + 90°$ for θ.

(2) *Shear strain* $\gamma_{x_1y_1}$. Now we turn to the shear strain $\gamma_{x_1y_1}$ associated with the x_1y_1 axes. This strain is equal to the decrease in angle between lines in the material that were initially along the x_1 and y_1 axes. To clarify this idea, consider Fig. 7-33, which shows both the xy and x_1y_1 axes, with the angle θ between them. Let line Oa represent a line in the material that *initially* was along the x_1 axis (that is, along the diagonal of the element in Fig. 7-32). The deformations caused by the strains ϵ_x, ϵ_y, and γ_{xy} (Fig. 7-32) cause line Oa to rotate through a counterclockwise angle α from the x_1 axis to the position shown in Fig. 7-33. Similarly, line Ob was originally along the y_1 axis, but because of the deformations it rotates through a clockwise angle β. The shear strain $\gamma_{x_1y_1}$ is the decrease in angle between the two lines that originally were at right angles; therefore,

$$\gamma_{x_1y_1} = \alpha + \beta \tag{7-67}$$

Thus, in order to find the shear strain $\gamma_{x_1y_1}$, we must determine the angles α and β.

The angle α can be found from the deformations pictured in Fig. 7-32 as follows. The strain ϵ_x (Fig. 7-32a) produces a clockwise rotation of the diagonal of the element. Let us denote this angle of rotation as α_1.

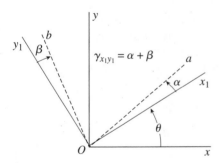

Fig. 7-33 Shear strain $\gamma_{x_1y_1}$ associated with the x_1y_1 axes

The angle α_1 is equal to the distance $\epsilon_x \, dx \sin \theta$ divided by the length ds of the diagonal:

$$\alpha_1 = \epsilon_x \frac{dx}{ds} \sin \theta \tag{f}$$

Similarly, the strain ϵ_y produces a counterclockwise rotation of the diagonal through an angle α_2 (Fig. 7-32b). This angle is equal to the distance $\epsilon_y \, dy \cos \theta$ divided by ds:

$$\alpha_2 = \epsilon_y \frac{dy}{ds} \cos \theta \tag{g}$$

Finally, the strain γ_{xy} produces a clockwise rotation through an angle α_3 (Fig. 7-32c) equal to the distance $\gamma_{xy} \, dy \sin \theta$ divided by ds:

$$\alpha_3 = \gamma_{xy} \frac{dy}{ds} \sin \theta \tag{h}$$

Therefore, the resultant counterclockwise rotation of the diagonal (Fig. 7-32), equal to the angle α shown in Fig. 7-33, is

$$\alpha = -\alpha_1 + \alpha_2 - \alpha_3$$

$$= -\epsilon_x \frac{dx}{ds} \sin \theta + \epsilon_y \frac{dy}{ds} \cos \theta - \gamma_{xy} \frac{dy}{ds} \sin \theta \tag{i}$$

Again observing that $dx/ds = \cos \theta$ and $dy/ds = \sin \theta$, we obtain

$$\alpha = -(\epsilon_x - \epsilon_y) \sin \theta \cos \theta - \gamma_{xy} \sin^2 \theta \tag{7-68}$$

The rotation of line Ob (Fig. 7-33), which initially was at 90° to line Oa, can be found by substituting $\theta + 90°$ for θ in the expression for α. The resulting expression is counterclockwise when positive (because α is counterclockwise when positive), hence it is equal to the negative of the angle β (because β is positive when clockwise). Thus,

$$\beta = (\epsilon_x - \epsilon_y) \sin (\theta + 90°) \cos (\theta + 90°) + \gamma_{xy} \sin^2 (\theta + 90°)$$

$$= -(\epsilon_x - \epsilon_y) \sin \theta \cos \theta + \gamma_{xy} \cos^2 \theta \tag{7-69}$$

Adding α and β gives the shear strain $\gamma_{x_1 y_1}$ (see Eq. 7-67):

$$\gamma_{x_1 y_1} = -2(\epsilon_x - \epsilon_y) \sin \theta \cos \theta + \gamma_{xy} (\cos^2 \theta - \sin^2 \theta) \tag{j}$$

To put the equation in a more useful form, we divide each term by 2:

$$\frac{\gamma_{x_1 y_1}}{2} = -(\epsilon_x - \epsilon_y) \sin \theta \cos \theta + \frac{\gamma_{xy}}{2} (\cos^2 \theta - \sin^2 \theta) \tag{7-70}$$

We have now obtained an expression for the shear strain $\gamma_{x_1 y_1}$ associated with the $x_1 y_1$ axes in terms of the strains ϵ_x, ϵ_y, and γ_{xy} associated with the xy axes.

Transformation equations. The equations for plane strain (Eqs. 7-66 and 7-70) can be expressed in terms of the angle 2θ by using the following trigonometric identities:

$$\cos^2\theta = \frac{1}{2}(1 + \cos 2\theta) \qquad \sin^2\theta = \frac{1}{2}(1 - \cos 2\theta)$$

$$\sin\theta\cos\theta = \frac{1}{2}\sin 2\theta$$

Thus, the **transformation equations for plane strain** are

$$\epsilon_{x_1} = \frac{\epsilon_x + \epsilon_y}{2} + \frac{\epsilon_x - \epsilon_y}{2}\cos 2\theta + \frac{\gamma_{xy}}{2}\sin 2\theta \qquad (7\text{-}71a)$$

$$\frac{\gamma_{x_1y_1}}{2} = -\frac{\epsilon_x - \epsilon_y}{2}\sin 2\theta + \frac{\gamma_{xy}}{2}\cos 2\theta \qquad (7\text{-}71b)$$

These equations are the counterparts of Eqs. (7-4a) and (7-4b) for plane stress.

When comparing the two sets of equations, note that ϵ_{x_1} corresponds to σ_{x_1}, $\gamma_{x_1y_1}/2$ corresponds to $\tau_{x_1y_1}$, ϵ_x corresponds to σ_x, ϵ_y corresponds to σ_y, and $\gamma_{xy}/2$ corresponds to τ_{xy}. The corresponding variables in the two sets of transformation equations are listed in Table 7-1.

The analogy between the transformation equations for plane stress and those for plane strain shows that all of the observations made in Sections 7.2, 7.3, and 7.4 concerning plane stress, principal stresses, maximum shear stresses, and Mohr's circle have their counterparts in plane strain. For instance, the sum of the normal strains in perpendicular directions is a constant (compare with Eq. 7-6):

$$\epsilon_{x_1} + \epsilon_{y_1} = \epsilon_x + \epsilon_y \qquad (7\text{-}72)$$

This equality can be verified easily by substituting the expressions for ϵ_{x_1} (from Eq. 7-71a) and ϵ_{y_1} (from Eq. 7-71a with θ replaced by $\theta + 90°$).

Table 7-1 Corresponding variables in the transformation equations for plane stress (Eqs. 7-4a and b) and plane strain (Eqs. 7-71a and b)

Stresses	Strains
σ_x	ϵ_x
σ_y	ϵ_y
τ_{xy}	$\gamma_{xy}/2$
σ_{x_1}	ϵ_{x_1}
$\tau_{x_1y_1}$	$\gamma_{x_1y_1}/2$

■ **Principal Strains**

Principal strains exist on perpendicular planes with the principal angles θ_p calculated from the following equation (compare with Eq. 7-11):

$$\tan 2\theta_p = \frac{\gamma_{xy}}{\epsilon_x - \epsilon_y} \qquad (7\text{-}73)$$

The principal strains can be calculated from the equation

$$\epsilon_{1,2} = \frac{\epsilon_x + \epsilon_y}{2} \pm \sqrt{\left(\frac{\epsilon_x - \epsilon_y}{2}\right)^2 + \left(\frac{\gamma_{xy}}{2}\right)^2} \qquad (7\text{-}74)$$

which corresponds to Eq. (7-17) for the principal stresses. The two principal strains (in the xy plane) can be correlated with the two principal directions using the technique described in Section 7.3 for the principal stresses. (This technique is illustrated later in Example 7-7.) Finally, note that in plane strain the third principal strain is $\epsilon_z = 0$. Also, the shear strains are zero on the principal planes.

■ **Maximum Shear Strains**

The maximum shear strains in the xy plane are associated with axes at 45° to the directions of the principal strains. The algebraically maximum shear strain (in the xy plane) is given by the following equation (compare with Eq. 7-25):

$$\frac{\gamma_{\max}}{2} = \sqrt{\left(\frac{\epsilon_x - \epsilon_y}{2}\right)^2 + \left(\frac{\gamma_{xy}}{2}\right)^2} \qquad (7\text{-}75)$$

The algebraically minimum shear strain has the same magnitude but is negative. In the directions of maximum shear strain, the normal strains are

$$\epsilon_{\text{aver}} = \frac{\epsilon_x + \epsilon_y}{2} \qquad (7\text{-}76)$$

which is analogous to Eq. (7-27) for stresses. The maximum out-of-plane shear strains, that is, the shear strains in the xz and yz planes, can be obtained from equations analogous to Eq. (7-75).

An element in plane stress that is oriented to the principal directions of stress (see Fig. 7-12b) has no shear stresses acting on its faces. Therefore, the shear strain $\gamma_{x_1y_1}$ for this element is zero. It follows that the normal strains in this element are the principal strains. Thus, at a given point in a stressed body, *the principal strains and principal stresses occur in the same directions.*

■ Mohr's Circle for Plane Strain

Mohr's circle for plane strain is constructed in the same manner as the circle for plane stress, as illustrated in Fig. 7-34. Normal strain ϵ_{x_1} is plotted as the abscissa (positive to the right) and one-half the shear strain $(\gamma_{x_1y_1}/2)$ is plotted as the ordinate (positive downward). The center C of the circle has an abscissa equal to ϵ_{aver} (Eq. 7-76).

Fig. 7-34 Mohr's circle for plane strain

Point A, representing the strains associated with the x direction ($\theta = 0$), has coordinates ϵ_x and $\gamma_{xy}/2$. Point B, at the opposite end of a diameter from A, has coordinates ϵ_y and $-\gamma_{xy}/2$, representing the strains associated with a pair of axes rotated through an angle $\theta = 90°$.

The strains associated with axes rotated through an angle θ are given by point D, which is located on the circle by measuring an angle 2θ counterclockwise from radius CA. The principal strains are represented by points P_1 and P_2, and the maximum shear strains by points S_1 and S_2. All of these strains can be determined from the geometry of the circle or from the transformation equations.

■ Strain Measurements

An electrical-resistance **strain gage** is a device for measuring normal strains on the surface of a stressed object. These gages are quite small, with lengths typically in the range from one-eighth to one-half of an inch. The gages are bonded securely to the surface of the object so that they change in length in proportion to the strains in the object itself. Each gage consists of a fine metal grid that is stretched or shortened when the object is strained at the point where the gage is attached. The grid is equivalent to a continuous wire that goes back and forth from one end of the grid to the other, thereby effectively increasing its length (Fig. 7-35). The electrical resistance of the wire is altered when it stretches or shortens—then this change in resistance is converted into a measurement of strain. The gages are extremely sensitive and can measure strains as small as 1×10^{-6}.

Since each gage measures the normal strain in only one direction, and since the directions of the principal stresses are usually unknown, it is necessary to use three gages in combination, with each gage measuring the strain in a different direction. From three such measurements, it is possible to calculate the strains in any direction, as illustrated in Example 7-8. A group of three gages arranged in a particular pattern is called a **strain rosette**. Because the rosette is mounted on the surface of the body, where the material is in plane stress, we can use the transformation equations for plane strain to calculate the strains in various directions. (As explained earlier in this section, the transformation equations for plane strain can also be used for the strains in plane stress.)

Fig. 7-35 Three electrical-resistance strain gages arranged as a 45° strain rosette (magnified view). (Courtesy of Micro-Measurements Division, Measurements Group, Inc., Raleigh, NC, USA.)

■ Calculation of Stresses

The strain equations presented in this section are derived solely from geometry, as already pointed out. Therefore, the equations apply to any material, whether linear or nonlinear, elastic or inelastic. However, if it is desired to determine the stresses from the strains, the material properties must be taken into account.

If the material follows Hooke's law, we can find the stresses using the appropriate stress-strain equations from either Section 7.5 (for plane stress) or Section 7.6 (for triaxial stress).

As a first example, suppose that the material is in plane stress and that we know the strains ϵ_x, ϵ_y, and γ_{xy}, perhaps from strain-gage measurements. Then we can use the stress-strain equations for plane stress (Eqs. 7-36 and 7-37) to obtain the stresses in the material.

Now consider a second example. Suppose we have determined the three principal strains ϵ_1, ϵ_2, and ϵ_3 for an element of material (if the element is in plane strain, then $\epsilon_3 = 0$). Knowing these strains, we can find the principal stresses using Hooke's law for triaxial stress (see Eqs.

7-54a, b, and c). Once the principal stresses are known, we can find the stresses on inclined planes using the transformation equations for plane stress (see the discussion at the beginning of Section 7.6).

WORK SHEET 7   Mathcad   **Example 7-7**

An element of material in plane strain undergoes the following strains:

$$\epsilon_x = 340 \times 10^{-6} \qquad \epsilon_y = 110 \times 10^{-6} \qquad \gamma_{xy} = 180 \times 10^{-6}$$

These strains are shown highly exaggerated in Fig. 7-36a, which shows the deformations of an element of unit dimensions. Since the edges of the element have unit lengths, the changes in linear dimensions have the same magnitudes as the normal strains ϵ_x and ϵ_y. The shear strain γ_{xy} is the decrease in angle at the lower-left corner of the element.

Determine the following quantities: (a) the strains for an element oriented at an angle $\theta = 30°$, (b) the principal strains, and (c) the maximum shear strains. (Consider only the in-plane strains, and show all results on sketches of properly oriented elements.)

Fig. 7-36 Example 7-7. Element of material in plane strain: (a) element oriented to the x and y axes, (b) element oriented at an angle $\theta = 30°$, (c) principal strains, and (d) maximum shear strains. (*Note:* The edges of the elements have unit lengths.)

Solution   (a) *Element oriented at an angle $\theta = 30°$.* The strains for an element oriented at an angle θ to the x axis can be found from the transformation

(a)

(b)

(c)

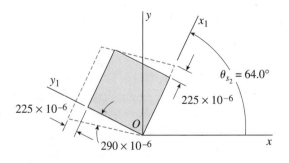

(d)

equations (Eqs. 7-71a and 7-71b). As a preliminary matter, we make the following calculations:

$$\frac{\epsilon_x + \epsilon_y}{2} = \frac{(340 + 110)10^{-6}}{2} = 225 \times 10^{-6}$$

$$\frac{\epsilon_x - \epsilon_y}{2} = \frac{(340 - 110)10^{-6}}{2} = 115 \times 10^{-6}$$

$$\frac{\gamma_{xy}}{2} = 90 \times 10^{-6}$$

Now substituting into Eqs. (7-71a) and (7-71b), we get

$$\epsilon_{x_1} = \frac{\epsilon_x + \epsilon_y}{2} + \frac{\epsilon_x - \epsilon_y}{2} \cos 2\theta + \frac{\gamma_{xy}}{2} \sin 2\theta$$

$$= (225 \times 10^{-6}) + (115 \times 10^{-6})(\cos 60°) + (90 \times 10^{-6})(\sin 60°)$$

$$= 360 \times 10^{-6}$$

$$\frac{\gamma_{x_1y_1}}{2} = -\frac{\epsilon_x - \epsilon_y}{2} \sin 2\theta + \frac{\gamma_{xy}}{2} \cos 2\theta$$

$$= -(115 \times 10^{-6})(\sin 60°) + (90 \times 10^{-6})(\cos 60°)$$

$$= -55 \times 10^{-6}$$

Therefore, the shear strain is

$$\gamma_{x_1y_1} = -110 \times 10^{-6}$$

The strain ϵ_{y_1} can be obtained from Eq. (7-72), as follows:

$$\epsilon_{y_1} = \epsilon_x + \epsilon_y - \epsilon_{x_1} = (340 + 110 - 360)10^{-6} = 90 \times 10^{-6}$$

The strains ϵ_{x_1}, ϵ_{y_1}, and $\gamma_{x_1y_1}$ are shown in Fig. 7-36b for an element oriented at $\theta = 30°$. Note that the angle at the lower-left corner of the element increases because $\gamma_{x_1y_1}$ is negative.

(b) *Principal strains.* The principal strains are readily determined from Eq. (7-74), as follows:

$$\epsilon_{1,2} = \frac{\epsilon_x + \epsilon_y}{2} \pm \sqrt{\left(\frac{\epsilon_x - \epsilon_y}{2}\right)^2 + \left(\frac{\gamma_{xy}}{2}\right)^2}$$

$$= 225 \times 10^{-6} \pm \sqrt{(115 \times 10^{-6})^2 + (90 \times 10^{-6})^2}$$

$$= 225 \times 10^{-6} \pm 146 \times 10^{-6}$$

Thus, the principal strains are

$$\epsilon_1 = 370 \times 10^{-6} \qquad \epsilon_2 = 80 \times 10^{-6}$$

in which ϵ_1 denotes the algebraically larger principal strain and ϵ_2 denotes the algebraically smaller principal strain. (Recall that we are considering only in-plane strains in this example.)

The angles to the principal directions can be obtained from Eq. (7-73):

$$\tan 2\theta_p = \frac{\gamma_{xy}}{\epsilon_x - \epsilon_y} = \frac{180}{340 - 110} = 0.7826$$

The values of $2\theta_p$ between 0 and 360° are 38.0° and 218.0°, and therefore the angles to the principal directions are

$$\theta_p = 19.0° \text{ and } 109.0°$$

To determine the value of θ_p associated with each principal strain, we substitute $\theta_p = 19.0°$ into the first transformation equation (Eq. 7-71a) and solve for the strain:

$$\epsilon_{x_1} = \frac{\epsilon_x + \epsilon_y}{2} + \frac{\epsilon_x + \epsilon_y}{2} \cos 2\theta + \frac{\gamma_{xy}}{2} \sin 2\theta$$

$$= (225 \times 10^{-6}) + (115 \times 10^{-6})(\cos 38.0°) + (90 \times 10^{-6})(\sin 38.0°)$$

$$= 370 \times 10^{-6}$$

This result shows that the larger principal strain ϵ_1 is at the angle $\theta_{p_1} = 19.0°$. The smaller strain ϵ_2 acts at 90° from that direction ($\theta_{p_2} = 109.0°$). Thus,

$$\epsilon_1 = 370 \times 10^{-6} \quad \text{and} \quad \theta_{p_1} = 19.0° \qquad \Longleftarrow$$

$$\epsilon_2 = 80 \times 10^{-6} \quad \text{and} \quad \theta_{p_2} = 109.0° \qquad \Longleftarrow$$

Note that $\epsilon_1 + \epsilon_2 = \epsilon_x + \epsilon_y$.

The principal strains are portrayed in Fig. 7-36c. There are, of course, no shear strains on the principal planes.

(c) *Maximum shear strain.* The maximum shear strain is calculated from Eq. (7-75):

$$\frac{\gamma_{max}}{2} = \sqrt{\left(\frac{\epsilon_x - \epsilon_y}{2}\right)^2 + \left(\frac{\gamma_{xy}}{2}\right)^2} = 146 \times 10^{-6} \qquad \gamma_{max} = 290 \times 10^{-6} \qquad \Longleftarrow$$

The element having the maximum shear strains is oriented at 45° to the principal directions; therefore, $\theta_s = 19.0° + 45° = 64.0°$ and $2\theta_s = 128.0°$. By substituting this value of $2\theta_s$ into the second transformation equation (Eq. 7-71b), we can determine the sign of the shear strain associated with this direction. The calculations are as follows:

$$\frac{\gamma_{x_1 y_1}}{2} = -\frac{\epsilon_x - \epsilon_y}{2} \sin 2\theta + \frac{\gamma_{xy}}{2} \cos 2\theta$$

$$= -(115 \times 10^{-6})(\sin 128.0°) + (90 \times 10^{-6})(\cos 128.0°)$$

$$= -146 \times 10^{-6}$$

This result shows that an element oriented at an angle $\theta_{s_2} = 64.0°$ has the maximum negative shear strain.

We can arrive at the same result by observing that the angle θ_{s_1} to the direction of maximum positive shear strain is always 45° less than θ_{p_1}. Hence,

$$\theta_{s_1} = \theta_{p_1} - 45° = 19.0° - 45° = -26.0° \qquad \Longleftarrow$$

$$\theta_{s_2} = \theta_{s_1} + 90° = 64.0° \qquad \Longleftarrow$$

The shear strains corresponding to θ_{s_1} and θ_{s_2} are $\gamma_{max} = 290 \times 10^{-6}$ and $\gamma_{min} = -290 \times 10^{-6}$, respectively.

The normal strains on the element having the maximum and minimum shear strains are

$$\epsilon_{aver} = \frac{\epsilon_x + \epsilon_y}{2} = 225 \times 10^{-6} \qquad \longleftarrow$$

A sketch of the element having the maximum in-plane shear strains is shown in Fig. 7-36d.

In this example, we solved for the strains by using the transformation equations. However, all of the results can be obtained just as easily from Mohr's circle.

■ Example 7-8

A 45° strain rosette (also called a *rectangular rosette*) consists of three electrical-resistance strain gages arranged to measure strains in two perpendicular directions and also at a 45° angle between them, as shown in Fig. 7-37a. The rosette is bonded to the surface of the structure before it is loaded. Gages A, B, and C measure the normal strains ϵ_a, ϵ_b, and ϵ_c in the directions of lines Oa, Ob, and Oc, respectively.

Explain how to obtain the strains ϵ_{x_1}, ϵ_{y_1}, and $\gamma_{x_1y_1}$ associated with an element oriented at an angle θ to the xy axes (Fig. 7-37b).

Fig. 7-37 Example 7-8. (a) 45° strain rosette, and (b) element oriented at an angle θ to the xy axes

(a)   (b)

Solution   At the surface of the stressed object, the material is in plane stress. Since the strain-transformation equations (Eqs. 7-71a and 7-71b) apply to plane stress as well as to plane strain, we can use those equations to determine the strains in any desired direction.

Strains associated with the xy axes. We begin by determining the strains associated with the xy axes. Because gages A and C are aligned with the x and y axes, respectively, they give the strains ϵ_x and ϵ_y directly:

$$\epsilon_x = \epsilon_a \qquad \epsilon_y = \epsilon_c \tag{7-77a,b}$$

To obtain the shear strain γ_{xy}, we use the transformation equation for normal strains (Eq. 7-71a):

$$\epsilon_{x_1} = \frac{\epsilon_x + \epsilon_y}{2} + \frac{\epsilon_x - \epsilon_y}{2} \cos 2\theta + \frac{\gamma_{xy}}{2} \sin 2\theta$$

For an angle $\theta = 45°$, we know that $\epsilon_{x_1} = \epsilon_b$ (Fig. 7-37a); therefore, the preceding equation gives

$$\epsilon_b = \frac{\epsilon_a + \epsilon_c}{2} + \frac{\epsilon_a - \epsilon_c}{2} (\cos 90°) + \frac{\gamma_{xy}}{2} (\sin 90°)$$

Solving for γ_{xy}, we get

$$\gamma_{xy} = 2\epsilon_b - \epsilon_a - \epsilon_c \qquad (7\text{-}78)$$

Thus, the strains ϵ_x, ϵ_y, and γ_{xy} are easily determined from the given strain-gage readings.

Strains associated with the x_1y_1 axes. Knowing the strains ϵ_x, ϵ_y, and γ_{xy}, we can calculate the strains for an element oriented at any angle θ (Fig. 7-37b) from the strain-transformation equations (Eqs. 7-71a and 7-71b) or from Mohr's circle. We can also calculate the principal strains and the maximum shear strains from Eqs. (7-74) and (7-75), respectively.

PROBLEMS/CHAPTER 7

 PLANE STRESS

7.2-1 An element in plane stress is subjected to stresses $\sigma_x = 7500$ psi, $\sigma_y = 4500$ psi, and $\tau_{xy} = 3000$ psi, as shown in the figure. Determine the stresses acting on an element oriented at an angle $\theta = 50°$ from the x axis, where the angle θ is positive when counterclockwise. Show these stresses on a sketch of an element oriented at the angle θ.

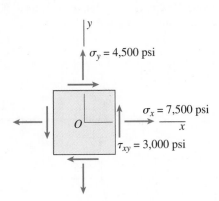

Prob. 7.2-1

7.2-2 Solve the preceding problem for $\sigma_x = 110$ MPa, $\sigma_y = 40$ MPa, $\tau_{xy} = 30$ MPa, and $\theta = 30°$ (see figure).

Prob. 7.2-2

7.2-3 The stresses at point A in the web of a train rail are found to be 9,300 psi tension in the horizontal direction and 22,100 psi compression in the vertical direction (see figure). Also, the shear stresses are 4,800 psi acting in the directions shown. Determine the stresses acting on an element oriented at a counterclockwise angle of 60° from the horizontal. (Show the stresses on a sketch of an element oriented at this angle.)

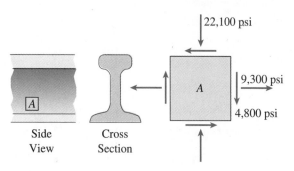

Prob. 7.2-3

7.2-4 The stresses at point A in the web of a wide-flange beam are found to be 52 MPa tension in the horizontal direction and 16 MPa compression in the vertical direction (see figure). Also, the shear stresses are 31 MPa acting in the directions shown. Determine the stresses acting on an element oriented at a counterclockwise angle of 50° to the horizontal. (Show the stresses on a sketch of an element oriented at this angle.)

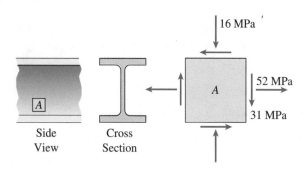

Prob. 7.2-4

7.2-5 The polyethylene liner of a settling pond is subjected to stresses $\sigma_x = 800$ psi, $\sigma_y = 400$ psi, and $\tau_{xy} = -200$ psi, as shown by the plane-stress element in part (a) of the figure. Determine the normal and shear stresses acting on a seam inclined at an angle of 30° to the element, as shown in part (b) of the figure. (Show the stresses on a sketch of an element having its sides parallel and perpendicular to the seam.)

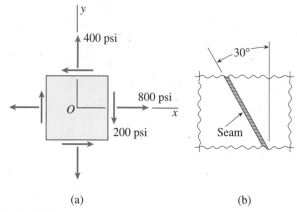

(a)   (b)

Prob. 7.2-5

7.2-6 A thin rectangular plate of dimensions 75 mm × 200 mm is formed by welding two triangular plates (see figure). The plate is subjected to a compressive stress of 3.0 MPa in the long direction and a tensile stress of 10.0 MPa in the short direction. Determine the normal stress σ_w acting perpendicular to the line of the weld and the shear stress τ_w acting parallel to the weld.

Prob. 7.2-6

7.2-7 Solve the preceding problem for a plate of dimensions 3.0 in. × 6.0 in. subjected to a tensile stress of 2000 psi in the long direction and a compressive stress of 800 psi in the short direction (see figure).

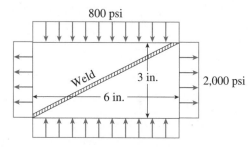

Prob. 7.2-7

7.2-8 At a point on the surface of a machine the material is in biaxial stress with $\sigma_x = 120$ MPa and $\sigma_y = -40$ MPa, as shown in part (a) of the figure. Part (b) of the figure shows an inclined plane aa through the same point in the material but oriented at an angle θ. Determine all values of the angle θ such that no normal stresses act on plane aa. For each angle θ, sketch a stress element having plane aa as one of its sides and show all stresses acting on the element.

stresses have the magnitudes and directions shown in the figure. Determine the normal and shear stresses acting on an element whose sides are parallel to the xy axes; that is, determine σ_x, σ_y, and τ_{xy}.

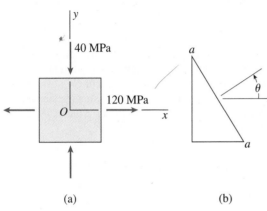

(a)   (b)

Prob. 7.2-8

Prob. 7.2-10

7.2-9 Solve the preceding problem for $\sigma_x = 2000$ psi and $\sigma_y = -8000$ psi (see figure).

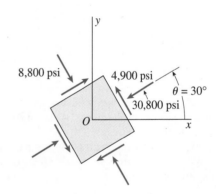

Prob. 7.2-11

(a)   (b)

Prob. 7.2-9

7.2-10 and 7.2-11 An element in plane stress from the frame of a racing car is oriented at a known angle θ (see figure). On this inclined element, the normal and shear

7.2-12 A thin plate in plane stress is subjected to normal stresses σ_x and σ_y and a shear stress τ_{xy}, as shown in part (a) of the figure. At angles $\theta = 40°$ and $80°$ from the x axis the normal stress is 50 MPa tension [see parts (b) and (c) of the figure]. If the stress σ_x equals 20 MPa tension, what are the stresses σ_y and τ_{xy}?

(a)

(b)   (c)

Prob. 7.2-12

7.2-13 The surface of an airplane wing is subjected to plane stress with normal stresses σ_x and σ_y and shear stress τ_{xy}, as shown in part (a) of the figure. At an angle $\theta = 30°$ from the x axis the normal stress is 350 psi tension, and at an angle $\theta = 50°$ it is 100 psi compression [see parts (b) and (c) of the figure]. If the stress σ_x equals 1000 psi tension, what are the stresses σ_y and τ_{xy}?

(a)

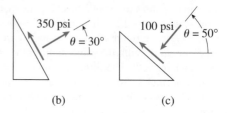

(b)   (c)

Prob. 7.2-13

*7.2-14 At a point in a structure subjected to plane stress, the stresses have the magnitudes and directions shown acting on element A in the first part of the figure. Element B, located at the same point in the structure, is oriented at an angle θ_1 (between zero and 90°) such that the stresses are as shown in the second part of the figure. Calculate the normal stress σ_b, the shear stress τ_b, and the angle θ_1.

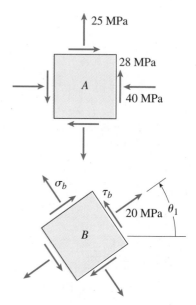

Prob. 7.2-14

*7.2-15 At a certain point in a structure the material is subjected to plane stress such that on a vertical plane through the point, the normal stress is 14 ksi tension and the shear stress is 12 ksi counterclockwise [see part (a) of the figure], and on a plane oriented at an angle θ_1, the normal stress is 21 ksi tension and the shear stress is 5 ksi counterclockwise [see part (b) of the figure]. What is the angle θ_1?

(a)   (b)

Prob. 7.2-15

*7.2-16 The stresses on the surface of a crane hook are σ_x = 12 MPa, σ_y = 36 MPa, and τ_{xy} = 27 MPa, as shown by the stress element in the figure. For what angles θ between 0 and 90° (measured counterclockwise) is the normal stress equal numerically to twice the shear stress? For each such angle, sketch a properly oriented stress element and show all stresses acting on the element.

Prob. 7.2-16

  PRINCIPAL STRESSES AND MAXIMUM SHEAR STRESSES

WORK SHEETS 5 & 6   Mathcad

When solving the problems for Section 7.3, consider only the in-plane stresses (the stresses in the xy plane).

7.3-1 An element in plane stress is subjected to stresses σ_x = 7500 psi, σ_y = 4500 psi, and τ_{xy} = 3000 psi (see the figure for Problem 7.2-1). Determine the principal stresses and show them on a sketch of a properly oriented element.

7.3-2 An element in plane stress is subjected to stresses σ_x = 110 MPa, σ_y = 40 MPa, and τ_{xy} = 30 MPa (see the figure for Problem 7.2-2). Determine the principal stresses and show them on a sketch of a properly oriented element.

7.3-3 An element in plane stress is subjected to stresses σ_x = 9,300 psi, σ_y = −22,100 psi, and τ_{xy} = −4,800 psi (see the figure for Problem 7.2-3). Determine the maximum shear stresses and associated normal stresses and show them on a sketch of a properly oriented element.

7.3-4 An element in plane stress is subjected to stresses σ_x = 52 MPa, σ_y = −16 MPa, and τ_{xy} = −31 MPa (see the figure for Problem 7.2-4). Determine the maximum shear stresses and associated normal stresses and show them on a sketch of a properly oriented element.

7.3-5 A shear wall in a reinforced concrete building is subjected to a vertical uniform load q and a horizontal force H, as shown in the first part of the figure. As a consequence, the stresses at point A on the surface of the wall have the values shown in the second part of the figure (compressive stress equal to 1400 psi and shear stress equal to 240 psi). (a) Determine the principal stresses and show them on a sketch of a properly oriented element. (b) Determine the maximum shear stresses and associated normal stresses and show them on a sketch of a properly oriented element.

Prob. 7.3-5

7.3-6 A propeller shaft subjected to combined torsion and axial thrust is designed to resist a shear stress of 60 MPa and a normal stress of 90 MPa (see figure). (a) Determine the principal stresses and show them on a sketch of a properly oriented element. (b) Determine the maximum shear stresses and associated normal stresses and show them on a sketch of a properly oriented element.

Prob. 7.3-6

7.3-7 At a point on the surface of a machine component the stresses acting on the x face of a stress element are $\sigma_x = 6500$ psi and $\tau_{xy} = 2100$ psi (see figure). What is the allowable range of values for the stress σ_y if the maximum shear stress is limited to $\tau_0 = 2900$ psi?

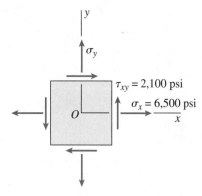

Prob. 7.3-7

7.3-8 Solve the preceding problem if $\sigma_x = 45$ MPa, $\tau_{xy} = 30$ MPa, and $\tau_0 = 34$ MPa (see figure).

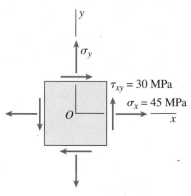

Prob. 7.3-8

7.3-9 At a point in the web of a steel girder the stresses consist of a normal stress σ_x and a shear stress τ_{xy}, as shown in part (a) of the figure. At this same point, the principal stresses are 1.0 ksi tension at an angle $\theta = 76.0°$ and 16.0 ksi compression [see part (b) of the figure]. Determine the stresses σ_x and τ_{xy} acting on the xy element.

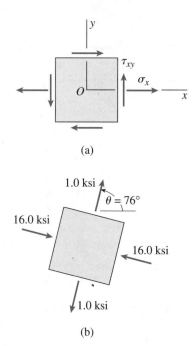

(a)

(b)

Prob. 7.3-9

7.3-10 to 7.3-15 An element in plane stress (see figure) is subjected to stresses σ_x, σ_y, and τ_{xy}. (a) Determine the principal stresses and show them on a sketch of a properly oriented element. (b) Determine the maximum shear stresses and associated normal stresses and show them on a sketch of a properly oriented element.

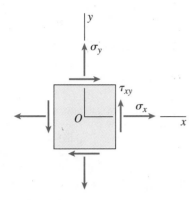

Probs. 7.3-10 through 7.3-15

7.3-10 $\sigma_x = 72$ MPa, $\sigma_y = 30$ MPa, $\tau_{xy} = 28$ MPa

7.3-11 $\sigma_x = 6,500$ psi, $\sigma_y = -27,000$ psi, $\tau_{xy} = -8,000$ psi

7.3-12 $\sigma_x = -81$ MPa, $\sigma_y = 129$ MPa, $\tau_{xy} = -36$ MPa

7.3-13 $\sigma_x = -11,000$ psi, $\sigma_y = 31,000$ psi, $\tau_{xy} = -29,000$ psi

7.3-14 $\sigma_x = -71$ MPa, $\sigma_y = -33$ MPa, $\tau_{xy} = 29$ MPa

7.3-15 $\sigma_x = -21,200$ psi, $\sigma_y = -4,200$ psi, $\tau_{xy} = 3,900$ psi

7.3-16 and 7.3-17 At a point on the surface of a machine, the stresses acting on an element oriented at an angle θ to the x axis have the magnitudes and directions shown in the figure. (a) Determine the principal stresses and show them on a sketch of a properly oriented element. (b) Determine the maximum shear stresses and associated normal stresses and show them on a sketch of a properly oriented element.

Prob. 7.3-16

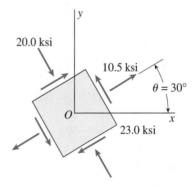

Prob. 7.3-17

7.3-18 An element in plane stress is subjected to stresses $\sigma_x = -68.5$ MPa and $\tau_{xy} = 39.2$ MPa (see figure). It is known that one of the principal stresses equals 26.1 MPa in tension. (a) Determine the stress σ_y. (b) Determine the other principal stress and the orientation of the principal planes; then show the principal stresses on a sketch of a properly oriented element.

Prob. 7.3-18

7.3-19 Solve the preceding problem if $\sigma_x = 7620$ psi and $\tau_{xy} = -2910$ psi (see figure) and one of the principal stresses equals 8400 psi in tension.

Prob. 7.3-19

 WORK SHEETS 6 & 7 Mathcad **MOHR'S CIRCLE**

The problems for Section 7.4 are to be solved using Mohr's circle. Consider only the in-plane stresses (the stresses in the xy plane).

7.4-1 An element in uniaxial stress is subjected to tensile stresses $\sigma_x = 11,600$ psi, as shown in the figure. Using Mohr's circle, determine: (a) the stresses acting on an element oriented at an angle $\theta = 21.8°$ from the x axis, and (b) the maximum shear stresses and associated normal stresses. Show all results on sketches of properly oriented elements.

Prob. 7.4-1

7.4-2 Solve the preceding problem for an element subjected to stresses $\sigma_x = 68$ MPa (see figure) if the angle $\theta = 31°$.

Prob. 7.4-2

7.4-3 An element in uniaxial stress is subjected to a compressive stress of 6000 psi, as shown in the figure. Using Mohr's circle, determine: (a) the stresses acting on an element oriented at a slope of 1 on 2 (see figure), and (b) the maximum shear stresses and associated normal stresses. Show all results on sketches of properly oriented elements.

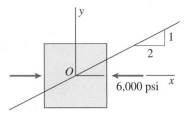

Prob. 7.4-3

7.4-4 An element in pure shear is subjected to stresses $\tau_{xy} = -32$ MPa, as shown in the figure. Using Mohr's circle, determine: (a) the stresses acting on an element oriented at an angle $\theta = 20°$ from the x axis, and (b) the principal stresses. Show all results on sketches of properly oriented elements.

Prob. 7.4-4

7.4-5 Solve the preceding problem for an element subjected to stresses $\tau_{xy} = 4500$ psi (see figure) if the angle $\theta = 75°$.

Prob. 7.4-5

7.4-6 An element in biaxial stress is subjected to stresses $\sigma_x = -80$ MPa and $\sigma_y = -20$ MPa, as shown in the figure. Using Mohr's circle, determine: (a) the stresses acting on an element oriented at an angle $\theta = 22.5°$ from the x axis, and (b) the maximum shear stresses and associated normal stresses. Show all results on sketches of properly oriented elements.

Prob. 7.4-6

7.4-7 Solve the preceding problem for an element subjected to stresses $\sigma_x = 4000$ psi and $\sigma_y = -1000$ psi (see figure) if the angle $\theta = 60°$.

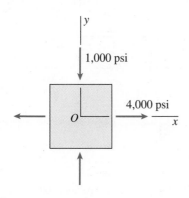

Prob. 7.4-7

7.4-8 An element in biaxial stress is subjected to stresses $\sigma_x = -16$ MPa and $\sigma_y = 42$ MPa, as shown in the figure. Using Mohr's circle, determine: (a) the stresses acting on an element oriented at a slope of 1 on 2.5 (see figure), and (b) the maximum shear stresses and associated normal stresses. Show all results on sketches of properly oriented elements.

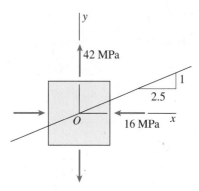

Prob. 7.4-8

7.4-9 For an element in plane stress the normal stresses are $\sigma_x = -1800$ psi and $\sigma_y = 7800$ psi (see figure). Using Mohr's circle, determine the maximum permissible value of the shear stress τ_{xy} if the allowable shear stress in the material is 8000 psi.

Prob. 7.4-9

7.4-10 For an element in plane stress, the normal stresses are $\sigma_x = 26$ MPa and $\sigma_y = -70$ MPa (see figure). Using Mohr's circle, determine the maximum permissible value of the shear stress τ_{xy} if the allowable shear stress in the material is 52 MPa.

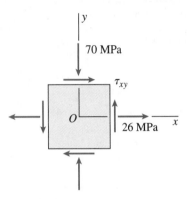

Prob. 7.4-10

7.4-11 An element in plane stress is subjected to stresses $\sigma_x = -1300$ psi, $\sigma_y = 7100$ psi, and $\tau_{xy} = 4000$ psi, as shown in the figure. Using Mohr's circle, determine the stresses acting on an element oriented at an angle $\theta = 45°$ from the x axis. Show these stresses on a sketch of an element oriented at the angle θ.

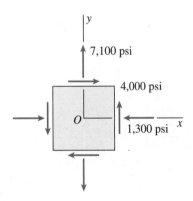

Prob. 7.4-11

7.4-12 Solve the preceding problem for $\sigma_x = -6.2$ MPa, $\sigma_y = -20.4$ MPa, $\tau_{xy} = -9.0$ MPa, and $\theta = 40°$ (see figure).

Prob. 7.4-12

7.4-13 Solve Problem 7.2-1 by using Mohr's circle.

7.4-14 Solve Problem 7.2-2 by using Mohr's circle.

7.4-15 Solve Problem 7.2-3 by using Mohr's circle.

7.4-16 Solve Problem 7.2-4 by using Mohr's circle.

7.4-17 Solve Problem 7.2-5 by using Mohr's circle.

7.4-18 An element in plane stress is subjected to stresses $\sigma_x = 71$ MPa, $\sigma_y = 16$ MPa, and $\tau_{xy} = 24$ MPa (see figure). (a) Determine the principal stresses and show them on a sketch of a properly oriented element. (b) Determine the maximum shear stresses and associated normal stresses and show them on a sketch of a properly oriented element.

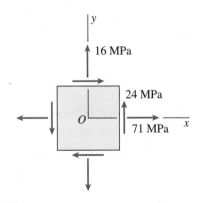

Prob. 7.4-18

7.4-19 Solve the preceding problem for $\sigma_x = -4200$ psi, $\sigma_y = 1400$ psi, and $\tau_{xy} = 4500$ psi (see figure).

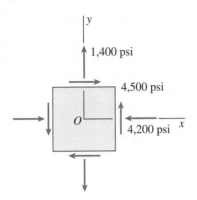

1,400 psi

4,500 psi

O

4,200 psi x

Prob. 7.4-19

7.4-20 Solve Problem 7.3-10 by using Mohr's circle.

7.4-21 Solve Problem 7.3-11 by using Mohr's circle.

7.4-22 Solve Problem 7.3-12 by using Mohr's circle.

7.4-23 Solve Problem 7.3-13 by using Mohr's circle.

7.4-24 Solve Problem 7.3-14 by using Mohr's circle.

7.4-25 Solve Problem 7.3-15 by using Mohr's circle.

HOOKE'S LAW FOR PLANE STRESS

When solving the problems for Section 7.5, assume that the material is linearly elastic with modulus of elasticity E and Poisson's ratio v.

7.5-1 A rectangular steel plate with thickness $t = 0.25$ in. is subjected to uniform normal stresses σ_x and σ_y, as shown in the figure. Strain gages A and B, oriented in the x and y directions, respectively, are attached to the plate. The gage readings give normal strains $\epsilon_x = 0.00062$ (elongation) and $\epsilon_y = -0.00045$ (shortening). Knowing that $E = 30 \times 10^6$ psi and $v = 0.3$, determine the stresses σ_x and σ_y and the change Δt in the thickness of the plate.

σ_y

B A

O x

σ_x

Probs. 7.5-1 and 7.5-2

7.5-2 Solve the preceding problem if the thickness of the steel plate is $t = 10$ mm, the gage readings are $\epsilon_x = 350 \times 10^{-6}$ (elongation) and $\epsilon_y = 85 \times 10^{-6}$ (elongation), the modulus is $E = 200$ GPa, and Poisson's ratio is $v = 0.30$.

7.5-3 Assume that the normal strains ϵ_x and ϵ_y for an element in plane stress (see figure) are measured with strain gages. (a) Obtain a formula for the normal strain ϵ_z in the z direction in terms of ϵ_x, ϵ_y, and Poisson's ratio v. (b) Obtain a formula for the dilatation e in terms of ϵ_x, ϵ_y, and Poisson's ratio v.

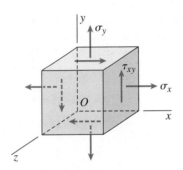

σ_y

τ_{xy}

σ_x

O

x

z

Prob. 7.5-3

7.5-4 A magnesium plate in biaxial stress is subjected to tensile stresses $\sigma_x = 30$ MPa and $\sigma_y = 15$ MPa (see figure). The corresponding strains in the plate are $\epsilon_x = 550 \times 10^{-6}$ and $\epsilon_y = 100 \times 10^{-6}$. Determine Poisson's ratio v and the modulus of elasticity E for the material.

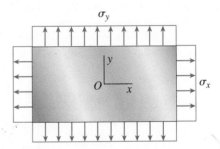

σ_y

σ_x

O x

Probs. 7.5-4 through 7.5-7

7.5-5 Solve the preceding problem for a steel plate with $\sigma_x = 18,000$ psi (tension), $\sigma_y = -9,000$ psi (compression), $\epsilon_x = 700 \times 10^{-6}$ (elongation), and $\epsilon_y = -500 \times 10^{-6}$ (shortening).

7.5-6 A rectangular plate in biaxial stress (see figure) is subjected to normal stresses $\sigma_x = 65$ MPa (tension) and $\sigma_y = -20$ MPa (compression). The plate has dimensions

$200 \times 300 \times 15$ mm and is made of aluminum with $E = 75$ GPa and $\nu = 0.33$. (a) Determine the maximum in-plane shear strain γ_{max} in the plate. (b) Determine the change Δt in the thickness of the plate. (c) Determine the change ΔV in the volume of the plate.

7.5-7 Solve the preceding problem for a steel plate with $\sigma_x = 9500$ psi (tension), $\sigma_y = -3200$ psi (compression), dimensions $10 \times 12 \times 1.0$ in., $E = 32 \times 10^6$ psi, and $\nu = 0.29$.

7.5-8 A brass cube 40 mm on each edge is compressed in two perpendicular directions by forces $P = 120$ kN (see figure). Calculate the decrease ΔV in the volume of the cube and the strain energy U stored in the cube, assuming $E = 100$ GPa and $\nu = 0.34$.

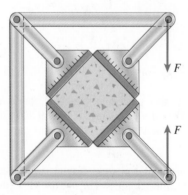

$P = 120$ kN

$P = 120$ kN

Prob. 7.5-8

7.5-9 A 4.0-inch cube of concrete ($E = 3.0 \times 10^6$ psi, $\nu = 0.1$) is compressed in biaxial stress by means of a framework that is loaded as shown in the figure. Assuming that each load F equals 20 k, determine the decrease ΔV in the volume of the cube and the strain energy U stored in the cube.

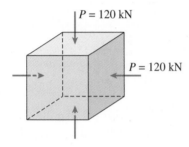

F

F

Prob. 7.5-9

7.5-10 A square plate of width b and thickness t is loaded by normal forces P_x and P_y, and by shear forces V, as shown in the figure. These forces produce uniformly distributed stresses acting on the side faces of the plate. Calculate the change ΔV in the volume of the plate and the strain energy U stored in the plate if the dimensions are $b = 500$ mm and $t = 40$ mm, the plate is made of magnesium with $E = 45$ GPa and $\nu = 0.35$, and the forces are $P_x = 450$ kN, $P_y = 150$ kN, and $V = 110$ kN.

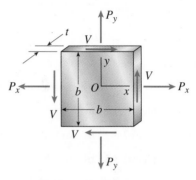

Probs. 7.5-10 and 7.5-11

7.5-11 Solve the preceding problem for an aluminum plate with $b = 10$ in., $t = 1$ in., $E = 10,600$ ksi, $\nu = 0.33$, $P_x = 120$ k, $P_y = 30$ k, and $V = 20$ k.

7.5-12 A circle of diameter $d = 200$ mm is etched on a brass plate (see figure). The plate has dimensions $400 \times 400 \times 20$ mm. Forces are applied to the plate, producing uniformly distributed normal stresses $\sigma_x = 42$ MPa and $\sigma_y = 14$ MPa. Calculate the following quantities: (a) the change in length Δac of diameter ac; (b) the change in length Δbd of diameter bd; (c) the change Δt in the thickness of the plate; (d) the change ΔV in the volume of the plate, and (e) the strain energy U stored in the plate. (Assume $E = 100$ GPa and $\nu = 0.34$.)

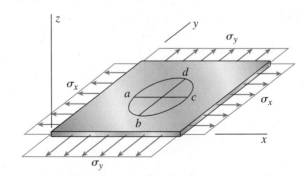

Prob. 7.5-12

*7.5-13 A cantilever beam of length L and rectangular cross section (width b and height h) supports a concentrated load P at the free end (see figure). (a) Derive a formula for the increase ΔV_t in the volume of the tension half of the beam when the load P is applied. (b) Derive a formula for the decrease ΔV_c in the volume of the compression half of the beam. (c) What is the net change in volume of the entire beam?

Prob. 7.5-13

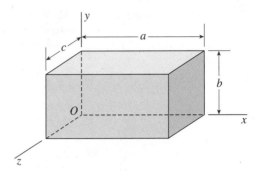

Probs. 7.6-1 and 7.6-2

7.6-2 Solve the preceding problem if the element is steel ($E = 200$ GPa, $v = 0.30$) with dimensions $a = 250$ mm, $b = 150$ mm, and $c = 150$ mm and the stresses are $\sigma_x = -50$ MPa, $\sigma_y = -40$ MPa, and $\sigma_z = -36$ MPa.

7.6-3 A cube of cast iron with sides of length $a = 3$ in. (see figure) is tested in a laboratory under triaxial stress. Gages mounted on the testing machine show that the compressive strains in the material are $\epsilon_x = -350 \times 10^{-6}$ and $\epsilon_y = \epsilon_z = -65 \times 10^{-6}$. Determine the following quantities: (a) the normal stresses σ_x, σ_y, and σ_z acting on the x, y, and z faces of the cube; (b) the maximum shear stress τ_{max} in the material; (c) the change ΔV in the volume of the cube; and (d) the strain energy U stored in the cube. (Assume $E = 14,000$ ksi and $v = 0.25$.)

TRIAXIAL STRESS

When solving the problems for Section 7.6, assume that the material is linearly elastic with modulus of elasticity E and Poisson's ratio v.

7.6-1 An element of aluminum in the form of a rectangular parallelepiped (see figure) of dimensions $a = 5$ in., $b = 4$ in., and $c = 3$ in. is subjected to triaxial stresses $\sigma_x = 11,000$ psi, $\sigma_y = -5,000$ psi, and $\sigma_z = -1,500$ psi acting on the x, y, and z faces, respectively. Determine the following quantities: (a) the maximum shear stress τ_{max} in the material; (b) the changes Δa, Δb, and Δc in the dimensions of the element; (c) the change ΔV in the volume; and (d) the strain energy U stored in the element. (Assume $E = 10,400$ ksi and $v = 0.33$.)

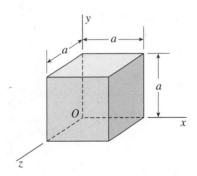

Probs. 7.6-3 and 7.6-4

7.6-4 Solve the preceding problem if the cube is granite ($E = 60$ GPa, $v = 0.25$) with dimensions $a = 50$ mm and compressive strains $\epsilon_x = -620 \times 10^{-6}$ and $\epsilon_y = \epsilon_z = -250 \times 10^{-6}$.

7.6-5 An element of aluminum in triaxial stress (see figure) is subjected to stresses $\sigma_x = 5200$ psi (tension), $\sigma_y = -4750$ psi (compression), and $\sigma_z = -3090$ psi (compression). It is also known that the normal strains in the x and y directions are $\epsilon_x = 713.8 \times 10^{-6}$ (elongation) and $\epsilon_y = -502.3 \times 10^{-6}$ (shortening). What is the bulk modulus K for the aluminum?

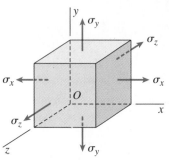

Probs. 7.6-5 and 7.6-6

7.6-6 Solve the preceding problem if the material is nylon subjected to compressive stresses $\sigma_x = -4.5$ MPa, $\sigma_y = -3.6$ MPa, and $\sigma_z = -2.1$ MPa, and the normal strains are $\epsilon_x = -740 \times 10^{-6}$ and $\epsilon_y = -320 \times 10^{-6}$ (shortenings).

7.6-7 A rubber cylinder R of length L and cross-sectional area A is compressed inside a steel cylinder S by a force F that applies a uniformly distributed pressure to the rubber (see figure). (a) Derive a formula for the lateral pressure p between the rubber and the steel. (Disregard friction between the rubber and the steel, and assume that the steel cylinder is rigid when compared to the rubber.) (b) Derive a formula for the shortening δ of the rubber cylinder.

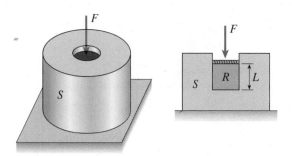

Prob. 7.6-7

7.6-8 A block R of rubber is confined between plane parallel walls of a steel block S (see figure). A uniformly distributed pressure p_0 is applied to the top of the rubber block by a force F. (a) Derive a formula for the lateral pressure p between the rubber and the steel. (Disregard friction between the rubber and the steel, and assume that the steel block is rigid when compared to the rubber.) (b) Derive a

formula for the dilatation e of the rubber. (c) Derive a formula for the strain-energy density u of the rubber.

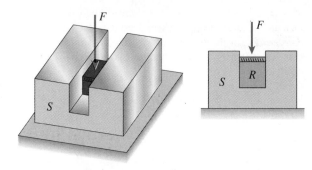

Prob. 7.6-8

7.6-9 A solid spherical ball of brass ($E = 15 \times 10^6$ psi, $\nu = 0.34$) is lowered into the ocean to a depth of 10,000 ft. The diameter of the ball is 11.0 in. Determine the decrease Δd in diameter, the decrease ΔV in volume, and the strain energy U of the ball.

7.6-10 A solid steel sphere ($E = 210$ GPa, $\nu = 0.3$) is subjected to hydrostatic pressure p such that its volume is reduced by 0.4%. (a) Calculate the pressure p. (b) Calculate the volume modulus of elasticity K for the steel. (c) Calculate the strain energy U stored in the sphere if its diameter is $d = 150$ mm.

7.6-11 A solid bronze sphere (volume modulus of elasticity $K = 14.5 \times 10^6$ psi) is suddenly heated around its outer surface. The tendency of the heated part of the sphere to expand produces uniform tension in all directions at the center of the sphere. If the stress at the center is 12,000 psi, what is the strain? Also, calculate the unit volume change e and the strain-energy density u at the center.

7.6-12 A cube of magnesium 100 mm on each side is lowered into the ocean to a depth such that the length of each side shortens by 0.018 mm. Assuming that $E = 45$ GPa and $\nu = 0.35$, calculate the following quantities: (a) the depth d to which the cube is lowered, and (b) the percent increase in the density of the magnesium.

PLANE STRAIN   **WORK SHEET 7**

When solving the problems for Section 7.7, consider only the in-plane strains (the strains in the xy plane) unless stated otherwise. Use the transformation equations of plane strain unless Mohr's circle is specified (Problems 7.7-23 through 7.7-28).

7.7-1 A thin rectangular plate in biaxial stress is subjected to stresses σ_x and σ_y, as shown in part (a) of the figure. The width and height of the plate are $b = 8.0$ in. and $h = 4.0$ in., respectively. Measurements show that the normal strains in the x and y directions are $\epsilon_x = 195 \times 10^{-6}$ and $\epsilon_y = -125 \times 10^{-6}$, respectively. With reference to part (b) of the figure, which shows a two-dimensional view of the plate, determine the following quantities: (a) the increase Δd in the length of diagonal Od; (b) the change $\Delta\phi$ in the angle ϕ between diagonal Od and the x axis; and (c) the change $\Delta\psi$ in the angle ψ between diagonal Od and the y axis.

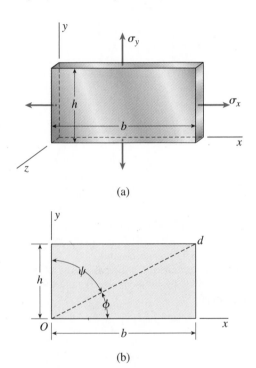

(a)

(b)

Probs. 7.7-1 and 7.7-2

7.7-2 Solve the preceding problem if $b = 160$ mm, $h = 60$ mm, $\epsilon_x = 410 \times 10^{-6}$, and $\epsilon_y = -320 \times 10^{-6}$.

7.7-3 A thin square plate in biaxial stress is subjected to stresses σ_x and σ_y, as shown in part (a) of the figure. The width of the plate is $b = 12.0$ in. Measurements show that the normal strains in the x and y directions are $\epsilon_x = 427 \times 10^{-6}$ and $\epsilon_y = 113 \times 10^{-6}$, respectively. With reference to part (b) of the figure, which shows a two-dimensional view of the plate, determine the following quantities: (a) the increase Δd in the length of diagonal Od; (b) the change $\Delta\phi$ in the angle ϕ between diagonal Od and the x axis; and

(c) the shear strain γ associated with diagonals Od and cf (that is, find the decrease in angle ced).

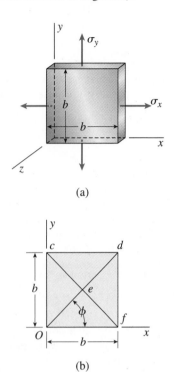

(a)

(b)

Probs. 7.7-3 and 7.7-4

7.7-4 Solve the preceding problem if $b = 225$ mm, $\epsilon_x = 845 \times 10^{-6}$, and $\epsilon_y = 211 \times 10^{-6}$.

7.7-5 An element of material subjected to plane strain (see figure) has strains as follows: $\epsilon_x = 220 \times 10^{-6}$, $\epsilon_y = 480 \times 10^{-6}$, and $\gamma_{xy} = 180 \times 10^{-6}$. Calculate the strains for an element oriented at an angle $\theta = 50°$ and show these strains on a sketch of a properly oriented element.

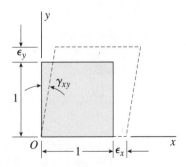

Probs. 7.7-5 through 7.7-10

7.7-6 Solve the preceding problem for the following data: $\epsilon_x = 420 \times 10^{-6}$, $\epsilon_y = -170 \times 10^{-6}$, $\gamma_{xy} = 310 \times 10^{-6}$, and $\theta = 37.5°$.

7.7-7 The strains for an element of material in plane strain (see figure) are as follows: $\epsilon_x = 480 \times 10^{-6}$, $\epsilon_y = 140 \times 10^{-6}$, and $\gamma_{xy} = -350 \times 10^{-6}$. Determine the principal strains and maximum shear strains, and show these strains on sketches of properly oriented elements.

7.7-8 Solve the preceding problem for the following strains: $\epsilon_x = 120 \times 10^{-6}$, $\epsilon_y = -450 \times 10^{-6}$, and $\gamma_{xy} = -360 \times 10^{-6}$.

7.7-9 An element of material in plane strain (see figure) is subjected to strains $\epsilon_x = 480 \times 10^{-6}$, $\epsilon_y = 70 \times 10^{-6}$, and $\gamma_{xy} = 420 \times 10^{-6}$. Determine the following quantities: (a) the strains for an element oriented at an angle $\theta = 75°$, (b) the principal strains, and (c) the maximum shear strains. Show the results on sketches of properly oriented elements.

7.7-10 Solve the preceding problem for the following data: $\epsilon_x = -1120 \times 10^{-6}$, $\epsilon_y = -430 \times 10^{-6}$, $\gamma_{xy} = 780 \times 10^{-6}$, and $\theta = 45°$.

7.7-11 A steel plate with modulus of elasticity $E = 30 \times 10^6$ psi and Poisson's ratio $\nu = 0.30$ is loaded in biaxial stress by normal stresses σ_x and σ_y (see figure). A strain gage is bonded to the plate at an angle $\phi = 30°$. If the stress σ_x is 18,000 psi and the strain measured by the gage is $\epsilon = 407 \times 10^{-6}$, what is the maximum in-plane shear stress $(\tau_{max})_{xy}$ and shear strain $(\gamma_{max})_{xy}$? What is the maximum shear strain $(\gamma_{max})_{xz}$ in the xz plane? What is the maximum shear strain $(\gamma_{max})_{yz}$ in the yz plane?

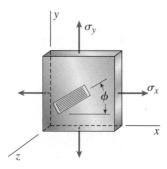

Probs. 7.7-11 and 7.7-12

7.7-12 Solve the preceding problem if the plate is made of aluminum with $E = 72$ GPa and $\nu = 1/3$, the stress σ_x is 86.4 MPa, the angle ϕ is 21°, and the strain ϵ is 946×10^{-6}.

7.7-13 An element in plane stress is subjected to stresses $\sigma_x = -8400$ psi, $\sigma_y = 1100$ psi, and $\tau_{xy} = -1700$ psi (see figure). The material is aluminum with modulus of elasticity $E = 10,000$ ksi and Poisson's ratio $\nu = 0.33$. Determine the following quantities: (a) the strains for an element oriented at an angle $\theta = 30°$, (b) the principal strains, and (c) the maximum shear strains. Show the results on sketches of properly oriented elements.

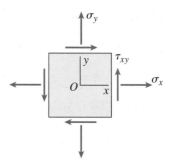

Probs. 7.7-13 and 7.7-14

7.7-14 Solve the preceding problem for the following data: $\sigma_x = -150$ MPa, $\sigma_y = -210$ MPa, $\tau_{xy} = -16$ MPa, and $\theta = 50°$. The material is brass with $E = 100$ GPa and $\nu = 0.34$.

7.7-15 During a test of an airplane wing, the strain gage readings from a 45° rosette (see figure) are as follows: gage A, 520×10^{-6}; gage B, 360×10^{-6}; and gage C, -80×10^{-6}. Determine the principal strains and maximum shear strains, and show them on sketches of properly oriented elements.

Probs. 7.7-15 and 7.7-16

7.7-16 A 45° strain rosette (see figure) mounted on the surface of an automobile frame gives the following readings: gage A, 310×10^{-6}; gage B, 180×10^{-6}; and gage C, -160×10^{-6}. Determine the principal strains and maximum shear strains, and show them on sketches of properly oriented elements.

7.7-17 A solid circular bar of diameter $d = 1.5$ in. is subjected to an axial force P and a torque T (see figure). Strain gages A and B mounted on the surface of the bar give readings $\epsilon_a = 100 \times 10^{-6}$ and $\epsilon_b = -55 \times 10^{-6}$. The bar is made of steel having $E = 30 \times 10^6$ psi and $\nu = 0.29$. (a) Determine the axial force P and the torque T. (b) Determine the maximum shear strain γ_{max} and the maximum shear stress τ_{max} in the bar.

Prob. 7.7-17

7.7-18 A cantilever beam of rectangular cross section (width $b = 25$ mm, height $h = 100$ mm) is loaded by a force P that acts at the midheight of the beam and is inclined at an angle α to the vertical (see figure). Two strain gages are placed at point C, which also is at the midheight of the beam. Gage A measures the strain in the horizontal direction and gage B measures the strain at an angle $\beta = 60°$ to the horizontal. The measured strains are $\epsilon_a = 125 \times 10^{-6}$ and $\epsilon_b = -375 \times 10^{-6}$. Determine the force P and the angle α, assuming the material is steel with $E = 200$ GPa and $\nu = 1/3$.

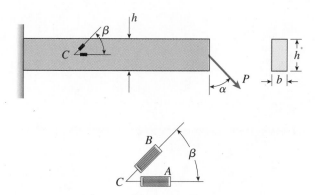

Probs. 7.7-18 and 7.7-19

7.7-19 Solve the preceding problem if the cross-sectional dimensions are $b = 1.0$ in. and $h = 3.0$ in., the gage angle is $\beta = 75°$, the measured strains are $\epsilon_a = 171 \times 10^{-6}$ and $\epsilon_b = -266 \times 10^{-6}$, and the material is a magnesium alloy with modulus $E = 6.0 \times 10^6$ psi and Poisson's ratio $\nu = 0.35$.

7.7-20 A $60°$ strain rosette, or *delta rosette*, consists of three electrical-resistance strain gages arranged as shown in the figure. Gage A measures the normal strain ϵ_a in the direction of the x axis. Gages B and C measure the strains ϵ_b and ϵ_c in the inclined directions shown. Obtain the equations for the strains ϵ_x, ϵ_y, and γ_{xy} associated with the xy axes.

Prob. 7.7-20

7.7-21 On the surface of a structural component in a space vehicle, the strains are monitored by means of three strain gages arranged as shown in the figure. During a certain maneuver, the following strains were recorded: $\epsilon_a = 1100 \times 10^{-6}$, $\epsilon_b = 200 \times 10^{-6}$, and $\epsilon_c = 200 \times 10^{-6}$. Determine the principal strains and principal stresses in the material, which is a magnesium alloy for which $E = 6000$ ksi and $\nu = 0.35$. (Show the principal strains and principal stresses on sketches of properly oriented elements.)

Prob. 7.7-21

7.7-22 The strains on the surface of an experimental device made of pure aluminum ($E = 70$ GPa, $\nu = 0.33$) and tested in a space shuttle were measured by means of strain gages. The gages were oriented as shown in the figure, and the measured strains were $\epsilon_a = 1100 \times 10^{-6}$, $\epsilon_b = 1496 \times 10^{-6}$, and $\epsilon_c = -39.44 \times 10^{-6}$. What is the stress σ_x in the x direction?

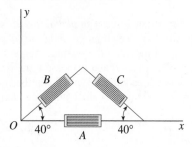

Prob. 7.7-22

7.7-23 Solve Problem 7.7-5 by using Mohr's circle for plane strain.

7.7-24 Solve Problem 7.7-6 by using Mohr's circle for plane strain.

7.7-25 Solve Problem 7.7-7 by using Mohr's circle for plane strain.

7.7-26 Solve Problem 7.7-8 by using Mohr's circle for plane strain.

7.7-27 Solve Problem 7.7-9 by using Mohr's circle for plane strain.

7.7-28 Solve Problem 7.7-10 by using Mohr's circle for plane strain.

8

APPLICATIONS OF PLANE STRESS
(PRESSURE VESSELS, BEAMS, AND COMBINED LOADINGS)

8.1 INTRODUCTION

In the preceding chapter we analyzed the stresses and strains at a point in a structure subjected to *plane stress* (see Sections 7.2 through 7.5). Plane stress is a common stress condition that exists in all ordinary structures, including buildings, machines, vehicles, and aircraft. In this chapter we examine some practical applications involving plane stress, and therefore the material presented in Sections 7.2 through 7.5 should be thoroughly understood before proceeding.

The first structures to be discussed are *pressure vessels*, such as compressed-air tanks and water pipes (Sections 8.2 and 8.3). We will determine the stresses and strains in the walls of these structures due to the internal pressures from the compressed gases or liquids. Next, we return to the subject of *stresses in beams* and show how to determine the principal stresses and maximum shear stresses at various points (Section 8.4). Finally, in Section 8.5, we analyze structures subjected to *combined loadings*, that is, combinations of axial loads, torsional loads, bending loads, and internal pressure. Our objective is to determine the maximum normal and shear stresses at various points in these structures.

8.2 SPHERICAL PRESSURE VESSELS

Pressure vessels are closed structures containing liquids or gases under pressure. Familiar examples include tanks, pipes, pressurized cabins in aircraft, and space vehicles. When pressure vessels have walls that are

Fig. 8-1 Spherical pressure vessel

Fig. 8-2 Cross section of spherical pressure vessel showing inner radius r, wall thickness t, and internal pressure p

thin in comparison to their radii and lengths, they are classified as **shell structures**. Examples of shell structures are roof domes, boilers, airplane wings, and submarine hulls.

In this section we consider thin-walled pressure vessels of spherical shape, like the compressed-air tank shown in Fig. 8-1. The term *thin-walled* is not precise, but as a general rule, pressure vessels are considered to be thin-walled when the ratio of radius r to wall thickness t (Fig. 8-2) is greater than 10. When this condition is met, we can determine the stresses in the walls with reasonable accuracy using statics alone.

We assume in the following discussions that the internal pressure p (Fig. 8-2) exceeds the pressure acting on the outside of the shell. Otherwise, the vessel may collapse inward due to buckling.

A sphere is the theoretically ideal shape for a vessel that resists internal pressure. We only need to contemplate the familiar soap bubble to recognize that a sphere is the "natural" shape for this purpose. To determine the stresses in a spherical vessel, let us cut through the sphere on a vertical diametral plane (Fig. 8-3a) and isolate half of the shell *and its fluid contents* as a single free body (Fig. 8-3b). Acting on this free body are the tensile stresses σ in the wall of the vessel and the fluid pressure p. The pressure acts horizontally against the plane circular area of fluid remaining inside the hemisphere. Since the pressure is uniform, the resultant pressure force is

$$P = p(\pi r^2) \tag{a}$$

where r is the inner radius of the sphere. Note that the pressure p is not the absolute pressure inside the vessel but is the net internal pressure, or the *gage pressure*. Gage pressure is the internal pressure *above* the pressure acting on the outside of the vessel. If the internal and external pressures are the same, no stresses are developed in the wall of the vessel—only the excess of internal pressure over external pressure has any effect on these stresses.

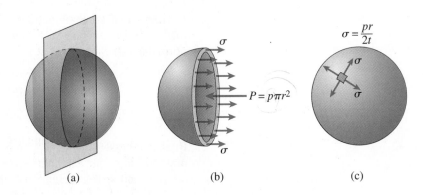

Fig. 8-3 Tensile stresses σ in the wall of a spherical pressure vessel

Because of the symmetry of the vessel and its loading (Fig. 8-3b), the tensile stress σ is uniform around the circumference. Furthermore, since the wall is thin, we can assume with good accuracy that the stress is uniformly distributed across the thickness t. The accuracy of this approximation increases as the shell becomes thinner and decreases as it becomes thicker.

The resultant of the tensile stresses σ in the wall is a horizontal force equal to the stress σ times the area over which it acts, or

$$\sigma(2\pi r_m t)$$

where t is the thickness of the wall and r_m is its mean radius:

$$r_m = r + \frac{t}{2} \tag{b}$$

Consequently, equilibrium of forces in the horizontal direction (Fig. 8-3b) gives

$$\Sigma F_{\text{horiz}} = 0 \qquad \sigma(2\pi r_m t) - p(\pi r^2) = 0 \tag{c}$$

from which we obtain the tensile stresses in the wall of the vessel:

$$\sigma = \frac{pr^2}{2r_m t} \tag{d}$$

Since our analysis is valid only for thin shells, we can disregard the small difference between the two radii and replace r by r_m or replace r_m by r. While either choice is satisfactory from the standpoint of accuracy, it turns out that the stresses are closer to the theoretically exact stresses if we use the inner radius r instead of the mean radius r_m. Therefore, we adopt the following formula for calculating the tensile stresses in the wall of a spherical shell:

$$\sigma = \frac{pr}{2t} \tag{8-1}$$

As is evident from the symmetry of a spherical shell, we obtain the same equation for the tensile stresses when we cut through the center of the sphere in any direction whatsoever. Thus, we reach the following conclusion: *A pressurized sphere is subjected to uniform tensile stresses σ in all directions.* This stress condition is represented in Fig. 8-3c by the small stress element with stresses σ acting in mutually perpendicular directions.

Stresses that act tangentially to the curved surface of a shell, such as the stresses σ shown in Fig. 8-3c, are known as **membrane stresses.** The name arises from the fact that these are the only stresses that exist in true membranes, such as soap films and thin sheets of rubber.

■ Stresses at the Outer Surface

The outer surface of a spherical pressure vessel is usually free of any loads. Therefore, the element shown in Fig. 8-3c is in *biaxial stress*. To aid in analyzing the stresses acting on this element, we show it again in Fig. 8-4a, where a set of coordinate axes is oriented parallel to the sides of the element. The x and y axes are tangential to the surface of the sphere, and the z axis is perpendicular to the surface. Thus, the normal stresses σ_x and σ_y are the same as the membrane stresses σ, and the normal stress σ_z is zero. No shear stresses act on the sides of this element.

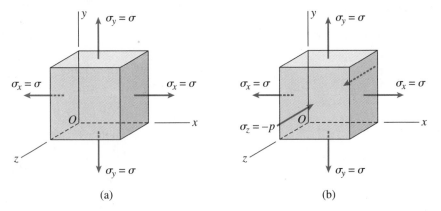

Fig. 8-4 Stresses in a spherical pressure vessel at: (a) the outer surface, and (b) the inner surface

(a)   (b)

If we analyze the element of Fig. 8-4a by using the transformation equations for plane stress (see Fig. 7-1 and Eqs. 7-4a and 7-4b of Section 7.2), we find

$$\sigma_{x_1} = \sigma \quad \text{and} \quad \tau_{x_1 y_1} = 0$$

as expected. In other words, when we consider elements obtained by rotating the axes about the z axis, the normal stresses remain constant and there are no shear stresses. Every plane is a principal plane and every direction is a principal direction. Thus, the principal stresses for the element are

$$\sigma_1 = \sigma_2 = \frac{pr}{2t} \qquad \sigma_3 = 0 \tag{8-2a,b}$$

The stresses σ_1 and σ_2 lie in the xy plane and the stress σ_3 pertains to the z direction.

To obtain the maximum shear stresses, we must consider out-of-plane rotations, that is, rotations about the x and y axes (because all in-plane shear stresses are zero). Elements oriented by making 45° rotations about the x and y axes have maximum shear stresses equal to $\sigma/2$ and normal stresses equal to $\sigma/2$. Therefore,

$$\tau_{max} = \frac{\sigma}{2} = \frac{pr}{4t} \tag{8-3}$$

These stresses are the largest shear stresses in the element.

■ **Stresses at the Inner Surface**

At the inner surface of the wall of a spherical vessel, a stress element (Fig. 8-4b) has the same membrane stresses σ_x and σ_y as does an element at the outer surface (Fig. 8-4a). In addition, a compressive stress σ_z equal to the pressure p acts in the z direction (Fig. 8-4b). This compressive stress decreases from p at the inner surface to zero at the outer surface.

The element shown in Fig. 8-4b is in triaxial stress with principal stresses

$$\sigma_1 = \sigma_2 = \frac{pr}{2t} \qquad \sigma_3 = -p \tag{e,f}$$

The in-plane shear stresses are zero, but the maximum out-of-plane shear stress (obtained by a 45° rotation about either the x or y axis) is

$$\tau_{max} = \frac{\sigma + p}{2} = \frac{pr}{4t} + \frac{p}{2} = \frac{p}{2}\left(\frac{r}{2t} + 1\right) \tag{g}$$

When the vessel is thin-walled and the ratio r/t is large, we can disregard the number 1 in comparison with the term $r/2t$. In other words, we can disregard the existence of the principal stress σ_3 in the z direction because it is small when compared with the principal stresses σ_1 and σ_2. Consequently, we can consider the stress state at the inner surface to be the same as at the outer surface (biaxial stress). This approximation is consistent with the approximate nature of thin-shell theory, and therefore we will use Eqs. (8-1), (8-2), and (8-3) to obtain the stresses anywhere in the wall of a spherical pressure vessel, including at the inner surface.

■ **General Comments**

Pressure vessels usually have openings in the wall (to serve as inlets and outlets for the fluid contents) as well as fittings and supports that exert forces on the shell (Fig. 8-1). These features result in nonuniformities in the stress distribution, or *stress concentrations*, that cannot be analyzed by the elementary formulas given here. Instead, more advanced methods of analysis are needed. Other factors that affect the design of pressure vessels include corrosion, accidental impacts, and temperature changes.

Some of the limitations of thin-shell theory as applied to pressure vessels are listed here:

1. The wall thickness must be small in comparison to the other dimensions (the ratio r/t should be 10 or more).

2. The internal pressure must exceed the external pressure (to avoid inward buckling).
3. The analysis presented in this section is based only on the effects of internal pressure (the effects of external loads, reactions, and the weight of the structure are not considered).
4. The formulas derived in this section are valid throughout the wall of the vessel *except* near points of stress concentrations.

The following example illustrates how the principal stresses and maximum shear stresses are used in the analysis of a spherical shell.

Fig. 8-5 Example 8-1. Spherical pressure vessel

■ Example 8-1

A compressed-air tank having an inner diameter of 18 inches and a wall thickness of 1/4 inch is formed by welding two steel hemispheres (Fig. 8-5).

a. If the allowable tensile stress in the steel is 14,000 psi, what is the maximum permissible air pressure p_a in the tank?
b. If the allowable shear stress in the steel is 6,000 psi, what is the maximum permissible pressure p_b?
c. If the normal strain at the outer surface of the tank is not to exceed 0.0003, what is the maximum permissible pressure p_c? (Assume that Hooke's law is valid and that the modulus of elasticity for the steel is 29×10^6 psi and Poisson's ratio is 0.28.)
d. Tests on the welded seam show that failure occurs when the tensile load on the welds exceeds 8.1 kips per inch of weld. If the required factor of safety against failure of the weld is 2.5, what is the maximum permissible pressure p_d?
e. Considering the four preceding factors, what is the allowable pressure p_{allow} in the tank?

Solution   (a) *Allowable pressure based upon the tensile stress in the steel.* The maximum tensile stress in the wall of the tank is given by the formula $\sigma = pr/2t$ (see Eq. 8-1). Solving this equation for the pressure in terms of the allowable stress, we get

$$p_a = \frac{2t\sigma_{\text{allow}}}{r} = \frac{2(0.25 \text{ in.})(14{,}000 \text{ psi})}{9.0 \text{ in.}} = 777.8 \text{ psi} \quad \longleftarrow$$

Thus, the maximum allowable pressure based upon tension in the wall of the tank is $p_a = 777$ psi. (Note that in a calculation of this kind, we round downward, not upward.)

(b) *Allowable pressure based upon the shear stress in the steel.* The maximum shear stress in the wall of the tank is given by Eq. (8-3), from which we get the following equation for the pressure:

$$p_b = \frac{4t\tau_{\text{allow}}}{r} = \frac{4(0.25 \text{ in.})(6{,}000 \text{ psi})}{9.0 \text{ in.}} = 666.7 \text{ psi} \quad \longleftarrow$$

Therefore, the allowable pressure based upon shear is $p_b = 666$ psi.

(c) *Allowable pressure based upon the normal strain in the steel.* The normal strain is obtained from Hooke's law for biaxial stress (Eq. 7-39a):

$$\epsilon_x = \frac{1}{E}(\sigma_x - \nu\sigma_y) \tag{h}$$

Substituting $\sigma_x = \sigma_y = \sigma = pr/2t$ (see Fig. 8-4a), we obtain

$$\epsilon_x = \frac{\sigma}{E}(1 - \nu) = \frac{pr}{2tE}(1 - \nu) \tag{8-4}$$

This equation can be solved for the pressure p_c:

$$p_c = \frac{2tE\epsilon_{\text{allow}}}{r(1 - \nu)} = \frac{2(0.25 \text{ in.})(29 \times 10^6 \text{ psi})(0.0003)}{(9.0 \text{ in.})(1 - 0.28)} = 671.3 \text{ psi} \quad \longleftarrow$$

Thus, the allowable pressure based upon the normal strain in the wall is $p_c = 671$ psi.

(d) *Allowable pressure based upon the tension in the welded seam.* The allowable tensile load on the welded seam is equal to the failure load divided by the factor of safety:

$$T_{\text{allow}} = \frac{T_{\text{failure}}}{n} = \frac{8.1 \text{ k/in.}}{2.5} = 3.24 \text{ k/in.} = 3240 \text{ lb/in.}$$

The corresponding allowable tensile stress is equal to the allowable load on a one-inch length of weld divided by the cross-sectional area of a one-inch length of weld:

$$\sigma_{\text{allow}} = \frac{T_{\text{allow}}(1.0 \text{ in.})}{(1.0 \text{ in.})(t)} = \frac{(3240 \text{ lb/in.})(1.0 \text{ in})}{(1.0 \text{ in.})(0.25 \text{ in.})} = 12,960 \text{ psi}$$

Finally, we solve for the internal pressure by using Eq. (8-1):

$$p_d = \frac{2t\sigma_{\text{allow}}}{r} = \frac{2(0.25 \text{ in.})(12,960 \text{ psi})}{9.0 \text{ in.}} = 720.0 \text{ psi} \quad \longleftarrow$$

This result gives the allowable pressure based upon tension in the welded seam.

(e) *Allowable pressure.* Comparing the preceding results for p_a, p_b, p_c, and p_d, we see that shear stress in the wall governs and the allowable pressure in the tank is

$$p_{\text{allow}} = 666 \text{ psi} \quad \longleftarrow$$

This example illustrates how various stresses and strains enter into the design of a spherical pressure vessel.

Note: When the internal pressure is at its maximum allowable value (666 psi), the tensile stresses in the shell are

$$\sigma = \frac{pr}{2t} = \frac{(666 \text{ psi})(9.0 \text{ in.})}{2(0.25 \text{ in.})} = 12,000 \text{ psi}$$

Thus, at the inner surface of the shell (Fig. 8-4b), the ratio of the principal stress in the z direction (666 psi) to the in-plane principal stresses (12,000 psi) is only 0.056. Therefore, our assumption that we can disregard the principal stress σ_3 in the z direction and consider the entire shell to be in biaxial stress is justified.

(a)

(b)

Fig. 8-6 Cylindrical pressure vessels with circular cross sections

8.3 CYLINDRICAL PRESSURE VESSELS

Cylindrical pressure vessels with a circular cross section (Fig. 8-6) are found in industrial settings (compressed air tanks and rocket motors), in homes (fire extinguishers and spray cans), and in the countryside (propane tanks and grain silos). Pressurized pipes, such as water-supply pipes and penstocks, are also classified as cylindrical pressure vessels.

We begin our analysis of cylindrical vessels by determining the normal stresses in a thin-walled circular tank AB subjected to internal pressure (Fig. 8-7a). A stress element with its faces parallel and perpendicular to the axis of the tank is shown on the wall of the tank. The normal stresses σ_1 and σ_2 acting on the side faces of this element are the membrane stresses in the wall. No shear stresses act on these faces because of the symmetry of the vessel and its loading. Therefore, the stresses σ_1 and σ_2 are principal stresses. Because of their directions, the stress σ_1 is called the **circumferential stress** or the **hoop stress**, and the stress σ_2 is called the **longitudinal stress** or the **axial stress**. Each of these stresses can be calculated from equilibrium by using appropriate free-body diagrams.

To determine the circumferential stress σ_1, we make two cuts (*mn* and *pq*) perpendicular to the longitudinal axis and distance b apart (Fig. 8-7a). Then we make a third cut in a vertical plane through the longitudinal axis of the tank, resulting in the free body shown in Fig. 8-7b. This free body consists not only of the half-circular piece of the tank but also of the fluid contained within the cuts. Acting on the longitudinal cut (plane *mpqn*) are the circumferential stresses σ_1 and the internal pressure p. Stresses and pressures also act on the left-hand and right-hand faces of the free body. However, these stresses and pressures are not shown in the figure because they do not enter the equation of equilibrium that we will use. As in our analysis of a spherical vessel, we will disregard the weight of the tank and its contents.

The circumferential stresses σ_1 acting in the wall of the vessel have a resultant equal to $\sigma_1(2bt)$, where t is the thickness of the wall. Also, the resultant force P_1 of the internal pressure is equal to $2pbr$, where r is the inner radius of the cylinder. Hence, we have the following equation of equilibrium:

$$\sigma_1(2bt) - 2pbr = 0$$

From this equation we obtain the following formula for the *circumferential stress* in the cylinder:

$$\sigma_1 = \frac{pr}{t} \tag{8-5}$$

(a)

(b)

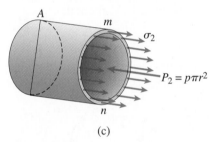

(c)

Fig. 8-7 Stresses in a circular cylindrical pressure vessel

This stress is uniformly distributed over the thickness of the wall, provided the thickness is small compared to the radius.

The longitudinal stress σ_2 is obtained from the equilibrium of a free body of the part of the vessel to the left of cross section mn (Fig. 8-7c). Again, the free body includes not only part of the tank but also its contents. The stresses σ_2 act longitudinally and have a resultant force equal to $\sigma_2(2\pi rt)$. Note that we are using the inner radius of the shell in place of the mean radius, as explained in Section 8.2.

The resultant force P_2 of the internal pressure is a force equal to $p\pi r^2$. Thus, the equation of equilibrium for the free body is

$$\sigma_2(2\pi rt) - p\pi r^2 = 0$$

Solving this equation for σ_2, we obtain the following formula for the *longitudinal stress* in a cylindrical pressure vessel:

$$\sigma_2 = \frac{pr}{2t} \tag{8-6}$$

This stress is equal to the membrane stress in a spherical vessel (Eq. 8-1).

Comparing Eqs. (8-5) and (8-6), we see that the circumferential stress in a cylindrical vessel is equal to twice the longitudinal stress:

$$\sigma_1 = 2\sigma_2 \tag{8-7}$$

From this result we note that a longitudinal welded seam in a pressurized tank must be twice as strong as a circumferential seam.

■ **Stresses at the Outer Surface**

The principal stresses σ_1 and σ_2 at the outer surface of a cylindrical vessel are shown on the stress element of Fig. 8-8a. Since the third principal stress (acting in the z direction) is zero, the element is in *biaxial stress*.

The maximum *in-plane shear stresses* occur on planes that are rotated 45° about the z axis; these stresses are

$$(\tau_{\max})_z = \frac{\sigma_1 - \sigma_2}{2} = \frac{\sigma_1}{4} = \frac{pr}{4t} \tag{8-8}$$

The maximum *out-of-plane shear stresses* are obtained by 45° rotations about the x and y axes, respectively; thus,

$$(\tau_{\max})_x = \frac{\sigma_1}{2} = \frac{pr}{2t} \qquad (\tau_{\max})_y = \frac{\sigma_2}{2} = \frac{pr}{4t} \tag{8-9a,b}$$

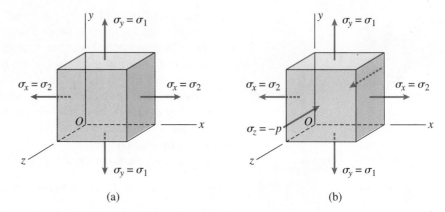

Fig. 8-8 Stresses in a circular cylindrical pressure vessel at: (a) the outer surface, and (b) the inner surface

(a)   (b)

Comparing the preceding results, we see that the *absolute maximum shear stress* is

$$\tau_{\text{max}} = \frac{\sigma_1}{2} = \frac{pr}{2t} \tag{8-10}$$

This stress occurs on a plane that has been rotated 45° about the *x* axis.

■ **Stresses at the Inner Surface**

The stress conditions at the inner surface of the wall of the vessel are shown in Fig. 8-8b. The principal stresses are

$$\sigma_1 = \frac{pr}{t} \qquad \sigma_2 = \frac{pr}{2t} \qquad \sigma_3 = -p \tag{a,b,c}$$

The three maximum shear stresses, obtained by 45° rotations about the *x*, *y*, and *z* axes, are

$$(\tau_{\text{max}})_x = \frac{\sigma_1 - \sigma_3}{2} = \frac{pr}{2t} + \frac{p}{2} \qquad (\tau_{\text{max}})_y = \frac{\sigma_2 - \sigma_3}{2} = \frac{pr}{4t} + \frac{p}{2} \tag{d,e}$$

$$(\tau_{\text{max}})_z = \frac{\sigma_1 - \sigma_2}{2} = \frac{pr}{4t} \tag{f}$$

The first of these three stresses is the largest. However, as explained in the discussion of shear stresses in a spherical shell, we may disregard the additional term $p/2$ in Eqs. (d) and (e) when the shell is thin-walled. Equations (d), (e), and (f) then become the same as Eqs. (8-9) and (8-8). Therefore, in all of our examples and problems pertaining to cylindrical pressure vessels, we will disregard the presence of the compressive stress in the *z* direction. (This compressive stress varies from *p* at the inner surface to zero at the outer surface.) With this approximation, the stresses at the inner surface become the same as the stresses at the outer surface (biaxial stress). As explained in the discussion of spherical pressure vessels, this procedure is satisfactory when we consider the numerous other approximations in our theory.

■ General Comments

The preceding formulas for stresses in a circular cylinder are valid in parts of the cylinder away from any discontinuities that cause stress concentrations, as discussed previously for spherical shells. An obvious discontinuity exists at the ends of the cylinder where the heads are attached, because the geometry of the structure changes abruptly. Other stress concentrations occur at openings, at points of support, and wherever objects or fittings are attached to the cylinder. The stresses at such points cannot be determined solely from equilibrium equations; instead, more advanced methods of analysis (such as shell theory and finite-element analysis) must be used.

Some of the limitations of the elementary theory for thin-walled shells are listed in Section 8.2.

Fig. 8-9 Example 8-2. Cylindrical pressure vessel with a helical weld

■ Example 8-2

A cylindrical pressure vessel is constructed from a long, narrow steel plate by wrapping the plate around a mandrel and then welding along the edges of the plate to make a helical joint (Fig. 8-9). The helical weld makes an angle $\alpha = 55°$ with the longitudinal axis. The vessel has inner radius $r = 1.8$ m and wall thickness $t = 20$ mm. The material is steel with modulus $E = 200$ GPa and Poisson's ratio $\nu = 0.30$. The internal pressure p is 800 kPa.

Calculate the following quantities for the cylindrical part of the vessel: (a) the circumferential and longitudinal stresses σ_1 and σ_2, respectively; (b) the maximum in-plane and out-of-plane shear stresses; (c) the circumferential and longitudinal strains ϵ_1 and ϵ_2, respectively; and (d) the normal stress σ_w and shear stress τ_w acting perpendicular and parallel, respectively, to the welded seam.

Solution (a) *Circumferential and longitudinal stresses.* The circumferential and longitudinal stresses σ_1 and σ_2, respectively, are pictured in Fig. 8-10a, where they are shown acting on a stress element at point A on the wall of the vessel. The magnitudes of the stresses can be calculated from Eqs. (8-5) and (8-6):

$$\sigma_1 = \frac{pr}{t} = \frac{(800 \text{ kPa})(1.8 \text{ m})}{20 \text{ mm}} = 72 \text{ MPa} \qquad \sigma_2 = \frac{pr}{2t} = \frac{\sigma_1}{2} = 36 \text{ MPa} \quad \Longleftarrow$$

The stress element at point A is shown again in Fig. 8-10b, where the x axis is in the longitudinal direction of the cylinder and the y axis is in the circumferential direction. Since there is no stress in the z direction ($\epsilon_3 = 0$), the element is in biaxial stress.

Note that the ratio of the internal pressure (800 kPa) to the smaller in-plane principal stress (36 MPa) is 0.022. Therefore, our assumption that we may disregard any stresses in the z direction and consider all elements in the cylindrical shell, even those at the inner surface, to be in biaxial stress is justified.

(b) *Maximum shear stresses.* The largest in-plane shear stress is obtained from Eq. (8-8):

$$(\tau_{max})_z = \frac{\sigma_1 - \sigma_2}{2} = \frac{\sigma_1}{4} = \frac{pr}{4t} = 18 \text{ MPa} \quad \Longleftarrow$$

(a)

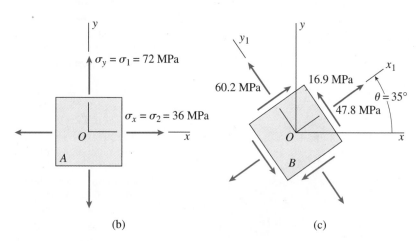

Fig. 8-10 Solution to Example 8-2

(b)   (c)

Because we are disregarding the normal stress in the z direction, the largest out-of-plane shear stress is obtained from Eq. (8-9a):

$$\tau_{max} = \frac{\sigma_1}{2} = \frac{pr}{2t} = 36 \text{ MPa} \qquad \longleftarrow$$

This last stress is the absolute maximum shear stress in the wall of the vessel.

(c) *Circumferential and longitudinal strains.* Since the largest stresses are well below the yield stress of steel (see Table H-3, Appendix H), we may assume that Hooke's law applies to the wall of the vessel. Then we can obtain the strains in the x and y directions (Fig. 8-10b) from Eqs. (7-39a) and (7-39b) for biaxial stress:

$$\epsilon_x = \frac{1}{E}(\sigma_x - \nu\sigma_y) \qquad \epsilon_y = \frac{1}{E}(\sigma_y - \nu\sigma_x) \qquad \text{(g,h)}$$

We note that the strain ϵ_x is the same as the principal strain ϵ_2 in the longitudinal direction and that the strain ϵ_y is the same as the principal strain ϵ_1 in the circumferential direction. Also, the stress σ_x is the same as the stress σ_2, and the stress σ_y is the same as the stress σ_1. Therefore, the preceding two equations can be written in the following forms:

$$\epsilon_2 = \frac{\sigma_2}{E}(1 - 2\nu) = \frac{pr}{2tE}(1 - 2\nu) \qquad \text{(8-11a)}$$

$$\epsilon_1 = \frac{\sigma_1}{2E}(2 - \nu) = \frac{pr}{2tE}(2 - 2\nu) \qquad \text{(8-11b)}$$

Substituting numerical values, we find

$$\epsilon_2 = \frac{\sigma_2}{E}(1 - 2\nu) = \frac{(36 \text{ MPa})[1 - 2(0.30)]}{200 \text{ GPa}} = 72 \times 10^{-6} \qquad \longleftarrow$$

$$\epsilon_1 = \frac{\sigma_1}{2E}(2 - \nu) = \frac{(72 \text{ MPa})(2 - 0.30)}{2(200 \text{ GPa})} = 306 \times 10^{-6} \qquad \longleftarrow$$

These are the longitudinal and circumferential strains, respectively, in the cylinder.

(d) *Normal and shear stresses acting on the welded seam.* The stress element at point B in the wall of the cylinder (Fig. 8-10a) is oriented so that its sides are parallel and perpendicular to the weld. The angle θ for the element is

$$\theta = 90° - \alpha = 35°$$

as shown in Fig. 8-10c. Either the stress-transformation equations or Mohr's circle may be used to obtain the normal and shear stresses acting on the side faces of this element.

Stress-transformation equations. The normal stress σ_{x_1} and the shear stress $\tau_{x_1y_1}$ acting on the x_1 face of the element (Fig. 8-10c) are obtained from Eqs. (7-4a) and (7-4b), which are repeated here:

$$\sigma_{x_1} = \frac{\sigma_x + \sigma_y}{2} + \frac{\sigma_x - \sigma_y}{2}\cos 2\theta + \tau_{xy}\sin 2\theta \qquad \text{(8-12a)}$$

$$\tau_{x_1y_1} = -\frac{\sigma_x - \sigma_y}{2}\sin 2\theta + \tau_{xy}\cos 2\theta \qquad \text{(8-12b)}$$

Substituting $\sigma_x = \sigma_2 = pr/2t$, $\sigma_y = \sigma_1 = pr/t$, and $\tau_{xy} = 0$, we obtain

$$\sigma_{x_1} = \frac{pr}{4t}(3 - \cos 2\theta) \qquad \tau_{x_1y_1} = \frac{pr}{4t}\sin 2\theta \qquad \text{(8-13a,b)}$$

These equations give the normal and shear stresses acting on an inclined plane oriented at an angle θ with the longitudinal axis of the cylinder.

Substituting $pr/4t = 18$ MPa and $\theta = 35°$ into Eqs. (8-13a) and (8-13b), we obtain

$$\sigma_{x_1} = 47.8 \text{ MPa} \qquad \tau_{x_1y_1} = 16.9 \text{ MPa}$$

These stresses are shown on the stress element of Fig. 8-10c.

To complete the stress element, we can calculate the normal stress σ_{y_1} acting on the y_1 face of the element from the sum of the normal stresses on perpendicular faces (Eq. 7-6):

$$\sigma_1 + \sigma_2 = \sigma_{x_1} + \sigma_{y_1} \qquad \text{(8-14)}$$

Substituting numerical values, we get

$$\sigma_{y_1} = \sigma_1 + \sigma_2 - \sigma_{x_1} = 72 \text{ MPa} + 36 \text{ MPa} - 47.8 \text{ MPa} = 60.2 \text{ MPa}$$

as shown in Fig. 8-10c.

From the figure, we see that the normal and shear stresses acting perpendicular and parallel, respectively, to the welded seam are

$$\sigma_w = 47.8 \text{ MPa} \qquad \tau_w = 16.9 \text{ MPa} \qquad \longleftarrow$$

Mohr's circle. The Mohr's circle construction for the biaxial stress element of Fig. 8-10b is shown in Fig. 8-11. Point A represents the stress $\sigma_2 = 36$ MPa on the x face ($\theta = 0$) of the element, and point B represents the stress $\sigma_1 = 72$ MPa on the y face ($\theta = 90°$). The center C of the circle is at a stress of 54 MPa, and the radius of the circle is

$$R = \frac{72 \text{ MPa} - 36 \text{ MPa}}{2} = 18 \text{ MPa}$$

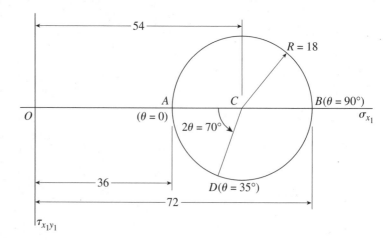

Fig. 8-11 Mohr's circle for the biaxial stress element of Fig. 8-10b. (*Note:* All stresses on the circle have units of MPa.)

A counterclockwise angle $2\theta = 70°$ (measured on the circle from point A) locates point D, which corresponds to the stresses on the x_1 face ($\theta = 35°$) of the element. The coordinates of point D (from the geometry of the circle) are

$$\sigma_{x_1} = 54 \text{ MPa} - R\cos 70° = 54 \text{ MPa} - (18 \text{ MPa})(\cos 70°) = 47.8 \text{ MPa}$$

$$\tau_{x_1 y_1} = R \sin 70° = (18 \text{ MPa})(\sin 70°) = 16.9 \text{ MPa}$$

These results agree with those found earlier, as expected.

Note: When seen in a side view, a **helix** follows the shape of a sine curve (Fig. 8-12). The pitch of the helix is

$$p = \pi d \tan \theta \tag{8-15}$$

where d is the diameter of the circular cylinder and θ is the angle between a normal to the helix and a longitudinal line. The width of the plate that wraps into the cylindrical shape is

$$w = \pi d \sin \theta \tag{8-16}$$

Thus, if the diameter of the cylinder and the angle θ are given, both the pitch and the plate width are established. For practical reasons, the angle θ is usually in the range from 20° to 35°.

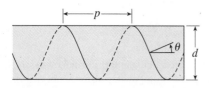

Fig. 8-12 Side view of a helix

8.4 ■ MAXIMUM STRESSES IN BEAMS

The stress analysis of a beam usually begins by finding the normal and shear stresses acting on cross sections. For instance, when Hooke's law holds, we can obtain the normal and shear stresses from the flexure and shear formulas (Eqs. 5-13 and 5-38, respectively, of Chapter 5):

$$\sigma = -\frac{My}{I} \qquad \tau = \frac{VQ}{Ib} \qquad (8\text{-}17\text{a,b})$$

In the flexure formula, σ is the normal stress acting on the cross section, M is the bending moment, y is the distance from the neutral axis, and I is the moment of inertia of the cross-sectional area with respect to the neutral axis. (The sign conventions for M and y in the flexure formula are shown in Figs. 5-9 and 5-10 of Chapter 5).

In the case of the shear formula, τ is the shear stress at any point in the cross section, V is the shear force, Q is the first moment of the cross-sectional area outside of the point in the cross section where the stress is being found, and b is the width of the cross section. (The shear formula is usually written without regard to signs because the directions of the shear stresses are apparent from the directions of the loads.)

The normal stresses obtained from the flexure formula have their maximum values at the farthest distances from the neutral axis, whereas the shear stresses obtained from the shear formula usually have their highest values at the neutral axis. In most circumstances, these are the only stresses that are needed for design purposes. The normal stresses are calculated at the cross section of maximum bending moment, and the shear stresses are calculated at the cross section of maximum shear force.

However, to obtain a more complete picture of the stresses in a beam, we need to determine the principal stresses and maximum shear stresses at various points in the beam. We will begin by discussing the stresses in a rectangular beam.

■ Beams of Rectangular Cross Section

We can gain an understanding of how the stresses in a beam vary by considering the simple beam of rectangular cross section shown in Fig. 8-13a. For the purposes of this discussion, we choose a cross section to the left of the load and then select five points (A, B, C, D, and E) on the side of the beam. Points A and E are at the top and bottom of the beam, respectively, point C is at the midheight of the beam, and points B and D are in between. If Hooke's law applies, the normal and shear stresses at each of these points can be readily calculated from the flexure and shear formulas. Since these stresses act on the cross section, we can picture them on stress elements having vertical and horizontal faces, as shown in

Fig. 8-13b. Note that all elements are in plane stress, because there are no stresses acting perpendicular to the plane of the figure.

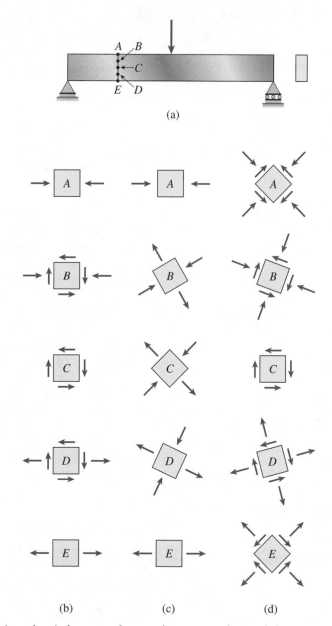

Fig. 8-13 Stresses in a beam of rectangular cross section: (a) simple beam with points A, B, C, D, and E on the side of the beam, (b) normal and shear stresses acting on stress elements at points A, B, C, D, and E, (c) principal stresses, and (d) maximum shear stresses

At point A the normal stress is compressive and there are no shear stresses. Similarly, at point E the normal stress is tensile and again there are no shear stresses. Thus, the elements at these locations are in uniaxial stress. At the neutral axis (point C) the element is in pure shear. At the other two locations (points B and D), both normal and shear stresses act on the stress elements.

To find the principal stresses and maximum shear stresses at each point, we may use either the transformation equations of plane stress or Mohr's circle. The directions of the principal stresses are shown in Fig. 8-13c, and the directions of the maximum shear stresses are shown in Fig. 8-13d. (Note that we are considering only the in-plane stresses.)

Now let us examine the principal stresses in more detail. From the sketches in Fig. 8-13c, we can observe how the principal stresses change as we go from top to bottom of the beam. Let us begin with the compressive principal stress. At point *A* the compressive stress acts in the horizontal direction and the other principal stress is zero. As we move toward the neutral axis the compressive principal stress becomes inclined, and at the neutral axis (point *C*) it acts at 45° to the horizontal. At point *D* the compressive principal stress is further inclined from the horizontal, and at the bottom of the beam its direction becomes vertical (except that its magnitude is now zero). Thus, the direction and magnitude of the compressive principal stress vary continuously from top to bottom of the beam. If the chosen cross section is located in a region of large bending moment, the largest compressive principal stress occurs at the top of the beam (point *A*), and the smallest compressive principal stress (zero) occurs at the bottom of the beam (point *E*). If the cross section is located in a region of small bending moment and large shear force, then the largest compressive principal stress is at the neutral axis.

Analogous comments apply to the tensile principal stress, which also varies in both magnitude and direction as we move from point *A* to point *E*. At point *A* the tensile stress is zero and at point *E* it has its maximum value. (Graphs showing how the principal stresses vary in magnitude for a particular beam and particular cross section are given later in Fig. 8-19 of Example 8-3.)

The maximum shear stresses (Fig. 8-13d) at the top and bottom of the beam occur on 45° planes (because the elements are in uniaxial stress). At the neutral axis, the maximum shear stresses occur on horizontal and vertical planes (because the element is in pure shear). At all points, the maximum shear stresses occur on planes oriented at 45° to the principal planes. In regions of high bending moment, the largest shear stresses occur at the top and bottom of the beam; in regions of low bending moment and high shear force, the largest shear stresses occur at the neutral axis.

By investigating the stresses at many cross sections of the beam, we can determine how the principal stresses vary throughout the beam. Then we can construct two systems of orthogonal curves, called *stress trajectories*, that give the directions of the principal stresses. Examples of stress trajectories for rectangular beams are shown in Fig. 8-14. Part (a) of the figure shows a cantilever beam with a load acting at the free end, and part (b) shows a simple beam with a uniform load. Solid lines are used for tensile principal stresses and dashed lines for compressive principal stresses. The curves for tensile and compressive principal stresses always intersect at right angles, and every trajectory crosses the longitudinal axis at 45°. At the top and bottom surfaces of the beam, where the shear stress is zero, the trajectories are either horizontal or vertical.*

Fig. 8-14 Principal-stress trajectories for beams of rectangular cross section: (a) cantilever beam, and (b) simple beam. (Solid lines represent tensile principal stresses and dashed lines represent compressive principal stresses.)

(a)   (b)

Fig. 8-15 Stress contours for a cantilever beam (tensile principal stresses only)

Another type of curve that may be plotted from the principal stresses is a *stress contour*, which is a curve connecting points of equal principal stress. Stress contours for a cantilever beam of rectangular cross section are shown in Fig. 8-15 (for tensile principal stresses only). The contour of largest stress is at the upper left part of the figure. As we move downward in the figure, the tensile stresses represented by the contours become smaller and smaller. The contour line of zero tensile stress is at the lower edge of the beam. Thus, the largest tensile stress occurs at the support, where the bending moment has its largest value.

Note that stress trajectories (Fig. 8-14) give the directions of the principal stresses but give no information about the magnitudes of the stresses. In general, the magnitudes of the principal stresses vary as we move along a trajectory. In contrast, the magnitudes of the principal stresses are constant as we move along a stress contour (Fig. 8-15), but the contours give no information about the directions of the stresses. In particular, the principal stresses are neither parallel nor perpendicular to a stress contour.

The stress trajectories and contours of Figs. 8-14 and 8-15 were plotted from the flexure and shear formulas (Eqs. 8-17a and b). Stress concentrations near the supports and near the concentrated loads, as well as the direct compressive stresses caused by the uniform load bearing on the top of the beam (Fig. 8-14b), were disregarded in plotting these figures.

■ Wide-flange Beams

Beams having other cross-sectional shapes, such as wide-flange beams, can be analyzed for the principal stresses in a manner similar to that described above for rectangular beams. For instance, consider the simply supported wide-flange beam shown in Fig. 8-16a. Proceeding as for a

---

*Stress trajectories were originated by the German engineer Karl Culmann (1821–1881); see Refs. 8-1 and 8-2.

rectangular beam, we identify points A, B, C, D, and E from top to bottom of the beam (Fig. 8-16b). Points B and D are in the web where it meets the flange, and point C is at the neutral axis. We can think of these points as being located either on the side of the beam (Fig. 8-16c) or along the axis of symmetry (Fig. 8-16d). The stresses determined from the flexure and shear formulas are the same at both sets of points.

Fig. 8-16 Stresses in a wide-flange beam

Stress elements at points A, B, C, D, and E (as seen in a side view of the beam) are shown in parts (e) through (i) of Fig. 8-16. These elements have the same general appearance as those for a rectangular beam (Fig. 8-13b), but of course the stresses are different.

The largest principal stresses usually occur at the top and bottom of the beam (points A and E) where the stresses obtained from the flexure formula have their largest values. However, depending upon the relative magnitudes of the bending moment and shear force, the largest stresses sometimes occur in the web where it meets the flange (points B and D). The explanation lies in the fact that the normal stresses at points B and D are only slightly smaller than those at points A and E, whereas the shear stresses (which are zero at points A and E) may be significant at points B and D because of the thin web. (*Note:* Figure 5-37 in Chapter 5 shows how the shear stresses vary in the web of a wide-flange beam).

The maximum shear stresses acting directly on a cross section of a wide-flange beam always occur at the neutral axis, as shown by the shear formula (Eq. 8-17b). However, the maximum shear stresses acting on inclined planes usually occur either at the top and bottom of the beam (points A and E) or in the web where it meets the flange (points B and D) because of the presence of normal stresses.

When analyzing a wide-flange beam for the maximum stresses, remember that high stresses may exist near supports, points of loading, fillets, and holes. Such stress concentrations are confined to the region very close to the discontinuity and cannot be calculated by elementary beam formulas.

The following example illustrates the procedure for determining the principal stresses and maximum shear stresses at a selected cross-section in a rectangular beam. The procedures for a wide-flange beam are similar.

WORK
SHEET
8
Mathcad

Example 8-3

A simple beam AB with span length $L = 6$ ft supports a concentrated load $P = 10,800$ lb acting at distance $c = 2$ ft from the right-hand support (Fig. 8-17). The beam is made of steel and has a rectangular cross section of width $b = 2$ in. and height $h = 6$ in. Investigate the principal stresses and maximum shear stresses at cross section mn, located at distance $x = 9$ in. from end A of the beam. (Consider only the in-plane stresses.)

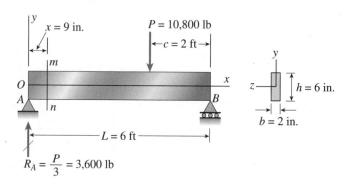

Fig. 8-17 Example 8-3. Beam of rectangular cross section

Solution   We begin by using the flexure and shear formulas to calculate the stresses acting on cross section mn. Once those stresses are known, we can determine the principal stresses and maximum shear stresses from the equations of plane stress. Finally, we can plot graphs of these stresses to show how they vary over the height of the beam.

As a preliminary matter, we note that the reaction of the beam at support A is $R_A = P/3 = 3600$ lb, and therefore the bending moment and shear force at section mn are

$$M = R_A x = (3600 \text{ lb})(9 \text{ in.}) = 32,400 \text{ lb-in.} \qquad V = R_A = 3600 \text{ lb}$$

Normal stresses on cross section mn. These stresses are found from the flexure formula (Eq. 8-17a), as follows:

$$\sigma_x = -\frac{My}{I} = -\frac{12My}{bh^3} = -\frac{12(32,400 \text{ lb-in.})y}{(2 \text{ in.})(6 \text{ in.})^3} = -900y \qquad \text{(a)}$$

in which y has units of inches (in.) and σ_x has units of pounds per square inch (psi). The stresses calculated from Eq. (a) are positive when in tension and negative when in compression. For instance, note that a positive value of y (upper half of the beam) gives a negative stress, as expected.

A stress element cut from the side of the beam at cross section mn (Fig. 8-17) is shown in Fig. 8-18. For reference purposes, a set of xy axes is associated with the element. The normal stress σ_x and the shear stress τ_{xy} are shown acting on the element in their positive directions. (Note that in this example there is no normal stress σ_y acting on the element.)

Shear stresses on cross section mn. The shear stresses are given by the shear formula (Eq. 8-17b) in which the first moment Q for a rectangular cross section is

$$Q = b\left(\frac{h}{2} - y\right)\left(y + \frac{h/2 - y}{2}\right) = \frac{b}{2}\left(\frac{h^2}{4} - y^2\right) \tag{8-18}$$

Thus, the shear formula becomes

$$\tau = \frac{VQ}{Ib} = \frac{12V}{(bh^3)(b)}\left(\frac{b}{2}\right)\left(\frac{h^2}{4} - y^2\right) = \frac{6V}{bh^3}\left(\frac{h^2}{4} - y^2\right) \tag{8-19}$$

The shear stresses τ_{xy} acting on the x face of the stress element (Fig. 8-18) are positive in the upward direction, whereas the actual shear stresses τ (Eq. 8-19) act downward. Therefore, the shear stresses τ_{xy} are given by the following formula:

$$\tau_{xy} = -\frac{6V}{bh^3}\left(\frac{h^2}{4} - y^2\right) \tag{8-20}$$

Substituting numerical values into this equation gives

$$\tau_{xy} = -\frac{6(3600 \text{ lb})}{(2 \text{ in.})(6 \text{ in.})^3}\left(\frac{(6 \text{ in.})^2}{4} - y^2\right) = -50(9 - y^2) \tag{b}$$

in which y (as before) has units of inches (in.) and τ_{xy} has units of pounds per square inch (psi).

Calculation of stresses. For the purpose of calculating the stresses at cross section mn, let us divide the height of the beam into six equal intervals and label the corresponding points from A to G, as shown in the side view of the beam (Fig. 8-19a). The y coordinates of these points are listed in column 2 of Table 8-1, and the corresponding stresses σ_x and τ_{xy} (calculated from Eqs. a and b, respectively) are listed in columns 3 and 4. These stresses are plotted in Figs. 8-19b and 8-19c. The normal stresses vary linearly from a compressive stress of -2700 psi at the top of the beam (point A) to a tensile stress of 2700 psi at the bottom of the beam (point G). The shear stresses have a parabolic distribution with the maximum stress at the neutral axis (point D).

Principal stresses and maximum shear stresses. The principal stresses at each of the seven points A through G may be determined from Eq. (7-17), which is repeated here:

$$\sigma_{1,2} = \frac{\sigma_x + \sigma_y}{2} \pm \sqrt{\left(\frac{\sigma_x - \sigma_y}{2}\right)^2 + \tau_{xy}^2} \tag{8-21}$$

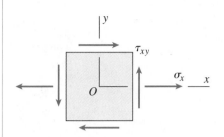

Fig. 8-18 Plane-stress element at cross section *mn* of the beam of Fig. 8-17 (Example 8-3)

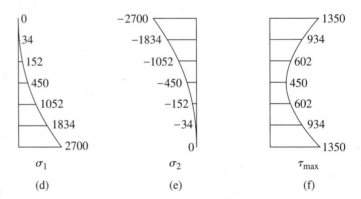

Fig. 8-19 Stresses in the beam of Fig. 8-17 (Example 8-3). (a) Points A, B, C, D, E, F, and G at cross section mn, (b) normal stresses σ_x acting on cross section mn, (c) shear stresses τ_{xy} acting on cross section mn, (d) principal tensile stresses σ_1, (e) principal compressive stresses σ_2, and (f) maximum shear stresses τ_{max}. (*Note:* All stresses have units of psi.)

Table 8-1 Stresses at cross section mn in the beam of Fig. 8-17 (Example 8-3)

(1) Point	(2) y (in.)	(3) σ_x (psi)	(4) τ_{xy} (psi)	(5) σ_1 (psi)	(6) σ_2 (psi)	(7) τ_{max} (psi)
A	3	−2700	0	0	−2700	1350
B	2	−1800	−250	34	−1834	934
C	1	−900	−400	152	−1052	602
D	0	0	−450	450	−450	450
E	−1	900	−400	1052	−152	602
F	−2	1800	−250	1834	−34	934
G	−3	2700	0	2700	0	1350

Since there is no normal stress in the y direction (Fig. 8-18), this equation simplifies to

$$\sigma_{1,2} = \frac{\sigma_x}{2} \pm \sqrt{\left(\frac{\sigma_x}{2}\right)^2 + \tau_{xy}^2} \tag{8-22}$$

Also, the maximum shear stresses (from Eq. 7-25) are

$$\tau_{max} = \sqrt{\left(\frac{\sigma_x - \sigma_y}{2}\right)^2 + \tau_{xy}^2} \tag{8-23}$$

which simplifies to

$$\tau_{max} = \sqrt{\left(\frac{\sigma_x}{2}\right)^2 + \tau_{xy}^2} \tag{8-24}$$

Thus, by substituting the values of σ_x and τ_{xy} (from Table 8-1) into Eqs. (8-22) and (8-24), we can calculate the principal stresses σ_1 and σ_2 and the maximum shear stress τ_{max}. These quantities are listed in the last three columns of Table 8-1 and are plotted in Figs. 8-19d, e, and f.

The tensile principal stresses σ_1 increases from zero at the top of the beam to a maximum of 2700 psi at the bottom (Fig. 8-19d). The directions of the stresses also change, varying from vertical at the top to horizontal at the bottom. At midheight, the stress σ_1 acts on a 45° plane. Similar comments apply to the compressive principal stress σ_2, except in reverse. For instance, the stress is largest at the top of the beam and zero at the bottom (Fig. 8-19e).

The maximum shear stresses at cross section mn occur on 45° planes at the top and bottom of the beam. These stresses are equal to one-half of the normal stresses σ_x at the same points. At the neutral axis, where the normal stress σ_x is zero, the maximum shear stresses occur on the horizontal and vertical planes.

Note 1: If we consider other cross sections of the beam, the maximum normal and shear stresses will be different from those shown in Fig. 8-19. For instance, at a cross section between section mn and the concentrated load (Fig. 8-17), the normal stresses σ_x are larger than shown in Fig. 8-19b because the bending moment is larger. However, the shear stresses τ_{xy} are the same as those shown in Fig. 8-19c because the shear force doesn't change in that region of the beam. Consequently, the principal stresses σ_1 and σ_2 and maximum shear stresses τ_{max} will vary from top to bottom of the beam in the same general manner as shown in Figs. 8-19d, e, and f but with different numerical values for the stresses.

The largest tensile stress anywhere in the beam is the normal stress at the bottom of the beam at the cross section of maximum bending moment. This stress is

$$(\sigma_{tens})_{max} = 14{,}400 \text{ psi}$$

The largest compressive stress has the same numerical value and occurs at the top of the beam at the same cross section:

$$(\sigma_{comp})_{max} = -14{,}400 \text{ psi}$$

The largest shear stress τ_{xy} acting on a cross section of the beam occurs to the right of the load P (Fig. 8-17) because the shear force is larger in that region of the beam:

$$V = R_B = 7200 \text{ lb}$$

Therefore, the largest value of τ_{xy}, which occurs at the neutral axis, is

$$(\tau_{xy})_{\text{max}} = 900 \text{ psi}$$

The largest shear stress anywhere in the beam occurs on 45° planes at either the top or bottom of the beam at the cross section of maximum bending moment:

$$\tau_{\text{max}} = \frac{14,400 \text{ psi}}{2} = 7200 \text{ psi}$$

Note 2: In this example, all stresses are calculated from the flexure and shear formulas without regard to stress concentrations.

Note 3: In the practical design of ordinary beams, the principal stresses and maximum shear stresses are rarely calculated. Instead, the tensile and compressive stresses to be used in design are calculated from the flexure formula at the cross section of maximum bending moment, and the shear stress to be used in design is calculated from the shear formula at the cross section of maximum shear force.

8.5 COMBINED LOADINGS

In previous chapters we analyzed structural members subjected to a single type of loading. For instance, we analyzed axially loaded bars in Chapters 1 and 2, shafts in torsion in Chapter 3, and beams in bending in Chapters 4, 5, and 6. We also analyzed pressure vessels earlier in this chapter. For each type of loading, we developed methods for finding stresses, strains, and deformations.

However, in many structures the members are required to resist more than one kind of loading. For example, a beam may be subjected to the simultaneous action of bending moments and axial forces (Fig. 8-20a), a pressure vessel may be supported so that it also functions as a beam (Fig. 8-20b), or a shaft in torsion may carry a bending load (Fig. 8-20c). Known as **combined loadings**, situations similar to those shown in Fig. 8-20 occur in a great variety of machines, buildings, vehicles, tools, equipment, and many other kinds of structures.

A structural member subjected to combined loadings can often be analyzed by superimposing the stresses and strains caused by each load acting separately. However, superposition of both stresses and strains is permissible only under certain conditions, as explained in earlier chapters. One requirement is that the stresses and strains must be linear functions of the applied loads, which in turn requires that the material follow

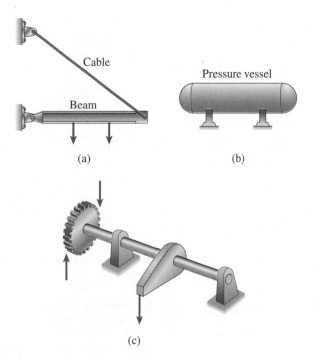

Fig. 8-20 Examples of structures subjected to combined loadings: (a) wide-flange beam supported by a cable (combined bending and axial load), (b) cylindrical pressure vessel supported as a beam, and (c) shaft in combined torsion and bending

Hooke's law and the displacements remain small. Another requirement is that there must be no interaction between the various loads, that is, the stresses and strains due to one load must not be affected by the presence of the other loads. Most ordinary structures satisfy these conditions, and therefore the use of superposition is very common in engineering work.

■ Method of Analysis

While there are many ways to analyze a structure subjected to more than one type of load, the procedure usually includes the following steps:

1. Select a point in the structure where the stresses and strains are to be determined. (The point is usually selected at a cross section where the stresses are large, such as at a cross section where the bending moment has its maximum value.)
2. For each load on the structure, determine the stress resultants at the cross section containing the selected point. (The possible stress resultants are an axial force, a twisting moment, a bending moment, and a shear force.)
3. Calculate the normal and shear stresses at the selected point due to each of the stress resultants. Also, if the structure is a pressure vessel, determine the stresses due to the internal pressure. (The stresses are found from the stress formulas derived previously; for instance, $\sigma = P/A$, $\tau = T\rho/I_p$, $\sigma = My/I$, $\tau = VQ/Ib$, and $\sigma = pr/t$.)

4. Combine the individual stresses to obtain the resultant stresses at the selected point. In other words, obtain the stresses σ_x, σ_y, and τ_{xy} acting on a stress element at the point. (Note that in this chapter we are dealing only with elements in plane stress.)

5. Determine the principal stresses and maximum shear stresses at the selected point, using either the stress-transformation equations or Mohr's circle. If required, determine the stresses acting on other inclined planes.

6. Determine the strains at the point with the aid of Hooke's law for plane stress.

7. Select additional points and repeat the process. Continue until enough stress and strain information is available to satisfy the purposes of the analysis.

(a)

(b)

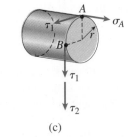

(c)

Fig. 8-21 Cantilever bar subjected to combined torsion and bending: (a) loads acting on the bar, (b) stress resultants at a cross section, and (c) stresses at points *A* and *B*

■ Illustration of the Method

To illustrate the procedure for analyzing a member subjected to combined loadings, we will discuss in general terms the stresses in the cantilever bar of circular cross section shown in Fig. 8-21a. This bar is subjected to two types of loads—a torque *T* and a vertical load *P*, both acting at the free end of the bar.

Let us begin by arbitrarily selecting two points *A* and *B* for investigation (Fig. 8-21a). Point *A* is located at the top of the bar and point *B* is located on the side. Both points are located at the same cross section.

The stress resultants acting at the cross section (Fig. 8-21b) are a twisting moment equal to the torque *T*, a bending moment *M* equal to the load *P* times the distance *b* from the free end of the bar to the cross section, and a shear force *V* equal to the load *P*.

The stresses acting at points *A* and *B* are shown in Fig. 8-21c. The twisting moment *T* produces torsional shear stresses

$$\tau_1 = \frac{Tr}{I_p} = \frac{2T}{\pi r^3} \qquad \text{(a)}$$

in which *r* is the radius of the bar and $I_p = \pi r^4/2$ is the polar moment of inertia of the cross-sectional area. The stress τ_1 acts horizontally to the left at point *A* and vertically downward at point *B*, as shown in the figure.

The bending moment *M* produces a tensile stress at point *A*:

$$\sigma_A = \frac{Mr}{I} = \frac{4M}{\pi r^3} \qquad \text{(b)}$$

in which $I = \pi r^4/4$ is the moment of inertia about the neutral axis. However, the bending moment produces no stress at point *B*, because *B* is located on the neutral axis.

The shear force *V* produces no shear stress at the top of the bar (point *A*), but at point *B* the shear stress is (see Eq. 5-42 in Chapter 5):

$$\tau_2 = \frac{4V}{3A} = \frac{4V}{3\pi r^2} \qquad\qquad (c)$$

in which $A = \pi r^2$ is the cross-sectional area.

The stresses σ_A and τ_1 acting at point A (Fig. 8-21c) are shown acting on a stress element in Fig. 8-22a. This element is cut from the top of the bar at point A. A two-dimensional view of the element, obtained by looking vertically downward on the element, is shown in Fig. 8-22b. For the purpose of determining the principal stresses and maximum shear stresses, we construct x and y axes through the element. The x axis is parallel to the longitudinal axis of the circular bar (Fig. 8-21a) and the y axis is horizontal. Note that the element is in plane stress with $\sigma_x = \sigma_A$, $\sigma_y = 0$, and $\tau_{xy} = -\tau_1$.

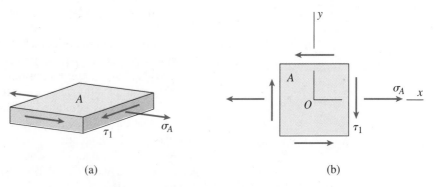

Fig. 8-22 Stress element at point A (a)   (b)

A stress element at point B (also in plane stress) is shown in Fig. 8-23a. The only stresses acting on this element are the shear stresses, equal to $\tau_1 + \tau_2$ (see Fig. 8-21c). A two-dimensional view of the stress element is shown in Fig. 8-23b, with the x axis parallel to the longitudinal axis of the bar and the y axis in the vertical direction. The stresses acting on the element are $\sigma_x = \sigma_y = 0$ and $\tau_{xy} = -(\tau_1 + \tau_2)$.

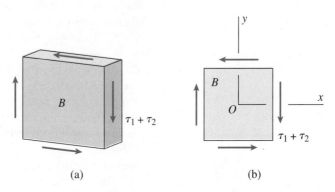

Fig. 8-23 Stress element at point B (a)   (b)

Now that we have determined the stresses acting at points A and B and constructed the corresponding stress elements, we can use the transformation equations of plane stress (Sections 7.2 and 7.3) or Mohr's circle (Section 7.4) to determine principal stresses, maximum shear stresses, and stresses acting in inclined directions. We can also use Hooke's law (Section 7.5) to determine the strains at points *A* and *B*.

The procedure described above for analyzing the stresses at points *A* and *B* (Fig. 8-21a) can be used at other points in the bar. Of particular interest are the points where the stresses calculated from the flexure and shear formulas have maximum or minimum values, called **critical points**. For instance, the normal stresses due to bending are largest at the cross section of maximum bending moment, which is at the support. Therefore, points *C* and *D* at the top and bottom of the beam at the fixed end (Fig. 8-21a) are critical points where the stresses should be calculated. Another critical point is point *B* itself, because the shear stresses are a maximum at this point. (Note that in this example the shear stresses do not change if point *B* is moved along the bar in the longitudinal direction.)

As a final step, the principal stresses and maximum shear stresses at the critical points can be compared with one another in order to determine the absolute maximum normal and shear stresses in the bar.

This example illustrates the general procedure for determining the stresses produced by combined loadings. No new theories are involved—only applications of previously derived formulas and concepts. Since the variety of practical situations seems to be endless, it is not worthwhile to derive specific formulas for calculating the maximum stresses. Instead, each structure is usually treated as a special case.

■ Selection of Critical Points

If the objective of the analysis is to determine the largest stresses anywhere in the structure, then the critical points should be selected at cross sections where the stress resultants have their largest values. Furthermore, within those cross sections, the points should be selected where either the normal stresses or the shear stresses have their largest values. By using good judgment in the selection of the points, we often can be reasonably certain of obtaining the absolute maximum stresses in the structure. However, it is sometimes difficult to recognize in advance where the maximum stresses in the member are to be found. Then it may be necessary to investigate the stresses at a large number of points, perhaps even using trial-and-error in the selection of points. Other strategies may also prove fruitful—such as deriving equations specific to the problem at hand or making simplifying assumptions to facilitate an otherwise difficult analysis.

The following examples illustrate the methods used to calculate stresses in structures subjected to combined loadings.

■ **Example 8-4**

The rotor shaft of a helicopter drives the rotor blades that provide the lifting force to support the helicopter in the air (Fig. 8-24a). As a consequence, the shaft is subjected to a combination of torsion and axial loading (Fig. 8-24b).

For a 50-mm diameter shaft transmitting a torque $T = 2.4$ kN·m and a tensile force $P = 125$ kN, determine the maximum tensile stress, maximum compressive stress, and maximum shear stress in the shaft.

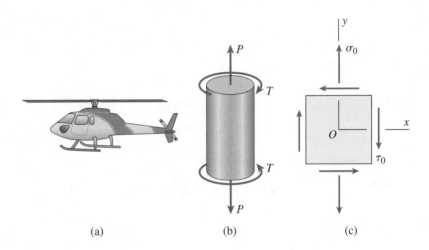

Fig. 8-24 Example 8-4. Rotor shaft of a helicopter (combined torsion and axial force)

(a)   (b)   (c)

Solution   The stresses in the rotor shaft are produced by the combined action of the axial force P and the torque T (Fig. 8-24b). Therefore, the stresses at any point on the surface of the shaft consist of a tensile stress σ_0 and shear stresses τ_0, as shown on the stress element of Fig. 8-24c. Note that the y axis is parallel to the longitudinal axis of the shaft.

The tensile stress σ_0 equals the axial force divided by the cross-sectional area:

$$\sigma_0 = \frac{P}{A} = \frac{4P}{\pi d^2} = \frac{4(125 \text{ kN})}{\pi (50 \text{ mm})^2} = 63.66 \text{ MPa}$$

The shear stress τ_0 is obtained from the torsion formula (Eq. 3-11 of Section 3.3):

$$\tau_0 = \frac{Tr}{I_p} = \frac{16T}{\pi d^3} = \frac{16(2.4 \text{ kN·m})}{\pi (50 \text{ mm})^3} = 97.78 \text{ MPa}$$

The stresses σ_0 and τ_0 act directly on cross sections of the shaft.

Knowing the stresses σ_0 and τ_0, we can now obtain the principal stresses and maximum shear stresses by the methods described in Section 7.3. The principal stresses are obtained from Eq. (7-17):

$$\sigma_{1,2} = \frac{\sigma_x + \sigma_y}{2} \pm \sqrt{\left(\frac{\sigma_x - \sigma_y}{2}\right)^2 + \tau_{xy}^2} \qquad \text{(d)}$$

Substituting $\sigma_x = 0$, $\sigma_y = \sigma_0 = 63.66$ MPa, and $\tau_{xy} = -\tau_0 = -97.78$ MPa, we get

$$\sigma_{1,2} = 32 \text{ MPa} \pm 103 \text{ MPa}$$

or $\qquad\qquad \sigma_1 = 135 \text{ MPa} \qquad \sigma_2 = -71 \text{ MPa}$ ⬅

These are the maximum tensile and compressive stresses in the rotor shaft.

The maximum in-plane shear stresses (Eq. 7-25) are

$$\tau_{max} = \sqrt{\left(\frac{\sigma_x - \sigma_y}{2}\right)^2 + \tau_{xy}^2} \tag{e}$$

This term was evaluated above, so we see immediately that

$$\tau_{max} = 103 \text{ MPa} ⬅$$

Because the principal stresses σ_1 and σ_2 have opposite signs, the maximum in-plane shear stresses are larger than the maximum out-of-plane shear stresses (see Eqs. 7-28a, b, and c and the accompanying discussion). Therefore, the maximum shear stress in the shaft is 103 MPa.

■ **Example 8-5**

A thin-walled cylindrical pressure vessel with a circular cross section is subjected to internal gas pressure p and simultaneously compressed by an axial load $P = 12$ k (Fig. 8-25a). The cylinder has inner radius $r = 2.1$ in. and wall thickness $t = 0.15$ in. Determine the maximum allowable internal pressure p_{allow} based upon an allowable shear stress of 6500 psi in the wall of the vessel.

Solution   The stresses in the wall of the pressure vessel are caused by the combined action of the internal pressure and the axial force. Since both actions produce uniform normal stresses throughout the wall, we can select any point on the surface for investigation. At a typical point, such as point A (Fig. 8-25a), we isolate a stress element as shown in Fig. 8-25b. The x axis is parallel to the longitudinal axis of the pressure vessel and the y axis is circumferential. Note that there are no shear stresses acting on the element.

Principal stresses. The longitudinal stress σ_x is equal to the tensile stress σ_2 produced by the internal pressure (see Fig. 8-7a and Eq. 8-6) minus the compressive stress produced by the axial force; thus,

$$\sigma_x = \frac{pr}{2t} - \frac{P}{A} = \frac{pr}{2t} - \frac{P}{2\pi rt} \tag{f}$$

in which $A = 2\pi rt$ is the cross-sectional area of the cylinder. (Note that for convenience we are using the inner radius r in all calculations.)

The circumferential stress σ_y is equal to the tensile stress σ_1 produced by the internal pressure (Fig. 8-7a and Eq. 8-5):

$$\sigma_y = \frac{pr}{t} \tag{g}$$

Note that σ_y is algebraically larger than σ_x.

(a)

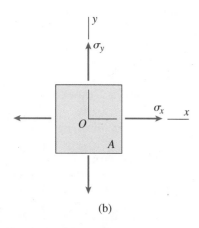

(b)

Fig. 8-25 Example 8-5. Pressure vessel subjected to combined internal pressure and axial force

Since no shear stresses act on the element (Fig. 8-25), the normal stresses σ_x and σ_y are also the principal stresses:

$$\sigma_1 = \sigma_y = \frac{pr}{t} \qquad \sigma_2 = \sigma_x = \frac{pr}{2t} - \frac{P}{2\pi rt} \qquad\qquad \text{(h,i)}$$

Now substituting numerical values, we obtain

$$\sigma_1 = \frac{pr}{t} = \frac{p(2.1 \text{ in.})}{0.15 \text{ in.}} = 14.0p$$

$$\sigma_2 = \frac{pr}{2t} - \frac{P}{2\pi rt} = \frac{p(2.1 \text{ in.})}{2(0.15 \text{ in.})} - \frac{12 \text{ k}}{2\pi(2.1 \text{ in.})(0.15 \text{ in.})}$$

$$= 7.0p - 6063 \text{ psi}$$

in which σ_1, σ_2, and p have units of pounds per square inch (psi).

In-plane shear stresses. The maximum in-plane shear stress (Eq. 7-26) is

$$\tau_{\max} = \frac{\sigma_1 - \sigma_2}{2} = \frac{1}{2}(14.0p - 7.0p + 6063 \text{ psi}) = 3.5p + 3032 \text{ psi}$$

Since $\tau_{\max}$ is limited to 6500 psi, the preceding equation becomes

$$6500 \text{ psi} = 3.5p + 3032 \text{ psi}$$

from which we get

$$p = \frac{3468 \text{ psi}}{3.5} = 990.9 \text{ psi} \quad \text{or} \quad (p_{\text{allow}})_1 = 990 \text{ psi}$$

because we round downward.

Out-of-plane shear stresses. The maximum out-of-plane shear stress (see Eqs. 7-28a and 7-28b) is either

$$\tau_{\max} = \frac{\sigma_2}{2} \quad \text{or} \quad \tau_{\max} = \frac{\sigma_1}{2}$$

From the first of these two equations we get

$$6500 \text{ psi} = 3.5p - 3032 \text{ psi} \quad \text{or} \quad (p_{\text{allow}})_2 = 2720 \text{ psi}$$

From the second equation we get

$$6500 \text{ psi} = 7.0p \quad \text{or} \quad (p_{\text{allow}})_3 = 928 \text{ psi}$$

Allowable internal pressure. Comparing the three calculated values for the allowable pressure, we see that $(p_{\text{allow}})_3$ governs, and therefore the allowable internal pressure is

$$p_{\text{allow}} = 928 \text{ psi} \qquad\qquad \longleftarrow$$

At this pressure the principal stresses are $\sigma_1 = 13,000$ psi and $\sigma_2 = 430$ psi. These stresses have the same signs, thus confirming that one of the out-of-plane shear stresses must be the largest shear stress (see the discussion following Eqs. 7-28a, b, and c).

Note: In this example, we determined the allowable pressure in the vessel assuming that the axial load was equal to 12 k. A more complete analysis would include the possibility that the axial force may not be present. (As it turns out, the allowable pressure does not change if the axial force is removed.)

■ Example 8-6

0.5 m

2.0 m

Sign

1.2 m

6.0 m

A *B*

C *B*

A

180 mm

220 mm

Fig. 8-26 Example 8-6. Wind pressure against a sign (combined bending, torsion, and shear of the pole)

A sign of dimensions 2.0 m × 1.2 m is supported by a hollow circular pole having outer diameter 220 mm and inner diameter 180 mm (Fig. 8-26). The sign is offset 0.5 m from the centerline of the pole and its lower edge is 6.0 m above the ground.

Determine the principal stresses and maximum shear stresses at points *A* and *B* at the base of the pole due to a wind pressure of 2.0 kPa against the sign.

Solution   *Stress resultants.* The wind pressure against the sign produces a resultant force *W* that acts at the midpoint of the sign (Fig. 8-27a) and is equal to the pressure *p* times the area *A* over which it acts:

$$W = pA = (2.0 \text{ kPa})(2.0 \text{ m} \times 1.2 \text{ m}) = 4.8 \text{ kN}$$

The line of action of this force is at height *h* = 6.6 m above the ground and at distance *b* = 1.5 m from the centerline of the pole.

The wind force acting on the sign is statically equivalent to a lateral force *W* and a torque *T* acting on the pole (Fig. 8-27b). The torque is equal to the force *W* times the distance *b*:

$$T = Wb = (4.8 \text{ kN})(1.5 \text{ m}) = 7.2 \text{ kN·m}$$

The stress resultants at the base of the pole (Fig. 8-27c) consist of a bending moment *M*, a torque *T*, and a shear force *V*. Their magnitudes are

$$M = Wh = (4.8 \text{ kN})(6.6 \text{ m}) = 31.68 \text{ kN·m}$$

$$T = 7.2 \text{ kN·m} \qquad V = W = 4.8 \text{ kN}$$

Examination of these stress resultants shows that maximum bending stresses occur at point *A* and maximum shear stresses at point *B*. Therefore, *A* and *B* are critical points where the stresses should be determined. (Another critical point is diametrically opposite point *A*, as explained in the *Note* below.)

Stresses at points A and B. The bending moment *M* produces a tensile stress σ_A at point *A* (Fig. 8-27d) but no stress at point *B* (which is located on the neutral axis). The stress σ_A is obtained from the flexure formula:

$$\sigma_A = \frac{M(d_2/2)}{I}$$

in which d_2 is the outer diameter (220 mm) and *I* is the moment of inertia of the cross section. The moment of inertia is

$$I = \frac{\pi}{64}\left(d_2^4 - d_1^4\right) = \frac{\pi}{64}\left[(220 \text{ mm})^4 - (180 \text{ mm})^4\right] = 63.46 \times 10^{-6} \text{ m}^4$$

in which d_1 is the inner diameter. Therefore,

$$\sigma_A = \frac{Md_2}{2I} = \frac{(31.68 \text{ kN·m})(220 \text{ mm})}{2(63.46 \times 10^{-6} \text{ m}^4)} = 54.91 \text{ MPa}$$

(a)   (b)

(c)   (d)

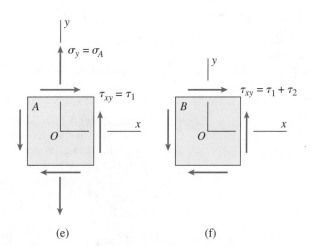

Fig. 8-27 Solution to Example 8-6   (e)   (f)

The torque T produces shear stresses τ_1 at points A and B (Fig. 8-27d). We can calculate these stresses from the torsion formula:

$$\tau_1 = \frac{T(d_2/2)}{I_p}$$

in which I_p is the polar moment of inertia:

$$I_p = \frac{\pi}{32}\left(d_2^4 - d_1^4\right) = 2I = 126.92 \times 10^{-6} \text{ m}^4$$

Thus,

$$\tau_1 = \frac{Td_2}{2I_p} = \frac{(7.2 \text{ kN·m})(220 \text{ mm})}{2(126.92 \times 10^{-6} \text{ m}^4)} = 6.24 \text{ MPa}$$

Finally, we calculate the shear stresses at points A and B due to the shear force V. The shear stress at point A is zero, and the shear stress at point B (denoted τ_2 in Fig. 8-27d) is obtained from the shear formula for a circular tube (Eq. 5-44 of Section 5.9):

$$\tau_2 = \frac{4V}{3A}\left(\frac{r_2^2 + r_2 r_1 + r_1^2}{r_2^2 + r_1^2}\right) \tag{j}$$

in which r_2 and r_1 are the outer and inner radii, respectively, and A is the cross-sectional area:

$$r_2 = \frac{d_2}{2} = 110 \text{ mm} \qquad r_1 = \frac{d_1}{2} = 90 \text{ mm}$$

$$A = \pi(r_2^2 - r_1^2) = 12{,}570 \text{ mm}^2$$

Substituting numerical values into Eq. (j), we obtain

$$\tau_2 = 0.76 \text{ MPa}$$

All stresses acting at points A and B have now been calculated.

Stress elements. The next step is to show these stresses on stress elements (Figs. 8-27e and f). For both elements, the y axis is parallel to the longitudinal axis of the pole and the x axis is horizontal. At point A the stresses acting on the element are

$$\sigma_x = 0 \qquad \sigma_y = \sigma_A = 54.91 \text{ MPa} \qquad \tau_{xy} = \tau_1 = 6.24 \text{ MPa}$$

At point B the stresses are

$$\sigma_x = \sigma_y = 0 \qquad \tau_{xy} = \tau_1 + \tau_2 = 6.24 \text{ MPa} + 0.76 \text{ MPa} = 7.00 \text{ MPa}$$

Since there are no normal stresses acting on the element, point B is in pure shear.

Now that all stresses acting on the stress elements (Figs. 8-27e and f) are known, we can use the equations given in Section 7.3 to determine the principal stresses and maximum shear stresses.

Principal stresses and maximum shear stresses at point A. The principal stresses are obtained from Eq. (7-17), which is repeated here:

$$\sigma_{1,2} = \frac{\sigma_x + \sigma_y}{2} \pm \sqrt{\left(\frac{\sigma_x - \sigma_y}{2}\right)^2 + \tau_{xy}^2} \tag{k}$$

Substituting $\sigma_x = 0$, $\sigma_y = 54.91$ MPa, and $\tau_{xy} = 6.24$ MPa, we get

$$\sigma_{1,2} = 27.5 \text{ MPa} \pm 28.2 \text{ MPa}$$

or $\qquad\qquad \sigma_1 = 55.7 \text{ MPa} \qquad \sigma_2 = -0.7 \text{ MPa}$

The maximum in-plane shear stresses may be obtained from Eq. (7-25):

$$\tau_{max} = \sqrt{\left(\frac{\sigma_x - \sigma_y}{2}\right)^2 + \tau_{xy}^2} \tag{l}$$

This term was evaluated above, so we see immediately that

$$\tau_{max} = 28.2 \text{ MPa}$$

Because the principal stresses σ_1 and σ_2 have opposite signs, the maximum in-plane shear stresses are larger than the maximum out-of-plane shear stresses (see Eqs. 7-28a, b, and c and the accompanying discussion). Therefore, the maximum shear stress at point A is 28.2 MPa.

Principal stresses and maximum shear stresses at point B. The stresses at this point are $\sigma_x = 0$, $\sigma_y = 0$, and $\tau_{xy} = 7.0$ MPa. Since the element is in pure shear, the principal stresses are

$$\sigma_1 = 7.0 \text{ MPa} \qquad \sigma_2 = -7.0 \text{ MPa} \qquad \longleftarrow$$

and the maximum in-plane shear stress is

$$\tau_{max} = 7.0 \text{ MPa} \qquad \longleftarrow$$

The maximum out-of-plane shear stresses are half this value.

Note: If the largest stresses anywhere in the pole are needed, then we must also determine the stresses at the critical point diametrically opposite point A, because at that point the compressive stress due to bending has its largest value. The principal stresses at that point are

$$\sigma_1 = 0.7 \text{ MPa} \qquad \sigma_2 = -55.7 \text{ MPa}$$

and the maximum shear stress is 28.2 MPa. Therefore, the largest tensile stress in the pole is 55.7 MPa, the largest compressive stress is -55.7 MPa, and the largest shear stress is 28.2 MPa. (Keep in mind that only the effects of the wind pressure are considered in this analysis. Other loads, such as the weight of the structure, also produce stresses at the base of the pole.)

■ Example 8-7

A tubular post of square cross section supports a horizontal platform (Fig. 8-28). The tube has outer dimension $b = 6$ in. and wall thickness $t = 0.5$ in. The platform has dimensions 6.75 in. × 24.0 in. and supports a uniformly distributed load of 20 psi acting over its upper surface. The resultant of the distributed load is a vertical force P_1:

$$P_1 = (20 \text{ psi})(6.75 \text{ in.} \times 24.0 \text{ in.}) = 3240 \text{ lb}$$

This force acts at the midpoint of the platform, which is at distance $d = 9$ in. from the longitudinal axis of the post. A second load $P_2 = 800$ lb acts horizontally on the post at height $h = 52$ in. above the base.

Determine the principal stresses and maximum shear stresses at points A and B at the base of the post due to the loads P_1 and P_2.

Solution   *Stress resultants.* The force P_1 acting on the platform (Fig. 8-28) is statically equivalent to a force P_1 and a moment $M_1 = P_1 d$ acting at the midpoint of the post (Fig. 8-29a). The stress resultants at the base of the post due to the loads P_1 and P_2 and the moment M_1 are shown in Fig. 8-29b. These stress resultants are the following:

1. An axial compressive force $P_1 = 3240$ lb
2. A bending moment M_1 produced by the force P_1:

$$M_1 = P_1 d = (3240 \text{ lb})(9 \text{ in.}) = 29{,}160 \text{ lb-in.}$$

$d = 9$ in. $P_1 = 3,240$ lb

$P_2 = 800$ lb

b b

$h = 52$ in.

A B

$b = 6$ in.

$t = 0.5$ in.

B $\dfrac{b}{2} = 3$ in.

$\dfrac{b}{2} = 3$ in.

A

$t = 0.5$ in.

Fig. 8-28 Example 8-7. Loads on a post (combined axial load, bending, and shear)

3. A shear force $P_2 = 800$ lb
4. A bending moment M_2 produced by the force P_2:

$$M_2 = P_2 h = (800\text{ lb})(52\text{ in.}) = 41,600\text{ lb-in.}$$

Examination of these stress resultants (Fig. 8-29b) shows that both M_1 and M_2 produce maximum compressive stresses at point A and the shear force produces maximum shear stresses at point B. Therefore, A and B are critical points where the stresses should be determined. (Another critical point is diagonally opposite point A, as explained in the *Note* below.)

Stresses at points A and B. (1) The axial force P_1 (Fig. 8-29b) produces uniform compressive stresses throughout the post. These stresses are

$$\sigma_{P_1} = \frac{P_1}{A}$$

in which A is the cross-sectional area of the post:

$$A = b^2 - (b - 2t)^2 = 4t(b - t)$$

$$= 4(0.5\text{ in.})(6\text{ in.} - 0.5\text{ in.}) = 11.00\text{ in.}^2$$

Therefore, the axial compressive stress is

$$\sigma_{P_1} = \frac{P_1}{A} = \frac{3240\text{ lb}}{11.00\text{ in.}^2} = 295\text{ psi}$$

The stress σ_{P_1} is shown acting at points A and B in Fig. 8-29c.

(2) The bending moment M_1 (Fig. 8-29b) produces compressive stresses σ_{M_1} at points A and B (Fig. 8-29c). These stresses are obtained from the flexure formula:

$$\sigma_{M_1} = \frac{M_1(b/2)}{I} = \frac{M_1 b}{2I}$$

in which I is the moment of inertia of the cross-sectional area:

$$I = \frac{b^4}{12} - \frac{(b - 2t)^4}{12} = \frac{1}{12}\left[(6\text{ in.})^4 - (5\text{ in.})^4\right] = 55.92\text{ in.}^4$$

Thus, the stress σ_{M_1} is

$$\sigma_{M_1} = \frac{M_1 b}{2I} = \frac{(29,160\text{ lb-in.})(6\text{ in.})}{2(55.92\text{ in.}^4)} = 1564\text{ psi}$$

(3) The shear force P_2 (Fig. 8-29b) produces a shear stress at point B but not at point A. From the discussion of shear stresses in the webs of beams with flanges (Section 5.10), we know that an approximate value of the shear stress can be obtained by dividing the shear force by the web area (see Eq. 5-50 in Section 5.10). Thus, the shear stress produced at point B by the force P_2 is

$$\tau_{P_2} = \frac{P_2}{A_{\text{web}}} = \frac{P_2}{2t(b - 2t)} = \frac{800\text{ lb}}{2(0.5\text{ in.})(6\text{ in.} - 1\text{ in.})} = 160\text{ psi}$$

The stress τ_{P_2} acts at point B in the direction shown in Fig. 8-29c.

Fig. 8-29 Solution to Example 8-7

If desired, we can calculate the shear stress τ_{P_2} from the more accurate formula of Eq. (5-48a) in Section 5.10. The result of that calculation is $\tau_{P_2} = 163$ psi, which shows that the shear stress obtained from the approximate formula is satisfactory.

(4) The bending moment M_2 (Fig. 8-29b) produces a compressive stress at point A but no stress at point B. The stress at A is

$$\sigma_{M_2} = \frac{M_2(b/2)}{I} = \frac{M_2 b}{2I} = \frac{(41{,}600\ \text{lb-in.})(6\ \text{in.})}{2(55.92\ \text{in.}^4)} = 2232\ \text{psi}$$

This stress is also shown in Fig. 8-29c.

Stress elements. The next step is to show the stresses acting on stress elements at points A and B (Figs. 8-29d and e). Each element is oriented so that the y axis is vertical (that is, parallel to the longitudinal axis of the post) and the x axis is horizontal. At point A the only stress is a compressive stress σ_A in the y direction (Fig. 8-29d):

$$\sigma_A = \sigma_{P_1} + \sigma_{M_1} + \sigma_{M_2}$$
$$= 295\ \text{psi} + 1564\ \text{psi} + 2232\ \text{psi} = 4090\ \text{psi (compression)}$$

Since this is the only stress acting on the element, the element is in uniaxial stress.

At point B the compressive stress in the y direction (Fig. 8-29e) is

$$\sigma_B = \sigma_{P_1} + \sigma_{M_1} = 295\ \text{psi} + 1564\ \text{psi} = 1860\ \text{psi (compression)}$$

and the shear stress is

$$\tau_{P_2} = 160\ \text{psi}$$

The shear stress acts leftward on the top face of the element and downward on the x face of the element.

Principal stresses and maximum shear stresses at point A. Using the standard notation for an element in plane stress (Fig. 8-30), we write the stresses for element A (Fig. 8-29d) as follows:

$$\sigma_x = 0 \qquad \sigma_y = -\sigma_A = -4090\ \text{psi} \qquad \tau_{xy} = 0$$

Since the element is in uniaxial stress, the principal stresses are

$$\sigma_1 = 0 \qquad \sigma_2 = -4090\ \text{psi} \qquad \longleftarrow$$

and the maximum in-plane shear stress (Eq. 7-26) is

$$\tau_{max} = \frac{\sigma_1 - \sigma_2}{2} = \frac{4090\ \text{psi}}{2} = 2050\ \text{psi} \qquad \longleftarrow$$

The maximum out-of-plane shear stress (Eq. 7-28a) has the same magnitude.

Principal stresses and maximum shear stresses at point B. Again using the standard notation for plane stress, we see that the stresses at point B (Fig. 8-29e) are

$$\sigma_x = 0 \qquad \sigma_y = -\sigma_B = -1860\ \text{psi} \qquad \tau_{xy} = -\tau_{P_2} = -160\ \text{psi}$$

To obtain the principal stresses, we use Eq. (7-17), which is repeated here:

$$\sigma_{1,2} = \frac{\sigma_x + \sigma_y}{2} \pm \sqrt{\left(\frac{\sigma_x - \sigma_y}{2}\right)^2 + \tau_{xy}^2} \tag{m}$$

Substituting for σ_x, σ_y, and τ_{xy}, we get

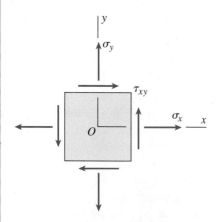

Fig. 8-30 Notation for an element in plane stress

$$\sigma_{1,2} = -930 \text{ psi} \pm 944 \text{ psi}$$

or $$\sigma_1 = 14 \text{ psi} \qquad \sigma_2 = -1870 \text{ psi} \qquad \leftarrow$$

The maximum in-plane shear stresses may be obtained from Eq. (7-25):

$$\tau_{max} = \sqrt{\left(\frac{\sigma_x - \sigma_y}{2}\right)^2 + \tau_{xy}^2} \qquad (n)$$

This term was evaluated above, so we see immediately that

$$\tau_{max} = 944 \text{ psi} \qquad \leftarrow$$

Because the principal stresses σ_1 and σ_2 have opposite signs, the maximum in-plane shear stresses are larger than the maximum out-of-plane shear stresses (see Eqs. 7-28a, b, and c and the accompanying discussion). Therefore, the maximum shear stress at point B is 944 psi.

Note: If the largest stresses anywhere at the base of the post are needed, then we must also determine the stresses at the critical point diagonally opposite point A (Fig. 8-29c), because at that point each bending moment produces the maximum tensile stress. Thus, the tensile stress acting at that point is

$$\sigma_y = -\sigma_{P_1} + \sigma_{M_1} + \sigma_{M_2} = -295 \text{ psi} + 1564 \text{ psi} + 2232 \text{ psi} = 3500 \text{ psi}$$

The stresses acting on a stress element at that point (see Fig. 8-30) are

$$\sigma_x = 0 \qquad \sigma_y = 3500 \text{ psi} \qquad \tau_{xy} = 0$$

and therefore the principal stresses and maximum shear stress are

$$\sigma_1 = 3500 \text{ psi} \qquad \sigma_2 = 0 \qquad \tau_{max} = 1750 \text{ psi}$$

Thus, the largest tensile stress anywhere at the base of the post is 3500 psi, the largest compressive stress is 4090 psi, and the largest shear stress is 2050 psi. (Keep in mind that only the effects of the loads P_1 and P_2 are considered in this analysis. Other loads, such as the weight of the structure, also produce stresses at the base of the post.)

PROBLEMS/CHAPTER 8

SPHERICAL PRESSURE VESSELS

When solving problems for Section 8.2, assume that the given radius or diameter is an inside dimension.

8.2-1 A large spherical tank (see figure) contains gas at a pressure of 500 psi. The tank is 60 ft in diameter and is constructed of high-strength steel having a yield stress in tension of 80 ksi. Determine the minimum thickness (to the nearest 1/4 inch) of the wall of the tank if a factor of safety of 3.2 with respect to yielding is required.

Probs. 8.2-1 and 8.2-2

8.2-2 Solve the preceding problem if the internal pressure is 3.0 MPa, the diameter is 16 m, the yield stress is 490 MPa, and the factor of safety is 2.5. Determine the required thickness to the nearest millimeter.

8.2-3 A hemispherical window (or *viewport*) in a decompression chamber (see figure) is subjected to an internal air pressure of 90 psi. The port is attached to the wall of the chamber by 18 bolts. Find the tensile force F in each bolt and the tensile stress σ in the viewport if the radius of the hemisphere is 15 in. and its thickness is 1.0 in.

Prob. 8.2-3

8.2-4 A rubber ball is inflated to a pressure of 80 kPa. At that pressure the diameter of the ball is 208 mm and the wall thickness is 1.2 mm. The rubber has modulus of elasticity $E = 3.5$ MPa and Poisson's ratio $\nu = 0.45$. Determine the maximum stress and strain in the ball.

8.2-5 Solve the preceding problem if the pressure is 12 psi, the diameter is 8.20 in., the wall thickness is 0.048 in., the modulus of elasticity is 500 psi, and Poisson's ratio is 0.45.

8.2-6 A spherical steel pressure vessel (diameter 600 mm, thickness 10 mm) is coated with brittle lacquer that cracks when the strain exceeds 200×10^{-6}. What internal pressure p will cause the lacquer to develop cracks? (Assume $E = 205$ GPa and $\nu = 0.30$.)

8.2-7 A spherical tank of diameter 48 in. and wall thickness 2 in. contains compressed air at a pressure of 2500 psi. The tank is constructed of two hemispheres joined by a welded seam. (a) What is the tensile load f (lb per in. of length of weld) carried by the weld? (b) What is the maximum shear stress τ_{max} in the wall of the tank? (c) What is the maximum normal strain ϵ in the wall? (For steel, assume $E = 31 \times 10^6$ psi and $\nu = 0.29$.)

8.2-8 Solve the preceding problem for the following data: diameter 1200 mm, thickness 50 mm, pressure 20 MPa, modulus 210 GPa, and Poisson's ratio 0.29.

8.2-9 A spherical stainless-steel tank having a diameter of 16 in. is used to store propane gas at a pressure of 3000 psi. The properties of the steel are as follows: yield stress in tension, 140,000 psi; yield stress in shear, 65,000 psi; modulus of elasticity, 30×10^6 psi; and Poisson's ratio, 0.28. The desired factor of safety with respect to yielding is 2.75. Also, the normal strain must not exceed 1000×10^{-6}. Determine the minimum permissible thickness t_{min} of the tank.

8.2-10 Solve the preceding problem if the diameter is 400 mm, the pressure is 20 MPa, the yield stress in tension is 950 MPa, the yield stress in shear is 450 MPa, the factor of safety is 2.4, the modulus of elasticity is 200 GPa, Poisson's ratio is 0.28, and the normal strain must not exceed 1200×10^{-6}.

8.2-11 A hollow pressurized sphere having radius $r = 6.0$ in. and wall thickness $t = 0.5$ in. is lowered into a lake (see figure). The compressed air in the tank is at a pressure of 20 psi (gage pressure when the tank is out of the water). At what depth d will the wall of the tank be subjected to a compressive stress of 100 psi?

Prob. 8.2-11

CYLINDRICAL PRESSURE VESSELS

When solving problems for Section 8.3, assume that the given radius or diameter is an inside dimension.

8.3-1 A scuba tank is being designed for an internal pressure of 1800 psi with a factor of safety of 2.0 with respect to yielding. The yield stress of the steel is 42,000 psi in tension and 20,000 psi in shear. If the diameter of the tank is 6.0 in., what is the required wall thickness?

8.3-2 A tall standpipe with an open top (see figure) has diameter $d = 2$ m and wall thickness $t = 5$ mm. (a) What height h of water will produce a circumferential stress of 32 MPa in the wall of the standpipe? (b) What is the axial stress in the wall of the tank due to the water pressure?

Prob. 8.3-2

8.3-3 An inflatable structure used by a traveling circus has the shape of a half-circular cylinder with closed ends (see figure). The fabric and plastic structure is inflated by a small blower and has a radius of 40 ft when fully inflated. A longitudinal seam runs the entire length of the "ridge" of the structure. If the seam tears open when it is subjected to a tensile load of 540 pounds per inch of seam, what is the factor of safety n against tearing when the internal pressure is 0.5 psi and the structure is fully inflated?

Longitudinal seam

Prob. 8.3-3

8.3-4 A thin-walled cylindrical pressure vessel of radius r is subjected simultaneously to internal gas pressure p and a compressive force F acting at the ends (see figure). What should be the magnitude of the force F in order to produce pure shear in the wall of the cylinder?

$F \rightarrow$ $\leftarrow F$

Prob. 8.3-4

8.3-5 A strain gage is installed in the longitudinal direction on the surface of an aluminum beverage can (see figure). The radius-to-thickness ratio of the can is 200. When the lid of the can is popped open, the strain changes by $\epsilon_0 = 170 \times 10^{-6}$. What was the internal pressure p in the can? (Assume $E = 10 \times 10^6$ psi and $\nu = 0.33$.)

Soda Pop

Prob. 8.3-5

8.3-6 A circular cylindrical steel tank (see figure) contains a volatile fuel under pressure. A strain gage at point A records the longitudinal strain in the tank and transmits this information to a control room. The ultimate shear stress in the wall of the tank is 82 MPa and a factor of safety of 2 is required. At what value of the strain should the operators take action to reduce the pressure in the tank? (Data for the steel are as follows: modulus of elasticity $E = 205$ GPa and Poisson's ratio $\nu = 0.30$.)

Pressure relief valve

Cylindrical tank

A

Prob. 8.3-6

8.3-7 A cylinder filled with oil is under pressure from a piston, as shown in the figure. The diameter d of the piston is 1.91 in. and the compressive force F is 3600 lb. The maximum allowable shear stress τ_{allow} in the wall of the cylinder is 6000 psi. What is the minimum permissible thickness t_{min} of the cylinder wall?

Probs. 8.3-7 and 8.3-8

8.3-8 Solve the preceding problem if $d = 90.4$ mm, $F = 42.6$ kN, and $\tau_{allow} = 50$ MPa.

8.3-9 A standpipe in a water-supply system (see figure) is 12 ft in diameter and 6 inches thick. Two horizontal pipes carry water out of the standpipe; each is 2 ft in diameter and 1 inch thick. When the system is shut down and water fills the pipes but is not moving, the hoop stress at the bottom of the standpipe is 130 psi. (a) What is the height h of the water in the standpipe? (b) If the bottoms of the pipes are at the same elevation as the bottom of the standpipe, what is the hoop stress in the pipes?

Prob. 8.3-9

8.3-10 A cylindrical tank with hemispherical heads is constructed of steel sections that are welded circumferentially (see figure). The tank diameter is 1.2 m, the wall thickness is 20 mm, and the maximum internal pressure is 1800 kPa. (a) Determine the maximum tensile stress σ_h in the heads of the tank. (b) Determine the maximum tensile stress σ_c in the cylindrical part of the tank. (c) Determine the maximum tensile stress σ_w acting perpendicular to the welded joints. (d) Determine the maximum shear stress τ_h in the heads of the tank. (e) Determine the maximum shear stress τ_c in the cylindrical part of the tank.

Probs. 8.3-10 and 8.3-11

8.3-11 A cylindrical tank with diameter $d = 12$ in. is subjected to internal gas pressure $p = 320$ psi. The tank is constructed of steel sections that are welded circumferentially (see figure). The heads of the tank are hemispherical. The allowable tensile and shear stresses are 8000 psi and 3200 psi, respectively. Also, the allowable tensile stress perpendicular to a weld is 6400 psi. Determine the minimum required thickness t_{min} of (a) the cylindrical part of the tank, and (b) the hemispherical heads.

8.3-12 A pressurized steel tank is constructed with a helical weld that makes an angle $\alpha = 60°$ with the longitudinal axis (see figure). The tank has radius $r = 0.5$ m, wall thickness $t = 15$ mm, and internal pressure $p = 2.4$ MPa. Also, the steel has modulus of elasticity $E = 200$ GPa and Poisson's ratio $\nu = 0.30$. Determine the following quantities for the cylindrical part of the tank: (a) the circumferential and longitudinal stresses, (b) the maximum in-plane and out-of-plane shear stresses, (c) the circumferential and longitudinal strains, and (d) the normal and shear stresses acting on planes parallel and perpendicular to the weld (show these stresses on a properly oriented stress element).

Probs. 8.3-12 and 8.3-13

8.3-13 Solve the preceding problem for a welded tank with $\alpha = 65°$, $r = 18$ in., $t = 0.6$ in., $p = 200$ psi, $E = 30 \times 10^6$ psi, and $\nu = 0.30$.

8.3-14 A 60° strain-gage rosette (or delta rosette) is mounted on the outside of a cylindrical compressed air tank (see figure). The recorded strains are $\epsilon_a = 80 \times 10^{-6}$ and $\epsilon_b = \epsilon_c = 275 \times 10^{-6}$. If the tank has an r/t ratio of 25 and the modulus of elasticity is 200 GPa, what is the air pressure p in the tank?

Delta rosette

Probs. 8.3-14 and 8.3-15

8.3-15 Solve the preceding problem for a tank with an r/t ratio of 80 and modulus of elasticity $E = 30 \times 10^6$ psi if the measured strains are $\epsilon_a = 64 \times 10^{-6}$ and $\epsilon_b = \epsilon_c = 220 \times 10^{-6}$.

WORK SHEET 8 · · · **MAXIMUM STRESSES IN BEAMS**

Mathcad *When solving the problems for Section 8.4, consider only the in-plane stresses and disregard the weights of the beams.*

8.4-1 A cantilever beam of rectangular cross section is subjected to a concentrated load $P = 15$ k acting at the free end (see figure). The beam has width $b = 4$ in. and height $h = 10$ in. Point A is located at distance $c = 2$ ft from the free end and distance $d = 3$ in. from the bottom of the beam. Calculate the principal stresses σ_1 and σ_2 and the maximum shear stress τ_{max} at point A. Show these stresses on sketches of properly oriented elements.

Probs. 8.4-1 and 8.4-2

8.4-2 Solve the preceding problem for the following values: $P = 120$ kN, $b = 100$ mm, $h = 200$ mm, $c = 0.5$ m, and $d = 150$ mm.

8.4-3 A simple beam of rectangular cross section (width 4 in., height 8 in.) carries a uniform load of 1000 lb/ft on a span of 10 ft (see figure). Find the principal stresses σ_1 and

σ_2 and the maximum shear stress τ_{max} at a cross section 1 ft from the left-hand support at each of the following locations: (a) the neutral axis, (b) 2 in. above the neutral axis, and (c) the top of the beam. (Disregard the direct compressive stresses produced by the uniform load bearing against the top of the beam.)

Prob. 8.4-3

8.4-4 An overhanging beam ABC of rectangular cross section supports a concentrated load P at the free end (see figure). The span length from A to B is L, and the length of the overhang is $L/2$. The cross section has width b and height h. Point D is located midway between the supports at a distance d from the top face of the beam. Knowing that the maximum tensile stress (principal stress) at point D is $\sigma_1 = 49.0$ MPa, determine the magnitude of the load P. Data for the beam are as follows: $L = 1.25$ m, $b = 25$ mm, $h = 150$ mm, and $d = 25$ mm.

Probs. 8.4-4 and 8.4-5

8.4-5 Solve the preceding problem if the stress and dimensions are as follows: $\sigma_1 = 3620$ psi, $L = 50$ in., $b = 1$ in., $h = 6$ in., and $d = 1$ in.

8.4-6 A beam of wide-flange cross section (see figure) has the following dimensions: $b = 120$ mm, $t = 10$ mm, $h = 300$ mm, and $h_1 = 260$ mm. The beam is simply supported with span length $L = 3.0$ m. A concentrated load $P = 120$ kN acts at the midpoint of the span. At a cross section located 1.0 m from the left-hand support, calculate the principal stresses σ_1 and σ_2 and the maximum shear stress τ_{max} at each of the following locations: (a) the top of the beam, (b) the top of the web, and (c) the neutral axis.

Probs. 8.4-6 and 8.4-7

8.4-7 A beam of wide-flange cross section (see figure) has the following dimensions: $b = 5$ in., $t = 0.5$ in., $h = 12$ in., and $h_1 = 10.5$ in. The beam is simply supported with span length $L = 10$ ft and supports a uniform load $q = 6$ k/ft. Calculate the principal stresses σ_1 and σ_2 and the maximum shear stress τ_{max} at a cross section located 3 ft from the left-hand support at each of the following locations: (a) the bottom of the beam, (b) the bottom of the web, and (c) the neutral axis.

8.4-8 A W 10×30 wide-flange beam (see Table E-l, Appendix E) is simply supported with a span length of 8 ft (see figure). The beam supports a concentrated load of 30 kips at midspan. At a cross section located 2 ft from the left-hand support, determine the principal stresses σ_1 and σ_2 and the maximum shear stress τ_{max} at each of the following locations: (a) the top of the beam, (b) the top of the web, and (c) the neutral axis.

Prob. 8.4-8

8.4-9 A W 8×28 wide-flange beam (see Table E-l, Appendix E) is simply supported with a span length of 120 in. (see figure). The beam supports two symmetrically placed concentrated loads of 5 k each. At a cross section located 20 in. from the right-hand support, determine the principal stresses σ_1 and σ_2 and the maximum shear stress τ_{max} at each of the following locations: (a) the top of the beam, (b) the top of the web, and (c) the neutral axis.

Prob. 8.4-9

8.4-10 A cantilever beam of T-section is loaded by an inclined force of magnitude 10 kN (see figure). The line of action of the force is inclined at an angle of 60° to the horizontal and intersects the top of the beam at the end cross section. The beam is 1.5 m long and the cross section has the dimensions shown. Determine the principal stresses σ_1 and σ_2 and the maximum shear stress τ_{max} at points A and B in the web of the beam.

Prob. 8.4-10

8.4-11 A simple beam of rectangular cross section has span length $L = 60$ in. and supports a concentrated load $P = 18$ k at the midpoint (see figure). The height of the beam is $h = 6$ in. and the width is $b = 2$ in. Plot graphs of the principal stresses σ_1 and σ_2 and the maximum shear stress τ_{max}, showing how they vary over the height of the beam at cross section mn, which is located 20 in. from the left-hand support.

Probs. 8.4-11 and 8.4-12

8.4-12 Solve the preceding problem for a cross section mn located 0.15 m from the support if $L = 0.7$ m, $P = 144$ kN, $h = 120$ mm, and $b = 20$ mm.

*8.4-13 A cantilever beam of rectangular cross section supports a concentrated load P at the free end (see figure). The beam has length L, width b, and height h. Construct a diagram of stress contours for the maximum shear stress τ_{max}. Data for the beam are as follows: $P = 200$ lb, $b = 1$ in., $h = 3$ in., and $L = 18$ in. (*Note:* Draw contour lines for $\tau_{max} = 100, 200, 400, 600, 800,$ and 1000 psi.)

Probs. 8.4-13 and 8.4-14

*8.4-14 Solve the preceding problem if the data for the beam are as follows: $P = 1800$ N, $b = 30$ mm, $h = 90$ mm, and $L = 540$ mm. (*Note:* Draw contour lines for $\tau_{max} = 1, 2, 4, 6, 8,$ and 10 MPa.)

COMBINED LOADINGS

The problems for Section 8.5 are to be solved assuming that the structure behaves linearly elastically and that the stresses caused by two or more loads may be superimposed to obtain the resultant stresses acting at a point. Consider both in-plane and out-of-plane shear stresses unless otherwise specified.

8.5-1 A bracket $ABCD$ having a hollow circular cross section consists of a vertical arm AB, a horizontal arm BC parallel to the x_0 axis, and a horizontal arm CD parallel to the z_0 axis (see figure). The arms BC and CD have lengths $b_1 = 4$ ft and $b_2 = 3$ ft, respectively. The outer and inner diameters of the bracket are $d_2 = 8.0$ in. and $d_1 = 7.0$ in. A vertical load $P = 2.0$ k acts at point D. Determine the maximum tensile, compressive, and shear stresses in the bracket.

Prob. 8.5-1

8.5-2 The hollow drill pipe for an oil well (see figure) is 150 mm in outer diameter and 15 mm in thickness. Just above the bit, the compressive force in the pipe (due to the weight of the pipe) is 265 kN and the torque (due to drilling) is 19 kN·m. Determine the maximum tensile, compressive, and shear stresses in the drill pipe.

Prob. 8.5-2

8.5-3 A segment of a generator shaft is subjected to a torque T and an axial force P, as shown in the figure. The shaft is hollow (outer diameter $d_2 = 11$ in. and inner diameter $d_1 = 9$ in.) and delivers 2500 hp at 250 rpm. If the compressive force $P = 120$ k, what are the maximum tensile, compressive, and shear stresses in the shaft?

8.5-4 A generator shaft of hollow circular cross section is subjected to a torque $T = 25$ kN·m (see figure). The outer and inner diameters of the shaft are 200 mm and 160 mm, respectively. What is the maximum permissible compressive load P that can be applied to the shaft if the allowable in-plane shear stress is $\tau_{allow} = 45$ MPa?

Probs. 8.5-3 and 8.5-4

8.5-5 A gondola on a ski lift is supported by two bent arms, as shown in part (a) of the figure. Each arm is offset by the distance $b = 7.0$ in. from the line of action of the weight force W [see part (b) of the figure]. The allowable stresses in the arms are 15,000 psi in tension and 7,500 psi in shear. If the loaded gondola weighs 1300 lb, what is the required diameter d of the arms?

(a)   (b)

Prob. 8.5-5

8.5-6 A machine shaft of solid circular cross section is being designed for a tensile load $P = 80$ kN and a torque $T = 1.1$ kN·m (see figure). The allowable stresses are 60 MPa in tension and 30 MPa in shear. What is the required diameter of the shaft?

Prob. 8.5-6

8.5-7 A cylindrical tank having diameter $d = 2.5$ in. is subjected to internal gas pressure $p = 600$ psi and an external tensile load $P = 1000$ lb (see figure). Determine the required thickness t of the wall of the tank based upon an allowable shear stress of 3000 psi.

Prob. 8.5-7

8.5-8 A thin circular cylindrical shell subjected to internal pressure p is simultaneously compressed by an axial load $P = 72$ kN (see figure). The cylinder has diameter $d = 100$ mm and wall thickness $t = 4$ mm. Calculate the maximum allowable internal pressure p_{max} based upon an allowable shear stress in the wall of the shell of 60 MPa.

8.5-9 A cylindrical pressure vessel must resist an internal pressure $p = 1000$ psi and a compressive load $P = 12,000$ lb (see figure). The allowable shear stress in the wall of the cylinder is 6000 psi. It is desired to have the cross-sectional area of the vessel as small as possible while keeping the r/t ratio (radius-to-thickness ratio) between 10 and 20. Determine the optimum radius and thickness. (*Hints:* Express the principal stresses σ_1 and σ_2 and the maximum in-plane shear stresses τ_{max} in terms of r and t. Then set τ_{max} equal to the allowable stress and solve for t in terms of r. From this equation, find t, r/t, and the cross-sectional area A for various assumed values of r. From a plot of these results, decide on the value of r/t that makes the area A the smallest. Then calculate the corresponding values of r and t. As a final check, verify that the out-of-plane shear stresses are less than τ_{max}.)

Probs. 8.5-8 and 8.5-9

8.5-10 The torsional pendulum shown in the figure consists of a horizontal circular disk of mass $M = 50$ kg suspended by a vertical steel wire ($G = 80$ GPa) of length $L = 1.0$ m and diameter $d = 4$ mm. Calculate the maximum permissible angle of rotation ϕ_{max} of the disk (that is, the maximum amplitude of torsional vibrations) so that the stresses in the wire do not exceed 80 MPa in tension or 50 MPa in shear.

Prob. 8.5-10

8.5-11 Determine the maximum tensile, compressive, and shear stresses at point A on the bicycle pedal crank shown in the figure. The pedal and crank are in a horizontal plane and point A is located on the top of the crank. The load $P = 160$ lb acts in the vertical direction and the distances (in the horizontal plane) between the line of action of the load and point A are $b_1 = 5.0$ in. and $b_2 = 2.5$ in. Assume that the crank has a circular cross section with diameter $d = 0.6$ in.

Prob. 8.5-11

8.5-12 An L-shaped bracket lying in a horizontal plane supports a load $P = 600$ N (see figure). The bracket has a hollow rectangular cross section with thickness $t = 4$ mm and outer dimensions $b = 50$ mm and $h = 90$ mm. The centerline lengths of the arms are $b_1 = 500$ mm and $b_2 = 750$ mm. Considering only the load P, calculate the maximum tensile stress σ_t, maximum compressive stress σ_c, and maximum shear stress τ_{max} at point A, which is located on the top of the bracket at the support.

Prob. 8.5-12

8.5-13 A cylindrical pressure vessel having radius $r = 12$ in. and wall thickness $t = 0.6$ in. is subjected to internal pressure $p = 360$ psi. In addition, a torque $T = 90$ k-ft acts at each end of the cylinder (see figure). (a) Determine the maximum tensile stress σ_{max} and the maximum in-plane shear stress τ_{max} in the wall of the cylinder. (b) If the allowable in-plane shear stress is 3000 psi, what is the maximum allowable torque T?

Prob. 8.5-13

8.5-14 For purposes of analysis, a segment of the crankshaft in a vehicle is represented as shown in the figure. The load P equals 1.0 kN, and the dimensions are $b_1 = 80$ mm, $b_2 = 120$ mm, and $b_3 = 40$ mm. The diameter of the upper shaft is $d = 20$ mm. (a) Determine the maximum tensile, compressive, and shear stresses at point A, which is located on the surface of the shaft at the z_0 axis. (b) Determine the maximum tensile, compressive, and shear stresses at point B, which is located on the surface of the shaft at the y_0 axis.

Prob. 8.5-14

8.5-15 A post having a hollow circular cross section supports a horizontal load $P = 360$ lb acting at the end of an arm that is 4 ft long (see figure). The height of the post is 25 ft, and its section modulus is $S = 10$ in.[3] (a) Calculate the maximum tensile stress σ_{max} and maximum in-plane shear stress τ_{max} at point A due to the load P. Point A is located on the "front" of the post, that is, at the point where the tensile stress due to bending alone is a maximum. (b) If the maximum tensile stress and maximum in-plane shear stress at point A are limited to 16,000 psi and 6,000 psi, respectively, what is the largest permissible value of the load P?

Prob. 8.5-15

8.5-16 A sign is supported by a pipe (see figure) having outer diameter 100 mm and inner diameter 80 mm. The dimensions of the sign are 2.0 m × 0.75 m, and its lower edge is 3.2 m above the base. The wind pressure against the sign is 1.8 kPa. Determine the maximum in-plane shear stresses due to the wind pressure on the sign at points A, B, and C, located on the outer surface at the base of the pipe.

Prob. 8.5-16

8.5-17 A sign is supported by a pole of hollow circular cross section, as shown in the figure. The outer and inner diameters of the pole are 10.0 in. and 8.0 in., respectively. The pole is 40 ft high and weighs 3.8 k. The sign has dimensions 6 ft × 3 ft and weighs 400 lb; its center of gravity is 41 in. from the axis of the pole. The wind pressure against the sign is 30 lb/ft². (a) Determine the stresses acting on a stress element at point A, which is on the outer surface of the pole at the "front" of the pole, that is, the part of the pole nearest to the viewer. (b) Determine the maximum tensile, compressive, and shear stresses at point A.

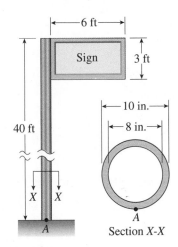

Prob. 8.5-17

8.5-18 A horizontal bracket ABC (see figure) consists of two perpendicular arms AB and BC, the latter having a length of 0.4 m. Arm AB has a solid circular cross section with diameter equal to 60 mm. At point C a load $P_1 = 2.02$ kN acts vertically and a load $P_2 = 3.07$ kN acts horizontally and parallel to arm AB. Considering only the forces P_1 and P_2, calculate the maximum tensile stress σ_t, the maximum compressive stress σ_c, and the maximum in-plane shear stress τ_{max} at point p, which is located at support A on the side of the bracket at midheight.

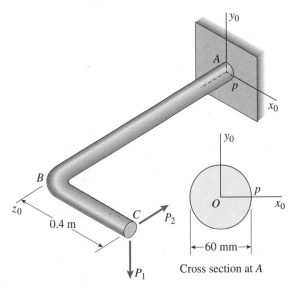

Cross section at A

Prob. 8.5-18

8.5-19 A segment of a cylindrical pressure vessel is subjected to a torque T and a bending moment M (see figure). The outer radius is 12.0 in. and the wall thickness is 1.0 in. The loads are as follows: $T = 800$ k-in., $M = 1000$ k-in., and the internal pressure $p = 900$ psi. Determine the maximum tensile stress σ_t, maximum compressive stress σ_c, and maximum shear stress τ_{max} in the wall of the cylinder.

Prob. 8.5-19

8.5-20 A pressurized cylindrical tank is loaded by torques T and tensile forces P (see figure). The tank has radius $r = 50$ mm and wall thickness $t = 3$ mm. The internal pressure $p = 3.5$ MPa and the torque $T = 500$ N·m. What is the maximum permissible value of the forces P if the allowable tensile stress in the wall of the cylinder is 70 MPa?

Prob. 8.5-20

9

DEFLECTIONS
OF BEAMS

9.1 INTRODUCTION

When a beam with a straight longitudinal axis is loaded by lateral forces, the axis is deformed into a curve, called the **deflection curve** of the beam. In Chapter 5 we used the curvature of the deflection curve to determine the normal strains and stresses in a beam. However, we did not develop a method for finding the deflections themselves. In this chapter, we will determine the equation of the deflection curve and also find deflections at specific points along the axis of the beam.

The calculation of deflections is an important part of structural analysis and design. For example, finding deflections is an essential ingredient in the analysis of statically indeterminate structures (Chapter 10). Deflections are also important in dynamic analysis, as when investigating the vibrations of aircraft or the response of buildings to earthquakes.

Deflections are sometimes calculated in order to verify that they are within tolerable limits. For instance, specifications for the design of buildings usually place upper limits on the deflections. Large deflections in buildings are unsightly (and even unnerving) and can cause cracks in ceilings and walls. In the design of machines and aircraft, specifications may limit deflections in order to prevent undesirable vibrations.

9.2 DIFFERENTIAL EQUATIONS OF THE DEFLECTION CURVE

All procedures for finding beam deflections are based on the differential equations of the deflection curve and their associated relationships.

Consequently, we will begin by deriving the basic equation for the deflection curve of a beam.

For discussion purposes, consider a cantilever beam with a concentrated load acting upward at the free end (Fig. 9-1a). Under the action of this load, the axis of the beam deforms into a curve, as shown in Fig. 9-1b. The reference axes have their origin at the fixed end of the beam, with the x axis directed to the right and the y axis directed upward. The z axis is directed outward from the figure (toward the viewer).

As in our previous discussions of beam bending in Chapter 5, we assume that the xy plane is a plane of symmetry of the beam, and we assume that all loads act in this plane (the *plane of bending*).

The **deflection** v is the displacement in the y direction of any point on the axis of the beam (Fig. 9-1b). Because the y axis is positive upward, the deflections are also positive upward. To obtain the equation of the deflection curve, we must express v as a function of x.*

Now let us consider the deflection curve in more detail. The deflection v at any point m_1 on the deflection curve is shown in Fig. 9-2a. Point m_1 is located at distance x from the origin (measured along the x axis). A second point m_2, located at distance $x + dx$ from the origin, is also shown. The deflection at this second point is $v + dv$, where dv is the increment in deflection as we move along the curve from m_1 to m_2.

When the beam is bent, there is not only a deflection at each point along the axis but also a rotation. The **angle of rotation** θ of the axis of the beam is the angle between the x axis and the tangent to the deflection curve, as shown for point m_1 in the enlarged view of Fig. 9-2b. For our choice of axes (x positive to the right and y positive upward), the angle of rotation is positive when counterclockwise. (Other names for the angle of rotation are *angle of inclination* and *angle of slope*.)

Fig. 9-1 Deflection curve of a cantilever beam

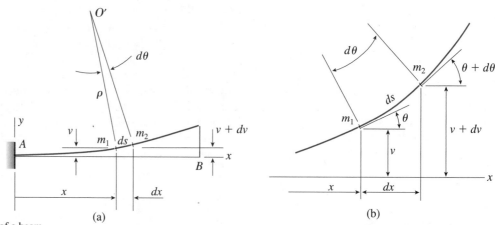

Fig. 9-2 Deflection curve of a beam

---

*As mentioned in Section 5.1, the traditional symbols for displacements in the x, y, and z directions are u, v, and w, respectively. The advantage of this notation is that it emphasizes the distinction between a *coordinate* and a *displacement*.

The angle of rotation at point m_2 is $\theta + d\theta$, where $d\theta$ is the increase in angle as we move from point m_1 to point m_2. It follows that if we construct lines normal to the tangents (Figs. 9-2a and b), the angle between these normals is $d\theta$. Also, as discussed earlier in Section 5.3, the point of intersection of these normals is the **center of curvature** O' (Fig. 9-2a) and the distance from O' to the curve is the **radius of curvature** ρ. From Fig. 9-2a we see that

$$\rho \, d\theta = ds \tag{a}$$

in which $d\theta$ is in radians and ds is the distance along the deflection curve between points m_1 and m_2. Therefore, the **curvature** κ (equal to the reciprocal of the radius of curvature) is given by the equation

$$\kappa = \frac{1}{\rho} = \frac{d\theta}{ds} \tag{9-1}$$

The **sign convention** for curvature is pictured in Fig. 9-3, which is repeated from Fig. 5-6 of Section 5.3. Note that curvature is positive when the angle of rotation increases as we move along the beam in the positive x direction.

The **slope of the deflection curve** is the first derivative dv/dx of the expression for the deflection v. In geometric terms, the slope is the increment dv in the deflection (as we go from point m_1 to point m_2 in Fig. 9-2) divided by the increment dx in the distance along the x axis. Since dv and dx are infinitesimally small, the slope dv/dx is equal to the tangent of the angle of rotation θ (Fig. 9-2b). Thus,

$$\frac{dv}{dx} = \tan \theta \qquad \theta = \arctan \frac{dv}{dx} \tag{9-2a,b}$$

In a similar manner, we also obtain the following relationships:

$$\cos \theta = \frac{dx}{ds} \qquad \sin \theta = \frac{dv}{ds} \tag{9-3a,b}$$

Note that when the x and y axes have the directions shown in Fig. 9-2a, the slope dv/dx is positive when the tangent to the curve slopes upward to the right.

Equations (9-1) through (9-3) are based only upon geometric considerations, and therefore they are valid for beams of any material. Furthermore, there are no restrictions on the magnitudes of the slopes and deflections.

Positive
curvature

(a)

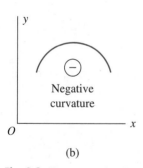

Negative
curvature

(b)

Fig. 9-3 Sign convention for curvature

■ **Beams with Small Rotations**

Most beams undergo only very small deflections and angles of rotation under service conditions, and so their deflection curves have extremely small curvatures. Therefore, we can make some mathematical approximations that greatly simplify the analysis. For instance, since $\cos \theta \approx 1$ when θ is small, Eq. (9-3a) gives

$$ds \approx dx \tag{b}$$

and Eq. (9-1) becomes

$$\kappa = \frac{1}{\rho} = \frac{d\theta}{dx} \tag{9-4}$$

Also, since $\tan \theta \approx \theta$ when θ is small, we can make the following approximation to Eq. (9-2a):

$$\theta \approx \tan\theta = \frac{dv}{dx} \tag{c}$$

Thus, if the rotations of a beam are small, we can assume that the angle of rotation θ and the slope dv/dx are equal. (Note that the angle of rotation must be measured in radians.)

Taking the derivative of θ with respect to x in Eq. (c), we get

$$\frac{d\theta}{dx} = \frac{d^2v}{dx^2} \tag{d}$$

Combining this equation with Eq. (9-4), we obtain a relation between the curvature of a beam and its deflection:

$$\kappa = \frac{1}{\rho} = \frac{d^2v}{dx^2} \tag{9-5}$$

This equation is valid for a beam of any material, provided the rotations are small quantities.

If the material of a beam is linearly elastic and follows Hooke's law, the curvature (from Eq. 5-12, Chapter 5) is:

$$\kappa = \frac{1}{\rho} = \frac{M}{EI} \tag{9-6}$$

in which M is the bending moment and EI is the flexural rigidity of the beam. Equation (9-6) shows that a positive bending moment produces positive curvature and a negative bending moment produces negative curvature, as shown earlier in Fig. 5-10.

Combining Eq. (9-5) with Eq. (9-6) yields the basic **differential equation of the deflection curve** of a beam:

$$\frac{d^2v}{dx^2} = \frac{M}{EI} \tag{9-7}$$

This equation can be integrated in each particular case to find the deflection v, provided the bending moment M and flexural rigidity EI are known as functions of x.

As a reminder, the **sign conventions** to be used with the preceding equations are repeated here: (1) the x and y axes are positive to the right and upward, respectively; (2) the deflection v is positive upward; (3) the slope dv/dx and angle of rotation θ are positive when counterclockwise with respect to the positive x axis; (4) the curvature κ is positive when the beam is bent concave upward; and (5) the bending moment M is positive when it produces compression in the upper part of the beam.

Other equations can be obtained by replacing the bending moment M with the shear force V and the intensity q of the distributed load. In Chapter 4 we derived the following relations between M, V, and q (see Eqs. 4-4 and 4-6):

$$\frac{dV}{dx} = -q \qquad \frac{dM}{dx} = V \tag{9-8a,b}$$

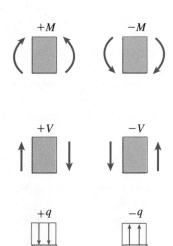

The sign conventions for these quantities are shown in Fig. 9-4. By differentiating Eq. (9-7) with respect to x and then substituting the preceding equations for shear force and load, we can obtain the additional equations. In so doing, we will consider two cases, nonprismatic beams and prismatic beams.

In the case of a **nonprismatic beam**, the flexural rigidity EI is variable, and therefore we write Eq. (9-7) in the form

$$EI_x \frac{d^2v}{dx^2} = M \tag{9-9a}$$

where the subscript x is inserted as a reminder that the flexural rigidity may vary with x. Differentiating both sides of this equation and using Eqs. (9-8a) and (9-8b), we obtain

$$\frac{d}{dx}\left(EI_x \frac{d^2v}{dx^2}\right) = \frac{dM}{dx} = V \tag{9-9b}$$

$$\frac{d^2}{dx^2}\left(EI_x \frac{d^2v}{dx^2}\right) = \frac{dV}{dx} = -q \tag{9-9c}$$

Fig. 9-4 Sign conventions for bending moment M, shear force V, and intensity q of distributed load

The deflection of a nonprismatic beam can be found by solving (either analytically or numerically) any one of the three preceding differential equations. The choice usually depends upon which equation provides the most efficient solution.

In the case of a **prismatic beam** (constant EI), the differential equations become

$$EI\frac{d^2v}{dx^2} = M \qquad EI\frac{d^3v}{dx^3} = V \qquad EI\frac{d^4v}{dx^4} = -q \qquad \text{(9-10a,b,c)}$$

To simplify the writing of these and other equations, **primes** are often used to denote differentiation:

$$v' \equiv \frac{dv}{dx} \qquad v'' \equiv \frac{d^2v}{dx^2} \qquad v''' \equiv \frac{d^3v}{dx^3} \qquad v'''' \equiv \frac{d^4v}{dx^4} \qquad \text{(9-11)}$$

Using this notation, we can express the differential equations for a prismatic beam in the following forms:

$$EIv'' = M \qquad EIv''' = V \qquad EIv'''' = -q \qquad \text{(9-12a,b,c)}$$

We will refer to these equations as the *bending-moment equation*, the *shear-force equation*, and the *load equation*, respectively.

In the next two sections we will use the preceding equations to find deflections of beams. The general procedure consists of integrating the equations and then evaluating the constants of integration from boundary and other conditions pertaining to the beam.

When deriving the differential equations (Eqs. 9-9, 9-10, and 9-12), we assumed that the material followed Hooke's law and that the slopes of the deflection curve were very small. We also assumed that any shear deformations were negligible; consequently, we considered only the deformations due to pure bending. All of these assumptions are satisfied by most beams in common use.

■ Exact Expression for Curvature

If the deflection curve of a beam has large slopes, we cannot use the approximations given by Eqs. (b) and (c). Instead, we must resort to the exact expressions for curvature and angle of rotation (see Eqs. 9-1 and 9-2b). Combining those expressions, we get

$$\kappa = \frac{1}{\rho} = \frac{d\theta}{ds} = \frac{d(\arctan v')}{dx}\frac{dx}{ds} \qquad \text{(e)}$$

From Fig. 9-2 we see that

$$ds^2 = dx^2 + dv^2 \quad \text{or} \quad ds = [dx^2 + dv^2]^{1/2} \qquad \text{(f,g)}$$

Dividing both sides of Eq. (g) by dx gives

$$\frac{ds}{dx} = \left[1 + \left(\frac{dv}{dx}\right)^2\right]^{1/2} = [1 + (v')^2]^{1/2} \quad \text{or} \quad \frac{dx}{ds} = \frac{1}{[1 + (v')^2]^{1/2}} \qquad \text{(h,i)}$$

Also, differentiation of the arctangent function (see Appendix C) gives

$$\frac{d}{dx}(\arctan v') = \frac{v''}{1 + (v')^2} \tag{j}$$

Substitution of expressions (i) and (j) into the equation for curvature (Eq. e) yields

$$\kappa = \frac{1}{\rho} = \frac{v''}{[1 + (v')^2]^{3/2}} \tag{9-13}$$

Comparing this equation with Eq. (9-5), we see that the assumption of small rotations is equivalent to disregarding $(v')^2$ in comparison to one. Equation (9-13) should be used for the curvature whenever the slopes are large.*

9.3 DEFLECTIONS BY INTEGRATION OF THE BENDING-MOMENT EQUATION

We are now ready to solve the differential equations of the deflection curve and obtain deflections of beams. The first equation we will use is the bending-moment equation (Eq. 9-12a). Since this equation is of second order, two integrations are required. The first integration produces the slope $v' = dv/dx$, and the second produces the deflection v.

We begin the analysis by writing the equation (or equations) for the bending moments in the beam. Since only statically determinate beams are considered in this chapter, we can obtain the bending moments from free-body diagrams and equations of equilibrium, using the procedures described in Chapter 4. In some cases a single bending-moment expression holds for the entire length of the beam, as illustrated below in Examples 9-1 and 9-2. In other cases the bending moment changes abruptly at one or more points along the axis of the beam. Then we must write separate bending-moment expressions for each region of the beam between points where changes occur, as illustrated in Example 9-3.

Regardless of the number of bending-moment expressions, the general procedure for solving the differential equations is as follows. For each region of the beam, we substitute the expression for M into the differential equation and integrate to obtain the slope v'. Each such integration produces one constant of integration. Next, we integrate each slope equation to obtain the corresponding deflection v. Again, each integration produces a new constant. Thus, there are two constants of

---

*The basic relationship stating that the curvature of a beam is proportional to the bending moment (Eq. 9-6) was first obtained by Jacob Bernoulli, although he obtained an incorrect value for the constant of proportionality. The relationship was used later by Euler, who solved the differential equation of the deflection curve for both large deflections (using Eq. 9-13) and small deflections (using Eq. 9-7). For the history of deflection curves, see Ref. 9-1.

$v_A = 0$ $v_B = 0$

Fig. 9-5 Boundary conditions at simple supports

$v_A = 0$

$v'_A = 0$

Fig. 9-6 Boundary conditions at a fixed support

integration for each region of the beam. These constants are evaluated from known conditions pertaining to the slopes and deflections. The conditions fall into three categories: (1) boundary conditions, (2) continuity conditions, and (3) symmetry conditions.

Boundary conditions pertain to the deflections and slopes at the supports of a beam. For example, at a simple support (either a pin or a roller) the deflection is zero (Fig. 9-5), and at a fixed support both the deflection and the slope are zero (Fig. 9-6). Each such boundary condition supplies one equation that can be used to evaluate the constants of integration.

Continuity conditions occur at points where the regions of integration meet, such as at point C in the beam of Fig. 9-7. The deflection curve of this beam is physically continuous at point C, and therefore the deflection at point C as determined for the left-hand part of the beam must be equal to the deflection at point C as determined for the right-hand part. Similarly, the slopes found for each part of the beam must be equal at point C. Each of these continuity conditions supplies an equation for evaluating the constants of integration.

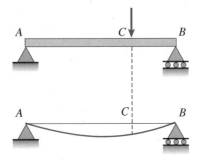

At point C: $(v)_{AC} = (v)_{CB}$

$(v')_{AC} = (v')_{CB}$

Fig. 9-7 Continuity conditions at point C

Symmetry conditions may also be available. For instance, if a simple beam supports a uniform load throughout its length, we know in advance that the slope of the deflection curve at the midpoint must be zero. This condition supplies an additional equation, as illustrated in Example 9-1.

Each boundary, continuity, and symmetry condition leads to an equation containing one or more of the constants of integration. Since the number of *independent* conditions always matches the number of constants of integration, we can always solve these equations for the constants. (The boundary and continuity conditions alone are always sufficient to determine the constants. Any symmetry conditions provide · additional equations, but they are not independent of the other equations. The choice of which conditions to use is a matter of convenience.)

Once the constants are evaluated, they can be substituted back into the expressions for slopes and deflections, thus yielding the final equations of the deflection curve. These equations can then be used to obtain the deflections and angles of rotation at particular points along the axis of the beam.

The preceding method for finding deflections is sometimes called the **method of successive integrations**. The following examples illustrate the method in detail.

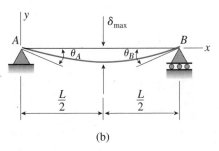

(a)

(b)

Fig. 9-8 Example 9-1. Deflections of a simple beam with a uniform load

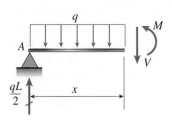

Fig. 9-9 Free-body diagram used in determining the bending moment M (Example 9-1)

■ Example 9-1

Determine the equation of the deflection curve for a simple beam AB supporting a uniform load of intensity q acting throughout the span of the beam (Fig. 9-8a). Also, determine the maximum deflection δ_{max} at the midpoint of the beam and the angles of rotation θ_A and θ_B at the supports (Fig. 9-8b). (*Note:* The beam has length L and constant flexural rigidity EI.)

Solution   *Bending moment in the beam.* The bending moment at a cross section distance x from the left-hand support is obtained from the free-body diagram of Fig. 9-9. The reaction at the support is $qL/2$, and therefore the equation for the bending moment is

$$M = \frac{qLx}{2} - \frac{qx^2}{2} \tag{9-14}$$

Differential equation of the deflection curve. By substituting the expression for the bending moment (Eq. 9-14) into the differential equation (Eq. 9-12a), we obtain

$$EIv'' = \frac{qLx}{2} - \frac{qx^2}{2} \tag{9-15}$$

This equation can now be integrated to obtain the slopes and deflections of the beam.

Slope of the beam. Multiplying both sides of the differential equation by dx and integrating, we get the following equation for the slope:

$$EIv' = \frac{qLx^2}{4} - \frac{qx^3}{6} + C_1 \tag{a}$$

in which C_1 is a constant of integration. To evaluate the constant C_1, we observe from the symmetry of the beam and its load that the slope of the deflection curve at midspan is equal to zero. Thus, we have the following symmetry condition:

$$v' = 0 \quad \text{when} \quad x = \frac{L}{2}$$

This condition may be expressed more succinctly as

$$v'\left(\frac{L}{2}\right) = 0$$

Applying this condition to Eq. (a) gives

$$0 = \frac{qL}{4}\left(\frac{L}{2}\right)^2 - \frac{q}{6}\left(\frac{L}{2}\right)^3 + C_1 \quad \text{or} \quad C_1 = -\frac{qL^3}{24}$$

The equation for the slope of the beam (Eq. a) then becomes

$$EIv' = \frac{qLx^2}{4} - \frac{qx^3}{6} - \frac{qL^3}{24} \tag{b}$$

or

$$v' = -\frac{q}{24EI}(L^3 - 6Lx^2 + 4x^3) \tag{9-16} \quad \longleftarrow$$

As expected, the slope is negative (i.e., clockwise) at the left-hand end of the beam ($x = 0$), positive at the right-hand end ($x = L$), and equal to zero at the midpoint ($x = L/2$).

Deflection of the beam. The deflection is obtained by integrating the equation for the slope. Thus, upon multiplying both sides of Eq. (b) by dx and integrating, we obtain

$$EIv = \frac{qLx^3}{12} - \frac{qx^4}{24} - \frac{qL^3x}{24} + C_2 \tag{c}$$

The constant of integration C_2 may be evaluated from the condition that the deflection of the beam at the left-hand support is equal to zero; that is, $v = 0$ when $x = 0$, or

$$v(0) = 0$$

Applying this condition to Eq. (c) yields $C_2 = 0$; hence the equation for the deflection curve is

$$EIv = \frac{qLx^3}{12} - \frac{qx^4}{24} - \frac{qL^3x}{24} \tag{d}$$

or

$$v = -\frac{qx}{24EI}(L^3 - 2Lx^2 + x^3) \tag{9-17} \quad \longleftarrow$$

This equation gives the deflection at any point along the axis of the beam. Note that the deflection is zero at both ends of the beam ($x = 0$ and $x = L$) and negative elsewhere (recall that downward deflections are negative).

Maximum deflection. The maximum deflection δ_{max} occurs at the midpoint of the span (Fig. 9-8b) and is obtained by setting x equal to $L/2$ in Eq. (9-17). However, since δ_{max} represents the magnitude of the maximum deflection, and since the deflection v is negative when downward, we must insert a minus sign in the equation, as follows:

$$\delta_{max} = \left| v\left(\frac{L}{2}\right) \right| = -v\left(\frac{L}{2}\right) = \frac{5qL^4}{384EI} \tag{9-18} \quad \longleftarrow$$

Angles of rotation. The maximum angles of rotation occur at the supports of the beam. At the left-hand end of the beam, the angle θ_A, which is a clockwise angle (Fig. 9-8b), is equal to the negative of the slope v'. Thus, by substituting $x = 0$ into Eq. (9-16), we find

$$\theta_A = -v'(0) = \frac{qL^3}{24EI} \qquad (9\text{-}19) \quad \blacktriangleleft$$

In a similar manner, we can obtain the angle of rotation θ_B at the right-hand end of the beam. Since θ_B is a counterclockwise angle, it is equal to the slope at the end:

$$\theta_B = v'(L) = \frac{qL^3}{24EI} \qquad (9\text{-}20) \quad \blacktriangleleft$$

Because the beam and loading are symmetric about the midpoint, the angles of rotation at the ends are equal.

This example illustrates the process of setting up and solving the differential equation of the deflection curve. It also illustrates the process of finding slopes and deflections at selected points along the axis of a beam.

Note: Now that we have derived formulas for the maximum deflection and maximum angles of rotation (see Eqs. 9-18, 9-19, and 9-20), we can evaluate those quantities numerically and observe that the deflections and angles are indeed small, as the theory requires.

Consider a steel beam on simple supports with a span length $L = 6$ ft. The cross section is rectangular with width $b = 3$ in. and height $h = 6$ in. The intensity of uniform load is $q = 8000$ lb/ft, which is relatively large because it produces a stress in the beam of 24,000 psi. (Thus, the deflections and slopes are larger than would normally be expected.)

Substituting into Eq. (9-18), and using $E = 30 \times 10^6$ psi, we find that the maximum deflection is $\delta_{max} = 0.144$ in., which is only 1/500 of the span length. Also, from Eq. (9-19), we find that the maximum angle of rotation is $\theta_A = 0.0064$ radians, or $0.37°$, which is a very small angle.

Thus, our assumption that the slopes and deflections are small is validated.

■ Example 9-2

Determine the equation of the deflection curve for a cantilever beam AB subjected to a uniform load of intensity q (Fig. 9-10a). Also, determine the angle of rotation θ_B and the deflection δ_B at the free end (Fig. 9-10b). (*Note:* The beam has length L and constant flexural rigidity EI.)

Solution   *Bending moment in the beam.* The bending moment at distance x from the fixed support is obtained from the free-body diagram of Fig. 9-11. Note that the vertical reaction at the support is equal to qL and the moment reaction is equal to $qL^2/2$. Consequently, the expression for the bending moment M is

$$M = -\frac{qL^2}{2} + qLx - \frac{qx^2}{2} \qquad (9\text{-}$$

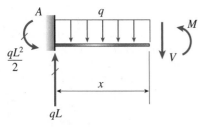

Fig. 9-10 Example 9-2. Deflections of a cantilever beam with a uniform load

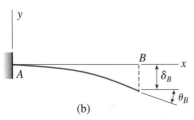

Fig. 9-11 Free-body diagram used in determining the bending moment M (Example 9-2)

Differential equation of the deflection curve. When the preceding expression for the bending moment is substituted into the differential equation (Eq. 9-12a), we obtain

$$EIv'' = -\frac{qL^2}{2} + qLx - \frac{qx^2}{2} \tag{9-22}$$

We now integrate both sides of this equation to obtain the slopes and deflections.

Slope of the beam. The first integration of Eq. (9-22) gives the following equation for the slope:

$$EIv' = -\frac{qL^2x}{2} + \frac{qLx^2}{2} - \frac{qx^3}{6} + C_1 \tag{e}$$

The constant of integration C_1 can be found from the boundary condition that the slope of the beam is zero at the support; thus, we have the following condition:

$$v'(0) = 0$$

When this condition is applied to Eq. (e) we get $C_1 = 0$. Therefore, Eq. (e) becomes

$$EIv' = -\frac{qL^2x}{2} + \frac{qLx^2}{2} - \frac{qx^3}{6} \tag{f}$$

and the slope is

$$v' = -\frac{qx}{6EI}(3L^2 - 3Lx + x^2) \tag{9-23}$$

As expected, the slope is zero at the support ($x = 0$) and negative (i.e., clockwise) throughout the length of the beam.

Deflection of the beam. Integration of the slope equation (Eq. f) yields

$$EIv = -\frac{qL^2x^2}{4} + \frac{qLx^3}{6} - \frac{qx^4}{24} + C_2 \tag{g}$$

The constant C_2 is found from the boundary condition that the deflection of the beam is zero at the support:

$$v(0) = 0$$

When this condition is applied to Eq. (g), we see immediately that $C_2 = 0$. Therefore, the equation for the deflection v is

$$v = -\frac{qx^2}{24EI}(6L^2 - 4Lx + x^2) \tag{9-24}$$

As expected, the deflection v is zero at the support ($x = 0$) and negative (that is, downward) elsewhere.

Angle of rotation at the free end of the beam. The clockwise angle of rotation θ_B at end B of the beam (Fig. 9-10b) is equal to the negative of the slope at that point. Thus, using Eq. (9-23), we get

$$\theta_B = -v'(L) = \frac{qL^3}{6EI} \tag{9-25}$$

This angle is the maximum angle of rotation for the beam.

Deflection at the free end of the beam. Since the deflection δ_B is downward (Fig. 9-10b), it is equal to the negative of the deflection obtained from Eq. (9-24):

$$\delta_B = -v(L) = \frac{qL^4}{8EI} \qquad (9\text{-}26) \quad \longleftarrow$$

This deflection is the maximum deflection of the beam.

(b)

Fig. 9-12 Example 9-3. Deflections of a simple beam with a concentrated load

■ **Example 9-3**

A simple beam AB supports a concentrated load P acting at distances a and b from the left-hand and right-hand supports, respectively (Fig. 9-12a). Determine the equations of the deflection curve, the angles of rotation θ_A and θ_B at the supports, the maximum deflection $\delta_{\max}$, and the deflection δ_C at the midpoint C of the beam (Fig. 9-12b). (*Note:* The beam has length L and constant flexural rigidity EI.)

Solution   *Bending moments in the beam.* In this example the bending moments are expressed by two equations, one for each part of the beam. Using the free-body diagrams of Fig. 9-13, we arrive at the following equations:

$$M = \frac{Pbx}{L} \qquad (0 \leq x \leq a) \qquad (9\text{-}27a)$$

$$M = \frac{Pbx}{L} - P(x - a) \qquad (a \leq x \leq L) \qquad (9\text{-}27b)$$

Differential equations of the deflection curve. The differential equations for the two parts of the beam are obtained by substituting the bending-moment expressions (Eqs. 9-27a and b) into Eq. (9-12a). The results are

$$EIv'' = \frac{Pbx}{L} \qquad (0 \leq x \leq a) \qquad (9\text{-}28a)$$

$$EIv'' = \frac{Pbx}{L} - P(x - a) \qquad (a \leq x \leq L) \qquad (9\text{-}28b)$$

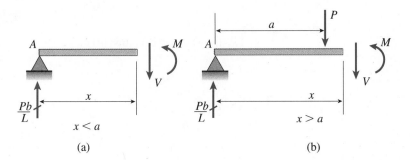

Fig. 9-13 Free-body diagrams used in determining the bending moments (Example 9-3)

Slopes and deflections of the beam. The first integrations of the two differential equations yield the following expressions for the slopes:

$$EIv' = \frac{Pbx^2}{2L} + C_1 \qquad (0 \leq x \leq a) \tag{h}$$

$$EIv' = \frac{Pbx^2}{2L} - \frac{P(x-a)^2}{2} + C_2 \qquad (a \leq x \leq L) \tag{i}$$

in which C_1 and C_2 are constants of integration. A second pair of integrations gives the deflections:

$$EIv = \frac{Pbx^3}{6L} + C_1 x + C_3 \qquad (0 \leq x \leq a) \tag{j}$$

$$EIv = \frac{Pbx^3}{6L} - \frac{P(x-a)^3}{6} + C_2 x + C_4 \qquad (a \leq x \leq L) \tag{k}$$

These equations contain two additional constants of integration, making a total of four constants to be evaluated.

Constants of integration. The four constants of integration can be found from the following four conditions:

(1) At $x = a$, the slopes v' for the two parts of the beam are the same
(2) At $x = a$, the deflections v for the two parts of the beam are the same
(3) At $x = 0$, the deflection v is zero
(4) At $x = L$, the deflection v is zero

The first two conditions are continuity conditions based upon the fact that the axis of the beam is a continuous curve. Conditions (3) and (4) are boundary conditions that must be satisfied at the supports.

Condition (1) means that the slopes determined from Eqs. (h) and (i) must be equal when $x = a$; therefore,

$$\frac{Pba^2}{2L} + C_1 = \frac{Pba^2}{2L} + C_2 \quad \text{or} \quad C_1 = C_2$$

Condition (2) means that the deflections found from Eqs. (j) and (k) must be equal when $x = a$; therefore,

$$\frac{Pba^3}{6L} + C_1 a + C_3 = \frac{Pba^3}{6L} + C_2 a + C_4$$

Inasmuch as $C_1 = C_2$, this equation gives $C_3 = C_4$.

Next, we apply condition (3) to Eq. (j) and obtain $C_3 = 0$; therefore,

$$C_3 = C_4 = 0 \tag{l}$$

Finally, we apply condition (4) to Eq. (k) and obtain

$$\frac{PbL^2}{6} - \frac{Pb^3}{6} + C_2 L = 0$$

Therefore,

$$C_1 = C_2 = -\frac{Pb(L^2 - b^2)}{6L} \tag{m}$$

Equations of the deflection curve. We now substitute the constants of integration (Eqs. l and m) into the equations for the deflections (Eqs. j and k) and obtain the deflection equations for the two parts of the beam. The resulting equations, after a slight rearrangement, are

$$v = -\frac{Pbx}{6LEI}(L^2 - b^2 - x^2) \qquad (0 \le x \le a) \qquad\qquad (9\text{-}29a) \quad\Longleftarrow$$

$$v = -\frac{Pbx}{6LEI}(L^2 - b^2 - x^2) - \frac{P(x-a)^3}{6EI} \qquad (a \le x \le L) \quad (9\text{-}29b) \quad\Longleftarrow$$

The first of these equations gives the deflection curve for the part of the beam to the left of the load P, and the second gives the deflection curve for the part of the beam to the right of the load.

The slopes for the two parts of the beam can be found either by substituting the values of C_1 and C_2 into Eqs. (h) and (i) or by taking the first derivatives of the deflection equations (Eqs. 9-29a and b). The resulting equations are

$$v' = -\frac{Pb}{6LEI}(L^2 - b^2 - 3x^2) \qquad (0 \le x \le a) \qquad\qquad (9\text{-}30a) \quad\Longleftarrow$$

$$v' = -\frac{Pb}{6LEI}(L^2 - b^2 - 3x^2) - \frac{P(x-a)^2}{2EI} \qquad (a \le x \le L) \quad (9\text{-}30b) \quad\Longleftarrow$$

The deflection and slope at any point along the axis of the beam can be calculated from Eqs. (9-29) and (9-30).

Angles of rotation at the supports. To obtain the angles of rotation θ_A and θ_B at the ends of the beam (Fig. 9-12b), we substitute $x = 0$ into Eq. (9-30a) and $x = L$ into Eq. (9-30b):

$$\theta_A = -v'(0) = \frac{Pb(L^2 - b^2)}{6LEI} = \frac{Pab(L + b)}{6LEI} \qquad\qquad (9\text{-}31a) \quad\Longleftarrow$$

$$\theta_B = v'(L) = \frac{Pb(2L^2 - 3bL + b^2)}{6LEI} = \frac{Pab(L + a)}{6LEI} \qquad (9\text{-}31b) \quad\Longleftarrow$$

Note that the angle θ_A is clockwise and the angle θ_B is counterclockwise, as shown in Fig. 9-12b.

The angles of rotation are functions of the position of the load and reach their largest values when the load is located near the midpoint of the beam. In the case of the angle of rotation θ_A, the maximum value of the angle is

$$(\theta_A)_{\max} = \frac{PL^2\sqrt{3}}{27EI} \qquad\qquad (9\text{-}32)$$

and occurs when $b = L/\sqrt{3} = 0.577L$ (or $a = 0.423L$). This result is obtained by taking the derivative of θ_A (Eq. 9-31a) with respect to b and setting it equal to zero.

Maximum deflection of the beam. The maximum deflection $\delta_{\max}$ occurs at point D (Fig. 9-12b) where the deflection curve has a horizontal tangent. If the load is to the right of the midpoint, that is, if $a > b$, point D is in the part of the beam to the left of the load. We can locate this point by equating the slope v' from Eq. (9-30a) to zero and solving for the distance x, which we now denote as x_1. In this manner we obtain the following formula for x_1:

$$x_1 = \sqrt{\frac{L^2 - b^2}{3}} \qquad (a \geq b) \qquad (9\text{-}33)$$

From this equation we see that as the load P moves from the middle of the beam ($b = L/2$) to the right-hand end ($b = 0$), the distance x_1 varies from $L/2$ to $L/\sqrt{3} = 0.577L$. Thus, the maximum deflection occurs at a point very close to the midpoint of the beam, and this point is always between the midpoint of the beam and the load.

The maximum deflection δ_{max} is found by substituting x_1 (from Eq. 9-33) into the deflection equation (Eq. 9-29a) and then inserting a minus sign:

$$\delta_{max} = -(v)_{x=x_1} = \frac{Pb(L^2 - b^2)^{3/2}}{9\sqrt{3}\,LEI} \qquad (a \geq b) \qquad (9\text{-}34) \quad \longleftarrow$$

The minus sign is needed because the maximum deflection is downward (Fig. 9-12b) whereas the deflection v is positive upward.

The maximum deflection of the beam depends on the position of the load P, that is, on the distance b. The maximum value of the maximum deflection (the "max-max" deflection) occurs when $b = L/2$ and the load is at the midpoint of the beam. This maximum deflection is equal to $PL^3/48EI$.

Deflection at the midpoint of the beam. The deflection δ_C at the midpoint C when the load is acting to the right of the midpoint (Fig. 9-12b) is obtained by substituting $x = L/2$ into Eq. (9-29a), as follows:

$$\delta_C = -v\left(\frac{L}{2}\right) = \frac{Pb(3L^2 - 4b^2)}{48EI} \qquad (a \geq b) \qquad (9\text{-}35) \quad \longleftarrow$$

Because the maximum deflection always occurs near the midpoint of the beam, Eq. (9-35) yields a close approximation to the maximum deflection. In the most unfavorable case (when b approaches zero), the difference between the maximum deflection and the deflection at the midpoint is less than 3% of the maximum deflection, as demonstrated in Problem 9.3-10.

Special case (load at the midpoint of the beam). An important special case occurs when the load P acts at the midpoint of the beam ($a = b = L/2$). Then we obtain the following results from Eqs. (9-30a), (9-29a), (9-31), and (9-34), respectively:

$$v' = -\frac{P}{16EI}(L^2 - 4x^2) \qquad \left(0 \leq x \leq \frac{L}{2}\right) \qquad (9\text{-}36)$$

$$v = -\frac{Px}{48EI}(3L^2 - 4x^2) \qquad \left(0 \leq x \leq \frac{L}{2}\right) \qquad (9\text{-}37)$$

$$\theta_A = \theta_B = \frac{PL^2}{16EI} \qquad (9\text{-}38)$$

$$\delta_{max} = \delta_C = \frac{PL^3}{48EI} \qquad (9\text{-}39)$$

Since the deflection curve is symmetric about the midpoint of the beam, the equations for v' and v are given only for the left-hand half of the beam (Eqs. 9-36 and 9-37). If needed, the equations for the right-hand half can be obtained from Eqs. (9-30b) and (9-29b) by substituting $a = b = L/2$.

9.4 DEFLECTIONS BY INTEGRATION OF THE SHEAR-FORCE AND LOAD EQUATIONS

The equations of the deflection curve in terms of the shear force V and the load q (Eqs. 9-12b and c, respectively) may also be integrated to obtain slopes and deflections. Since the loads are usually given quantities, whereas the bending moments must be determined from free-body diagrams and equations of equilibrium, many analysts prefer to start with the load equation. For this same reason, most computer programs for finding deflections begin with the load equation and then perform numerical integrations to obtain the shear forces, bending moments, slopes, and deflections.

The procedure for solving either the load equation or the shear-force equation is similar to that for solving the bending-moment equation, except that more integrations are required. For instance, if we begin with the load equation, four integrations are needed in order to arrive at the deflections. Thus, four constants of integration are introduced for each load equation that is integrated. As before, these constants are found from boundary, continuity, and symmetry conditions. However, these conditions now include conditions on the shear forces and bending moments as well as conditions on the slopes and deflections.

Conditions on the shear forces are equivalent to conditions on the third derivative (because $EIv''' = V$). In a similar manner, conditions on the bending moments are equivalent to conditions on the second derivative (because $EIv'' = M$). When the shear force and bending moment conditions are added to those for the slopes and deflections, we always have enough independent conditions to solve for the constants of integration.

The following examples illustrate the techniques of analysis in detail. The first example begins with the load equation and the second begins with the shear-force equation.

■ **Example 9-4**

Determine the equation of the deflection curve for a cantilever beam AB supporting a triangularly distributed load of maximum intensity q_0 (Fig. 9-14a). Also, determine the deflection δ_B and angle of rotation θ_B at the free end (Fig. 9-14b). Use the fourth-order differential equation of the deflection curve (the load equation). (*Note:* The beam has length L and constant flexural rigidity EI.)

(a)

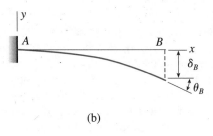

(b)

Fig. 9-14 Example 9-4. Deflections of a cantilever beam with a triangular load

Solution   *Differential equation of the deflection curve.* The intensity of the distributed load is given by the following equation (see Fig. 9-14a):

$$q = \frac{q_0(L - x)}{L} \tag{9-40}$$

Consequently, the fourth-order differential equation (Eq. 9-12c) becomes

$$EIv'''' = -q = -\frac{q_0(L - x)}{L} \tag{a}$$

Shear force in the beam. The first integration of Eq. (a) gives

$$EIv''' = \frac{q_0}{2L}(L - x)^2 + C_1 \tag{b}$$

The right-hand side of this equation represents the shear force V (see Eq. 9-12b). Because the shear force is zero at $x = L$, we have the following boundary condition:

$$v'''(L) = 0$$

Using this condition with Eq. (b), we get $C_1 = 0$. Therefore, Eq. (b) simplifies to

$$EIv''' = \frac{q_0}{2L}(L - x)^2 \tag{c}$$

and the shear force in the beam is

$$V = \frac{q_0}{2L}(L - x)^2 \tag{9-41}$$

Bending moment in the beam. Integrating a second time, we obtain the following equation from Eq. (c):

$$EIv'' = -\frac{q_0}{6L}(L - x)^3 + C_2 \tag{d}$$

This equation is equal to the bending moment M (see Eq. 9-12a). Since the bending moment is zero at the free end of the beam, we have the following boundary condition:

$$v''(L) = 0$$

Applying this condition to Eq. (d), we obtain $C_2 = 0$, and therefore the bending moment is

$$M = EIv'' = -\frac{q_0}{6L}(L - x)^3 \tag{9-42}$$

Slope and deflection of the beam. The third and fourth integrations yield

$$EIv' = \frac{q_0}{24L}(L - x)^4 + C_3 \tag{e}$$

$$EIv = -\frac{q_0}{120L}(L - x)^5 + C_3 x + C_4 \tag{f}$$

The boundary conditions at the fixed support, where both the slope and deflection equal zero, are

$$v'(0) = 0 \qquad v(0) = 0$$

Applying these conditions to Eqs. (e) and (f), respectively, we find

$$C_3 = -\frac{q_0 L^3}{24} \qquad C_4 = \frac{q_0 L^4}{120}$$

Substituting these expressions for the constants into Eqs. (e) and (f), we obtain the following equations for the slope and deflection of the beam:

$$v' = -\frac{q_0 x}{24LEI}(4L^3 - 6L^2 x + 4Lx^2 - x^3) \qquad (9\text{-}43) \quad \Longleftarrow$$

$$v = -\frac{q_0 x^2}{120LEI}(10L^3 - 10L^2 x + 5Lx^2 - x^3) \qquad (9\text{-}44) \quad \Longleftarrow$$

Angle of rotation and deflection at the free end of the beam. The angle of rotation θ_B and deflection δ_B at the free end of the beam (Fig. 9-14b) are obtained from Eqs. (9-43) and (9-44), respectively, by substituting $x = L$. The results are

$$\theta_B = -v'(L) = \frac{q_0 L^3}{24EI} \qquad \delta_B = -v(L) = \frac{q_0 L^4}{30EI} \qquad (9\text{-}45\text{a,b}) \quad \Longleftarrow$$

Thus, we have determined the required slopes and deflections of the beam by solving the fourth-order differential equation of the deflection curve.

■ **Example 9-5**

A simple beam AB with an overhang BC supports a concentrated load P at the end of the overhang (Fig. 9-15a). The main span of the beam has length L and the overhang has length $L/2$. Determine the equations of the deflection curve and the deflection δ_C at the end of the overhang (Fig. 9-15b). Use the third-order differential equation of the deflection curve (the shear-force equation). (*Note:* The beam has constant flexural rigidity EI.)

Solution   *Differential equations of the deflection curve.* Because reactive forces act at supports A and B, we must write separate differential equations for parts AB and BC of the beam. Therefore, we begin by finding the shear forces in each part of the beam.

The downward reaction at support A is equal to $P/2$, and the upward reaction at support B is equal to $3P/2$ (see Fig. 9-15a). It follows that the shear forces in parts AB and BC are

$$V = -\frac{P}{2} \qquad (0 < x < L) \qquad (9\text{-}46\text{a})$$

$$V = P \qquad \left(L < x < \frac{3L}{2}\right) \qquad (9\text{-}46\text{b})$$

in which x is measured from end A of the beam (Fig. 9-12b).

(a)

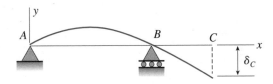

Fig. 9-15 Example 9-5. Deflections of a
beam with an overhang

(b)

The third-order differential equations for the beam now become (see
Eq. 9-12b):

$$EIv''' = -\frac{P}{2} \qquad (0 < x < L) \tag{g}$$

$$EIv''' = P \qquad \left(L < x < \frac{3L}{2}\right) \tag{h}$$

Bending moments in the beam. Integration of the preceding two equations
yields the bending-moment equations:

$$M = EIv'' = -\frac{Px}{2} + C_1 \qquad (0 \leq x \leq L) \tag{i}$$

$$M = EIv'' = Px + C_2 \qquad \left(L \leq x \leq \frac{3L}{2}\right) \tag{j}$$

The bending moments at points A and C are zero; hence we have the following
boundary conditions:

$$v''(0) = 0 \qquad v''\left(\frac{3L}{2}\right) = 0$$

Using these conditions with Eqs. (i) and (j), we get

$$C_1 = 0 \qquad C_2 = -\frac{3PL}{2}$$

Therefore, the bending moments are

$$M = EIv'' = -\frac{Px}{2} \qquad (0 \leq x \leq L) \tag{9-47a}$$

$$M = EIv'' = -\frac{P(3L - 2x)}{2} \qquad \left(L \le x \le \frac{3L}{2}\right) \qquad \text{(9-47b)}$$

These equations can be verified by determining the bending moments from free-body diagrams and equations of equilibrium.

Slopes and deflections of the beam. The next integrations yield the slopes:

$$EIv' = -\frac{Px^2}{4} + C_3 \qquad (0 \le x \le L)$$

$$EIv' = -\frac{Px(3L - x)}{2} + C_4 \qquad \left(L \le x \le \frac{3L}{2}\right)$$

The only condition on the slopes is the continuity condition at support B. According to this condition, the slope at point B as found for part AB of the beam is equal to the slope at the same point as found for part BC of the beam. Therefore, we substitute $x = L$ into each of the two preceding equations for the slopes and obtain

$$-\frac{PL^2}{4} + C_3 = -PL^2 + C_4$$

This equation eliminates one constant of integration because we can express C_4 in terms of C_3:

$$C_4 = C_3 + \frac{3PL^2}{4} \qquad \text{(k)}$$

The third and last integrations give

$$EIv = -\frac{Px^3}{12} + C_3x + C_5 \qquad (0 \le x \le L) \qquad \text{(l)}$$

$$EIv = -\frac{Px^2(9L - 2x)}{12} + C_4x + C_6 \qquad \left(L \le x \le \frac{3L}{2}\right) \qquad \text{(m)}$$

For part AB of the beam (Fig. 9-15a), we have two boundary conditions on the deflections, namely, the deflection is zero at points A and B:

$$v(0) = 0 \quad \text{and} \quad v(L) = 0$$

Applying these conditions to Eq. (l), we obtain

$$C_5 = 0 \qquad C_3 = \frac{PL^2}{12} \qquad \text{(n,o)}$$

Substituting the preceding expression for C_3 in Eq. (k), we get

$$C_4 = \frac{5PL^2}{6} \qquad \text{(p)}$$

For part BC of the beam, the deflection is zero at point B. Therefore, the boundary condition is

$$v(L) = 0$$

Applying this condition to Eq. (m), and also substituting Eq. (p) for C_4, we get

$$C_6 = -\frac{PL^3}{4} \qquad \text{(q)}$$

All constants of integration have now been evaluated.

The deflection equations are obtained by substituting the constants of integration (Eqs. n, o, p, and q) into Eqs. (l) and (m). The results are

$$v = \frac{Px}{12EI}(L^2 - x^2) \qquad (0 \le x \le L) \qquad \text{(9-48a)} \quad \Longleftarrow$$

$$v = -\frac{P}{12EI}(3L^3 - 10L^2x + 9Lx^2 - 2x^3) \qquad \left(L \le x \le \frac{3L}{2}\right) \text{ (9-48b)} \quad \Longleftarrow$$

Note that the deflection is always positive (upward) in part AB of the beam (Eq. 9-48a) and always negative (downward) in the overhang BC (Eq. 9-48b).

Deflection at the end of the overhang. We can find the deflection δ_C at the end of the overhang (Fig. 9-15b) by substituting $x = 3L/2$ in Eq. (9-48b):

$$\delta_C = -v\left(\frac{3L}{2}\right) = \frac{PL^3}{8EI} \qquad \text{(9-49)} \quad \Longleftarrow$$

Thus, we have determined the deflections of the overhanging beam (Eqs. 9-48 and 9-49) by solving the third-order differential equation of the deflection curve.

9.5 METHOD OF SUPERPOSITION

The **method of superposition** is a practical and commonly used technique for obtaining deflections and angles of rotation of beams. The underlying concept is quite simple and may be stated as follows: *The deflection of a beam produced by several different loads acting simultaneously can be found by superposing the deflections produced by the same loads acting separately.* For instance, if v_1 represents the deflection at a particular point on the axis of a beam due to a load q_1, and if v_2 represents the deflection at that same point due to a different load q_2, then the deflection at that point due to loads q_1 and q_2 acting simultaneously is $v_1 + v_2$. (Note that the loads q_1 and q_2 may act anywhere along the axis of the beam; in particular, they do not have to act at the point where the deflection is determined.)

The justification for superposing deflections lies in the nature of the differential equations of the deflection curve (Eqs. 9-12a, b, and c). These equations are *linear* differential equations, because all terms containing the deflection v and its derivatives are raised to the first power. Therefore, the solutions of these equations for several loading conditions may be added algebraically, or *superposed.*

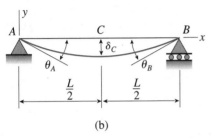

Fig. 9-16 Simple beam with two loads

As an illustration of the superposition method, consider the simple beam *ACB* shown in Fig. 9-16a. This beam supports two loads: (1) a uniform load of intensity *q* acting throughout the span, and (2) a concentrated load *P* acting at the midpoint. Suppose we wish to find the deflection δ_C at the midpoint and the angles of rotation θ_A and θ_B at the ends (Fig. 9-16b). Using superposition, we obtain the effects of each load acting separately and then combine the results. For the uniform load acting alone, the deflection at the midpoint and the angles of rotation are obtained from the formulas of Example 9-1 (see Eqs. 9-18, 9-19, and 9-20):

$$(\delta_C)_1 = \frac{5qL^4}{384EI} \qquad (\theta_A)_1 = (\theta_B)_1 = \frac{qL^3}{24EI}$$

in which *EI* is the flexural rigidity of the beam and *L* is its length. For the load *P* acting alone, the corresponding quantities are obtained from the formulas of Example 9-3 (see Eqs. 9-38 and 9-39):

$$(\delta_C)_2 = \frac{PL^3}{48EI} \qquad (\theta_A)_2 = (\theta_B)_2 = \frac{PL^2}{16EI}$$

Therefore, the deflection and angles of rotation due to the combined loading (Fig. 9-16) are

$$\delta_C = (\delta_C)_1 + (\delta_C)_2 = \frac{5qL^4}{384EI} + \frac{PL^3}{48EI} \qquad (a)$$

$$\theta_A = \theta_B = (\theta_A)_1 + (\theta_A)_2 = \frac{qL^3}{24EI} + \frac{PL^2}{16EI} \qquad (b)$$

The deflection and angle of rotation at any point on the beam axis can be found by this same procedure. However, the method of superposition is not limited to finding deflections and angles of rotation at a single point. The method may also be used to obtain the general equations for the slopes and deflections of a beam.

■ **Tables of Beam Deflections**

The method of superposition is useful only when formulas for deflections and slopes are readily available. To provide convenient access to such formulas, tables for both cantilever and simple beams are given in Appendix G at the back of the book. Similar tables can be found in engineering handbooks. Using these tables and the method of superposition, we can find deflections and angles of rotation for many different loading conditions, as illustrated in the examples at the end of this section.

■ **Distributed Loads**

Sometimes we encounter a distributed load that is not included in a table of beam deflections. In such cases, superposition may still be useful. We can consider an element of the distributed load as though it were a concentrated load, and then we can find the required deflection by integrating throughout the region of the beam where the load is applied.

To illustrate this process of integration, consider a simple beam ACB with a triangular load acting on the left-hand half (Fig. 9-17a). We wish to obtain the deflection δ_C at the midpoint C and the angle of rotation θ_A at the left-hand support (Fig. 9-17c).

We begin by visualizing an element $q\,dx$ of the distributed load as a concentrated load (Fig. 9-17b). Note that the load acts to the left of the midpoint of the beam. The deflection at the midpoint due to this concentrated load is obtained from Case 5 of Table G-2, Appendix G. The formula given there for the midpoint deflection (for the case in which $a \leq b$) is

$$\frac{Pa}{48EI}(3L^2 - 4a^2)$$

In our example (Fig. 9-17b), we substitute $q\,dx$ for P and x for a:

$$\frac{(q\,dx)(x)}{48EI}(3L^2 - 4x^2) \tag{c}$$

Next, we note that the intensity of the uniform load is

$$q = \frac{2q_0 x}{L} \tag{d}$$

where q_0 is the maximum intensity of the load (Fig. 9-17a). With this substitution for q, the formula for the deflection (Eq. c) becomes

$$\frac{q_0 x^2}{24LEI}(3L^2 - 4x^2)dx$$

Finally, we integrate throughout the region of the load to obtain the deflection δ_C at the midpoint of the beam due to the entire triangular load:

$$\delta_C = \int_0^{L/2} \frac{q_0 x^2}{24LEI}(3L^2 - 4x^2)dx$$

$$= \frac{q_0}{24LEI}\int_0^{L/2}(3L^2 - 4x^2)x^2\,dx = \frac{q_0 L^4}{240EI} \tag{9-50}$$

By a similar procedure, we can calculate the angle of rotation θ_A at the left-hand end of the beam (Fig. 9-17c). The expression for this angle due to a concentrated load P (see Case 5 of Table G-2) is

$$\frac{Pab(L + b)}{6LEI}$$

(a)

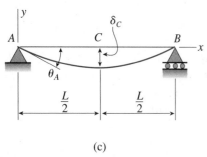

(b)

(c)

Fig. 9-17 Simple beam with a triangular load

Replacing P with $2q_0x\,dx/L$, a with x, and b with $L - x$, we obtain

$$\frac{2q_0x^2(L - x)(L + L - x)}{6L^2EI}dx \quad \text{or} \quad \frac{q_0}{3L^2EI}(L - x)(2L - x)x^2dx$$

Finally, we integrate throughout the region of the load:

$$\theta_A = \int_0^{L/2} \frac{q_0}{3L^2EI}(L - x)(2L - x)x^2dx = \frac{41q_0L^3}{2880EI} \tag{9-51}$$

This is the angle of rotation produced by the triangular load.

This example illustrates how we can use superposition and integration to find deflections and angles of rotation produced by distributed loads of almost any kind. If the integration cannot be performed easily by analytical means, numerical methods can be used.

■ **Principle of Superposition**

The method of superposition for finding beam deflections is an example of a more general concept known in mechanics as the **principle of superposition**. This principle is valid whenever the quantity to be determined is a linear function of the applied loads. When that is the case, the desired quantity may be found due to each load acting separately, and then these results may be superposed to obtain the desired quantity due to all loads acting simultaneously. In ordinary structures, the principle is usually valid for stresses, strains, bending moments, and many other quantities besides deflections.

In the particular case of beam deflections, the principle of superposition is valid under the following conditions: (1) Hooke's law holds for the material, (2) the deflections and rotations are small, and (3) the presence of the deflections does not alter the actions of the applied loads. These requirements ensure that the differential equations of the deflection curve are linear.

The following examples provide additional illustrations in which the principle of superposition is used to calculate deflections and angles of rotation of beams.

■ **Example 9-6**

A cantilever beam AB supports a uniform load of intensity q acting over part of the span and a concentrated load P acting at the free end (Fig. 9-18a). Determine the deflection δ_B and angle of rotation θ_B at end B of the beam (Fig. 9-18b). (*Note:* The beam has length L and constant flexural rigidity EI.)

Solution   We can obtain the deflection and angle of rotation at end B of the beam by combining the effects of the loads acting separately. If the uniform load

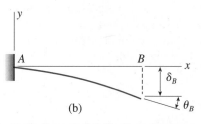

(a)

(b)

Fig. 9-18 Example 9-6. Cantilever beam with a uniform load and a concentrated load

acts alone, the deflection and angle of rotation (obtained from Case 2 of Table G-1, Appendix G) are

$$(\delta_B)_1 = \frac{qa^3}{24EI}(4L - a) \qquad (\theta_B)_1 = \frac{qa^3}{6EI}$$

If the load P acts alone, the corresponding quantities (from Case 4, Table G-1) are

$$(\delta_B)_2 = \frac{PL^3}{3EI} \qquad (\theta_B)_2 = \frac{PL^2}{2EI}$$

Therefore, the deflection and angle of rotation due to the combined loading (Fig. 9-18a) are

$$\delta_B = (\delta_B)_1 + (\delta_B)_2 = \frac{qa^3}{24EI}(4L - a) + \frac{PL^3}{3EI} \qquad (9\text{-}52) \quad \Longleftarrow$$

$$\theta_B = (\theta_B)_1 + (\theta_B)_2 = \frac{qa^3}{6EI} + \frac{PL^2}{2EI} \qquad (9\text{-}53) \quad \Longleftarrow$$

Thus, we have found the required quantities by using the method of superposition and tabulated formulas.

■ **Example 9-7**

A cantilever beam AB with a uniform load of intensity q acting on the right-hand half of the beam is shown in Fig. 9-19a. Obtain formulas for the deflection δ_B and angle of rotation θ_B at the free end (Fig. 9-19c). (*Note:* The beam has length L and constant flexural rigidity EI.)

Solution   In this example we will determine the deflection and angle of rotation by treating an element of the uniform load as a concentrated load and then integrating (see Fig. 9-19b). The element of load has magnitude $q\,dx$ and is located at distance x from the support. The resulting differential deflection $d\delta_B$ and differential angle of rotation $d\theta_B$ at the free end are found from the corresponding formulas in Case 5 of Table G-1, Appendix G, by replacing P with $q\,dx$ and a with x; thus,

$$d\delta_B = \frac{(q\,dx)(x^2)(3L - x)}{6EI} \qquad d\theta_B = \frac{(q\,dx)(x^2)}{2EI}$$

By integrating over the loaded region, we get

$$\delta_B = \int d\delta_B = \frac{q}{6EI}\int_{L/2}^{L} x^2(3L - x)\,dx = \frac{41qL^4}{384EI} \qquad (9\text{-}54) \quad \Longleftarrow$$

$$\theta_B = \int d\theta_B = \frac{q}{2EI}\int_{L/2}^{L} x^2\,dx = \frac{7qL^3}{48EI} \qquad (9\text{-}55) \quad \Longleftarrow$$

Note: These same results can be obtained by using the formulas in Case 3 of Table G-1 and substituting $a = b = L/2$.

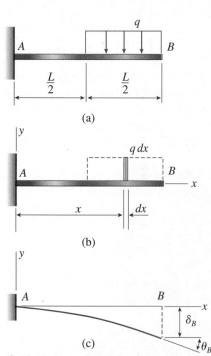

(a)

(b)

(c)

Fig. 9-19 Example 9-7. Cantilever beam with a uniform load acting on the right-hand half of the beam

■ **Example 9-8**

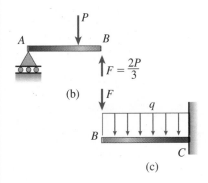

(a)

A compound beam ABC has a roller support at A, an internal hinge at B, and a fixed support at C (Fig. 9-20a). A concentrated load P acts at distance $2a/3$ from support A and a uniform load of intensity q acts between points B and C. Determine the deflection δ_B at the hinge and the angle of rotation θ_A at support A (Fig. 9-20d). (*Note:* The beam has constant flexural rigidity EI.)

Solution   For purposes of analysis, we may consider the beam as composed of two parts: (1) a simple beam AB of length a, and (2) a cantilever beam BC of length b. The two beams are linked together by a pin connection at B.

If we separate beam AB from the rest of the structure (Fig. 9-20b), we see that there is a vertical force F at end B equal to $2P/3$. This same force acts downward at end B of the cantilever (Fig. 9-20c). Consequently, the cantilever beam BC is subjected to two loads: a uniform load and a concentrated load. The deflection at the end of this cantilever (which is the same as the deflection δ_B of the hinge) is readily found from Cases 1 and 4 of Table G-1, Appendix G:

$$\delta_B = \frac{qb^4}{8EI} + \frac{Fb^3}{3EI}$$

or, since $F = 2P/3$,

$$\delta_B = \frac{qb^4}{8EI} + \frac{2Pb^3}{9EI} \tag{9-56}$$ ⬅

The angle of rotation θ_A at support A (Fig. 9-20d) consists of two parts: (1) an angle BAB' produced by the downward displacement of the hinge, and (2) an additional angle of rotation produced by the bending of beam AB (or beam AB') as a simple beam. The angle BAB' is

$$(\theta_A)_1 = \frac{\delta_B}{a} = \frac{qb^4}{8aEI} + \frac{2Pb^3}{9aEI}$$

The angle of rotation at the end of a simple beam with a concentrated load is obtained from Case 5 of Table G-2. The formula given there is

$$\frac{Pab(L + b)}{6LEI}$$

in which L is the length of the beam, a is the distance from the left-hand support to the load, and b is the distance from the right-hand support to the load. Thus, in the notation of our example (Fig. 9-20a), the angle of rotation is

$$(\theta_A)_2 = \frac{P\left(\dfrac{2a}{3}\right)\left(\dfrac{a}{3}\right)\left(a + \dfrac{a}{3}\right)}{6aEI} = \frac{4Pa^2}{81EI}$$

Combining the two angles, we obtain the total angle of rotation at support A:

$$\theta_A = (\theta_A)_1 + (\theta_A)_2 = \frac{qb^4}{8aEI} + \frac{2Pb^3}{9aEI} + \frac{4Pa^2}{81EI} \tag{9-57}$$ ⬅

This example illustrates how the method of superposition can be adapted to handle a seemingly complex situation in a relatively simple manner.

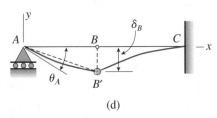

(b)

(c)

(d)

Fig. 9-20 Example 9-8. Compound beam with a hinge

■ **Example 9-9**

A simple beam AB of span length L has an overhang BC of length a (Fig. 9-21a). The beam supports a uniform load of intensity q throughout its length. Obtain a formula for the deflection δ_C at the end of the overhang (Fig. 9-21c). (*Note:* The beam has constant flexural rigidity EI.)

Solution   We can find the deflection of point C by imagining the overhang BC (Fig. 9-21a) to be a cantilever beam subjected to two actions. The first action is the rotation of the support of the cantilever through an angle θ_B, which is the angle of rotation of the original beam at support B (Fig. 9-21c). This angle of rotation causes a rigid-body rotation of the overhang BC, resulting in a downward displacement δ_1 of point C. The second action is the bending of BC as a cantilever beam supporting a uniform load. This bending produces an additional downward displacement δ_2 (Fig. 9-21c). The superposition of these two displacements gives the total displacement δ_C at point C.

(a)

(b)

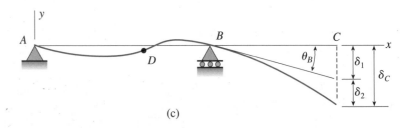

Fig. 9-21 Example 9-9. Simple beam with an overhang

(c)

Let us begin by finding the deflection δ_1 caused by the angle of rotation θ_B at point B. To find this angle, we observe that part AB of the beam is in the same condition as a simple beam (Fig. 9-21b) subjected to the following loads: (1) a uniform load of intensity q, (2) a couple M_B (equal to $qa^2/2$), and (3) a vertical load P (equal to qa). Only the loads q and M_B produce angles of rotation at end B of this simple beam. These angles are found from Cases 1 and 7 of Table G-2, Appendix G. Thus, the angle θ_B is

$$\theta_B = -\frac{qL^3}{24EI} + \frac{M_B L}{3EI} = -\frac{qL^3}{24EI} + \frac{qa^2 L}{6EI} = \frac{qL(4a^2 - L^2)}{24EI} \tag{9-58}$$

in which a clockwise angle is positive, as shown in Fig. 9-21c.

The downward deflection δ_1 of point C, due solely to the angle of rotation θ_B, is equal to the length of the overhang times the angle (Fig. 9-21c):

$$\delta_1 = a\theta_B = \frac{qaL(4a^2 - L^2)}{24EI} \tag{e}$$

Bending of the overhang BC produces an additional downward deflection δ_2 at point C. This deflection is equal to the deflection of a cantilever beam of length a subjected to a uniform load of intensity q (see Case 1 of Table G-1):

$$\delta_2 = \frac{qa^4}{8EI} \tag{f}$$

Therefore, the total downward deflection of point C is

$$\delta_C = \delta_1 + \delta_2 = \frac{qaL(4a^2 - L^2)}{24EI} + \frac{qa^4}{8EI} = \frac{qa}{24EI}[L(4a^2 - L^2) + 3a^3]$$

or

$$\delta_C = \frac{qa}{24EI}(a + L)(3a^2 + aL - L^2) \tag{9-59}$$ ⬅

From the preceding equation we see that the deflection δ_C may be upward or downward, depending upon the relative magnitudes of the lengths L and a. If a is relatively large, the last term in the equation (the three-term expression in parentheses) is positive and the deflection δ_C is downward. If a is relatively small, the last term is negative and the deflection is upward. The deflection is zero when the last term is equal to zero:

$$3a^2 + aL - L^2 = 0$$

or

$$a = \frac{L(\sqrt{13} - 1)}{6} = 0.4343L \tag{g}$$

From this result, we see that if a is greater than $0.4343L$, the deflection of point C is downward; if a is less than $0.4343L$, the deflection is upward.

The shape of the deflection curve for the beam in this example is shown in Fig. 9-21c for the case where a is large enough $(a > 0.4343L)$ to produce a downward deflection at C and small enough $(a < L)$ to ensure that the reaction at A is upward. Under these conditions the beam has a positive bending moment between support A and a point such as D. The deflection curve in region AD is concave upward (positive curvature). From D to C, the bending moment is negative, and therefore the deflection curve is concave downward (negative curvature).

At point D the curvature of the deflection curve is zero because the bending moment is zero. A point such as D where the curvature and bending moment *change signs* is called a **point of inflection** (or *point of contraflexure*). The bending moment M and the second derivative d^2v/dx^2 always vanish at an inflection point. (However, a point where M and d^2v/dx^2 equal zero is not *always* an inflection point because it is possible for those quantities to be zero without changing signs at that point; for example, they could have maximum or minimum values.)

9.6 MOMENT-AREA METHOD

In this section we will describe another method for finding deflections and angles of rotation of beams. Because the method is based upon two theorems related to the *area* of the bending-moment diagram, it is called the **moment-area method**. The assumptions used in deriving the two theorems are the same as those used in deriving the differential equations of the deflection curve. Therefore, the moment-area method is valid only for linearly elastic beams with small slopes. From a practical standpoint, the method is limited to finding deflections and angles of rotation at specific points on the axis of a beam.

■ First Moment-area Theorem

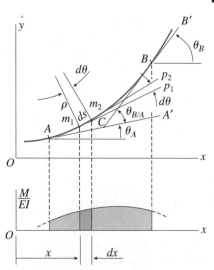

Fig. 9-22 Derivation of the first moment-area theorem

To derive the first theorem, consider a segment AB of the deflection curve of a beam in a region where the curvature is positive (Fig. 9-22). Of course, the deflections and slopes shown in the figure are highly exaggerated for clarity. At point A the tangent AA' to the deflection curve is at an angle θ_A to the x axis, and at point B the tangent BB' is at an angle θ_B. These two tangents meet at point C. The angle between the tangents, denoted $\theta_{B/A}$, is equal to the difference between θ_B and θ_A:

$$\theta_{B/A} = \theta_B - \theta_A \qquad (9\text{-}60)$$

Thus, the angle $\theta_{B/A}$ may be described as the angle to the tangent at B measured relative to, or with respect to, the tangent at A. Note that the angles θ_A and θ_B, which are the angles of rotation of the beam axis at points A and B, respectively, are also equal to the slopes at those points, because in reality the slopes and angles are very small quantities.

Next, consider two points m_1 and m_2 on the deflected axis of the beam (Fig. 9-22). These points are a small distance ds apart. The tangents to the deflection curve at these points are shown in the figure as lines m_1p_1 and m_2p_2. The normals to these tangents intersect at the center of curvature (not shown in the figure). The angle $d\theta$ between the normals is given by the following equation:

$$d\theta = \frac{ds}{\rho} \tag{a}$$

in which ρ is the radius of curvature and $d\theta$ is measured in radians (see Eq. 9-1). Because the normals and the tangents (m_1p_1 and m_2p_2) are perpendicular, it follows that the angle between the tangents is also equal to $d\theta$.

For a beam with small angles of rotation, we can replace ds with dx, as explained in Section 9.2. Thus,

$$d\theta = \frac{dx}{\rho} \tag{b}$$

Also, from Eq. (9-6) we know that

$$\frac{1}{\rho} = \frac{M}{EI} \tag{c}$$

and therefore

$$d\theta = \frac{M\,dx}{EI} \tag{9-61}$$

in which M is the bending moment and EI is the flexural rigidity of the beam.

The quantity $M\,dx/EI$ has a simple geometric interpretation. To see this, refer to Fig. 9-22 where we have drawn the M/EI diagram directly below the beam. At any point along the x axis, the height of this diagram is equal to the bending moment M at that point divided by the flexural rigidity EI at that point. Thus, the M/EI diagram has the same shape as the bending-moment diagram provided EI is constant. The term $M\,dx/EI$ is the area of the shaded strip of width dx within the M/EI diagram. (Note that since the curvature of the deflection curve in Fig. 9-22 is positive, the bending moment M and the area of the M/EI diagram are also positive.)

Let us now integrate $d\theta$ (Eq. 9-61) between points A and B of the deflection curve:

$$\int_A^B d\theta = \int_A^B \frac{M\,dx}{EI} \tag{d}$$

When evaluated, the integral on the left-hand side becomes $\theta_B - \theta_A$, which is equal to the angle $\theta_{B/A}$ between the tangents at B and A (Eq. 9-60). The integral on the right-hand side is equal to the area of the M/EI diagram between points A and B. (Note that the area of the M/EI diagram is an algebraic quantity and may be positive or negative, depending upon whether the bending moment is positive or negative.)

Now we can write Eq. (d) as follows:

$$\theta_{B/A} = \int_A^B \frac{M\,dx}{EI}$$

$$= \text{Area of the } M/EI \text{ diagram between points } A \text{ and } B \qquad (9\text{-}62)$$

This equation may be stated as a theorem:

> **First moment-area theorem:** The angle $\theta_{B/A}$ between the tangents to the deflection curve at two points A and B is equal to the area of the M/EI diagram between those points.

The sign conventions used in deriving this theorem are as follows: (1) The angles θ_A and θ_B are positive when counterclockwise. (2) The angle $\theta_{B/A}$ between the tangents is positive when the angle θ_B is algebraically larger than the angle θ_A. Note that point B must be to the right of point A; that is, it must be further along the axis of the beam as we move in the x direction. (3) The bending moment M is positive according to our usual sign convention; that is, M is positive when it produces compression in the upper part of the beam. (4) The area of the M/EI diagram is given a positive or negative sign according to whether the bending moment is positive or negative. If part of the bending-moment diagram is positive and part is negative, then the corresponding parts of the M/EI diagram are given those same signs.

The preceding sign conventions for θ_A, θ_B, and $\theta_{B/A}$ are often ignored in practice because (as explained later) the directions of the angles of rotation are usually obvious from an inspection of the beam and its loading. When this is the case, we can simplify the calculations by ignoring signs and using only absolute values when applying the first moment-area theorem.

■ **Second Moment-area Theorem**

Now we turn to the second theorem, which is related primarily to deflections rather than to angles of rotation. Consider again the deflection curve between points A and B (Fig. 9-23). We draw the tangent at point A and note that its intersection with a vertical line through point B is at point B_1. The vertical distance between points B and B_1 is denoted $t_{B/A}$ in the figure. This distance is referred to as the *tangential deviation of B with respect to A*; more precisely, it is the vertical deviation of point B on the deflection curve from the tangent at A. The tangential deviation is positive when point B is above the tangent at A.

To determine the tangential deviation, we again select two points m_1 and m_2 a small distance apart on the deflection curve (Fig. 9-23). The angle between the tangents at these two points is $d\theta$, and the segment on line BB_1 between these tangents is dt. Since the angles between the tangents and the x axis are actually very small, we see that the vertical distance dt is equal to $x_1\,d\theta$, where x_1 is the horizontal distance from point B to the small element m_1m_2. Since $d\theta = M\,dx/EI$ (Eq. 9-61), we obtain

$$dt = x_1\,d\theta = x_1\frac{M\,dx}{EI} \tag{e}$$

The distance dt represents the contribution made by the bending of element m_1m_2 to the tangential deviation $t_{B/A}$. The expression $x_1M\,dx/EI$ may be interpreted geometrically as the first moment of the area of the shaded strip of width dx within the M/EI diagram. This first moment is evaluated with respect to a vertical line through point B.

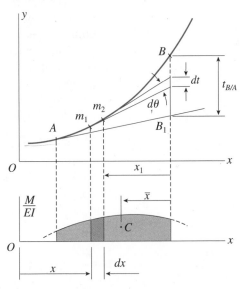

Fig. 9-23 Derivation of the second moment-area theorem

Integrating Eq. (e) between points A and B, we get

$$\int_A^B dt = \int_A^B x_1\frac{M\,dx}{EI} \tag{f}$$

The integral on the left-hand side is equal to $t_{B/A}$, that is, it is equal to the deviation of point B from the tangent at A. The integral on the right-hand side represents the first moment with respect to point B of the area of the M/EI diagram between A and B. Therefore, we can write Eq. (f) as follows:

$$t_{B/A} = \int_A^B x_1\frac{M\,dx}{EI}$$

$\qquad$ = First moment of the area of the M/EI diagram
$\qquad\qquad$ between points A and B, evaluated with respect to B $\qquad$ (9-63)

This equation represents the second theorem:

Second moment-area theorem: The tangential deviation $t_{B/A}$ of point B from the tangent at point A is equal to the first moment of the area of the *M/EI* diagram between *A* and *B*, evaluated with respect to *B*.

If the bending moment is positive, then the first moment of the *M/EI* diagram is also positive, provided point *B* is to the right of point *A*. Under these conditions the tangential deviation $t_{B/A}$ is positive and point *B* is above the tangent at *A* (as shown in Fig. 9-23). If, as we move from *A* to *B* in the *x* direction, the area of the *M/EI* diagram is negative, then the first moment is also negative and the tangential deviation is negative, which means that point *B* is below the tangent at *A*.

The first moment of the area of the *M/EI* diagram can be obtained by taking the product of the area of the diagram and the distance $\bar{x}$ from point *B* to the centroid *C* of the area (Fig. 9-23). This procedure is usually more convenient than integrating, because the *M/EI* diagram usually consists of familiar geometric figures such as rectangles, triangles, and parabolic segments. The areas and centroidal distances of such figures are tabulated in Appendix D.

As a method of analysis, the moment-area method is feasible only for relatively simple kinds of beams. Therefore, it is usually obvious whether the beam deflects upward or downward and whether an angle of rotation is clockwise or counterclockwise. Consequently, it is seldom necessary to follow the formal (and somewhat awkward) sign conventions described above for the tangential deviation. Instead, we can determine the directions by inspection and use only absolute values when applying the moment-area theorems.*

■ **Example 9-10**

Determine the angle of rotation θ_B and deflection δ_B at the free end *B* of a cantilever beam *AB* supporting a concentrated load *P* (Fig. 9-24). (*Note:* The beam has length *L* and constant flexural rigidity *EI*.)

Solution   By inspection of the beam and its loading, we know that the angle of rotation θ_B is clockwise and the deflection δ_B is downward (Fig. 9-24). Therefore, we can use absolute values when applying the moment-area theorems.

M/EI diagram. The bending-moment diagram is triangular in shape with the moment at the support equal to $-PL$. Since the flexural rigidity *EI* is constant, the *M/EI* diagram has the same shape as the bending-moment diagram, as shown in the last part of Fig. 9-24.

Angle of rotation. From the first moment-area theorem, we know that the angle $\theta_{B/A}$ between the tangents at points *B* and *A* is equal to the area of the *M/EI* diagram between those points. This area, which we will denote as A_1, is determined as follows:

$$A_1 = \frac{1}{2}(L)\left(\frac{PL}{EI}\right) = \frac{PL^2}{2EI}$$

---

*The moment-area method was introduced by Saint-Venant (Ref. 9-6). The method was developed more fully by Mohr (Ref. 9-7) and Greene (Ref. 9-8).

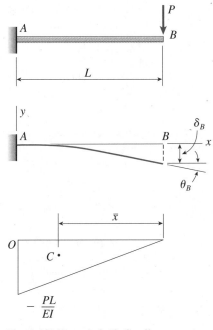

Fig. 9-24 Example 9-10. Cantilever beam with a concentrated load

Note that we are using only the absolute value of the area.

The relative angle of rotation between points A and B (from the first theorem) is

$$\theta_{B/A} = \theta_B - \theta_A = A_1 = \frac{PL^2}{2EI}$$

Since the tangent to the deflection curve at support A is horizontal ($\theta_A = 0$), we obtain

$$\theta_B = \frac{PL^2}{2EI} \qquad (9\text{-}64)$$

This result agrees with the formula for θ_B given in Case 4 of Table G-1, Appendix G.

Deflection. The deflection δ_B at the free end can be obtained from the second moment-area theorem. In this case, the tangential deviation $t_{B/A}$ of point B from the tangent at A is equal to the deflection δ_B itself (see Fig. 9-24). The first moment of the area of the M/EI diagram, evaluated with respect to point B, is

$$Q_1 = A_1 \bar{x} = \left(\frac{PL^2}{2EI}\right)\left(\frac{2L}{3}\right) = \frac{PL^3}{3EI}$$

Note again that we are disregarding signs and using only absolute values.

From the second moment-area theorem, we know that the deflection δ_B is equal to the first moment Q_1. Therefore,

$$\delta_B = \frac{PL^3}{3EI} \qquad (9\text{-}65)$$

This result also appears in Case 4 of Table G-1.

■ **Example 9-11**

Find the angle of rotation θ_B and deflection δ_B at the free end B of a cantilever beam ACB supporting a uniform load of intensity q acting over the right-hand half of the beam (Fig. 9-25). (*Note:* The beam has length L and constant flexural rigidity EI.)

Solution The deflection and angle of rotation at end B of the beam have the directions shown in Fig. 9-25. Since we know these directions in advance, we can write the moment-area expressions using only absolute values.

M/EI diagram. The bending-moment diagram consists of a parabolic curve in the region of the uniform load and a straight line in the left-hand half of the beam. Since EI is constant, the M/EI diagram has the same shape (see the last part of Fig. 9-25). The values of M/EI at points A and C are $-3qL^2/8EI$ and $-qL^2/8EI$, respectively.

Angle of rotation. For the purpose of evaluating the area of the M/EI diagram, it is convenient to divide the diagram into three parts: (1) a parabolic

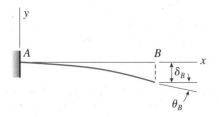

Fig. 9-25 Example 9-11. Cantilever beam supporting a uniform load on the right-hand half of the beam

spandrel of area A_1, (2) a rectangle of area A_2, and (3) a triangle of area A_3. These areas are

$$A_1 = \frac{1}{3}\left(\frac{L}{2}\right)\left(\frac{qL^2}{8EI}\right) = \frac{qL^3}{48EI} \qquad A_2 = \frac{L}{2}\left(\frac{qL^2}{8EI}\right) = \frac{qL^3}{16EI}$$

$$A_3 = \frac{1}{2}\left(\frac{L}{2}\right)\left(\frac{3qL^2}{8EI} - \frac{qL^2}{8EI}\right) = \frac{qL^3}{16EI}$$

According to the first moment-area theorem, the angle between the tangents at A and B is equal to the area of the M/EI diagram between those points. Since the angle at A is zero, it follows that the angle of rotation θ_B is equal to the area of the diagram; thus,

$$\theta_B = A_1 + A_2 + A_3 = \frac{7qL^3}{48EI} \tag{9-66} \quad \Longleftarrow$$

Deflection. The deflection δ_B is the tangential deviation of point B with respect to a tangent at point A (Fig. 9-25). Therefore, from the second moment-area theorem, δ_B is equal to the first moment of the M/EI diagram, evaluated with respect to point B:

$$\delta_B = A_1\bar{x}_1 + A_2\bar{x}_2 + A_3\bar{x}_3 \tag{g}$$

in which $\bar{x}_1$, $\bar{x}_2$, and $\bar{x}_3$, are the distances from point B to the centroids of the respective areas. These distances are

$$\bar{x}_1 = \frac{3}{4}\left(\frac{L}{2}\right) = \frac{3L}{8} \qquad \bar{x}_2 = \frac{L}{2} + \frac{L}{4} = \frac{3L}{4} \qquad \bar{x}_3 = \frac{L}{2} + \frac{2}{3}\left(\frac{L}{2}\right) = \frac{5L}{6}$$

Substituting into Eq. (g), we find

$$\delta_B = \frac{qL^3}{48EI}\left(\frac{3L}{8}\right) + \frac{qL^3}{16EI}\left(\frac{3L}{4}\right) + \frac{qL^3}{16EI}\left(\frac{5L}{6}\right) = \frac{41qL^4}{384EI} \tag{9-67} \quad \Longleftarrow$$

This example illustrates how the area and first moment of a complex M/EI diagram can be determined by dividing the area into parts having known properties. The results of this analysis (Eqs. 9-66 and 9-67) can be verified by using the formulas of Case 3, Table G-1, Appendix G, and substituting $a = b = L/2$.

■ **Example 9-12**

A simple beam ADB supports a concentrated load P acting at the position shown in Fig. 9-26. Determine the angle of rotation θ_A at support A and the deflection δ_D under the load P. (*Note:* The beam has length L and constant flexural rigidity EI.)

Solution   The deflection curve, showing the angle of rotation θ_A and the deflection δ_D, is sketched in the second part of Fig. 9-26. Because we can determine the directions of θ_A and δ_D by inspection, we can write the moment-area expressions using only absolute values.

M/EI diagram. The bending-moment diagram is triangular, with the maximum moment (equal to Pab/L) occurring under the load. Since EI is constant, the *M/EI* diagram has the same shape as the moment diagram (see the third part of Fig. 9-26).

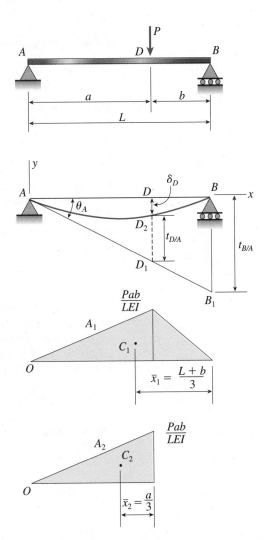

Fig. 9-26 Example 9-12. Simple beam with a concentrated load

Angle of rotation at support A. To find this angle, we construct the tangent AB_1 at support *A*. We then note that the distance BB_1 is the tangential deviation $t_{B/A}$ of point *B* from the tangent at *A*. We can calculate this distance by evaluating the first moment of the area of the *M/EI* diagram with respect to point *B* and then applying the second moment-area theorem.

The area of the *M/EI* diagram is

$$A_1 = \frac{1}{2}(L)\left(\frac{Pab}{LEI}\right) = \frac{Pab}{2EI}$$

The centroid C_1 of this area is at distance $\bar{x}_1$ from point B (see Fig. 9-26). This distance, obtained from Case 3 of Appendix D, is

$$\bar{x}_1 = \frac{L + b}{3}$$

Therefore, the tangential deviation is

$$t_{B/A} = A_1 \bar{x}_1 = \frac{Pab}{2EI}\left(\frac{L + b}{3}\right) = \frac{Pab}{6EI}(L + b)$$

The angle θ_A is equal to the tangential deviation divided by the length of the beam:

$$\theta_A = \frac{t_{B/A}}{L} = \frac{Pab}{6LEI}(L + b) \tag{9-68}$$ ⬅

Thus, the angle of rotation at support A has been found.

Deflection under the load. As shown in the second part of Fig. 9-26, the deflection δ_D under the load P is equal to the distance DD_1 minus the distance D_2D_1. The distance DD_1 is equal to the angle of rotation θ_A times the distance a; thus,

$$DD_1 = a\theta_A = \frac{Pa^2b}{6LEI}(L + b) \tag{h}$$

The distance D_2D_1 is the tangential deviation $t_{D/A}$ at point D; that is, it is the deviation of point D from the tangent at A. This distance can be found from the second moment-area theorem by taking the first moment of the area of the M/EI diagram between points A and D with respect to D (see the last part of Fig. 9-26). The area of this part of the M/EI diagram is

$$A_2 = \frac{1}{2}(a)\left(\frac{Pab}{LEI}\right) = \frac{Pa^2b}{2LEI}$$

and its centroidal distance from point D is

$$\bar{x}_2 = \frac{a}{3}$$

Thus, the first moment of this area with respect to point D is

$$t_{D/A} = A_2 \bar{x}_2 = \left(\frac{Pa^2b}{2LEI}\right)\left(\frac{a}{3}\right) = \frac{Pa^3b}{6LEI} \tag{i}$$

The deflection at point D is

$$\delta_D = DD_1 - D_2D_1 = DD_1 - t_{D/A}$$

Upon substituting from Eqs. (h) and (i), we find

$$\delta_D = \frac{Pa^2b}{6LEI}(L + b) - \frac{Pa^3b}{6LEI} = \frac{Pa^2b^2}{3LEI} \tag{9-69}$$ ⬅

The preceding formulas for θ_A and δ_D (Eqs. 9-68 and 9-69) can be verified by using the formulas of Case 5, Table G-2, Appendix G.

9.7 ┃ NONPRISMATIC BEAMS

The methods presented in the preceding sections for finding deflections of prismatic beams can also be used to find deflections of beams having varying moments of inertia. Two examples of nonprismatic beams are shown in Fig. 9-27. The first beam has two different moments of inertia, and the second is a tapered beam having continuously varying moment of inertia. In both cases the objective is to save material by increasing the moment of inertia in regions where the bending moment is largest.

Although no new concepts are involved, the analysis of a nonprismatic beam is more complex than the analysis of a beam with constant moment of inertia. Some of the procedures that can be used are illustrated in the examples that follow (Examples 9-13 and 9-14). In the first example (a simple beam having two different moments of inertia), the deflections are found by solving the differential equation of the deflection curve. In the second example (a cantilever beam having two different moments of inertia), the method of superposition is used. These examples, as well as the problems for this section, involve relatively simple and idealized beams. When more complex beams (such as tapered beams) are encountered, numerical methods of analysis are often required. (Computer programs for the numerical calculation of beam deflections are readily available.)

Fig. 9-27 Beams with varying moments of inertia (see also Fig. 5-23)

■ Example 9-13

A beam *ABCDE* on simple supports is constructed from a wide-flange beam by welding cover plates over the middle half of the beam (Fig. 9-28a). The effect of the cover plates is to double the moment of inertia (Fig. 9-28b). A concentrated load *P* acts at the midpoint *C* of the beam. Determine the equations of the deflection curve, the angle of rotation θ_A at the left-hand support, and the deflection δ_C at the midpoint (Fig. 9-28c).

Solution   *Differential equations of the deflection curve.* In this example we will determine the slopes and deflections of the beam by integrating the bending-moment equation, that is, the second-order differential equation of the deflection curve (Eq. 9-12a). Since the reaction at each support is *P*/2, the bending moment throughout the left-hand half of the beam is

$$M = \frac{Px}{2} \quad \left(0 \le x \le \frac{L}{2}\right) \tag{a}$$

Therefore, the differential equations for the left-hand half of the beam are

$$EIv'' = \frac{Px}{2} \quad \left(0 \le x \le \frac{L}{4}\right) \tag{b}$$

$$E(2I)v'' = \frac{Px}{2} \quad \left(\frac{L}{4} \le x \le \frac{L}{2}\right) \tag{c}$$

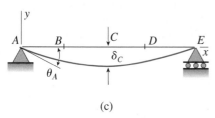

Fig. 9-28 Example 9-13. Simple beam with two different moments of inertia

Each of these equations can be integrated twice to obtain expressions for the slopes and deflections in their respective regions. These integrations produce four constants of integration that can be found from the following four conditions:

(1) Boundary condition: At support A ($x = 0$), the deflection is zero ($v = 0$).
(2) Symmetry condition: At point C ($x = L/2$), the slope is zero ($v' = 0$).
(3) Continuity condition: At point B ($x = L/4$), the slope obtained from part AB of the beam is equal to the slope obtained from part BC of the beam.
(4) Continuity condition: At point B ($x = L/4$), the deflection obtained from part AB of the beam is equal to the deflection obtained from part BC of the beam.

Slopes of the beam. Integrating each of the differential equations (Eqs. b and c), we obtain the following equations for the slopes in the left-hand half of the beam:

$$v' = \frac{Px^2}{4EI} + C_1 \qquad \left(0 \le x \le \frac{L}{4}\right) \tag{d}$$

$$v' = \frac{Px^2}{8EI} + C_2 \qquad \left(\frac{L}{4} \le x \le \frac{L}{2}\right) \tag{e}$$

Applying the symmetry condition (2) to Eq. (e), we obtain the constant C_2:

$$C_2 = -\frac{PL^2}{32EI}$$

Therefore, the slope of the beam between points B and C (from Eq. e) is

$$v' = -\frac{P}{32EI}(L^2 - 4x^2) \qquad \left(\frac{L}{4} \le x \le \frac{L}{2}\right) \tag{9-70}$$

From this equation we can find the slope of the deflection curve at point B where the moment of inertia changes from I to $2I$:

$$v'\left(\frac{L}{4}\right) = -\frac{3PL^2}{128EI} \tag{f}$$

Because the deflection curve is continuous at point B, we can use the continuity condition (3) and equate the slope at point B as obtained from Eq. (d) to the slope at the same point given by Eq. (f). In this manner we find the constant C_1:

$$\frac{P}{4EI}\left(\frac{L}{4}\right)^2 + C_1 = -\frac{3PL^2}{128EI} \quad \text{or} \quad C_1 = -\frac{5PL^2}{128EI}$$

Therefore, the slope between points A and B (see Eq. d) is

$$v' = -\frac{P}{128EI}(5L^2 - 32x^2) \qquad \left(0 \le x \le \frac{L}{4}\right) \tag{9-71}$$

At support A, where $x = 0$, the angle of rotation (Fig. 9-28c) is

$$\theta_A = -v'(0) = \frac{5PL^2}{128EI} \tag{9-72}$$

Deflections of the beam. Integrating the equations for the slopes (Eqs. 9-71 and 9-70), we get

$$v = -\frac{P}{128EI}\left(5L^2x - \frac{32x^3}{3}\right) + C_3 \qquad \left(0 \le x \le \frac{L}{4}\right) \tag{g}$$

$$v = -\frac{P}{32EI}\left(L^2x - \frac{4x^3}{3}\right) + C_4 \qquad \left(\frac{L}{4} \le x \le \frac{L}{2}\right) \tag{h}$$

Applying the boundary condition at the support (condition 1) to Eq. (g), we get $C_3 = 0$. Therefore, the deflection between points A and B (from Eq. g) is

$$v = -\frac{Px}{384EI}(15L^2 - 32x^2) \qquad (0 \le x \le \frac{L}{4}) \tag{9-73}$$ ⬅

From this equation we can find the deflection at point B:

$$v\left(\frac{L}{4}\right) = -\frac{13PL^3}{1536EI} \tag{i}$$

Since the deflection curve is continuous at point B, we can use the continuity condition (4) and equate the deflection at point B as obtained from Eq. (h) to the deflection given by Eq. (i):

$$-\frac{P}{32EI}\left[L^2\left(\frac{L}{4}\right) - \frac{4}{3}\left(\frac{L}{4}\right)^3\right] + C_4 = -\frac{13PL^3}{1536EI}$$

from which

$$C_4 = -\frac{PL^3}{768EI}$$

Therefore, the deflection between points B and C (from Eq. h) is

$$v = -\frac{P}{768EI}(L^3 + 24L^2x - 32x^3) \qquad \left(\frac{L}{4} \le x \le \frac{L}{2}\right) \tag{9-74}$$ ⬅

Thus, we have obtained the equations of the deflection curve for the left-hand half of the beam. (The deflections in the right-hand half of the beam can be obtained from symmetry.)

Finally, we obtain the deflection at the midpoint C by substituting $x = L/2$ into Eq. (9-74):

$$\delta_C = -v\left(\frac{L}{2}\right) = \frac{3PL^3}{256EI} \tag{9-75}$$ ⬅

All required quantities have now been found, and the analysis of the nonprismatic beam is completed.

Notes: Using the differential equation for finding deflections is practical only if the number of equations to be solved is limited to one or two and only if the integrations are easily performed, as in this example. In the case of a tapered beam (Fig. 9-27), it may be difficult to solve the differential equation analytically because the moment of inertia is a continuous function of x. In such a case, the differential equation has variable coefficients instead of constant coefficients.

When a beam has abrupt changes in cross-sectional dimensions, as in this example, there are stress concentrations at the points where changes occur. However, because the stress concentrations affect only a small region of the beam, they have no noticeable effect on the deflections.

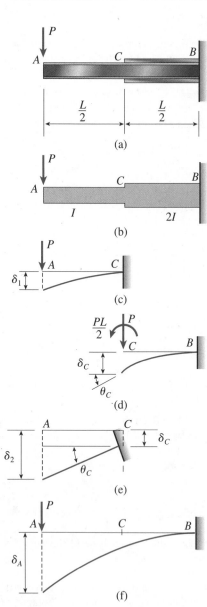

Fig. 9-29 Example 9-14. Cantilever beam with two different moments of inertia

■ **Example 9-14**

A cantilever beam ACB with two different moments of inertia supports a concentrated load P at the free end A (Figs. 9-29a and b). Determine the deflection δ_A at the free end.

Solution   In this example we will use the method of superposition to determine the deflection δ_A at the end of the beam. We begin by recognizing that the deflection consists of two parts: the deflection due to bending of part AC of the beam and the deflection due to bending of part CB. We can determine these deflections separately and then superpose them to obtain the total deflection.

Deflection due to bending of part AC of the beam. Imagine that the beam is held rigidly at point C, so that the beam neither deflects nor rotates at that point (Fig. 9-29c). We can easily calculate the deflection δ_1 of point A in this beam. Since the beam has length $L/2$ and moment of inertia I, its deflection (see Case 4 of Table G-1, Appendix G) is

$$\delta_1 = \frac{P(L/2)^3}{3EI} = \frac{PL^3}{24EI} \tag{j}$$

Deflection due to bending of part CB of the beam. Part CB of the beam also behaves as a cantilever beam (Fig. 9-29d) and contributes to the deflection of point A. The end of this cantilever is subjected to a concentrated load P and a moment $PL/2$. Therefore, the deflection δ_C and angle of rotation θ_C at the free end (Fig. 9-29d) are as follows (see Cases 4 and 6 of Table G-1):

$$\delta_C = \frac{P(L/2)^3}{3(2EI)} + \frac{(PL/2)(L/2)^2}{2(2EI)} = \frac{5PL^3}{96EI}$$

$$\theta_C = \frac{P(L/2)^2}{2(2EI)} + \frac{(PL/2)(L/2)}{2EI} = \frac{3PL^2}{16EI}$$

This deflection and angle of rotation make an additional contribution δ_2 to the deflection at end A (Fig. 9-29e). We again visualize part AC as a cantilever beam, but now its support (at point C) moves downward by the amount δ_C and rotates counterclockwise through the angle θ_C (Fig. 9-29e). These rigid-body displacements produce a downward displacement at end A equal to the following:

$$\delta_2 = \delta_C + \theta_C\left(\frac{L}{2}\right) = \frac{5PL^3}{96EI} + \frac{3PL^2}{16EI}\left(\frac{L}{2}\right) = \frac{7PL^3}{48EI} \tag{k}$$

Therefore, the total deflection at the free end A of the original cantilever beam (Fig. 9-29f) is

$$\delta_A = \delta_1 + \delta_2 = \frac{PL^3}{24EI} + \frac{7PL^3}{48EI} = \frac{3PL^3}{16EI} \tag{9-76}$$ ◀

This example illustrates how the techniques of superposition may be used to find deflections of nonprismatic beams.

9.8 STRAIN ENERGY OF BENDING

The general concepts pertaining to strain energy were explained previously in our discussions of bars subjected to axial loads and shafts subjected to torsion (Sections 2.7 and 3.9, respectively). In this section, we will apply the same concepts to beams. Since we will be using the equations for curvature and deflection derived earlier in this chapter, our discussion of strain energy applies only to beams that behave in a linearly elastic manner. This requirement means that the material must follow Hooke's law and the deflections and rotations must be small.

Let us begin with a simple beam AB in pure bending under the action of two couples, each having a moment M (Fig. 9-30a). The deflection curve (Fig. 9-30b) is a nearly flat circular arc of constant curvature $\kappa = M/EI$ (see Eq. 9-6). The angle θ subtended by this arc equals L/ρ, where L is the length of the beam and ρ is the radius of curvature. Therefore,

$$\theta = \frac{L}{\rho} = \kappa L = \frac{ML}{EI} \tag{9-77}$$

This linear relationship between the moments M and the angle θ is shown graphically by line OA in Fig. 9-31. As the bending couples gradually increase in magnitude from zero to their maximum values, they perform work W represented by the shaded area below line OA. This work, equal to the strain energy U stored in the beam, is

$$W = U = \frac{M\theta}{2} \tag{9-78}$$

This equation is analogous to Eq. (2-29) for the strain energy of an axially loaded bar.

By combining Eqs. (9-77) and (9-78), we can express the strain energy stored in a beam in pure bending in either of the following forms:

$$U = \frac{M^2 L}{2EI} \qquad U = \frac{EI\theta^2}{2L} \tag{9-79a,b}$$

The first of these equations expresses the strain energy in terms of the applied moments M, and the second equation expresses it in terms of the

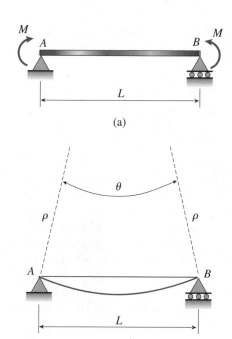

Fig. 9-30 Beam in pure bending by couples of moment M

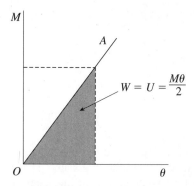

Fig. 9-31 Diagram showing linear relationship between moments M and the angle θ

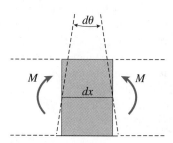

Fig. 9-32 Side view of an element of a beam subjected to bending moments M

angle θ. The equations are similar in form to those for strain energy in an axially loaded bar (Eqs. 2-31a and b).

If the bending moment in a beam varies along its length (nonuniform bending), then we may obtain the strain energy by applying Eqs. (9-79a) and (9-79b) to an element of the beam (Fig. 9-32) and integrating along the length. The length of the element is dx and the angle $d\theta$ between its side faces can be obtained from Eqs. (9-4) and (9-5), as follows:

$$d\theta = \kappa\,dx = \frac{d^2v}{dx^2}dx \qquad\text{(a)}$$

Therefore, the strain energy dU of the element is given by either of the following equations (see Eqs. 9-79a and b):

$$dU = \frac{M^2\,dx}{2EI} \qquad dU = \frac{EI(d\theta)^2}{2dx} = \frac{EI}{2dx}\left(\frac{d^2v}{dx^2}dx\right)^2 = \frac{EI}{2}\left(\frac{d^2v}{dx^2}\right)^2 dx \qquad\text{(b,c)}$$

By integrating the preceding equations throughout the length of a beam, we can express the strain energy stored in a beam in either of the following forms:

$$U = \int \frac{M^2\,dx}{2EI} \qquad U = \int \frac{EI}{2}\left(\frac{d^2v}{dx^2}\right)^2 dx \qquad\text{(9-80a,b)}$$

Note that M is the bending moment in the beam and may vary as a function of x. We use the first equation when the bending moment is known, and the second equation when the equation of the deflection curve is known. (Examples 9-15 and 9-16 illustrate the use of these equations.)

In the derivation of Eqs. (9-80a) and (9-80b), we considered only the effects of the bending moments. If shear forces are also present, additional strain energy will be stored in the beam. However, the strain energy of shear is relatively small (in comparison with the strain energy of bending) for beams of usual proportions, that is, beams in which the lengths are much greater than the depths (say, $L/d > 8$). Therefore, the strain energy of shear may safely be disregarded.

■ Deflections Caused by a Single Load

If a beam supports a single load, either a concentrated load P or a couple M_0, the corresponding deflection δ or angle of rotation θ, respectively, can be determined from the strain energy of the beam. In the case of a beam supporting a concentrated load, the *corresponding deflection* δ is the deflection of the beam axis at the point where the load is applied. The deflection must be measured along the line of action of the load, and is positive in the direction of the load. In the case of a beam supporting a couple as a load, the *corresponding angle of rotation* θ is the angle of rotation of the beam axis at the point where the couple is applied.

Since the strain energy of a beam is equal to the work done by the load, and since δ and θ correspond to P and M_0, respectively, we obtain the following equations:

$$U = W = \frac{P\delta}{2} \qquad U = W = \frac{M_0\theta}{2} \tag{9-81a,b}$$

The first equation applies to a beam loaded *only* by a force P, and the second equation applies to a beam loaded *only* by a couple M_0. It follows from Eqs. (9-81a) and (9-81b) that

$$\delta = \frac{2U}{P} \qquad \theta = \frac{2U}{M_0} \tag{9-82a,b}$$

As explained in Section 2.7, this method for finding deflections and angles of rotation is extremely limited in its application because only one deflection (or one angle) can be found. Furthermore, the only deflection (or angle) that can be found is the one corresponding to the load (or couple). Nevertheless, the method is illustrated later in Example 9-16.

■ Example 9-15

A simple beam AB of length L supports a uniform load of intensity q (Fig. 9-33). (a) Evaluate the strain energy of the beam from the bending moment in the beam. (b) Evaluate the strain energy of the beam from the equation of the deflection curve.

Solution   (a) *Strain energy from the bending moment.* The reaction of the beam at support A is $qL/2$, and therefore the expression for the bending moment in the beam is

$$M = \frac{qLx}{2} - \frac{qx^2}{2} = \frac{q}{2}(Lx - x^2) \tag{d}$$

The strain energy of the beam (from Eq. 9-80a) is

$$U = \int_0^L \frac{M^2 dx}{2EI} = \frac{1}{2EI} \int_0^L \left[\frac{q}{2}(Lx - x^2) \right]^2 dx = \frac{q^2}{8EI} \int_0^L (L^2x^2 - 2Lx^3 + x^4)dx \tag{e}$$

from which we get

$$U = \frac{q^2L^5}{240EI} \tag{9-83} \quad \longleftarrow$$

Note that the load q appears to the second power, which is consistent with the fact that strain energy is always positive. Furthermore, Eq. (9-83) shows that strain energy is *not* a linear function of the loads, even though the beam itself behaves in a linearly elastic manner.

Fig. 9-33 Example 9-15. Strain energy of a beam

(b) *Strain energy from the deflection curve.* The equation of the deflection curve for a simple beam with a uniform load is given in Case 1 of Table G-2, Appendix G, as follows:

$$v = -\frac{qx}{24EI}(L^3 - 2Lx^2 + x^3) \tag{f}$$

Taking two derivatives of this equation, we get

$$\frac{dv}{dx} = -\frac{q}{24EI}(L^3 - 6Lx^2 + 4x^3) \qquad \frac{d^2v}{dx^2} = \frac{q}{2EI}(Lx - x^2)$$

Substituting the latter expression into the equation for strain energy (Eq. 9-80b), we obtain

$$U = \int_0^L \frac{EI}{2}\left(\frac{d^2v}{dx^2}\right)^2 dx = \frac{EI}{2}\int_0^L \left[\frac{q}{2EI}(Lx - x^2)\right]^2 dx$$

$$= \frac{q^2}{8EI}\int_0^L (L^2x^2 - 2Lx^3 + x^4)dx \tag{g}$$

Since the final integral in this equation is the same as the final integral in Eq. (e), we get the same result as before (Eq. 9-83).

■ **Example 9-16**

A cantilever beam *AB* (Fig. 9-34) is subjected to three different loading conditions: (a) a concentrated load *P* at its free end, (b) a couple M_0 at its free end, and (c) both loads acting simultaneously. For each loading condition, determine the strain energy of the beam. Also, determine the vertical deflection δ_A at end *A* of the beam due to the load *P* acting alone (Fig. 9-34a), and determine the angle of rotation θ_A at end *A* due to the moment M_0 acting alone (Fig. 9-34b).

Solution   (a) *Beam with concentrated load P (Fig. 9-34a).* The bending moment in the beam at distance *x* from the free end is $M = -Px$. Substituting this expression for *M* into Eq. (9-80a), we get the following expression for the strain energy of the beam:

$$U = \int_0^L \frac{M^2 dx}{2EI} = \int_0^L \frac{(-PX)^2 dx}{2EI} = \frac{P^2L^3}{6EI} \tag{9-84}$$ ⬅

To obtain the vertical deflection δ_A under the load *P*, we equate the work done by the load to the strain energy:

$$W = U \quad \text{or} \quad \frac{P\delta_A}{2} = \frac{P^2L^3}{6EI}$$

from which

$$\delta_A = \frac{PL^3}{3EI}$$ ⬅

The deflection δ_A is the only deflection we can find by this procedure.

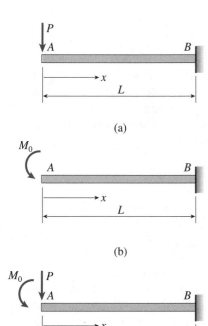

(a)

(b)

(c)

Fig. 9-34 Example 9-16. Strain energy of a beam

(b) *Beam with moment M_0 (Fig. 9-34b).* In this case the bending moment is constant and equal to $-M_0$. Therefore, the strain energy (from Eq. 9-80a) is

$$U = \int_0^L \frac{M^2 \, dx}{2EI} = \int_0^L \frac{(-M_0)^2 \, dx}{2EI} = \frac{M_0^2 L}{2EI} \qquad (9\text{-}85) \quad \Longleftarrow$$

The work W done by the couple M_0 during loading of the beam is $M_0 \theta_A / 2$, where θ_A is the angle of rotation at end A. Therefore,

$$W = U \quad \text{or} \quad \frac{M_0 \theta_A}{2} = \frac{M_0^2 L}{2EI}$$

and

$$\theta_A = \frac{M_0 L}{EI} \quad \Longleftarrow$$

The angle of rotation has the same sense as the moment (counterclockwise in this example).

(c) *Beam with both loads acting simultaneously (Fig. 9-34c).* When both loads act on the beam, the bending moment in the beam is

$$M = -Px - M_0$$

Therefore, the strain energy is

$$U = \int_0^L \frac{M^2 \, dx}{2EI} = \frac{1}{2EI} \int_0^L (-Px - M_0)^2 \, dx$$

$$= \frac{P^2 L^3}{6EI} + \frac{PM_0 L^2}{2EI} + \frac{M_0^2 L}{2EI} \qquad (9\text{-}86) \quad \Longleftarrow$$

The first term in this result gives the strain energy due to P acting alone (Eq. 9-84), and the last term gives the strain energy due to M_0 alone (Eq. 9-85). However, when both loads act simultaneously, an additional term appears in the expression for strain energy.

Therefore, we conclude that *the strain energy in a structure due to two or more loads acting simultaneously cannot be obtained by adding the strain energies due to the loads acting separately.* The reason is that strain energy is a quadratic function of the loads, not a linear function. Therefore, the principle of superposition does not apply to strain energy.

We also observe that we cannot calculate a deflection for a beam with two or more loads by equating the work done by the loads to the strain energy. For instance, if we equate work and energy for the beam of Fig. 9-34c, we get

$$W = U \quad \text{or} \quad \frac{P \delta_{A2}}{2} + \frac{M_0 \theta_{A2}}{2} = \frac{P^2 L^3}{6EI} + \frac{PM_0 L^2}{2EI} + \frac{M_0^2 L}{2EI} \qquad (\text{h})$$

in which δ_{A2} and θ_{A2} represent the deflection and angle of rotation at end A of the beam with two loads (Fig. 9-34c). Although the work done by the loads is indeed equal to the strain energy, and Eq. (h) is quite correct, we cannot solve for either δ_{A2} or θ_{A2} because there are two unknowns and only one equation.

CASTIGLIANO'S THEOREM

Castigliano's theorem provides a means for finding the deflections of a structure from the strain energy of the structure. To illustrate what we mean by that statement, consider a cantilever beam with a concentrated load P acting at the free end (Fig. 9-35a). The strain energy of this beam is obtained from Eq. (9-84) of Example 9-16:

$$U = \frac{P^2 L^3}{6EI} \tag{a}$$

Now take the derivative of this expression with respect to the load P:

$$\frac{dU}{dP} = \frac{d}{dP}\left(\frac{P^2 L^3}{6EI}\right) = \frac{PL^3}{3EI} \tag{b}$$

We immediately recognize this result as the deflection δ_A at the free end A of the beam (see Fig. 9-35b). Note especially that the deflection δ_A *corresponds* to the load P itself. (Recall that a deflection corresponding to a concentrated load is the deflection at the point where the concentrated load is applied. Furthermore, the deflection is in the direction of the load.) Thus, Eq. (b) shows that *the derivative of the strain energy with respect to the load is equal to the deflection corresponding to the load.* Castigliano's theorem is a generalized statement of this observation, and we will now derive it in more general terms.

(a)

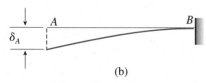

(b)

Fig. 9-35 Beam supporting a single load P

■ **Derivation of Castigliano's Theorem**

Let us consider a beam subjected to any number of loads, say n loads P_1, $P_2, \ldots, P_i, \ldots, P_n$ (Fig. 9-36a). The deflections of the beam corresponding to the various loads are denoted $\delta_1, \delta_2, \ldots, \delta_i, \ldots, \delta_n$, as shown in Fig. 9-36b. As in our earlier discussions of deflections and strain energy, we assume that the principle of superposition is applicable to the beam and its loads.

Now we will determine the strain energy of this beam. When the loads are applied to the beam, they gradually increase in magnitude from zero to their maximum values. At the same time, each load moves through its corresponding displacement and does work. The total work W done by the loads is equal to the strain energy U stored in the beam:

$$W = U \tag{c}$$

Note that W (and hence U) is a function of the loads $P_1, P_2, \ldots, P_n$ acting on the beam.

Next, let us suppose that one of the loads, say the ith load, is increased slightly by the amount dP_i while the other loads are held constant. This increase in load will cause a small increase dU in the strain energy of the beam. This increase in strain energy may be expressed as

(a)

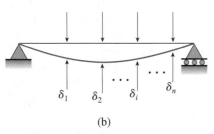

(b)

Fig. 9-36 Beam supporting n loads

the rate of change of U with respect to P_i times the small increase in P_i. Thus, the increase in strain energy is

$$dU = \frac{\partial U}{\partial P_i} dP_i \qquad \text{(d)}$$

where $\partial U / \partial P_i$ is the rate of change of U with respect to P_i. (Since U is a function of *all* the loads, the derivative with respect to any one of the loads is a partial derivative.) The final strain energy of the beam is

$$U + dU = U + \frac{\partial U}{\partial P_i} dP_i \qquad \text{(e)}$$

in which U is the strain energy referred to in Eq. (c).

Because the principal of superposition holds for this beam, the total strain energy is independent of the order in which the loads are applied. That is, the final displacements of the beam (and the work done by the loads in reaching those displacements) are the same regardless of the order in which the loads are applied. In arriving at the strain energy given by Eq. (e), we first applied the n loads $P_1, P_2, \ldots, P_n$, and then we applied the load dP_i. However, we can reverse the order of application and apply the load dP_i first, followed by the loads $P_1, P_2, \ldots, P_n$. The final amount of strain energy is the same in either case.

When the load dP_i is applied first, it produces strain energy equal to one-half the product of the load dP_i and its corresponding displacement $d\delta_i$. Thus, the amount of strain energy due to the load dP_i is

$$\frac{dP_i\, d\delta_i}{2} \qquad \text{(f)}$$

When the loads $P_1, P_2, \ldots, P_n$ are applied, they produce the same displacements as before ($\delta_1, \delta_2, \ldots, \delta_n$) and do the same amount of work as before (Eq. c). However, during the application of these loads, the force dP_i automatically moves through the displacement δ_i. In so doing, it produces additional work equal to the product of the force and the distance through which it moves. (Note that the work does not have a factor 1/2, because the force dP_i acts at full value throughout this displacement.) Thus, the additional work, equal to the additional strain energy, is

$$dP_i\, \delta_i \qquad \text{(g)}$$

Therefore, the final strain energy for the second loading sequence is

$$\frac{dP_i\, d\delta_i}{2} + U + dP_i\, \delta_i \qquad \text{(h)}$$

Equating this expression for the final strain energy to the earlier expression (e), which was obtained for the first loading sequence, we get

$$\frac{dP_i\, d\delta_i}{2} + U + dP_i\, \delta_i = U + \frac{\partial U}{\partial P_i} dP_i \qquad \text{(i)}$$

We can discard the first term because it contains the product of two differentials and is infinitesimally small compared to the other terms. We then obtain the following relationship:

$$\delta_i = \frac{\partial U}{\partial P_i}$$

(9-87)

This equation is known as **Castigliano's theorem.**[*]

Although we derived Castigliano's theorem by using a beam as an illustration, we could have used any other type of structure (for example, a truss) and any other kinds of loads (for example, loads in the form of couples). The important requirements are that the structure be linearly elastic and that the principal of superposition be applicable. Also, note that the strain energy must be expressed as a function of the loads (and not as a function of the displacements), a condition which is implied in the theorem itself, since the partial derivative is taken with respect to a load. With these limitations in mind, we can state Castigliano's theorem in general terms as follows:

The partial derivative of the strain energy of a structure with respect to any load is equal to the displacement corresponding to that load.

The strain energy of a linearly elastic structure is a *quadratic* function of the loads (for instance, see Eq. a), and therefore the partial derivatives and the displacements (Eq. 9-87) are *linear* functions of the loads (as expected).

When using the terms *load* and *corresponding displacement* in connection with Castigliano's theorem, it is understood that these terms are used in a generalized sense. The load P_i and corresponding displacement δ_i may be a force and a corresponding translation, or a couple and a corresponding rotation, or some other set of corresponding quantities.

■ **Application of Castigliano's Theorem**

As an application of Castigliano's theorem, let us consider a cantilever beam AB carrying a concentrated load P and a couple of moment M_0 acting at the free end (Fig. 9-37a). We wish to determine the vertical deflection δ_A and angle of rotation θ_A at the end of the beam (Fig. 9-37b). Note that δ_A is the deflection corresponding to the load P, and θ_A is the angle of rotation corresponding to the moment M_0.

---

[*]Castigliano's theorem, one of the most famous theorems in structural analysis, was discovered by Carlos Alberto Pio Castigliano (1847–1884), an Italian engineer (Ref. 9-9). The theorem quoted here (Eq. 9-87) is actually the second of two theorems presented by Castigliano and is properly called *Castigliano's second theorem*. The first theorem is the reverse of the second theorem, in the sense that it gives the *loads* on a structure in terms of the partial derivatives of the strain energy with respect to the *displacements*.

Fig. 9-37 Application of Castigliano's theorem to a beam

The first step in the analysis is to determine the strain energy of the beam. For that purpose, we write the equation for the bending moment as follows:

$$M = -Px - M_0 \tag{j}$$

in which x is the distance from the free end (Fig. 9-37a). The strain energy is found by substituting this expression for M into Eq. (9-80a):

$$U = \int_0^L \frac{M^2 dx}{2EI} = \frac{1}{2EI} \int_0^L (-Px - M_0)^2 dx$$

$$= \frac{P^2 L^3}{6EI} + \frac{P M_0 L^2}{2EI} + \frac{M_0^2 L}{2EI} \tag{k}$$

in which L is the length of the beam and EI is its flexural rigidity. Note that the strain energy is a quadratic function of the loads P and M_0.

To obtain the vertical deflection δ_A at the end of the beam, we use Castigliano's theorem (Eq. 9-87) and take the partial derivative of the strain energy with respect to P:

$$\delta_A = \frac{\partial U}{\partial P} = \frac{PL^3}{3EI} + \frac{M_0 L^2}{2EI} \tag{l}$$

This expression for the deflection can be verified by comparing it with the formulas of Cases 4 and 6 of Table G-1, Appendix G.

In a similar manner, we can find the angle of rotation θ_A at the end of the beam by taking the partial derivative with respect to M_0:

$$\theta_A = \frac{\partial U}{\partial M_0} = \frac{PL^2}{2EI} + \frac{M_0 L}{EI} \tag{m}$$

which also can be checked with Cases 4 and 6 of Table G-1.

■ Use of a Fictitious Load

The only displacements that can be found from Castigliano's theorem are those that correspond to loads acting on the structure. If we wish to calculate a displacement at a point on a structure where there is no load, then a fictitious load *corresponding to the desired displacement* must be applied to the structure. We can then determine the displacement by evaluating the strain energy and taking the partial derivative with respect to the fictitious load. The result is the displacement produced by the actual loads and the fictitious load acting simultaneously. By setting the fictitious load equal to zero, we obtain the displacement produced only by the actual loads.

To illustrate this concept, suppose we wish to find the vertical deflection δ_C at the midpoint C of the cantilever beam shown in Fig. 9-38a. Since the deflection δ_C is downward (Fig. 9-38b), the load corre-

(a)

(b)

Fig. 9-38 Beam supporting loads P and M_0

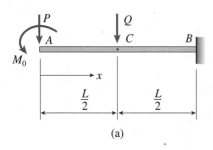

(a)

(b)

Fig. 9-39 Beam with a fictitious load Q

sponding to that deflection is a downward vertical force acting at the same point. Therefore, we must supply a fictitious load Q acting at point C in the downward direction (Fig. 9-39a). Then we can use Castigliano's theorem to determine the deflection $(\delta_C)_0$ at the midpoint of this beam (Fig. 9-39b). From that deflection, we can obtain the deflection δ_C in the beam of Fig. 9-38 by setting Q equal to zero.

We begin by finding the bending moments in the beam of Fig. 9-39a:

$$M = -Px - M_0 \qquad \left(0 \le x \le \frac{L}{2}\right) \tag{n}$$

$$M = -Px - M_0 - Q\left(x - \frac{L}{2}\right) \qquad \left(\frac{L}{2} \le x \le L\right) \tag{o}$$

Next, we determine the strain energy of the beam by applying Eq. (9-80a) to each half of the beam. For the left-hand half of the beam (from point A to point C), the strain energy is

$$U_{AC} = \int_0^{L/2} \frac{M^2 \, dx}{2EI} = \frac{1}{2EI} \int_0^{L/2} (-Px - M_0)^2 \, dx$$

$$= \frac{P^2 L^3}{48EI} + \frac{PM_0 L^2}{8EI} + \frac{M_0^2 L}{4EI} \tag{p}$$

For the right-hand half, the strain energy is

$$U_{CB} = \int_{L/2}^{L} \frac{M^2 \, dx}{2EI} = \frac{1}{2EI} \int_{L/2}^{L} \left[-Px - M_0 - Q\left(x - \frac{L}{2}\right) \right]^2 \, dx$$

$$= \frac{7P^2 L^3}{48EI} + \frac{3PM_0 L^2}{8EI} + \frac{5PQL^3}{48EI} + \frac{M_0^2 L}{4EI} + \frac{M_0 QL^2}{8EI} + \frac{Q^2 L^3}{48EI} \tag{q}$$

which requires a very lengthy process of integration. Adding the strain energies for the two parts of the beam, we obtain the strain energy for the entire beam (Fig. 9-39a):

$$U = U_{AC} + U_{CB}$$

$$= \frac{P^2 L^3}{6EI} + \frac{PM_0 L^2}{2EI} + \frac{5PQL^3}{48EI} + \frac{M_0^2 L}{2EI} + \frac{M_0 QL^2}{8EI} + \frac{Q^2 L^3}{48EI} \tag{r}$$

The deflection at the midpoint of the beam shown in Fig. 9-39a can now be obtained from Castigliano's theorem:

$$(\delta_C)_0 = \frac{\partial U}{\partial Q} = \frac{5PL^3}{48EI} + \frac{M_0 L^2}{8EI} + \frac{QL^3}{24EI} \tag{s}$$

This equation gives the deflection at point C produced by all three loads acting on the beam. To obtain the deflection produced by the loads P and

M_0 only, we set the load Q equal to zero in the preceding equation. The result is the deflection at point C for the beam with two loads (Fig. 9-38a):

$$\delta_C = \frac{5PL^3}{48EI} + \frac{M_0L^2}{8EI} \qquad \text{(t)}$$

Thus, the deflection in the original beam has been obtained.

This method is often called the *dummy-load method*, because of the introduction of a fictitious, or dummy, load.

■ Differentiation Under the Integral Sign

As we saw in the preceding example, the use of Castigliano's theorem for determining beam deflections may lead to lengthy integrations, especially when more than two loads act on the beam. The reason is clear—finding the strain energy requires the integration of the *square* of the bending moment (Eq. 9-80a). For instance, if the bending moment expression has three terms, its square may have as many as six terms, each of which must be integrated.

After the integrations are completed and the strain energy has been determined, we differentiate the strain energy to obtain the deflections. However, we can bypass the step of finding the strain energy by differentiating *before* integrating. This procedure does not eliminate the integrations, but it does make them much simpler.

To derive this method, we begin with the equation for the strain energy (Eq. 9-80a) and apply Castigliano's theorem (Eq. 9-87):

$$\delta_i = \frac{\partial U}{\partial P_i} = \frac{\partial}{\partial P_i} \int \frac{M^2 dx}{2EI} \qquad \text{(u)}$$

Following the rules of calculus, we can differentiate the integral by differentiating under the integral sign:

$$\delta_i = \frac{\partial}{\partial P_i} \int \frac{M^2 dx}{2EI} = \int \left(\frac{M}{EI}\right)\left(\frac{\partial M}{\partial P_i}\right) dx \qquad \text{(9-88)}$$

We will refer to this equation as the **modified Castigliano's theorem**.

When using the modified theorem, we integrate the product of the bending moment and its derivative. By contrast, when using the standard Castigliano's theorem (see Eq. u), we integrate the square of the bending moment. Since the derivative is a shorter expression than the moment itself, this new procedure is much simpler. To show this, we will now solve the preceding examples using the modified theorem (Eq. 9-88).

Let us begin with the beam shown in Fig. 9-37 and recall that we wish to find the deflection and angle of rotation at the free end. The bending moment and its derivatives (see Eq. j) are

$$M = -Px - M_0$$

$$\frac{\partial M}{\partial P} = -x \qquad \frac{\partial M}{\partial M_0} = -1$$

Therefore, the deflection δ_A and angle of rotation θ_A (from Eq. 9-88) are

$$\delta_A = \frac{1}{EI} \int_0^L (-Px - M_0)(-x)dx = \frac{PL^3}{3EI} + \frac{M_0 L^2}{2EI} \qquad \text{(v)}$$

$$\theta_A = \frac{1}{EI} \int_0^L (-Px - M_0)(-1)dx = \frac{PL^2}{2EI} + \frac{M_0 L}{EI} \qquad \text{(w)}$$

These equations agree with the earlier results (Eqs. l and m). However, the calculations are shorter than those performed earlier, because we did not have to integrate the square of the bending moment (see Eq. k).

The advantages of differentiating under the integral sign are even more apparent when there are more than two loads acting on the structure, as in the example of Fig. 9-38. In that example, we wished to determine the deflection δ_C at the midpoint C of the beam due to the loads P and M_0. To do so, we added a fictitious load Q at the midpoint (Fig. 9-39). We then proceeded to find the deflection $(\delta_C)_0$ at the midpoint of the beam when all three loads (P, M_0, and Q) were acting. Finally, we set $Q = 0$ to obtain the deflection δ_C due to P and M_0 alone. The solution was time-consuming, because the integrations were extremely long. However, if we use the modified theorem and differentiate first, the calculations are much shorter.

With all three loads acting (Fig. 9-39), the bending moments and their derivatives are as follows (see Eqs. n and o):

$$M = -Px - M_0 \qquad \frac{\partial M}{\partial Q} = 0 \qquad \left(0 \leq x \leq \frac{L}{2}\right)$$

$$M = -Px - M_0 - Q\left(x - \frac{L}{2}\right) \qquad \frac{\partial M}{\partial Q} = -\left(x - \frac{L}{2}\right) \qquad \left(\frac{L}{2} \leq x \leq L\right)$$

Therefore, the deflection $(\delta_C)_0$, from Eq. (9-88), is

$$(\delta_C)_0 = \frac{1}{EI} \int_0^{L/2} (-Px - M_0)(0)dx$$

$$+ \frac{1}{EI} \int_{L/2}^{L} \left[-Px - M_0 - Q\left(x - \frac{L}{2}\right)\right]\left[-\left(x - \frac{L}{2}\right)\right]dx$$

Since Q is a fictitious load, and since we have already taken the partial derivatives, we can set Q equal to zero before integrating and obtain the deflection δ_C due to the two loads P and M_0 as follows:

$$\delta_C = \frac{1}{EI} \int_{L/2}^{L} [-Px - M_0]\left[-\left(x - \frac{L}{2}\right)\right] dx = \frac{5PL^3}{48EI} + \frac{M_0 L^2}{8EI}$$

which agrees with the earlier result (Eq. t). Again, the integrations are greatly simplified by differentiating under the integral sign and using the modified theorem.

The partial derivative that appears under the integral sign in Eq. (9-88) has a simple physical interpretation. It represents the rate of change of the bending moment M with respect to the load P_i, that is, it is equal to the bending moment M produced by a load P_i of unit value. This observation leads to a method of finding deflections known as the *unit-load method*. Castigliano's theorem also leads to a method of structural analysis known as the *flexibility method*. Both the unit-load method and the flexibility method are widely used in structural analysis and are described in textbooks on that subject.

The following examples provide additional illustrations of the use of Castigliano's theorem for finding deflections of beams. However, it should be remembered that the theorem is not limited to finding beam deflections—it applies to any kind of linearly elastic structure for which the principle of superposition is valid.

■ Example 9-17

A simple beam AB supports a uniform load of intensity $q = 1.5$ k/ft and a concentrated load $P = 5$ k (Fig. 9-40). The load P acts at the midpoint C of the beam. The beam has length $L = 8.0$ ft, modulus of elasticity $E = 30 \times 10^6$ psi, and moment of inertia $I = 75.0$ in.[4] Determine the downward deflection δ_C at the midpoint of the beam by the following methods: (1) Obtain the strain energy of the beam and use Castigliano's theorem, and (2) use the modified form of Castigliano's theorem (differentiation under the integral sign).

Solution   *Method (1).* Because the beam and its loading are symmetrical about the midpoint, the strain energy for the entire beam is equal to twice the strain energy for the left-hand half of the beam. Therefore, we begin by finding the bending moment in the left-hand half of the beam.

The reaction at support A (Fig. 9-41) is

$$R_A = \frac{P}{2} + \frac{qL}{2}$$

and therefore the bending moment M is

$$M = R_A x - \frac{qx^2}{2} = \frac{Px}{2} + \frac{qLx}{2} - \frac{qx^2}{2} \tag{x}$$

Fig. 9-40 Example 9-17. Simple beam with two loads

in which x is measured from support A. The strain energy of the beam (from Eq. 9-80a) is

$$U = \int \frac{M^2\, dx}{2EI} = 2 \int_0^{L/2} \frac{1}{2EI}\left(\frac{Px}{2} + \frac{qLx}{2} - \frac{qx^2}{2}\right)^2 dx$$

After squaring the term in parentheses and performing a lengthy integration, we find

$$U = \frac{P^2 L^3}{96EI} + \frac{5PqL^4}{384EI} + \frac{q^2 L^5}{240EI}$$

Since the deflection at the midpoint C (Fig. 9-40) corresponds to the load P, we can find the deflection by using Castigliano's theorem:

$$\delta_C = \frac{\partial U}{\partial P} = \frac{\partial}{\partial P}\left(\frac{P^2 L^3}{96EI} + \frac{5PqL^4}{384EI} + \frac{q^2 L^5}{240EI}\right) = \frac{PL^3}{48EI} + \frac{5qL^4}{384EI} \quad \text{(y)} \quad \longleftarrow$$

Method (2). By using the modified form of Castigliano's theorem (Eq. 9-88), we avoid the lengthy integration for finding the strain energy. The bending moment in the left-hand half of the beam has already been determined (see Eq. x), and its partial derivative with respect to the load P is

$$\frac{\partial M}{\partial P} = \frac{x}{2}$$

Therefore, the modified Castigliano's theorem becomes

$$\delta_C = \int \left(\frac{M}{EI}\right)\left(\frac{\partial M}{\partial P}\right) dx$$

$$= 2 \int_0^{L/2} \frac{1}{EI}\left(\frac{Px}{2} + \frac{qLx}{2} - \frac{qx^2}{2}\right)\left(\frac{x}{2}\right) dx = \frac{PL^3}{48EI} + \frac{5qL^4}{384EI} \quad \text{(z)} \quad \longleftarrow$$

which agrees with the earlier result (Eq. y) but requires a much simpler integration.

Numerical solution. Now that we have an expression for the deflection at point C, we can substitute numerical values, as follows:

$$\delta_C = \frac{PL^3}{48EI} + \frac{5qL^4}{384EI}$$

$$= \frac{(5\ \text{k})(96\ \text{in.})^3}{48(30 \times 10^6\,\text{psi})(75.0\ \text{in.}^4)} + \frac{5(1.5\ \text{k/ft})(1/12\ \text{ft/in.})(96\ \text{in.})^4}{384(30 \times 10^6\ \text{psi})(75.0\ \text{in.}^4)}$$

$$= 0.0410\ \text{in.} + 0.0614\ \text{in.} = 0.1024\ \text{in.} \quad \longleftarrow$$

Note that numerical values cannot be substituted until *after* the partial derivative is obtained. If numerical values are substituted prematurely, either in the expression for the bending moment or the expression for the strain energy, it may be impossible to take the derivative.

$$R_A = \frac{P}{2} + \frac{qL}{2}$$

Fig. 9-41 Free-body diagram for determining the bending moment M (Example 9-17)

■ **Example 9-18**

A simple beam with an overhang supports a uniform load of intensity q on span AB and a concentrated load P at end C of the overhang (Fig. 9-42). Determine the deflection δ_C and angle of rotation θ_C at point C. (Use the modified form of Castigliano's theorem.)

(a)

(b)

Fig. 9-42 Example 9-18. Beam with an overhang

Solution   *Deflection δ_C at the end of the overhang (Fig. 9-42b).* Since the load P corresponds to this deflection, we do not need to supply a fictitious load. Instead, we can begin immediately to find the bending moments throughout the length of the beam. The reaction at support A is

$$R_A = \frac{qL}{2} - \frac{P}{2}$$

as shown in Fig. 9-43. Therefore, the bending moment in span AB is

$$M_{AB} = R_A x_1 - \frac{qx_1^2}{2} = \frac{qLx_1}{2} - \frac{Px_1}{2} - \frac{qx_1^2}{2} \qquad (0 \le x_1 \le L)$$

where x_1 is measured to the right from support A (Fig. 9-43). The bending moment in the overhang is

$$M_{BC} = -Px_2 \qquad \left(0 \le x_2 \le \frac{L}{2}\right)$$

where x_2 is measured to the left from point C (Fig. 9-43).

Next, we determine the partial derivatives with respect to the load P:

$$\frac{\partial M_{AB}}{\partial P} = -\frac{x_1}{2} \qquad (0 \le x_1 \le L)$$

$$\frac{\partial M_{BC}}{\partial P} = -x_2 \qquad \left(0 \le x_2 \le \frac{L}{2}\right)$$

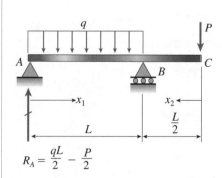

Fig. 9-43 Reaction at support A and coordinates x_1 and x_2 for the beam of Example 9-18

Now that we have obtained expressions for the bending moments and partial derivatives, we can use the modified form of Castigliano's theorem (Eq. 9-88) to obtain the deflection at point C:

$$\delta_C = \int \left(\frac{M}{EI}\right)\left(\frac{\partial M}{\partial P}\right)dx$$

$$= \frac{1}{EI}\int_0^L M_{AB}\left(\frac{\partial M_{AB}}{\partial P}\right)dx + \frac{1}{EI}\int_0^{L/2} M_{BC}\left(\frac{\partial M_{BC}}{\partial P}\right)dx$$

Substituting the expressions for the bending moments and partial derivatives, we get

$$\delta_C = \frac{1}{EI}\int_0^L \left(\frac{qLx_1}{2} - \frac{Px_1}{2} - \frac{qx_1^2}{2}\right)\left(-\frac{x_1}{2}\right)dx_1 + \frac{1}{EI}\int_0^{L/2}(-Px_2)(-x_2)dx_2$$

By performing the integrations and combining terms, we get

$$\delta_C = \frac{PL^3}{8EI} - \frac{qL^4}{48EI} \qquad\qquad (9\text{-}89) \quad \Longleftarrow$$

Since the load P acts downward, the deflection δ_C is positive downward. In other words, if the preceding equation produces a positive result, the deflection is downward. If the result is negative, the deflection is upward.

Comparing the two terms in Eq. (9-89), we see that the deflection at the end of the overhang is downward when $P > qL/6$ and upward when $qL > 6P$.

Angle of rotation θ_C at the end of the overhang (Fig. 9-42b). Since there is no load on the original beam (Fig. 9-42a) corresponding to this angle of rotation, we must supply a fictitious load. Therefore, we place a couple of moment M_C at point C (Fig. 9-44). Note that the couple M_C acts at the point on the beam where the angle of rotation is to be determined. Furthermore, it has the same clockwise direction as the angle of rotation.

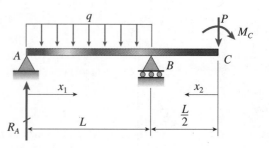

Fig. 9-44 Fictitious moment M_C acting on the beam of Example 9-18

We now follow the same steps as when determining the deflection at C. First, we note that the reaction at support A (Fig. 9-44) is

$$R_A = \frac{qL}{2} - \frac{P}{2} - \frac{M_C}{L}$$

Consequently, the bending moment in span AB becomes

$$M_{AB} = R_A x_1 - \frac{qx_1^2}{2} = \frac{qLx_1}{2} - \frac{Px_1}{2} - \frac{M_C x_1}{L} - \frac{qx_1^2}{2} \qquad (0 \le x_1 \le L)$$

and the bending moment in the overhang becomes

$$M_{BC} = -Px_2 - M_C \qquad \left(0 \le x_2 \le \frac{L}{2}\right)$$

The partial derivatives are taken with respect to the moment M_C, which is the load corresponding to the angle of rotation. Therefore,

$$\frac{\partial M_{AB}}{\partial M_C} = -\frac{x_1}{L} \qquad \left(0 \le x_1 \le L\right)$$

$$\frac{\partial M_{BC}}{\partial M_C} = -1 \qquad \left(0 \le x_2 \le \frac{L}{2}\right)$$

Now we use the modified form of Castigliano's theorem (Eq. 9-88) to obtain the angle of rotation at point C:

$$\theta_C = \int \left(\frac{M}{EI}\right)\left(\frac{\partial M}{\partial M_C}\right) dx$$

$$= \frac{1}{EI}\int_0^L M_{AB}\left(\frac{\partial M_{AB}}{\partial M_C}\right) dx + \frac{1}{EI}\int_0^{L/2} M_{BC}\left(\frac{\partial M_{BC}}{\partial M_C}\right) dx$$

Substituting the expressions for the bending moments and partial derivatives, we obtain

$$\theta_C = \frac{1}{EI}\int_0^L \left(\frac{qLx_1}{2} - \frac{Px_1}{2} - \frac{M_C x_1}{L} - \frac{qx_1^2}{2}\right)\left(-\frac{x_1}{L}\right) dx_1$$

$$+ \frac{1}{EI}\int_0^{L/2} (-Px_2 - M_C)(-1) dx_2$$

Since M_C is a fictitious load, and since we have already taken the partial derivatives, we can set M_C equal to zero at this stage of the calculations and simplify the integrations:

$$\theta_C = \frac{1}{EI}\int_0^L \left(\frac{qLx_1}{2} - \frac{Px_1}{2} - \frac{qx_1^2}{2}\right)\left(-\frac{x_1}{L}\right) dx_1 + \frac{1}{EI}\int_0^{L/2} (-Px_2)(-1) dx_2$$

After carrying out the integrations and combining terms, we obtain

$$\theta_C = \frac{7PL^2}{24EI} - \frac{qL^3}{24EI} \qquad (9\text{-}90) \quad \Longleftarrow$$

If this equation produces a positive result, the angle of rotation is clockwise. If the result is negative, the angle is counterclockwise.

Comparing the two terms in Eq. (9-90), we see that the angle of rotation is clockwise when $P > qL/7$ and counterclockwise when $qL > 7P$.

If numerical data are available, it is now a routine matter to substitute numbers into Eqs. (9-89) and (9-90) and calculate the deflection and angle of rotation at the end of the overhang.

*9.10 DEFLECTIONS PRODUCED BY IMPACT

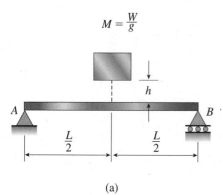

$$M = \frac{W}{g}$$

h

A B

$$\frac{L}{2} \qquad \frac{L}{2}$$

(a)

A B

δ_{max}

(b)

Fig. 9-45 Deflection of a beam produced by a falling body

In this section we will discuss the impact of an object falling onto a beam (Fig. 9-45a). We will determine the dynamic deflection of the beam by equating the potential energy lost by the falling mass to the strain energy acquired by the beam. This approximate method was described in detail in Section 2.8 for a mass striking an axially loaded bar; consequently, Section 2.8 should be fully understood before proceeding.

Most of the assumptions described in Section 2.8 apply to beams as well as to axially loaded bars. Some of these assumptions are: (1) The falling weight sticks to the beam and moves with it, (2) no energy losses occur, (3) the beam behaves in a linearly elastic manner, (4) the deflected shape of the beam is the same under a dynamic load as under a static load, and (5) the potential energy of the beam due to its change in position is relatively small and may be disregarded. In general, these assumptions are reasonable if the mass of the falling object is very large compared to the mass of the beam. Otherwise, the approximate analysis given here is not valid and a more advanced analysis is required.

As an example, we will consider the simple beam AB shown in Fig. 9-45. The beam is struck at its midpoint by a falling body of mass M and weight W. Based upon the preceding idealizations, we may assume that all of the potential energy lost by the body during its fall is transformed into elastic strain energy that is stored in the beam. Since the distance through which the body falls is $h + \delta_{max}$, where h is the initial height above the beam (Fig. 9-45a) and δ_{max} is the maximum dynamic deflection of the beam (Fig. 9-45b), the potential energy lost is

$$\text{Potential energy} = W(h + \delta_{max}) \qquad \text{(a)}$$

The strain energy acquired by the beam can be determined from the deflection curve by using Eq. (9-80b), which is repeated here:

$$U = \int \frac{EI}{2}\left(\frac{d^2v}{dx^2}\right)^2 dx \qquad \text{(b)}$$

The deflection curve for a simple beam subjected to a concentrated load acting at the midpoint (see Case 4 of Table G-2, Appendix G) is

$$v = -\frac{Px}{48EI}(3L^2 - 4x^2) \qquad \left(0 \le x \le \frac{L}{2}\right) \qquad \text{(c)}$$

Also, the maximum deflection of the beam is

$$\delta_{max} = \frac{PL^3}{48EI} \qquad \text{(d)}$$

Eliminating the load P between Eqs. (c) and (d), we get the equation of the deflection curve in terms of the maximum deflection:

$$v = -\frac{\delta_{max}x}{L^3}(3L^2 - 4x^2) \qquad \left(0 \le x \le \frac{L}{2}\right) \qquad \text{(e)}$$

Taking two derivatives, we find

$$\frac{d^2v}{dx^2} = \frac{24\delta_{max}x}{L^3} \qquad \text{(f)}$$

Finally, we substitute the second derivative into Eq. (b) and obtain the following expression for the strain energy of the beam in terms of the maximum deflection:

$$U = 2\int_0^{L/2} \frac{EI}{2}\left(\frac{d^2v}{dx^2}\right)^2 dx = EI\int_0^{L/2}\left(\frac{24\delta_{max}x}{L^3}\right)^2 dx = \frac{24EI\delta_{max}^2}{L^3} \qquad \text{(g)}$$

Equating the potential energy lost by the falling mass (Eq. a) to the strain energy acquired by the beam (Eq. g), we get

$$W(h + \delta_{max}) = \frac{24EI\delta_{max}^2}{L^3} \qquad \text{(9-91)}$$

This equation is quadratic in δ_{max} and can be solved for its positive root:

$$\delta_{max} = \frac{WL^3}{48EI} + \left[\left(\frac{WL^3}{48EI}\right)^2 + 2h\left(\frac{WL^3}{48EI}\right)\right]^{1/2} \qquad \text{(9-92)}$$

We see that the maximum deflection increases if either the weight of the falling object or the height of fall is increased, and the maximum deflection decreases if the stiffness EI/L^3 of the beam is increased.

To simplify the preceding equation, we will denote the static deflection of the beam due to the weight W as δ_{st}:

$$\delta_{st} = \frac{WL^3}{48EI} \qquad \text{(9-93)}$$

Then Eq. (9-92) for the maximum dynamic deflection can be written as follows:

$$\delta_{max} = \delta_{st} + (\delta_{st}^2 + 2h\delta_{st})^{1/2} \qquad \text{(9-94)}$$

This equation shows that the dynamic deflection is always larger than the static deflection. If the height h equals 0, which means that the load is applied suddenly but without any free fall, the dynamic deflection is twice the static deflection. If h is very large compared to the deflection, then the term containing h in Eq. (9-94) predominates, and the equation can be simplified to

$$\delta_{max} = \sqrt{2h\delta_{st}} \qquad (9\text{-}95)$$

These observations are analogous to those discussed previously in Section 2.8 for impact on a bar in tension or compression.

The deflection δ_{max} calculated from Eq. (9-94) generally represents an upper limit, because we assumed there were no energy losses during impact. Several other factors also tend to reduce the deflection, including localized deformation of the contact surfaces, the tendency of the falling mass to bounce upward, and inertia effects of the mass of the beam.

*9.11 ■ TEMPERATURE EFFECTS

In the preceding sections of this chapter we considered the deflections of beams due to lateral loads. In this section, we will consider the deflections caused by nonuniform temperature changes. As a preliminary matter, recall that the effects of *uniform* temperature changes have already been described in Section 2.5, where it was shown that a uniform temperature increase causes an unconstrained bar or beam to have its length increased by the amount

$$\delta_T = \alpha(\Delta T)L \qquad (9\text{-}96)$$

In this equation, α is the coefficient of thermal expansion,* ΔT is the uniform increase in temperature, and L is the length of the bar (see Fig. 2-21 and Eq. 2-14 in Chapter 2).

If a beam is supported in such a manner that longitudinal expansion is free to occur, as is the case for all of the statically determinate beams considered in this chapter, then a uniform temperature change will not produce any stresses in the beam. Also, there will be no lateral deflections of such a beam, because there is no tendency for the beam to bend.

The behavior of a beam is quite different if the temperature is not constant across its height. For example, assume that a simple beam, initially straight and at a uniform temperature T_0, has its temperature changed to T_1 on its upper surface and T_2 on its lower surface, as pictured in Fig. 9-46a. If we assume that the variation in temperature is linear between the top and bottom of the beam, then the *average temperature* of the beam is

$$T_{aver} = \frac{T_1 + T_2}{2} \qquad (9\text{-}97)$$

---

*Values of the thermal coefficient α are given in Table H-4, Appendix H, for various materials.

and occurs at midheight. Any difference between this average temperature and the initial temperature T_0 results in a change in length of the beam, given by Eq. (9-96), as follows:

$$\delta_T = \alpha(T_{aver} - T_0)L = \alpha\left(\frac{T_1 + T_2}{2} - T_0\right)L \qquad (9\text{-}98)$$

In addition, the temperature differential $T_2 - T_1$ between the bottom and top of the beam produces a *curvature* of the axis of the beam, with the accompanying lateral deflections (Fig. 9-46b).

To investigate the deflections due to a temperature differential, consider an element of length dx cut out from the beam (Figs. 9-46a and c). The changes in length of the element at the bottom and top are $\alpha(T_2 - T_0)dx$ and $\alpha(T_1 - T_0)dx$, respectively. If T_2 is greater than T_1, the sides of the element will rotate with respect to each other through an angle $d\theta$, as shown in Fig. 9-46c. The angle $d\theta$ is related to the changes in dimension by the following equation, obtained from the geometry of the figure:

$$h\,d\theta = \alpha(T_2 - T_0)dx - \alpha(T_1 - T_0)dx$$

from which we get

$$\frac{d\theta}{dx} = \frac{\alpha(T_2 - T_1)}{h} \qquad (9\text{-}99)$$

in which h is the height of the beam.

We have already seen that the quantity $d\theta/dx$ represents the curvature of the deflection curve of the beam (see Eq. 9-4). Since the curvature is equal to d^2v/dx^2 (Eq. 9-5), we may write the following **differential equation of the deflection curve**:

$$\frac{d^2v}{dx^2} = \frac{\alpha(T_2 - T_1)}{h} \qquad (9\text{-}100)$$

Note that when T_2 is greater than T_1, the curvature is positive and the beam is bent concave upward, as shown in Fig. 9-46b. The quantity $\alpha(T_2 - T_1)/h$ in Eq. (9-100) is the counterpart of the quantity M/EI, which appears in the differential equation that we used previously (Eq. 9-7).

We can solve Eq. (9-100) by the same integration techniques described earlier for the effects of bending moments (see Section 9.3). We can integrate the differential equation to obtain dv/dx and v, and we can use boundary or other conditions to evaluate the constants of integration. In this manner we can obtain the equations for the slopes and deflections of the beam, as illustrated by Problems 9.11-1 through 9.11-4 at the end of this chapter.

(a)

(b)

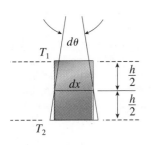

(c)

Fig. 9-46 Temperature effects in a beam

If the beam is able to change in length and deflect freely, there will be no stresses associated with the temperature changes described in this section. However, if the beam is restrained against longitudinal expansion or lateral deflection, or if the temperature changes do not vary linearly from top to bottom of the beam, internal temperature stresses will develop. The determination of such stresses requires the use of more advanced methods of analysis.

PROBLEMS/CHAPTER 9

DEFLECTIONS OF BEAMS

The beams described in Problems 9.3-1 through 9.3-10 have constant flexural rigidity EI.

9.3-1 The deflection curve for a simple beam *AB* (see figure) is given by the following equation:

$$v = -\frac{q_0 x}{360LEI}(7L^4 - 10L^2x^2 + 3x^4)$$

Describe the load acting on the beam.

Probs. 9.3-1 and 9.3-2

9.3-2 The deflection curve for a simple beam *AB* (see figure) is given by the following equation:

$$v = -\frac{q_0 x}{90L^2 EI}(3L^5 - 5L^3x^2 + 3Lx^4 - x^5)$$

Describe the load acting on the beam.

9.3-3 A simple beam *AB* with span length *L* and overhangs of length *d* is loaded by forces P_A and P_B as shown in the figure. The deflection curve in the region from *A* to *B* is given by the equation

$$v = -\frac{Px}{6EI}(L - x)^2$$

in which the origin of coordinates is at support *A*. Determine the forces P_A and P_B.

Prob. 9.3-3

9.3-4 A uniformly loaded steel beam with simple supports (see Fig. 9-8 and Example 9-1) has a downward deflection of 12 mm at the midpoint and angles of rotation equal to 0.01 radians at the ends. Calculate the height *h* of the beam if the maximum bending stress is 75 MPa and the modulus of elasticity is 200 GPa.

9.3-5 A wide-flange beam (W 12 × 35) supports a uniform load on a simple span of length *L* = 14 ft (see Fig. 9-8 and Example 9-1). (a) Calculate the maximum deflection δ_{max} at the midpoint and the angles of rotation θ at the supports if *q* = 1.75 k/ft and *E* = 30 × 10⁶ psi. (b) What is the radius of curvature ρ at the midpoint of the beam?

9.3-6 A gold-alloy microbeam attached to a silicon wafer behaves like a cantilever beam subjected to a uniform load (see figure). The beam has length *L* = 25 μm and rectangular cross section of width *b* = 15 μm and thickness *t* = 0.87 μm. The total load on the beam is 44 μN. If the deflection at the end of the beam is 1.3 μm, what is the modulus of elasticity E_g of the gold alloy? (*Hint:* See Fig. 9-10 and Example 9-2.)

Prob. 9.3-6

9.3-7 What is the span length L of a uniformly loaded simple beam (see Fig. 9-8 and Example 9-1) of wide-flange cross section if the maximum bending stress is 12,000 psi, the maximum deflection is 0.1 in., the height of the beam is 12 in., and the modulus of elasticity is 30×10^6 psi?

9.3-8 A cantilever beam with a uniform load (see Fig. 9-10 and Example 9-2) has a height h equal to 1/10 of the length L. The beam is a steel wide-flange section with $E = 208$ GPa and an allowable bending stress of 130 MPa in both tension and compression. Calculate the ratio δ/L of the deflection at the free end to the length, assuming that the beam carries the maximum allowable load.

9.3-9 (a) Calculate the maximum deflection δ_{max} of a uniformly loaded simple beam (see Fig. 9-8 and Example 9-1) if the span length $L = 7$ ft, the intensity of the uniform load $q = 150$ lb/ft, and the maximum bending stress $\sigma = 9000$ psi. The cross section of the beam is square, and the material is aluminum having modulus of elasticity $E = 10 \times 10^6$ psi. (b) Calculate the radius of curvature ρ at midspan of the beam.

9.3-10 Obtain a formula for the ratio δ_C/δ_{max} of the deflection at the midpoint to the maximum deflection for a simple beam supporting a concentrated load P (see Fig. 9-12 and Example 9-3). From the formula, plot a graph of δ_C/δ_{max} versus the ratio a/L that defines the position of the load ($0.5 < a/L < 1$). What conclusion do you draw from the graph?

DEFLECTIONS BY INTEGRATION OF THE BENDING-MOMENT EQUATION

Problems 9.3-11 through 9.3-19 are to be solved by integrating the second-order differential equation of the deflection curve (the bending-moment equation). All beams have constant flexural rigidity EI.

9.3-11 Derive the equation of the deflection curve for a cantilever beam AB supporting a load P at the free end (see figure). Also, determine the deflection δ_B and angle of rota-

tion θ_B at the free end. (*Note:* Use the second-order differential equation of the deflection curve.)

Prob. 9.3-11

9.3-12 Derive the equation of the deflection curve for a simple beam AB loaded by a couple M_0 at the left-hand support (see figure). Also, determine the maximum deflection δ_{max}. (*Note:* Use the second-order differential equation of the deflection curve.)

Prob. 9.3-12

9.3-13 A cantilever beam AB supporting a triangularly distributed load of maximum intensity q_0 is shown in the figure. Derive the equation of the deflection curve and then obtain formulas for the deflection δ_B and angle of rotation θ_B at the free end. (*Note:* Use the second-order differential equation of the deflection curve.)

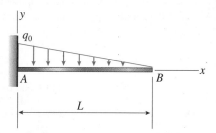

Prob. 9.3-13

9.3-14 A cantilever beam AB is acted upon by a uniformly distributed moment (bending moment, not torque) of intensity m per unit distance along the axis of the beam (see figure). Derive the equation of the deflection curve and then obtain formulas for the deflection δ_B and angle of rotation θ_B at the free end. (*Note:* Use the second-order differential equation of the deflection curve.)

Prob. 9.3-14

9.3-15 The beam shown in the figure has a roller support at A and a guided support at B. The guided support permits vertical movement but no rotation. Derive the equation of the deflection curve and determine the deflection δ_B at end B due to the uniform load of intensity q. (*Note:* Use the second-order differential equation of the deflection curve.)

Prob. 9.3-15

9.3-16 Derive the equations of the deflection curve for a simple beam AB loaded by a couple M_0 acting at distance a from the left-hand support (see figure). Also, determine the deflection δ_0 at the point where the load is applied. (*Note:* Use the second-order differential equation of the deflection curve.)

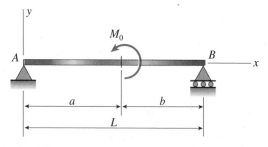

Prob. 9.3-16

9.3-17 Derive the equations of the deflection curve for a cantilever beam AB carrying a uniform load of intensity q over part of the span (see figure). Also, determine the deflection δ_B at the end of the beam. (*Note:* Use the second-order differential equation of the deflection curve.)

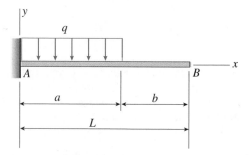

Prob. 9.3-17

9.3-18 Derive the equations of the deflection curve for a cantilever beam AB supporting a uniform load of intensity q acting over one-half of the length (see figure). Also, obtain formulas for the deflections δ_B and δ_C at points B and C, respectively. (*Note:* Use the second-order differential equation of the deflection curve.)

Prob. 9.3-18

9.3-19 Derive the equations of the deflection curve for a simple beam AB with a uniform load of intensity q acting over the left-hand half of the span (see figure). Also, determine the deflection δ_C at the midpoint of the beam. (*Note:* Use the second-order differential equation of the deflection curve.)

Prob. 9.3-19

DEFLECTIONS BY INTEGRATION OF THE SHEAR-FORCE AND LOAD EQUATIONS

The beams described in the problems for Section 9.4 have constant flexural rigidity EI.

9.4-1 Derive the equation of the deflection curve for a cantilever beam AB when a couple M_0 acts counterclockwise at the free end (see figure). Also, determine the deflection δ_B and slope θ_B at the free end. Use the third-order differential equation of the deflection curve (the shear-force equation).

Prob. 9.4-1

9.4-2 A simple beam AB is subjected to a distributed load of intensity $q = q_0 \sin \pi x/L$, where q_0 is the maximum intensity of the load (see figure). Derive the equation of the deflection curve, and then determine the deflection δ_{max} at the midpoint of the beam. Use the fourth-order differential equation of the deflection curve (the load equation).

Prob. 9.4-2

9.4-3 The simple beam AB shown in the figure has moments $2M_0$ and M_0 acting at the ends. Derive the equation of the deflection curve, and then determine the maximum deflection δ_{max}. Use the third-order differential equation of the deflection curve (the shear-force equation).

Prob. 9.4-3

9.4-4 A simple beam with a uniform load is pin supported at one end and spring supported at the other (see figure). The spring has stiffness $k = 48EI/L^3$. Derive the equation of the deflection curve by starting with the third-order differential equation (the shear-force equation). Also, determine the angle of rotation θ_A at support A.

Prob. 9.4-4

9.4-5 The distributed load acting on a cantilever beam AB has an intensity q given by the expression $q_0 \cos \pi x /2L$, where q_0 is the maximum intensity of the load (see figure). Derive the equation of the deflection curve, and then determine the deflection δ_B at the free end. Use the fourth-order differential equation of the deflection curve (the load equation).

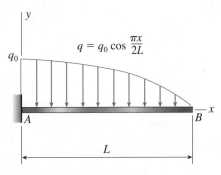

Prob. 9.4-5

9.4-6 A cantilever beam AB is subjected to a parabolically varying load of intensity $q = q_0(L^2 - x^2)/L^2$, where q_0 is the maximum intensity of the load (see figure). Derive the equation of the deflection curve, and then determine the deflection δ_B and angle of rotation θ_B at the free end. Use the fourth-order differential equation of the deflection curve (the load equation).

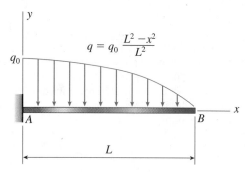

$$q = q_0 \frac{L^2 - x^2}{L^2}$$

Prob. 9.4-6

9.4-7 A beam on simple supports is subjected to a parabolically distributed load of intensity $q = 4q_0 x(L - x)/L^2$, where q_0 is the maximum intensity of the load (see figure). Derive the equation of the deflection curve, and then determine the maximum deflection δ_{max}. Use the fourth-order differential equation of the deflection curve (the load equation).

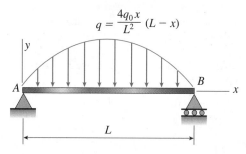

$$q = \frac{4q_0 x}{L^2}(L - x)$$

Prob. 9.4-7

9.4-8 Derive the equation of the deflection curve for a simple beam AB carrying a triangularly distributed load of maximum intensity q_0 (see figure). Also, determine the maximum deflection δ_{max} of the beam. Use the fourth-order differential equation of the deflection curve (the load equation).

Prob. 9.4-8

*9.4-9 Derive the equations of the deflection curve for an overhanging beam ABC subjected to a uniform load of intensity q acting on the overhang (see figure). Also, obtain formulas for the deflection δ_C and angle of rotation θ_C at the end of the overhang. Use the fourth-order differential equation of the deflection curve (the load equation).

Prob. 9.4-9

*9.4-10 Derive the equations of the deflection curve for a simple beam AB supporting a triangularly distributed load of maximum intensity q_0 acting on the right-hand half of the beam (see figure). Also, determine the angles of rotation θ_A and θ_B at the ends and the deflection δ_C at the midpoint. Use the fourth-order differential equation of the deflection curve (the load equation).

Prob. 9.4-10

METHOD OF SUPERPOSITION

The problems for Section 9.5 are to be solved by the method of superposition. All beams have constant flexural rigidity EI.

9.5-1 A cantilever beam AB carries three equally spaced concentrated loads, as shown in the figure. Obtain formulas for the angle of rotation θ_B and deflection δ_B at the free end of the beam.

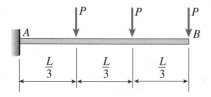

Prob. 9.5-1

9.5-2 A simple beam AB supports five equally spaced loads P (see figure). (a) Determine the deflection δ_1 at the midpoint of the beam. (b) If the same total load ($5P$) is distributed as a uniform load on the beam, what is the deflection δ_2 at the midpoint? (c) Calculate the ratio of δ_1 to δ_2.

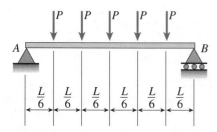

Prob. 9.5-2

9.5-3 The cantilever beam AB shown in the figure has an extension BCD attached to its free end. A force P acts at the end of the extension. (a) Find the ratio a/L so that the vertical deflection of point B will be zero. (b) Find the ratio a/L so that the angle of rotation at point B will be zero.

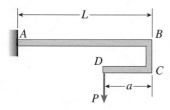

Prob. 9.5-3

9.5-4 Beam ACB hangs from two springs, as shown in the figure. The springs have stiffnesses k_1 and k_2 and the beam has flexural rigidity EI. What is the downward displacement of point C, which is at the midpoint of the beam, when the load P is applied? Data for the structure are as follows: $P = 8.0$ kN, $L = 1.8$ m, $EI = 216$ kN·m², $k_1 = 250$ kN/m, and $k_2 = 160$ kN/m.

Prob. 9.5-4

9.5-5 What must be the equation $y = f(x)$ of the axis of the slightly curved beam AB (see figure) *before* the load is applied in order that the load P, moving along the bar, always stays at the same level?

Prob. 9.5-5

9.5-6 Determine the angle of rotation θ_B and deflection δ_B at the free end of a cantilever beam AB having a uniform load of intensity q acting over the middle third of its length (see figure).

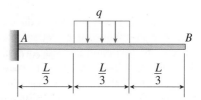

Prob. 9.5-6

9.5-7 The cantilever beam *ACB* shown in the figure has flexural rigidity $EI = 2.1 \times 10^6$ k-in.^{2} Calculate the deflections δ_C and δ_B at points *C* and *B*, respectively, due to the simultaneous action of the moment of 35 k-in. applied at point *C* and the concentrated load of 3.6 k applied at the free end *B*.

Prob. 9.5-7

9.5-8 A beam *ABCD* consisting of a simple span *BD* and an overhang *AB* is loaded by a force *P* acting at the end of the bracket *CEF* (see figure). (a) Determine the deflection δ_A at the end of the overhang. (b) Under what conditions is this deflection upward? Under what conditions is it downward?

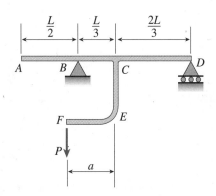

Prob. 9.5-8

9.5-9 A horizontal load *P* acts at end *C* of the bracket *ABC* shown in the figure. (a) Determine the deflection δ_C of point *C*. (b) Determine the maximum upward deflection δ_{max} of member *AB*. (Assume that the flexural rigidity *EI* is constant throughout the frame. Also, disregard the effects of axial deformations and consider only the effects of bending due to the load *P*.)

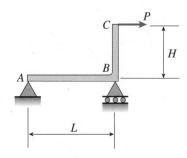

Prob. 9.5-9

9.5-10 A beam *ABC* having flexural rigidity $EI = 75$ kN·m^2 is loaded by a force $P = 480$ N at end *C* and tied down at end *A* by a wire having axial rigidity $EA = 900$ kN (see figure). What is the deflection at point *C* when the load *P* is applied?

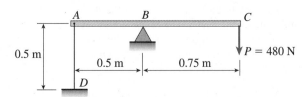

Prob. 9.5-10

9.5-11 Determine the angle of rotation θ_B and deflection δ_B at the free end of a cantilever beam *AB* supporting a parabolic load defined by the equation $q = q_0 x^2/L^2$ (see figure).

Prob. 9.5-11

9.5-12 A simple beam *AB* supports a uniform load of intensity *q* acting over the middle region of the span (see figure). Determine the angle of rotation θ_A at the left-hand support and the deflection δ_{max} at the midpoint.

Prob. 9.5-12

Prob. 9.5-15

9.5-13 The overhanging beam *ABCD* supports two concentrated loads *P* and *Q* (see figure). (a) For what ratio *P/Q* will the deflection at point *B* be zero? (b) For what ratio will the deflection at point *D* be zero?

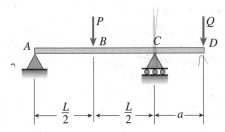

Prob. 9.5-13

9.5-14 A thin metal strip of total weight *W* and length *L* is placed across the top of a flat table of width *L*/3 as shown in the figure. What is the clearance δ between the strip and the middle of the table? (The strip of metal has flexural rigidity *EI*.)

Prob. 9.5-14

9.5-15 An overhanging beam *ABC* with flexural rigidity $EI = 21$ k-in.^{2} is supported by a pin support at *A* and by a spring of stiffness *k* at point *B* (see figure). Span *AB* has length *L* = 30 in. and carries a uniformly distributed load. The overhang *BC* has length *b* = 15 in. For what stiffness *k* of the spring will the uniform load produce no deflection at the free end *C*?

9.5-16 A beam *ABCD* rests on simple supports at *B* and *C* (see figure). The beam has a slight initial curvature so that end *A* is 15 mm above the elevation of the supports and end *D* is 10 mm above. What loads *P* and *Q*, acting at points *A* and *D*, respectively, will move points *A* and *D* downward to the level of the supports? (The flexural rigidity *EI* of the beam is 5.0×10^6 N·m^2.)

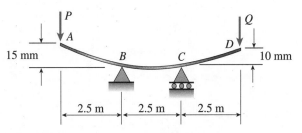

Prob. 9.5-16

9.5-17 The compound beam *ABCD* shown in the figure has fixed supports at ends *A* and *D* and consists of three members joined by pin connections at *B* and *C*. Find the deflection δ under the load *P*.

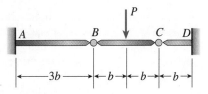

Prob. 9.5-17

9.5-18 A compound beam *ABCDE* (see figure) consists of two parts (*ABC* and *CDE*) connected by a hinge at *C*. Determine the deflection δ_E at the free end *E* due to the load *P* acting at that point.

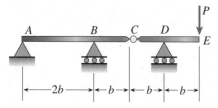

Prob. 9.5-18

9.5-19 A steel beam *ABC* is simply supported at *A* and held by a high-strength steel wire at *B* (see figure). A load $P = 240$ lb acts at the free end *C*. The wire has axial rigidity $EA = 300 \times 10^3$ lb, and the beam has flexural rigidity $EI = 30 \times 10^6$ lb-in.^{2} What is the deflection δ_C of point *C* due to the load *P*?

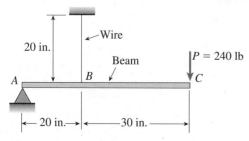

Prob. 9.5-19

9.5-20 The compound beam shown in the figure consists of a cantilever beam *AB* (length *L*) that is pin-connected to a simple beam *BD* (length *2L*). After the beam is constructed, a clearance *c* exists between the beam and a support at *C*, midway between points *B* and *D*. Subsequently, a uniform load is placed along the entire length of the beam. What intensity *q* of the load is needed to close the gap at *C* and bring the beam into contact with the support?

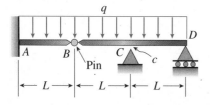

Prob. 9.5-20

*9.5-21 A beam *ABCDE* has simple supports at *B* and *D* and symmetrical overhangs at each end (see figure). The center span has length *L* and each overhang has length *b*. A uniform load of intensity *q* acts on the beam. (a) Determine the ratio *b/L* so that the deflection δ_C at the midpoint of the beam is equal to the deflections δ_A and δ_E at the ends. (b) For this value of *b/L*, what is the deflection δ_C at the midpoint?

Prob. 9.5-21

*9.5-22 Two equal wheel loads *P*, distance *L/4* apart, move slowly across a simple beam *AB* of span length *L* (see figure). Determine the maximum value of the deflection δ_C at the midpoint of the beam.

Prob. 9.5-22

*9.5-23 Find the horizontal deflection δ_h and vertical deflection δ_v at the free end *C* of the frame *ABC* shown in the figure. (The flexural rigidity *EI* is constant throughout the frame.) *Note:* Disregard the effects of axial deformations and consider only the effects of bending due to the load *P*.

Prob. 9.5-23

*9.5-24 The frame *ABCD* shown in the figure is squeezed by two collinear forces *P* acting at points *A* and *D*. What is the decrease δ in the distance between points A and D when the loads P are applied? (The flexural rigidity EI is constant throughout the frame.) *Note:* Disregard the effects of axial deformations and consider only the effects of bending due to the loads P.

Prob. 9.5-24

*9.5-25 A frame ABC is loaded at point C by a force P acting at an angle α to the horizontal (see figure). Both members of the frame have the same length and the same flexural rigidity. Determine the angle α so that the deflection of point C is in the same direction as the load. (Disregard the effects of axial deformations and consider only the effects of bending due to the load P.) *Note:* A direction of loading such that the resulting deflection is in the same direction as the load is called a *principal direction*. For a given load on a planar structure, there are two principal directions, perpendicular to each other.

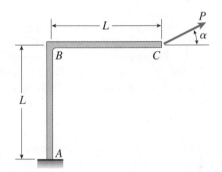

Prob. 9.5-25

MOMENT-AREA METHOD

The problems for Section 9.6 are to be solved by the moment-area method. All beams have constant flexural rigidity EI.

9.6-1 A cantilever beam AB is subjected to a uniform load of intensity q acting throughout its length (see figure). Determine the angle of rotation θ_B and the deflection δ_B at the free end.

Prob. 9.6-1

9.6-2 The load on a cantilever beam AB has a triangular distribution with maximum intensity q_0 (see figure). Determine the angle of rotation θ_B and the deflection δ_B at the free end.

Prob. 9.6-2

9.6-3 A cantilever beam AB is subjected to a concentrated load P and a couple M_0 acting at the free end (see figure). Obtain formulas for the angle of rotation θ_B and the deflection δ_B at end B.

Prob. 9.6-3

9.6-4 Determine the angle of rotation θ_B and the deflection δ_B at the free end of a cantilever beam AB with a uniform load of intensity q acting over the middle third of the length (see figure).

Prob. 9.6-4

9.6-5 Calculate the deflections δ_B and δ_C at points B and C, respectively, of the cantilever beam ACB shown in the figure. Assume $M_0 = 36$ k-in., $P = 3.8$ k, $L = 8$ ft, and $EI = 2.25 \times 10^9$ lb-in.^{2}

Prob. 9.6-5

9.6-6 A cantilever beam ACB supports two concentrated loads P_1 and P_2 as shown in the figure. Calculate the deflections δ_B and δ_C at points B and C, respectively. Assume $P_1 = 10$ kN, $P_2 = 5$ kN, $L = 2.6$ m, $E = 200$ GPa, and $I = 20.1 \times 10^6$ mm^4.

Prob. 9.6-6

9.6-7 Obtain formulas for the angle of rotation θ_A at support A and the deflection δ at the midpoint for a simple beam AB with a uniform load of intensity q (see figure).

Prob. 9.6-7

9.6-8 A simple beam AB supports two concentrated loads P at the positions shown in the figure. A support C at the midpoint of the beam is positioned at distance d below the beam before the loads are applied. Assuming that $d = 10$ mm, $L = 6$ m, $E = 200$ GPa, and $I = 198 \times 10^6$ mm^4, calculate the magnitude of the loads P so that the beam just touches the support at C.

Prob. 9.6-8

9.6-9 A simple beam AB is subjected to a load in the form of a couple M_0 acting at end B (see figure). Determine the angles of rotation θ_A and θ_B at the supports and the deflection δ at the midpoint.

Prob. 9.6-9

9.6-10 The simple beam AB shown in the figure supports two equal concentrated loads P, one acting downward and the other upward. Determine the angle of rotation θ_A at the left-hand end, the deflection δ_1 under the downward load, and the deflection δ_2 at the midpoint of the beam.

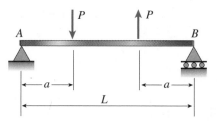

Prob. 9.6-10

9.6-11 A simple beam AB is subjected to couples M_0 and $2M_0$ acting as shown in the figure. Determine the angles of rotation θ_A and θ_B at the ends of the beam and the deflection δ at the point where the load M_0 is applied.

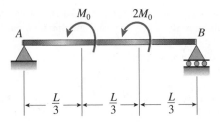

Prob. 9.6-11

9.6-12 A simple beam AB supports two concentrated loads P_1 and P_2 acting as shown in the figure. Calculate the angles of rotation θ_A and θ_B at the supports and the deflections δ_1 and δ_2 at the loads, assuming $P_1 = 100$ kN, $P_2 = 200$ kN, $L = 10$ m, $E = 200$ GPa, and $I = 1.20 \times 10^9$ mm.[4]

Prob. 9.6-12

NONPRISMATIC BEAMS

9.7-1 The cantilever beam ACB shown in the figure has moments of inertia I_2 and I_1 in parts AC and CB, respectively. (a) Using the method of superposition, determine the deflection δ_B at the free end due to the load P. (b) Determine the ratio r of the deflection δ_B to the deflection δ_1 at the free end of a prismatic cantilever with moment of inertia I_1 carrying the same load. (c) Plot a graph of the deflection ratio r versus the ratio I_2/I_1 of the moments of inertia. (Let I_2/I_1 vary from 1 to 5.)

Prob. 9.7-1

9.7-2 The cantilever beam ACB shown in the figure supports a uniform load of intensity q throughout its length. The beam has moments of inertia I_2 and I_1 in parts AC and CB, respectively. (a) Using the method of superposition, determine the deflection δ_B at the free end due to the uniform load. (b) Determine the ratio r of the deflection δ_B to the deflection δ_1 at the free end of a prismatic cantilever with moment of inertia I_1 carrying the same load. (c) Plot a graph of the deflection ratio r versus the ratio I_2/I_1 of the moments of inertia. (Let I_2/I_1 vary from 1 to 5.)

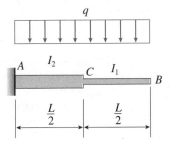

Prob. 9.7-2

***9.7-3** A simple beam $ABCD$ has moment of inertia I near the supports and moment of inertia $2I$ in the middle region, as shown in the figure. A uniform load of intensity q acts over the entire length of the beam. Determine the equations of the deflection curve for the left-hand half of the beam. Also, find the angle of rotation θ_A at the left-hand support and the deflection δ_{max} at the midpoint.

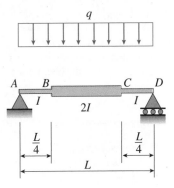

Prob. 9.7-3

**9.7-4 A beam ABC has a rigid segment from A to B and a flexible segment with moment of inertia I from B to C (see figure). A concentrated load P acts at point B. Determine the angle of rotation θ_A of the rigid segment, the deflection δ_B at point B, and the maximum deflection δ_{max}.

Prob. 9.7-4

**9.7-5 A simple beam ABC has moment of inertia $1.5I$ from A to B and I from B to C (see figure). A concentrated load P acts at point B. Obtain the equations of the deflection curves for both parts of the beam. From the equations, determine the angles of rotation θ_A and θ_C at the supports and the deflection δ_B at point B.

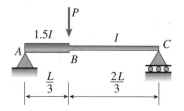

Prob. 9.7-5

**9.7-6 The tapered cantilever beam AB shown in the figure has thin-walled, hollow circular cross sections of constant thickness t. The diameters at the ends A and B are d_A and $d_B = 2d_A$, respectively. Thus, the diameter d and moment of inertia I at distance x from the free end are, respectively,

$$d = \frac{d_A}{L}(L + x) \qquad I = \frac{\pi t d^3}{8} = \frac{\pi t d_A^3}{8L^3}(L + x)^3 = \frac{I_A}{L^3}(L + x)^3$$

in which I_A is the moment of inertia at end A of the beam. Determine the equation of the deflection curve and the deflection δ_A at the free end of the beam due to the load P.

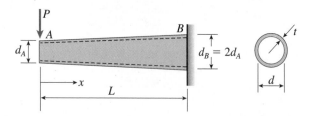

Prob. 9.7-6

**9.7-7 The tapered cantilever beam AB shown in the figure has a solid circular cross section. The diameters at the ends A and B are d_A and $d_B = 2d_A$, respectively. Thus, the diameter d and moment of inertia I at distance x from the free end are, respectively,

$$d = \frac{d_A}{L}(L + x) \qquad I = \frac{\pi d^4}{64} = \frac{\pi d_A^4}{64L^4}(L + x)^4 = \frac{I_A}{L^4}(L + x)^4$$

in which I_A is the moment of inertia at end A of the beam. Determine the equation of the deflection curve and the deflection δ_A at the free end of the beam due to the load P.

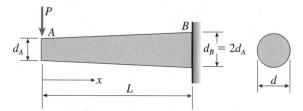

Prob. 9.7-7

**9.7-8 A tapered cantilever beam AB supports a concentrated load P at the free end (see figure). The cross sections of the beam are rectangular with constant width b, depth d_A at support A, and depth $d_B = 3d_A/2$ at the support. Thus, the depth d and moment of inertia I at distance x from the free end are, respectively,

$$d = \frac{d_A}{2L}(2L + x)$$

$$I = \frac{bd^3}{12} = \frac{bd_A^3}{96L^3}(2L + x)^3 = \frac{I_A}{8L^3}(2L + x)^3$$

in which I_A is the moment of inertia at end A of the beam. Determine the equation of the deflection curve and the deflection δ_A at the free end of the beam due to the load P.

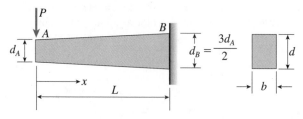

Prob. 9.7-8

**9.7-9 A simple beam ACB is constructed with square cross sections and a double taper (see figure). The depth of the beam at the supports is d_A and at the midpoint is $d_C = 2d_A$. Each half of the beam has length L. Thus, the depth d and moment of inertia I at distance x from the left-hand end are, respectively,

$$d = \frac{d_A}{L}(L + x) \qquad I = \frac{d^4}{12} = \frac{d_A^4}{12L^4}(L + x)^4 = \frac{I_A}{L^4}(L + x)^4$$

in which I_A is the moment of inertia at end A of the beam. (These equations are valid for x between 0 and L, that is, for the left-hand half of the beam.) (a) Obtain equations for the slope and deflection of the left-hand half of the beam due to the uniform load. (b) From those equations obtain formulas for the angle of rotation θ_A at support A and the deflection δ_C at the midpoint.

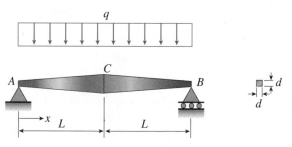

Prob. 9.7-9

STRAIN ENERGY

The beams described in the problems for Section 9.8 have constant flexural rigidity EI.

9.8-1 A simple beam AB of length L supports a concentrated load P at the midpoint (see figure). (a) Evaluate the strain energy of the beam from the bending moment in the beam. (b) Evaluate the strain energy of the beam from the equation of the deflection curve. (c) From the strain energy, determine the deflection δ under the load P.

Prob. 9.8-1

9.8-2 A uniformly loaded simple beam AB (see figure) of span length L and rectangular cross section (b = width, h = height) has a maximum bending stress σ_{max} due to the uniform load. Determine the strain energy U stored in the beam.

Prob. 9.8-2

9.8-3 A cantilever beam AB of length L supports a uniform load of intensity q (see figure). (a) Evaluate the strain energy of the beam from the bending moment in the beam. (b) Evaluate the strain energy of the beam from the equation of the deflection curve. (c) Calculate the numerical value of U if the beam is a W 8 × 15 steel wide-flange section 6 ft long loaded to a maximum bending stress of 16,000 psi. (Assume $E = 30 × 10^6$ psi.)

Prob. 9.8-3

9.8-4 A simple beam AB of length L is subjected to loads that produce a symmetric deflection curve with maximum deflection δ at the midpoint of the span (see figure). How much strain energy U is stored in the beam if the deflection curve is: (a) a parabola, and (b) a half wave of a sine curve?

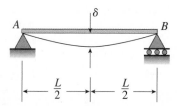

Prob. 9.8-4

9.8-5 A beam ABC with simple supports at A and B and an overhang BC supports a concentrated load P at the free end C (see figure). (a) Determine the strain energy U stored in the beam due to the load P. (b) From the strain energy, find the deflection δ_C under the load P. (c) Calculate the numerical values of U and δ_C if the length L is 8 ft, the overhang length a is 3 ft, the beam is a W 10 × 12 steel wide-flange section, and the load P produces a maximum stress of 12,000 psi in the beam. (Use $E = 29 \times 10^6$ psi.)

Prob. 9.8-5

9.8-6 A simple beam AB supporting a concentrated load P at the midpoint and a couple of moment M_0 at one end is shown in the figure. Determine the strain energy U stored in the beam due to the force P and the couple M_0 acting simultaneously.

Prob. 9.8-6

9.8-7 The frame shown in the figure consists of a beam ACB supported by a strut CD. The beam has length $2L$ and is continuous through joint C. A concentrated load P acts at the free end B. Determine the vertical deflection δ_B at point B due to the load P. (Let EI denote the flexural rigidity of the beam, and let EA denote the axial rigidity of the strut. Disregard axial and shearing effects in the beam, and disregard any bending effects in the strut.)

Prob. 9.8-7

CASTIGLIANO'S THEOREM

The beams described in the problems for Section 9.9 have constant flexural rigidity EI.

9.9-1 A simple beam AB of length L is loaded at the left-hand end by a couple of moment M_0 (see figure). Determine the angle of rotation θ_A at support A. (Obtain the solution by determining the strain energy of the beam and then using Castigliano's theorem.)

Prob. 9.9-1

9.9-2 The simple beam shown in the figure supports a concentrated load P acting at distance a from the left-hand support and distance b from the right-hand support. Determine the deflection δ_D at point D where the load is applied. (Obtain the solution by determining the strain energy of the beam and then using Castigliano's theorem.)

Prob. 9.9-2

9.9-3 An overhanging beam ABC supports a concentrated load P at the end of the overhang (see figure). Span AB has length L and the overhang has length a. Determine the deflection δ_C at the end of the overhang. (Obtain the solution by determining the strain energy of the beam and then using Castigliano's theorem.)

Prob. 9.9-3

9.9-4 The cantilever beam shown in the figure supports a triangularly distributed load of maximum intensity q_0. Determine the deflection δ_B at the free end B. (Obtain the solution by determining the strain energy of the beam and then using Castigliano's theorem.)

Prob. 9.9-4

9.9-5 A simple beam ACB supports a uniform load of intensity q on the left-hand half of the span (see figure). Determine the angle of rotation θ_B at support B. (Obtain the solution by using the modified form of Castigliano's theorem.)

Prob. 9.9-5

9.9-6 A cantilever beam ACB supports two concentrated loads P_1 and P_2, as shown in the figure. Determine the deflections δ_C and δ_B at points C and B, respectively. (Obtain the solution by using the modified form of Castigliano's theorem.)

Prob. 9.9-6

9.9-7 The cantilever beam ACB shown in the figure is subjected to a uniform load of intensity q acting between

points A and C. Determine the angle of rotation θ_A at the free end A. (Obtain the solution by using the modified form of Castigliano's theorem.)

Prob. 9.9-7

9.9-8 The frame ABC supports a concentrated load P at point C (see figure). Members AB and BC have lengths h and b, respectively. Determine the vertical deflection δ_C and angle of rotation θ_C at end C of the frame. (Obtain the solution by using the modified form of Castigliano's theorem.)

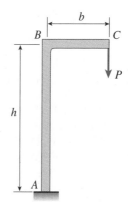

Prob. 9.9-8

9.9-9 An overhanging beam ABC is subjected to a couple M_A at the free end (see figure). The lengths of the overhang and the main span are a and L, respectively. Determine the angle of rotation θ_A and deflection δ_A at end A. (Obtain the solution by using the modified form of Castigliano's theorem.)

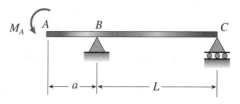

Prob. 9.9-9

9.9-10 A simple beam *ABCDE* supports a uniform load of intensity *q* (see figure). The moment of inertia in the central part of the beam (*BCD*) is twice the moment of inertia in the end parts (*AB* and *DE*). Find the deflection δ_C at the midpoint *C* of the beam. (Obtain the solution by using the modified form of Castigliano's theorem.)

Prob. 9.9-10

Prob. 9.9-12

***9.9-11** An overhanging beam *ABC* rests on a simple support at *A* and a spring support at *B* (see figure). A concentrated load *P* acts at the end of the overhang. Span *AB* has length *L*, the overhang has length *a*, and the spring has stiffness *k*. Determine the downward displacement δ_C of the end of the overhang. (Obtain the solution by using the modified form of Castigliano's theorem.)

Prob. 9.9-11

****9.9-12** A symmetric beam *ABCD* with overhangs at both ends supports a uniform load of intensity *q* (see figure). Determine the deflection δ_D at the end of the overhang. (Obtain the solution by using the modified form of Castigliano's theorem.)

DEFLECTIONS PRODUCED BY IMPACT

The beams described in the problems for Section 9.10 have constant flexural rigidity EI. Disregard the weights of the beams themselves, and consider only the effects of the given loads.

9.10-1 A heavy object of weight *W* is dropped onto the midpoint of a simple beam *AB* from a height *h* (see figure). Obtain a formula for the maximum bending stress σ_{max} due to the falling weight in terms of *h*, σ_{st}, and δ_{st}, where σ_{st} is the maximum bending stress and δ_{st} is the deflection at the midpoint when the weight *W* acts on the beam as a statically applied load. Plot a graph of the ratio σ_{max}/σ_{st} (that is, the ratio of the dynamic stress to the static stress) versus the ratio h/δ_{st}. (Let h/δ_{st} vary from 0 to 10.)

Prob. 9.10-1

9.10-2 An object of weight *W* is dropped onto the midpoint of a simple beam *AB* from a height *h* (see figure). The beam has a rectangular cross section of area *A*. Assuming that *h* is very large compared to the deflection of the beam when the weight *W* is applied statically, obtain a formula for the maximum bending stress σ_{max} in the beam due to the falling weight.

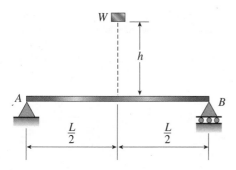

Prob. 9.10-2

9.10-3 A cantilever beam AB of length $L = 6$ ft is constructed of a W 8 × 21 wide-flange section (see figure). A weight $W = 1500$ lb falls through a height $h = 0.25$ in. onto the end of the beam. Calculate the maximum deflection δ_{max} of the end of the beam and the maximum bending stress σ_{max} due to the falling weight. (Assume $E = 30 \times 10^6$ psi.)

Prob. 9.10-3

9.10-4 A weight $W = 20$ kN falls through a height $h = 1.0$ mm onto the midpoint of a simple beam of length $L = 3$ m (see figure). The beam is made of wood with square cross section (dimension d on each side) and $E = 12$ GPa. If the allowable bending stress in the wood is $\sigma_{allow} = 10$ MPa, what is the minimum required dimension d?

Probs. 9.10-4 and 9.10-5

9.10-5 A weight $W = 4000$ lb falls through a height $h = 0.5$ in. onto the midpoint of a simple beam of length $L = 10$ ft (see figure). Assuming that the allowable bending stress in the beam is $\sigma_{allow} = 18,000$ psi and $E = 30 \times 10^6$ psi, select the lightest wide-flange beam listed in Table E-1 in Appendix E that will be satisfactory.

9.10-6 An overhanging beam ABC of rectangular cross section has the dimensions shown in the figure. A weight $W = 750$ N drops onto end C of the beam. If the allowable normal stress in bending is 45 MPa, what is the maximum height h from which the weight may be dropped? (Assume $E = 12$ GPa.)

Prob. 9.10-6

9.10-7 A heavy flywheel rotating at an angular speed ω (radians per second) is attached through bearings to the end of a simply supported beam of flexural rigidity EI and length L (see figure). The flywheel has mass moment of inertia I_m about its axis of rotation. If the bearings suddenly freeze, what will be the reaction R at support A of the beam?

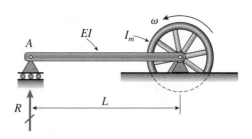

Prob. 9.10-7

TEMPERATURE EFFECTS

The beams described in the problems for Section 9.11 have constant flexural rigidity EI. In every problem, the temperature varies linearly between the top and bottom of the beam.

9.11-1 A simple beam AB of length L and height h undergoes a temperature change such that the bottom of the beam is at temperature T_2 and the top of the beam is at temperature T_1 (see figure). Determine the equation of the deflection curve of the beam, the angle of rotation θ_A at the left-hand support, and the deflection δ_{max} at the midpoint.

Prob. 9.11-1

9.11-2 A cantilever beam AB of length L and height h (see figure) is subjected to a temperature change such that the temperature at the top is T_1 and at the bottom is T_2. Determine the equation of the deflection curve of the beam, the angle of rotation θ_B at end B, and the deflection δ_B at end B.

Prob. 9.11-2

9.11-3 An overhanging beam ABC of height h is heated to a temperature T_1 on the top and T_2 on the bottom (see figure). Determine the equation of the deflection curve of the beam, the angle of rotation θ_C at end C, and the deflection δ_C at end C.

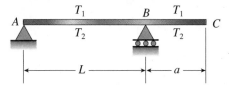

Prob. 9.11-3

9.11-4 A simple beam AB of length L and height h (see figure) is heated in such a manner that the temperature difference $T_2 - T_1$ between the bottom and top of the beam is proportional to the distance from support A; that is,

$$T_2 - T_1 = T_0 x$$

where T_0 is a constant. Determine the maximum deflection δ_{max} of the beam.

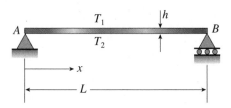

Prob. 9.11-4

10
STATICALLY INDETERMINATE BEAMS

INTRODUCTION

In this chapter we will analyze beams in which the number of reactions exceeds the number of independent equations of equilibrium. Since the reactions of such beams cannot be determined by statics alone, the beams are said to be **statically indeterminate**.

The analysis of statically indeterminate beams is quite different from that of statically determinate beams. When a beam is statically determinate, we can obtain all reactions, shear forces, and bending moments from free-body diagrams and equations of equilibrium. Then, knowing the shear forces and bending moments, we can obtain the stresses and deflections.

However, when a beam is statically indeterminate, the equilibrium equations are not sufficient and additional equations are needed. The most fundamental method for analyzing a statically indeterminate beam is to solve the differential equations of the deflection curve, as described later in Section 10.3. Although this method serves as a good starting point in our analysis, it is practical for only the simplest types of statically indeterminate beams. Therefore, we also discuss the method of superposition (Section 10.4), a method that is applicable to a wide variety of structures. In the method of superposition, we supplement the equilibrium equations with compatibility equations and force-displacement equations. (The same method was described earlier in Section 2.4, where we analyzed statically indeterminate bars subjected to tension and compression.)

In the last half of this chapter we discuss three specialized topics pertaining to statically indeterminate beams, namely, continuous beams (Section 10.5), beams with temperature changes (Section 10.6), and longitudinal displacements at the ends of beams (Section 10.7). Throughout this chapter, we assume that the beams are made of linearly elastic materials (this assumption underlies both the differential equations and the method of superposition).

Although only statically indeterminate beams are discussed in this chapter, the fundamental ideas have much wider application. Most of the structures we encounter in everyday life, including automobile frames, buildings, and aircraft, are statically indeterminate. However, they are much more complex than beams and must be designed by very sophisticated analytical techniques. Many of these techniques rely on the concepts described in this chapter, and therefore this chapter may be viewed as an introduction to the analysis of statically indeterminate structures of all kinds.

10.2   TYPES OF STATICALLY INDETERMINATE BEAMS

Statically indeterminate beams are usually identified by the arrangement of their supports. For instance, a beam that is fixed at one end and simply supported at the other (Fig. 10-1a) is called a **propped cantilever beam**. The reactions of the beam shown in the figure consist of horizontal and vertical forces at end A, a moment at end A, and a vertical force at end B. Because there are only three independent equations of equilibrium for this beam, it is not possible to calculate all four of the reactions from equilibrium alone. The number of reactions in *excess* of the number of equilibrium equations is called the **degree of static indeterminacy**. Thus, a propped cantilever beam is statically indeterminate to the first degree.

The excess reactions are called **static redundants** and must be selected in each particular case. For example, the reaction R_B of the propped cantilever beam shown in Fig. 10-1a may be selected as the

Fig. 10-1 Propped cantilever beam: (a) beam with load and reactions; (b) released structure when the reaction at end B is selected as the redundant; and (c) released structure when the moment reaction at end A is selected as the redundant

(a)

(b)

(c)

redundant reaction. Since this reaction is in excess of those needed to maintain equilibrium, it can be released from the structure by removing the support at *B*. When support *B* is removed, we are left with a cantilever beam (Fig. 10-1b). The structure that remains when the redundants are released is called the **released structure** or the **primary structure**. The released structure must be stable (so that it is capable of carrying loads), and it must be statically determinate (so that all force quantities can be determined by equilibrium alone).

Another possibility for the analysis of the propped cantilever beam of Fig. 10-1a is to select the reactive moment M_A as the redundant. Then, when the moment restraint at support *A* is removed, the released structure is a simple beam with a pin support at one end and a roller support at the other (Fig. 10-1c).

A special case arises if all loads acting on the beam are vertical (Fig. 10-2). Then the horizontal reaction at support *A* vanishes, and three reactions remain. However, only two independent equations of equilibrium are now available, and therefore the beam is still statically indeterminate to the first degree. If the reaction R_B is chosen as the redundant, the released structure is a cantilever beam; if the moment M_A is chosen, the released structure is a simple beam.

Another type of statically indeterminate beam, known as a **fixed-end beam**, is shown in Fig. 10-3a. This beam has fixed supports at both ends, resulting in a total of six unknown reactions (two forces and a moment at each support). Because there are only three equations of equilibrium, the beam is statically indeterminate to the third degree. (Other names for this type of beam are *clamped beam* and *built-in beam*.)

If we select the three reactions at end *B* of the beam as the redundants, and if we remove the corresponding restraints, we are left with a cantilever beam as the released structure (Fig. 10-3b). If we release the two fixed-end moments and one horizontal reaction, the released structure is a simple beam (Fig. 10-3c).

Again considering the special case of vertical loads only (Fig. 10-4), we find that the fixed-end beam now has only four nonzero reactions (one force and one moment at each support). The number of available equilibrium equations is two, and therefore the beam is statically indeterminate to the second degree. If the two reactions at end *B* are selected as

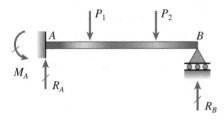

Fig. 10-2 Propped cantilever beam with vertical loads only

Fig. 10-3 Fixed-end beam: (a) beam with load and reactions; (b) released structure when the three reactions at end *B* are selected as the redundants; and (c) released structure when the two moment reactions and the horizontal reaction at end *B* are selected as the redundants

(a)

(b)

(c)

Fig. 10-4 Fixed-end beam with vertical loads only

the redundants, the released structure is a cantilever beam; if the two moment reactions are selected, the released structure is a simple beam.

The beam shown in Fig. 10-5a is an example of a **continuous beam**, so called because it has more than one span and is continuous over an interior support. This particular beam is statically indeterminate to the first degree because there are four reactive forces and only three equations of equilibrium. If the reaction R_B at the interior support is selected as the redundant, and if we remove the corresponding support from the beam, then there remains a released structure in the form of a statically determinate simple beam (Fig. 10-5b). If the reaction R_C is selected as the redundant, the released structure is a simple beam with an overhang (Fig. 10-5c).

In the following sections, we will discuss methods for analyzing statically indeterminate beams similar to the beams described above. The objective in each case is to determine the redundant reactions. Once they are known, all remaining reactions (plus the shear forces and bending moments) can be found from equations of equilibrium. In effect, the structure has become statically determinate. As the final step in the analysis, the stresses and deflections can be found by the methods described in preceding chapters.

10.3

ANALYSIS BY THE DIFFERENTIAL EQUATIONS OF THE DEFLECTION CURVE

Statically indeterminate beams may be analyzed by solving any one of the three differential equations of the deflection curve: (1) the second-order equation in terms of the bending moment (Eq. 9-12a), (2) the third-order equation in terms of the shear force (Eq. 9-12b), or (3) the fourth-order equation in terms of the intensity of distributed load (Eq. 9-12c). The procedure is essentially the same as that for a statically determinate beam (see Sections 9.2, 9.3, and 9.4) and consists of writing the differential equation, integrating to obtain its general solution, and then applying boundary and other conditions to evaluate the unknown

Fig. 10-5 Example of a continuous beam: (a) beam with loads and reactions; (b) released structure when the reaction at support B is selected as the redundant; and (c) released structure when the reaction at end C is selected as the redundant

(a)

(b)

(c)

quantities. The unknowns consist of the redundant reactions as well as the constants of integration.

The differential equation for a beam may be solved in symbolic terms only when the beam and its loading are relatively simple and uncomplicated. The resulting solutions are in the form of general purpose formulas. However, in more complex situations the differential equations must be solved numerically, using computer programs intended for that purpose. In such cases the results apply only to specific numerical problems.

The following examples illustrate the analysis of statically indeterminate beams by solving the differential equations in symbolic terms.*

Example 10-1

A propped cantilever beam AB of length L supports a uniform load of intensity q (Fig. 10-6). Analyze this beam by solving the second-order differential equation of the deflection curve (the bending-moment equation). Determine the reactions, shear forces, bending moments, slopes, and deflections of the beam.

Solution   Because the load on this beam acts in the vertical direction, we conclude that there is no horizontal reaction at the fixed support. Therefore, the beam has three unknown reactions (M_A, R_A, and R_B). Only two equations of equilibrium are available for determining these reactions, and therefore the beam is statically indeterminate to the first degree.

Since we will be analyzing this beam by solving the bending-moment equation, we must begin with a general expression for the bending moment. This expression will be in terms of both the load and the redundant reaction.

Redundant reaction. Let us choose the reaction R_B at the simple support as the redundant. Then, by considering the equilibrium of the entire beam, we can express the other two reactions in terms of R_B:

$$R_A = qL - R_B \qquad M_A = \frac{qL^2}{2} - R_B L \qquad \text{(a,b)}$$

Bending moment. The bending moment M at distance x from the fixed support can be expressed in terms of the reactions as follows:

$$M = R_A x - M_A - \frac{qx^2}{2} \qquad \text{(c)}$$

This equation can be obtained by the customary technique of constructing a free-body diagram of part of the beam and solving an equation of equilibrium.

Substituting into Eq. (c) from Eqs. (a) and (b), we obtain the bending moment in terms of the load and the redundant reaction:

$$M = qLx - R_B x - \frac{qL^2}{2} + R_B L - \frac{qx^2}{2} \qquad \text{(d)}$$

Fig. 10-6 Example 10-1. Propped cantilever beam with a uniform load

---

*The earliest use of the differential equation method for analyzing statically indeterminate beams appears in Navier's book on strength of materials (Ref. 10-1).

Differential equation. The second-order differential equation of the deflection curve (Eq. 9-12a) now becomes

$$EIv'' = M = qLx - R_Bx - \frac{qL^2}{2} + R_BL - \frac{qx^2}{2} \qquad \text{(e)}$$

After two successive integrations, we obtain the following equations for the slopes and deflections of the beam:

$$EIv' = \frac{qLx^2}{2} - \frac{R_Bx^2}{2} - \frac{qL^2x}{2} + R_BLx - \frac{qx^3}{6} + C_1 \qquad \text{(f)}$$

$$EIv = \frac{qLx^3}{6} - \frac{R_Bx^3}{6} - \frac{qL^2x^2}{4} + \frac{R_BLx^2}{2} - \frac{qx^4}{24} + C_1x + C_2 \qquad \text{(g)}$$

These equations contain three unknown quantities (C_1, C_2, and R_B).

Boundary conditions. Three boundary conditions pertaining to the deflections and slopes of the beam are apparent from an inspection of Fig. 10-6. These conditions are as follows: (1) the deflection at the fixed support is zero, (2) the slope at the fixed support is zero, and (3) the deflection at the simple support is zero. Thus,

$$v(0) = 0 \qquad v'(0) = 0 \qquad v(L) = 0$$

Applying these conditions to the equations for slopes and deflections (Eqs. f and g), we find $C_1 = 0$, $C_2 = 0$, and

$$R_B = \frac{3qL}{8} \qquad \text{(10-1)} \quad \Longleftarrow$$

Thus, the redundant reaction R_B is now known.

Reactions. With the value of the redundant established, we can find the remaining reactions from Eqs. (a) and (b). The results are

$$R_A = \frac{5qL}{8} \qquad M_A = \frac{qL^2}{8} \qquad \text{(10-2a,b)} \quad \Longleftarrow$$

Knowing these reactions, we can find the shear forces and bending moments in the beam.

Shear forces and bending moments. These quantities can be obtained by the usual techniques involving free-body diagrams and equations of equilibrium. The results are

$$V = R_A - qx = \frac{5qL}{8} - qx \qquad \text{(10-3)} \quad \Longleftarrow$$

$$M = R_Ax - M_A - \frac{qx^2}{2} = \frac{5qLx}{8} - \frac{qL^2}{8} - \frac{qx^2}{2} \qquad \text{(10-4)} \quad \Longleftarrow$$

Shear-force and bending-moment diagrams for the beam can be drawn with the aid of these equations (see Fig. 10-7).

Fig. 10-7 Shear-force and bending-moment diagrams for the propped cantilever beam of Fig. 10-6

From the diagrams, we see that the maximum shear force occurs at the fixed support and is equal to

$$V_{max} = \frac{5qL}{8} \tag{10-5}$$

Also, the maximum positive and negative bending moments are

$$M_{pos} = \frac{9qL^2}{128} \qquad M_{neg} = -\frac{qL^2}{8} \tag{10-6a,b}$$

Finally, we note that the bending moment is equal to zero at distance $x = L/4$ from the fixed support.

Slopes and deflections of the beam. Returning to Eqs. (f) and (g) for the slopes and deflections, we now substitute the values of the constants of integration ($C_1 = 0$ and $C_2 = 0$) as well as the expression for the redundant (Eq. 10-1) and obtain

$$v' = \frac{qx}{48EI}(-6L^2 + 15Lx - 8x^2) \tag{10-7} \quad \Longleftarrow$$

$$v = -\frac{qx^2}{48EI}(3L^2 - 5Lx + 2x^2) \tag{10-8} \quad \Longleftarrow$$

The deflected shape of the beam as obtained from Eq. (10-8) is shown in Fig. 10-8.

To determine the maximum deflection δ_{max} of the beam, we set the slope (Eq. 10-7) equal to zero and solve for the distance x_1 to the point of maximum deflection:

$$v' = 0 \quad \text{or} \quad -6L^2 + 15Lx - 8x^2 = 0$$

from which

$$x_1 = \frac{15 - \sqrt{33}}{16}L = 0.5785L \tag{10-9}$$

Substituting this value of x into the equation for the deflection (Eq. 10-8), we get

$$\delta_{max} = -(v)_{x=x_1} = \frac{qL^4}{65{,}536EI}(39 + 55\sqrt{33})$$

$$= \frac{qL^4}{184.6EI} = 0.005416\frac{qL^4}{EI} \tag{10-10}$$

The point of inflection is located where the bending moment is equal to zero, that is, where $x = L/4$. The corresponding deflection of the beam (from Eq. 10-8) is

$$v_0 = (v)_{x=L/4} = -\frac{5qL^4}{2048EI} = -0.002441\frac{qL^4}{EI} \tag{10-11}$$

Fig. 10-8 Deflection curve for the propped cantilever beam of Fig. 10-6

The negative sign means that the deflection is downward, as expected. Also, we note that when $x < L/4$, both the curvature and the bending moment are negative, and when $x > L/4$, the curvature and bending moment are positive.

To determine the angle of rotation θ_B at the simply supported end of the beam, we use Eq. (10-7), as follows:

$$\theta_B = (v')_{x=L} = \frac{qL^3}{48EI} \tag{10-12}$$

Slopes and deflections at other points along the axis of the beam can be obtained by similar procedures.

Note: In this example, we analyzed the beam by taking the reaction R_B (Fig. 10-6) as the redundant reaction. An alternative approach is to take the reactive moment M_A as the redundant. Then we can express the bending moment M in terms of M_A, substitute the resulting expression into the second-order differential equation, and solve as before. Still another approach is to begin with the fourth-order differential equation, as illustrated in the next example.

WORK
SHEET
11 Mathcad

Example 10-2

The fixed-end beam *ACB* shown in Fig. 10-9 supports a concentrated load *P* at the midpoint. Analyze this beam by solving the fourth-order differential equation of the deflection curve (the load equation). Determine the reactions, shear forces, bending moments, slopes, and deflections of the beam.

Solution   Because the load on this beam acts only in the vertical direction, we know that there are no horizontal reactions at the supports. Therefore, the beam has four unknown reactions, two at each support. Since only two equations of equilibrium are available, the beam is statically indeterminate to the second degree.

However, we can simplify the analysis by observing from the symmetry of the beam and its loading that the forces and moments at supports *A* and *B* are equal, that is,

$$R_A = R_B \quad \text{and} \quad M_A = M_B$$

Since the vertical reactions at the supports are equal, we know from equilibrium of forces in the vertical direction that each force is equal to $P/2$:

$$R_A = R_B = \frac{P}{2} \tag{10-13} \quad \longleftarrow$$

Thus, the only unknown quantities that remain are the moment reactions M_A and M_B. For convenience, we will select the moment M_A as the redundant quantity.

Differential equation. The fourth-order differential equation (Eq. 9-12c) for the left-hand half of the beam is

$$EIv'''' = -q = 0 \qquad (0 < x < L/2) \tag{h}$$

because there is no load acting on the beam in this region.

Successive integrations of Eq. (h) yield the following equations, which are valid for the left-hand half of the beam:

Fig. 10-9 Example 10-2. Fixed-end beam with a concentrated load at the midpoint

$$EIv''' = C_1 \tag{i}$$

$$EIv'' = C_1 x + C_2 \tag{j}$$

$$EIv' = \frac{C_1 x^2}{2} + C_2 x + C_3 \tag{k}$$

$$EIv = \frac{C_1 x^3}{6} + \frac{C_2 x^2}{2} + C_3 x + C_4 \tag{l}$$

These equations contain four unknown constants of integration. Since we now have five unknowns (C_1, C_2, C_3, C_4, and M_A), we need five boundary conditions.

Boundary conditions. The boundary conditions applicable to the left-hand half of the beam are as follows:

(1) The shear force in the left-hand segment of the beam is equal to R_A, or $P/2$. Therefore, from Eq. (9-12b) we find

$$EIv''' = V = \frac{P}{2}$$

Combining this equation with Eq. (i), we obtain $C_1 = P/2$.

(2) The bending moment at the left-hand support is equal to $-M_A$. Therefore, from Eq. (9-12a) we get

$$EIv'' = M = -M_A \quad \text{at } x = 0$$

Combining this equation with Eq. (j), we obtain $C_2 = -M_A$.

(3) The slope of the beam at the left-hand support ($x = 0$) is equal to zero. Therefore, Eq. (k) yields $C_3 = 0$.

(4) The slope of the beam at the midpoint ($x = L/2$) is also equal to zero (from symmetry). Therefore, from Eq. (k) we find

$$M_A = M_B = \frac{PL}{8} \tag{10-14}$$

Thus, the reactive moments at the ends of the beam have been determined.

(5) The deflection of the beam at the left-hand support ($x = 0$) is equal to zero. Therefore, from Eq. (l) we find $C_4 = 0$.

In summary, the four constants of integration are

$$C_1 = \frac{P}{2} \qquad C_2 = -M_A = -\frac{PL}{8} \qquad C_3 = 0 \qquad C_4 = 0$$

Shear forces and bending moments. The shear forces and bending moments can be found by substituting the appropriate constants of integration into Eqs. (i) and (j). The results are

$$EIv''' = V = \frac{P}{2} \qquad (0 < x < L/2) \tag{10-15}$$

$$EIv'' = M = \frac{Px}{2} - \frac{PL}{8} \qquad (0 \le x \le L/2) \tag{10-16}$$

Since we know the reactions of the beam, we can also obtain these expressions directly from free-body diagrams and equations of equilibrium.

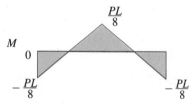

Fig. 10-10 Shear-force and bending-moment diagrams for the fixed-end beam of Fig. 10-9

The shear-force and bending moment diagrams are shown in Fig. 10-10.

Slopes and deflections. The slopes and deflections in the left-hand half of the beam can be found from Eqs. (k) and (l) by substituting the expressions for the constants of integration. In this manner, we find

$$v' = -\frac{Px}{8EI}(L - 2x) \qquad (0 \le x \le L/2) \qquad (10\text{-}17) \quad \Longleftarrow$$

$$v = -\frac{Px^2}{48EI}(3L - 4x) \qquad (0 \le x \le L/2) \qquad (10\text{-}18) \quad \Longleftarrow$$

The deflection curve of the beam is shown in Fig. 10-11.

To find the maximum deflection δ_{max} we set x equal to $L/2$ in Eq. (10-18) and change the sign; thus,

$$\delta_{max} = -(v)_{x=L/2} = \frac{PL^3}{192EI} \qquad (10\text{-}19)$$

The point of inflection in the left-hand half of the beam occurs where the bending moment M is equal to zero, that is, where $x = L/4$ (see Eq. 10-16). The corresponding deflection (from Eq. 10-18) is

$$v_0 = (v)_{x=L/4} = -\frac{PL^3}{384EI} \qquad (10\text{-}20)$$

which is equal numerically to one-half of the maximum deflection. A second point of inflection occurs in the right-hand half of the beam at distance $L/4$ from end B.

Notes: As we observed in this example, the number of boundary and other conditions is always sufficient to evaluate not only the constants of integration but also the redundant reactions.

Sometimes it is necessary to set up differential equations for more than one region of the beam and use conditions of continuity between regions, as illustrated in Chapter 9 for statically determinate beams. Such analyses are likely to be long and tedious because of the large number of conditions that must be satisfied. However, if deflections and angles of rotation are needed at only one or two specific points, the method of superposition may be used (see the next section).

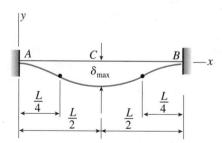

Fig. 10-11 Deflection curve for the fixed-end beam of Fig. 10-9

10.4 ❙ METHOD OF SUPERPOSITION

The method of superposition is of fundamental importance in the analysis of statically indeterminate bars, trusses, beams, frames, and many other kinds of structures. We have already used the superposition method to analyze statically indeterminate structures composed of bars in tension and compression (Section 2.4) and shafts in torsion (Section 3.8). In this section, we will apply the method to beams.

We begin the analysis by noting the degree of static indeterminacy and selecting the redundant reactions. Then, having identified the redundants, we can write **equations of equilibrium** that relate the other unknown reactions to the redundants and the loads.

Next, we assume that both the original loads and the redundants act upon the released structure. Then we find the deflections in the released structure by superposing the separate deflections due to the loads and the redundants. The sum of these deflections must match the deflections in the original beam. However, the deflections in the original beam (at the points where restraints were removed) are either zero or have known values. Therefore, we can write **equations of compatibility** (or *equations of superposition*) expressing the fact that the deflections of the released structure (at the points where restraints were removed) are the same as the deflections in the original beam (at those same points).

Since the released structure is statically determinate, we can easily determine its deflections by using the techniques described in Chapter 9. The relationships between the loads and the deflections of the released structure are called **force-displacement relations**. When these relations are substituted into the equations of compatibility, we obtain equations in which the redundants are the unknown quantities. Therefore, we can solve those equations for the redundant reactions. Then, with the redundants known, we can determine all other reactions from the equations of equilibrium. Furthermore, we can also determine the shear forces and bending moments from equilibrium.

The steps described in the preceding paragraphs can be made clearer by discussing a particular case. For this purpose, let us again analyze a propped cantilever beam supporting a uniform load (Fig. 10-12a), and let us again select the reaction R_B at the simple support as the redundant. Then the *equations of equilibrium* that express the other unknown reactions in terms of the redundant are as follows:

$$R_A = qL - R_B \qquad M_A = \frac{qL^2}{2} - R_B L \qquad \text{(a,b)}$$

These equations are obtained from equations of equilibrium that apply to the entire beam taken as a free body.

(a)

(b)

(c)

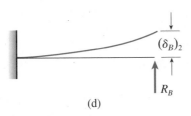

(d)

Fig. 10-12 Analysis of a propped cantilever beam by the method of superposition with the reaction R_B selected as the redundant

The next step is to remove the restraint corresponding to the redundant (in this example, we remove the support at end B). The released structure that remains is a cantilever beam (Fig. 10-12b). The uniform load q and the redundant force R_B are now applied as loads on the released structure (Figs. 10-12c and d). The deflection at end B of the released structure due solely to the uniform load is denoted $(\delta_B)_1$, and the deflection at the same point due solely to the redundant is denoted $(\delta_B)_2$. The deflection δ_B at point B in the original structure is obtained by superposing these two deflections. Since the deflection in the original beam is equal to zero, we obtain the following *equation of compatibility*:

$$\delta_B = (\delta_B)_1 - (\delta_B)_2 = 0 \tag{c}$$

The minus sign appears in this equation because $(\delta_B)_1$ is positive downward whereas $(\delta_B)_2$ is positive upward.

The *force-displacement relations* that give the deflections $(\delta_B)_1$ and $(\delta_B)_2$ in terms of the uniform load q and the redundant R_B, respectively, are found with the aid of Table G-1 in Appendix G (see Cases 1 and 4). Using the formulas given there, we find

$$(\delta_B)_1 = \frac{qL^4}{8EI} \qquad (\delta_B)_2 = \frac{R_B L^3}{3EI} \tag{d,e}$$

Substituting these force-displacement relations into the equation of compatibility yields

$$\delta_B = \frac{qL^4}{8EI} - \frac{R_B L^3}{3EI} = 0 \tag{f}$$

which can be solved for the redundant reaction:

$$R_B = \frac{3qL}{8} \tag{10-21}$$

Note that this equation gives the redundant in terms of the loads acting on the original beam.

The remaining reactions (R_A and M_A) can be found from the equilibrium equations (Eqs. a and b); the results are

$$R_A = \frac{5qL}{8} \qquad M_A = \frac{qL^2}{8} \tag{10-22a,b}$$

Knowing all reactions, we can now obtain the shear forces and bending moments throughout the beam and plot the corresponding diagrams (see Fig. 10-7 for these diagrams).

The same propped cantilever beam (Fig. 10-13a) can be analyzed in a different way by selecting the moment reaction M_A as the redundant, in which case the released structure is a simple beam (Fig. 10-13b). The *equations of equilibrium* for the reactions R_A and R_B in the original beam are

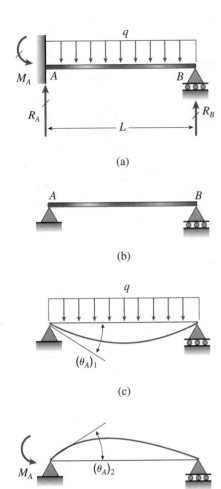

(a)

(b)

(c)

(d)

Fig. 10-13 Analysis of a propped cantilever beam by the method of superposition with the moment reaction M_A selected as the redundant

$$R_A = \frac{qL}{2} + \frac{M_A}{L} \qquad R_B = \frac{qL}{2} - \frac{M_A}{L} \qquad \text{(g,h)}$$

The equation of compatibility expresses the fact that the angle of rotation θ_A at the fixed end of the original beam is equal to zero. Since this angle is obtained by superposing the angles of rotation $(\theta_A)_1$ and $(\theta_A)_2$ in the released structure (Figs. 10-13c and d), the *compatibility equation* becomes

$$\theta_A = (\theta_A)_1 - (\theta_A)_2 = 0 \qquad \text{(i)}$$

In this equation, the angle $(\theta_A)_1$ is assumed to be positive when clockwise and the angle $(\theta_A)_2$ is assumed to be positive when counterclockwise.

The angles of rotation in the released structure are obtained from the formulas given in Table G-2 of Appendix G (see Cases 1 and 7). Thus, the *force-displacement relations* are

$$(\theta_A)_1 = \frac{qL^3}{24EI} \qquad (\theta_A)_2 = \frac{M_A L}{3EI}$$

Substituting into the compatibility equation (Eq. i), we get

$$\theta_A = \frac{qL^3}{24EI} - \frac{M_A L}{3EI} = 0 \qquad \text{(j)}$$

Solving this equation for the redundant, we get $M_A = qL^2/8$, which agrees with the previous result (Eq. 10-22b). Also, the equations of equilibrium (Eqs. g and h) yield the same results as before for the reactions R_A and R_B (see Eqs. 10-22a and 10-21).

As already mentioned, we can readily determine the shear forces and bending moments once we know the reactions. Furthermore, we can also determine the deflections and slopes of the original beam. One method is to solve the differential equation of the deflection curve, which is a straightforward matter since the reactions are already known. Another method is to determine the deflections in the released structure (Figs. 10-13c and d) from the formulas in Appendix G, and then use the principle of superposition to obtain the deflections in the original beam.

In the following examples, and also in the problems at the end of the chapter, we are concerned primarily with finding the reactions, since this is the key step in the solutions.

The method of superposition described in this section is also called the *flexibility method* or the *force method* of analysis. The latter name arises from the use of force quantities (forces and moments) as the redundants; the former name is used because the coefficients of the unknown quantities in the compatibility equation (terms such as $L^3/3EI$ in Eq. f and $L/3EI$ in Eq. j) are *flexibilities* (that is, deflections or angles produced by a unit load).

Since the method of superposition involves the superposition of deflections, it is applicable only to linearly elastic structures. (Recall that this same limitation applies to all topics discussed in this chapter.)

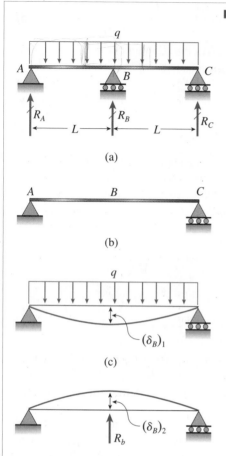

(a)

(b)

(c)

(d)

Fig. 10-14 Example 10-3. Two-span continuous beam with a uniform load

■ **Example 10-3**

A two-span continuous beam *ABC* supports a uniform load of intensity q, as shown in Fig. 10-14a. Each span of the beam has length L. Using the method of superposition, determine all reactions for this beam.

Solution   This beam has three unknown reactions (R_A, R_B, and R_C). Since there are two equations of equilibrium for the beam as a whole, it is statically indeterminate to the first degree. For convenience, let us select the reaction R_B at the middle support as the redundant.

Equations of equilibrium. We can express the reactions R_A and R_C in terms of the redundant R_B by means of two equations of equilibrium. The first equation, which is for equilibrium of moments about point B, shows that R_A and R_C are equal. The second equation, which is for equilibrium in the vertical direction, yields the following result:

$$R_A = R_C = qL - \frac{R_B}{2} \tag{k}$$

Equation of compatibility. Because the reaction R_B is selected as the redundant, the released structure is a simple beam with supports at A and C (Fig. 10-14b). The deflections at point B in the released structure due to the uniform load q and the redundant R_B are shown in Figs. 10-14c and d, respectively. Note that the deflections are denoted $(\delta_B)_1$ and $(\delta_B)_2$. The superposition of these deflections must produce the deflection δ_B in the original beam at point B. Since the latter deflection is equal to zero, the equation of compatibility is

$$\delta_B = (\delta_B)_1 - (\delta_B)_2 = 0 \tag{l}$$

in which the deflection $(\delta_B)_1$ is positive downward and the deflection $(\delta_B)_2$ is positive upward.

Force-displacement relations. The deflection $(\delta_B)_1$ caused by the uniform load acting on the released structure (Fig. 10-14c) is obtained from Table G-2, Case 1, as follows:

$$(\delta_B)_1 = \frac{5q(2L)^4}{384EI} = \frac{5qL^4}{24EI}$$

where $2L$ is the length of the released structure. The deflection $(\delta_B)_2$ produced by the redundant (Fig. 10-14d) is

$$(\delta_B)_2 = \frac{R_B(2L)^3}{48EI} = \frac{R_BL^3}{6EI}$$

as obtained from Table G-2, Case 4.

The equation of compatibility pertaining to the vertical deflection at point B (Eq. l) now becomes

$$\delta_B = \frac{5qL^4}{24EI} - \frac{R_BL^3}{6EI} = 0 \tag{m}$$

from which we find the reaction at the middle support:

$$R_B = \frac{5qL}{4} \qquad (10\text{-}23) \quad \Longleftarrow$$

The other reactions are obtained from Eq. (k):

$$R_A = R_C = \frac{3qL}{8} \qquad (10\text{-}24) \quad \Longleftarrow$$

With the reactions known, we can find the shear forces, bending moments, stresses, and deflections without difficulty.

■ Example 10-4

A fixed-end beam AB (Fig. 10-15a) is loaded by a force P acting at an intermediate point D. Find the reactive forces and moments at the ends of the beam using the method of superposition. Also, determine the deflection at point D where the load is applied.

Solution   This beam has four unknown reactions (a force and a moment at each support), but only two independent equations of equilibrium are available. Therefore, the beam is statically indeterminate to the second degree. In this example, we will select the reactive moments M_A and M_B as the redundants.

Equations of equilibrium. The two unknown force reactions (R_A and R_B) can be expressed in terms of the redundants (M_A and M_B) with the aid of two equations of equilibrium. The first equation is for moments about point B, and the second is for moments about point A. The resulting expressions are

$$R_A = \frac{Pb}{L} + \frac{M_A}{L} - \frac{M_B}{L} \qquad R_B = \frac{Pa}{L} - \frac{M_A}{L} + \frac{M_B}{L} \qquad (\text{n,o})$$

Equations of compatibility. When both redundants are released by removing the rotational restraints at the ends of the beam, we are left with a simple beam as the released structure (Figs. 10-15b, c, and d). The angles of rotation at the ends of the released structure due to the concentrated load P are denoted $(\theta_A)_1$ and $(\theta_B)_1$, as shown in Fig. 10-15b. In a similar manner, the angles at the ends due to the redundant M_A are denoted $(\theta_A)_2$ and $(\theta_B)_2$, and the angles due to the redundant M_B are denoted $(\theta_A)_3$ and $(\theta_B)_3$.

Since the angles of rotation at the supports of the original beam are equal to zero, the two equations of compatibility are

$$\theta_A = (\theta_A)_1 - (\theta_A)_2 - (\theta_A)_3 = 0 \qquad (\text{p})$$

$$\theta_B = (\theta_B)_1 - (\theta_B)_2 - (\theta_B)_3 = 0 \qquad (\text{q})$$

in which the signs of the various terms are determined by inspection from the figures.

Force-displacement relations. The angles at the ends of the beam due to the load P (Fig. 10-15b) are obtained from Case 5 of Table G-2:

$$(\theta_A)_1 = \frac{Pab(L + b)}{6LEI} \qquad (\theta_B)_1 = \frac{Pab(L + a)}{6LEI}$$

(a)

(b)

(c)

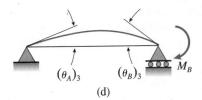

(d)

Fig. 10-15 Example 10-4. Fixed-end beam with a concentrated load in which a and b are the distances from the supports to point D where the load is applied.

Also, the angles at the ends due to the redundant moment M_A are (see Case 7 of Table G-2):

$$(\theta_A)_2 = \frac{M_A L}{3EI} \qquad (\theta_B)_2 = \frac{M_A L}{6EI}$$

Similarly, the angles due to the moment M_B are

$$(\theta_A)_3 = \frac{M_B L}{6EI} \qquad (\theta_B)_3 = \frac{M_B L}{3EI}$$

When the preceding expressions for the angles are substituted into the equations of compatibility (Eqs. p and q), we arrive at two simultaneous equations containing M_A and M_B as unknowns:

$$\frac{M_A L}{3EI} + \frac{M_B L}{6EI} = \frac{Pab(L + b)}{6LEI} \tag{r}$$

$$\frac{M_A L}{6EI} + \frac{M_B L}{3EI} = \frac{Pab(L + a)}{6LEI} \tag{s}$$

Solving these equations for the redundants, we obtain

$$M_A = \frac{Pab^2}{L^2} \qquad M_B = \frac{Pa^2 b}{L^2} \qquad \text{(10-25a,b)} \quad \Longleftarrow$$

Substituting these expressions for M_A and M_B into the equations of equilibrium (Eqs. n and o), we obtain the vertical reactions:

$$R_A = \frac{Pb^2}{L^3}(L + 2a) \qquad R_B = \frac{Pa^2}{L^3}(L + 2b) \quad \text{(10-26a,b)} \quad \Longleftarrow$$

Thus, all reactions for the fixed-end beam have been determined.

The reactions at the supports of a beam with fixed ends are commonly referred to as **fixed-end moments** and **fixed-end forces**. They are widely used in structural analysis, and formulas for these quantities are occasionally listed in engineering handbooks.

Deflection at point D. To obtain the deflection at point D in the original fixed-end beam (Fig. 10-15a), we again use the principle of superposition. The deflection at point D is equal to the sum of three deflections: (1) the downward deflection $(\delta_D)_1$ at point D in the released structure due to the load P (Fig. 10-15b); (2) the upward deflection $(\delta_D)_2$ at the same point in the released structure due to the redundant M_A (Fig. 10-15c), and (3) the upward deflection $(\delta_D)_3$ at the same point in the released structure due to the redundant M_B (Fig. 10-15d). This superposition of deflections is expressed by the following equation:

$$\delta_D = (\delta_D)_1 - (\delta_D)_2 - (\delta_D)_3 \tag{t}$$

in which δ_D is the downward deflection in the original beam.

The deflections appearing in Eq. (t) are obtained from the formulas given in Table G-2 of Appendix G (see Cases 5 and 7) by substituting $x = a$ and simplifying. The results are

$$(\delta_D)_1 = \frac{Pa^2b^2}{3LEI} \qquad (\delta_D)_2 = \frac{M_Aab}{6LEI}(L + b) \qquad (\delta_D)_3 = \frac{M_Bab}{6LEI}(L + a)$$

Substituting the expressions for M_A and M_B from Eqs. (10-25a and b) into the last two expressions, we get

$$(\delta_D)_2 = \frac{Pa^2b^3}{6L^3EI}(L + b) \qquad (\delta_D)_3 = \frac{Pa^3b^2}{6L^3EI}(L + a)$$

Therefore, the deflection at point D in the original beam, obtained by substituting $(\delta_D)_1$, $(\delta_D)_2$, and $(\delta_D)_3$ into Eq. (t) and simplifying, is

$$\delta_D = \frac{Pa^3b^3}{3L^3EI} \qquad\qquad (10\text{-}27) \quad \Longleftarrow$$

The method described above for finding the deflection δ_D can be used not only to find deflections at individual points but also to find the equations of the deflection curve.

Concentrated load acting at the midpoint of the beam (Fig. 10-16). When the load P acts at the midpoint C, the reactions of the beam (from Eqs. 10-25 and 10-26 with $a = b = L/2$) are

Fig. 10-16 Fixed-end beam with a concentrated load acting at the midpoint

$$M_A = M_B = \frac{PL}{8} \qquad R_A = R_B = \frac{P}{2} \qquad (10\text{-}28a,b)$$

Also, the deflection at the midpoint (from Eq. 10-27) is

$$\delta_C = \frac{PL^3}{192EI} \qquad\qquad (10\text{-}29)$$

This deflection is only one-fourth of the deflection at the midpoint of a simple beam with the same load, which shows the stiffening effect of clamping the ends of the beam.

The above results for the reactions at the ends and the deflection at the middle (Eqs. 10-28 and 10-29) agree with those found in Example 10-2 by solving the differential equation of the deflection curve (see Eqs. 10-13, 10-14, and 10-19).

■ **Example 10-5**

A fixed-end beam AB supports a uniform load of intensity q acting over part of the span (Fig. 10-17a). Determine the reactions of this beam (that is, find the fixed-end moments and fixed-end forces).

Solution   We can find the reactions of this beam by using superposition together with the results obtained in the preceding example (Example 10-4). In that example we found the reactions of a fixed-end beam subjected to a concentrated load P acting at distance a from the left-hand end (see Fig. 10-15a and Eqs. 10-25 and 10-26). In order to apply those results to the uniform load of Fig. 10-17a, we will treat an element of the uniform load as a concentrated load of magnitude $q\,dx$ acting at distance x from the left-hand end (Fig. 10-17b). Then, using the formulas derived in Example 10-4, we can obtain the reactions caused

Fig. 10-17 Example 10-5. (a) Fixed-end beam with a uniform load over part of the span, and (b) reactions produced by an element $q\,dx$ of the uniform load

Fig. 10-18 Fixed-end beam with a uniform load

by this element of load. Finally, by integrating over the length a of the uniform load, we can obtain the reactions due to the entire uniform load.

Let us begin with the moment reactions, which are given by Eqs. (10-25a and b). To obtain the moments caused by the element $q\,dx$ of the uniform load (compare Fig. 10-17b with Fig. 10-15a), we replace P with $q\,dx$, a with x, and b with $L - x$. Thus, the fixed-end moments due to the element of load are

$$dM_A = \frac{qx(L-x)^2dx}{L^2} \qquad dM_B = \frac{qx^2(L-x)dx}{L^2}$$

Integrating over the loaded part of the beam, we get the fixed-end moments due to the entire uniform load:

$$M_A = \int dM_A = \frac{q}{L^2}\int_0^a x(L-x)^2dx = \frac{qa^2}{12L^2}(6L^2 - 8aL + 3a^2) \quad \text{(10-30a)} \ \blacktriangleleft$$

$$M_B = \int dM_B = \frac{q}{L^2}\int_0^a x^2(L-x)dx = \frac{qa^3}{12L^2}(4L - 3a) \quad \text{(10-30b)} \ \blacktriangleleft$$

Proceeding in a similar manner but using Eqs. (10-26a and b), we obtain the following expressions for the fixed-end forces due to the element of load:

$$dR_A = \frac{q(L-x)^2(L+2x)dx}{L^3} \qquad dR_B = \frac{qx^2(3L-2x)dx}{L^3}$$

Integration gives

$$R_A = \int dR_A = \frac{q}{L^3}\int_0^a (L-x)^2(L+2x)dx = \frac{qa}{2L^3}(2L^3 - 2a^2L + a^3) \quad \text{(10-31a)} \ \blacktriangleleft$$

$$R_B = \int dR_B = \frac{q}{L^3}\int_0^a x^2(3L-2x)dx = \frac{qa^3}{2L^3}(2L - a) \quad \text{(10-31b)} \ \blacktriangleleft$$

Thus, all reactions (fixed-end moments and fixed-end forces) have been found.

Uniform load acting over the entire length of the beam (Fig. 10-18). When the load acts over the entire span, we can obtain the reactions by substituting $a = L$ into the preceding equations, yielding

$$M_A = M_B = \frac{qL^2}{12} \qquad R_A = R_B = \frac{qL}{2} \qquad \text{(10-32a,b)}$$

The deflection at the midpoint of a uniformly loaded beam is also of interest. The simplest procedure for obtaining this deflection is to use the method of superposition. The first step is to remove the moment restraints at the supports and obtain a released structure in the form of a simple beam. The downward deflection at the midpoint of a simple beam due to a uniform load (from Case 1, Table G-2) is

$$(\delta_C)_1 = \frac{5qL^4}{384EI} \qquad \text{(u)}$$

and the upward deflection at the midpoint due to the end moments (from Case 10, Table G-2) is

$$(\delta_C)_2 = \frac{M_A L^2}{8EI} = \frac{(qL^2/12)L^2}{8EI} = \frac{qL^4}{96EI} \tag{v}$$

Thus, the final downward deflection of the original fixed-end beam (Fig. 10-18) is

$$\delta_C = (\delta_C)_1 - (\delta_C)_2$$

Substituting for the deflections from Eqs. (u) and (v), we get

$$\delta_C = \frac{qL^4}{384EI} \tag{10-33}$$

This deflection is one-fifth of the deflection at the midpoint of a simple beam with a uniform load, again illustrating the stiffening effect of fixity at the ends of the beam.

■ Example 10-6

A beam ABC (Fig. 10-19a) rests on simple supports at points A and B and is supported by a cable at point C. The beam has total length $2L$ and supports a uniform load of intensity q. Prior to the application of the load, there is no force in the cable nor is there any slack in the cable. When the load is applied, the beam deflects downward at point C and a tensile force T develops in the cable. Find the magnitude of this force.

Solution   The structure $ABCD$, consisting of the beam and cable, has three vertical reactions (at points A, B, and D). However, only two equations of equilibrium are available from a free-body diagram of the entire structure. Therefore, the structure is statically indeterminate to the first degree, and we must select one redundant quantity for purposes of analysis.

The tensile force T in the cable is a suitable choice for the redundant. We can release this force by removing the connection at point C, thereby cutting the

Fig. 10-19 Example 10-6. Beam ABC with one end supported by a cable

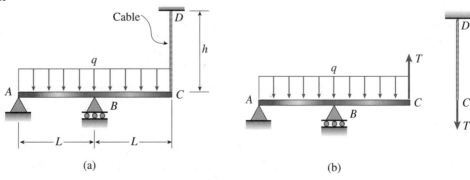

(a)   (b)

structure into two parts (Fig. 10-19b). The released structure consists of the beam ABC and the cable CD as separate elements, with the redundant force T acting upward on the beam and downward on the cable.

The deflection at point C of beam ABC (Fig. 10-19b) consists of two parts, a downward deflection $(\delta_C)_1$ due to the uniform load and an upward deflection $(\delta_C)_2$ due to the force T. At the same time, the lower end C of cable CD displaces downward by an amount $(\delta_C)_3$, equal to the elongation of the cable due to the force T. Therefore, the *equation of compatibility*, which expresses the fact that the downward deflection of end C of the beam is equal to the elongation of the cable, is

$$(\delta_C)_1 - (\delta_C)_2 = (\delta_C)_3 \tag{w}$$

Having formulated this equation, we now turn to the task of evaluating all three displacements.

The deflection $(\delta_C)_1$ at the end of the overhang (point C in beam ABC) due to the uniform load can be found from the results given in Example 9-9 of Section 9.5 (see Fig. 9-21). Using Eq. (9-59) of that example, and substituting $a = L$, we get

$$(\delta_C)_1 = \frac{qL^4}{4E_b I_b} \tag{x}$$

where $E_b I_b$ is the flexural rigidity of the beam.

The deflection of the beam at point C due to the force T can be taken from the answer to Problem 9.8-5, which gives the deflection $(\delta_C)_2$ at the end of the overhang when the length of the overhang is a:

$$(\delta_C)_2 = \frac{Ta^2(L + a)}{3E_b I_b}$$

Again substituting $a = L$, we find

$$(\delta_C)_2 = \frac{2TL^3}{3E_b I_b} \tag{y}$$

Finally, the elongation of the cable is

$$(\delta_C)_3 = \frac{Th}{E_c A_c} \tag{z}$$

where h is the length of the cable and $E_c A_c$ is its axial rigidity.

By substituting the three displacements (Eqs. x, y, and z) into the equation of compatibility (Eq. w), we get

$$\frac{qL^4}{4E_b I_b} - \frac{2TL^3}{3E_b I_b} = \frac{Th}{E_c A_c}$$

Now solving for the force T, we get

$$T = \frac{3qL^4 E_c A_c}{8L^3 E_c A_c + 12hE_b I_b} \tag{10-34} \quad \Longleftarrow$$

With the force T known, we can find all reactions, shear forces, and bending moments by means of free-body diagrams and equations of equilibrium.

This example illustrates how an internal force quantity (instead of an external reaction) can be used as the redundant.

*10.5 ▌ CONTINUOUS BEAMS

Beams that have more than one span and that are continuous throughout their lengths (Fig. 10-20) are known as **continuous beams**. They are commonly encountered in aircraft, bridges, buildings, pipelines, and various kinds of specialized structures. Continuous beams are statically indeterminate and may be analyzed by the method of superposition. In this section, we describe a particular form of the superposition method, called the **method of three moments**, that is especially useful in the analysis of continuous beams.

Fig. 10-20 Continuous beam

For purposes of this discussion, we assume that all loads acting on the beam are vertical (Fig. 10-20) and that there are no restraints against rotation at the supports (that is, none of the supports is fixed, or clamped). In addition, we assume that there are no axial deformations due to restraints against horizontal movement. Under these conditions, all reactions at the supports are vertical forces. We can represent this behavior schematically by considering one of the supports to be a pin support and all others to be roller supports, as shown in the figure. Since there is one reaction for each support, and since only two equations of equilibrium are available for the structure as a whole, the degree of indeterminacy is two less than the number of reactions. Thus, the beam shown in Fig. 10-20 has five vertical reactions, of which three are redundants.

Analyzing a continuous beam by solving the differential equation of the deflection curve is likely to be extremely long and tedious. Therefore, the method of superposition is generally preferred. When using the superposition method, the first step is to select the redundant forces or moments. One possibility is to select the reactions at the intermediate supports (that is, the reactions at all supports except the two end supports). When these reactions are released from the structure by removing the corresponding supports, we are left with a released structure in the form of a simple beam. This choice for the released structure was used in Example 10-3 of Section 10.4 (see Fig. 10-14).

If a beam has more than two spans, it is usually advantageous to select the bending moments at the intermediate supports as the redundants. This choice simplifies the calculations because it leads to a set of simultaneous equations in which no more than three unknowns appear in any one equation, regardless of the number of redundants. Furthermore, the deflection calculations are simpler because only simple beams are involved, as we will see later. Because each equation contains three redundant bending moments, this variation of the method of superposition is known as the *method of three moments*.

■ Method of Three Moments

Let us begin the discussion by describing the method in very general terms. After that, we will examine the procedure in more detail and derive the required equations.

As in any analysis that uses the method of superposition, we begin by releasing the redundants and obtaining the released structure. Since the redundants are the bending moments at the intermediate supports, we must cut through the beam at those points in order to remove the moments. Thus, the continuity of the beam at the supports is broken, and the released structure is a series of simple beams. Each simple beam is subjected to two sets of loads: (1) the external loads that act on the corresponding portion of the continuous beam, and (2) the redundant moments acting at the ends of the simple beam. These loads produce angles of rotation at the ends of each simple beam. The equations of compatibility express the fact that at each support the two adjacent beams (in the original structure) have the same angle of rotation. By solving these equations, we obtain the unknown bending moments. Since each equation contains only three of the unknown bending moments, the equations of compatibility are called *three-moment equations*.

To illustrate the method described above, we will refer to a segment of a continuous beam, as shown in Fig. 10-21a. This segment of the beam includes three consecutive supports, identified as supports A, B, and C. The span of the beam to the right of support A has length L_A and moment of inertia I_A. The span to the right of support B has length L_B and moment of inertia I_B. Also, we will denote the unknown bending moments at the three supports as M_A, M_B, and M_C. The true directions of these moments depend upon the loads on the beam, but for purposes of our derivation we will assume that the bending moments are positive (recall that a positive bending moment produces compression on the upper part of a beam and tension on the lower part).

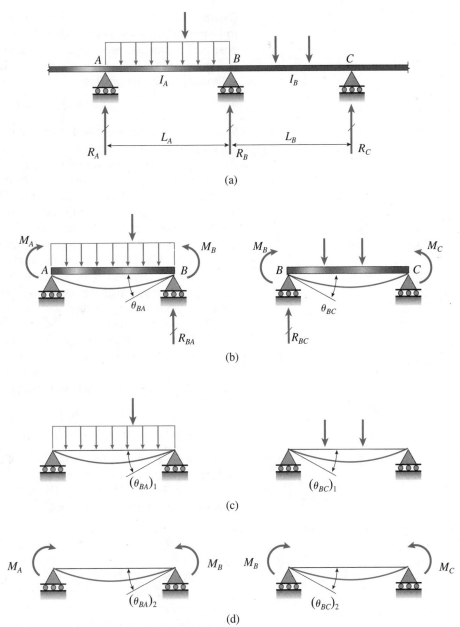

Fig. 10-21 Derivation of the three-moment equation

The released structure, consisting of a series of simple beams, is shown in Fig. 10-21b for the two adjacent spans *AB* and *BC*. Each simple beam is loaded by the external forces plus the redundant bending moments. The angle of rotation of the left-hand beam (beam *AB*) at support *B* is denoted θ_{BA}, and the angle of rotation of the right-hand beam

(beam BC) at this same support is denoted θ_{BC}. These angles are assumed to be positive as shown in the figure (that is, in the same directions as the positive bending moments M_B). Because the longitudinal axis of the original beam (Fig. 10-21a) is continuous across support B, the *equation of compatibility* at support B is

$$\theta_{BA} = -\theta_{BC} \qquad (10\text{-}35)$$

This equation expresses the fact that the angles of rotation of the two adjacent simple beams in the released structure must be consistent with the condition of continuity at support B.

The next step in the analysis is to develop expressions for the angles θ_{BA} and θ_{BC} for substitution into Eq. (10-35). We note that each of these angles is equal to the superposition of two angles: (1) the angle of rotation in a simple beam due to the external loads, and (2) the angle of rotation in a simple beam due to the unknown bending moments. The former are shown in Fig. 10-21c, where the angle at B in member AB is denoted $(\theta_{BA})_1$ and the angle at B in member BC is denoted $(\theta_{BC})_1$. Thus, the subscript 1 denotes that the angles are produced by the external loads acting on the released structure.

The angles of rotation due to the bending moments are shown in Fig. 10-21d, where the subscript 2 is used as the identifier. The angle at B in member AB is denoted $(\theta_{BA})_2$, and the angle at B in member BC is denoted $(\theta_{BC})_2$. These angles are readily found with the aid of the formulas in Table G-2, Appendix G (see Case 7). The results are:

$$(\theta_{BA})_2 = \frac{M_A L_A}{6EI_A} + \frac{M_B L_A}{3EI_A} \qquad (\theta_{BC})_2 = \frac{M_B L_B}{3EI_B} + \frac{M_C L_B}{6EI_B}$$

Note that we are assuming the modulus of elasticity E to be the same for both spans of the beam.

The angles of rotation due to both the external loads and the bending moments are obtained by superposition, as follows:

$$\theta_{BA} = (\theta_{BA})_1 + (\theta_{BA})_2 = (\theta_{BA})_1 + \frac{M_A L_A}{6EI_A} + \frac{M_B L_A}{3EI_A} \qquad (10\text{-}36a)$$

$$\theta_{BC} = (\theta_{BC})_1 + (\theta_{BC})_2 = (\theta_{BC})_1 + \frac{M_B L_B}{3EI_B} + \frac{M_C L_B}{6EI_B} \qquad (10\text{-}36b)$$

These equations give the angles of rotation in the released structure at support B (Fig. 10-21b). In the context of the method of superposition, Eqs. (10-36a and b) are the *force-displacement relations*, although in this case we might better refer to them as *force-rotation relations*.

The final step in the analysis is to substitute the expressions for θ_{BA} and θ_{BC} into the equation of compatibility (Eq. 10-35) and rearrange the terms. In this manner, we obtain the following equation:

$$M_A\left(\frac{L_A}{I_A}\right) + 2M_B\left(\frac{L_A}{I_A} + \frac{L_B}{I_B}\right) + M_C\left(\frac{L_B}{I_B}\right) = -6E(\theta_{BA})_1 - 6E(\theta_{BC})_1$$

(10-37)

This equation is called the **three-moment equation** because it relates three consecutive bending moments in the beam.

If the two adjoining spans of the beam have the same moment of inertia I, the three-moment equation simplifies to the following:

$$M_A L_A + 2M_B(L_A + L_B) + M_C L_B = -6EI(\theta_{BA})_1 - 6EI(\theta_{BC})_1 \qquad (10\text{-}38)$$

If both spans have the same length L, the equation becomes even simpler:

$$M_A + 4M_B + M_C = -\frac{6EI}{L}(\theta_{BA})_1 - \frac{6EI}{L}(\theta_{BC})_1 \qquad (10\text{-}39)$$

As previously mentioned, we are assuming that the modulus of elasticity is the same throughout all spans of the continuous beam. Otherwise, we must make alterations in the equations, beginning with Eqs. (10-36a and b).

When performing calculations with the three-moment equations, the angles $(\theta_{BA})_1$ and $(\theta_{BC})_1$ must be expressed in units of radians.

■ **Procedure**

The procedure for analyzing a continuous beam by the method of three moments may be summarized as follows:

(1) Write a three-moment equation for each intermediate support of the beam. In each of these equations, the moment M_A is the bending moment at the support to the left of the selected intermediate support, the moment M_B is the moment at the intermediate support itself, and the moment M_C is the moment at the support to the right of the selected intermediate support. The angles of rotation $(\theta_{BA})_1$ and $(\theta_{BC})_1$ are the rotations at the ends of simple beams subjected to the external loads (Fig. 10-21c). These angles can be found with the aid of tables such as Table G-2 in the appendix (see the examples given below).

(2) Solve the three-moment equations and obtain the bending moments at the supports. (Note that the number of equations is the same as the number of intermediate supports, which in turn is the same as the number of redundant moments.)

(3) Knowing the bending moments at all supports, use free-body diagrams and equations of equilibrium to find the reactions. Then, knowing the reactions, determine the shear forces and bending moments in all members and draw the corresponding shear-force and bending-moment diagrams.

■ Angles of Rotation

In order to write the three-moment equations, we need to determine the angles of rotation $(\theta_{BA})_1$ and $(\theta_{BC})_1$ appearing on the right-hand sides of the equations (see step 1 in the procedure described above). These angles can be obtained from a table giving deflections and slopes of simple beams, such as Table G-2 in the appendix. For instance, if a uniform load of intensity q acts on span AB (see Fig. 10-21a and c), the angle of rotation at support B is

$$(\theta_{BA})_1 = \frac{qL_A^3}{24EI_A} \tag{10-40a}$$

Similarly, if a uniform load acts on span BC, the angle of rotation at B is

$$(\theta_{BC})_1 = \frac{qL_B^3}{24EI_B} \tag{10-40b}$$

These expressions are obtained from Case 1 of Table G-2.

If a concentrated load P acts at the center of each span, the corresponding angles of rotation are

$$(\theta_{BA})_1 = \frac{PL_A^2}{16EI_A} \qquad (\theta_{BC})_1 = \frac{PL_B^2}{16EI_B} \tag{10-41a,b}$$

as obtained from Case 4 of Table G-2.

These two examples show that there is no major difficulty in evaluating the angles of rotation $(\theta_{BA})_1$ and $(\theta_{BC})_1$ that appear in the three-moment equations. Once the angles are determined, the equations can be formulated merely by writing them down. Then the equations can be solved for the unknown bending moments.*

■ Beam with a Fixed Support

Throughout the foregoing discussion we assumed that the two extreme ends of the continuous beam were simply supported. If either or both of these ends is fixed, the number of redundant bending moments will increase (Fig. 10-22a). An easy way to handle this situation is to replace the fixed support by an additional span having an infinite moment of inertia (Fig. 10-22b). The effect of this extra span having infinitely large

---

*The three-moment equation was developed by the French engineers B. P. E. Clapeyron and H. Bertot in the middle of the 19th century (see Ref. 10-2).

(a)

(b)

Fig. 10-22 Replacement of a fixed support by a span having infinite moment of inertia

stiffness is to prevent rotation at support 1, which is identical to the condition imposed by the fixed support. The bending moments found at points 1, 2, and 3 in the continuous beam shown in Fig. 10-22b are identical to those in the original beam. The length assigned to the extra span is not relevant (except that it must be greater than zero) because it always disappears from the three-moment equation.

■ Reactions

After the bending moments at the supports of a continuous beam have been found, we can use equations of equilibrium to find the reactions. To illustrate the methodology, consider again the two adjacent spans shown in Fig. 10-21. Let R_{BA} and R_{BC} represent the reactions at support B for the two simple beams AB and BC shown in part (b) of the figure. The sum of these two reactions gives the total reaction R_B at support B of the original beam (Fig. 10-21a).

The reaction R_{BA} of beam AB is made up of three parts: (1) the simple-beam reaction due to the external loads, (2) the reaction due to the moment M_A (equal to M_A/L_A), and (3) the reaction due to the moment M_B (equal to $-M_B/L_A$).

In a similar manner, the reaction R_{BC} of beam BC is equal to the simple-beam reaction due to the external loads, plus M_C/L_B, minus M_B/L_B.

Combining all six of the preceding terms gives the total reaction R_B in the original beam:

R_B = Simple-beam reaction of member AB due to the loads

+ Simple-beam reaction of member BC due to the loads

$$+ \frac{M_A}{L_A} - \frac{M_B}{L_A} - \frac{M_B}{L_B} + \frac{M_C}{L_B} \tag{10-42}$$

This equation may be used at each support until all reactions are calculated.

A special case occurs when a concentrated load acts directly over a support. A load of this kind is transmitted directly into the reaction and need not appear in the released structure. Of course, it must be added to the reaction calculated from Eq. (10-42).

■ **Example 10-7**

The continuous beam shown in Fig. 10-23a has three spans of equal length L and constant moment of inertia I. The first span is subjected to a uniform load of intensity q and the third span supports a concentrated load P of magnitude qL. The concentrated load acts at distance $3L/4$ from support 3. Determine the reactions of the beam using the three-moment equation, and then construct the shear-force and bending-moment diagrams for the beam.

Solution   This beam rests on four simple supports (supports 1, 2, 3, and 4), and therefore it is statically indeterminate to the second degree. The bending moments M_2 and M_3 at supports 2 and 3, respectively, are the redundant quantities.

Since the moments of inertia and span lengths are the same for all three spans, we can use the three-moment equation in the form given by Eq. (10-39):

$$M_A + 4M_B + M_C = -\frac{6EI}{L}(\theta_{BA})_1 - \frac{6EI}{L}(\theta_{BC})_1 \tag{a}$$

Now we will write this equation at supports 2 and 3.

Three-moment equation at support 2. The adjoining spans are 1-2 and 2-3, and therefore the moments M_A, M_B, and M_C in Eq. (a) become M_1, M_2, and M_3, respectively. Since $M_1 = 0$, only the moments M_2 and M_3 will appear in the equation.

The angle of rotation $(\theta_{BA})_1$ is the rotation at the right-hand end of member 1-2 when it is treated as a simple beam. Thus, from Eq. (10-40a), we get

$$(\theta_{BA})_1 = (\theta_{21})_1 = \frac{qL^3}{24EI} \tag{b}$$

Similarly, the angle $(\theta_{BC})_1$ is the simple-beam rotation at the left-hand end of member 2-3. Since there is no load on member 2-3, we obtain

$$(\theta_{BC})_1 = (\theta_{23})_1 = 0 \tag{c}$$

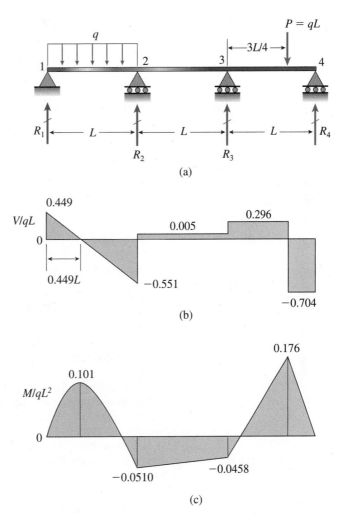

Fig. 10-23 Example 10-7.
(a) Continuous beam, (b) shear-force
diagram, and (c) bending-moment
diagram

Thus, the three-moment equation for support 2 (see Eq. a) becomes

$$4M_2 + M_3 = -\frac{6EI}{L}\left(\frac{qL^3}{24EI}\right) = -\frac{qL^2}{4} \tag{d}$$

Three-moment equation at support 3. At support 3 the moments M_A, M_B,
and M_C in Eq. (a) become M_2, M_3, and M_4, respectively. Since $M_4 = 0$, only the
moments M_2 and M_3 will appear in the equation.

The angle of rotation $(\theta_{BA})_1$ is the angle at the right-hand end of member
2-3 when it is treated as a simple beam. Thus,

$$(\theta_{BA})_1 = (\theta_{32})_1 = 0 \tag{e}$$

The angle $(\theta_{BC})_1$ is the rotation at the left-hand end of member 3-4.
Since the load on member 3-4 is a concentrated load, we determine the angle of

rotation from Case 5, Table G-2. Using the formula for the angle θ_A in that table, we obtain

$$(\theta_{BC})_1 = (\theta_{34})_1 = \frac{Pab(L + b)}{6LEI} = \frac{P(3L/4)(L/4)(L + L/4)}{6LEI} = \frac{5PL^2}{128EI}$$

Since the load P is equal to qL, this equation becomes

$$(\theta_{BC})_1 = (\theta_{34})_1 = \frac{5qL^3}{128EI} \qquad (f)$$

Thus, the three-moment equation for support 3 (Eq. a) becomes

$$M_2 + 4M_3 = -\frac{6EI}{L}\left(\frac{5qL^3}{128EI}\right) = -\frac{15qL^2}{64} \qquad (g)$$

Solution of equations. To find the redundant bending moments, we solve simultaneously the three-moment equations (Eqs. d and g). The results are

$$M_2 = -\frac{49qL^2}{960} \qquad M_3 = -\frac{11qL^2}{240} \qquad (h,i)$$

Reactions. The reactions at each support can now be obtained by using free-body diagrams and equations of equilibrium, or by substituting directly into Eq. (10-42).

At support 1, the terms in Eq. (10-42) are as follows: (1) The simple-beam reaction for member AB is zero because span AB is nonexistent; (2) the simple-beam reaction for member BC, which is now member 1-2, is $qL/2$; (3) the moment M_A is zero because it is nonexistent; (4) the moment M_B is zero because it is the bending moment at support 1; and (5) the moment M_C is the bending moment M_2, which is given by Eq. (h). Thus, substituting term-by-term into Eq. (10-42), we get

$$R_1 = 0 + \frac{qL}{2} + 0 - 0 - 0 - \frac{49qL^2}{960L} = \frac{431qL}{960}$$

At support 2, the terms in Eq. (10-42) are as follows: (1) The simple-beam reaction for member AB, which is now member 1-2, is $qL/2$; (2) the simple-beam reaction for member BC, which is now member 2-3, is zero; (3) the moment M_A is the bending moment at support 1 and is equal to zero; (4) the moment M_B is the bending moment M_2, which is given by Eq. (h); and (5) the moment M_C is the bending moment M_3, which is given by Eq. (i). Thus, substituting term-by-term into Eq. (10-42), we get

$$R_2 = \frac{qL}{2} + 0 + 0 + \frac{49qL^2}{960L} + \frac{49qL^2}{960L} - \frac{11qL^2}{240L} = \frac{89qL}{160}$$

Proceeding in this manner for the remaining two reactions, we are able to obtain all reactions of the beam from Eq. (10-42). As an alternative to using this equation, we can simply draw free-body diagrams of each member and solve the equations of equilibrium.

The final results are

$$R_1 = \frac{431qL}{960} = 0.449qL \qquad R_2 = \frac{89qL}{160} = 0.556qL \qquad \text{(j,k)} \quad \Longleftarrow$$

$$R_3 = \frac{93qL}{320} = 0.291qL \qquad R_4 = \frac{169qL}{240} = 0.704qL \qquad \text{(l,m)} \quad \Longleftarrow$$

Shear-force and bending-moment diagrams. Now that the reactions have been found, we can isolate parts of the beam as free bodies and use equations of equilibrium to determine the shear forces and bending moments throughout the beam. The resulting diagrams are shown in Figs. 10-23b and c.

*10.6 ∎ TEMPERATURE EFFECTS

Temperature changes may produce changes in lengths of bars (see Section 2.5) and lateral deflections of beams (see Section 9.11). If free expansion of the bar or beam is prevented, stresses will be produced. For instance, in Section 2.5 we determined the stresses in statically indeterminate bars subjected to uniform temperature changes. Now, in this section, we will analyze statically indeterminate beams subjected to nonuniform temperature changes.

The stresses and deflections produced by temperature changes in a statically indeterminate beam can be analyzed by methods that are similar to those already described for loads. In order to illustrate some of the possibilities, we will use a fixed-end beam AB as an example (Fig. 10-24a). We assume that the beam was originally at a uniform temperature T_0, but later its temperature is increased to T_1 on the upper surface and T_2 on the lower surface. The variation of temperature over the height h of the beam is assumed to be linear.

Because the temperature varies linearly, the *average* temperature of the beam is

$$T_{\text{aver}} = \frac{T_1 + T_2}{2} \tag{10-43}$$

and occurs at midheight of the beam. The difference between this average temperature and the initial temperature T_0 results in a tendency for the beam to change in length. If the beam is free to expand longitudinally, its length will increase by an amount δ_T given by Eq. (9-98), which is repeated here:

$$\delta_T = \alpha(T_{\text{aver}} - T_0)L = \alpha\left(\frac{T_1 + T_2}{2} - T_0\right)L \tag{10-44}$$

In this equation, α is the coefficient of thermal expansion of the material and L is the length of the beam. If longitudinal expansion is free to occur, no axial stresses will be produced by the temperature changes. However, if longitudinal expansion is restrained, axial stresses will develop, as described in Section 2.5.

Now consider the effects of the temperature differential $T_2 - T_1$, which tends to produce a *curvature* of the beam but no change in length. Curvature due to temperature changes is described in Section 9.11, where the following differential equation of the deflection curve is derived (see Eq. 9-100):

$$\frac{d^2v}{dx^2} = \frac{\alpha(T_2 - T_1)}{h} \tag{10-45}$$

This equation applies to a beam that is unrestrained by supports and therefore is free to deflect and rotate. Note that when T_2 is greater than T_1, the curvature is positive and the beam tends to bend concave upward. Deflections and rotations of simple beams and cantilever beams due to a temperature differential can be determined with the aid of Eq. (10-45), as discussed in Section 9.11. We can use those results when analyzing statically indeterminate beams by the method of superposition.

■ **Method of Superposition**

To illustrate the use of superposition, let us determine the reactions of the fixed-end beam of Fig. 10-24a due to the temperature differential. As usual, we begin the analysis by selecting the redundant reactions. Although other choices may result in more efficient calculations, we will select the reactive force R_B and reactive moment M_B as the redundants in order to illustrate the general methodology.

When the supports corresponding to the redundants are removed, we obtain the released structure shown in Fig. 10-24b (a cantilever beam). The deflection and angle of rotation at end B of this cantilever (due to the temperature differential) are (see Fig. 10-24c):

$$(\delta_B)_1 = \frac{\alpha(T_2 - T_1)L^2}{2h} \qquad (\theta_B)_1 = \frac{\alpha(T_2 - T_1)L}{h}$$

These equations are obtained from the solution to Problem 9.11-2 in the preceding chapter. Note that when T_2 is greater than T_1, the deflection $(\delta_B)_1$ is upward and the angle of rotation $(\theta_B)_1$ is counterclockwise.

Next, we need to find the deflections and angles of rotation in the released structure (Fig. 10-24b) due to the redundants R_B and M_B. These quantities are obtained from Cases 4 and 6, respectively, of Table G-1:

$$(\delta_B)_2 = \frac{R_B L^3}{3EI} \qquad (\theta_B)_2 = \frac{R_B L^2}{2EI}$$

$$(\delta_B)_3 = -\frac{M_B L^2}{2EI} \qquad (\theta_B)_3 = -\frac{M_B L}{EI}$$

Fig. 10-24 (a) Fixed-end beam with a temperature differential, (b) released structure, and (c) deflection curve for the released structure

In these expressions, upward deflection and counterclockwise rotation are positive (as in Fig. 10-24c).

We can now write the equations of compatibility for the deflection and angle of rotation at support B as follows:

$$\delta_B = (\delta_B)_1 + (\delta_B)_2 + (\delta_B)_3 = 0 \qquad \text{(a)}$$

$$\theta_B = (\theta_B)_1 + (\theta_B)_2 + (\theta_B)_3 = 0 \qquad \text{(b)}$$

or, upon substituting the appropriate expressions,

$$\frac{\alpha(T_2 - T_1)L^2}{2h} + \frac{R_B L^3}{3EI} - \frac{M_B L^2}{2EI} = 0 \qquad \text{(c)}$$

$$\frac{\alpha(T_2 - T_1)L}{h} + \frac{R_B L^2}{2EI} - \frac{M_B L}{EI} = 0 \qquad \text{(d)}$$

These equations can be solved simultaneously for the two redundants:

$$R_B = 0 \qquad M_B = \frac{\alpha EI(T_2 - T_1)}{h}$$

The fact that R_B is zero could have been anticipated initially from the symmetry of the fixed-end beam. If we had utilized this fact from the outset, the preceding solution would have been simplified because only one equation of compatibility would have been required.

We also know from symmetry (or from equations of equilibrium) that the reaction R_B is equal to the reaction R_A and that the moment M_A is equal to the moment M_B. Therefore, the reactions for the fixed-end beam shown in Fig. 10-24a are as follows:

$$R_A = R_B = 0 \qquad M_A = M_B = \frac{\alpha EI(T_2 - T_1)}{h} \qquad \text{(10-46a,b)}$$

From these results we see that the entire beam is subjected to a constant bending moment due to the temperature changes.

■ **Differential Equation of the Deflection Curve**

We can also analyze the fixed-end beam of Fig. 10-24a by solving the differential equation of the deflection curve. When a beam is subjected to both a bending moment M and a temperature differential $T_2 - T_1$, the differential equation becomes (see Eqs. 9-7 and 10-45):

$$\frac{d^2v}{dx^2} = \frac{M}{EI} + \frac{\alpha(T_2 - T_1)}{h} \qquad \text{(10-47a)}$$

or
$$EIv'' = M + \frac{\alpha EI(T_2 - T_1)}{h} \qquad \text{(10-47b)}$$

For the fixed-end beam of Fig. 10-24a, the expression for the bending moment in the beam is

$$M = R_A x - M_A \tag{e}$$

where x is measured from support A. Substituting into the differential equation and integrating, we obtain the following equation for the slope of the beam:

$$EIv' = \frac{R_A x^2}{2} - M_A x + \frac{\alpha EI(T_2 - T_1)x}{h} + C_1 \tag{f}$$

The two boundary conditions on the slope ($v' = 0$ when $x = 0$ and $x = L$) give $C_1 = 0$ and

$$\frac{R_A L}{2} - M_A = -\frac{\alpha EI(T_2 - T_1)}{h} \tag{g}$$

A second integration gives the deflection of the beam:

$$EIv = \frac{R_A x^3}{6} - \frac{M_A x^2}{2} + \frac{\alpha EI(T_2 - T_1)x^2}{2h} + C_2 \tag{h}$$

The boundary conditions on the deflection ($v = 0$ when $x = 0$ and $x = L$) give $C_2 = 0$ and

$$\frac{R_A L}{3} - M_A = -\frac{\alpha EI(T_2 - T_1)}{h} \tag{i}$$

Solving simultaneously Eqs. (g) and (i), we find

$$R_A = 0 \qquad M_A = \frac{\alpha EI(T_2 - T_1)}{h}$$

From the equilibrium of the beam, we obtain $R_B = 0$ and $M_B = M_A$. Thus, these results agree with those found by the method of superposition (see Eqs. 10-46a and b).

Note that we carried out the preceding solution without taking advantage of symmetry in order to show the general approach of the integration method.

Knowing the reactions of the beam, we can now find the shear forces, bending moments, slopes, and deflections. The simplicity of the results may surprise you.

*10.7   LONGITUDINAL DISPLACEMENTS AT THE ENDS OF A BEAM

When a beam is bent by lateral loads, the ends of the beam move closer together. It is common practice to disregard these longitudinal displacements because usually they have no noticeable effect on the behavior of the beam. In this section, we will show how to evaluate these displacements and determine whether or not they are important.

Consider a simple beam AB that is pin-supported at one end and free to displace longitudinally at the other (Fig. 10-25a). When this beam is bent by lateral loads, the deflection curve has the shape shown in part (b) of the figure. In addition to the lateral deflections, there is a longitudinal displacement at end B of the beam. End B moves horizontally from point B to point B' through a small distance λ, called the **curvature shortening** of the beam. As the name implies, curvature shortening is due to bending of the axis of the beam and is *not* due to axial strains produced by tensile or compressive forces. As we see from Fig. 10-25b, the curvature shortening is equal to the difference between the initial length L of the straight beam and the length of the chord AB' of the bent beam. Of course, both the lateral deflections and the curvature shortening are highly exaggerated in the figure.

(a)

(b)

(c)

Fig. 10-25 (a) Simple beam with lateral loads, (b) horizontal displacement λ at the end of the beam, and (c) horizontal reactions H for a beam with immovable supports

■ Curvature Shortening

To determine the curvature shortening, we begin by considering an element of length ds measured along the curved axis of the beam (Fig. 10-25b). The projection of this element on the horizontal axis has length dx. The relationship between the length of the element and the length of its horizontal projection is obtained from the Pythagorean theorem:

$$(ds)^2 = (dx)^2 + (dv)^2$$

where dv is the increment in the deflection v of the beam as we move through the distance dx. Thus,

$$ds = \sqrt{(dx)^2 + (dv)^2} = dx\sqrt{1 + \left(\frac{dv}{dx}\right)^2} \tag{a}$$

The difference between the length of the element and the length of its horizontal projection is

$$ds - dx = dx\sqrt{1 + \left(\frac{dv}{dx}\right)^2} - dx = dx\left[\sqrt{1 + \left(\frac{dv}{dx}\right)^2} - 1\right] \tag{b}$$

Now let us introduce the following binomial series (see Appendix C):

$$\sqrt{1 + t} = 1 + \frac{t}{2} - \frac{t^2}{8} + \frac{t^3}{16} - \cdots \tag{10-48}$$

which converges when t is numerically less than 1. If t is very small compared to 1, we can disregard the terms involving t^2, t^3, and so on, in comparison with the first two terms. Then we obtain

$$\sqrt{1 + t} \approx 1 + \frac{t}{2} \tag{10-49}$$

The term $(dv/dx)^2$ in Eq. (b) is ordinarily very small compared to 1. Therefore, we can use Eq. (10-49) with $t = (dv/dx)^2$ and rewrite Eq. (b) as

$$ds - dx = dx\left[1 + \frac{1}{2}\left(\frac{dv}{dx}\right)^2 - 1\right] = \frac{1}{2}\left(\frac{dv}{dx}\right)^2 dx \qquad (c)$$

If the left- and right-hand sides of this expression are integrated over the length of the beam, we obtain an expression for the difference between the length of the beam and the length of the chord AB' (Fig. 10-25b):

$$L - \overline{AB'} = \int_0^L \frac{1}{2}\left(\frac{dv}{dx}\right)^2 dx$$

Thus, the curvature shortening is

$$\lambda = \frac{1}{2}\int_0^L \left(\frac{dv}{dx}\right)^2 dx \qquad (10\text{-}50)$$

This equation is valid provided the deflections and slopes are small.

Note that when the equation of the deflection curve is known, we can substitute into Eq. (10-50) and determine the shortening λ.

■ Horizontal Reactions

Now suppose that the ends of the beam are prevented from translating longitudinally by immovable supports (Fig. 10-25c). Because the ends cannot move toward each other, a horizontal reaction H will develop at each end. This force will cause the axis of the beam to elongate as bending occurs. In addition, the force H itself will have an effect upon the bending moments in the beam, because an additional bending moment (equal to H times the deflection) will exist at every cross section. Thus, the deflection curve of the beam depends not only upon the lateral loads but also upon the reaction H, which in turn depends upon the shape of the deflection curve, as shown by Eq. (10-50).

Rather than attempt an exact analysis of this complicated problem, let us obtain an approximate expression for the force H in order to ascertain its importance. For that purpose, we can use any reasonable approximation to the deflection curve. In the case of a pin-ended beam with downward loads (Fig. 10-25c), a good approximation is a parabola having the equation

$$v = -\frac{4\delta x(L - x)}{L^2} \tag{10-51}$$

where δ is the downward deflection at the midpoint of the beam. The curvature shortening λ corresponding to this assumed deflected shape can be found by substituting the expression for the deflection v into Eq. (10-50) and integrating; the result is

$$\lambda = \frac{8\delta^2}{3L} \tag{10-52}$$

The horizontal force H required to elongate the beam by this amount is

$$H = \frac{EA\lambda}{L} = \frac{8EA\delta^2}{3L^2} \tag{10-53}$$

in which EA is the axial rigidity of the beam. The corresponding axial tensile stress in the beam is

$$\sigma_t = \frac{H}{A} = \frac{8E\delta^2}{3L^2} \tag{10-54}$$

Now let us substitute some numerical values so that we can assess the significance of the curvature shortening. The deflection δ at the midpoint of the beam is usually very small compared to the length; for example, the ratio δ/L might be 1/500 or smaller. Using this value, and also assuming that the material is steel with $E = 30 \times 10^6$ psi, we find from Eq. (10-54) that the tensile stress is only 320 psi. Since the allowable tensile stress in the steel is typically 15,000 psi or larger, it becomes clear that the axial stress due to the horizontal force H may be disregarded when compared to the ordinary working stresses in the beam.

Furthermore, in the derivation of Eq. (10-54) we assumed that the ends of the beam were held rigidly against horizontal displacements, which is not physically possible. In reality, small longitudinal displacements always occur, thereby reducing the axial stress calculated from Eq. (10-54).*

From the above discussions, we conclude that the customary practice of disregarding the effects of any longitudinal restraints and assuming that one end of the beam is on a roller support (regardless of the actual construction) is justified. The stiffening effect of longitudinal restraints is significant only when the beam is very long and slender and supports large loads. This behavior is sometimes referred to as "string action," because it is analogous to the action of a cable, or string, supporting a load.

---

*For a more complete analysis of beams with immovable supports, see Ref. 10-5.

PROBLEMS/CHAPTER 10

DIFFERENTIAL EQUATIONS

The problems for Section 10.3 are to be solved by integrating the differential equations of the deflection curve. All beams have constant flexural rigidity EI. When drawing shear-force and bending-moment diagrams, be sure to label all critical ordinates, including maximum and minimum values.

10.3-1 A propped cantilever beam AB of length L is loaded by a counterclockwise moment M_0 acting at support B (see figure). Beginning with the second-order differential equation of the deflection curve (the bending-moment equation), obtain the reactions, shear forces, bending moments, slopes, and deflections of the beam. Construct the shear-force and bending-moment diagrams, labeling all critical ordinates.

Prob. 10.3-1

10.3-2 A fixed-end beam AB of length L supports a uniform load of intensity q (see figure). Beginning with the second-order differential equation of the deflection curve (the bending-moment equation), obtain the reactions, shear forces, bending moments, slopes, and deflections of the beam. Construct the shear-force and bending-moment diagrams, labeling all critical ordinates.

Prob. 10.3-2

10.3-3 A cantilever beam AB of length L has a fixed support at A and a roller support at B (see figure). The support at B is moved downward through a distance δ_B. Using the fourth-order differential equation of the deflection curve (the load equation), determine the reactions of the beam and the equation of the deflection curve. (*Note:* Express all results in terms of the imposed displacement δ_B.)

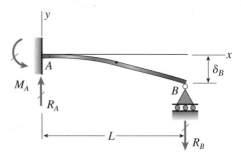

Prob. 10.3-3

10.3-4 A cantilever beam AB of length L has a fixed support at A and a spring support at B (see figure). The spring behaves in a linearly elastic manner with stiffness k. If a uniform load of intensity q acts on the beam, what is the downward displacement δ_B of end B of the beam? (Use the second-order differential equation of the deflection curve, that is, the bending-moment equation.)

Prob. 10.3-4

10.3-5 A propped cantilever beam AB of length L supports a triangularly distributed load of maximum intensity q_0 (see figure). Beginning with the fourth-order differential equation of the deflection curve (the load equation), obtain the reactions of the beam and the equation of the deflection curve.

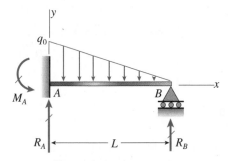

Prob. 10.3-5

10.3-6 The load on a propped cantilever beam AB of length L is parabolically distributed according to the equation $q = q_0(1 - x^2/L^2)$, as shown in the figure. Beginning with the fourth-order differential equation of the deflection curve (the load equation), obtain the reactions of the beam and the equation of the deflection curve.

Prob. 10.3-6

10.3-7 The load on a fixed-end beam AB of length L is distributed in the form of a sine curve (see figure). The intensity of the load is given by the equation $q = q_0 \sin \pi x/L$.

Prob. 10.3-7

Beginning with the fourth-order differential equation of the deflection curve (the load equation), obtain the reactions of the beam and the equation of the deflection curve.

10.3-8 A fixed-end beam AB of length L supports a triangularly distributed load of maximum intensity q_0 (see figure). Beginning with the fourth-order differential equation of the deflection curve (the load equation), obtain the reactions of the beam and the equation of the deflection curve.

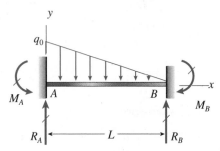

Prob. 10.3-8

* **10.3-9** A counterclockwise moment M_0 acts at the midpoint of a fixed-end beam ACB of length L (see figure). Beginning with the second-order differential equation of the deflection curve (the bending-moment equation), determine all reactions of the beam and obtain the equation of the deflection curve for the left-hand half of the beam. Then construct the shear-force and bending-moment diagrams for the entire beam, labeling all critical ordinates. Also, draw the deflection curve for the entire beam.

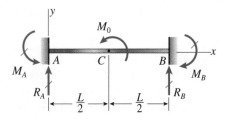

Prob. 10.3-9

** **10.3-10** A propped cantilever beam AB supports a concentrated load P acting at the midpoint C (see figure). Beginning with the second-order differential equation of the deflection curve (the bending-moment equation), determine all reactions of the beam and draw the shear-force and bending-moment diagrams. Also, obtain the equations of the deflection curves for both halves of the beam, and draw the complete deflection curve of the beam.

Prob. 10.3-10

METHOD OF SUPERPOSITION

The problems for Section 10.4 are to be solved by the method of superposition. All beams have constant flexural rigidity EI unless otherwise stated. When drawing shear-force and bending-moment diagrams, be sure to label all critical ordinates, including maximum and minimum values.

10.4-1 A propped cantilever beam AB of length L carries a concentrated load P acting at the position shown in the figure. Determine the reactions R_A, R_B, and M_A for this beam. Also, draw the shear-force and bending-moment diagrams, labeling all critical ordinates.

Prob. 10.4-1

10.4-2 The propped cantilever beam shown in the figure supports a uniform load of intensity q on the left-hand half of the beam. Find the reactions R_A, R_B, and M_A, and then

draw the shear-force and bending-moment diagrams, labeling all critical ordinates.

Prob. 10.4-2

10.4-3 The figure shows a propped cantilever beam ABC having span length L and an overhang of length a. A concentrated load P acts at the end of the overhang. Determine the reactions R_A, R_B, and M_A for this beam. Also, draw the shear-force and bending-moment diagrams, labeling all critical ordinates.

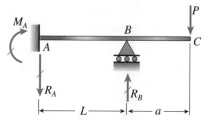

Prob. 10.4-3

10.4-4 Determine the fixed-end moments (M_A and M_B) and fixed-end forces (R_A and R_B) for a beam of length L supporting a triangular load of maximum intensity q_0 (see figure). Then draw the shear-force and bending-moment diagrams, labeling all critical ordinates.

Prob. 10.4-4

10.4-5 Two flat beams AB and CD, lying in horizontal planes, cross at right angles and jointly support a vertical load P at their midpoints (see figure). Before the load P is applied, the beams just touch each other. Both beams are made of the same material and have the same widths. Also, the ends of both beams are simply supported, but beam AB is longer than beam CD. What should be the ratio t_{AB}/t_{CD} of their thicknesses if all four reactions are to be the same?

Prob. 10.4-5

10.4-6 A continuous beam ABC with two unequal spans, one of length L and one of length $2L$, supports a uniform load of intensity q (see figure). Determine the reactions R_A, R_B, and R_C for this beam. Also, draw the shear-force and bending-moment diagrams, labeling all critical ordinates.

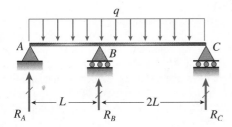

Prob. 10.4-6

10.4-7 Beam ABC is fixed at support A and rests (at point B) upon the midpoint of beam DE (see the first part of the figure). Thus, beam ABC may be represented as a propped cantilever beam with an overhang BC and a linearly elastic support of stiffness k at point B (see the second part of the figure). The distance from A to B is $L = 10$ ft, the distance from B to C is $L/2 = 5$ ft, and the length of beam DE is $L = 10$ ft. Both beams have the same flexural rigidity EI. A concentrated load $P = 1700$ lb acts at the free end of beam ABC. Determine the reactions R_A, R_B, and M_A for beam ABC. Also, draw the shear-force and bending-moment diagrams for beam ABC, showing all critical ordinates.

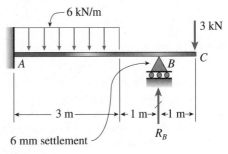

Prob. 10.4-7

10.4-8 The beam ABC shown in the figure has flexural rigidity $EI = 4.0$ MN·m². When the loads are applied to the beam, the support at B settles vertically downward through a distance of 6.0 mm. Calculate the reaction R_B at support B.

Prob. 10.4-8

10.4-9 A beam AB is cantilevered from a wall at one end and held by a tie rod at the other end (see figure). The beam is an S 6 × 12.5 I-beam section with length $L = 6$ ft. The tie rod has a diameter of 1/4 inch and length $H = 3$ ft. Both members are made of steel with $E = 30 \times 10^6$ psi. A uniform load of intensity $q = 200$ lb/ft acts along the length of the beam. Before the load q is applied, the tie rod just meets the end of the cable. (a) Determine the tensile force T

in the tie rod due to the load q. (b) Draw the shear-force and bending-moment diagrams, showing all critical ordinates, for the beam.

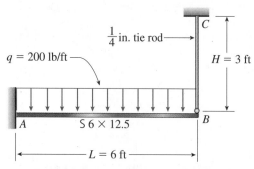

Prob. 10.4-9

10.4-10 The figure shows a nonprismatic, propped cantilever beam AB with flexural rigidity $2EI$ from A to C and EI from C to B. Determine all reactions of the beam due to the uniform load of intensity q. (*Hint:* Use the results of Problems 9.7-1 and 9.7-2.)

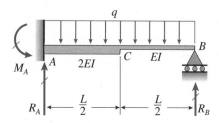

Prob. 10.4-10

10.4-11 A beam ABC is fixed at end A and supported by beam DE at point B (see figure). Both beams have the same cross section and are made of the same material. (a) Determine all reactions due to the load P. (b) What is the numerically largest bending moment in either beam?

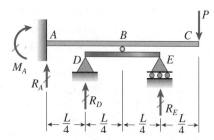

Prob. 10.4-11

10.4-12 A three-span continuous beam $ABCD$ with three equal spans supports a uniform load of intensity q (see figure). Determine all reactions of this beam and draw the shear-force and bending-moment diagrams, labeling all critical ordinates.

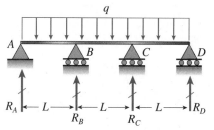

Prob. 10.4-12

10.4-13 A beam AC rests on simple supports at points A and C (see figure). A small gap $\Delta = 0.4$ in. exists between the unloaded beam and a support at point B, which is midway between the ends of the beam. The beam has total length $2L = 80$ in. and flexural rigidity $EI = 0.4 \times 10^9$ lb-in.^{2} Plot a graph of the bending moment M_B at the midpoint of the beam as a function of the intensity q of the uniform load.

Hints: Begin by determining the intensity q_0 of the load that will just close the gap. Then determine the corresponding bending moment $(M_B)_0$. Next, determine the bending moment M_B (in terms of q) for the case where $q < q_0$. Finally, make a statically indeterminate analysis and determine the moment M_B (in terms of q) for the case where $q > q_0$. Plot M_B (units of lb-in.) versus q (units of lb/in.) with q varying from 0 to 2500 lb/in.

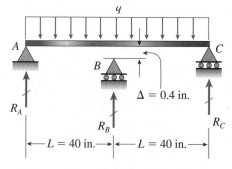

Prob. 10.4-13

10.4-14 A fixed-end beam AB of length L is subjected to a moment M_0 acting at the position shown in the figure. (a) Determine all reactions for this beam. (b) Draw shear-force and bending-moment diagrams for the special case in which $a = b = L/2$.

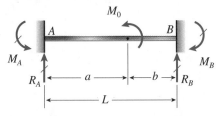

Prob. 10.4-14

10.4-15 A temporary wood flume serving as a channel for irrigation water is shown in the figure. The vertical boards forming the sides of the flume are sunk in the ground, which provides a fixed support. The top of the flume is held by tie rods that are tightened so that there is no deflection of the boards at that point. Thus, the vertical boards may be modeled as a beam AB, supported and loaded as shown in the last part of the figure. Assuming that the thickness t of the boards is 1.5 in., the depth d of the water is 40 in., and the height h to the tie rods is 50 in., what is the maximum bending stress σ in the boards? (*Hint:* The numerically largest bending moment occurs at the fixed support.)

Prob. 10.4-15

10.4-16 Two identical, simply supported beams AB and CD are placed so that they cross each other at their midpoints (see figure). Before the uniform load is applied, the beams just touch each other at the crossing point. Determine the maximum bending moments $(M_{AB})_{max}$ and $(M_{CD})_{max}$ in beams AB and CD, respectively, due to the uniform load if the intensity of the load is $q = 6.4$ kN/m and the length of each beam is $L = 4$ m.

Prob. 10.4-16

* **10.4-17** The cantilever beam AB shown in the figure is an S 6 × 12.5 steel I-beam with $E = 30 \times 10^6$ psi. The simple beam DE is a wood beam 4 in. × 12 in. (nominal dimensions) in cross section with $E = 1.5 \times 10^6$ psi. A steel rod AC of diameter 0.25 in., length 10 ft, and $E = 30 \times 10^6$ psi serves as a hanger joining the two beams. The hanger fits snugly between the beams before the uniform load is applied to beam DE. Determine the force F in the hanger and the maximum bending moments M_{AB} and M_{DE} in the two beams due to the uniform load, which has intensity $q = 400$ lb/ft. (*Hint:* To aid in obtaining the maximum bending moment in beam DE, draw the shear-force and bending-moment diagrams.)

Prob. 10.4-17

*10.4-18 The beam *AB* shown in the figure is simply supported at *A* and *B* and supported on a spring of stiffness *k* (force per unit of shortening) at its midpoint *C*. The beam has flexural rigidity *EI* and length 2*L*. What should be the stiffness *k* of the spring in order that the maximum bending moment in the beam (due to the uniform load) will have the smallest possible value?

Prob. 10.4-18

*10.4-19 The continuous frame *ABC* has a fixed support at *A*, a roller support at *C*, and a rigid corner connection at *B* (see figure). Members *AB* and *CD* each have length *L* and flexural rigidity *EI*. A horizontal force *P* acts at midheight of member *AB*. (a) Find all reactions of the frame. (b) What is the largest bending moment M_{max} in the frame? (*Note:* Disregard axial deformations in member *AB* and consider only the effects of bending.)

Prob. 10.4-19

*10.4-20 The continuous frame *ABC* has a pinned support at *A*, a pinned support at *C*, and a rigid corner connection at *B* (see figure). Members *AB* and *CD* each have length *L* and flexural rigidity *EI*. A horizontal force *P* acts at midheight of member *AB*. (a) Find all reactions of the frame. (b) What is the largest bending moment M_{max} in the frame? (*Note:* Disregard axial deformations in members *AB* and *BC* and consider only the effects of bending.)

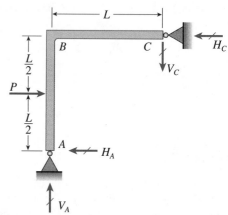

Prob. 10.4-20

*10.4-21 A wide-flange beam *ABC* rests on three identical spring supports at points *A*, *B*, and *C* (see figure). The flexural rigidity of the beam is $EI = 6912 \times 10^6$ lb-in.², and each spring has stiffness $k = 62,500$ lb/in. The length of the beam is $L = 16$ ft. If the load *P* is 6000 lb, what are the reactions R_A, R_B, and R_C? Also, draw the shear-force and bending-moment diagrams for the beam, labeling all critical ordinates.

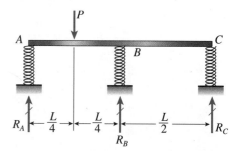

Probs. 10.4-21 and 10.4-22

*10.4-22 Solve the preceding problem if the flexural rigidity *EI* of the beam is 9.216×10^6 N·m², the stiffness *k* of each spring is $93,750 \times 10^3$ N/m, the length *L* of the beam is 3.84 m, and the load *P* is 27 kN.

*10.4-23 A fixed-end beam *AB* of length *L* is subjected to a uniform load of intensity *q* acting over the middle region of the beam (see figure). (a) Obtain a formula for the fixed-end moments M_A and M_B in terms of the load *q*, the length *L*, and the length *b* of the loaded part of the beam. (b) Plot a graph of the fixed-end moment M_A versus the length *b* of the loaded part of the beam. For convenience, plot the graph in the following nondimensional form:

$$\frac{M_A}{qL^2/12} \quad \text{versus} \quad \frac{b}{L}$$

with the ratio b/L varying between its extreme values of 0 and 1. (c) For the special case in which $a = b = L/3$, draw the shear-force and bending-moment diagrams for the beam, labeling all critical ordinates.

Prob. 10.4-23

*10.4-24 A beam supporting a uniform load of intensity q throughout its length rests on pistons at points A, C, and B (see figure). The cylinders are filled with oil and are connected by a tube so that the oil pressure on each piston is the same. The pistons at A and B have diameter d_1, and the piston at C has diameter d_2. (a) Determine the ratio of d_2 to d_1 so that the largest bending moment in the beam is as small as possible. (b) Under these optimum conditions, what is the largest bending moment M_{max} in the beam? (c) What is the difference in elevation between point C and the end supports?

Prob. 10.4-24

**10.4-25 A *trussed beam* consisting of a beam ACB, a strut at the midpoint C, and four tie rods (one on each side of the beam from A to D and one on each side from B to D) is shown in the figure. The beam is a steel wide-flange member (W 18 × 119) of length $L = 36$ ft. The vertical distance between the longitudinal axis of the beam and the pin at D is $H = 3$ ft. Each tie rod has a cross-sectional area of 2.0 in.[2], so that the total area for member AD (or for

member BD) is $A_t = 4.0$ in.[2] All members have the same modulus of elasticity. For purposes of this analysis, the strut and tie rods may be idealized as pinned-end members, the shortening of the strut may be disregarded, and the effects of the axial compressive force in the beam may be disregarded. Thus, the only effects to be included in the analysis are the bending of the beam and the lengthening of the tie rods. (a) Determine the compressive force F in the strut due to a concentrated load $P = 20$ k acting at the midpoint of the beam. (b) Determine the maximum bending stress σ_b in the beam and the tensile stress σ_t in the tie rods.

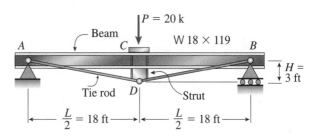

Prob. 10.4-25

** 10.4-26 A thin steel beam AB used in conjunction with an electromagnet in a high-energy physics experiment is securely bolted to rigid supports (see figure). A magnetic field produced by coils C results in a force acting on the beam. The force is trapezoidally distributed with maximum intensity $q_0 = 18$ kN/m. The length of the beam between supports is $L = 200$ mm and the dimension c of the trapezoidal load is 50 mm. The beam has a rectangular cross section with width $b = 60$ mm and height $h = 20$ mm. Determine the maximum bending stress σ_{max} and the maximum deflection δ_{max} for the beam. (Disregard any effects of axial deformations and consider only the effects of bending. Use $E = 200$ GPa.)

Prob. 10.4-26

**10.4-27 A beam BCD having weight q per unit length is placed over three supports, distance L apart (see the first part of the figure). Under these conditions the largest bending moment in the beam occurs at the midpoint C and is equal to $qL^2/8$. In order to reduce the bending moment in the beam, it is decided to lengthen the beam by the addition of overhangs at each end (see the second part of the figure). What should be the length a of each overhang (in terms of the span length L) in order that the largest bending moment in the beam will be as small as possible? Draw the bending-moment diagram for this optimum condition, labeling all critical values.

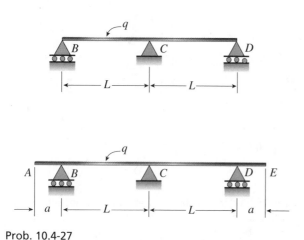

Prob. 10.4-27

CONTINUOUS BEAMS (THREE-MOMENT EQUATION)

The problems for Section 10.5 are to be solved by using the three-moment equation. All beams have constant flexural rigidity EI unless otherwise stated. When drawing shear-force and bending-moment diagrams, be sure to label all critical ordinates, including maximum and minimum values.

10.5-1 A continuous beam with three spans, each of length L, supports a concentrated load P at the midpoint of each span (see figure). (a) Determine all reactions for this beam. (b) Draw the shear-force and bending-moment diagrams, labeling all critical ordinates.

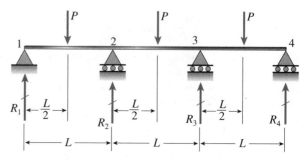

Prob. 10.5-1

10.5-2 The three-span beam shown in the figure supports a uniform load of intensity q acting on both end spans (see figure). All spans have length L. (a) Determine all reactions for this beam. (b) Draw the shear-force and bending-moment diagrams, labeling all critical ordinates.

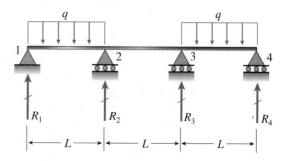

Prob. 10.5-2

10.5-3 A continuous beam with three equal spans (each span has length L) has a bracket attached to each end span (see figure). A concentrated load P acts at the end of each bracket. (a) Determine all reactions for this beam. (b) Draw the shear-force and bending-moment diagrams, labeling all critical ordinates.

Prob. 10.5-3

10.5-4 The two-span continuous beam shown in the figure has an overhang at one end. Partial uniform loads act on the left-hand span, and a concentrated load acts at the end of the overhang. (a) Calculate the reactions R_1, R_2, and R_3 due to these loads. (b) Draw the shear-force and bending-moment diagrams, labeling all critical ordinates.

Note: The presence of an overhang may be handled in the following manner. Remove the overhang from the beam and replace its action on the rest of the beam by a statically equivalent force and moment acting at the end support. Observe that the vertical component of the force is transmitted directly into the reaction at the end support (and therefore does not enter the three-moment equation, although it does affect the reaction itself) and the moment becomes the bending moment at the end support (which is a known quantity). Now analyze the continuous beam using the three-moment equation in the customary manner.

Prob. 10.5-4

10.5-5 A continuous beam with three equal spans has an overhang at each end (see figure). A uniform load acts on the middle span, and concentrated loads act at the ends of the overhangs. (a) Calculate all reactions for this beam. (b) Draw the shear-force and bending-moment diagrams, labeling all critical ordinates. (*Note:* The method for handling an overhang is described in the note to Problem 10.5-4.)

Prob. 10.5-5

10.5-6 A continuous beam with three spans, each of length L, supports a uniform load of intensity q on the end span (see figure). (a) Determine all reactions for this beam. (b) Draw the shear-force and bending-moment diagrams, labeling all critical ordinates.

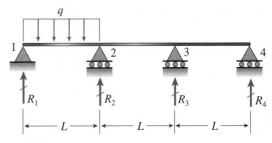

Prob. 10.5-6

10.5-7 A two-span continuous beam has a fixed support at the left-hand end (see figure). A uniform load of intensity q acts on the rightmost span. Both spans have length L. (a) Determine all reactions for this beam. (b) Draw the shear-force and bending-moment diagrams, labeling all critical ordinates.

Prob. 10.5-7

10.5-8 The symmetrical five-span continuous beam shown in the figure has a uniform load of intensity q acting on the center span. Each span has length L. (a) Determine all reactions for this beam. (b) Draw the shear-force and bending-moment diagrams, labeling all critical ordinates.

Prob. 10.5-8

10.5-9 A four-span continuous beam has a uniform load of intensity q throughout its entire length (see figure). Each span has length L. (a) Determine all reactions for this beam. (b) Draw the shear-force and bending-moment diagrams, labeling all critical ordinates.

Prob. 10.5-9

10.5-10 A three-span continuous beam has a uniform load of intensity q acting on two adjoining spans (see figure). Each span has length L. (a) Determine all reactions for this beam. (b) Draw the shear-force and bending-moment diagrams, labeling all critical ordinates.

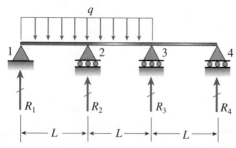

Prob. 10.5-10

10.5-11 Visualize a continuous beam with seven spans of equal length L with the two end spans supporting uniform loads of intensity q. Determine the largest positive bending moment M_{pos} and largest negative bending moment M_{neg} in the beam.

10.5-12 A two-span beam with a fixed support at one end is shown in the figure. The beam is subjected to a uniform load of 7.0 kN/m acting on one span and a concentrated load of 22.4 kN acting on the other span. Calculate the bending moments M_1 and M_2 at the supports.

Prob. 10.5-12

10.5-13 A two-span beam with a fixed support has spans of 8 ft and 6 ft, as shown in the figure. The beam is subjected to a uniform load of 240 lb/ft acting on one span and a concentrated load of 1600 lb acting at the midpoint of the other span. Calculate the bending moments M_1 and M_2 at the supports.

Prob. 10.5-13

10.5-14 A continuous beam on four supports has an overhang at the left-hand end, as shown in the figure. The moment of inertia I_1 for spans 1-2 and 3-4 is 46.6×10^6 mm⁴, and the moment of inertia I_2 for span 2-3 is 43.7×10^6 mm⁴. Determine the bending moments M_1, M_2, and M_3

at the supports of the beam. (*Note:* The method for handling an overhang is described in the note to Problem 10.5-4.)

Prob. 10.5-14

10.5-15 A two-span beam with a fixed support at one end and an overhang at the other (see figure) supports a uniform load of intensity q over its entire length. Determine the bending moments M_1, M_2, and M_3 at supports 1, 2, and 3, respectively. Also, draw the shear-force and bending-moment diagrams for the beam, labeling all critical ordinates. (*Note:* The method for handling an overhang is described in the note to Problem 10.5-4.)

Prob. 10.5-15

TEMPERATURE EFFECTS

10.6-1 A cable CD of length H is attached to the midpoint of a simple beam AB of length L (see figure). The moment of inertia of the beam is I, and the effective cross-sectional area of the cable is A. The cable is initially taut but without any initial tension. Obtain a formula for the tensile force S

Prob. 10.6-1

in the cable when the temperature drops uniformly by ΔT degrees, assuming that the beam and cable are made of the same material (modulus of elasticity E and coefficient of thermal expansion α). (Use the method of superposition in the solution.)

10.6-2 A propped cantilever beam, fixed at the left-hand end A and simply supported at the right-hand end B, is subjected to a temperature differential with temperature T_1 on its upper surface and T_2 on its lower surface (see figure). Find all reactions for this beam. (Use the method of superposition in the solution. Also, if desired, use the results from Problem 9.11-1.)

Probs. 10.6-2 and 10.6-3

10.6-3 Solve the preceding problem by integrating the differential equation of the deflection curve.

10.6-4 A two-span beam with spans of lengths L and $L/2$ is subjected to a temperature differential with temperature T_1 on its upper surface and T_2 on its lower surface (see figure). Determine all reactions for this beam. (Use the method of superposition in the solution. Also, if desired, use the results from Problems 9.8-5 and 9.11-3.)

Probs. 10.6-4 and 10.6-5

10.6-5 Solve the preceding problem by integrating the differential equation of the deflection curve.

LONGITUDINAL DISPLACEMENTS AT THE ENDS OF BEAMS

10.7-1 Assume that the deflected shape of a beam AB with *immovable* pinned supports (see figure) is given by the equation $v = -\delta \sin \pi x/L$, where δ is the deflection at the midpoint of the beam and L is the length. Also, assume that the beam has axial rigidity EA. (a) Obtain formulas for the longitudinal force H at the ends of the beam and the corresponding axial tensile stress σ_t. (b) For an aluminum-alloy beam with $E = 10 \times 10^6$ psi, calculate the tensile stress σ_t when the ratio of the deflection δ to the length L equals 1/200, 1/400, and 1/600.

Prob. 10.7-1

10.7-2 A simple beam AB with length L and height h supports a uniform load of intensity q (see the first part of the figure). (a) Obtain a formula for the curvature shortening λ of this beam. (b) Obtain a formula for the maximum bending stress σ_b in the beam due to the load q. (c) Now assume that the ends of the beam are pinned so that curvature shortening is prevented and a horizontal force H develops at the supports (see the second part of the figure).

Obtain a formula for the corresponding axial tensile stress σ_t. (d) Calculate the curvature shortening λ, the maximum bending stress σ_b, and the tensile stress σ_t for the following steel beam: length $L = 3$ m, height $h = 300$ mm, modulus of elasticity $E = 200$ GPa, and moment of inertia $I = 36 \times 10^6$ mm^4. Also, the load on the beam has intensity $q = 25$ kN/m. Compare the tensile stress σ_t produced by the force H with the maximum bending stress σ_b produced by the uniform load.

Prob. 10.7-2

COLUMNS

INTRODUCTION

Load-carrying structures may fail in a variety of ways, depending upon the type of structure, the conditions of support, the kinds of loads, and the materials used. For instance, an axle in a vehicle may fracture suddenly from repeated cycles of loading, or a tension member may stretch excessively, so that the structure is unable to perform its intended functions. These kinds of failures are prevented by designing structures so that the maximum stresses and maximum displacements remain within tolerable limits. Thus, **strength** and **stiffness** are important factors in design, as discussed throughout the preceding chapters.

Another type of failure is **buckling**, which is the subject matter of this chapter. We will consider specifically the buckling of **columns**, which are long, slender structural members loaded axially in compression (Fig. 11-1a). If a compression member is relatively slender, it may fail by bending or deflecting laterally (Fig. 11-1b) rather than by direct compression of the material. You can demonstrate this behavior by compressing a plastic ruler or other slender object. When lateral bending occurs, we say that the column has *buckled*. Under an increasing axial load, the lateral deflections will increase too, and eventually the column will collapse completely.

The phenomenon of buckling is not limited to columns. Buckling can occur in many kinds of structures and can take many forms. When you step on the top of an empty aluminum can, the thin cylindrical walls buckle under your weight and the can collapses. When a large bridge collapsed a few years ago, investigators found that failure was caused by the buckling of a thin steel plate that wrinkled under compressive

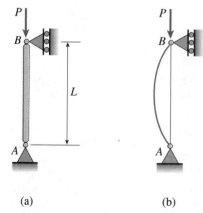

Fig. 11-1 Buckling of a column due to an axial compressive load P

stresses. Buckling is one of the major causes of failures in structures, and therefore the possibility of buckling should always be considered in design.

BUCKLING AND STABILITY

To illustrate the fundamental concepts of buckling and stability, we will analyze the *idealized structure*, or *buckling model*, shown in Fig. 11-2a. This hypothetical structure consists of two rigid bars AB and BC, each of length $L/2$. They are joined at B by a pin connection and held in a vertical position by a rotational spring having stiffness β_R.* This structure is analogous to the column of Fig. 11-1a, because both structures have simple supports at the ends and are compressed by an axial load P. However, the elasticity of the idealized structure is "concentrated" in the rotational spring, whereas a real column can bend throughout its length (Fig. 11-1b).

In the idealized structure, the two bars are perfectly aligned and the axial load P has its line of action along the longitudinal axis (Fig. 11-2a). Consequently, the spring is unstressed and the bars are in direct compres-

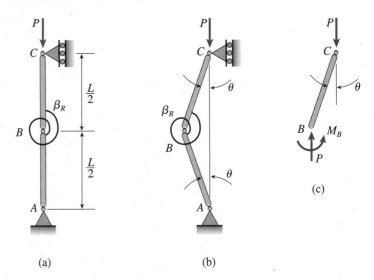

Fig. 11-2 Buckling of an idealized structure consisting of two rigid bars and a rotational spring

(a)   (b)

*The general relationship for a rotational spring is $M = \beta_R \theta$, where M is the moment acting on the spring, β_R is the rotational stiffness of the spring, and θ is the angle through which the spring rotates. Thus, rotational stiffness has units of moment divided by angle, such as lb-in./rad or N·m/rad. The analogous relationship for a translational spring is $F = \beta\delta$, where F is the force acting on the spring, β is the translational stiffness of the spring (or spring constant), and δ is the change in length of the spring. Thus, translational stiffness has units of force divided by length, such as lb/in. or N/m.

sion. Now suppose that the structure is disturbed by some external force that causes point B to move a small distance laterally (Fig. 11-2b). The rigid bars rotate through small angles θ and a moment develops in the spring. The direction of this moment is such that it tends to return the structure to its original straight position, and therefore it is called a *restoring moment*. At the same time, however, the tendency of the axial compressive force is to increase the lateral displacement. Thus, these two actions have opposite effects—the restoring moment tends to *decrease* the displacement and the axial force tends to *increase* it.

Now consider what happens when the disturbing force is removed. If the axial force P is relatively small, the action of the restoring moment will predominate over the action of the axial force and the structure will return to its initial straight position. Under these conditions, the structure is said to be **stable**. However, if the axial force P is large, the lateral displacement of point B will increase and the bars will rotate through larger and larger angles until the structure collapses. Under these conditions, the structure is **unstable** and fails by lateral buckling.

■ Critical Load

The transition between the stable and unstable conditions occurs at a special value of the axial force known as the **critical load** (denoted by the symbol P_{cr}). We can determine the critical load by considering the structure in the disturbed position (Fig. 11-2b) and investigating its equilibrium. First, we consider the entire structure as a free body and sum moments about support A. This step leads to the conclusion that there is no horizontal reaction at support C. Second, we consider bar BC as a free body (Fig. 11-2c) and note that it is subjected to the action of the axial forces P and the moment M_B in the spring. The moment M_B is equal to the rotational stiffness β_R times the angle of rotation 2θ of the spring; thus,

$$M_B = 2\beta_R\theta \tag{a}$$

Since the angle θ is a small quantity, the lateral displacement of point B is $\theta L/2$. Therefore, we obtain the following equation of equilibrium by summing moments about point B (Fig. 11-2c):

$$M_B - P\left(\frac{\theta L}{2}\right) = 0 \tag{b}$$

or, upon substituting from Eq. (a),

$$\left(2\beta_R - \frac{PL}{2}\right)\theta = 0 \tag{11-1}$$

One solution of this equation is $\theta = 0$, which means that the structure is in equilibrium when it is perfectly straight, regardless of the magnitude of the force P. A second solution is obtained by setting the term in parentheses equal to zero and solving for the load P, which is the critical load:

$$P_{cr} = \frac{4\beta_R}{L} \qquad (11\text{-}2)$$

At this value of the load the structure is in equilibrium regardless of the magnitude of the angle θ (provided the angle remains small, because we made that assumption when deriving Eq. b).

From the preceding analysis we see that the critical load is the *only* load for which the structure will be in equilibrium in the disturbed position. At this value of the load, the restoring effect of the moment in the spring just matches the buckling effect of the axial load. Therefore, the critical load represents the boundary between the stable and unstable conditions. If the axial load is less than P_{cr}, the effect of the moment in the spring predominates and the structure returns to the vertical position after a slight disturbance; if the axial load is larger than P_{cr}, the effect of the axial force predominates and the structure buckles:

If $P < P_{cr}$, the structure is stable

If $P > P_{cr}$, the structure is unstable

From Eq. (11-2) we see that the stability of the structure is increased either by *increasing its stiffness* or by *decreasing its length*. Later in this chapter, when we determine critical loads for various types of columns, we will see that these same observations apply.

■ **Summary**

Let us now summarize the behavior of the idealized structure (Fig. 11-2a) as the axial load P increases from zero to a large value.

When the axial load is less than the critical load ($0 < P < P_{cr}$), the structure is in equilibrium when it is perfectly straight. Because the equilibrium is *stable*, the structure returns to its initial position after being disturbed. Thus, the structure is in equilibrium *only* when it is perfectly straight ($\theta = 0$).

When the axial load is greater than the critical load ($P > P_{cr}$), the structure is still in equilibrium when $\theta = 0$ (because it is in direct compression and there is no moment in the spring), but the equilibrium is *unstable* and cannot be maintained. The slightest disturbance will cause the structure to buckle.

At the critical load ($P = P_{cr}$), the structure is in equilibrium even when point B is displaced laterally by a small amount. In other words, the structure is in equilibrium for *any* small angle θ, including $\theta = 0$. However, the structure is neither stable nor unstable—it is at the boundary between stability and instability. This condition is referred to as **neutral equilibrium**.

The equilibrium conditions for the idealized structure are shown in

Fig. 11-3 Equilibrium diagram for buckling of an idealized structure

the graph of axial load P versus angle of rotation θ (Fig. 11-3). The two heavy lines, one vertical and one horizontal, represent the equilibrium conditions. Point B, where the equilibrium diagram branches, is called a *bifurcation point*. The horizontal line for neutral equilibrium extends to the left and right of the vertical axis because the angle θ may be clockwise or counterclockwise. The line extends only a short distance, however, because our analysis is based upon the assumption that θ is a small angle. (This assumption is quite valid, because θ is indeed small when the structure first departs from its vertical position. If buckling continues and θ becomes large, the line labeled "Neutral equilibrium" curves upward, as explained later in Section 11.3.)

The three equilibrium conditions represented by the diagram of Fig. 11-3 are analogous to those of a ball placed upon a smooth surface (Fig. 11-4). If the surface is concave upward, like the inside of a dish, the equilibrium is stable and the ball always returns to the low point when disturbed. If the surface is convex upward, like a dome, the ball can theoretically be in equilibrium on top of the surface, but the equilibrium is unstable and in reality the ball rolls away. If the surface is perfectly flat, the ball is in neutral equilibrium and remains wherever it is placed.

Fig. 11-4 Ball in stable, unstable, and neutral equilibrium

As we will see in the next section, the behavior of an ideal elastic column is analogous to that of the buckling model shown in Fig. 11-2. Furthermore, many other kinds of structural and mechanical systems fit this model. (For a more detailed discussion of buckling models, see Ref. 11-1.)

11.3 | COLUMNS WITH PINNED ENDS

We begin our investigation of the stability behavior of columns by analyzing a slender column with pinned ends (Fig. 11-5a). The column is loaded by a vertical force P that is applied through the centroid of the end cross section. The column itself is perfectly straight and is made of a linearly elastic material that follows Hooke's law. Since the column is assumed to have no imperfections, it is referred to as an **ideal column**.

For purposes of analysis, we construct a coordinate system with its origin at support A and with the x axis along the longitudinal axis of the column. The y axis is directed to the left in the figure, and the z axis (not shown) comes out of the plane of the figure toward the viewer. We assume that the xy plane is a plane of symmetry of the column and that

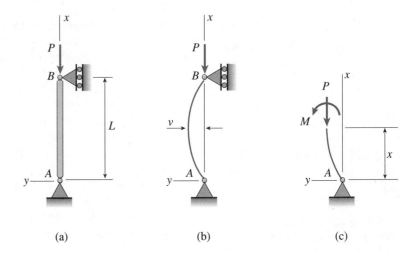

Fig. 11-5 Column with pinned ends:
(a) ideal column, (b) buckled shape, and
(c) axial force P and bending moment M
acting at a cross section

(a)   (b)   (c)

any bending takes place in that plane (Fig. 11-5b). The coordinate system is identical to the one used in our previous discussions of beams, as can be seen by rotating the column clockwise through an angle of 90°.

When the axial load P has a small value, the column remains perfectly straight and undergoes direct axial compression. The only stresses are the uniform compressive stresses obtained from the equation $\sigma = P/A$. The column is in **stable equilibrium**, which means that it returns to the straight position after a disturbance. For instance, if we apply a small lateral load and cause the column to bend, the deflection will disappear and the column will return to its original position when the lateral load is removed.

As the axial load P is gradually increased, we reach a condition of **neutral equilibrium** in which the column may have a bent shape. The corresponding value of the load is the **critical load** P_{cr}. At this load the column may undergo small lateral deflections with no change in the axial force. For instance, a small lateral load will produce a bent shape that does not disappear when the lateral load is removed. Thus, the critical load can maintain the column in equilibrium *either* in the straight position or in a slightly bent position.

At higher values of the load, the column is **unstable** and may collapse by buckling, that is, by excessive bending. For the ideal case that we are discussing, the column will be in equilibrium in the straight position even when the axial force P is greater than the critical load. However, since the equilibrium is unstable, the smallest imaginable disturbance will cause the column to deflect sideways. Once that happens, the deflections will immediately increase and the column will fail by buckling. The behavior is similar to that described in the preceding section for the idealized buckling model (Fig. 11-2).

The behavior of an ideal column compressed by an axial load P (Figs. 11-5a and b) may be summarized as follows:

If $P < P_{cr}$, the column is in stable equilibrium in the straight position.

If $P = P_{cr}$, the column is in neutral equilibrium in either the straight or a slightly bent position.

If $P > P_{cr}$, the column is in unstable equilibrium in the straight position and will buckle under the slightest disturbance.

Of course, a real column does not behave in this idealized manner because imperfections are always present. For instance, the column is not *perfectly* straight, and the load is not *exactly* at the centroid. Nevertheless, we begin by studying ideal columns because they provide insight into the behavior of real columns.

▪ Differential Equation for Column Buckling

To determine the critical loads and corresponding deflected shapes for an ideal pin-ended column (Fig. 11-5a), we use one of the differential equations of the deflection curve of a beam (see Eqs. 9-12a, b, and c in Section 9.2). These equations are applicable to a buckled column because the column bends as though it were a beam (Fig. 11-5b). Although both the fourth-order differential equation (the load equation) and the third-order differential equation (the shear-force equation) are suitable for analyzing columns, we will elect to use the second-order equation (the bending-moment equation) because its general solution is usually the simplest. The bending-moment equation (Eq. 9-12a) is

$$EIv'' = M \tag{11-3}$$

in which M is the bending moment at any cross section, v is the lateral deflection in the y direction, and EI is the flexural rigidity for bending in the xy plane.

The bending moment M at distance x from end A of the buckled column is shown acting in its positive direction in Fig. 11-5c. Note that the bending moment sign convention is the same as that used in earlier chapters, namely, positive bending moment produces positive curvature (see Figs. 9-3 and 9-4.) The axial force P acting at the cross section is also shown in Fig. 11-5c. Since there are no horizontal forces acting at the supports, there are no shear forces in the column. Therefore, from equilibrium of moments about point A, we obtain

$$M + Pv = 0 \quad \text{or} \quad M = -Pv \tag{11-4}$$

where v is the deflection at the cross section.

This same expression for the bending moment is obtained if we assume that the column buckles to the right instead of to the left (Fig. 11-6a). When the column deflects to the right, the deflection itself is $-v$

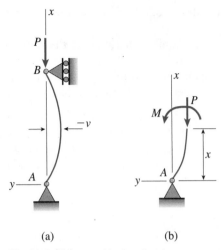

(a)   (b)

Fig. 11-6 Column with pinned ends
(alternative direction of buckling)

but the moment of the axial force about point A also changes sign. Thus, the equilibrium equation for moments about point A (see Fig. 11-6b) is

$$M - P(-v) = 0$$

which gives the same expression for the bending moment M as before.

The differential equation of the deflection curve (Eq. 11-3) now becomes

$$EIv'' + Pv = 0 \tag{11-5}$$

By solving this equation, which is a *homogeneous, linear, differential equation of second order with constant coefficients*, we can determine the magnitude of the critical load and the deflected shape of the buckled column.

Note that we are analyzing the buckling of columns by solving the same basic differential equation as the one we solved in Chapters 9 and 10 when finding beam deflections. However, there is a fundamental difference in the two types of analysis. In the case of beam deflections, the bending moment M appearing in Eq. (11-3) is a function of the loads only—it does not depend upon the deflections of the beam. In the case of buckling, the bending moment is a function of the deflections themselves (Eq. 11-4). Thus, we now encounter a new aspect of bending analysis. In our previous work, the deflected shape of the structure was not considered, and the equations of equilibrium were based upon the geometry of the *undeformed* structure. Now, however, the geometry of the *deformed* structure is taken into account when writing equations of equilibrium.

■ Solution of the Differential Equation

For convenience in writing the general solution of the differential equation (Eq. 11-5), we introduce the notation

$$k^2 = \frac{P}{EI} \quad \text{or} \quad k = \sqrt{\frac{P}{EI}} \tag{11-6a,b}$$

in which k is always taken as a positive quantity. Note that k has units of the reciprocal of length, and therefore quantities such as kx and kL are nondimensional.

Using this notation, we can rewrite Eq. (11-5) in the form

$$v'' + k^2 v = 0 \tag{11-7}$$

From mathematics we know that the solution of this equation is

$$v = C_1 \sin kx + C_2 \cos kx \tag{11-8}$$

in which C_1 and C_2 are constants of integration (to be evaluated from the boundary conditions, or end conditions, of the column). Note that the number of arbitrary constants in the solution (two in this case) agrees with the order of the differential equation. Also, note that we can verify the solution by substituting the expression for v (Eq. 11-8) into the differential equation (Eq. 11-7) and reducing it to an identity.

To evaluate the constants of integration appearing in the solution, we use the boundary conditions at the ends of the column; namely, the deflection is zero when $x = 0$ and $x = L$ (see Fig. 11-5b):

$$v(0) = 0 \quad \text{and} \quad v(L) = 0 \tag{a,b}$$

The first condition gives $C_2 = 0$, and therefore

$$v = C_1 \sin kx \tag{c}$$

The second condition gives

$$C_1 \sin kL = 0 \tag{d}$$

From this equation we conclude that either $C_1 = 0$ or $\sin kL = 0$. We will consider both of these possibilities.

Case 1. If the constant C_1 equals zero, the deflection v is also zero (see Eq. c), and therefore the column remains straight. In addition, we see that when C_1 equals zero, Eq. (d) is satisfied for *any* value of the quantity kL. Consequently, the axial load P may also have any value (see Eq. 11-6b). This solution of the differential equation (known in mathematics as the *trivial solution*) is represented by the vertical axis of the load-deflection diagram (Fig. 11-7). It gives the behavior of an ideal column that is in equilibrium (either stable or unstable) in the straight position (no deflection) under the action of the compressive load P.

Case 2. The second possibility for satisfying Eq. (d) is given by the following equation, known as the **buckling equation**:

$$\sin kL = 0 \tag{11-9}$$

This equation is satisfied when $kL = 0, \pi, 2\pi, \ldots$. However, since $kL = 0$ means that $P = 0$, this solution is not of interest. Therefore, the solutions we will consider are:

$$kL = n\pi \quad n = 1, 2, 3, \ldots \tag{e}$$

or (see Eq. 11-6a):

$$P = \frac{n^2 \pi^2 EI}{L^2} \quad n = 1, 2, 3, \ldots \tag{11-10}$$

This formula gives the values of P that satisfy the buckling equation and provide solutions (other than the trivial solution) to the differential equation.

The equation of the deflection curve (from Eqs. c and e) is

$$v = C_1 \sin kx = C_1 \sin \frac{n\pi x}{L} \quad n = 1, 2, 3, \ldots \tag{11-11}$$

Fig. 11-7 Load-deflection diagram for an ideal, linearly elastic column

Only when P has one of the values given by Eq. (11-10) is it theoretically possible for the column to have a bent shape (given by Eq. 11-11). For all other values of P, the column is in equilibrium only if it remains straight. Therefore, the values of P given by Eq. (11-10) are the **critical loads** for this column.

■ Critical Loads

The lowest critical load for a column with pinned ends (Fig. 11-8a) is obtained when $n = 1$:

$$P_{cr} = \frac{\pi^2 EI}{L^2} \qquad (11\text{-}12)$$

The corresponding buckled shape (sometimes called a *mode shape*) is

$$v = C_1 \sin \frac{\pi x}{L} \qquad (11\text{-}13)$$

as shown in Fig. 11-8b. The constant C_1 represents the deflection at the midpoint of the column and may have any small value, either positive or negative. Therefore, the part of the load-deflection diagram corresponding to P_{cr} is a horizontal straight line (Fig. 11-7). Thus, the deflection at the critical load is *undefined*, although it must remain small for our equations to be valid. Above the bifurcation point B the equilibrium is unstable, and below point B it is stable. Buckling of a pinned-end column in the first mode is called the **fundamental case** of column buckling.

Fig. 11-8 Buckled shapes for an ideal column with pinned ends: (a) initially straight column, (b) buckled shape for $n = 1$, and (c) buckled shape for $n = 2$

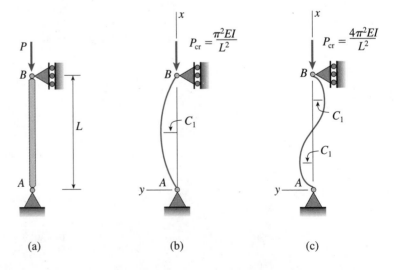

The type of buckling described in this section is called **Euler buckling**, and the critical load for an ideal elastic column is often called the **Euler load**. The famous mathematician Leonhard Euler (1707–1783), generally recognized as the greatest mathematician of all time, was the first person to investigate the buckling of a slender column and determine its critical load (Euler published his results in 1744); see Ref. 11-2.

By taking higher values of the index n in Eqs. (11-10) and (11-11), we obtain an infinite number of critical loads and corresponding mode shapes. The mode shape for $n = 2$ has two half-waves, as pictured in Fig. 11-8c. The corresponding critical load is four times larger than the critical load for the fundamental case. The magnitudes of the critical loads are proportional to the square of n, and the number of half-waves in the buckled shape is equal to n. Buckled shapes for the higher modes are often of no practical interest because the column buckles when the axial load P reaches its lowest critical value. The only way to obtain modes of buckling higher than the first is to provide lateral support of the column at intermediate points, such as at the midpoint of the column shown in Fig. 11-8 (see Example 11-1 at the end of this section).

From Eq. (11-12) we see that the critical load of a column is proportional to the flexural rigidity EI and inversely proportional to the square of the length. Of particular interest is the fact that the strength of the material itself, as represented by a quantity such as the proportional limit or the yield stress, does not appear in the equation for the critical load. Therefore, increasing a strength property does not raise the critical load of a slender column. It can only be raised by increasing the flexural rigidity, reducing the length, or providing additional lateral support.

The flexural rigidity can be increased by using a "stiffer" material (that is, a material with larger modulus of elasticity E) or by distributing the material in such a way as to increase the moment of inertia I of the cross section, just as a beam can be made stiffer by increasing the moment of inertia. The moment of inertia is increased by distributing the material farther from the centroid of the cross section. Hence, a hollow tubular member is generally more economical for use as a column than a solid member having the same cross-sectional area. Reducing the wall thickness of a tubular member and increasing its lateral dimensions (while keeping the cross-sectional area constant) also increases the critical load because the moment of inertia is increased. This process has a practical limit, however, because eventually the wall itself will become unstable. When that happens, localized buckling occurs in the form of small corrugations or wrinkles in the walls of the column. Thus, we must distinguish between *overall buckling* of a column, which is illustrated in Fig. 11-8, and *local buckling* of its parts. The latter requires more detailed investigations, such as those found in books on buckling and stability (Refs. 11-5, 11-6, 11-7, and current textbooks). In this chapter we are considering only overall buckling of columns.

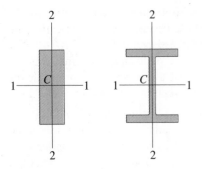

Fig. 11-9 Cross sections of columns showing principal centroidal axes with $I_1 > I_2$

In the preceding analysis we assumed that the xy plane was a plane of symmetry of the column and that buckling took place in that plane (Fig. 11-8). The latter assumption will be met if the column has lateral supports perpendicular to the plane of the figure, so that the column is constrained to buckle in the xy plane. If the column is supported only at its ends and is free to buckle in *any* direction, then bending will occur about the principal centroidal axis having the smaller moment of inertia. For instance, consider the rectangular and wide-flange cross sections shown in Fig. 11-9. In each case, the moment of inertia I_1 is greater than the moment of inertia I_2; hence the column will buckle in the 1-1 plane, and the smaller moment of inertia I_2 should be used in the formula for the critical load. If the cross section is square or circular, all centroidal axes have the same moment of inertia and buckling may occur in any longitudinal plane.

■ Critical Stress

After finding the critical load for a column, we can calculate the corresponding **critical stress** by dividing the load by the cross-sectional area. For the fundamental case of buckling (Fig. 11-8b), the critical stress is

$$\sigma_{cr} = \frac{P_{cr}}{A} = \frac{\pi^2 EI}{AL^2} \tag{11-14}$$

in which I is the moment of inertia for the principal axis about which buckling occurs. This equation can be written in a more useful form by introducing the notation

$$r = \sqrt{\frac{I}{A}} \tag{11-15}$$

in which r is the **radius of gyration** of the cross section in the plane of bending.* Then the equation for the critical stress becomes

$$\sigma_{cr} = \frac{\pi^2 E}{(L/r)^2} \tag{11-16}$$

in which L/r is a nondimensional ratio called the **slenderness ratio**:

$$\text{Slenderness ratio} = \frac{L}{r} \tag{11-17}$$

---

*Radius of gyration is described in Section 12.4.

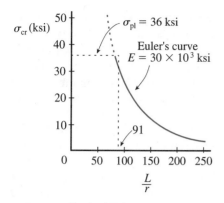

Fig. 11-10 Graph of Euler's curve (from Eq. 11-16) for structural steel with $E = 30 \times 10^3$ ksi and $\sigma_{pl} = 36$ ksi

Note that the slenderness ratio depends only on the dimensions of the column. A column that is long and slender will have a high slenderness ratio and therefore a low critical stress. A column that is short and stubby will have a low slenderness ratio and will buckle at a high stress. Typical values of slenderness ratio for actual columns are between 30 and 150.

The critical stress is the average compressive stress on the cross section at the instant the load reaches its critical value. We can a plot a graph of this stress as a function of the slenderness ratio and obtain a curve known as **Euler's curve** (Fig. 11-10). The curve shown in the figure is plotted for a structural steel with $E = 30 \times 10^3$ ksi. The curve is valid only when the critical stress is less than the proportional limit of the steel, because the equations were derived using Hooke's law. Therefore, we draw a horizontal line on the graph at the proportional limit of the steel (assumed to be 36 ksi) and terminate Euler's curve at that level of stress.*

■ Effects of Large Deflections, Imperfections, and Inelastic Behavior

The equations for critical loads were derived for ideal columns, that is, columns for which the loads are precisely applied, the construction is perfect, and the material follows Hooke's law. As a consequence, we found that the magnitudes of the small deflections at buckling were undefined.** Thus, when $P = P_{cr}$, the column may have any small deflection, a condition represented by the horizontal line labeled A in the load-deflection diagram of Fig. 11-11. (In this figure, we show only the right-hand half of the diagram, but the two halves are symmetric about the vertical axis.) The theory for ideal columns is limited to small deflections because we used the second derivative v'' for the curvature. A more exact analysis, based upon the exact expression for curvature (Eq. 9-13 in Section 9.2), shows that there is no indefiniteness in the magnitudes of the deflections at buckling. Instead, for an ideal, linearly elastic column, the load-deflection diagram goes upward in accord with curve B of Fig. 11-11. Thus, after a linearly elastic column begins to buckle, an increasing load is required to cause an increase in the deflections.

Now suppose that the column is not constructed perfectly; for instance, the column might have an imperfection in the form of a small initial curvature, so that the unloaded column is not perfectly straight. Such imperfections produce deflections from the onset of loading, as shown by curve C in Fig. 11-11. For small deflections, curve C approaches line A as an asymptote. However, as the deflections become

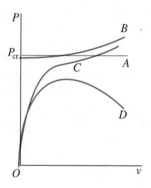

Fig. 11-11 Load-deflection diagram for columns: Line A, ideal elastic column with small deflections; Curve B, ideal elastic column with large deflections; Curve C, elastic column with imperfections; and Curve D, inelastic column with imperfections

---

*Euler's curve is not a common geometric shape. It is sometimes mistakenly called a hyperbola, but hyperbolas are plots of polynomial equations of the second degree in two variables, whereas Euler's curve is a plot of an equation of the third degree in two variables.

**In mathematical terminology, we solved a *linear eigenvalue problem*. The critical load is an *eigenvalue* and the corresponding buckled mode shape is an *eigenfunction*.

large, it approaches curve *B*. The larger the imperfections, the further curve *C* moves to the right, away from the vertical line. Conversely, if the column is constructed with considerable accuracy, curve *C* approaches the vertical axis and the horizontal line labeled *A*. By comparing lines *A*, *B*, and *C*, we see that for practical purposes the critical load represents the maximum load-carrying capacity of an elastic column, because large deflections are not acceptable in most applications.

Finally, consider what happens when the stresses exceed the proportional limit and the material no longer follows Hooke's law. Of course, the load-deflection diagram is unchanged up to the level of load at which the proportional limit is reached. Then the curve for inelastic behavior (curve *D*) departs from the elastic curve, continues upward, reaches a maximum, and turns downward.

The precise shapes of the curves in Fig. 11-11 depend upon the material properties and column dimensions, but the general nature of the behavior is typified by the curves shown.

Only extremely slender columns remain elastic up to the critical load. Stockier columns behave inelastically and follow a curve such as *D*. Thus, the maximum load that can be supported by an inelastic column may be considerably less than the Euler load for that same column. Furthermore, the descending part of curve *D* represents sudden and catastrophic collapse, because it takes smaller and smaller loads to maintain larger and larger deflections. By contrast, the curves for elastic columns are quite stable, because they continue upward as the deflections increase, and therefore it takes larger and larger loads to cause an increase in deflection. (Inelastic buckling is described in more detail in Sections 11.7 and 11.8.)

■ Optimum Shapes of Columns

Compression members usually have the same cross sections throughout their lengths, and therefore only prismatic columns are analyzed in this chapter. However, prismatic columns are not the optimum shape if minimum weight is desired. The critical load of a column consisting of a given amount of material may be increased by varying the shape so that the column has larger cross sections in those regions where the bending moments are larger. Consider, for instance, a column of solid circular cross section with pinned ends. A column shaped as shown in Fig. 11-12a will have a larger critical load than a prismatic column made from the same volume of material. As a means of approximating this optimum shape, prismatic columns are sometimes reinforced over part of their lengths (Fig. 11-12b).

Now consider a prismatic column with pinned ends that is free to buckle in *any* lateral direction (Fig. 11-13a). Also, assume that the column has a solid cross section, such as a circle, square, triangle, rectangle, or hexagon (Fig. 11-13b). An interesting question arises: For a given cross-sectional area, which of these shapes makes the most effi-

(a)   (b)

Fig. 11-12 Nonprismatic columns

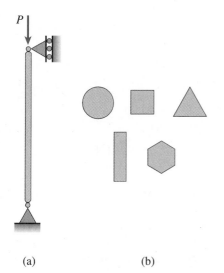

(a)   (b)

Fig. 11-13 Which cross-sectional shape is the optimum shape for a prismatic column?

cient column? Or, in more precise terms, which cross section gives the largest critical load? Of course, we are assuming that the critical load is calculated from the Euler formula $P_{cr} = \pi^2 EI/L^2$ using the smallest moment of inertia for the cross section.

While a common answer to this question is "the circular shape," you can readily demonstrate that a cross section in the shape of an equilateral triangle gives a 21% higher critical load than does a circular cross section of the same area (see Problem 11.3-10). The critical load for an equilateral triangle is also higher than the loads obtained for the other shapes; hence, an equilateral triangle is the optimum cross section (based only upon theoretical considerations). For a mathematical analysis of optimum column shapes, including columns with varying cross sections, see Ref. 11-8.

WORK SHEET 12 Mathcad

Example 11-1

A long, slender column *ABC* is pin-supported at the ends and compressed by an axial load *P* (Fig. 11-14). Lateral support is provided at the midpoint *B* in the plane of the figure. However, lateral support perpendicular to the plane of the figure is provided only at the ends. The column is constructed of a steel wide-flange section (W 8 × 28) having modulus of elasticity $E = 29 \times 10^3$ ksi and proportional limit $\sigma_{pl} = 42$ ksi. The total length of the column is $L = 25$ ft. Determine the allowable load P_{allow} using a factor of safety $n = 2.5$ with respect to Euler buckling of the column.

Solution   Because of the manner in which it is supported, this column may buckle in either of the two principal planes of bending. First, it may buckle in the plane of the figure, in which case the distance between lateral supports is $L/2 = 12.5$ ft and bending occurs about axis 2-2 (see Fig. 11-8c for the mode shape of buckling). Second, it may buckle perpendicular to the plane of the figure with bending about axis 1-1. Because the only lateral support in this direction is at the ends, the distance between lateral supports is $L = 25$ ft (see Fig. 11-8b for the mode shape of buckling).

Column properties. From Table E-1, Appendix E, we obtain the following moments of inertia and cross-sectional area for a W 8 × 28 column:

$$I_1 = 98.0 \text{ in.}^4 \qquad I_2 = 21.7 \text{ in.}^4 \qquad A = 8.25 \text{ in.}^2$$

Critical loads. If the column buckles in the plane of the figure, the critical load is

$$P_{cr} = \frac{\pi^2 EI_2}{(L/2)^2} = \frac{4\pi^2 EI_2}{L^2}$$

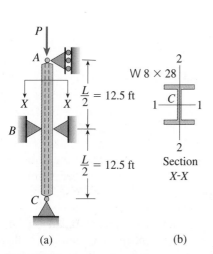

(a)   (b)

Fig. 11-14 Example 11-1. Euler buckling of a slender column

Substituting numerical values, we obtain

$$P_{cr} = \frac{4\pi^2 EI_2}{L^2} = \frac{4\pi^2(29 \times 10^3 \text{ ksi})(21.7 \text{ in.}^4)}{[(25 \text{ ft})(12 \text{ in./ft})]^2} = 276 \text{ k}$$

If the column buckles perpendicular to the plane of the figure, the critical load is

$$P_{cr} = \frac{\pi^2 EI_1}{L^2} = \frac{\pi^2(29 \times 10^3 \text{ ksi})(98.0 \text{ in.}^4)}{[(25 \text{ ft}))(12 \text{ in./ft})]^2} = 312 \text{ k}$$

Therefore, the critical load for the column (the smaller of the two preceding values) is:

$$P_{cr} = 276 \text{ k}$$

and buckling occurs in the plane of the figure.

Critical stresses. Since the calculations for the critical loads are valid only if the material follows Hooke's law, we need to verify that the critical stresses do not exceed the proportional limit of the material. In the case of the larger critical load, we obtain the following critical stress:

$$\sigma_{cr} = \frac{P_{cr}}{A} = \frac{312 \text{ k}}{8.25 \text{ in.}^2} = 37.8 \text{ ksi}$$

Since this stress is less than the proportional limit ($\sigma_{pl} = 42$ ksi), both critical-load calculations are satisfactory.

Allowable load. The allowable axial load for the column, based on Euler buckling, is

$$P_{\text{allow}} = \frac{P_{cr}}{n} = \frac{276 \text{ k}}{2.5} = 110 \text{ k} \qquad \longleftarrow$$

in which $n = 2.5$ is the desired factor of safety.

11.4 COLUMNS WITH OTHER SUPPORT CONDITIONS

Buckling of a column with pinned ends (described in the preceding section) is usually considered as the most basic case of buckling. However, in practice we encounter many other end conditions, such as fixed ends, free ends, and elastic supports. The critical loads for columns with various kinds of support conditions can be determined from the differential equation of the deflection curve by following the same procedure that we used when analyzing a pinned-end column.

The procedure is as follows. First, with the column assumed to be in the buckled state, we obtain an expression for the bending moment in the column. Second, we set up the differential equation of the deflection curve, using the bending-moment equation ($EIv'' = M$). Third, we solve the equation and obtain its general solution, which contains two constants of integration plus other unknown quantities. Fourth, we apply boundary conditions pertaining to the deflection v and the slope v' and obtain a set of simultaneous equations. Finally, we solve those equations to obtain the critical load and the deflected shape of the buckled column.

This straightforward mathematical procedure is illustrated in the following discussion of three different types of columns.

■ Column Fixed at the Base and Free at the Top

The first case we will consider is an ideal column that is fixed at the base, free at the top, and subjected to an axial load P (Fig. 11-15a).* The deflected shape of the buckled column is shown in Fig. 11-15b. From this figure we see that the bending moment at distance x from the base is

$$M = P(\delta - v) \tag{11-18}$$

where δ is the deflection at the free end of the column. The differential equation of the deflection curve then becomes

$$EIv'' = M = P(\delta - v) \tag{11-19}$$

in which I is the moment of inertia for buckling in the xy plane.

Using the notation $k^2 = P/EI$ (Eq. 11-6a), we can rearrange Eq. (11-19) into the form

$$v'' + k^2v = k^2\delta \tag{11-20}$$

which is a linear differential equation of second order with constant coefficients. However, it is a more complicated equation than the equation for a column with pinned ends (see Eq. 11-7) because it has a nonzero term on the right-hand side. The general solution of Eq. (11-20) consists of

Fig. 11-15 Ideal column fixed at the base and free at the top: (a) initially straight column, (b) buckled shape for $n = 1$, (c) buckled shape for $n = 3$, and (d) buckled shape for $n = 5$

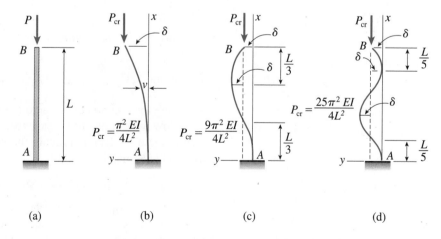

(a)   (b)   (c)   (d)

---

*This particular column is of special interest because it is the one first analyzed by Euler in 1744.

two parts: (1) the *homogeneous solution*, which is the solution of the homogeneous equation obtained by replacing the right-hand side with zero, and (2) the *particular solution*, which is the solution of Eq. (11-20) that produces the term on the right-hand side.

The homogeneous solution (also called the *complementary solution*) is the same as the solution of Eq. (11-7); hence

$$v_H = C_1 \sin kx + C_2 \cos kx \tag{a}$$

where C_1 and C_2 are constants of integration. Note that when v_H is substituted into the left-hand side of the differential equation (Eq. 11-20), it produces zero.

The particular solution of the differential equation is

$$v_P = \delta \tag{b}$$

When v_P is substituted into the left-hand side of the differential equation, it produces the right-hand side, that is, it produces the term $k^2\delta$. Consequently, the general solution of the equation, equal to the sum of v_H and v_P, is

$$v = C_1 \sin kx + C_2 \cos kx + \delta \tag{11-21}$$

This equation contains three unknown quantities (C_1, C_2, and δ), and therefore three boundary conditions are needed to complete the solution.

At the base of the column, the deflection and slope are each equal to zero. Therefore, we obtain the following boundary conditions:

$$v(0) = 0 \qquad v'(0) = 0$$

Applying the first condition to Eq. (11-21), we find

$$C_2 = -\delta \tag{c}$$

To apply the second condition, we first differentiate Eq. (11-21) to obtain the slope:

$$v' = C_1 k \cos kx - C_2 k \sin kx \tag{d}$$

Applying the second condition to this equation, we find $C_1 = 0$.

Now we can substitute the expressions for C_1 and C_2 into the general solution (Eq. 11-21) and obtain the equation of the deflection curve for the buckled column:

$$v = \delta(1 - \cos kx) \tag{11-22}$$

Note that this equation gives only the *shape* of the deflection curve—the amplitude δ remains undefined. Thus, when the column buckles, the deflection given by Eq. (11-22) may have any arbitrary magnitude, except that it must remain small (because the differential equation is based upon small deflections).

The third boundary condition applies to the upper end of the column, where the deflection v is equal to δ:

$$v(L) = \delta$$

Using this condition with Eq. (11-22), we get

$$\delta \cos kL = 0 \tag{11-23}$$

From this equation we conclude that either $\delta = 0$ or $\cos kL = 0$. If $\delta = 0$, there is no deflection of the bar (see Eq. 11-22) and we have the trivial solution—the column remains straight and buckling does not occur. In that case, Eq. (11-23) will be satisfied for any value of the quantity kL, that is, for any value of the load P. This conclusion is represented by the vertical line in the load-deflection diagram of Fig. 11-7.

The other possibility for solving Eq. (11-23) is

$$\cos kL = 0 \tag{11-24}$$

in which case Eq. (11-23) is satisfied regardless of the value of the deflection δ. Thus, as already observed, δ is undefined and may have any small value. The equation $\cos kL = 0$, which is the *buckling equation*, is satisfied when

$$kL = \frac{n\pi}{2} \qquad n = 1, 3, 5, \ldots \tag{11-25}$$

Using the expression $k^2 = P/EI$, we obtain the following formula for the critical loads:

$$P_{cr} = \frac{n^2 \pi^2 EI}{4L^2} \qquad n = 1, 3, 5, \ldots \tag{11-26}$$

Also, the buckled mode shapes are obtained from Eq. (11-22):

$$v = \delta\left(1 - \cos \frac{n\pi x}{2L}\right) \qquad n = 1, 3, 5, \ldots \tag{11-27}$$

The lowest critical load is obtained by substituting $n = 1$ in Eq. (11-26):

$$P_{cr} = \frac{\pi^2 EI}{4L^2} \tag{11-28}$$

The corresponding buckled shape (from Eq. 11-27) is

$$v = \delta\left(1 - \cos \frac{\pi x}{2L}\right) \tag{11-29}$$

and is shown in Fig. 11-15b.

By taking higher values of the index n, we can theoretically obtain an infinite number of critical loads from Eq. (11-26). The corresponding buckled mode shapes have additional waves in them. For instance, when $n = 3$ the buckled column has the shape shown in Fig. 11-15c and P_{cr} is nine times larger than for $n = 1$. Similarly, the buckled shape for $n = 5$ has even more waves (Fig. 11-15d) and the critical load is twenty-five times larger.

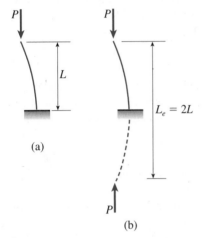

(a)

(b)

Fig. 11-16 Deflection curves showing the effective length L_e for a column fixed at the base and free at the top

■ Effective Lengths of Columns

The critical loads for columns with various support conditions can be related to the critical load of a pinned-end column through the concept of an **effective length**. To demonstrate this idea, consider the deflected shape of a column fixed at the base and free at the top (Fig. 11-16a). This column buckles in a curve that is one-quarter of a complete sine (or cosine) wave. If we extend the deflection curve (Fig. 11-16b), it becomes one-half of a complete sine wave, which is the deflection curve for a pinned-end column.

The effective length L_e for any column is the length of the equivalent pinned-end column, that is, it is the length of a pinned-end column having a deflection curve that exactly matches all or part of the deflection curve of the original column. Another way of expressing this idea is to say that the effective length of a column is the distance between points of inflection (that is, points of zero moment) in its deflection curve, assuming that the curve is extended (if necessary) until points of inflection are reached. Thus, for a fixed-free column (Fig. 11-16), the effective length is

$$L_e = 2L \tag{11-30}$$

Since the effective length is the length of an equivalent pinned-end column, we can write a general formula for critical loads as follows:

$$P_{cr} = \frac{\pi^2 EI}{L_e^2} \tag{11-31}$$

Thus, if we know the effective length of a column (no matter how complex the end conditions may be), we can substitute into this equation and determine the critical load. For instance, in the case of a fixed-free column, we can substitute $L_e = 2L$ and obtain Eq. (11-28).

The effective length is often expressed in terms of an **effective-length factor** K:

$$L_e = KL \tag{11-32}$$

where L is the actual length of the column. Thus, the critical load is

$$P_{cr} = \frac{\pi^2 EI}{(KL)^2}$$

(11-33)

The factor K equals 2 for a column fixed at the base and free at the top and equals 1 for a pinned-end column. The effective-length factor is often included in design formulas for columns, as illustrated later in Section 11.9.

■ Column with Both Ends Fixed against Rotation

Next, let us consider a column with both ends fixed against rotation (Fig. 11-17a). Note that in this figure we use the standard symbol for the fixed support at the base of the column. However, since the column is free to shorten under an axial load, we must introduce a new symbol at the top of the column. This new symbol shows a rigid block that is constrained in such a manner that rotation and horizontal displacement are prevented but vertical movement can occur. (As a convenience when drawing sketches, we often replace this more accurate symbol with the standard symbol for a fixed support—see Fig. 11-17b—with the understanding that the column is free to shorten.)

The buckled shape of the column in the first mode is shown in Fig. 11-17c. Note that the deflection curve is symmetrical (with zero slope at the midpoint) and has zero slope at the ends. Because rotation at the ends is prevented, reactive moments M_0 develop at the supports. These moments, as well as the axial forces, are shown in the figure.

Fig. 11-17 Buckling of a column with both ends fixed against rotation

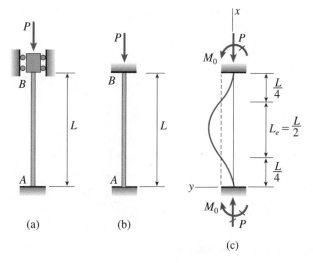

(a)   (b)

(c)

From our previous solutions of the differential equation, we know that the equation of the deflection curve involves sine and cosine functions. Also, we know that the curve is symmetric about the midpoint. Therefore, we see immediately that the curve must have inflection points at distances $L/4$ from the ends. It follows that the middle portion of the deflection curve has the same shape as the deflection curve for a pinned-end column. Thus, the effective length of a column with fixed ends, equal to the distance between inflection points, is

$$L_e = \frac{L}{2} \tag{11-34}$$

Substituting into Eq. (11-31) gives the critical load:

$$P_{cr} = \frac{4\pi^2 EI}{L^2} \tag{11-35}$$

This formula shows that the critical load for a column with fixed ends is four times that for a column with pinned ends. As a check, this result may be verified by solving the differential equation of the deflection curve (see Problem 11.4-10).

■ **Column Fixed at the Base and Pinned at the Top**

The critical load and buckled mode shape for a column that is fixed at the base and pinned at the top (Fig. 11-18a) can be determined by solving the

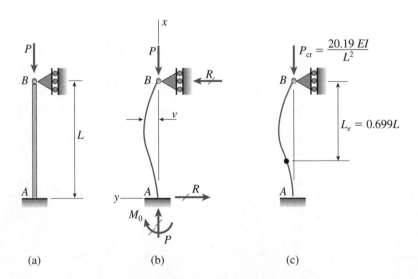

Fig. 11-18 Column fixed at the base and pinned at the top

(a) (b) (c)

differential equation of the deflection curve. When the column buckles (Fig. 11-18b), a reactive moment M_0 develops at the base because there can be no rotation at that point. Then, from the equilibrium of the entire column, we know that there must be horizontal reactions R at each end such that

$$M_0 = RL \tag{e}$$

The bending moment in the buckled column, at distance x from the base, is

$$M = M_0 - Pv - Rx = -Pv + R(L - x) \tag{11-36}$$

and therefore the differential equation is

$$EIv'' = M = -Pv + R(L - x) \tag{11-37}$$

Again substituting $k^2 = P/EI$ and rearranging, we get

$$v'' + k^2 v = \frac{R}{EI}(L - x) \tag{11-38}$$

The general solution of this equation is

$$v = C_1 \sin kx + C_2 \cos kx + \frac{R}{P}(L - x) \tag{11-39}$$

in which the first two terms on the right-hand side constitute the homogeneous solution and the last term is the particular solution. (The general solution can be verified by substitution into the differential equation.)

Since the solution contains three unknown quantities (C_1, C_2, and R), we need three boundary conditions. They are

$$v(0) = 0 \qquad v'(0) = 0 \qquad v(L) = 0$$

Applying these conditions to Eq. (11-39) yields

$$C_2 + \frac{RL}{P} = 0 \qquad C_1 k - \frac{R}{P} = 0 \qquad C_1 \tan kL + C_2 = 0 \tag{f,g,h}$$

All three equations are satisfied if $C_1 = C_2 = R = 0$, in which case we have the trivial solution and the deflection is zero. To obtain the solution for buckling, we must solve Eqs. (f), (g), and (h) in a more general manner. One method of solution is to eliminate R from the first two equations, which yields

$$C_1 kL + C_2 = 0 \quad \text{or} \quad C_2 = -C_1 kL \tag{i}$$

Next, we substitute this expression for C_2 into Eq. (h) and obtain the *buckling equation*:

$$kL = \tan kL \tag{11-40}$$

The solution of this equation gives the critical load.

Since the buckling equation is a transcendental equation,* it cannot be solved explicitly. Nevertheless, the values of kL that satisfy the equation can be determined numerically with the aid of a computer program for finding roots of equations. The smallest nonzero value of kL that satisfies Eq. (11-40) is

$$kL = 4.4934 \qquad (11\text{-}41)$$

The corresponding critical load is

$$P_{cr} = \frac{20.19EI}{L^2} = \frac{2.046\pi^2 EI}{L^2} \qquad (11\text{-}42)$$

which (as expected) is higher than the critical load for a column with pinned ends and lower than the critical load for a column with fixed ends (see Eqs. 11-12 and 11-35).

The effective length of the column may be obtained by comparing Eqs. (11-42) and (11-31); thus,

$$L_e = 0.699L \approx 0.7L \qquad (11\text{-}43)$$

This length is the distance from the pinned end of the column to the point of inflection in the buckled shape (Fig. 11-18c).

The equation of the buckled mode shape is obtained by substituting $C_2 = -C_1 kL$ (Eq. i) and $R/P = C_1 k$ (Eq. g) into the general solution (Eq. 11-39):

$$v = C_1[\sin kx - kL \cos kx + k(L - x)] \qquad (11\text{-}44)$$

in which $k = 4.4934/L$. The term in brackets gives the mode shape for the deflection of the buckled column. However, the amplitude of the deflection curve is undefined because C_1 may have any value (within the usual limitation that the deflections must remain small).

In addition to the requirement of small deflections, the Euler buckling theory used in this section is valid only if the column is perfectly straight and the material follows Hooke's law, as previously explained in Section 11.3.

The lowest critical loads and corresponding effective lengths for the four columns we have analyzed are summarized in Fig. 11-19.

---

*In a transcendental equation, the variables are contained within transcendental functions. A transcendental function cannot be expressed by a finite number of algebraic operations; hence trigonometric, logarithmic, exponential, and other such functions are transcendental.

(a) Pinned - pinned column	(b) Fixed - free column	(c) Fixed - fixed column	(d) Fixed - pinned column
$P_{cr} = \dfrac{\pi^2 EI}{L^2}$	$P_{cr} = \dfrac{\pi^2 EI}{4L^2}$	$P_{cr} = \dfrac{4\pi^2 EI}{L^2}$	$P_{cr} = \dfrac{2.046\,\pi^2 EI}{L^2}$
$L_e = L$	$L_e = 2L$	$L_e = 0.5L$	$L_e = 0.699L$
$K = 1$	$K = 2$	$K = 0.5$	$K = 0.699$

Fig. 11-19 Critical loads, effective lengths, and effective-length factors for ideal columns

---

■ **Example 11-2**

A viewing platform in a wild-animal park (Fig. 11-20a) is supported by a row of aluminum pipe columns having length $L = 3.25$ m and outer diameter $d = 100$ mm. The bases of the columns are set in concrete footings and the tops of the columns are supported laterally by the platform. The columns are being designed to support compressive loads $P = 100$ kN. Determine the minimum required thickness t of the columns (Fig. 11-20b) if a factor of safety $n = 3$ is required with respect to Euler buckling. (For the aluminum, use $E = 72$ GPa for the modulus of elasticity and use $\sigma_{pl} = 480$ MPa for the proportional limit.)

Solution   Because of the manner in which the columns are constructed, we will model each column as a fixed-pinned column (see Fig. 11-19d). Therefore, the critical load is

$$P_{cr} = \frac{2.046\,\pi^2 EI}{L^2} \qquad (j)$$

(a)

(b)

Fig. 11-20 Example 11-2

in which I is the moment of inertia of the tubular cross section:

$$I = \frac{\pi}{64}\Big[d^4 - (d - 2t)^4\Big] \tag{k}$$

Substituting $d = 100$ mm (or 0.1 m), we get

$$I = \frac{\pi}{64}\Big[(0.1 \text{ m})^4 - (0.1 \text{ m} - 2t)^4\Big] \tag{l}$$

in which t is expressed in meters.

Required thickness of the columns. Since the load per column is 100 kN and the factor of safety is 3, each column must be designed for the following critical load:

$$P_{\text{cr}} = nP = 3(100 \text{ kN}) = 300 \text{ kN}$$

Substituting this value for P_{cr} in Eq. (j), and also replacing I with its expression from Eq. (l), we obtain

$$300,000 \text{ N} = \frac{2.046\pi^2(72 \times 10^9 \text{ Pa})}{(3.25 \text{ m})^2}\left(\frac{\pi}{64}\right)\Big[(0.1 \text{ m})^4 - (0.1 \text{ m} - 2t)^4\Big]$$

Note that all terms in this equation are expressed in units of newtons and meters. After multiplying and dividing, the preceding equation simplifies to

$$44.40 \times 10^{-6} \text{ m}^4 = (0.1 \text{ m})^4 - (0.1 \text{ m} - 2t)^4$$

or $(0.1 \text{ m} - 2t)^4 = (0.1 \text{ m})^4 - 44.40 \times 10^{-6} \text{ m}^4 = 55.60 \times 10^{-6} \text{ m}^4$

from which we obtain

$$0.1 \text{ m} - 2t = 0.08635 \text{ m} \quad \text{and} \quad t = 0.006825 \text{ m}$$

Therefore, the minimum required thickness of the column to meet the specified conditions is

$$t_{\min} = 6.83 \text{ mm} \qquad \longleftarrow$$

Notes: Knowing the diameter and thickness of the column, we can now calculate its moment of inertia, cross-sectional area, and radius of gyration. Using the minimum thickness of 6.83 mm, we obtain

$$I = \frac{\pi}{64}\Big[d^4 - (d - 2t)^4\Big] = 2.18 \times 10^6 \text{ mm}^4$$

$$A = \frac{\pi}{4}\Big[d^2 - (d - 2t)^2\Big] = 1999 \text{ mm}^2 \qquad r = \sqrt{\frac{I}{A}} = 33.0 \text{ mm}$$

The slenderness ratio L/r of the column is approximately 98, which is in the customary range for slender columns, and the diameter-to-thickness ratio d/t is approximately 15, which should be adequate to prevent local buckling of the walls of the column.

The critical stress in the column must be less than the proportional limit of the aluminum if the formula for the critical load (Eq. j) is to be valid. The critical stress is

$$\sigma_{cr} = \frac{P_{cr}}{A} = \frac{300 \text{ kN}}{1999 \text{ mm}^2} = 150 \text{ MPa}$$

which is less than the proportional limit (480 MPa). Therefore, our calculation for the critical load using Euler buckling theory is satisfactory.

11.5 COLUMNS WITH ECCENTRIC AXIAL LOADS

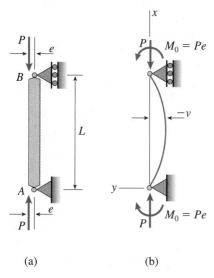

Fig. 11-21 Column with eccentric axial loads

In Sections 11.3 and 11.4 we analyzed ideal columns in which the axial loads acted through the centroids of the cross sections. Under these conditions, the columns remain straight until the critical load is reached, after which bending may occur. Now we will assume that the load is applied with a small eccentricity e measured from the axis of the column (Fig. 11-21a). The eccentric axial load is equivalent to a centric load P and a couple of moment $M_0 = Pe$ (Fig. 11-21b). This moment exists from the instant the load is first applied, and therefore the column begins to deflect at the onset of loading. The deflection then becomes steadily larger as the load increases.

To analyze the pin-ended column shown in Fig. 11-21, we make the same assumptions as in previous sections; namely, the column is initially perfectly straight, the material is linearly elastic, and the xy plane is a plane of symmetry. The bending moment in the column at distance x from the lower end (see Fig. 11-21b) is

$$M = M_0 + P(-v) = Pe - Pv \tag{11-45}$$

where v is the deflection of the column (positive when in the positive direction of the y axis). Note that the deflections of the column are negative when the eccentricity of the load is positive.

The differential equation of the deflection curve is

$$EIv'' = M = Pe - Pv \tag{11-46}$$

or $$v'' + k^2 v = k^2 e \tag{11-47}$$

in which $k^2 = P/EI$, as before. The general solution of this equation is

$$v = C_1 \sin kx + C_2 \cos kx + e \tag{11-48}$$

in which C_1 and C_2 are constants of integration in the homogeneous solution and e is the particular solution. As usual, we can verify the solution by substituting it into the differential equation.

The boundary conditions for determining the constants C_1 and C_2 are obtained from the deflections at the ends of the column (Fig. 11-21b):

$$v(0) = 0 \qquad v(L) = 0$$

These conditions yield

$$C_2 = -e \qquad C_1 = -\frac{e(1 - \cos kL)}{\sin kL} = -e \tan \frac{kL}{2}$$

Therefore, the equation of the deflection curve is

$$v = -e\left(\tan \frac{kL}{2} \sin kx + \cos kx - 1\right) \qquad (11\text{-}49)$$

For a column with a known load P and known eccentricity e, we can use this equation to calculate the deflection at any point along the x axis.

The behavior of a column with an eccentric load is quite different from that of a centrally loaded column, as can be seen by comparing Eq. (11-49) with Eqs. (11-13), (11-27), and (11-44). Equation (11-49) shows that each value of the eccentric load P produces a definite value of the deflection, just as each value of the load on a beam produces a definite deflection. In contrast, the deflection equations for centrally loaded columns give the buckled mode shape (when $P = P_{cr}$) but with the amplitude undefined.

Because the column shown in Fig. 11-21 has pinned ends, its critical load (when centrally loaded) is

$$P_{cr} = \frac{\pi^2 EI}{L^2} \qquad (11\text{-}50)$$

We will use this formula in the equations for the maximum deflection and maximum bending moment in the eccentrically loaded column.

■ Maximum Deflection

The maximum deflection δ produced by the eccentric loads occurs at the midpoint of the column (Fig. 11-22) and is obtained by setting x equal to $L/2$ in Eq. (11-49):

$$\delta = -v_{x=L/2} = e\left(\tan \frac{kL}{2} \sin \frac{kL}{2} + \cos \frac{kL}{2} - 1\right)$$

or, after simplifying,

$$\delta = e\left(\sec \frac{kL}{2} - 1\right) \qquad (11\text{-}51)$$

This equation can be written in a slightly different form by replacing the quantity k with its equivalent in terms of the critical load (see Eq. 11-50):

$$k = \sqrt{\frac{P}{EI}} = \sqrt{\frac{P\pi^2}{P_{cr}L^2}} = \frac{\pi}{L}\sqrt{\frac{P}{P_{cr}}} \qquad (11\text{-}52)$$

Thus, the nondimensional term kL becomes

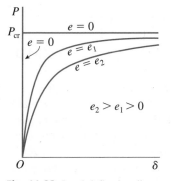

Fig. 11-22 Maximum deflection δ of a column with eccentric axial loads

Fig. 11-23 Load-deflection diagram for a column with eccentric axial loads (see Fig. 11-22 and Eq. (11-54))

$$kL = \pi \sqrt{\frac{P}{P_{cr}}} \qquad (11\text{-}53)$$

and Eq. (11-51) for the maximum deflection becomes

$$\delta = e\left[\sec\left(\frac{\pi}{2}\sqrt{\frac{P}{P_{cr}}}\right) - 1\right] \qquad (11\text{-}54)$$

As special cases, we note the following: (1) The deflection δ is zero when the eccentricity e is zero and P is not equal to P_{cr}, (2) the deflection is zero when the axial load P is zero, and (3) the deflection becomes infinitely large as P approaches P_{cr}.

A load-deflection diagram plotted from Eq. (11-54) is shown in Fig. 11-23. To plot the diagram, we select a particular value e_1 of the eccentricity and then calculate δ for various values of the load P. The resulting curve is labeled $e = e_1$ in Fig. 11-23. We note immediately that the deflection δ increases as P increases, but the relationship is nonlinear. Therefore, we cannot use the principle of superposition for calculating deflections due to more than one load, even though the material of the column is linearly elastic. In simple terms, the deflection due to an axial load $2P$ is *not* equal to twice the deflection caused by an axial load P.

Additional curves, such as the curve labeled $e = e_2$, are plotted in a similar manner. Since the deflection δ is linear with e (Eq. 11-54), the curve for $e = e_2$ has the same *shape* as the curve for $e = e_1$ but the abscissas are larger by the ratio e_2/e_1.

As the load P approaches the critical load, the deflection δ increases without limit and the horizontal line corresponding to $P = P_{cr}$ becomes an asymptote for the curves. In the limit, as e approaches zero, the curves on the diagram approach two straight lines, one vertical and one horizontal (compare with Fig. 11-7). Thus, as expected, an ideal column with a centrally applied load ($e = 0$) is the limiting case of a column with an eccentric load ($e > 0$).

Although the curves plotted in Fig. 11-23 are mathematically correct, we must keep in mind that the differential equation is valid only for small deflections. Therefore, when the deflections become large, the curves are no longer physically valid and must be modified to take into account the presence of large deflections and (if the proportional limit of the material is exceeded) inelastic bending effects (see Fig. 11-11).

The reason for the nonlinear relationship between loads and deflections, even when the deflections are small and Hooke's law holds, can be understood if we observe once again that the axial loads P are equivalent to centrally applied loads P plus couples Pe acting at the ends of the column (Fig. 11-21b). The couples Pe, if acting alone, would produce bending deflections of the column in the same manner as for a beam. In a beam, the presence of the deflections does not change the action of the loads, and the bending moments are the same whether the deflections exist or not. However, when an axial load is applied to the member, the existence of deflections increases the bending moments (the increases are equal to the product of the axial load and the deflections). When the bending moments increase, the deflections are further increased—hence the moments increase even more, and so on. Thus, the bending moments in a column depend upon the deflections, which in turn depend upon the bending moments. This type of behavior results in a nonlinear relationship between the axial loads and the deflections.

In general, a straight structural member subjected to both bending loads and axial compressive loads is called a **beam-column**. In the case of a column with eccentric loads (Fig. 11-21), the bending loads are the moments $M_0 = Pe$ and the axial loads are the forces P. (Beam-columns with various loading conditions are discussed in detail in Ref. 11-5 and in current textbooks on buckling and stability.)

■ Maximum Bending Moment

The maximum bending moment in an eccentrically loaded column occurs at the midpoint where the deflection is a maximum (Fig. 11-22):

$$M_{max} = P(e + \delta) \tag{11-55}$$

Substituting for δ from Eqs. (11-51) and (11-54), we obtain

$$M_{max} = Pe \sec \frac{kL}{2} = Pe \sec\left(\frac{\pi}{2}\sqrt{\frac{P}{P_{cr}}}\right) \tag{11-56}$$

The manner in which M_{max} varies as a function of the axial load P is shown in Fig. 11-24. When P is small, the maximum moment is equal to Pe, which means that the effect of the deflections is negligible. As P increases, the bending moment grows nonlinearly and theoretically becomes infinitely large as P approaches the critical load. However, as explained before, our equations are valid only when the deflections are small and cannot be used when the axial load approaches the critical load. Nevertheless, the preceding equations and accompanying graphs indicate the general behavior of beam-columns.

The equations given in this section were derived for a pinned-end column, as shown in Figs. 11-21 and 11-22. If a column is fixed at the base and free at the top (Fig. 11-19b), we can use all of the equations derived in this section by replacing the actual length L with the equivalent length $2L$ (see Problem 11.5-9). However, the equations do not apply to a column that is fixed at the base and pinned at the top (Fig. 11-19d). The use of an equivalent length equal to $0.699L$ gives erroneous results; instead, we must return to the differential equation and derive a new set of equations. In the case of a column with both ends fixed against rotation (Fig. 11-19c), the concept of an eccentric axial load acting at the end of the column has no meaning. Any moment applied at the end of the column is resisted directly by the supports and produces no bending of the column itself.

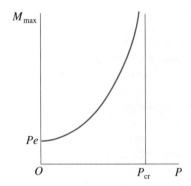

Fig. 11-24 Maximum bending moment in a column with eccentric axial loads (see Fig. 11-22 and Eq. 11-56)

■ **Example 11-3**

A brass bar AB projecting from the side of a large machine is loaded at end B by a force $P = 1500$ lb acting with an eccentricity $e = 0.45$ in. (Fig. 11-25). The bar has a rectangular cross section with dimensions 1.2 in. × 0.6 in. What is the longest permissible length L_{max} of the bar if the deflection at the end is limited to 0.12 in.? (For the brass, use $E = 16 \times 10^6$ psi.)

Solution   We will model this bar as a slender column that is fixed at end A and free at end B. Therefore, its critical load is given by the following equation (see Fig. 11-19b):

$$P_{cr} = \frac{\pi^2 EI}{4L^2} \tag{a}$$

The moment of inertia for the axis about which bending occurs is

$$I = \frac{(1.2 \text{ in.})(0.6 \text{ in.})^3}{12} = 0.02160 \text{ in.}^4$$

Therefore, the expression for the critical load becomes

$$P_{cr} = \frac{\pi^2(16,000,000 \text{ psi})(0.02160 \text{ in.}^4)}{4L^2} = \frac{852,700 \text{ lb-in.}^2}{L^2} \tag{b}$$

in which P_{cr} has units of pounds and L has units of inches.
The deflection at the end of the brass bar is given by Eq. (11-54), which applies to a fixed-free column as well as a pinned-end column:

$$\delta = e\left[\sec\left(\frac{\pi}{2}\sqrt{\frac{P}{P_{cr}}}\right) - 1\right] \tag{c}$$

in which P_{cr} is given by Eq.(a). To find the maximum permissible length of the bar, we substitute for δ its limiting value of 0.12 in. Also, we substitute numerical values for the eccentricity e and the load P, and finally, we substitute the expression for the critical load (Eq. b). Thus,

$$0.12 \text{ in.} = (0.45 \text{ in.})\left[\sec\left(\frac{\pi}{2}\sqrt{\frac{1500 \text{ lb}}{852,700/L^2}}\right) - 1\right]$$

This equation can be solved for the length L (inches). We begin by carrying out the multiplications and divisions: $0.2667 = \sec(0.06588L) - 1$, from which we get

$$\sec(0.06588L) = 1.2667 \quad \text{and} \quad \cos(0.06588L) = 0.7895$$

Therefore,

$$0.06588L = 0.6608 \quad \text{and} \quad L = 10.03 \text{ in.}$$

Thus, the maximum permissible length of the bar is

$$L_{max} = 10.0 \text{ in.} \qquad \longleftarrow$$

If a longer bar is used, the deflection will exceed the allowable value of 0.12 in.

Fig. 11-25 Example 11-3

THE SECANT FORMULA FOR COLUMNS

In the preceding section we determined the maximum deflection and maximum bending moment in a pin-ended column subjected to eccentric axial loads. In this section, we will investigate the maximum stresses in the column and obtain a special formula for calculating them.

The maximum stresses in a column with eccentric axial loads occur at the cross section where the deflection and bending moment have their largest values, that is, at the midpoint (Fig. 11-26a). Acting at this cross section are the compressive force P and the bending moment M_{max} (Fig. 11-26b). The stresses due to the force P are equal to P/A, where A is the cross-sectional area of the column, and the stresses due to the bending moment M_{max} are obtained from the flexure formula. Thus, the maximum compressive stress, which occurs on the concave side of the column, is

$$\sigma_{\text{max}} = \frac{P}{A} + \frac{M_{\text{max}}c}{I} \tag{11-57}$$

in which I is the moment of inertia in the plane of bending and c is the distance from the centroidal axis to the extreme point on the concave side of the column. Note that in this equation we consider compressive stresses to be positive.

The bending moment M_{max} is obtained from Eq. (11-56), which is repeated here:

$$M_{\text{max}} = Pe \sec\left(\frac{\pi}{2}\sqrt{\frac{P}{P_{\text{cr}}}}\right)$$

Since $P_{\text{cr}} = \pi^2 EI/L^2$ for a pinned-end column, and since $I = Ar^2$, where r is the radius of gyration in the plane of bending, the preceding equation becomes

$$M_{\text{max}} = Pe \sec\left(\frac{L}{2r}\sqrt{\frac{P}{EA}}\right) \tag{11-58}$$

Substituting into Eq. (11-57), we obtain the following formula for the maximum compressive stress:

$$\sigma_{\text{max}} = \frac{P}{A} + \frac{Pec}{I}\sec\left(\frac{L}{2r}\sqrt{\frac{P}{EA}}\right)$$

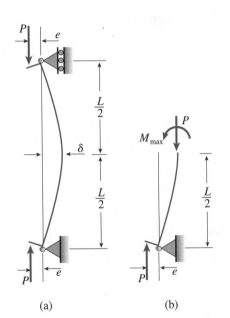

Fig. 11-26 Column with eccentric axial loads

or
$$\sigma_{max} = \frac{P}{A}\left[1 + \frac{ec}{r^2}\sec\left(\frac{L}{2r}\sqrt{\frac{P}{EA}}\right)\right] \qquad (11\text{-}59)$$

This equation is known as the **secant formula** for an eccentrically loaded column with pinned ends.

The secant formula gives the maximum compressive stress in the column as a function of the average compressive stress P/A, the modulus of elasticity E, and two nondimensional ratios—the slenderness ratio L/r (Eq. 11-17) and the **eccentricity ratio**:

$$\text{Eccentricity ratio} = \frac{ec}{r^2} \qquad (11\text{-}60)$$

As the name implies, the eccentricity ratio is a measure of the eccentricity of the load as compared to the dimensions of the cross section. Its numerical value depends upon the position of the load, but typical values are in the range from 0 to 3, with the most common values being less than 1.

When analyzing a column, we can use the secant formula to calculate the maximum compressive stress whenever the axial load P and its eccentricity e are known. Then the maximum stress can be compared with the allowable stress to determine if the column is adequate to support the load.

We can also use the secant formula in the reverse manner, that is, if we know the allowable stress, we can calculate the corresponding value of the load P. However, because the secant formula is transcendental, it is not practical to derive a formula for the load P. Instead, we can solve Eq. (11-59) numerically in each individual case.

A graph of the secant formula is shown in Fig. 11-27. The abscissa is the slenderness ratio L/r, and the ordinate is the average compressive stress P/A. The graph is plotted for a steel column with modulus of elasticity $E = 30 \times 10^3$ ksi and maximum stress $\sigma_{max} = 36$ ksi. Curves are plotted for several values of the eccentricity ratio ec/r^2. These curves are valid only when the maximum stress is less than the proportional limit of the material, because the secant formula was derived using Hooke's law.

A special case arises when the eccentricity of the load disappears ($e = 0$), because then we have an ideal column with a centrally applied load. Under these conditions the maximum load is the critical load ($P_{cr} = \pi^2 EI/L^2$) and the corresponding maximum stress is the critical stress (see Eqs. 11-14 and 11-16):

$$\sigma_{cr} = \frac{P_{cr}}{A} = \frac{\pi^2 EI}{AL^2} = \frac{\pi^2 E}{(L/r)^2} \qquad (11\text{-}61)$$

Since this equation gives the stress P/A in terms of the slenderness ratio L/r, we can plot it on the graph of the secant formula as Euler's curve

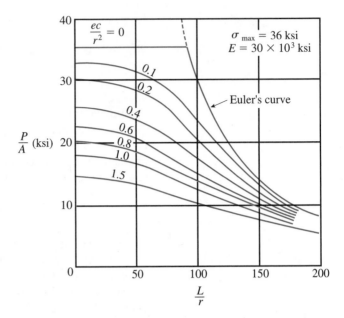

Fig. 11-27 Graph of the secant formula (Eq. 11-59) for $\sigma_{max} = 36$ ksi and $E = 30 \times 10^3$ ksi

(Fig. 11-27). Let us now assume that the proportional limit of the material is the same as the selected maximum stress, that is, 36 ksi. Then we construct a horizontal line on the graph at a value of 36 ksi, and we terminate Euler's curve at that stress. The horizontal line and Euler's curve represent the limits of the secant-formula curves as the eccentricity e approaches zero.

■ Discussion of the Secant Formula

The graph of the secant formula shows that the load-carrying capacity of a column decreases significantly as the slenderness ratio L/r increases, especially in the middle region of L/r values. Thus, long slender columns are much less stable than short, stocky columns. The graph also shows that the load-carrying capacity decreases with increasing eccentricity e; furthermore, this effect is relatively greater for short columns than for long ones.

The secant formula was derived for a column with pinned ends, but it can also be used for a column that is fixed at the base and free at the top. All that is required is to replace the length L in the formula with the equivalent length $2L$. However, because it is based upon Eq. (11-56), the secant formula is not valid for the other end conditions that we discussed.

Now let us consider an actual column, which inevitably differs from an ideal column because of imperfections, such as initial curvature of the longitudinal axis, imperfect support conditions, and nonhomogeneity of the material. Furthermore, even when the load is supposed to be centrally applied, there will be unavoidable eccentricities in its direction and point of application. The extent of these imperfections varies from one column to another, and therefore there is considerable scatter in the results of laboratory tests performed with actual columns.

All imperfections have the effect of producing bending in addition to direct compression. Therefore, it is reasonable to assume that the behavior of an imperfect, centrally loaded column is similar to that of an ideal, eccentrically loaded column. In such cases, the secant formula can be used by choosing an approximate value of the eccentricity ratio ec/r^2 to account for the combined effects of the various imperfections. For instance, a commonly used value of the eccentricity ratio for pinned-end columns in structural-steel design is $ec/r^2 = 0.25$. The use of the secant formula in this manner for columns with centrally applied loads provides a rational means of accounting for the effects of imperfections, rather than accounting for them simply by increasing the factor of safety. (For further discussions of the effects of imperfections, see Refs. 11-5, 11-6, 11-7, 11-9, and current textbooks on buckling and stability.)

The procedure for analyzing a centrally loaded column by means of the secant formula depends upon the particular conditions. For instance, if the objective is to determine the allowable load, the procedure is as follows. Assume a value of the eccentricity ratio ec/r^2 based upon test results, code values, or practical experience. Substitute this value into the secant formula, along with the values of L/r, A, and E for the actual column. Assign a value to σ_{max}, such as the yield stress σ_y or the proportional limit σ_{pl} of the material. Then solve the secant formula for the load P_{max} that produces the maximum stress. (This load will always be less than the critical load P_{cr} for the column.) The allowable load on the column equals the load P_{max} divided by the factor of safety n.

The following example illustrates how the secant formula may be used to determine the maximum stress in a column when the load is known, and also how to determine the load when the maximum stress is given.

■ **Example 11-4**

A steel wide-flange column of W 14 × 82 shape (Fig. 11-28a) is pin-supported at the ends and has a length of 25 ft. The column supports a centrally applied load $P_1 = 320$ k and an eccentrically applied load $P_2 = 40$ k (Fig. 11-28b). Bending takes place about axis 1-1 of the cross section, and the eccentric load acts on axis 2-2 at a distance of 13.5 in. from the centroid C.

(a) Using the secant formula, and assuming $E = 30,000$ ksi, calculate the maximum compressive stress in the column. (b) If the yield stress for the steel is $\sigma_y = 42$ ksi, what is the factor of safety with respect to yielding?

Solution   (a) *Maximum compressive stress.* The two loads P_1 and P_2 acting as shown in Fig. 11-28b are statically equivalent to a single load $P = 360$ k acting with an eccentricity $e = 1.5$ in. (Fig. 11-28c). Since the column is now

(a)

(b)

(c)

Fig. 11-28 Example 11-4. Column with an eccentrically applied axial load

loaded by a single force P having an eccentricity e, we can use the secant formula to find the maximum stress.

The required properties of the W 14 × 82 wide-flange shape are obtained from Table E-1 in Appendix E:

$$A = 24.1 \text{ in.}^2 \qquad r = 6.05 \text{ in.} \qquad c = \frac{14.31 \text{ in.}}{2} = 7.155 \text{ in.}$$

The required terms in the secant formula (Eq. 11-59) are calculated as follows:

$$\frac{P}{A} = \frac{360 \text{ k}}{24.1 \text{ in.}^2} = 14.94 \text{ ksi} \qquad \frac{ec}{r^2} = \frac{(1.5 \text{ in.})(7.155 \text{ in.})}{(6.05 \text{ in.})^2} = 0.2932$$

$$\frac{L}{r} = \frac{(25 \text{ ft})(12 \text{ in./ft})}{6.05 \text{ in.}} = 49.59$$

$$\frac{P}{EA} = \frac{360 \text{ k}}{(30,000 \text{ ksi})(24.1 \text{ in.}^2)} = 497.9 \times 10^{-6}$$

Substituting these values into the secant formula, we get

$$\sigma_{\max} = \frac{P}{A}\left[1 + \frac{ec}{r^2}\sec\left(\frac{L}{2r}\sqrt{\frac{P}{EA}}\right)\right]$$

$$= (14.94 \text{ ksi})(1 + 0.345) = 20.1 \text{ ksi} \qquad \longleftarrow$$

This stress occurs at midheight of the column on the concave side (the right-hand side in Fig. 11-28b).

(b) *Factor of safety with respect to yielding.* To find the factor of safety, we need to determine the value of the load P, acting at the eccentricity e, that will produce a maximum stress equal to the yield stress $\sigma_y = 42$ ksi. Since this value of the load is just sufficient to produce initial yielding of the material, we will denote it as P_y.

Note that we cannot determine P_y by multiplying the load P (equal to 360 k) by the ratio $\sigma_y/\sigma_{\max}$. The reason is that we are dealing with a nonlinear relationship between load and stress. Instead, we must substitute $\sigma_{\max} = \sigma_y = 42$ ksi in the secant formula and then solve for the corresponding load P, which becomes P_y. In other words, we must find the value of P_y that satisfies the following equation:

$$\sigma_y = \frac{P_y}{A}\left[1 + \frac{ec}{r^2}\sec\left(\frac{L}{2r}\sqrt{\frac{P_y}{EA}}\right)\right] \qquad (11\text{-}62)$$

Substituting numerical values, we obtain

$$42 \text{ ksi} = \frac{P_y}{24.1 \text{ in.}^2}\left[1 + 0.2939 \sec\left(\frac{49.59}{2}\sqrt{\frac{P_y}{(30,000 \text{ ksi})(24.1 \text{ in.}^2)}}\right)\right]$$

or

$$1012 \text{ k} = P_y\left[1 + 0.2939 \sec\left(0.02916\sqrt{P_y}\right)\right]$$

in which P_y has units of kips. Solving this equation numerically, we get

$$P_y = 716 \text{ k}$$

This load will produce yielding of the material (in compression) at the cross section of maximum bending moment.

Since the actual load is $P = 360$ k, the factor of safety against yielding is

$$n = \frac{P_y}{P} = \frac{716 \text{ k}}{360 \text{ k}} = 1.99 \qquad \longleftarrow$$

This example illustrates two of the many ways in which the secant formula may be used. Other types of analysis are illustrated in the problems at the end of the chapter.

11.7 ELASTIC AND INELASTIC COLUMN BEHAVIOR

In the preceding sections we described the behavior of columns when the material is stressed below the proportional limit. We began by considering an ideal column subjected to a centrally applied load (Euler buckling), and we arrived at the concept of a critical load P_{cr}. Then we considered columns with eccentric axial loads and derived the secant formula. We portrayed the results of these analyses on a diagram of average compressive stress P/A versus the slenderness ratio L/r (see Fig. 11-27). The behavior of an ideal column is represented in the figure by Euler's curve, and the behavior of columns with eccentric loads is represented by the family of curves having various values of the eccentricity ratio ec/r^2.

We will now extend our discussion to include **inelastic buckling**, that is, the buckling of columns when the proportional limit is exceeded. We will portray the behavior on the same kind of diagram as before, namely, a diagram of average compressive stress P/A versus slenderness ratio L/r (Fig. 11-29). Note that Euler's curve is shown on this diagram as curve ECD. This curve is valid only in the region CD where the stress is below the proportional limit σ_{pl} of the material. Therefore, the part of the curve above the proportional limit is shown by a dashed line.

The value of slenderness ratio above which Euler's curve is valid is obtained by setting the critical stress (Eq. 11-61) equal to the proportional limit σ_{pl} and solving for the slenderness ratio. Thus, letting $(L/r)_c$ represent the **critical slenderness ratio** (Fig. 11-29), we get

$$\left(\frac{L}{r}\right)_c = \sqrt{\frac{\pi^2 E}{\sigma_{pl}}} \qquad (11\text{-}63)$$

Fig. 11-29 Diagram of average compressive stress P/A versus slenderness ratio L/r

As an example, consider structural steel with $\sigma_{pl} = 36$ ksi and $E = 30,000$ ksi. Then the critical slenderness ratio $(L/r)_c$ is equal to 90.7.

Above this value, an ideal column buckles elastically and the Euler load is valid. Below this value, the stress in the column exceeds the proportional limit and the column buckles inelastically.

If we take into account the effects of eccentricities in loading or imperfections in construction, but still assume that the material follows Hooke's law, we obtain a curve such as the one labeled "Secant formula" in Fig. 11-29. This curve is plotted for a maximum stress σ_{max} equal to the proportional limit σ_{pl}.

When comparing the secant-formula curve with Euler's curve, we must keep in mind an important distinction. In the case of Euler's curve, the stress P/A not only is proportional to the applied load P but also is the actual maximum stress in the column when buckling occurs. Consequently, as we move from C to D along Euler's curve, both the maximum stress P/A (equal to the critical stress) and the axial load P decrease. However, in the case of the secant-formula curve, the *average* stress P/A decreases as we move from left to right along the curve (and therefore the axial load P also decreases) but the maximum stress (equal to the proportional limit) remains constant.

From Euler's curve, we see that long columns with large slenderness ratios buckle at low values of the average compressive stress P/A. This condition cannot be improved by using a higher-strength material, because collapse results from instability of the column as a whole and not from failure of the material itself. The stress can only be raised by reducing the slenderness ratio L/r or by using a material with higher modulus of elasticity E.

When a compression member is very short, it fails by yielding and crushing of the material, and no buckling or stability considerations are involved. In such a case, we can define an ultimate compressive stress σ_{ult} as the failure stress for the material. This stress establishes a **strength limit** for the column, represented by the horizontal line AB in Fig. 11-29. The strength limit is much higher than the proportional limit, since it represents the ultimate stress in compression.

Between the regions of short and long columns, there is a range of intermediate slenderness ratios too small for elastic stability to govern and too large for strength considerations alone to govern. Such an intermediate-length column fails by inelastic buckling, which means that the maximum stresses are above the proportional limit when buckling occurs. Because the proportional limit is exceeded, the slope of the stress-strain curve for the material is less than the modulus of elasticity; hence the critical load for inelastic buckling is always less than the Euler load (see the next section).

The dividing lines between short, intermediate, and long columns are not precise. Nevertheless, it is useful to make these distinctions because the maximum load-carrying capacity of columns in each category is based upon quite different types of behavior. The maximum load-carrying capacity of a particular column (as a function of its length) is represented by curve *ABCD* in Fig. 11-29. If the length is very small (region *AB*), the column fails by direct compression; if the column is longer (region *BC*), it fails by inelastic buckling; and if it is even longer (region *CD*), it fails by elastic buckling (that is, Euler buckling). Curve *ABCD* applies to columns with various support conditions if the length *L* in the slenderness ratio is replaced by the effective length L_e.

The results of load tests on columns are in reasonably good agreement with curve *ABCD*. When test results are plotted on the diagram, they generally form a band that lies just below this curve. Considerable scatter of test results is to be expected, because column performance is sensitive to such matters as the accuracy of construction, the alignment of loads, and the details of support conditions. To account for these variables, we usually obtain the allowable stress for a column by dividing the maximum stress (from curve *ABCD*) by a suitable *factor of safety*, which often has a value of about 2. Because imperfections are apt to increase with increase in length, a variable factor of safety (increasing as *L/r* increases) is sometimes used. In Section 11.9 we will give some typical formulas for allowable stresses.

11.8 INELASTIC BUCKLING

The critical load for elastic buckling is valid only for relatively long columns, as explained previously (see curve *CD* in Fig. 11-29). If a column is of intermediate length, the stress in the column will reach the proportional limit before buckling begins (curve *BC* in Fig. 11-29). To calculate critical loads in this intermediate range, we need a theory of **inelastic buckling**. Three such theories are described in this section: the tangent-modulus theory, the reduced-modulus theory, and the Shanley theory.

■ Tangent-modulus Theory

Let us again consider an ideal, pinned-end column subjected to an axial force *P* (Fig. 11-30a). The column is assumed to have a slenderness ratio *L/r* that is less than the critical slenderness ratio (Eq. 11-63), and therefore the axial stress *P/A* reaches the proportional limit before the critical load is reached.

The compressive stress-strain diagram for the material of the column is shown in Fig. 11-31. The proportional limit of the material is indicated as σ_{pl}, and the actual stress σ_A in the column (equal to *P/A*) is represented by point *A* (which is above the proportional limit). If the load is increased, so that a small increase in stress occurs, the relationship between the increment of stress and the corresponding increment of strain is given by the *slope* of the stress-strain diagram at point *A*. This slope, equal to the slope of the tangent line at *A*, is called the **tangent modulus** and is denoted by E_t; thus,

$$E_t = \frac{d\sigma}{d\epsilon} \tag{11-64}$$

Note that the tangent modulus *decreases* as the stress increases beyond the proportional limit. When the stress is below the proportional limit, the tangent modulus is the same as the ordinary elastic modulus *E*.

According to the tangent-modulus theory of inelastic buckling, the column shown in Fig. 11-30a remains straight until the inelastic critical load is reached. At that value of load, the column may undergo a small lateral deflection (Fig. 11-30b). The resulting bending stresses are superimposed upon the axial compressive stresses σ_A. Since the column starts bending from a straight position, the initial bending stresses represent only a small increment of stress. Therefore, the relationship between the bending stresses and the resulting strains is given by the tangent modulus E_t. Since the strains vary linearly across the cross section of the column, the initial bending stresses also vary linearly, and therefore the expressions for curvature are the same as those for linearly elastic bending except that E_t replaces *E*:

$$\kappa = \frac{1}{\rho} = \frac{d^2v}{dx^2} = \frac{M}{E_t I} \tag{11-65}$$

(compare with Eqs. 9-5 and 9-7).

Because the bending moment $M = -Pv$ (see Fig. 11-30b), the differential equation of the deflection curve is

$$E_t I v'' + Pv = 0 \tag{11-66}$$

This equation has the same form as the equation for elastic buckling (Eq. 11-5) except that E_t appears in place of *E*. Therefore, we can solve the equation in the same manner as before and obtain the following equation for the **tangent-modulus load**:

$$P_t = \frac{\pi^2 E_t I}{L^2} \tag{11-67}$$

This load represents the critical load for the column according to the tangent-modulus theory. The corresponding critical stress is

$$\sigma_t = \frac{P_t}{A} = \frac{\pi^2 E_t}{(L/r)^2} \tag{11-68}$$

which is similar in form to Eq. (11-61) for the Euler critical stress.

Since the tangent modulus E_t varies with the compressive stress $\sigma = P/A$ (Fig. 11-31), we usually obtain the tangent-modulus load by an itera-

Fig. 11-30 Ideal column of intermediate length that buckles inelastically

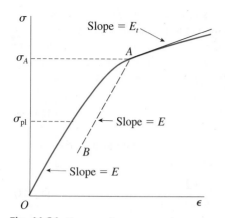

Fig. 11-31 Compression stress-strain diagram for the material of the column shown in Fig. 11-30

tive procedure. We begin by estimating the value of P_t. This trial value, call it P_1, should be slightly larger than $\sigma_{pl}A$, which is the axial load when the stress just reaches the proportional limit. Knowing P_1, we can calculate the corresponding axial stress $\sigma_1 = P_1/A$ and determine the tangent modulus E_t from the stress-strain diagram. Next, we use Eq. (11-67) to obtain a second estimate of P_t. Let us call this value P_2. If P_2 is very close to P_1, we may accept the load P_2 as the tangent-modulus load. However, it is more likely that additional cycles of iteration will be required until we reach a load that is in close agreement with the preceding trial load. This value is the tangent-modulus load.

A diagram showing how the critical stress σ_t varies with the slenderness ratio is given in Fig. 11-32 for a typical metal column with pinned ends. Note that the curve is above the proportional limit and below Euler's curve.

The tangent-modulus formulas may be used for columns with various support conditions by using the effective length L_e in place of the actual length L.

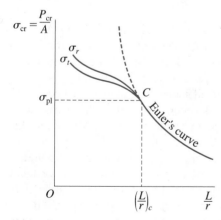

Fig. 11-32 Diagram of critical stress versus slenderness ratio

■ Reduced-modulus Theory

The tangent-modulus theory is distinguished by its simplicity and ease of use. However, it is conceptually deficient because it does not account for the complete behavior of the column. To explain the difficulty, we will consider again the column shown in Fig. 11-30a. When this column first departs from the straight position (Fig. 11-30b), bending stresses are added to the existing compressive stresses P/A. These additional stresses are compressive on the concave side of the column and tensile on the convex side. Therefore, the compressive stresses in the column become larger on the concave side and smaller on the other side.

Now imagine that the axial stress P/A is represented by point A on the stress-strain curve (Fig. 11-31). On the concave side of the column (where the compressive stress is increased), the material follows the tangent modulus E_t. However, on the convex side (where the compressive stress is decreased), the material follows the unloading line AB on the stress-strain diagram. This line is parallel to the initial linear part of the diagram, and therefore its slope is equal to the elastic modulus E. Thus, at the onset of bending, the column behaves as if it were made of two different materials, a material of modulus E_t on the concave side and a material of modulus E on the convex side.

A bending analysis of such a column can be made using the bending theories for a beam of two materials (Sections 6.2 and 6.3). The results of such analyses show that the column bends as though the material had a modulus of elasticity between the values of E and E_t. This "effective modulus" is known as the **reduced modulus** E_r, and its value depends not only upon the magnitude of the stress (because E_t depends upon the magnitude of the stress) but also upon the shape of the cross section of the column. Thus, the reduced modulus E_r is more difficult to determine than is the tangent modulus E_t. In the case of a column having a *rectangular cross section*, the equation for the reduced modulus is

$$E_r = \frac{4EE_t}{(\sqrt{E} + \sqrt{E_t})^2} \tag{11-69}$$

For a *wide-flange beam* with the area of the web disregarded, the reduced modulus for bending about the strong axis is

$$E_r = \frac{2EE_t}{E + E_t} \tag{11-70}$$

Further discussion of the reduced modulus, which is also called the *double modulus*, is given in Refs. 11-5 and 11-6 as well as in current textbooks on buckling and stability.

Since the reduced modulus represents an effective modulus that governs the bending of the column when it first departs from the straight position, we can formulate a reduced-modulus theory of inelastic buckling. Proceeding in the same manner as for the tangent-modulus theory, we begin with an equation for the curvature and then we write the differential equation of the deflection curve. These equations are the same as Eqs. (11-65) and (11-66) except that E_r appears instead of E_t. Thus, we arrive at the following equation for the reduced-modulus load:

$$P_r = \frac{\pi^2 E_r I}{L^2} \tag{11-71}$$

The corresponding equation for the critical stress is

$$\sigma_r = \frac{\pi^2 E_r}{(L/r)^2} \tag{11-72}$$

To find the reduced-modulus load P_r, we again must use an iterative procedure, because E_r depends upon E_t. The critical stress according to the reduced-modulus theory is shown in Fig. 11-32. Note that the curve for σ_r is above that for σ_t, because E_r is always greater than E_t.

The reduced-modulus theory is difficult to use in practice because E_r depends upon the shape of the cross section as well as the stress-strain curve and must be evaluated for each particular column. Moreover, this theory also has a conceptual defect. In order for the reduced modulus E_r to apply, the material on the convex side of the column must be undergoing a reduction in stress. However, such a reduction in stress cannot occur until bending actually takes place. Therefore, the axial load P, applied to an ideal straight column, can never actually reach the reduced-modulus load P_r. To reach that load would require that bending already exist, which is a contradiction.

■ Shanley Theory

From the preceding discussions we see that neither the tangent-modulus theory nor the reduced-modulus theory is entirely rational in explaining the phenomenon of inelastic buckling. Nevertheless, an understanding of both theories is necessary in order to develop a more complete and logically consistent theory. Such a theory was developed by F. R. Shanley in 1946 (see the historical note that follows) and today is called the *Shanley theory of inelastic buckling.*

The Shanley theory overcomes the difficulties with both the tangent-modulus and reduced-modulus theories by recognizing that it is not possible for a column to buckle inelastically in a manner that is analogous to Euler buckling. In Euler buckling, a critical load is reached at which the column is in neutral equilibrium, represented by a horizontal line on the load-deflection diagram (Fig. 11-33). As already explained, neither the tangent-modulus load P_t nor the reduced-modulus load P_r can represent this type of behavior. In both cases, we are led to a contradiction if we try to associate the load with a condition of neutral equilibrium.

Instead of neutral equilibrium, wherein a deflected shape suddenly becomes possible with no change in load, we must think of a column that has an ever-increasing axial load. When the load reaches the tangent-modulus load (which is less than the reduced-modulus load), bending can begin only if the load continues to increase. Under these conditions, bending occurs simultaneously with an increase in load, resulting in a decrease in strain on the convex side of the column. Thus, the effective modulus of the material throughout the cross section becomes greater than E_t, and therefore an increase in load is possible. However, the effective modulus is not as great as E_r, because E_r is based upon full strain reversal on the convex side of the column. In other words, E_r is based upon the amount of strain reversal that exists if the column bends without a change in the axial force, whereas the presence of an increasing axial force means that the reduction in strain is not as great.

Thus, instead of neutral equilibrium, where the relationship between load and deflection is undefined, we now have a definite relationship between each value of the load and the corresponding deflection. This behavior is shown by the curve labeled "Shanley theory" in Fig. 11-33. Note that buckling begins at the tangent-modulus load; then the load increases but does not reach the reduced-modulus load until the deflection becomes infinitely large (theoretically). However, other effects become important as the deflection increases, and in reality the curve eventually goes downward, as shown by the dashed line.

The Shanley concept of inelastic buckling has been verified by numerous investigators and by many tests and is now fully accepted. However, the maximum load attained by real columns (see the dashed curve trending downward in Fig. 11-33) is only slightly above the

Fig. 11-33 Load-deflection diagram for elastic and inelastic buckling

tangent-modulus load P_t. In addition, the tangent-modulus load is very simple to calculate. Therefore, for practical purposes it is reasonable to adopt the tangent-modulus load as the critical load for inelastic buckling of columns.

The preceding discussions of elastic and inelastic buckling are based upon idealized conditions. Although theoretical concepts are important in understanding column behavior, the actual design of columns must take into account additional factors not considered in the theory. For instance, steel columns always contain residual stresses produced by the rolling process. These stresses vary greatly in different parts of the cross section, and therefore the stress level required to produce yielding varies throughout the cross section. For such reasons, a variety of empirical design formulas have been developed for use in designing columns. Some of the commonly used formulas are given in the next section.

HISTORICAL NOTE   Over 200 years elapsed between the first calculation of a buckling load by Euler (in 1744) and the final development of the theory by Shanley (in 1946). Several famous investigators in the field of mechanics contributed to this development, and their work is described in this note.

After Euler's pioneering studies (Ref. 11-2), little progress was made until 1845, when the French engineer A. H. E. Lamarle pointed out that Euler's formula should be used only for slenderness ratios beyond a certain limit and that experimental data should be relied upon for columns with smaller ratios (Ref. 11-10). Then, in 1889, another French engineer, A. G. Considère, published the results of the first comprehensive tests on columns (Ref. 11-11). He pointed out that the stresses on the concave side of the column increased with E_t and the stresses on the convex side decreased with E. Thus, he showed why the Euler formula was not applicable to inelastic buckling, and he stated that the effective modulus was between E and E_t. Although he made no attempt to evaluate the effective modulus, Considère was responsible for beginning the reduced-modulus theory.

In the same year, and quite independently, the German engineer F. Engesser suggested the tangent-modulus theory (Ref. 11-12). He denoted the tangent modulus by the symbol T (equal to $d\sigma/d\epsilon$) and proposed that T be substituted for E in Euler's formula for the critical load. Later, in March 1895, Engesser again presented the tangent-modulus theory (Ref. 11-13), obviously without knowledge of Considère's work. Today, the tangent-modulus theory is often called the *Engesser theory*.

Three months later, Polish-born F. S. Jasinsky, then a professor in St. Petersburg, pointed out that Engesser's tangent-modulus theory was incorrect, called attention to Considère's work, and presented the reduced-modulus theory (Ref. 11-14). He also stated that the reduced modulus could not be calculated theoretically. In response, and only one month later, Engesser acknowledged the error in the tangent-modulus approach and showed how to calculate the reduced modulus for any cross section (Ref. 11-15). Thus, the reduced-modulus theory is also known as the *Considère-Engesser theory*.

The reduced-modulus theory was also presented by the famous scientist Theodore von Kármán in 1908 and 1910 (Refs. 11-16, 11-17, and 11-18), apparently independently of the earlier investigations. In Ref. 11-17 he derived the formulas for E_r for both rectangular and idealized wide-flange sections (that is, wide-flange sections without a web). He extended the theory to include the effects of eccentricities of the buckling load, and he showed that the maximum load decreases rapidly as the eccentricity increases.

The reduced-modulus theory was the accepted theory of inelastic buckling until 1946, when the American aeronautical-engineering professor F. R. Shanley pointed out the logical paradoxes in both the tangent-modulus and reduced-modulus theories. In a remarkable one-page paper (Ref. 11-19), Shanley not only explained what was wrong with the generally accepted theories but also proposed his own theory that resolved the paradoxes. In a second paper, five months later, he gave further analyses to support his earlier theory and gave results from tests on columns (Ref. 11-20). Since that time, many other investigators have confirmed and expanded Shanley's concept.

The background of the inelastic-buckling theories is described in Refs. 11-5, 11-6, and 11-7. For excellent discussions of the column-buckling problem, see the comprehensive papers by Hoff (Refs. 11-21 and 11-22), and for a historical account, see the paper by Johnston (Ref. 11-23).

11.9 DESIGN FORMULAS FOR COLUMNS

In the preceding sections of this chapter we discussed the theoretical load-carrying capacity of columns for both elastic and inelastic buckling. With that background in mind, we are now ready to examine some practical formulas that are used in the design of columns. These design formulas are based not only upon the theoretical analysis but also upon the behavior of real columns as observed in laboratory tests.

The theoretical results are represented by the column curves shown in Figs. 11-29 and 11-32. A common design approach is to approximate these curves in the inelastic buckling range (low values of slenderness ratio) by empirical formulas and to use Euler's formula in the elastic range (high values of slenderness ratio). Of course, a factor of safety must be applied to obtain the allowable loads from the maximum loads (or to obtain the allowable stresses from the maximum stresses).

The following examples of column design formulas are applicable to centrally loaded columns of structural steel, aluminum, and wood. The formulas give the allowable stresses in terms of the column properties, such as length, cross-sectional dimensions, and conditions of support. Thus, for a given column, the allowable stress can be readily obtained.

Once the allowable stress is known, we can determine the allowable load by multiplying by the cross-sectional area:

$$P_{\text{allow}} = \sigma_{\text{allow}} A \qquad (11\text{-}73)$$

The allowable load must be larger than the actual load if the allowable stress is not to be exceeded.

The selection of a column often requires an iterative or trial-and-error procedure. Such a procedure is necessary whenever we don't know in advance which design formula to use. Since each formula is valid only for a certain range of slenderness ratios, and since the slenderness ratio is unknown until after the column is selected, we usually don't know which formula is applicable until we have made at least one trial.

A common trial-and-error procedure for selecting a column to support a given axial load is the following:

(1) Estimate the allowable stress σ_{allow}. (Note that an upper limit for σ_{allow} is the allowable stress for a column of zero length. This stress is readily obtained from the design formulas, and the estimated stress should be equal to or less than this upper limit.)

(2) Calculate an approximate value of the cross-sectional area A by dividing the given axial load P by the estimated allowable stress.

(3) Determine a column size and/or shape that supplies the required area, either by calculating a required dimension or by selecting a column from a table of available shapes.

(4) Knowing the dimensions of a trial column from step (3), determine the allowable stress σ_{allow} in the column from the appropriate design formula.

(5) Using Eq. (11-73), calculate the allowable load P_{allow} and compare it with the actual load P.

(6) If the column is not adequate to support the given load, select a larger column and repeat the process. If the column appears to be overdesigned (because the allowable load is much larger than the given load), select a smaller column and repeat the process. A suitable column can usually be obtained with only two or three trials.

Many variations of this procedure are possible, depending upon the type of column and what quantities are known in advance. Sometimes a direct design procedure, bypassing the trial-and-error steps, can be devised.

■ **Structural Steel**

Let us begin with the design formulas for centrally loaded, structural-steel columns. The formulas used today in the United States are based upon research conducted many years ago by the Structural Stability Research Council (SSRC), a professional engineering organization (Ref. 11-7). The formulas proposed by the SSRC give the maximum stress in the column, that is, the stress obtained by dividing the maximum load the column can carry (without a safety factor) by the cross-sectional area.

When the slenderness ratio L/r is large, the maximum stress is based upon the Euler load:

$$\sigma_{max} = \frac{\pi^2 E}{(KL/r)^2} \qquad (11\text{-}74)$$

where the effective length KL is used so that the formula may be applied to a variety of support conditions.

Equation (11-74) is valid only when the stresses in the column are less than the proportional limit σ_{pl}. Under ordinary conditions, we assume that the proportional limit of steel is equal to the yield stress σ_y. However, rolled steel sections (such as wide-flange sections) contain significant residual stresses—stresses that may be as large as one-half the yield stress. For such a column, the proportional limit is reached when the axial stress σ_{max} due to the compressive load equals one-half the yield stress:

$$\sigma_{max} = 0.5\sigma_y \qquad (11\text{-}75)$$

To determine the smallest slenderness ratio for which Eq. (11-74) is applicable, we set σ_{max} equal to $0.5\sigma_y$ and solve for the corresponding value of KL/r, which is known as the *critical slenderness ratio* (compare with Eq. 11-63):

$$\left(\frac{KL}{r}\right)_c = \sqrt{\frac{2\pi^2 E}{\sigma_y}} \qquad (11\text{-}76)$$

If the actual slenderness ratio is equal to or larger than $(KL/r)_c$, the Euler formula for the maximum stress (Eq. 11-74) may be used. Thus, the critical slenderness ratio given by Eq. (11-76) determines the boundary between elastic and inelastic buckling for rolled steel columns.

Equation (11-74) may be expressed in nondimensional form by dividing by the yield stress σ_y and then substituting from Eq. (11-76):

$$\frac{\sigma_{max}}{\sigma_y} = \frac{\pi^2 E}{\sigma_y\left(\dfrac{KL}{r}\right)^2} = \frac{\left(\dfrac{KL}{r}\right)_c^2}{2\left(\dfrac{KL}{r}\right)^2} \qquad \frac{KL}{r} \geq \left(\frac{KL}{r}\right)_c \qquad (11\text{-}77)$$

This equation is plotted in Fig. 11-34 and labeled *Euler's curve*.

For the region of inelastic buckling, where $KL/r \leq (KL/r)_c$, the maximum stress, as proposed by the SSRC, is given by a parabolic formula:

$$\frac{\sigma_{max}}{\sigma_y} = 1 - \frac{\left(\dfrac{KL}{r}\right)^2}{2\left(\dfrac{KL}{r}\right)_c^2} \qquad \frac{KL}{r} \leq \left(\frac{KL}{r}\right)_c \qquad (11\text{-}78)$$

This empirical formula is also plotted in Fig. 11-34. Note that the curve is a parabola with a horizontal tangent at $KL/r = 0$, where the maximum stress is equal to σ_y. At the critical slenderness ratio $(KL/r)_c$ the curve merges smoothly with Euler's curve (both curves have the same slope at the point where they meet). Thus, the empirical formula provides a design curve that fits the general shape of the theoretical curves (Figs. 11-29 and 11-32) while also being simple to use. The validity of the formula for use in design has been verified by numerous tests.

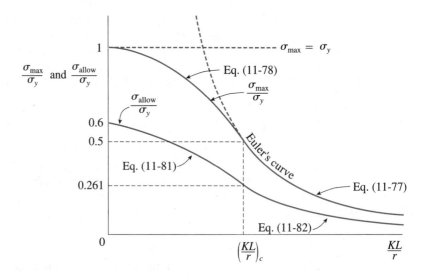

Fig. 11-34 Design formulas for structural-steel columns

The American Institute of Steel Construction (AISC), a technical organization that prepares specifications for structural-steel design, adopted the formulas proposed by the SSRC for the maximum stresses (Eqs. 11-77 and 11-78). In addition, the AISC proposed the following formulas for the factors of safety to be applied to the maximum stresses:

$$n_1 = \frac{5}{3} + \frac{3\left(\dfrac{KL}{r}\right)}{8\left(\dfrac{KL}{r}\right)_c} - \frac{\left(\dfrac{KL}{r}\right)^3}{8\left(\dfrac{KL}{r}\right)_c^3} \qquad \frac{KL}{r} \leq \left(\frac{KL}{r}\right)_c \qquad (11\text{-}79)$$

$$n_2 = \frac{23}{12} \approx 1.92 \qquad \frac{KL}{r} \geq \left(\frac{KL}{r}\right)_c \qquad (11\text{-}80)$$

Thus, the factor of safety is 5/3 when $KL/r = 0$ and gradually increases to 23/12 when $KL/r = (KL/r)_c$. For higher slenderness ratios, the factor of safety remains constant at that value.

The AISC formulas for the allowable stresses are obtained by dividing the maximum stresses σ_{max} by the appropriate factors of safety; thus,

$$\frac{\sigma_{\text{allow}}}{\sigma_y} = \frac{1}{n_1}\left[1 - \frac{\left(\dfrac{KL}{r}\right)^2}{2\left(\dfrac{KL}{r}\right)_c^2}\right] \qquad \frac{KL}{r} \le \left(\frac{KL}{r}\right)_c \qquad (11\text{-}81)$$

$$\frac{\sigma_{\text{allow}}}{\sigma_y} = \frac{\left(\dfrac{KL}{r}\right)_c^2}{2n_2\left(\dfrac{KL}{r}\right)^2} \qquad \frac{KL}{r} \ge \left(\frac{KL}{r}\right)_c \qquad (11\text{-}82)$$

These equations for the allowable stresses are also plotted in Fig. 11-34. The AISC specifications (Ref. 5-4) place an upper limit of 200 on the slenderness ratio KL/r.

All of the preceding design formulas for structural steel may be used with either USCS or SI units. The formulas are applicable to wide-flange and other rolled shapes, as well as to columns with rectangular and circular cross sections.

The symbols used in the AISC specifications differ slightly from those in the above formulas. For instance, the critical slenderness ratio is denoted C_c:

$$C_c = \left(\frac{KL}{r}\right)_c = \sqrt{\frac{2\pi^2 E}{\sigma_y}} \qquad (11\text{-}83\text{a})$$

Also, the allowable stress is denoted F_a and the yield stress is denoted F_y:

$$F_a = \sigma_{\text{allow}} \qquad F_y = \sigma_y \qquad (11\text{-}83\text{b,c})$$

With these changes in notation, and also by substituting for n_1 and n_2, we obtain the design formulas (Eqs. 11-81 and 11-82) in the style presented in the AISC specifications:

$$F_a = \frac{\left[1 - \dfrac{(KL/r)^2}{2C_c^2}\right]F_y}{\dfrac{5}{3} + \dfrac{3(KL/r)}{8C_c} - \dfrac{(KL/r)^3}{8C_c^3}} \qquad \frac{KL}{r} \le C_c \qquad (11\text{-}84\text{a})$$

$$F_a = \frac{12\pi^2 E}{23(KL/r)^2} \qquad \frac{KL}{r} \ge C_c \qquad (11\text{-}84\text{b})$$

In addition, the AISC specifies the modulus of elasticity E as 29,000 ksi.

▪ Aluminum

The design formulas for aluminum columns are presented in the specifications of the Aluminum Association (Ref. 5-5). These formulas are also based upon the curves given in Figs. 11-29 and 11-32. For instance, Euler's curve is used for large values of slenderness ratio. For smaller

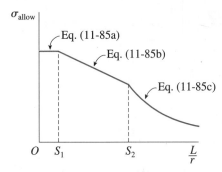

Fig. 11-35 Design formulas for aluminum columns

slenderness ratios, two straight lines (one horizontal and one inclined) are used to represent the inelastic buckling curves (Fig. 11-35). The horizontal line is for short columns, and the sloping line is for columns of intermediate length. The slenderness ratios that separate short, intermediate, and long columns are denoted S_1 and S_2.

The general formulas for the allowable stresses are as follows:

$$\sigma_{allow} = \frac{\sigma_y}{k_c n_y} \qquad 0 \le \frac{L}{r} \le S_1 \qquad (11\text{-}85a)$$

$$\sigma_{allow} = \frac{1}{n_u}\left(B_c - D_c\frac{L}{r}\right) \qquad S_1 \le \frac{L}{r} \le S_2 \qquad (11\text{-}85b)$$

$$\sigma_{allow} = \frac{\pi^2 E}{n_u\left(\dfrac{L}{r}\right)^2} \qquad S_2 \le \frac{L}{r} \qquad (11\text{-}85c)$$

In these equations, the length L is defined as the distance between points of lateral support, or twice the length of a cantilever column. Thus, L may be interpreted as the effective length. The stress σ_y is the compressive yield stress (0.2% offset), k_c is a constant, n_y is the factor of safety with respect to the yield stress, n_u is the factor of safety with respect to the ultimate stress, and B_c, D_c, S_1, and S_2 are constants.

The values of the quantities appearing in Eqs. (11-85a, b, and c) vary according to the particular aluminum alloy, the temper of the finished product, and the use to which it will be put. Numerous alloys and tempers are available, so the Aluminum Association gives tables of values based upon the material and usage. As examples, the following formulas apply to two alloys often used in buildings and aircraft structures.

(1) Alloy 2014-T6

$$\sigma_{allow} = 28 \text{ ksi} \qquad 0 \le \frac{L}{r} \le 12 \qquad (11\text{-}86a)$$

$$\sigma_{allow} = 30.7 - 0.23\left(\frac{L}{r}\right) \text{ ksi} \qquad 12 \le \frac{L}{r} \le 55 \qquad (11\text{-}86b)$$

$$\sigma_{allow} = \frac{54{,}000 \text{ ksi}}{(L/r)^2} \qquad 55 \le \frac{L}{r} \qquad (11\text{-}86c)$$

(2) Alloy 6061-T6

$$\sigma_{allow} = 19 \text{ ksi} \qquad 0 \le \frac{L}{r} \le 9.5 \qquad (11\text{-}87a)$$

$$\sigma_{allow} = 20.2 - 0.126\left(\frac{L}{r}\right) \text{ ksi} \qquad 9.5 \le \frac{L}{r} \le 66 \qquad (11\text{-}87b)$$

$$\sigma_{allow} = \frac{51{,}000 \text{ ksi}}{(L/r)^2} \qquad 66 \le \frac{L}{r} \qquad (11\text{-}87c)$$

Note that these formulas give the allowable stresses, hence they already incorporate the factors of safety, which vary from 1.65 to 2.20, depending upon the alloy.

■ Wood

The design of wood structural members is usually governed by the *National Design Specification for Wood Construction,* published by the American Forest and Paper Association (Ref. 5-6). The AFPA formulas for the allowable stresses in *rectangular* wood columns for short, intermediate, and long columns are as follows:

$$\sigma_{\text{allow}} = F_c \qquad 0 \le \frac{L}{d} \le 11 \tag{11-88a}$$

$$\sigma_{\text{allow}} = F_c\left[1 - \frac{1}{3}\left(\frac{L/d}{K_c}\right)^4\right] \qquad 11 \le \frac{L}{d} \le K_c \tag{11-88b}$$

$$\sigma_{\text{allow}} = \frac{0.3E}{(L/d)^2} = \frac{2F_c}{3}\left(\frac{K_c}{L/d}\right)^2 \qquad K_c \le \frac{L}{d} \le 50 \tag{11-88c}$$

The stress F_c is the design stress in direct compression parallel to the grain. Thus, this stress depends upon the species and grade of wood, the duration of loading, and other service conditions. The ratio L/d is the effective length L of the column for buckling in a principal plane divided by the cross-sectional dimension d in that same plane. Finally, the factor K_c is the critical L/d ratio that separates intermediate from long columns:

$$K_c = \sqrt{\frac{0.45E}{F_c}} \tag{11-89}$$

Note especially that the symbol K_c in the formulas for wood columns represents an L/d ratio, not an effective-length factor. Also, note that the equations for wood columns may be used with either USCS or SI units.

A graph of Eqs. (11-88a, b, and c) is shown in Fig. 11-36 for the ratio $E/F_c = 1500$. For short columns the graph has a horizontal line and for intermediate columns it has a fourth-degree curve. For long columns, the graph is based upon Euler's curve.

Typical values of the design stress F_c for structural lumber are in the range 700 to 2000 psi, depending upon the particular wood being used and the conditions of service. The modulus of elasticity E also varies with species, grade, and service conditions, but typical values are in the range of 1.0×10^6 psi to 1.8×10^6 psi. Values of the ratio K_c are usually in the range from 18 to 30.

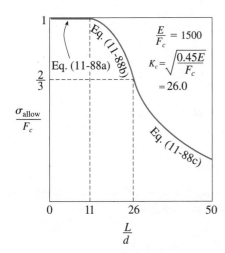

Fig. 11-36 Design formulas for rectangular wood columns with $E/F_c = 1500$

■ **Limitations**

The preceding formulas for the design of steel, aluminum, and wood columns are intended solely for use in solving problems. They should not be used for the design of actual columns, because they represent only a small part of the complete design process. Many factors besides those discussed here enter into the design of columns, and therefore textbooks or other references on structural design should be consulted before designing a column for a specific application.

Furthermore, all design formulas presented in specifications and codes, such as the formulas given in this section, require informed judgment in their use. There are many cases of structures that "met the code" but nevertheless collapsed or failed to perform adequately. Meeting the code requirements is not enough for a safe design—practical design experience is also essential.

Fig. 11-37 Example 11-5. Steel wide-flange column

■ **Example 11-5**

A steel column is constructed from a W 10 × 60 wide-flange section (Fig. 11-37). Assume that the column has pin supports and may buckle in any direction. Also, assume that the steel has modulus of elasticity $E = 29,000$ ksi and yield stress $\sigma_y = 36$ ksi. (a) If the length of the column is $L = 20$ ft, what is the allowable axial load? (b) If the column is subjected to an axial load $P = 200$ k, what is the maximum permissible length?

Solution   We will use the AISC formulas (Eqs. 11-79 through 11-82) when analyzing this column. Since the column has pin supports, the effective-length factor $K = 1$. Also, since the column will buckle about the weak axis of bending, we will use the smaller radius of gyration:

$$r = 2.57 \text{ in.}$$

as obtained from Table E-1, Appendix E. The critical slenderness ratio (Eq. 11-76) is

$$\left(\frac{KL}{r}\right)_c = \sqrt{\frac{2\pi^2 E}{\sigma_y}} = \sqrt{\frac{2\pi^2(29,000 \text{ ksi})}{36 \text{ ksi}}} = 126.1 \quad\quad \text{(a)}$$

(a) *Allowable axial load.* If the length $L = 20$ ft, the slenderness ratio of the column is

$$\frac{L}{r} = \frac{(20 \text{ ft})(12 \text{ in./ft})}{2.57 \text{ in.}} = 93.4$$

which is less than the critical ratio (Eq. a). Therefore, we will use Eqs. (11-79) and (11-81) to obtain the factor of safety and allowable stress, respectively:

$$n_1 = \frac{5}{3} + \frac{3\left(\dfrac{KL}{r}\right)}{8\left(\dfrac{KL}{r}\right)_c} - \frac{\left(\dfrac{KL}{r}\right)^3}{8\left(\dfrac{KL}{r}\right)_c^3} = \frac{5}{3} + \frac{3(93.4)}{8(126.1)} - \frac{(93.4)^3}{8(126.1)^3} = 1.89$$

$$\frac{\sigma_{allow}}{\sigma_y} = \frac{1}{n_1}\left[1 - \frac{\left(\frac{KL}{r}\right)^2}{2\left(\frac{KL}{r}\right)_c^2}\right] = \frac{1}{1.89}\left[1 - \frac{(93.4)^2}{2(126.1)^2}\right] = 0.384$$

$$\sigma_{allow} = 0.384\sigma_y = 0.384(36 \text{ ksi}) = 13.8 \text{ ksi}$$

Since the cross-sectional area of the column is $A = 17.6 \text{ in.}^2$ (from Table E-1), the allowable axial load is

$$P_{allow} = \sigma_{allow}A = (13.8 \text{ ksi})(17.6 \text{ in.}^2) = 243 \text{ k} \quad \longleftarrow$$

(b) *Maximum permissible length.* To determine the maximum length when the axial load $P = 200$ k, we begin with an estimated value of the length and then use a trial-and-error procedure. Note that when the load $P = 200$ k, the maximum length is greater than 20 ft (because a length of 20 ft corresponds to an axial load of 243 k). Therefore, as a trial value, we will assume $L = 25$ ft. The corresponding slenderness ratio is

$$\frac{L}{r} = \frac{(25 \text{ ft})(12 \text{ in./ft})}{2.57 \text{ in.}} = 116.7$$

which is less than the critical ratio. Therefore, we again use Eqs. (11-79) and (11-81) to obtain the factor of safety and allowable stress:

$$n_1 = \frac{5}{3} + \frac{3\left(\frac{KL}{r}\right)}{8\left(\frac{KL}{r}\right)_c} - \frac{\left(\frac{KL}{r}\right)^3}{8\left(\frac{KL}{r}\right)_c^3} = \frac{5}{3} + \frac{3(116.7)}{8(126.1)} - \frac{(116.7)^3}{8(126.1)^3} = 1.915$$

$$\frac{\sigma_{allow}}{\sigma_y} = \frac{1}{n_1}\left[1 - \frac{\left(\frac{KL}{r}\right)^2}{2\left(\frac{KL}{r}\right)_c^2}\right] = \frac{1}{1.915}\left[1 - \frac{(116.7)^2}{2(126.1)^2}\right] = 0.299$$

$$\sigma_{allow} = 0.299\sigma_y = 0.299(36 \text{ ksi}) = 10.8 \text{ ksi}$$

Thus, the allowable axial load corresponding to a length $L = 25$ ft is

$$P_{allow} = \sigma_{allow}A = (10.8 \text{ ksi})(17.6 \text{ in.}^2) = 190 \text{ k}$$

which is less than the given load of 200 k. Therefore, the permissible length is less than 25 ft.

Performing similar calculations for $L = 24.0$ ft and $L = 24.5$ ft, we obtain the following results:

$$L = 24.0 \text{ ft} \qquad P_{allow} = 201 \text{ k}$$
$$L = 24.5 \text{ ft} \qquad P_{allow} = 194 \text{ k}$$
$$L = 25.0 \text{ ft} \qquad P_{allow} = 190 \text{ k}$$

Interpolating between these results, we see that a load of 200 k corresponds to a length of 24.1 ft. Thus, the maximum permissible length of the column is

$$L_{max} = 24.1 \text{ ft} \quad \longleftarrow$$

Fig. 11-38 Example 11-6. Steel pipe column

■ **Example 11-6**

Find the minimum required thickness t_{min} for a steel pipe column of length $L = 3.6$ m and outer diameter $d = 160$ mm supporting an axial load $P = 240$ kN (Fig. 11-38). The column is fixed at the base and free at the top. (Use $E = 200$ GPa and $\sigma_y = 250$ MPa.)

Solution   We will use the AISC formulas (Eqs. 11-79 through 11-82) when analyzing this column. Since the column has fixed-free end conditions, the effective length is

$$KL = 2(3.6 \text{ m}) = 7.2 \text{ m}$$

Also, the critical slenderness ratio (Eq. 11-76) is

$$\left(\frac{KL}{r}\right)_c = \sqrt{\frac{2\pi^2 E}{\sigma_y}} = \sqrt{\frac{2\pi^2(200 \text{ GPa})}{250 \text{ MPa}}} = 125.7 \qquad \text{(b)}$$

To determine the required thickness of the column, we will use a trial-and-error method. Let us start by assuming a trial value $t = 7.0$ mm. Then the moment of inertia of the cross-sectional area is

$$I = \frac{\pi}{64}\left[d^4 - (d - 2t)^4\right] = \frac{\pi}{64}\left[(160 \text{ mm})^4 - (146 \text{ mm})^4\right] = 9.866 \times 10^6 \text{ mm}^4$$

Also, the cross-sectional area and radius of gyration are

$$A = \frac{\pi}{4}\left[d^2 - (d - 2t)^2\right] = \frac{\pi}{4}\left[(160 \text{ mm})^2 - (146 \text{ mm})^2\right] = 3365 \text{ mm}^2$$

$$r = \sqrt{\frac{I}{A}} = \sqrt{\frac{9.866 \times 10^6 \text{ mm}^4}{3365 \text{ mm}^2}} = 54.15 \text{ mm}$$

Therefore, the slenderness ratio of the column is

$$\frac{KL}{r} = \frac{2(3.6 \text{ m})}{54.15 \text{ mm}} = 133.0$$

Since this ratio is larger than the critical slenderness ratio (Eq. b), we obtain the factor of safety and the allowable stress from Eqs. (11-80) and (11-82):

$$n_2 = 1.92$$

$$\frac{\sigma_{allow}}{\sigma_y} = \frac{\left(\dfrac{KL}{r}\right)_c^2}{2n_2\left(\dfrac{KL}{r}\right)^2} = \frac{(125.7)^2}{2(1.92)(133.0)^2} = 0.2326$$

$$\sigma_{allow} = 0.2326\sigma_y = 0.2326(250 \text{ MPa}) = 58.15 \text{ MPa}$$

Thus, the allowable axial load is

$$P_{allow} = \sigma_{allow} A = (58.15 \text{ MPa})(3365 \text{ mm}^2) = 196 \text{ kN}$$

Since this load is less than the required load of 240 kN, we must try a larger value of the thickness t.

Performing calculations similar to those above, but for $t = 8$ mm and $t = 9$ mm, we get the following results:

$$t = 7.0 \text{ mm} \qquad P_{allow} = 196 \text{ kN}$$

$$t = 8.0 \text{ mm} \qquad P_{allow} = 220 \text{ kN}$$

$$t = 9.0 \text{ mm} \qquad P_{allow} = 243 \text{ kN}$$

By interpolation, we see that $t = 8.9$ mm corresponds to a load of 240 kN. Therefore, the required thickness of the pipe column is

$$t_{min} = 8.9 \text{ mm} \qquad\qquad \longleftarrow$$

Fig. 11-39 Example 11-7. Aluminum tube

■ Example 11-7

An aluminum tube (alloy 2014-T6) with an effective length $L = 16.0$ in. is compressed by an axial force $P = 5.0$ k (Fig. 11-39). Determine the minimum required outer diameter d if the thickness t equals one-tenth the outer diameter.

Solution   We will use the Aluminum Association formulas for alloy 2014-T6 (Eqs. 11-86a, b, and c) for analyzing this column. However, we must make an initial guess as to which formula is applicable, because each formula applies to a different range of slenderness ratios. Let us assume that the slenderness ratio of the tube is between 12 and 55, in which case we use Eq. (11-86b):

$$\sigma_{allow} = 30.7 - 0.23\left(\frac{L}{r}\right) \text{ ksi} \qquad\qquad (c)$$

In this equation, we can replace the allowable stress by the actual stress P/A, that is, by the axial load divided by the cross-sectional area. The axial load $P = 5.0$ k, and the cross-sectional area is

$$A = \frac{\pi}{4}\left[d^2 - (d - 2t)^2\right] = \frac{\pi}{4}\left[d^2 - (0.8d)^2\right] = 0.2827d^2 \qquad (d)$$

Therefore, the stress P/A is

$$\frac{P}{A} = \frac{5.0 \text{ k}}{0.2827d^2} = \frac{17.69}{d^2}$$

in which P/A has units of kips per square inch (ksi) and d has units of inches (in.). Substituting into Eq. (c), we get

$$\frac{17.69}{d^2} = 30.7 - 0.23\left(\frac{L}{r}\right) \text{ ksi} \qquad\qquad (e)$$

The slenderness ratio L/r can also be expressed in terms of the diameter d. First, we find the moment of inertia and radius of gyration of the cross section:

$$I = \frac{\pi}{64}\left[d^4 - (d - 2t)^4\right] = \frac{\pi}{64}\left[d^4 - (0.8d)^4\right] = 0.02898d^4$$

$$r = \sqrt{\frac{I}{A}} = \sqrt{\frac{0.02898d^4}{0.2827d^2}} = 0.3202d$$

Therefore, the slenderness ratio is

$$\frac{L}{r} = \frac{16.0 \text{ in.}}{0.3202d} = \frac{49.97 \text{ in.}}{d}$$

where (as before) the diameter d has units of inches. Substituting into Eq. (e), we obtain the following equation, in which d is the only unknown quantity:

$$\frac{17.69}{d^2} = 30.7 - 0.23\left(\frac{49.97}{d}\right)$$

With a little rearranging, this equation becomes

$$30.7d^2 - 11.49d - 17.69 = 0$$

from which we find

$$d = 0.97 \text{ in.}$$

This result is satisfactory provided the slenderness ratio is between 12 and 55, as we assumed when we selected Eq. (c). To verify that this is the case, we calculate the slenderness ratio:

$$\frac{L}{r} = \frac{49.97 \text{ in.}}{d} = \frac{49.97 \text{ in.}}{0.97 \text{ in.}} = 51.5$$

Therefore, the solution is valid, and the minimum required diameter is

$$d_{min} = 0.97 \text{ in.}$$

←

■ **Example 11-8**

A wood post of rectangular cross-section (Fig. 11-40) is constructed of structural grade, Douglas fir lumber ($F_c = 15$ MPa, $E = 14$ GPa). The cross-sectional dimensions are $b = 120$ mm and $h = 160$ mm. Assume that the supports provide pinned-end conditions. (a) Determine the allowable axial load P_{allow} if the length $L = 1.8$ m. (b) Determine the maximum allowable length L_{max} if the axial load $P = 140$ kN.

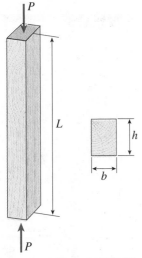

Fig. 11-40 Example 11-8. Wood post

Solution   We will use the AFPA formulas (Eqs. 11-88a, b, and c) when analyzing this post. The critical L/d ratio (Eq. 11-89) is

$$K_c = \sqrt{\frac{0.45E}{F_c}} = \sqrt{\frac{0.45(14\ \text{GPa})}{15\ \text{MPa}}} = 20.5$$

Since buckling may occur in any direction, the actual L/d ratio is

$$\frac{L}{d} = \frac{L}{0.120\ \text{m}}$$

in which L has units of meters and d is the least dimension of the cross section, that is, $d = b$.

(a) *Allowable axial load.* Since the length $L = 1.8$ m, the L/d ratio is

$$\frac{L}{d} = \frac{1.8\ \text{m}}{0.120\ \text{m}} = 15$$

which is between 11 and K_c. Therefore, we use Eq. (11-88b) to obtain the allowable stress:

$$\sigma_{\text{allow}} = F_c\left[1 - \frac{1}{3}\left(\frac{L/d}{K_c}\right)^4\right] = (15\ \text{MPa})\left[1 - \frac{1}{3}\left(\frac{15}{20.5}\right)^4\right] = 13.6\ \text{MPa}$$

Knowing the allowable stress, we can now calculate the allowable axial load:

$$P_{\text{allow}} = \sigma_{\text{allow}}A = \sigma_{\text{allow}}bh = (13.6\ \text{MPa})(120\ \text{mm} \times 160\ \text{mm}) = 261\ \text{kN} \quad \longleftarrow$$

(b) *Maximum allowable length.* Since the length of the column is unknown at this stage, we do not know the L/d ratio. Therefore, we do not know which formula to use for the allowable stress. As a trial, we will use Eq. (11-88c), for which L/d is between K_c and 50:

$$\sigma_{\text{allow}} = \frac{0.3E}{(L/d)^2} = \frac{0.3(14\ \text{GPa})}{(L/d)^2} = \frac{4200\ \text{MPa}}{(L/d)^2}$$

Now we replace σ_{allow} by the actual stress P/A, which is

$$\frac{P}{A} = \frac{140\ \text{kN}}{(120\ \text{mm})(160\ \text{mm})} = 7.29\ \text{MPa}$$

and obtain

$$7.29\ \text{MPa} = \frac{4200\ \text{MPa}}{(L/d)^2}$$

Thus, the L/d ratio is

$$\frac{L}{d} = 24$$

which is between K_c and 50, as assumed. Therefore, the maximum allowable length is

$$L_{\text{max}} = 24d = 24(120\ \text{mm}) = 2.88\ \text{m} \quad \longleftarrow$$

PROBLEMS/CHAPTER 11

IDEALIZED BUCKLING MODELS

11.2-1 through 11.2-5. The figure shows an idealized structure consisting of one or more rigid bars with pinned connections and linearly elastic springs. Each spring has rotational stiffness β_R. Determine the critical load P_{cr} for the structure. *Hints:* (1) When writing equations, assume that the displacements and angles of rotation are small. (2) When two critical loads are found (Problems 11.2-4 and 11.2-5), use the lower load.

Prob. 11.2-1   Prob. 11.2-2

Prob. 11.2-3   Prob. 11.2-4

Prob. 11.2-5

CRITICAL LOADS OF COLUMNS WITH PINNED SUPPORTS

The problems for Section 11.3 are to be solved using the assumptions of ideal, slender, prismatic, linearly elastic columns (Euler buckling). Buckling occurs in the plane of the figure unless otherwise stated.

11.3-1 Calculate the critical load P_{cr} for a W 8 × 35 steel column (see figure) having length $L = 24$ ft and $E = 30 \times 10^6$ psi under the following conditions: (a) The column buckles by bending about its strong axis (axis 1-1), and (b) the column buckles by bending about its weak axis (axis 2-2). In both cases, assume that the column has pinned ends.

Probs. 11.3-1 through 11.3-3

11.3-2 Solve the preceding problem for a W 10 × 60 steel column having length L = 30 ft.

11.3-3 Solve Problem 11.3-1 for a W 10 × 45 steel column having length L = 28 ft.

11.3-4 A horizontal beam AB is supported by a pinned-end column CD, as shown in the figure. The column is a solid steel bar (E = 200 GPa) of square cross section having length L = 1.8 m and side dimensions b = 50 mm. Based upon the critical load of the column, determine the allowable load Q if the factor of safety with respect to buckling is n = 2.0.

Probs. 11.3-4 and 11.3-5

11.3-5 Solve the preceding problem if the column is aluminum (E = 10 × 10^6 psi), the length L = 25 in., the side dimension b = 1.5 in., and the factor of safety n = 1.8.

11.3-6 A horizontal beam AB is pin-supported at end A and carries a load Q at end B, as shown in the figure. The beam is supported at C and D by two identical pinned-end columns of length L. Each column has flexural rigidity EI. What is the critical load Q_{cr}? (In other words, at what load Q_{cr} does the system collapse because of Euler buckling of the columns?)

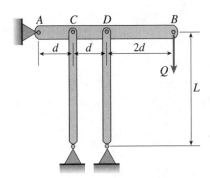

Prob. 11.3-6

11.3-7 A slender bar AB with pinned ends and length L is held between immovable supports (see figure). What increase ΔT in the temperature of the bar will produce buckling at the Euler load?

Prob. 11.3-7

11.3-8 A rectangular column with cross-sectional dimensions b and h is pin-supported at ends A and C (see figure). At midheight, the column is restrained in the plane of the figure but is free to deflect perpendicular to the plane of the figure. Determine the ratio h/b such that the critical load is the same for buckling in the two principal planes of the column.

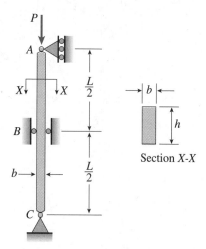

Section X-X

Prob. 11.3-8

11.3-9 Three identical, solid circular rods, each of radius r and length L, are placed together to form a compression member (see the cross section shown in the figure). Assuming pinned-end conditions, determine the critical load P_{cr} as follows: (a) The rods act independently as individual columns, and (b) the rods are bonded by epoxy throughout their lengths so that they function as a single member.

Prob. 11.3-9

11.3-10 Three pinned-end columns of the same material have the same length and the same cross-sectional area (see figure). The columns are free to buckle in any direction. The columns have cross sections as follows: (1) a circle, (2) a square, and (3) an equilateral triangle. Determine the ratios $P_1 : P_2 : P_3$ of the critical loads for these columns.

Prob. 11.3-10

11.3-11 A long slender column ABC is pinned at ends A and C and compressed by an axial force P (see figure). At the midpoint B, lateral support is provided to prevent deflection in the plane of the figure. The column is a steel wide-flange section (W 10 × 45) with $E = 30 × 10^6$ psi. The distance between lateral supports is $L = 18$ ft. Calculate the allowable load P using a factor of safety $n = 2.5$, taking into account the possibility of Euler buckling about either principal centroidal axis 1-1 or 2-2.

Prob. 11.3-11

11.3-12 The multifaceted glass roof over the lobby of a museum building is supported by the use of pretensioned cables. At a typical joint in the roof structure, a strut AB is compressed by the action of tensile forces F in a cable that makes an angle $\alpha = 75°$ with the strut (see figure). The strut is a circular tube of aluminum ($E = 72$ GPa) with outer diameter $d_2 = 50$ mm and inner diameter $d_1 = 40$ mm. The strut is 1.0 m long and is assumed to be pin-connected at both ends. Using a factor of safety $n = 2.25$ with respect to the critical load, determine the allowable force F in the cable.

Prob. 11.3-12

11.3-13 The hoisting arrangement for lifting a large pipe is shown in the figure. The spreader is a steel tubular section with outer diameter 2.75 in. and inner diameter 2.25 in. Its length is 8.5 ft and its modulus of elasticity is 29×10^6 psi. Based upon a factor of safety of 2.0 with respect to Euler buckling of the spreader, what is the maximum weight of pipe that can be lifted? (Assume pinned conditions at the ends of the spreader.)

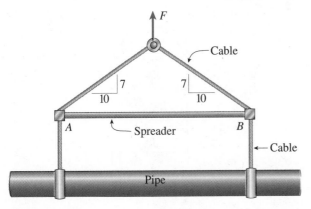

Prob. 11.3-13

11.3-14 A pinned-end strut of aluminum ($E = 72$ GPa) with length $L = 1.8$ m is constructed of circular tubing with outside diameter $d = 50$ mm (see figure). The strut must resist an axial load $P = 18$ kN with a factor of safety $n = 2.0$ with respect to the critical load. Determine the required thickness t of the tube.

Prob. 11.3-14

11.3-15 The cross section of a column built up of two steel I-beams (S 6 × 17.25 sections) is shown in the figure. The beams are connected by spacer bars (represented by the dashed lines) to ensure that they act together as a single column. The column is assumed to have pinned ends and may buckle in any direction. Assuming $E = 30 \times 10^6$ psi and $L = 27.5$ ft, calculate the critical load P_{cr} for the column.

Prob. 11.3-15

11.3-16 The truss ABC shown in the figure supports a vertical load W. Each member is a slender circular steel pipe ($E = 200$ GPa) with outside diameter 100 mm and wall thickness 6.0 mm. The distance between supports is $L = 7$ m. Determine the critical value W_{cr} of the load.

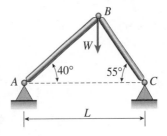

Prob. 11.3-16

11.3-17 The three-dimensional framework shown in the figure consists of a steel strut AB (modulus of elasticity $E = 30 \times 10^6$ psi) supported by cables BC and BD. A vertical load $W = 2000$ lb is supported at joint B. The strut has a hollow circular cross section (outer diameter 2.0 in. and inner diameter 1.8 in.) and is pinned at both ends. Triangle BCD is equilateral (side dimension $2b = 90$ in.) and lies in a horizontal plane parallel to the xz plane. A factor of safety of 2.0 against Euler buckling of the strut is required. Considering only the critical load of the strut, determine the smallest allowable value of the height H of the framework.

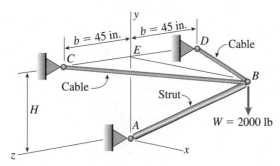

Prob. 11.3-17

*11.3-18 A pin-connected truss ABC (see figure) is composed of two bars of the same material with identical cross sections. The angle between the two bars is 90°, and the angle between bar AB and the horizontal is β. A load P is applied at joint B at an angle θ from the prolongation of line AB. The angle θ may be varied from 0 to 90°. (a) Assuming that failure occurs by Euler buckling of the bars, obtain a formula for the angle θ (in terms of the angle β) such that the load P will have its largest value. (b) From the formula, calculate the angle θ for a truss with $\beta = 60°$.

Prob. 11.3-18

*11.3-19 A truss ABC supports a load W at joint B, as shown in the figure. The length L_1 of member AB is fixed, but the length of strut BC varies as the angle θ is changed. Strut BC has a solid circular cross section. Assuming that collapse occurs by Euler buckling of the strut, determine the angle θ for minimum weight of the strut.

Prob. 11.3-19

COLUMNS WITH OTHER SUPPORT CONDITIONS

WORK
SHEET
12

The problems for Section 11.4 are to be solved using the assumptions of ideal, slender, prismatic, linearly elastic columns (Euler buckling). Buckling occurs in the plane of the figure unless otherwise stated.

11.4-1 An aluminum pipe column ($E = 10,400$ ksi) with length $L = 10.5$ ft has inside and outside diameters $d_1 = 5.0$ in. and $d_2 = 6.0$ in., respectively (see figure). The column is supported only at the ends and may buckle in any direction. Calculate the critical load P_{cr} for the following end conditions: (1) pinned-pinned, (2) fixed-free, (3) fixed-pinned, and (4) fixed-fixed.

Probs. 11.4-1 and 11.4-2

11.4-2 Solve the preceding problem for a steel pipe column ($E = 210$ GPa) with length $L = 1500$ mm, inner diameter $d_1 = 36$ mm, and outer diameter $d_2 = 40$ mm.

11.4-3 A wide-flange steel column ($E = 30 \times 10^6$ psi) of W 12×87 shape (see figure) has length $L = 30$ ft. It is supported only at the ends and may buckle in any direction. Calculate the allowable load P_{allow} based upon the critical load and using a factor of safety $n = 2.25$. Consider the following end conditions: (1) pinned-pinned, (2) fixed-free, (3) fixed-pinned, and (4) fixed-fixed.

Probs. 11.4-3 and 11.4-4

11.4-4 Solve the preceding problem for a W 10×60 shape with length $L = 25$ ft.

11.4-5 The upper end of a wide-flange steel column ($E = 30 \times 10^3$ ksi) is supported laterally between two pipes (see figure). The pipes are not attached to the column, and friction between the pipes and the column is unreliable. The base of the column provides a fixed support. The column is a W 8×21 section, 14 ft long. Determine the critical load for the column, considering Euler buckling in the plane of the web and also perpendicular to the plane of the web.

W8 × 21

Prob. 11.4-5

11.4-6 A vertical post AB is embedded in a concrete foundation and held at the top by two cables (see figure). The post is a hollow steel tube with modulus of elasticity 200 GPa, outer diameter 40 mm, and thickness 5 mm. The cables are tightened equally by turnbuckles. If a factor of safety of 3.0 against Euler buckling in the plane of the figure is desired, what is the maximum allowable tensile force T_{allow} in the cables?

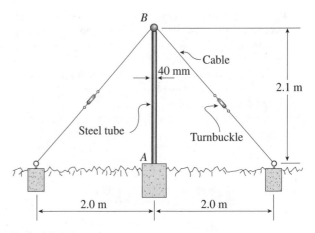

Prob. 11.4-6

11.4-7 The horizontal beam ABC shown in the figure is supported by columns BD and CE. The beam is prevented from moving horizontally by the roller support at end A, but vertical displacement at end A is free to occur. Each column is pinned at its upper end to the beam, but at the lower ends, support D is fixed and support E is pinned. Both columns are solid steel bars ($E = 30 \times 10^6$ psi) of square cross section with width equal to 0.625 in. A load Q acts at distance a from column BD. (a) If the distance $a = 10$ in., what is the critical value Q_{cr} of the load? (b) If the distance a can be varied between 0 and 40 in., what is the maximum possible value of Q_{cr}? What is the corresponding value of the distance a?

Prob. 11.4-7

11.4-8 The roof beams of a warehouse are supported by pipe columns (see figure) having outer diameter $d_2 = 100$ mm and inner diameter $d_1 = 90$ mm. The columns have length $L = 4.9$ m, modulus $E = 210$ GPa, and fixed supports at the base. Calculate the critical load P_{cr} of one of the columns using the following assumptions: (1) the upper end is pinned and the beam prevents horizontal displacement; (2) the upper end is fixed against rotation and the beam prevents horizontal displacement; (3) the upper end is pinned but the beam is free to move horizontally; and (4) the upper end is fixed against rotation but the beam is free to move horizontally.

Prob. 11.4-8

11.4-9 An aluminum tube AB of circular cross section is fixed at the base and pinned at the top to a horizontal beam supporting a load $Q = 45$ k (see figure). Determine the required thickness t of the tube if its outside diameter d is 4.5 in. and the desired factor of safety with respect to Euler buckling is $n = 2.5$. (Assume $E = 10,500$ ksi.)

Prob. 11.4-9

11.4-10 Determine the critical load P_{cr} and the equation of the buckled shape for an ideal column with fixed ends (see Fig. 11-17) by solving the differential equation of the deflection curve.

* 11.4-11 The frame ABC consists of two members AB and BC that are rigidly connected at joint B, as shown in part (a) of the figure. The frame has pin supports at A and C. A concentrated load P acts at joint B, thereby placing member AB in direct compression. To assist in determining the buckling load for member AB, we represent it as a pinned-end column, as shown in part (b) of the figure. At the top of the column, a rotational spring of stiffness β_R represents the restraining action of the horizontal beam BC on the column (note that the horizontal beam provides resistance to rotation of joint B when the column buckles). (a) By solving the differential equation of the deflection curve, derive the following buckling equation for this column:

$$\frac{\beta_R L}{EI}(kL \cot kL - 1) - k^2 L^2 = 0$$

in which L is the length of the column and EI is its flexural rigidity. (b) For the particular case when member BC is identical to member AB, the rotational stiffness β_R equals $3EI/L$ (see Case 7, Table G-2, Appendix G). For this special case, calculate the critical load P_{cr}.

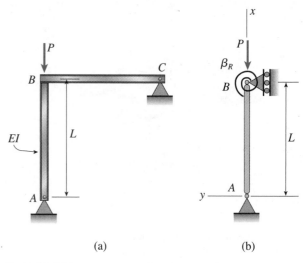

(a)   (b)

Prob. 11.4-11

* 11.4-12 Column AB is fixed at the base and pin-supported at the top by a linearly elastic spring of stiffness b (see figure). (a) By solving the differential equation of the deflection curve, derive the following buckling equation for this column:

$$\frac{\beta L^3}{EI}(\tan kL - kL) + k^3L^3 = 0$$

in which L is the length of the column and EI is its flexural rigidity. (b) For the particular case when $\beta = 2EI/L^3$, calculate the critical load P_{cr}.

Prob. 11.4-12

COLUMNS WITH ECCENTRIC AXIAL LOADS

When solving the problems for Section 11.5, assume that bending occurs in the principal plane containing the eccentric axial load.

11.5-1 Determine the bending moment M in the pinned-end column with eccentric axial loads shown in Fig. 11-21. Then plot the bending-moment diagram for an axial load $P = 0.3P_{cr}$. (*Note:* Express the moment as a function of the distance x from the end of the column, and plot the diagram in nondimensional form with M/Pe as ordinate and x/L as abscissa.)

11.5-2 Plot the load-deflection diagram for a pinned-end column with eccentric axial loads (Fig. 11-21) if the eccentricity e of the load is 5 mm and the column has length $L = 3.6$ m, moment of inertia $I = 9.0 \times 10^6$ mm⁴, and modulus of elasticity $E = 210$ GPa. (*Note:* Plot the axial load as ordinate and the deflection at the midpoint as abscissa.)

11.5-3 Solve the preceding problem for a column with $e = 0.20$ in., $L = 12$ ft, $I = 21.7$ in.⁴, and $E = 30 \times 10^6$ psi.

11.5-4 A steel bar having a square cross section (50 mm $\times$ 50 mm) and length $L = 2.0$ m is simply supported at the ends and compressed by axial loads that have a resultant $P = 60$ kN acting at the midpoint of one side of the cross

section (see figure). Assuming that the modulus of elasticity E is equal to 210 GPa, calculate the maximum deflection δ of the bar and the maximum bending moment M_{max}.

Prob. 11.5-4   Prob. 11.5-5

11.5-5 An aluminum bar having a rectangular cross section (2.0 in. $\times$ 1.0 in.) and length $L = 30$ in. is simply supported at the ends and compressed by axial loads that have a resultant $P = 2800$ lb acting at the midpoint of the long side of the cross section (see figure). Assuming that the modulus of elasticity E is equal to 10×10^6 psi, calculate the maximum deflection δ of the bar and the maximum bending moment M_{max}.

11.5-6 A wide-flange member (W 8 $\times$ 15) supports a compressive load that has its resultant P acting at the point shown in the figure. The member has modulus of elasticity $E = 29,000$ ksi and is simply supported at the ends. Lateral supports prevent any bending about the weak axis of the cross section. If the length of the member is 20 ft and the deflection is limited to 1/4 inch, what is the maximum allowable load P_{allow}?

Prob. 11.5-6

11.5-7 A wide-flange member (W 10 × 30) is compressed by axial loads that have their resultant $P = 20$ k acting at the point shown in the figure. The material is steel with modulus of elasticity $E = 29{,}000$ ksi. Assuming that the member is simply supported at the ends, determine the maximum permissible length L_{max} if the deflection is not to exceed 1/400th of the length.

$P = 20$ k

W 10 × 30

Prob. 11.5-7

11.5-8 Solve the preceding problem (W 10 × 30) if the resultant force P equals 25 k.

11.5-9 The column shown in the figure is fixed at the base and free at the upper end. A compressive load P acts at the top of the column with an eccentricity e from the axis of the column. Beginning with the differential equation of the deflection curve, derive formulas for the maximum deflection δ of the column and the maximum bending moment M_{max} in the column.

(a)   (b)

Prob. 11.5-9

11.5-10 An aluminum box column of square cross section is fixed at the base and free at the top (see figure). The outside dimension b of each side is 100 mm and the thickness t of the wall is 8 mm. The resultant of the compressive loads acting on the top of the column is a force $P = 50$ kN acting at the outer edge of the column at the midpoint of one side. What is the longest permissible length L_{max} of the column if the deflection at the top is not to exceed 30 mm? (Assume $E = 73$ GPa.)

Section A-A

Probs. 11.5-10 and 11.5-11

11.5-11 Solve the preceding problem for an aluminum column with $b = 6.0$ in., $t = 0.5$ in., $P = 30$ k, and $E = 10.6 \times 10^3$ ksi. The deflection at the top is limited to 2.0 in.

11.5-12 A steel post AB of hollow circular cross section is fixed at the base and free at the top (see figure). The inner and outer diameters are $d_1 = 96$ mm and $d_2 = 110$ mm, respectively, and the length $L = 4.0$ m. A cable CBD passes through a fitting that is welded to the side of the post. The distance between the plane of the cable (plane CBD) and the axis of the post is $e = 100$ mm, and the angles between the cable and the ground are $\alpha = 53.13°$. The cable is pretensioned by tightening the turnbuckles. If the deflection at the top of the post is limited to $\delta = 20$ mm, what is the maximum allowable tensile force T in the cable? (Assume $E = 205$ GPa.)

$e = 100$ mm

$L = 4.0$ m

B

Cable

d_1

d_2

$\alpha = 53.13°$

$-d_2$

$\alpha = 53.13°$

C

A

D

Prob. 11.5-12

11.5-13 A steel I-beam (S 8 × 23 shape) used as a column supports a load P applied on axis 1-1 at the position shown in the figure (that is, the eccentricity of the load equals half the width of the flange). The column is fixed at the base, free at the top, and 14 ft long. Assuming $E = 30 \times 10^3$ ksi, determine the load P that will produce a maximum bending moment of 40 k-in. in the column.

P

2

1———1

2

Section A-A

Probs. 11.5-13 and 11.5-14

11.5-14 Solve the preceding problem for a 16-ft long column of S 10 × 35 shape if the maximum bending moment is 70 k-in.

11.5-15 A frame *ABCD* is constructed of steel wide-flange members (W 8 × 21; $E = 30 \times 10^6$ psi) and subjected to triangularly distributed loads of maximum intensity q_0 acting along the vertical members (see figure). The distance between supports is $L = 20$ ft and the height of the frame is $h = 4$ ft. (a) Calculate the intensity of load q_0 required to produce a maximum bending moment of 80 k-in. in the horizontal member *BC*. (b) If the load q_0 is reduced to one-half of the value calculated in part (a), what is the maximum bending moment in member *BC*? What is the ratio of this moment to the moment of 80 k-in. in part (a)?

A

D

h

E

B

C

E

q_0

q_0

E

L

Section E-E

Prob. 11.5-15

THE SECANT FORMULA

When solving the problems for Section 11.6, assume that bending occurs in the principal plane containing the eccentric axial load.

11.6-1 A steel bar has a square cross section of width $b = 2.0$ in. (see figure). The bar has pinned supports at the ends and is 3.0 ft long. The axial forces acting at the end of the bar have a resultant $P = 20$ k located at distance $e = 0.75$ in. from the center of the cross section. Also, the modulus of elasticity of the steel is 29,000 ksi. (a) Determine the maximum compressive stress σ_{max} in the bar. (b) If the allowable stress in the steel is 18,000 psi, what is the maximum permissible length L_{max} of the bar?

Probs. 11.6-1 through 11.6-3

Probs. 11.6-4 through 11.6-6

11.6-2 A brass bar ($E = 100$ GPa) with a square cross section is subjected to axial forces having their resultant P acting at distance e from the center (see figure). The bar is simply supported at the ends and is 0.6 m in length. The side dimension b of the bar is 30 mm and the eccentricity e of the load is 10 mm. If the allowable stress in the brass is 150 MPa, what is the allowable axial force P_{allow}?

11.6-3 A square aluminum bar with simply supported ends carries a load $P = 25$ k acting at distance $e = 2.0$ in. from the center (see figure). The bar has length $L = 54$ in. and modulus of elasticity $E = 10,600$ ksi. If the stress in the bar is not to exceed 6 ksi, what is the minimum permissible width b_{min} of the bar?

11.6-4 A pinned-end column of length $L = 2.1$ m is constructed of steel pipe ($E = 210$ GPa) having inside diameter $d_1 = 60$ mm and outside diameter $d_2 = 68$ mm (see figure). A compressive load $P = 10$ kN acts with eccentricity $e = 30$ mm. (a) What is the maximum compressive stress σ_{max} in the column? (b) If the allowable stress in the steel is 50 MPa, what is the maximum permissible length L_{max} of the column?

11.6-5 A pinned-end strut of length $L = 5.2$ ft is constructed of steel pipe ($E = 30 \times 10^3$ ksi) having inside diameter $d_1 = 2.0$ in. and outside diameter $d_2 = 2.2$ in. (see figure). A compressive load $P = 2.0$ k is applied with eccentricity $e = 1.0$ in. (a) What is the maximum compressive stress σ_{max} in the strut? (b) What is the allowable load P_{allow} if a factor of safety $n = 2$ with respect to yielding is required? (Assume that the yield stress σ_y of the steel is 42 ksi.)

11.6-6 A circular aluminum tube with pinned ends supports a load $P = 18$ kN acting at distance $e = 50$ mm from the center (see figure). The length of the tube is 3.5 m and its modulus of elasticity is 73 GPa. If the maximum permissible stress in the tube is 20 MPa, what is the required outer diameter d_2 if the ratio of diameters is to be $d_1/d_2 = 0.9$?

11.6-7 A copper tube AB is loaded by forces P acting at an eccentricity e, as shown in the figure. Data for the tube are as follows: $L = 30$ in., $d_2 = 2.0$ in., $t = 0.1$ in., and $E = 18,000$ ksi. (a) Assuming that the compressive force P is equal to 2.0 k, construct a graph showing the maximum compressive stress σ_{max} in the tube versus the eccentricity e of the load. (Plot σ_{max} as ordinate and e as abscissa, and let e vary from 0 to 3.0 in.) (b) Assuming that the eccentricity e is equal to 0.5 in., construct a graph showing the maximum compressive stress σ_{max} versus the force P. (Plot σ_{max} as ordinate and P as abscissa, and let P vary from 0 to 20 k.)

Prob. 11.6-7

Prob. 11.6-9

11.6-8 A steel column ($E = 30 \times 10^3$ ksi) with pinned ends is constructed of a W 10 × 60 wide-flange shape (see figure). The column is 24 ft long. The resultant of the axial loads acting on the column is a force P acting with an eccentricity $e = 2.0$ in. (a) If $P = 120$ k, determine the maximum compressive stress σ_{max} in the column. (b) Determine the allowable load P_{allow} if the yield stress is $\sigma_y = 42$ ksi and the factor of safety with respect to yielding of the material is $n = 2.5$.

11.6-10 A steel column ($E = 30 \times 10^3$ ksi) that is fixed at the base and free at the top is constructed of a W 8 × 35 wide-flange member (see figure). The column is 9.0 ft long. The force P acting at the top of the column has an eccentricity $e = 1.25$ in. (a) If $P = 40$ k, what is the maximum compressive stress in the column? (b) If the yield stress is 36 ksi and the required factor of safety with respect to yielding is 2.1, what is the allowable load P_{allow}?

Prob. 11.6-8

Probs. 11.6-10 and 11.6-11

11.6-9 A W 16 × 57 steel column is compressed by a force $P = 75$ k acting with an eccentricity $e = 1.5$ in., as shown in the figure. The column is simply supported at the ends and has length L. Also, the steel has modulus of elasticity $E = 30 \times 10^3$ ksi and yield stress $\sigma_y = 36$ ksi. (a) If the length $L = 10$ ft, what is the maximum compressive stress σ_{max} in the column? (b) If a factor of safety $n = 2.0$ is required with respect to yielding, what is the longest permissible length L_{max} of the column?

11.6-11 A W 12 × 50 wide-flange steel column with length $L = 12.5$ ft is fixed at the base and free at the top (see figure). The load P acting on the column is intended to be centrally applied, but because of unavoidable discrepancies in construction, an eccentricity ratio of 0.25 is specified.

Also, the following data are supplied: $E = 30 \times 10^3$ ksi, $\sigma_y = 42$ ksi, and $P = 70$ k. (a) What is the maximum compressive stress σ_{max} in the column? (b) What is the factor of safety n with respect to yielding of the steel?

11.6-12 A vertical post constructed from a C 8 × 11.5 channel section is fixed at the base and free at the top (see figure). The channel has length $L = 5.0$ ft and modulus of elasticity $E = 30 \times 10^6$ psi. A load $P = 4000$ lb acts on the axis of symmetry of the cross section and at the back of the channel. Determine the maximum compressive and tensile stresses in the post.

Section A-A

Probs. 11.6-12 and 11.6-13

11.6-13 Solve the preceding problem for a C 6 × 13 channel section with $L = 4.0$ ft, $E = 29 \times 10^6$ psi, and $P = 6000$ lb.

11.6-14 A pinned-end column with length $L = 18$ ft is constructed from a W 12 × 87 wide-flange shape (see figure). The column is subjected to a centrally applied load $P_1 = 180$ k and an eccentrically applied load $P_2 = 75$ k. The load P_2 acts at distance $s = 5.0$ in. from the centroid of the cross section. The properties of the steel are $E = 29,000$ ksi and $\sigma_y = 36$ ksi. (a) Calculate the maximum compressive stress in the column. (b) Determine the factor of safety with respect to yielding.

Wide-flange column

Probs. 11.6-14 and 11.6-15

11.6-15 The wide-flange column shown in the figure is simply supported at the ends and carries two loads, a force $P_1 = 100$ k acting at the centroid and a force $P_2 = 60$ k acting at distance $s = 4.0$ in. from the centroid. The column is a W 10 × 45 shape with $L = 13.5$ ft, $E = 29 \times 10^3$ ksi, and $\sigma_y = 42$ ksi. (a) What is the maximum compressive stress in the column? (b) If the load P_1 remains at 100 k, what is the largest permissible value of the load P_2 in order to maintain a factor of safety of 2.0 with respect to yielding?

11.6-16 A W 14 × 53 wide-flange column of length $L = 15$ ft is fixed at the base and free at the top (see figure). The column supports a centrally applied load $P_1 = 120$ k and a load $P_2 = 40$ k supported on a bracket. The distance from the centroid of the column to the load P_2 is $s = 12$ in. Also, the modulus of elasticity is $E = 29,000$ ksi and the yield stress is $\sigma_y = 36$ ksi. (a) Calculate the maximum compressive stress in the column. (b) Determine the factor of safety with respect to yielding.

Section A-A

Probs. 11.6-16 and 11.6-17

11.6-17 A wide-flange column with a bracket is fixed at the base and free at the top (see figure). The column supports a load $P_1 = 75$ k acting at the centroid and a load $P_2 = 25$ k acting on the bracket at distance $s = 10.0$ in. from the load P_1. The column is a W 12 × 35 shape with $L = 16$ ft, $E = 29 \times 10^3$ ksi, and $\sigma_y = 42$ ksi. (a) What is the maximum compressive stress in the column? (b) If the load P_1 remains at 75 k, what is the largest permissible value of the load P_2 in order to maintain a factor of safety of 1.8 with respect to yielding?

DESIGN FORMULAS FOR COLUMNS

The problems for Section 11.9 are to be solved assuming that the axial loads are centrally applied at the ends of the columns. Unless otherwise stated, the columns may buckle in any direction.

STEEL COLUMNS

11.9-1 Determine the allowable axial load P_{allow} for a W 8 × 35 steel wide-flange column with pinned ends (see figure) for each of the following lengths: $L = 16$ ft, 20 ft, 24 ft, and 28 ft. (Assume $E = 29,000$ ksi and $\sigma_y = 36$ ksi.)

Section A-A

Probs. 11.9-1 through 11.9-6

11.9-2 Determine the allowable axial load P_{allow} for a W 10 × 30 steel wide-flange column with pinned ends (see figure) for each of the following lengths: $L = 12$ ft, 14 ft, 16 ft, and 18 ft. (Assume $E = 29,000$ ksi and $\sigma_y = 36$ ksi.)

11.9-3 Determine the allowable axial load P_{allow} for a W 12 × 50 steel wide-flange column with pinned ends (see figure) for each of the following lengths: $L = 12$ ft, 15 ft, 18 ft, and 21 ft. (Assume $E = 29,000$ ksi and $\sigma_y = 50$ ksi.)

11.9-4 Select a steel wide-flange column of nominal depth 10 in. (W 10 shape) to support an axial load $P = 180$ k (see figure). The column has pinned ends and length $L = 14$ ft. Assume $E = 29,000$ ksi and $\sigma_y = 36$ ksi. (*Note:* The selection of columns is limited to those listed in Table E-1, Appendix E.)

11.9-5 Select a steel wide-flange column of nominal depth 12 in. (W 12 shape) to support an axial load $P = 175$ k (see figure). The column has pinned ends and length $L = 35$ ft. Assume $E = 29,000$ ksi and $\sigma_y = 36$ ksi. (*Note:* The selection of columns is limited to those listed in Table E-1, Appendix E.)

11.9-6 Select a steel wide-flange column of nominal depth 14 in. (W 14 shape) to support an axial load $P = 250$ k (see figure). The column has pinned ends and length $L = 20$ ft. Assume $E = 29,000$ ksi and $\sigma_y = 50$ ksi. (*Note:* The selection of columns is limited to those listed in Table E-1, Appendix E.)

11.9-7 Determine the allowable axial load P_{allow} for a steel pipe column that is fixed at the base and free at the top (see figure) for each of the following lengths: $L = 6$ ft, 9 ft, 12 ft, and 15 ft. The column has outside diameter $d = 6.625$ in. and wall thickness $t = 0.280$ in. (Assume $E = 29,000$ ksi and $\sigma_y = 36$ ksi.)

Section A-A

Probs. 11.9-7 through 11.9-12

11.9-8 Determine the allowable axial load P_{allow} for a steel pipe column that is fixed at the base and free at the top (see figure) for each of the following lengths: $L = 2.6$ m, 2.8 m, 3.0 m, and 3.2 m. The column has outside diameter $d = 140$ mm and wall thickness $t = 7$ mm. (Assume $E = 200$ GPa and $\sigma_y = 250$ MPa.)

11.9-9 Determine the allowable axial load P_{allow} for a steel pipe column that is fixed at the base and free at the top (see figure) for each of the following lengths: L = 4 ft, 6 ft, 8 ft, and 10 ft. The column has outside diameter d = 4.50 in. and wall thickness t = 0.237 in. (Assume E = 29,000 ksi and σ_y = 36 ksi.)

11.9-10 Determine the maximum permissible length L_{max} for a steel pipe column that is fixed at the base and free at the top and must support an axial load P = 500 kN (see figure). The column has outside diameter d = 200 mm, wall thickness t = 10 mm, E = 200 GPa, and σ_y = 250 MPa.

11.9-11 Determine the maximum permissible length L_{max} for a steel pipe column that is fixed at the base and free at the top and must support an axial load P = 40 k (see figure). The column has outside diameter d = 4.0 in., wall thickness t = 0.226 in., E = 29,000 ksi, and σ_y = 42 ksi.

11.9-12 Determine the maximum permissible length L_{max} for a steel pipe column that is fixed at the base and free at the top and must support an axial load P = 400 kN (see figure). The column has outside diameter d = 160 mm, wall thickness t = 8 mm, E = 210 GPa, and σ_y = 290 MPa.

11.9-13 A frame ABC consists of a tension member AB and a compression member BC (see figure). Member BC is a steel pipe section that is fixed at the base and pinned at the top. Joint B is laterally supported so that sideways displacement is prevented. The outside and inside diameters of the pipe section are 3.5 in. and 3.0 in., respectively. A horizontal force H acts at joint B. (a) If the length L of member BC equals 12 ft, what is the maximum allowable force H_{max}? (b) If the force H equals 16 k, what is the maximum allowable length L_{max}? (Assume E = 29,000 ksi and σ_y = 50 ksi.)

Prob. 11.9-13

11.9-14 The steel columns used in a college recreation center are 55 ft long and are formed by welding three wide-flange sections (see figure). The columns are pin-supported at the ends and may buckle in any direction. Calculate the allowable load P_{allow} for one column, assuming E = 29,000 ksi and σ_y = 36 ksi.

W 12 × 87

W 24 × 162

W 12 × 87

Prob. 11.9-14

11.9-15 A W 8 × 28 steel wide-flange column with pinned ends carries an axial load P. What is the maximum permissible length L_{max} of the column if: (a) P = 60 k, and (b) P = 120 k? (Assume E = 29,000 ksi and σ_y = 36 ksi.)

Probs. 11.9-15 and 11.9-16

11.9-16 A W 10 × 45 steel wide-flange column with pinned ends carries an axial load P. What is the maximum permissible length L_{max} of the column if: (a) P = 125 k, and (b) P = 200 k? (Assume E = 29,000 ksi and σ_y = 42 ksi.)

11.9-17 Find the required outside diameter d for a steel pipe column (see figure) of length L = 20 ft that is pinned at both ends and must support an axial load P = 25 k. Assume that the wall thickness t is equal to $d/20$. (Use E = 29,000 ksi and σ_y = 36 ksi.)

Probs. 11.9-17 through 11.9-20

11.9-18 Find the required outside diameter d for a steel pipe column (see figure) of length $L = 3.5$ m that is pinned at both ends and must support an axial load $P = 130$ kN. Assume that the wall thickness t is equal to $d/20$. (Use $E = 200$ GPa and $\sigma_y = 275$ MPa.)

11.9-19 Find the required outside diameter d for a steel pipe column (see figure) of length $L = 11.5$ ft that is pinned at both ends and must support an axial load $P = 80$ k. Assume that the wall thickness t is 0.30 in. (Use $E = 29,000$ ksi and $\sigma_y = 42$ ksi.)

11.9-20 Find the required outside diameter d for a steel pipe column (see figure) of length $L = 3.0$ m that is pinned at both ends and must support an axial load $P = 800$ kN. Assume that the wall thickness t is 9 mm. (Use $E = 200$ GPa and $\sigma_y = 300$ MPa.)

ALUMINUM COLUMNS

11.9-21 An aluminum pipe column (alloy 2014-T6) with pinned ends has outside diameter $d_2 = 5.60$ in. and inside diameter $d_1 = 4.80$ in. (see figure). Determine the allowable axial load P_{allow} for each of the following lengths: $L = 6$ ft, 8 ft, 10 ft, and 12 ft.

Probs. 11.9-21 through 11.9-24

11.9-22 An aluminum pipe column (alloy 2014-T6) with pinned ends has outside diameter $d_2 = 120$ mm and inside diameter $d_1 = 110$ mm (see figure). Determine the allowable axial load P_{allow} for each of the following lengths: $L = 1.0$ m, 2.0 m, 3.0 m, and 4.0 m. (*Hint:* Convert given data to USCS units, determine the required quantities, and then convert back to SI units.)

11.9-23 An aluminum pipe column (alloy 6061-T6) that is fixed at the base and free at the top has outside diameter $d_2 = 3.25$ in. and inside diameter $d_1 = 3.00$ in. (see figure). Determine the allowable axial load P_{allow} for each of the following lengths: $L = 2$ ft, 3 ft, 4 ft, and 5 ft.

11.9-24 An aluminum pipe column (alloy 6061-T6) that is fixed at the base and free at the top has outside diameter

$d_2 = 80$ mm and inside diameter $d_1 = 72$ mm (see figure). Determine the allowable axial load P_{allow} for each of the following lengths: $L = 0.6$ m, 0.8 m, 1.0 m, and 1.2 m. (*Hint:* Convert the given data to USCS units, determine the required quantities, and then convert back to SI units.)

11.9-25 A solid round bar of aluminum having diameter d (see figure) is compressed by an axial force $P = 60$ k. The bar has simple supports and is made of alloy 2014-T6. (a) If the diameter $d = 2.0$ in., what is the maximum allowable length L_{max} of the bar? (b) If the length $L = 30$ in., what is the minimum required diameter d_{min}?

Probs. 11.9-25 through 11.9-28

11.9-26 A solid round bar of aluminum having diameter d (see figure) is compressed by an axial force $P = 175$ kN. The bar has simple supports and is made of alloy 2014-T6. (a) If the diameter $d = 40$ mm, what is the maximum allowable length L_{max} of the bar? (b) If the length $L = 0.6$ m, what is the minimum required diameter d_{min}? (*Hint:* Convert the given data to USCS units, determine the required quantities, and then convert back to SI units.)

11.9-27 A solid round bar of aluminum having diameter d (see figure) is compressed by an axial force $P = 10$ k. The bar has simple supports and is made of alloy 6061-T6. (a) If the diameter $d = 1.0$ in., what is the maximum allowable length L_{max} of the bar? (b) If the length $L = 20$ in., what is the minimum required diameter d_{min}?

11.9-28 A solid round bar of aluminum having diameter d (see figure) is compressed by an axial force $P = 60$ kN. The bar has simple supports and is made of alloy 6061-T6. (a) If the diameter $d = 30$ mm, what is the maximum allowable length L_{max} of the bar? (b) If the length $L = 0.6$ m, what is the minimum required diameter d_{min}? (*Hint:* Convert the given data to USCS units, determine the required quantities, and then convert back to SI units.)

WOOD COLUMNS

11.9-29 A wood post of rectangular cross section (see figure) is constructed of 4 in. × 6 in. structural grade, Douglas fir lumber ($F_c = 2,000$ psi, $E = 1,800,00$ psi). The net cross-sectional dimensions of the post are $b = 3.5$ in.

and $h = 5.5$ in. (see Appendix F). Assuming pinned-end conditions, calculate the allowable axial load P_{allow} for each of the following lengths: $L = 2.5$ ft, 5.0 ft, 7.5 ft, and 10.0 ft.

Probs. 11.9-29 through 11.9-32

11.9-30 A wood post of rectangular cross section (see figure) is constructed of structural grade, southern pine lumber ($F_c = 14$ MPa, $E = 12$ GPa). The cross-sectional dimensions of the post (actual dimensions) are $b = 100$ mm and $h = 150$ mm. Assuming pinned-end conditions, calculate the allowable axial load P_{allow} for each of the following lengths: $L = 1.0$ m, 1.5 m, 2.0 m, and 2.5 m.

11.9-31 A wood post of rectangular cross section (see figure) is constructed of 4 in. × 8 in. construction grade, western hemlock lumber ($F_c = 1,000$ psi, $E = 1,300,000$ psi). The net cross-sectional dimensions of the post are $b = 3.5$ in. and $h = 7.25$ in. (see Appendix F). Assuming pinned-end conditions, calculate the allowable axial load P_{allow} for each of the following lengths: $L = 4$ ft, 6 ft, 8 ft, and 10 ft.

11.9-32 A wood post of rectangular cross section (see figure) is constructed of structural grade, Douglas fir lumber ($F_c = 12$ MPa, $E = 10$ GPa). The cross-sectional dimensions of the post (actual dimensions) are $b = 140$ mm and $h = 210$ mm. Assuming pinned-end conditions, calculate the allowable axial load P_{allow} for each of the following lengths: $L = 1.5$ m, 2.5 m, 3.5 m, and 4.5 m.

11.9-33 A square wood column with side dimensions b (see figure) is constructed of a structural grade of Douglas fir for which $F_c = 1,700$ psi and $E = 1,400,000$ psi. The

ends of the column are assumed to be simply supported. An axial force $P = 40$ k acts on the column. (a) If the dimension $b = 5.5$ in., what is the maximum allowable length L_{max} of the column? (b) If the length $L = 11$ ft, what is the minimum required dimension b?

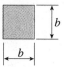

Probs. 11.9-33 through 11.9-36

11.9-34 A square wood column with side dimensions b (see figure) is constructed of a structural grade of southern pine for which $F_c = 10.5$ MPa and $E = 12$ GPa. The ends of the column are assumed to be simply supported. An axial force $P = 200$ kN acts on the column. (a) If the dimension $b = 150$ mm, what is the maximum allowable length L_{max} of the column? (b) If the length $L = 4.0$ m, what is the minimum required dimension b?

11.9-35 A square wood column with side dimensions b (see figure) is constructed of a structural grade of spruce for which $F_c = 900$ psi and $E = 1,500,000$ psi. The ends of the column are assumed to be simply supported. An axial force $P = 8.0$ k acts on the column. (a) If the dimension $b = 3.5$ in., what is the maximum allowable length L_{max} of the column? (b) If the length $L = 10$ ft, what is the minimum required dimension b?

11.9-36 A square wood column with side dimensions b (see figure) is constructed of a structural grade of eastern white pine for which $F_c = 8.0$ MPa and $E = 8.5$ GPa. The ends of the column are assumed to be simply supported. An axial force $P = 100$ kN acts on the column. (a) If the dimension $b = 120$ mm, what is the maximum allowable length L_{max} of the column? (b) If the length $L = 4.0$ m, what is the minimum required dimension b?

12 REVIEW OF CENTROIDS AND MOMENTS OF INERTIA

INTRODUCTION

This chapter is a review of the definitions and formulas pertaining to centroids and moments of inertia of plane areas. The word "review" is appropriate because these topics are usually covered in mathematics or statics courses, and most readers will already have been exposed to the material. However, since centroids and moments of inertia are used repeatedly throughout the preceding chapters, they must be clearly understood by the reader and the relevant definitions and formulas must be readily accessible.

Topics covered in this chapter include centroids and how to locate them, moments of inertia, polar moments of inertia, products of inertia, parallel-axis theorems, rotation of axes, and principal axes. Only plane areas are considered. A table of centroids and moments of inertia for a variety of common shapes is given in Appendix D for convenient reference.

The terminology used in this and earlier chapters may appear puzzling to some readers. For instance, the term "moment of inertia" is clearly a misnomer when referring to properties of an area, since no mass is involved. Even the word "area" is used inappropriately in the preceding discussions. When we say "plane area," we really mean "plane surface." Strictly speaking, area is a measure of the *size* of a surface and is not the same thing as the surface itself. In spite of its deficiencies, the terminology used in this book is so entrenched in the engineering literature that it rarely causes confusion.

12.2 CENTROIDS OF PLANE AREAS

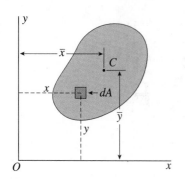

Fig. 12-1 Plane area of arbitrary shape with centroid C

The position of the centroid of a plane area is an important geometric property. To obtain formulas for locating centroids, we will refer to Fig. 12-1, which shows a plane area of irregular shape with its centroid at point C. The xy coordinate system is oriented arbitrarily with its origin at any point O. The **area** of the geometric figure* is defined by the following integral:

$$A = \int dA \tag{12-1}$$

in which dA is a differential element of area having coordinates x and y (Fig. 12-1) and A is the total area of the figure.

The **first moments** of the area with respect to the x and y axes are defined, respectively, as follows:

$$Q_x = \int y\, dA \qquad Q_y = \int x\, dA \tag{12-2a,b}$$

Thus, the first moments represent the sums of the products of the differential areas and their coordinates. First moments may be positive or negative, depending upon the position of the xy axes. Also, first moments have units of length raised to the third power; for instance, in.³ or mm³.

The coordinates $\bar{x}$ and $\bar{y}$ of the **centroid** C (Fig. 12-1) are equal to the first moments divided by the area:

$$\bar{x} = \frac{Q_y}{A} = \frac{\int x\, dA}{\int dA} \qquad \bar{y} = \frac{Q_x}{A} = \frac{\int y\, dA}{\int dA} \tag{12-3a,b}$$

If the boundaries of the area are defined by simple mathematical expressions, we can evaluate the integrals appearing in Eqs. (12-3a) and (12-3b) in closed form and thereby obtain formulas for $\bar{x}$ and $\bar{y}$. The formulas listed in Appendix D were obtained in this manner. In general, the coordinates $\bar{x}$ and $\bar{y}$ may be positive or negative, depending upon the position of the centroid with respect to the reference axes.

---

*As explained in the preceding paragraph, it would be awkward to say, "The area of the area." Therefore, in this instance we substitute the term "geometric figure."

Fig. 12-2 Area with one axis of symmetry

Fig. 12-3 Area with two axes of symmetry

Fig. 12-4 Area that is symmetric about a point

The use of Eqs. (12-3a) and (12-3b) for locating the centroid of an area is illustrated later in Example 12-1.

If an area is **symmetric about an axis**, the centroid must lie on that axis because the first moment about an axis of symmetry equals zero. For example, the centroid of the singly symmetric area shown in Fig. 12-2 must lie on the x axis, which is the axis of symmetry. Therefore, only one coordinate must be calculated in order to locate the centroid C.

If an area has **two axes of symmetry**, as illustrated in Fig. 12-3, the position of the centroid can be determined by inspection because it lies at the intersection of the axes of symmetry.

An area of the type shown in Fig. 12-4 is **symmetric about a point**. It has no axes of symmetry, but there is a point (called the **center of symmetry**) such that every line drawn through that point contacts the area in a symmetrical manner. The centroid of such an area coincides with the center of symmetry, and therefore the centroid can be located by inspection.

If an area has **irregular boundaries** not defined by simple mathematical expressions, we can locate the centroid by numerically evaluating the integrals in Eqs. (12-3a) and (12-3b). The simplest procedure is to divide the geometric figure into small finite elements and replace the integrations with summations. If we denote the area of the ith element by ΔA_i, then the expressions for the summations are

$$A = \sum_{i=1}^{n} \Delta A_i \qquad Q_x = \sum_{i=1}^{n} \bar{y}_i \Delta A_i \qquad Q_y = \sum_{i=1}^{n} \bar{x}_i \Delta A_i \qquad \text{(12-4a,b,c)}$$

in which n is the total number of elements, $\bar{y}_i$ is the y coordinate of the centroid of the ith element, and $\bar{x}_i$ is the x coordinate of the centroid of the ith element. Replacing the integrals in Eqs. (12-3a) and (12-3b) by the corresponding summations, we obtain the following formulas for the coordinates of the centroid:

$$\bar{x} = \frac{Q_y}{A} = \frac{\displaystyle\sum_{i=1}^{n} \bar{x}_i \Delta A_i}{\displaystyle\sum_{i=1}^{n} \Delta A_i} \qquad \bar{y} = \frac{Q_x}{A} = \frac{\displaystyle\sum_{i=1}^{n} \bar{y}_i \Delta A_i}{\displaystyle\sum_{i=1}^{n} \Delta A_i} \qquad \text{(12-5a,b)}$$

The accuracy of the calculations for $\bar{x}$ and $\bar{y}$ depends upon how closely the selected elements fit the actual area. If they fit exactly, the results are exact. Many computer programs for locating centroids use a numerical scheme similar to the one expressed by Eqs. (12-5a) and (12-5b).

■ **Example 12-1**

A parabolic semisegment *OAB* is bounded by the *x* axis, the *y* axis, and a parabolic curve having its vertex at *A* (Fig. 12-5). The equation of the curve is

$$y = f(x) = h\left(1 - \frac{x^2}{b^2}\right) \tag{a}$$

in which *b* is the base and *h* is the height of the semisegment. Locate the centroid *C* of the semisegment.

Solution   To determine the coordinates $\bar{x}$ and $\bar{y}$ of the centroid *C* (Fig. 12-5), we will use Eqs. (12-3a) and (12-3b). We begin by selecting an element of area *dA* in the form of a thin vertical strip of width *dx* and height *y*. The area of this differential element is

$$dA = ydx = h\left(1 - \frac{x^2}{b^2}\right)dx \tag{b}$$

Therefore, the area of the parabolic semisegment is

$$A = \int dA = \int_0^b h\left(1 - \frac{x^2}{b^2}\right)dx = \frac{2bh}{3} \tag{c}$$

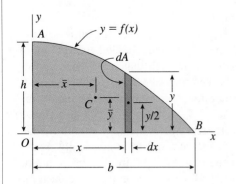

Fig. 12-5 Example 12-1. Centroid of a parabolic semisegment

Note that this area is 2/3 of the area of the surrounding rectangle.

The first moment of an element of area *dA* with respect to an axis is obtained by multiplying the area of the element by the distance from its centroid to the axis. Since the *x* and *y* coordinates of the centroid of the element shown in Fig. 12-5 are *x* and *y*/2, respectively, the first moments of the element with respect to the *x* and *y* axes are

$$Q_x = \int \frac{y}{2}dA = \int_0^b \frac{h^2}{2}\left(1 - \frac{x^2}{b^2}\right)^2 dx = \frac{4bh^2}{15} \tag{d}$$

$$Q_y = \int xdA = \int_0^b hx\left(1 - \frac{x^2}{b^2}\right)dx = \frac{b^2 h}{4} \tag{e}$$

in which we have substituted for *dA* from Eq. (b).

We can now determine the coordinates of the centroid *C*:

$$\bar{x} = \frac{Q_y}{A} = \frac{3b}{8} \qquad \bar{y} = \frac{Q_x}{A} = \frac{2h}{5} \tag{f,g} \quad \Longleftarrow$$

These results agree with the formulas listed in Appendix D, Case 17.

Notes: The centroid *C* of the parabolic semisegment may also be located by taking the element of area *dA* as a horizontal strip of height *dy* and width

$$x = b\sqrt{1 - \frac{y}{h}} \tag{h}$$

This expression is obtained by solving Eq. (a) for *x* in terms of *y*.

Another possibility is to take the differential element as a rectangle of width *dx* and height *dy*. Then the expressions for *A*, Q_x, and Q_y are in the form of double integrals instead of single integrals.

12.3 CENTROIDS OF COMPOSITE AREAS

In engineering work we rarely need to locate centroids by integration, because the centroids of common geometric figures are already known and tabulated (see Appendix D). However, we frequently need to locate the centroids of areas composed of several parts, each part having a familiar geometric shape, such as a rectangle or a circle. Examples of such **composite areas** are the cross sections of beams and columns, which usually consist of rectangular elements (for instance, see Figs. 12-2, 12-3, and 12-4).

The areas and first moments of composite areas may be calculated by summing the corresponding properties of the component parts. Let us assume that a composite area is divided into a total of n parts, and let us denote the area of the ith part as A_i. Then we can obtain the area and first moments by the following summations:

$$A = \sum_{i=1}^{n} A_i \qquad Q_x = \sum_{i=1}^{n} \bar{y}_i A_i \qquad Q_y = \sum_{i=1}^{n} \bar{x}_i A_i \qquad \text{(12-6a,b,c)}$$

in which $\bar{x}_i$ and $\bar{y}_i$ are the coordinates of the centroid of the ith part.

The coordinates of the centroid of the composite area are

$$\bar{x} = \frac{Q_y}{A} = \frac{\sum_{i=1}^{n} \bar{x}_i A_i}{\sum_{i=1}^{n} A_i} \qquad \bar{y} = \frac{Q_x}{A} = \frac{\sum_{i=1}^{n} \bar{y}_i A_i}{\sum_{i=1}^{n} A_i} \qquad \text{(12-7a,b)}$$

Since the composite area is represented exactly by the n parts, the preceding equations give exact results for the coordinates of the centroid.

To illustrate the use of Eqs. (12-7a) and (12-7b), consider the L-shaped area (or angle section) shown in Fig. 12-6a. This area has side dimensions b and c and thickness t. The area can be divided into two rectangles of areas A_1 and A_2 with centroids C_1 and C_2, respectively (Fig. 12-6b). The areas and centroidal coordinates of these two parts are

$$A_1 = bt \qquad \bar{x}_1 = \frac{t}{2} \qquad \bar{y}_1 = \frac{b}{2}$$

$$A_2 = (c - t)t \qquad \bar{x}_2 = \frac{c + t}{2} \qquad \bar{y}_2 = \frac{t}{2}$$

Therefore, the area and first moments of the composite area (from Eqs. 12-6a, b, and c) are

$$A = A_1 + A_2 = t(b + c - t)$$

(a)

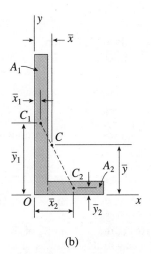

(b)

Fig. 12-6 Centroid of a composite area consisting of two parts

$$Q_x = \bar{y}_1 A_1 + \bar{y}_2 A_2 = \frac{t}{2}(b^2 + ct - t^2)$$

$$Q_y = \bar{x}_1 A_1 + \bar{x}_2 A_2 = \frac{t}{2}(bt + c^2 - t^2)$$

Finally, we can obtain the coordinates $\bar{x}$ and $\bar{y}$ of the centroid C of the composite area (Fig. 12-6b) from Eqs. (12-7a) and (12-7b):

$$\bar{x} = \frac{Q_y}{A} = \frac{bt + c^2 - t^2}{2(b + c - t)} \qquad \bar{y} = \frac{Q_x}{A} = \frac{b^2 + ct - t^2}{2(b + c - t)} \qquad \text{(12-8a,b)}$$

A similar procedure can be used for more complex areas, as illustrated below in Example 12-2.

Note 1: When a composite area is divided into only two parts, the centroid C of the entire area lies on the line joining the centroids C_1 and C_2 of the two parts (as shown in Fig. 12-6b for the L-shaped area).

Note 2: When using the formulas for composite areas (Eqs. 12-6 and 12-7), we can handle the *absence* of an area by subtraction. This procedure is useful when there are cutouts or holes in the figure. For instance, consider the area shown in Fig. 12-7a. We can analyze this figure as a composite area by subtracting the properties of the inner rectangle *efgh* from the corresponding properties of the outer rectangle *abcd*. (From another viewpoint, we can think of the outer rectangle as a "positive area" and the inner rectangle as a "negative area.") Similarly, if an area has a hole (Fig. 12-7b), we can subtract the properties of the area of the hole from those of the outer rectangle. (Again, the same effect is achieved if we treat the outer rectangle as a "positive area" and the hole as a "negative area.")

(a)

(b)

Fig. 12-7 Composite areas with a cutout and a hole

WORK SHEET 14 ∫ƒ=ˣ Mathcad

Example 12-2

The cross-section of a steel beam is constructed of a W 18 × 71 wide-flange section with a 6 in. × 1/2 in. cover plate welded to the top flange and a C 10 × 30 channel section welded to the bottom flange (Fig. 12-8). Locate the centroid C of the cross-sectional area.

Solution   Let us denote the areas of the cover plate, the wide-flange section, and the channel section as areas A_1, A_2, and A_3, respectively. The centroids of these three areas are labeled C_1, C_2, and C_3, respectively, in Fig. 12-8. Note that the composite area has an axis of symmetry, and therefore all centroids lie on that axis. The three partial areas are

$$A_1 = (6 \text{ in.})(0.5 \text{ in.}) = 3.0 \text{ in.}^2 \qquad A_2 = 20.8 \text{ in.}^2 \qquad A_3 = 8.82 \text{ in.}^2$$

in which the areas A_2 and A_3 are obtained from Tables E-1 and E-3 of Appendix E.

Let us place the origin of the x and y axes at the centroid C_2 of the wide-flange section. Then the distances from the x axis to the centroids of the three areas are as follows:

$$\bar{y}_1 = \frac{18.47 \text{ in.}}{2} + \frac{0.5 \text{ in.}}{2} = 9.485 \text{ in.}$$

$$\bar{y}_2 = 0 \qquad \bar{y}_3 = \frac{18.47 \text{ in.}}{2} + 0.649 \text{ in.} = 9.884 \text{ in.}$$

Plate
6 in. $\times \frac{1}{2}$ in.

W 18 $\times$ 71

C 10 $\times$ 30

Fig. 12-8 Example 12-2. Centroid of a composite area

in which the pertinent dimensions of the wide-flange and channel sections are obtained from Tables E-1 and E-3.

The area A and first moment Q_x of the entire cross section are obtained from Eqs. (12-6a) and (12-6b) as follows:

$$A = \sum_{i=1}^{n} A_i = A_1 + A_2 + A_3$$

$$= 3.0 \text{ in.}^2 + 20.8 \text{ in.}^2 + 8.82 \text{ in.}^2 = 32.62 \text{ in.}^2$$

$$Q_x = \sum_{i=1}^{n} \bar{y}_i A_i = \bar{y}_1 A_1 + \bar{y}_2 A_2 + \bar{y}_3 A_3$$

$$= (9.485 \text{ in.})(3.0 \text{ in.}^2) + 0 - (9.884 \text{ in.})(8.82 \text{ in.}^2) = -58.72 \text{ in.}^3$$

Now we can obtain the coordinate $\bar{y}$ to the centroid C of the composite area from Eq. (12-7b):

$$\bar{y} = \frac{Q_x}{A} = \frac{-58.72 \text{ in.}^3}{32.62 \text{ in.}^2} = -1.80 \text{ in.}$$

Since $\bar{y}$ is positive in the positive direction of the y axis, the minus sign means that the centroid C of the composite area is located below the x axis, as shown in Fig. 12-8. Thus, the distance $\bar{c}$ between the x axis and the centroid C is

$$\bar{c} = 1.80 \text{ in.} \qquad \longleftarrow$$

Note that the position of the reference axis (the x axis) is arbitrary; however, in this example we placed it through the centroid of the wide-flange section because it slightly simplifies the calculations.

12.4 MOMENTS OF INERTIA OF PLANE AREAS

The **moments of inertia** of a plane area (Fig. 12-9) with respect to the x and y axes, respectively, are defined by the integrals

$$I_x = \int y^2 dA \qquad I_y = \int x^2 dA \qquad (12\text{-}9\text{a,b})$$

in which x and y are the coordinates of the differential element of area dA. Because the element dA is multiplied by the square of the distance

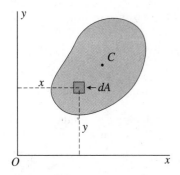

Fig. 12-9 Plane area of arbitrary shape

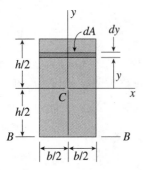

Fig. 12-10 Moments of inertia of a rectangle

from the reference axis, moments of inertia are also called **second moments of area**. Also, we see that moments of inertia of areas·(unlike first moments) are always positive quantities.

To illustrate how moments of inertia are obtained by integration, we will consider a rectangle having width b and height h (Fig. 12-10). The x and y axes have their origin at the centroid C. For convenience, we use a differential element of area dA in the form of a thin horizontal strip of width b and height dy (therefore, $dA = b\,dy$). Since all parts of the elemental strip are the same distance from the x axis, we can express the moment of inertia I_x with respect to the x axis as follows:

$$I_x = \int y^2 dA = \int_{-h/2}^{h/2} y^2 b\, dy = \frac{bh^3}{12} \tag{a}$$

In a similar manner, we can use an element of area in the form of a vertical strip with area $dA = h\,dx$ and obtain the moment of inertia with respect to the y axis:

$$I_y = \int x^2 dA = \int_{-b/2}^{b/2} x^2 h\, dx = \frac{hb^3}{12} \tag{b}$$

If a different set of axes is selected, the moments of inertia will have different values. For instance, consider axis BB at the base of the rectangle (Fig. 12-10). If this axis is selected as the reference, we must define y as the coordinate distance from that axis to the element of area dA. Then the calculations for the moment of inertia become

$$I_{BB} = \int y^2 dA = \int_0^h y^2 b\, dy = \frac{bh^3}{3} \tag{c}$$

Note that the moment of inertia with respect to axis BB is larger than the moment of inertia with respect to the centroidal x axis. In general, the moment of inertia increases as the reference axis is moved parallel to itself farther from the centroid.

The moment of inertia of a **composite area** with respect to any particular axis is the sum of the moments of inertia of its parts with respect to that same axis. An example is the hollow box section shown in Fig. 12-11a, where the x and y axes are axes of symmetry through the centroid C. The moment of inertia I_x with respect to the x axis is equal to the algebraic sum of the moments of inertia of the outer and inner rectangles. (As explained earlier, we can think of the inner rectangle as a "negative area" and the outer rectangle as a "positive area.") Therefore,

$$I_x = \frac{bh^3}{12} - \frac{b_1 h_1^3}{12} \tag{d}$$

This same formula applies to the channel section shown in Fig. 12-11b, where we may consider the cutout as a "negative area."

(a)

(b)

Fig. 12-11 Composite areas

For the hollow box section, we can use a similar technique to obtain the moment of inertia I_y with respect to the vertical axis. However, in the case of the channel section, the determination of the moment of inertia I_y requires the use of the parallel-axis theorem, which is described in the next section (Section 12.5).

Formulas for moments of inertia are listed in Appendix D. For shapes not shown, the moments of inertia can usually be obtained by using the listed formulas in conjunction with the parallel-axis theorem. If an area is of such irregular shape that its moments of inertia cannot be obtained in this manner, then we can use numerical methods. The procedure is to divide the area into small elements of area ΔA_i, multiply each such area by the square of its distance from the reference axis, and then sum the products.

■ **Radius of Gyration**

A distance known as the **radius of gyration** is occasionally encountered in mechanics. Radius of gyration of a plane area is defined as the square root of the moment of inertia of the area divided by the area itself; thus,

$$r_x = \sqrt{\frac{I_x}{A}} \qquad r_y = \sqrt{\frac{I_y}{A}} \qquad \text{(12-10a,b)}$$

in which r_x and r_y denote the radii of gyration with respect to the x and y axes, respectively. Since moment of inertia has units of length to the fourth power and area has units of length to the second power, radius of gyration has units of length.

Although the radius of gyration of an area does not have an obvious physical meaning, we may consider it to be the distance (from the reference axis) at which the entire area could be concentrated and still have the same moment of inertia as the original area.

WORK
SHEET
13
Mathcad

Example 12-3

Determine the moments of inertia I_x and I_y for the parabolic semisegment OAB shown in Fig. 12-12. The equation of the parabolic boundary is

$$y = f(x) = h\left(1 - \frac{x^2}{b^2}\right) \qquad \text{(e)}$$

(This same area was considered previously in Example 12-1.)

Solution   To determine the moments of inertia by integration, we will use Eqs. (12-9a) and (12-9b). The differential element of area dA is selected as a vertical strip of width dx and height y, as shown in Fig. 12-12. The area of this element is

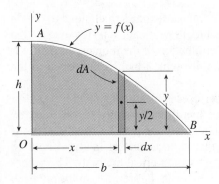

Fig. 12-12 Example 12-3. Moments of inertia of a parabolic semisegment

$$dA = y\, dx = h\left(1 - \frac{x^2}{b^2}\right)dx \tag{f}$$

Since every point in this element is at the same distance from the y axis, the moment of inertia of the element with respect to the y axis is $x^2 dA$. Therefore, the moment of inertia of the entire area with respect to the y axis is obtained as follows:

$$I_y = \int x^2 dA = \int_0^b x^2 h\left(1 - \frac{x^2}{b^2}\right)dx = \frac{2hb^3}{15} \tag{g} \quad \Longleftarrow$$

To obtain the moment of inertia with respect to the x axis, we note that the differential element of area dA has a moment of inertia dI_x with respect to the x axis equal to

$$dI_x = \frac{1}{3}(dx)y^3 = \frac{y^3}{3}dx$$

as obtained from Eq. (c). Hence, the moment of inertia of the entire area with respect to the x axis is

$$I_x = \int_0^b \frac{y^3}{3}dx = \int_0^b \frac{h^3}{3}\left(1 - \frac{x^2}{b^2}\right)^3 dx = \frac{16bh^3}{105} \tag{h} \quad \Longleftarrow$$

These same results for I_x and I_y can be obtained by using an element in the form of a horizontal strip of area $dA = x\, dy$ or by using a rectangular element of area $dA = dx\, dy$ and performing a double integration. Also, note that the preceding formulas for I_x and I_y agree with those given in Case 17 of Appendix D.

12.5 PARALLEL-AXIS THEOREM FOR MOMENTS OF INERTIA

In this section we will derive a very important and useful theorem pertaining to moments of inertia of plane areas. Known as the *parallel-axis theorem*, it gives the relationship between the moment of inertia with respect to a centroidal axis and the moment of inertia with respect to any parallel axis.

To derive the theorem, we consider an area of arbitrary shape with centroid C (Fig. 12-13). We also consider two sets of coordinate axes: (1) the $x_c y_c$ axes with origin at the centroid, and (2) a set of parallel xy axes with origin at any point O. The distances between the two sets of parallel axes are denoted d_1 and d_2. Also, we identify a differential element of area dA having coordinates x and y with respect to the centroidal axes.

From the definition of moment of inertia, we can write the following equation for the moment of inertia I_x with respect to the x axis:

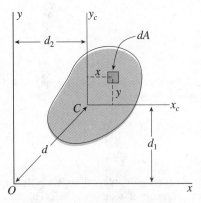

Fig. 12-13 Derivation of parallel-axis theorems

$$I_x = \int (y + d_1)^2 dA = \int y^2 dA + 2d_1 \int y\, dA + d_1^2 \int dA \qquad \text{(a)}$$

The first integral on the right-hand side is the moment of inertia I_{x_c} with respect to the x_c axis. The second integral is the first moment of the area with respect to the x_c axis (this integral equals zero because the x_c axis passes through the centroid). The third integral is the area A itself. Therefore, the preceding equation reduces to

$$I_x = I_{x_c} + Ad_1^2 \qquad \text{(12-11a)}$$

Proceeding in the same manner for the moment of inertia with respect to the y axis, we obtain

$$I_y = I_{y_c} + Ad_2^2 \qquad \text{(12-11b)}$$

Equations (12-11a) and (12-11b) represent the **parallel-axis theorem for moments of inertia**: *The moment of inertia of an area with respect to any axis in its plane is equal to the moment of inertia with respect to a parallel centroidal axis plus the product of the area and the square of the distance between the two axes.*

To illustrate the use of the theorem, consider again the rectangle shown in Fig. 12-10. Knowing that the moment of inertia about the x axis, which is through the centroid, is equal to $bh^3/12$ (see Eq. a of Section 12.4), we can determine the moment of inertia I_{BB} about the base of the rectangle by using the parallel-axis theorem:

$$I_{BB} = I_x + Ad^2 = \frac{bh^3}{12} + bh\left(\frac{h}{2}\right)^2 = \frac{bh^3}{3}$$

This result agrees with the moment of inertia obtained previously by integration (Eq. c of Section 12.4).

From the parallel-axis theorem, we see that the moment of inertia increases as the axis is moved parallel to itself farther from the centroid. Therefore, the moment of inertia about a centroidal axis is the least moment of inertia for an area (for a given direction of the axes).

When using the parallel-axis theorem, it is essential to remember that one of the two parallel axes *must* be a centroidal axis. If it is necessary to find the moment of inertia I_2 about a noncentroidal axis 2-2 (Fig. 12-14) when the moment of inertia I_1 about another noncentroidal (and parallel) axis 1-1 is known, we must apply the parallel-axis theorem twice. First, we find the centroidal moment of inertia I_{x_c} from the known moment of inertia I_1:

$$I_{x_c} = I_1 - Ad_1^2$$

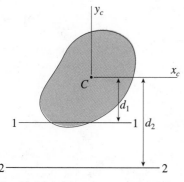

Fig. 12-14 Plane area with two parallel noncentroidal axes

Then we find the moment of inertia I_2 from the centroidal moment of inertia:

$$I_2 = I_{x_c} + Ad_2^2 = I_1 + A(d_2^2 - d_1^2) \qquad (12\text{-}12)$$

This equation shows again that the moment of inertia increases with increasing distance from the centroid of the area.

In the case of a composite area, we can find the moment of inertia (with respect to a particular axis) for each individual part of the area, and then we can add these moments of inertia to obtain the moment of inertia for the entire area (see Example 12-5).

WORK SHEET 14   Mathcad

Example 12-4

The parabolic semisegment OAB shown in Fig. 12-15 has base b and height h. Using the parallel-axis theorem, determine the moments of inertia I_{x_c} and I_{y_c} with respect to the centroidal axes x_c and y_c.

Solution   We can use the parallel-axis theorem (rather than integration) to find the centroidal moments of inertia because we already know the area A, the centroidal coordinates $\bar{x}$ and $\bar{y}$, and the moments of inertia I_x and I_y with respect to the x and y axes. These quantities were obtained earlier in Examples 12-1 and 12-3. They also are listed in Case 17 of Appendix D and are repeated here:

$$A = \frac{2bh}{3} \qquad \bar{x} = \frac{3b}{8} \qquad \bar{y} = \frac{2h}{5} \qquad I_x = \frac{16bh^3}{105} \qquad I_y = \frac{2hb^3}{15}$$

To obtain the moment of inertia with respect to the x_c axis, we write the parallel-axis theorem as follows:

$$I_x = I_{x_c} + A\bar{y}^2$$

Therefore, the moment of inertia I_{x_c} is

$$I_{x_c} = I_x - A\bar{y}^2 = \frac{16bh^3}{105} - \frac{2bh}{3}\left(\frac{2h}{5}\right)^2 = \frac{8bh^3}{175} \qquad (12\text{-}13a) \;\Longleftarrow$$

In a similar manner, we obtain the moment of inertia with respect to the y_c axis:

$$I_{y_c} = I_y - A\bar{x}^2 = \frac{2hb^3}{15} - \frac{2bh}{3}\left(\frac{3b}{8}\right)^2 = \frac{19hb^3}{480} \qquad (12\text{-}13b) \;\Longleftarrow$$

Thus, we have found the centroidal moments of inertia of the semisegment.

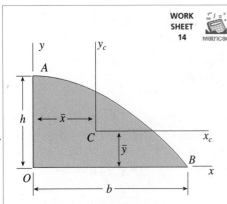

Fig. 12-15 Example 12-4. Parallel-axis theorem

WORK SHEET 14   Mathcad

Example 12-5

Determine the moment of inertia I_c with respect to the horizontal axis C-C through the centroid C of the beam cross section shown in Fig. 12-16. (The position of the centroid C was determined previously in Example 12-2 of Section 12.3.)

Plate 6 in. × $\frac{1}{2}$ in.

W 18 × 71

C 10 × 30

Fig. 12-16 Example 12-5. Moment of inertia of a composite area

Note: From beam theory (Chapter 5), we know that axis $C\text{-}C$ is the neutral axis for bending of this beam, and therefore the moment of inertia I_c must be determined in order to calculate the stresses and deflections of this beam.

Solution   We will determine the moment of inertia I_c with respect to axis $C\text{-}C$ by applying the parallel-axis theorem to each individual part of the composite area. The area divides naturally into three parts: (1) the cover plate, (2) the wide-flange section, and (3) the channel section. The following areas and centroidal distances were obtained previously in Example 12-2:

$$A_1 = 3.0 \text{ in.}^2 \qquad A_2 = 20.8 \text{ in.}^2 \qquad A_3 = 8.82 \text{ in.}^2$$

$$\bar{y}_1 = 9.485 \text{ in.} \qquad \bar{y}_2 = 0 \qquad \bar{y}_3 = 9.884 \text{ in.} \qquad \bar{c} = 1.80 \text{ in.}$$

The moments of inertia of the three parts with respect to horizontal axes through their own centroids C_1, C_2, and C_3 are as follows:

$$I_1 = \frac{bh^3}{12} = \frac{1}{12}(6.0 \text{ in.})(0.5 \text{ in.})^3 = 0.063 \text{ in.}^4$$

$$I_2 = 1170 \text{ in.}^4 \qquad I_3 = 3.94 \text{ in.}^4$$

The moments of inertia I_2 and I_3 are obtained from Tables E-1 and E-3, respectively, of Appendix E.

Now we can use the parallel-axis theorem to calculate the moments of inertia about axis $C\text{-}C$ for each of the three parts of the composite area:

$$I_{c1} = I_1 + A_1(\bar{y}_1 + \bar{c})^2 = 0.063 \text{ in.}^4 + (3.0 \text{ in.}^2)(11.28 \text{ in.})^2 = 382 \text{ in.}^4$$

$$I_{c2} = I_2 + A_2\bar{c}^2 = 1170 \text{ in.}^4 + (20.8 \text{ in.}^2)(1.80 \text{ in.})^2 = 1240 \text{ in.}^4$$

$$I_{c3} = I_3 + A_3(\bar{y}_3 - \bar{c})^2 = 3.94 \text{ in.}^4 + (8.82 \text{ in.}^2)(8.084 \text{ in.})^2 = 580 \text{ in.}^4$$

The sum of these individual moments of inertia gives the moment of inertia of the entire cross-sectional area about its centroidal axis $C\text{-}C$:

$$I_c = I_{c1} + I_{c2} + I_{c3} = 2200 \text{ in.}^4 \qquad \longleftarrow$$

This example shows how to calculate moments of inertia of composite areas by using the parallel-axis theorem.

12.6

POLAR MOMENTS OF INERTIA

The moments of inertia discussed in the preceding sections are defined with respect to axes lying in the plane of the area itself, such as the x and y axes in Fig. 12-17. Now we will consider an axis *perpendicular* to the plane of the area and intersecting the plane at the origin O. The moment of inertia with respect to this perpendicular axis is called the **polar moment of inertia** and is denoted by the symbol I_p.

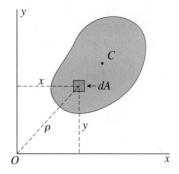

Fig. 12-17 Plane area of arbitrary shape

The polar moment of inertia with respect to an axis through O perpendicular to the plane of the figure is defined by the integral

$$I_p = \int \rho^2 dA \qquad (12\text{-}14)$$

in which ρ is the distance from point O to the differential element of area dA (Fig. 12-17). This integral is similar in form to those for moments of inertia I_x and I_y (see Eqs. 12-9a and 12-9b).

Inasmuch as $\rho^2 = x^2 + y^2$, where x and y are the rectangular coordinates of the element dA, we obtain the following expression for I_p:

$$I_p = \int \rho^2 dA = \int (x^2 + y^2)dA = \int x^2 dA + \int y^2 dA$$

Thus, we obtain the important relationship

$$I_p = I_x + I_y \qquad (12\text{-}15)$$

This equation shows that the polar moment of inertia with respect to an axis perpendicular to the plane of the figure at any point O is equal to the sum of the moments of inertia with respect to *any* two perpendicular axes x and y passing through that same point and lying in the plane of the figure.

For convenience, we usually refer to I_p simply as the polar moment of inertia with respect to point O, without mentioning the axis perpendicular to the plane of the figure. Also, to distinguish them from *polar* moments of inertia, we sometimes refer to I_x and I_y as *rectangular* moments of inertia.

Polar moments of inertia with respect to various points in the plane of an area are related by the **parallel-axis theorem for polar moments of inertia**. We can derive this theorem by referring again to Fig. 12-13. Let us denote the polar moments of inertia with respect to the origin O and the centroid C by $(I_p)_O$ and $(I_p)_C$, respectively. Then, using Eq. (12-15), we can write the following equations:

$$(I_p)_O = I_x + I_y \qquad (I_p)_C = I_{x_c} + I_{y_c} \qquad (a)$$

Now refer to the parallel-axis theorems derived in Section 12.5 for rectangular moments of inertia (Eqs. 12-11a and 12-11b). Adding those two equations, we get

$$I_x + I_y = I_{x_c} + I_{y_c} + A(d_1^2 + d_2^2)$$

Substituting from Eqs. (a), and also noting that $d^2 = d_1^2 + d_2^2$ (Fig. 12-13), we obtain

$$(I_p)_O = (I_p)_C + Ad^2 \qquad (12\text{-}16)$$

This equation represents the parallel-axis theorem for polar moments of inertia: *The polar moment of inertia of an area with respect to any point O in its plane is equal to the polar moment of inertia with respect to the centroid C plus the product of the area and the square of the distance between points O and C.*

To illustrate the determination of polar moments of inertia and the use of the parallel-axis theorem, consider a circle of radius r (Fig. 12-18). Let us take a differential element of area dA in the form of a thin ring of radius ρ and thickness $d\rho$ (thus, $dA = 2\pi\rho\, d\rho$). Since every point in the element is at the same distance ρ from the center of the circle, the polar moment of inertia of the circle with respect to the center is

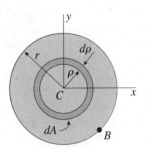

Fig. 12-18 Polar moment of inertia of a circle

$$(I_p)_C = \int \rho^2 dA = \int_0^r 2\pi\rho^3 d\rho = \frac{\pi r^4}{2} \qquad (12\text{-}17)$$

This result is listed in Case 9 of Appendix D.

The polar moment of inertia of the circle with respect to any point B on its circumference (Fig. 12-18) can be obtained from the parallel-axis theorem:

$$(I_p)_B = (I_p)_C + Ad^2 = \frac{\pi r^4}{2} + \pi r^2(r^2) = \frac{3\pi r^4}{2} \qquad (12\text{-}18)$$

As an incidental matter, note that the polar moment of inertia has its smallest value when the reference point is the centroid of the area.

A circle is a special case in which the polar moment of inertia can be determined by integration. However, most of the shapes encountered in engineering work do not lend themselves to this technique. Instead, polar moments of inertia are usually obtained by summing the rectangular moments of inertia for two perpendicular axes (Eq. 12-15).

12.7 ▌ PRODUCTS OF INERTIA

The product of inertia of a plane area is defined with respect to a set of perpendicular axes lying in the plane of the area. Thus, referring to the area shown in Fig. 12-19, we define the **product of inertia** with respect to the x and y axes as follows:

$$I_{xy} = \int xy\, dA \qquad (12\text{-}19)$$

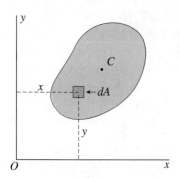

Fig. 12-19 Plane area of arbitrary shape

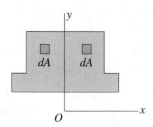

Fig. 12-20 The product of inertia equals zero when one axis is an axis of symmetry

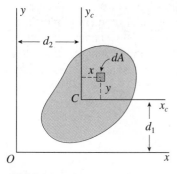

Fig. 12-21 Plane area of arbitrary shape

From this definition we see that each differential element of area dA is multiplied by the product of its coordinates. As a consequence, products of inertia may be positive, negative, or zero, depending upon the position of the xy axes with respect to the area. If the area lies entirely in the first quadrant of the axes (as in Fig. 12-19), then the product of inertia is positive because every element dA has positive coordinates x and y. If the area lies entirely in the second quadrant, the product of inertia is negative because every element has a positive y coordinate and a negative x coordinate. Similarly, areas entirely within the third and fourth quadrants have positive and negative products of inertia, respectively. When the area is located in more than one quadrant, the sign of the product of inertia depends upon the distribution of the area within the quadrants.

A special case arises when one of the axes is an **axis of symmetry** of the area. For instance, consider the area shown in Fig. 12-20, which is symmetric about the y axis. For every element dA having coordinates x and y, there exists an equal and symmetrically located element dA having the same y coordinate but an x coordinate of opposite sign. Therefore, the products $xy\,dA$ cancel each other and the integral in Eq. (12-19) vanishes. Thus, *the product of inertia of an area is zero with respect to any pair of axes in which at least one axis is an axis of symmetry of the area.*

As examples of the preceding rule, the product of inertia I_{xy} equals zero for the areas shown in Figs. 12-10, 12-11, 12-16, and 12-18. In contrast, the product of inertia I_{xy} has a positive nonzero value for the area shown in Fig. 12-15. (These observations are valid for products of inertia with respect to the particular xy axes shown in the figures. If the axes are shifted to another position, the product of inertia may change.)

Products of inertia of an area with respect to parallel sets of axes are related by a **parallel-axis theorem** that is analogous to the corresponding theorems for rectangular moments of inertia and polar moments of inertia. To obtain this theorem, consider the area shown in Fig. 12-21, which has centroid C and centroidal $x_c y_c$ axes. The product of inertia I_{xy} with respect to any other set of axes, parallel to the $x_c y_c$ axes, is

$$I_{xy} = \int (x + d_2)(y + d_1)dA = \int xy\,dA + d_1\int x\,dA + d_2\int y\,dA + d_1 d_2\int dA \quad \text{(a)}$$

in which d_1 and d_2 are the coordinates of the centroid C with respect to the xy axes (thus, d_1 and d_2 may have positive or negative values). The first integral in the last expression is the product of inertia $I_{x_c y_c}$ with respect to the centroidal axes; the second and third integrals equal zero because they are the first moments of the area with respect to the centroidal axes; and the last integral is the area A. Therefore, the preceding equation reduces to

$$I_{xy} = I_{x_c y_c} + Ad_1 d_2 \quad (12\text{-}20)$$

This equation represents the parallel-axis theorem for products of inertia: *The product of inertia of an area with respect to any pair of axes in its plane is equal to the product of inertia with respect to parallel centroidal axes plus the product of the area and the coordinates of the centroid with respect to the pair of axes.*

To demonstrate the use of this parallel-axis theorem, let us determine the product of inertia of a rectangle with respect to *xy* axes having their origin at point *O* at the lower left-hand corner of the rectangle (Fig. 12-22). The product of inertia with respect to the centroidal $x_c y_c$ axes is zero because of symmetry. Also, the coordinates of the centroid with respect to the *xy* axes are

$$d_1 = \frac{h}{2} \qquad d_2 = \frac{b}{2}$$

Substituting into Eq. (12-20), we obtain

$$I_{xy} = I_{x_c y_c} + Ad_1 d_2 = 0 + bh\left(\frac{h}{2}\right)\left(\frac{b}{2}\right) = \frac{b^2 h^2}{4} \qquad (12\text{-}21)$$

Fig. 12-22 Parallel-axis theorem for products of inertia

This product of inertia is positive because the entire area lies in the first quadrant. If the *xy* axes are translated horizontally so that the origin moves to point *B* at the lower right-hand corner of the rectangle (Fig. 12-22), the entire area lies in the second quadrant and the product of inertia becomes $-b^2 h^2/4$.

The following examples illustrate the use of both integration and the parallel-axis theorem for determining products of inertia.

■ **Example 12-6**

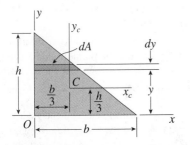

A right triangle with base *b* and height *h* is shown in Fig. 12-23. (a) Determine the product of inertia I_{xy} with respect to the *xy* axes having their origin *O* at the 90° vertex of the triangle. (b) Determine the product of inertia $I_{x_c y_c}$ with respect to the centroidal axes $x_c y_c$.

Solution   (a) *Product of inertia with respect to the xy axes.* We will use the method of integration to evaluate this product of inertia. We begin by considering a differential element of area *dA* (Fig. 12-23) in the form of a thin, horizontal strip of height *dy* and width equal to

Fig. 12-23 Example 12-6. Products of inertia of a triangle

$$\frac{(h - y)b}{h}$$

The area of this elemental strip is

$$dA = \frac{(h - y)b}{h} dy$$

and the coordinates of its centroid with respect to the *xy* axes are

$$d_1 = y \qquad d_2 = \left(\frac{1}{2}\right)\frac{(h-y)b}{h}$$

in which d_1 and d_2 are defined in Fig. 12-21. The product of inertia of the strip with respect to axes through its own centroid and parallel to the xy axes is zero (from symmetry). Therefore, its product of inertia dI_{xy} with respect to the xy axes (from the parallel-axis theorem, Eq. 12-20) is

$$dI_{xy} = 0 + (dA)d_1 d_2 = \left[\frac{(h-y)b}{h}dy\right]\left[y\left(\frac{1}{2}\right)\frac{(h-y)b}{h}\right] = \frac{b^2}{2h^2}(h-y)^2 y \, dy$$

The product of inertia I_{xy} of the entire triangle is obtained by integration:

$$I_{xy} = \int dI_{xy} = \frac{b^2}{2h^2}\int_0^h (h-y)^2 y \, dy = \frac{b^2 h^2}{24} \qquad (12\text{-}22) \quad \blacktriangleleft$$

Note that the entire area lies in the first quadrant, and therefore the product of inertia is positive.

(b) *Product of inertia with respect to the $x_c y_c$ axes.* The product of inertia with respect to axes through the centroid may be determined from the parallel-axis theorem (Eq. 12-20):

$$I_{x_c y_c} = I_{xy} - A\left(\frac{h}{3}\right)\left(\frac{b}{3}\right) = \frac{b^2 h^2}{24} - \frac{bh}{2}\left(\frac{h}{3}\right)\left(\frac{b}{3}\right) = -\frac{b^2 h^2}{72} \qquad (12\text{-}23) \quad \blacktriangleleft$$

in which $h/3$ and $b/3$ are the coordinates of point C with respect to the xy axes. Since most of the area is located in the second and fourth quadrants, the product of inertia turns out to be negative.

The products of inertia given by Eqs. (12-22) and (12-23) are listed in Cases 7 and 6, respectively, of Appendix D.

■ **Example 12-7**

Fig. 12-24 Example 12-7. Product of inertia of a Z-section

Determine the product of inertia I_{xy} of the Z-section shown in Fig. 12-24. The section has width b, height h, and thickness t.

Solution   To obtain the product of inertia with respect to the xy axes through the centroid, we divide the area into three parts and use the parallel-axis theorem. The parts are as follows: (1) a rectangle of width $b - t$ and thickness t in the upper flange, (2) a similar rectangle in the lower flange, and (3) a web rectangle with height h and thickness t.

The product of inertia of the web rectangle with respect to the xy axes is zero (from symmetry). The product of inertia $(I_{xy})_1$ of the upper flange rectangle (with respect to the xy axes) is determined by using the parallel-axis theorem:

$$(I_{xy})_1 = I_{x_c y_c} + Ad_1 d_2$$

in which $I_{x_c y_c}$ is the product of inertia of the rectangle with respect to its own centroid, A is the area of the rectangle, d_1 is the y coordinate of the centroid of the rectangle, and d_2 is the x coordinate of the centroid of the rectangle. Thus,

$$I_{x_c y_c} = 0 \qquad A = (b-t)(t) \qquad d_1 = \frac{h}{2} - \frac{t}{2} \qquad d_2 = \frac{b}{2}$$

and the product of inertia of the upper flange rectangle is

$$(I_{xy})_1 = I_{x_c y_c} + A d_1 d_2 = 0 + (b - t)(t)\left(\frac{h}{2} - \frac{t}{2}\right)\left(\frac{b}{2}\right) = \frac{bt}{4}(h - t)(b - t)$$

The product of inertia of the lower flange rectangle is the same. Therefore, the product of inertia of the entire Z-section is twice $(I_{xy})_1$, or

$$I_{xy} = \frac{bt}{2}(h - t)(b - t) \qquad (12\text{-}24) \quad \longleftarrow$$

Note that this product of inertia is positive because the flanges lie in the first and third quadrants.

12.8 ROTATION OF AXES

The moments of inertia of a plane area depend upon the position of the origin and the orientation of the reference axes. For a given origin, the moments and product of inertia vary as the axes are rotated about that origin. The manner in which they vary, and the magnitudes of the maximum and minimum values, are discussed in this and the following section.

Let us consider the plane area shown in Fig. 12-25, and let us assume that the xy axes are a pair of arbitrarily located reference axes. The moments and products of inertia with respect to those axes are

$$I_x = \int y^2 dA \qquad I_y = \int x^2 \, dA \qquad I_{xy} = \int xy \, dA \qquad (a,b,c)$$

in which x and y are the coordinates of a differential element of area dA.

The $x_1 y_1$ axes have the same origin as the xy axes but are rotated through a counterclockwise angle θ with respect to those axes. The moments and product of inertia with respect to the $x_1 y_1$ axes are denoted I_{x_1}, I_{y_1}, and $I_{x_1 y_1}$, respectively. To obtain these quantities, we need the coordinates of the element of area dA with respect to the $x_1 y_1$ axes. These coordinates may be expressed in terms of the xy coordinates and the angle θ by geometry, as follows:

$$x_1 = x \cos \theta + y \sin \theta \qquad y_1 = y \cos \theta - x \sin \theta \qquad (12\text{-}25)$$

Then the moment of inertia with respect to the x_1 axis is

$$I_{x_1} = \int y_1^2 \, dA = \int (y \cos \theta - x \sin \theta)^2 \, dA$$

$$= \cos^2 \theta \int y^2 \, dA + \sin^2 \theta \int x^2 \, dA - 2 \sin \theta \cos \theta \int xy \, dA$$

or, by using Eqs. (a), (b), and (c),

$$I_{x_1} = I_x \cos^2 \theta + I_y \sin^2 \theta - 2I_{xy} \sin \theta \cos \theta \qquad (d)$$

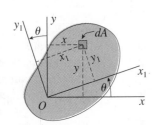

Fig. 12-25 Rotation of axes

Now we introduce the following trigonometric identities:

$$\cos^2 \theta = \frac{1}{2}(1 + \cos 2\theta) \qquad \sin^2 \theta = \frac{1}{2}(1 - \cos 2\theta)$$

$$2 \sin \theta \cos \theta = \sin 2\theta$$

Then Eq. (d) becomes

$$I_{x_1} = \frac{I_x + I_y}{2} + \frac{I_x - I_y}{2} \cos 2\theta - I_{xy} \sin 2\theta \qquad (12\text{-}26)$$

In a similar manner, we can obtain the product of inertia with respect to the $x_1 y_1$ axes:

$$I_{x_1 y_1} = \int x_1 y_1 \, dA = \int (x \cos \theta + y \sin \theta)(y \cos \theta - x \sin \theta) dA$$
$$= (I_x - I_y)\sin \theta \cos \theta + I_{xy}(\cos^2 \theta - \sin^2 \theta)$$

Again using the trigonometric identities, we obtain

$$I_{x_1 y_1} = \frac{I_x - I_y}{2} \sin 2\theta + I_{xy} \cos 2\theta \qquad (12\text{-}27)$$

Equations (12-26) and (12-27) give the moment of inertia I_{x_1} and the product of inertia $I_{x_1 y_1}$ with respect to the rotated axes in terms of the moments and product of inertia for the original axes. These equations are called the **transformation equations for moments and products of inertia**.

Note that these transformation equations have the same form as the transformation equations for plane stress (Eqs. 7-4a and 7-4b of Section 7.2). Upon comparing the two sets of equations, we see that I_{x_1} corresponds to σ_{x_1}, $I_{x_1 y_1}$ corresponds to $-\tau_{x_1 y_1}$, I_x corresponds to σ_x, I_y corresponds to σ_y, and I_{xy} corresponds to $-\tau_{xy}$. Therefore, we can also analyze moments and products of inertia by the use of **Mohr's circle** (see Section 7.4).

The moment of inertia I_{y_1} may be obtained by the same procedure that we used for finding I_{x_1} and $I_{x_1 y_1}$. However, a simpler procedure is to replace θ with $\theta + 90°$ in Eq. (12-26). The result is

$$I_{y_1} = \frac{I_x + I_y}{2} - \frac{I_x - I_y}{2} \cos 2\theta + I_{xy} \sin 2\theta \qquad (12\text{-}28)$$

This equation shows how the moment of inertia I_{y_1} varies as the axes are rotated about the origin.

A useful equation related to moments of inertia is obtained by taking the sum of I_{x_1} and I_{y_1} (Eqs. 12-26 and 12-28). The result is

$$I_{x_1} + I_{y_1} = I_x + I_y \qquad (12\text{-}29)$$

This equation shows that the sum of the moments of inertia with respect to a pair of axes remains constant as the axes are rotated about the origin. This sum is the polar moment of inertia of the area with respect to the origin. Note that Eq. (12-29) is analogous to Eq. (7-6) for stresses and Eq. (7-72) for strains.

12.9 | PRINCIPAL AXES AND PRINCIPAL MOMENTS OF INERTIA

The transformation equations for moments and products of inertia (Eqs. 12-26, 12-27, and 12-28) show how the moments and products of inertia vary as the angle of rotation θ varies. Of special interest are the maximum and minimum values of the moment of inertia. These values are known as the **principal moments of inertia**, and the corresponding axes are known as **principal axes**.

■ Principal Axes

To find the values of the angle θ that make the moment of inertia I_{x_1} a maximum or a minimum, we take the derivative with respect to θ of the expression on the right-hand side of Eq. (12-26) and set it equal to zero:

$$(I_x - I_y) \sin 2\theta + 2I_{xy} \cos 2\theta = 0 \qquad (a)$$

Solving for θ from this equation, we get

$$\tan 2\theta_p = -\frac{2I_{xy}}{I_x - I_y} \qquad (12\text{-}30)$$

in which θ_p denotes the angle defining a principal axis. This same result is obtained if we take the derivative of I_{y_1} (Eq. 12-28).

Equation (12-30) yields two values of the angle $2\theta_p$ in the range from 0 to 360°; these values differ by 180°. The corresponding values of θ_p differ by 90° and define the two perpendicular principal axes. One of these axes corresponds to the maximum moment of inertia and the other corresponds to the minimum moment of inertia.

Now let us examine the variation in the product of inertia $I_{x_1y_1}$ as θ changes (see Eq. 12-27). If $\theta = 0$, we get $I_{x_1y_1} = I_{xy}$, as expected. If $\theta = 90°$, we obtain $I_{x_1y_1} = -I_{xy}$. Thus, during a 90° rotation the product of inertia changes sign, which means that for an intermediate orientation of the axes, the product of inertia must equal zero. To determine this orientation, we set $I_{x_1y_1}$ (Eq. 12-27) equal to zero:

$$(I_x - I_y) \sin 2\theta + 2I_{xy} \cos 2\theta = 0$$

This equation is the same as Eq. (a), which defines the angle θ_p to the principal axes. Therefore, we conclude that *the product of inertia is zero for the principal axes*.

In Section 12.7 we showed that the product of inertia of an area with respect to a pair of axes equals zero if at least one of the axes is an axis of symmetry. It follows that if an area has an axis of symmetry, that axis and any axis perpendicular to it constitute a set of principal axes.

The preceding observations may be summarized as follows: (1) principal axes through an origin O are a pair of orthogonal axes for which the moments of inertia are a maximum and a minimum; (2) the orientation of the principal axes is given by the angle θ_p obtained from Eq. (12-30); (3) the product of inertia is zero for principal axes; and (4) an axis of symmetry is always a principal axis.

■ **Principal Points**

Now consider a pair of principal axes with origin at a given point O. If there exists a different pair of principal axes through that same point, then *every* pair of axes through that point is a set of principal axes. Furthermore, the moment of inertia must be constant as the angle θ is varied. These conclusions follow from the nature of the transformation equation for I_{x_1} (Eq. 12-26). Because this equation contains trigonometric functions of the angle 2θ, there is one maximum value and one minimum value of I_{x_1} as 2θ varies through a range of $360°$ (or as θ varies through a range of $180°$). If a second maximum exists, then the only possibility is that I_{x_1} remains constant, which means that every pair of axes is a set of principal axes and all moments of inertia are the same. A point so located that every axis through the point is a principal axis, and hence the moments of inertia are the same for all axes through the point, is called a **principal point**.

An illustration of this situation is the rectangle of width $2b$ and height b shown in Fig. 12-26. The xy axes, with origin at point O, are principal axes of the rectangle because the y axis is an axis of symmetry. The $x'y'$ axes, with the same origin, are also principal axes because the product of inertia $I_{x'y'}$ equals zero (because the triangles are symmetrically located with respect to the x' and y' axes). It follows that every pair of axes through O is a set of principal axes and every moment of inertia is the same (and equal to $2b^4/3$). Therefore, point O is a principal point for the rectangle. (A second principal point is located where the y axis intersects the upper side of the rectangle.)

A useful corollary of the concepts described in the preceding two paragraphs applies to axes through the centroid of an area. Consider an area having *two different pairs* of centroidal axes such that at least one

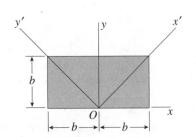

Fig. 12-26 Rectangle for which every axis (in the plane of the area) through point O is a principal axis

(a)

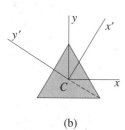

(b)

Fig. 12-27 Examples of areas for which every centroidal axis is a principal axis and the centroid C is a principal point (a square and an equilateral triangle)

axis in each pair is an axis of symmetry. In other words, there exist two different axes of symmetry that are not perpendicular to each other. Then it follows that the centroid is a principal point.

Two examples, a square and an equilateral triangle, are shown in Fig. 12-27. In each case the xy axes are principal centroidal axes because their origin is at the centroid C and at least one of the two axes is an axis of symmetry. In addition, a second pair of centroidal axes (the $x'y'$ axes) has at least one axis of symmetry. It follows that both the xy and $x'y'$ axes are principal axes. Therefore, every axis through the centroid C is a principal axis, and every such axis has the same moment of inertia.

If an area has three different axes of symmetry, even if two of them are perpendicular, the conditions described in the preceding paragraph are automatically fulfilled. Therefore, if an area has three or more axes of symmetry, the centroid is a principal point and every axis through the centroid is a principal axis and has the same moment of inertia. These conditions are fulfilled for a circle, for all regular polygons (equilateral triangle, square, regular pentagon, regular hexagon, and so on), and for many other symmetric shapes.

In general, every plane area has two principal points. These points lie equidistant from the centroid on the principal centroidal axis having the larger principal moment of inertia. A special case occurs when the two principal centroidal moments of inertia are equal; then the two principal points merge at the centroid, which becomes the sole principal point.

■ **Principal Moments of Inertia**

Let us now determine the principal moments of inertia, assuming that I_x, I_y, and I_{xy} are known. One method is to determine the two values of θ_p (differing by 90°) from Eq. (12-30) and then substitute these values into Eq. (12-26) for I_{x_1}. The resulting two values are the principal moments of inertia, denoted by I_1 and I_2. The advantage of this method is that we know which of the two principal angles θ_p corresponds to each principal moment of inertia.

It is also possible to obtain general formulas for the principal moments of inertia. We note from Eq. (12-30) and Fig. 12-28 (which is a geometric representation of Eq. 12-30) that

$$\cos 2\theta_p = \frac{I_x - I_y}{2R} \qquad \sin 2\theta_p = \frac{-I_{xy}}{R} \qquad \text{(12-31a,b)}$$

in which

$$R = \sqrt{\left(\frac{I_x - I_y}{2}\right)^2 + I_{xy}^2} \qquad \text{(12-32)}$$

is the hypotenuse of the triangle. When evaluating R, we always take the positive square root.

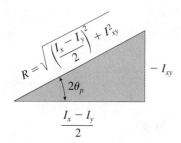

Fig. 12-28 Geometric representation of Eq. (12-30)

Now we substitute the expressions for $\cos 2\theta_p$ and $\sin 2\theta_p$ (from Eqs. 12-31a and b) into Eq. (12-26) for I_{x_1} and obtain the algebraically larger of the two principal moments of inertia, denoted by the symbol I_1:

$$I_1 = \frac{I_x + I_y}{2} + \sqrt{\left(\frac{I_x - I_y}{2}\right)^2 + I_{xy}^2} \qquad (12\text{-}33\text{a})$$

The smaller principal moment of inertia, denoted as I_2, may be obtained from the equation

$$I_1 + I_2 = I_x + I_y$$

(see Eq. 12-29). Substituting the expression for I_1 into this equation and solving for I_2, we get

$$I_2 = \frac{I_x + I_y}{2} - \sqrt{\left(\frac{I_x - I_y}{2}\right)^2 + I_{xy}^2} \qquad (12\text{-}33\text{b})$$

Equations (12-33a) and (12-33b) provide a convenient way to calculate the principal moments of inertia.

The following example illustrates the method for locating the principal axes and determining the principal moments of inertia.

■ **Example 12-8**

Determine the orientations of the principal centroidal axes and the magnitudes of the principal centroidal moments of inertia for the cross-sectional area of the Z-section shown in Fig. 12-29. Use the following numerical data: height $h = 200$ mm, width $b = 90$ mm, and thickness $t = 15$ mm.

Solution   Let us use the xy axes (Fig. 12-29) as the reference axes through the centroid C. The moments and product of inertia with respect to these axes can be obtained by dividing the area into three rectangles and using the parallel-axis theorems. The results of such calculations are as follows:

$$I_x = 29.29 \times 10^6 \text{ mm}^4 \qquad I_y = 5.667 \times 10^6 \text{ mm}^4 \qquad I_{xy} = -9.366 \times 10^6 \text{ mm}^4$$

Substituting these values into the equation for the angle θ_p (Eq. 12-30), we get

$$\tan 2\theta_p = -\frac{2I_{xy}}{I_x - I_y} = 0.7930 \qquad 2\theta_p = 38.4° \text{ and } 218.4°$$

Thus, the two values of θ_p are

$$\theta_p = 19.2° \text{ and } 109.2°$$

Using these values of θ_p in the transformation equation for I_{x_1} (Eq. 12-26), we find $I_{x_1} = 32.6 \times 10^6 \text{ mm}^4$ and $2.4 \times 10^6 \text{ mm}^4$, respectively. These same values

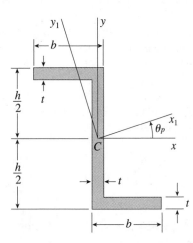

Fig. 12-29 Example 12-8. Principal axes and principal moments of inertia for a Z-section

are obtained if we substitute into Eqs. (12-33a) and (12-33b). Thus, the principal moments of inertia and the angles to the corresponding principal axes are:

$$I_1 = 32.6 \times 10^6 \text{ mm}^4 \qquad \theta_{p_1} = 19.2°$$

$$I_2 = 2.4 \times 10^6 \text{ mm}^4 \qquad \theta_{p_2} = 109.2°$$

The principal axes are shown in Fig. 12-29 as the $x_1 y_1$ axes.

PROBLEMS/CHAPTER 12

  CENTROIDS OF AREAS

The problems for Section 12.2 are to be solved by integration.

12.2-1 Determine the distances $\bar{x}$ and $\bar{y}$ to the centroid C of a right triangle having base b and altitude h (see Case 6, Appendix D).

12.2-2 Determine the distance $\bar{y}$ to the centroid C of a trapezoid having bases a and b and altitude h (see Case 8, Appendix D).

12.2-3 Determine the distance $\bar{y}$ to the centroid C of a semicircle of radius r (see Case 10, Appendix D).

12.2-4 Determine the distances $\bar{x}$ and $\bar{y}$ to the centroid C of a parabolic spandrel of base b and height h (see Case 18, Appendix D).

12.2-5 Determine the distances $\bar{x}$ and $\bar{y}$ to the centroid C of a semisegment of nth degree having base b and height h (see Case 19, Appendix D).

   CENTROIDS OF COMPOSITE AREAS

The problems for Section 12.3 are to be solved by using the formulas for composite areas.

12.3-1 Determine the distance $\bar{y}$ to the centroid C of a trapezoid having bases a and b and altitude h (see Case 8, Appendix D) by dividing the trapezoid into two triangles.

12.3-2 One quarter of a square of side a is removed (see figure). What are the coordinates $\bar{x}$ and $\bar{y}$ of the centroid C of the remaining area?

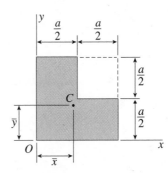

Probs. 12.3-2 and 12.5-2

12.3-3 Calculate the distance $\bar{y}$ to the centroid C of the channel section shown in the figure if $a = 6$ in., $b = 1$ in., and $c = 2$ in.

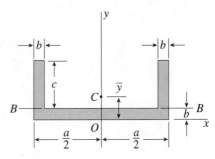

Probs. 12.3-3, 12.3-4, and 12.5-3

12.3-4 What must be the relationship between the dimensions a, b, and c of the channel section shown in the figure in order that the centroid C will lie on line BB?

12.3-5 The cross section of a beam constructed of a W 24 × 162 wide-flange section with an 8 in. × 3/4 in. cover plate welded to the top flange is shown in the figure. Determine the distance $\bar{y}$ from the base of the beam to the centroid C of the cross-sectional area.

Probs. 12.3-5 and 12.5-5

12.3-6 Determine the distance $\bar{y}$ to the centroid C of the composite area shown in the figure.

Probs. 12.3-6, 12.5-6, and 12.7-6

12.3-7 Determine the coordinates $\bar{x}$ and $\bar{y}$ of the centroid C of the L-shaped area shown in the figure.

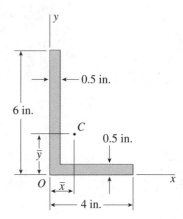

Probs. 12.3-7, 12.4-7, 12.5-7, and 12.7-7

12.3-8 Determine the coordinates $\bar{x}$ and $\bar{y}$ of the centroid C of the area shown in the figure.

Prob. 12.3-8

MOMENTS OF INERTIA WORK SHEETS 13 & 14

Problems 12.4-1 through 12.4-4 are to be solved by integration.

12.4-1 Determine the moment of inertia I_x of a triangle of base b and altitude h with respect to its base (see Case 4, Appendix D).

12.4-2 Determine the moment of inertia I_{BB} of a trapezoid having bases a and b and altitude h with respect to its base (see Case 8, Appendix D).

12.4-3 Determine the moment of inertia I_x of a parabolic spandrel of base b and height h with respect to its base (see Case 18, Appendix D).

12.4-4 Determine the moment of inertia I_x of a circle of radius r with respect to a diameter (see Case 9, Appendix D).

Problems 12.4-5 through 12.4-9 are to be solved by considering the area to be a composite area.

12.4-5 Determine the moment of inertia I_{BB} of a rectangle having sides of lengths b and h with respect to a diagonal of the rectangle (see the figure for Case 2, Appendix D).

12.4-6 Calculate the moment of inertia I_x for the composite circular area shown in the figure. The origin of the axes is at the center of the concentric circles, and the three diameters are 50, 100, and 150 mm.

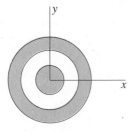

Prob. 12.4-6

12.4-7 Calculate the moments of inertia I_x and I_y with respect to the x and y axes for the L-shaped area shown in the figure for Prob. 12.3-7.

12.4-8 A semicircular area of radius 150 mm has a rectangular cutout of dimensions 50 mm × 100 mm (see figure). Calculate the moments of inertia I_x and I_y with respect to the x and y axes. Also, calculate the corresponding radii of gyration r_x and r_y.

Prob. 12.4-8

12.4-9 Calculate the moments of inertia I_1 and I_2 of a W 16 × 100 wide-flange section using the cross-sectional dimensions given in Table E-l, Appendix E. (Disregard the cross-sectional areas of the fillets.) Also, calculate the corresponding radii of gyration r_1 and r_2, respectively.

WORK SHEET 14

PARALLEL-AXIS THEOREM

12.5-1 Calculate the moment of inertia I_b of a W 12 × 50 wide-flange section with respect to its base. (Use data from Table E-l, Appendix E.)

12.5-2 Determine the moment of inertia I_c with respect to an axis through the centroid C and parallel to the x axis for the figure described in Prob. 12.3-2.

12.5-3 For the channel section described in Prob. 12.3-3, calculate the moment of inertia I_c with respect to an axis through the centroid C and parallel to the x axis.

12.5-4 The moment of inertia with respect to axis 1-1 of the scalene triangle shown in the figure is 90×10^3 mm⁴. Calculate its moment of inertia I_2 with respect to axis 2-2.

Prob. 12.5-4

12.5-5 For the beam cross section described in Prob. 12.3-5, calculate the centroidal moments of inertia I_{x_c} and I_{y_c} with respect to axes through the centroid C such that the x_c axis is parallel to the x axis and the y_c axis coincides with the y axis.

12.5-6 Calculate the moment of inertia I_c with respect to an axis through the centroid C and parallel to the x axis for the composite area shown in Prob. 12.3-6.

12.5-7 Calculate the centroidal moments of inertia I_{x_c} and I_{y_c} with respect to axes through the centroid C and parallel to the x and y axes, respectively, for the L-shaped area shown in Prob. 12.3-7.

12.5-8 The wide-flange beam section shown in the figure has a total height of 250 mm and a constant thickness of 15 mm. Determine the flange width b if it is required that the centroidal moments of inertia I_x and I_y be in the ratio 3 to 1, respectively.

Prob. 12.5-8

POLAR MOMENTS OF INERTIA

12.6-1 Determine the polar moment of inertia I_p of an isosceles triangle of base b and altitude h with respect to its apex (see Case 5, Appendix D).

12.6-2 Determine the polar moment of inertia I_{p_c} with respect to the centroid C for a circular sector (see Case 13, Appendix D).

12.6-3 Determine the polar moment of inertia I_p for a W 8 × 21 wide-flange section with respect to one of its outermost corners.

12.6-4 Obtain a formula for the polar moment of inertia I_p with respect to the midpoint of the hypotenuse for a right triangle of base b and height h (see Case 6, Appendix D).

12.6-5 Determine the polar moment of inertia I_{p_c} with respect to the centroid C for a quarter-circular spandrel (see Case 12, Appendix D).

∗∗12.6-6 Calculate the polar moment of inertia I_{p_c} with respect to the centroid C for the hollow section of constant thickness shown in the figure.

Probs. 12.6-6 and 12.7-8

PRODUCTS OF INERTIA

12.7-1 Using integration, determine the product of inertia I_{xy} for the parabolic semisegment shown in Fig. 12-5 (see Example 12-1 in Section 12.2 and Case 17 in Appendix D).

12.7-2 Using integration, determine the product of inertia I_{xy} for the quarter-circular spandrel shown in Case 12, Appendix D.

12.7-3 Find the relationship between the radius r and the distance b for the composite area shown in the figure in order that the product of inertia I_{xy} will be zero.

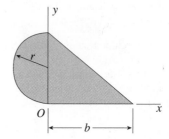

Prob. 12.7-3

12.7-4 Obtain a formula for the product of inertia I_{xy} of the symmetrical L-shaped area shown in the figure.

Prob. 12.7-4

12.7-5 Calculate the product of inertia I_{12} with respect to the centroidal axes 1-1 and 2-2 for an L 6 × 6 × 1 in. angle section (see Table E-4, Appendix E). (Disregard the cross-sectional areas of the fillet and rounded corners.)

12.7-6 Calculate the product of inertia I_{xy} for the composite area shown in Prob. 12.3-6.

12.7-7 Determine the product of inertia $I_{x_c y_c}$ with respect to centroidal axes x_c and y_c parallel to the x and y axes, respectively, for the L-shaped area shown in Prob. 12.3-7.

** **12.7-8** Calculate the product of inertia I_{xy} for the hollow section of constant thickness shown in the figure for Prob. 12.6-6.

ROTATION OF AXES

The problems for Section 12.8 are to be solved by using the transformation equations for moments and products of inertia.

12.8-1 Determine the moments of inertia I_{x_1} and I_{y_1} and the product of inertia $I_{x_1 y_1}$ for a square with sides b, as shown in the figure. (Note that the $x_1 y_1$ axes are centroidal axes rotated through an angle θ with respect to the xy axes.)

Prob. 12.8-1

12.8-2 Determine the moments and product of inertia with respect to the $x_1 y_1$ axes for the rectangle shown in the figure. (Note that the x_1 axis is a diagonal of the rectangle.)

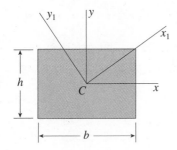

Prob. 12.8-2

12.8-3 Calculate the moment of inertia I_d for a W 12 × 50 wide-flange section with respect to a diagonal passing through the centroid and two outside corners of the flanges. (Use the dimensions and properties given in Table E-1.)

12.8-4 Calculate the moments of inertia I_{x_1} and I_{y_1} and the product of inertia $I_{x_1 y_1}$ with respect to the $x_1 y_1$ axes for the L-shaped area shown in the figure if $a = 150$ mm, $b = 100$ mm, $t = 15$ mm, and $\theta = 30°$.

Probs. 12.8-4 and 12.9-4

12.8-5 Calculate the moments of inertia I_{x_1} and I_{y_1} and the product of inertia $I_{x_1 y_1}$ with respect to the $x_1 y_1$ axes for the Z-section shown in the figure if $b = 3$ in., $h = 4$ in., $t = 0.5$ in., and $\theta = 60°$.

Probs. 12.8-5, 12.8-6, 12.9-5, and 12.9-6

12.8-6 Solve the preceding problem if $b = 80$ mm, $h = 120$ mm, $t = 12$ mm, and $\theta = 30°$.

PRINCIPAL AXES, PRINCIPAL POINTS, AND PRINCIPAL MOMENTS OF INERTIA

12.9-1 An ellipse with major axis of length $2a$ and minor axis of length $2b$ is shown in the figure. (a) Determine the distance c from the centroid C of the ellipse to the principal points P on the minor axis (y axis). (b) For what ratio a/b do the principal points lie on the circumference of the ellipse? (c) For what ratios do they lie inside the ellipse?

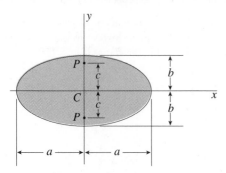

Prob. 12.9-1

12.9-2 Demonstrate that the two points P_1 and P_2, located as shown in the figure, are the principal points of the isosceles right triangle.

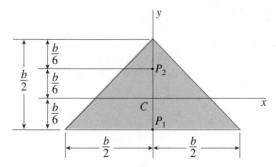

Prob. 12.9-2

12.9-3 Determine the angles θ_{p_1} and θ_{p_2} defining the orientations of the principal axes through the origin O for the right triangle shown in Fig. 12-23 of Example 12-6 (Section 12.7) if $b = 6$ in. and $h = 8$ in. Also, calculate the corresponding principal moments of inertia I_1 and I_2.

12.9-4 Determine the angles θ_{p_1} and θ_{p_2} defining the orientations of the principal axes through the origin O and the corresponding principal moments of inertia I_1 and I_2 for the L-shaped area described in Prob. 12.8-4 ($a = 150$ mm, $b = 100$ mm, and $t = 15$ mm).

12.9-5 Determine the angles θ_{p_1} and θ_{p_2} defining the orientations of the principal axes through the centroid C and the corresponding principal centroidal moments of inertia I_1 and I_2 for the Z-section described in Prob. 12.8-5 ($b = 3$ in., $h = 4$ in., and $t = 0.5$ in.).

12.9-6 Solve the preceding problem for the Z-section described in Prob. 12.8-6 ($b = 80$ mm, $h = 120$ mm, and $t = 12$ mm).

12.9-7 Determine the angles θ_{p_1} and θ_{p_2} defining the orientations of the principal axes through the centroid C for the right triangle shown in Fig. 12-23 of Example 12-6 (Section 12.7) if $h = 2b$. Also, determine the corresponding principal centroidal moments of inertia I_1 and I_2.

12.9-8 Determine the angles θ_{p_1} and θ_{p_2} defining the orientations of the principal axes through the centroid C and the corresponding principal centroidal moments of inertia I_1 and I_2 for the Z-section shown in Fig. 12-24 of Example 12-7 (Section 12.7) if $b = 80$ mm, $h = 150$ mm, and $t = 10$ mm.

12.9-9 Solve the preceding problem if $b = 3$ in., $h = 6$ in., and $t = 3/8$ in.

12.9-10 Determine the angles θ_{p_1} and θ_{p_2} defining the orientations of the principal centroidal axes and the corresponding principal moments of inertia I_1 and I_2 for the L-shaped area shown in the figure if $a = 80$ mm, $b = 150$ mm, and $t = 16$ mm.

Probs. 12.9-10 and 12.9-11

12.9-11 Solve the preceding problem if $a = 3$ in., $b = 6$ in., and $t = 5/8$ in.

REFERENCES AND HISTORICAL NOTES

1-1 Timoshenko, S. P., *History of Strength of Materials*, Dover Publications, Inc., New York, 1983 (originally published by McGraw-Hill Book Co., Inc., New York, 1953).

Note: Stephen P. Timoshenko (1878–1972) was a famous scientist, engineer, and teacher. Born in Russia, he came to the United States in 1922. He was a researcher with the Westinghouse Research Laboratory, a professor at the University of Michigan, and later a professor at Stanford University, where he retired in 1944. Timoshenko made many original contributions, both theoretical and experimental, to the field of applied mechanics, and he wrote twelve pioneering textbooks that revolutionized the teaching of mechanics in the United States. These books, which were published in as many as five editions and translated into as many as 35 languages, covered the subjects of statics, dynamics, mechanics of materials, vibrations, structural theory, stability, elasticity, plates, and shells.

1-2 Todhunter, I., and Pearson, K., *A History of the Theory of Elasticity and of the Strength of Materials*, Vols. I and II, Dover Publications, Inc., New York, 1960 (originally published by the Cambridge University Press in 1886 and 1893). *Note:* Isaac Todhunter (1820–1884) and Karl Pearson (1857–1936) were English mathematicians and educators. Pearson was especially noteworthy for his original contributions to statistics.

1-3 Love, A. E. H., *A Treatise on the Mathematical Theory of Elasticity*, 4th Ed., Dover Publications, Inc., New York, 1944 (originally published by the Cambridge University Press in 1927); see "Historical Introduction," pp. 1–31. *Note:* Augustus Edward Hough Love (1863–1940) was an outstanding English elastician who taught at Oxford University. His many important investigations included the analysis of seismic surface waves, now called *Love waves* by geophysicists.

1-4 Jacob Bernoulli (1654–1705), also known by the names James, Jacques, and Jakob, was a member of the famous family of mathematicians and scientists of Basel, Switzerland. He did important work in connection with elastic curves of beams. Bernoulli also developed polar coordinates and became famous for his work in theory of probability, analytic geometry, and other fields. Jean Victor Poncelet (1788–1867) was a Frenchman who fought in Napoleon's campaign against Russia and was given up for dead on the battlefield. He survived, was taken prisoner, and later returned to France to continue his work in mathematics. His major contributions to mathematics are in geometry, and in mechanics he is best known for his work on properties of materials and dynamics. (For the work of Bernoulli and Poncelet in connection with stress-strain diagrams, see Ref. 1-1, p. 88, and Ref. 1-2, Vol. I, pp. 10, 533, and 873.)

1-5 James & James, *Mathematics Dictionary*, Van Nostrand Reinhold, New York (latest edition).

1-6 Robert Hooke (1635–1703) was an English scientist who performed many experiments with elastic bodies and developed improvements in timepieces. He also formulated the laws of gravitation independently of Newton, of whom he was a contemporary. Upon the founding of the Royal Society of London in 1662, Hooke was appointed its first curator. (For the origins of Hooke's law, see Ref. 1-1, pp. 17–20, and Ref. 1-2, Vol. I, p. 5.)

1-7 Thomas Young (1773–1829) was an outstanding English scientist who did pioneering work in optics, sound, impact, and other subjects. (For information about his work with materials, see Ref. 1-1, pp. 90–98, and Ref. 1-2, Vol. I, pp. 80–86.)

1-8 Siméon Denis Poisson (1781–1840) was a great French mathematician. He made many contributions to both mathematics and mechanics, and his name is preserved in numerous ways besides Poisson's ratio. For instance, we have Poisson's equation in partial differential equations and the Poisson distribution in theory of probability. Based upon his own theory for the behavior of materials, he calculated the lateral strain of a bar in tension and found it to be one-fourth of the longitudinal strain. (For information about his theories of material behavior, see Ref. 1-1, pp. 111–114; Ref. 1-2, Vol. I, pp. 208–318; and Ref. 1-3, p. 13.)

2-1 Timoshenko, S. P., and Goodier, J. N., *Theory of Elasticity*, 3rd Ed., McGraw-Hill Book Co., Inc., New York, 1970 (see p. 110). *Note:* James Norman Goodier (1905–1969) was well known for his research contributions to theory of elasticity, stability, wave propagation in solids, and other branches of applied mechanics. Born in England, he studied at Cambridge University and the University of Michigan. He was a professor at Cornell University and later at Stanford University, where he headed the program in applied mechanics.

2-2 Weaver, W., Jr., and Gere, J. M., *Matrix Analysis of Framed Structures*, 3rd Ed., Van Nostrand Reinhold Co., New York, 1990. *Note:* The flexibility and stiffness methods for analyzing statically indeterminate structures by computer methods are presented in this book.

2-3 Leonhard Euler (1707–1783) was a famous Swiss mathematician, perhaps the greatest mathematician of all time. Ref. 11-2 gives information about his life and works. (See Ref. 1-1, p. 36, and Ref. 2-4, p. 650, regarding his work on statically indeterminate structures.)

2-4 Oravas, G. A., and McLean, L., "Historical development of energetical principles in elastomechanics," *Applied Mechanics Reviews,* Part I, Vol. 19, No. 8, August 1966, pp. 647–658, and Part II, Vol. 19, No. 11, November 1966, pp. 919–933.

2-5 Louis Marie Henri Navier (1785–1836), a famous French mathematician and engineer, was one of the founders of the mathematical theory of elasticity. He contributed to beam, plate, and shell theory, to theory of vibrations, and to the theory of viscous fluids. (See Ref. 1-1, p. 75; Ref. 1-2, Vol. I, p. 146; and Ref. 2-4, p. 652, for his analysis of a statically indeterminate structure.)

2-6 Piobert, G., Morin, A.-J., and Didion, I., "Commission des Principes du Tir," *Mémorial de l'Artillerie*, Vol. 5, 1842, pp. 501–552. *Note:* This paper describes experiments made by firing artillery projectiles against iron plating. On page 505 appears the description of the markings that are the slip bands. The description is quite brief, and there is no indication that the authors attributed the markings to inherent material characteristics. Guillaume Piobert (1793–1871) was a French general and mathematician who made many studies of ballistics; when this paper was written, he was a captain in the artillery.

2-7 Lüders, W., "Ueber die Äusserung der elasticität an stahlartigen Eisenstäben und Stahlstäben, und über eine beim Biegen solcher Stäbe beobachtete Molecularbewegung," *Dingler's Polytechnisches Journal*, Vol. 155, 1860, pp. 18–22. *Note:* This paper clearly describes and illustrates the bands that appear on the surface of a polished steel specimen during yielding. Of course, these bands are only the surface manifestation of three-dimensional zones of deformation; hence, the zones should probably be characterized as "wedges" rather than bands. Nevertheless, the name *Lüders' bands* is usually given to the markings, although sometimes they are called *Piobert's bands*.

2-8 Benoit Paul Emile Clapeyron (1799–1864) was a famous French structural engineer and bridge designer; he taught engineering at the École des Ponts et Chaussées in Paris. It appears that Clapeyron's theorem, which states that the work of the external loads acting on a linearly elastic body is equal to the strain energy, was first published in 1833. (See Ref. 1-1, pp. 118 and 288; Ref. 1-2, Vol. I, p. 578; and Ref. 1-2, Vol. II, p. 418.)

2-9 Poncelet investigated longitudinal vibrations of a bar due to impact loads (see Ref. 1-1, p. 88). Ref. 1-4 gives additional information about his life and works.

2-10 Roark, R. J., and Young, W. C., *Formulas for Stress and Strain*, 6th Ed., McGraw-Hill Book Co., Inc., New York, 1989.

2-11 Peterson, R. E., *Stress Concentration Factors*, John Wiley and Sons, Inc., New York, 1974.

2-12 Barré de Saint-Venant (1797–1886) is generally recognized as the most outstanding elastician of all time. Born near Paris, he studied briefly at the École Polytechnique and later graduated from the École des Ponts et Chaussées. His later professional career suffered greatly from his refusal, as a matter of conscience and politics, to join his schoolmates in preparing for the defense of Paris in March 1814, just prior to Napoleon's abdication. As a consequence, his achievements received greater recognition in other countries than they did in France. Some of his most famous contributions are the formulation of the fundamental equations of elasticity and the development of the exact theories of bending and torsion. He also developed theories for plastic deformations and vibrations. His full name was Adéhmar Jean Claude Barré, Count de Saint-Venant. (See Ref. 1-1, pp. 229–242; Ref. 1-2, Vol. I, pp. 833–872, Vol. II,

Part I, pp. 1–286, Vol. II, Part II, pp. 1–51; and Ref. 2-1, pp. 39–40.)

2-13 Zaslavsky, A., "A note on Saint-Venant's principle," *Israel Journal of Technology*, Vol. 20, 1982, pp. 143–144.

2-14 Ramberg, W. A., and Osgood, W. R., "Description of stress-strain curves by three parameters," *National Advisory Committee for Aeronautics*, Technical Note No. 902, July 1943.

3-1 The relationship between torque and angle of twist in the case of a circular bar was correctly established in 1784 by Charles Augustin de Coulomb (1736–1806), a famous French scientist (see Ref. 1-1, pp. 51–53, 82, and 92, and Ref. 1-2, Vol. I, p. 69). Coulomb made contributions in electricity and magnetism, viscosity of fluids, friction, beam bending, retaining walls and arches, torsion and torsional vibrations, and other subjects (see Ref. 1-1, pp. 47–54). Thomas Young (Ref. 1-6) observed that the applied torque is balanced by the shear stresses on the cross section and that the shear stresses are proportional to the distance from the axis. The French engineer Alphonse J. C. B. Duleau (1789–1832) performed tests on bars in torsion and also developed a theory for circular bars (see Ref. 1-1, p. 82).

3-2 Saint-Venant's conclusive memoir on torsion, published in 1855, is described in Ref. 1-1, pp. 229–237, and Ref. 1-2, Vol. II, Part II, pp. 1-51. (See Ref. 2-12 for information about Saint-Venant's life and works.)

3-3 For additional information about Poisson, see Ref. 1-1, p. 216, and Ref. 1-7.

3-4 Bredt, R., "Kritische Bemerkungen zur Drehungselastizität," *Zeitschrift des Vereines Deutscher Ingenieure*, Vol. 40, 1896, pp. 785–790, and 813–817. *Note:* Rudolph Bredt (1842–1900) was a German engineer who studied in Karlsruhe and Zürich. Then he worked for a while in Crewe, England, at a train factory, where he learned about the design and construction of cranes. This experience formed the basis for his later work as a crane manufacturer in Germany. His theory of torsion was developed in connection with the design of box-girder cranes.

5-1 A proof of the theorem that cross sections of a beam in pure bending remain plane can be found in the paper by Fazekas, G. A., "A note on the bending of Euler beams," *Journal of Engineering Education*, Vol. 57, No. 5, January 1967. Proofs can also be found in well-known textbooks on mechanics of materials, such as the books by F. P. Beer and E. R. Johnston, Jr.; S. H. Crandall, N. C. Dahl, and T. J. Lardner; and I. H. Shames. The validity of the theorem has long been recognized, and it was used by early investigators such as Jacob Bernoulli (Ref. 1-4) and L. M. H. Navier

(Ref. 2-5). For a discussion of the work done by Bernoulli and Navier in connection with bending of beams, see Ref. 1-1, pp. 25–27 and 70–75.

5-2 Galilei, Galileo, *Dialogues Concerning Two New Sciences*, translated from the Italian and Latin into English by Henry Crew and Alfonso De Salvio, The Macmillan Company, New York, 1933 (translation first published in 1914).

Note: This book was published in 1638 by Louis Elzevir in Leida, now Leiden, Netherlands. *Two New Sciences* represents the culmination of Galileo's work on dynamics and mechanics of materials. It can truly be said that these two subjects, as we know them today, began with Galileo and the publication of this famous book.

Galileo Galilei was born in Pisa in 1564. He made many famous experiments and discoveries, including those on falling bodies and pendulums that initiated the science of dynamics. Galileo was an eloquent lecturer and attracted students from many countries. He pioneered in astronomy and developed a telescope with which he made many astronomical discoveries, including the mountainous character of the moon, Jupiter's satellites, the phases of Venus, and sunspots. Because his scientific views of the solar system were contrary to theology, he was condemned by the church in Rome and spent the last years of his life in seclusion in Florence; during this period he wrote *Two New Sciences*. Galileo died in 1642 and was buried in Florence.

5-3 The history of beam theory is described in Ref. 1-1, pp. 11–47 and 135–141, and Ref. 1-2. Edme Mariotte (1620–1684) was a French physicist who made developments in dynamics, hydrostatics, optics, and mechanics. He made tests on beams and developed a theory for calculating load-carrying capacity; his theory was an improvement on Galileo's work, but still not correct. Jacob Bernoulli (1654–1705), who is described in Ref. 1-4, first determined that the curvature is proportional to the bending moment. However, his constant of proportionality was incorrect. Leonhard Euler (1707–1783) obtained the differential equation of the deflection curve of a beam and used it to solve many problems of both large and small deflections (Euler's life and work are described in Ref. 11-2). The first person to obtain the distribution of stresses in a beam and correctly relate the stresses to the bending moment probably was Antoine Parent (1666–1716), a French physicist and mathematician. Later, a rigorous investigation of strains and stresses in beams was made by Saint-Venant (1797–1886); see Ref. 2-12.

5-4 *Manual of Steel Construction*, published by the American Institute of Steel Construction, Inc., One East Wacker Drive, Suite 3100, Chicago, Illinois 60601-2001.

5-5 *Aluminum Construction Manual*, published by The Aluminum Association, Inc., 900 19th Street, N.W., Washington, D.C. 20006.

5-6 *National Design Specification for Wood Construction* and *Wood Structural Design Data*, published by the American Forest and Paper Association, 1250 Connecticut Avenue, N.W., Washington, D.C., 20036.

5-7 D. J. Jourawski (1821–1891) was a Russian bridge and railway engineer who developed the now widely used approximate theory for shear stresses in beams (see Ref. 1-1, pp. 141–144, and Ref. 1-2, Vol. II, Part I, pp. 641–642). In 1844, only two years after graduating from the Institute of Engineers of Ways of Communication in St. Petersburg, he was assigned the task of designing and constructing a major bridge on the first railway line from Moscow to St. Petersburg. He noticed that some of the large timber beams split longitudinally in the centers of the cross sections, where he knew the bending stresses were zero. Jourawski drew free-body diagrams and quickly discovered the existence of horizontal shear stresses in the beams. He derived the shear formula and applied his theory to various shapes of beams. Although the exact theory for shear stresses in beams was derived by Saint-Venant, it is useful in very few practical cases, so it appears that the critical remarks about Jourawski's theory made by Todhunter and Pearson on p. 642 of Ref. 1-2 are unjustified. Jourawski's paper on shear in beams is cited in Ref. 5-8. His name is also transliterated as Dimitrii Ivanovich Zhuravskii.

5-8 Jourawski, D. J., "Sur la résistance d'un corps prismatique . . .," *Annales des Ponts et Chaussées*, Mémoires et Documents, 3rd Series, Vol. 12, Part 2, 1856, pp. 328–351.

5-9 Zaslavsky, A., "On the limitations of the shearing stress formula," *International Journal of Mechanical Engineering Education*, Vol. 8, No. 1, 1980, pp. 13–19.

5-10 Maki, A. C., and Kuenzi, E. W., "Deflection and stresses of tapered wood beams," Research Paper FPL 34, U. S. Forest Service, Forest Products Laboratory, Madison, Wisconsin, September 1965, 54 pages.

5-11 See Ref. 2-1, pp. 358–359.

5-12 The French engineer Jacques Antoine Charles Bresse (1822-1883) is best known for his work with curved bars and arches. For details concerning his analysis of the core of a section, see Ref. 1-1, p. 147.

6-1 Timoshenko, S. P., "Use of stress functions to study flexure and torsion of prismatic bars," (in Russian), St. Petersburg, 1913 (reprinted in Vol. 82 of the *Memoirs of the Institute of Ways of Communication*, pp. 1–21).

Note: In this paper, the point in the cross section of a beam through which a concentrated force should act in order to eliminate rotation was found. Thus, this work contains the first determination of a shear center. The particular beam under investigation had a solid semicircular cross section (see Ref. 2-1, pp. 371–373). In 1909, C. Bach tested channel beams and observed the twisting that occurs when a load parallel to the plane of the web is applied to the beam (Ref. 6-2). He also observed that the rotation changed as the load was translated laterally, but apparently he did not locate the shear center. In 1917, A. A. Griffith and G. I. Taylor used the soap-film method to investigate bending; they located the shear center, which they called the *flexural center*, for several shapes of structural sections (Ref. 6-3). The general approximate solution for the shear center of a thin-walled bar of open cross section was developed by R. Maillart, who explained the practical significance of the shear center for structural shapes (Ref. 6-4). It also appears that Maillart introduced the name *shear center*. A comprehensive discussion of the shear center, as we'ı as a general presentation of the problem of bending and torsion of beams, is given in Ref. 6-5, and some historical comments concerning the shear center can be found in Ref. 1-1, p. 401, and Ref. 6-6.

6-2 Bach, C., "Versuche über die tatsächliche Widerstandsfähigkeit von Balkan mit [-förmigem Querschnitt," *Zeitschrift des Vereines Deutscher Ingenieure*, Vol. 53, No. 44, 1909, pp. 1790–1795, and "Versuche über die tatsächliche Widerstandsfähigkeit von Trägern mit [-förmigem Querschnitt," *ibid.*, Vol. 54, No. 10, 1910, pp. 382–387.

6-3 Griffith, A. A., and Taylor, G. I., "The problem of flexure and its solution by the soap-film method," *Technical Report of the Advisory Committee for Aeronautics*, 1917–1918, Vol. 3, published in 1921 by His Majesty's Stationery Office, London (Reports and Memoranda, No. 399, November 1917), pp. 950–969.

6-4 Maillart, R., "Zur Frage der Biegung," *Schweizerische Bauzeitung*, Vol. 77, 1921, pp. 195–197; "Bemerkungen zur Frage der Biegung," *ibid.*, Vol. 78, 1921, pp. 18–19; "Ueber Drehung und Biegung," *ibid.*, Vol. 79, 1922, pp. 254–257; "Der Schubmittelpunkt," *ibid.*, Vol. 83, 1924, pp. 109–111; and "Zur Frage des Schubmittelpunktes," *ibid.*, Vol. 83, 1924, pp. 176–177.

6-5 Timoshenko, S. P., "Theory of bending, torsion, and buckling of thin-walled members of open cross section," *Journal of the Franklin Institute*, Vol. 239, No. 3, March 1945, pp. 201–219; No. 4, April 1945, pp. 249–268; and No. 5, May 1945, pp. 343–361.

6-6 Nowinski, J. L., "Theory of thin-walled bars," *Applied Mechanics Surveys*, edited by Abramson, H. N., et al., Spartan Books, Washington, D.C., 1966, pp. 325–338.

7-1 Augustin Louis Cauchy (1789–1857) was one of the greatest mathematicians. Born in Paris, he entered the École Polytechnique at the age of 16, where he studied under Lagrange, Laplace, Fourier, and Poisson. He was quickly recognized for his mathematical prowess, and at age 27 he became a professor at the École and a member of the Academy of Sciences. His major works in pure mathematics were in group theory, number theory, series, integration, differential equations, and analytical functions. In applied mathematics, he introduced the concept of stress as we know it today, developed the equations of theory of elasticity, and introduced the notion of principal stresses and principal strains (see Ref. 1-1, pp. 107–111). An entire chapter is devoted to his work on theory of elasticity in Ref. 1-2 (see Vol. I, pp. 319–376).

7-2 See Ref. 1-1, pp. 229–242. *Note:* Saint-Venant was a pioneer in many aspects of theory of elasticity, and Todhunter and Pearson dedicated their book, *A History of the Theory of Elasticity* (Ref. 1-2), to him. For further information about Saint-Venant, see Ref. 2-12.

7-3 William John Macquorn Rankine (1820–1872) was born in Edinburgh, Scotland, and taught engineering at Glasgow University. He derived the stress transformation equations in 1852 and made many other contributions to theory of elasticity and applied mechanics (see Ref. 1-1, pp. 197–202, and Ref. 1-2, Vol. II, Part I, pp. 86 and 287–322). His engineering subjects included arches, retaining walls, and structural theory. He also achieved scientific fame for his work with fluids, light, sound, and behavior of crystals.

7-4 The famous German civil engineer Otto Christian Mohr (1835–1918) was both a theoretician and a practical designer. He was a professor at the Stuttgart Polytechnikum and later at the Dresden Polytechnikum. He developed the circle of stress in 1882 (Ref. 7-5 and Ref. 1-1, pp. 283–288). The circle is also described in Mohr's book (Ref. 7-6, pp. 187–219). Mohr made numerous contributions to the theory of structures, including the Williot-Mohr diagram for truss displacements, the moment-area method for beam deflections, and the Maxwell-Mohr method for analyzing statically indeterminate structures. (*Note:* Joseph Victor Williot, 1843–1907, was a French engineer, and James Clerk Maxwell, 1831–1879, was a famous British scientist.)

7-5 Mohr, O., "Über die Darstellung des Spannungszustandes und des Deformationszustandes eines Körperelementes," *Zivilingenieur*, 1882, p. 113.

7-6 Mohr, O., *Abhandlungen aus dem Gebiete der technischen Mechanik*, Wilhelm Ernst and Sohn, Berlin, 1906.

8-1 Karl Culmann (1821–1881) was a famous German bridge and railway engineer. In 1849–1850 he spent two years traveling in England and the United States to study bridges, which he later wrote about in Germany. He designed numerous bridge structures in Europe, and in 1855 he became professor of structures at the newly organized Zürich Polytechnicum. Culmann made many developments in graphical methods and wrote the first book on graphic statics, published in Zürich in 1866. Stress trajectories are one of the original topics presented in this book (see Ref. 1-1, pp. 190–197). For additional information about Culmann and the history of bridges, see Ref. 8-2.

8-2 Kuzmanovic, B. O., "History of the theory of bridge structures," *Proceedings of the American Society of Civil Engineers, Journal of the Structural Division*, Vol. 103, No. ST5, May 1977, pp. 1095–1111.

9-1 The work of Jacob Bernoulli, Euler, and many others with respect to elastic curves is described in Ref. 1-1, pp. 27 and 30–36, and Ref. 1-2. Another member of the Bernoulli family, Daniel Bernoulli (1700–1782), proposed to Euler that he obtain the differential equation of the deflection curve by minimizing the strain energy, which Euler did. Daniel Bernoulli, a nephew of Jacob Bernoulli, is renowned for his work in hydrodynamics, kinetic theory of gases, beam vibrations, and other subjects. His father, John Bernoulli (1667–1748), a younger brother of Jacob, was an equally famous mathematician and scientist who first formulated the principle of virtual displacements, solved the problem of the brachystochrone, and established the rule for obtaining the limiting value of a fraction when both the numerator and denominator tend to zero. He communicated this last rule to G. F. A. de l'Hôpital (1661–1704), a French nobleman who wrote the first book on calculus (1696) and included this theorem, which consequently became known as *L'Hôpital's rule* (see Ref. 9-2). Daniel's nephew, Jacob Bernoulli (1759–1789), also known as James or Jacques, was a pioneer in the theory of plate bending and plate vibrations. Much interesting information about the many prominent members of the Bernoulli family, as well as other pioneers in mechanics and mathematics, can be found in books on the history of mathematics (for instance, see Refs. 9-3, 9-4, and 9-5).

9-2 Struik, D. J., "The origin of L'Hôpital's rule," *The Mathematics Teacher*, Vol. 56, No. 4, April 1963, pp. 257–260.

9-3 Newman, J. R., *The World of Mathematics*, Vols. 1–4, Simon and Schuster, New York, 1956, 2469 pages.

9-4 Struik, D. J., *A Concise History of Mathematics*, 4th Ed., Dover Publications, Inc., New York, 1987.

9-5 Cajori, F., *A History of Mathematics*, 4th Ed., Chelsea Publishing Co., New York, 1985.

9-6 See Ref. 1-1, p. 137, and Saint-Venant, Barré de, notes and appendices to the third edition of the book by Navier, *Résumé des Leçons données à l'école des ponts et chaussées sur l'application de la mécanique à l'établissement des constructions et des machines*, 1st Part, "De la Résistance des corps solides," Paris, 1864, p. 72.

9-7 See Ref. 1-1, p. 284, and Mohr, O., "Beitrag zur Theorie der Holz- und Eisen-Constructionen," *Zeitschrift des Architekten- und Ingenieur-Vereins zu Hannover*, Vol. 14, 1868, pp. 19–51.

9-8 Greene, C. E., *Graphical Method for the Analysis of Bridge Trusses*, D. Van Nostrand Co., Inc., New York, 1875. *Note:* The moment-area method was discovered independently by Greene, a professor at the University of Michigan, in 1873. He began teaching it to students in that year. The method is described under the name *area-moment method* on pp. 35–40 of this book.

9-9 Castigliano, A., *Théorie de l'équilibre des systèmes élastiques et ses applications*, A. F. Negro, Turin, 1879, 480 pages.

Note: In this book Castigliano presented in very complete form many fundamental concepts and principles of structural analysis. Although Castigliano was Italian, he wrote this book in French in order to gain a wider audience for his work. It was translated into both German and English (Refs. 9-10 and 9-11). The English translation was republished in 1966 by Dover Publications and is especially valuable because of the introductory material by Gunhard A. Oravas (Refs. 9-12 and 9-13).

Castigliano's first and second theorems appear on pp. 15–16 of the 1966 edition of his book. He identified them as Part 1 and Part 2 of the "Theorem of the Differential Coefficients of the Internal Work." In mathematical form, they appear in his book as

$$F_p = \frac{dW_i}{dr_p} \quad \text{and} \quad r_p = \frac{dW_i}{dF_p}$$

where W_i is the internal work (or strain energy), F_p represents any one of the external forces, and r_p is the displacement of the point of application of F_p.

Castigliano did not claim complete originality for the first theorem, although he stated in the Preface to his book that his presentation and proof were more general than anything published previously. The second theorem was original with him and was part of his thesis for the civil engineering degree at the Polytechnic Institute of Turin in 1873.

Carlo Alberto Pio Castigliano was born of a poor family in Asti in 1847 and died of pneumonia in 1884, while at the height of his productivity. The story of his life is told by Oravas in the introduction to the 1966 edition, and a bibliography of Castigliano's works and a list of his honors and awards are also given there. His contributions are also documented in Refs. 2-4 and 1-1. He used the name Alberto Castigliano when signing his writings.

9-10 Hauff, E., *Theorie des Gleichgewichtes elastischer Systeme und deren Anwendung*, Carl Gerold's Sohn, Vienna, 1886. (A translation of Castigliano's book, Ref. 9-9.)

9-11 Andrews, E. S., *Elastic Stresses in Structures*, Scott, Greenwood and Son, London, 1919. (A translation of Castigliano's book, Ref. 9-9.)

9-12 Castigliano, C. A. P., *The Theory of Equilibrium of Elastic Systems and Its Applications*, translated by E. S. Andrews, with a new introduction and biographical portrait section by G. A. Oravas, Dover Publications, Inc., New York, 1966. (A republication of Ref. 9-11 but with the addition of historical material by Oravas.)

9-13 Oravas, G. A., "Historical Review of Extremum Principles in Elastomechanics," an introductory section (pp. xx–xlvi) of the book, *The Theory of Equilibrium of Elastic Systems and Its Applications*, by C. A. P. Castigliano, translated by E. S. Andrews, Dover Publications, Inc., New York, 1966 (Ref. 9-12).

10-1 Navier, L. M. H., *Résumé des Leçons données à l'école des ponts et chaussées sur l'application de la mécanique à l'établissement des constructions et des machines*, 1st Ed., 1826; 2nd Ed., 1833; 3rd Ed. (with notes and appendices by Saint-Venant), Paris, 1864. *Note:* Navier prepared the first two editions; the third edition was prepared by Saint-Venant after Navier's death. Saint-Venant's additions are ten times the volume of the original book. For discussions of this famous book, see Ref. 1-1, pp. 73–77 and 232–233, and Ref. 1-2, Vol. I, pp. 144–146, Vol. II, Part I, pp. 105–135.

10-2 B. P. E. Clapeyron (1799–1864) was a French engineer who developed the three-moment equation in connection with his work on bridges (see Ref. 1-1, pp. 144–146). He is also known for his theorems on work and strain energy (see Ref. 2-8). Clapeyron's paper giving the three-moment equation is cited in Ref. 10-3. Bertot, another French engineer, was the first to publish the three-moment equation in its present-day form (Ref. 10-4). While Bertot's paper was published prior to Clapeyron's, it seems that Clapeyron used the method several years earlier and is the true originator of the equation, as pointed out in Ref. 1-1.

10-3 Clapeyron, B. P. E., "Calcul d'une poutre élastique reposant librement sur des appuis inégalement espacés," *Comptes Rendus*, Vol. 45, 1857, pp. 1076–1080.

10-4 Bertot, H., *Mémoires et compte-rendu des travaux de la Société des Ingénieurs Civils*, Vol. 8, 1855, p. 278.

10-5 Zaslavsky, A., "Beams on immovable supports," *Publications of the International Association for Bridge and Structural Engineering*, Vol. 25, 1965, pp. 353–362.

11-1 Zaslavsky, A., "Models for elastic buckling behaviour," *Israel Journal of Technology*, Vol. 6, No. 4, 1968, pp. 212–226.

11-2 Euler, L., "Methodus inveniendi lineas curvas maximi minimive proprietate gaudentes . . . ," Appendix I, "De curvis elasticis," Bousquet, Lausanne and Geneva, 1744. (English translation: Oldfather, W. A., Ellis, C. A., and Brown, D. M., *Isis*, Vol. 20, 1933, pp. 72–160. Also, republished in *Leonhardi Euleri Opera Omnia*, series 1, Vol. 24, 1952.)

Note: Leonhard Euler (1707–1783) made many remarkable contributions to mathematics and mechanics. He was the most productive mathematician of all time (Ref. 9-4, p. 120) and Newman refers to him as "a hero for mathematicians" (Ref. 9-3, p. 150). His name, which is pronounced "oiler," appears repeatedly in present-day textbooks; for instance, in mechanics we have Euler's equations of motion of a rigid body, Euler's angles, Euler's equations of fluid flow, the Euler load in column buckling, and much more; and in mathematics we encounter the famous Euler constant, as well as Euler's numbers, the Euler identity ($e^{i\theta} = \cos\theta + i\sin\theta$), Euler's formula ($e^{i\pi} + 1 = 0$), Euler's differential equation, Euler's equation of a variational problem, Euler's quadrature formula, the Euler summation formula, Euler's theorem on homogeneous functions, Euler's integrals, and even Euler squares (square arrays of numbers possessing special properties).

In applied mechanics, Euler was the first to derive the formula for the critical buckling load of an ideal, slender column and the first to solve the problem of the elastica. This work was published in 1744, as cited above. He dealt with a column that is fixed at the base and free at the top. Later, he extended his work on columns (Ref. 11-3). Euler's numerous books include treatises on celestial mechanics, dynamics, and hydromechanics, and his papers include subjects such as vibrations of beams and plates and statically indeterminate structures.

In the field of mathematics, Euler made outstanding contributions to trigonometry, algebra, number theory, differential and integral calculus, infinite series, analytic geometry, differential equations, calculus of variations, and many other subjects. He was the first to conceive of trigonometric values as the ratios of numbers and the first to present the famous equation $e^{i\theta} = \cos\theta + i\sin\theta$. Within his books on mathematics, all of which were classical references for many generations, we find the first development of the calculus of variations as well as such intriguing items as the proof of Fermat's "last theorem" for $n = 3$ and $n = 4$. Euler also solved the famous problem of the seven bridges of Königsberg, a problem of topology, another field in which he pioneered.

Euler was born near Basel, Switzerland, and attended the University of Basel, where he studied under John Bernoulli (1667–1748). From 1727 to 1741 he lived and worked in St. Petersburg, where he established a great reputation as a mathematician. In 1741 he moved to Berlin upon the invitation of Frederick the Great, King of Prussia. He continued his mathematical research in Berlin until the year 1766, when he returned to St. Petersburg at the request of Catherine II, Empress of Russia. Euler continued to be prolific until his death in St. Petersburg at the age of 76; during this final period of his life he wrote more than 400 papers. In his entire lifetime, the number of books and papers written by Euler totaled 886; he left many manuscripts at his death and they continued to be published by the Russian Academy of Sciences in St. Petersburg for 47 years afterward. All this in spite of the fact that one of his eyes went blind in 1735 and the other in 1766! The story of Euler's life is told in Ref. 1-1, pp. 28–30, and Ref. 9-3, pp. 148–151; some of his contributions to mechanics are described in Ref. 1-1, pp. 30–36 (see also Refs. 1-2, 1-3, 2-3, and 5-3).

11-3 Euler, L., "Sur la force des colonnes," *Histoire de L'Académie Royale des Sciences et Belles Lettres*, 1757, published in *Mémoires* of the Academie, Vol. 13, Berlin, 1759, pp. 252–282. (See Ref. 11-4 for a translation and discussion of this paper.)

11-4 Van den Broek, J. A., "Euler's classic paper 'On the strength of columns,'" *American Journal of Physics*, Vol. 15, No. 4, July–August 1947, pp. 309–318.

11-5 Timoshenko, S. P., and Gere, J. M., *Theory of Elastic Stability*, 2nd Ed., McGraw-Hill Book Company, Inc., New York, 1961.

11-6 Bleich, F., *Buckling Strength of Metal Structures*, McGraw-Hill Book Company, Inc., New York, 1952.

11-7 Johnston, B. G., editor, *Guide to Stability Design Criteria for Metal Structures*, 3rd Ed., John Wiley and Sons, New York, 1976.

11-8 Keller, J. B., "The shape of the strongest column," *Archive for Rational Mechanics and Analysis*, Vol. 5, No. 4, 1960, pp. 275–285.

11-9 Young, D. H., "Rational design of steel columns," *Transactions of the American Society of Civil Engineers*, Vol. 101, 1936, pp. 422–451. *Note:* Donovan Harold Young (1904–1980) was a well-known engineering educator. He was a professor at the University of Michigan and later at Stanford University. His five textbooks in the field of applied mechanics, written with S. P. Timoshenko, were translated into many languages and used throughout the world.

11-10 Lamarle, A. H. E., "Mémoire sur la flexion du bois," *Annales des Travaux Publiques de Belgique*, Part 1, Vol. 3, 1845, pp. 1–64, and Part 2, Vol. 4, 1846, pp. 1–36. *Note:* Anatole Henri Ernest Lamarle (1806–1875) was an engineer and professor. He was born in Calais, studied in Paris, and became a professor at the University of Gand, or Ghent, Belgium. For his work on columns, see Ref. 1-1, p. 208.

11-11 Considère, A., "Résistance des pièces comprimées," *Congrès International des Procédés de Construction*, Paris, September 9–14, 1889, proceedings published by Librairie Polytechnique, Paris, Vol. 3, 1891, p. 371. *Note:* Armand Gabriel Considère (1841–1914) was a French engineer.

11-12 Engesser, F., "Ueber die Knickfestigkeit gerader Stäbe," *Zeitschrift für Architektur und Ingenieurwesen*, Vol. 35, No. 4, 1889, pp. 455–462. *Note:* Friedrich Engesser (1848–1931) was a German railway and bridge engineer. Later, he became a professor at the Karlsruhe Polytechnical Institute, where he made important advances in the theory of structures, especially in buckling and energy methods. For his work on columns, see Ref. 1-1, pp. 292 and 297–299.

11-13 Engesser, F., "Knickfragen," *Schweizerische Bauzeitung*, Vol. 25, No. 13, March 30, 1895, pp. 88–90.

11-14 Jasinski, F., "Noch ein Wort zu den 'Knickfragen,'" *Schweizerische Bauzeitung*, Vol. 25, No. 25, June 22, 1895, pp. 172–175. *Note:* Félix S. Jasinski (1856–1899) was born in Warsaw and studied in Russia. He became a professor at the Institute of Engineers of Ways of Communication in St. Petersburg, now Leningrad.

11-15 Engesser, F., "Ueber Knickfragen," *Schweizerische Bauzeitung*, Vol. 26, No. 4, July 27, 1895, pp. 24–26.

11-16 von Kármán, T., "Die Knickfestigkeit gerader Stäbe," *Physikalische Zeitschrift*, Vol. 9, No. 4, 1908, pp. 136–140 (this paper also appears in Vol. I of Ref. 11-18). *Note:* Theodore von Kármán (1881–1963) was born in Hungary and later worked at the University of Göttingen in the field of aerodynamics. After coming to the United States in 1929, he founded the Jet Propulsion Laboratory and pioneered in aircraft and rocket problems. His research also included inelastic buckling of columns and stability of shells.

11-17 von Kármán, T., "Untersuchungen über Knickfestigkeit," *Mitteilungen über Forschungsarbeiten auf dem Gebiete des Ingenieurwesens, Verein Deutscher Ingenieure*, Berlin, Heft 81, 1910 (this paper also appears in Ref. 11-18).

11-18 *Collected Works of Theodore von Kármán*, Vols. I–IV, Butterworths Scientific Publications, London, 1956.

11-19 Shanley, F. R., "The column paradox," *Journal of the Aeronautical Sciences*, Vol. 13, No. 12, December 1946, p. 678. *Note:* Francis Reynolds Shanley (1904–1968) was a professor of aeronautical engineering at the University of California, Los Angeles.

11-20 Shanley, F. R., "Inelastic column theory," *ibid.*, Vol. 14, No. 5, May 1947, pp. 261–267.

11-21 Hoff, N. J., "Buckling and Stability," The Forty-First Wilbur Wright Memorial Lecture, *Journal of the Royal Aeronautical Society*, Vol. 58, January 1954, pp. 3–52.

11-22 Hoff, N. J., "The idealized column," *Ingenieur-Archiv*, Vol. 28, 1959 (Festschrift Richard Grammel), pp. 89–98.

11-23 Johnston, B. G., "Column buckling theory: Historical highlights," *Journal of Structural Engineering*, Structural Division, American Society of Civil Engineers, Vol. 109, No. 9, September 1983, pp. 2086–2096.

A

SYSTEMS OF UNITS AND CONVERSION FACTORS

SYSTEMS OF UNITS

Measurement systems have been a necessity since people first began to build and barter, and every ancient culture developed some sort of measurement system to serve its needs. Standardization of units took place gradually over the centuries, often through royal edicts. Development of the **British Imperial System** from earlier measurement standards began in the 13th century and was well established by the 18th century. The British system spread to many parts of the world, including the United States, through commerce and colonization. In the United States the system gradually evolved into the **U.S. Customary System (USCS)** that is in common use today.

The concept of the **metric system** originated in France about 300 years ago and was formalized in the 1790's, at the time of the French Revolution. France mandated the use of the metric system in 1840, and since then many other countries have done the same. In 1866 the United States Congress legalized the metric system without making it compulsory.

A new system of units was created when the metric system underwent a major revision in the 1950's. Officially adopted in 1960 and named the **International System of Units** (Système International d'Unités), this newer system is commonly referred to as **SI**. Although some SI units are the same as in the old metric system, SI has many new features and simplifications. Thus, SI is an improved metric system.

Length, time, mass, and force are the basic concepts of mechanics for which units of measurement are needed. However, only three of these quantities are independent since all four of them are related by Newton's second law of motion:

$$F = ma \tag{A-1}$$

843

in which F is the force acting on a particle, m is the mass of the particle, and a is its acceleration. Since acceleration has units of length divided by time squared, all four quantities are involved in the second law.

The International System of Units, like the metric system, is based upon length, time, and mass as fundamental quantities. In these systems, force is derived from Newton's second law. Therefore, the unit of force is expressed in terms of the basic units of length, time, and mass, as shown in the next section. SI is classified as an **absolute system of units** because measurements of the three fundamental quantities are independent of the locations at which the measurements are made; that is, the measurements do not depend upon the effects of gravity. Therefore, the SI units for length, time, and mass may be used anywhere on earth, in space, on the moon, or even on another planet. This is one of the reasons why the metric system has always been preferred for scientific work.

The British Imperial System and the U.S. Customary System are based upon length, time, and force as the fundamental quantities with mass being derived from the second law. Therefore, in these systems the unit of mass is expressed in terms of the units of length, time, and force. The unit of force is defined as the force required to give a certain standard mass an acceleration equal to the acceleration of gravity, which means that the unit of force varies with location and altitude. For this reason, these systems are called **gravitational systems of units**. Such systems were the first to evolve, probably because weight is such a readily discernible property and because variations in gravitational attraction were not noticeable. It is clear, however, that in the modern technological world an absolute system is preferable.

A.2 SI UNITS

The International System of Units has seven **base units** from which all other units are derived. The base units of importance in mechanics are the meter (m) for length, second (s) for time, and kilogram (kg) for mass. Other SI base units pertain to temperature, electric current, amount of substance, and luminous intensity.

The **meter** was originally defined as one ten-millionth of the distance from the North Pole to the equator. Later, this distance was converted to a physical standard, and for many years the standard for the meter was the distance between two marks on a platinum-iridium bar stored at the headquarters of the International Bureau of Weights and Measures (Bureau International des Poids et Mesures) in Sèvres, a suburb on the western edge of Paris, France. Because of the inaccuracies inherent in the use of a physical bar as a standard, the definition of the meter was changed in 1983 to the length of the path traveled by light in a vacuum during a time interval of 1/299792458 of a second.* The advantages of this "natural" standard are that it is not subject to physical damage and is reproducible at laboratories anywhere in the world.

---

*Taking the reciprocal of this number gives the speed of light in a vacuum (299,792,458 meters per second).

The **second** was originally defined as 1/86400 of a mean solar day (24 hours equals 86,400 seconds). However, since 1967 a highly accurate atomic clock has set the standard, and a second is now defined to be the duration of 9,192,631,770 periods of the radiation corresponding to the transition between the two hyperfine levels of the ground state of the cesium-133 atom. (Most engineers would probably prefer the original definition over the new one, which hasn't noticeably changed the second but which is necessary because the earth's rotation rate is gradually slowing down.)

Of the seven base units in SI, the **kilogram** is the only one that is still defined by a physical object. Since the mass of an object can only be determined by comparing it experimentally with the mass of some other object, a physical standard is needed. For this purpose, a one-kilogram cylinder of platinum-iridium, called the International Prototype Kilogram, is kept by the International Bureau of Weights and Measures at Sèvres.

Other units used in mechanics, called **derived units**, are expressed in terms of the base units of meter, second, and kilogram. For instance, the unit of **force** is the **newton**, which is defined as the force required to impart an acceleration of one meter per second squared to a mass of one kilogram.* From Newton's second law ($F = ma$), we can derive the unit of force in terms of base units:

$$1 \text{ newton} = (1 \text{ kilogram})(1 \text{ meter per second squared})$$

Thus, the newton (N) is given in terms of base units by the formula

$$1 \text{ N} = 1 \text{kg·m/s}^2 \tag{A-2}$$

To provide a point of reference, we note that a small apple weighs approximately one newton.

The unit of **work** and **energy** is the **joule**, defined as the work done when the point of application of a force of one newton is displaced a distance of one meter in the direction of the force.** Therefore,

$$1 \text{ joule} = (1 \text{ newton})(1 \text{ meter}) = 1 \text{ newton meter}$$

or
$$1 \text{ J} = 1 \text{ N·m} \tag{A-3}$$

When you raise this book from desktop to eye level, you do about one joule of work, and when you walk up one flight of stairs, you do about 200 joules of work.

The names, symbols, and formulas for SI units of importance in mechanics are listed in Table A-1. Some of the derived units have special names, such as newton, joule, hertz, watt, and pascal. These units are named for notable persons in science and engineering and have symbols (N, J, Hz, W, and Pa) that are capitalized, although the unit names themselves are written in lowercase letters. Other derived units have no special names (for example, the units of acceleration, area, and density) and must be expressed in terms of base units and other derived units.

---

*Sir Isaac Newton (1642–1727) was an English mathematician, physicist, and astronomer. He invented calculus and discovered the laws of motion and gravitation.

**James Prescott Joule (1818–1889) was an English physicist who developed a method for determining the mechanical equivalent of heat. His last name is pronounced "jool."

Table A-1 Principal Units Used in Mechanics

Quantity	International System (SI)			U.S. Customary System (USCS)		
	Unit	Symbol	Formula	Unit	Symbol	Formula
Acceleration (angular)	radian per second squared		rad/s^2	radian per second squared		rad/s^2
Acceleration (linear)	meter per second squared		m/s^2	foot per second squared		ft/s^2
Area	square meter		m^2	square foot		ft^2
Density (mass) (Specific mass)	kilogram per cubic meter		kg/m^3	slug per cubic foot		$slug/ft^3$
Density (weight) (Specific weight)	newton per cubic meter		N/m^3	pound per cubic foot	pcf	lb/ft^3
Energy; work	joule	J	$N \cdot m$	foot-pound		ft-lb
Force	newton	N	$kg \cdot m/s^2$	pound	lb	(base unit)
Force per unit length (Intensity of force)	newton per meter		N/m	pound per foot		lb/ft
Frequency	hertz	Hz	s^{-1}	hertz	Hz	s^{-1}
Length	meter	m	(base unit)	foot	ft	(base unit)
Mass	kilogram	kg	(base unit)	slug		$lb\text{-}s^2/ft$
Moment of a force; torque	newton meter		$N \cdot m$	pound-foot		lb-ft
Moment of inertia (area)	meter to fourth power		m^4	inch to fourth power		$in.^4$
Moment of inertia (mass)	kilogram meter squared		$kg \cdot m^2$	slug foot squared		$slug\text{-}ft^2$
Power	watt	W	J/s $(N \cdot m/s)$	foot-pound per second		ft-lb/s
Pressure	pascal	Pa	N/m^2	pound per square foot	psf	lb/ft^2
Section modulus	meter to third power		m^3	inch to third power		$in.^3$
Stress	pascal	Pa	N/m^2	pound per square inch	psi	$lb/in.^2$
Time	second	s	(base unit)	second	s	(base unit)
Velocity (angular)	radian per second		rad/s	radian per second		rad/s
Velocity (linear)	meter per second		m/s	foot per second	fps	ft/s
Volume (liquids)	liter	L	$10^{-3} m^3$	gallon	gal.	$231\ in.^3$
Volume (solids)	cubic meter		m^3	cubic foot	cf	ft^3

Notes: 1 joule (J) = 1 newton meter (N·m) = 1 watt second (W·s)

1 hertz (Hz) = 1 cycle per second (cps) or 1 revolution per second (rev/s)

1 watt (W) = 1 joule per second (J/s) = 1 newton meter per second (N·m/s)

1 pascal (Pa) = 1 newton per meter squared (N/m²)

1 liter (L) = 0.001 cubic meter (m³) = 1000 cubic centimeters (cm³)

The relationships between various SI units and some commonly used metric units are given in the upper part of Table A-2. Metric units such as dyne, erg, gal, and micron are no longer recommended for engineering or scientific use.

The **weight** of an object is the **force of gravity** acting on that object, and therefore weight is measured in newtons. Since the force of gravity depends upon altitude and position on the earth, weight is not an invariant property of a body. Furthermore, the weight of a body as measured by a spring scale is affected not only by the gravitational pull of the earth but also by the centrifugal effects associated with the rotation of the earth. As a consequence, we must recognize two kinds of weight, *absolute weight* and *apparent weight*. The former is based upon the force of gravity alone, and the latter includes the effects of rotation. Thus, apparent weight is always less than absolute weight (except at the poles). Apparent weight, which is the weight of an object as measured with a spring scale, is the weight we customarily use in business and everyday life; absolute weight is used in astroengineering and certain kinds of scientific work. In this book, the term "weight" will always mean "apparent weight."

Table A-2   Additional Units in Common Use

SI and Metric Units	
1 gal = 1 centimeter per second squared (cm/s^2) for example, $g \approx 981$ gals)	1 micron = 1 micrometer (μm) = 10^{-6} meters (m)
1 are (a) = 100 square meters (m^2)	1 gram (g) = 10^{-3} kilograms (kg)
1 hectare (ha) = 10,000 square meters (m^2)	1 metric ton (t) = 1 megagram (Mg) = 1000 kilograms (kg)
1 erg = 10^{-7} joules (J)	1 watt (W) = 10^7 ergs per second (erg/s)
1 kilowatt-hour (kWh) = 3.6 megajoules (MJ)	1 dyne per square centimeter ($dyne/cm^2$) = 10^{-1} pascals (Pa)
1 dyne = 10^{-5} newtons (N)	1 bar = 10^5 pascals (Pa)
1 kilogram-force (kgf) = 1 kilopond (kp) = 9.80665 newtons (N)	1 stere = 1 cubic meter (m^3)
1 centimeter (cm) = 10^{-2} meters (m)	

USCS and Imperial Units	
1 kilowatt-hour (kWh) = 2,655,220 foot-pounds (ft-lb)	1 horsepower (hp) = 550 foot-pounds per second (ft-lb/s)
1 British thermal unit (Btu) = 778.171 foot-pounds (ft-lb)	1 kilowatt (kW) = 737.562 foot-pounds per second (ft-lb/s)
1 kip (k) = 1000 pounds (lb)	= 1.34102 horsepower (hp)
1 ounce (oz) = 1/16 pound (lb)	1 pound per square inch (psi)
1 ton = 2000 pounds (lb)	= 144 pounds per square foot (psf)
1 Imperial ton (or long ton) = 2240 pounds (lb)	1 revolution per minute (rpm)
1 poundal (pdl) = 0.0310810 pounds (lb) = 0.138255 newtons (N)	= $2\pi/60$ radians per second (rad/s)
1 inch (in.) = 1/12 foot (ft)	1 mile per hour (mph)
1 mil = 0.001 inch (in.)	= 22/15 feet per second (fps)
1 yard (yd) = 3 feet (ft)	1 quart (qt) = 2 pints = 1/4 gallon (gal.)
1 mile = 5280 feet (ft)	1 cubic foot (cf) = 576/77 gallons = 7.48052 gallons (gal.)
	1 Imperial gallon = 277.420 cubic inches (in.^{3})

The **acceleration of gravity**, denoted by the letter g, is directly proportional to the force of gravity, and therefore it too depends upon position. In contrast, **mass** is a measure of the amount of material in a body and does not change with location.

The fundamental relationship between weight, mass, and acceleration of gravity can be obtained from Newton's second law ($F = ma$), which in this case becomes

$$W = mg \qquad\qquad (A\text{-}4)$$

In this equation, W is the weight in newtons (N), m is the mass in kilograms (kg), and g is the acceleration of gravity in meters per second squared (m/s^2). Equation (A-4) shows that *a body having a mass of one kilogram has a weight in newtons numerically equal to g*. The values of the weight W and the acceleration g depend upon many factors, including latitude and elevation. However, for scientific calculations a standard international value of g has been established as

$$g = 9.806650 \text{ m/s}^2 \qquad\qquad (A\text{-}5)$$

This value is intended for use under standard conditions of elevation and latitude (sea level at a latitude of approximately 45°). The recommended value of g for ordinary engineering purposes on or near the surface of the earth is

$$g = 9.81 \text{ m/s}^2 \qquad\qquad (A\text{-}6)$$

Thus, a body having a mass of one kilogram has a weight of 9.81 newtons.

Atmospheric pressure varies considerably with weather conditions, location, altitude, and other factors. Consequently, a standard international value for the pressure at the earth's surface has been defined:

$$1 \text{ standard atmosphere} = 101.325 \text{ kilopascals} \qquad\qquad (A\text{-}7)$$

The following simplified value is recommended for ordinary engineering work:

$$1 \text{ standard atmosphere} = 101 \text{ kPa} \qquad\qquad (A\text{-}8)$$

Of course, the values given in Eqs. (A-7) and (A-8) are intended for use in calculations and do not represent the actual ambient pressure at any given location.

A basic concept in mechanics is **moment** or **torque**, especially the moment of a force and the moment of a couple. Moment is expressed in units of force times length, or newton meters (N·m). Other important concepts in mechanics are **work** and **energy**, both of which are expressed in joules, a derived unit that happens to have the same units (newton meters) as the units of moment. However, moment is a distinctly different quantity from work or energy, and the joule should *never* be used for moment or torque.

Frequency is measured in units of **hertz** (Hz), a derived unit equal to the reciprocal of seconds (1/s or s^{-1}). The hertz is defined as the frequency of a periodic phenomenon for which the period is one second; thus, it is equivalent to one cycle per second (cps) or one revolution per second (rev/s). It is customarily used for the frequencies of vibrations, sound waves, and electromagnetic waves, and occasionally it is used for rotational frequency in place of the more traditional units of revolution per minute (rpm) and revolution per second (rev/s).*

---

*Heinrich Rudolf Hertz (1857–1894) was a German physicist who discovered electromagnetic waves and showed that light waves and electromagnetic waves are identical.

Two other derived units that have special names in SI are the **watt** (W) and the **pascal** (Pa). The watt is the unit of power, which is work per unit of time, and one watt is equal to one joule per second (J/s) or one newton meter per second (N · m/s). The pascal is the unit of pressure and stress, or force per unit area, and is equal to one newton per square meter (N/m^2).*

The **liter** is not an accepted SI unit, yet it is so commonly used that it cannot easily be discarded. Therefore, SI permits its use under limited conditions for volumetric capacity, dry measure, and liquid measure. Both uppercase L and lowercase l are permitted as symbols for liter in SI, but in the United States only L is permitted (to avoid confusion with the numeral 1). The only prefixes permitted with liter are milli and micro.

Loads on structures, whether due to gravity or other actions, are usually expressed in force units, such as newtons, newtons per meter, or pascals (newtons per square meter). Examples of such loads are a concentrated load of 25 kN acting on an axle, a uniformly distributed load of intensity 800 N/m acting on a small beam, and air pressure of intensity 2.1 kPa acting on an airplane wing. However, there is one circumstance in SI in which it is permissible to express a load in mass units. If the load acting on a structure is produced by gravity acting on a mass, then that load may be expressed in mass units (kilograms, kilograms per meter, or kilograms per square meter). The usual procedure in such cases is to convert the load to force units by multiplying by the acceleration of gravity ($g = 9.81$ m/s^2).

■ SI Prefixes

Multiples and submultiples of SI units (both base units and derived units) are created by attaching prefixes to the units (see Table A-3 for a list of prefixes). The use of a prefix avoids unusually large or small numbers. The general rule is that prefixes should be used to keep numbers in the range 0.1 to 1000.

All of the recommended prefixes change the size of the quantity by a multiple or submultiple of three. Similarly, when powers of 10 are used as multipliers, the exponents of 10 should be multiples of three (for example, 40×10^3 N is satisfactory but 400×10^2 N is not). Also, the exponent on a unit with a prefix refers to the entire unit; for instance, the symbol mm^2 means (mm)^2 and not m(m)^2.

■ Styles for Writing SI Units

Rules for writing SI units have been established by international agreement, and the most pertinent ones are described here. Examples of the rules are shown in parentheses.

(1) Units are always written as symbols (kg) in equations and numerical calculations. In text, units are written as words (kilograms) unless numerical values are being reported, in which case either words or symbols may be used (12 kg or 12 kilograms).

(2) Multiplication is shown in a compound unit by a raised dot (kN·m). When the unit is written in words, no dot is required (kilonewton meter).

---

*James Watt (1736–1819) was a Scottish inventor and engineer who developed a practical steam engine and discovered the composition of water. Watt also originated the term "horsepower."

Blaise Pascal (1623–1662) was a French mathematician and philosopher. He founded probability theory, constructed the first calculating machine, and proved experimentally that atmospheric pressure varies with altitude.

Table A-3 SI Prefixes

Prefix	Symbol	Multiplication factor	
tera	T	10^{12} =	1 000 000 000 000
giga	G	10^{9} =	1 000 000 000
mega	M	10^{6} =	1 000 000
kilo	k	10^{3} =	1 000
hecto	h	10^{2} =	100
deka	da	10^{1} =	10
deci	d	10^{-1} =	0.1
centi	c	10^{-2} =	0.01
milli	m	10^{-3} =	0.001
micro	μ	10^{-6} =	0.000 001
nano	n	10^{-9} =	0.000 000 001
pico	p	10^{-12} =	0.000 000 000 001

Note: The use of the prefixes hecto, deka, deci, and centi is not recommended in SI.

(3) Division is shown in a compound unit by a slash (or *solidus*) or by multiplication using a negative exponent (m/s or m·s^{-1}). When the unit is written in words, the slash is always replaced by "per" (meter per second).

(4) A space is always used between a number and its units (200 Pa or 200 pascals) with the exception of the degree symbol (either angle or temperature), where no space is used between the number and the symbol (45°, 20°C).

(5) Units and their prefixes are always printed in roman type (that is, upright or vertical type) and never in italic type (slanted type), even when the surrounding text is in italic type.

(6) When written as words, units are not capitalized (newton) except at the beginning of a sentence or in capitalized material such as a title. When written as a symbol, units are capitalized when they are derived from the name of a person (N). An exception is the symbol for liter, which may be either L or l, but the use of uppercase L is preferred to avoid confusion with the numeral 1. Also, some prefixes are written with capital letters when used in symbols (MPa) but not when used in words (megapascal).

(7) When written as words, units are singular or plural as appropriate to the context (1 kilometer, 20 kilometers, 6 seconds). When written as symbols, units are always singular (1 km, 20 km, 6 s). The plural of hertz is hertz; the plurals of other units are formed in the customary manner (newtons, watts).

(8) Unit symbols never have periods as part of the symbol (kg, not kg.). Of course, a period may follow a unit at the end of a sentence.

(9) In SI, spaces rather than commas are used to separate long numbers into groups of three digits, counting from the decimal point both to the right and to the left:

$$12\ 345\ 678.876\ 54$$

The difficulty with this scheme is that spaces are somewhat vague even when printed, but especially so when written by hand. This awkwardness can be minimized by using prefixes and powers of 10 to reduce or eliminate the need for spaces.

(10) Prefixes are not used in the denominator of a compound unit. An exception is the kilogram (kg),which is a base unit and therefore the letter "k" is not considered as a prefix. For example, we can write kN/m but not N/mm, and we can write J/kg but not mJ/g.

■ Calculations in SI Units

One of the advantages of SI is the convenience of not having to convert units through the use of conversion factors such as 12 (feet to inches) and 144 (square feet to square inches). However, it is still necessary to convert between different powers of 10, and it is surprisingly easy to make errors in doing so. To reduce the likelihood of errors, it is strongly recommended that all calculations be performed in base units (meters, seconds, kilograms). Then the final results are converted to convenient size numbers by adjusting the prefixes or powers of ten and moving the decimal points to correspond.

■ Pronunciation of SI Prefixes and Units

A guide to the pronunciation of a few SI names that are sometimes mispronounced is given in Table A-4. For instance, kilometer is pronounced *kill-oh-meter*, not *kil-om-eter*. The only prefix that generates arguments is giga—the official pronunciation is *jig-uh*, but most engineers in the United States say *gig-uh*.

A.3 U.S. CUSTOMARY UNITS

The units of measurement traditionally used in the United States have never been made mandatory by the government; hence for lack of a better name they are called the "customary" units. In this system the **base units** of relevance to mechanics are the foot (ft) for length, second (s) for time, and pound (lb) for force. The **foot** is defined as

$$1 \text{ ft} = 0.3048 \text{ m (exactly)} \tag{A-9}$$

The **second** is the same as in SI and is described in the preceding section.

The **pound** is defined as the **force** that will give to a certain standard mass an acceleration equal to the acceleration of gravity. In other words, the pound is the weight of the standard mass, which is defined as 0.45359237 kg (exactly). The weight of this amount of mass (see Eq. A-4) is

$$W = (0.45359237 \text{ kg})(9.806650 \text{ m/s}^2) = 4.448222 \text{ N}$$

in which the standard international value of g is used (see Eq. A-5). Thus, the pound is defined as follows:

$$1 \text{ lb} = 4.448222 \text{ N} \tag{A-10}$$

which shows that the pound (like the foot) is actually defined in terms of SI units.

The unit of **mass** in USCS, called the **slug**, is a derived unit defined as the mass that will be accelerated one foot per second squared when acted upon by a

Table A-4   Pronunciation of SI Prefixes and Units

Prefix	Pronunciation
tera	same as *terra*, as in *terra firma*
giga	prononuced *jig-uh*; with *a* pronounced as in *about* (Alternate pronunciation: *gig-uh*)
mega	same as *mega* in *megaphone*
kilo	pronounced *kill-oh*; rhymes with *pillow*
milli	pronounced *mill-eh*, as in *military*
micro	same as *micro* in *microphone*
nano	pronounced *nan-oh*; rhymes with *man-oh*
pico	pronounced *pea-ko*
	Note: The first syllable of every prefix is accented.

Unit	Pronunciation
joule	pronounced *jool*; rhymes with *cool* and *pool*
kilogram	pronounced *kill-oh-gram*
kilometer	pronounced *kill-oh-meter*
pascal	pronounced *pas-kal*, with the accent on *kal*

force of one pound. Writing Newton's second law in the form $m = F/a$, we get

$$1 \text{ slug} = \frac{1 \text{ pound}}{1 \text{ ft/s}^2}$$

which shows that the slug is expressed in terms of base units by the formula

$$1 \text{ slug} = 1 \text{ lb} \cdot \text{s}^2/\text{ft} \tag{A-11}$$

To obtain the mass of an object of known weight, we use the second law in the form

$$m = \frac{W}{g} \tag{A-12}$$

where m is the mass in slugs, W is the weight in pounds, and g is the acceleration of gravity in feet per second squared. As discussed previously, the value of g depends upon the location, but in calculations where location is not relevant, the standard international value of g may be used:

$$g = 32.1740 \text{ ft/s}^2 \tag{A-13}$$

For ordinary purposes, the recommended value is

$$g = 32.2 \text{ ft/s}^2 \tag{A-14}$$

From the preceding equations we conclude that an object having a mass of 1 slug will weigh 32.2 pounds at the earth's surface.

Another unit of mass in USCS is the pound-mass (lbm), which is the mass of an object weighing 1 pound, that is, 1 lbm = 1/32.2 slug.

As mentioned previously, **atmospheric pressure** varies considerably with local conditions; however, for many purposes the standard international value may be used:

$$1 \text{ standard atmosphere} = 14.6959 \text{ pounds per square inch} \qquad (A\text{-}15)$$

or, for ordinary engineering work:

$$1 \text{ standard atmosphere} = 14.7 \text{ psi} \qquad (A\text{-}16)$$

These values are intended for use in calculations and obviously do not represent the actual atmospheric pressure.

The unit of **work** and **energy** in USCS is the **foot-pound** (ft-lb), defined as the work done when the point of application of a force of one pound is displaced a distance of one foot in the direction of the force. The unit of **moment** or **torque** is the **pound-foot** (lb-ft), which comes from the fact that moment is expressed in units of force times length. Although in reality the same units (ft-lb) apply to work, energy, and moment, it is common practice to use the pound-foot for moment and the foot-pound for work and energy.

The symbols and formulas for the most important USCS units used in mechanics are listed in Table A-1.

Many additional units from the U. S. Customary and Imperial systems appear in the mechanics literature; a few of these units are listed in the lower part of Table A-2.

A.4 | TEMPERATURE UNITS

Temperature is measured in SI by a unit called the kelvin (K), and the corresponding scale is the **Kelvin temperature scale**. The Kelvin scale is an absolute scale, which means that its origin (zero kelvins, or 0 K) is at absolute zero temperature, a theoretical temperature characterized by the complete absence of heat. On the Kelvin scale, water freezes at approximately 273 K and boils at approximately 373 K.

For nonscientific purposes the **Celsius temperature scale** is normally used. The corresponding unit of temperature is the degree Celsius (°C), which is equal to one kelvin. On this scale, water freezes at approximately zero degrees (0°C) and boils at approximately 100 degrees (100°C) under certain standard conditions. The Celsius scale is also known as the *centigrade temperature scale*.

The relationship between Kelvin temperature and Celsius temperature is given by the following equations:

$$\text{Temperature in degrees Celsius} = \text{temperature in kelvins} - 273.15$$

or
$$T(°C) = T(K) - 273.15 \qquad (A\text{-}17)$$

where T denotes the temperature. When working with *changes* in temperature, or *temperature intervals*, as is usually the case in mechanics, either unit can be used because the intervals are the same.*

---

*Lord Kelvin (1824–1907), William Thomson, was a British physicist who made many scientific discoveries, developed theories of heat, and proposed the absolute scale of temperature.

Anders Celsius (1701–1744) was a Swedish scientist and astronomer. In 1742 he developed the temperature scale in which 0 and 100 correspond respectively to the freezing and boiling points of water.

The U.S. Customary unit for temperature is the degree Fahrenheit (°F). On the **Fahrenheit temperature scale**, water freezes at approximately 32 degrees (32°F) and boils at approximately 212 degrees (212°F). Each Fahrenheit degree is exactly 5/9 of one kelvin or one degree Celsius. The corresponding absolute scale is the **Rankine temperature scale**, related to the Fahrenheit scale by the equation

$$T(°F) = T(°R) - 459.67 \tag{A-18}$$

Thus, absolute zero corresponds to −459.67°F.*

The conversion formulas between the Fahrenheit and Celsius scales are as follows:

$$T(°C) = \frac{5}{9}[T(°F) - 32] \qquad T(°F) = \frac{9}{5}T(°C) + 32 \tag{A-19a,b}$$

As before, T denotes the temperature on the indicated scale.

A.5 ▮ CONVERSIONS BETWEEN UNITS

Quantities given in either USCS or SI units can be converted quickly to the other system by using the **conversion factors** listed in Table A-5.

If the given quantity is expressed in USCS units, it can be converted to SI units by *multiplying* by the conversion factor. To illustrate this process, assume that the stress in a beam is given as 10,600 psi and we wish to convert this quantity to SI units. From Table A-5 we see that a stress of 1 psi converts to 6894.76 Pa. Therefore, the conversion of the given value is performed in the following manner:

$$(10,600 \text{ psi})(6894.76) = 73100000 \text{ Pa} = 73.1 \text{ MPa}$$

Because the original value is given to three significant digits, we have rounded the final result to three significant digits also (see Appendix B for a discussion of significant digits). Note that the conversion factor of 6894.76 has units of pascals divided by pounds per square inch, and therefore the equation is dimensionally correct.

To reverse the conversion process (that is, to convert from SI units to USCS units), the quantity in SI units is *divided* by the conversion factor. For instance, suppose that the moment of inertia of the cross-sectional area of a beam is given as $94.73 \times 10^6 \text{ mm}^4$. Then the moment of inertia in USCS units is

$$\frac{94.73 \times 10^6 \text{ mm}^4}{416,231} = 228 \text{ in.}^4$$

in which the term 416,231 is the appropriate conversion factor for moment of inertia.

---

*William John Macquorn Rankine (1820–1872) was a Scottish engineer and physicist. He made important contributions in such diverse fields as thermodynamics, light, sound, stress analysis, and bridge engineering.

Gabriel Daniel Fahrenheit (1686–1736) was a German physicist who experimented with thermometers and made them more accurate by using mercury in the tube. He set the origin (0°) of his temperature scale at the freezing point of a mixture of ice, salt, and water.

Table A-5 Conversions Between U.S. Customary Units and SI Units

Customary unit		Times conversion factor		Equals SI unit	
		Accurate	Practical		
Acceleration (linear)					
foot per second squared	ft/s^2	0.3048*	0.305	meter per second squared	m/s^2
inch per second squared	$in./s^2$	0.0254*	0.0254	meter per second squared	m/s^2
Area					
square foot	ft^2	0.09290304*	0.0929	square meter	m^2
square inch	$in.^2$	645.16*	645	square millimeter	mm^2
Density (mass)					
(Specific mass)					
slug per cubic foot	$slug/ft^3$	515.379	515	kilogram per cubic meter	kg/m^3
Density (weight)					
(Specific weight)					
pound per cubic foot	lb/ft^3	157.087	157	newton per cubic meter	N/m^3
pound per cubic inch	$lb/in.^3$	271.447	271	kilonewton per cubic meter	kN/m^3
Energy; work					
foot-pound	ft-lb	1.35582	1.36	joule	J
kilowatt-hour	kWh	3.6*	3.6	megajoule	MJ
British thermal unit	Btu	1055.06	1055	joule	J
Force					
pound	lb	4.44822	4.45	newton	N
kip (1000 pounds)	k	4.44822	4.45	kilonewton	kN
Force per unit length					
(Intensity of force)					
pound per foot	lb/ft	14.5939	14.6	newton per meter	N/m
pound per inch	lb/in.	175.127	175	newton per meter	N/m
kip per foot	k/ft	14.5939	14.6	kilonewton per meter	kN/m
kip per inch	k/in.	175.127	175	kilonewton per meter	kN/m
Length					
foot	ft	0.3048*	0.305	meter	m
inch	in.	25.4*	25.4	millimeter	mm
mile	mi	1.609344*	1.61	kilometer	km
Mass					
slug		14.5939	14.6	kilogram	kg
Moment of a force; torque					
pound-foot	lb-ft	1.35582	1.36	newton meter	N·m
pound-inch	lb-in.	0.112985	0.113	newton meter	N·m
kip-foot	k-ft	1.35582	1.36	kilonewton meter	kN·m
kip-inch	k-in.	0.112985	0.113	kilonewton meter	kN·m

(Continued)

Table A-5   (Continued)

Customary unit		Times conversion factor		Equals SI unit	
		Accurate	Practical		
Moment of inertia (area)					
inch to fourth power	in.^{4}	416,231	416,000	millimeter to fourth power	mm^4
inch to fourth power	in.^{4}	0.416231×10^{-6}	0.416×10^{-6}	meter to fourth power	m^4
Moment of inertia (mass)					
slug foot squared	slug-ft^2	1.35582	1.36	kilogram meter squared	kg·m^2
Power					
foot-pound per second	ft-lb/s	1.35582	1.36	watt	W
foot-pound per minute	ft-lb/min	0.0225970	0.0226	watt	W
horsepower		745.701	746	watt	W
(550 ft-lb/s)	hp				
Pressure; stress					
pound per square foot	psf	47.8803	47.9	pascal (N/m^2)	Pa
pound per square inch	psi	6894.76	6890	pascal	Pa
kip per square foot	ksf	47.8803	47.9	kilopascal	kPa
kip per square inch	ksi	6.89476	6.89	megapascal	MPa
Section modulus					
inch to third power	in.^{3}	16,387.1	16,400	millimeter to third power	mm^3
inch to third power	in.^{3}	16.3871×10^{-6}	16.4×10^{-6}	meter to third power	m^3
Velocity (linear)					
foot per second	ft/s	0.3048*	0.305	meter per second	m/s
inch per second	in./s	0.0254*	0.0254	meter per second	m/s
mile per hour	mph	0.44704*	0.447	meter per second	m/s
mile per hour	mph	1.609344*	1.61	kilometer per hour	km/h
Volume					
cubic foot	ft^3	0.0283168	0.0283	cubic meter	m^3
cubic inch	in.^{3}	16.3871×10^{-6}	16.4×10^{-6}	cubic meter	m^3
cubic inch	in.^{3}	16.3871	16.4	cubic centimeter (cc)	cm^3
gallon (231 in.^{3})	gal.	3.78541	3.79	liter	L
gallon (231 in.^{3})	gal.	0.00378541	0.00379	cubic meter	m^3

*An asterisk denotes an *exact* conversion factor
Note: To convert from SI units to USCS units, divide by the conversion factor

B
PROBLEM SOLVING

B.1 TYPES OF PROBLEMS

The study of mechanics of materials divides naturally into two parts: first, *understanding* the general concepts and principles, and second, *applying* those concepts and principles to physical situations. An understanding of the general concepts is obtained by studying the discussions and derivations presented in books such as this one. Skill in applying the concepts is accomplished by solving problems on your own. Of course, these two aspects of mechanics are closely related, and many experts in mechanics will argue that you don't really understand the concepts if you can't apply them. It is easy to recite the principles, but applying them to real situations requires an in-depth understanding. That is why teachers of mechanics place so much emphasis on problems. Problem solving gives meaning to the concepts and also provides an opportunity to gain experience and develop judgment.

Some of the homework problems in this book require symbolic solutions and others require numerical solutions. In the case of **symbolic problems** (also called *analytical*, *algebraic*, or *literal problems*), the data are supplied in the form of symbols for the various quantities, such as P for load, L for length, and E for modulus of elasticity. Such problems are solved in terms of algebraic variables, and the results are expressed as formulas or mathematical expressions. Symbolic problems usually do not involve numerical calculations, except when numerical data are substituted into the final symbolic result in order to obtain a numerical value. However, this final substitution of numerical data should not obscure the fact that the problem was solved in symbolic terms.

In contrast, **numerical problems** are those in which the data are given in the form of numbers (with appropriate units); for example, a load might be given as 12 kN, a length as 3 m, and a dimension as 0.25 m. The solution of a numerical problem is carried out by performing calculations from the beginning, and the results, both intermediate and final, are in the form of numbers.

An advantage of a numerical problem is that the magnitudes of all quantities are evident at every stage of the solution, thereby providing an opportunity to observe whether the calculations are producing reasonable results. If common sense suggests the results are unreasonable, you can stop and look for an error in the calculations. Also, a numerical solution makes it possible to keep the magnitudes of quantities within prescribed limits. For instance, suppose the stress at a particular point in a beam must not exceed a certain allowable value. If this stress is calculated as an intermediate step in the numerical solution, you can verify immediately whether or not it exceeds the limit. But if the solution is performed in symbolic terms, the magnitude of the stress would not be known at that stage of the solution.

Symbolic problems have several advantages too. Because the results are algebraic formulas or expressions, you can see immediately how the variables affect the answers. For instance, if a load appears to the first power in the numerator of the final result, you know that doubling the load will double the result. Or if a dimension appears in the denominator and is squared, you know that doubling that dimension will divide the result by four. Equally important is the fact that a symbolic solution shows what variables do *not* affect the result. For instance, a certain quantity may cancel out of the solution, a fact that might not even be noticed in a numerical solution. Furthermore, a symbolic solution makes it convenient to check the dimensional homogeneity of all terms in the solution. And most important, a symbolic solution provides a general formula that is applicable to many different problems, each with a different set of numerical data. In contrast, a numerical solution is good for only one set of circumstances, and a complete new solution is required if the data are changed. Of course, symbolic solutions are not feasible when the formulas become too complex to manipulate; when that happens, a numerical solution is the best possibility.

In more advanced work in mechanics, problem solving requires the use of **numerical methods**. This term refers to a wide variety of computational methods, including standard mathematical procedures (such as numerical integration and numerical solution of differential equations), and advanced methods of analysis (such as the finite-element method). Numerical methods involve lengthy and repetitive calculations, so computer programs are used. Programs for performing mathematical procedures, such as those just mentioned, are readily available. More specialized computer programs are also available for performing routine tasks, such as finding deflections of beams and finding principal stresses. However, when studying mechanics of materials, we usually concentrate on the concepts rather than on the use of particular computer programs.

B.2   STEPS IN SOLVING PROBLEMS

The procedures used in solving problems will vary among individuals and will vary according to the type of problem. Nevertheless, the following suggestions will help in reducing mistakes.

1. Make a clear statement of the problem and draw a figure portraying the mechanical or structural system to be investigated. An important part of this step is identifying what is known and what is to be found.

2. Simplify the mechanical or structural system by making assumptions about its physical nature. This step is called *modeling*, because it involves

creating (on paper) an idealized model of the real system. The objective is to create a model that represents the real system to a sufficient degree of accuracy that the results obtained from the model can be applied to the real system.

Here are a few examples of idealizations used in modeling mechanical systems. (a) Finite objects are sometimes modeled as particles, as when determining the forces acting on a joint of a truss. (b) Deformable bodies are sometimes represented as rigid bodies, as when finding the reactions of a statically determinate beam or the forces in the members of a statically determinate truss. (c) The geometry and shapes of objects may be simplified, as when we consider the earth to be a sphere or a beam to be perfectly straight. (d) Distributed forces acting on machines and structures may be represented as concentrated forces or line forces. (e) Forces that are small compared to other forces, or forces that are known to have only a minor effect on the results, may be disregarded in the analysis (friction forces are sometimes in this category). (f) Supports of structures often may be considered as immovable.

3. Draw large and clear sketches as you solve problems. Sketches always aid in understanding the physical situation and often bring out aspects of the problem that would otherwise be overlooked.

4. Apply the principles of mechanics to the idealized model to obtain the governing equations. In statics, the equations usually are equations of equilibrium obtained from Newton's first law; in dynamics, they usually are equations of motion obtained from Newton's second law. In mechanics of materials, the equations are associated with stresses, strains, deformations, and displacements.

5. Use mathematical and computational techniques to solve the equations and obtain results, either in the form of mathematical formulas or numerical values.

6. Interpret the results in terms of the physical behavior of the mechanical or structural system; that is, give meaning or significance to the results, and draw conclusions about the behavior of the system.

7. Check the results in as many ways as you can. Because errors can be disastrous and expensive, engineers should never rely on a single solution.

8. Finally, as preparation for later engineering work, present your solution in clear, neat fashion so that it can be easily reviewed and checked by others. In an academic setting, your instructor will most likely do the reviewing and checking, but if you are participating in a design project in industry or private practice, your analysis and calculations will be thoroughly reviewed by other engineers working on the same project. Your engineering reputation will depend upon the quality of the work you do. Furthermore, you want the archives of the project to contain an accurate, high quality, professional-looking record of your work. Like any other engineering document, your analysis of a problem may someday appear in court or in the public media—and since this document will have *your* name on it, you want it to stand up to public scrutiny.

B.3 DIMENSIONAL HOMOGENEITY

The basic concepts in mechanics are length, time, mass, and force. Each of these physical quantities has a *dimension*, that is, a generalized unit of measurement. For example, consider the concept of length. There are many units of length, such as the meter, kilometer, yard, foot, and inch, yet all of these units have something in common—each one represents a distinct length and not some other quantity such as volume or force. Therefore, we can refer to the *dimension of length* without being specific as to the particular unit of measurement. Similar comments can be made for the dimensions of time, mass, and force. These four dimensions are customarily denoted by the symbols L, T, M, and F, respectively.

Every equation, whether in numeric form or symbolic form, must be *dimensionally homogeneous*, that is, the dimensions of all terms in the equation must be the same. To check the dimensional correctness of an equation, we disregard all numbers and write only the dimensions of each quantity in the equation. The resulting equation must have identical dimensions in all terms.

As an example, consider the following equation for the deflection δ at the midpoint of a simple beam with a uniformly distributed load:

$$\delta = \frac{5qL^4}{384EI}$$

The corresponding dimensional equation is obtained by replacing each quantity by its dimensions; thus, the deflection δ is replaced by the dimension L, the intensity of uniform load q is replaced by F/L (force per unit of length), the length L of the beam is replaced by the dimension L, the modulus of elasticity E is replaced by F/L^2 (force per unit of area), and the moment of inertia I is replaced by L^4. Therefore, the dimensional equation is

$$L = \frac{(F/L)L^4}{(F/L^2)L^4}$$

When simplified, this equation reduces to the dimensional equation L = L, as expected.

Dimensional equations can be written either in generalized terms using the LTMF notation or in terms of the actual units being used in the problem. For instance, if we are making calculations for the beam deflection using USCS units, we can write the dimensional equation as follows:

$$\text{in.} = \frac{(\text{lb/in.})\text{in.}^4}{(\text{lb/in.}^2)\text{in.}^4}$$

which reduces to in. = in. and is dimensionally correct. Frequent checks for dimensional homogeneity (or *consistency of units*) help to eliminate errors when performing derivations and calculations.

Another important way to reduce errors when making numerical calculations is to write the units after *every* numerical value, including the units of conversion factors. By displaying the units of every term in an equation, we can readily verify that the dimensions are consistent.

B.4   SIGNIFICANT DIGITS

Engineering calculations are performed by calculators and computers that operate with great precision. For instance, some computers routinely perform calculations with more than 25 digits in every numerical value, and output values with 10 or more digits are available in even the most inexpensive hand-held calculators. Under these conditions it is important to realize that the accuracy of the results obtained from an engineering analysis is determined not only by the calculations but also by factors such as the accuracy of the given data, the approximations inherent in the analytical models, and the validity of the assumptions used in the theories. In many engineering situations, these considerations mean that the results are valid to only two or three significant digits. Because of the large number of digits available from calculators, there is a strong possibility that numerical results will be reported to an accuracy much greater than is justified, with the possibility of misleading anyone who uses those results.

As an example, suppose that a computation yields the result $R = 6287.46$ lb for the reaction of a statically indeterminate beam. To state the result in this manner is misleading, because it implies that the reaction is known to the nearest 1/100 of a pound even though its magnitude is over 6000 pounds. Thus, it implies an accuracy of approximately 1/600,000 and a precision of 0.01 lb, neither of which is justified. Instead, the accuracy of the calculated reaction depends upon matters such as the following: (1) how accurately the loads, dimensions, and other data used in the analysis are known, and (2) the approximations inherent in the theories of beam behavior. Most likely, the reaction R in this example would be known only to the nearest 10 pounds, or perhaps only to the nearest 100 pounds. Consequently, the result of the computation should be stated as either $R = 6290$ lb or $R = 6300$ lb.

To make clear the accuracy of a given numerical value, it is common practice to use **significant digits**. This practice is widespread in engineering and is a very convenient approach to the matter of accuracy, although it does not correctly portray the accuracy of a number, as explained later.

A *significant digit* is a digit from 1 to 9 or any zero not used to show the position of the decimal point; for instance, the numbers 417, 8.29, 7.30, and 0.00254 each have three significant digits. However, the number of significant digits in a number such as 29,000 is not apparent. It may have two significant digits, with the three zeros serving only to locate the decimal point, or it may have three, four, or five significant digits if one or more of the zeros is valid. By using powers of ten, the accuracy of a number such as 29,000 can be made clearer. When written as 29×10^3 or 0.029×10^6, the number is understood to have two significant digits; when written as 29.0×10^3 or 0.0290×10^6, it has three significant digits.

When a number is obtained by calculation, its accuracy depends upon the accuracy of the numbers used in performing the calculations. A rule of thumb that serves for multiplication and division is the following: The number of significant digits in the calculated result is the same as the least number of significant digits in any of the numbers used in the calculation. As an illustration, consider the product of 2339.3 and 35.4. The calculated result is 82,811.220 when recorded to eight digits. However, stating the result in this manner is misleading because it implies much greater accuracy than is warranted by either of the original numbers. Inasmuch as the number 35.4 has only three significant digits, the proper way to write the result is 82.8×10^3.

For calculations involving addition or subtraction of a column of numbers, the last significant digit in the result is found in the last column of digits that has significant digits in all of the numbers being added or subtracted. To make this notion clearer, consider the following three examples:

	459.637	838.49	856,400
	+ 7.2	− 7	− 847,900
Result from calculator:	466.837	831.49	8,500
Write the result as:	466.8	831	8,500

In the first example, the number 459.637 has six significant digits and the number 7.2 has two. When added, the result has four significant digits because all digits in the result to the right of the column containing the 2 are meaningless. In the second example, the number 7 is accurate to one significant digit (that is, it is not an exact number). Therefore, the final result is accurate only as far as the column containing the 7, which means it has three significant digits and is recorded as 831. In the third example, the numbers 856,400 and 847,900 are assumed to be accurate to four significant digits, but the result of the subtraction is accurate to only two significant digits since none of the zeros is significant. In general, subtraction results in reduced accuracy, and the loss of accuracy is especially noticeable when the difference between the original numbers is relatively small.

These three examples show that numbers obtained by calculation may contain superfluous digits having no physical meaning. Therefore, when reporting such numbers as final results, you should give only those digits that are significant.

In mechanics of materials, the data for problems are usually accurate to about 1%, or perhaps 0.1% in some cases, and therefore the final results should be reported to a comparable accuracy. When greater accuracy is warranted, it will be obvious from the statement of the problem.

Although the use of significant digits provides a handy way to deal with the matter of numerical accuracy, it should be recognized that significant digits are not valid indicators of accuracy. To illustrate this fact, consider the numbers 999 and 101. Three significant digits in the number 999 correspond to an accuracy of 1/999, or 0.1%, whereas the same number of significant digits in the number 101 corresponds to an accuracy of only 1/101, or 1.0%. This disparity in accuracy can be reduced by always using one additional significant digit for numbers beginning with the digit 1. Thus, four significant digits in the number 101.1 gives about the same accuracy as three significant digits in the number 999.

In this book we generally will follow the common engineering rule that *final* numerical results beginning with the digits 2 through 9 should be recorded to three significant digits and those beginning with the digit 1 should be recorded to four significant digits. However, to preserve numerical accuracy and avoid round-off errors during the calculation process, the results of *intermediate* calculations will usually be recorded with additional digits.

Many of the numbers entering into our calculations are exact, for example, the number π, fractions such as 1/2, integers such as the number 48 in the formula $PL^3/48EI$ for a beam deflection, and some of the conversion factors for units. Exact numbers are significant to an infinite number of digits and therefore have no role in determining the accuracy of a calculated result.

B.5 | ROUNDING OF NUMBERS

The process of discarding the insignificant digits and keeping only the significant ones is called *rounding*. To illustrate the process, assume that a number is to be rounded to three significant digits. Then the following rules apply:

(a) If the fourth digit is less than 5, the first three digits are left unchanged and all succeeding digits are dropped or replaced by zeros. For example, 37.44 rounds to 37.4 and 673,289 rounds to 673,000.

(b) If the fourth digit is greater than 5, or if the fourth digit is 5 and is followed by at least one digit other than zero, then the third digit is increased by 1 and all following digits are dropped or replaced by zeros. For example, 26.37 rounds to 26.4 and 3.245002 rounds to 3.25.

(c) Finally, if the fourth digit is 5 and all following digits (if any) are zeros, then the third digit is unchanged if it is an even number and increased by 1 if it is an odd number, and the 5 is replaced by a zero. (Trailing and leading zeros are retained only if they are needed to locate the decimal point.) This process is usually described as "rounding to the even digit." Since the occurrence of even and odd digits is more or less random, the use of this rule means that numbers are rounded upward about as often as downward, thereby reducing the chances of accumulating round-off errors. (Unfortunately, instead of following the "even" rule, most calculators round upward when the last nonzero digit is 5.)

The rules described in the preceding paragraphs for rounding to three significant digits apply in the same general manner when rounding to any other number of significant digits.

C

MATHEMATICAL FORMULAS

■ **Mathematical Constants**

$\pi = 3.14159\ldots \qquad e = 2.71828\ldots \qquad 2\pi \text{ radians} = 360 \text{ degrees}$

$1 \text{ radian} = \dfrac{180}{\pi} \text{ degrees} = 57.2958° \quad 1 \text{ degree} = \dfrac{\pi}{180} \text{ radians} = 0.0174533 \text{ rad}$

Conversions: Multiply degrees by $\dfrac{\pi}{180}$ to obtain radians

Multiply radians by $\dfrac{180}{\pi}$ to obtain degrees

■ **Exponents**

$A^n A^m = A^{n+m} \qquad \dfrac{A^m}{A^n} = A^{m-n} \qquad (A^m)^n = A^{mn} \qquad A^{-m} = \dfrac{1}{A^m}$

$(AB)^n = A^n B^n \qquad \left(\dfrac{A}{B}\right)^n = \dfrac{A^n}{B^n} \qquad A^{m/n} = \sqrt[n]{A^m} \qquad A^0 = 1 \ (A \neq 0)$

■ **Logarithms**

$\log \equiv$ common logarithm (logarithm to the base 10) $\qquad 10^x = y \qquad \log y = x$

$\ln \equiv$ natural logarithm (logarithm to the base e) $\qquad e^x = y \qquad \ln y = x$

$e^{\ln A} = A \qquad 10^{\log A} = A \qquad \ln e^A = A \qquad \log 10^A = A$

$\log AB = \log A + \log B \qquad \log \dfrac{A}{B} = \log A - \log B$

$\log \dfrac{1}{A} = -\log A \qquad \log A^n = n \log A$

$\log 1 = \ln 1 = 0 \qquad \log 10 = 1 \qquad \ln e = 1$

$\ln A = (\ln 10)(\log A) = 2.30259 \log A \qquad \log A = (\log e)(\ln A) = 0.434294 \ln A$

■ Trigonometric Functions

$$\tan x = \frac{\sin x}{\cos x} \qquad \cot x = \frac{\cos x}{\sin x} \qquad \sec x = \frac{1}{\cos x} \qquad \csc x = \frac{1}{\sin x}$$

$$\sin^2 x + \cos^2 x = 1 \qquad \tan^2 x + 1 = \sec^2 x \qquad \cot^2 x + 1 = \csc^2 x$$

$$\sin(-x) = -\sin x \qquad \cos(-x) = \cos x \qquad \tan(-x) = -\tan x$$

$$\cot(-x) = -\cot x$$

$$\sin(x \pm y) = \sin x \cos y \pm \cos x \sin y \qquad \cos(x \pm y) = \cos x \cos y \mp \sin x \sin y$$

$$\sin 2x = 2 \sin x \cos x \qquad \cos 2x = \cos^2 x - \sin^2 x \qquad \tan 2x = \frac{2 \tan x}{1 - \tan^2 x}$$

$$\tan x = \frac{1 - \cos 2x}{\sin 2x} = \frac{\sin 2x}{1 + \cos 2x}$$

$$\sin^2 x = \frac{1}{2}(1 - \cos 2x) \qquad \cos^2 x = \frac{1}{2}(1 + \cos 2x)$$

For any triangle with sides a, b, c and opposite angles A, B, C:

Law of sines $\qquad \dfrac{a}{\sin A} = \dfrac{b}{\sin B} = \dfrac{c}{\sin C}$

Law of cosines $\qquad c^2 = a^2 + b^2 - 2ab \cos C$

■ Quadratic Equation and Quadratic Formula

$$ax^2 + bx + c = 0 \qquad x = \frac{-b \pm \sqrt{b^2 - 4ac}}{2a}$$

■ Infinite Series

$$\frac{1}{1 + x} = 1 - x + x^2 - x^3 + \dots \qquad (-1 < x < 1)$$

$$\sqrt{1 + x} = 1 + \frac{x}{2} - \frac{x^2}{8} + \frac{x^3}{16} - \dots \qquad (-1 < x < 1)$$

$$\frac{1}{\sqrt{1 + x}} = 1 - \frac{x}{2} + \frac{3x^2}{8} - \frac{5x^3}{16} + \dots \qquad (-1 < x < 1)$$

$$e^x = 1 + x + \frac{x^2}{2!} + \frac{x^3}{3!} + \dots \qquad (-\infty < x < \infty)$$

$$\sin x = x - \frac{x^3}{3!} + \frac{x^5}{5!} - \frac{x^7}{7!} + \dots \qquad (-\infty < x < \infty)$$

$$\cos x = 1 - \frac{x^2}{2!} + \frac{x^4}{4!} - \frac{x^6}{6!} + \dots \qquad (-\infty < x < \infty)$$

Note: If x is very small compared to 1, only the first few terms in the series are needed.

■ Derivatives

$$\frac{d}{dx}(ax) = a \qquad \frac{d}{dx}(x^n) = nx^{n-1} \qquad \frac{d}{dx}(au) = a\frac{du}{dx}$$

$$\frac{d}{dx}(uv) = u\frac{dv}{dx} + v\frac{du}{dx} \qquad \frac{d}{dx}\left(\frac{u}{v}\right) = \frac{v(du/dx) - u(dv/dx)}{v^2}$$

$$\frac{d}{dx}(u^n) = nu^{n-1}\frac{du}{dx} \qquad \frac{dy}{dx} = \frac{dy}{du}\frac{du}{dx} \qquad \frac{du}{dx} = \frac{1}{dx/du}$$

$$\frac{d}{dx}(\sin u) = \cos u\frac{du}{dx} \qquad \frac{d}{dx}(\cos u) = -\sin u\frac{du}{dx}$$

$$\frac{d}{dx}(\tan u) = \sec^2 u\frac{du}{dx} \qquad \frac{d}{dx}(\cot u) = -\csc^2 u\frac{du}{dx}$$

$$\frac{d}{dx}(\sec u) = \sec u \tan u\frac{du}{dx} \qquad \frac{d}{dx}(\csc u) = -\csc u \cot u\frac{du}{dx}$$

$$\frac{d}{dx}(\arctan u) = \frac{1}{1 + u^2}\frac{du}{dx} \qquad \frac{d}{dx}(\log u) = \frac{\log e}{u}\frac{du}{dx} \qquad \frac{d}{dx}(\ln u) = \frac{1}{u}\frac{du}{dx}$$

$$\frac{d}{dx}(a^u) = a^u \ln a\frac{du}{dx} \qquad \frac{d}{dx}(e^u) = e^u\frac{du}{dx}$$

■ Indefinite Integrals

Note: A constant must be added to the result of every integration

$$\int a\,dx = ax \qquad \int u\,dv = uv - \int v\,du \quad \text{(integration by parts)}$$

$$\int x^n\,dx = \frac{x^{n+1}}{n + 1} \quad (n \neq -1) \qquad \int\frac{dx}{x} = \ln|x| \quad (x \neq 0)$$

$$\int\frac{dx}{x^n} = \frac{x^{1-n}}{1 - n} \quad (n \neq 1) \qquad \int(a + bx)^n\,dx = \frac{(a + bx)^{n+1}}{b(n + 1)} \quad (n \neq -1)$$

$$\int\frac{dx}{a + bx} = \frac{1}{b}\ln(a + bx) \qquad \int\frac{dx}{(a + bx)^2} = -\frac{1}{b(a + bx)}$$

$$\int\frac{dx}{(a + bx)^3} = -\frac{1}{2b(a + bx)^2} \qquad \int\frac{dx}{(a + bx)^4} = -\frac{1}{3b(a + bx)^3}$$

$$\int\frac{dx}{(a + bx)^n} = -\frac{1}{(n - 1)(b)(a + bx)^{n-1}} \quad (n \neq 1)$$

$$\int\frac{dx}{a^2 + b^2x^2} = \frac{1}{ab}\tan^{-1}\frac{bx}{a} \quad (x \text{ in radians}) \quad (a > 0, b > 0)$$

$$\int\frac{dx}{a^2 - b^2x^2} = \frac{1}{ab}\tanh^{-1}\frac{bx}{a} = \frac{1}{2ab}\ln\left(\frac{a + bx}{a - bx}\right) \quad (x \text{ in radians}) \quad (a > 0, b > 0)$$

$$\int\frac{x\,dx}{a + bx} = \frac{1}{b^2}[bx - a\ln(a + bx)] \qquad \int\frac{x\,dx}{(a + bx)^2} = \frac{1}{b^2}\left[\frac{a}{a + bx} + \ln(a + bx)\right]$$

$$\int \frac{x\,dx}{(a + bx)^3} = -\frac{a + 2bx}{2b^2(a + bx)^2} \qquad \int \frac{x\,dx}{(a + bx)^4} = -\frac{a + 3bx}{6b^2(a + bx)^3}$$

$$\int \frac{x^2 dx}{a + bx} = \frac{1}{2b^3}\left[(a + bx)(-3a + bx) + 2a^2 \ln(a + bx)\right]$$

$$\int \frac{x^2 dx}{(a + bx)^2} = \frac{1}{b^3}\left[\frac{bx(2a + bx)}{a + bx} - 2a \ln (a + bx)\right]$$

$$\int \frac{x^2 dx}{(a + bx)^3} = \frac{1}{b^3}\left[\frac{a(3a + 4bx)}{2(a + bx)^2} + \ln (a + bx)\right]$$

$$\int \frac{x^2 dx}{(a + bx)^4} = -\frac{a^2 + 3abx + 3b^2x^2}{3b^3(a + bx)^3}$$

$$\int \sin ax\,dx = -\frac{\cos ax}{a} \qquad \int \cos ax\,dx = \frac{\sin ax}{a}$$

$$\int \tan ax\,dx = \frac{1}{a} \ln (\sec ax) \qquad \int \cot ax\,dx = \frac{1}{a} \ln (\sin ax)$$

$$\int \sec ax\,dx = \frac{1}{a} \ln (\sec ax + \tan ax) \qquad \int \csc ax\,dx = \frac{1}{a} \ln (\csc ax - \cot ax)$$

$$\int \sin^2 ax\,dx = \frac{x}{2} - \frac{\sin 2ax}{4a} \qquad \int \cos^2 ax\,dx = \frac{x}{2} + \frac{\sin 2ax}{4a} \quad (x \text{ in radians})$$

$$\int x \sin ax\,dx = \frac{\sin ax}{a^2} - \frac{x \cos ax}{a} \quad (x \text{ in radians})$$

$$\int x \cos ax\,dx = \frac{\cos ax}{a^2} + \frac{x \sin ax}{a} \quad (x \text{ in radians})$$

$$\int \frac{dx}{1 + \sin ax} = -\frac{1}{a} \tan \left(\frac{\pi}{4} - \frac{ax}{2}\right)$$

$$\int e^{ax}\,dx = \frac{e^{ax}}{a} \qquad \int xe^{ax}\,dx = \frac{e^{ax}}{a^2}(ax - 1) \qquad \int \ln ax\,dx = x(\ln ax - 1)$$

$$\int \sqrt{a + bx}\,dx = \frac{2}{3b}(a + bx)^{3/2}$$

$$\int \sqrt{a^2 + b^2x^2}\,dx = \frac{x}{2}\sqrt{a^2 + b^2x^2} + \frac{a^2}{2b} \ln \left(\frac{bx}{a} + \sqrt{1 + \frac{b^2x^2}{a^2}}\right)$$

$$\int \frac{dx}{\sqrt{a^2 + b^2x^2}} = \frac{1}{b} \ln \left(\frac{bx}{a} + \sqrt{1 + \frac{b^2x^2}{a^2}}\right)$$

$$\int \sqrt{a^2 - b^2x^2}\,dx = \frac{x}{2}\sqrt{a^2 - b^2x^2} + \frac{a^2}{2b} \sin^{-1}\frac{bx}{a}$$

$$\int \frac{dx}{\sqrt{a^2 - b^2x^2}} = \frac{1}{b} \sin^{-1}\frac{bx}{a}$$

■ **Definite Integrals**

$$\int_a^b f(x)\,dx = -\int_b^a f(x)\,dx \qquad \int_a^b f(x)\,dx = \int_a^c f(x)\,dx + \int_c^b f(x)\,dx$$

PROPERTIES OF PLANE AREAS

Notation: A = area

$\bar{x}, \bar{y}$ = distances to centroid C

I_x, I_y = moments of inertia with respect to the x and y axes, respectively

I_{xy} = product of inertia with respect to the x and y axes

$I_p = I_x + I_y$ = polar moment of inertia with respect to the origin of the x and y axes

I_{BB} = moment of inertia with respect to axis B-B

1

Rectangle   (Origin of axes at centroid)

$$A = bh \qquad \bar{x} = \frac{b}{2} \qquad \bar{y} = \frac{h}{2}$$

$$I_x = \frac{bh^3}{12} \qquad I_y = \frac{hb^3}{12} \qquad I_{xy} = 0 \qquad I_p = \frac{bh}{12}(h^2 + b^2)$$

2

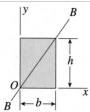

Rectangle   (Origin of axes at corner)

$$I_x = \frac{bh^3}{3} \qquad I_y = \frac{hb^3}{3} \qquad I_{xy} = \frac{b^2h^2}{4} \qquad I_p = \frac{bh}{3}(h^2 + b^2)$$

$$I_{BB} = \frac{b^3h^3}{6(b^2 + h^2)}$$

3

Triangle   (Origin of axes at centroid)

$$A = \frac{bh}{2} \qquad \bar{x} = \frac{b + c}{3} \qquad \bar{y} = \frac{h}{3}$$

$$I_x = \frac{bh^3}{36} \qquad I_y = \frac{bh}{36}(b^2 - bc + c^2)$$

$$I_{xy} = \frac{bh^2}{72}(b - 2c) \qquad I_p = \frac{bh}{36}(h^2 + b^2 - bc + c^2)$$

4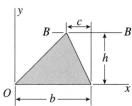

Triangle   (Origin of axes at vertex)

$$I_x = \frac{bh^3}{12} \qquad I_y = \frac{bh}{12}(3b^2 - 3bc + c^2)$$

$$I_{xy} = \frac{bh^2}{24}(3b - 2c) \qquad I_{BB} = \frac{bh^3}{4}$$

5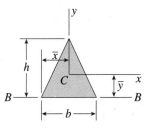

Isosceles triangle   (Origin of axes at centroid)

$$A = \frac{bh}{2} \qquad \bar{x} = \frac{b}{2} \qquad \bar{y} = \frac{h}{3}$$

$$I_x = \frac{bh^3}{36} \qquad I_y = \frac{hb^3}{48} \qquad I_{xy} = 0$$

$$I_p = \frac{bh}{144}(4h^2 + 3b^2) \qquad I_{BB} = \frac{bh^3}{12}$$

(*Note:* For an equilateral triangle, $h = \sqrt{3}\ b/2$.)

6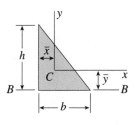

Right triangle   (Origin of axes at centroid)

$$A = \frac{bh}{2} \qquad \bar{x} = \frac{b}{3} \qquad \bar{y} = \frac{h}{3}$$

$$I_x = \frac{bh^3}{36} \qquad I_y = \frac{hb^3}{36} \qquad I_{xy} = -\frac{b^2h^2}{72}$$

$$I_p = \frac{bh}{36}(h^2 + b^2) \qquad I_{BB} = \frac{bh^3}{12}$$

7

Right triangle   (Origin of axes at vertex)

$$I_x = \frac{bh^3}{12} \qquad I_y = \frac{hb^3}{12} \qquad I_{xy} = \frac{b^2h^2}{24}$$

$$I_p = \frac{bh}{12}(h^2 + b^2) \qquad I_{BB} = \frac{bh^3}{4}$$

8

Trapezoid   (Origin of axes at centroid)

$$A = \frac{h(a + b)}{2} \qquad \bar{y} = \frac{h(2a + b)}{3(a + b)}$$

$$I_x = \frac{h^3(a^2 + 4ab + b^2)}{36(a + b)} \qquad I_{BB} = \frac{h^3(3a + b)}{12}$$

9

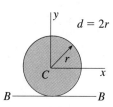

Circle   (Origin of axes at center)

$$A = \pi r^2 = \frac{\pi d^2}{4} \qquad I_x = I_y = \frac{\pi r^4}{4} = \frac{\pi d^4}{64}$$

$$I_{xy} = 0 \qquad I_p = \frac{\pi r^4}{2} = \frac{\pi d^4}{32} \qquad I_{BB} = \frac{5\pi r^4}{4} = \frac{5\pi d^4}{64}$$

10

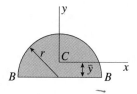

Semicircle   (Origin of axes at centroid)

$$A = \frac{\pi r^2}{2} \qquad \bar{y} = \frac{4r}{3\pi}$$

$$I_x = \frac{(9\pi^2 - 64)r^4}{72\pi} \approx 0.1098 r^4 \qquad I_y = \frac{\pi r^4}{8} \qquad I_{xy} = 0 \qquad I_{BB} = \frac{\pi r^4}{8}$$

11

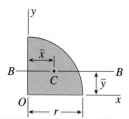

Quarter circle   (Origin of axes at center of circle)

$$A = \frac{\pi r^2}{4} \qquad \bar{x} = \bar{y} = \frac{4r}{3\pi}$$

$$I_x = I_y = \frac{\pi r^4}{16} \qquad I_{xy} = \frac{r^4}{8} \qquad I_{BB} = \frac{(9\pi^2 - 64)r^4}{144\pi} \approx 0.05488 r^4$$

12

Quarter-circular spandrel   (Origin of axes at point of tangency)

$$A = \left(1 - \frac{\pi}{4}\right)r^2 \qquad \bar{x} = \frac{2r}{3(4 - \pi)} \approx 0.7766r \qquad \bar{y} = \frac{(10 - 3\pi)r}{3(4 - \pi)} \approx 0.2234r$$

$$I_x = \left(1 - \frac{5\pi}{16}\right)r^4 \approx 0.01825 r^4 \qquad I_y = I_{BB} = \left(\frac{1}{3} - \frac{\pi}{16}\right)r^4 \approx 0.1370 r^4$$

13

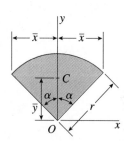

Circular sector   (Origin of axes at center of circle)

$$\alpha = \text{angle in radians} \qquad (\alpha \leq \pi/2)$$

$$A = \alpha r^2 \qquad \bar{x} = r \sin \alpha \qquad \bar{y} = \frac{2r \sin \alpha}{3\alpha}$$

$$I_x = \frac{r^4}{4}(\alpha + \sin \alpha \cos \alpha) \qquad I_y = \frac{r^4}{4}(\alpha - \sin \alpha \cos \alpha) \qquad I_{xy} = 0 \qquad I_p = \frac{\alpha r^4}{2}$$

14

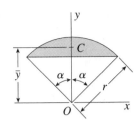

Circular segment   (Origin of axes at center of circle)

α = angle in radians $(\alpha \le \pi/2)$

$$A = r^2(\alpha - \sin \alpha \cos \alpha) \qquad \bar{y} = \frac{2r}{3}\left(\frac{\sin^3 \alpha}{\alpha - \sin \alpha \cos \alpha}\right)$$

$$I_x = \frac{r^4}{4}(\alpha - \sin \alpha \cos \alpha + 2 \sin^3 \alpha \cos \alpha) \qquad I_{xy} = 0$$

$$I_y = \frac{r^4}{12}(3\alpha - 3 \sin \alpha \cos \alpha - 2 \sin^3 \alpha \cos \alpha)$$

15

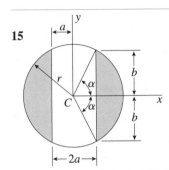

Circle with core removed   (Origin of axes at center of circle)

α = angle in radians $(\alpha \le \pi/2)$

$$\alpha = \arccos \frac{a}{r} \qquad b = \sqrt{r^2 - a^2} \qquad A = 2r^2\left(\alpha - \frac{ab}{r^2}\right)$$

$$I_x = \frac{r^4}{6}\left(3\alpha - \frac{3ab}{r^2} - \frac{2ab^3}{r^4}\right) \qquad I_y = \frac{r^4}{2}\left(\alpha - \frac{ab}{r^2} + \frac{2ab^3}{r^4}\right) \qquad I_{xy} = 0$$

16

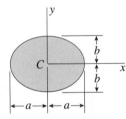

Ellipse   (Origin of axes at centroid)

$$A = \pi ab \qquad I_x = \frac{\pi ab^3}{4} \qquad I_y = \frac{\pi ba^3}{4}$$

$$I_{xy} = 0 \qquad I_p = \frac{\pi ab}{4}(b^2 + a^2)$$

$$\text{Circumference} \approx \pi[1.5(a + b) - \sqrt{ab}] \qquad (a/3 \le b \le a)$$

$$\approx 4.17b^2/a + 4a \qquad (0 \le b \le a/3)$$

17

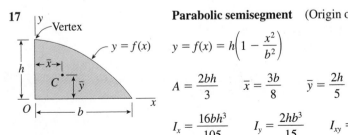

Parabolic semisegment   (Origin of axes at corner)

$$y = f(x) = h\left(1 - \frac{x^2}{b^2}\right)$$

$$A = \frac{2bh}{3} \qquad \bar{x} = \frac{3b}{8} \qquad \bar{y} = \frac{2h}{5}$$

$$I_x = \frac{16bh^3}{105} \qquad I_y = \frac{2hb^3}{15} \qquad I_{xy} = \frac{b^2h^2}{12}$$

18

Parabolic spandrel   (Origin of axes at vertex)

$$y = f(x) = \frac{hx^2}{b^2}$$

$$A = \frac{bh}{3} \qquad \bar{x} = \frac{3b}{4} \qquad \bar{y} = \frac{3h}{10}$$

$$I_x = \frac{bh^3}{21} \qquad I_y = \frac{hb^3}{5} \qquad I_{xy} = \frac{b^2h^2}{12}$$

19

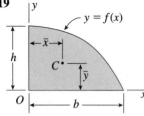

Semisegment of nth degree   (Origin of axes at corner)

$$y = f(x) = h\left(1 - \frac{x^n}{b^n}\right) \qquad (n > 0)$$

$$A = bh\left(\frac{n}{n+1}\right) \qquad \bar{x} = \frac{b(n+1)}{2(n+2)} \qquad \bar{y} = \frac{hn}{2n+1}$$

$$I_x = \frac{2bh^3n^3}{(n+1)(2n+1)(3n+1)} \qquad I_y = \frac{hb^3n}{3(n+3)} \qquad I_{xy} = \frac{b^2h^2n^2}{4(n+1)(n+2)}$$

20

Spandrel of nth degree   (Origin of axes at point of tangency)

$$y = f(x) = \frac{hx^n}{b^n} \qquad (n > 0)$$

$$A = \frac{bh}{n+1} \qquad \bar{x} = \frac{b(n+1)}{n+2} \qquad \bar{y} = \frac{h(n+1)}{2(2n+1)}$$

$$I_x = \frac{bh^3}{3(3n+1)} \qquad I_y = \frac{hb^3}{n+3} \qquad I_{xy} = \frac{b^2h^2}{4(n+1)}$$

21

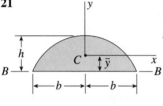

Sine wave   (Origin of axes at centroid)

$$A = \frac{4bh}{\pi} \qquad \bar{y} = \frac{\pi h}{8}$$

$$I_x = \left(\frac{8}{9\pi} - \frac{\pi}{16}\right)bh^3 \approx 0.08659bh^3 \qquad I_y = \left(\frac{4}{\pi} - \frac{32}{\pi^3}\right)hb^3 \approx 0.2412hb^3$$

$$I_{xy} = 0 \qquad I_{BB} = \frac{8bh^3}{9\pi}$$

22

Thin circular ring   (Origin of axes at center)
Approximate formulas for case when t is small

$$A = 2\pi rt = \pi dt \qquad I_x = I_y = \pi r^3 t = \frac{\pi d^3 t}{8}$$

$$I_{xy} = 0 \qquad I_p = 2\pi r^3 t = \frac{\pi d^3 t}{4}$$

23

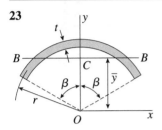

Thin circular arc   (Origin of axes at center of circle)
Approximate formulas for case when t is small

$$\beta = \text{angle in radians} \qquad (\beta \le \pi/2)$$

$$A = 2\beta rt \qquad \overline{y} = \frac{r \sin \beta}{\beta}$$

$$I_x = r^3 t(\beta + \sin \beta \cos \beta) \qquad I_y = r^3 t(\beta - \sin \beta \cos \beta)$$

$$I_{xy} = 0 \qquad I_{BB} = r^3 t\left(\frac{2\beta + \sin 2\beta}{2} - \frac{1 - \cos 2\beta}{\beta}\right)$$

(*Note:* For a semicircular arc, $\beta = \pi/2$.)

24

Thin rectangle   (Origin of axes at centroid)
Approximate formulas for case when t is small

$$A = bt$$

$$I_x = \frac{tb^3}{12} \sin^2 \beta \qquad I_y = \frac{tb^3}{12} \cos^2 \beta \qquad I_{BB} = \frac{tb^3}{3} \sin^2 \beta$$

25

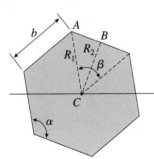

Regular polygon with n sides   (Origin of axes at centroid)

C = centroid (at center of polygon)

n = number of sides ($n \ge 3$) b = length of a side

β = central angle for a side α = interior angle (or vertex angle)

$$\beta = \frac{360°}{n} \qquad \alpha = \left(\frac{n-2}{n}\right)180° \qquad \alpha + \beta = 180°$$

R_1 = radius of circumscribed circle (line CA) R_2 = radius of inscribed circle (line CB)

$$R_1 = \frac{b}{2} \csc \frac{\beta}{2} \qquad R_2 = \frac{b}{2} \cot \frac{\beta}{2} \qquad A = \frac{nb^2}{4} \cot \frac{\beta}{2}$$

I_c = moment of inertia about any axis through C (the centroid C is a principal point and
    every axis through C is a principal axis)

$$I_c = \frac{nb^4}{192}\left(\cot \frac{\beta}{2}\right)\left(3\cot^2 \frac{\beta}{2} + 1\right) \qquad I_p = 2I_c$$

E

PROPERTIES OF STRUCTURAL-STEEL SHAPES

In the following tables, the properties of a few structural-steel shapes are presented as an aid to the reader in solving problems in the text. These tables were compiled from the extensive tables in the *Manual of Steel Construction*, published by the American Institute of Steel Construction, Inc. (Ref. 5-4).

Notation:

I = moment of inertia
S = section modulus
$r = \sqrt{I/A}$ = radius of gyration

Table E-1   Properties of **W** shapes (wide-flange sections)
(abridged list)

Designation	Weight per foot	Area	Depth	Web thickness	Flange		Axis 1-1			Axis 2-2		
					Width	Thickness	I	S	r	I	S	r
	lb	in.²	in.	in.	in.	in.	in.⁴	in.³	in.	in.⁴	in.³	in.
W 30 × 211	211	62.0	30.94	0.775	15.105	1.315	10300	663	12.9	757	100	3.49
W 30 × 132	132	38.9	30.31	0.615	10.545	1.000	5770	380	12.2	196	37.2	2.25
W 24 × 162	162	47.7	25.00	0.705	12.955	1.220	5170	414	10.4	443	68.4	3.05
W 24 × 94	94	27.7	24.31	0.515	9.065	0.875	2700	222	9.87	109	24.0	1.98
W 18 × 119	119	35.1	18.97	0.655	11.265	1.060	2190	231	7.90	253	44.9	2.69
W 18 × 71	71	20.8	18.47	0.495	7.635	0.810	1170	127	7.50	60.3	15.8	1.70
W 16 × 100	100	29.4	16.97	0.585	10.425	0.985	1490	175	7.10	186	35.7	2.51
W 16 × 77	77	22.6	16.52	0.455	10.295	0.760	1110	134	7.00	138	26.9	2.47
W 16 × 57	57	16.8	16.43	0.430	7.120	0.715	758	92.2	6.72	43.1	12.1	1.60
W 16 × 31	31	9.12	15.88	0.275	5.525	0.440	375	47.2	6.41	12.4	4.49	1.17
W 14 × 120	120	35.3	14.48	0.590	14.670	0.940	1380	190	6.24	495	67.5	3.74
W 14 × 82	82	24.1	14.31	0.510	10.130	0.855	882	123	6.05	148	29.3	2.48
W 14 × 53	53	15.6	13.92	0.370	8.060	0.660	541	77.8	5.89	57.7	14.3	1.92
W 14 × 26	26	7.69	13.91	0.255	5.025	0.420	245	35.3	5.65	8.91	3.54	1.08
W 12 × 87	87	25.6	12.53	0.515	12.125	0.810	740	118	5.38	241	39.7	3.07
W 12 × 50	50	14.7	12.19	0.370	8.080	0.640	394	64.7	5.18	56.3	13.9	1.96
W 12 × 35	35	10.3	12.50	0.300	6.560	0.520	285	45.6	5.25	24.5	7.47	1.54
W 12 × 14	14	4.16	11.91	0.200	3.970	0.225	88.6	14.9	4.62	2.36	1.19	0.753
W 10 × 60	60	17.6	10.22	0.420	10.080	0.680	341	66.7	4.39	116	23.0	2.57
W 10 × 45	45	13.3	10.10	0.350	8.020	0.620	248	49.1	4.32	53.4	13.3	2.01
W 10 × 30	30	8.84	10.47	0.300	5.810	0.510	170	32.4	4.38	16.7	5.75	1.37
W 10 × 12	12	3.54	9.87	0.190	3.960	0.210	53.8	10.9	3.90	2.18	1.10	0.785
W 8 × 35	35	10.3	8.12	0.310	8.020	0.495	127	31.2	3.51	42.6	10.6	2.03
W 8 × 28	28	8.25	8.06	0.285	6.535	0.465	98.0	24.3	3.45	21.7	6.63	1.62
W 8 × 21	21	6.16	8.28	0.250	5.270	0.400	75.3	18.2	3.49	9.77	3.71	1.26
W 8 × 15	15	4.44	8.11	0.245	4.015	0.315	48.0	11.8	3.29	3.41	1.70	0.876

Note: Axes 1-1 and 2-2 are principal centroidal axes.

Table E-2   Properties of S shapes (I-beam sections)
(abridged list)

Designation	Weight per foot	Area	Depth	Web thickness	Flange		Axis 1-1			Axis 2-2		
					Width	Average thickness	I	S	r	I	S	r
	lb	in.^{2}	in.	in.	in.	in.	in.^{4}	in.^{3}	in.	in.^{4}	in.^{3}	in.
S 24 × 100	100	29.3	24.00	0.745	7.245	0.870	2390	199	9.02	47.7	13.2	1.27
S 24 × 80	80	23.5	24.00	0.500	7.000	0.870	2100	175	9.47	42.2	12.1	1.34
S 20 × 96	96	28.2	20.30	0.800	7.200	0.920	1670	165	7.71	50.2	13.9	1.33
S 20 × 75	75	22.0	20.00	0.635	6.385	0.795	1280	128	7.62	29.8	9.32	1.16
S 18 × 70	70	20.6	18.00	0.711	6.251	0.691	926	103	6.71	24.1	7.72	1.08
S 18 × 54.7	54.7	16.1	18.00	0.461	6.001	0.691	804	89.4	7.07	20.8	6.94	1.14
S 15 × 50	50	14.7	15.00	0.550	5.640	0.622	486	64.8	5.75	15.7	5.57	1.03
S 15 × 42.9	42.9	12.6	15.00	0.411	5.501	0.622	447	59.6	5.95	14.4	5.23	1.07
S 12 × 50	50	14.7	12.00	0.687	5.477	0.659	305	50.8	4.55	15.7	5.74	1.03
S 12 × 35	35	10.3	12.00	0.428	5.078	0.544	229	38.2	4.72	9.87	3.89	0.980
S 10 × 35	35	10.3	10.00	0.594	4.944	0.491	147	29.4	3.78	8.36	3.38	0.901
S 10 × 25.4	25.4	7.46	10.00	0.311	4.661	0.491	124	24.7	4.07	6.79	2.91	0.954
S 8 × 23	23	6.77	8.00	0.441	4.171	0.426	64.9	16.2	3.10	4.31	2.07	0.798
S 8 × 18.4	18.4	5.41	8.00	0.271	4.001	0.426	57.6	14.4	3.26	3.73	1.86	0.831
S 6 × 17.25	17.25	5.07	6.00	0.465	3.565	0.359	26.3	8.77	2.28	2.31	1.30	0.675
S 6 × 12.5	12.5	3.67	6.00	0.232	3.332	0.359	22.1	7.37	2.45	1.82	1.09	0.705
S 4 × 9.5	9.5	2.79	4.00	0.326	2.796	0.293	6.79	3.39	1.56	0.903	0.646	0.569
S 4 × 7.7	7.7	2.26	4.00	0.193	2.663	0.293	6.08	3.04	1.64	0.764	0.574	0.581

Note: Axes 1-1 and 2-2 are principal centroidal axes.

Table E-3 Properties of channel sections (C shapes)
(abridged list)

Designation	Weight per foot	Area	Depth	Web thickness	Flange		Axis 1-1			Axis 2-2			
					Width	Average thickness	I	S	r	I	S	r	c
	lb	in.²	in.	in.	in.	in.	in.⁴	in.³	in.	in.⁴	in.³	in.	in.
C 15 × 50	50.0	14.7	15.00	0.716	3.716	0.650	404	53.8	5.24	11.0	3.78	0.867	0.798
C 15 × 40	40.0	11.8	15.00	0.520	3.520	0.650	349	46.5	5.44	9.23	3.37	0.886	0.777
C 15 × 33.9	33.9	9.96	15.00	0.400	3.400	0.650	315	42.0	5.62	8.13	3.11	0.904	0.787
C 12 × 30	30.0	8.82	12.00	0.510	3.170	0.501	162	27.0	4.29	5.14	2.06	0.763	0.674
C 12 × 25	25	7.35	12.00	0.387	3.047	0.501	144	24.1	4.43	4.47	1.88	0.780	0.674
C 12 × 20.7	20.7	6.09	12.00	0.282	2.942	0.501	129	21.5	4.61	3.88	1.73	0.799	0.698
C 10 × 30	30.0	8.82	10.00	0.673	3.033	0.436	103	20.7	3.42	3.94	1.65	0.669	0.649
C 10 × 25	25	7.35	10.00	0.526	2.886	0.436	91.2	18.2	3.52	3.36	1.48	0.676	0.617
C 10 × 20	20.0	5.88	10.00	0.379	2.739	0.436	78.9	15.8	3.66	2.81	1.32	0.692	0.606
C 10 × 15.3	15.3	4.49	10.00	0.240	2.600	0.436	67.4	13.5	3.87	2.28	1.16	0.713	0.634
C 8 × 18.75	18.75	5.51	8.00	0.487	2.527	0.390	44.0	11.0	2.82	1.98	1.01	0.599	0.565
C 8 × 13.75	13.75	4.04	8.00	0.303	2.343	0.390	36.1	9.03	2.99	1.53	0.854	0.615	0.553
C 8 × 11.5	11.5	3.38	8.00	0.220	2.260	0.390	32.6	8.14	3.11	1.32	0.781	0.625	0.571
C 6 × 13	13.0	3.83	6.00	0.437	2.157	0.343	17.4	5.80	2.13	1.05	0.642	0.525	0.514
C 6 × 10.5	10.5	3.09	6.00	0.314	2.034	0.343	15.2	5.06	2.22	0.866	0.564	0.529	0.499
C 6 × 8.2	8.2	2.40	6.00	0.200	1.920	0.343	13.1	4.38	2.34	0.693	0.492	0.537	0.511
C 4 × 7.25	7.25	2.13	4.00	0.321	1.721	0.296	4.59	2.29	1.47	0.433	0.343	0.450	0.459
C 4 × 5.4	5.4	1.59	4.00	0.184	1.584	0.296	3.85	1.93	1.56	0.319	0.283	0.449	0.457

Notes: 1. Axes 1-1 and 2-2 are principal centroidal axes.

2. The distance c is measured from the centroid to the back of the web.

3. For axis 2-2, the tabulated value of S is the smaller of the two section moduli for this axis.

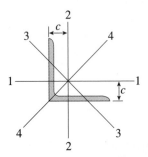

Table E-4 Properties of angle sections with equal legs (L shapes)
(abridged list)

Designation	Weight per foot	Area	Axis 1-1 and Axis 2-2				Axis 3-3
			I	S	r	c	r_{min}
in.	lb	in.^{2}	in.^{4}	in.^{3}	in.	in.	in.
L 8 × 8 × 1	51.0	15.0	89.0	15.8	2.44	2.37	1.56
L 8 × 8 × 3/4	38.9	11.4	69.7	12.2	2.47	2.28	1.58
L 8 × 8 × 1/2	26.4	7.75	48.6	8.36	2.50	2.19	1.59
L 6 × 6 × 1	37.4	11.0	35.5	8.57	1.80	1.86	1.17
L 6 × 6 × 3/4	28.7	8.44	28.2	6.66	1.83	1.78	1.17
L 6 × 6 × 1/2	19.6	5.75	19.9	4.61	1.86	1.68	1.18
L 5 × 5 × 7/8	27.2	7.98	17.8	5.17	1.49	1.57	0.973
L 5 × 5 × 1/2	16.2	4.75	11.3	3.16	1.54	1.43	0.983
L 5 × 5 × 3/8	12.3	3.61	8.74	2.42	1.56	1.39	0.990
L 4 × 4 × 3/4	18.5	5.44	7.67	2.81	1.19	1.27	0.778
L 4 × 4 × 1/2	12.8	3.75	5.56	1.97	1.22	1.18	0.782
L 4 × 4 × 3/8	9.8	2.86	4.36	1.52	1.23	1.14	0.788
L 3 1/2 × 3 1/2 × 3/8	8.5	2.48	2.87	1.15	1.07	1.01	0.687
L 3 1/2 × 3 1/2 × 1/4	5.8	1.69	2.01	0.794	1.09	0.968	0.694
L 3 × 3 × 1/2	9.4	2.75	2.22	1.07	0.898	0.932	0.584
L 3 × 3 × 1/4	4.9	1.44	1.24	0.577	0.930	0.842	0.592

Notes: 1. Axes 1-1 and 2-2 are centroidal axes parallel to the legs.

2. The distance c is measured from the centroid to the back of the legs.

3. For axes 1-1 and 2-2, the tabulated value of S is the smaller of the two section moduli for those axes.

4. Axes 3-3 and 4-4 are principal centroidal axes.

5. The moment of inertia for axis 3-3, which is the smaller of the two principal moments of inertia, can be found from the equation $I_{33} = Ar_{min}^2$.

6. The moment of inertia for axis 4-4, which is the larger of the two principal moments of inertia, can be found from the equation $I_{44} + I_{33} = I_{11} + I_{22}$.

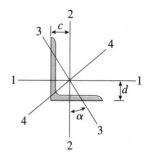

Table E-5 Properties of angle sections with unequal legs (L shapes) (abridged list)

Designation	Weight per foot	Area	Axis 1-1				Axis 2-2				Axis 3-3	
			I	S	r	d	I	S	r	c	r_{min}	$\tan \alpha$
in.	lb	in.^{2}	in.^{4}	in.^{3}	in.	in.	in.^{4}	in.^{3}	in.	in.	in.	
L 8 × 6 × 1	44.2	13.00	80.8	15.1	2.49	2.65	38.8	8.92	1.73	1.65	1.28	0.543
L 8 × 6 × 1/2	23.0	6.75	44.3	8.02	2.56	2.47	21.7	4.79	1.79	1.47	1.30	0.558
L 7 × 4 × 3/4	26.2	7.69	37.8	8.42	2.22	2.51	9.05	3.03	1.09	1.01	0.860	0.324
L 7 × 4 × 1/2	17.9	5.25	26.7	5.81	2.25	2.42	6.53	2.12	1.11	0.917	0.872	0.335
L 6 × 4 × 3/4	23.6	6.94	24.5	6.25	1.88	2.08	8.68	2.97	1.12	1.08	0.860	0.428
L 6 × 4 × 1/2	16.2	4.75	17.4	4.33	1.91	1.99	6.27	2.08	1.15	0.987	0.870	0.440
L 5 × 3 1/2 × 3/4	19.8	5.81	13.9	4.28	1.55	1.75	5.55	2.22	0.977	0.996	0.748	0.464
L 5 × 3 1/2 × 1/2	13.6	4.00	9.99	2.99	1.58	1.66	4.05	1.56	1.01	0.906	0.755	0.479
L 5 × 3 × 1/2	12.8	3.75	9.45	2.91	1.59	1.75	2.58	1.15	0.829	0.750	0.648	0.357
L 5 × 3 × 1/4	6.6	1.94	5.11	1.53	1.62	1.66	1.44	0.614	0.861	0.657	0.663	0.371
L 4 × 3 1/2 × 1/2	11.9	3.50	5.32	1.94	1.23	1.25	3.79	1.52	1.04	1.00	0.722	0.750
L 4 × 3 1/2 × 1/4	6.2	1.81	2.91	1.03	1.27	1.16	2.09	0.808	1.07	0.909	0.734	0.759
L 4 × 3 × 1/2	11.1	3.25	5.05	1.89	1.25	1.33	2.42	1.12	0.864	0.827	0.639	0.543
L 4 × 3 × 3/8	8.5	2.48	3.96	1.46	1.26	1.28	1.92	0.866	0.879	0.782	0.644	0.551
L 4 × 3 × 1/4	5.8	1.69	2.77	1.00	1.28	1.24	1.36	0.599	0.896	0.736	0.651	0.558

Notes: 1. Axes 1-1 and 2-2 are centroidal axes parallel to the legs.

2. The distances c and d are measured from the centroid to the backs of the legs.

3. For axes 1-1 and 2-2, the tabulated value of S is the smaller of the two section moduli for those axes.

4. Axes 3-3 and 4-4 are principal centroidal axes.

5. The moment of inertia for axis 3-3, which is the smaller of the two principal moments of inertia, can be found from the equation $I_{33} = Ar_{min}^2$.

6. The moment of inertia for axis 4-4, which is the larger of the two principal moments of inertia, can be found from the equation $I_{44} + I_{33} = I_{11} + I_{22}$.

F

PROPERTIES OF STRUCTURAL LUMBER

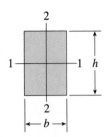

Properties of surfaced lumber (abridged list)

Nominal dimensions $b \times h$	Net dimensions $b \times h$	Area $A = bh$	Axis 1-1 Moment of inertia $I_1 = \dfrac{bh^3}{12}$	Axis 1-1 Section modulus $S_1 = \dfrac{bh^2}{6}$	Axis 2-2 Moment of inertia $I_2 = \dfrac{hb^3}{12}$	Axis 2-2 Section modulus $S_2 = \dfrac{hb^2}{6}$	Weight per linear foot (weight density = 35 lb/ft³)
in.	in.	in.²	in.⁴	in.³	in.⁴	in.³	lb
2 × 4	1.5 × 3.5	5.25	5.36	3.06	0.98	1.31	1.3
2 × 6	1.5 × 5.5	8.25	20.80	7.56	1.55	2.06	2.0
2 × 8	1.5 × 7.25	10.88	47.63	13.14	2.04	2.72	2.6
2 × 10	1.5 × 9.25	13.88	98.93	21.39	2.60	3.47	3.4
2 × 12	1.5 × 11.25	16.88	177.98	31.64	3.16	4.22	4.1
3 × 4	2.5 × 3.5	8.75	8.93	5.10	4.56	3.65	2.1
3 × 6	2.5 × 5.5	13.75	34.66	12.60	7.16	5.73	3.3
3 × 8	2.5 × 7.25	18.13	79.39	21.90	9.44	7.55	4.4
3 × 10	2.5 × 9.25	23.13	164.89	35.65	12.04	9.64	5.6
3 × 12	2.5 × 11.25	28.13	296.63	52.73	14.65	11.72	6.8
4 × 4	3.5 × 3.5	12.25	12.51	7.15	12.51	7.15	3.0
4 × 6	3.5 × 5.5	19.25	48.53	17.65	19.65	11.23	4.7
4 × 8	3.5 × 7.25	25.38	111.15	30.66	25.90	14.80	6.2
4 × 10	3.5 × 9.25	32.38	230.84	49.91	33.05	18.89	7.9
4 × 12	3.5 × 11.25	39.38	415.28	73.83	40.20	22.97	9.6
6 × 6	5.5 × 5.5	30.25	76.3	27.7	76.3	27.7	7.4
6 × 8	5.5 × 7.5	41.25	193.4	51.6	104.0	37.8	10.0
6 × 10	5.5 × 9.5	52.25	393.0	82.7	131.7	47.9	12.7
6 × 12	5.5 × 11.5	63.25	697.1	121.2	159.4	58.0	15.4
8 × 8	7.5 × 7.5	56.25	263.7	70.3	263.7	70.3	13.7
8 × 10	7.5 × 9.5	71.25	535.9	112.8	334.0	89.1	17.3
8 × 12	7.5 × 11.5	86.25	950.5	165.3	404.3	107.8	21.0

Note: Axes 1-1 and 2-2 are principal centroidal axes.

DEFLECTIONS AND SLOPES OF BEAMS

Table G-1 Deflections and slopes of cantilever beams

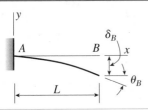

v = deflection in the y direction (positive upward)
$v' = dv/dx$ = slope of the deflection curve
$\delta_B = -v(L)$ = deflection at end B of the beam (downward)
$\theta_B = -v'(L)$ = angle of rotation at end B of the beam (clockwise)
EI = constant

1

$$v = -\frac{qx^2}{24EI}(6L^2 - 4Lx + x^2) \qquad v' = -\frac{qx}{6EI}(3L^2 - 3Lx + x^2)$$

$$\delta_B = \frac{qL^4}{8EI} \qquad \theta_B = \frac{qL^3}{6EI}$$

2

$$v = -\frac{qx^2}{24EI}(6a^2 - 4ax + x^2) \qquad (0 \le x \le a)$$

$$v' = -\frac{qx}{6EI}(3a^2 - 3ax + x^2) \qquad (0 \le x \le a)$$

$$v = -\frac{qa^3}{24EI}(4x - a) \qquad v' = -\frac{qa^3}{6EI} \qquad (a \le x \le L)$$

$$\text{At } x = a: \quad v = -\frac{qa^4}{8EI} \qquad v' = -\frac{qa^3}{6EI}$$

$$\delta_B = \frac{qa^3}{24EI}(4L - a) \qquad \theta_B = \frac{qa^3}{6EI}$$

(Continued)

3

$$v = -\frac{qbx^2}{12EI}(3L + 3a - 2x) \qquad (0 \le x \le a)$$

$$v' = -\frac{qbx}{2EI}(L + a - x) \qquad (0 \le x \le a)$$

$$v = -\frac{q}{24EI}(x^4 - 4Lx^3 + 6L^2x^2 - 4a^3x + a^4) \qquad (a \le x \le L)$$

$$v' = -\frac{q}{6EI}(x^3 - 3Lx^2 + 3L^2x - a^3) \qquad (a \le x \le L)$$

At $x = a$: $v = -\dfrac{qa^2b}{12EI}(3L + a) \qquad v' = -\dfrac{qabL}{2EI}$

$$\delta_B = \frac{q}{24EI}(3L^4 - 4a^3L + a^4) \qquad \theta_B = \frac{q}{6EI}(L^3 - a^3)$$

4

$$v = -\frac{Px^2}{6EI}(3L - x) \qquad v' = -\frac{Px}{2EI}(2L - x)$$

$$\delta_B = \frac{PL^3}{3EI} \qquad \theta_B = \frac{PL^2}{2EI}$$

5

$$v = -\frac{Px^2}{6EI}(3a - x) \qquad v' = -\frac{Px}{2EI}(2a - x) \qquad (0 \le x \le a)$$

$$v = -\frac{Pa^2}{6EI}(3x - a) \qquad v' = -\frac{Pa^2}{2EI} \qquad (a \le x \le L)$$

At $x = a$: $v = -\dfrac{Pa^3}{3EI} \qquad v' = -\dfrac{Pa^2}{2EI}$

$$\delta_B = \frac{Pa^2}{6EI}(3L - a) \qquad \theta_B = \frac{Pa^2}{2EI}$$

6

$$v = -\frac{M_0x^2}{2EI} \qquad v' = -\frac{M_0x}{EI}$$

$$\delta_B = \frac{M_0L^2}{2EI} \qquad \theta_B = \frac{M_0L}{EI}$$

(Continued)

7

$$v = -\frac{M_0 x^2}{2EI} \qquad v' = -\frac{M_0 x}{EI} \qquad (0 \le x \le a)$$

$$v = -\frac{M_0 a}{2EI}(2x - a) \qquad v' = -\frac{M_0 a}{EI} \qquad (a \le x \le L)$$

$$At\ x = a: \quad v = -\frac{M_0 a^2}{2EI} \qquad v' = -\frac{M_0 a}{EI}$$

$$\delta_B = \frac{M_0 a}{2EI}(2L - a) \qquad \theta_B = \frac{M_0 a}{EI}$$

8

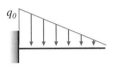

$$v = -\frac{q_0 x^2}{120LEI}(10L^3 - 10L^2 x + 5Lx^2 - x^3)$$

$$v' = -\frac{q_0 x}{24LEI}(4L^3 - 6L^2 x + 4Lx^2 - x^3)$$

$$\delta_B = \frac{q_0 L^4}{30EI} \qquad \theta_B = \frac{q_0 L^3}{24EI}$$

9

$$v = -\frac{q_0 x^2}{120LEI}(20L^3 - 10L^2 x + x^3)$$

$$v' = -\frac{q_0 x}{24LEI}(8L^3 - 6L^2 x + x^3)$$

$$\delta_B = \frac{11q_0 L^4}{120EI} \qquad \theta_B = \frac{q_0 L^3}{8EI}$$

10

$$v = -\frac{q_0 L}{3\pi^4 EI}\left(48L^3 \cos\frac{\pi x}{2L} - 48L^3 + 3\pi^3 Lx^2 - \pi^3 x^3\right)$$

$$v' = -\frac{q_0 L}{\pi^3 EI}\left(2\pi^2 Lx - \pi^2 x^2 - 8L^2 \sin\frac{\pi x}{2L}\right)$$

$$\delta_B = \frac{2q_0 L^4}{3\pi^4 EI}(\pi^3 - 24) \qquad \theta_B = \frac{q_0 L^3}{\pi^3 EI}(\pi^2 - 8)$$

Table G-2 Deflections and slopes of simple beams

v = deflection in the y direction (positive upward)
v' = dv/dx = slope of the deflection curve
δ_C = $-v(L/2)$ = deflection at midpoint C of the beam (downward)
x_1 = distance from support A to point of maximum deflection
δ_{max} = $-v_{max}$ = maximum deflection (downward)
θ_A = $-v'(0)$ = angle of rotation at left-hand end of the beam (clockwise)
θ_B = $v'(L)$ = angle of rotation at right-hand end of the beam (counterclockwise)

EI = constant

1

$$v = -\frac{qx}{24EI}(L^3 - 2Lx^2 + x^3)$$

$$v' = -\frac{q}{24EI}(L^3 - 6Lx^2 + 4x^3)$$

$$\delta_C = \delta_{max} = \frac{5qL^4}{384EI} \qquad \theta_A = \theta_B = \frac{qL^3}{24EI}$$

2

$$v = -\frac{qx}{384EI}(9L^3 - 24Lx^2 + 16x^3) \qquad \left(0 \leq x \leq \frac{L}{2}\right)$$

$$v' = -\frac{q}{384EI}(9L^3 - 72Lx^2 + 64x^3) \qquad \left(0 \leq x \leq \frac{L}{2}\right)$$

$$v = -\frac{qL}{384EI}(8x^3 - 24Lx^2 + 17L^2x - L^3) \qquad \left(\frac{L}{2} \leq x \leq L\right)$$

$$v' = -\frac{qL}{384EI}(24x^2 - 48Lx + 17L^2) \qquad \left(\frac{L}{2} \leq x \leq L\right)$$

$$\delta_C = \frac{5qL^4}{768EI} \qquad \theta_A = \frac{3qL^3}{128EI} \qquad \theta_B = \frac{7qL^3}{384EI}$$

3

$$v = -\frac{qx}{24LEI}(a^4 - 4a^3L + 4a^2L^2 + 2a^2x^2 - 4aLx^2 + Lx^3) \qquad (0 \leq x \leq a)$$

$$v' = -\frac{q}{24LEI}(a^4 - 4a^3L + 4a^2L^2 + 6a^2x^2 - 12aLx^2 + 4Lx^3) \qquad (0 \leq x \leq a)$$

$$v = -\frac{qa^2}{24LEI}(-a^2L + 4L^2x + a^2x - 6Lx^2 + 2x^3) \qquad (a \leq x \leq L)$$

$$v' = -\frac{qa^2}{24LEI}(4L^2 + a^2 - 12Lx + 6x^2) \qquad (a \leq x \leq L)$$

$$\theta_A = \frac{qa^2}{24LEI}(2L - a)^2 \qquad \theta_B = \frac{qa^2}{24LEI}(2L^2 - a^2)$$

(Continued)

4

$$v = -\frac{Px}{48EI}(3L^2 - 4x^2) \qquad v' = -\frac{P}{16EI}(L^2 - 4x^2) \qquad \left(0 \le x \le \frac{L}{2}\right)$$

$$\delta_C = \delta_{max} = \frac{PL^3}{48EI} \qquad \theta_A = \theta_B = \frac{PL^2}{16EI}$$

5

$$v = -\frac{Pbx}{6LEI}(L^2 - b^2 - x^2) \qquad v' = -\frac{Pb}{6LEI}(L^2 - b^2 - 3x^2) \qquad (0 \le x \le a)$$

$$\theta_A = \frac{Pab(L + b)}{6LEI} \qquad \theta_B = \frac{Pab(L + a)}{6LEI}$$

$$\text{If } a \ge b, \quad \delta_C = \frac{Pb(3L^2 - 4b^2)}{48EI} \qquad \text{If } a \le b, \quad \delta_C = \frac{Pa(3L^2 - 4a^2)}{48EI}$$

$$\text{If } a \ge b, \quad x_1 = \sqrt{\frac{L^2 - b^2}{3}} \quad \text{and} \quad \delta_{max} = \frac{Pb(L^2 - b^2)^{3/2}}{9\sqrt{3}\,LEI}$$

6

$$v = -\frac{Px}{6EI}(3aL - 3a^2 - x^2) \qquad v' = -\frac{P}{2EI}(aL - a^2 - x^2) \qquad (0 \le x \le a)$$

$$v = -\frac{Pa}{6EI}(3Lx - 3x^2 - a^2) \qquad v' = -\frac{Pa}{2EI}(L - 2x) \qquad (a \le x \le L - a)$$

$$\delta_C = \delta_{max} = \frac{Pa}{24EI}(3L^2 - 4a^2) \qquad \theta_A = \theta_B = \frac{Pa(L - a)}{2EI}$$

7

$$v = -\frac{M_0 x}{6LEI}(2L^2 - 3Lx + x^2) \qquad v' = -\frac{M_0}{6LEI}(2L^2 - 6Lx + 3x^2)$$

$$\delta_C = \frac{M_0 L^2}{16EI} \qquad \theta_A = \frac{M_0 L}{3EI} \qquad \theta_B = \frac{M_0 L}{6EI}$$

$$x_1 = L\left(1 - \frac{\sqrt{3}}{3}\right) \quad \text{and} \quad \delta_{max} = \frac{M_0 L^2}{9\sqrt{3}\,EI}$$

8

$$v = -\frac{M_0 x}{24LEI}(L^2 - 4x^2) \qquad v' = -\frac{M_0}{24LEI}(L^2 - 12x^2) \qquad \left(0 \le x \le \frac{L}{2}\right)$$

$$\delta_C = 0 \qquad \theta_A = \frac{M_0 L}{24EI} \qquad \theta_B = -\frac{M_0 L}{24EI}$$

(Continued)

$$v = -\frac{M_0 x}{6LEI}(6aL - 3a^2 - 2L^2 - x^2) \qquad (0 \le x \le a)$$

$$v' = -\frac{M_0}{6LEI}(6aL - 3a^2 - 2L^2 - 3x^2) \qquad (0 \le x \le a)$$

At $x = a$: $v = -\frac{M_0 ab}{3LEI}(2a - L)$ $v' = -\frac{M_0}{3LEI}(3aL - 3a^2 - L^2)$

$$\theta_A = \frac{M_0}{6LEI}(6aL - 3a^2 - 2L^2) \qquad \theta_B = \frac{M_0}{6LEI}(3a^2 - L^2)$$

10

$$v = -\frac{M_0 x}{2EI}(L - x) \qquad v' = -\frac{M_0}{2EI}(L - 2x)$$

$$\delta_C = \delta_{max} = \frac{M_0 L^2}{8EI} \qquad \theta_A = \theta_B = \frac{M_0 L}{2EI}$$

11

$$v = -\frac{q_0 x}{360LEI}(7L^4 - 10L^2 x^2 + 3x^4)$$

$$v' = -\frac{q_0}{360LEI}(7L^4 - 30L^2 x^2 + 15x^4)$$

$$\delta_C = \frac{5q_0 L^4}{768EI} \qquad \theta_A = \frac{7q_0 L^3}{360EI} \qquad \theta_B = \frac{q_0 L^3}{45EI}$$

$$x_1 = 0.5193L \qquad \delta_{max} = 0.00652\frac{q_0 L^4}{EI}$$

12

$$v = -\frac{q_0 x}{960LEI}(5L^2 - 4x^2)^2 \qquad \left(0 \le x \le \frac{L}{2}\right)$$

$$v' = -\frac{q_0}{192LEI}(5L^2 - 4x^2)(L^2 - 4x^2) \qquad \left(0 \le x \le \frac{L}{2}\right)$$

$$\delta_C = \delta_{max} = \frac{q_0 L^4}{120EI} \qquad \theta_A = \theta_B = \frac{5q_0 L^3}{192EI}$$

13

$q = q_0 \sin \dfrac{\pi x}{L}$

$$v = -\frac{q_0 L^4}{\pi^4 EI} \sin \frac{\pi x}{L} \qquad v' = -\frac{q_0 L^3}{\pi^3 EI} \cos \frac{\pi x}{L}$$

$$\delta_C = \delta_{max} = \frac{q_0 L^4}{\pi^4 EI} \qquad \theta_A = \theta_B = \frac{q_0 L^3}{\pi^3 EI}$$

H
PROPERTIES
OF MATERIALS

1. Properties of materials vary greatly depending upon manufacturing processes, chemical composition, internal defects, temperature, previous loading history, age, dimensions of test specimens, and other factors. The tabulated values are typical but should never be used for specific engineering or design purposes. Manufacturers and materials suppliers should be consulted for information about a particular product.

2. Except when compression or bending is indicated, the modulus of elasticity E, yield stress σ_y, and ultimate stress σ_u are for materials in tension.

nd mass densities

ial	Weight density γ		Mass density ρ	
	lb/ft³	kN/m³	slugs/ft³	kg/m³
alloys	160–180	26–28	5.2–5.4	2,600–2,800
, 7075-T6	175	28	5.4	2,800
.6	170	26	5.2	2,700
	520–540	82–85	16–17	8,400–8,600
ɔnze	510–550	80–86	16–17	8,200–8,800
Cast iron	435–460	68–72	13–14	7,000–7,400
Concrete				
Plain	145	23	4.5	2,300
Reinforced	150	24	4.7	2,400
Lightweight	70–115	11–18	2.2–3.6	1,100–1,800
Copper	556	87	17	8,900
Glass	150–180	24–28	4.7–5.4	2,400–2,800
Magnesium alloys	110–114	17–18	3.4–3.5	1,760–1,830
Monel (67% Ni, 30% Cu)	550	87	17	8,800
Nickel	550	87	17	8,800
Plastics				
Nylon	55–70	8.6–11	1.7–2.2	880–1,100
Polyethylene	60–90	9.4–14	1.9–2.8	960–1,400
Rock				
Granite, marble, quartz	165–180	26–28	5.1–5.6	2,600–2,900
Limestone, sandstone	125–180	20–28	3.9–5.6	2,000–2,900
Rubber	60–80	9–13	1.9–2.5	960–1,300
Sand, soil, gravel	75–135	12–21	2.3–4.2	1,200–2,200
Steel	490	77.0	15.2	7,850
Titanium	280	44	8.7	4,500
Tungsten	1,200	190	37	1,900
Water, fresh	62.4	9.81	1.94	1,000
sea	63.8	10.0	1.98	1,020
Wood (air dry)				
Douglas fir	30–35	4.7–5.5	0.9–1.1	480–560
Oak	40–45	6.3–7.1	1.2–1.4	640–720
Southern pine	35–40	5.5–6.3	1.1–1.2	560–640

Table H-2 Moduli of elasticity and Poisson's ratios

Material	Modulus of elasticity E		Shear modulus of elasticity G		Poisson's ratio ν
	ksi	GPa	ksi	GPa	
Aluminum alloys	10,000–11,400	70–79	3,800–4,300	26–30	0.33
2014-T6	10,600	73	4,000	28	0.33
6061-T6	10,000	70	3,800	26	0.33
7075-T6	10,400	72	3,900	27	0.33
Brass	14,000–16,000	96–110	5,200–6,000	36–41	0.34
Bronze	14,000–17,000	96–120	5,200–6,300	36–44	0.34
Cast iron	12,000–25,000	83–170	4,600–10,000	32–69	0.2–0.3
Concrete (compression)	2,500–4,500	17–31			0.1–0.2
Copper and copper alloys	16,000–18,000	110–120	5,800–6,800	40–47	0.33–0.36
Glass	7,000–12,000	48–83	2,700–5,100	19–35	0.17–0.27
Magnesium alloys	6,000–6,500	41–45	2,200–2,400	15–17	0.35
Monel (67% Ni, 30% Cu)	25,000	170	9,500	66	0.32
Nickel	30,000	210	11,400	80	0.31
Plastics					
Nylon	300–500	2.1–3.4			0.4
Polyethylene	100–200	0.7–1.4			0.4
Rock (compression)					
Granite, marble, quartz	6,000–14,000	40–100			0.2–0.3
Limestone, sandstone	3,000–10,000	20–70			0.2–0.3
Rubber	0.1–0.6	0.0007–0.004	0.03–0.2	0.0002–0.001	0.45–0.50
Steel	28,000–30,000	190–210	10,800–11,800	75–80	0.27–0.30
Titanium alloys	15,000–17,000	100–120	5,600–6,400	39–44	0.33
Tungsten	50,000–55,000	340–380	21,000–23,000	140–160	0.2
Wood (bending)					
Douglas fir	1,600–1,900	11–13			
Oak	1,600–1,800	11–12			
Southern pine	1,600–2,000	11–14			

Table H-3   Mechanical properties

Material	Yield stress σ_y		Ultimate stress σ_u		Percent elongation (2 in. gage length)
	ksi	MPa	ksi	MPa	
Aluminum alloys	5–70	35–500	15–80	100–550	1–45
2014-T6	60	410	70	480	13
6061-T6	40	270	45	310	17
7075-T6	70	480	80	550	11
Brass	10–80	70–550	30–90	200–620	4–60
Bronze	12–100	82–690	30–120	200–830	5–60
Cast iron (tension)	17–42	120–290	10–70	69–480	0–1
Cast iron (compression)			50–200	340–1,400	
Concrete (compression)			1.5–10	10–70	
Copper and copper alloys	8–110	55–760	33–120	230–830	4–50
Glass			5–150	30–1,000	0
Plate glass			10	70	
Glass fibers			1,000–3,000	7,000–20,000	
Magnesium alloys	12–40	80–280	20–50	140–340	2–20
Monel (67% Ni, 30% Cu)	25–160	170–1,100	65–170	450–1,200	2–50
Nickel	15–90	100–620	45–110	310–760	2–50
Plastics					
Nylon			6–12	40–80	20–100
Polyethylene			1–4	7–28	15–300
Rock (compression)					
Granite, marble, quartz			8–40	50–280	
Limestone, sandstone			3–30	20–200	
Rubber	0.2–1.0	1–7	1–3	7–20	100–800
Steel					
High-strength	50–150	340–1,000	80–180	550–1,200	5–25
Machine	50–100	340–700	80–125	550–860	5–25
Spring	60–240	400–1,600	100–270	700–1,900	3–15
Stainless	40–100	280–700	60–150	400–1,000	5–40
Tool	75	520	130	900	8
Steel, structural	30–100	200–700	50–120	340–830	10–40
ASTM-A36	36	250	60	400	30
ASTM-A572	50	340	70	500	20
ASTM-A514	100	700	120	830	15

(Continued)

Table H-3 Mechanical properties (Continued)

Material	Yield stress σ_y		Ultimate stress σ_u		Percent elongation (2 in. gage length)
	ksi	MPa	ksi	MPa	
Steel wire	40–150	280–1,000	80–200	550–1,400	5–40
Titanium alloys	110–150	760–1,000	130–170	900–1,200	10
Tungsten			200–600	1,400–4,000	0–4
Wood (bending)					
Douglas fir	5–8	30–50	8–12	50–80	
Oak	6–9	40–60	8–14	50–100	
Southern pine	6–9	40–60	8–14	50–100	
Wood (compression parallel to grain)					
Douglas fir	4–8	30–50	6–10	40–70	
Oak	4–6	30–40	5–8	30–50	
Southern pine	4–8	30–50	6–10	40–70	

Table H-4 Coefficients of thermal expansion

Material	Coefficient of thermal expansion α		Material	Coefficient of thermal expansion α	
	$10^{-6}/°F$	$10^{-6}/°C$		$10^{-6}/°F$	$10^{-6}/°C$
Aluminum alloys	13	23	Plastics		
			Nylon	40–80	70–140
Brass	10.6–11.8	19.1–21.2	Polyethylene	80–160	140–290
Bronze	9.9–11.6	18–21	Rock	3–5	5–9
Cast iron	5.5–6.6	9.9–12			
			Rubber	70–110	130–200
Concrete	4–8	7–14	Steel	5.5–9.9	10–18
Copper and copper alloys	9.2–9.8	16.6–17.6	High-strength	8.0	14
Glass	3–6	5–11	Stainless	9.6	17
			Structural	6.5	12
Magnesium alloys	14.5–16.0	26.1–28.8	Titanium alloys	4.5–6.0	8.1–11
Monel (67% Ni, 30% Cu)	7.7	14	Tungsten	2.4	4.3
Nickel	7.2	13			

ANSWERS TO PROBLEMS

1.2-1 $\sigma_{BC} = 9800$ psi **1.2-2** $\sigma_c = 11.2$ MPa
1.2-3 (a) $\delta = 0.0114$ in.; (b) $P = 8290$ lb
1.2-4 (a) $\sigma_c = 12.5$ MPa; (b) $\bar{x} = 0.395$ m, $\bar{y} = 0.6$ m
1.2-5 $\sigma_t = 37,400$ psi **1.2-6** $\sigma_c = 5.21$ MPa
1.2-7 $\sigma_t = 17.5$ ksi
1.2-8 (a) $\sigma_{AB} = 62.5$ MPa (tension), $\sigma_{BC} = 29.4$ MPa
(compression); (b) $\epsilon_{AB} = 440 \times 10^{-6}$;
(c) $\epsilon_{BC} = 148 \times 10^{-6}$
1.2-9 $P = 8120$ lb, $\sigma_{BC} = 68,600$ psi, $\sigma_{DE} = 7250$ psi
1.2-10 $\sigma_t = 42.8$ MPa, $\sigma_c = 2.9$ MPa
1.2-11 $\sigma_t = 22,600$ psi
1.2-12 (a) $\sigma_x = \gamma\omega^2(L^2 - x^2)/2g$, (b) $\sigma_{max} = \gamma\omega^2L^2/2g$
1.3-1 (a) $L = 10,600$ ft; (b) $L = 12,200$ ft
1.3-2 Percent elongation = 8.2, 26.0, 41.2;
Percent reduction = 8.2, 37.9, 74.9; Brittle, ductile, ductile
1.3-3 $\sigma_t \approx 30$ ksi
1.3-4 $\sigma_{pl} \approx 47$ MPa, Slope ≈ 2.4 GPa, $\sigma_y \approx 53$ MPa,
Brittle
1.3-5 $\sigma_{pl} \approx 65$ ksi, Slope $\approx 30 \times 10^3$ ksi, $\sigma_y \approx 69$ ksi,
$\sigma_u \approx 113$ ksi, Elongation = 6%, Reduction = 31%
1.3-6 Aluminum, 11.9×10^3 m; Douglas fir,
12.7×10^3 m; nylon, 6.1×10^3 m; steel, 6.5×10^3 m;
titanium, 23.9×10^3 m
1.4-1 0.13 in. longer **1.4-2** 5.6 mm longer
1.4-3 (a) 0.0032; (b) 0.037; (c) 1.8 in.; (d) 32 ksi
1.4-4 2.8 mm **1.4-5** 0.088 in.
1.5-1 $P = 157$ k
1.5-2 $\delta = 6.98$ mm, $\Delta d = 0.0233$ mm, $\Delta V = 1640$ mm^3
1.5-3 (a) $E = 29.5 \times 10^6$ psi; (b) $\nu = 0.29$;
(c) $e = 990 \times 10^{-6}$ (increase)
1.5-4 $P = 27.4$ kN (tension), $e = 1.65 \times 10^{-3}$ (increase
in volume)
1.5-5 (a) $\delta = 0.00978$ in., $\Delta d = 97.8 \times 10^{-6}$ in.;

(b) $\Delta V = 0.000173$ in.^{3}, $e = 0.00044$
1.5-6 (a) $E = 104$ GPa; (b) $\nu = 0.34$; (c) $e = 0.00078$
1.5-7 $\nu = 3/10$
1.5-8 (a) Slope = $(b/L)(1 - \nu\sigma/E)/(1 + \sigma/E)$;
(b) $e_{face} = \sigma(1 - \nu)/E$; (c) $e_{area} = -2\nu\sigma/E$
1.5-9 $\Delta V = WL(1 - 2\nu)/2E$
1.6-1 $\tau_{aver} = 5300$ psi **1.6-2** $\tau_{aver} = 3.2$ MPa
1.6-3 $\gamma = 4.36 \times 10^{-3}$; $\tau = 17,300$ psi
1.6-4 $\sigma_b = 48.6$ MPa, $\tau_{aver} = 69.6$ MPa
1.6-5 (a) $\sigma_b = 17.1$ ksi, (b) $P_{ult} = 39.3$ k
1.6-6 $\tau_{aver} = 52.7$ MPa
1.6-7 (a) $\sigma_b = 4560$ psi, (b) $\tau_{aver} = 2550$ psi
1.6-8 $\tau_{aver} = 26.8$ MPa, $\sigma_b = 7.57$ MPa
1.6-9 $\tau_{aver} = 14.1$ ksi **1.6-10** $G = 2.0$ MPa
1.6-11 (a) $\gamma_{aver} = 0.295$; (b) $\delta = 0.15$ in.
1.6-12 If the ratio of the length of the crank arm to the
radius of the chain wheel is 1.80 to 1, then: (a) $T = 1440$ N;
(b) $\tau_{aver} = 147$ MPa
1.6-13 (a) $\tau_{aver} = 212$ psi; (b) $\tau_{max} = 340$ psi
1.6-14 $\tau_{aver} = 4PL/[bc(2d + b)]$
1.6-15 (a) $V_{max} = 250$ lb; (b) $\tau_{aver} = 141$ psi
1.6-16 (a) $\tau = \dfrac{P}{2\pi rh}$; (b) $\delta = \dfrac{P}{2\pi hG}\ln\dfrac{b}{d}$
1.7-1 $P_{allow} = 607$ lb **1.7-2** $P_{allow} = 30.2$ kN
1.7-3 $T_{allow} = 295$ lb **1.7-4** $P_{allow} = 49.1$ kN
1.7-5 $W_{max} = 58.7$ lb
1.7-6 $C_{ult} = 4.28$ kN; $P_{allow} = 601$ N
1.7-7 $P_{allow} = 6.14$ k **1.7-8** $T_{max} = 9.14$ kN
1.7-9 (a) $P_{allow} = \sigma_c A\sqrt{1 - (R/L)^2}$; (b) $P_{allow} = 2.11$ k
1.7-10 $W_{max} = 1370$ N **1.7-11** $P = 41.2$ k
1.7-12 (a) $P_{allow} = 0.5380d^2\sigma_{allow}$; (b) $P_{allow} = 36.3$ kN
1.7-13 $P_{allow} = 16.9$ k **1.7-14** $P_{allow} = 65.9$ N
1.8-1 $d_{min} = 3.58$ in. **1.8-2** $d_{min} = 112$ mm
1.8-3 $d_{max} = 1.01$ in. **1.8-4** $d_{min} = 20.7$ mm

1.8-5 $d_{min} = 0.639$ in.   **1.8-6** $d_{min} = 14.8$ mm

1.8-7 $n = 11.2$; use 12 bolts   **1.8-8** $A_c = 82.9$ mm^2

1.8-9 (a) $d_m = 0.618$ in.; (b) $P_{max} = 6910$ lb

1.8-10 (a) $t = 18.8$ mm, use 20 mm;

(b) $D = 296.6$ mm, use 298 mm

1.8-11 $A = 2M_1L\omega^2/(2\sigma_t - \rho L^2\omega^2)$

1.8-12 (a) $d_m = 22.3$ mm; (b) $P_{max} = 18.6$ kN

1.8-13 $\theta = 45°$ **1.8-14** $\theta = 54.7°$

1.8-15 $A_s = 2.00$ in.^{2}

CHAPTER 2

2.2-1 (a) $\delta_c/\delta_s = 1.61$; (b) $d_c/d_s = 1.27$

2.2-2 $\delta = 186$ mm

2.2-3 (a) $\delta = 1.38$ in.; (b) $n = 3.04$

2.2-4 $h = L - p\pi d^2/4k$

2.2-5 $x = 5.96$ in.   **2.2-6** $x = 135$ mm

2.2-7 (a) 14.52 ft; (b) 966; (c) $P_{max} = 5080$ lb

2.2-8 (a) $L = 3.03$ m; (b) $P_{max} = 186$ N

2.2-9 (a) $\delta = 0.278$ in.; (b) $P_{max} = 91.7$ k

2.2-10 (a) $\delta_B = 1.31$ mm; (b) $P_{max} = 472$ kN

2.2-11 $t = 0.623$ in.   **2.2-12** $t = 16.9$ mm

2.2-13 $\delta_A = 0.0133$ in., $\delta_D = 0.0278$ in.

2.2-14 $x = 0.729$ m

2.3-1 $\dfrac{P_2}{P_1} = \dfrac{a}{b}\left(1 + \dfrac{A_1L_2}{A_2L_1}\right)$

2.3-2 (a) $\delta = 0.308$ mm elongation; (b) $P = 5.6$ kN

2.3-3 (a) $\delta_B = 0.00171$ in. (to the right), $\delta_C = 0.00171$ in. (to the right), $\delta_D = -0.000857$ in. (to the left); (b) $x = 10$ ft

2.3-4 (a) $\delta_C = 3.72$ mm; (b) $P_0 = 44.2$ kN

2.3-5 (a) $\delta_D = 0.0302$ in.; (b) $P_3 = 5$ k

2.3-6 (a) $L_1 = 0.75$ m, $L_2 = 0.25$ m; (b) $\delta = 0.400$ mm

2.3-7 (a) $\delta = 0.0589$ in.; (b) $\delta = 0.0501$ in.

2.3-8 $d = 23.9$ mm

2.3-9 (a) $\delta = PL/2EA$; (b) $\sigma_c = Py/AL$

2.3-10 (a) $\epsilon_A = 0.00236$; (b) $\epsilon_B = 0.00210$; (c) $\delta = 1004$ mm

2.3-11 (a) $\delta_C = W(L^2 - c^2)/2EAL$; (b) $\delta_B = WL/2EA$; (c) $r = 3$

2.3-12 (a) $\delta = PL/2Eb^2$; (b) $\delta = 0.234$ mm

2.3-13 (a) $\delta = P(L_1 + 3L_2)/3EA$; (b) Lower part of pile: $\sigma_c = Py^2/AL_1^2$

2.3-14 (a) $\delta_c = 2WL/\pi d^2E$; (b) $\delta = 6WL/\pi d^2E$

2.3-15 $\delta = \rho\omega^2L^2(3R + 2L)/6E$

2.3-17 (a) $\delta = \dfrac{qL^3}{8hEA}\left(1 + \dfrac{16h^2}{3L^2}\right)$; (b) $\delta = 11.14$ ft

2.4-1 (a) $P = 24,650$ lb; (b) $P_{max} = 28,200$ lb

2.4-2 (a) $P = 9.24$ kN; (b) $P_{allow} = 7.07$ kN

2.4-3 (a) $\sigma_s = 9100$ psi, $\sigma_c = 1130$ psi; (b) $\delta = 0.045$ in.

2.4-4 (a) 3/11; (b) 1/2; (c) Ratio = 1

2.4-5 (a) 41.7%; (b) $\sigma_m = 32.7$ ksi, $\sigma_o = 51.4$ ksi

2.4-6 (b) $\delta_C = b_1b_2P/[E(b_1A_2 + b_2A_1)]$

2.4-7 (b) $\sigma_a = 1610$ psi (compression), $\sigma_s = 9350$ psi (tension)

2.4-8 (a) $R_A = 2P/3$, $R_D = P/3$; (b) $\delta_B = PL/6EA_0$, $\delta_C = PL/12EA_0$

2.4-9 (a) $R_A = R_B = PbA_1/(bA_1 + 2aA_2)$; (b) $F = 2PaA_2/(bA_1 + 2aA_2)$; (c) $b = 2a$

2.4-10 (a) $P_1 = \dfrac{PE_1}{E_1 + E_2}$, $P_2 = \dfrac{PE_2}{E_1 + E_2}$;

(b) $e = \dfrac{b(E_2 - E_1)}{2(E_2 + E_1)}$; (c) $\dfrac{\sigma_1}{\sigma_2} = \dfrac{E_1}{E_2}$

2.4-11 $x = L/4$ **2.4-12** $P = 1.50$ kN

2.4-13 $d_2 = 0.3382$ in., $L_2 = 48$ in.

2.4-14 $\delta_C = 0.176$ mm

2.4-15 (a) $T_C = \dfrac{2aPL}{2a^2 + b^2}$, $T_D = \dfrac{bPL}{2a^2 + b^2}$;

(b) $\delta_B = \dfrac{2PL^2h}{EA(2a^2 + b^2)}$

2.4-16 $P_{max} = 1.69$ kN

2.4-17 (b) $L/3 < x < 5L/3$; (c) $x > 5L/3$

2.4-18 $T_{BD} = 81P/116$, $T_{CE} = 9P/58$

2.4-19 (a) $T_{AC} = 1.294P$, $T_{AD} = 1.079P$;

(b) $\delta_D = 8.089Pb/EA$

2.4-20 $P_{max} = 1.80$ kN

2.4-21 $T_{AB} = T_{CD} = P$, $\theta = 25P/12EA$

2.4-22 $\sigma_s = 58.9$ MPa, $\sigma_b = 28.0$ MPa, $\sigma_c = 33.6$ MPa

2.5-1 $\sigma = 13,200$ psi

2.5-2 (a) $\Delta d = 0.193$ mm; (b) $\sigma = 171$ MPa

2.5-3 115.6°F and 48.1°F

2.5-4 $\delta = (\Delta T)(L)[\alpha_m + (b/a)(\alpha_m - \alpha_t)]$

2.5-5 $\Delta T = P/[E_cA_c(\alpha_c - \alpha_s)]$

2.5-6 $\sigma_c = E\alpha(\Delta T_1)/3$ **2.5-7** $\Delta T = 34.2°$F

2.5-8 $\delta = 5.00$ mm

2.5-9 (a) $P = 11.88$ k; (b) $\sigma_c = 3.78$ ksi; (c) $\delta_C = 0.0126$ in. (to the left)

2.5-10 (a) $P = \dfrac{2E\alpha(\Delta T)A_1A_2}{A_1 + A_2}$;

(b) $\delta_C = \dfrac{\alpha(\Delta T)(L)(A_2 - A_1)}{2(A_1 + A_2)}$

2.5-11 $\Delta T = 93.4°$F   **2.5-12** $\tau_{aver} = 36.6$ MPa

2.5-13 $P_{allow} = 8920$ lb

2.5-14 (a) $T_A = 1760$ N, $T_B = 880$ N, (b) $T_A = 2008$ N, $T_B = 383$ N; (c) $\Delta T = 177°$C

2.6-1 $P_{max} = 40,500$ lb   **2.6-2** $d_{min} = 30.4$ mm

2.6-3 $P_{max} = 22,000$ lb   **2.6-4** $\Delta T = 56.4°$C

2.6-5 (a) $\tau_{max} = 12,210$ psi; (b) $\Delta T = 109.4°F$

2.6-6 (a) $\sigma_{max} = 79.6$ MPa; (b) $\tau_{max} = 39.8$ MPa

2.6-7 (a) $\sigma_{max} = 21,750$ psi; (b) $\tau_{max} = 10,880$ psi

2.6-8 (a) $\sigma_x = -61.7$ MPa; (b) $\sigma_\theta = -52.7$ MPa, $\tau_\theta = 21.8$ MPa; (c) $\sigma_\theta = -30.9$ MPa, $\tau_\theta = 30.9$ MPa

2.6-9 (a) $\sigma_x = -127.3$ psi; (b) $\sigma_\theta = -108.7$ psi, $\tau_\theta = 45.0$ psi; (c) $\sigma_\theta = -63.7$ psi, $\tau_\theta = 63.7$ psi

2.6-10 (a) $\sigma_\theta = 0.57$ MPa, $\tau_\theta = -1.58$ MPa; (b) $\alpha = 33.3°$; (c) $\alpha = 26.6°$

2.6-11 (a) $\Delta T = 55.4°F$; (b) $\sigma_\theta = -3380$ psi, $\tau_\theta = 4500$ psi

2.6-12 (a) $\tau_\theta = 6.00$ MPa; (b) For $\theta = 26.57°$, $\sigma_\theta = -12.0$ MPa, $\tau_\theta = 6.00$ MPa

2.6-13 $d_{min} = 1.19$ in.

2.6-14 For $\theta = 60°$: $\sigma_\theta = -18.7$ MPa, $\tau_\theta = 32.3$ MPa

2.6-15 For $\theta = 50°$: $\sigma_\theta = 5.37$ ksi, $\tau_\theta = -6.40$ ksi

2.6-16 (a) $\theta = 30°$, $\tau_\theta = -34.6$ MPa; (b) $\sigma_{max} = 80$ MPa, $\tau_{max} = 40$ MPa

2.6-17 $\sigma_{max} = 10,000$ psi, $\tau_{max} = 5,000$ psi

2.6-18 (a) $\theta = 33.69°$; (b) $P_{max} = 11.7$ kN

2.7-1 $U = 5570$ in.-lb

2.7-2 (a) $U = 5P^2L/4\pi Ed^2$; (b) $U = 1.036$ J

2.7-3 (a) $U = 23P^2L/12EA$; (b) $U = 80.1$ in.-lb

2.7-4 (c) $U = P^2L/2EA + PQL/2EA + Q^2L/4EA$

2.7-5 (a) $U_1 = 2.07$ in.-lb; (b) $U_2 = 71.4$ in.-lb; (c) $U = 58.0$ in.-lb

2.7-6 $U = \pi\gamma^2 d^2L^3/360E$

2.7-7 (a) $U = P^2L/EA$; (b) $\delta_B = 2PL/EA$

2.7-8 (a) $U = 1.457P^2L/EA$; (b) $\delta_D = 2.914PL/EA$

2.7-9 (a) $U = 7k\delta^2/2$; (b) $\delta = W/7k$; (c) $F_1 = W/7$, $F_2 = 2W/7$

2.7-10 (a) $U = 2P^2L/\pi Ed_1d_2$; (b) $\delta = 4PL/\pi Ed_1d_2$

2.7-11 (a) $U = \dfrac{P^2L}{2Et(b_2 - b_1)}\ln\dfrac{b_2}{b_1}$

2.7-12 $U = 2\rho^2 A\omega^4 L^5/15E$

2.7-13 (a) $P_1 = 106.4$ k; (b) $\delta = 0.0637$ in.; (c) $U = 4420$ in.-lb

2.7-14 (a) $x = 2s$, $P = 2(k_1 + k_2)s$; (b) $U_1 = (2k_1 + k_2)s^2$

2.7-15 (a) $U = P^2L/(4EA \sin\beta \cos^2\beta)$; $\beta = 35.26°$; (c) $U_{min} = 3\sqrt{3}P^2L/8EA$; (d) $\delta_B = 3\sqrt{3}PL/4EA$

2.7-16 (a) $U = EA\delta_D^2(1 + 2\cos^3\beta)/2H$; (b) $\delta_D = PH/[EA(1 + 2\cos^3\beta)]$; (c) $F_{BD} = P/(1 + 2\cos^3\beta)$

2.7-17 (a) $U = 58.0$ in.-lb; (b) $\delta_C = 6.65$ in.

2.8-1 (a) $\delta_{max} = 0.0361$ in.; (b) $\sigma_{max} = 22,600$ psi; (c) Impact factor $= 113$

2.8-2 (a) $\delta_{max} = 6.33$ mm; (b) $\sigma_{max} = 359$ MPa; (c) Impact factor $= 160$

2.8-3 $v = 45.9$ ft/s

2.8-4 (a) $\delta_{max} = 215$ mm; (b) Impact factor $= 3.86$

2.8-5 (a) $\delta_{max} = 9.21$ in.; (b) Impact factor $= 4.61$

2.8-6 $\delta_{max} = 200$ mm **2.8-7** $v = 11.6$ mph

2.8-8 $h = 1.13$ m **2.8-9** $h = 45.7$ in.

2.8-10 (a) Impact factor $= 1 + (1 + 2EA/W)^{1/2}$; (b) Impact factor $= 10$

2.8-11 $\delta_{max} = 0.377$ in., $\sigma_{max} = 2620$ psi, Impact factor $= 65.7$

2.8-12 $\sigma_2/\sigma_1 = 0.447$ **2.8-13** $L = 83.8$ ft

2.8-15 $\sigma_{max} = 4410$ psi

2.8-16 $\delta_{max} = 0.413$ mm, $\sigma_{max} = 43.4$ MPa, Impact factor $= 39.1$

2.10-1 (a) $\sigma_{max} \approx 5.5$ ksi and 6.2 ksi; (b) $\sigma_{max} \approx 9.8$ ksi and 8.0 ksi

2.10-2 (a) $\sigma_{max} \approx 70$ MPa and 77 MPa; (b) $\sigma_{max} \approx 67$ MPa and 58 MPa

2.10-3 $P_{max} = \sigma_t bt/3$

2.10-4 $\sigma_{max} \approx 46$ MPa **2.10-5** $\sigma_{max} \approx 6100$ psi

2.10-6 (a) No, it makes it weaker, $P_1 = 34.6$ kN; $P_2 = 20.3$ kN; (b) $d_0 = 15.3$ mm

2.10-7 (a) $P_{max} = 850$ lb; (b) Make d smaller

2.10-8 (a) $P_{max} = 18$ kN; (b) Make d smaller

2.10-9 $d_{max} = 0.51$ in.

2.11-2 (a) $\delta_C = 1.67$ mm; (b) $\delta_C = 5.13$ mm; (c) $\delta_C = 11.88$ mm

2.11-3 (b) $P = 123$ lb **2.11-4** $\delta_B = 18$ mm

2.11-5 For $P = 24$ k, $\delta_B = 0.36$ in.; for $P = 40$ k, $\delta_B = 1.36$ in.

2.11-6 For $P = 3.2$ kN, $\delta_B = 4.85$ mm; for $P = 4.8$ kN, $\delta_B = 17.32$ mm

2.11-7 $n = 204$ rpm **2.11-8** $\delta = 528$ mm

2.11-9 $F_A = 30$ lb, $F_C = 25$ lb, $\delta_A = 1.25$ in., $\delta_C = 0.833$ in., $\theta = 4.77°$

2.11-10 $F_A = 150$ N, $F_C = 135$ N, $\delta_A = 41.67$ mm, $\delta_C = 31.25$ mm, $\theta = 5.97°$

2.11-11 $\delta = 0.1966$ in., $F_1 = 238.6$ lb, $F_2 = 272.8$ lb, $\sigma_1 = 30,380$ psi, $\sigma_2 = 34,740$ psi, $\epsilon_1 = 0.007561$, $\epsilon_2 = 0.009829$

2.11-12 $\delta = 4.194$ mm, $F_1 = 947.3$ N, $F_2 = 1105.5$ N, $\sigma_1 = 193.0$ MPa, $\sigma_2 = 225.2$ MPa, $\epsilon_1 = 0.006453$, $\epsilon_2 = 0.008388$

2.11-13 $F_1 = 1119$ lb, $F_2 = 2441$ lb, $\delta_1 = 0.2245$ in., $\delta_2 = 0.4490$ in., $\sigma_1 = 63,300$ psi, $\sigma_2 = 77,690$ psi, $\epsilon_1 = 0.006414$, $\epsilon_2 = 0.01283$

2.11-14 $F_1 = 5.238$ kN, $F_2 = 10.381$ kN, $\delta_1 = 4.795$ mm, $\delta_2 = 9.590$ mm, $\sigma_1 = 416.9$ MPa, $\sigma_2 = 528.7$ MPa, $\epsilon_1 = 0.005993$, $\epsilon_2 = 0.01199$

2.12-1 $P_p = 5\sigma_y A$ **2.12-2** $P_p = 221$ kN

2.12-3 $P = 3080$ lb **2.12-4** $P_p = 56.5$ kN

2.12-5 $P_p = 2\sigma_y A(1 + \sin\alpha)$ **2.12-6** $P_p = 95.7$ kN

2.12-7   (a) $P_y = \sigma_y A$, $\delta_y = 3\sigma_y L/2E$; (b) $P_p = 4\sigma_y A/3$, $\delta_p = 3\sigma_y L/E$

2.12-8 $P_{\text{allow}} = 2.155\sigma_y A/n$

2.12-9   (a) $P_y = 17\sigma_y A/16$, $\delta_y = 2\sigma_y L/E$; (b) $P_p = 5\sigma_y A/4$, $\delta_p = 8\sigma_y L/3E$

2.12-10   (a) $W_y = 28.8$ kN, $\delta_y = 125$ mm; (b) $W_p = 48$ kN, $\delta_p = 225$ mm

2.12-11   (a) $P_y = 70{,}100$ lb, $\delta_y = 0.01862$ in.; (b) $P_p = 104{,}300$ lb, $\delta_p = 0.02862$ in.

2.12-12 $\beta = 60°$

CHAPTER 3

3.2-1   (a) $\gamma_1 = 200 \times 10^{-6}$ rad; (b) $(r_2)_{\min} = 1.83$ in.

3.2-2   (a) $\gamma_1 = 388 \times 10^{-6}$ rad; (b) $(r_2)_{\max} = 51.6$ mm

3.2-3   (a) $\gamma_1 = 314 \times 10^{-6}$ rad; (b) $(r_2)_{\max} = 1.91$ in.

3.3-1   (a) $\tau_{\max} = 18{,}300$ psi; $\phi = 3.32°$

3.3-2   (a) $k_T = 1860$ N·m; (b) $\tau_{\max} = 24.4$ MPa, $\gamma_{\max} = 872 \times 10^{-6}$ rad

3.3-3   (a) $L = 25.0$ in.; (b) $k_T = 9.96$ k-in.

3.3-4 $T_{\max} = 6.03$ N·m; $\phi = 2.20°$

3.3-5 $\tau_{\max} = 7760$ psi, $G = 4.0 \times 10^6$ psi, $\gamma_{\max} = 1940 \times 10^{-6}$ rad

3.3-6 $\tau_{\max} = 69.8$ MPa, $G = 28.0$ GPa, $\gamma_{\max} = 2490 \times 10^{-6}$ rad

3.3-7 $\tau_{\max} = 4840$ psi

3.3-8   (a) $\tau_2 = 35.1$ MPa; (b) $\tau_1 = 21.1$ MPa; (c) $\theta = 0.503°/$m

3.3-9 $T_{\max} = 81{,}700$ lb-in.   **3.3-10** $d_{\min} = 71.4$ mm

3.3-11 $d = 0.75$ in., $P = 149$ lb

3.3-12   (a) $T_1 = 424$ N·m; (b) $T_2 = 398$ N·m; (c) 6.25% and 25%

3.3-13   (a) $\phi = 5.50°$; (b) $d = 2.98$ in.; (c) 0.422

3.3-14   (a) $d_2/d = 1.279$; (b) $\beta_1 = 0.454$; (c) $\beta_2 = 1.279$

3.3-15   (a) $d_1 = 1.365$ in.; (b) $d_0 = 1.25$ in., $\beta_1 = 0.667$, $\beta_2 = 1.370$

3.3-16   (a) $T_2/T = 1 - (r_1/r_2)^4$; (c) 93.75%; (d) 75%

3.3-17 $r_2 = 1.234$ in.

3.4-1   (a) $\tau_{\max} = 7600$ psi; (b) $\phi_C = 0.16°$

3.4-2   (a) $\tau_{\text{bar}} = 79.6$ MPa, $\tau_{\text{tube}} = 32.3$ MPa; (b) $\phi_A = 9.43°$

3.4-3   (a) $\tau_{\max} = 5870$ psi, (b) $\phi_D = 1.68°$

3.4-4 $T_{\text{allow}} = 348$ N·m   **3.4-5** $d_1 = 0.818$ in.

3.4-6   (a) $d = 25.5$ mm; (b) $d = 22.6$ mm

3.4-7   (a) $d = 1.78$ in.; (b) $d = 1.83$ in.

3.4-8 $d = 77.5$ mm   **3.4-9** $d_B/d_A = 1.45$

3.4-10 $d_A = 60.3$ mm   **3.4-11** $\phi = 3TL / 2\pi Gt d_A^3$

3.4-12   (a) $\tau_{\max} = 16tL/\pi d^3$; (b) $\phi = 16tL^2/\pi Gd^4$

3.4-13   (a) $\tau_{\max} = 8t_A L/\pi d^3$; (b) $\phi_B = 16t_A L^2/3\pi Gd^4$

3.4-14   (a) $\tau_{\max} = 16WH_A/3\pi d^3$; (b) $\phi_B = 8WH_A L/3\pi Gd^4$

3.5-1 $G = 11.5 \times 10^6$ psi

3.5-2 $\gamma_{\max} = 1740 \times 10^{-6}$ rad, $\epsilon_{\max} = 870 \times 10^{-6}$

3.5-3 $T = 4200$ lb-in.   **3.5-4** $T = 6700$ N·m

3.5-5 $d = 1.43$ in.

3.5-6 $\epsilon_{\max} = 320 \times 10^{-6}$, $\sigma_{\max} = 51.1$ MPa, $T = 20.0$ kN·m

3.5-7 $d_1 = 0.60$ in.

3.7-1   (a) $\tau_{\max} = 4950$ psi; (b) $d = 3.22$ in.

3.7-2   (a) $\tau_{\max} = 50$ MPa; (b) $d = 32.3$ mm

3.7-3   (a) 28,500 hp; (b) Shear stress is halved

3.7-4   (a) $\tau_{\max} = 16.8$ MPa; (b) $P_{\max} = 312$ kW

3.7-5 $d = 3.56$ in.   **3.7-6** $d = 68.1$ mm

3.7-7 $d_1 = 1.221d$ **3.7-8** $P = 91.0$ kW

3.7-9 $d = 2.75$ in.   **3.7-10** $d = 53.4$ mm

3.8-1 $\phi_{\max} = 11T_0 L/16GI_p$

3.8-2 $T_A = T_D = T_0/3$, $\phi_{\max} = T_0 L/9GI_p$, $\phi_0 = 0$

3.8-3 $P = 716$ lb   **3.8-4** $\phi_{\text{allow}} = 2b\tau_{\text{allow}}/Gd$

3.8-5 $T_{\text{allow}} = 879$ lb-in.   **3.8-6** $T_{\text{allow}} = 639$ N·m

3.8-7 $x = 0.579L$

3.8-8 $T = 2bckGI_p/(2c^2 Lk + GI_p)$

3.8-9 $T_A = t_0 L/6$, $T_B = t_0 L/3$

3.8-10   (a) $\tau_t = 30.7$ MPa, $\tau_b = 20.4$ MPa; (b) $\phi = 0.44°$; (c) $k_T = 260 \times 10^3$ N·m

3.8-11   (a) $\tau_t = 5270$ psi, $\tau_b = 3520$ psi; (b) $\phi = 0.46°$; (c) $k_T = 2.52 \times 10^6$ lb-in.

3.8-12   (a) $\tau_{s2} = 77.3$ MPa; (b) $\gamma_{s2} = 1030 \times 10^{-6}$ rad

3.8-13   (a) $\tau_s = 9790$ psi, $\tau_b = 3970$ psi; (b) $\phi = 1.22°$; (c) $k_T = 939 \times 10^3$ lb-in.

3.8-14 $T_{\text{allow}} = 4170$ N·m   **3.8-15** $T_{\text{allow}} = 253$ k-in.

3.8-16 $\tau_{\max} = \dfrac{16\,T_0 d_3}{\pi(d_1^4 + d_3^4 - d_2^4)}$

3.9-1   (a) $U = 15.9$ in.-lb; (b) $\phi = 0.92°$

3.9-2   (a) $U = 75.2$ J; (b) $\phi = 1.89°$

3.9-3 $U = 122$ in.-lb   **3.9-4** $U = 3.10$ J

3.9-5   (c) $U_3 = \dfrac{T^2 L}{2GI_p} + \dfrac{TtL^2}{2GI_p} + \dfrac{t^2 L^3}{6GI_p}$

3.9-6 $U = t_0^2 L^3/40GI_p$ **3.9-7** $U = 19T_0^2 L/32GI_p$

3.9-8 $\phi = \dfrac{L_A L_B T}{G(L_B I_{pA} + L_A I_{pB})}$

3.9-9   (b) $\phi = \dfrac{2TL(d_A + d_B)}{\pi Gt\, d_A^2\, d_B^2}$

3.9-10 $U = \dfrac{\beta^2 GI_{pA} I_{pB}}{2L(I_{pA} + I_{pB})}$

3.9-11 $\phi = \dfrac{2n}{15d^2}\sqrt{\dfrac{2\pi I_m L}{G}}$; $\tau_{\max} = \dfrac{n}{15d}\sqrt{\dfrac{2\pi GI_m}{L}}$

3.10-1   (a) $\tau_{\text{approx}} = 6310$ psi; (b) $\tau_{\text{exact}} = 6830$ psi

3.10-2 $\tau_{\min} = \pi d/64$ **3.10-3** $U_1/U_2 = 2$

3.10-4 (a) $\tau = 10.0$ MPa; (b) $\phi = 0.386°$
3.10-5 $\tau = 3320$ psi; $\theta = 0.0067°/$in.
3.10-6 $\tau = 35.0$ MPa, (b) $\phi = 0.38°$
3.10-7 $\tau = T\sqrt{3}/9b^2t$, $\theta = 2T/9Gb^3t$
3.10-8 $t = 4.57$ mm   **3.10-9** $\phi_1/\phi_2 = 1 + 1/4\beta^2$
3.10-10 $\tau_{max} = T/(2\pi r^2 t_0)$, $\phi = TL/(G\pi^2 r^3 t_0)$
3.10-11 (a) $t = 0.263$ in.; (b) $t = 0.277$ in.
3.10-12 $\tau = 2T(1 + \beta)^2/tL_m^2\beta$
3.10-13 $\theta = 4T(1 + \beta)^4/GtL_m^3\beta^2$
3.10-14 $\phi = 2TL(d_A + d_B)/\pi Gtd_A^2 d_B^2$
3.11-1 $T_{max} \approx 12{,}000$ lb-in.   **3.11-2** $R_{min} \approx 4$ mm
3.11-3 For $D_1 = 0.8.$: $\tau_{max} \approx 6400$ psi
3.11-4 $D_2 \approx 116$ mm; lower limit
3.11-5 $D_1 \approx 1.31$ in.
3.12-2 $\tau_{max} = Tr(3n + 1)/4I_p n$
3.12-3 $\dfrac{T_p}{T_y} = \dfrac{4r_2(r_2^3 - r_1^3)}{3(r_2^4 - r_1^4)}$
3.12-4 $\theta = 3.71°/$m   **3.12-5** $\theta = 0.142°/$in.
3.12-7 (a) $T = 1161$ k-in.; (b) $T_p = 1188$ k-in.;
(c) $\gamma_p = 0.004$ rad
3.12-8 $\tau_{max} = 50.8$ MPa

CHAPTER 4

4.3-1 $V = 4.0$ k, $M = 60.0$ k-ft
4.3-2 $V = 7.0$ kN, $M = -9.5$ kN·m
4.3-3 $V = 2Pa/b$, $M = 0$
4.3-4 $V = 1.0$ kN, $M = 9.0$ kN·m
4.3-5 $V = -1540$ lb, $M = -4640$ lb-ft
4.3-6 $V = -1.0$ kN, $M = -7.0$ kN·m
4.3-7 $M_{max} = 450$ lb-ft
4.3-8 $V = -2.5$ kN, $M = 45$ kN·m
4.3-9 $N = P\sin\theta$, $V = P\cos\theta$, $M = Pr\sin\theta$
4.3-10 (b) $M_{max} = 2qb^2/9$ **4.3-11** $P = 7.5$ k
4.3-12 $N = 21.6$ kN (comp.), $V = 7.2$ kN,
$M = 50.4$ kN·m
4.3-13 (a) $V_B = 6000$ lb, $M_B = 9000$ lb-ft; (b) $V_m = 0$,
$M_m = 21{,}000$ lb-ft
4.3-14 $a/L = 1/4$
4.3-15 $V = 13.3$ k, $M = 117$ k-ft
4.3-16 $V_{max} = 131wL^2\alpha/36g$, $M_{max} = 1339wL^3\alpha/360g$
4.5-1 $V_{max} = P$, $M_{max} = Pa$
4.5-2 $V_{max} = qL/2$, $M_{max} = -3qL^2/8$
4.5-3 $V_{max} = M_0/L$, $M_{pos} = M_0a/L$
4.5-4 $V_{max} = -7P/12$, $M_{max} = 7PL/36$
4.5-5 $V_{max} = P/2$, $M_{max} = 3PL/8$
4.5-6 $V_{max} = -12.0$ kN, $M_{max} = 38.4$ kN·m
4.5-7 $V_{max} = -3M_1/L$, $M_{max} = M_1$
4.5-8 $V_{max} = P$, $M_{max} = -Pa$
4.5-9 $V_{max} = 1.5$ k, $M_{max} = -12.0$ k-ft

4.5-10 $V_{max} = -10$ kN, $M_{max} = -16$ kN·m
4.5-11 $V_{pos} = 8$ k, $M_{max} = 64$ k-ft
4.5-12 $V_{max} = 7$ kN, $M_{max} = -13$ kN·m
4.5-13 $V_{pos} = 2$ Pa/b, $M_{max} = Pa$
4.5-14 $V_{pos} = 5.0$ kN, $M_{max} = 10.5$ kN·m
4.5-15 $V_{max} = 2460$ lb, $M_{pos} = 7560$ lb-ft
4.5-16 $V_{max} = 8.0$ kN, $M_{max} = -8.0$ kN·m
4.5-17 $V_{max} = -q_0L/2$, $M_{max} = -q_0L^2/6$
4.5-18 $a = 0.5858L$, $V_{max} = 0.2929qL$,
$M_{max} = 0.02145qL^2$
4.5-19 $V_{max} = -7q_0L/18$, $M_{pos} = 2\sqrt{2}\,q_0L^2/81$
4.5-20 $V_{max} = 13.0$ kN, $M_{max} = -28.0$ kN·m
4.5-21 $V_{max} = -6.2$ k, $M_{max} = -34.5$ k-ft
4.5-22 $V_{max} = -q_0L/3$, $M_{max} = q_0L^2/9\sqrt{3}$
4.5-23 $V_{max} = -7000$ lb, $M_{max} = 22{,}360$ lb-ft
4.5-24 $V_{neg} = -q_0L/12$, $M_{max} = 0.01604q_0L^2$
4.5-25 $V_{max} = -165$ lb, $M_{max} = 2722$ lb-in.
4.5-26 $V_{max} = -12.5$ kN, $M_{max} = 20.0$ kN·m
4.5-27 $V_{max} = -1560$ lb, $M_{max} = -2880$ lb-ft
4.5-28 $V_{max} = 32.97$ kN, $M_{max} = -61.15$ kN·m
4.5-29 $V_{max} = 12.5$ k, $M_{pos} = 78.12$ k-ft
4.5-30 $V_{max} = -12.0$ kN, $M_{max} = -24.0$ kN·m
4.5-31 $b = L/3$, $V_{max} = 5q_0L/36$, $M_{max} = -q_0L^2/81$
4.5-32 $V_{max} = qL$, $M_{max} = -3qL^2/8$
4.5-33 $V_{max} = -q_0L/3$, $M_{max} = 3q_0L^2/64$
4.5-34 $V_{max} = 2.5$ kN, $M_{max} = 5.0$ kN·m
4.5-35 $M_{max} = 20{,}000$ lb-ft
4.5-36 $M_{max} = 30$ kN·m
4.5-37 $V_{max} = -9q_0L/32$, $M_{max} = 3q_0L^2/64$
4.5-38 $q_A = 2P/3L$, $V_{max} = 11P/9$, $M_{max} = -3PL/16$
4.5-39 (a) $x = 0$ and $x = L - d$, $V_{max} = P(2 - d/L)$;
(b) $x = (2L - d)/4$, $M_{max} = P(L - d/2)^2/2L$
4.5-40 (a) $x = 6.4$ m, $V_{max} = 21$ kN; (b) $x = 2.667$ m,
$M_{max} = 39.2$ kN·m
4.5-41 $x = 24$ ft, $M_{max} = 535$ k-ft
4.5-42 $M_{max} = 560$ N·m
4.5-43 $M_{max} = PL(n + 2)/[8(n + 1)$ for n even;
$M_{max} = PL(n + 1)/8n$ for n odd
4.5-44 (a) $a/L = 0.4749$, (b) $M_{max} = 0.01785q_0L^2$

CHAPTER 5

5.4-1 $\epsilon_{max} = 1300 \times 10^{-6}$ **5.4-2** $L_{min} = 3.81$ m
5.4-3 $\epsilon_{max} = 8.18 \times 10^{-3}$ **5.4-4** $\delta = 11.5$ mm
5.4-5 $\epsilon = 358 \times 10^{-6}$ (shortening)
5.4-6 $\epsilon = 533 \times 10^{-6}$
5.5-1 (a) $\sigma_{max} = 48{,}000$ psi; (b) Stress decreases
5.5-2 (a) $\sigma_{max} = 355$ MPA; (b) Stress increases
5.5-3 (a) $\sigma_{max} = 71{,}000$ psi; (b) Stress increases

5.5-4 $\sigma_{max} = 5.0$ MPa   **5.5-5** $\sigma_{max} = 3420$ psi

5.5-6 $\sigma_{max} = 37.0$ MPa   **5.5-7** $\sigma_{max} = 18.6$ ksi

5.5-8 $\sigma_{max} = 68.8$ MPa   **5.5-9** $\sigma_{max} = 22,500$ psi

5.5-10 $\sigma_{max} = 5.26$ MPa   **5.5-11** $\sigma_{max} = 900$ psi

5.5-12 $\sigma_{max} = 2.04$ MPa

5.5-13   (a) $\sigma_{max} = 32\sqrt{2}\,M/\pi d^3$; (b) $\sigma_{max} = 8M/5b^3$

5.5-14   (a) $\sigma_t = 768Ml/(9\pi^2 - 64)d^3$;

(b) $\sigma_t = 168M/37bh^2$

5.5-15 $\sigma_{max} = 21.4$ ksi   **5.5-16** $\sigma_{max} = 55.4$ MPa

5.5-17 $\sigma_t = 2{,}890$ psi, $\sigma_c = 10{,}190$ psi

5.5-18   (a) $\sigma_t = 1.36$ MPa, $\sigma_c = 2.72$ MPa;

(b) No change; (c) Reduced by half

5.5-19 $\sigma_t = 8{,}190$ psi, $\sigma_c = 14{,}600$ psi

5.5-20 $\sigma_{max} = 3\rho L^2 a_0/t$

5.5-21 $\sigma_t = 18{,}400$ psi, $\sigma_c = 13{,}800$ psi

5.5-22 $\sigma = 25.1$ MPa, 17.8 MPa, -23.5 MPa

5.5-23 $d = 3$ ft, $\sigma_{max} = 171$ psi; $d = 6$ ft, $\sigma_{max} = 830$ psi

5.6-1 $d_{min} = 0.471$ in.   **5.6-2** $d_{min} = 94.9$ mm

5.6-3   W 8×21 **5.6-4**   W 12×35

5.6-5   S 8×18.4 **5.6-6** $b_{min} = 188$ mm

5.6-7 $S = 18.38$ in.^{3}; 2×10 in. joists

5.6-8 $s = 382$ mm

5.6-9 $q_{allow} = 314$ lb/ft   **5.6-10** $s = 32.1$ mm

5.6-11   (a) $S_{min} = 15.41$ in.^{3}; (b) S 8×23

5.6-12 $d_{min} = 23.7$ mm   **5.6-13** $q = 711$ lb/ft

5.6-14 $b = 152$ mm   **5.6-15** $b = 10.25$ in.

5.6-16 $t = 15$ mm   **5.6-17** $1 : 1.260 : 1.408$

5.6-18   Increase when $d/h > 0.6861$; decrease when $d/h < 0.6861$

5.6-19 $q = 21.5$ lb/in.

5.6-20   (a) $b = 9.95$ mm; (b) $b = 9.96$ mm

5.6-21   6.8%

5.6-22   (a) Stronger when bent about axis 1-1;

(b) $\beta = 1.184$

5.6-23 $t_{min} = \sqrt{8\gamma h^3/\sigma_{allow}\sin^2\alpha}$

5.6-24   (a) $\beta = 1/9$; (b) 5.35%

5.6-25 $s = 72.0$ in.

5.7-1 $x = 2L/5$; $\sigma_{max} = 512PL/135\pi d_A^3$;

$\sigma_{max}/\sigma_B = 27/20 = 1.35$

5.7-2 $x = L/3$; $\sigma_{max} = 16PL/27h_A^3$;

$\sigma_{max}/\sigma_B = 125/81 = 1.543$

5.7-3 $x = 12.5$ ft; $\sigma_{max} = 5197$ psi;

$\sigma_{max}/\sigma_B = 9/8 = 1.125$

5.7-4 $x = 200$ mm; $\sigma_{max} = 16$ MPa;

$\sigma_{max}/\sigma_B = 25/24 = 1.042$

5.7-5   (a) $\sigma_A = 8488$ psi; (b) $\sigma_B = 7427$ psi;

(c) $x = 15$ in.; (d) $\sigma_{max} = 10{,}865$ psi

5.7-6   (a) $d_B/d_A = 1.906$; (b) $V_{min} = 2.405L(PL/\sigma_{allow})^{2/3}$;

(c) $V_{min}/V_{prismatic} = 0.6518$

5.7-7 $h_x = h_Bx/L$ **5.7-8** $h_x = h_B\sqrt{x/L}$

5.7-9 $b_x = b_Bx/L$ **5.7-10** $h_x = h_1\sqrt{1 - 4x^2/L^2}$

5.8-2 $\tau_{max} = 656$ kPa; $\sigma_{max} = 4.59$ MPa

5.8-3 $P_{max} = 6500$ lb   **5.8-4** $\tau_{max} = 500$ kPa

5.8-5 $\tau_{max} = 2400$ psi

5.8-6   (a) $L_0 = h(\sigma_{allow}/\tau_{allow})$; (b) $L_0 = (h/2)(\sigma_{allow}/\tau_{allow})$

5.8-7 $P_{allow} = 1800$ lb   **5.8-8** $b = 162$ mm

5.8-9   (a) 8×10 in. beam; (b) Satisfactory

5.8-10   (a) $P = 402$ N; (b) $P = 448$ N

5.8-11   (a) $P = 8590$ lb; (b) $P = 9700$ lb

5.8-12   (a) $b = 99.8$ mm; (b) $b = 71.8$ mm

5.8-13 $w_1 = 118$ lb/ft^2, $w_2 = 320$ lb/ft^2, $w_3 = 121$ lb/ft^2,

$w_4 = 86.9$ lb/ft^2

5.9-1   (a) $d_{min} = 5.59$ in.; (b) $d_{min} = 2.38$ in.

5.9-2   (a) $W = 30.7$ kN; (b) $W = 41.3$ kN

5.9-3   (a) $d = 10.5$ in.; (b) $d = 2.56$ in.

5.9-4   (a) $d = 266$ mm; (b) $d = 64$ mm

5.10-1   (a) $\tau_{max} = 4680$ psi; (b) $\tau_{min} = 3920$ psi;

(c) $\tau_{aver} = 4670$ psi; (d) $V_{web} = 18{,}960$ lb

5.10-2   (a) $\tau_{max} = 23.5$ MPa; (b) $\tau_{min} = 18.8$ MPa;

(c) $\tau_{aver} = 22.8$ MPa; (d) $V_{web} = 48.0$ kN

5.10-3   (a) $\tau_{max} = 4860$ psi; (b) $\tau_{min} = 4200$ psi;

(c) $\tau_{aver} = 4920$ psi; (d) $V_{web} = 9.43$ k

5.10-4   (a) $\tau_{max} = 32.3$ MPa; (b) $\tau_{min} = 21.5$ MPa;

(c) $\tau_{aver} = 29.2$ MPa; (d) $V_{web} = 196$ kN

5.10-5   (a) $\tau_{max} = 4570$ psi; (b) $\tau_{min} = 3380$ psi;

(c) $\tau_{aver} = 4300$ psi; (d) $V_{web} = 48.5$ k

5.10-6   (a) $\tau_{max} = 28.4$ MPa; (b) $\tau_{min} = 19.4$ MPa;

(c) $\tau_{aver} = 26.0$ MPa; (d) $V_{web} = 58.6$ kN

5.10-7   (a) $q_{allow} = 2700$ lb/ft; (b) $q_{allow} = 4230$ lb/ft

5.10-8   (a) $q_{allow} = 141$ kN/m; (b) $q_{allow} = 199$ kN/m

5.10-9   S 8×23 **5.10-10** $V_{allow} = 303$ kN

5.10-11 $V_{allow} = 42{,}200$ lb   **5.10-12** $\tau_{max} = 22.0$ MPa

5.10-13 $\tau_{max} = 1830$ psi

5.10-14   (a) $\tau_{max} = 24.8$ MPa; $\tau_{min} = 19.6$ MPa;

(b) $\tau_{aver} = 25.0$ MPa; (c) $V_{web} = 65.7$ kN

5.11-1 $V_{max} = 728$ lb   **5.11-2** $V_{max} = 763$ kN

5.11-3 $F = 1670$ lb/in.   **5.11-4** $V_{max} = 5.95$ kN

5.11-5 $s_{max} = 3.20$ in.

5.11-6   (a) $s_A = 78.3$ mm; (b) $s_B = 97.9$ mm

5.11-7 $V_{max} = 25.4$ k   **5.11-8** $s = 50$ mm

5.11-9 $s_{max} = 6.03$ in.   **5.11-10** $V = 12.5$ k

5.12-1 $\sigma_t = 14{,}660$ psi, $\sigma_c = -14{,}990$ psi

5.12-2 $\sigma_t = 480$ kPa, $\sigma_c = -930$ kPa

5.12-3 $\sigma_t = 2860$ psi, $\sigma_c = -2980$ psi

5.12-4 $t_{min} = 12.38$ mm

5.12-5 $\sigma_t = 302$ psi, $\sigma_c = -314$ psi

5.12-6 $T_{allow} = 66.2$ kN

5.12-7 $\alpha = \arctan[(d_2^2 + d_1^2)/(4hd_2)]$

5.12-8 $d = 61.8$ mm   **5.12-9** $H_{max} = 32.2$ ft

5.12-10   (a) $\sigma_t = 826$ kPa, $\sigma_c = -918$ kPa;

$d_{max} = 0.750$ m

5.12-11 $W = 7560$ lb   **5.12-12** $\sigma_{max} = 720$ kPa

5.12-13 (a) $\sigma_t = 1900$ psi, $\sigma_c = -1100$ psi; (b) Both stresses increase in magnitude

5.12-14 (a) $b = \pi d/6$; (b) Rectangular post

5.12-15 (a) $\sigma_t = 8P/b^2$, $\sigma_c = -4P/b^2$; (b) $\sigma_t = 9.11P/b^2$, $\sigma_c = -6.36P/b^2$

5.12-16 (a) $\sigma_t = 3.57$ MPa, $\sigma_c = -26.4$ MPa; (b) $y_0 = -76.2$ mm

5.12-17 (a) $\sigma_t = 480$ psi, $\sigma_c = -3200$ psi; (b) $y_0 = -3.86$ in.

5.12-18 (a) $y_0 = -75.1$ mm, (b) $P_{max} = 170$ kN

5.12-19 (a) $y_0 = -0.847$ in.; (b) $P_{max} = 16.2$ k

5.12-20 (a) $r_c = (r_2^2 + r_1^2)/4r_2$

5.12-21 Rhombus with diagonals of lengths 7.35 in. and 1.30 in.

5.12-22 Rhombus with diagonals of lengths 9.23 in. and 3.11 in.

5.12-23 Square inclined at 45° with diagonals of length $(b/3)(1 + c^2/b^2)$

5.12-24 $d = 0.3692b$

5.13-1 (a) $d = 0.50$ in., $\sigma_{max} = 15,500$ psi; (b) $R = 0.10$ in., $\sigma_{max} \approx 49,000$ psi

5.13-2 (a) $d = 16$ mm, $\sigma_{max} = 81$ MPa; (b) $R = 4$ mm, $\sigma_{max} \approx 200$ MPa

5.13-3 $b_{min} \approx 0.24$ in.   **5.13-4** $b_{min} \approx 33$ mm

5.13-5 (a) $R_{min} \approx 0.45$ in.; (b) $d_{max} \approx 4.13$ in.

CHAPTER 6

6.2-1 $\sigma_{face} = 3960$ psi, $\sigma_{core} = 1060$ psi

6.2-2 $M_{max} = 67.9$ kN·m   **6.2-3** $M_{max} = 207$ k-in.

6.2-4 $M = \pi d^3 \sigma_s (15 + E_a/E_s)/512$

6.2-5 $\sigma_s = 13,730$ psi, $\sigma_w = 658$ psi

6.2-6 $M_{allow} = 768$ N·m

6.2-7 (a) $\sigma_{face} = 3610$ psi, $\sigma_{core} = 4$ psi; (b) $\sigma_{face} = 3630$ psi

6.2-8 (a) $\sigma_{face} = 14.12$ MPa, $\sigma_{core} = 0.21$ MPa; (b) $\sigma_{face} = 14.9$ MPa

6.2-9 $\sigma_a = 4120$ psi, $\sigma_c = 5230$ psi

6.2-10 $\sigma_w = 5.4$ MPa (compression), $\sigma_s = 40.2$ MPa (tension)

6.3-1 $M_{max} = 1025$ k-in.   **6.3-2** $t = 15.0$ mm

6.3-3 $q_{allow} = 811$ lb/ft

6.3-4 $\sigma_s = 62.3$ MPa, $\sigma_w = 2.3$ MPa

6.3-5 $\sigma_a = 3100$ psi, $\sigma_p = 120$ psi

6.3-6 $\sigma_a = 12.14$ MPa, $\sigma_p = 0.472$ MPa

6.3-7 $\sigma_s = 8490$ psi (tension), $\sigma_w = 1510$ psi (compression)

6.3-8 $\sigma_s = 93.5$ MPa   **6.3-9** $M_{max} = 179$ k-in.

6.3-10 $S_A = 50.6$ mm³; Metal A

6.3-11 $\sigma_s = 14,110$ psi, $\sigma_c = -855$ psi

6.3-12 $M_{max} = 16.2$ kN·m   **6.3-13** $M_{max} = 307$ k-in.

6.4-2 $\beta = 66.6°$, $\sigma_{max} = 14.93$ MPa

6.4-3 $\beta = 30.7°$, $\sigma_{max} = 642$ psi

6.4-4 $\beta = 78.9°$, $\sigma_A = -\sigma_E = 13,400$ psi, $\sigma_D = -\sigma_B = 6,410$ psi

6.4-5 $\beta = 70.4°$, $\sigma_A = -\sigma_E = 13,100$ psi, $\sigma_D = -\sigma_B = 3,700$ psi

6.4-6 $\beta = -79.71°$, $\sigma_A = \sigma_{max} = 9.41$ MPa

6.4-7 $\beta = -75.96°$, $\sigma_A = \sigma_{max} = 1790$ psi

6.4-8 $\beta = -83.6°$, $\sigma_{max} = 9850$ psi

6.4-9 $\beta = 72.9°$, $\sigma_{max} = 10,750$ psi

6.4-10 $\beta = 59.8°$, $\sigma_{max} = 10,760$ psi

6.4-11 (a) $\sigma_A = 45,420 \sin\alpha + 3630 \cos\alpha$ (psi); (b) $\tan\beta = 37.54 \tan\alpha$

6.5-1 $\beta = 83.1°$, $\sigma_t = 5,060$ psi, $\sigma_c = -10,420$ psi

6.5-2 $\beta = 77.3°$, $\sigma_t = 1470$ psi, $\sigma_c = -2860$ psi

6.5-3 $\beta = 75.6°$, $\sigma_t = 3080$ psi, $\sigma_c = -3450$ psi

6.5-4 $\beta = 75.4°$, $\sigma_t = 3090$ psi, $\sigma_c = -3510$ psi

6.5-5 (a) $\beta = \theta$; (b) $\sigma_A = (32M \sin\theta)/b^3$, $\sigma_B = -16M(\sqrt{3} \cos\theta + \sin\theta)/b^3$, $\sigma_D = 16M(\sqrt{3} \cos\theta - \sin\theta)/b^3$

6.5-6 $\theta = 0$: $\sigma_t = -\sigma_c = 2.546M/r^3$; $\theta = 45°$: $\sigma_t = 4.535M/r^3$, $\sigma_c = -3.955M/r^3$; $\theta = 90°$: $\sigma_t = 3.867M/r^3$; $\sigma_c = -5.244M/r^3$

6.5-7 $\beta = 68.2°$, $\sigma_t = 1870$ psi, $\sigma_c = -1840$ psi

6.5-8 $\beta = 67.8°$, $\sigma_t = 2930$ psi, $\sigma_c = -2950$ psi

6.7-1 (a) $\tau_{max} = 2360$ psi; (b) $\tau_B = 490$ psi

6.7-2 (a) $\tau_{max} = 15.1$ MPa; (b) $\tau_B = 3.4$ MPa

6.7-3 (a) $\tau_{max} = 500$ psi; (b) 3/10, 2/10, 2/10, and 3/10

6.7-4 (a) $\tau_{max} = 3P/2bt$

6.7-5 $\tau_{max} = 3Pb_1^2/[2(t_1 b_1^3 + t_2 b_2^3)]$

6.7-6 (a) $\tau_{max} = 3P/2\sqrt{2}\, tb$ **6.7-7** (a) $\tau_{max} = 1370$ psi

6.7-8 (a) $\tau_{max} = 7.56$ MPa   **6.7-9** (a) $\tau_{max} = 1890$ psi

6.7-10 (a) $\tau_{max} = 31.3$ MPa

6.8-1 $e = 1.011$ in.   **6.8-2** $e = 0.674$ in.

6.9-1 $f = 2(2b_1 + b_2)/(3b_1 + b_2)$

6.9-2 (a) $f = 16r_2(r_2^3 - r_1^3)/[3\pi(r_2^4 - r_1^4)]$; (b) $f = 4/\pi$

6.9-3 $q = 800$ lb/in.

6.9-4 (a) 82.4%; (b) 15.4 kN·m

6.9-5 $f = 1.15$ **6.9-6** $f = 1.14$

6.9-7 $f = 1.12$ **6.9-8** $f = 1.12$

6.9-9 $M_y = 4800$ k-in., $M_p = 5360$ k-in.

6.9-10 $M_y = 645$ k-in., $M_p = 724$ k-in.

6.9-11 $M_y = 4320$ k-in., $M_p = 5450$ k-in.

6.9-12 $M_y = 511$ kN·m, $M_p = 648$ kN·m

6.9-13 $M_y = 1196$ k-in., $M_p = 1485$ k-in.

6.9-14 $M_y = 122$ kN·m, $M_p = 147$ kN·m

6.9-15 (a) $M = 4430$ k-in.; (b) 25%

6.9-16 (a) $M = 524$ kN·m; (b) 36%
6.9-17 (a) $M = 2410$ k-in.; (b) 11%
6.9-18 (a) $M = 59.8$ kN·m; (b) 13%
6.9-19 (a) $Z = 53.25$ in.^{3}; (b) $f = 1.77$
6.9-20 (a) $Z = 440.6 \times 10^3$ mm^3; (b) $f = 1.81$
6.9-21 $M_p = 1120$ k-in.
6.9-22 $M_p = 295$ kN·m
6.10-1 $M_{allow} = 18.3$ k-in.
6.10-2 $\sigma_t = 24.4$ MPa, $\sigma_c = -41.2$ MPa
6.10-3 $\sigma_t = 3020$ psi, $\sigma_c = -3490$ psi
6.10-4 $M = 21.8$ kN·m **6.10-5** $M = 414$ k-in.
6.10-6 $M = bh^2\epsilon_1(4B_1 - 3B_2\epsilon_1)/24$
6.10-9 $M = 951$ k-in. **6.10-10** $\sigma_{max} = 174$ MPa
6.10-11 $\sigma_{max} = 25,200$ psi **6.10-12** $M = 566$ N·m

CHAPTER 7

7.2-1 For $\theta = 50°$: $\sigma_{x1} = 8690$ psi, $\tau_{x1y1} = -2000$ psi
7.2-2 For $\theta = 30°$: $\sigma_{x1} = 118.5$ MPa, $\tau_{x1y1} = -15.3$ MPa
7.2-3 For $\theta = 60°$: $\sigma_{x1} = -18,410$ psi, $\tau_{x1y1} = -11,200$ psi
7.2-4 For $\theta = 50°$: $\sigma_{x1} = -18.4$ MPa, $\tau_{x1y1} = -28.1$ MPa
7.2-5 For $\theta = 30°$: $\sigma_{x1} = 527$ psi, $\tau_{x1y1} = -273$ psi
7.2-6 $\sigma_w = 8.40$ MPa, $\tau_w = -4.27$ MPa
7.2-7 $\sigma_w = -240$ psi, $\tau_w = 1120$ psi
7.2-8 $\theta = \pm60°$ **7.2-9** $\theta = \pm26.57°$
7.2-10 $\sigma_x = 29.5$ MPa, $\sigma_y = -9.5$ MPa, $\tau_{xy} = -11.8$ MPa
7.2-11 $\sigma_x = -29,540$ psi, $\sigma_y = -10,060$ psi, $\tau_{xy} = -7,080$ psi
7.2-12 $\sigma_y = 43.7$ MPa, $\tau_{xy} = 20.5$ MPa
7.2-13 $\sigma_y = -193$ psi, $\tau_{xy} = -406$ psi
7.2-14 $\sigma_b = -35.0$ MPa, $\tau_b = 32.9$ MPa, $\theta_1 = 44.56°$
7.2-15 $\theta_1 = 22.38°$ **7.2-16** $\theta = 33.05°$ and $80.91°$
7.3-1 $\sigma_1 = 9350$ psi, $\theta_{p1} = 31.72°$
7.3-2 $\sigma_1 = 121$ MPa, $\theta_{p1} = 20.30°$
7.3-3 $\tau_{max} = 16,420$ psi, $\theta_{s1} = -53.5°$
7.3-4 $\tau_{max} = 46.0$ MPa, $\theta_{s1} = -66.18°$
7.3-5 (a) $\sigma_1 = 40$ psi, $\theta_{p1} = -9.46°$;
(b) $\tau_{max} = 740$ psi, $\theta_{s1} = -54.46°$
7.3-6 (a) $\sigma_1 = 30$ MPa, $\theta_{p1} = -63.43°$;
(b) $\tau_{max} = 75$ MPa, $\theta_{s1} = 71.57°$
7.3-7 $2,500$ psi $\le \sigma_y \le 10,500$ psi
7.3-8 13 MPa $\le \sigma_y \le 77$ MPa
7.3-9 $\sigma_x = -15.01$ ksi, $\tau_{xy} = 3.99$ ksi
7.3-10 (a) $\sigma_1 = 86.0$ MPa, $\theta_{p1} = 26.57°$;
(b) $\tau_{max} = 35.0$ MPa, $\theta_{s1} = -18.43°$

7.3-11 (a) $\sigma_1 = 8,310$ psi, $\theta_{p1} = -12.76°$;
(b) $\tau_{max} = 18,560$ psi, $\theta_{s1} = -57.76°$
7.3-12 (a) $\sigma_1 = 135$ MPa, $\theta_{p1} = 99.46°$;
(b) $\tau_{max} = 111$ MPa, $\theta_{s1} = 54.46°$
7.3-13 (a) $\sigma_1 = 45,800$ psi, $\theta_{p1} = 117.0°$;
(b) $\tau_{max} = 35,800$ psi, $\theta_{s1} = 72.0°$
7.3-14 (a) $\sigma_1 = -17.3$ MPa, $\theta_{p1} = 61.62°$;
(b) $\tau_{max} = 34.7$ MPa, $\theta_{s1} = 16.62°$
7.3-15 (a) $\sigma_1 = -3350$ psi, $\theta_{p1} = 77.68°$;
(b) $\tau_{max} = 9350$ psi, $\theta_{s1} = 32.68°$
7.3-16 (a) $\sigma_1 = 74.5$ MPa, $\theta_{p1} = 10.02°$;
(b) $\tau_{max} = 55.5$ MPa, $\theta_{s1} = -34.98°$
7.3-17 (a) $\sigma_1 = 22.8$ ksi, $\theta_{p1} = 58.23°$;
(b) $\tau_{max} = 27.6$ ksi, $\theta_{s1} = 13.23°$
7.3-18 (a) $\sigma_y = 9.86$ MPa; (b) $\sigma_2 = -84.74$ MPa, $\theta_{p2} = -22.51°$
7.3-19 (a) $\sigma_y = -2460$ psi; (b) $\sigma_2 = -3240$ psi, $\theta_{p2} = 75.00°$
7.4-1 (a) For $\theta = 21.8°$, $\sigma_{x1} = 10,000$ psi, $\tau_{x1y1} = -4,000$ psi; (b) $\tau_{max} = 5,800$ psi, $\theta_{s1} = -45°$
7.4-2 (a) For $\theta = 31°$, $\sigma_{x1} = 50.0$ MPa, $\tau_{x1y1} = -30.0$ MPa; (b) $\tau_{max} = 34.0$ MPa, $\theta_{s1} = -45°$
7.4-3 (a) For $\theta = 26.57°$, $\sigma_{x1} = -4800$ psi, $\tau_{x1y1} = 2400$ psi; (b) $\tau_{max} = 3000$ psi, $\theta_{s1} = 45°$
7.4-4 (a) For $\theta = 20°$, $\sigma_{x1} = -20.6$ MPa, $\tau_{x1y1} = -24.5$ MPa; (b) $\sigma_1 = 32.0$ MPa, $\theta_{p1} = -45°$
7.4-5 (a) For $\theta = 75°$: $\sigma_{x1} = 2250$ psi, $\tau_{x1y1} = -3900$ psi; (b) $\sigma_1 = 4500$ psi, $\theta_{p1} = 45°$
7.4-6 (a) For $\theta = 22.5°$: $\sigma_{x1} = -71.2$ MPa, $\tau_{x1y1} = 21.2$ MPa; (b) $\tau_{max} = 30.0$ MPa, $\theta_{s1} = 45°$
7.4-7 (a) For $\theta = 60°$: $\sigma_{x1} = 250$ psi, $\tau_{x1y1} = -2170$ psi; (b) $\tau_{max} = 2500$ psi, $\theta_{s1} = -45°$
7.4-8 (a) For $\theta = 21.80°$: $\sigma_{x1} = -8.0$ MPa, $\tau_{x1y1} = 20.0$ MPa; (b) $\tau_{max} = 29.0$ MPa, $\theta_{s1} = 45°$
7.4-9 $(\tau_{xy})_{max} = 6400$ psi **7.4-10** $(\tau_{xy})_{max} = 20$ MPa
7.4-11 For $\theta = 45°$: $\sigma_{x1} = 6900$ psi, $\tau_{x1y1} = 4200$ psi
7.4-12 For $\theta = 40°$: $\sigma_{x1} = -20.9$ MPa, $\tau_{x1y1} = -8.6$ MPa
7.4-13 For $\theta = 50°$: $\sigma_{x1} = 8690$ psi, $\tau_{x1y1} = -2000$ psi
7.4-14 For $\theta = 30°$: $\sigma_{x1} = 118.5$ MPa, $\tau_{x1y1} = -15.3$ MPa
7.4-15 For $\theta = 60°$: $\sigma_{x1} = -18,410$ psi, $\tau_{x1y1} = -11,200$ psi
7.4-16 For $\theta = 50°$: $\sigma_{x1} = -18.4$ MPa, $\tau_{x1y1} = -28.1$ MPa
7.4-17 For $\theta = 30°$: $\sigma_{x1} = 527$ psi, $\tau_{x1y1} = -273$ psi
7.4-18 (a) $\sigma_1 = 80.0$ MPa, $\theta_{p1} = 20.56°$;
(b) $\tau_{max} = 36.5$ MPa, $\theta_{s1} = -24.44°$

7.4-19 (a) $\sigma_1 = 3900$ psi, $\theta_{p1} = 60.95°$;
(b) $\tau_{max} = 5300$ psi, $\theta_{s1} = 15.95°$
7.4-20 (a) $\sigma_1 = 86.0$ MPa, $\theta_{p1} = 26.57°$;
(b) $\tau_{max} = 35.0$ MPa, $\theta_{s1} = -18.43°$
7.4-21 (a) $\sigma_1 = 8,310$ psi, $\theta_{p1} = -12.76°$;
(b) $\tau_{max} = 18,560$ psi, $\theta_{s1} = -57.76°$
7.4-22 (a) $\sigma_1 = 135$ MPa, $\theta_{p1} = 99.46°$;
(b) $\tau_{max} = 111$ MPa, $\theta_{s1} = 54.46°$
7.4-23 (a) $\sigma_1 = 45,800$ psi, $\theta_{p1} = 117.0°$;
(b) $\tau_{max} = 35,800$ psi, $\theta_{s1} = 72.0°$
7.4-24 (a) $\sigma_1 = -17.3$ MPa, $\theta_{p1} = 61.62°$;
(b) $\tau_{max} = 34.7$ MPa, $\theta_{s1} = 16.62°$
7.4-25 (a) $\sigma_1 = -3350$ psi, $\theta_{p1} = 77.68°$;
(b) $\tau_{max} = 9350$ psi, $\theta_{s1} = 32.68°$
7.5-1 $\Delta t = -18.2 \times 10^{-6}$ in. (decrease)
7.5-2 $\Delta t = -1860 \times 10^{-6}$ mm (decrease)
7.5-3 (a) $\epsilon_z = -\nu(\epsilon_x + \epsilon_y)/(1 - \nu)$;
(b) $e = (1 - 2\nu)(\epsilon_x + \epsilon_y)/(1 - \nu)$
7.5-4 $\nu = 0.35$, $E = 45$ GPa
7.5-5 $\nu = 1/3$, $E = 30 \times 10^6$ psi
7.5-6 (a) $\gamma_{max} = 1510 \times 10^{-6}$;
(b) $\Delta t = 2970 \times 10^{-6}$ mm (decrease);
(c) $\Delta V = 184$ mm^3 (increase)
7.5-7 (a) $\gamma_{max} = 512 \times 10^{-6}$;
(b) $\Delta t = 57.1 \times 10^{-6}$ in. (decrease);
(c) $\Delta V = 0.00992$ in.^{3} (increase)
7.5-8 $\Delta V = 30.7$ mm^3 (decrease); $U = 2.38$ J
7.5-9 $\Delta V = 0.060$ in.^{3} (decrease); $U = 60$ in.-lb
7.5-10 $\Delta V = 2000$ mm^3 (increase); $U = 58.5$ J
7.5-11 $\Delta V = 0.0481$ in.^{3} (increase); $U = 660$ in.-lb
7.5-12 (a) $\Delta ac = 0.0745$ mm (increase);
(b) $\Delta bd = 0.000560$ mm (decrease); (c) $\Delta t = 0.00381$ mm
(decrease); (d) $\Delta V = 573$ mm^3 (increase) ;
(e) $U = 25.0$ J
7.5-13 (a) $\Delta V_t = 3PL^2(1 - 2\nu)/4Eh$; (b) $\Delta V_c = \Delta V_t$;
(c) Zero
7.6-1 (a) $\tau_{max} = 8000$ psi; (b) $\Delta a = 0.00632$ in.,
$\Delta b = -0.00313$ in., $\Delta c = -0.00100$ in.;
(c) $\Delta V = 0.00883$ in.^{3}; (d) $U = 549$ in.-lb
7.6-2 (a) $\tau_{max} = 7.0$ MPa; (b) $\Delta a = -0.03400$ mm,
$\Delta b = -0.01065$ mm, $\Delta c = -0.00675$ mm;
(c) $\Delta V = -1418$ mm^3; (d) $U = 31.7$ J
7.6-3 (a) $\sigma_x = -6610$ psi, $\sigma_y = \sigma_z = -3420$ psi;
(b) $\tau_{max} = 1600$ psi; (c) $\Delta V = -0.0130$ in.^{3};
(d) $U = 37.2$ in.-lb
7.6-4 (a) $\sigma_x = -56.6$ MPa, $\sigma_y = \sigma_z = -38.9$ MPa;
(b) $\tau_{max} = 8.9$ MPa; (c) $\Delta V = -140$ mm^3; (d) $U = 3.41$ J
7.6-5 $K = 10.0 \times 10^6$ psi   **7.6-6** $K = 5.0$ GPa

7.6-7 (a) $p = \nu F/[A(1 - \nu)]$;
(b) $\delta = FL(1 + \nu)(1 - 2\nu)/[EA(1 - \nu)]$
7.6-8 (a) $p = \nu p_0$; (b) $e = -p_0(1 + \nu)(1 - 2\nu)/E$;
(c) $u = p_0^2 (1 - \nu^2)/2E$
7.6-9 $\Delta d = 0.00104$ in. (decrease); $\Delta V = 0.198$ in.^{3}
(decrease); $U = 438$ in.-lb
7.6-10 (a) $p = 700$ MPa; (b) $K = 175$ GPa;
(c) $U = 2470$ J
7.6-11 $\epsilon_0 = 276 \times 10^{-6}$, $e = 828 \times 10^{-6}$, $u = 4.97$ psi
7.6-12 (a) $d = 2700$ m; (b) Increase $= 0.05403\%$
7.7-1 (a) $\Delta d = 0.00117$ in. (increase);
(b) $\Delta \phi = 128 \times 10^{-6}$ rad (decrease);
(c) $\Delta \psi = 128 \times 10^{-6}$ rad (increase)
7.7-2 (a) $\Delta d = 0.0547$ mm (increase);
(b) $\Delta \phi = 240 \times 10^{-6}$ rad (decrease);
(c) $\Delta \psi = 240 \times 10^{-6}$ rad (increase)
7.7-3 (a) $\Delta d = 0.00458$ in. (increase);
(b) $\Delta \phi = 157 \times 10^{-6}$ rad (decrease);
(c) $\gamma = -314 \times 10^{-6}$ rad (angle *ced* increases)
7.7-4 (a) $\Delta d = 0.168$ mm (increase);
(b) $\Delta \phi = 317 \times 10^{-6}$ rad (decrease);
(c) $\gamma = -634 \times 10^{-6}$ rad (angle *ced* increases)
7.7-5 For $\theta = 50°$: $\epsilon_{x1} = 461 \times 10^{-6}$, $\gamma_{x1y1} = 225 \times 10^{-6}$
7.7-6 For $\theta = 37.5°$: $\epsilon_{x1} = 351 \times 10^{-6}$,
$\gamma_{x1y1} = -490 \times 10^{-6}$
7.7-7 $\epsilon_1 = 554 \times 10^{-6}$, $\theta_{p1} = -22.9°$;
$\gamma_{max} = 488 \times 10^{-6}$
7.7-8 $\epsilon_1 = 172 \times 10^{-6}$, $\theta_{p1} = 163.9°$;
$\gamma_{max} = 674 \times 10^{-6}$
7.7-9 (a) For $\theta = 75°$: $\epsilon_{x1} = 202 \times 10^{-6}$,
$\gamma_{x1y1} = -569 \times 10^{-6}$; (b) $\epsilon_1 = 568 \times 10^{-6}$, $\theta_{p1} = 22.8°$;
$\gamma_{max} = 587 \times 10^{-6}$
7.7-10 (a) For $\theta = 45°$: $\epsilon_{x1} = -385 \times 10^{-6}$,
$\gamma_{x1y1} = 690 \times 10^{-6}$; (b) $\epsilon_1 = -254 \times 10^{-6}$, $\theta_{p1} = 65.7°$;
(c) $\gamma_{max} = 1041 \times 10^{-6}$
7.7-11 $(\tau_{max})_{xy} = 7800$ psi, $(\gamma_{max})_{xy} = 676 \times 10^{-6}$,
$(\gamma_{max})_{xz} = 780 \times 10^{-6}$, $(\gamma_{max})_{yz} = 104 \times 10^{-6}$
7.7-12 $(\tau_{max})_{xy} = 32.4$ MPa, $(\gamma_{max})_{xy} = 1200 \times 10^{-6}$,
$(\gamma_{max})_{xz} = 1600 \times 10^{-6}$, $(\gamma_{max})_{yz} = 399 \times 10^{-6}$
7.7-13 (a) For $\theta = 30°$: $\epsilon_{x1} = -756 \times 10^{-6}$,
$\gamma_{x1y1} = 868 \times 10^{-6}$; (b) $\epsilon_1 = 426 \times 10^{-6}$, $\theta_{p1} = 99.8°$;
(c) $\gamma_{max} = 1342 \times 10^{-6}$
7.7-14 (a) For $\theta = 50°$: $\epsilon_{x1} = -1469 \times 10^{-6}$,
$\gamma_{x1y1} = -717 \times 10^{-6}$; (b) $\epsilon_1 = -732 \times 10^{-6}$,
$\theta_{p1} = 166.0°$; (c) $\gamma_{max} = 911 \times 10^{-6}$
7.7-15 $\epsilon_1 = 551 \times 10^{-6}$, $\theta_{p1} = 12.5°$; $\gamma_{max} = 662 \times 10^{-6}$
7.7-16 $\epsilon_1 = 332 \times 10^{-6}$, $\theta_{p1} = 12.0°$; $\gamma_{max} = 515 \times 10^{-6}$
7.7-17 (a) $P = 5300$ lb, $T = 1390$ lb-in.;
(b) $\gamma_{max} = 222 \times 10^{-6}$, $\tau_{max} = 2580$ psi

7.7-18 $P = 125$ kN, $\alpha = 30°$

7.7-19 $P = 5000$ lb, $\alpha = 38°$

7.7-20 $\epsilon_x = \epsilon_a$, $\epsilon_y = (2\epsilon_b + 2\epsilon_c - \epsilon_a)/3$, $\gamma_{xy} = 2(\epsilon_b - \epsilon_c)/\sqrt{3}$

7.7-21 For $\theta_{p1} = 30°$: $\epsilon_1 = 1550 \times 10^{-6}$, $\epsilon_2 = -250 \times 10^{-6}$, $\sigma_1 = 10,000$ psi, $\sigma_2 = 2,000$ psi

7.7-22 $\sigma_x = 91.6$ MPa

7.7-23 For $\theta = 50°$: $\epsilon_{x1} = 461 \times 10^{-6}$, $\gamma_{x1y1} = 225 \times 10^{-6}$

7.7-24 For $\theta = 37.5°$: $\epsilon_{x1} = 351 \times 10^{-6}$, $\gamma_{x1y1} = -490 \times 10^{-6}$

7.7-25 $\epsilon_1 = 554 \times 10^{-6}$, $\theta_{p1} = -22.9°$; $\gamma_{max} = 488 \times 10^{-6}$

7.7-26 $\epsilon_1 = 172 \times 10^{-6}$, $\theta_{p1} = 163.9°$; $\gamma_{max} = 674 \times 10^{-6}$

7.7-27 For $\theta = 75°$: $\epsilon_{x1} = 202 \times 10^{-6}$, $\gamma_{x1y1} = -569 \times 10^{-6}$; (b) $\epsilon_1 = 568 \times 10^{-6}$, $\theta_{p1} = 22.8°$; (c) $\gamma_{max} = 587 \times 10^{-6}$

7.7-28 (a) For $\theta = 45°$: $\epsilon_{x1} = -385 \times 10^{-6}$, $\gamma_{x1y1} = 690 \times 10^{-6}$; (b) $\epsilon_1 = -254 \times 10^{-6}$, $\theta_{p1} = 65.7°$; (c) $\gamma_{max} = 1041 \times 10^{-6}$

CHAPTER 8

8.2-1 $t_{min} = 3.75$ in. **8.2-2** $t_{min} = 62$ mm

8.2-3 $F = 3530$ lb, $\sigma = 675$ psi

8.2-4 $\sigma_{max} = 3.47$ MPa, $\epsilon_{max} = 0.545$

8.2-5 $\sigma_{max} = 513$ psi, $\epsilon_{max} = 0.564$

8.2-6 $p = 3.90$ MPa

8.2-7 (a) $f = 30.0$ k/in.; (b) $\tau_{max} = 7500$ psi; (c) $\epsilon_{max} = 3.44 \times 10^{-6}$

8.2-8 (a) $f = 6.0$ MN/m; (b) $\tau_{max} = 60.0$ MPa; (c) $\epsilon_{max} = 406 \times 10^{-6}$

8.2-9 $t_{min} = 0.29$ in. **8.2-10** $t_{min} = 6.0$ mm

8.2-11 $d = 84.6$ ft **8.3-1** $t_{min} = 0.27$ in.

8.3-2 (a) $h = 16.3$ m; (b) Zero **8.3-3** $n = 2.25$

8.3-4 $F = 3\pi pr^2$ **8.3-5** $p = 50$ psi

8.3-6 $\epsilon_{max} = 80 \times 10^{-6}$ **8.3-7** $t_{min} = 0.10$ in.

8.3-8 $t_{min} = 3.0$ mm

8.3-9 (a) $h = 25$ ft; (b) $\sigma = 125$ psi

8.3-10 (a) $\sigma_h = 27.0$ MPa; (b) $\sigma_c = 54.0$ MPa; (c) $\sigma_w = 27.0$ MPa; (d) $\tau_h = 13.5$ MPa; (e) $\tau_c = 27.0$ MPa

8.3-11 (a) $t_{min} = 0.30$ in.; (b) $t_{min} = 0.15$ in.

8.3-12 (a) $\sigma_1 = 80$ MPa, $\sigma_2 = 40$ MPa; (b) $\tau_1 = 20$ MPa, $\tau_2 = 40$ MPa; (c) $\epsilon_1 = 340 \times 10^{-6}$, $\epsilon_2 = 80 \times 10^{-6}$; (d) For $\theta = 30°$: $\theta_{x1} = 50$ MPa, $\tau_{x1y1} = 17.3$ MPa

8.3-13 (a) $\sigma_1 = 6000$ psi, $\sigma_2 = 3000$ psi; (b) $\tau_1 = 1500$ psi, $\tau_2 = 3000$ psi; (c) $\epsilon_1 = 170 \times 10^{-6}$, $\epsilon_2 = 40 \times 10^{-6}$; (d) For $\theta = 25°$: $\sigma_{x1} = 3540$ psi, $\tau_{x1y1} = 1150$ psi

8.3-14 $p = 3200$ kPa **8.3-15** $p = 120$ psi

8.4-1 $\sigma_1 = 99$ psi, $\sigma_2 = -2260$ psi, $\tau_{max} = 1180$ psi

8.4-2 $\sigma_1 = 46.0$ MPa, $\sigma_2 = -1.0$ MPa, $\tau_{max} = 23.5$ MPa

8.4-3 (b) $\sigma_1 = 30$ psi, $\sigma_2 = -663$ psi, $\tau_{max} = 346$ psi

8.4-4 $P = 22.0$ kN **8.4-5** $P = 2600$ lb

8.4-6 (b) $\sigma_1 = 4.5$ MPa, $\sigma_2 = -76.1$ MPa, $\tau_{max} = 40.3$ MPa

8.4-7 (b) $\sigma_1 = 14,100$ psi, $\sigma_2 = -220$ psi, $\tau_{max} = 7,160$ psi

8.4-8 (b) $\sigma_1 = 1,620$ psi, $\sigma_2 = -11,630$ psi, $\tau_{max} = 6,620$ psi

8.4-9 (b) $\sigma_1 = 930$ psi, $\sigma_2 = -4570$ psi, $\tau_{max} = 2750$ psi

8.4-10 At point B: $\sigma_1 = 20.2$ MPa, $\sigma_2 = -0.3$ MPa, $\tau_{max} = 10.3$ MPa

8.4-11 Top of beam: $\sigma_1 = 0$, $\sigma_2 = -15,000$ psi, $\tau_{max} = 7500$ psi

8.4-12 Top of beam: $\sigma_1 = 0$, $\sigma_2 = -225$ MPa, $\tau_{max} = 112$ MPa

8.4-13 $\tau_{NA} = 100$ psi **8.4-14** $\tau_{NA} = 1.0$ MPa

8.5-1 $\sigma_t = 5600$ psi, $\sigma_c = -5940$ psi, $\tau_{max} = 2970$ psi

8.5-2 $\sigma_t = 32.0$ MPa, $\sigma_c = -73.7$ MPa, $\tau_{max} = 52.8$ MPa

8.5-3 $\sigma_t = 2860$ psi, $\sigma_c = -6680$ psi, $\tau_{max} = 4770$ psi

8.5-4 $P = 815$ kN **8.5-5** $d = 1.47$ in.

8.5-6 $d = 59.7$ mm **8.5-7** $t = 0.125$ in.

8.5-8 $p_{max} = 9.6$ MPa

8.5-9 $r = 1.65$ in., $t = 0.165$ in.

8.5-10 $\phi_{max} = 0.288$ rad

8.5-11 $\sigma_t = 39,950$ psi, $\sigma_c = -2,230$ psi, $\tau_{max} = 21,090$ psi

8.5-12 $\sigma_t = 21.6$ MPa, $\sigma_c = -9.4$ MPa, $\tau_{max} = 15.5$ MPa

8.5-13 (a) $\sigma_{max} = 8080$ psi, $\tau_{max} = 2680$ psi; (b) $T = 109$ k-ft

8.5-14 (a) $\sigma_t = 32$ MPa, $\sigma_c = -184$ MPa, $\tau_{max} = 108$ MPa

8.5-15 (a) $\sigma_{max} = 10,870$ psi, $\tau_{max} = 5,470$ psi; (b) $P_{allow} = 395$ lb

8.5-16 $\tau_A = 86.8$ MPa, $\tau_B = 22.6$ MPa, $\tau_C = 26.4$ MPa

8.5-17 (a) $\sigma_x = 0$, $\sigma_y = 4156$ psi, $\tau_{xy} = 191$ psi; (b) $\sigma_t = 4160$ psi, $\sigma_c = -10$ psi, $\tau_{max} = 2090$ psi

8.5-18 $\sigma_t = 6.1$ MPa, $\sigma_c = -65.1$ MPa, $\tau_{max} = 35.6$ MPa

8.5-19 $\sigma_t = 10,680$ psi; No compressive stresses; $\tau_{max} = 5,340$ psi

8.5-20 $P = 29.4$ kN

CHAPTER 9

9.3-1 $q = q_0 x/L$; Triangularly distributed load of maximum intensity q_0, acting downward

9.3-2 $q = 4q_0x(L - x)/L^2$; Symmetrical, parabolically distributed load of maximum intensity q_0, acting downward

9.3-3 $P_A = 2PL/3d$, $P_B = PL/3d$ **9.3-4** $h = 96$ mm

9.3-5 (a) $\delta_{max} = 0.177$ in., $\theta = 0.193°$; (b) $\rho = 1385$ ft

9.3-6 $E_g = 80.3$ GPa   **9.3-7** $L = 10$ ft

9.3-8 $\delta/L = 1/320$

9.3-9 (a) $\delta_{max} = 0.680$ in.; (b) $\rho = 90.0$ ft

9.3-10 If $a/L = 3/4$, $\delta_C/\delta_{max} = 0.9839$; Deflection at the midpoint is within 3% of the maximum deflection

9.3-14 $v = -mx^2(3L - x)/6EI$, $\delta_B = mL^3/3EI$, $\theta_B = mL^2/2EI$

9.3-15 $v = -qx(8L^3 - 4Lx^2 + x^3)/24EI$, $\delta_B = 5qL^4/24EI$

9.3-16 $v = -M_0x(6aL - 3a^2 - 2L^2 - x^2)/6LEI$ for $0 \leq x \leq a$; $v = -M_0(3a^2L - 3a^2x - 2L^2x + 3Lx^2 - x^3)/6LEI$ for $a \leq x \leq L$; $\delta_0 = M_0ab(2a - L)/3LEI$

9.3-18 $v = -qLx^2(9L - 4x)/48EI$ for $0 \leq x \leq L/2$; $v = -q(16x^4 - 64Lx^3 + 96L^2x^2 - 8L^3x + L^4)/384EI$ for $L/2 \leq x \leq L$; $\delta_B = 41qL^4/384EI$; $\delta_C = 7qL^4/192EI$

9.4-3 $v = -M_0x(L - x)^2/2LEI$; $\delta_{max} = 2M_0L^2/27EI$

9.4-4 $v = -qx(5L^3 - 8Lx^2 + 4x^3)/96EI$; $\theta_A = 5qL^3/96EI$

9.4-6 $v = -q_0x^2(45L^4 - 40L^3x + 15L^2x^2 - x^4)/360L^2EI$; $\delta_B = 19q_0L^4/360EI$; $\theta_B = q_0L^3/15EI$

9.4-7 $v = -q_0x(3L^5 - 5L^3x^2 + 3Lx^4 - x^5)/90L^2EI$; $\delta_{max} = 61q_0L^4/5760EI$

9.4-9 $v = qLx(L^2 - x^2)/48EI$ for $0 \leq x \leq L$; $v = -q(L - x)(7L^3 - 17L^2x + 10Lx^2 - 2x^3)/48EI$ for $L \leq x \leq 3L/2$; $\delta_C = 11qL^4/384EI$; $\theta_C = qL^3/16EI$

9.4-10 $v = -q_0Lx(37L^2 - 40x^2)/5760EI$ for $0 \leq x \leq L/2$; $v = -q_0[L^2x(37L^2 - 40x^2) + 3(2x - L)^5]/5760LEI$ for $L/2 \leq x \leq L$; $\theta_A = 37q_0L^3/5760EI$; $\theta_B = 53q_0L^3/5760EI$; $\delta_C = 3q_0L^4/1280EI$

9.5-1 $\theta_B = 7PL^2/9EI$; $\delta_B = 5PL^3/9EI$

9.5-2 (a) $\delta_1 = 11PL^3/144EI$; (b) $\delta_2 = 25PL^3/384EI$; (c) $\delta_1/\delta_2 = 88/75 = 1.173$

9.5-3 (a) $a/L = 2/3$; (b) $a/L = 1/2$

9.5-4 $\delta_C = 25$ mm   **9.5-5** $y = Px^2(L - x)^2/3LEI$

9.5-6 $\theta_B = 7qL^3/162EI$; $\delta_B = 23qL^4/648EI$

9.5-7 $\delta_C = 0.139$ in., $\delta_B = 0.448$ in.

9.5-8 (a) $\delta_A = PL^2(10L - 9a)/324EI$ (positive upward); (b) Upward when $a/L < 10/9$, downward when $a/L > 10/9$

9.5-9 (a) $\delta_C = PH^2(L + H)/3EI$; (b) $\delta_{max} = PHL^2/9\sqrt{3}EI$

9.5-10 $\delta_C = 2.1$ mm

9.5-11 $\theta_B = q_0L^3/10EI$, $\delta_B = 13q_0L^4/180EI$

9.5-12 $\theta_A = q(L^3 - 6La^2 + 4a^3)/24EI$; $\delta_{max} = q(5L^4 - 24L^2a^2 + 16a^4)/384EI$

9.5-13 (a) $P/Q = 3a/L$; (b) $P/Q = 16a(L + a)/3L^2$

9.5-14 $\delta = 19WL^3/31,104EI$ **9.5-15** $k = 28$ lb/in.

9.5-16 $P = 6.4$ kN, $Q = 3.2$ kN

9.5-17 $\delta = 5Pb^3/2EI$ **9.5-18** $\delta_E = 5Pb^3/3EI$

9.5-19 $\delta_C = 0.22$ in.   **9.5-20** $q = 16cEI/7L^4$

9.5-21 (a) $b/L = 0.4030$; (b) $\delta_C = 0.002870qL^4/EI$

9.5-22 $(\delta_C)_{max} = 39PL^3/1024EI$

9.5-23 $\delta_h = Pcb^2/2EI$, $\delta_v = Pc^2(c + 3b)/3EI$

9.5-24 $\delta = PL^2(2L + 3a)/3EI$

9.5-25 $\alpha = 22.5°, 112.5°, -67.5°,$ or $-157.5°$

9.6-4 $\theta_B = 7qL^3/162EI$, $\delta_B = 23qL^4/648EI$

9.6-5 $\delta_B = 0.443$ in., $\delta_C = 0.137$ in.

9.6-6 $\delta_B = 11.8$ mm, $\delta_C = 4.10$ mm

9.6-8 $P = 64$ kN

9.6-9 $\theta_A = M_0L/6EI$, $\theta_B = M_0L/3EI$, $\delta = M_0L^2/16EI$

9.6-10 $\theta_A = Pa(L - a)(L - 2a)/6LEI$, $\delta_1 = Pa^2(L - 2a)^2/6LEI$, $\delta_2 = 0$

9.6-11 $\theta_A = M_0L/6EI$, $\theta_B = 0$, $\delta = M_0L^2/27EI$ (downward)

9.6-12 $\theta_A = 0.00586$ rad, $\theta_B = 0.00716$ rad, $\delta_1 = 20.6$ mm, $\delta_2 = 15.7$ mm

9.7-1 (a) $\delta_B = PL^3(1 + 7I_1/I_2)/24EI_1$; (b) $r = (1 + 7I_1/I_2)/8$

9.7-2 (a) $\delta_B = qL^4(1 + 15I_1/I_2)/128EI_1$; (b) $r = (1 + 15I_1/I_2)/16$

9.7-3 $v = -qx(21L^3 - 64Lx^2 + 32x^3)/768EI$ for $0 \leq x \leq L/4$; $v = -q(13L^4 + 256L^3x - 512Lx^3 + 256x^4)/12,288EI$ for $L/4 \leq x \leq L/2$; $\theta_A = 7qL^3/256EI$; $\delta_{max} = 31qL^4/4096EI$

9.7-4 $\theta_A = 8PL^2/243EI$, $\delta_B = 8PL^3/729EI$, $\delta_{max} = 0.01363PL^3/EI$

9.7-5 $v = -2Px(19L^2 - 27x^2)/729EI$ for $0 \leq x \leq L/3$; $v = P(13L^3 - 175L^2x + 243Lx^2 - 81x^3)/1458EI$ for $L/3 \leq x \leq L$; $\theta_A = 38PL^2/729EI$, $\theta_C = 34PL^2/729EI$, $\delta_B = 32PL^3/2187EI$

9.7-6 $v = \dfrac{PL^3}{EI_A}\left[\dfrac{L}{2(L + x)} - \dfrac{3x}{8L} + \dfrac{1}{8} + \ln\left(\dfrac{L + x}{2L}\right)\right]$;

$\delta_A = \dfrac{PL^3}{8EI_A}(8 \ln 2 - 5)$

9.7-7 $v = \dfrac{PL^3}{24EI_A}\left[7 - \dfrac{4L(2L + 3x)}{(L + x)^2} - \dfrac{2x}{L}\right]$; $\delta_A = \dfrac{PL^3}{24EI_A}$

9.7-8 $v = \dfrac{8PL^3}{EI_A}\left[\dfrac{L}{2L + x} - \dfrac{2x}{9L} - \dfrac{1}{9} + \ln\left(\dfrac{2L + x}{3L}\right)\right]$;

$\delta_A = \dfrac{8PL^3}{EI_A}\left(\ln\dfrac{3}{2} - \dfrac{7}{18}\right)$

9.7-9 (a) $v' = -\dfrac{qL^3}{16EI_A}\left[1 - \dfrac{8Lx^2}{(L + x)^3}\right]$ for $0 \leq x \leq L$,

$v = -\dfrac{qL^4}{2EI_A}\left[\dfrac{(9L^2 + 14Lx + x^2)x}{8L(L + x)^2} - \ln\left(1 + \dfrac{x}{L}\right)\right]$ for

$0 \leq x \leq L$; (b) $\theta_A = \dfrac{qL^3}{16EI_A}$, $\delta_C = \dfrac{qL^4(3 - 4 \ln 2)}{8EI_A}$

9.8-1 (a) and (b) $U = P^2L^3/96EI$; (c) $\delta = PL^3/48EI$

9.8-2 $U = 4bhL\sigma_{max}^2/45E$

9.8-3 (a) and (b) $U = q^2L^5/40EI$; (c) $U = 179$ in.-lb

9.8-4 (a) $U = 32EI\delta^2/L^3$; (b) $U = \pi^4EI\delta^2/4L^3$

9.8-5 (a) $U = P^2a^2(L + a)/6EI$; (b) $\delta_C = Pa^2(L + a)/3EI$;
(c) $U = 241$ in.-lb, $\delta_C = 0.133$ in.

9.8-6 $U = P^2L^3/96EI + PM_0L^2/16EI + M_0^2L/6EI$

9.8-7 $\delta_B = 2PL^3/3EI + 8\sqrt{2}PL/EA$

9.9-2 $\delta_D = Pa^2b^2/3LEI$ **9.9-3** $\delta_C = Pa^2(L + a)/3EI$

9.9-6 $\delta_C = L^3(2P_1 + 5P_2)/48EI$,
$\delta_B = L^3(5P_1 + 16P_2)/48EI$

9.9-7 $\theta_A = 7qL^3/48EI$

9.9-8 $\delta_C = Pb^2(b + 3h)/3EI$, $\theta_C = Pb(b + 2h)/2EI$

9.9-9 $\theta_A = M_A(L + 3a)/3EI$, $\delta_A = M_Aa(2L + 3a)/6EI$

9.9-10 $\delta_C = 31qL^4/4096EI$

9.9-11 $\delta_C = Pa^2(L + a)/3EI + P(L + a)^2/kL^2$

9.9-12 $\delta_D = 37qL^4/6144EI$ (upward)

9.10-1 $\sigma_{max} = \sigma_{st}[1 + (1 + 2h/\delta_{st})^{1/2}]$

9.10-2 $\sigma_{max} = \sqrt{18WEh/AL}$

9.10-3 $\delta_{max} = 0.302$ in., $\sigma_{max} = 21,700$ psi

9.10-4 $d = 281$ mm **9.10-5** W 14×53

9.10-6 $h = 360$ mm **9.10-7** $R = \sqrt{3EII_m\omega^2 / L^3}$

9.11-1 $v = -\alpha(T_2 - T_1)(x)(L - x)/2h$ (positive upward);
$\theta_A = \alpha L(T_2 - T_1)/2h$ (positive clockwise);
$\delta_{max} = \alpha L^2(T_2 - T_1)/8h$ (positive downward)

9.11-2 $v = \alpha(T_2 - T_1)(x^2)/2h$ (positive upward);
$\theta_B = \alpha L(T_2 - T_1)/h$ (positive counterclockwise);
$\delta_B = \alpha L^2(T_2 - T_1)/2h$ (positive upward)

9.11-3 $v = \alpha(T_2 - T_1)(x^2 - Lx)/2h$ (positive upward);
$\theta_C = \alpha(T_2 - T_1)(L + 2a)/2h$ (positive counterclockwise);
$\delta_C = \alpha(T_2 - T_1)(L + a)(a)/2h$ (positive upward)

9.11-4 $\delta_{max} = \alpha T_0L^3/9\sqrt{3}h$ (positive downward)

CHAPTER 10

10.3-1 $R_A = -R_B = 3M_0/2L$, $M_A = M_0/2$,
$v = -M_0x^2(L - x)/4LEI$

10.3-2 $R_A = R_B = qL/2$, $M_A = M_B = qL^2/12$,
$v = -qx^2(L - x)^2/24EI$

10.3-3 $R_A = R_B = 3EI\delta_B/L^3$, $M_A = 3EI\delta_B/L^2$,
$v = -\delta_Bx^2(3L - x)/2L^3$

10.3-4 $\delta_B = 3qL^4/(24EI + 8kL^3)$

10.3-5 $R_A = 2q_0L/5$, $R_B = q_0L/10$, $M_A = q_0L^2/15$,
$v = -q_0x^2(4L^3 - 8L^2x + 5Lx^2 - x^3)/120LEI$

10.3-6 $R_A = 61q_0L/120$, $R_B = 19q_0L/120$,
$M_A = 11q_0L^2/120$,
$v = -q_0x^2(L - x)(33L^3 - 28L^2x + 2Lx^2 + 2x^3)/720L^2EI$

10.3-7 $R_A = R_B = q_0L/\pi$, $M_A = M_B = 2q_0L^2/\pi^3$,
$v = -q_0L^2(L^2 \sin \pi x/L - \pi Lx + \pi x^2)/\pi^4EI$

10.3-8 $R_A = 7q_0L/20$, $R_B = 3q_0L/20$,
$M_A = q_0L^2/20$, $M_B = q_0L^2/30$,
$v = -q_0x^2(L - x)^2(3L - x)/120LEI$

10.3-9 $R_A = -R_B = 3M_0/2L$, $M_A = -M_B = M_0/4$,
$v = -M_0x^2(L - 2x)/8LEI$ for $0 \le x \le L/2$

10.3-10 $R_A = 11P/16$, $R_B = 5P/16$, $M_A = 3PL/16$;
For $0 \le x \le L/2$, $v = -Px^2(9L - 11x)/96EI$;
For $L/2 \le x \le L$,
$v = -P(L - x)(-2L^2 + 10Lx - 5x^2)/96EI$

10.4-1 $R_A = Pb(3L^2 - b^2)/2L^3$, $R_B = Pa^2(3L - a)/2L^3$,
$M_A = Pab(L + b)/2L^2$

10.4-2 $R_A = 57qL/128$, $R_B = 7qL/128$,
$M_A = 9qL^2/128$

10.4-3 $R_A = 3Pa/2L$, $R_B = P(2L + 3a)/2L$, $M_A = Pa/2$

10.4-4 $R_A = R_B = q_0L/4$, $M_A = M_B = 5q_0L^2/96$

10.4-5 $t_{AB}/t_{CD} = L_{AB}/L_{CD}$

10.4-6 $R_A = qL/8$, $R_B = 33qL/16$, $R_C = 13qL/16$

10.4-7 $R_A = 1100$ lb (downward), $R_B = 2800$ lb
(upward), $M_A = 30,000$ lb-in. (clockwise)

10.4-8 $R_B = 7.11$ kN **10.4-9** $T = 398$ lb

10.4-10 $R_A = 31qL/48$, $R_B = 17qL/48$, $M_A = 7qL^2/48$

10.4-11 (a) $R_A = -23P/17$, $R_D = R_E = 20P/17$,
$M_A = 3PL/17$; (b) $M_{max} = -PL/2$

10.4-12 $R_A = R_D = 2qL/5$, $R_B = R_C = 11qL/10$

10.4-13 $q_0 = 300$ lb/in.; $(M_B)_0 = 240,000$ lb-in.;
$M_B = 800q$ for $q < q_0$; $M_B = 300,000 - 200q$ for $q > q_0$

10.4-14 $M_A = M_0b(3a - L)/L^2$, $M_B = -M_0a(3b - L)/L^2$,
$R_A = -R_B = 6M_0ab/L^3$

10.4-15 $\sigma = 509$ psi

10.4-16 $(M_{AB})_{max} = 121qL^2/2048 = 6.05$ kN·m,
$(M_{CD})_{max} = 5qL^2/64 = 8.0$ kN·m

10.4-17 $F = 3,160$ lb, $M_{AB} = 18,960$ lb-ft,
$M_{DE} = 7,320$ lb-ft

10.4-18 $k = 48EI(6 + 5\sqrt{2})/7L^3 = 89.63EI/L^3$

10.4-19 $V_A = V_C = 3P/32$, $H_A = P$, $M_A = 13PL/32$,
$M_{max} = 13PL/32$

10.4-20 $H_A = 13P/32$, $V_A = 3P/32$, $H_C = 19P/32$,
$V_C = 3P/32$, $M_{max} = 13PL/64$

10.4-21 $R_A = R_B = 3000$ lb, $R_C = 0$

10.4-22 $R_A = 11.5$ kN, $R_B = 17.5$ kN, $R_C = -2.0$ kN

10.4-23 (a) $M_A = M_B = qb(3L^2 - b^2)/24L$;
(c) For $a = b = L/3$, $M_{pos} = 19qL^2/648$

10.4-24 (a) $d_2/d_1 = \sqrt[4]{8} = 1.682$;
(b) $M_{max} = qL^2(3 - 2\sqrt{2})/2 = 0.08579qL^2$;
(c) Point C is below points A and B by the amount
0.01307 qL^4/EI

10.4-25 (a) $F = 8.62$ k; (b) $\sigma_b = 5320$ psi, $\sigma_t = 6550$ psi

10.4-26 $M_{max} = 19q_0L^2/256$, $\sigma_{max} = 13.4$ MPa,
$\delta_{max} = 19q_0L^4/7680EI = 0.00891$ mm

10.4-27 $a = L/\sqrt{6}$, $M_{max} = qL^2/12$

10.5-1 $R_1 = R_4 = 7P/20$, $R_2 = R_3 = 23P/20$, $M_{max} = 7PL/40$

10.5-2 At $x_1 = 9L/20$, $M_{max} = 81qL^2/800$

10.5-3 $R_1 = R_4 = 31P/135$, $R_2 = R_3 = 104P/135$, $M_{max} = 166PL/405$

10.5-4 $R_1 = 17.63$ kN, $R_2 = 24.44$ kN, $R_3 = 5.93$ kN, $M_{max} = 15.54$ kN·m

10.5-5 (a) $R_1 = R_4 = 10.68$ k, $R_2 = R_3 = -3.48$ k; (b) $M_{pos} = 5.88$ k-ft

10.5-6 $R_1 = 13qL/30$, $R_2 = 13qL/20$, $R_3 = -qL/10$, $R_4 = qL/60$, $M_{max} = 169qL^2/1800$

10.5-7 (a) $R_1 = -3qL/28$, $R_2 = 19qL/28$, $R_3 = 3qL/7$, $M_1 = qL^2/28$; (b) $M_{max} = 9qL^2/98$

10.5-8 (a) $R_1 = qL/76$, $R_2 = -3qL/38$, $R_3 = 43qL/76$; (b) $M_{max} = 11qL^2/152$

10.5-9 (a) $R_1 = 11qL/28$, $R_2 = 8qL/7$, $R_3 = 13qL/14$; (b) $M_{max} = M_2 = -3qL^2/28$

10.5-10 (a) $R_1 = 23qL/60$, $R_2 = 6qL/5$, $R_3 = 9qL/20$, $R_4 = -qL/30$; (b) $M_{max} = M_2 = -7qL^2/60$

10.5-11 $M_{pos} = 15,129qL^2/161,312$; $M_{neg} = -19qL^2/284$

10.5-12 $M_1 = -3840$ N·m, $M_2 = -10,240$ N·m

10.5-13 $M_1 = -1150$ lb-ft, $M_2 = -1540$ lb-ft

10.5-14 $M_1 = -15$ kN·m, $M_2 = -25.52$ kN·m, $M_3 = -36.48$ kN·m

10.5-15 $M_1 = -5qL^2/56$, $M_2 = -qL^2/14$, $M_3 = -qL^2/8$, $M_{pos} = 281qL^2/6272$, $M_{neg} = -qL^2/8$

10.6-1 $S = 48EIAH\alpha(\Delta T)/(AL^3 + 48IH)$

10.6-2 $R_A = -R_B = 3\alpha EI(T_2 - T_1)/2hL$, $M_A = 3\alpha EI(T_2 - T_1)/2h$

10.6-3 $R_A = -R_B = 3\alpha EI(T_2 - T_1)/2hL$, $M_A = 3\alpha EI(T_2 - T_1)/2h$

10.6-4 $R_A = -3\alpha EI(T_2 - T_1)/2hL$, $R_B = -3R_A$, $R_C = 2R_A$

10.6-5 $R_A = -3\alpha EI(T_2 - T_1)/2hL$, $R_B = -3R_A$, $R_C = 2R_A$

10.7-1 (a) $H = \pi^2 EA\delta^2/4L^2$, $\sigma_t = \pi^2 E\delta^2/4L^2$; (b) $\sigma_t = 617$, 154, and 69 psi

10.7-2 (a) $\lambda = 17q^2L^7/40,320E^2I^2$; (b) $\sigma_b = qhL^2/16I$; (c) $\sigma_t = 17q^2L^6/40,320EI^2$; (d) $\lambda = 0.01112$ mm, $\sigma_b = 117.2$ MPa, $\sigma_t = 0.741$ MPa

CHAPTER 11

11.2-1 $P_{cr} = \beta_R/L$ **11.2-2** $P_{cr} = 6\beta_R/L$

11.2-3 $P_{cr} = \beta_R/L$

11.2-4 $P_{cr} = (3 - \sqrt{5})\beta_R/L = 0.764\beta_R/L$

11.2-5 $P_{cr} = 3\beta_R/L$

11.3-1 (a) $P_{cr} = 453$ k; (b) $P_{cr} = 152$ k

11.3-2 (a) $P_{cr} = 779$ k; (b) $P_{cr} = 265$ k

11.3-3 (a) $P_{cr} = 650$ k; (b) $P_{cr} = 140$ k

11.3-4 $Q_{allow} = 52.9$ kN

11.3-5 $Q_{allow} = 12.3$ k   **11.3-6** $Q_{cr} = 3\pi^2 EI/4L^2$

11.3-7 $\Delta T = \pi^2 I/\alpha AL^2$ **11.3-8** $h/b = 2$

11.3-9 (a) $P_{cr} = 3\pi^3 Er^4/4L^2$; (b) $P_{cr} = 11\pi^3 Er^4/4L^2$

11.3-10 $P_1 : P_2 : P_3 = 1 : 1.047 : 1.209$

11.3-11 $P_{allow} = 136$ k   **11.3-12** $F_{allow} = 110$ kN

11.3-13 $W_{max} = 29.8$ k   **11.3-14** $t_{min} = 4.36$ mm

11.3-15 $P_{cr} = 123$ k

11.3-16 $W_{cr} = 203$ kN   **11.3-17** $H_{min} = 28.6$ in.

11.3-18 (a) $\theta = \arctan(\cot^2 \beta)$; (b) $\theta = 18.43°$

11.3-19 $\theta = \arctan 0.5 = 26.57°$

11.4-1 $P_{cr} = 213$ k, 53 k, 436 k, 852 k

11.4-2 $P_{cr} = 39.8$ kN, 10.0 kN, 81.5 kN, 159 kN

11.4-3 $P_{allow} = 245$ k, 61 k, 501 k, 979 k

11.4-4 $P_{allow} = 170$ k, 42 k, 347 k, 678 k

11.4-5 $P_{cr} = 197$ k   **11.4-6** $T_{allow} = 18.1$ kN

11.4-7 (a) $Q_{cr} = 7440$ lb; (b) $Q_{cr} = 8150$ lb, $a = 9.13$ in.

11.4-8 $P_{cr} = 298$ kN, 583 kN, 36 kN, 146 kN

11.4-9 $t_{min} = 0.674$ in.

11.4-10 $P_{cr} = 4\pi^2 EI/L^2$, $v = \delta(1 - \cos 2\pi x/L)/2$

11.4-11 (b) $P_{cr} = 13.89EI/L^2$

11.4-12 (b) $P_{cr} = 4.070EI/L^2$

11.5-1 For $P = 0.3P_{cr}$: $M/Pe = 1.162(\sin 1.721x/L) + \cos 1.721x/L$

11.5-2 $\delta = 14.3$ mm when $P = 1000$ kN; $P = 884$ kN when $\delta = 10$ mm

11.5-3 $\delta = 0.458$ in. when $P = 200$ k; $P = 190$ k when $\delta = 0.4$ in.

11.5-4 $\delta = 8.87$ mm, $M_{max} = 2.03$ kN·m

11.5-5 $\delta = 0.112$ in., $M_{max} = 1710$ lb-in.

11.5-6 $P_{allow} = 11,300$ lb   **11.5-7** $L_{max} = 12.5$ ft

11.5-8 $L_{max} = 10.2$ ft

11.5-9 $\delta = e(\sec kL - 1)$ and $M_{max} = Pe \sec kL$

11.5-10 $L_{max} = 2.21$ m   **11.5-11** $L_{max} = 10.9$ ft

11.5-12 $T_{max} = 8.29$ kN   **11.5-13** $P = 6.73$ k

11.5-14 $P = 9.97$ k

11.5-15 (a) $q_0 = 2230$ lb/ft; (b) $M_{max} = 37.7$ k-in.

11.6-1 (a) $\sigma_{max} = 17.3$ ksi; (b) $L_{max} = 46.2$ in.

11.6-2 $P_{allow} = 37.2$ kN   **11.6-3** $b_{min} = 4.10$ in.

11.6-4 (a) $\sigma_{max} = 38.8$ MPa; (b) $L_{max} = 5.03$ m

11.6-5 (a) $\sigma_{max} = 9.65$ ksi; (b) $P_{allow} = 3.59$ k

11.6-6 $d_2 = 131$ mm

11.6-7 (a) $\sigma_{max} = 18.9$ ksi when $e = 2$ in.; (b) $\sigma_{max} = 40.6$ ksi when $P = 10$ k

11.6-8 (a) $\sigma_{max} = 10.9$ ksi; (b) $P_{allow} = 160$ k

11.6-9 (a) $\sigma_{max} = 14.8$ ksi; (b) $L_{max} = 12.6$ ft

11.6-10 (a) $\sigma_{max} = 9.60$ ksi; (b) $P_{allow} = 53.6$ k

11.6-11 (a) $\sigma_{max} = 6.85$ ksi; (b) $n = 2.35$

11.6-12 $\sigma_c = 2.38$ ksi; $\sigma_t = 2.37$ ksi
11.6-13 $\sigma_c = 3.50$ ksi; $\sigma_t = 4.61$ ksi
11.6-14 (a) $\sigma_{max} = 13.4$ ksi; (b) $n = 2.61$
11.6-15 (a) $\sigma_{max} = 17.3$ ksi; (b) $P_2 = 78.4$ k
11.6-16 (a) $\sigma_{max} = 17.6$ ksi; (b) $n = 1.89$
11.6-17 (a) $\sigma_{max} = 16.7$ ksi; (b) $P_2 = 34.3$ k
11.9-1 $P_{allow} = 141$ k, 109 k, 76 k, 56 k
11.9-2 $P_{allow} = 109$ k, 88 k, 67 k, 53 k
11.9-3 $P_{allow} = 298$ k, 243 k, 181 k, 133 k
11.9-4 W 10×45 **11.9-5** W 12×87
11.9-6 W 14×82
11.9-7 $P_{allow} = 95$ k, 75 k, 51 k, 32 k
11.9-8 $P_{allow} = 235$ kN, 211 kN, 186 kN, 163 kN
11.9-9 $P_{allow} = 54$ k, 43 k, 29 k, 19 k
11.9-10 $L_{max} = 3.59$ m **11.9-11** $L_{max} = 5.23$ ft
11.9-12 $L_{max} = 2.55$ m
11.9-13 (a) $H_{max} = 25.8$ k; (b) $L_{max} = 16.1$ ft
11.9-14 $P_{allow} = 1260$ k
11.9-15 (a) $L_{max} = 19.4$ ft; (b) $L_{max} = 11.8$ ft
11.9-16 (a) $L_{max} = 21.0$ ft; (b) $L_{max} = 15.6$ ft
11.9-17 $d = 4.89$ in. **11.9-18** $d = 99$ mm
11.9-19 $d = 5.23$ in. **11.9-20** $d = 194$ mm
11.9-21 $P_{allow} = 142$ k, 122 k, 83 k, 58 k
11.9-22 $P_{allow} = 312$ kN, 242 kN, 124 kN, 70 kN
11.9-23 $P_{allow} = 18.1$ k, 14.7 k, 8.3 k, 5.3 k
11.9-24 $P_{allow} = 96$ kN, 84 kN, 61 kN, 42 kN
11.9-25 (a) $L_{max} = 25.2$ in.; (b) $d_{min} = 2.12$ in.
11.9-26 (a) $L_{max} = 457$ mm; (b) $d_{min} = 43.1$ mm
11.9-27 (a) $L_{max} = 14.8$ in.; (b) $d_{min} = 1.12$ in.
11.9-28 (a) $L_{max} = 469$ mm; (b) $d_{min} = 33.4$ mm
11.9-29 $P_{allow} = 38.5$ k, 31.7 k, 15.7 k, 8.8 k
11.9-30 $P_{allow} = 210$ kN, 186 kN, 135 kN, 86 kN
11.9-31 $P_{allow} = 24.5$ k, 21.0 k, 13.2 k, 8.4 k
11.9-32 $P_{allow} = 353$ kN, 268 kN, 141 kN, 85 kN
11.9-33 (a) $L_{max} = 7.97$ ft; (b) $b_{min} = 6.38$ in.
11.9-34 (a) $L_{max} = 2.80$ m; (b) $b_{min} = 173$ mm
11.9-35 (a) $L_{max} = 7.61$ ft; (b) $b_{min} = 4.00$ in.
11.9-36 (a) $L_{max} = 2.08$ m; (b) $b_{min} = 158$ mm

CHAPTER 12

12.3-2 $\bar{x} = \bar{y} = 5a/12$ **12.3-3** $\bar{y} = 1.1$ in.
12.3-4 $2c^2 = ab$ **12.3-5** $\bar{y} = 13.9$ in.
12.3-6 $\bar{y} = 52.5$ mm **12.3-7** $\bar{x} = 0.99$ in., $\bar{y} = 1.99$ in.
12.3-8 $\bar{x} = 137$ mm, $\bar{y} = 132$ mm
12.4-6 $I_x = 20.2 \times 10^6$ mm^4
12.4-7 $I_x = 36.1$ in.^{4}, $I_y = 10.9$ in.^{4}
12.4-8 $I_x = I_y = 195 \times 10^6$ mm^4,
$r_x = r_y = 80.1$ mm
12.4-9 $I_1 = 1480$ in.^{4}, $I_2 = 186$ in.^{4},
$r_1 = 7.10$ in., $r_2 = 2.52$ in.

12.5-1 $I_b = 940$ in.^{4} **12.5-2** $I_c = 11a^4/192$
12.5-3 $I_c = 7.23$ in.^{4} **12.5-4** $I_2 = 405 \times 10^3$ mm^4
12.5-5 $I_{x_c} = 6050$ in.^{4}, $I_{y_c} = 475$ in.^{4}
12.5-6 $I_c = 106 \times 10^6$ mm^4
12.5-7 $I_{x_c} = 17.40$ in.^{4}, $I_{y_c} = 6.27$ in.^{4}
12.5-8 $b = 250$ mm **12.6-1** $I_p = bh(b^2 + 12h^2)/48$
12.6-2 $I_{p_c} = r^4(9\alpha^2 - 8 \sin^2 \alpha)/18\alpha$
12.6-3 $I_p = 233$ in.^{4} **12.6-4** $I_p = bh(b^2 + h^2)/24$
12.6-5 $I_{p_c} = r^4(176 - 84\pi + 9\pi^2)/[72(4 - \pi)]$
12.6-6 $I_{p_c} = 132 \times 10^6$ mm^4 **12.7-2** $I_{xy} = r^4/24$
12.7-3 $b = 2r$ **12.7-4** $I_{xy} = t^2(2b^2 - t^2)/4$
12.7-5 $I_{12} = -20.5$ in.^{4} **12.7-6** $I_{xy} = 24.3 \times 10^6$ mm^4
12.7-7 $I_{x_c y_c} = -6.079$ in.^{4}
12.7-8 $I_{xy} = 99.9 \times 10^6$ mm^4
12.8-1 $I_{x1} = I_{y1} = b^4/12, I_{x1y1} = 0$
12.8-2 $I_{x1} = \dfrac{b^3 h^3}{6(b^2 + h^2)}, I_{y1} = \dfrac{bh(b^4 + h^4)}{12(b^2 + h^2)},$
$I_{x1y1} = \dfrac{b^2 h^2 (h^2 - b^2)}{12(b^2 + h^2)}$
12.8-3 $I_d = 159$ in.^{4}
12.8-4 $I_{x1} = 12.4 \times 10^6$ mm^4,
$I_{y1} = 9.68 \times 10^6$ mm^4, $I_{x1y1} = 6.02 \times 10^6$ mm^4
12.8-5 $I_{x1} = 13.5$ in.^{4}, $I_{y1} = 3.84$ in.^{4}, $I_{x1y1} = 4.76$ in.^{4}
12.8-6 $I_{x1} = 8.75 \times 10^6$ mm^4,
$I_{y1} = 1.02 \times 10^6$ mm^4, $I_{x1y1} = -0.356 \times 10^6$ mm^4
12.9-1 (a) $c = \sqrt{a^2 - b^2}/2$; (b) $a/b = \sqrt{5}$;
(c) $1 \le a/b < \sqrt{5}$
12.9-2 Show that two different sets of principal axes exist at each point
12.9-3 $\theta_{p1} = -29.9°, \theta_{p2} = 60.1°, I_1 = 311.1$ in.^{4},
$I_2 = 88.9$ in.^{4}
12.9-4 $\theta_{p1} = -8.54°, \theta_{p2} = 81.46°$,
$I_1 = 17.24 \times 10^6$ mm^4, $I_2 = 4.88 \times 10^6$ mm^4
12.9-5 $\theta_{p1} = 37.7°, \theta_{p2} = 127.7°, I_1 = 15.5$ in.^{4},
$I_2 = 1.9$ in.^{4}
12.9-6 $\theta_{p1} = 32.6°, \theta_{p2} = 122.6°, I_1 = 8.76 \times 10^6$ mm^4,
$I_2 = 1.00 \times 10^6$ mm^4
12.9-7 $\theta_{p1} = 16.8°, \theta_{p2} = 106.8°, I_1 = 0.2390b^4$,
$I_2 = 0.03873b^4$
12.9-8 $\theta_{p1} = -24.4°, \theta_{p2} = 65.6°, I_1 = 11.46 \times 10^6$ mm^4,
$I_2 = 1.05 \times 10^6$ mm^4
12.9-9 $\theta_{p1} = -22.4°, \theta_{p2} = 67.6°, I_1 = 25.8$ in.^{4},
$I_2 = 2.2$ in.^{4}
12.9-10 $\theta_{p1} = 74.1°, \theta_{p2} = -15.9°, I_1 = 8.29 \times 10^6$ mm^4,
$I_2 = 1.00 \times 10^6$ mm^4
12.9-11 $\theta_{p1} = 75.7°, \theta_{p2} = -14.3°, I_1 = 20.1$ in.^{4},
$I_2 = 2.12$ in.^{4}

NAME INDEX

Note: Pages identified with an asterisk contain biographical information.

SUBJECT INDEX

Principal Units Used in Mechanics

Quantity	International System (SI)			U.S. Customary System (USCS)		
	Unit	Symbol	Formula	Unit	Symbol	Formula
Acceleration (angular)	radian per second squared		rad/s^2	radian per second squared		rad/s^2
Acceleration (linear)	meter per second squared		m/s^2	foot per second squared		ft/s^2
Area	square meter		m^2	square foot		ft^2
Density (mass) (Specific mass)	kilogram per cubic meter		kg/m^3	slug per cubic foot		slug/ft^3
Density (weight) (Specific weight)	newton per cubic meter		N/m^3	pound per cubic foot	pcf	lb/ft^3
Energy; work	joule	J	N·m	foot-pound		ft-lb
Force	newton	N	kg·m/s^2	pound	lb	(base unit)
Force per unit length (Intensity of force)	newton per meter		N/m	pound per foot		lb/ft
Frequency	hertz	Hz	s^{-1}	hertz	Hz	s^{-1}
Length	meter	m	(base unit)	foot	ft	(base unit)
Mass	kilogram	kg	(base unit)	slug		$\text{lb-s}^2\text{/ft}$
Moment of a force; torque	newton meter		N·m	pound-foot		lb-ft
Moment of inertia (area)	meter to fourth power		m^4	inch to fourth power		in.^4
Moment of inertia (mass)	kilogram meter squared		kg·m^2	slug foot squared		slug-ft^2
Power	watt	W	J/s (N·m/s)	foot-pound per second		ft-lb/s
Pressure	pascal	Pa	N/m^2	pound per square foot	psf	lb/ft^2
Section modulus	meter to third power		m^3	inch to third power		in.^3
Stress	pascal	Pa	N/m^2	pound per square inch	psi	lb/in.^2
Time	second	s	(base unit)	second	s	(base unit)
Velocity (angular)	radian per second		rad/s	radian per second		rad/s
Velocity (linear)	meter per second		m/s	foot per second	fps	ft/s
Volume (liquids)	liter	L	$10^{-3}\ \text{m}^3$	gallon	gal.	$231\ \text{in.}^3$
Volume (solids)	cubic meter		m^3	cubic foot	cf	ft^3

Physical Properties in SI and USCS Units

Property	SI	USCS
Water (fresh)		
weight density	9.81 kN/m^3	62.4 lb/ft^3
mass density	1000 kg/m^3	1.94 slugs/ft^3
Sea water		
weight density	10.0 kN/m^3	63.8 lb/ft^3
mass density	1020 kg/m^3	1.98 slugs/ft^3
Aluminum (structural alloys)		
weight density	27.5 kN/m^3	175 lb/ft^3
mass density	2800 kg/m^3	5.43 slugs/ft^3
Steel		
weight density	77.0 kN/m^3	490 lb/ft^3
mass density	7850 kg/m^3	15.2 slugs/ft^3
Reinforced concrete		
weight density	23.6 kN/m^3	150 lb/ft^3
mass density	2400 kg/m^3	4.66 slugs/ft^3
Atmospheric pressure (sea level)		
Recommended value	101 kPa	14.7 psi
Standard international value	101.325 kPa	14.6959 psi
Acceleration of gravity (sea level, approx. 45° latitude)		
Recommended value	9.81 m/s^2	32.2 ft/s^2
Standard international value	9.80665 m/s^2	32.1740 ft/s^2

Note: Properties of materials vary with chemical composition, temperature, and other factors. The values given here are typical.

SI Prefixes

Prefix	Symbol	Multiplication factor	
tera	T	10^{12}	= 1 000 000 000 000
giga	G	10^{9}	= 1 000 000 000
mega	M	10^{6}	= 1 000 000
kilo	k	10^{3}	= 1 000
hecto	h	10^{2}	= 100
deka	da	10^{1}	= 10
deci	d	10^{-1}	= 0.1
centi	c	10^{-2}	= 0.01
milli	m	10^{-3}	= 0.001
micro	μ	10^{-6}	= 0.000 001
nano	n	10^{-9}	= 0.000 000 001
pico	p	10^{-12}	= 0.000 000 000 001

Note: The use of the prefixes hecto, deka, deci, and centi is not recommended in SI.